Conversion of U.S. Customary Units to SI Units (continued)

Customary unit		*Times conversion factor*		*Equals SI unit*	
		Accurate	*Practical*		
Power					
foot-pound per second	ft-lb/s	1.35582	1.36	watt	W
foot-pound per minute	ft-lb/min	0.0225970	0.0226	watt	W
horsepower (550 foot-pounds per second)	hp	745.701	746	watt	W
Pressure; stress					
pound per square foot	psf	47.8803	47.9	pascal (N/m^2)	Pa
pound per square inch	psi	6894.76	6890	pascal	Pa
kip per square foot	ksf	47.8803	47.9	kilopascal	kPa
kip per square inch	ksi	6894.76	6890	kilopascal	kPa
Section modulus					
inch to third power	$in.^3$	16,387.1	16,400	millimeter to third power	mm^3
inch to third power	$in.^3$	16.3871×10^{-6}	16.4×10^{-6}	meter to third power	m^3
Specific weight (weight density)					
pound per cubic foot	lb/ft^3	157.087	157	newton per cubic meter	N/m^3
pound per cubic inch	$lb/in.^3$	271.447	271	kilonewton per cubic meter	kN/m^3
Velocity					
foot per second	ft/s	0.3048*	0.305	meter per second	m/s
inch per second	in./s	0.0254*	0.0254	meter per second	m/s
mile per hour	mph	0.44704*	0.447	meter per second	m/s
mile per hour	mph	1.609344*	1.61	kilometer per hour	km/h
Volume					
cubic foot	ft^3	0.0283168	0.0283	cubic meter	m^3
cubic inch	$in.^3$	16.3871×10^{-6}	16.4×10^{-6}	cubic meter	m^3
cubic inch	$in.^3$	16.3871	16.4	cubic centimeter	cm^3
gallon	gal.	3.78541	3.79	liter	L
gallon	gal.	0.00378541	0.00379	cubic meter	m^3

* Exact conversion factor

Note: To convert from SI units to U.S. Customary units, *divide* by the conversion factor.

Temperature Conversions

$$T(\mathrm{K}) = \frac{5}{9}[T(^\circ\mathrm{F}) - 32] + 273.15 = T(^\circ\mathrm{C}) + 273.15$$

$$T(^\circ\mathrm{F}) = \frac{9}{5}T(^\circ\mathrm{C}) + 32 = \frac{9}{5}T(\mathrm{K}) - 459.67$$

Mechanics of Materials

THIRD EDITION

PWS SERIES IN ENGINEERING

Askeland, *The Science and Engineering of Materials*, Second Edition
Askeland, *The Science and Engineering of Materials*, Alternate Edition
Bauld, *Mechanics of Materials*, Second Edition
Borse, *Fortran 77 and Numerical Methods for Engineers*
Das, *Principles of Foundation Engineering*, Second Edition
Das, *Principles of Geotechnical Engineering*, Second Edition
Duff, *Concepts of Technical Graphics*
Gere and Timoshenko, *Mechanics of Materials*, Third Edition
Keedy, *An Introduction to CAD Using CADKey*
McGill and King, *Engineering Mechanics: Statics*, Second Edition
McGill and King, *Engineering Mechanics: An Introduction to Dynamics*, Second Edition
McGill and King, *Engineering Mechanics: Statics and An Introduction to Dynamics*, Second Edition
Reynolds, *Unit Operations and Processes in Environmental Engineering*
Sack, *Matrix Structural Analysis*
Schmidt and Wong, *Fundamentals of Surveying*, Third Edition
Segui, *Fundamentals of Structural Steel Design*
Weinman, *Fortran for Scientists and Engineers*

Mechanics of Materials

THIRD EDITION

James M. Gere
STANFORD UNIVERSITY

Stephen P. Timoshenko (1878–1972)
Formerly, **STANFORD UNIVERSITY**

PWS Publishing Company
BOSTON

PWS PUBLISHING COMPANY
20 Park Plaza, Boston, MA 02116-4324

I(T)P ™

International Thomson Publishing
The trademark ITP is used under license

PWS Publishing Company is a division of Wadsworth, Inc.

Printed in the United States of America

7 8 9 - 95 94

Library of Congress Cataloging-in-Publication Data

Gere, James M.
Mechanics of materials/James M. Gere, Stephen P. Timoshenko.—3rd ed.
p. cm.
Includes bibliographical references.
ISBN 0-534-92174-4
1. Strength of materials. I. Timoshenko, Stephen P., 1878-1972. II. Title.
TA405.G44 1990
620.1′123—dc20 89-27415
CIP

Sponsoring Editor: Jonathan Plant
Assistant Editor: Mary Thomas
Production Coordinator: Helen Walden
Manufacturing Coordinator: Peter Leatherwood
Interior and Cover Design: Helen Walden
Interior Illustrations: Scientific Illustrators and Mary Burkhardt, Carol Johnston, Georg Klatt, Judith McCarty, Pat Rogondino
Typesetting: Polyglot Pte Ltd.
Printing and Binding: R. R. Donnelley & Sons
Cover Printing: Henry H. Sawyer Company

Cover Illustration: Construction of the Forth Railway Bridge over the Firth of Forth in Scotland, courtesy of The Bettmann Archive.

For many years the longest bridge in the world, the Forth Railway Bridge on the rail line between Edinburgh and Dundee was completed in 1890. Of trussed cantilever design, it has a total length of 8300 ft with two main spans of 1710 ft each. The compression members are steel tubes and the tension members are open lattice steel girders.

Preface

A course in mechanics of materials provides an opportunity to teach students a basic engineering subject and, at the same time, to develop their analytical and problem-solving abilities. While preparing this new edition, I have kept both of these important goals in mind. The facts, theories, and methodologies of mechanics are presented in a teachable and easy-to learn manner, with ample discussions and numerous illustrative examples, so that undergraduate students can readily master the subject matter. In addition, considerable emphasis is placed on how to analyze mechanical and structural systems, and most of the examples and problems require that students do some original thinking.

This book covers all of the standard topics of mechanics of materials and presents them at a level suitable for sophomore and junior engineering students. Subject matter of a more advanced and specialized nature is also included, so that the book serves both as a classroom text and as a permanent engineering reference.

A glance at the table of contents shows the topics covered and the way in which they are organized. The topics include the analysis and design of structural members subjected to axial loads, torsion, and bending, as well as such fundamental concepts as stress, strain, elastic and inelastic behavior, and strain energy. Other topics of general interest are the transformations of stress and strain, stress concentrations, deflections of beams, behavior of columns, and energy methods. More specialized topics are thermal and prestrain effects, pressure vessels, nonprismatic members, discontinuity functions, shear center, and inelastic bending. (An asterisk indicates a section dealing with an advanced or specialized topic.)

Much more material than can be covered in a single course is included in the book, and therefore teachers have the opportunity to select the topics that they feel are the most fundamental and relevant. Teachers will also appreciate the hundreds of new problems (over 1,000 problems total) that are available for homework assignments and classroom discussions. Difficult and lengthy problems are indicated by asterisks.

Both the International System of Units (SI) and the U.S. Customary System (USCS) are used in the numerical examples and problems. Discussions of both systems and a table of conversion factors are given in the appendix. For problems involving numerical solutions, odd-numbered problems are in USCS units and even-numbered problems are in SI units. (The only exceptions are problems involving the tabulated properties of structural-steel shapes, because the tables are available only in USCS units.) Answers to most problems are listed at the back of the book, so that problem-solvers can have the satisfaction of checking their results.

References and historical notes are also collected at the back of the book. They include the original sources of the subject matter and biographical notes about the pioneering engineers, scientists, and mathematicians who created the subject. The book concludes with an extensive and carefully prepared index, which makes it very easy to locate any desired subject matter.

This third edition of *Mechanics of Materials* has been thoroughly rewritten with improved discussions, many new problems, and some changes in organization to make the book more useful in the classroom. Discussions of fatigue and stress concentrations in axial load, torsion, and bending have been added, and a few specialized topics have been dropped. All changes in the subject-matter coverage were prompted by the suggestions of teachers who were familiar with the second edition.

Every effort has been made to eliminate errors, but if you find any, please jot them down and mail them to the author (Department of Civil Engineering, Stanford University, Stanford, California 94305-4020); then we can correct them immediately for the next printing of the book.

To acknowledge everyone who contributed to this book in some manner is clearly impossible, but a major debt is owed my former Stanford teachers (those giants of mechanics, including Timoshenko himself, Wilhelm Flügge, James Norman Goodier, Miklós Hetényi, Nicholas J. Hoff, and Donovan H. Young) from whom I learned so much. My current Stanford colleagues, especially Tom Kane, Anne Kiremidjian, Helmut Krawinkler, Haresh Shah, Sheri Sheppard, and Bill Weaver, made suggestions for the book and cooperated during its writing. Professor Thalia Anagnos of San Jose State University gave me many valuable ideas for improving the presentation, and the following reviewers contributed both general ideas and detailed comments: Robert R. Archer, University of Massachusetts–Amherst; Peter Chang, University of Maryland; J. S. DeNatale, California Polytechnic State University; Brian D. Harper, Ohio State University; and Terry Kohutek, Texas A&M University.

Typing and word processing of the manuscript were done with considerable care by Christina Doering, Sharon Garner, Emily Gere,

Gretchen Howard, Richard Koch, Duc Wong, and Laurie Yadon. Several graduate students provided expert assistance by proofreading and preparing solutions to problems: Upal Basu, Narayanan Doraswamy, Mohsen Rahnama, and Rajan Vasudevan. Editing and production were handled skillfully and efficiently by the staff of PWS-KENT, including David Geggis, Eileen Pfeffer, Jonathan Plant, Kathy Smith, Wendy Snyder, Mary Thomas, and Helen Walden. Their friendly and cooperative spirit made my role a pleasure. Finally, my wife, Janice, offered encouragement and exercised great patience throughout this project. So also did other family members—Susan Gere and DeWitt Durham, Bill Gere, and David Gere. To all of these wonderful people I am pleased to express my gratitude.

James M. Gere

Contents

CHAPTER 3
Torsion 158

CHAPTER 4
Shear Force and Bending Moment 220

CHAPTER 5
Stresses in Beams 250

CHAPTER 8

Statically Indeterminate Beams 535

CHAPTER 9

Columns 577

CHAPTER 10

Energy Methods 627

List of Symbols

A	area, action (force or couple), constant
a, b, c	dimensions, distances, constants
C	centroid, constant of integration, compressive force
c	distance from neutral axis to outer surface of a beam
D	displacement (translation or rotation), diameter
d	diameter, dimension, distance
E	modulus of elasticity, elliptic integral of the second kind
E_r	reduced modulus of elasticity
E_t	tangent modulus of elasticity
e	eccentricity, dimension, distance, unit volume change (dilatation, volumetric strain)
F	force, discontinuity function, elliptic integral of the first kind, flexibility
f	shear flow, shape factor for plastic bending, flexibility, frequency (Hz)
f_s	form factor for shear
G	modulus of elasticity in shear
g	acceleration of gravity
H	distance, force, reaction, horsepower
h	height, dimension
I	moment of inertia (or second moment) of a plane area
I_x, I_y, I_z	moments of inertia with respect to x, y, and z axes
I_{xy}	product of inertia with respect to the x and y axes
I_p	polar moment of inertia
I_1, I_2	principal moments of inertia
J	torsion constant

K stress-concentration factor, bulk modulus of elasticity, effective length factor for a column
K_L kinetic energy loss factor
k spring constant, stiffness, symbol for $\sqrt{P/EI}$
L length, distance, span length
L_e effective length of a column
ln natural logarithm
log common logarithm
M bending moment, couple, mass
M_p plastic moment for a beam
M_y yield moment for a beam
m moment per unit length, mass per unit length
N axial force
n factor of safety, number, ratio, integer, revolutions per minute (rpm)
O origin of coordinates
O' center of curvature
P force, concentrated load, axial force, power
P_{allow} allowable load (or working load)
P_{cr} critical load for a column
P_r reduced-modulus load for a column
P_t tangent-modulus load for a column
P_u ultimate load
P_y yield load
p pressure (force per unit area)
Q force, concentrated load, first moment (or static moment) of a plane area
q intensity of distributed load (force per unit distance), intensity of distributed torque (torque per unit distance)
q_u ultimate load intensity
q_y yield load intensity
R reaction, radius, force
r radius, distance, radius of gyration ($r = \sqrt{I/A}$)
S section modulus of the cross section of a beam, shear center, stiffness, force
s distance, length along a curved line
T twisting couple or torque, temperature, tensile force
T_u ultimate torque
T_y yield torque
t thickness, time
U strain energy
u strain energy density (strain energy per unit volume)
u_r modulus of resilience
u_t modulus of toughness

$U^{\star}$	complementary energy
$u^{\star}$	complementary energy density (complementary energy per unit volume)
V	shear force, volume
v	deflection of a beam, velocity
v', v'', etc.	dv/dx, d^2v/dx^2, etc.
W	weight, work
$W^{\star}$	complementary work
w	load per unit of area (force per unit area)
X	statical redundant
x, y, z	rectangular coordinates, distances
$\bar{x}$, $\bar{y}$, $\bar{z}$	coordinates of centroid
Z	plastic modulus of the cross section of a beam
α	angle, coefficient of thermal expansion, nondimensional ratio, spring constant, stiffness
α_s	shear coefficient
β	angle, nondimensional ratio, spring constant, stiffness
γ	shear strain, specific weight (weight per unit volume)
γ_{xy}, γ_{yz}, γ_{zx}	shear strains in the xy, yz, and zx planes
γ_{θ}	shear strain for inclined axes
$\gamma_{x_1y_1}$	shear strain in the x_1y_1 plane
δ, Δ	deflection, displacement, elongation
ϵ	normal strain
ϵ_x, ϵ_y, ϵ_z	normal strains in the x, y, and z directions
ϵ_{θ}	normal strain for inclined axes
ϵ_{x_1}, ϵ_{y_1}	normal strains in the x_1 and y_1 directions
ϵ_1, ϵ_2, ϵ_3	principal normal strains
ϵ_y	yield strain
θ	angle, angle of twist per unit length, angle of rotation of beam axis
θ_p	angle to a principal plane or to a principal axis
θ_s	angle to a plane of maximum shear stress
κ	curvature ($\kappa = 1/\rho$)
κ_y	yield curvature
λ	distance
ρ	radius, radius of curvature, radial distance in polar coordinates, mass density (mass per unit volume, specific mass)
ν	Poisson's ratio
σ	normal stress
σ_x, σ_y, σ_z	normal stresses on planes perpendicular to the x, y, and z axes
σ_{θ}	normal stress on inclined plane
σ_{x_1}, σ_{y_1}	normal stresses on planes perpendicular to the rotated x_1y_1 axes
σ_1, σ_2, σ_3	principal stresses

σ_{allow}	allowable stress (or working stress)
σ_{cr}	critical stress for a column ($\sigma_{\text{cr}} = P_{\text{cr}}/A$)
σ_{pl}	proportional-limit stress
σ_r	residual stress
σ_u	ultimate stress
σ_y	yield stress
τ	shear stress
$\tau_{xy}, \tau_{yz}, \tau_{zx}$	shear stresses on planes perpendicular to the x, y, and z axes and parallel to the y, z, and x axes
τ_θ	shear stress on inclined plane
$\tau_{x_1y_1}$	shear stress on plane perpendicular to the rotated x_1 axis and parallel to the y_1 axis
τ_{allow}	allowable stress (or working stress) in shear
τ_u	ultimate stress in shear
τ_y	yield stress in shear
ϕ	angle, angle of twist
ψ	nondimensional ratio
ω	angular velocity, angular frequency ($\omega = 2\pi f$)

* An asterisk denotes a difficult problem or a section dealing with an advanced or specialized topic

Greek Alphabet

Α	α	Alpha	Ν	ν	Nu
Β	β	Beta	Ξ	ξ	Xi
Γ	γ	Gamma	Ο	ο	Omicron
Δ	δ	Delta	Π	π	Pi
Ε	ϵ	Epsilon	Ρ	ρ	Rho
Ζ	ζ	Zeta	Σ	σ	Sigma
Η	η	Eta	Τ	τ	Tau
Θ	θ	Theta	Υ	υ	Upsilon
Ι	ι	Iota	Φ	ϕ	Phi
Κ	κ	Kappa	Χ	χ	Chi
Λ	λ	Lambda	Ψ	ψ	Psi
Μ	μ	Mu	Ω	ω	Omega

CHAPTER 1

Tension, Compression, and Shear

1.1 INTRODUCTION

Mechanics of materials is a branch of applied mechanics that deals with the behavior of solid bodies subjected to various types of loading. This field of study is known by several names, including "strength of materials" and "mechanics of deformable bodies." The solid bodies considered in this book include axially loaded members, shafts in torsion, thin shells, beams, and columns, as well as structures that are assemblies of these components. Usually the objectives of our analysis will be the determination of the stresses, strains, and displacements produced by the loads. If these quantities can be found for all values of load up to the failure load, we will have a complete picture of the mechanical behavior of the body.

A thorough understanding of mechanical behavior is essential for the safe design of all structures, whether buildings and bridges, machines and motors, submarines and ships, or airplanes and antennas. That is why mechanics of materials is a basic subject in so many engineering fields. Of course, statics and dynamics are also essential, but they deal primarily with the forces and motions associated with particles and rigid bodies. In mechanics of materials, we go one step further by examining the stresses and strains that occur inside real bodies that deform under loads. We use the physical properties of the materials (obtained from experiments) as well as numerous theoretical laws and concepts, many of which are explained in succeeding sections of this book.

Theoretical analyses and experimental results have equally important roles in the study of mechanics of materials. Sometimes we will make logical derivations to obtain formulas and equations for predicting mechanical behavior, but we must recognize that these formulas cannot be used unless certain properties of the materials are known. These properties are available to us only after suitable experiments have been carried out in the laboratory. Also, because many practical problems of great importance in engineering cannot be handled efficiently by theoretical means, experimental measurements are a necessity.

The historical development of mechanics of materials is a fascinating blend of both theory and experiment; experiments have pointed the way to useful results in some instances, and theory has done so in others. Such famous men as Leonardo da Vinci (1452–1519) and Galileo Galilei (1564–1642) performed experiments to determine the strength of wires, bars, and beams, although they did not develop any adequate theories (by today's standards) to explain their test results. Such theories came much later. By contrast, the famous mathematician Leonhard Euler (1707–1783) developed the mathematical theory of columns and calculated the theoretical critical load of a column in 1744, long before any experimental evidence existed to show the significance of his results. Without appropriate tests, Euler's results remained unused for many years, although today they form the basis of column theory.*

When studying mechanics of materials, you will find that your efforts are divided naturally into two parts: first, understanding the logical development of the concepts, and second, applying those concepts to practical situations. The former is accomplished by studying the derivations, discussions, and examples, and the latter by solving problems. Some of the examples and problems are numerical in character, and the others are algebraic (or symbolic). An advantage of numerical problems is that the magnitudes of all quantities are evident at every stage of the calculations. Sometimes these values are needed to ensure that practical limits (such as allowable stresses) are not exceeded. Algebraic solutions have certain advantages, too. Because they lead to formulas, algebraic solutions make clear the variables that affect the final result. For instance, a certain quantity may actually cancel out of the solution, a fact that would not be evident from a numerical problem. Also apparent in algebraic solutions is the manner in which variables affect the results, such as the appearance of one variable in the numerator and another in the denominator. Furthermore, a symbolic solution provides the opportunity to check the dimensions at any stage of the work. Finally, the most important reason for deriving an algebraic solution is to obtain a general formula that can be used for many different problems. In contrast, a numerical solution applies to only one set of circumstances. You must be adept at both kinds of solutions, and therefore you will find a mixture of numerical and algebraic problems throughout this book.

Numerical problems require that you work with specific units for measurements. This book utilizes both the International System of Units

* The history of mechanics of materials, beginning with Leonardo and Galileo, is given in Refs. 1-1, 1-2, and 1-3.

(SI) and the U.S. Customary System (USCS). A discussion of both systems appears in Appendix A, which also provides useful tables and conversion factors. The matter of significant digits, which is very important in engineering, is discussed in Appendix B. As explained there, numerical examples and problems in this book are solved to a final accuracy of three or four significant digits.

All problems appear at the ends of the chapters, with the problem numbers identifying the sections to which they pertain. In the case of problems requiring numerical solutions, odd-numbered problems are in USCS units and even-numbered problems are in SI units. The only exceptions are a few problems in some of the later chapters where structural-steel shapes are used, because the properties of these shapes are given only in USCS units (see Appendix E).

1.2 NORMAL STRESS AND STRAIN

The fundamental concepts of stress and strain can be illustrated by considering a prismatic bar that is loaded by axial forces P at the ends, as shown in Fig. 1-1a. A **prismatic bar** is a straight structural member having constant cross section throughout its length. In this illustration, the axial forces produce a uniform stretching of the bar, and the bar is said to be in **tension**.

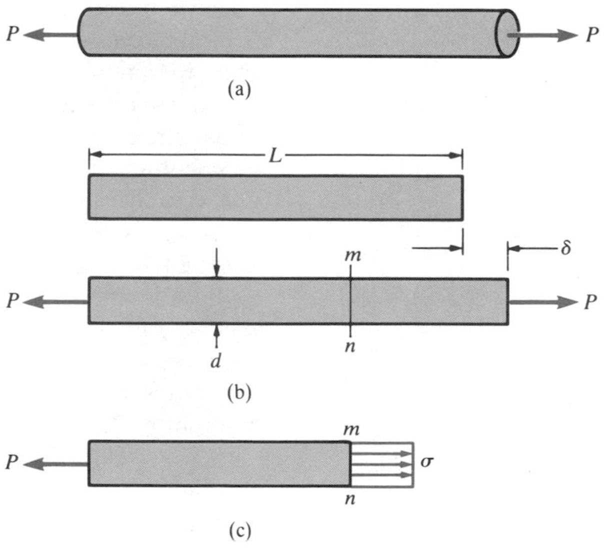

Fig. 1-1 Prismatic bar in tension

To investigate the **stresses and strains** in this bar, we consider in Fig. 1-1b two side views of the bar, one showing the bar (length L) before the loads are applied and the other showing the elongated bar after the loads are applied. The internal stresses produced in the bar by the axial forces are exposed if we make an imaginary cut at section *mn* (Fig. 1-1b). Because this section is taken perpendicular to the longitudinal axis of the bar, it is called a **cross section**. We now isolate the part of the bar to the left of the cut as a free body (Fig. 1-1c). The tensile load *P* acts at the left-hand end of this free body; at the other end are forces representing the action of the removed part of the bar upon the part that remains. These forces are continuously distributed over the cross section, analogous to the continuous distribution of hydrostatic pressure over a submerged horizontal surface. The intensity of force (that is, the force per unit area) is called the **stress** and is commonly denoted by the Greek letter σ (sigma).

Assuming that the stress has a uniform distribution over the cross section (Fig. 1-1c), we see that its resultant is equal to the intensity σ times the cross-sectional area A of the bar. Furthermore, from the equilibrium of the body shown in Fig. 1-1c, it is also evident that the resultant must be equal in magnitude and opposite in direction to the applied load P. From these observations we obtain

$$\sigma = \frac{P}{A} \tag{1-1}$$

as the equation for the uniform stress in an axially loaded, prismatic bar of arbitrary cross-sectional shape. When the bar is stretched by the forces P, as shown in the figure, the resulting stresses are **tensile stresses**; if the forces are reversed in direction, causing the bar to be compressed, we obtain **compressive stresses**. Inasmuch as the stress σ acts in a direction perpendicular to the cut surface, it is referred to as a **normal stress**. Thus, normal stresses may be either tensile or compressive stresses. Later, we will encounter another type of stress, called a **shear stress**, that acts parallel to the surface.

When a **sign convention** for normal stresses is required, it is customary to define tensile stresses as positive and compressive stresses as negative.

Because the normal stress σ is obtained by dividing the axial force by the cross-sectional area, it has **units** of force per unit of area. When USCS units are used, stress is customarily expressed in pounds per square inch (psi) or kips per square inch (ksi).* For instance, a typical stress in a steel bar might be 20,000 psi or 20 ksi.

When using SI units, force is expressed in newtons (N) and area in square meters (m^2). Hence, stress has units of newtons per square meter (N/m^2), or pascals (Pa). However, the pascal is such a small unit of stress that it is necessary to work with large multiples. To illustrate this point, we have only to note that it takes almost 7000 pascals to make 1 psi.** As an

* One kip, or kilopound, equals 1000 lb.

** See Table A-3 in Appendix A for conversion factors between USCS and SI units.

example, a typical tensile stress in a steel bar might have a magnitude of 140 megapascals (140 MPa), which is 140×10^6 pascals. Other common units are the kilopascal (kPa) and gigapascal (GPa); the former equals 10^3 pascals and the latter equals 10^9 pascals. Although it is not recommended in SI, you will sometimes find stress given in newtons per square millimeter (N/mm^2), which is a unit equal to the megapascal (MPa).

If the equation $\sigma = P/A$ is to be valid, the stress σ must be uniformly distributed over the cross section of the bar. This condition is realized if the axial force P acts through the centroid of the cross-sectional area, as demonstrated later in Example 1. When the load P does not act at the centroid, bending of the bar will result, and a more complicated analysis is necessary (see Section 5.9). However, we will assume throughout this book that all axial forces are applied at the centroid of the cross section unless specifically stated otherwise.

The uniform stress condition pictured in Fig. 1-1c exists throughout the length of the member except near the ends. The stress distribution at the ends of the bar depends upon the details of how the axial load P is actually applied. If the load itself is distributed uniformly over the end, then the stress pattern at the end will be the same as elsewhere. However, the load usually is concentrated over a small area, resulting in high localized stresses (called **stress concentrations**) and nonuniform stress distributions over cross sections in the vicinity of the load. As we move away from the ends, the stress distribution gradually approaches the uniform distribution shown in Fig. 1-1c. It is usually safe to assume that the formula $\sigma = P/A$ may be used with good accuracy at any point within the bar that is at least a distance d away from the ends, where d is the largest transverse dimension of the bar (see Fig. 1-1b). Of course, even when the stress is not uniformly distributed, the equation $\sigma = P/A$ will give the **average normal stress** on the cross section.*

An axially loaded bar undergoes a change in length, becoming longer when in tension and shorter when in compression. The change in length, denoted by the Greek letter δ (delta), is pictured in Fig. 1-1b for the bar in tension. This elongation is the cumulative result of the stretching of all elements of the material throughout the length L of the bar. Now let us assume that the material is the same everywhere in the bar. Then, if we consider half of the bar, it will have an elongation equal to $\delta/2$; similarly, if we consider a unit length of the bar, it will have an elongation equal to $1/L$ times the total elongation δ. In this manner, we arrive at the concept of elongation per unit length, or **strain**, denoted by the Greek letter ϵ (epsilon) and given by the equation

$$\epsilon = \frac{\delta}{L} \qquad (1\text{-}2)$$

* Stress concentrations are discussed in Section 2.10.

If the bar is in tension, the strain is called a **tensile strain**, representing an elongation or stretching of the material. If the bar is in compression, the strain is a **compressive strain** and the bar shortens. Tensile strain is usually taken as positive and compressive strain as negative. The strain ϵ is called a **normal strain** because it is associated with normal stresses.

Because normal strain ϵ is the ratio of two lengths, it is a **dimensionless quantity**; that is, it has no units. Thus, strain is expressed as a pure number, independent of any system of units. Numerical values of strain are usually very small, especially for structural materials, which ordinarily undergo only small changes in dimensions. As an example, consider a steel bar having length L equal to 2.0 m. When loaded in tension, the bar might elongate by an amount δ equal to 1.4 mm. The corresponding strain is

$$\epsilon = \frac{\delta}{L} = \frac{1.4 \text{ mm}}{2.0 \text{ m}} = 0.0007 = 700 \times 10^{-6}$$

In practice, the original units of δ and L are sometimes attached to the strain itself, and then the strain is recorded in forms such as mm/m, μm/m, and in./in. For instance, the strain ϵ in the preceding illustration could be given as 700 μm/m or 700×10^{-6} in./in.

The definitions of normal stress and strain are based upon purely statical and geometrical considerations, which means that Eqs. (1-1) and (1-2) can be used for loads of any magnitude and for any material. The principal requirement is that the deformation of the bar be uniform, which in turn requires that the bar be prismatic, the loads act through the centroids of the cross sections, and the material be **homogeneous** (that is, the same throughout all parts of the bar). The resulting state of stress and strain is called **uniaxial stress and strain**. Further discussions of uniaxial stress, including stresses and strains in other than the longitudinal direction of the bar, are given in later sections. We will also encounter more complicated stress states, such as biaxial stress and plane stress, in later chapters.

Example 1

Show that the axial forces P producing uniform tension or compression in a prismatic member (see Fig. 1-1) must act through the centroid of the cross section.

The uniform stress distribution acting at cross section *mn* of the bar of Fig. 1-1 is shown again in Fig. 1-2a. The resultant of this stress distribution is the axial force P. Now let us assume that the cross section has the arbitrary shape shown in Fig. 1-2b, and for reference let us take a set of xy axes in the plane of the cross section. Then the z axis is parallel to the long direction of the bar (Fig. 1-2a).

The x and y coordinates of the line of action of the force P are denoted by $\bar{x}$ and $\bar{y}$ in the figure. To determine these coordinates, we observe that the moments M_x and M_y of the force P about the x and y axes, respectively, must be equal to the

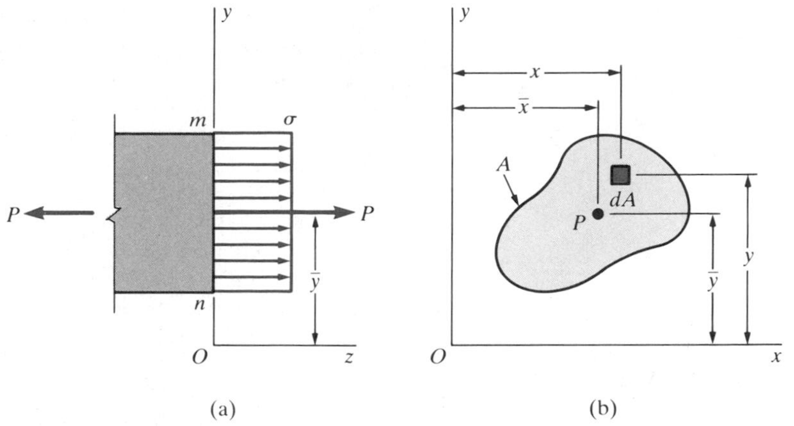

Fig. 1-2 Example 1. Axial force P acting at the centroid of the cross section

corresponding moments of the uniformly distributed stress σ. The moments of the force P are

$$M_x = P\bar{y} \qquad M_y = -P\bar{x} \tag{a}$$

in which a moment is considered positive when its vector (using the right-hand rule) acts in the positive direction of the axis. To obtain the moments of the distributed stresses, we consider an element of area dA in the cross section (Fig. 1-2b) and note that the differential force acting on this element is $\sigma\,dA$. The moments of this elemental force about the x and y axes are $\sigma y\,dA$ and $-\sigma x\,dA$, respectively, in which x and y denote the coordinates of the element dA. The total moments can be obtained by integrating over the entire cross-sectional area A; thus, the moments of the stress σ about the x and y axes are

$$M_x = \int \sigma y\,dA \qquad M_y = -\int \sigma x\,dA \tag{b}$$

Next, we equate the moments given by expressions (a) and (b):

$$P\bar{y} = \int \sigma y\,dA \qquad P\bar{x} = \int \sigma x\,dA$$

Noting that the force P equals σA and that the stress σ is a constant, we obtain from the preceding equations the following formulas for the coordinates $\bar{y}$ and $\bar{x}$:

$$\bar{y} = \frac{\int y\,dA}{A} \qquad \bar{x} = \frac{\int x\,dA}{A} \tag{c}$$

These equations show that the coordinates of the line of action of the resultant force P are equal to the first moments of the cross-sectional area divided by the area itself. Therefore, these equations are identical to the equations defining the coordinates of the centroid of the area.*

* Centroids of an area are explained in Section C.1 of Appendix C.

We have now arrived at an important general conclusion. In order to have uniform tension or compression in a prismatic bar, *the axial force must act through the centroid of the cross-sectional area.*

Example 2

A prismatic bar with rectangular cross section (20×40 mm) and length $L = 2.8$ m is subjected to an axial tensile force of 70 kN (Fig. 1-3). The measured elongation of the bar is $\delta = 1.2$ mm. Calculate the tensile stress and strain in the bar.

Assuming that the axial forces act at the centroids of the end cross sections, we can use Eq. (1-1) to calculate the normal stress:

$$\sigma = \frac{P}{A} = \frac{70\text{ kN}}{(20\text{ mm})(40\text{ mm})} = 87.5\text{ MPa}$$

The normal strain (from Eq. 1-2) is

$$\epsilon = \frac{\delta}{L} = \frac{1.2\text{ mm}}{2.8\text{ m}} = 429 \times 10^{-6}$$

The quantities σ and ϵ are the tensile stress and strain, respectively, in the longitudinal direction of the bar.

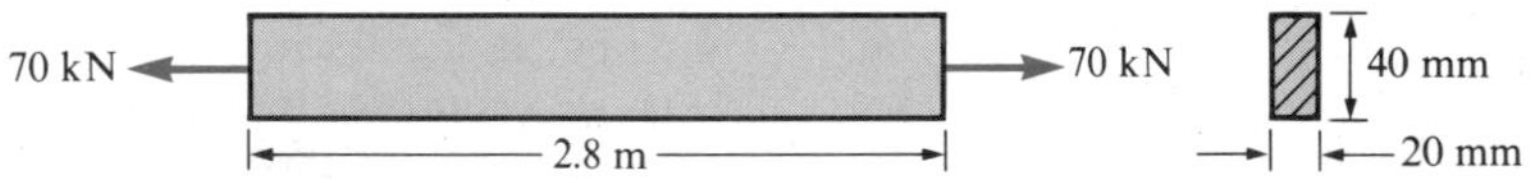

Fig. 1-3 Example 2. Prismatic bar of rectangular cross section

Example 3

A deep-well pump is operated by a crank that rotates in fixed bearings and moves a piston up and down in the well (Fig. 1-4). The pump rod has a diameter $d = 0.6$ in. and a length $L = 320$ ft. It is made of steel having specific weight $\gamma = 490\text{ lb/ft}^3$. The resistance encountered by the piston during the downstroke is 200 lb and during the upstroke is 2000 lb. Determine the maximum tensile and compressive stresses in the pump rod due to the combined effects of the resistance forces and the weight of the rod.

During the downstroke, the resistance of the piston creates a compressive force $C = 200$ lb throughout the length of the rod, and during the upstroke it creates a tensile force $T = 2000$ lb. The weight of the rod produces a tensile force that varies from zero at the lower end of the rod to a maximum at the upper end. The maximum force equals the weight of the entire rod, given by the expression

$$W = \gamma L A \tag{d}$$

where γ is the specific weight of the material, L is the length of the rod, and A is the cross-sectional area. Substituting into this equation, we get

$$W = (490\text{ lb/ft}^3)(320\text{ ft})\left(\frac{\pi}{4}\right)\left(\frac{0.6\text{ in.}}{12\text{ in./ft}}\right)^2 = 308\text{ lb}$$

as the tensile force at the upper end due to the weight.

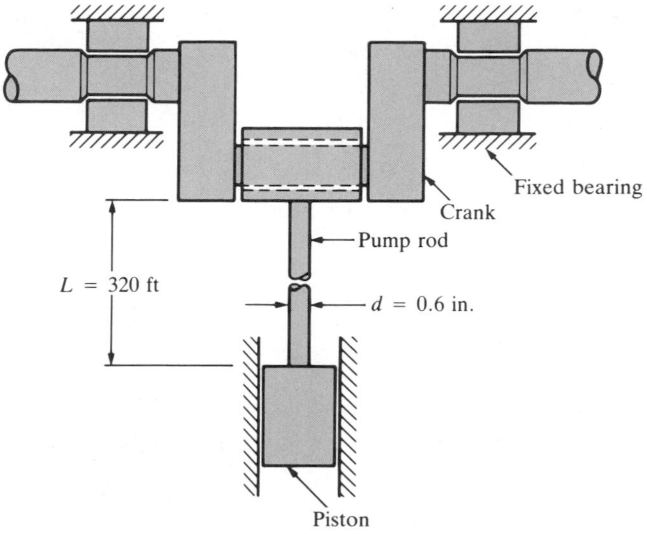

Fig. 1-4 Example 3. Deep-well pump rod

The maximum tensile force in the pump rod occurs during the upstroke at the upper end and is equal to $T + W$, or 2308 lb. The corresponding maximum tensile stress is

$$\sigma_t = \frac{P}{A} = \frac{2308 \text{ lb}}{\pi(0.6 \text{ in.})^2/4} = 8160 \text{ psi}$$

The maximum compressive stress occurs at the lower end during the downstroke:

$$\sigma_t = \frac{P}{A} = \frac{200 \text{ lb}}{\pi(0.6 \text{ in.})^2/4} = 710 \text{ psi}$$

These calculations give the axial stresses in the pump rod due only to the specified loads. Other considerations, such as bending of the pump rod and dynamic effects, have not been taken into account in this example.

1.3 STRESS–STRAIN DIAGRAMS

The mechanical properties of materials used in engineering are determined by tests performed on small specimens of the material. The tests are conducted in materials-testing laboratories equipped with testing machines capable of loading the specimens in a variety of ways, including static and dynamic loading in tension and compression. A **tensile-test machine** is shown in Fig. 1-5. The test specimen is installed between the two large *grips* of the testing machine and then loaded in tension. The automatic control and data-processing systems are at the left.

In order that test results will be comparable, the dimensions of test specimens and the methods of applying loads have been standardized.

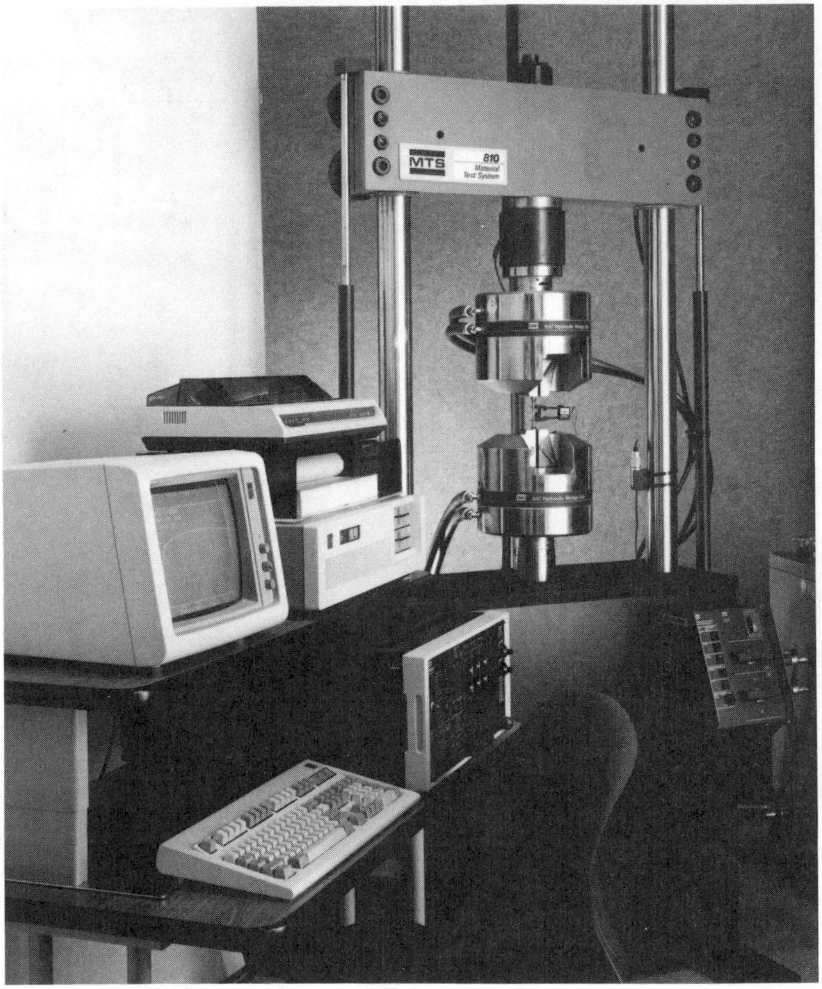

Fig. 1-5 Tensile-test machine with automatic data-processing system. (Courtesy of MTS Systems Corporation)

One of the major standards organizations is the *American Society for Testing and Materials* (*ASTM*), a national technical society that publishes specifications and standards for materials and testing. Other standardizing organizations are the *American Standards Association* (*ASA*) and the *National Bureau of Standards* (*NBS*).

The most common materials test is the **tension test**, in which tensile loads are applied to a cylindrical specimen like the one shown in Fig. 1-6. The ends of the specimen are enlarged where they fit in the grips so that failure will not occur near the grips themselves. (Failure at the ends would not produce the desired information about the material, because the stress distribution near the grips is not uniform and is affected by the stress concentrations.) Instead, failure occurs in the prismatic portion of the specimen where the stress distribution is uniform and the bar is subjected only to pure tension. The figure shows a steel specimen that has just

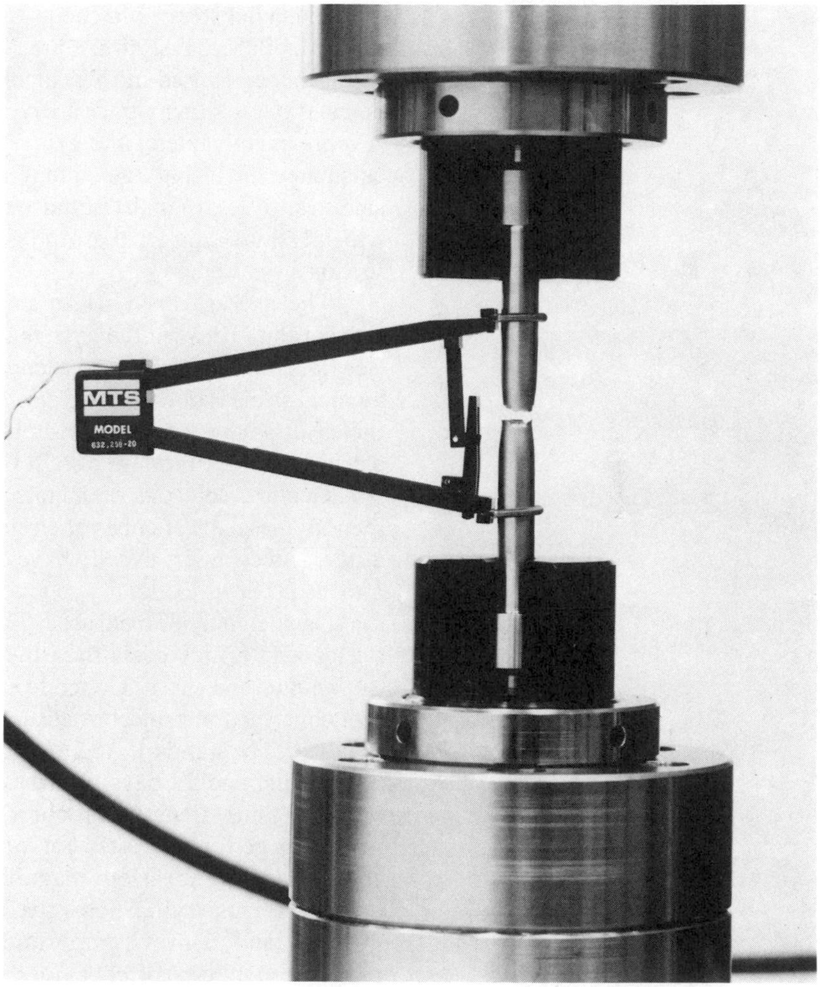

Fig. 1-6 Typical tensile-test specimen with extensometer attached; the specimen has just fractured in tension. (Courtesy of MTS Systems Corporation)

fractured under load. The device at the left, which is attached by two arms to the specimen, is an **extensometer** that measures the elongation during loading.

The ASTM standard tension specimen has a diameter of 0.505 in. and a **gage length** of 2.0 in. between the gage marks, which are the points where the extensometer arms are attached to the specimen (see Fig. 1-6). As the specimen is pulled, the load P is measured and recorded, either automatically or by reading from a dial. The elongation over the gage length is measured simultaneously, either by mechanical gages of the kind shown in Fig. 1-6 or by electrical-resistance strain gages. In a **static test**, the load is applied very slowly and the *rate* of loading need not be measured. However, in a **dynamic test** the rate of loading may be very high and must be measured because it affects the properties of the materials.

The axial stress σ in the test specimen is calculated by dividing the load P by the cross-sectional area A (see Eq. 1-1). When the initial area of the specimen is used in this calculation, the resulting stress is called the **nominal stress** (other names are *conventional stress* and *engineering stress*). A more exact value of the axial stress, known as the **true stress**, can be calculated by using the actual area of the bar, which can become significantly less than the initial area (as shown in Fig. 1-6). The true stress, which is always larger than the nominal stress, is discussed later in this section.

The average axial strain in the bar is found from the measured elongation δ between the gage marks by dividing δ by the gage length L (see Eq. 1-2). If the initial gage length is used (for instance, 2.0 in.), then the **nominal strain** is obtained. Of course, the distance between the gage marks increases as the tensile load is applied. If the actual distance is used in calculating the strain, we obtain the **true strain**, or **natural strain**.

Compression tests of metals are customarily made on small specimens in the shape of cubes or circular cylinders. Cubes are often 2.0 in. on a side, and cylinders usually have diameters of about 1 in. and lengths of 1 to 12 in. Both the load applied by the machine and the shortening of the specimen may be measured. The shortening should be measured over a gage length that is less than the total length of the specimen in order to eliminate end effects. Concrete is tested in compression on every important construction project to ensure that the required strengths have been obtained. The standard ASTM concrete test specimen is 6 in. in diameter, 12 in. long, and 28 days old (the age of concrete is important because concrete gains strength as it cures).

After performing a tension or compression test and determining the stress and strain at various magnitudes of the load, we can plot a diagram of stress versus strain. Such a **stress-strain diagram** is characteristic of the material and conveys important information about the mechanical properties and type of behavior.*

The first material we will discuss is **structural steel**, also known as mild steel or low-carbon steel. Structural steel is one of the most widely used metals, and it is found in buildings, bridges, towers, cranes, machines, and many other types of construction. A stress-strain diagram for a typical structural steel in tension is shown in Fig. 1-7 (not to scale). Strains are plotted on the horizontal axis and stresses on the vertical axis. The diagram begins with a straight line from the origin O to point A, which means that the stress and strain are **proportional**.** Beyond point A, the

* Stress-strain diagrams were originated by Jacob Bernoulli (1654–1705) and J. V. Poncelet (1788–1867); see Ref. 1-4.

** Two variables are said to be **proportional** if their ratio remains constant. Therefore, a proportional relationship may always be represented by a straight line through the origin. A proportional relationship is also a **linear** relationship, but the converse is not true, because a relationship represented by a straight line that does not pass through the origin is linear but not proportional.

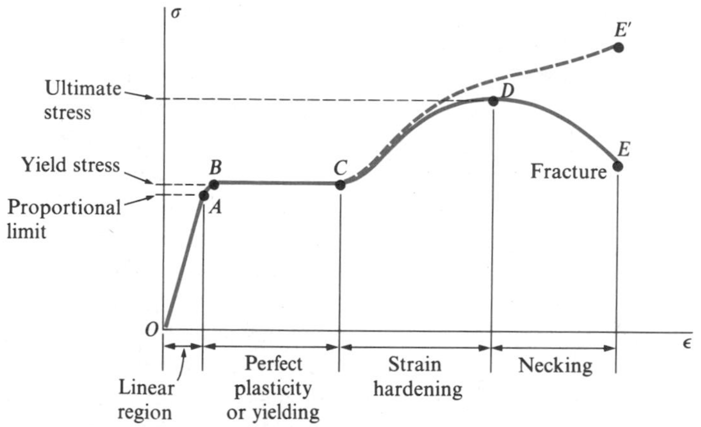

Fig. 1-7 Stress-strain diagram for a typical structural steel in tension (not to scale)

proportionality between stress and strain no longer exists; hence the stress at *A* is called the **proportional limit**. For low-carbon steels, this limit is in the range 30 to 40 ksi, but high-strength steels (with higher carbon content plus other alloys) can have proportional limits of 80 ksi and more. The slope of the straight line from *O* to *A* is called the **modulus of elasticity**. Because strain is nondimensional, this slope has the same units as stress. (Modulus of elasticity is discussed later in Section 1.5.)

With an increase in the load beyond the proportional limit, the strain begins to increase more rapidly for each increment in stress. The stress-strain curve then has a smaller and smaller slope, until, at point *B*, the curve becomes horizontal. Beginning at this point, considerable elongation occurs, with no noticeable increase in the tensile force (from *B* to *C* in Fig. 1-7). This phenomenon is known as **yielding** of the material, and point *B* is called the **yield point**. The corresponding stress is known as the **yield stress** of the steel. In the region from *B* to *C*, the material becomes **perfectly plastic**, which means that it can deform without an increase in the applied load. The elongation of a mild-steel specimen in the perfectly plastic region is typically 10 to 15 times the elongation that occurs between the onset of loading and the proportional limit.

After undergoing the large strains that occur during yielding in the region *BC*, the steel begins to **strain harden**. During strain hardening, the material undergoes changes in its atomic and crystalline structure, resulting in increased resistance of the material to further deformation. Additional elongation now requires an increase in the tensile load, and the stress-strain diagram has a positive slope from *C* to *D*. The load eventually reaches its maximum value, and the corresponding stress (at point *D*) is called the **ultimate stress**. Further stretching of the bar is actually accompanied by a reduction in the load, and **fracture** finally occurs at a point such as *E* in Fig. 1-7.

Lateral contraction of the specimen occurs when it is stretched, resulting in a decrease in the cross-sectional area, as previously mentioned.

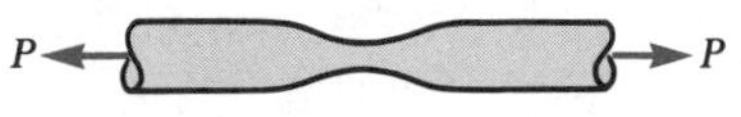

Fig. 1-8 Necking of a bar in tension

The reduction in area is too small to have a noticeable effect on the calculated value of stress up to about point C, but beyond that point the reduction begins to alter the shape of the diagram. As explained before, the true stress is larger than the nominal stress because it is calculated with a smaller area. In the vicinity of the ultimate stress, the reduction in area of the bar becomes clearly visible and a pronounced **necking** of the bar occurs (see Figs. 1-6 and 1-8). If the actual cross-sectional area at the narrow part of the neck is used to calculate the stress, the **true stress-strain curve** will follow the dashed line CE' in Fig. 1-7. The total load the bar can carry does indeed diminish after the ultimate stress is reached (curve DE), but this reduction is due to the decrease in area of the bar and not to a loss in strength of the material itself. In reality, the material withstands an increase in stress up to failure (point E'). For most practical purposes, however, the conventional stress-strain curve $OABCDE$, which is based upon the original cross-sectional area of the specimen and is easy to calculate, provides satisfactory information for use in design.

The diagram in Fig. 1-7 shows the general characteristics of the stress-strain curve for mild steel, but its proportions are not realistic because, as already mentioned, the strain that occurs from B to C may be 15 times the strain occurring from O to A. Furthermore, the strains from C to E are many times greater than those from B to C. Figure 1-9 shows a stress-strain diagram for mild steel drawn to scale. In this figure, the strains from O to A are so small in comparison to the strains from A to E that they cannot be seen, and the initial linear part of the diagram appears to be a vertical line.

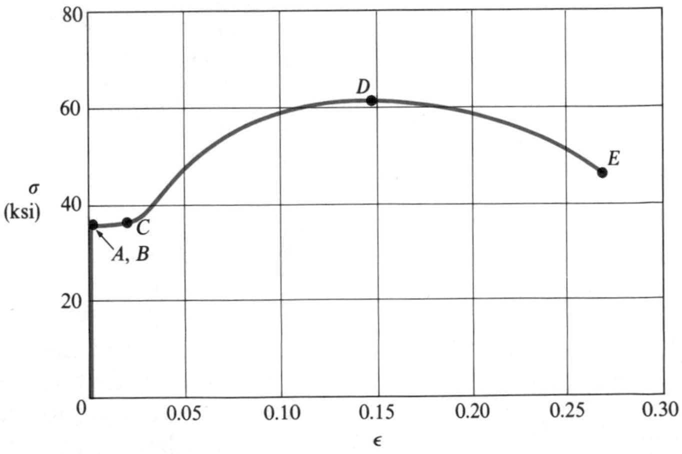

Fig. 1-9 Stress-strain diagram for a typical structural steel in tension (drawn to scale)

The presence of a clearly defined yield point followed by large plastic strains is an important characteristic of mild steel that is sometimes utilized in practical design (see, for instance, the discussion of plastic bending in Section 5.17). Materials that undergo large strains before failure are classified as **ductile**. An advantage of ductility is that visible distortions may occur if the loads become too large, thus providing an opportunity to take remedial action before an actual fracture occurs. Also, ductile materials are capable of absorbing large amounts of energy prior to fracture, as explained in Sections 2.6 and 2.7. Ductile materials include mild steel, aluminum and some of its alloys, copper, magnesium, lead, molybdenum, nickel, brass, bronze, monel metal, nylon, teflon, and many others.

Structural steel is an alloy containing about 0.2% carbon and is classified as a low-carbon steel. With increasing carbon content, steel becomes less ductile but has a higher yield stress and higher ultimate stress. The physical properties of steel are also affected by heat treating, the presence of other metals, and manufacturing processes such as rolling.

Many **aluminum alloys** possess considerable ductility, although they do not have a clearly definable yield point. A typical stress-strain diagram for an aluminum alloy is shown in Fig. 1-10. Aluminum alloys suitable for structural purposes are available with proportional limits in the range 10 to 60 ksi and ultimate stresses in the range 20 to 80 ksi.

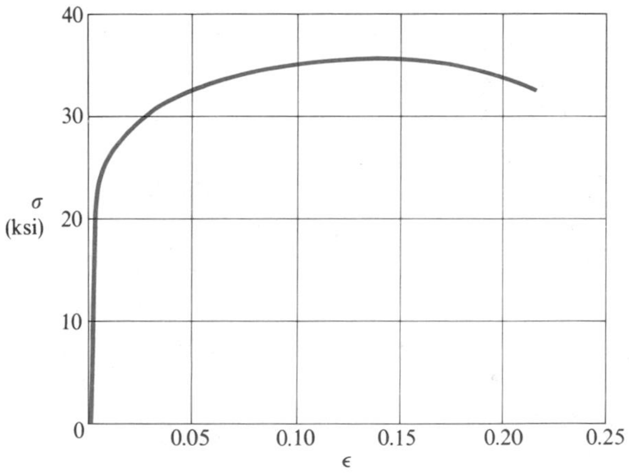

Fig. 1-10 Typical stress-strain diagram for an aluminum alloy

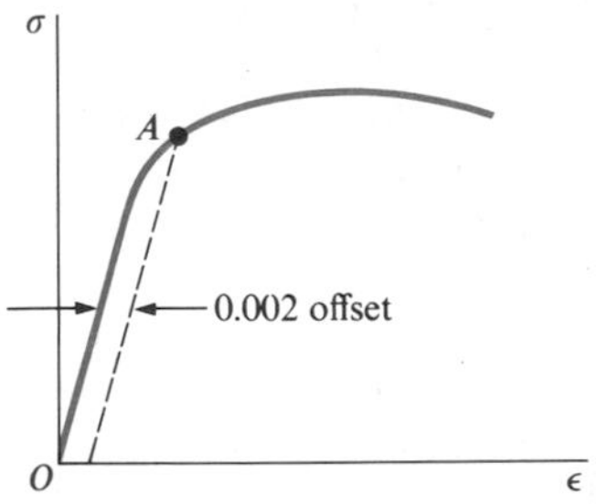

Fig. 1-11 Arbitrary yield stress determined by the offset method

When a material such as aluminum does not have an obvious yield point and yet undergoes large strains after the proportional limit is exceeded, an arbitrary yield stress may be determined by the **offset method**. A line is drawn on the stress-strain diagram parallel to the initial linear part of the curve (see Fig. 1-11) but offset by some standard amount of strain, such as 0.002 (or 0.2%). The intersection of the offset line and the stress-strain curve (point A in the figure) defines the yield stress. Because this stress is determined by an arbitrary rule and is not an inherent physical property of the material, it should be referred to as the **offset yield stress**. For a material such as aluminum, the offset yield stress is slightly above the proportional limit. In the case of structural steel, with its abrupt transition from the linear region to the region of plastic stretching, the offset stress is essentially the same as both the yield stress and the proportional limit.

Rubber maintains a linear relationship between stress and strain up to very large strains in the vicinity of 0.1 or 0.2. The behavior after the proportional limit is exceeded depends upon the type of rubber (see Fig. 1-12). Some kinds of soft rubber continue to stretch enormously without failure. The material eventually offers increasing resistance to the load, and the stress-strain curve turns markedly upward prior to failure. You can easily sense this characteristic behavior by stretching a rubber band.

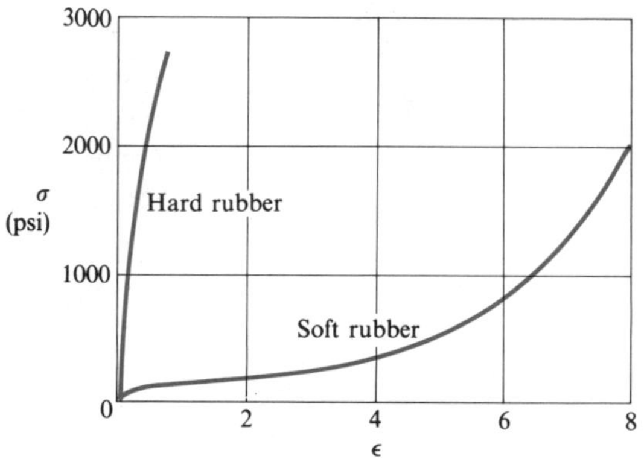

Fig. 1-12 Stress-strain diagrams for two kinds of rubber in tension

The ductility of a material in tension can be characterized by its elongation and by the reduction in area at the cross-section where fracture occurs. The **percent elongation** is defined as follows:

$$\text{Percent elongation} = \frac{L_f - L_o}{L_o}(100) \tag{1-3}$$

in which L_o is the original gage length and L_f is the distance between the gage marks at fracture. Because the elongation is not uniform over the length of the specimen but is concentrated in the region of necking, the percent elongation depends upon the gage length. Therefore, when stating the percent elongation, the gage length should also be given. For a 2 in. gage length, steel may have an elongation in the range of 10% to 40%, depending upon composition; for structural steel, values of 25% or 30% are common. In the case of aluminum alloys, the elongation varies from 1% to 45%, depending upon composition and treatment.

The **percent reduction in area** measures the amount of necking that occurs and is defined as follows:

$$\text{Percent reduction in area} = \frac{A_o - A_f}{A_o}(100) \tag{1-4}$$

in which A_o is the original cross-sectional area and A_f is the final area at the fracture section. For ductile steels, the reduction is about 50%.

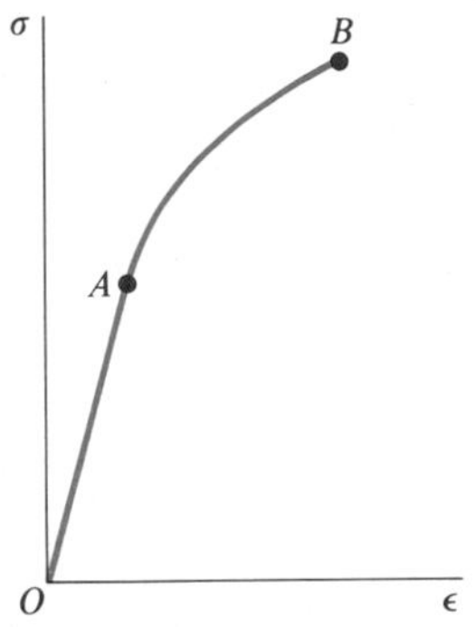

Fig. 1-13 Typical stress-strain diagram for a brittle material

Materials that fail in tension at relatively low values of strain are classified as **brittle** materials. Examples are concrete, stone, cast iron, glass, ceramic materials, and many common metallic alloys. These materials fail with only little elongation after the proportional limit (point A in Fig. 1-13) is exceeded, and the fracture stress (point B) is the same as the ultimate stress. High-carbon steels behave in a brittle manner; they may have a very high yield stress (over 100 ksi in some cases), but fracture occurs at an elongation of only a few percent.

Ordinary **glass** is a nearly ideal brittle material, because it exhibits almost no ductility whatsoever. The stress-strain curve for glass in tension is essentially a straight line, with failure occurring before any yielding takes place. The ultimate stress is about 10,000 psi for certain kinds of plate glass, but great variation exists, depending upon the type of glass, the size of the specimen, and the presence of microscopic defects. **Glass fibers** can develop enormous strengths, and ultimate stresses over 1,000,000 psi have been attained.

Plastics are being used increasingly for structural purposes and thousands of different kinds are available. Light weight, resistance to corrosion, and good electrical insulation properties have made plastics important in many applications. Their mechanical properties vary tremendously, with some plastics being brittle and others ductile. When designing with plastics it is important to realize that their properties are greatly affected by both temperature changes and the passage of time. For instance, the ultimate tensile stress of some plastics is cut in half by raising the temperature from 50 °F to 120 °F. Also, a loaded plastic may stretch gradually over time until it is no longer serviceable. For example, a specimen of polyvinyl chloride subjected to a load that initially produces a strain ϵ equal to 0.005 may have that strain doubled after a week, even though the load has remained constant. (This phenomenon, known as *creep*, is discussed in the next section.) Ultimate tensile stresses for plastics are generally in the range 2,000 to 50,000 psi and specific weights vary from 55 to 90 lb/ft^3. Nylon, a familiar plastic, has an ultimate stress in the 6,000 to 10,000 psi range and weighs 70 lb/ft^3, about 12% heavier than water. Because of its light weight, the strength-to-weight ratio for nylon is about the same as for structural steel.

A **filament-reinforced material** consists of a base material (or matrix) in which are embedded high-strength filaments, fibers, or whiskers. The resulting composite material has much greater strength than the base material. As an example, the use of glass fibers can more than double the strength of a plastic matrix. Composites are widely used in aircraft, rockets, boats, and automobiles where high strength and light weight are needed.

Stress-strain diagrams for **compression** have different shapes from those for tension. Ductile metals such as steel, aluminum, and copper have proportional limits in compression very close to those in tension; and therefore the initial regions of their compression stress-strain diagrams are very similar to the tension diagrams. However, when yielding begins, the behavior is quite different. In a tension test, the specimen is stretched, necking may occur, and fracture ultimately takes place. When a small specimen of ductile material is compressed, it begins to bulge outward on the sides and become barrel shaped. With increasing load, the specimen is flattened out, thus offering increased resistance to further shortening (which means the stress-strain curve goes upward). These characteristics are illustrated in Fig. 1-14, which shows a compression stress-strain diagram for copper.

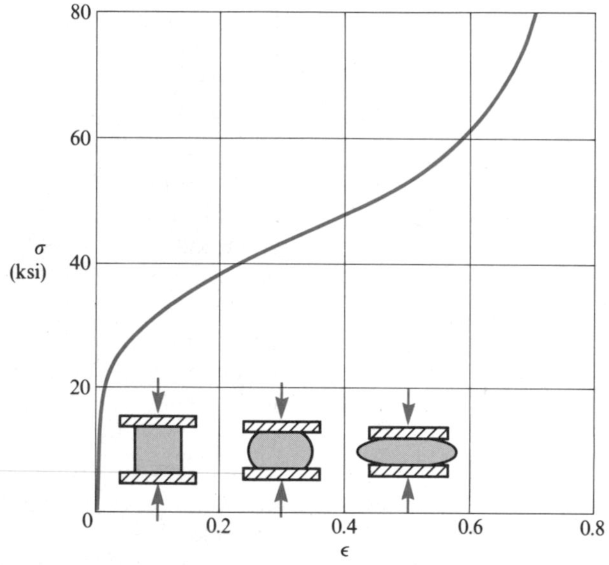

Fig. 1-14 Compression stress-strain diagram for copper

Brittle materials in compression typically have an initial linear region followed by a region in which the shortening increases at a higher rate than does the load. Thus, the compression stress-strain diagram has a shape that is similar to the shape of the tensile diagram. However, brittle materials usually reach much higher ultimate stresses in compression than in tension. Also, unlike ductile materials in compression, brittle materials actually fracture or break at the maximum load. The tension and compression stress-strain diagrams for a particular type of cast iron are given in Fig. 1-15. Curves for other brittle materials, such as concrete and stone, have similar shapes but quite different numerical values.

A table of important **mechanical properties** for various materials is given in Appendix H. These data are suitable for solving problems in this book. However, properties and stress-strain curves vary greatly, even for the same material, because of different manufacturing processes, chemical composition, internal defects, temperature, and many other factors. Hence, any data obtained from general tables such as those in Appendix H should be considered as typical, but not necessarily suitable for a specific application.

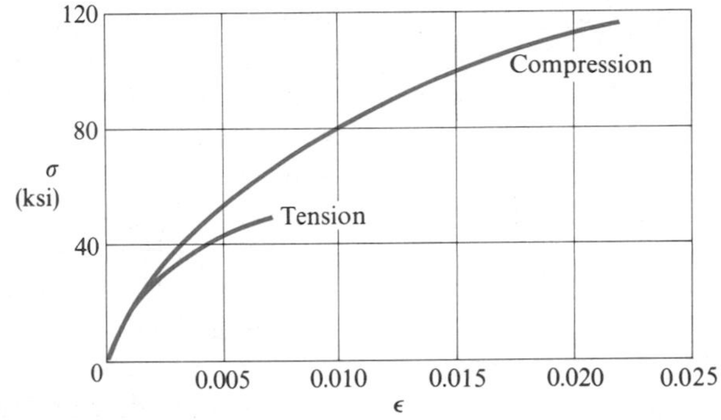

Fig. 1-15 Stress-strain diagrams for a cast iron in tension and compression

1.4 ELASTICITY AND PLASTICITY

The stress-strain diagrams described in the preceding section illustrate the behavior of various materials as they are **loaded** statically in tension or compression. Now let us consider what happens when the load is slowly removed and the material is **unloaded**. Assume, for instance, that we apply a load to a tensile specimen so that the stress and strain go from O to A on the stress-strain curve in Fig. 1-16a. Suppose further that when the load is removed, the material follows exactly the same curve back to the origin O. This property of a material, by which it returns to its original dimensions during unloading, is called **elasticity**, and the material itself is said to be **elastic**. Note that the stress-strain curve from O to A need not be linear in order for the material to be elastic.

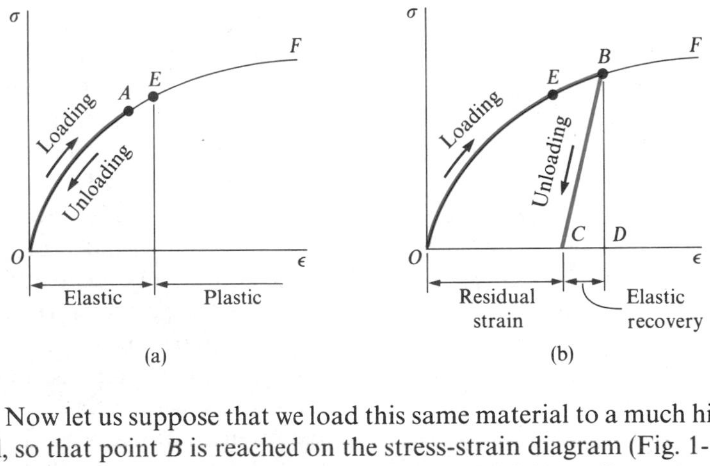

Fig. 1-16 Stress-strain diagrams illustrating: (a) elastic behavior, and (b) partially elastic behavior

Now let us suppose that we load this same material to a much higher level, so that point B is reached on the stress-strain diagram (Fig. 1-16b). In this case, when unloading occurs, the material follows line BC on the diagram. This unloading line typically is parallel to the initial portion of the loading curve; that is, line BC is parallel to a tangent to the stress-strain curve at O. When point C is reached, the load has been entirely removed, but a **residual strain**, or **permanent strain**, OC remains in the material. The corresponding residual elongation of the bar is called the **permanent set**. Of the total strain OD developed during loading from O to B, the strain CD has been recovered elastically and the strain OC remains as a permanent strain. Thus, during unloading the bar returns partially to its original shape, and the material is said to be **partially elastic**.

When a bar is being tested, the load can be increased from zero to some small selected value and then removed. If there is no permanent set (that is, if the elongation of the bar returns to zero), then the material is elastic up to the stress represented by the selected value of the load. This process of loading and unloading can be repeated for successively higher values of load. Eventually, a stress will be reached such that not all the strain is recovered during unloading. By this procedure, it is possible to determine the stress at the upper limit of the elastic region; for instance, it could be the stress at point E in Figs. 1-16a and b. This stress is known as the **elastic limit** of the material.

Many materials, including most metals, have linear regions at the beginning of their stress-strain curves (see Figs. 1-7 and 1-10). As explained in Section 1.3, the upper limit of this linear region is defined by the proportional limit. Usually the elastic limit is slightly above, or nearly the same as, the proportional limit. Hence, for many materials the two limits are assigned the same numerical value. In the case of mild steel, the yield stress is also very close to the proportional limit, so that for practical purposes the yield stress, the elastic limit, and the proportional limit are assumed to be equal. Of course, this situation does not hold for all materials. Rubber provides the outstanding example of a material that is elastic far beyond the proportional limit.

The characteristic of a material by which it undergoes inelastic strains beyond those at the elastic limit is known as **plasticity**. Thus, on the stress-strain curve in Fig. 1-16a, we have an elastic region followed by a plastic region. When large deformations occur in a ductile material loaded into the plastic region, the material is said to undergo **plastic flow**.

If the material remains within the elastic range, it can be loaded, unloaded, and loaded again without significantly changing the behavior. However, when loaded into the plastic range, the internal structure of the material is altered and its properties change. For instance, we have already observed that a permanent strain exists in the specimen after unloading from the plastic region (Fig. 1-16b). Now suppose that the material is reloaded after such an unloading (Fig. 1-17). The new loading begins at point C on the diagram and continues upward to B, the point at which unloading began during the first loading cycle. The material then follows the original stress-strain diagram toward point F. During the second loading, the material behaves in a linearly elastic manner from C to B; hence the material has a higher proportional limit than before. Thus, by stretching a material such as steel or aluminum into the inelastic or plastic range, the linearly elastic region is increased. However, the ductility

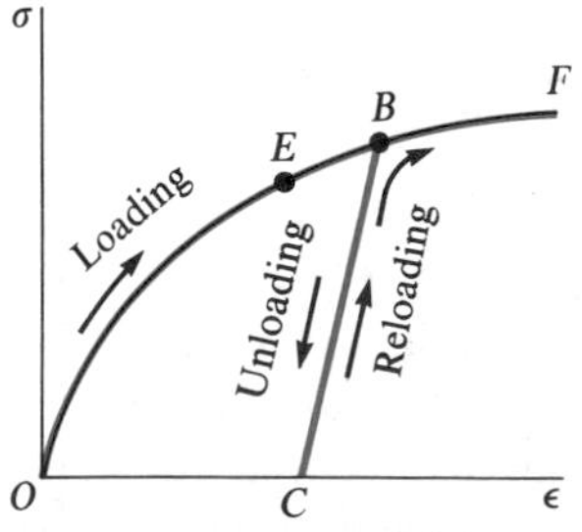

Fig. 1-17 Reloading of a material and raising of the elastic limit

is reduced because the amount of yielding from B to F is less than from E to F.*

CREEP

The stress-strain diagrams previously described were obtained from tension tests involving only static loading of the specimens, and so the passage of time did not enter into our discussions. However, some materials develop additional strains over long periods of time and are said to **creep**. This phenomenon can manifest itself in a variety of ways. For instance, let us suppose that a vertical bar (Fig. 1-18a) is loaded slowly by a force P, so that the bar elongates by an amount δ_0. Let us assume that this loading and the corresponding elongation take place during a time interval of duration t_0 (Fig. 1-18b). Subsequent to time t_0, the load remains constant. However, due to creep, the bar may gradually lengthen, as shown in Fig. 1-18b, even though the load does not change. This behavior occurs with many materials, although sometimes the change is too small to be of concern.

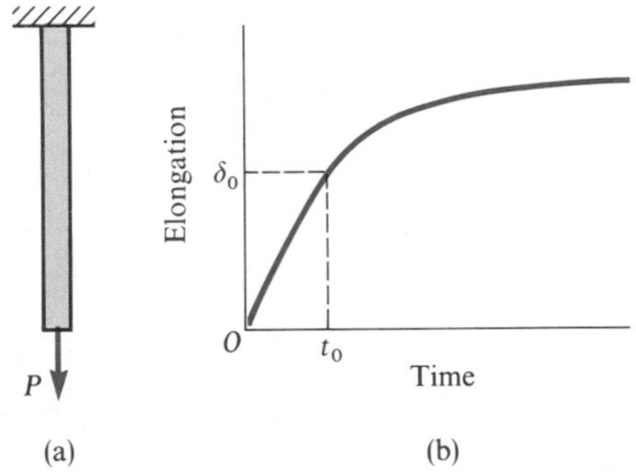

Fig. 1-18 Creep in a bar under constant load

As a second example of creep, consider a wire that is stretched between two immovable supports so that it has an initial tensile stress σ_0 (Fig. 1-19a). Again, we will denote the time during which the wire is loaded initially as t_0 (Fig. 1-19b). With the elapse of time, the stress in the wire gradually diminishes, eventually reaching a constant value, even though the supports at the ends of the wire do not move. This process, which is a manifestation of creep, is called **relaxation** of the material.

Creep is usually more important at high temperatures than at ordinary temperatures, and therefore it should always be considered in the design of engines, furnaces, and other structures that operate at elevated temperatures for long periods of time. However, materials such as steel, concrete, and wood will creep slightly even at atmospheric temperatures.

* The study of material behavior under various environmental and loading conditions is an important branch of applied mechanics. For more detailed engineering information about materials, consult a textbook devoted solely to this subject.

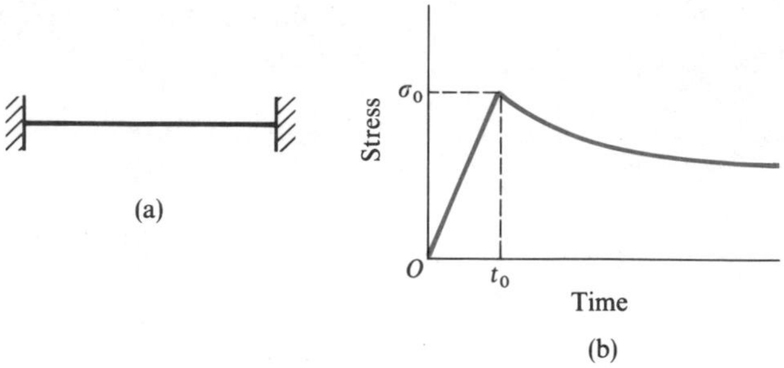

Fig. 1-19 Relaxation of stress in a wire under constant strain

For example, creep of concrete can create "waves" in bridge decks because of sagging between the supports over long periods of time. One remedy is to construct the deck with an upward **camber**, which is an initial displacement above the horizontal, so that, when creep occurs, the spans lower to the level position.

1.5 LINEAR ELASTICITY AND HOOKE'S LAW

Most structural materials have an initial region on the stress-strain diagram in which the material behaves both elastically and linearly. An example is the region from the origin O up to the proportional limit at point A on the stress-strain curve for structural steel (see Fig. 1-7). Other examples are the regions below *both* the proportional limits and the elastic limits on the diagrams of Figs. 1-10 through 1-15. When a material behaves elastically and also exhibits a linear relationship between stress and strain, it is said to be **linearly elastic**. This type of behavior is extremely important in engineering because many structures and machines are designed to function at low levels of stress in order to avoid permanent deformations from yielding or plastic flow. Linear elasticity in the initial region of the stress-strain diagram is a property of many solid materials, including metals, wood, concrete, plastics, and ceramics.

The linear relationship between stress and strain for a bar in simple tension or compression can be expressed by the equation

$$\sigma = E\epsilon \tag{1-5}$$

in which E is a constant of proportionality known as the **modulus of elasticity** for the material. The modulus of elasticity is the slope of the stress-strain diagram in the linearly elastic region, and its value depends upon the particular material being used. The units of E are the same as the units of stress, inasmuch as strain is dimensionless. Typical units of E are psi or ksi in USCS units and pascals (or multiples thereof) in SI units.

The equation $\sigma = E\epsilon$ is commonly known as **Hooke's law**, named for the famous English scientist Robert Hooke (1635–1703). Hooke was the first person to investigate the elastic properties of materials, and he tested such diverse materials as metal, wood, stone, bones, and sinews. He measured the stretching of long wires supporting weights and observed that the elongations "always bear the same proportions one to the other that the weights do that make them" (Refs. 1-5 and 1-6). Thus, Hooke established the linear relationship between the applied load and the resulting elongation.

Equation (1-5) is actually a very limited version of Hooke's law because it relates only to the normal stresses and strains developed in simple tension or compression of a bar. To deal with more complicated states of stress, such as those found in most structures and machines, the complete equations of Hooke's law are required (see Chapter 6).

The modulus of elasticity E has relatively large values for materials that are very stiff, such as structural metals. Steel has a modulus of approximately 30,000 ksi or 200 GPa; for aluminum, E equals approximately 10,600 ksi or 70 GPa. More flexible materials have a lower modulus; typical values for plastics range from 100 to 2000 ksi (or from 0.7 to 14 GPa). Some representative values of E are listed in Table H-2, Appendix H. For most materials, the value of E in compression is the same as in tension.

The modulus of elasticity is often called **Young's modulus**, after another English scientist, Thomas Young (1773–1829). In connection with an investigation of tension and compression of prismatic bars, Young introduced the idea of a "modulus of the elasticity." However, his modulus was not the same as the one in use today, because it involved properties of the bar as well as of the material (Refs. 1-7 and 1-8).

POISSON'S RATIO

When a prismatic bar is loaded in tension, the axial elongation is accompanied by **lateral contraction** (normal to the direction of the applied load). This change in shape is pictured in Fig. 1-20, in which the dashed lines represent the shape before loading and the solid lines give the shape after loading. Lateral contraction is readily seen in a stretched rubber band, but in metals the changes in lateral dimensions are usually too small to be visible. However, they can easily be detected with measuring devices.

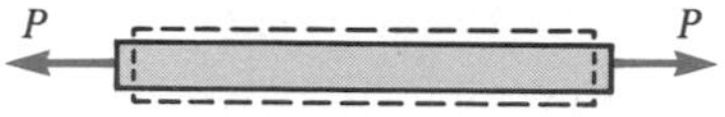

Fig. 1-20 Axial elongation and lateral contraction of a bar in tension

The **lateral strain** at a point in a bar is proportional to the axial strain at that same point if the material is linearly elastic. However, in order for the lateral strains to be the same throughout the bar, additional conditions must be met. First, the material must be **homogeneous**, which means it has the same composition throughout the body and hence the same elastic properties at every point. We already assumed that the material was homogeneous (see Section 1.2) in order that the stress and strain would be uniform throughout the bar. However, the elastic properties need not be the same in all directions for the material to be homogeneous. For instance, the modulus of elasticity could be different in the axial and lateral directions. Therefore, a second condition is that the elastic properties must be the same in all directions perpendicular to the longitudinal axis*. When these two conditions are met, as is normally the case with structural materials, the lateral strains in a bar subjected to uniform tension will be the same at every point in the bar and will be the same in all lateral directions.

The ratio of the strain in the lateral direction to the strain in the axial direction is known as **Poisson's ratio** and is denoted by the Greek letter ν (nu); thus,

$$\nu = -\frac{\text{lateral strain}}{\text{axial strain}} \tag{1-6}$$

For a bar in tension, the lateral strain represents a decrease in width (negative strain) and the axial strain represents elongation (positive strain). For compression we have the opposite situation, with the bar becoming shorter (negative axial strain) and wider (positive lateral strain). Therefore, Poisson's ratio has a positive value for most materials.

Poisson's ratio is named for the famous French mathematician Siméon Denis Poisson (1781–1840), who attempted to calculate this ratio by a molecular theory of materials (Ref. 1-9). For isotropic materials, Poisson found $\nu = \frac{1}{4}$. However, more recent calculations based upon a model of atomic structure give $\nu = \frac{1}{3}$. Both of these values are close to actual measured values, which are in the range 0.25 to 0.35 for many metals and other materials. Materials with an extremely low value of Poisson's ratio include cork, for which ν is practically zero,** and concrete, for which ν is about 0.1 or 0.2. A theoretical upper limit for Poisson's ratio is 0.5, as explained in the following subsection on volume changes. Rubber comes close to this limiting value. A table of Poisson's ratios for various materials in the elastic range is given in Appendix H (see Table H-2). For most practical purposes, the value of ν is assumed to be the same in both tension and compression.

The lateral contraction of a bar in tension, or the lateral expansion of a bar in compression, is an illustration of how a normal strain can exist in

* Materials having the same properties in *all* directions are said to be **isotropic**. If the properties differ in various directions, the material is **anisotropic** (or **aeolotropic**). A special case of anisotropy occurs when the properties in a particular direction are the same throughout the material and the properties in all directions perpendicular to that direction are the same (but different from the first properties); then the material is classified as **orthotropic**. Fiber-reinforced plastics and concrete reinforced with parallel steel bars are examples of orthotropic materials.

** Hence its suitability as a bottle stopper.

a certain direction without a normal stress acting in the same direction. Another instance in which this situation may occur is thermal strain (see Section 2.4).

VOLUME CHANGE

Because the dimensions of a bar in tension or compression are changed when the load is applied (see Fig. 1-20), the volume of the bar changes too. The change in volume can be calculated from the axial and lateral strains. Let us take a small element of isotropic material cut from a bar in tension (Fig. 1-21). The original shape of the element (shown by the dashed lines) is a rectangular parallelepiped having sides of lengths a, b, and c in the x, y, and z directions, respectively.* The x axis is taken in the longitudinal direction of the bar, which is the direction of the normal stresses σ produced by the axial forces. The final shape of the element is shown by solid lines. The elongation of the element in the direction of loading is $a\epsilon$, where ϵ is the axial strain. Because the lateral strains are $-\nu\epsilon$ (see Eq. 1-6), the lateral dimensions decrease by $b\nu\epsilon$ and $c\nu\epsilon$ in the y and z directions, respectively. Thus, the final dimensions of the element are $a(1+\epsilon)$, $b(1-\nu\epsilon)$, and $c(1-\nu\epsilon)$, and the final volume is

$$V_f = abc(1+\epsilon)(1-\nu\epsilon)(1-\nu\epsilon) \tag{a}$$

When this expression is expanded, we obtain terms involving the square and cube of ϵ. Because ϵ is very small compared to unity, its square and cube are negligible in comparison to ϵ itself and may be dropped from the

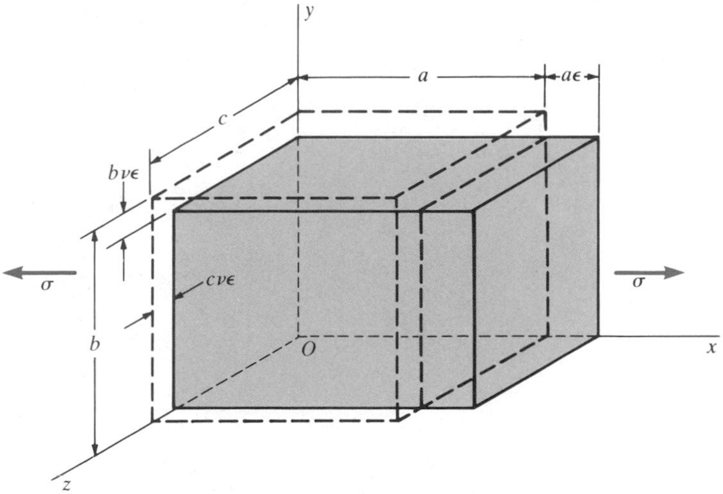

Fig. 1-21 Change of shape of an element from a bar in tension

* A **parallelepiped** is a prism whose bases are parallelograms; thus, a parallelepiped has six faces, each of which is a parallelogram. Opposite faces are parallel and identical parallelograms. A rectangular parallelepiped has all faces in the form of rectangles.

expression. Therefore, the final volume of the element is

$$V_f = abc(1 + \epsilon - 2\nu\epsilon) \tag{b}$$

and the change in volume is

$$\Delta V = V_f - V_o = abc\epsilon(1 - 2\nu)$$

where V_o is the original volume abc. The **unit volume change** e is defined as the change in volume divided by the original volume, or

$$e = \frac{\Delta V}{V_o} = \epsilon(1 - 2\nu) = \frac{\sigma}{E}(1 - 2\nu) \tag{1-7}$$

The quantity e is also known as the **dilatation**. Equation (1-7) can be used to calculate the increase in volume of a bar in tension provided the axial strain ϵ (or stress σ) and Poisson's ratio ν are known. This equation may also be used for compression, in which case ϵ is a negative strain and the volume of the bar decreases.

From Eq. (1-7) we see that the maximum possible value of ν for ordinary materials is 0.5, because any larger value means that the volume decreases when the material is stretched, which seems physically unlikely. As already pointed out, for most materials ν is about $\frac{1}{4}$ or $\frac{1}{3}$ in the linear elastic region, which means that the unit volume change is in the range 0.3ϵ to 0.5ϵ. In the perfectly plastic region of behavior, no volume changes occur, so Poisson's ratio may be taken as 0.5.

Example

A prismatic bar of circular cross section is loaded by tensile forces $P = 85$ kN (Fig. 1-22). The bar has length $L = 3.0$ m and diameter $d = 30$ mm. It is made of aluminum alloy (2014-T6) with modulus of elasticity $E = 70$ GPa and Poisson's ratio $\nu = \frac{1}{3}$. Calculate the elongation δ, the decrease in diameter Δd, and the increase in volume ΔV of the bar.

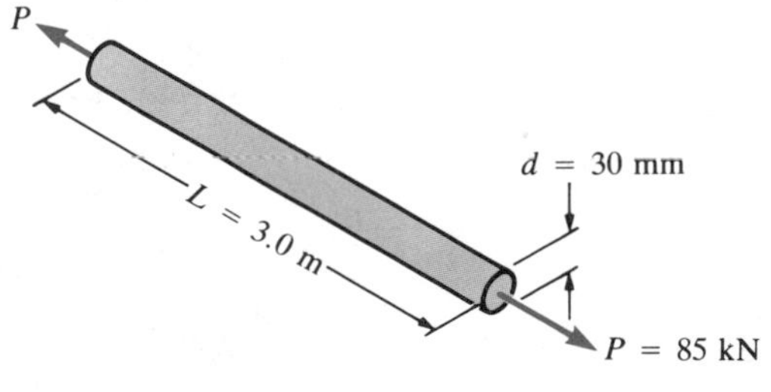

Fig. 1-22 Example

The longitudinal stress σ in the bar can be obtained from the equation

$$\sigma = \frac{P}{A} = \frac{85 \text{ kN}}{\pi(30 \text{ mm})^2/4} = 120.3 \text{ MPa}$$

Because this stress is well below the yield stress (see Table H-3, Appendix H), the material behaves linearly elastically and the axial strain may be found from Hooke's law:

$$\epsilon = \frac{\sigma}{E} = \frac{120.3 \text{ MPa}}{70 \text{ GPa}} = 0.001718$$

The total elongation is

$$\delta = \epsilon L = (0.001718)(3.0 \text{ m}) = 5.15 \text{ mm}$$

The lateral strain is obtained from Poisson's ratio:

$$\epsilon_{\text{lateral}} = -\nu\epsilon = -\frac{1}{3}(0.001718) = -0.0005726$$

The decrease in diameter equals numerically the lateral strain times the diameter:

$$\Delta d = \epsilon_{\text{lateral}} d = (0.0005726)(30 \text{ mm}) = 0.0172 \text{ mm}$$

Finally, the change in volume is calculated from Eq. (1-7):

$$\Delta V = V_o \epsilon(1 - 2\nu)$$

$$= \left(\frac{\pi}{4}\right)(30 \text{ mm})^2(3.0 \text{ m})(0.001718)\left(1 - \frac{2}{3}\right) = 1210 \text{ mm}^3$$

Because the bar is in tension, ΔV represents an increase in volume.

1.6 SHEAR STRESS AND STRAIN

In the preceding sections we dealt with the effects of normal stresses produced by axial loads acting on bars. We shall now consider a different kind of stress, known as a **shear stress**, that acts **parallel** or **tangential** to the surface of the material.

As an example of a practical situation in which shear stresses are present, consider the bolted connection shown in Fig. 1-23a. This connection consists of a flat bar *A*, a clevis *C*, and a bolt *B* that passes through holes in the bar and clevis. Under the action of the tensile loads *P*, the bar and clevis will press against the bolt in **bearing**, and contact stresses, called **bearing stresses**, will be developed against the bolt. In addition, the bar and clevis tend to shear off the bolt, and this tendency is resisted by **shear stresses** in the bolt.

To show more clearly the action of these stresses, let us look at the connection in a side view (Fig. 1-23b). Then a free-body diagram of the bolt (Fig. 1-23c) shows the bearing stresses exerted by the clevis and the bar against the bolt; the stresses on the left-hand side (labeled 1 and 3) are from the clevis and those on the right-hand side (labeled 2) are from the bar. The actual distribution of the bearing stresses is difficult to determine,

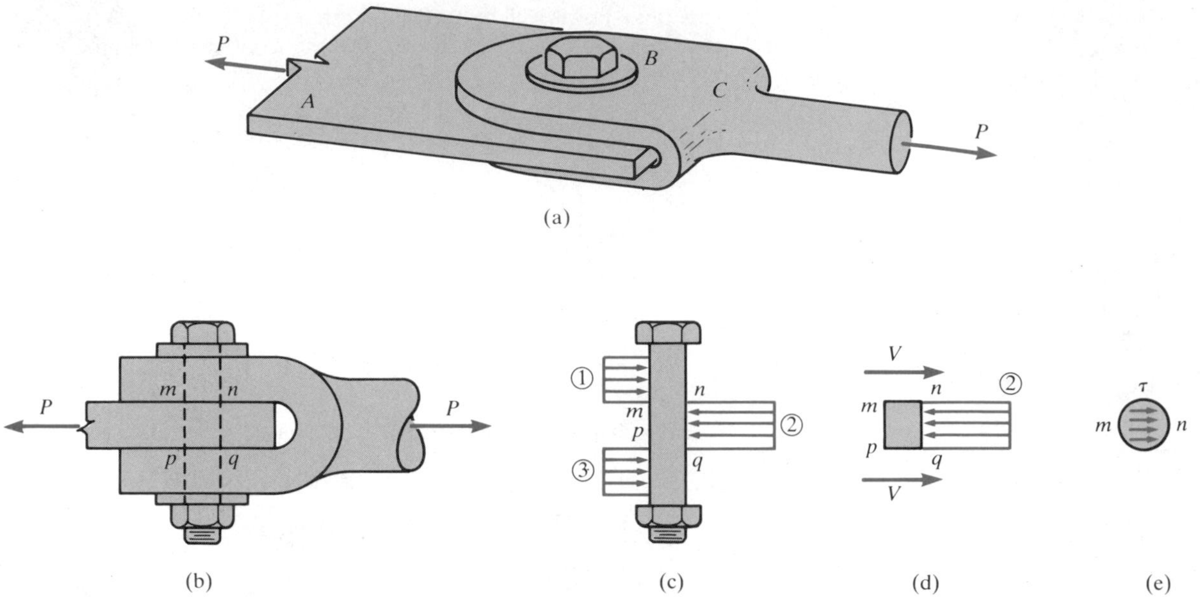

Fig. 1-23 Bolted connection with bolt subjected to bearing stresses and direct shear stresses

so it is usual to assume that they are uniformly distributed. Based upon the assumption of uniform distribution, we can calculate an average bearing stress by dividing the total bearing force by the bearing area. This area is taken as the projected area of the curved bearing surface, which in this case is a rectangle. For instance, consider the bearing stresses labeled 1. These stresses act against a projected area in the shape of a rectangle having a height equal to the thickness of the clevis and a width equal to the diameter of the bolt. The same applies to the stresses labeled 3. The bearing stresses between the flat bar A and the bolt (labeled 2) act against an area of height equal to the thickness of the bar and width equal to the bolt diameter.

The free-body diagram of Fig. 1-23c shows that there is a tendency to shear the bolt along cross sections mn and pq. From a free-body diagram of the portion $mnpq$ of the bolt (see Fig. 1-23d), we see that **shear forces** V must act over the cut surfaces of the bolt. In this particular example there are two planes of shear, and so the bolt is said to be in **double shear**. Each of the shear forces V is equal to $P/2$. These shear forces are actually the resultants of the shear stresses distributed over the cross-sectional areas of the bolt. The circular cross section at section mn is shown in Fig. 1-23e, and the shear stresses acting on that section are represented by small arrows. The exact distribution of these stresses is not known, but they are highest near the center and become zero at certain locations on the edges. Shear stresses are customarily denoted by the Greek letter τ (tau).

The average shear stress on the cross section of the bolt is obtained by dividing the total shear force V by the area A over which it acts:

$$\tau_{aver} = \frac{V}{A} \tag{1-8}$$

In the example shown in Fig. 1-23, the shear force V is $P/2$ and the area A is the cross-sectional area of the bolt. From Eq. (1-8) we see that shear stresses, like normal stresses, represent intensity of force, or force per unit of area. Thus, the units of shear stress are the same as those for normal stress, namely, psi or ksi in USCS units and pascals in SI units.

The loading arrangement shown in Fig. 1-23 is an example of **direct shear**, or **simple shear**, in which the shear stresses are created by a direct action of the forces in trying to cut through the material. Direct shear arises in the design of bolts, pins, rivets, keys, welds, and glued joints. Shear stresses also arise in an indirect manner when members are subjected to tension, torsion, and bending, as discussed in later chapters.

To obtain a more complete picture of the action of shear stresses, let us consider a small element of material in the form of a rectangular parallelepiped having sides of lengths a, b, and c in the x, y, and z directions, respectively (Fig. 1-24a). The front and rear faces of the element are assumed to be free of any stresses. Now suppose that a shear stress τ is distributed uniformly over the top face of the element. Then, in order for the element to be in equilibrium in the x direction, an equal but oppositely directed shear stress must act on the bottom face. Note that the total shear force on the top face is τac and that this force is balanced by the equal and

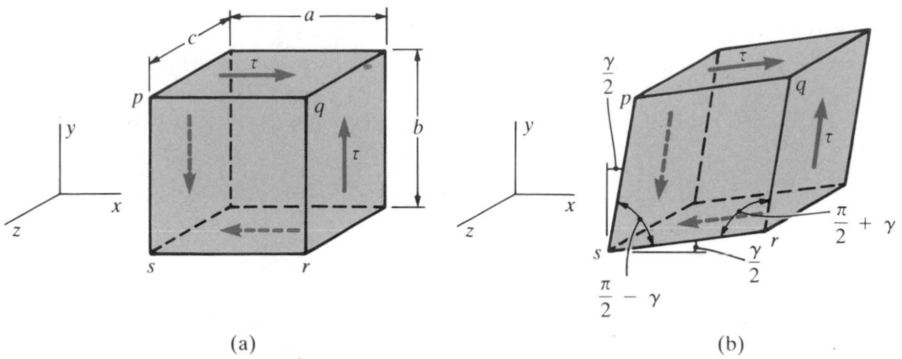

Fig. 1-24 Shear stress and strain

oppositely directed shear force on the bottom face. These two forces form a couple having a moment about the z axis of magnitude τabc, clockwise in the figure. Equilibrium of the element requires that this moment be balanced by an equal and opposite moment resulting from shear stresses acting on the side faces of the element. Denoting the stresses on the side faces as τ_1, we see that the vertical shear forces are $\tau_1 bc$ and that they form a counterclockwise couple of moment $\tau_1 abc$. It follows from moment equilibrium that $\tau_1 = \tau$, and hence the magnitudes of the shear stresses on the four faces of the element are equal, as shown in Fig. 1-24a. Thus, we have reached the following conclusions:

1. Shear stresses on opposite faces of an element are equal in magnitude and opposite in direction.
2. Shear stresses on perpendicular faces of an element are equal in magnitude and have directions such that both stresses point toward, or both point away from, the line of intersection of the faces.

These conclusions concerning the shear stresses are valid even when normal stresses also act on the faces of the element.

An element subjected to shear stresses only, as pictured in Fig. 1-24a, is said to be in **pure shear**.* Under the action of these shear stresses, the material is deformed, resulting in **shear strains**. In order to visualize these strains, we note first that the shear stresses have no tendency to elongate or to shorten the element in the x, y, and z directions; in other words, the lengths of the sides of the element do not change. Instead, the shear stresses produce a change in the shape of the element, as shown in Fig. 1-24b. The original element is deformed into an oblique parallelepiped,** and the front face $pqrs$ of the element becomes a rhomboid.*** The angles between the faces at points q and s, which were $\pi/2$ before deformation, are reduced by a small angle γ to $\pi/2 - \gamma$ (see Fig. 1-24b). At the same time, the angles at p and r are increased to $\pi/2 + \gamma$. The angle γ is a measure of the distortion, or change in shape, of the element and is called the **shear strain**. The unit of shear strain is the radian.

Shear stresses and strains having the directions shown in Fig. 1-24 are assumed to be positive. To make this sign convention clear, we will refer to the faces oriented toward the positive directions of the axes as the positive faces of the element. In other words, a positive face has its outward normal directed in the positive direction of a coordinate axis. The opposite faces are negative faces. Thus, in Fig. 1-24a, the right-hand, top, and front faces are the positive x, y, and z faces, respectively, and the opposite faces are the negative x, y, and z faces. Using this terminology, we may state the **sign convention for shear stresses** as follows: A shear stress acting on a positive

* Pure shear is discussed in greater detail in Section 3.4.

** An **oblique angle** can be either acute or obtuse, but it is *not* a right angle.

*** A **rhomboid** is a parallelogram with oblique angles and all four sides *not* equal. (If the sides a and b of the element are equal, then face $pqrs$ is a **rhombus**, which is a parallelogram with oblique angles and four equal sides.)

face of an element is positive if it acts in the positive direction of one of the coordinate axes and negative if it acts in the negative direction of the axis. A shear stress acting on a negative face of an element is positive if it acts in the negative direction of an axis and negative if it acts in the positive direction. Thus, all shear stresses shown in Fig. 1-24a are positive.

The **sign convention for shear strains** is related to that for shear stresses. Shear strain in an element is positive when the angle between two positive (or two negative) faces is reduced. The strain is negative when the angle between two positive (or two negative) faces is increased. Thus, the strains shown in Fig. 1-24b are positive, and we see that positive shear stresses produce positive shear strains.

The properties of a material in shear can be determined experimentally from direct-shear tests or torsion tests. The latter tests are performed by twisting hollow, circular tubes, thereby producing a state of pure shear stress, as explained in Chapter 3. From the results of these tests, **stress-strain diagrams in shear** may be plotted. These diagrams of τ versus γ are similar in shape to the tension-test diagrams (σ versus ϵ) for the same materials, although different in magnitudes. From the shear diagrams, we can obtain shear properties such as the proportional limit, the yield stress, and the ultimate stress. These properties in shear are usually about half as large as those in tension. For instance, the yield stress for structural steel in shear is 0.5 to 0.6 times the yield stress in tension.

For many materials, the initial part of the shear stress-strain diagram is a straight line, just as in tension. For this linearly elastic region, the shear stress and shear strain are directly proportional, and we have the following equation for **Hooke's law in shear**:

$$\tau = G\gamma \tag{1-9}$$

in which G is the **shear modulus of elasticity** (also called the *modulus of rigidity*). The shear modulus G has the same units as the tension modulus E, namely, psi or ksi in USCS units and pascals in SI units. For mild steel, typical values of G are 11,000 ksi or 75 GPa; for aluminum, typical values are 4000 ksi or 28 GPa. Additional values are listed in Table H-2, Appendix H.

The moduli of elasticity in tension and shear (E and G) are related by the following equation:

$$G = \frac{E}{2(1 + \nu)} \tag{1-10}$$

in which ν is Poisson's ratio. This relationship, which is derived in Section 3.5, shows that E, G, and ν are not independent elastic properties of the material. Because the value of Poisson's ratio for ordinary materials is between zero and one-half, we see from Eq. (1-10) that G must be from one-third to one-half of E.

Example 1

A punch with a diameter of 0.75 in. is used to punch a hole in a $\frac{1}{4}$ in. steel plate (Fig. 1-25). A force $P = 26{,}000$ lb is required. What is the average shear stress in the plate and the average compressive stress in the punch?

The average shear stress is obtained by dividing the force P by the area being sheared by the punch. This area is equal to the circumference of the hole times the thickness of the plate:

$$A_s = \pi(0.75 \text{ in.})(0.25 \text{ in.}) = 0.589 \text{ in.}^2$$

Therefore, the average shear stress is

$$\tau_{\text{aver}} = \frac{P}{A_s} = \frac{26{,}000 \text{ lb}}{0.589 \text{ in.}^2} = 44{,}100 \text{ psi}$$

Also, the average compressive stress in the punch is

$$\sigma_c = \frac{P}{A_c} = \frac{26{,}000 \text{ lb}}{\pi(0.75 \text{ in.})^2/4} = 58{,}900 \text{ psi}$$

in which A_c is the cross-sectional area of the punch.

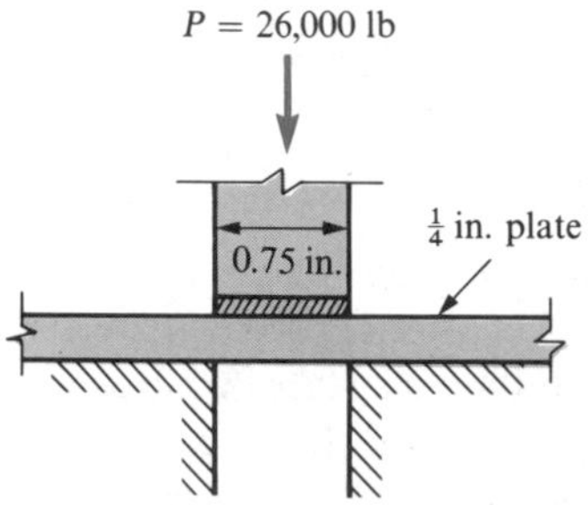

Fig. 1-25 Example 1

Example 2

A bearing pad consisting of a linearly elastic material of thickness h capped by a rigid steel plate of dimensions $a \times b$ (Fig. 1-26a) is subjected to a horizontal shear force V. Determine the average shear stress in the elastic material and the horizontal displacement d of the plate (Fig. 1-26b).

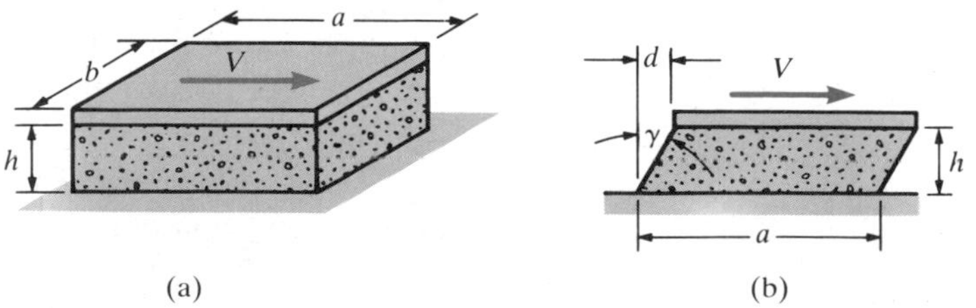

Fig. 1-26 Example 2

The average shear stress equals the force V divided by the area over which it acts:

$$\tau_{\text{aver}} = \frac{V}{ab} \qquad \text{(a)}$$

The corresponding shear strain is

$$\gamma = \frac{\tau_{\text{aver}}}{G} = \frac{V}{abG} \tag{b}$$

in which G is the shear modulus of the elastic material. Finally, the displacement d is equal to $h \tan \gamma$ (see Fig. 1-26b). However, if γ is a small angle, we may replace $\tan \gamma$ by γ and obtain

$$d = h\gamma = \frac{hV}{abG} \tag{1-11}$$

This expression for the horizontal displacement of the plate is approximate because it is based upon the assumption that the shear stress and strain are constant throughout the volume of the elastic material. In reality the stress is zero at the edges of the material (because there are no shear stresses on the free vertical faces) and the deformation of the material is more complex than pictured in Fig. 1-26b. However, if the dimension a of the plate is large compared with the thickness h, the preceding results are satisfactory.

The method described in this example is used in the design of elastomeric bearings subjected to shear forces. Elastomeric bearings are usually constructed of rubber (either natural or synthetic) compounded with other materials, and they are widely used to support machines, bridge girders, and other structures. In addition to supporting large vertical loads, they also must resist horizontal shear loads caused by actions such as creep, vibration, and seismic movement.

1.7 ALLOWABLE STRESSES AND ALLOWABLE LOADS

An important consideration in engineering design is the capacity of the object being designed to support or transmit loads. Objects that must sustain loads include building structures, machines, aircraft, vehicles, ships, and a seemingly endless list of other man-made things. For simplicity, we will refer to all such objects as **structures**; thus, a structure is any object that must support or transmit loads.

If structural failure is to be avoided, the loads that a structure actually can support must be greater than the loads it will be required to sustain when in service. The ability of a structure to resist loads is called **strength**; hence the preceding criterion can be restated as follows: The actual strength of a structure must exceed the required strength. The ratio of the actual strength to the required strength is called the **factor of safety** n:

$$\text{Factor of safety } n = \frac{\text{actual strength}}{\text{required strength}} \tag{1-12}$$

Of course, the factor of safety must be greater than 1.0 if failure is to be avoided. Depending upon the circumstances, factors of safety from slightly above 1.0 to as much as 10 are used.

The incorporation of factors of safety into design is not a simple matter, because both strength and failure have many different meanings. **Failure** can mean the fracture or complete collapse of a structure, or it can mean that the deformations have exceeded some limiting value so that the structure is no longer able to perform its intended functions. The latter kind of failure may occur at loads much smaller than those that cause actual collapse. The determination of a factor of safety must also take into account such matters as the following: probability of accidental overloading of the structure by loads that exceed the design loads; types of loads (static, dynamic, or repeated) and how accurately they are known; possibility of fatigue failure (see Section 2.9); inaccuracies in construction; quality of workmanship; variations in properties of materials; deterioration due to corrosion or other environmental effects; accuracy of the methods of analysis; whether failure is gradual (ample warning) or sudden (no warning); consequences of failure (minor damage or major catastrophe); and other such considerations. If the factor of safety is too low, the likelihood of failure will be high and hence the structure will be unacceptable; if the factor is too large, the structure will be wasteful of materials and perhaps unsuitable for its function (for instance, it may be too heavy). Because of these complexities, good engineering judgment is required when establishing factors of safety. They are usually determined by groups of experienced engineers who write the codes and specifications used by other designers.

In actual practice, there are several ways in which factors of safety are defined and implemented. For many structures, it is important that the material remain within the linearly elastic range in order to avoid permanent deformations when the loads are removed. Under these conditions, a common method of design is to use a factor of safety with respect to yielding of the structure. The structure begins to yield when the yield stress is reached at any point within the structure. By applying a factor of safety with respect to the yield stress, we obtain an **allowable stress**, or **working stress**, that must not be exceeded anywhere in the structure. Thus,

$$\text{Allowable stress} = \frac{\text{yield stress}}{\text{factor of safety}}$$

or

$$\sigma_{\text{allow}} = \frac{\sigma_y}{n} \tag{1-13}$$

in which we have introduced the notations σ_{allow} and σ_y for the allowable and yield stresses, respectively. In building design, a typical factor of safety n with respect to yielding is 1.67; thus, a mild steel having a yield stress σ_y of 36 ksi has an allowable stress σ_{allow} in tension of 21.6 ksi.

Another method of design is to establish the allowable stress by applying a factor of safety with respect to the **ultimate stress** instead of the yield stress. This method is suitable for brittle materials, such as concrete and some plastics, but it also is used for wood and high-strength steels.

The allowable stress is obtained from the equation

$$\sigma_{\text{allow}} = \frac{\sigma_u}{n} \tag{1-14}$$

in which σ_u is the ultimate stress. The factor of safety is normally much greater with respect to the ultimate stress than with respect to the yield stress. In the case of mild steel, a factor of safety of 1.67 with respect to yielding corresponds to a factor of approximately 2.8 with respect to the ultimate stress.

The last method we will describe involves the application of factors of safety to loads rather than to stresses. We will use the term **ultimate loads** to mean the loads that produce failure or collapse of the structure. The loads that the structure must support in service are called **service loads** or **working loads**. The factor of safety is the ratio of the former to the latter:

$$\text{Factor of safety } n = \frac{\text{ultimate load}}{\text{service load}} \tag{1-15}$$

Inasmuch as the service loads are assumed to be known quantities, the usual design procedure is to multiply them by the factor of safety to obtain the ultimate loads. Then the structure is designed so that it can just sustain the ultimate loads at failure. This method of design is known as **strength design**, or **ultimate-load design**, and the factor of safety is called the **load factor** because it is a multiplier of the service loads:

$$\text{Ultimate load} = (\text{service load})(\text{load factor}) \tag{1-16}$$

Typical load factors used in the design of reinforced concrete structures are 1.4 for **dead load**, which is the weight of the structure itself, and 1.7 for **live loads**, which are loads applied to the structure. The strength-design method is used regularly for reinforced concrete structures and occasionally for steel structures. Methods for determining ultimate loads for some simple structures are given in Section 2.8.

In aircraft design, it is customary to speak of the margin of safety rather than the factor of safety. The **margin of safety** is defined as the factor of safety minus one:

$$\text{Margin of safety} = n - 1 \tag{1-17}$$

Thus, a structure having an ultimate strength that is twice the required strength has a factor of safety of 2.0 and a margin of safety of 1.0. When the margin of safety is reduced to zero or less, the structure (presumably) will fail.

In addition to strength considerations, the design of a structure is likely to involve **stiffness** and **stability**. Stiffness refers to the ability of the structure to resist changes in shape, and stability refers to the ability of the structure to resist buckling under compressive stresses. These matters are discussed in detail in later chapters for various kinds of structural elements.

Example 1

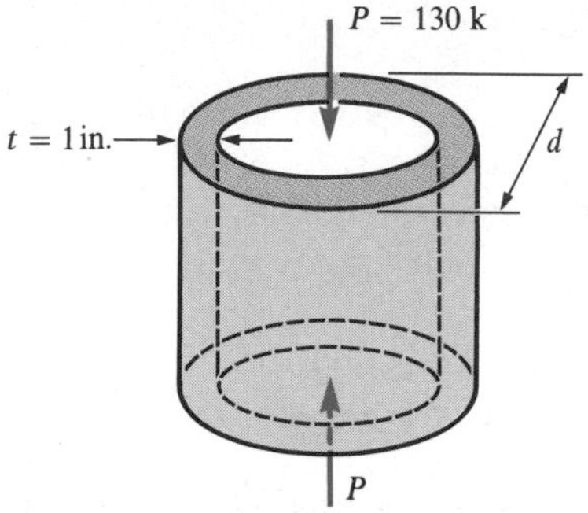

Fig. 1-27 Example 1

A short, hollow, circular, cast-iron cylinder (Fig. 1-27) is to support an axial compressive load $P = 130$ kips. The ultimate stress in compression for the material is $\sigma_u = 35{,}000$ psi. It is decided to design the cylinder with a wall thickness t of 1 in. and a factor of safety of 3.0 with respect to the ultimate strength. Compute the minimum required outside diameter d of the cylinder.

The allowable compressive stress is equal to the ultimate stress divided by the factor of safety (Eq. 1-14):

$$\sigma_{\text{allow}} = \frac{\sigma_u}{n} = \frac{35{,}000 \text{ psi}}{3.0} = 11{,}670 \text{ psi}$$

The required cross-sectional area can now be found:

$$A = \frac{P}{\sigma_{\text{allow}}} = \frac{130{,}000 \text{ lb}}{11{,}670 \text{ psi}} = 11.14 \text{ in.}^2$$

The actual cross-sectional area is

$$A = \frac{\pi d^2}{4} - \frac{\pi(d - 2t)^2}{4} = \pi t(d - t)$$

in which d is the outside diameter and $d - 2t$ is the inside diameter. Solving for d and then substituting $t = 1$ in. and $A = 11.14$ in.^{2}, we get

$$d = t + \frac{A}{\pi t} = 4.55 \text{ in.}$$

The outside diameter must be at least this large in order to have the desired factor of safety.

Example 2

A steel bar of rectangular cross section (10×40 mm) carries a tensile load P and is attached to a support by means of a round pin of diameter 15 mm (Fig. 1-28). The allowable stresses for the bar in tension and the pin in shear are $\sigma_{\text{allow}} = 120$ MPa and $\tau_{\text{allow}} = 60$ MPa, respectively. What is the maximum permissible value of the load P?

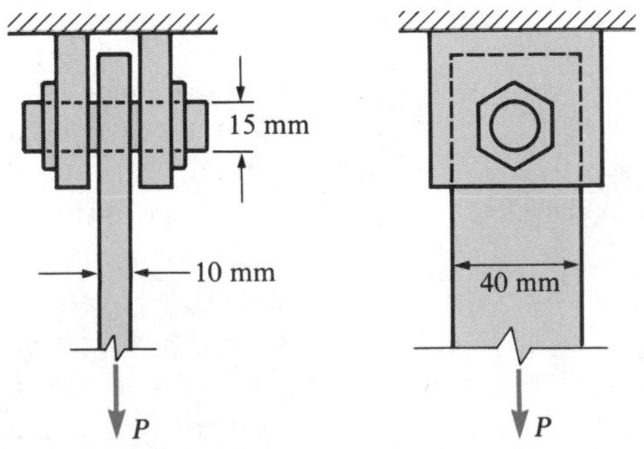

Fig. 1-28 Example 2

The tensile stress in the rectangular bar must be calculated using the net area at the cross section containing the hole for the pin. This area is

$$A_{net} = (40 \text{ mm} - 15 \text{ mm})(10 \text{ mm}) = 250 \text{ mm}^2$$

Hence, the allowable load P_1 based upon tension in the bar is

$$P_1 = \sigma_{allow} A_{net} = (120 \text{ MPa})(250 \text{ mm}^2) = 30 \text{ kN}$$

This calculation disregards the stress concentrations due to the presence of the hole (see Section 2.10).

Next, we calculate the allowable load based upon shear in the pin. The pin is in double shear and tends to shear on two cross sections; hence the total force that can be carried is

$$P_2 = \tau_{allow}(2A)$$

in which A is the cross-sectional area of the pin. Substitution of numerical values gives

$$P_2 = (60 \text{ MPa})(2)\left(\frac{\pi}{4}\right)(15 \text{ mm})^2 = 21.2 \text{ kN}$$

Comparing the two preceding values for P, we see that shear in the pin governs and

$$P_{allow} = 21.2 \text{ kN}$$

is the maximum permissible value of the load.

PROBLEMS / CHAPTER 1

1.2-1 A bar ABC having two different cross-sectional areas is loaded by an axial force P (see figure). Parts AB and BC are circular in cross section with diameters 2.0 in. and 1.5 in., respectively. If the normal stress in AB is 6000 psi, what is the normal stress σ in BC?

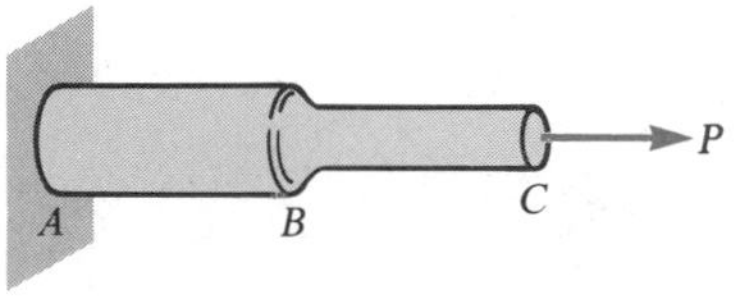

Prob. 1.2-1

1.2-2 A horizontal bar CBD having a length of 2.2 m is supported and loaded as shown in the figure. The vertical member AB has a cross-sectional area of 540 mm^2. Determine the magnitude of the load P so that it produces a normal stress equal to 50 MPa in member AB.

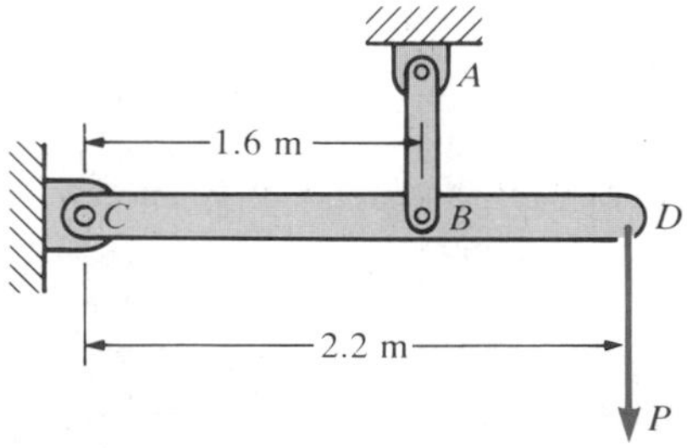

Prob. 1.2-2

1.2-3 Calculate the compressive stress σ_c in the piston rod (see figure on page 38) when a force $P = 8$ lb is applied to the brake pedal. (The diameter of the rod is $\frac{3}{16}$ in. The other dimensions shown in the figure are perpendicular to the line of action of the force P, which is parallel to the piston rod.)

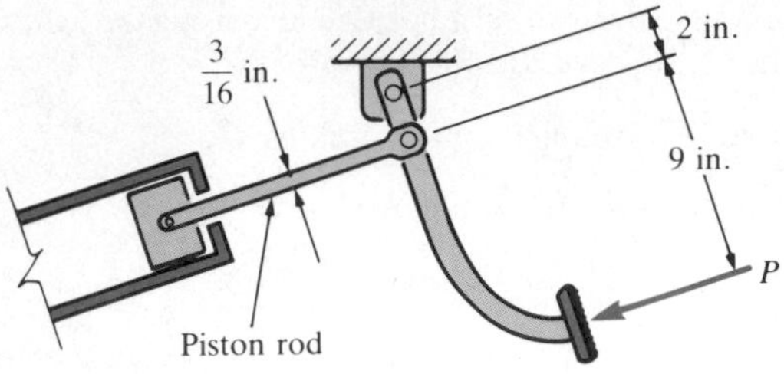

Prob. 1.2-3

1.2-4 A steel rod 30 m long hangs in a mine shaft and holds a 900 N weight at its lower end (see figure). If the diameter of the circular rod is 6 mm, calculate the maximum normal stress σ_{max} in the rod, taking into account the weight of the rod itself. (Obtain the specific weight of steel from Table H-1, Appendix H.)

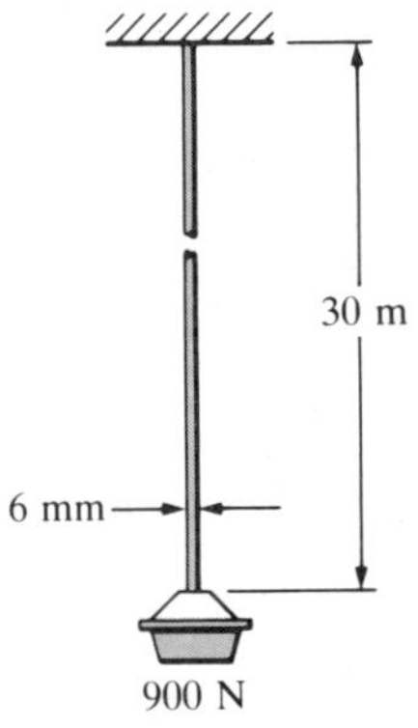

Prob. 1.2-4

1.2-5 A steel rod 3 ft long and 0.5 in. in diameter carries a tensile load of 3 k (see figure). The bar increases in length by 0.02 in. when the load is applied. Determine the normal stress and strain in the bar.

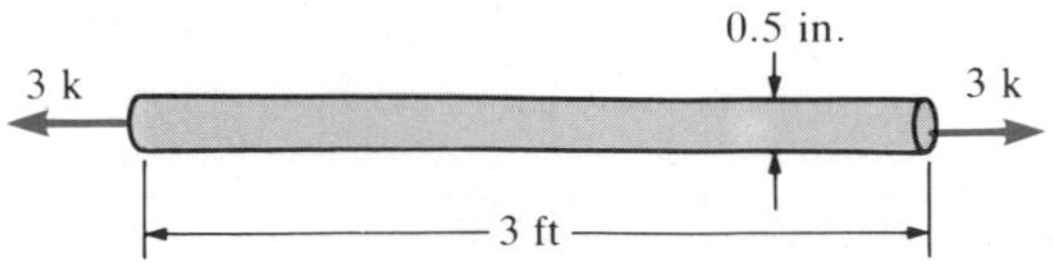

Prob. 1.2-5

1.2-6 A circular tube of length $L = 0.5$ m is loaded in compression by forces P (see figure). The outside and inside diameters are 30 mm and 25 mm, respectively. A strain gage is placed on the outside of the bar to measure normal strains in the longitudinal direction. (a) If the measured strain is $\epsilon = 900 \times 10^{-6}$, what is the shortening δ of the bar? (b) If the compressive stress in the bar is intended to be 60 MPa, what should be the load P?

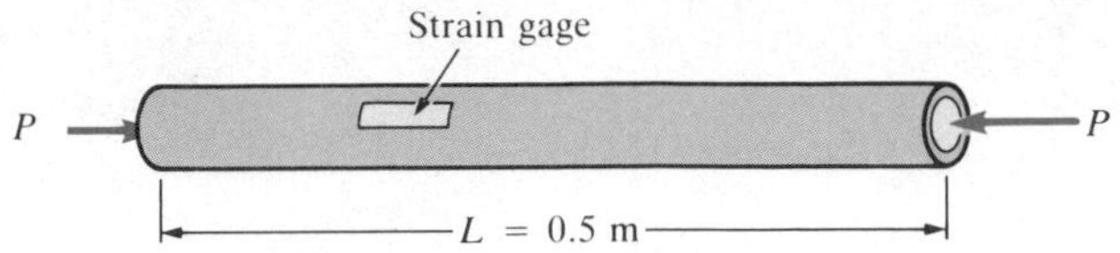

Prob. 1.2-6

1.2-7 A two-story column ABC in a building is constructed with a hollow, square box section (see figure). The outside dimensions are 8 in. × 8 in., and the wall thickness is $\frac{5}{8}$ in. The roof load applied at the top of the column is $P_1 = 70$ k, and the floor load applied at midheight is $P_2 = 100$ k. Obtain the compressive stresses σ_{ab} and σ_{bc} in the two parts of the column due to these loads.

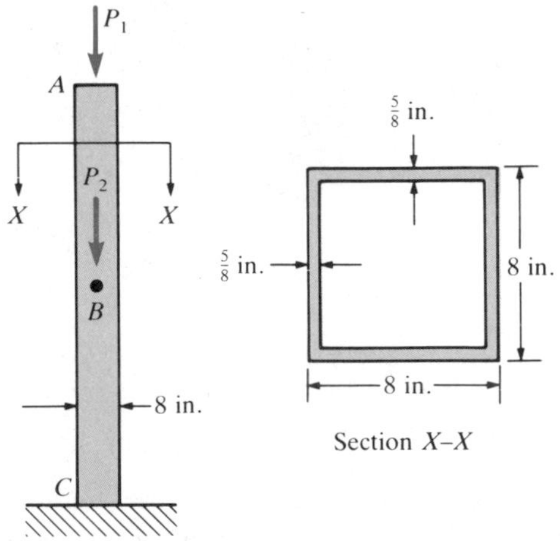

Prob. 1.2-7

1.2-8 The figure shows the cross section of a concrete pedestal that is loaded in compression. (a) Determine the

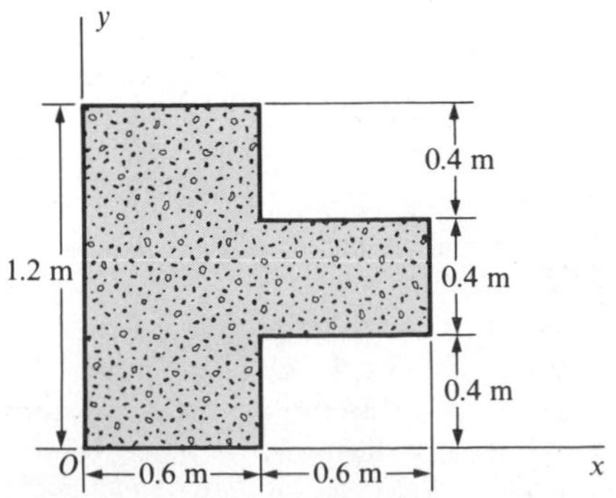

Prob. 1.2-8

coordinates $\bar{x}$ and $\bar{y}$ of the point where a concentrated load must act in order to produce uniform normal stress. (b) What is the magnitude of the compressive stress σ_c if the load is equal to 24 MN?

1.2-9 A steel wire *ABC* supporting a lamp at its midpoint is attached to supports that are 5 ft apart (see figure). The length of the wire is 6 ft and its diameter is 16 mils (it is common practice to specify wire diameters in mils; one mil equals 0.001 in.). If the lamp weighs 12 lb, what is the tensile stress σ_t in the wire *ABC*?

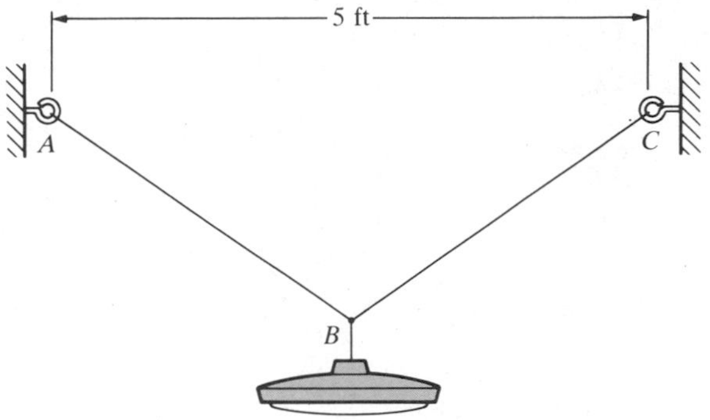

Prob. 1.2-9

1.2-10 A car having a mass of 5500 kg is pulled slowly up a steep inclined track by a steel cable (see figure). The cable has an effective cross-sectional area of 320 mm², and the angle of the incline is 31°. What is the tensile stress σ_t in the cable?

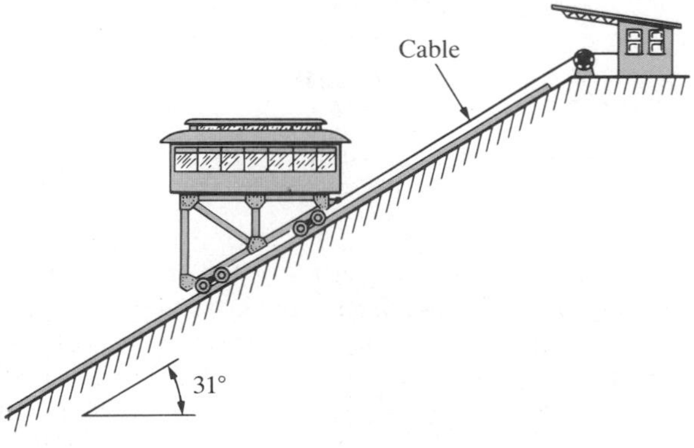

Prob. 1.2-10

1.2-11 An arrangement for pulling a pipe out of the ground is shown in the figure. A cable *ABC* is attached to the pipe at *A* and to the sturdy frame at *C*. A second cable *BDE* is then clamped to the first cable at *B* and attached to the frame at *E*. Finally, a man pulls vertically downward on the second cable at *D* with a force *Q*. The lengths of the cables are such that *AB* is vertical, *BD* is horizontal, *BC* is inclined by a small angle α to the vertical, and *DE* is inclined by a small angle β to the horizontal. The effective cross-sectional area of the cable is denoted A_c. (a) Obtain a formula for the tensile stress σ_{ab} in cable *AB*. (b) Calculate the stress σ_{ab} and also the lifting force *P* in cable *AB* for $Q = 100$ lb, $\alpha = 5°$, $\beta = 8°$, and $A_c = 0.119$ in.²

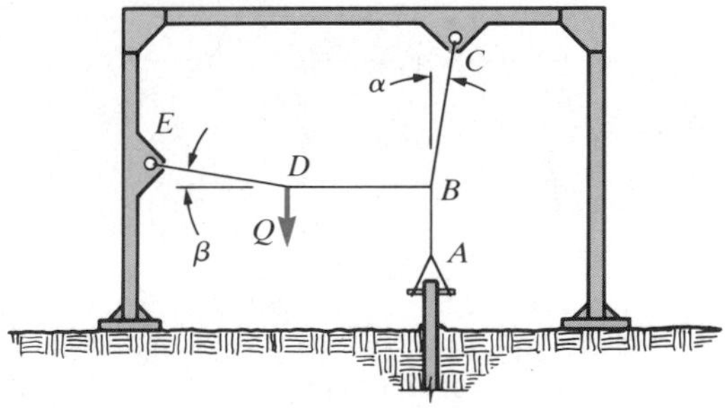

Prob. 1.2-11

1.2-12 A strut and cable assembly *ABC* (see figure) supports a vertical load $P = 18$ kN. The cable has an effective cross-sectional area of 120 mm², and the strut has an area of 250 mm². (a) Calculate the normal stresses σ_{ab} and σ_{bc} in the cable and strut, and indicate whether they are tension or compression. (b) If the cable elongates 1.3 mm, what is the strain? (c) If the strut shortens 0.62 mm, what is the strain?

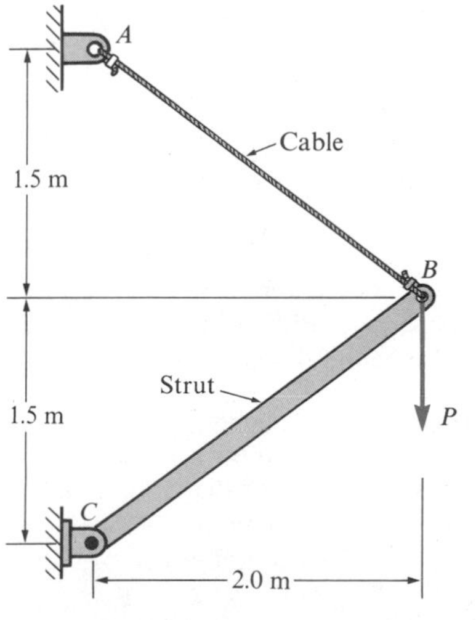

Prob. 1.2-12

***1.2-13** A round bar *ACB* of total length $2L$ (see figure) rotates about an axis through the midpoint *C* with constant angular speed ω (radians per second). The material of the bar has specific weight γ. Derive a formula for the tensile stress σ_x in the bar as a function of the distance x from point *C*. What is the maximum tensile stress?

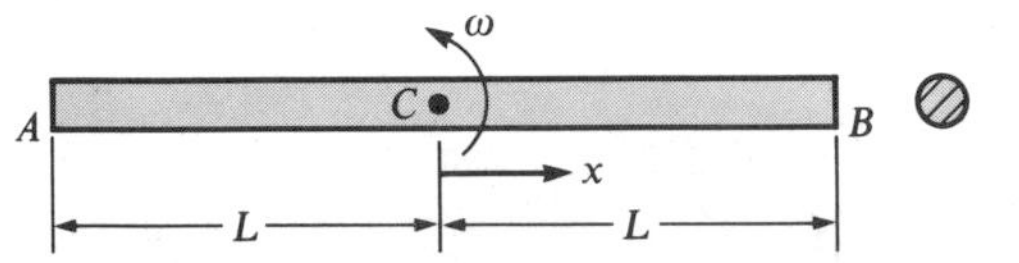

Prob. 1.2-13

1.3-1 A long steel wire hangs vertically in a deep well under its own weight. (a) What is the greatest length (feet) it can have without yielding if the yield stress is 40 ksi? (b) If the same wire hangs from a ship at sea, what is the greatest length? (The specific weights of sea water and steel are available in Table A-4, Appendix A.)

1.3-2 Three different materials *A*, *B*, and *C* are tested in tension using test specimens having diameters of 12 mm and gage lengths of 50 mm. After the specimens are fractured, the distances between the gage marks are found to be 54.1, 63.0, and 70.6 mm, respectively. Also, the diameters are 11.50, 9.46, and 6.01 mm, respectively, at the failure cross sections. Determine the percent elongation and percent reduction in area of each specimen. Also, classify the materials as brittle or ductile.

1.3-3 The data shown in the table were obtained from a tensile test of high-strength steel. The test specimen had a diameter of 0.505 in., and a gage length of 2.00 in. was used. The total elongation between the gage marks at fracture was 0.42 in., and the minimum diameter was 0.370 in. Plot the nominal stress-strain diagram for the steel and determine the proportional limit, modulus of elasticity, yield stress at 0.1% offset, ultimate stress, percent elongation in 2.00 in., and percent reduction in area.

1.3-4 A specimen of a methacrylate plastic is tested in tension at room temperature, producing the stress-strain data listed in the table. Plot the stress-strain diagram and determine the proportional limit, modulus of elasticity, and the yield stress at 0.2% offset. Do you classify this material as brittle or ductile?

Tensile-Test Data for Problem 1.3-3

Load (lb)	Elongation (in.)
1,000	0.0002
2,000	0.0006
6,000	0.0019
10,000	0.0033
12,000	0.0040
12,900	0.0043
13,400	0.0047
13,600	0.0054
13,800	0.0063
14,000	0.0090
14,400	0.0118
15,200	0.0167
16,800	0.0263
18,400	0.0380
20,000	0.0507
22,400	0.1108
25,400	fracture

Stress-Strain Data for Problem 1.3-4

Stress (MPa)	Strain
8.0	0.0032
17.5	0.0073
25.6	0.0111
31.1	0.0129
39.8	0.0163
44.0	0.0184
48.2	0.0209
53.9	0.0260
58.1	0.0331
62.0	0.0429
62.5	fracture

1.3-5 The mechanical properties of structural materials are available in various reference books and reports. Go to your technical library and find an actual stress-strain diagram for aluminum, copper, steel, nylon, rubber, fiberglass, or other structural material of your choice. The diagram may be in either system of units. (a) Make a photocopy of the diagram and use it to determine as many mechanical properties as you can. (b) Based upon the information from (a) plus supplementary reading, prepare a one-paragraph description of the material.

1.4-1 A 5 ft long bar is made of a structural steel having the stress-strain diagram shown in the figure. The yield stress is 42 ksi and the slope of the initial linear part of the diagram (modulus of elasticity) is 30×10^3 ksi. The bar is loaded axially until it elongates 0.3 in., and then the load is removed. How does the final length of the bar compare with its original length? (Hint: Use the concepts illustrated in Fig. 1-16.)

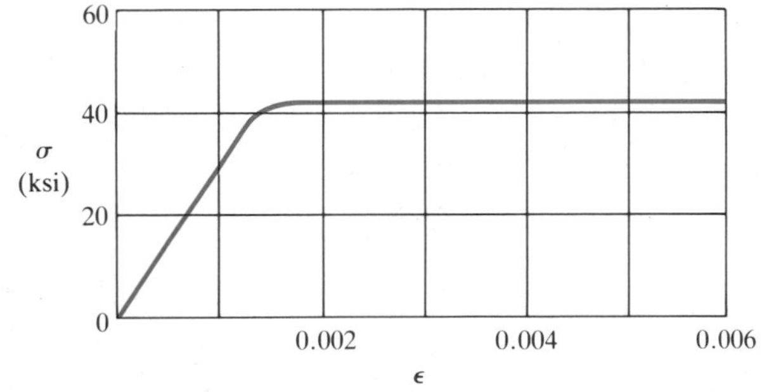

Prob. 1.4-1

1.4-2 A 2 m long bar is made of a structural steel having the stress-strain diagram shown in the figure. The yield stress is 250 MPa and the slope of the initial linear part of the diagram (modulus of elasticity) is 200 GPa. The bar is loaded axially until it elongates 6.5 mm, and then the load is removed. How does the final length of the bar compare with its original length? (Hint: Use the concepts illustrated in Fig. 1-16.)

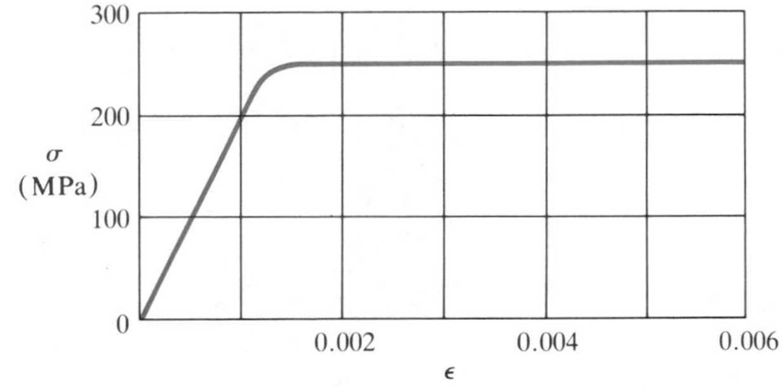

Prob. 1.4-2

***1.4-3** A round bar having a length of 4 ft and a diameter of 0.25 in. is made of an aluminum alloy for which the stress-strain diagram can be described mathematically by the following equation:

$$\epsilon = \frac{\sigma}{10{,}600}\left[1 + \frac{3}{7}\left(\frac{\sigma}{60}\right)^9\right]$$

in which σ has units of kips per square inch (ksi). The shape of this stress-strain diagram is shown in the figure. The bar is axially loaded by tensile forces $P = 2.8$ k, after which the loads are removed. What is the permanent set of the bar?

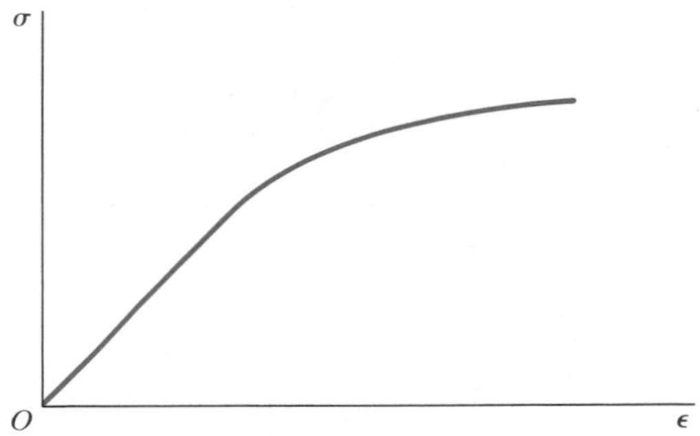

Probs. 1.4-3 and 1.4-4

***1.4-4** A round bar having a length of 2 m and a diameter of 10 mm is made of an aluminum alloy for which the stress-strain diagram can be described mathematically by the following equation:

$$\epsilon = \frac{\sigma}{70{,}000}\left[1 + \frac{3}{7}\left(\frac{\sigma}{270}\right)^9\right]$$

in which σ has units of megapascals (MPa). The shape of this stress-strain diagram is shown in the figure. The bar is axially loaded by tensile forces $P = 20$ kN, after which the loads are removed. What is the permanent set of the bar?

1.5-1 What is the modulus of elasticity E of the high-strength material used for a $\frac{1}{8}$ in. diameter wire if the wire stretches 1.15 in. when a 40 ft length of it is subjected to a tensile load of 900 lb?

1.5-2 A round bar of 6 mm diameter is made of pure aluminum. When the bar is stretched by an axial force P, its diameter decreases by 0.450 μm. What is the magnitude of the load P?

1.5-3 A high-strength steel rod ($E = 29 \times 10^6$ psi, $\nu = 0.3$) is compressed by an axial force P (see figure on page 42). When there is no axial load, the diameter of the rod is exactly 2 in. In order to maintain certain clearances, the diameter of the rod must not exceed 2.001 in. under load. What is the largest permissible load P?

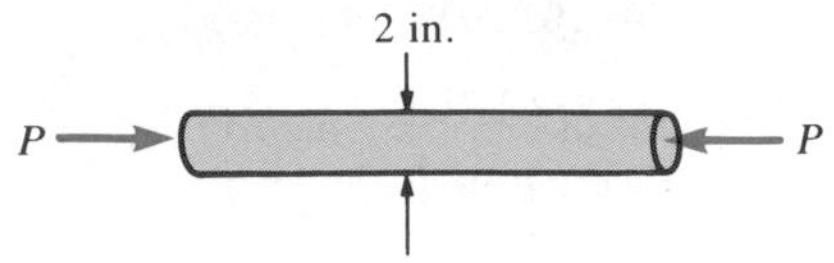

Prob. 1.5-3

1.5-4 A tensile test is performed on a brass specimen 10 mm in diameter using a gage length of 50 mm (see figure). When the tensile load P reaches a value of 25 kN, the distance between the gage marks has increased by 0.152 mm. (a) What is the modulus of elasticity E of the brass? (b) What is the decrease Δd in diameter of the bar? (c) Calculate the dilatation e of the bar.

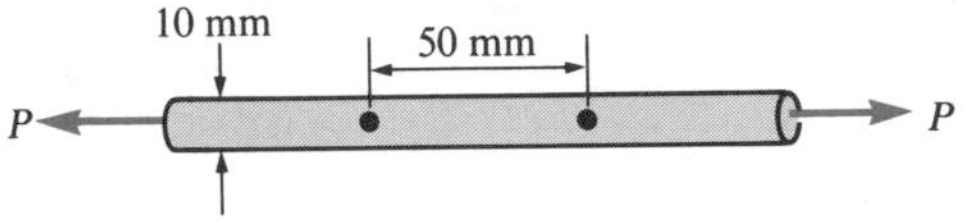

Prob. 1.5-4

1.5-5 A $\frac{1}{4}$ in. diameter bar that is 8 in. long is loaded axially in tension by an 800 lb force (see figure). The increase in length and decrease in diameter of the bar are measured as 0.00522 in. and 0.0000522 in., respectively. (a) Calculate the modulus of elasticity E and Poisson's ratio ν for the material. (b) Calculate the dilatation e of the bar. (c) Knowing that the material of the bar is listed in Table H-2 of the Appendix, determine the material.

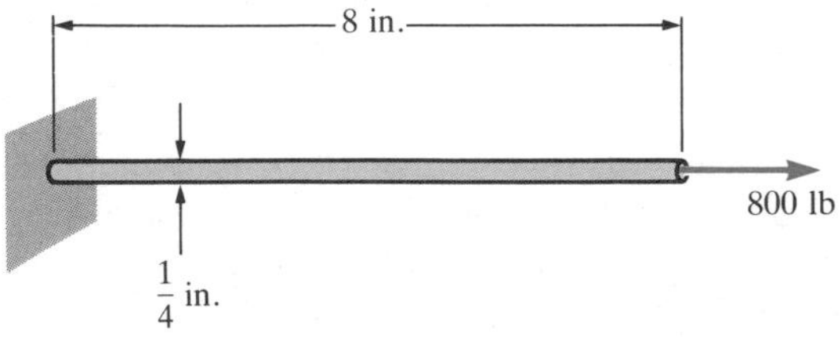

Prob. 1.5-5

1.5-6 A prismatic bar is loaded in tension by axial forces. Find Poisson's ratio ν for the material if the ratio of the unit volume change to the unit change in cross-sectional area is equal to $-\frac{3}{4}$.

1.5-7 An aluminum alloy bar of square cross section (0.5 in. × 0.5 in.) is loaded in tension by axial forces P. Stress-strain data obtained from a tension test of the same alloy are listed in the table. (a) What is the modulus of elasticity E of the aluminum? (b) What is the maximum load P if the stress in the bar is not to exceed the proportional limit?

Stress-Strain Data for Problem 1.5-7

Stress (ksi)	Strain
8	0.0006
17	0.0015
27	0.0024
35	0.0032
43	0.0040
50	0.0046
58	0.0052
62	0.0058
64	0.0062
65	0.0065
67	0.0073
68	0.0081

1.5-8 A steel pipe of length $L = 2$ m, outside diameter $d = 110$ mm, and wall thickness $t = 8$ mm is subjected to an axial compressive load $P = 160$ kN (see figure). Assuming that $E = 200$ GPa and $\nu = 0.3$, find (a) the shortening δ of the pipe, (b) the increase Δd in outside diameter, and (c) the increase Δt in wall thickness.

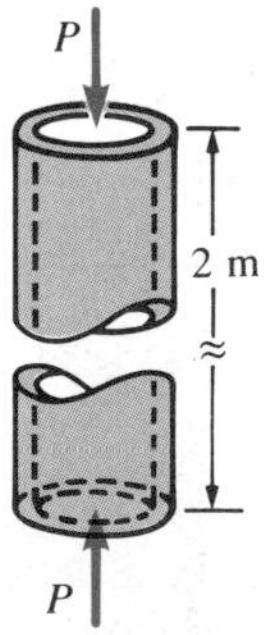

Prob. 1.5-8

1.5-9 A metal plate of length L and width b is subjected to a uniform tensile stress σ at the ends (see figure). Before loading, the slope of the diagonal line OA was b/L. What is the slope when the stress σ is acting?

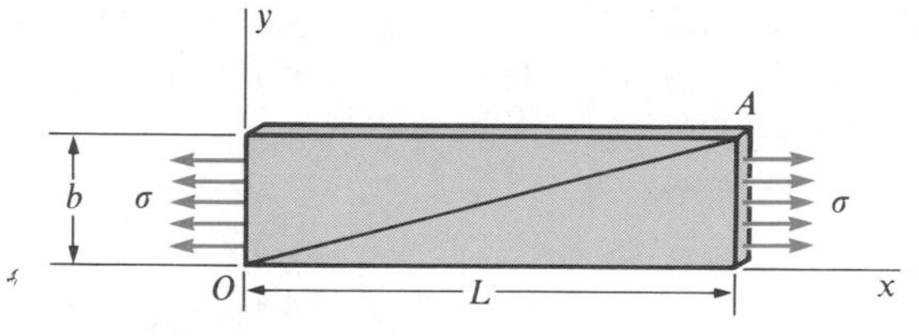

Prob. 1.5-9

***1.5-10** Derive a formula for the increase ΔV in the volume of a prismatic bar of length L hanging vertically under its own weight (W = total weight of the bar).

1.6-1 A block of wood is tested in direct shear using the loading frame and test specimen shown in the figure. The load P produces shear in the specimen along the plane AB. The width of the specimen (perpendicular to the plane of the paper) is 4 in., and the height h of plane AB is 2 in. For a load $P = 3600$ lb, what is the average shear stress τ_{aver} in the wood?

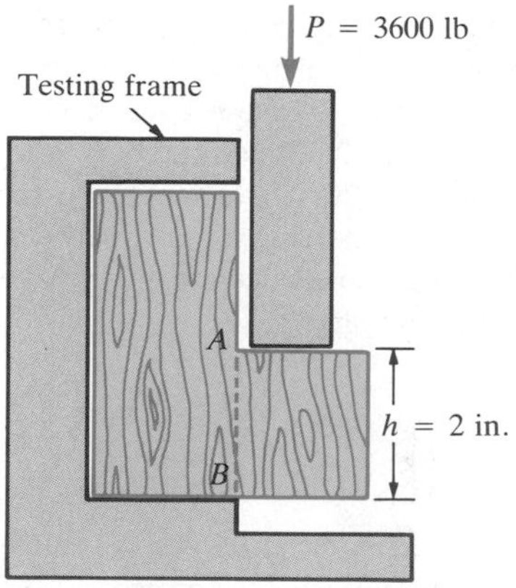

Prob. 1.6-1

1.6-2 An angle bracket is attached to a column with two 14 mm diameter bolts as shown in the figure. The bracket supports a load $P = 28$ kN. Calculate the average shear stress τ_{aver} in the bolts, disregarding friction between the bracket and the column. (Note that these bolts are loaded in *single shear*.)

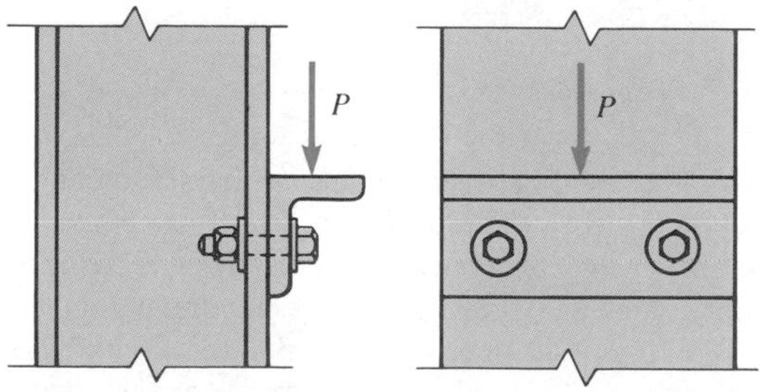

Prob. 1.6-2

1.6-3 The vertical load P acting on the wheel of a traveling crane is 12,000 lb (see figure). What is the average shear stress τ_{aver} in the 1.5 in. diameter axle?

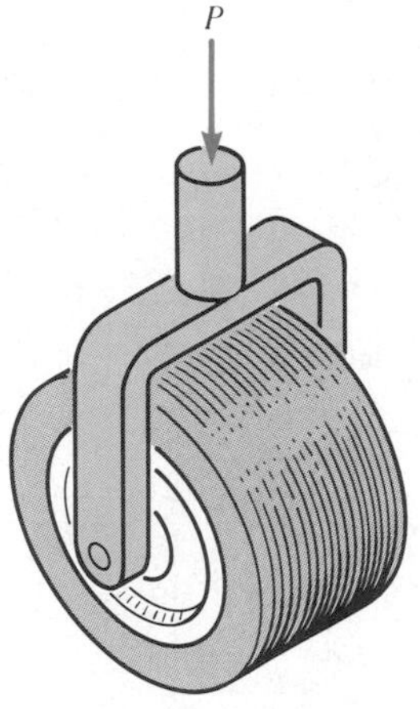

Prob. 1.6-3

1.6-4 Three steel plates are joined by two rivets as shown in the figure. If the rivets have diameters of 15 mm and the ultimate shear stress in the rivets is 210 MPa, what force P is required to cause the rivets to fail in shear?

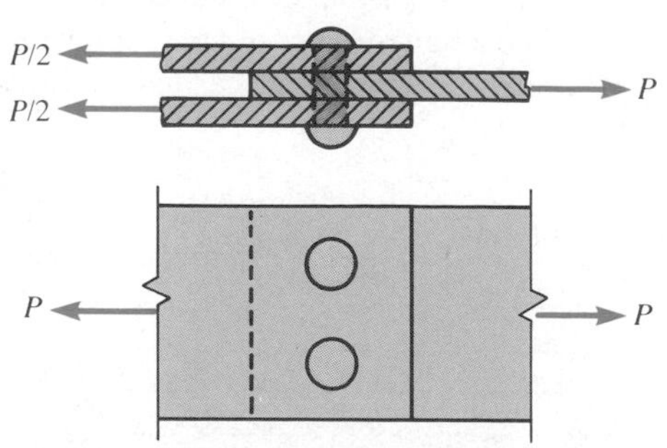

Prob. 1.6-4

1.6-5 Three pieces of wood are glued together and to rigid supports (see figure on page 44). A force $P = 3150$ lb pulls on the middle piece of wood. The cross section of each piece is 1.5×3.5 in., and the length of the glued surfaces is 6 in. What is the average shear stress τ_{aver} in the glued joints?

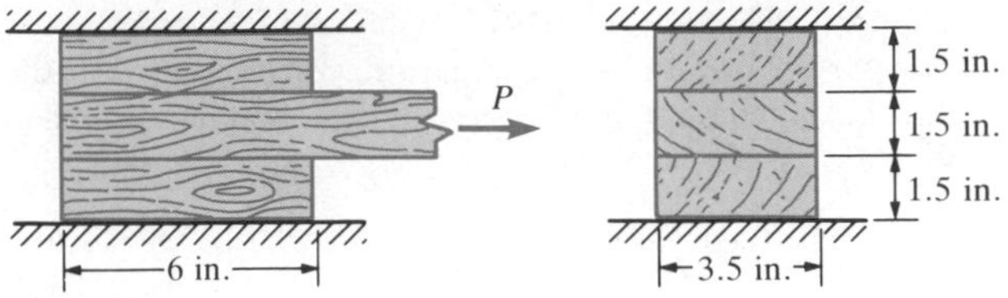

Prob. 1.6-5

1.6-6 A punch with a diameter $d = 20$ mm is used to punch a hole in an aluminum plate of thickness $t = 4$ mm (see figure). If the ultimate shear stress for the aluminum is 280 MPa, what force P is required to punch through the plate?

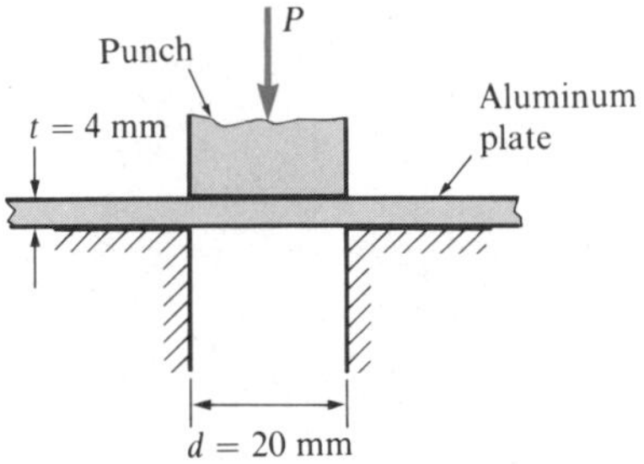

Prob. 1.6-6

1.6-7 The connection shown in the figure consists of five $\frac{3}{16}$ in. plates joined by a single $\frac{1}{4}$ in. diameter bolt. The total load transferred between the plates is 1600 lb, distributed between the plates as shown. (a) Calculate the maximum shear stress τ in the bolt, disregarding friction between the plates. (b) Calculate the maximum bearing stress σ_b between a plate and the bolt.

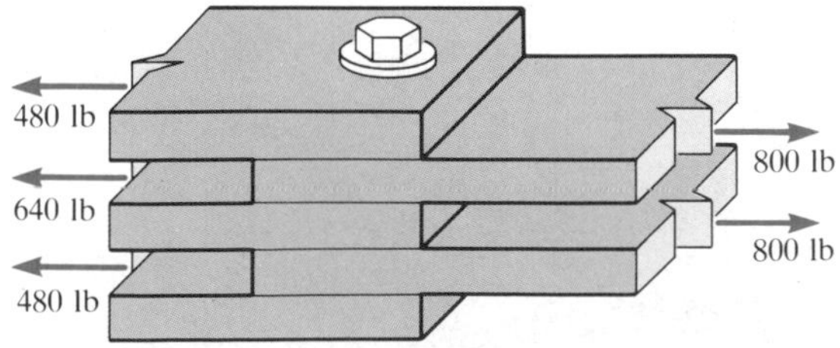

Prob. 1.6-7

1.6-8 A frame is made of a 2 m long vertical pipe CD and a brace AB formed from two flat bars (see figure). The frame is supported by bolted connections at points A and C, which are 2 m apart. The brace is fastened to the pipe at point B, which is 1 m above point C, by a 20 mm diameter bolt. If a load $P = 15$ kN acts horizontally at D, determine the average shear stress τ_{aver} in the bolt at B.

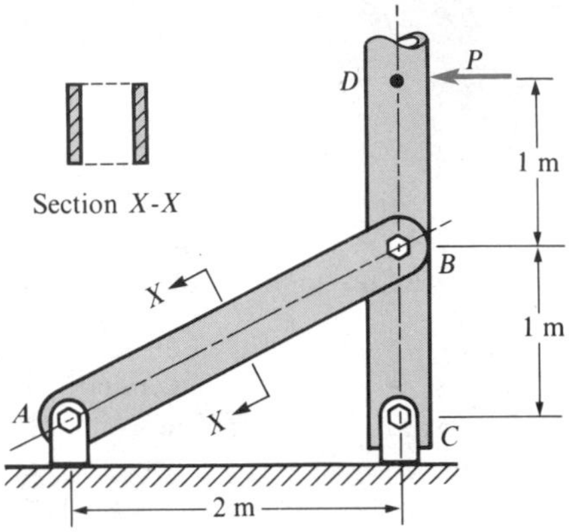

Prob. 1.6-8

1.6-9 A special-purpose bolt of diameter $d = 0.375$ in. passes through a hole in a steel plate (see figure). The head of the bolt bears directly against the steel plate when a tensile force $P = 2100$ lb acts on the bolt. The diameter of the bolt head is $D = 0.70$ in. and its thickness is $t = 0.20$ in. Calculate the average bearing stress σ_b between the head of the bolt and the plate, and calculate the average shear stress τ in the head of the bolt.

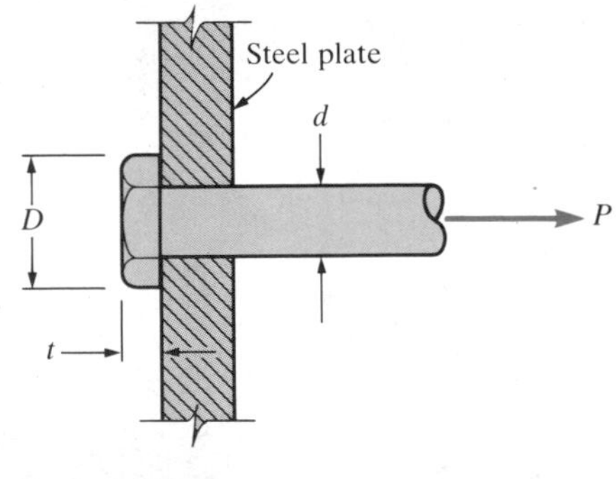

Prob. 1.6-9

1.6-10 The bond between reinforcing bars and concrete is tested by means of a "pull-out test" of a bar embedded in concrete (see figure). A tensile force P is applied to the end of the bar, which has diameter d and embedment length L. If $P = 18$ kN, $d = 12$ mm, and $L = 300$ mm, what average shear stress τ_{aver} is developed between the steel and concrete?

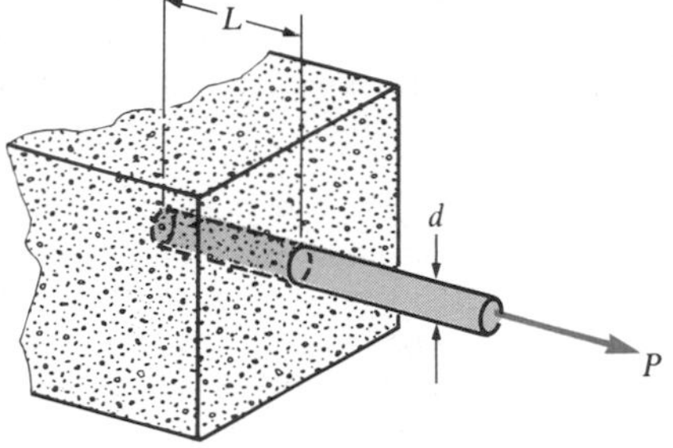

Prob. 1.6-10

1.6-11 A torque T_0 having a moment of 90 in.-k is transmitted between two flanged shafts by means of four $\frac{7}{8}$ in. bolts (see figure). What is the average shear stress τ_{aver} in each bolt if the diameter d of the bolt circle is 6 in.?

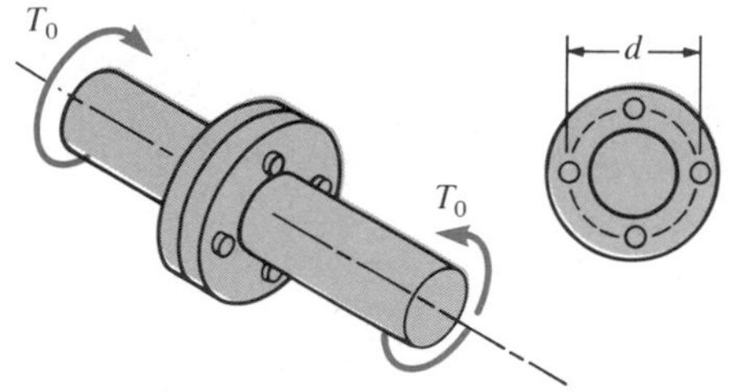

Prob. 1.6-11

1.6-12 An elastomeric bearing pad consisting of two steel plates bonded to a chloroprene elastomer (chloroprene is an artificial rubber) is subjected to a shear force V during a static test (see figure). The pad has dimensions $a = 150$ mm and $b = 250$ mm, and the elastomer has thickness $t = 50$ mm. When the force V equals 12 kN, the top plate is found to have displaced laterally 8.0 mm with respect to the bottom plate. What is the shear modulus of elasticity G of the elastomer?

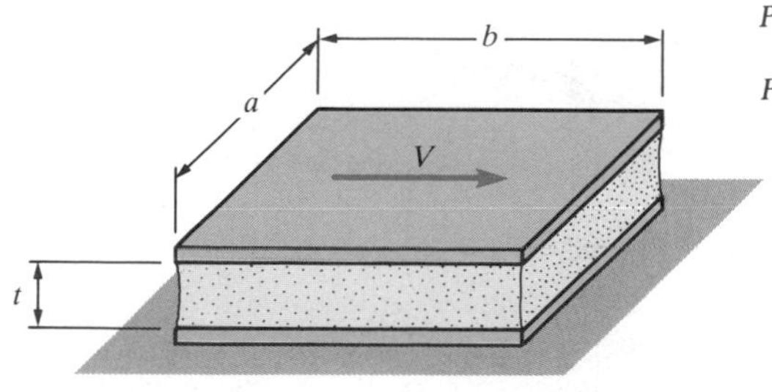

Prob. 1.6-12

1.6-13 A joint between two concrete slabs A and B is filled with a flexible epoxy that bonds securely to the concrete (see figure). The height of the joint is $b = 4.0$ in., its length perpendicular to the plane of the paper is $L = 40$ in., and its thickness is $t = 0.5$ in. Under the action of shear forces V, the slabs displace vertically through the distance $d = 0.002$ in. relative to each other. (a) What is the average shear strain γ_{aver} in the epoxy? (b) What is the magnitude of the forces V if $G = 140{,}000$ psi for the epoxy?

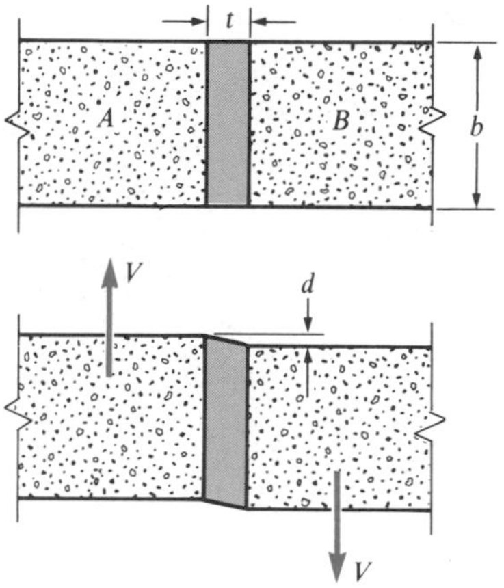

Prob. 1.6-13

1.6-14 A flexible connection consisting of rubber pads (thickness $t = 10$ mm) bonded to steel plates is shown in the figure. (a) Find the average shear strain γ in the rubber if the force $P = 15$ kN and the shear modulus for the rubber is $G = 800$ kPa. (b) Find the relative horizontal displacement δ between the interior plate and the outer plates.

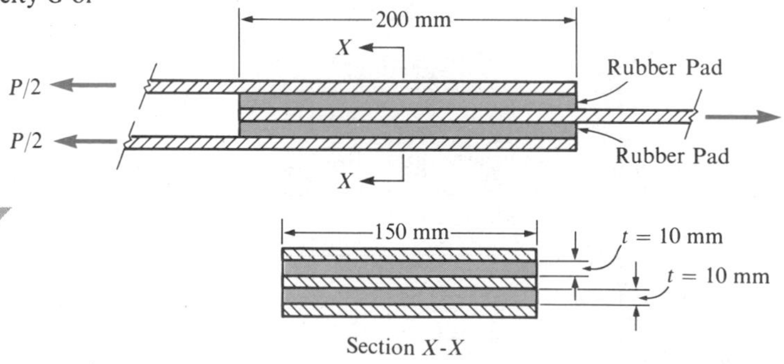

Prob. 1.6-14

***1.6-15** A special lever is used to twist a circular shaft by means of a square key that fits into slots (or keyways) in the shaft and lever, as shown in the figures. The shaft has diameter d, the key has a square cross section of dimensions $b \times b$, and the length of the key is c. The key fits half into the lever and half into the shaft (i.e., the keyways have a depth equal to $b/2$). Derive a formula for the average shear stress τ in the key when a load P is applied at distance L from the center of the shaft. (Hints: Disregard the effects of any friction, assume that the bearing pressure between the key and the lever is uniformly distributed, and be sure to draw a free-body diagram of the lever.)

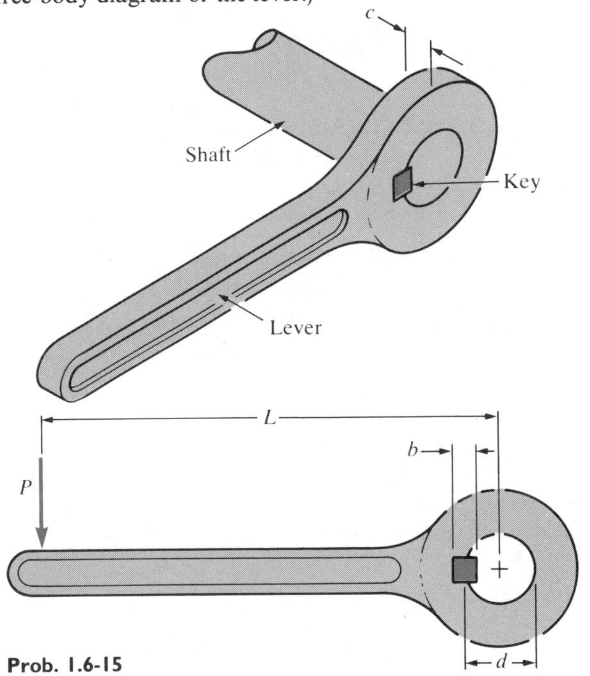

Prob. 1.6-15

***1.6-16** A thin rectangular plate of dimensions $b = 150$ mm and $c = 200$ mm (see figure) is made of a structural plastic having modulus of elasticity $E = 2.8$ GPa and Poisson's ratio $\nu = 0.40$. When the plate is subjected to shear stresses τ acting along the edge faces, the length of one of the diagonals increases by 1.2 mm. What is the magnitude of the shear stresses?

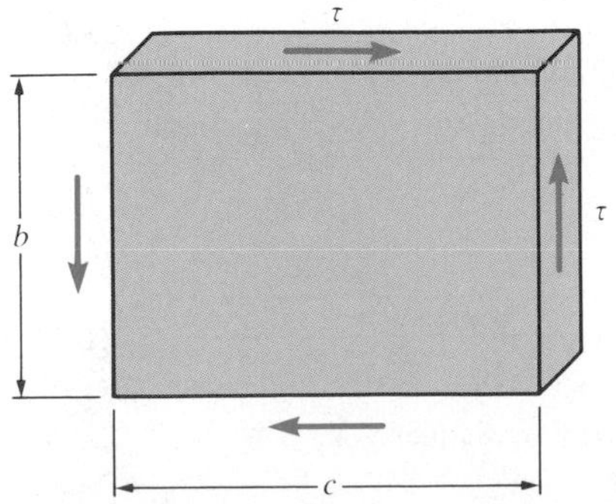

Prob. 1.6-16

***1.6-17** A shock mount designed as shown in the figure is used to support an instrument. The mount consists of a steel tube with inner diameter b, a central steel bar of diameter d that supports the load P, and a hollow rubber cylinder (height h) bonded to the steel tube and bar. (a) Obtain a formula for the shear stress τ in the rubber at distance r from the center of the mount. (b) Obtain a formula for the downward displacement δ of the bar, assuming that G is the shear modulus of elasticity of the rubber and that the steel tube and bar are rigid.

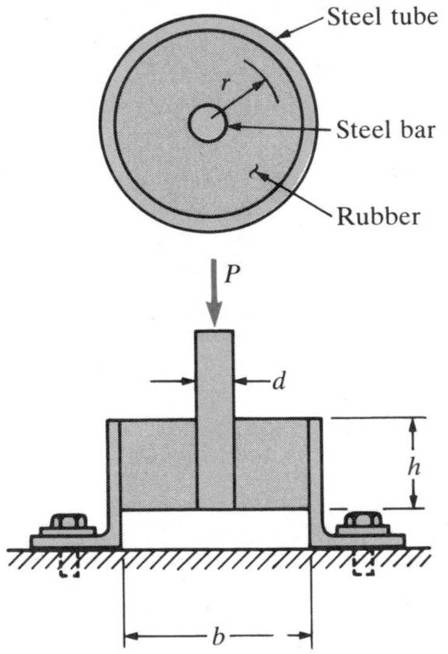

Prob. 1.6-17

1.7-1 A short piece of steel pipe having yield stress $\sigma_y = 42$ ksi is to carry an axial compressive load $P = 260$ k with a factor of safety of 1.8 against yielding (see figure). If

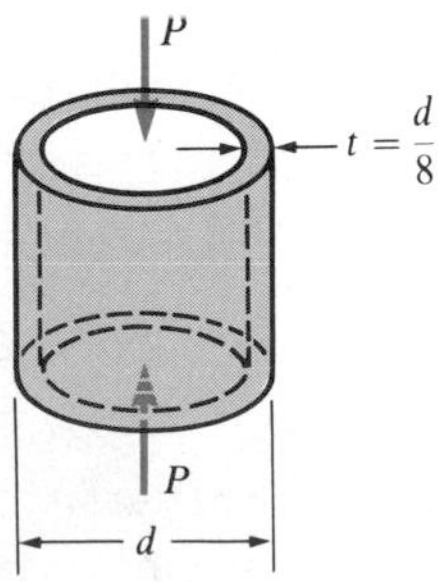

Prob. 1.7-1

the thickness t of the pipe is to be one-eighth of its outside diameter, find the minimum required outside diameter d.

1.7-2 Two members are connected by a bolt AB as shown in the figure. If the load $P = 28$ kN and the allowable shear stress in the bolt is $\tau_{allow} = 90$ MPa, find the minimum required diameter d of the bolt.

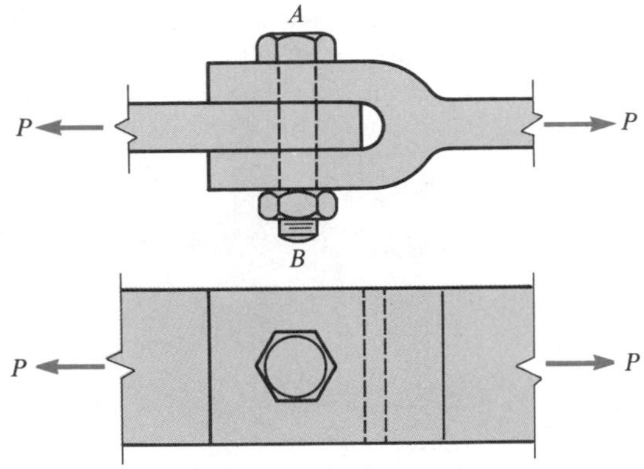

Prob. 1.7-2

1.7-3 A beam AB is supported by a strut CD and carries a load $P = 2500$ lb as shown in the figure. The strut, which consists of two bars, is connected to the beam by a bolt passing through each of the bars at joint C. If the allowable average shear stress in the bolt is 14,000 psi, what is the minimum required diameter d of the bolt?

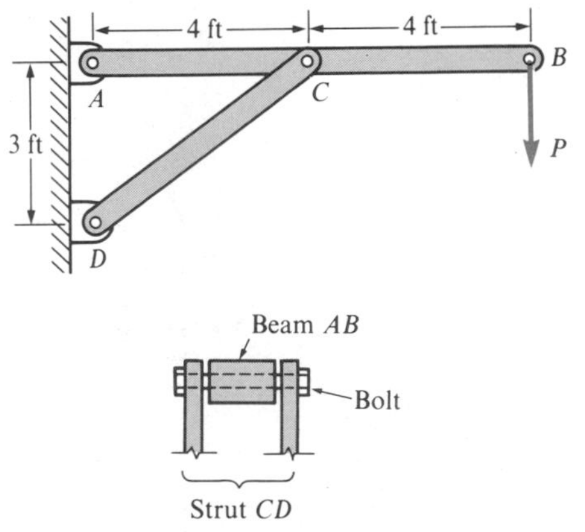

Prob. 1.7-3

1.7-4 A cylinder that has a sealed cover plate fastened with steel bolts contains a gas under pressure p (see figure). The diameter d_b of the bolts is 12 mm, and the allowable tensile stress in the bolts is 70 MPa. If the inside diameter D of the cylinder is 250 mm and the pressure p is 1.8 MPa, find the number n of bolts needed to fasten the cover.

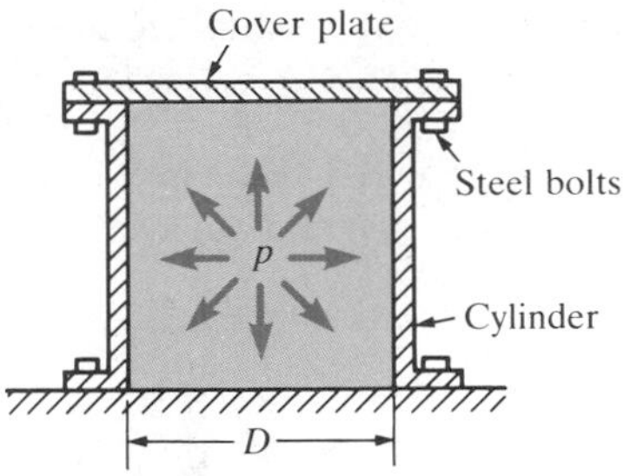

Prob. 1.7-4

1.7-5 A machine component consists of a piston attached to a connecting rod AB, which in turn is connected to a crank arm BC (see figure). The piston slides without friction in a cylinder and is subjected to a constant force P while moving to the right in the figure. The connecting rod, which has cross-sectional area A and length L, is attached at both ends by pins. The crank arm rotates about the axle at C with the pin at B moving in an arc of radius R. The axle at C is supported against any displacement and also exerts a resisting moment M against the crank arm. (a) Obtain a formula giving the maximum permissible force P based upon an allowable compressive stress σ_c in the connecting rod. (b) Calculate the maximum force P for the following data: $\sigma_c = 25$ ksi, $A = 0.10$ in.^{2}, and $R = 0.28L$.

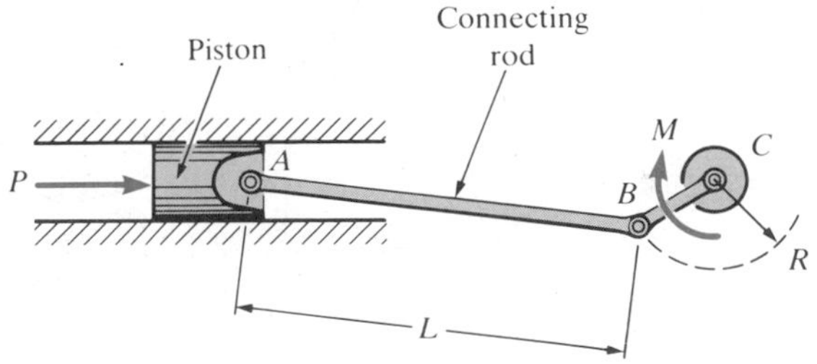

Prob. 1.7-5

1.7-6 A solid bar of circular cross section (diameter $d = 40$ mm) has a small hole drilled laterally through the center of the bar (see figure). The diameter of the hole is $d/4$. Assuming that the allowable average tensile stress on the net cross section of the bar at the hole is $\sigma_{allow} = 80$ MPa, find the allowable load P that the bar can carry in tension.

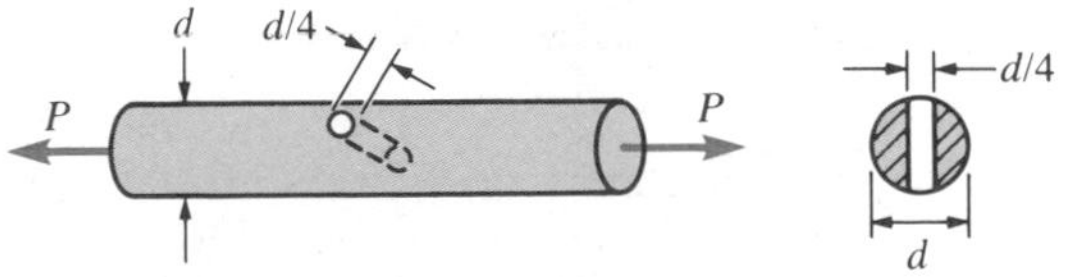

Prob. 1.7-6

1.7-7 An aluminum bar AB is attached to its support at A by a pin of diameter $d = 0.625$ in. (see figure). The bar has thickness $t = 0.6$ in. and width $b = 1.5$ in. If the allowable tensile stress in the bar is 22 ksi and the allowable shear stress in the pin is 12 ksi, find the allowable load P.

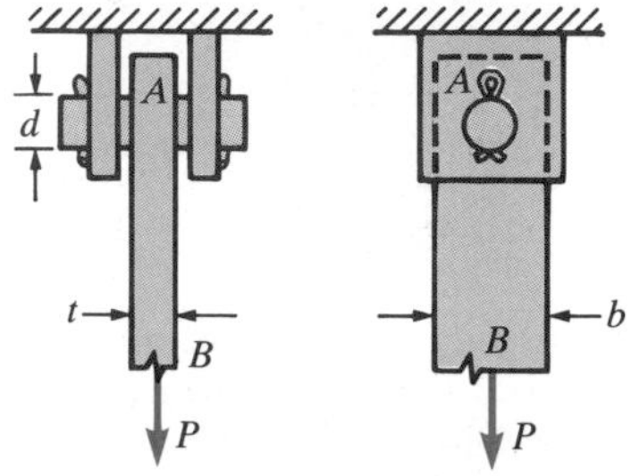

Prob. 1.7-7

1.7-8 Two flat bars loaded by a tensile force P are spliced using two rectangular splice plates and two 15 mm diameter rivets (see figure). The bars have width $b = 20$ mm (except at the splice, where the bars are wider) and thickness $t = 10$ mm. The bars are made of steel having an ultimate stress equal to 400 MPa. The ultimate shear stress for the rivet steel is 180 MPa. Determine the allowable load P if a safety factor of 2.5 is desired with respect to the ultimate load that can be carried. (Consider tension in the bars and shear in the rivets, and disregard friction between the plates.)

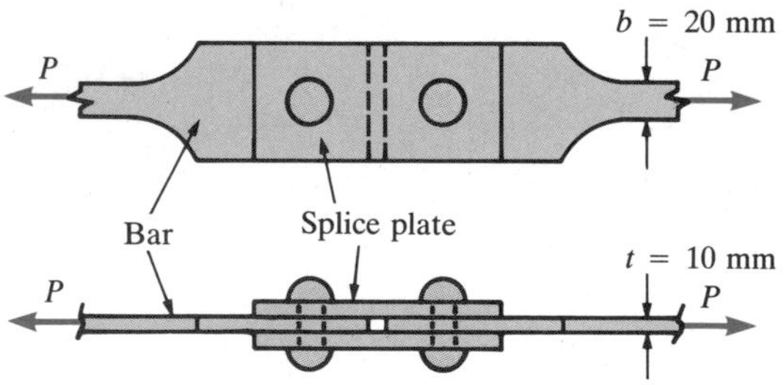

Prob. 1.7-8

1.7-9 A flat bar 2 in. wide and $\frac{1}{4}$ in. thick is subjected to a load P (see figure). A hole of diameter d is drilled through the bar to provide for a pin support. The allowable tensile stress on the net cross section of the bar is 21 ksi, and the allowable shear stress in the pin is 12 ksi. Find the pin diameter d for which the load P will be a maximum.

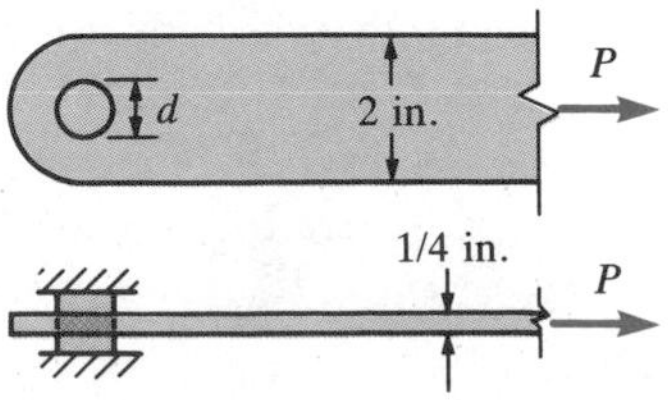

Prob. 1.7-9

1.7-10 A steel pipe column is supported on a circular base plate and a concrete pedestal (see figure). The pipe has an outside diameter of 250 mm and a wall thickness of 10 mm. The allowable average compressive stress for the concrete is 15 MPa and for the steel is 150 MPa. Find the minimum required diameter d of the base plate if it is to support the maximum load P that can be carried by the pipe.

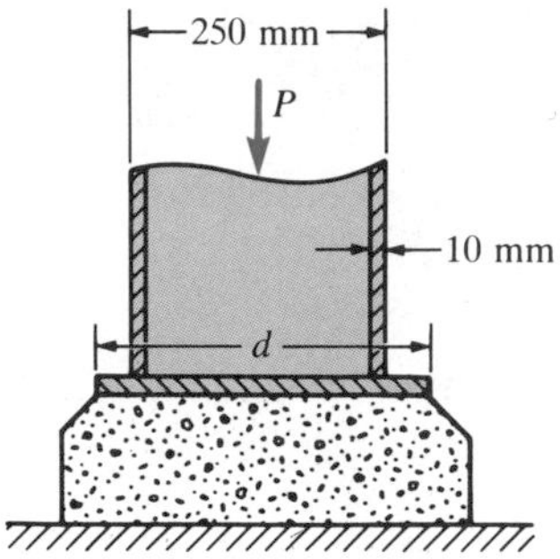

Prob. 1.7-10

1.7-11 A strut S used as bracing for a boat hoist is required to transmit a compressive force $P = 16{,}000$ lb to the deck of

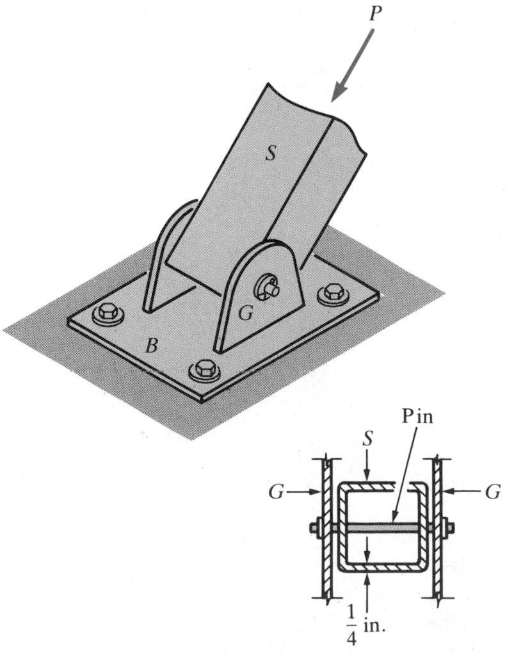

Prob. 1.7-11

a pier through the connection shown in the figure. The strut has a hollow square cross section with a wall thickness of $\frac{1}{4}$ in. A pin through the strut transmits the load to two gussets G that are welded to the base plate B. Four anchor bolts fasten the base plate to the deck. The allowable shear stress in the pin and anchor bolts is 6 ksi and the allowable bearing stress is 20 ksi. (a) Determine the minimum diameter d_P of the pin considering shear in the pin and bearing of the pin against the strut. (b) If the actual diameter of the pin is 1.5 in., determine the minimum thickness t_G of the gussets based upon bearing between the gussets and the pin. (c) If the base plate is $\frac{1}{4}$ in. thick and the strut is inclined at 40° to the horizontal, determine the minimum diameter d_B of the anchor bolts considering both shear in the bolts and bearing between the bolts and the base plate (neglect friction between the base plate and the deck).

1.7-12 A flat bar of width $b = 60$ mm and thickness $t = 10$ mm transmits an axial load P (see figure). A hole of diameter d is drilled through the bar for a pin support. The allowable tensile stress on the net cross section of the bar is $\sigma_t = 140$ MPa, the allowable shear stress in the pin is $\tau_s = 80$ MPa, and the allowable bearing stress between the pin and the bar is $\sigma_b = 200$ MPa. Determine the pin diameter d for which the load P will be a maximum.

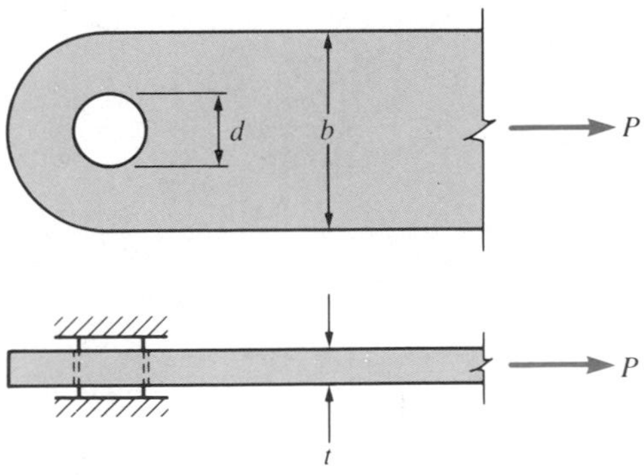

Prob. 1.7-12

1.7-13 A prismatic arm of length L rotates on a smooth horizontal surface about a vertical pivot C with constant angular speed ω (see figure). A mass M_1 is attached to the end of the arm. Denoting by ρ the mass density of the material of the arm and by σ_t the allowable stress in tension, derive a formula for the required cross-sectional area A of the arm.

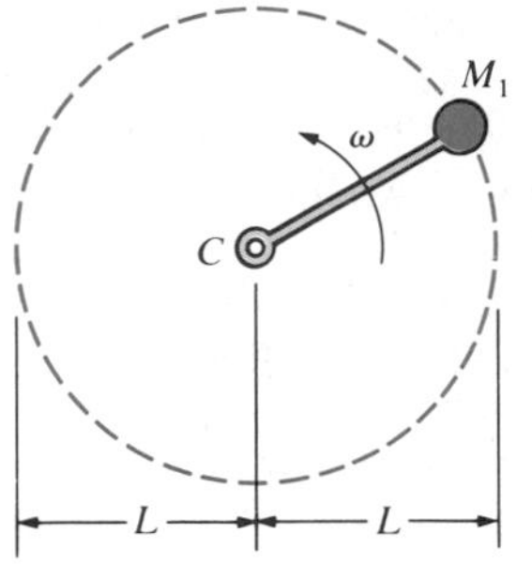

Prob. 1.7-13

1.7-14 Two identical bars AB and BC support a vertical load P (see figure). The distance L between supports is fixed, but the angle θ can be varied by changing the lengths of the bars. We observe that when the angle θ becomes smaller, the bars become shorter but their cross-sectional areas become larger (because the axial forces are larger). On the other hand, if the angle θ becomes larger, the bars become longer and the areas smaller. Thus, we see that the weight of the structure (which is proportional to the volume) is a function of the angle θ. Determine the angle θ so that the structure has minimum weight without exceeding the allowable stress in the bars. (Note: The weights of the bars are very small compared to the force P and may be disregarded.)

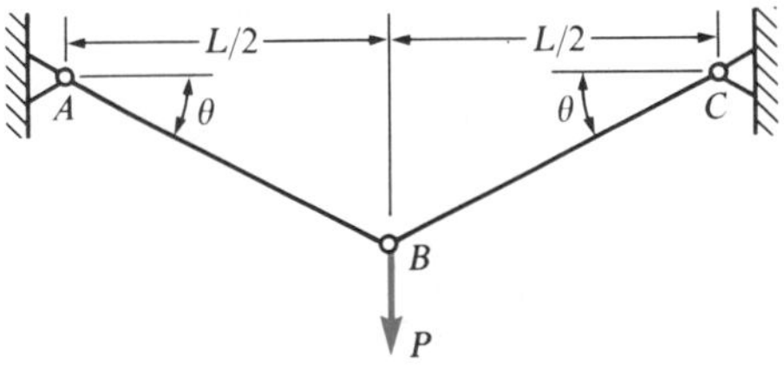

Prob. 1.7-14

***1.7-15** Two bars AB and BC of the same material support a vertical load P (see figure). The length L of the horizontal bar is fixed, but the angle θ can be varied by moving support A vertically and changing the length of bar AB to correspond with the new position of A. The allowable

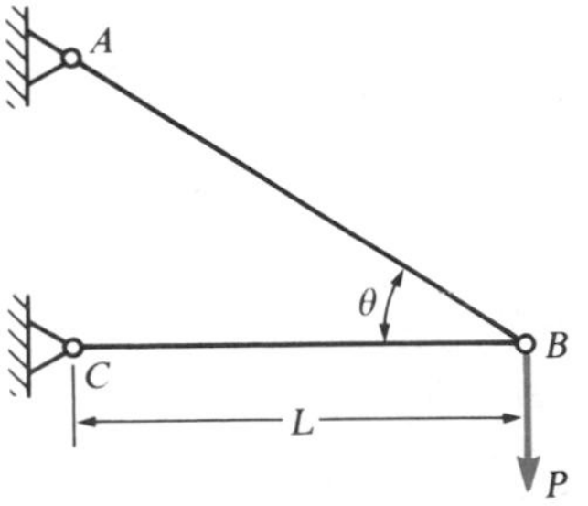

Prob. 1.7-15

stresses in the bars are the same in tension and compression. We observe that when the angle θ is reduced, bar AB becomes shorter but the cross-sectional areas of both bars increase (because the axial forces are larger). The opposite effects occur if the angle θ is increased. Thus, we see that the weight of the structure (which is proportional to the volume) depends upon the angle θ. Find the angle θ so that the structure has minimum weight without exceeding the allowable stresses in the bars. (Note: The weights of the bars are very small compared to the force P and may be disregarded.)

****1.7-16** A truss ABC is constructed of two bars AB and BC of equal lengths and made of the same material (see figure). The distance L between supports is fixed, but the angle θ can be varied by moving joint B horizontally and changing the lengths of the bars to correspond with the new location of B. The truss is subjected to three independent loading conditions: (1) the horizontal load P acts alone; (2) the vertical load P acts alone; and (3) both loads P act simultaneously. The allowable stress for compression in the bars is two-thirds of the allowable stress for tension. We observe that changing the angle θ produces changes in the lengths of the bars and changes in the cross-sectional areas (because the axial forces are changed); thus, the weight of the truss is a function of the angle θ. Determine the angle θ so that the truss has minimum weight without exceeding the allowable stresses in the bars. (Note: The weights of the bars are very small compared to the forces P and may be disregarded.)

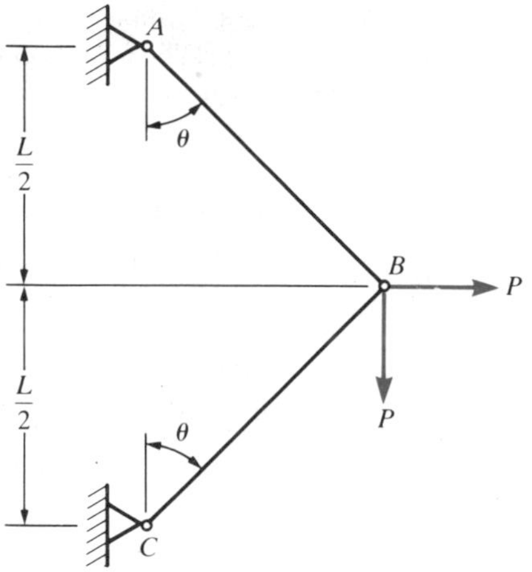

Prob. 1.7-16

****1.7-17** The ends of a rope of length $L = 6$ ft are fastened to supports at points A and B (see figure). The horizontal distance between the supports is $a = 60$ in. and point B is below the elevation of A. A sheave (or pulley) of diameter $d = 8$ in. supports a vertical load P and rolls along the rope until it comes to rest in the equilibrium position. If the rope has a breaking strength $S = 560$ lb, and if a factor of safety $n = 2$ with respect to breaking is required, what is the allowable load P?

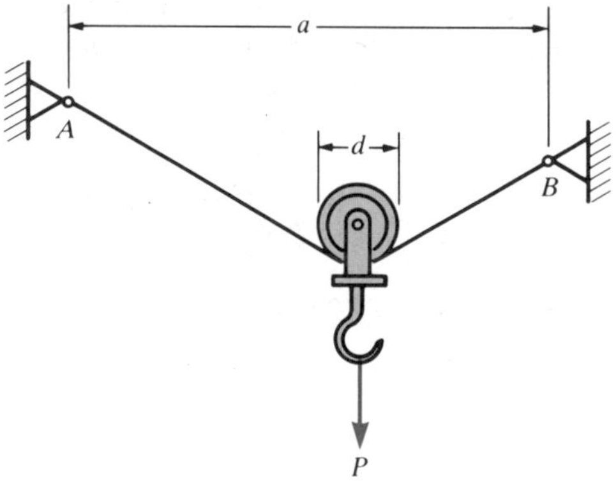

Prob. 1.7-17

CHAPTER 2

Axially Loaded Members

2.1 INTRODUCTION

This chapter is devoted to the behavior of **axially loaded members**, which are structural elements having straight longitudinal axes and carrying only axial forces (tensile or compressive). Members of this type appear in such diverse forms as bars in trusses, connecting rods in engines, spokes in bicycle wheels, columns in buildings, and struts in aircraft engine mounts. Their cross sections may be solid, hollow, or thin-walled and open (Fig. 2-1). Whether you are designing a new structure or analyzing an existing structure, it often is necessary to find not only the maximum stresses in the members (as discussed in Chapter 1) but also the displacements. For instance, the displacements may have to be kept within limits in order to maintain certain clearances. Displacements are also needed in the analysis of statically indeterminate structures, a vast subject that we will introduce in Section 2.3.

Other topics of this chapter include temperature effects, stresses on inclined sections, strain energy, dynamic loading, nonlinear behavior, fatigue, and stress concentrations. Although in this chapter we discuss only members with axial loads, we will see later that these same topics are important for all types of structural elements. In treating these subjects, we will make use of the material on tension, compression, and shear that we discussed in Chapter 1.

Throughout the remainder of this book, the words *analysis* and *design* appear many times. The term **analysis** is customarily used in mechanics to mean the calculation of such quantities as stresses, strains, displacements, and load-carrying capacity. When analyzing a structure, or part of a structure, we assume that the dimensions of the structure and

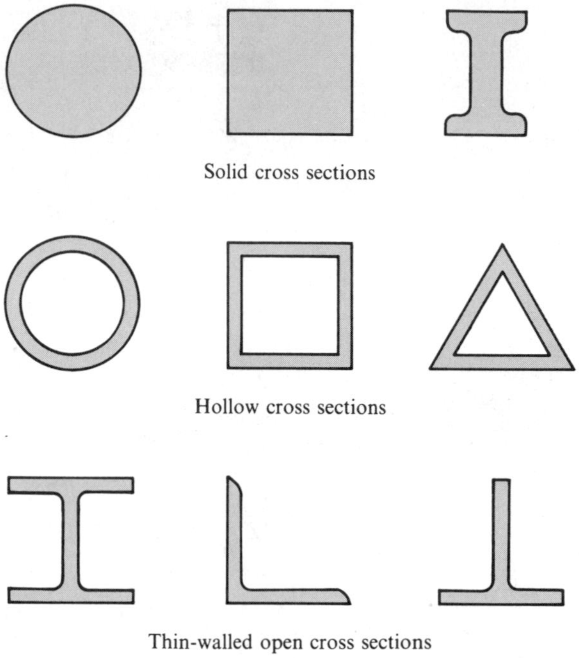

Fig. 2-1 Typical cross sections of structural members

the material of which it is composed are known. Then, using these properties and the various theories of mechanics of materials, we can determine the behavior of the structure when it is subjected to loads.

A more difficult task is that of **design**, which is the determination of the geometric configuration and other characteristics of a structure in order that it will fulfill all prescribed functions. For instance, we may design a structure to support certain given loads without exceeding the allowable stresses and allowable displacements. Designing a structure invariably requires that analyses be performed on that structure, often more than once.

Another part of the design process is **optimization**, which is the task of designing the "best" structure to meet a particular goal, such as minimum weight. Of course, the goal must be achieved within certain restrictions; for instance, the goal might be to design a structure having the least weight but with the displacements not to exceed a specified amount. Thus, analysis, design, and optimization are closely linked, as illustrated in the examples and problems that appear in this and later chapters.

2.2 DISPLACEMENTS OF AXIALLY LOADED MEMBERS

A prismatic bar of length L loaded in tension by axial forces P is shown in Fig. 2-2. If the forces P act at the centroid of the cross section, the uniform stress in the bar at sections away from the ends is given by the formula

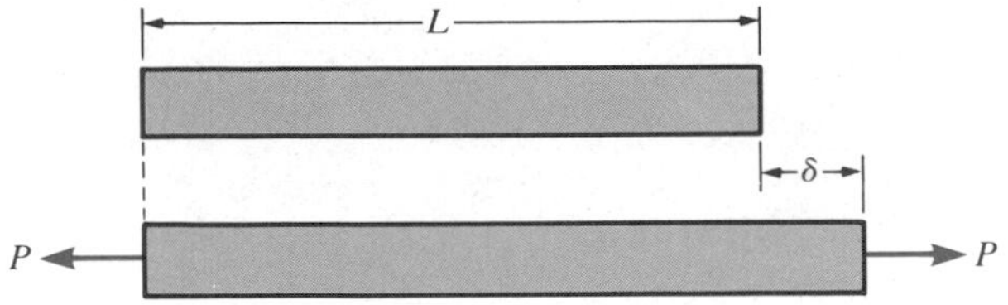

Fig. 2-2 Elongation of a prismatic bar in tension

$\sigma = P/A$, where A is the cross-sectional area. Also, if the bar is made of a homogeneous material, the axial strain is $\epsilon = \delta/L$, where δ is the total elongation produced by the axial forces. Let us also assume that the material is linearly elastic and follows Hooke's law, so that the longitudinal stress and strain are related by the equation $\sigma = E\epsilon$, where E is the modulus of elasticity. Then the preceding expressions can be combined to give the following equation for the **elongation** of the bar:

$$\delta = \frac{PL}{EA} \tag{2-1}$$

This equation shows that the elongation is directly proportional to the load P and the length L and inversely proportional to the modulus of elasticity E and the cross-sectional area A. The product EA is known as the **axial rigidity** of the bar. Of course, Eq. (2-1) also can be used for a member in compression, in which case δ represents the shortening of the bar. When a **sign convention** is needed, elongation is taken as positive and shortening as negative.

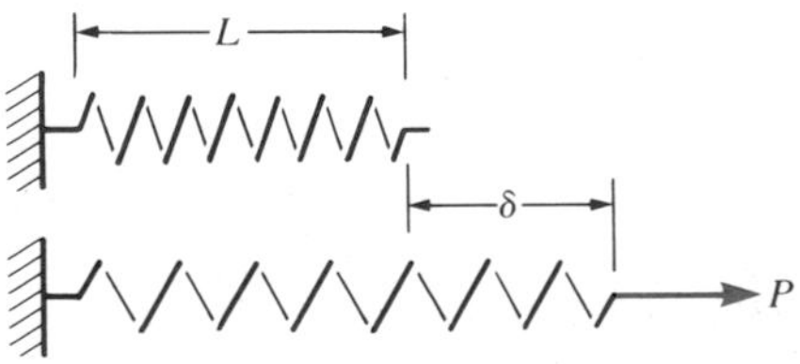

Fig. 2-3 Elongation of an axially loaded spring

From Eq. (2-1) we see that a bar in tension is analogous to an axially loaded spring (Fig. 2-3). Under the action of the force P, the spring elongates an amount δ, so that its total length becomes $L + \delta$, where L is the original length. For a linearly elastic spring, the **stiffness** k is defined as the force required to produce a unit elongation, that is, $k = P/\delta$. The **flexibility** f of the spring is the reciprocal of the stiffness, or the elongation produced by a load of unit value. Thus, we have $f = 1/k = \delta/P$. Sometimes the stiffness and flexibility of a spring are called the **spring constant** and **compliance**, respectively.

In the case of an axially loaded bar (Fig. 2-2), we define the stiffness in the same manner as for a spring, that is, the stiffness k is the force required to produce a unit elongation. From Eq. (2-1) we see that the stiffness of the bar in Fig. 2-2 is

$$k = \frac{EA}{L} \tag{2-2}$$

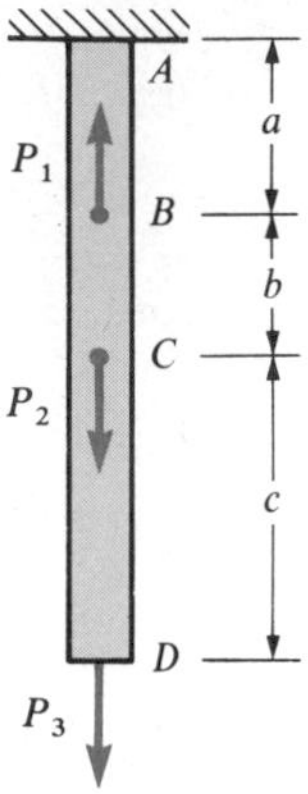

Fig. 2-4 Bar with intermediate axial loads

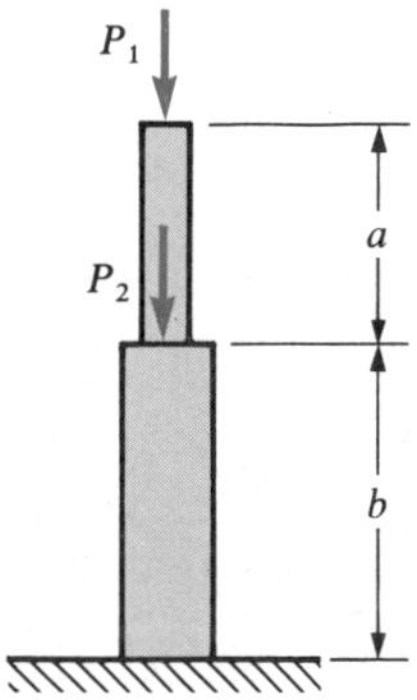

Fig. 2-5 Bar consisting of prismatic parts

In an analogous manner, the flexibility is defined as the elongation due to a unit load. Thus, from Eq. (2-1) we obtain

$$f = \frac{L}{EA} \tag{2-3}$$

which is the reciprocal of the stiffness. Stiffnesses and flexibilities have an important role in the analysis of many types of structures.

The change in length of a prismatic bar loaded only at the ends can be found without difficulty by using Eq. (2-1). Moreover, the equation can be readily adapted to handle more general situations. Suppose, for instance, that a prismatic bar is loaded by one or more intermediate axial loads (Fig. 2-4). Then we can determine the axial force in each part of the bar (that is, in parts AB, BC, and CD) by statics, and we can calculate the elongation or shortening of each part separately. Finally, these changes in lengths can be added algebraically to obtain the change in length of the entire bar. The same general method can be used when the bar consists of prismatic parts having different materials or different cross-sectional areas (Fig. 2-5).

In general, the total elongation δ of a bar consisting of several prismatic parts having different axial forces, dimensions, or materials may be obtained from the equation

$$\delta = \sum_{i=1}^{n} \frac{P_i L_i}{E_i A_i} \tag{2-4}$$

in which the subscript i is a numbering index for the various parts of the bar and n is the total number of parts. Note especially that P_i is not a load but is the axial force in part i. The use of Eq. (2-4) is illustrated in Example 1 at the end of this section.

If either the axial force or the cross-sectional area varies continuously along the axis of the bar, then Eq. (2-4) is not suitable. Instead, the elongation can be found by considering a differential element of the bar, obtaining an expression for its elongation, and then integrating over the length of the bar. To illustrate this idea, consider the tapered bar AB shown in Fig. 2-6. This bar is subjected to a continuously distributed axial load, as shown by the arrows in Fig. 2-6a. A continuous axial load of this kind may be produced by centrifugal forces, magnetic forces, or the weight of a bar hanging vertically. Now consider any intermediate cross section C, located at distance x from the left-hand end (Fig. 2-6a). The internal axial force $P(x)$ acting at this section (Fig. 2-6b) may be determined from static equilibrium using either part of the bar (AC or CB) as a free body. Thus, the force $P(x)$ is known in terms of x. Also, the cross-sectional area $A(x)$ at section C may be expressed as a function of x.

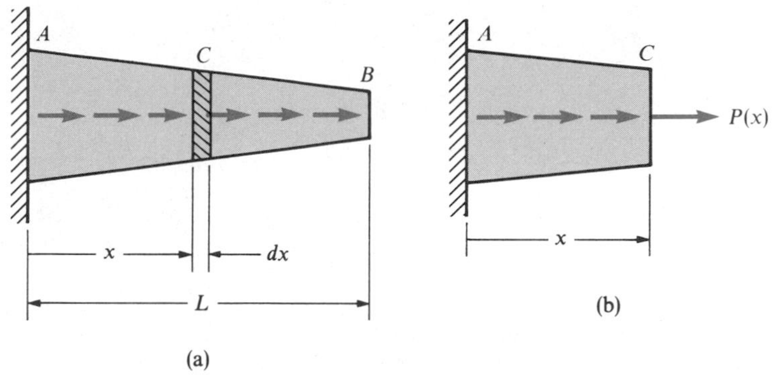

Fig. 2-6 Bar with varying cross-sectional area and varying axial force

Next we consider an element of the bar of length dx at section C (Fig. 2-6a). The elongation $d\delta$ of this element may be obtained from Eq. (2-1) by substituting $P(x)$ for P, dx for L, and $A(x)$ for A, as follows:

$$d\delta = \frac{P(x)\,dx}{EA(x)}$$

The elongation of the entire bar is obtained by integrating over the length:

$$\delta = \int_0^L d\delta = \int_0^L \frac{P(x)\,dx}{EA(x)} \qquad (2\text{-}5)$$

If the expressions for $P(x)$ and $A(x)$ are not too complicated, the integral can be evaluated analytically and a formula for δ can be obtained (see Example 3). However, if the integration is difficult or impossible, then a numerical method for evaluating the integral can be used.

The derivation of Eq. (2-5) was based upon the assumption that the stress distribution is uniform over the entire cross section ($\sigma = P/A$). This assumption is valid for prismatic bars but not for tapered bars, and Eq. (2-5) gives accurate results for a tapered bar only if the angle between the sides of the bar is small. For instance, if the angle between the sides is 20°, the normal stress calculated from the expression $\sigma = P/A$ is 3% less than the exact stress. For smaller angles, the error is less. If the taper of the bar is large, more accurate methods of analysis may be needed (see Ref. 2-1).

Example 1

A vertical steel bar ABC has length L_1 and cross-sectional area A_1 from A to B, and length L_2 and area A_2 from B to C (Fig. 2-7a). A load P_1 acts at point C. A horizontal arm BD is pinned to the vertical bar at B and carries a load P_2 at D. Calculate the vertical displacement δ at point C if $P_1 = 10$ kN, $P_2 = 26$ kN, $a = b$,

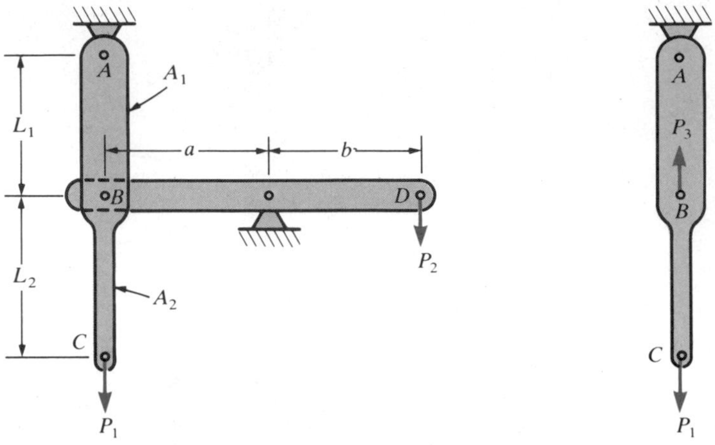

Fig. 2-7 Example 1 (a) (b)

$L_1 = 0.5$ m, $A_1 = 160$ mm^2, $L_2 = 0.8$ m, $A_2 = 100$ mm^2, and $E = 200$ GPa for steel. (Disregard the weights of the bars.)

From the moment equilibrium of arm BD, we see that a vertical force $P_3 = P_2 b/a$ is transmitted through the pin at B. This force acts upward on the vertical bar ABC (Fig. 2-7b). In this example, we have $a = b$; hence $P_3 = P_2 = 26$ kN. Part AB of the vertical bar is subjected to an axial compressive force equal to $P_3 - P_1$, or 16 kN. Part BC carries a tensile force equal to P_1, or 10 kN. Therefore, with tension considered positive, Eq. (2-4) yields

$$\delta = \sum \frac{P_i L_i}{E_i A_i} = -\frac{(16\ \text{kN})(0.5\ \text{m})}{(200\ \text{GPa})(160\ \text{mm}^2)} + \frac{(10\ \text{kN})(0.8\ \text{m})}{(200\ \text{GPa})(100\ \text{mm}^2)}$$

$$= -0.250 + 0.400 = 0.150\ \text{mm}$$

Since δ is positive, representing an elongation, the displacement of point C is downward.

Example 2

A collar of mass M slides without friction on a horizontal rod AB (Fig. 2-8a). The collar is restrained by a spring of stiffness k that is placed between the collar and the flange at end B of the rod. The rod rotates in a horizontal plane with constant angular speed ω about a vertical axis through A. The length of the rod is L and the natural length (or unloaded length) of the spring is b, with L being much larger than b (Fig. 2-8b). Also, the length of the collar is $2c$. (a) Determine the shortening δ of the spring. (b) Determine the angular speed ω in order for the spring to be compressed to half of its original length.

(a) When the bar rotates, the mass M moves outward due to centrifugal effects and compresses the spring by an amount δ (Fig. 2-8c). The radial distance r from the center of rotation A to the center of gravity of the collar is

$$r = L - b + \delta - c$$

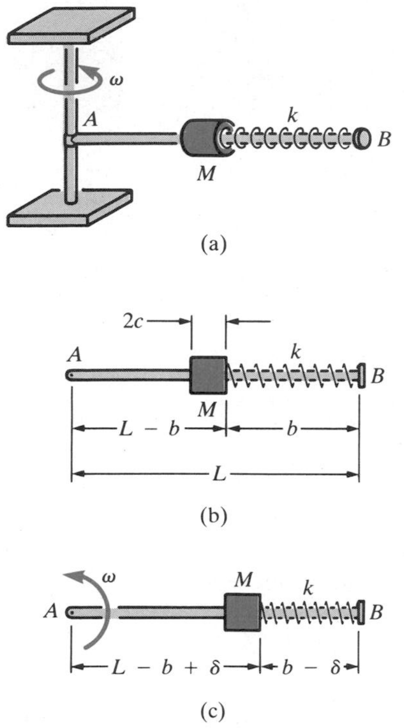

Fig. 2-8 Example 2. Collar of mass M sliding on a rotating rod AB

and the centrifugal force due to the mass M of the collar is

$$F = Mr\omega^2 = M(L - b + \delta - c)\omega^2$$

The shortening of the spring is equal to the force in the spring divided by its stiffness:

$$\delta = \frac{F}{k} = \frac{M\omega^2}{k}(L - b + \delta - c) \qquad \text{(a)}$$

Solving for δ from this equation, we get

$$\delta = \frac{(L - b - c)M\omega^2/k}{1 - M\omega^2/k} \qquad \text{(b)}$$

From this result we can calculate the shortening of the spring for any given values of the nondimensional parameter $M\omega^2/k$ and the dimensions L, b, and c.

(b) The angular speed ω can also be found from Eq. (a). Solving for the square of the frequency, we get

$$\omega^2 = \frac{k\delta}{M(L - b + \delta - c)} \qquad \text{(c)}$$

This formula can be used to calculate ω^2 for any desired value of the shortening δ. For instance, if $\delta = b/2$, we get

$$\omega^2 = \frac{kb}{M(2L - b - 2c)} \qquad \text{(d)}$$

which is the square of the angular frequency required to compress the spring to half of its natural length.

In this example we disregarded the elongation of the rod AB and considered only the shortening of the spring. This method is satisfactory provided the elongation of the rod is very small compared to the length of the rod itself, a condition that is met for metals and other ordinary structural materials. Under this condition, we can use the initial length L of the rod in the calculations. Also, we disregarded the effect of the mass of the spring on the shortening of the spring, a procedure that is satisfactory if the mass of the collar is much greater than the mass of the spring itself.

Example 3

A slightly tapered bar AB of circular cross section and length L is supported at end B and subjected to a tensile load P at the free end A (Fig. 2-9a). The diameters of the bar at ends A and B are d_1 and d_2, respectively. Derive a formula for the elongation δ of the bar due to the load P.

In this example, the bar has constant axial force P throughout its length, but the cross-sectional area varies. The first step in the solution is to obtain an expression for the area $A(x)$ at any cross section. To simplify the integration that must be performed, let us extend the sides of the tapered bar until they meet at point O (Fig. 2-9b). Then we will use this point as the origin for the distance x. The

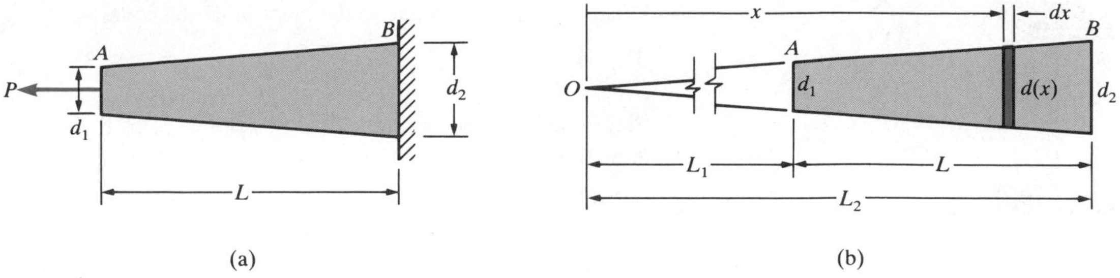

Fig. 2-9 Example 3. Tapered bar of circular cross section

distances L_1 and L_2 from the origin to ends A and B, respectively, are in the ratio

$$\frac{L_1}{L_2} = \frac{d_1}{d_2} \tag{e}$$

as obtained from the similar triangles in Fig. 2-9b.

The diameter $d(x)$ of the bar at distance x from the origin (Fig. 2-9b) is

$$d(x) = \frac{d_1 x}{L_1} \tag{f}$$

and the corresponding cross-sectional area is

$$A(x) = \frac{\pi[d(x)]^2}{4} = \frac{\pi d_1^2 x^2}{4L_1^2} \tag{g}$$

Now we substitute into Eq. (2-5) for δ and obtain

$$\delta = \int \frac{P(x)\,dx}{EA(x)} = \int_{L_1}^{L_2} \frac{P\,dx\,(4L_1^2)}{E(\pi d_1^2 x^2)} = \frac{4PL_1^2}{\pi E d_1^2} \int_{L_1}^{L_2} \frac{dx}{x^2}$$

By performing the integration and substituting the limits, we get

$$\delta = \frac{4PL_1^2}{\pi E d_1^2}\left[-\frac{1}{x}\right]_{L_1}^{L_2} = \frac{4PL_1^2}{\pi E d_1^2}\left(\frac{1}{L_1} - \frac{1}{L_2}\right)$$

This expression for δ can be simplified by noting that

$$\frac{1}{L_1} - \frac{1}{L_2} = \frac{L_2 - L_1}{L_1 L_2} = \frac{L}{L_1 L_2}$$

Thus, the equation for δ becomes

$$\delta = \frac{4PL}{\pi E d_1^2}\left(\frac{L_1}{L_2}\right)$$

Finally, we substitute $L_1/L_2 = d_1/d_2$ (see Eq. e) and obtain

$$\delta = \frac{4PL}{\pi E d_1 d_2} \tag{2-6}$$

This equation gives the required formula for the elongation of the tapered bar. As a special case, note that if the bar is prismatic with $d_1 = d_2 = d$, Eq. (2-6) simplifies to

$$\delta = \frac{4PL}{\pi E d^2} = \frac{PL}{EA}$$

in which $A = \pi d^2/4$.

A common mistake is to assume that the elongation of a tapered bar is equal to the elongation of a prismatic bar having the same cross-sectional area as the cross section located midway between the ends A and B of the tapered bar. Examination of Eq. (2-6) shows that such is not the case for the tapered bar in this example, nor is it generally the case.

2.3 STATICALLY INDETERMINATE STRUCTURES

In the preceding section we dealt with axially loaded bars and other simple structures that could be analyzed by static equilibrium. In every case it was possible to determine all axial forces in the members and all reactions at the supports by drawing free-body diagrams and solving equilibrium equations. Such structures are classified as **statically determinate**. For many structures, however, the equations of static equilibrium are not sufficient for the calculation of axial forces and reactions; these structures are called **statically indeterminate**. Structures of this type can be analyzed by supplementing the equilibrium equations with additional equations pertaining to the displacements of the structure.

Two important methods for analyzing statically indeterminate structures are the **flexibility method** and the **stiffness method**, both of which are described in this section as they apply to simple structures composed of axially loaded members. The flexibility method will also be used later when discussing statically indeterminate problems involving torsion (Section 3.7) and bending (Chapter 8).

To explain the **flexibility method**, let us consider the analysis of the statically indeterminate bar AB shown in Fig. 2-10a. This prismatic bar is attached at both ends to rigid supports and is axially loaded by the force P at an intermediate point C. As a consequence, reactions R_a and R_b will develop at the ends of the bar.* The reactions R_a and R_b cannot be found by statics alone, as is evident from Fig. 2-10a, which serves as a free-body

* **Reactions** are passive actions that are induced at the supports of a structure; their magnitudes and directions are determined by the nature of the structure, the materials, and the loads. **Loads** are active forces that are applied to the structure by some external cause. In most situations the loads are known quantities and the reactions must be found from an analysis of the structure.

To distinguish them from loads, reactive forces are indicated by drawing a slash, or slanted line, across the arrows, as in Fig. 2-10a.

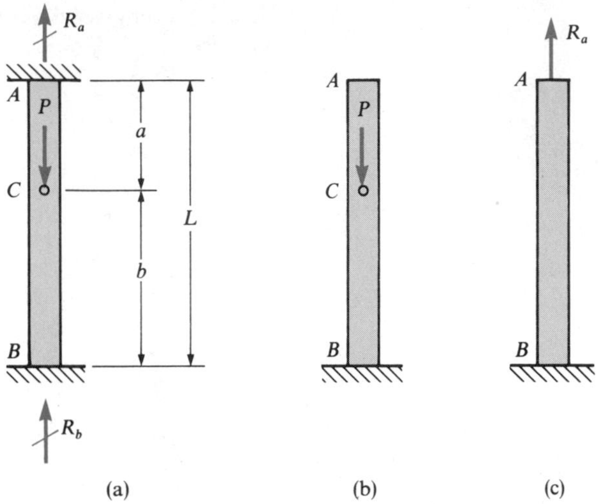

Fig. 2-10 Statically indeterminate bar (flexibility method of analysis)

diagram of the bar because all forces acting on the bar are shown. Only one independent equation of static equilibrium can be formulated from this free-body diagram, as follows:

$$R_a + R_b = P \tag{a}$$

Because this equation contains both unknown reactions, it is not sufficient for their calculation. A second equation must be obtained from the displacements of the bar.

To begin the analysis, we designate one of the unknown reactions as the **statical redundant**, or the force that is in excess of those that can be obtained by statics alone. Let us choose R_a as the redundant reaction in this example. If R_a can be found, then the other reaction R_b can be obtained by static equilibrium from Eq. (a). When the unknown reaction R_a is removed from the structure, the effect is to release the support at end A, thereby producing the statically determinate and stable structure shown in Fig. 2-10b. Thus, from the standpoint of having a structure that is capable of supporting loads, the reaction at end A is not needed; that is, it is redundant. The structure that remains after releasing the redundant is called the **released structure**, or the **primary structure**.

Now let us consider the effect of the load P on the displacement of point A in the released structure (Fig. 2-10b). This displacement is

$$\delta_P = \frac{Pb}{EA} \tag{b}$$

and is positive downward.

Next, let us consider the effect of the redundant force R_a on the displacement of point A (Fig. 2-10c). Note that, although it is an unknown quantity, R_a is now visualized as a load acting on the released structure.

The displacement of point A due to R_a is

$$\delta_R = \frac{R_a L}{EA} \tag{c}$$

and is positive upward.

The final displacement δ of point A due to both P and R_a acting simultaneously is found by combining δ_P and δ_R. Taking downward displacements as positive, we obtain

$$\delta = \delta_P - \delta_R$$

Because the actual displacement δ of point A is equal to zero (Fig. 2-10a), we get

$$\delta = 0 \qquad \text{or} \qquad \delta_R = \delta_P \tag{d}$$

Substituting from Eqs. (b) and (c), we now obtain

$$\frac{R_a L}{EA} = \frac{Pb}{EA} \tag{e}$$

from which

$$R_a = \frac{Pb}{L} \tag{2-7}$$

Thus, the redundant reaction has been calculated from an equation related to the displacements of the bar (Eq. d). Knowing the redundant, we can find R_b from equilibrium by using Eq. (a):

$$R_b = P - R_a = \frac{Pa}{L} \tag{2-8}$$

Both reactions for the bar have now been found.

The flexibility method for analyzing the statically indeterminate bar of Fig. 2-10a may be summarized as follows. First, one of the unknown reactions is selected as the redundant and then released from the structure by cutting through the bar and removing the support. The released structure, which is statically determinate and stable, is then loaded separately by the actual load P and by the redundant itself. Next, the displacements caused by these two quantities are calculated from the properties of the bar and the material (Eqs. b and c). Then the two displacements are combined into an **equation of compatibility** (Eq. d). The equation of compatibility expresses a condition pertaining to the original structure, namely, that the displacement δ at end A is zero. When the expressions for the displacements are substituted, the equation of compatibility takes the form of Eq. (e), which then can be solved for the redundant force R_a (Eq. 2-7). Finally, the remaining unknown force is found from an **equation of equilibrium**.

This method of analysis is called the **flexibility method** because flexibilities appear in the equation of compatibility. In this example, the equation of compatibility (Eq. e) contains the flexibility L/EA (see Eq. 2-3) as the coefficient of the unknown redundant R_a. Another name for the method is the **force method**, because forces are the unknown quantities in equations of compatibility. The flexibility method can be used for different types of structures and for structures having many redundant forces. However, in this section only elementary indeterminate structures with one or two redundant forces will be considered. Because the method requires adding the displacements caused by different forces, it is valid only when the material behaves in a linearly elastic manner and follows Hooke's law. This is an important limitation that will be discussed later when we explain the principle of superposition in more detail.

In general, the flexibility method is based upon the solution of two sets of equations: (1) equations of static equilibrium (such as Eq. a), and (2) compatibility equations (such as Eq. e). There will always be as many compatibility equations as redundants, so the two sets of equations are sufficient to find all of the unknown forces. When formulating the equations of compatibility, we need equations for the deformations of the members (such as Eqs. b and c). These equations are based upon the properties of the members, including dimensions and materials, and are known generally as *constitutive relations*. In our example, these relations are expressions of Hooke's law.

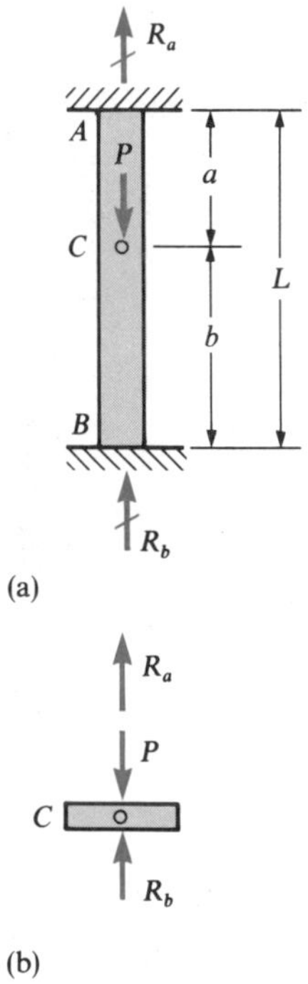

Fig. 2-11 Statically indeterminate bar (stiffness method of analysis)

The **stiffness method** for analyzing statically indeterminate structures differs from the flexibility method in that displacements (rather than forces) are taken as the unknown quantities. For this reason, the method is also called the **displacement method**. The unknown displacements are obtained by solving equations of equilibrium (rather than equations of compatibility) that contain coefficients in the form of stiffnesses (see Eq. 2-2). The stiffness method is quite general and can be used for a wide variety of structures. However, like the flexibility method, it is limited to structures that behave in a linearly elastic manner.

To explain the stiffness method, let us again analyze a prismatic bar AB held between rigid supports (Fig. 2-11a). The reactions at the supports are denoted R_a and R_b, as before. In this new solution, the vertical displacement δ_c of point C, which is the junction of the two parts of the bar, is taken as the unknown quantity. The axial forces R_a and R_b in the upper and lower parts of the bar can be expressed in terms of δ_c. To accomplish this step, we imagine that point C is moved downward by the distance δ_c. Then the upper part of the bar elongates and the lower part shortens by that amount. The axial forces in the two parts are equal to their respective stiffnesses multiplied by the displacement, as follows:

$$R_a = \frac{EA}{a}\delta_c \qquad R_b = \frac{EA}{b}\delta_c \tag{f}$$

In writing these equations, we have assumed that δ_c is positive downward, thereby producing tension in the upper part of the bar and compression in the lower part.

The next step is to isolate point C in the bar as a free body (Fig. 2-11b). Acting on the free body are the downward load P, the tensile force R_a in the upper part, and the compressive force R_b in the lower part. From static equilibrium, we then obtain

$$R_a + R_b = P \tag{g}$$

or, upon substituting from Eqs. (f),

$$\frac{EA}{a}\delta_c + \frac{EA}{b}\delta_c = P \tag{h}$$

which yields

$$\delta_c = \frac{Pab}{EA(a+b)} = \frac{Pab}{EAL} \tag{i}$$

Knowing the displacement δ_c, we now can find R_a and R_b from Eqs. (f):

$$R_a = \frac{Pb}{L} \qquad R_b = \frac{Pa}{L} \tag{j}$$

Of course, these results are the same as those already obtained (see Eqs. 2-7 and 2-8).

Let us now summarize the procedure for the stiffness method of analysis. The first step is to select a suitable displacement as the unknown quantity. A displacement will be suitable if the forces in the individual parts of the structure can be expressed in terms of that displacement (see Eqs. f). These expressions, which are based upon the properties of the bars and the material, are the constitutive relations. In all of our examples, the constitutive relations are based upon Hooke's law. Next, the forces are related by an **equation of equilibrium**, such as Eq. (g) in the preceding example. Then the expressions giving the forces in terms of the unknown displacement are substituted into the equation of equilibrium, thereby producing an equation with only the selected displacement as an unknown (Eq. h). Note that the coefficients of δ_c in this equation are stiffnesses. This equation is solved for the unknown displacement (Eq. i), and, finally, the forces are found from the displacement (Eqs. j). Thus, the stiffness method produces all of the desired results.

When comparing the flexibility and stiffness methods, we see that the flexibility method requires the solution of equations of compatibility for unknown forces whereas the stiffness method requires the solution of equations of equilibrium for unknown displacements. Both methods utilize constitutive relations, which are expressions relating forces and elongations through Hooke's law.

For the types of problems considered in this section, we do not need a highly formalized approach. It is sufficient to keep in mind that the analysis of a statically indeterminate structure requires the use of two basic concepts, static equilibrium and compatibility of displacements. The equations of equilibrium and the equations of compatibility are obtained from the nature of the structure. In setting up those equations, we use expressions (such as $\delta = PL/EA$ and $P = EA\delta/L$) to relate the forces to the changes in lengths of the members. Finally, we solve the equations of equilibrium and compatibility for the unknown forces and displacements. These techniques are illustrated in the following examples, all of which deal with simple structures in which there is one redundant force, that is, one force in excess of the number that can be found by statics alone. Of course, the flexibility and stiffness methods apply to much more complex structures in which there are many redundants. Such structures require the solution of many simultaneous equations (see Ref. 2-2).*

Example 1

A circular steel cylinder S and a hollow copper circular tube C having the same length L are compressed between the rigid plates of a testing machine by forces P (Fig. 2-12a). Determine the following quantities: (a) the compressive forces P_s and P_c in the steel cylinder and copper tube, respectively; (b) the corresponding compressive stresses σ_s and σ_c in the materials; and (c) the shortening δ of the assembly.

We begin by isolating the upper plate of the assembly as a free body (Fig. 2-12b). This plate is subjected to the force P and to the unknown compressive forces P_s and P_c; thus, the equation of equilibrium is

$$P_s + P_c = P \tag{k}$$

This is the only equilibrium equation available and, because it contains two unknowns, the structure is statically indeterminate.

When analyzing this structure by the *flexibility method*, we must select either P_s or P_c as the statical redundant; let us select P_s. Then we may remove the steel cylinder and obtain the released structure shown in Fig. 2-12c. This structure, which is statically determinate, is subjected to the loads P and the compressive forces P_s in the steel cylinder.

Under the action of the load P alone, the distance between the plates in the released structure (Fig. 2-12c) is reduced by the amount

$$\delta_P = \frac{PL}{E_c A_c}$$

in which $L/E_c A_c$ is the flexibility of the copper tube. In a similar way, we get the increase in distance caused by the redundant force acting alone:

$$\delta_R = \frac{P_s L}{E_c A_c}$$

* From a historical viewpoint, it appears that Euler in 1774 was the first to analyze a statically indeterminate system; he considered the problem of a rigid table with four legs supported on an elastic foundation (Refs. 2-3 and 2-4). The next work was done by the French mathematician and engineer L. M. H. Navier, who in 1825 pointed out that statically indeterminate reactions could be found only by taking into account the elasticity of the structure (Ref. 2-5). Navier solved a statically indeterminate three-bar truss.

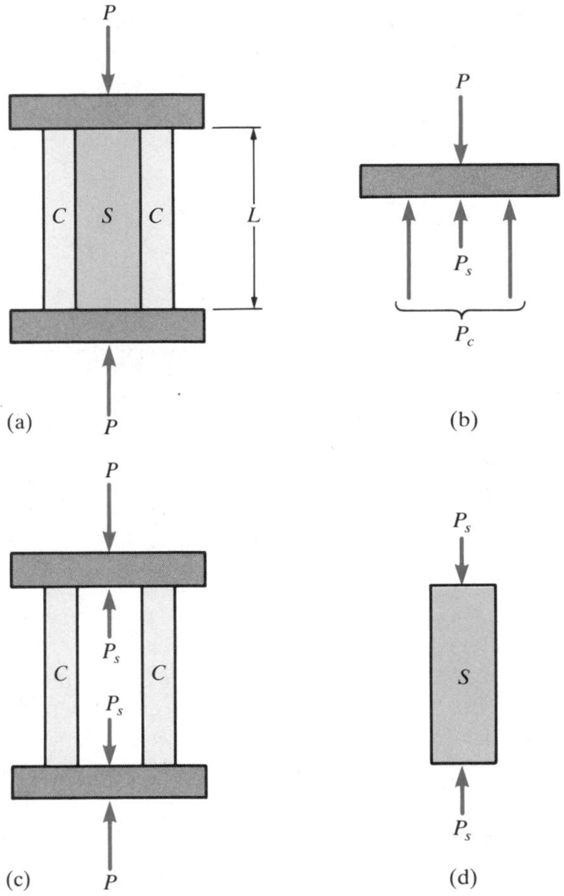

Fig. 2-12 Example 1. Statically indeterminate system (flexibility method)

The net decrease in distance between the plates, which is the same as the shortening δ of the copper tube, is $\delta_P - \delta_R$. Finally, we observe that the shortening of the steel bar due to the forces P_s (Fig. 2-12d) is

$$\delta_S = \frac{P_s L}{E_s A_s}$$

in which $L/E_s A_s$ is the flexibility of the steel bar. The preceding three equations for δ_P, δ_R, and δ_S are the constitutive relations that give the displacements in terms of the forces.

The equation of compatibility is obtained from the fact that the shortening of the copper tube $(\delta_P - \delta_R)$ is equal to the shortening of the steel cylinder (δ_S):

$$\delta_P - \delta_R = \delta_S$$

Substituting the constitutive relations yields

$$\frac{PL}{E_c A_c} - \frac{P_s L}{E_c A_c} = \frac{P_s L}{E_s A_s} \tag{1}$$

This equation can be solved for the redundant force P_s; the result is

$$P_s = \frac{E_s A_s}{E_s A_s + E_c A_c} P \tag{2-9a}$$

The force P_c in the copper tube is found from the equilibrium equation (Eq. k) by substituting the preceding value for P_s:

$$P_c = P - P_s = \frac{E_c A_c}{E_s A_s + E_c A_c} P \tag{2-9b}$$

Equations (2-9) show that the forces in the steel and copper parts are proportional to their axial rigidities.

Knowing the forces, we can now obtain the stresses in the materials:

$$\sigma_s = \frac{P_s}{A_s} = \frac{E_s P}{E_s A_s + E_c A_c} \qquad \sigma_c = \frac{P_c}{A_c} = \frac{E_c P}{E_s A_s + E_c A_c} \tag{2-10a,b}$$

Note that the stresses are proportional to the moduli of elasticity of the respective materials, and the "stiffer" material has the larger stress.

Lastly, we obtain the shortening δ of the assembly from either side of Eq. (l); using the right-hand side gives

$$\delta = \frac{P_s L}{E_s A_s} = \frac{PL}{E_s A_s + E_c A_c} \tag{2-11}$$

This result shows that the shortening of the assembly is equal to the total load divided by the sum of the stiffnesses of the two parts.

Let us now analyze this same structure by the *stiffness method* (Fig. 2-13a). We again isolate the upper plate as a free body (Fig. 2-13b) and obtain the same equation of equilibrium as before:

$$P_s + P_c = P \tag{m}$$

The unknown displacement in this analysis is the shortening δ of the assembly. For convenience, we may consider the lower plate as immovable and then δ becomes the downward displacement of the upper plate (Fig. 2-13b). The forces P_s and P_c can be expressed in terms of δ by the following constitutive relations:

$$P_s = \frac{E_s A_s}{L} \delta \qquad P_c = \frac{E_c A_c}{L} \delta \tag{n}$$

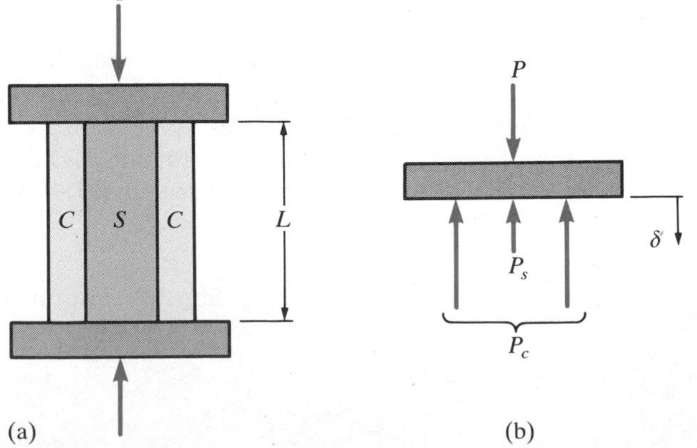

Fig. 2-13 Example 1. Statically indeterminate system (stiffness method)

Substituting these expressions for P_s and P_c into the equilibrium equation yields

$$\frac{E_s A_s}{L}\delta + \frac{E_c A_c}{L}\delta = P$$

This equation is now readily solved for the unknown displacement:

$$\delta = \frac{PL}{E_s A_s + E_c A_c}$$

Knowing the displacement, we can get the forces P_s and P_c by substituting into Eqs. (n). All of these results are the same as those given previously (see Eqs. 2-9 and 2-11). Thus, the forces and displacement have been found by the stiffness method.

This example illustrates the similarities and differences between the flexibility and stiffness methods. In the flexibility method we take a redundant force as the unknown quantity and obtain it from an equation of compatibility of displacements. The equation of compatibility incorporates constitutive relations giving the displacements in terms of the forces. In the stiffness method we take a displacement as the unknown and obtain it from an equilibrium equation that incorporates constitutive relations giving the forces in terms of the displacement. Both methods use equilibrium, compatibility, and constitutive relations. These are the fundamental conditions that all structures must satisfy.

Example 2

A horizontal rigid bar AB is supported by two wires CD and EF (Fig. 2-14a). Wire CD is made of aluminum ($E_1 = 10.4 \times 10^6$ psi), with diameter $d_1 = 0.16$ in. and length $L_1 = 15$ in. The other wire (EF) is made of magnesium ($E_2 = 6.5 \times 10^6$ psi), with diameter $d_2 = 0.12$ in. and length $L_2 = 12$ in. A vertical load P acts at the end of the rigid bar, which has length $3b$. Calculate the maximum permissible value of the load P if the allowable stresses in the aluminum and magnesium wires are $\sigma_1 = 30$ ksi and $\sigma_2 = 25$ ksi, respectively. (Disregard the weight of the bar.)

We begin by drawing a free-body diagram of bar AB (Fig. 2-14b). In this diagram T_1 and T_2 represent the unknown tensile forces in the wires and R is the reaction at the support. We see immediately that the structure is statically indeterminate because there are three unknown forces (T_1, T_2, and R) but only two equations of equilibrium (for instance, one force equation and one moment equation). Taking moments about point A yields

$$T_1 b + T_2(2b) - P(3b) = 0 \qquad \text{or} \qquad T_1 + 2T_2 = 3P \tag{o}$$

A second equilibrium equation will introduce the reaction R as an additional unknown and will be of no benefit in finding T_1 and T_2.

To solve this structure by the *flexibility method* we need an equation of compatibility to supplement the equation of equilibrium. We observe that the load P stretches the wires and causes bar AB to rotate about the support. The resulting displacements of the structure are shown in Fig. 2-14c, where line AB' represents the rotated position of the rigid bar and δ_1 and δ_2 represent the elongations of the wires. The equation of compatibility gives the following

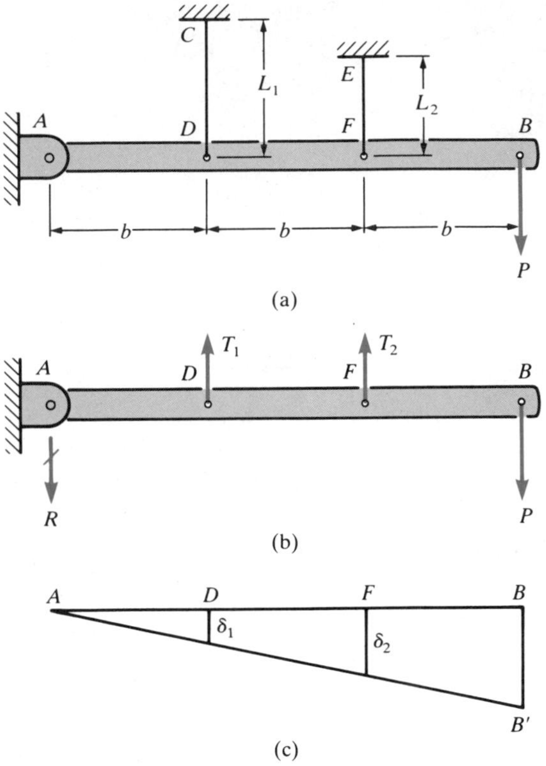

Fig. 2-14 Example 2. Statically indeterminate structure

relationship between the elongations:

$$\delta_2 = 2\delta_1 \tag{p}$$

as seen from the geometry of the figure.

The elongations of the wires can be expressed in terms of the unknown forces by means of the following constitutive relations:

$$\delta_1 = \frac{T_1 L_1}{E_1 A_1} \qquad \delta_2 = \frac{T_2 L_2}{E_2 A_2}$$

in which A_1 and A_2 are the cross-sectional areas of the wires. For convenience, let us introduce the following notation for the flexibilities of the wires (see Eq. 2-3):

$$f_1 = \frac{L_1}{E_1 A_1} \qquad f_2 = \frac{L_2}{E_2 A_2} \tag{2-12a,b}$$

Then the constitutive relations become

$$\delta_1 = f_1 T_1 \qquad \delta_2 = f_2 T_2 \tag{q}$$

Substituting these expressions into the equation of compatibility (Eq. p) gives

$$f_2 T_2 = 2 f_1 T_1 \tag{r}$$

This equation can be solved simultaneously with the equilibrium equation (Eq. o) to obtain the forces in the wires:

$$T_1 = \frac{3f_2P}{4f_1 + f_2} \qquad T_2 = \frac{6f_1P}{4f_1 + f_2} \tag{2-13a,b}$$

Knowing the forces T_1 and T_2, we can easily find the elongations δ_1 and δ_2 from Eqs. (q).

In the preceding solution by the flexibility method it was not necessary to formalize the procedure by specifically identifying a statical redundant and a corresponding released structure. Instead, we directly wrote an equation of equilibrium, an equation of compatibility, and two constitutive relations. Then we solved these equations in the most convenient manner. This direct and simple procedure is adequate in many situations.

As an alternative solution, we can solve the structure by the *stiffness method.* In this case we select a particular displacement as the unknown quantity; for instance, we will select δ_1. Then, using compatibility, we express the other displacement in terms of that unknown ($\delta_2 = 2\delta_1$). The constitutive relations are written in a form that gives the forces in terms of the unknown displacement:

$$T_1 = \frac{E_1A_1\delta_1}{L_1} \qquad T_2 = \frac{E_2A_2\delta_2}{L_2} = \frac{2E_2A_2\delta_1}{L_2}$$

or

$$T_1 = k_1\delta_1 \qquad T_2 = k_2\delta_2 = 2k_2\delta_1 \tag{s}$$

in which the stiffnesses of the wires (see Eq. 2-2) are

$$k_1 = \frac{E_1A_1}{L_1} \qquad k_2 = \frac{E_2A_2}{L_2} \tag{2-14a,b}$$

Next, we substitute the expressions for the forces (Eqs. s) into the equilibrium equation (Eq. o):

$$k_1\delta_1 + 4k_2\delta_1 = 3P$$

and solve for the unknown displacement:

$$\delta_1 = \frac{3P}{k_1 + 4k_2} \tag{2-15}$$

Finally, the forces in the wires are obtained by substituting for δ_1 in the constitutive relations (Eqs. s); the results are

$$T_1 = \frac{3k_1P}{k_1 + 4k_2} \qquad T_2 = \frac{6k_2P}{k_1 + 4k_2} \tag{2-16a,b}$$

These results agree with those obtained by the flexibility method (see Eqs. 2-13).

Now that the statically indeterminate analysis is completed and the forces in the wires are known, we can determine the permissible value of the load P. The stresses σ_1 and σ_2 in the wires CD and EF, respectively, are readily obtained from the forces (Eqs. 2-16):

$$\sigma_1 = \frac{T_1}{A_1} = \frac{3P}{A_1}\frac{k_1}{k_1 + 4k_2} \qquad \sigma_2 = \frac{T_2}{A_2} = \frac{6P}{A_2}\frac{k_2}{k_1 + 4k_2} \tag{2-17a,b}$$

From the first of these equations we can solve for the force P_1 based upon the allowable stress σ_1 in the aluminum bar:

$$P_1 = \frac{\sigma_1 A_1(k_1 + 4k_2)}{3k_1} \tag{t}$$

Similarly, from the second equation we get the force P_2 based upon the allowable stress σ_2 in the magnesium bar:

$$P_2 = \frac{\sigma_2 A_2(k_1 + 4k_2)}{6k_2} \tag{u}$$

The smaller of the two preceding values is the maximum allowable load P_{allow}.

Using the given data, we obtain the following numerical values:

$$\sigma_1 = 30 \text{ ksi} \qquad \sigma_2 = 25 \text{ ksi}$$

$$A_1 = \pi d_1^2/4 = \pi(0.16 \text{ in.})^2/4 = 0.02011 \text{ in.}^2$$

$$A_2 = \pi d_2^2/4 = \pi(0.12 \text{ in.})^2/4 = 0.01131 \text{ in.}^2$$

$$k_1 = E_1 A_1/L_1 = (10.4 \times 10^6 \text{ psi})(0.02011 \text{ in.}^2)/15 \text{ in.}$$

$$= 13{,}940 \text{ lb/in.}$$

$$k_2 = E_2 A_2/L_2 = (6.5 \times 10^6 \text{ psi})(0.01131 \text{ in.}^2)/12 \text{ in.}$$

$$= 6126 \text{ lb/in.}$$

Substituting into Eqs. (t) and (u) gives

$$P_1 = 554 \text{ lb} \qquad P_2 = 296 \text{ lb}$$

The first result is based upon the allowable stress σ_1 in the aluminum wire and the second is based upon the allowable stress σ_2 in the magnesium wire. The allowable load P is the smaller of the two values:

$$P_{allow} = 296 \text{ lb}$$

At this load the stress in the magnesium is 25 ksi (the allowable stress) and the stress in the aluminum is (296/554)(30 ksi) = 16.0 ksi, which is much less than the allowable stress of 30 ksi.

Example 3

A horizontal rigid bar ADB of length $2b$ is supported by two inclined wires CD and CB (Fig. 2-15a). The wires are attached to a support at C, which is located at distance b above point A. Both wires have axial rigidity EA, are made of the same material, and have the same diameter. Determine the tensile forces T_1 and T_2 in wires CD and CB, respectively, due to the vertical load P acting at the end of the rigid bar.

A free-body diagram of the bar (Fig. 2-15b) shows that there are four unknown forces, but only three equations of equilibrium; hence the structure is

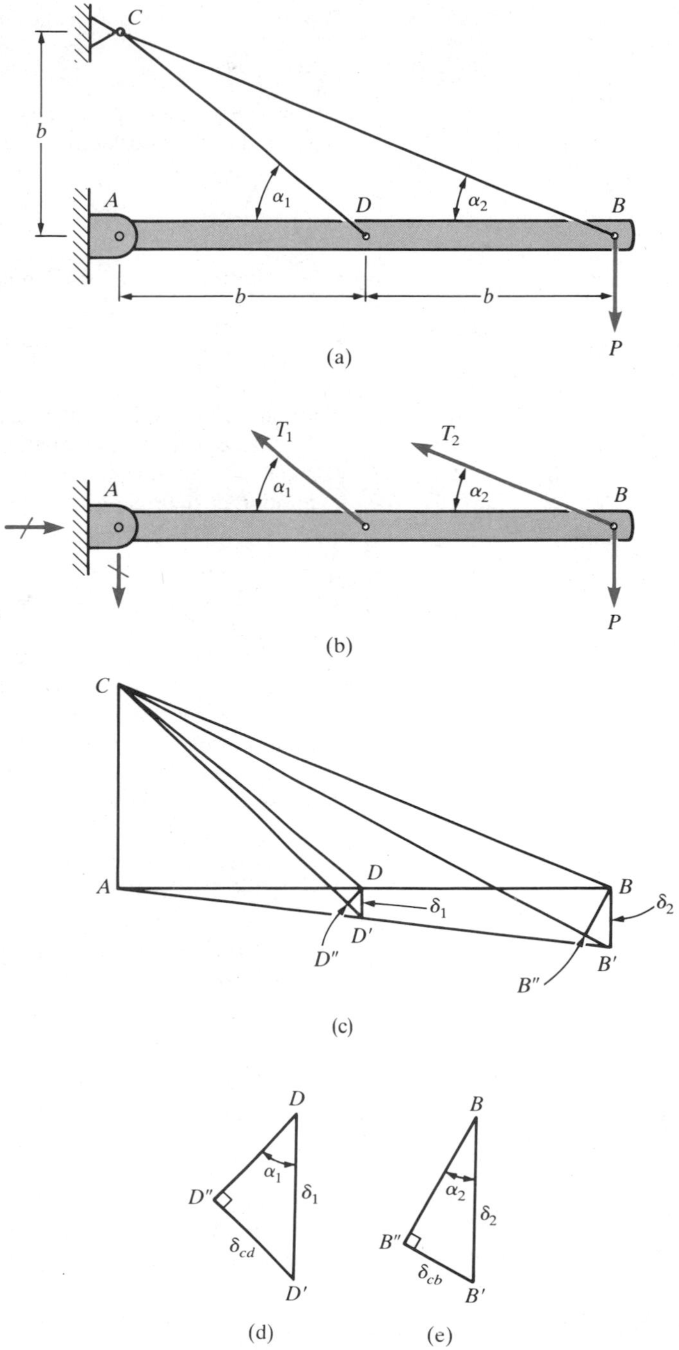

Fig. 2-15 Example 3. Statically indeterminate structure

statically indeterminate. The equation of moment equilibrium about point A is

$$bT_1 \sin \alpha_1 + 2bT_2 \sin \alpha_2 = 2bP$$

in which α_1 and α_2 are the angles between the wires and the bar. From the geometry of the structure we see that

$$\sin \alpha_1 = \frac{1}{\sqrt{2}} \qquad \sin \alpha_2 = \frac{1}{\sqrt{5}}$$

Substituting these values into the preceding equation, we get

$$\frac{T_1}{\sqrt{2}} + \frac{2T_2}{\sqrt{5}} = 2P \tag{v}$$

as the final form of the equation of equilibrium.

An equation of compatibility can be obtained by considering the displacements of points D and B. For this purpose, let line AB' represent the rotated position of the rigid bar AB (Fig. 2-15c). Then we see from the figure that

$$\delta_2 = 2\delta_1 \tag{w}$$

in which δ_1 is the vertical displacement of point D and δ_2 is the vertical displacement of point B.

To obtain the constitutive relations, we must relate the forces in the wires to the vertical displacements δ_1 and δ_2. Let us begin by considering wire CD (Fig. 2-15c) for which the final position is CD'. We now draw a perpendicular DD'' from D to line CD'. Because the displacements and angle changes are very small, we can take CD'' as equal in length to CD. (In other terms, a circular arc DD'' with point C as center is indistinguishable from a perpendicular line DD'' because the angle DCD'' is very small.) The triangle $DD'D''$ in Fig. 2-15c is redrawn for clarity in Fig. 2-15d. The distance DD' is the vertical displacement δ_1 and the distance $D'D''$ is the elongation δ_{cd} of wire CD. The angle α_1 appears in the triangle as angle $D'DD''$; to prove this, we note that DD' is perpendicular to AB, and DD'' is perpendicular to CD (for small angles of rotation). Therefore, from triangle $DD'D''$ we get

$$\delta_{cd} = \delta_1 \sin \alpha_1 = \frac{\delta_1}{\sqrt{2}}$$

But the elongation δ_{cd} is also equal to the following:

$$\delta_{cd} = \frac{T_1 L_{cd}}{EA} = \frac{T_1 b\sqrt{2}}{EA}$$

in which $L_{cd} = b\sqrt{2}$ is the length of wire CD. Combining the last two equations, we get

$$\delta_1 = \frac{2bT_1}{EA} \tag{x}$$

In a similar manner, we can relate the displacement δ_2 to the elongation δ_{cb} of wire CB (see Fig. 2-15e) and obtain

$$\delta_2 = \frac{5bT_2}{EA} \tag{y}$$

Equations (x) and (y) are the constitutive relations that give the displacements of the rigid bar in terms of the forces in the wires.

The final step in the solution is to solve the various equations we have assembled. Substituting δ_1 and δ_2 from the constitutive relations into the equation of compatibility (Eq. w) yields

$$5T_2 = 4T_1 \tag{z}$$

This equation can be solved simultaneously with the equation of equilibrium (Eq. v) to give

$$T_1 = 1.406P \qquad T_2 = 1.125P$$

Thus, we have found the tensile forces in the wires by solving the statically indeterminate structure. Knowing these forces, we can easily determine the displacements from Eqs. (x) and (y).

2.4 TEMPERATURE AND PRESTRAIN EFFECTS

A change in the temperature of an object tends to produce a change in its dimensions. A simple illustration of this effect is given in Fig. 2-16, which shows a block of homogeneous and isotropic material that is free to expand in all directions. If the block is heated uniformly, its sides will increase in length. Thus, with corner A taken as a reference point, the block will have the shape shown by the dashed lines. The material undergoes a **uniform thermal strain** ϵ_t given by the expression

$$\epsilon_t = \alpha(\Delta T) \tag{2-18}$$

in which α is the **coefficient of thermal expansion** and ΔT is the increase in temperature. The coefficient α is a property of the material* and has units equal to the reciprocal of temperature change. In SI units, α has dimensions of either 1/K (the reciprocal of kelvins) or 1/°C (the reciprocal of degrees Celsius), because a *change* in temperature is numerically the

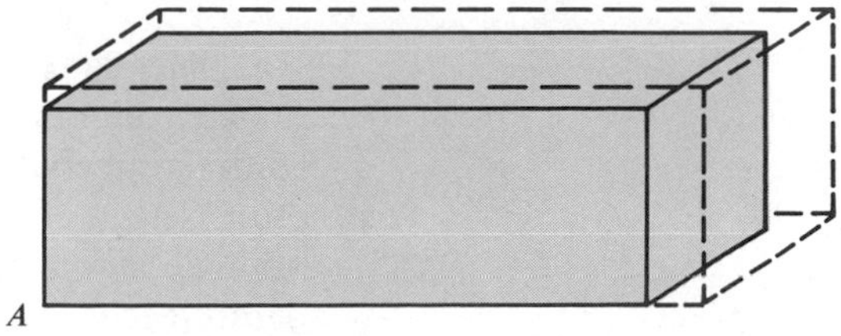

Fig. 2-16 Block with uniform increase in temperature

* Typical values of the coefficient of thermal expansion α are listed in Appendix H, Table H-4.

same in both kelvins and degrees Celsius. In USCS units, the dimensions of α are 1/°F (the reciprocal of degrees Fahrenheit).* Thermal strain ϵ_t is a dimensionless quantity that is considered positive when it represents expansion and negative when it represents contraction.

Ordinary materials expand when heated and contract when cooled, and therefore an increase in temperature produces a positive thermal strain. Thermal strains usually are reversible, in the sense that the member returns to its original shape when the temperature returns to its original value. However, some special metals have been developed recently that do not behave in the customary manner. Instead, over certain temperature ranges their dimensions decrease when heated and increase when cooled. These metals also differ from ordinary materials in that the strains are related nonlinearly to the temperature and sometimes are irreversible, because the internal structure of the material is altered. (Another unusual material is water, which expands when heated at temperatures above 4 °C but also expands when cooled below 4 °C. Thus, water has its maximum density at 4 °C.)

The changes in the dimensions of the block of material shown in Fig. 2-16 can be calculated by multiplying the original dimensions by the thermal strain. If, for instance, one of the dimensions is L, then that dimension will increase by the amount

$$\delta_t = \epsilon_t L = \alpha(\Delta T)L \tag{2-19}$$

in which δ_t represents the elongation due to the temperature increase ΔT. Equation (2-19) is used to calculate changes in lengths of structural members, such as the bar shown in Fig. 2-17. The transverse dimensions of this bar also change, but, since such changes usually have no effect on the forces being transmitted by the bar, they are not shown in the figure.

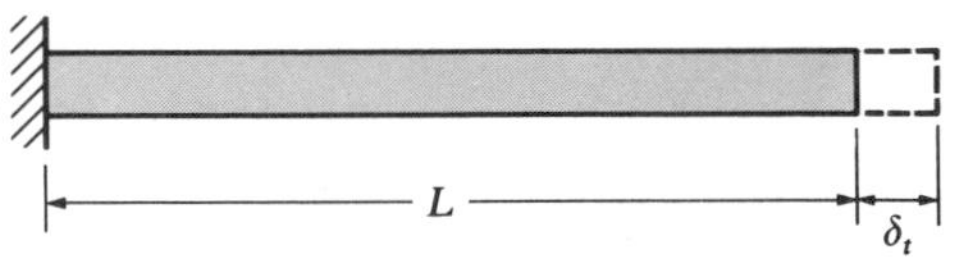

Fig. 2-17 Increase in length of a bar due to a uniform increase in temperature (Eq. 2-19)

In general, changes in lengths due to thermal strains may be calculated from Eq. (2-19) provided that the members are able to expand or contract freely, a situation that exists in statically determinate structures. As a consequence, no stresses are produced in a statically determinate structure when one or more members undergo a uniform temperature change. On the other hand, a temperature change in a statically indeterminate structure will usually produce stresses in the

* For a discussion of temperature units and scales, see Sections A.2 and A.3 of Appendix A.

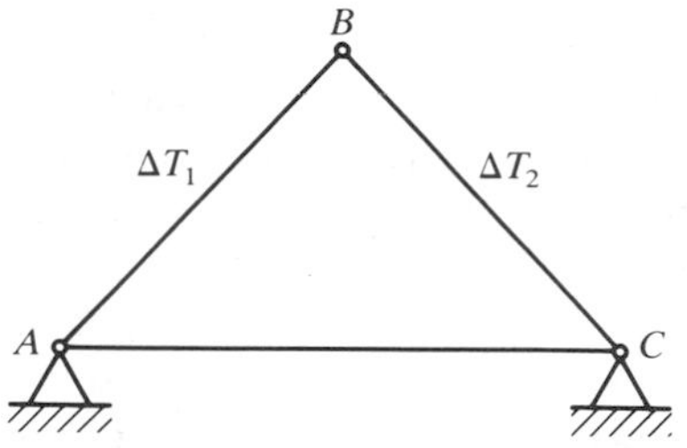

Fig. 2-18 Statically determinate truss with a uniform temperature change in each member

members, called **thermal stresses**. Such stresses also may occur when a member is heated in a nonuniform manner, irrespective of whether the structure is determinate or indeterminate.

To illustrate some of these ideas about thermal effects, consider the two-bar truss ABC of Fig. 2-18 and assume that the temperature of bar AB is changed by ΔT_1 and the temperature of bar BC is changed by ΔT_2. Because the truss is statically determinate, both bars are free to lengthen (or shorten), thereby producing a displacement of joint B. However, there are no stresses in either bar and no reactions at the supports. Thus, we observe that temperature changes can produce strains in a statically determinate structure without creating any corresponding stresses.

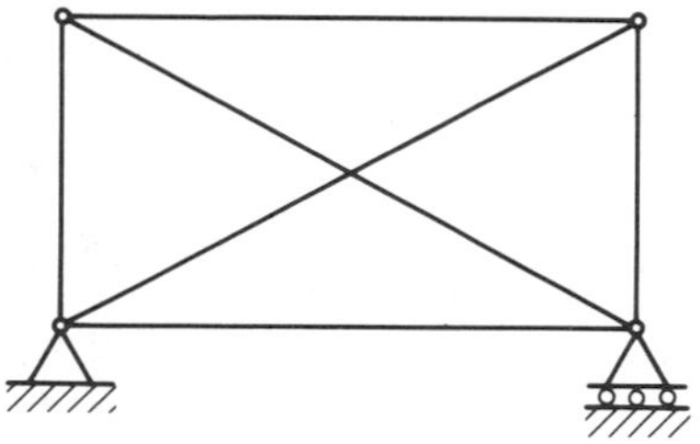

Fig. 2-19 Statically indeterminate truss with temperature changes

If a structure is statically indeterminate, it may or may not develop temperature stresses, depending upon the character of the structure and the nature of the temperature changes. To illustrate some of the possibilities, consider the statically indeterminate truss shown in Fig. 2-19. Because the supports of the structure permit unrestricted expansion to occur, no stresses are developed when the entire truss is heated uniformly. Instead, all members increase in length in proportion to their original lengths, and the entire truss is slightly larger. However, if some (but not all) of the bars are heated, stresses will develop because the statically indeterminate arrangement of the bars prevents free expansion. To visualize this, imagine that just one bar is heated; then elongation of that bar cannot occur without stressing the members.

The analysis of a statically indeterminate structure with temperature changes is based upon the same principles we discussed in the preceding section, namely equilibrium, compatibility, and constitutive relations. The principal difference is that some of the constitutive relations are now obtained from Eq. (2-19).

As an illustration of a temperature analysis, consider a bar AB held between immovable supports, as shown in Fig. 2-20a. If the temperature

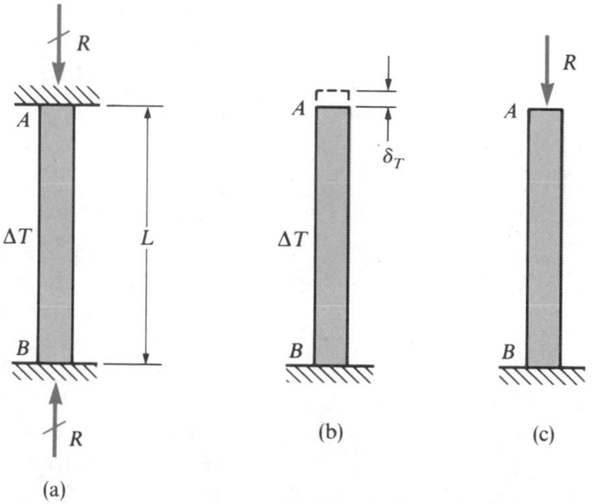

Fig. 2-20 Statically indeterminate bar with uniform temperature increase ΔT

increases uniformly by ΔT, reactions R will be developed at the supports and the bar will be subjected to uniform compressive stresses. To calculate the reactions R, we begin by removing the support at the upper end and allowing the temperature change ΔT to occur in the released structure (Fig. 2-20b). The resulting upward displacement of end A is

$$\delta_T = \alpha(\Delta T)L$$

When the force R acts on the released structure (Fig. 2-20c), it produces a downward displacement

$$\delta_R = \frac{RL}{EA}$$

The actual displacement of the end of the bar, equal to $\delta_T - \delta_R$, is zero (see Fig. 2-20a). Therefore, we obtain the following equation of compatibility:

$$\delta_T - \delta_R = \alpha(\Delta T)L - \frac{RL}{EA} = 0 \tag{a}$$

Solving this equation, we obtain the reaction R:

$$R = EA\alpha(\Delta T) \tag{2-20}$$

Note that R does not depend upon the length of the bar. Finally, we obtain the compressive stress in the bar:

$$\sigma = \frac{R}{A} = E\alpha(\Delta T) \tag{2-21}$$

which does not depend upon the cross-sectional area.

The preceding analysis shows how a temperature change can produce stresses in a statically indeterminate system even when there are no loads. In this particular example, the bar has zero longitudinal displacements, not only at the fixed ends but also at every cross section. Thus, there are no axial strains in this bar, and we have a situation involving longitudinal stresses without longitudinal strains. In a more general case, such as a bar consisting of two parts having different cross-sectional areas, there will be longitudinal strains as well as stresses when temperature changes occur (see Problems 2.4-12 and 2.4-13).

Example

A sleeve in the form of a circular tube of length L is placed around a bolt and fitted between washers at each end (Fig. 2-21a). The nut is then turned until it is just snug. The sleeve and bolt are made of different materials and have different coefficients of thermal expansion. If the temperature of the entire assembly is raised by an amount ΔT, what stresses σ_s and σ_b are developed in the sleeve and bolt, respectively?

Because the sleeve and bolt are of different materials, they will elongate by different amounts when they are able to expand freely. However, when they are

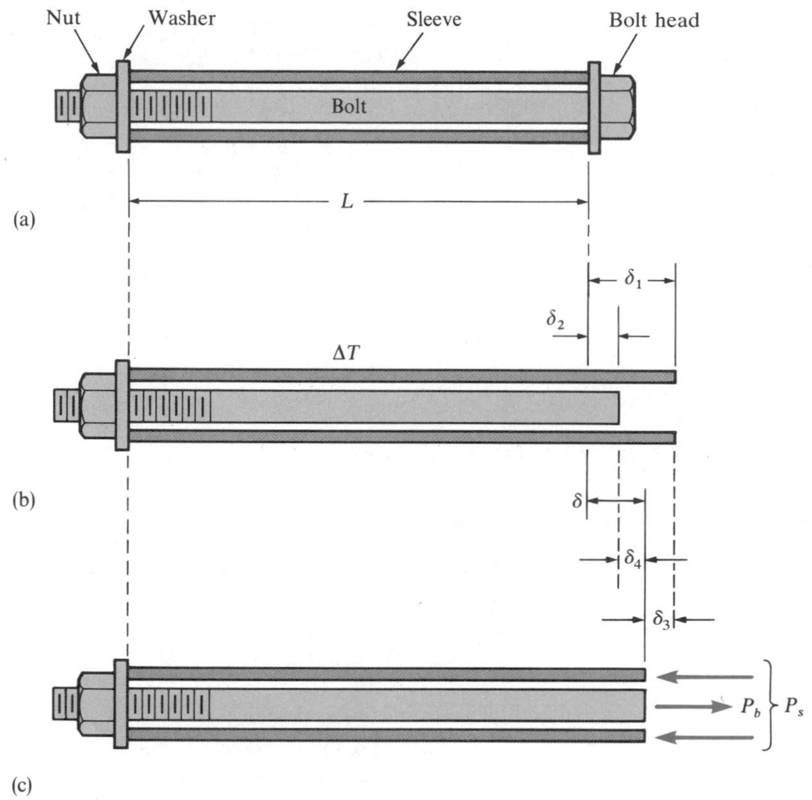

Fig. 2-21 Example. Sleeve and bolt assembly with uniform temperature increase ΔT

held together by the assembly, thermal stresses are developed. The assembly is statically indeterminate because the stresses cannot be determined by equilibrium alone. For illustration, we will obtain the stresses by both the flexibility and stiffness methods.

In the *flexibility method*, we begin by cutting the assembly apart so that a statically determinate released structure is obtained. A simple way to accomplish this result is to remove the head of the bolt, as shown in Fig. 2-21b. Then the temperature change ΔT is assumed to occur, producing elongations δ_1 and δ_2 of the sleeve and bolt, respectively:

$$\delta_1 = \alpha_s(\Delta T)L \qquad \delta_2 = \alpha_b(\Delta T)L$$

in which α_s and α_b are the coefficients of thermal expansion. In drawing the figure, it has been arbitrarily assumed that δ_1 is greater than δ_2 (that is, $\alpha_s > \alpha_b$).

The forces existing in the sleeve and bolt in the original assembly must be such that they shorten the sleeve and stretch the bolt until the final elongations of the sleeve and bolt are the same. These forces are shown in Fig. 2-21c, where P_s denotes the total compressive force in the sleeve and P_b denotes the tensile force in the bolt. The corresponding shortening δ_3 of the sleeve and elongation δ_4 of the bolt are

$$\delta_3 = \frac{P_s L}{E_s A_s} \qquad \delta_4 = \frac{P_b L}{E_b A_b}$$

in which E_sA_s and E_bA_b are the respective axial rigidities. The preceding expressions for δ_1, δ_2, δ_3, and δ_4 represent the constitutive relations in this analysis.

Now we can write an equation of compatibility expressing the fact that the final elongation δ is the same for both bars. The elongation of the sleeve is $\delta_1 - \delta_3$ and of the bolt is $\delta_2 + \delta_4$; therefore,

$$\delta = \delta_1 - \delta_3 = \delta_2 + \delta_4 \tag{b}$$

or

$$\delta = \alpha_s(\Delta T)L - \frac{P_sL}{E_sA_s} = \alpha_b(\Delta T)L + \frac{P_bL}{E_bA_b} \tag{c}$$

A second equation for the axial forces is obtained from static equilibrium (see Fig. 2-21c):

$$P_s = P_b \tag{d}$$

That is, the compressive force in the sleeve is equal to the tensile force in the bolt. Combining Eqs. (c) and (d) and solving, we obtain the forces in the original assembly (Fig. 2-21a) subjected to the temperature change:

$$P_s = P_b = \frac{(\alpha_s - \alpha_b)(\Delta T)E_sA_sE_bA_b}{E_sA_s + E_bA_b} \tag{2-22}$$

If α_s is greater than α_b, then P_s is compression and P_b is tension. We also note that the forces are independent of the length L.

Expressions for the stresses σ_s and σ_b in the sleeve and bolt, respectively, are obtained by dividing the corresponding forces by the appropriate areas:

$$\sigma_s = \frac{P_s}{A_s} = \frac{(\alpha_s - \alpha_b)(\Delta T)E_sE_bA_b}{E_sA_s + E_bA_b} \tag{2-23a}$$

$$\sigma_b = \frac{P_b}{A_b} = \frac{(\alpha_s - \alpha_b)(\Delta T)E_sA_sE_b}{E_sA_s + E_bA_b} \tag{2-23b}$$

These stresses are related by the relation $\sigma_s/\sigma_b = A_b/A_s$.

The final elongation δ of the system can be found by substituting from Eq. (2-22) into Eq. (c), yielding

$$\delta = \frac{(\alpha_sE_sA_s + \alpha_bE_bA_b)(\Delta T)L}{E_sA_s + E_bA_b} \tag{2-24}$$

A special case occurs when both bars are of the same material; then $\alpha_s = \alpha_b$, the forces and stresses are zero, and $\delta = \alpha_b(\Delta T)L$, as expected.

Now consider the analysis of this assembly by the *stiffness method*. In this method, we take the elongation δ of the assembly as the unknown quantity, and we express the forces in the sleeve and bolt in terms of that elongation. Then we write an equation of equilibrium for the forces and solve for δ.

Let us begin with the sleeve. Its total elongation is δ, but part of that elongation is produced by the temperature change, namely, $\alpha_s(\Delta T)L$. Therefore, the part of the elongation associated with the axial force P_s is $\delta - \alpha_s(\Delta T)L$, and the constitutive relation is

$$P_s = -\frac{E_s A_s}{L}[\delta - \alpha_s(\Delta T)L] \tag{e}$$

where the minus sign is used because we are taking P_s as positive in compression. In a similar manner, we get the following constitutive relation for the bolt:

$$P_b = \frac{E_b A_b}{L}[\delta - \alpha_b(\Delta T)L] \tag{f}$$

Note that we have implicitly used the compatibility condition which states that the final elongation of the sleeve and bolt are the same.

The preceding expressions for P_s and P_b are substituted into the equation of equilibrium ($P_s = P_b$) to give

$$-\frac{E_s A_s}{L}[\delta - \alpha_s(\Delta T)L] = \frac{E_b A_b}{L}[\delta - \alpha_b(\Delta T)L]$$

from which we solve for the displacement:

$$\delta = \frac{(\alpha_s E_s A_s + \alpha_b E_b A_b)(\Delta T)L}{E_s A_s + E_b A_b}$$

This result is the same as that obtained by the flexibility method (Eq. 2-24). Substituting this expression for δ into Eqs. (e) and (f) gives the forces P_s and P_b (see Eq. 2-22).

PRESTRAINS

Suppose that a member of a structure is manufactured with a length that is slightly different from its prescribed length. The effect on the structure is similar to that of a temperature change, which also produces a change from the intended length. If the structure is statically determinate, there will be departures from the theoretical configuration of the structure, but no strains or stresses will be created. However, in the case of a statically indeterminate structure that is not free to adjust to the change in length of the member, internal strains will develop. These strains exist at the time the structure is built and before any loads are applied; hence they are called **prestrains**. Such strains result in a **prestressed** condition in the structure. Sometimes structures are deliberately prestressed in order to obtain more favorable stress conditions under loads. Common examples of prestressing are shrink-fitted machine parts, spokes in bicycle wheels (which would collapse if not prestressed), prestressed concrete beams, and tennis racquets.

The analysis of a statically indeterminate structure with prestrains proceeds in the same manner as with temperature changes. To illustrate the methodology, consider an assembly consisting of a bar, rigid end plates, and two cables with turnbuckles (Fig. 2-22a). By rotating the turnbuckles, we tighten the cables and produce a prestressed condition in

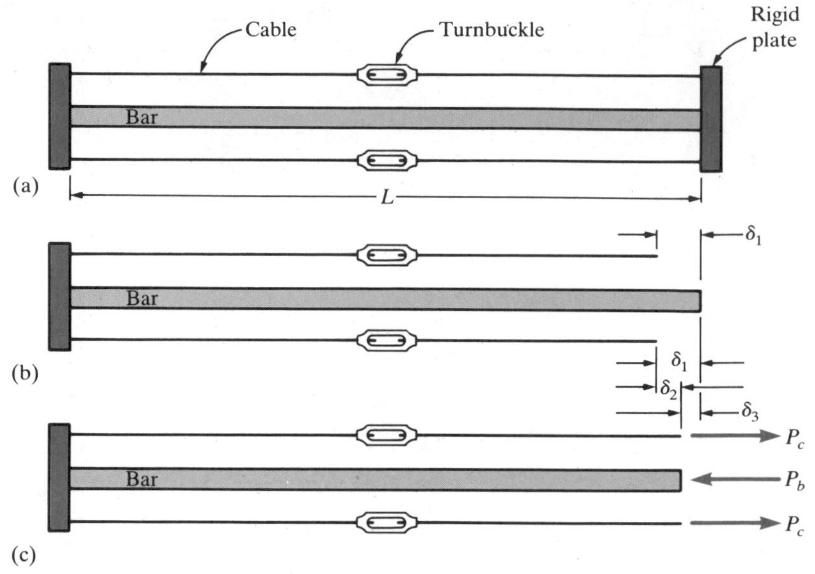

Fig. 2-22 Statically indeterminate assembly with a bar in compression and two cables in tension

which the cables are in tension and the bar is in compression. The turnbuckles shown in the figure are said to be "double-acting" because there are two end screws (Fig. 2-23). Each full turn of the turnbuckle causes it to travel a distance p along each screw, where p is the pitch of the threads (the distance from one thread to the next). Therefore, if the turnbuckle is tightened by one turn, the screws are drawn closer together by a distance $2p$ and the effect is to shorten the cable by $2p$.

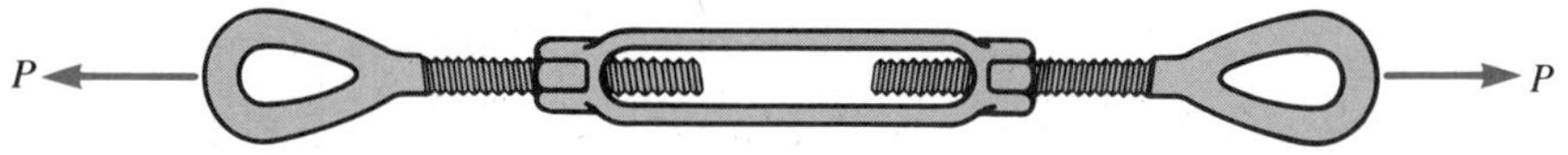

Fig. 2-23 Double-acting turnbuckle. (Each full turn of the turnbuckle shortens or lengthens the cable by $2p$, where p is the pitch of the screw threads.)

Let us determine the stresses in the bar and cables of Fig. 2-22a when the turnbuckles are given n turns (n does not have to be an integer; for instance, we could give each turnbuckle one-and-a-half turns). We begin by removing the plate at the right-hand end and observing that n turns of the turnbuckles will shorten the cables by a distance

$$\delta_1 = 2np$$

as shown in Fig. 2-22b. The tensile force P_c in each cable causes the cables to elongate by an amount (see Fig. 2-22c)

$$\delta_2 = \frac{P_c L}{E_c A_c}$$

where $E_c A_c$ is the axial rigidity of a cable. Also, the compressive force P_b in the bar causes it to shorten by

$$\delta_3 = \frac{P_b L}{E_b A_b}$$

where $E_b A_b$ is the axial rigidity of the bar. The final shortening of the cable $(\delta_1 - \delta_2)$ must equal the shortening δ_3 of the bar; therefore, the equation of compatibility is

$$\delta_1 - \delta_2 = \delta_3 \tag{g}$$

or, upon substituting the constitutive relations,

$$2np - \frac{P_c L}{E_c A_c} = \frac{P_b L}{E_b A_b} \tag{h}$$

Equilibrium of the released structure (Fig. 2-22c) gives

$$2P_c = P_b \tag{i}$$

Now we solve simultaneously the compatibility and equilibrium equations (Eqs. h and i) and obtain

$$P_b = 2P_c = \frac{4np E_b A_b E_c A_c}{L(E_b A_b + 2E_c A_c)} \tag{2-25}$$

The stresses in the bar and cables can now be obtained by dividing the forces P_b and P_c by the areas A_b and A_c, respectively.

This example illustrates that the methods of analysis of a statically indeterminate structure due to prestrains are essentially the same as for temperature changes. Furthermore, the results of such analyses can easily be converted from one case to the other. For instance, the analysis of the assembly in Fig. 2-22 when the turnbuckles are tightened is the same as when a temperature drop in the cables causes them to shorten (provided there is no temperature change in the bar). In such a case, the shortening $2np$ is replaced by $\alpha_c(\Delta T)L$, where α_c is the coefficient of thermal expansion for the cables and ΔT is the temperature drop in the cables.

2.5 STRESSES ON INCLINED SECTIONS

In the preceding discussions of tension and compression in bars, the only stresses we considered were the normal stresses acting on cross sections (such as the stress σ on cross section mn of bar AB shown in Fig. 2-24a). Let us now investigate the stresses acting on sections such as pq (Fig. 2-24a) that are inclined to the axis.

To begin the discussion, we recall that the normal stresses acting on a cross section may be calculated from the formula $\sigma = P/A$ provided that the stress distribution is uniform over the entire cross-sectional area. As explained previously, this condition exists if the bar is prismatic, the

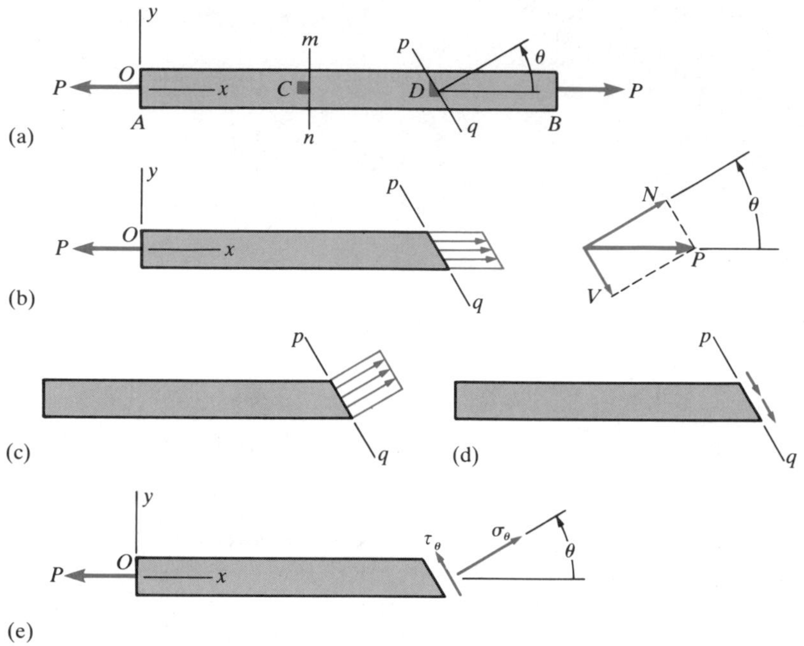

Fig. 2-24 Prismatic bar in tension with cross section *mn* and inclined section *pq*

material is homogeneous, the axial force P acts at the centroid of the cross-sectional area, and the cross section is away from the ends of the bar where high localized stresses may exist. Let us assume that the bar of Fig. 2-24a meets these conditions, so that the distribution of normal stresses on cross section *mn* is uniform. Of course, there are no shear stresses acting on this section, because it is cut at right angles to the longitudinal axis.

A convenient way to represent the stresses in the bar is to isolate a small element of material, such as the element labeled C in Fig. 2-24a, and then to show the stresses acting on all sides of this element. We shall refer to such an element as a **stress element**. The stress element at point C has the shape of a rectangular parallelepiped, and its right-hand face (or positive x face) is in cross section *mn*. The dimensions of a stress element are assumed to be infinitesimally small, but for clarity we can draw the element to a large scale, as in Fig. 2-25a. The edges of the element are parallel to the x, y, and z axes, and the only stresses acting on the element are the normal stresses on the x faces, denoted as σ_x. For convenience, we often will use only a two-dimensional sketch of the element, as shown in Fig. 2-25b.

The inclined section *pq* is cut through bar AB at an angle θ between the x axis and the normal to the inclined plane (Fig. 2-24a). Thus, cross section *mn* has an angle θ equal to zero, and a longitudinal section would have an angle θ equal to 90° (or $\pi/2$ radians). Because all parts of the bar have the same axial strains, the stresses acting over section *pq* must be uniformly distributed (Fig. 2-24b). The resultant of these stresses must be

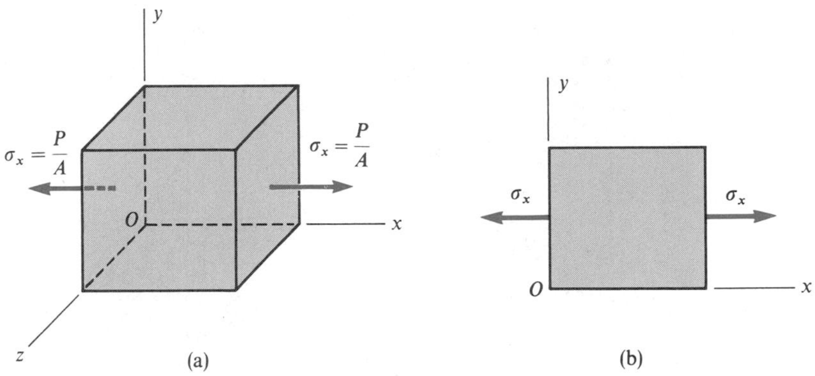

Fig. 2-25 Stress element at point C of the bar shown in Fig. 2-24a: (a) three-dimensional view of element, and (b) two-dimensional representation of element

a force that is equal in magnitude to the axial force P, in order to maintain equilibrium of the left-hand part of the bar. This resultant may be resolved into two components, a normal force N and a shear force V (Fig. 2-24b), that are normal and tangential, respectively, to the inclined plane pq. These components are

$$N = P\cos\theta \qquad V = P\sin\theta \tag{2-26a,b}$$

Associated with the forces N and V are normal stresses and shear stresses that are uniformly distributed over the inclined section (see Figs. 2-24c and d).

The sign convention and notation for normal and shear stresses on sections inclined at an angle θ are shown in Fig. 2-24e. Normal stresses σ_θ are positive in tension. Shear stresses τ_θ are positive when they tend to produce counterclockwise rotation of the material. (This convention is consistent with the sign convention for shear stresses described previously in Section 1.6.) For a bar in tension, the normal force N (see Fig. 2-24b) produces positive normal stresses σ_θ and the shear force V produces negative shear stresses τ_θ. These stresses can be evaluated by dividing the forces by the area over which they act. The area A_1 of the inclined section is $A/\cos\theta$, in which A is the cross-sectional area. Therefore, the stresses σ_θ and τ_θ are given by the following equations:

$$\sigma_\theta = \frac{N}{A_1} = \frac{P}{A}\cos^2\theta = \sigma_x\cos^2\theta \tag{2-27a}$$

$$\tau_\theta = -\frac{V}{A_1} = -\frac{P}{A}\sin\theta\cos\theta = -\sigma_x\sin\theta\cos\theta \tag{2-27b}$$

in which $\sigma_x = P/A$ is the normal stress on a cross section.

The preceding equations for σ_θ and τ_θ also can be obtained by considering the equilibrium of a stress element at point D in the bar of Fig. 2-24a. In this instance, the element is wedge shaped and has one face along the inclined section pq. The element is shown again in Fig. 2-26a, with the stresses σ_θ and τ_θ acting on the inclined face and the stress σ_x

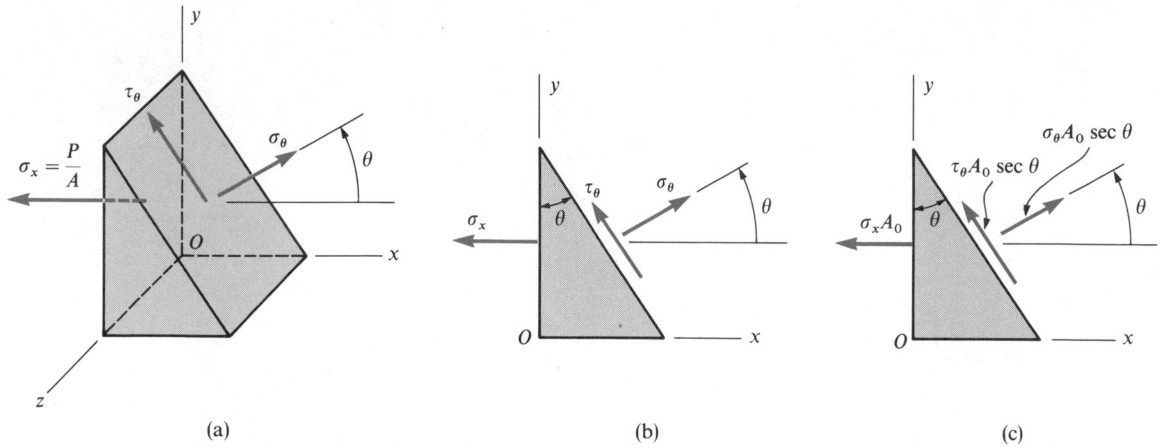

Fig. 2-26 Stress element at point D of the bar shown in Fig. 2-24a

acting on the left-hand face. There are no stresses on the other faces of the element. Again, a two-dimensional sketch of the element is useful for our purposes (Fig. 2-26b).

To obtain σ_θ and τ_θ, we consider the equilibrium of the element and draw a free-body diagram (Fig. 2-26c). The forces acting on the faces are obtained by multiplying the stresses by the areas over which they act (that is, by the areas of the faces). For instance, the force on the left-hand face is equal to $\sigma_x A_0$ (Fig. 2-26c), where A_0 is the area of the face. This force acts in the negative x direction. Because the thickness of the element is constant, the area of the inclined face is $A_0 \sec\theta$. Therefore, the normal and shear forces on this face are $\sigma_\theta A_0 \sec\theta$ and $\tau_\theta A_0 \sec\theta$, respectively. The force $\sigma_x A_0$ on the left-hand face can be resolved into components that are perpendicular and parallel to the inclined face; the perpendicular component is $\sigma_x A_0 \cos\theta$, and the parallel component is $\sigma_x A_0 \sin\theta$. Now we can write two equations of static equilibrium for the element, one in each of the directions mentioned. The first equation is obtained by summing forces perpendicular to the inclined face (that is, in the direction of σ_θ):

$$\sigma_\theta A_0 \sec\theta - \sigma_x A_0 \cos\theta = 0$$

or

$$\sigma_\theta = \sigma_x \cos^2\theta$$

which is the same as Eq. (2-27a). The second equation is obtained by summing forces in the direction of τ_θ:

$$\tau_\theta A_0 \sec\theta + \sigma_x A_0 \sin\theta = 0$$

or

$$\tau_\theta = -\sigma_x \sin\theta\cos\theta$$

which is the same as Eq. (2-27b). This method for finding stresses on inclined planes, in which we consider the equilibrium of a wedge-shaped element, will be used in discussions of pure shear (Section 3.4) and more general states of stress (Chapter 6).

Equations (2-27a and b) give the normal and shear stresses acting on any inclined section. Figure 2-27 shows the manner in which the stresses vary as the section is cut at angles varying from $\theta = -90°$ to $\theta = +90°$. Note that a positive angle θ is measured counterclockwise from the x axis and a negative angle is measured clockwise. When $\theta = 0$, plane pq becomes a cross section and the graph shows $\sigma_\theta = \sigma_x$, as expected. As θ increases or decreases, the stress σ_θ diminishes until at $\theta = \pm 90°$ it becomes zero, indicating that there are no normal stresses on a plane cut parallel to the longitudinal axis (as expected). The maximum normal stress occurs at $\theta = 0$ and is

$$\sigma_{max} = \sigma_x \tag{2-28}$$

At $\theta = \pm 45°$, the normal stress is one-half the maximum value.

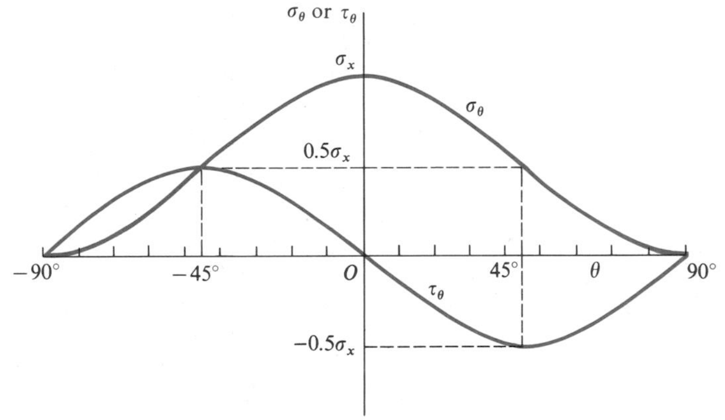

Fig. 2-27 Graph of normal stress σ_θ and shear stress τ_θ versus angle θ of the inclined section pq (see Fig. 2-24a)

The shear stress τ_θ is zero on cross sections ($\theta = 0$) and longitudinal sections ($\theta = \pm 90°$). Between these extremes, the stress varies as shown in Fig. 2-27, reaching the largest positive value when $\theta = -45°$ and the largest negative value when $\theta = 45°$. These maximum shear stresses have the same magnitude, but they tend to rotate the element in opposite directions. Thus, the numerically largest shear stresses are

$$\tau_{max} = \frac{\sigma_x}{2} \tag{2-29}$$

acting on planes at 45° to the axis.

The complete state of stress on sections cut at 45° to the axis is represented by the stress element shown in Fig. 2-28. On face ab ($\theta = 45°$), the normal and shear stresses (from Eqs. 2-27a and b) are $\sigma_x/2$ and $-\sigma_x/2$, respectively. Hence, the normal stress is tension and the shear stress acts clockwise against the element, as shown in the figure. The stresses on the

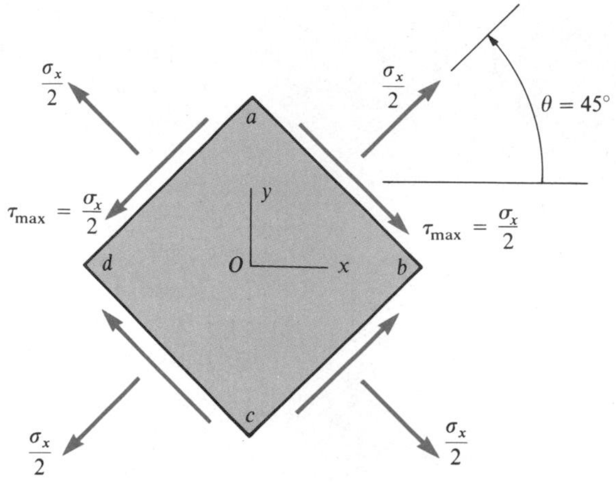

Fig. 2-28 Stress element at $\theta = 45°$ for a bar in tension

remaining faces *bc*, *cd*, and *ad* can be obtained in a similar manner by substituting $\theta = -45°$, $-135°$, and $135°$, respectively, into Eqs. (2-27a and b). In this special case, the normal stresses are the same on all four faces, and all shear stresses have the maximum value. Also, the shear stresses acting on perpendicular planes are equal in magnitude and have directions either toward or away from the line of intersection of the planes, as discussed in Section 1.6.

If a bar is loaded in compression instead of tension, the stress σ_x will have a negative value and the stresses acting on an element will have directions opposite to those for a bar in tension. Of course, Eqs. (2-27a and b) can still be used for calculations by substituting σ_x as a negative quantity.

Even though the maximum shear stress in an axially loaded bar is equal to one-half the maximum normal stress, the shear stress may be the controlling stress if the material is much weaker in shear than in tension. An example of such a shear failure is pictured in Fig. 2-29, which shows a short block of wood that was loaded in axial compression and that failed by shearing along a 45° plane. A related type of behavior occurs in mild steel loaded in tension. During a tensile test of a flat bar of low-carbon steel with polished surfaces, visible **slip bands** appear on the sides of the bar at approximately 45° to the axis (Fig. 2-30). These bands indicate that the material is failing in shear along planes on which the shear stress is a maximum. Such bands were first observed by G. Piobert in 1842 and W. Lüders in 1860 (see Refs. 2-6 through 2-9), and today they are called either *Lüders' bands* or *Piobert's bands.* They begin to appear when the yield stress is reached in the bar (point *B* in Fig. 1-7).

The state of stress described in this section is called **uniaxial stress**, because the bar is subjected to simple tension or compression and a stress element oriented to the axis of the bar is acted upon by stresses in only one direction (Fig. 2-25). It is apparent that the two most important

Fig. 2-29 Failure along a 45° plane of a wood block loaded in compression

Fig. 2-30 Slip bands (or Lüders' bands) in a polished steel specimen subjected to axial tension

orientations of stress elements for uniaxial stress are $\theta = 0$ (Fig. 2-25) and $\theta = 45°$ (Fig. 2-28); the former has the maximum normal stress, and the latter has the maximum shear stress. If sections are cut through the bar at other angles, the stresses acting on the faces of the corresponding stress element can be determined from Eqs. (2-27a and b), as illustrated in the following example. Uniaxial stress is a special case of a more general stress state known as *plane stress*, described in detail in Chapter 6.

Example

A prismatic bar has a cross-sectional area $A = 1200 \text{ mm}^2$ and is compressed by a load $P = 90$ kN (Fig. 2-31a). Determine the stresses acting on a plane cut through the bar at an angle $\theta = 25°$. Then show the complete state of stress for $\theta = 25°$ by determining the stresses on all faces of a stress element.

The normal stress σ_x acting on a cross section is

$$\sigma_x = -\frac{P}{A} = -\frac{90 \text{ kN}}{1200 \text{ mm}^2} = -75 \text{ MPa (compression)}$$

Substituting this value into Eqs. (2-27a and b) and using $\theta = 25°$, we obtain

$$\sigma_\theta = \sigma_x \cos^2\theta = (-75 \text{ MPa})(\cos 25°)^2 = -61.6 \text{ MPa}$$

$$\tau_\theta = -\sigma_x \sin\theta\cos\theta = (75 \text{ MPa})(\sin 25°)(\cos 25°) = 28.7 \text{ MPa}$$

Figure 2-31b shows these stresses acting on the inclined plane in their true directions.

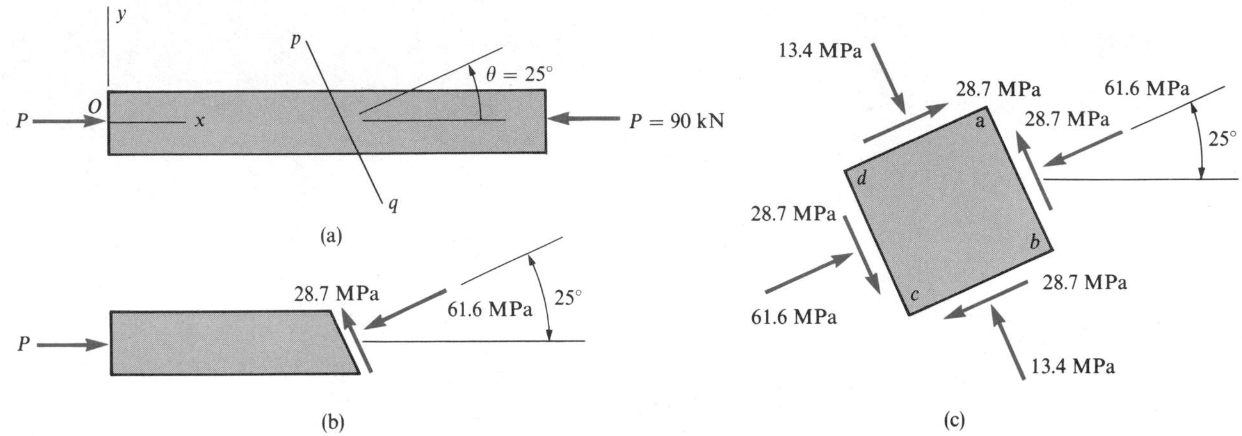

Fig. 2-31 Example

The stresses acting on the side faces of a stress element oriented at $\theta = 25^\circ$ are shown in Fig. 2-31c. Face *ab* has the same orientation as the inclined plane shown in Fig. 2-31b; hence the stresses are the same. The stresses on face *cd* are the same as those on face *ab*, which can be verified by substituting $\theta = 25^\circ + 180^\circ = 205^\circ$ into Eqs. (2-27a and b). For faces *bc* and *ad*, we substitute $\theta = -65^\circ$ and $\theta = 115^\circ$, respectively. Because these faces are on opposite sides of the element, the values of σ_θ and τ_θ are the same.

2.6 STRAIN ENERGY

The concept of **strain energy** is of fundamental importance in applied mechanics, and strain-energy principles are widely used when determining the response of machines and structures to both static and dynamic loads. This section introduces the subject in the simplest manner, by using only axially loaded members as examples. More complicated structural elements are discussed in later chapters.

As a means of illustrating the basic ideas, let us again consider a prismatic bar of length L subjected to a tensile force P (Fig. 2-32). We assume that the load is applied slowly, so that it gradually increases from zero to its maximum value P. Such a load is called a **static load** because

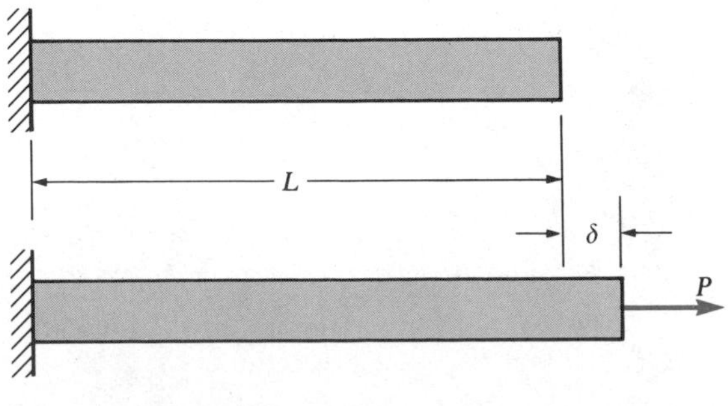

Fig. 2-32 Prismatic bar subjected to a statically applied load

there are no dynamic or inertial effects due to motion. The bar gradually elongates as the load is applied, eventually reaching its maximum elongation δ at the same time that the load reaches its full value P. Thereafter, the load remains at a constant value.

During the loading process, the load moves through the distance δ and does a certain amount of work. In order to evaluate the work done by the load, we use the **load-displacement diagram** plotted in Fig. 2-33. On this diagram the vertical axis represents the load and the horizontal axis represents the corresponding elongation of the bar. Of course, the shape of the diagram depends upon the material of the particular bar being analyzed. Let us denote by P_1 any value of the load between zero and the maximum value P; the corresponding elongation of the bar is denoted by δ_1. Then, an increment dP_1 in the load will produce an increment $d\delta_1$ in the elongation. The work done by P_1 during this incremental elongation is $P_1\, d\delta_1$, represented in the figure by the area of the shaded strip. The total work W done by the load as its value P_1 varies from zero to the maximum value P (and as δ_1 varies from zero to the maximum elongation δ) is the summation of all such elemental strips:

$$W = \int_0^\delta P_1\, d\delta_1 \tag{2-30}$$

In other words, the work done by the load is equal to the area below the load-displacement curve.

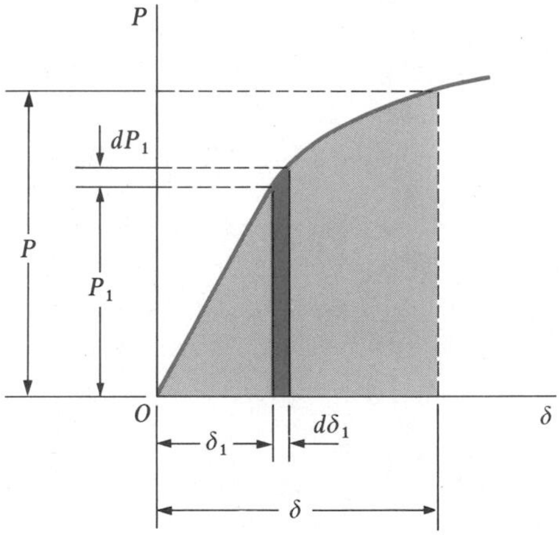

Fig. 2-33 Load-displacement diagram

The application of the load produces strains in the bar. The effect of these strains is to increase the energy level of the bar itself. Hence, a new quantity, called **strain energy**, is defined as the energy absorbed by the bar during the loading process. This strain energy, denoted by the letter U, is equal to the work done by the load provided no energy is added or subtracted in the form of heat. Therefore,

$$U = W = \int_0^{\delta} P_1 \, d\delta_1 \qquad (2\text{-}31)$$

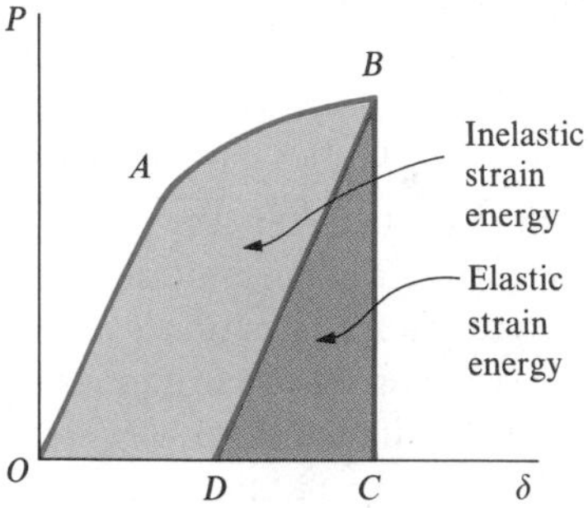

Fig. 2-34 Elastic and inelastic strain energy

Sometimes strain energy is referred to as **internal work** to distinguish it from external work W.

The unit of work and energy in SI is the joule (J), which is equal to one newton meter ($1 \text{ J} = 1 \text{ N} \cdot \text{m}$). In USCS units, work and energy are expressed in foot-pounds (ft-lb), foot-kips (ft-k), inch-pounds (in.-lb), and inch-kips (in.-k).*

If the force P is slowly removed from the bar (Fig. 2-32), the bar will shorten and either partially or fully return to its original length, depending upon whether the elastic limit was exceeded. Thus, during unloading, some or all of the strain energy of the bar may be recovered in the form of work. This behavior is portrayed in Fig. 2-34, which again shows a load-displacement diagram. During loading, the work done is equal to the area below the curve, or area $OABCDO$. When the load is removed, the load-displacement diagram follows line BD if point B is beyond the elastic limit, and a permanent elongation OD remains. Thus, the strain energy recovered during unloading is represented by the shaded triangle BCD; this recoverable energy is called **elastic strain energy**. Area $OABDO$ represents energy that is lost in the process of permanently deforming the bar; this energy is known as the **inelastic strain energy**.

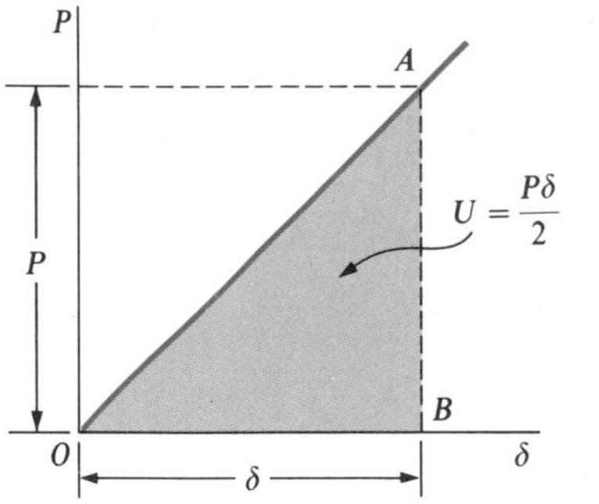

Fig. 2-35 Load-displacement diagram for a bar of linearly elastic material

Assume now that the load P acting on the bar is below the elastic-limit load (that is, below the load at which the stress in the material reaches the elastic limit). This load is represented by the ordinate at point A in the diagram of Fig. 2-34. As long as the load is below this value, all of the strain energy is recovered during unloading and no permanent set remains in the bar. Hence, the bar acts as an elastic spring that stores and releases energy as the load is applied and removed.

If the material of the bar is elastic and also follows Hooke's law, the load-displacement diagram is a straight line (Fig. 2-35). In this case, the strain energy U stored in the bar (equal to the work W done by the load P) is

$$U = W = \frac{P\delta}{2} \qquad (2\text{-}32)$$

which is the area of the shaded triangle OAB in the figure.**

* Conversion factors for work and energy are given in Appendix A, Table A-3.

** The principle that the work of the external loads is equal to the strain energy (for the case of linearly elastic behavior) was first stated by the French engineer B. P. E. Clapeyron (1799–1864) and is known as *Clapeyron's theorem* (Ref. 2-10).

In the case of a prismatic bar, we know that $\delta = PL/EA$, and therefore we can substitute into Eq. (2-32) and express the strain energy in either of the following forms:

$$U = \frac{P^2L}{2EA} \qquad U = \frac{EA\delta^2}{2L} \tag{2-33a,b}$$

The first of these equations expresses the strain energy of the bar as a function of the load P, and the second expresses it as a function of the elongation δ. Similar equations apply to a linearly elastic spring if the stiffness EA/L for a prismatic bar is replaced by the spring stiffness k:

$$U = \frac{P^2}{2k} \qquad U = \frac{k\delta^2}{2} \tag{2-34a,b}$$

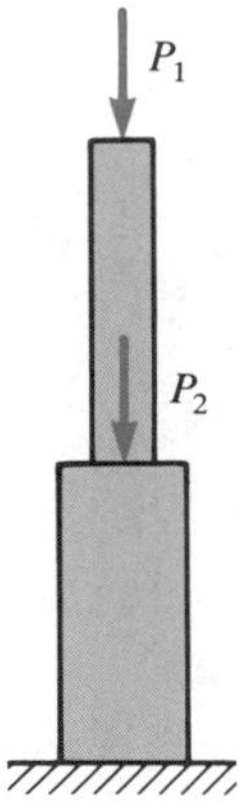

Fig. 2-36 Bar with changes in cross section

These expressions are equivalent inasmuch as $P = k\delta$ for a spring.

The strain energy of a bar consisting of two or more prismatic parts (Fig. 2-36) can be obtained by applying Eq. (2-33a) to each part of the bar and then summing to obtain the total strain energy:

$$U = \sum_{i=1}^{n} \frac{P_i^2 L_i}{2E_i A_i} \tag{2-35}$$

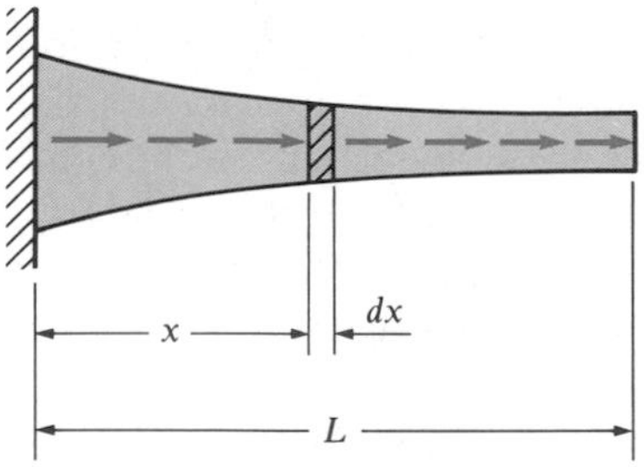

Fig. 2-37 Nonprismatic bar with varying axial force

In this equation, the quantity P_i is the axial force acting in part i of the bar and n is the number of parts.

In a similar manner, we can obtain the strain energy of a nonprismatic bar with varying axial force (Fig. 2-37) by applying Eq. (2-33a) to the differential element (shown shaded) and then integrating:

$$U = \int_0^L \frac{[P(x)]^2\, dx}{2EA(x)} \tag{2-36}$$

In this expression, $P(x)$ and $A(x)$ are the axial force and cross-sectional area at distance x from the end of the bar.

It is often convenient to use a quantity called the **strain energy density** u, which is the strain energy per unit volume of the material. Expressions for u in the linearly elastic case can be obtained by dividing the total strain energy U (Eqs. 2-33a and b) by the volume AL of the bar, because in this case the strain energy density is uniform throughout the volume; thus,

$$u = \frac{\sigma^2}{2E} \qquad u = \frac{E\epsilon^2}{2} \tag{2-37a,b}$$

in which $\sigma = P/A$ and $\epsilon = \delta/L$ are the normal stress and strain, respectively. In general, the strain energy density is equal to the area below the stress-strain curve from the origin to the point on the curve representing the stress σ and the strain ϵ.

Strain energy density has units of energy divided by volume. Thus, the SI units for u are joules per cubic meter (J/m^3), and the USCS units are foot-pounds per cubic foot, inch-pounds per cubic inch, or other similar units. These units are the same as the units of stress, so we can also use pascals or psi as the units for u.

The strain energy density when the material is stressed to the proportional limit is called the **modulus of resilience** u_r. It is found by substituting the proportional-limit stress σ_{pl} into Eq. (2-37a):

$$u_r = \frac{\sigma_{pl}^2}{2E} \tag{2-38}$$

As an example, a mild steel having $\sigma_{pl} = 30{,}000$ psi and $E = 30 \times 10^6$ psi has a modulus of resilience $u_r = 15$ psi (or 103 kPa). Note that the modulus of resilience is equal to the area below the stress-strain curve up to the proportional limit. **Resilience** represents the ability of the material to absorb energy within the elastic range. Another quantity, called **toughness**, refers to the ability of the material to absorb energy without fracturing. The corresponding modulus, called the **modulus of toughness** u_t, is the strain energy density when the material is stressed to the point of failure. It is equal to the area below the entire stress-strain curve.

The preceding discussions of strain energy for a member in tension apply also to a member in compression. Since the work done by an axial load is positive regardless of whether the load causes tension or compression, it follows that the strain energy is always a positive quantity. This conclusion is also apparent from the expressions for strain energy (Eqs. 2-33 through 2-37), all of which are positive because the algebraic quantities are squared.

Strain energy is a form of potential energy (or "energy of position") because it depends upon the relative locations of the particles or elements that make up the structure. For instance, when a spring is compressed, its particles are crowded more closely together; when stretched, the relative distances between particles are increased. In both cases the strain energy of the structure is increased as compared to the energy in the unloaded position.

Example 1

Three round bars having the same length L but different shapes are shown in Fig. 2-38. The first bar has diameter d over its entire length, the second has this diameter over one-fourth of its length, and the third has this diameter over one-eighth of its length. All three bars are subjected to the same load P. Compare the amounts of strain energy stored in the bars, assuming linearly elastic behavior.

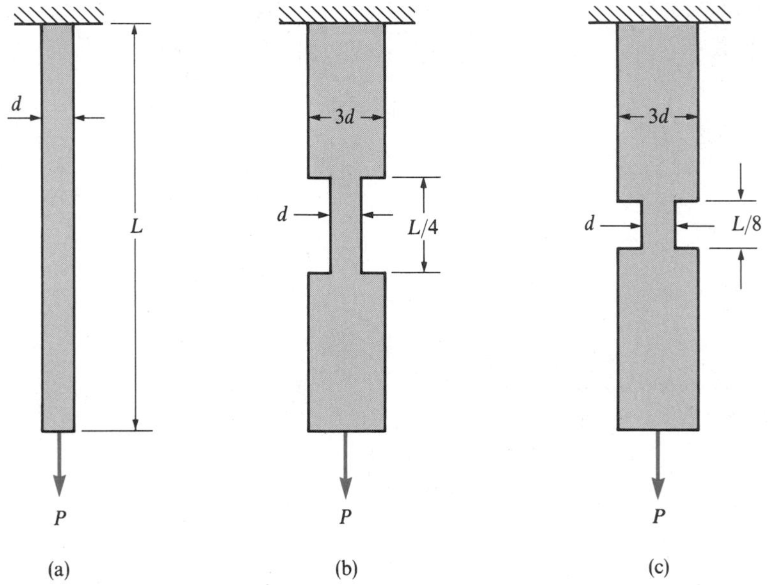

Fig. 2-38 Example 1. Calculation of strain energy

The strain energy of the first bar (from Eq. 2-33a) is

$$U_1 = \frac{P^2L}{2EA}$$

where $A = \pi d^2/4$. Assuming that the stress distribution is uniform over every cross section, we find the strain energy of the second bar to be

$$U_2 = \frac{P^2(L/4)}{2EA} + \frac{P^2(3L/4)}{2E(9A)} = \frac{P^2L}{6EA} = \frac{U_1}{3}$$

For the third bar we find

$$U_3 = \frac{P^2(L/8)}{2EA} + \frac{P^2(7L/8)}{2E(9A)} = \frac{P^2L}{9EA} = \frac{2U_1}{9}$$

Comparison of these results shows that the strain energy decreases as the volume of the bar increases, although all three bars have the same maximum stress. Thus, the third bar has less energy-absorbing capacity than the other two. Therefore, it takes only a small amount of work to bring the tensile stress to a high value in a bar with a groove, and the narrower the groove, the more severe the condition. When the loads are dynamic in character and the ability to absorb energy is important, the presence of grooves is very damaging. Of course, for static loads the maximum stresses, rather than the ability to absorb energy, are important in design.

Example 2

Determine the strain energy stored in a prismatic bar suspended from one end (Fig. 2-39) due to its own weight, assuming linearly elastic behavior.

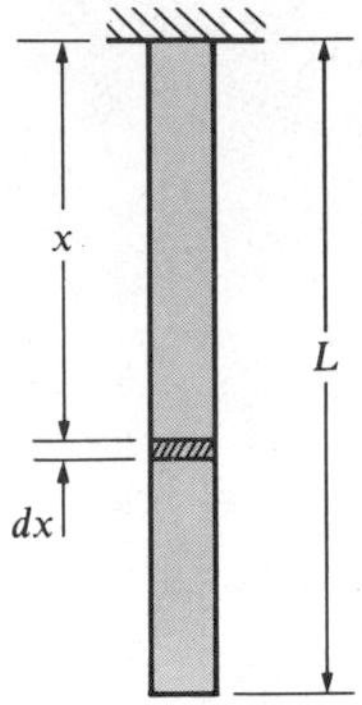

Fig. 2-39 Example 2. Bar hanging under its own weight

We begin by considering an element of the bar of length dx (shown shaded in the figure). The axial force $P(x)$ acting on this element is equal to the weight of the bar below the element:

$$P(x) = \gamma A(L - x) \tag{a}$$

in which γ is the specific weight of the material and A is the cross-sectional area of the bar. Substituting into Eq. (2-36) and integrating gives the total strain energy:

$$U = \int_0^L \frac{[\gamma A(L - x)]^2\,dx}{2EA} = \frac{\gamma^2 A L^3}{6E} \tag{2-39}$$

This same result can be obtained by integrating the strain energy density. At any distance x from the support, the stress is

$$\sigma = \frac{P(x)}{A} = \gamma(L - x)$$

and, therefore, the strain energy density (see Eq. 2-37a) is

$$u = \frac{\sigma^2}{2E} = \frac{\gamma^2(L - x)^2}{2E}$$

The total strain energy is now found by integrating u throughout the volume of the bar:

$$U = \int u\,dV = \int_0^L u(A\,dx) = \int_0^L \frac{\gamma^2 A(L - x)^2\,dx}{2E} = \frac{\gamma^2 A L^3}{6E}$$

which agrees with Eq. (2-39).

Example 3

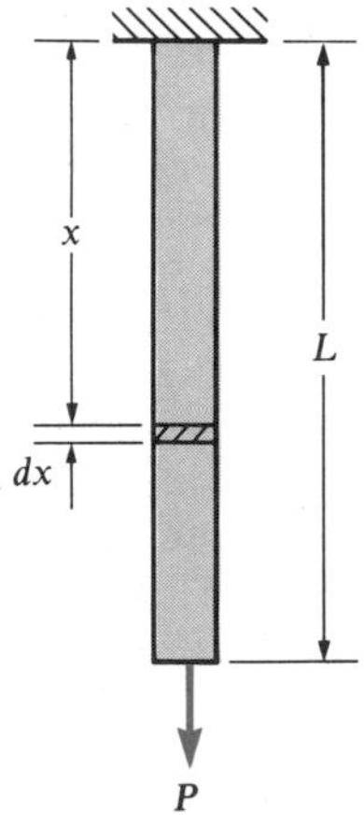

Fig. 2-40 Example 3. Bar hanging under its own weight and supporting a load P

Determine the strain energy in a prismatic bar suspended from one end if, in addition to its own weight, it supports a load P at the lower end (Fig. 2-40).

In this case, the axial force $P(x)$ acting on the element shown shaded in the figure is

$$P(x) = \gamma A(L - x) + P$$

(see Eq. a). From Eq. (2-36) we now obtain

$$U = \int_0^L \frac{[\gamma A(L - x) + P]^2\,dx}{2EA} = \frac{\gamma^2 A L^3}{6E} + \frac{\gamma P L^2}{2E} + \frac{P^2 L}{2EA} \tag{2-40}$$

The first term in this result is the same as that given in Eq. (2-39) for the strain energy of a bar hanging under its own weight. Also, the last term is the same as the strain energy of a bar subjected only to an axial force P (Eq. 2-33a). The middle term contains both γ and P, showing that it depends upon both the weight of the bar and the magnitude of the applied load.

The important conclusion from the preceding example is that we cannot obtain the strain energy of a structure due to more than one load merely by adding the strain energies obtained from the individual loads acting separately. The reason is that strain energy is a quadratic function of the loads (see Eq. 2-36), not a linear function.

DISPLACEMENTS CAUSED BY A SINGLE LOAD

The use of strain energy provides a convenient method for obtaining displacements of structures in certain elementary situations. Consider a linearly elastic structure that is subjected to a single concentrated load P. Then the work W done by this load, equal to the strain energy U stored in the structure, is given by Eq. (2-32):

$$U = W = \frac{P\delta}{2} \qquad \text{or} \qquad \delta = \frac{2U}{P} \tag{2-41a,b}$$

where δ is the displacement through which the force P moves. This equation provides a simple method for finding the displacement δ if the strain energy can be determined. However, the limitations of the method must be kept in mind: (1) only one load may act on the structure, and (2) the only displacement that can be determined by this method is the displacement corresponding to the load itself. The following example illustrates the procedure.

Example 4

Determine the vertical displacement δ_b of joint B for the truss of Fig. 2-41. Note that the only load acting on the truss is a vertical load P at joint B. Assume that both members have the same axial rigidity EA.

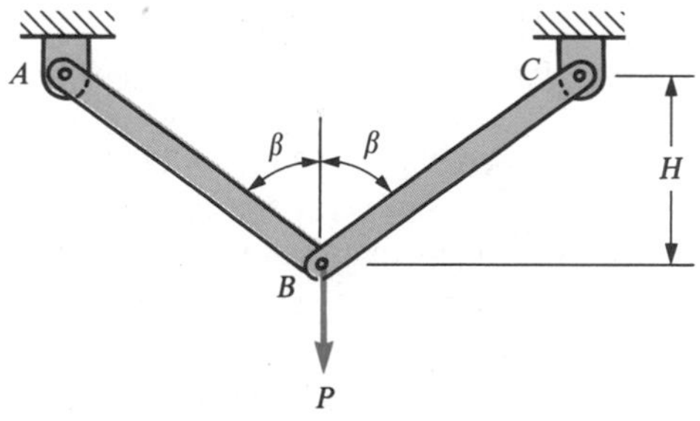

Fig. 2-41 Example 4. Displacement of a truss

In this example, the displacement produced by the load P will be found by equating the work of the load to the strain energy stored in the members. The force F in either bar is

$$F = \frac{P}{2\cos\beta}$$

as found from equilibrium of joint B. Also, the length of each bar is $L_1 = H/\cos\beta$. Then, using Eq. (2-33a), we get the strain energy of the two bars:

$$U = \frac{F^2 L_1}{2EA}(2) = \frac{P^2 H}{4EA\cos^3\beta}$$

The work of the load P is

$$W = \frac{P\delta_b}{2}$$

Equating U and W and solving for δ_b, we obtain

$$\delta_b = \frac{PH}{2EA\cos^3\beta} \tag{2-42}$$

This equation gives the vertical displacement of joint B of the truss.

*2.7 DYNAMIC LOADING

Dynamic loads differ from static loads in that they vary with time. A static load is applied slowly, gradually increasing from zero to its maximum value, and thereafter remaining constant. In contrast, dynamic loads may be applied very suddenly, thus causing vibrations of the structure, or they may change in magnitude as time elapses. Examples are impact loads, such as when two objects collide or when a falling object strikes a structure, and cyclical loads caused by rotating machinery. Other examples are the loads caused by traffic, wind gusts, water waves, earthquakes, and manufacturing processes, all of which are dynamic in character. To aid in understanding the nature of such loads, we will consider in this section the most basic type of dynamic load, an **impact load**.

As an example of an impact load, consider the simple arrangement shown in Fig. 2-42. A collar of mass M, initially at rest, falls from a height h onto a flange at the lower end of bar AB. When the collar strikes the flange, the bar begins to elongate, creating axial stresses and strains within the bar. In a very short interval of time, the flange will have moved downward to its position of maximum displacement; then the bar begins to shorten and longitudinal vibrations ensue. The situation is analogous to jumping onto a pogo stick or dropping a mass onto a spring. Soon the vibrations cease because of damping effects, and the bar will be at rest with the mass M supported on the flange. The behavior of this system under the action of the falling collar is obviously very complicated, and a complete analysis requires the use of advanced mathematical techniques. However, we can make an approximate analysis by using the concept of strain energy and making some simplifying assumptions.

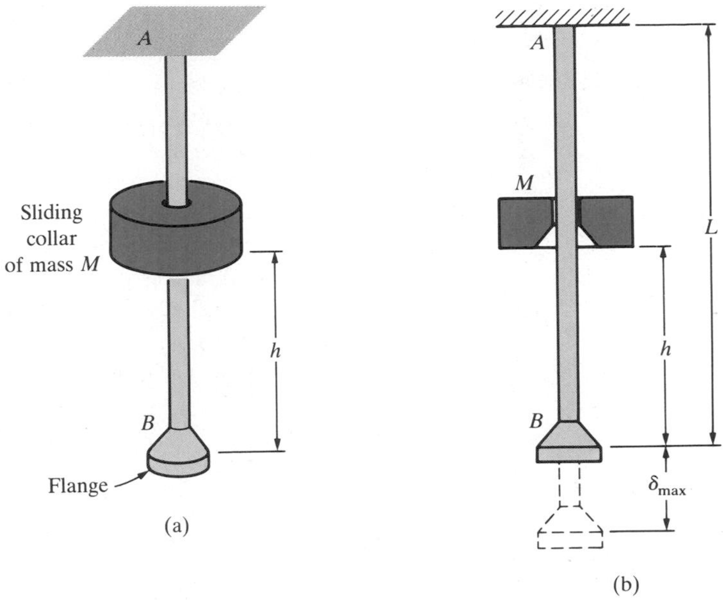

Fig. 2-42 Impact load on a prismatic bar *AB* due to a falling mass *M*

The potential energy of the mass M with respect to the elevation of the flange is Mgh, where g is the acceleration of gravity.* This potential energy is converted into kinetic energy as the mass falls. At the instant the mass strikes the flange, its potential energy (with respect to the elevation of the flange) has become zero and its kinetic energy is $Mv^2/2$, where $v = \sqrt{2gh}$ is the velocity of the falling mass. The kinetic energy of the mass is then transformed to other forms of energy. Part of it is transformed into strain energy of the stretched bar. Some of it is dissipated in the form of heat and some in causing localized plastic strains in the bar and flange. A small part may remain as kinetic energy of the mass, which either continues to move downward with the flange or bounces upward.

Let us assume that the collar and flange are designed so that the collar "sticks" to the flange and moves downward with it, a situation known as **perfectly plastic impact**. Also, we will disregard energy losses and assume that all of the kinetic energy of the falling mass is transformed into strain energy of the bar. This second assumption is conservative in the sense that it results in larger stresses in the bar than would result otherwise. However, disregarding energy losses is inconsistent with perfectly plastic impact, so later on we will show how to incorporate energy losses in the analysis. In addition, we will disregard any change in the potential energy of the bar itself, and we will ignore the existence of strain energy in the bar due to its own weight. Both of these effects are

* In SI units, $g = 9.81\ \text{m/s}^2$; in USCS units, $g = 32.2\ \text{ft/s}^2$. For more precise values of g, or for a discussion of mass and weight, see Appendix A.

extremely small. Finally, we will assume that the stresses in the bar remain within the linearly elastic range and that the stress distribution is the same as for static loading. On the basis of these assumptions, we can calculate the maximum elongation and maximum tensile stress in the bar due to the impact load.

The maximum elongation δ_{max} of the bar (Fig. 2-42b) can be obtained from the principle of **conservation of energy** by equating the potential energy lost by the falling mass to the strain energy acquired by the bar. The potential energy lost is $W(h + \delta_{max})$, where $W = Mg$ is the weight of the collar and $h + \delta_{max}$ is the distance through which it moves. The strain energy of the bar is $EA\delta_{max}^2/2L$, where EA is the axial rigidity and L is the length of the bar (see Eq. 2-33b). Thus, we obtain the following equation:

$$W(h + \delta_{max}) = \frac{EA\delta_{max}^2}{2L} \tag{2-43}$$

This equation is quadratic in δ_{max} and can be solved for its positive root; the result is

$$\delta_{max} = \frac{WL}{EA} + \left[\left(\frac{WL}{EA}\right)^2 + \frac{2WLh}{EA}\right]^{1/2} \tag{2-44}$$

Note that the elongation of the bar increases if either the mass or the height of fall is increased and diminishes if the stiffness EA/L is increased.

The preceding equation can be written in simpler form by introducing the notation

$$\delta_{st} = \frac{WL}{EA} \tag{2-45}$$

for the static elongation of the bar due to the weight W of the collar. Then Eq. (2-44) becomes

$$\delta_{max} = \delta_{st} + (\delta_{st}^2 + 2h\delta_{st})^{1/2} \tag{2-46}$$

If the static elongation is very small compared to the height h, we can further simplify the preceding equation to the following:

$$\delta_{max} \approx \sqrt{2h\delta_{st}} = \sqrt{\frac{Mv^2L}{EA}} \tag{2-47}$$

in which $v = \sqrt{2gh}$ is the velocity of the falling mass when it strikes the flange and $M = W/g$. This result can be obtained directly from Eq. (2-43) by omitting δ_{max} on the left-hand side of the equation. Values of δ_{max} calculated from Eq. (2-47) are always less than those obtained from Eq. (2-46). For instance, if $h = 40\delta_{st}$, the approximate equation gives $\delta_{max} = 8.9\delta_{st}$, which is 0.89 times the elongation of $10\delta_{st}$ obtained from Eq. (2-46).

The maximum stress in the bar can be calculated from the maximum elongation if we assume that the stress distribution is uniform throughout the length. Of course, this condition is only an approximation because, in reality, longitudinal stress waves exist in the bar. Nevertheless, based upon the assumption of a uniform stress distribution, we obtain the following equations for the maximum tensile stress (see Eq. 2-44):

$$\sigma_{\max} = \frac{E\delta_{\max}}{L} = \frac{W}{A} + \left[\left(\frac{W}{A}\right)^2 + \frac{2WhE}{AL}\right]^{1/2} \tag{2-48}$$

or

$$\sigma_{\max} = \sigma_{st} + \left(\sigma_{st}^2 + \frac{2hE}{L}\sigma_{st}\right)^{1/2} \tag{2-49}$$

in which $\sigma_{st} = W/A$ is the stress when the load acts statically. Again considering the case where the height h is large compared to the elongation (see Eq. 2-47), and also noting that $\sigma_{st} = Mg/A$, we obtain

$$\sigma_{\max} \approx \sqrt{\frac{2hE}{L}}\,\sigma_{st} = \sqrt{\frac{Mv^2E}{AL}} \tag{2-50}$$

From this result, we see that an increase in the kinetic energy $Mv^2/2$ of the falling mass will cause an increase in stress, whereas an increase in the volume AL of the bar will reduce the stress. This situation is quite different from static tension of the bar, in which case the stress is independent of the length L and the modulus of elasticity E. After the maximum elongation is reached in the bar, the bar will vibrate axially until it comes to rest at the static elongation.

The preceding equations for maximum elongation $\delta_{\max}$ and maximum stress $\sigma_{\max}$ were derived for a prismatic bar. If the bar is nonprismatic, the procedures for finding $\delta_{\max}$ and $\sigma_{\max}$ must be modified slightly. The simplest procedure involves determining an equivalent stiffness k of the bar by calculating the relationship between a load P acting on the bar and the corresponding static elongation δ (note that $k = P/\delta$). This stiffness is then used in Eqs. (2-44), (2-45), and (2-47) in place of EA/L, which is the stiffness of a prismatic bar. After the dynamic elongation $\delta_{\max}$ has been calculated, the static load $P_{\max}$ that would produce the same elongation can be found from the equation $P_{\max} = k\delta_{\max}$. Finally, the maximum stress can be obtained by dividing $P_{\max}$ by the minimum cross-sectional area of the bar. This technique is illustrated in Example 2.

Another example of an impact load is when an object that is moving horizontally strikes the end of a bar or spring (Fig. 2-43). Assuming that the mass M of the moving object is very large compared to the mass of the

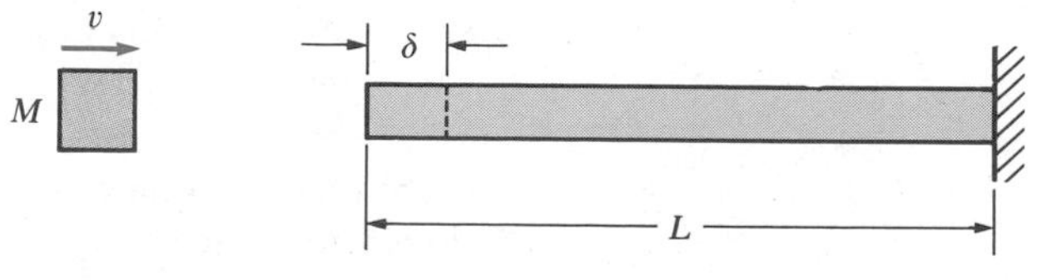

Fig. 2-43 Impact load on a horizontal bar due to a mass M that is moving with velocity v

bar, we can use the same approach as for a falling mass. At the instant of impact, the kinetic energy of the moving mass is $Mv^2/2$. If all of this energy is transformed into strain energy of the bar, we can write the following equation of conservation of energy:

$$\frac{Mv^2}{2} = \frac{EA\delta_{max}^2}{2L} \tag{2-51}$$

in which δ_{max} is the maximum displacement of the end of the bar. Therefore,

$$\delta_{max} = \sqrt{\frac{Mv^2L}{EA}} \tag{2-52}$$

The corresponding maximum compressive stress, assumed to be uniform throughout the bar, is

$$\sigma_{max} = \frac{E\delta_{max}}{L} = \sqrt{\frac{Mv^2E}{AL}} \tag{2-53}$$

These equations for δ_{max} and σ_{max} are the same as Eqs. (2-47) and (2-50), respectively.

SUDDENLY APPLIED LOAD

A special case of impact occurs when a load is suddenly applied to a bar, but without an initial velocity. To explain this kind of loading, consider again the vertical, prismatic bar shown in Fig. 2-42 and assume that the sliding collar is placed gently on the flange and then suddenly released. Although in this instance no kinetic energy exists at the beginning of extension of the bar, the problem is quite different from that of static loading of the bar. For static tension, we assume the load is gradually applied; consequently, equilibrium always exists between the applied load and the resisting force of the bar. The kinetic energy of the load does not enter into the analysis under such conditions. However, in a sudden application of the load, the elongation of the bar and the stress in the bar are initially zero, but then the suddenly applied load begins to move downward under the action of its own weight, elongating the bar as it does so. During this motion, the resisting force of the bar gradually increases until at some instant it just equals W, the weight of the collar. At this same instant, the elongation of the bar is δ_{st}. But the mass now has a certain kinetic energy, acquired during the displacement δ_{st}; hence it continues to move downward until its velocity is brought to zero by the resisting force in the bar. The maximum elongation for this condition is obtained from Eq. (2-46) by setting h equal to zero; thus,

$$\delta_{max} = 2\delta_{st} \tag{2-54}$$

From this equation we conclude that a suddenly applied load produces an elongation twice as large as the elongation caused by a static load. After the load is applied and the elongation $2\delta_{st}$ has occurred, the bar will vibrate up and down, eventually coming to rest at the static elongation, because the flange is still supporting the weight of the collar.*

ENERGY LOSSES

The preceding analyses were based upon the assumption that no energy losses occur during impact. In reality, energy losses always occur during perfectly plastic impact, with most of the energy loss being in the form of heat and localized deformation of the materials. Because of these losses, the kinetic energy of the system immediately after the impact is less than it was before the impact, which means that less energy is converted into strain energy of the bar. Therefore, the end of the bar does not displace as much as when energy is conserved.

Let us denote by KE_1 the kinetic energy immediately before impact and by KE_2 the kinetic energy after impact. Then we can relate these two quantities by the following equation:

$$KE_2 = K_L(KE_1) \tag{2-55}$$

The factor K_L, which is always between 0 and 1, is the *kinetic energy loss factor*, or *efficiency factor*.

The reduction in kinetic energy is taken into account by placing the factor K_L in the equations previously derived for no energy losses. For instance, Eq. (2-43) is modified by replacing Wh (which is equal to the kinetic energy before impact) with $K_L Wh$. The following equations (Eqs. 2-44 and 2-46 through 2-50) are altered by replacing h with $K_L h$. In this same manner, the equations derived for the case shown in Fig. 2-43 are modified by inserting the factor K_L on the left-hand side of Eq. (2-51). Then Eqs. (2-52) and (2-53) are changed by inserting K_L under the square root signs.

The loss factor K_L can sometimes be determined by advanced methods of analysis, or it can be estimated when no other information is available. One method for estimating K_L is based upon the principle of conservation of momentum. During an impact between two objects, the linear momentum of the objects is conserved even though energy is not. By using this principle and also by assuming that the velocity varies linearly along the length of the bar (maximum velocity at the free end and zero velocity at the supported end), it is possible to derive the following formula for the kinetic energy loss factor:

$$K_L = \frac{1 + \dfrac{M_1}{M} + \dfrac{M_2}{3M}}{\left(1 + \dfrac{M_1}{M} + \dfrac{M_2}{2M}\right)^2} \tag{2-56}$$

* Equation (2-54) was first obtained by the French mathematician and scientist J. V. Poncelet (1788–1867); see Ref. 2-11.

in which M is the mass of the striking object (see Figs. 2-42 and 2-43), M_1 is the mass of any object attached to the end of the bar (the flange in Fig. 2-42), and M_2 is the mass of the bar itself. The equation shows that K_L depends upon the relative masses of the objects. As the mass M of the striking object becomes very large compared to the masses of the flange and bar, the factor K_L approaches unity. This situation is equivalent to our earlier derivations where we disregarded energy losses and disregarded the masses of the flange and bar. The derivation of Eq. (2-56) for K_L is based upon the assumption of perfectly plastic impact, so it should be used only when the mass M is large as compared to M_1 and M_2.

Many considerations have been omitted from this discussion, including the effects of stress waves, inelastic deformations, damping, and imperfections at the contact surfaces. Therefore, all of the formulas in this section are based upon highly idealized conditions and give only a rough approximation, usually on the conservative side, of the true conditions.

INELASTIC EFFECTS AND CAUSES OF FAILURE

The preceding discussion of the effects of impact loads is based upon the assumption that the maximum stress in the bar remains within the proportional limit. Beyond this limit, the problem becomes more involved because the elongation of the bar is no longer proportional to the axial force. If the tensile-test diagram does not depend upon the rate of straining of the bar, the elongation beyond the elastic limit during impact can be determined from a static load-displacement diagram, such as the one shown in Fig. 2-44. For any assumed maximum elongation δ_1, the corresponding area $OABE$ gives the strain energy stored in the bar. This strain energy must be equal to the potential energy lost by the falling weight W when it drops through the distance $h + \delta$ (see Fig. 2-42). Therefore, when $W(h + \delta)$ is equal to or larger than the total area $OABCD$ of the load-displacement diagram, the falling body will fracture the bar. (In some materials, including ductile steel, the yield point is raised when the rate of straining of the bar is very large. Then the amount of work necessary to produce fracture is somewhat higher than in a static test.)

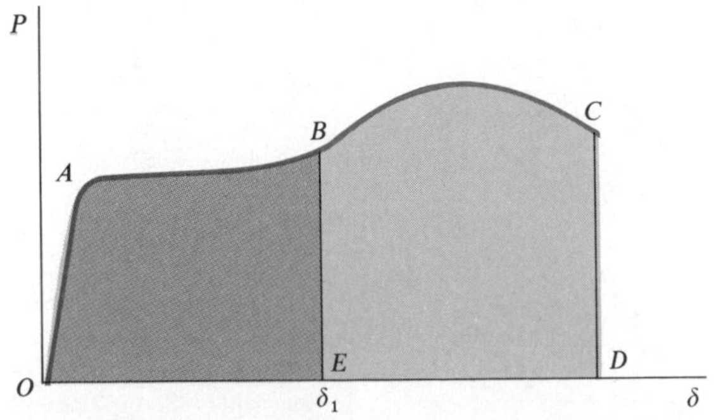

Fig. 2-44 Load-displacement diagram for a prismatic bar in tension

From this discussion, we see that any change in the form of the bar that diminishes the total area $OABCD$ of the load-displacement diagram (Fig. 2-44) also diminishes the resistance of the bar to impact. In the grooved specimens shown in Figs. 2-38b and c, for instance, the plastic flow of metal will be concentrated at the grooves; hence the total elongation and the work necessary to produce fracture will be much smaller in the grooved bars than in the cylindrical bar shown in Fig. 2-38a. Such grooved specimens are very weak against impact. A slight shock may produce fracture, although the material itself may be ductile. Members having holes or any sharp variation in cross section are similarly weak against impact (see Sections 2.9 and 2.10).

In general, ductile materials offer much greater resistance to impact loads than do brittle materials. The load-displacement curve for a bar of brittle material will have a much smaller area below it than does the curve for a bar of ductile material, even though the ultimate stresses for the two materials may be approximately equal.

Example 1

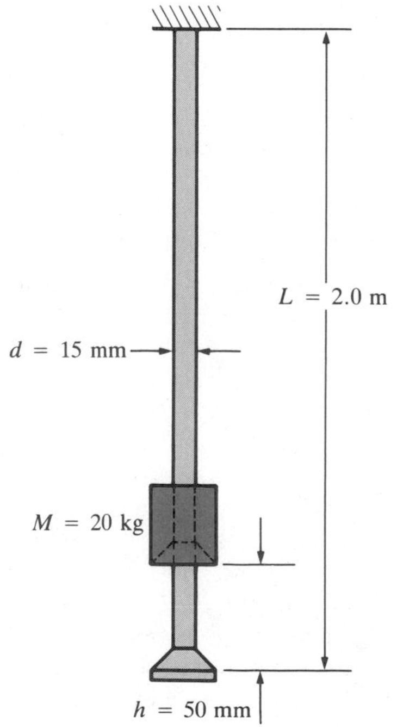

Fig. 2-45 Example 1

A round, prismatic steel bar ($E = 200$ GPa) of length $L = 2.0$ m and diameter $d = 15$ mm hangs vertically from a support at its upper end (Fig. 2-45). A sliding collar of mass $M = 20.0$ kg drops from a height $h = 50$ mm onto the flange at the lower end of the bar without rebounding. (a) Determine the maximum elongation δ_{max} and maximum tensile stress σ_{max} due to the impact, assuming no loss of energy during impact and disregarding the masses of the flange and bar. (b) Determine the maximum elongation δ_{max} and maximum tensile stress σ_{max} including the effect of energy losses. Assume the flange has mass $M_1 = 0.2$ kg and the bar has mass $M_2 = 2.8$ kg.

(a) When energy losses are disregarded, the maximum elongation δ_{max} produced by the falling mass M can be calculated directly from Eq. (2-46). We begin by determining the static elongation of the bar due to the mass (which has a weight $W = Mg$):

$$\delta_{st} = \frac{WL}{EA} = \frac{MgL}{EA} = \frac{(20.0 \text{ kg})(9.81 \text{ m/s}^2)(2.0 \text{ m})}{(200 \text{ GPa})(\pi/4)(15 \text{ mm})^2} = 0.0111 \text{ mm}$$

This elongation is now substituted into Eq. (2-46):

$$\delta_{max} = \delta_{st} + (\delta_{st}^2 + 2h\delta_{st})^{1/2}$$
$$= 0.0111 \text{ mm} + [(0.0111 \text{ mm})^2 + 2(50 \text{ mm})(0.0111 \text{ mm})]^{1/2} = 1.06 \text{ mm}$$

The ratio of δ_{max} to δ_{st}, called the **impact factor**, is 1.06/0.0111 or 95. This factor represents the amount by which the static elongation is amplified due to thc impact.

In this example the elongation may also be calculated from Eq. (2-47), as follows:

$$\delta_{max} = \sqrt{2h\delta_{st}} = [2(50 \text{ mm})(0.0111 \text{ mm})]^{1/2} = 1.05 \text{ mm}$$

This result is practically the same as the preceding one because δ_{st} is very small compared to h, the height of fall.

The maximum stress in the bar can be obtained from Eq. (2-48), as follows:

$$\sigma_{max} = \frac{E\delta_{max}}{L} = \frac{(200 \text{ GPa})(1.06 \text{ mm})}{2.0 \text{ m}} = 106 \text{ MPa}$$

This dynamic stress may be compared with the static stress:

$$\sigma_{st} = \frac{W}{A} = \frac{Mg}{A} = \frac{(20 \text{ kg})(9.81 \text{ m/s}^2)}{(\pi/4)(15 \text{ mm})^2} = 1.11 \text{ MPa}$$

The ratio of σ_{max} to σ_{st} is $106/1.11 = 95$, which is the same impact factor as for the elongations.

(b) When energy losses are taken into consideration, we may use Eq. (2-56) to obtain the kinetic energy loss factor. Substituting $M = 20.0$ kg, $M_1 = 0.2$ kg, and $M_2 = 2.8$ kg into that equation gives

$$K_L = 0.906$$

This factor gives the relative loss of kinetic energy during the impact, which is assumed to be perfectly plastic (see Eq. 2-55). To calculate the maximum elongation δ_{max}, we may use Eq. (2-46) with h replaced by $K_L h$ and $\delta_{st} = 0.0111$ mm as in part (a). In this manner we obtain

$$\delta_{max} = 1.01 \text{ mm}$$

which compares with 1.06 mm when energy losses were disregarded. The maximum tensile stress is

$$\sigma_{max} = \frac{E\delta_{max}}{L} = 101 \text{ MPa}$$

Note that the impact factor is now 91, which may be obtained either from the stresses (101 MPa/1.11 MPa) or from the elongations (1.01 mm/0.0111 mm).

Example 2

This example is similar to Example 1, except that the bar has a larger diameter $3d$ over the upper half of its length (Fig. 2-46). As in the preceding example, the bar has total length $L = 2.0$ m, the lower half of the bar has diameter $d = 15$ mm, and $E = 200$ GPa. The sliding collar has mass $M = 20.0$ kg and falls from a height $h = 50$ mm. We wish to determine the maximum elongation δ_{max} and the maximum tensile stress σ_{max} due to the impact, disregarding energy losses.

Because the bar is not prismatic, we cannot substitute directly into the formulas of this section, as we did in Example 1. Instead, we begin by calculating an equivalent stiffness k for the bar. If a static load P_1 acts on the free end of the bar, the resulting elongation δ_1 is

$$\delta_1 = \frac{P_1(L/2)}{EA} + \frac{P_1(L/2)}{E(9A)} = \frac{5P_1L}{9EA}$$

in which $A = \pi d^2/4$ is the cross-sectional area of the lower half of the bar. From

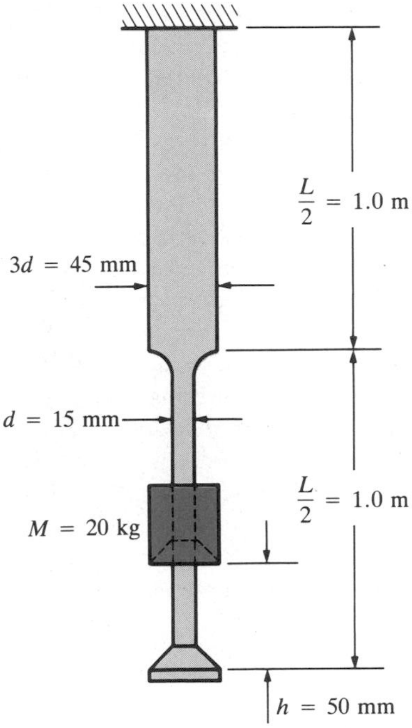

Fig. 2-46 Example 2

this equation, we can obtain the equivalent stiffness k:

$$k = \frac{P_1}{\delta_1} = \frac{9EA}{5L} = \frac{9(200\text{ GPa})(\pi/4)(15\text{ mm})^2}{5(2.0\text{ m})} = 31.8\text{ MN/m}$$

Next, the value of k is substituted into Eq. (2-45) in place of EA/L to obtain the static elongation caused by the mass M:

$$\delta_{st} = \frac{W}{k} = \frac{Mg}{k} = \frac{(20\text{ kg})(9.81\text{ m/s}^2)}{31.8\text{ MN/m}} = 0.00617\text{ mm}$$

Then we use Eq. (2-46) to obtain the dynamic elongation:

$$\begin{aligned}\delta_{max} &= \delta_{st} + (\delta_{st}^2 + 2h\delta_{st})^{1/2}\\ &= 0.00617\text{ mm} + [(0.00617\text{ mm})^2 + 2(50\text{ mm})(0.00617\text{ mm})]^{1/2}\\ &= 0.792\text{ mm}\end{aligned}$$

Note that this elongation is less than for the prismatic bar of Example 1. However, the impact factor is 0.792/0.00617 = 128, which is greater than the impact factor for the prismatic bar.

The static load P that would produce the same elongation δ is

$$P = k\delta = (31.8\text{ MN/m})(0.792\text{ mm}) = 25.2\text{ kN}$$

Therefore, the maximum stress σ is

$$\sigma = \frac{P}{A} = \frac{25.2\text{ kN}}{(\pi/4)(15\text{ mm})^2} = 143\text{ MPa}$$

This stress occurs in the lower half of the bar. Comparing it with the maximum stress of 106 MPa calculated in Example 1, we see that the stress in the nonprismatic bar is 35% greater than the stress in the prismatic bar. We conclude that enlarging the bar over part of its length is detrimental from the standpoint of the bar's resistance to an impact load, because the energy-absorbing capacity of the bar is reduced, as pointed out previously in Example 1 of Section 2.6. In general, bars that must resist dynamic loads should be prismatic so that the stress distributions are uniform throughout their lengths and so that stress concentrations are avoided (see Sections 2.9 and 2.10).

Example 3

An elevator car of weight W is supported by a cable that is moving downward with constant velocity v (Fig. 2-47). What maximum stress is produced in the cable when the drum is locked suddenly?

This analysis differs significantly from the preceding ones. In our earlier discussions the bars being struck were assumed to be unstressed prior to impact, hence they had no strain energy before impact. However, because the cable in this example supports the elevator, it has a significant amount of strain energy even before the drum is locked. Therefore, we cannot use any of the previously derived equations for maximum elongation or maximum stress. Instead, we must begin by writing a new equation of conservation of energy.

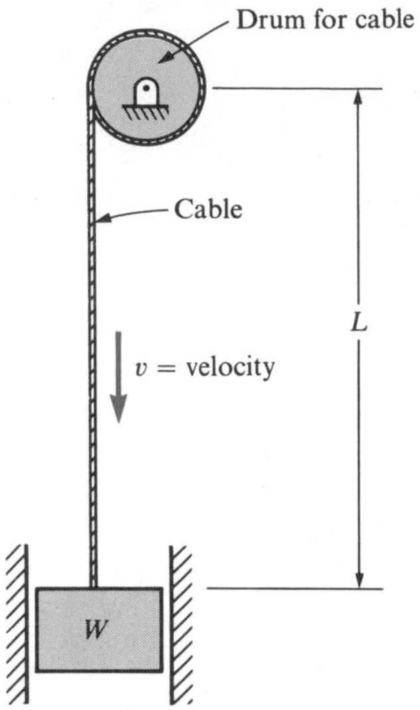

Fig. 2-47 Example 3

Let us be conservative and assume that there are no energy losses when the drum is locked. Then the total energy of the system (kinetic, potential, and strain energies) just prior to locking is equal to the total energy when the cable has its maximum elongation δ_{max}.

Before locking, the kinetic energy of the moving elevator is $Wv^2/2g$. The kinetic energies of the cable and rotating drum will be disregarded on the basis that they are relatively small compared to the kinetic energy of the elevator.

The potential energy of the elevator will be determined with respect to its lowest position (the position of the elevator when the cable has its greatest elongation). Let δ_1 be the distance that the elevator moves downward after the drum locks. This distance is not the same as the maximum elongation δ_{max} of the cable, because the cable already has a static elongation δ_{st} due to the weight W of the elevator. The additional elongation of the cable due to locking is $\delta_{max} - \delta_{st}$, and this elongation is the same as the downward displacement δ_1 of the elevator. Therefore, the potential energy of the elevator prior to locking is $W\delta_1$. The potential energy of the cable will be disregarded in comparison to that of the elevator.

The strain energy of the cable before locking is $EA\delta_{st}^2/2L$ (see Eq. 2-33b), where EA/L is the stiffness of the cable and δ_{st} is its static elongation:

$$\delta_{st} = \frac{WL}{EA} \tag{a}$$

Thus, the total energy of the system before locking is

$$\frac{Wv^2}{2g} + W(\delta_{max} - \delta_{st}) + \frac{EA\delta_{st}^2}{2L} \tag{b}$$

in which δ_{st} is given by Eq. (a).

After the drum is locked, and at the instant the cable has its maximum elongation, the velocity of the elevator is zero and it has no kinetic energy. We will assume that the same situation holds for the entire length of the cable. The potential energy is also zero because the elevator is at its lowest position (downward displacement equal to δ_1) and we are neglecting the potential energy of the cable. Finally, the strain energy of the cable at its maximum elongation is

$$\frac{EA\delta_{max}^2}{2L} \tag{c}$$

This quantity represents the total energy of the system after the cable has locked.

Using the principle of conservation of energy and equating the energies before and after locking of the drum (expressions b and c), we obtain

$$\frac{Wv^2}{2g} + W(\delta_{max} - \delta_{st}) + \frac{EA\delta_{st}^2}{2L} = \frac{EA\delta_{max}^2}{2L} \tag{d}$$

This equation can be solved for the maximum elongation δ_{max} of the cable in the following manner. First, we note from Eq. (a) that $W = EA\delta_{st}/L$. Then we substitute this expression for W into the second term of Eq. (d) and rearrange, thereby obtaining

$$\frac{Wv^2}{2g} + \frac{EA\delta_{st}}{L}(\delta_{max} - \delta_{st}) = \frac{EA}{2L}(\delta_{max}^2 - \delta_{st}^2)$$

Further rearrangement of terms gives

$$\frac{Wv^2}{2g} = \frac{EA}{2L}(\delta_{\text{max}} - \delta_{\text{st}})^2$$

Now we solve this equation for the maximum elongation of the cable:

$$\delta_{\text{max}} = \delta_{\text{st}} + \sqrt{\frac{Wv^2L}{gEA}} = \delta_{\text{st}}\left(1 + \sqrt{\frac{v^2EA}{gWL}}\right) \tag{e}$$

Finally, we obtain the maximum stress in the cable:

$$\sigma_{\text{max}} = \frac{E\delta_{\text{max}}}{L} = \frac{W}{A}\left(1 + \sqrt{\frac{v^2EA}{gWL}}\right) \tag{f}$$

The term in parentheses, which is the impact factor, may be many times greater than unity, showing that the dynamic elongation and dynamic stress in the cable may be much greater than the corresponding static quantities. If $v = 0$, the impact factor is 1 and Eqs. (e) and (f) give the static values.

*2.8 NONLINEAR BEHAVIOR

In earlier parts of this chapter we analyzed both statically determinate and statically indeterminate structures using the assumption that the materials followed Hooke's law. Let us now consider the behavior of axially loaded structures when the stresses exceed the proportional limit. In such cases it is necessary to make use of the stress-strain diagram for the material. The diagram may be available from an actual test of the material, or an **idealized stress-strain diagram** may be used (see Fig. 2-48).

The first diagram (Fig. 2-48a) consists of an initial linear region, in which Hooke's law holds and stress is proportional to strain, followed by a nonlinear region defined by an appropriate mathematical function. Aluminum alloys can be represented quite accurately by a diagram of this type (see Fig. 1-10). For purposes of analysis, a single mathematical expression is sometimes used for the entire stress-strain diagram, as illustrated later in Example 1.

Figure 2-48b portrays a diagram that is widely used for structural steel. Because this material has a linearly elastic region followed by a region of considerable yielding (see Fig. 1-7), it can be represented by two straight lines. The material is assumed to follow Hooke's law up to the yield stress, after which it yields under constant stress; the latter behavior is known as **perfect plasticity**. The yield stress and yield strain are denoted by σ_y and ϵ_y, respectively. The perfectly plastic region continues until the strains are many times larger than the yield strain. A material having a diagram of this kind is called an **elastic-plastic material**. As the strain becomes very large, the stress-strain curve rises above the yield stress due to strain hardening, as explained in Section 1.3. But, by the time strain hardening begins, the displacements are so large that the structure probably has lost its usefulness. Hence, it is common practice to analyze steel structures on the basis of the elastic-plastic diagram shown in Fig. 2-48b, with the same diagram being used for both tension and compression. An analysis made with these assumptions is usually called a **plastic analysis**, although the term **elastic-plastic analysis** is more accurate.

Figure 2-48c shows a diagram made up of two lines having different slopes, called a **bilinear diagram**. Note that in both parts of the diagram the relationship between stress and strain is linear, but only in the first part is the stress *proportional* to the strain (Hooke's law). This diagram is sometimes used to represent a material with strain hardening, or it can be used as an approximation to a diagram of the general shape shown in Fig. 2-48a.

Now let us consider the process for determining the elongation of a bar loaded into the nonlinear range. We will assume that both the cross-sectional area and the axial force vary along the length of the bar (Fig. 2-49a) and that the material has a general stress-strain diagram (Fig. 2-49b). Because the bar is statically determinate, the axial forces at all cross sections can be determined from static equilibrium without considering the material properties. Then the stresses can be found by dividing the axial forces by the cross-sectional areas, and the strains can be obtained from the stress-strain diagram. If the stresses and strains are expressed in mathematical form, the elongation of the bar can be determined by evaluating the following integral:

$$\delta = \int_0^L \epsilon \, dx \qquad (2\text{-}57)$$

where the strain ϵ is a function of the distance x from one end of the bar. If the stresses and strains are in numerical form, the bar can be divided into small segments of length Δx, the average stress and strain can be found for each segment, and the elongation of the entire bar can be calculated by summing the elongations for the individual segments. This latter process is equivalent to evaluating Eq. (2-57) by numerical means instead of by formal integration.

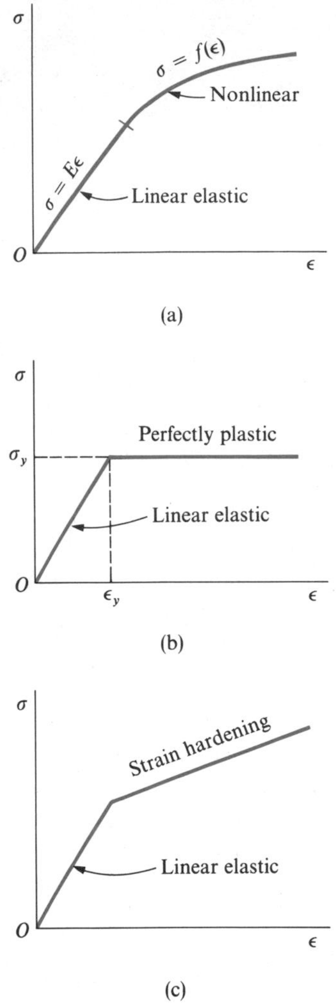

Fig. 2-48 Types of idealized nonlinear stress-strain diagrams: (a) elastic-nonlinear diagram, (b) elastic-plastic diagram, and (c) bilinear diagram

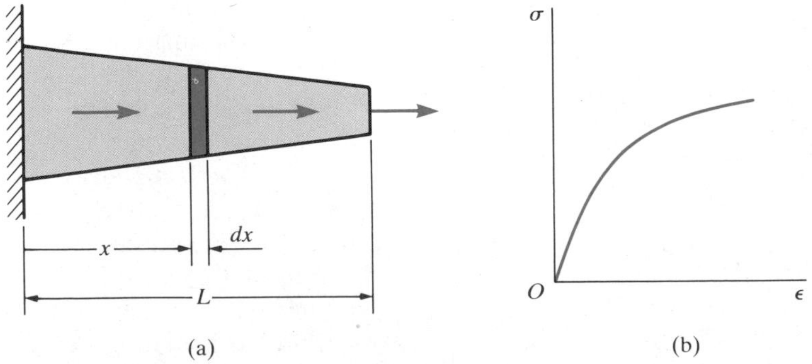

Fig. 2-49 Elongation of a bar consisting of a material having a nonlinear stress-strain diagram

Example 1

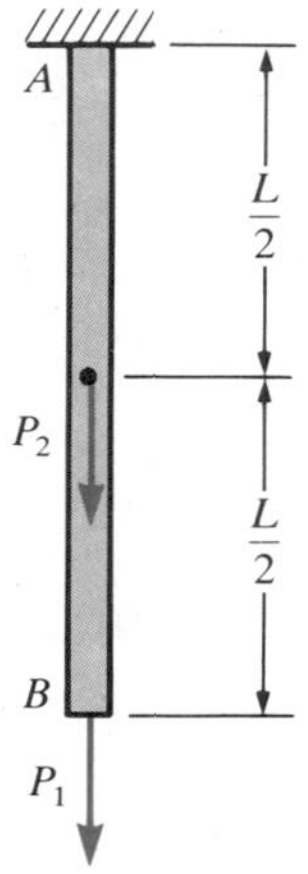

Fig. 2-50 Example 1

A prismatic bar AB of length $L = 10$ ft and cross-sectional area $A = 0.5$ in.^{2} supports two concentrated loads $P_1 = 15$ k and $P_2 = 5$ k, as shown in Fig. 2-50. The material of the bar has a nonlinear stress-strain diagram, as described in the next two paragraphs, and the objective of our analysis is to determine the displacement δ_b of the lower end of the bar under the following conditions: (a) the load P_1 acts alone, (b) the load P_2 acts alone, and (c) the loads P_1 and P_2 act simultaneously.

The stress-strain curve of the material of the bar is represented by the *Ramberg-Osgood equation*:

$$\frac{\epsilon}{\epsilon_0} = \frac{\sigma}{\sigma_0} + \alpha\left(\frac{\sigma}{\sigma_0}\right)^m \tag{2-58}$$

in which σ and ϵ are the stress and strain, respectively, and ϵ_0, σ_0, α, and m are constants of the material. When appropriate values of the constants are selected, this equation is especially useful for aluminum and magnesium. The Ramberg-Osgood stress-strain law also can be expressed in the form

$$\epsilon = \frac{\sigma}{E} + \frac{\sigma_0 \alpha}{E}\left(\frac{\sigma}{\sigma_0}\right)^m \tag{2-59}$$

in which $E = \sigma_0/\epsilon_0$ is the modulus of elasticity in the initial part of the stress-strain diagram.*

In this example, the material is an aluminum alloy for which $E = 10 \times 10^6$ psi, $\sigma_0 = 38{,}000$ psi, $\alpha = 3/7$, and $m = 10$. A graph of this stress-strain curve, for which the equation is

$$\epsilon = \frac{\sigma}{10 \times 10^6} + \frac{1}{614.0}\left(\frac{\sigma}{38{,}000}\right)^{10} \tag{2-60}$$

where σ has units of psi, is given in Fig. 2-51.

(a) *Displacement due to the load P_1 acting alone.* The load P_1 produces a uniform stress in the bar equal to P_1/A or 30,000 psi. Substituting this value into the stress-strain relation (Eq. 2-60) gives $\epsilon = 0.003153$. Therefore, the elongation of the bar, which is equal to the displacement at B, is

$$\delta_b = \epsilon L = (0.003153)(120 \text{ in.}) = 0.378 \text{ in.} \tag{a}$$

(b) *Displacement due to the load P_2 acting alone.* The stress in the upper half of the bar is P_2/A or 10,000 psi, and there is no stress in the lower half. Proceeding as before, we get the elongation as follows:

$$\delta_b = \epsilon L/2 = (0.00100)(60 \text{ in.}) = 0.060 \text{ in.} \tag{b}$$

(c) *Displacement due to both loads acting simultaneously.* The stress in the lower half of the bar is P_1/A and in the upper half is $(P_1 + P_2)/A$. The corresponding stresses are 30,000 psi and 40,000 psi, and the corresponding strains (from Eq. 2-60) are 0.003153 and 0.006720. Therefore, the elongation of the

* The Ramberg-Osgood stress-strain law was published in Ref. 2-12.

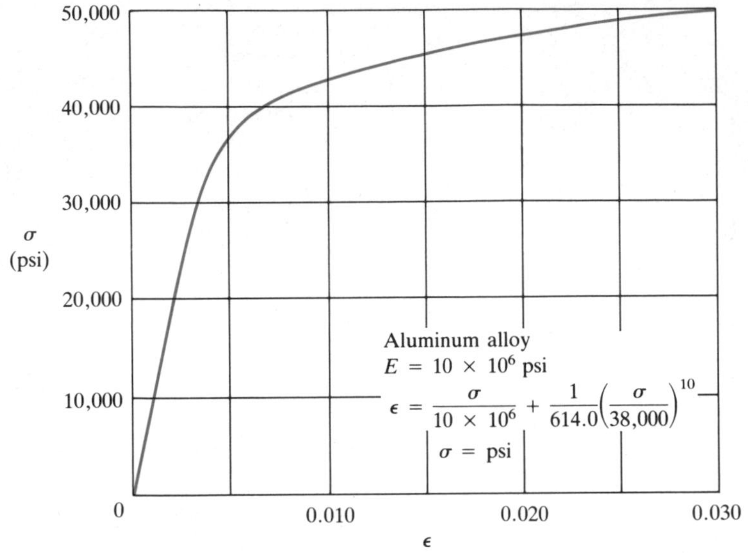

Fig. 2-51 Stress-strain diagram for the aluminum alloy of Example 1

bar is

$$\delta_b = (0.003153)(60 \text{ in.}) + (0.006720)(60 \text{ in.})$$

$$= 0.189 \text{ in.} + 0.403 \text{ in.} = 0.592 \text{ in.} \qquad \text{(c)}$$

These results demonstrate an important observation pertaining to a structure made of a material that behaves nonlinearly. In a nonlinear structure, the displacement produced by two (or more) loads acting simultaneously is *not* equal to the sum of the displacements produced by the loads acting separately.

PLASTIC ANALYSIS OF STATICALLY INDETERMINATE SYSTEMS

If a structure is statically indeterminate, the analysis is much more complicated than for a determinate system. In an indeterminate system, the forces cannot be found without knowing the displacements, but the displacements cannot be found without knowing the strains, which in turn depend upon the forces and the stress-strain diagram. As a consequence, a complete analysis usually requires a trial-and-error method or a method of successive approximations (described at the end of this section). An exception, however, is a material such as mild steel having an elastic-plastic stress-strain diagram (Fig. 2-48b). In this case the behavior of the structure is greatly simplified, and a plastic analysis can sometimes be performed without difficulty.

To illustrate the techniques of a plastic analysis, let us consider a statically indeterminate structure consisting of three bars supporting a load P acting through a rigid plate (Fig. 2-52a). Each outer bar has length L_1 and cross-sectional area A_1, and the inner bar has length L_2 and area

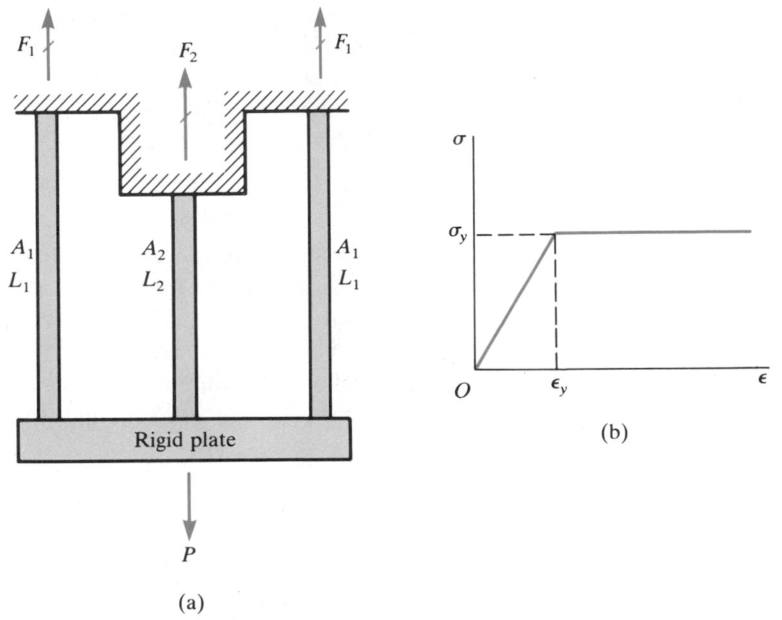

Fig. 2-52 Plastic analysis of a statically indeterminate structure

A_2. The bars are made of steel having an elastic-plastic stress-strain diagram (Fig. 2-52b) with σ_y representing the yield stress and ϵ_y representing the corresponding strain. In the elastic region the slope of the diagram is the modulus of elasticity $E = \sigma_y/\epsilon_y$.

When the load P is small, the stresses in the bars are less than the yield stress σ_y and the bar forces F_1 and F_2 (see Fig. 2-52a) can be determined by a linearly elastic analysis of the type described in Section 2.3. The analysis proceeds as follows. From static equilibrium we obtain

$$P = 2F_1 + F_2 \tag{d}$$

and from compatibility we know that $\delta_1 = \delta_2$, where δ_1 and δ_2 are the elongations of the outer and inner bars, respectively. Using the constitutive or stress-strain relation, which in this case is Hooke's law, we get

$$\frac{F_1 L_1}{EA_1} = \frac{F_2 L_2}{EA_2} \tag{e}$$

Solving these equations simultaneously yields

$$F_1 = \frac{A_1 L_2}{2A_1 L_2 + A_2 L_1} P \qquad F_2 = \frac{A_2 L_1}{2A_1 L_2 + A_2 L_1} P \tag{2-61a,b}$$

The corresponding stresses in the bars are

$$\sigma_1 = \frac{F_1}{A_1} = \frac{L_2}{2A_1 L_2 + A_2 L_1} P \qquad \sigma_2 = \frac{F_2}{A_2} = \frac{L_1}{2A_1 L_2 + A_2 L_1} P \tag{2-62a,b}$$

These equations for the forces and stresses are valid provided the stresses remain below the yield stress σ_y.

As the load P gradually increases, the stresses in the bars increase until the yield stress is reached either in the inner bar or in the two outer bars. Let us assume that the outer bars are longer than the inner bar:

$$L_1 > L_2$$

Then the inner bar is more highly stressed than the outer bars (see Eqs. 2-62a and b) and will reach the yield stress first. When that happens, the force in the inner bar is $F_2 = \sigma_y A_2$. The magnitude of the load P when the yield stress is first reached in any one of the bars is called the **yield load** P_y. We can determine P_y by setting F_2 equal to $\sigma_y A_2$ in Eq. (2-61b) and solving for the load:

$$P_y = \sigma_y(A_2 + 2A_1L_2/L_1) \tag{f}$$

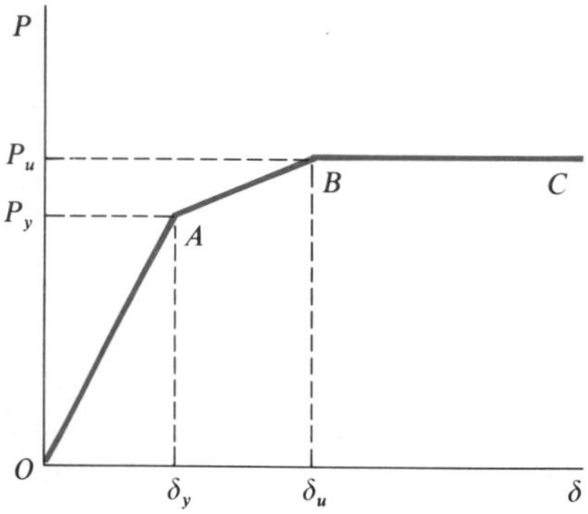

Fig. 2-53 Load-displacement diagram for the statically indeterminate structure shown in Fig. 2-52

As long as the load P is less than P_y, the structure behaves in a linearly elastic manner and the forces in the bars can be determined from Eqs. (2-61).

The downward displacement δ_y of the rigid bar at the yield load is equal to the elongation of the inner bar when its stress is σ_y:

$$\delta_y = \frac{F_2L_2}{EA_2} = \frac{\sigma_2L_2}{E} = \frac{\sigma_yL_2}{E} \tag{g}$$

The load-displacement behavior of the structure up to the yield load is represented by line OA in the diagram of Fig. 2-53.

With a further increase in the load P, the forces F_1 in the outer bars increase, but the force F_2 in the inner bar remains constant at the value $\sigma_y A_2$ because this bar is now perfectly plastic (see Fig. 2-52b). When the forces F_1 reach the value $\sigma_y A_1$, the outer bars also become plastic. When that happens, the structure cannot support any additional load. Instead, all three bars will elongate under this constant load, called the **ultimate load** P_u. The ultimate load is represented by point B on the load-displacement diagram (Fig. 2-53), and the horizontal line BC represents the region of continuous plastic deformation without any increase in the load.

The ultimate load P_u can be calculated from static equilibrium (Eq. d) knowing that

$$F_1 = \sigma_y A_1 \qquad F_2 = \sigma_y A_2 \tag{h}$$

Thus, from equilibrium we find

$$P_u = \sigma_y(2A_1 + A_2) \tag{i}$$

The displacement at the instant the load just attains its ultimate value P_u is denoted δ_u. This displacement is equal to the elongation of the outer bars at the instant they reach the yield stress. Therefore, we obtain

$$\delta_u = \frac{F_1 L_1}{EA_1} = \frac{\sigma_y L_1}{E} \tag{j}$$

Comparing δ_u with δ_y, we see that

$$\frac{\delta_u}{\delta_y} = \frac{L_1}{L_2} \tag{k}$$

Also, the ratio of the ultimate and yield loads is

$$\frac{P_u}{P_y} = \frac{2A_1L_1 + A_2L_1}{2A_1L_2 + A_2L_1} \tag{l}$$

For example, if $A_1 = A_2/2$ and $L_1 = 2L_2$, the ratios are $\delta_u/\delta_y = 2$ and $P_u/P_y = 4/3$.

When the load P is between P_y and P_u (region AB on the load-displacement diagram), the force F_2 is constant and equal to $\sigma_y A_2$, as already explained. Therefore, the force F_1 in the outer bars can be obtained from static equilibrium:

$$F_1 = \frac{P - F_2}{2} = \frac{P - \sigma_y A_2}{2} \tag{m}$$

This equation shows that the force F_1 is a linear function of the load P in region AB. It follows that the downward displacement δ of the rigid plate is also a linear function of P, because the outer bars still follow Hooke's law. However, the slope of the load-displacement diagram from A to B is less than in the initial linear region (from O to A) because only the outer bars remain linearly elastic and offer increasing resistance to the increasing load.

From this example we see that the calculation of the ultimate load P_u of a statically indeterminate structure requires only the use of statics, because all members have yielded and their axial forces are known. In contrast, the calculation of P_y requires a statically indeterminate analysis, which means that both static equilibrium and compatibility of displacements must be satisfied.

After the ultimate load is reached, the structure continues to deform. Strain hardening occurs eventually, and then the structure is able to support additional loads. However, the presence of very large displacements usually means that the structure has failed in a utilitarian sense, and so the load P_u is the ultimate load for most practical purposes.

This discussion has dealt with the behavior of a structure as the load P is applied for the first time. If the load is removed before the yield load P_y is reached, the structure will behave elastically and return to its original unstressed condition. However, if the load P_y is exceeded, some members of a statically indeterminate structure will retain a permanent set when the load is removed, thus creating a prestressed condition. Consequently, the structure will have **residual stresses** in it even though no external load is acting. Then, if the load is applied a second time, the structure will behave in a different manner.

Example 2

Determine the yield load P_y and ultimate load P_u for the structure shown in Fig. 2-54a if the horizontal bar AB is rigid and the two identical vertical wires are made of an elastic-plastic material. Also, construct a load-displacement diagram relating the load P to the displacement δ_b of point B, and determine the allowable load P_{allow} using a load factor of 1.85 with respect to the ultimate load.

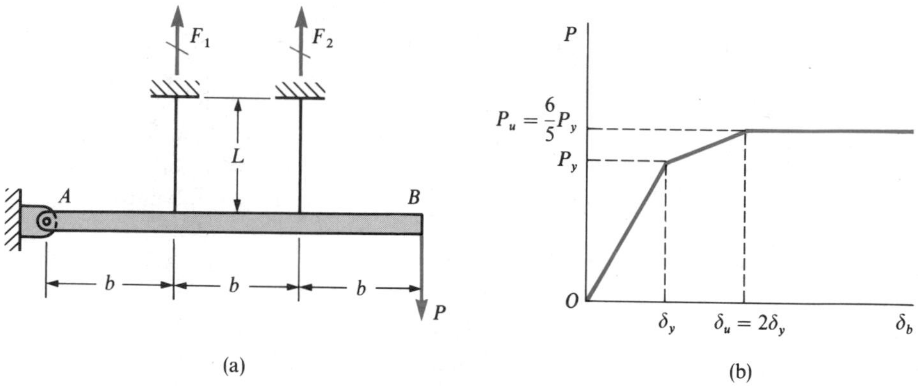

Fig. 2-54 Example 2

Because this structure is statically indeterminate, we must use both equilibrium and compatibility to obtain the yield load P_y. The equilibrium equation is obtained by taking moments about point A, which gives

$$3P = F_1 + 2F_2 \tag{n}$$

in which F_1 and F_2 are the forces in the wires. Because bar AB is rigid, the compatibility condition is $\delta_2 = 2\delta_1$, where δ_2 is the elongation of the right-hand wire and δ_1 is the elongation of the left-hand wire. Using Hooke's law to determine the elongations, we get the following equation from the compatibility condition:

$$\frac{F_2 L}{EA} = 2\frac{F_1 L}{EA} \quad \text{or} \quad F_2 = 2F_1 \tag{o}$$

in which L is the length and EA is the axial rigidity of each wire. Of course, the relation $F_2 = 2F_1$ holds only when the stresses are less than σ_y and P is less than P_y.

The force F_2 will be the first to reach the yield force $\sigma_y A$. At that instant the force F_1 will be equal to $\sigma_y A/2$. Therefore, the yield load P_y, found from equilibrium (Eq. n), is

$$P_y = \frac{5\sigma_y A}{6} \tag{p}$$

The corresponding elongation of the right-hand wire is $\delta_2 = \sigma_y L/E$, and the displacement at point B is

$$\delta_y = \frac{3\delta_2}{2} = \frac{3\sigma_y L}{2E} \tag{q}$$

Both P_y and δ_y are indicated on the load-displacement diagram (Fig. 2-54b).

When the ultimate load P_u is reached, both F_1 and F_2 will be equal to $\sigma_y A$. Thus, from equilibrium (Eq. n) we obtain

$$P_u = \sigma_y A \tag{r}$$

At this load, the left-hand wire has just reached the yield stress; therefore, its elongation is $\delta_1 = \sigma_y L/E$, and the displacement of B is

$$\delta_u = 3\delta_1 = \frac{3\sigma_y L}{E} \tag{s}$$

Note that the ratios δ_u/δ_y and P_u/P_y are 2 and 6/5, respectively, as shown on the load-displacement diagram.

The allowable load P_{allow} is found by dividing the ultimate load by the load factor, as explained in Section 1.7; thus,

$$P_{allow} = \frac{P_u}{\text{load factor}} = \frac{\sigma_y A}{1.85}$$

It happens that this load is less than P_y (see Eq. p), and therefore both wires are stressed within the elastic range under the allowable load.

NONLINEAR ANALYSIS OF STATICALLY INDETERMINATE SYSTEMS

If the material of the structure behaves nonlinearly but cannot be represented as an elastic-plastic material, then the analysis is more difficult to make. In the general case, a trial-and-error procedure is needed. To illustrate the method, let us refer again to the symmetric structure of Fig. 2-52a, but now we will assume that the material has a general stress-strain diagram such as the one shown in Fig. 2-48a.

We begin the analysis by assuming a trial value of the vertical displacement δ of the rigid plate. Compatibility is automatically satisfied because the elongations of all three bars are the same and equal to δ. Knowing δ, we can find the strains in the bars by dividing by their respective lengths. Then, using the stress-strain diagram (constitutive relation), we can get the stresses in the bars, and from the stresses we can get the forces by multiplying by the cross-sectional areas. Finally, we can check for static equilibrium of the structure. If the true value of δ was assumed initially, we will find that equilibrium of the forces is satisfied. If the forces are not in equilibrium, a second trial value of δ must be selected and the process repeated. Eventually, we will arrive at a value of δ such that both compatibility and equilibrium are satisfied. The corresponding forces in the bars, as well as the displacement δ, will have their true values.

An alternative trial-and-error procedure begins by assuming a value for one of the bar forces, say, the force F_2 in the middle bar (Fig. 2-52a). Then, from equilibrium, we can calculate the forces in the other bars. Next, we can determine the stresses (from the forces), then the strains (from the stress-strain diagram), and then the elongations (from the strains). Finally, we can determine whether the elongations of all three bars are compatible with one another. If they are, then the trial value of F_2 was the correct one, and the analysis is completed. Otherwise, a new trial value for F_2 must be selected and the process repeated until both equilibrium and compatibility are satisfied.

*2.9 FATIGUE

The behavior of a structure is highly dependent upon the character of the loads that act upon it. In some situations the loads are applied gradually, act for a long period of time, and change slowly. Such loads are essentially static in nature. When designing a structure to resist static loads, we usually apply a factor of safety to either the ultimate stress or the yield stress and then make certain that the calculated stresses remain below the allowable value.

Other loads are dynamic in character, such as loads that act very suddenly (for example, impact loads from falling objects) and loads that change continuously (for example, the load on a piston rod in an automobile engine). Some typical dynamic loading patterns are portrayed in Fig. 2-55. The first graph illustrates *repeated loading*, in which the load is applied, removed, and applied again. The second shows *reversed loading*, such as the load on an automobile axle, which reverses direction with every revolution. The third graph illustrates a *fluctuating load*, which is a load that varies about some average value. Loads in rotating machinery, turbines, airplanes, automobiles, and ships may go through millions (and sometimes billions) of cycles during the useful life of the structure.

A structure subjected to dynamic loads is likely to fail at a lower stress than when the loads are static in nature. This is especially true when the loads are repeated, reversed, or fluctuating. With loads of these kinds, failures are usually due to **fatigue**, or **progressive fracture**. A familiar example of a fatigue failure is the breaking of a metal paper clip by repeatedly bending it back and forth. If bent only once, the clip does not break. But if the load is reversed by bending the clip in the opposite direction, and if the entire loading cycle is repeated several times, the clip finally will break. Fatigue may be defined as the behavior of materials under repeated cycles of stress and strain, causing a deterioration of the material and resulting in progressive cracking that eventually produces fracture.

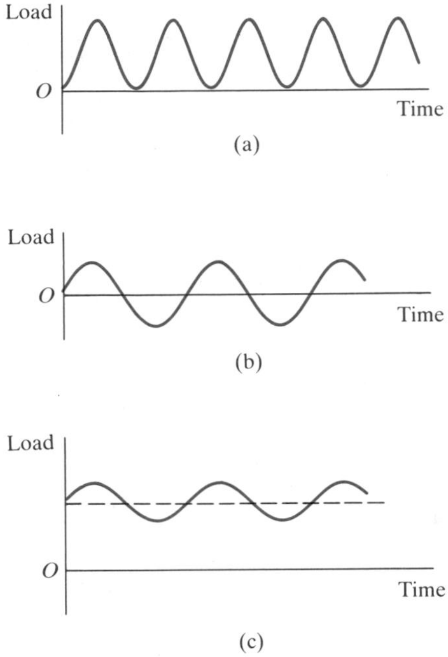

Fig. 2-55 Types of dynamic loads: (a) repeated, (b) reversed, and (c) fluctuating

In a typical fatigue failure, a microscopic crack forms at a point of high stress (such as a stress concentration, discussed in the next section) and gradually enlarges or spreads as the loads are applied repeatedly. When the crack becomes so large that the remaining material cannot resist the loads, a sudden fracture of the remaining material occurs (see Fig. 2-56). It may take thousands or even millions of load cycles to produce a fatigue failure.

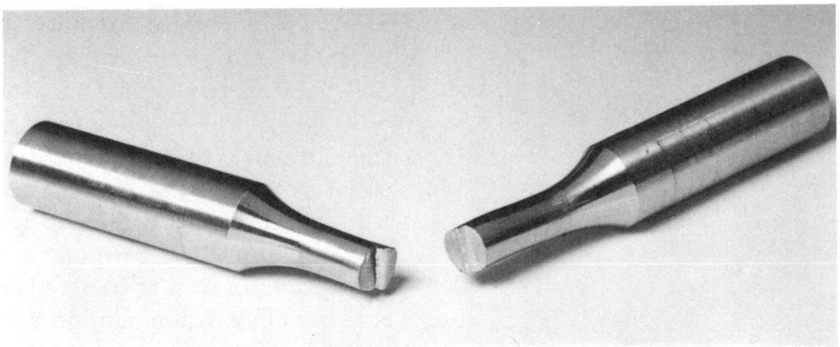

Fig. 2-56 Fatigue failure of a bar loaded cyclically in tension; the crack spread gradually over the cross section until fracture occurred suddenly. (Courtesy of MTS Systems Corporation)

The magnitude of the load acting on a structural component when a fatigue failure occurs is less than that which could be sustained as a static load. To determine the strength of a material under repeated loading, the material is tested at various stress levels. For instance, a specimen of the material is placed in a fatigue testing machine and loaded repeatedly to a certain stress, say σ_1. The loading cycles are continued until failure occurs, and the number n of loading cycles to failure is noted. Then the test is repeated for a different stress, say σ_2. If σ_2 is larger than σ_1, the number of cycles to failure will be less. If it is smaller, the number will be more. Eventually, enough data are accumulated to plot an **endurance curve**, or ***S-N* diagram** (stress versus number of cycles to failure). Such curves have the general shape shown in Fig. 2-57, where the vertical axis usually is a linear scale and the horizontal axis a log scale.

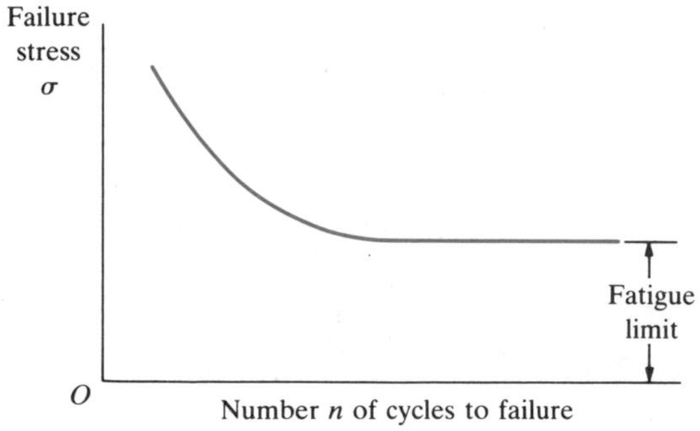

Fig. 2-57 Endurance curve, or *S-N* diagram, showing fatigue limit

The endurance curve shows that the smaller the stress, the larger the number of cycles to produce failure. For some materials, the diagram has a horizontal asymptote known as the **fatigue limit** or **endurance limit**. This limit (when it exists) is the stress below which a fatigue failure will not occur regardless of how many times the load is repeated. The precise shape of an endurance curve depends upon many factors, including the material, geometry of the test specimen, speed of testing, pattern of loading, and surface condition of the specimen. For this reason, the engineering literature contains the results of numerous fatigue tests made on a great variety of materials and structural components.

Typical *S-N* diagrams for steel and aluminum are shown in Fig. 2-58. The ordinate is the failure stress expressed as a percentage of the ultimate stress for the material, and the abscissa is the number of cycles at which failure occurred. Note that the number of cycles is plotted on a log scale. The curve for steel becomes horizontal at about 10^7 cycles, and the fatigue limit is seen to be about 50% of the ultimate tensile stress under static loading. Aluminum does not have a fatigue limit that is clearly defined, so it is often taken as the stress at 5×10^8 cycles.

Since fatigue failures usually begin with a microscopic crack at a point of high localized stress, the condition of the surface of the material is extremely important. Highly polished specimens give higher endurance limits. Rough surfaces, especially at stress concentrations around holes or grooves, greatly lower the endurance limit, as does corrosion, which creates tiny surface irregularities. For steel, ordinary corrosion may reduce the fatigue limit by as much as 50%.

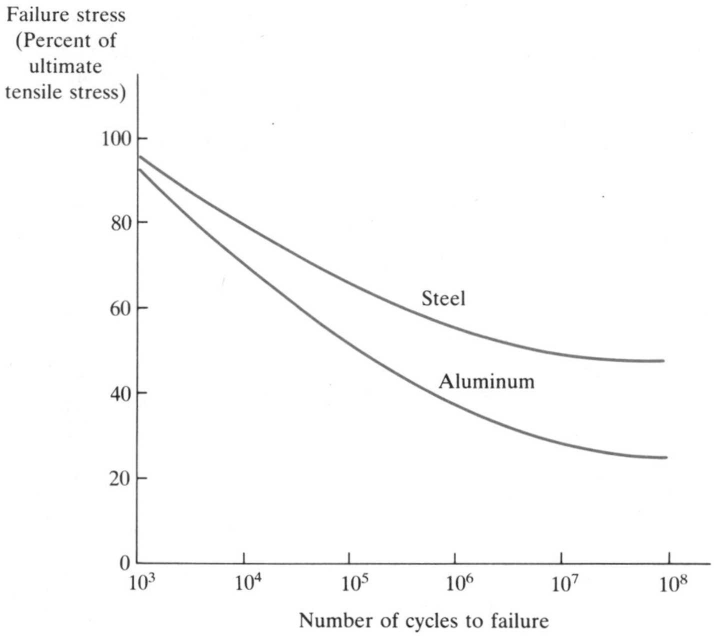

Fig. 2-58 Typical *S-N* diagrams for steel and aluminum in reversed loading

*2.10 STRESS CONCENTRATIONS

The stresses in a bar loaded in tension or compression are calculated by the formula $\sigma = P/A$ under ordinary circumstances. This formula is based upon a uniform stress distribution acting over the cross-sectional area of the bar, and therefore it is valid only for prismatic bars that are free of any changes in geometry (no holes, grooves, or other discontinuities in shape). The formula also requires that the loads applied at the ends be distributed over the cross section in the same manner as the calculated stress; that is, uniformly distributed over the area. In reality, concentrated loads often act over a small area and produce very high stresses in a very small region, such as when the load is applied through a pin connection. Also, bars often have holes or other changes in dimensions. These irregularities create a disruption in the stress pattern, again causing high stresses over a very small region. The high localized stresses created by holes, grooves, notches, keyways, shoulders, threads, and other abrupt changes in geometry, as well as the high stresses produced by concentrated loads and reactions, are known as **stress concentrations** or **stress raisers**.

The high stresses existing at stress concentrations can be determined by advanced methods, such as finite-element analysis and theory of elasticity. In some cases they are investigated by experimental methods. The results of such research are readily available in the engineering literature for many situations of practical interest, and a few results are given here and in Sections 3.10 and 5.10.

With respect to a concentrated load acting at the end of a bar (Fig. 2-59), the peak stress directly under the load may be several times the average stress P/bt, depending upon the area over which the load is applied. The maximum stress diminishes rapidly as we move along the axis of the bar, away from the point of load application. At a distance from the end of the bar approximately equal to the width b of the bar, the stress distribution is nearly uniform and the maximum stress is only a few percent larger than the average stress (see Fig. 2-59). This observation is true for most stress concentrations, so we can make the general statement that the formula $\sigma = P/A$ gives the stress in the bar with high accuracy provided the cross section of the bar is at least a distance b away from any concentrated load or discontinuity in shape, where b is the largest lateral dimension of the bar.

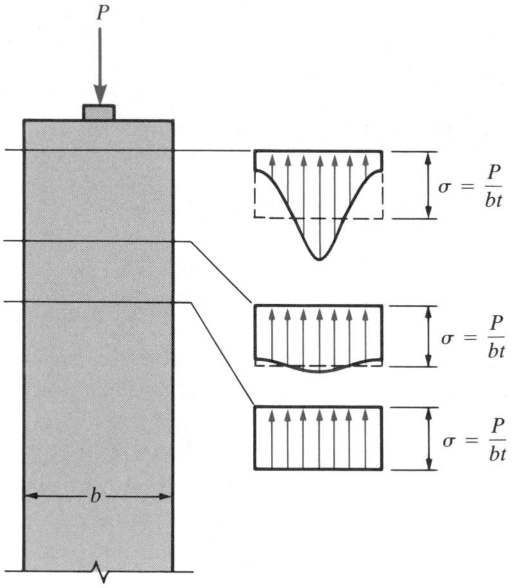

Fig. 2-59 Stress distribution near the end of a flat bar of rectangular cross section (width b, thickness t) subjected to a concentrated load P

The preceding statement about the stresses in a prismatic bar in tension or compression is part of a more general observation known as **Saint-Venant's principle**. With rare exceptions, this principle applies to linearly elastic bodies of all types. To explain the principle, consider a body with a system of loads acting over a small part of the surface. Then consider a different but statically equivalent load system acting over the same small region. (Statically equivalent means the load systems have the same force and moment resultants.) Saint-Venant's principle states that the stresses in the body caused by either of the two systems are the same, provided we move away from the loaded region a distance at least equal to the largest dimension of the loaded region. The practical meaning of this principle for structures of the type considered in mechanics of materials is as follows: If we move away by an amount equal to the width b of the bar from a point of loading, from a reaction, or from any other stress concentration, the effect of the stress concentration will become insignificant and the stresses can be calculated by the usual mechanics of materials formulas (such as $\sigma = P/A$). Within the distance b, the stresses depend upon the details of the loading and the geometry of the stress concentration. Since the stress concentration effects are highly localized, the formula for elongation of a bar ($\delta = PL/EA$) and other related displacement formulas give accurate results even when stress concentrations are present.*

With respect to stress concentrations caused by changes in geometry, let us begin with a flat bar having a circular hole through it (Fig. 2-60a). The bar has width b and constant thickness t, and the hole has diameter d. The normal stress acting on the cross section through the center of the hole has the distribution shown in Fig. 2-60b. The maximum stress σ_{max} occurs at the edges of the hole and may be significantly larger than the nominal stress $\sigma = P/ct$ at the same cross section. This stress-concentration effect is conveniently expressed by giving the ratio of the maximum stress to the nominal stress, called the **stress-concentration factor** K:

$$K = \frac{\sigma_{max}}{\sigma_{nom}} \tag{2-63}$$

For a bar in tension the nominal stress is the average stress $\sigma = P/ct$, based upon the net cross-sectional area, but in later chapters we will

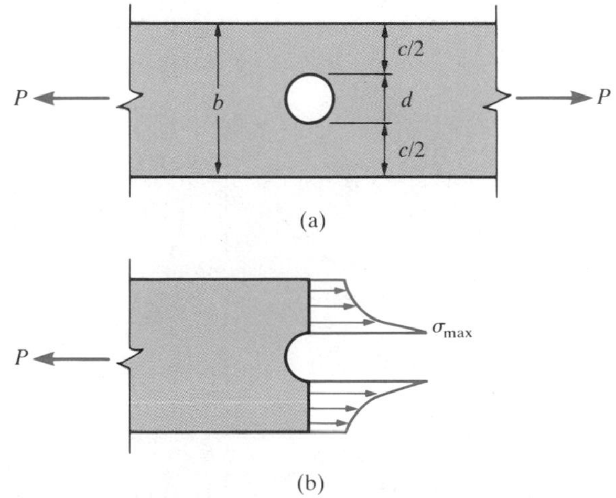

Fig. 2-60 Stress distribution in a flat bar with a circular hole (tension)

* Saint-Venant's principle is named for Barré de Saint-Venant (1797–1886), a famous French mathematician and elastician (Ref. 2-13).

encounter stress concentrations where the nominal stress is not an average stress but is some other quantity, such as a maximum stress.

A graph of the stress-concentration factor K for a bar with a hole is given in Fig. 2-61. If the hole is tiny, the factor K equals 3, which means that the maximum stress is three times the average stress. As the hole becomes larger in proportion to the width of the bar, K becomes smaller and the effect of the concentration is not as severe. From Saint-Venant's principle we know that at distances equal to the width b of the bar away from the hole in either axial direction, the stress distribution is practically uniform and equal to P divided by the gross cross-sectional area ($\sigma = P/bt$), regardless of the size of the hole.

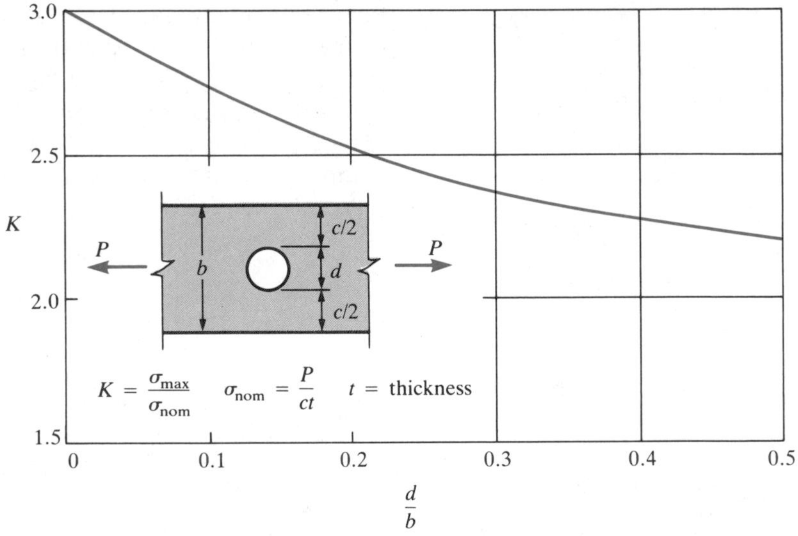

Fig. 2-61 Stress-concentration factor K for flat bars with circular holes (tension)

Stress-concentration factors for two other cases of practical interest are given in Figs. 2-62 and 2-63. These graphs are for flat bars and circular

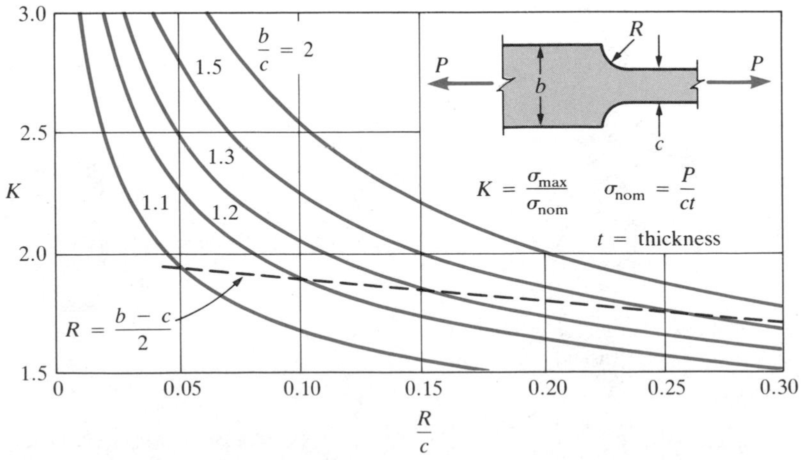

Fig. 2-62 Stress-concentration factor K for flat bars with shoulder fillets (tension). The dashed line is for a full quarter-circular fillet.

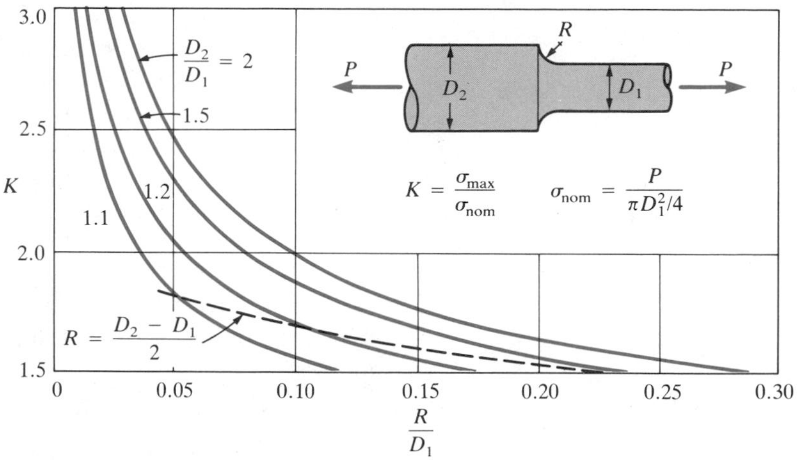

Fig. 2-63 Stress-concentration factor K for round bars with shoulder fillets (tension). The dashed line is for a full quarter-circular fillet.

bars, respectively, that are stepped down in size, forming a *shoulder*. To reduce the stress-concentration effects, fillets (radius R) are used to round off the re-entrant corners. Without the fillets, the stress-concentration factor would be extremely large, as indicated at the left-hand side of each graph where K approaches infinity as R approaches zero. In both cases the maximum stress occurs in the smaller part of the bar in the region of the fillet.*

Stress concentrations are especially important when the material is subjected to repeated loadings, because this creates the possibility of a fatigue failure. As explained in Section 2.9, cracks begin at the point of highest stress and then spread gradually through the material as the load is repeated. In practical design, the fatigue limit (see Fig. 2-57) is considered to be the ultimate stress for the material when the number of cycles is extremely large. A factor of safety is used with respect to this ultimate stress to obtain the allowable stress, and the peak stress at the stress concentration is compared with the allowable stress. The peak stress is obtained by multiplying the nominal stress by the stress-concentration factor K. Thus, the ultimate nominal stress is the fatigue limit divided by K, and the allowable nominal stress is that value divided by the factor of safety.

In many situations the procedure just described, which uses the full theoretical value of the stress-concentration factor, is too severe. Fatigue tests on specimens with stress concentrations usually produce failure at higher levels of the nominal stress than those obtained by dividing the fatigue limit by K. In other words, under repeated loading a structural member is not as sensitive to the stress concentration as the value of K indicates. In fatigue testing, this phenomenon is observed in the following

* The stress-concentration factors given in the graphs are theoretical factors for bars of linearly elastic material. The graphs are plotted from the formulas given in Ref. 2-14.

way. Suppose a material is tested in fatigue using bars of uniform cross section; then we obtain the fatigue limit (stress) in the conventional manner with the stress always being found by dividing the magnitude of the repeating load by the cross-sectional area. Now suppose we make tests of the same material using bars with stress concentrations, such as holes. Then we will obtain a new fatigue limit, with the stress being the nominal stress calculated by dividing the magnitude of the load by the net cross-sectional area. This new limit is smaller than the limit obtained from prismatic bars because of the stress concentration, but the reduction is not as great as the theoretical K value would predict. Consequently, we define a new quantity, called the **fatigue stress-concentration factor** K_f, as follows:

$$K_f = \frac{\text{fatigue limit for uniform bars}}{\text{fatigue limit for bars with stress concentrations}} \tag{2-64}$$

in which the stress in the denominator is calculated as the nominal stress. If the full stress-concentration factor is effective, then K_f will be the same as K. However, in most cases K_f (which must be greater than 1.0) is less than K, with typical values being in the range $0.6K$ to $0.9K$. The general phenomenon in which bars subjected to fatigue loading are not fully sensitive to the stress-concentration factor is called *notch sensitivity*. (The name refers specifically to stress concentrations caused by notches, but today the term is used in a more general sense to refer to all types of stress concentrations.) Note that using K_f equal to K is always on the safe side.

Other kinds of dynamic loads, such as impact loads, also require that stress concentration effects be taken into account. Unless better information is available, the full stress-concentration factor should be used. Members subjected to low temperatures also are highly susceptible to failures at stress concentrations, and so special precautions should be taken.

In the case of static loading of a member, the significance of a stress concentration depends upon the kind of material. With ductile materials, such as structural steel, it often is possible to ignore a stress concentration for the following reason. The material at the point of maximum stress will yield and plastic flow will occur, thereby reducing the intensity of the stress concentration and making the stress distribution more nearly uniform. On the other hand, with brittle materials (such as glass), the high stress concentration remains up to the point of fracture. Therefore, we can make the general observation that with static loads and ductile materials the stress-concentration effect is not likely to be important, but with static loads and brittle materials the full stress-concentration factor should be considered.

Stress concentrations can be reduced in intensity by properly proportioning the parts. Generous fillets help at re-entrant corners. The use of smooth surfaces at points of high stress, such as on the inside of a hole, is also beneficial because the formation of incipient cracks is inhibited. Proper reinforcing around holes can also be beneficial. There are many well-established techniques for smoothing out the stress distribution in a part and thereby reducing the stress-concentration factor. These techniques, which are usually studied in design courses, are extremely important in engineering because many unnecessary failures have occurred when designers failed to recognize the effects of stress concentrations and fatigue.

PROBLEMS / CHAPTER 2

2.2-1 A 1 in. diameter high-strength steel rod ($E = 30 \times 10^6$ psi) must carry a load in tension of 30,000 lb (see figure). If the initial length of the stressed portion of the rod is 42.50 in., what is its final length?

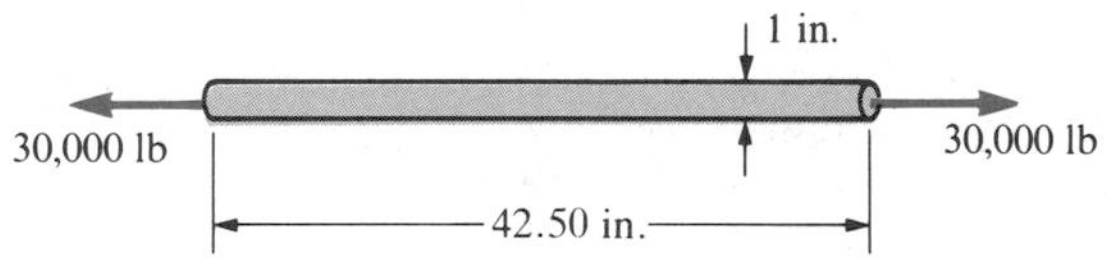

Prob. 2.2-1

2.2-2 An 8 m long hollow circular bar (see figure) of aluminum carries a tensile load of 120 kN. The outside diameter of the bar is 50 mm and the modulus of elasticity of the material is 70 GPa. If the allowable stress in the bar is 100 MPa and the allowable elongation of the bar is 10 mm, what is the minimum required wall thickness t?

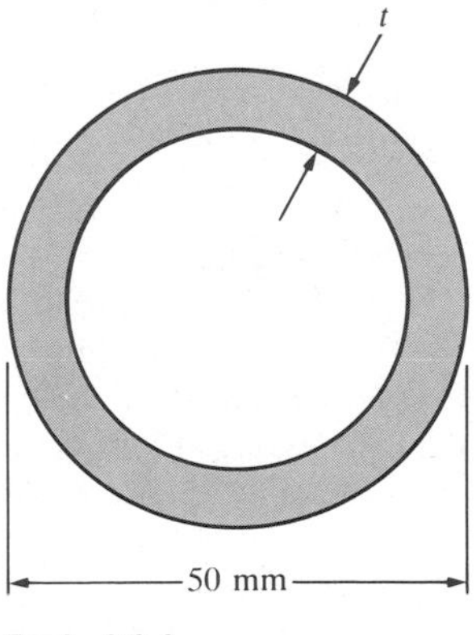

Prob. 2.2-2

2.2-3 A 1.0 in. diameter steel cable is used in a shipyard to lift a ship section weighing 17 tons. The cable has a metallic cross-sectional area $A = 0.471$ in.^{2} and an effective modulus of elasticity $E = 21 \times 10^6$ psi. If the cable is 40 ft long, how much will it stretch when the load is picked up?

(Note: A **cable** consists of a large number of wires wound in a special manner, as shown in the figure. While many variations are used, a typical cable is formed by six or more *strands* wound helically, with each strand consisting of many wires, also wound helically. For this reason, cables are often referred to as **wire rope**. The cross-sectional area of a cable is the total area of the individual wires, called the *metallic area.* This area is less than the area of a circle having the same diameter as the cable because of the spaces between the individual wires. For example, in this problem the cross-sectional area of the wires is 0.471 in.^{2}, whereas the area of a 1.0 in. diameter circle is 0.785 in.^{2} Furthermore, a cable stretches more than does a solid bar of the same material and same metallic cross-sectional area, because the wires "tighten up" in the same manner as the fibers in a rope. Therefore, a cable has a smaller modulus of elasticity, called the *effective modulus*, than does the material of which it is made. The value of E used in this problem is typical of steel cables for which the steel itself has a modulus of 30×10^6 psi. Dimensions and properties of cables are customarily obtained from the manufacturers.)

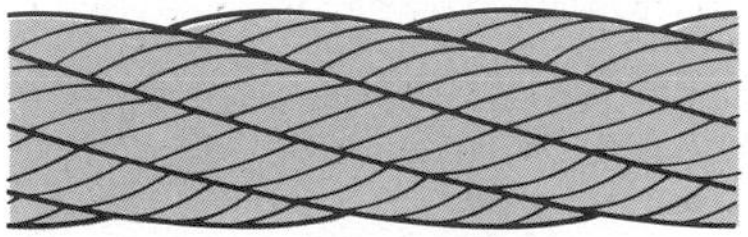

Prob. 2.2-3

2.2-4 A heavy object is being raised from the sea floor by a tug equipped with a crane and winch (see figure). The steel cable has an effective modulus of elasticity equal to 140 GPa, and the load produces a stress in the cable equal to 350 MPa. If the loaded part of the cable is 160 m long, how much does the cable stretch?

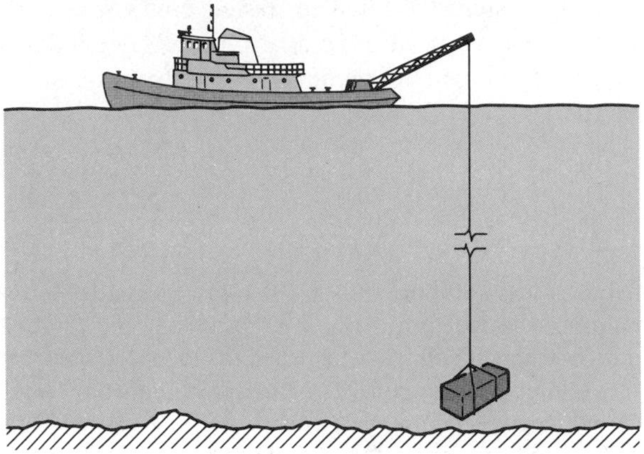

Prob. 2.2-4

2.2-5 What is the effective modulus of elasticity E of a $\frac{3}{8}$ in. diameter steel cable that stretches $\frac{5}{16}$ in. when a 10 ft length of it is subjected to a tensile load of 4 k? (The metallic cross-sectional area of a $\frac{3}{8}$ in. cable is 0.065 in.^{2})

2.2-6 The T-shaped arm ABC shown in the figure lies in a vertical plane and pivots about a horizontal pin at A. The arm has constant cross-sectional area and total weight W, and it is supported by a vertical spring of stiffness k at B. What is the elongation δ of the spring due to the weight of the arm?

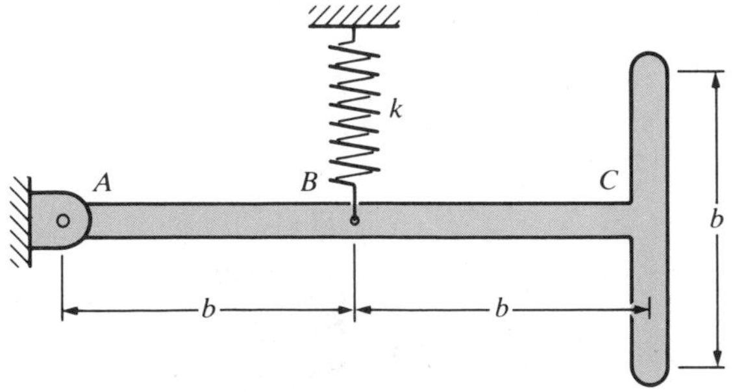

Prob. 2.2-6

2.2-7 A uniform bar AB of length L is suspended in a horizontal position under its own weight by two vertical wires attached to its ends (see figure). Both wires are made of the same material and have the same cross-sectional area, but the lengths are L_1 and L_2. Derive a formula for the distance x (from end A) to the point on the bar where a vertical load P should be applied if the bar is to remain horizontal.

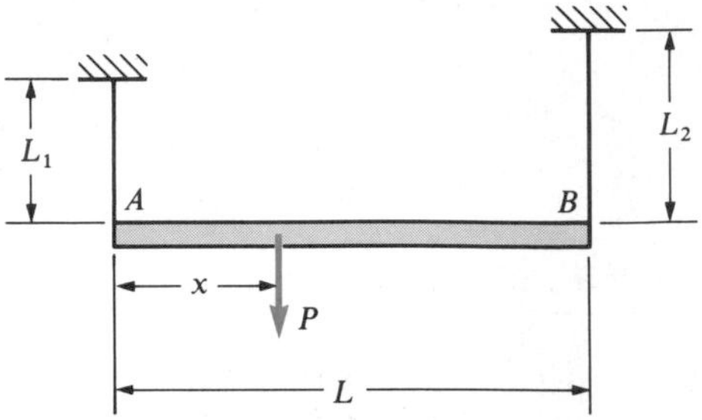

Prob. 2.2-7

2.2-8 Two horizontal rigid bars AB and CD are connected by wires as shown in the figure. The wires have length L, modulus of elasticity E, and diameters d_1 and d_2. Vertical loads P act at the midpoints E and F of the bars. What is the increase δ in the distance between points E and F when the loads are applied?

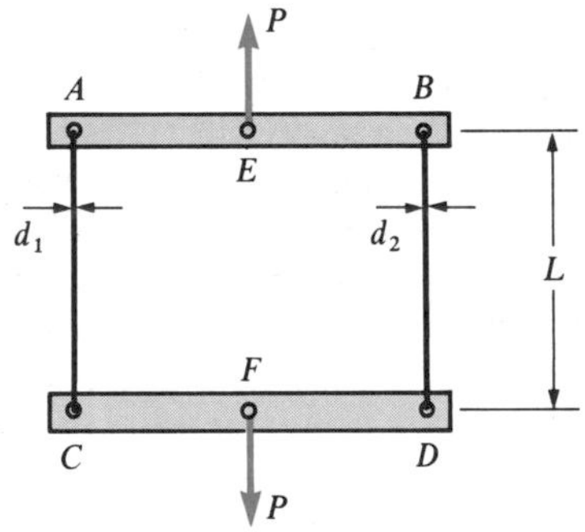

Prob. 2.2-8

2.2-9 A two-story building has columns AB at the first floor and BC at the second floor (see figure on p. 127). The columns are loaded as shown in the figure, with the roof load P_1 equal to 100 kips and the load P_2 applied at the second floor equal to 180 kips. The cross-sectional areas of the upper and lower columns are 6.1 in.^{2} and 17.1 in.^{2}, respectively, and each column has length $a = 12$ ft. Assuming that $E = 30 \times 10^6$ psi, determine the downward displacement δ_c at point C due to the applied loads.

2.2-10 A prismatic bar $ABCD$ is subjected to loads P_1, P_2, and P_3 acting at points B, C, and D, respectively (see figure). The bar is made of steel with modulus of elasticity $E =$

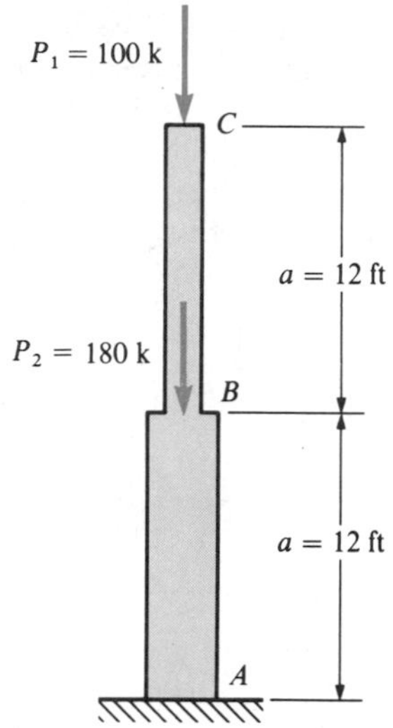

Prob. 2.2-9

200 GPa and cross-sectional area $A = 256\ \text{mm}^2$. Determine the displacement δ at the lower end of the bar due to the loads P_1, P_2, and P_3. Does the bar elongate or shorten?

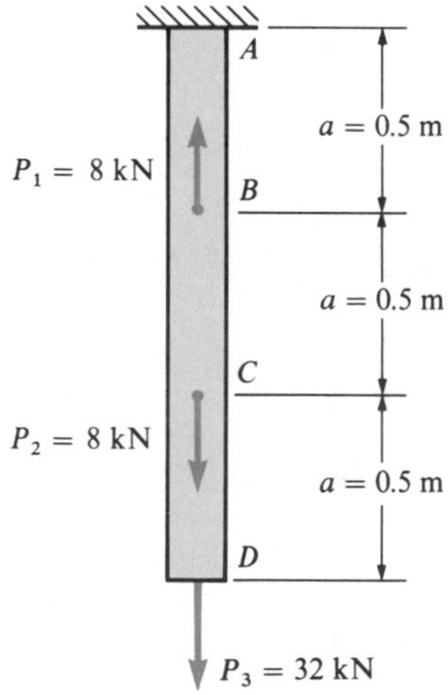

Prob. 2.2-10

2.2-11 The steel bar AD (see figure) has a length of 12 ft and a square cross section that is 0.625 in. on each side. The bar is loaded by axial forces acting at points A, B, C, and D. Assuming $E = 30 \times 10^6$ psi, calculate the change in length δ of the bar due to the loads. Does the bar elongate or shorten?

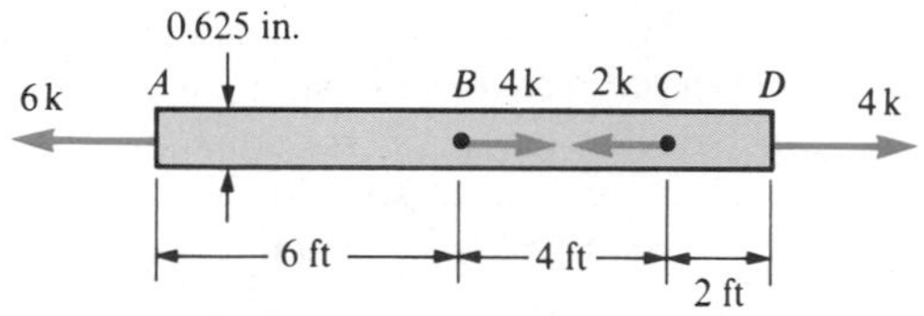

Prob. 2.2-11

2.2-12 A homogeneous prismatic bar is loaded by forces P_1, P_2, and P_3 acting at points B, C, and D, respectively (see figure). The dimensions of the bar are as follows: $a = 1.0$ m, $b = 0.4$ m, and $c = 0.6$ m. Assuming that $P_2 = 15$ kN and $P_3 = 9$ kN, determine the force P_1 so that the lower end D of the bar does not move vertically when the loads are applied.

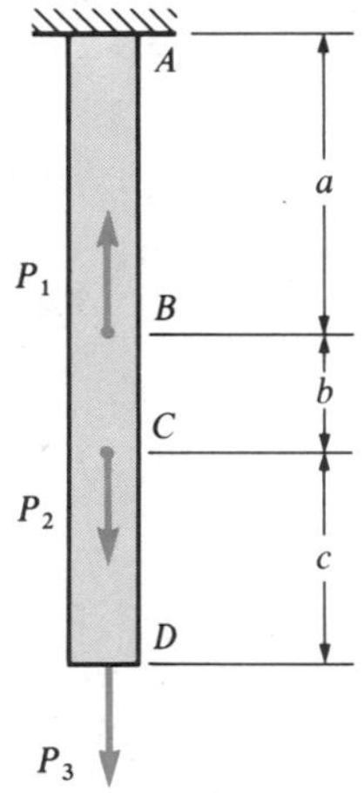

Prob. 2.2-12

2.2-13 A steel bar 10 ft long has a circular cross section of diameter $d_1 = 0.75$ in. over one-half of its length and diameter $d_2 = 0.50$ in. over the other half (see figure). (a) How much will the bar elongate under a tensile load $P = 5000$ lb? (b) If the same volume of material is made into a bar of constant diameter d and length 10 ft, what will be the elongation under the same load P? (Assume $E = 30 \times 10^6$ psi.)

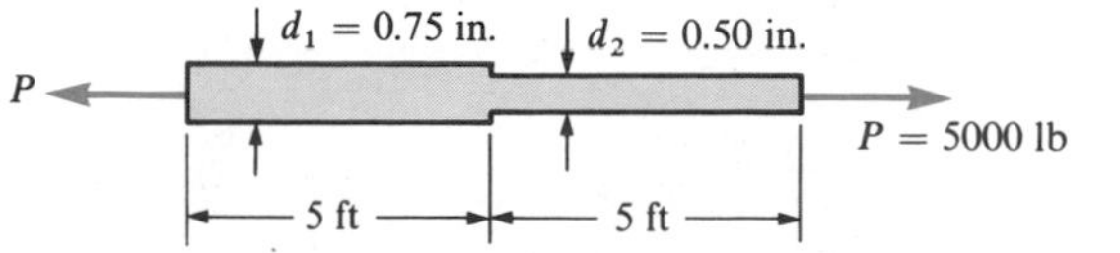

Prob. 2.2-13

2.2-14 A copper bar AB under a tensile load $P = 450$ kN hangs from a pin supported by two steel pillars (see figure). The copper bar has length 10 m, cross-sectional area 8100 mm^2, and modulus of elasticity $E_c = 103$ GPa. Each steel pillar has height 1 m, cross-sectional area 7500 mm^2, and modulus of elasticity $E_s = 200$ GPa. Determine the displacement δ of point A caused by the load P.

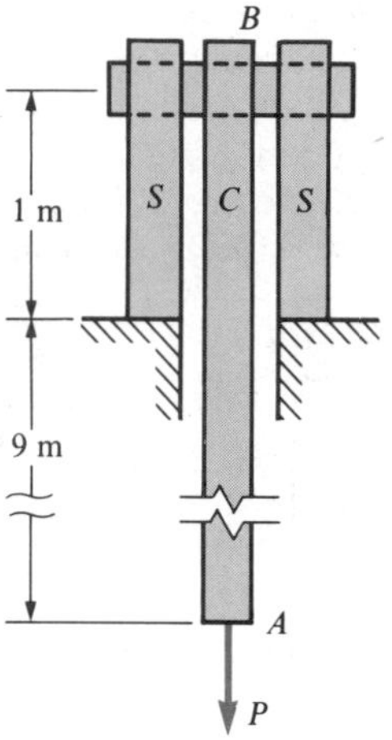

Prob. 2.2-14

2.2-15 The assembly shown in the figure is loaded by forces P_1 and P_2. Assuming that both parts of the vertical bar ABC are made of the same material, obtain a formula for the ratio P_2/P_1 so that the vertical displacement of point C will be zero. (Express the result in terms of the cross-sectional areas A_1 and A_2 and dimensions L_1, L_2, L_3, and L_4 shown in the figure.)

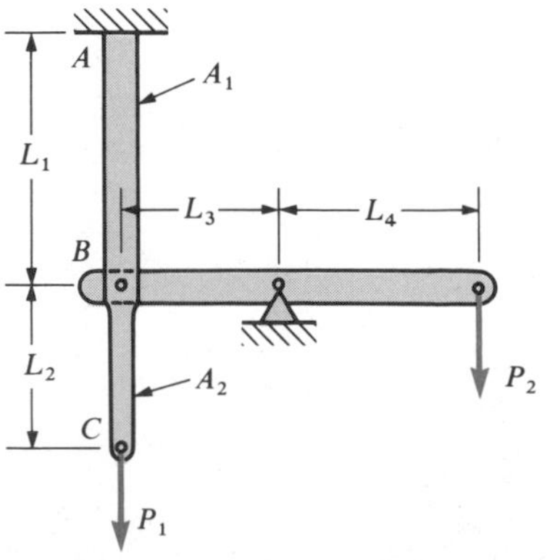

Prob. 2.2-15

2.2-16 Bar ABC is composed of two materials and has a total length of 1 m and a diameter of 50 mm (see figure). Part AB is steel ($E_s = 210$ GPa), and part BC is aluminum ($E_a = 70$ GPa). The bar is subjected to a tensile force of 130 kN. (a) Determine the lengths L_1 and L_2 for the steel and aluminum parts, respectively, in order that both parts have the same elongation. (b) What is the total elongation of the bar?

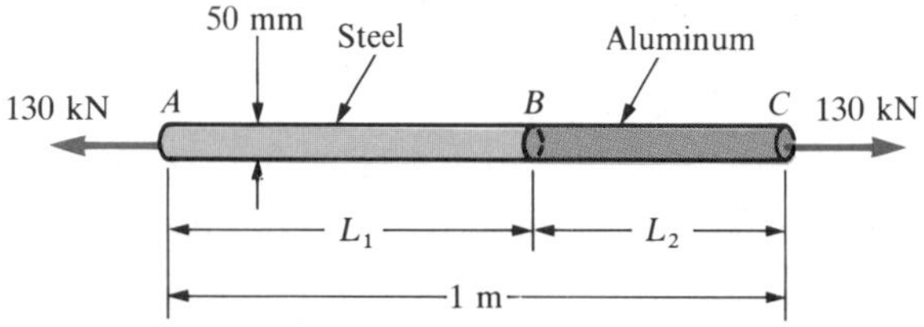

Prob. 2.2-16

2.2-17 A composite bar is made of two copper sections C with cross-sectional area $A_c = 2$ in.^{2} and a steel section S with cross-sectional area $A_s = 1$ in.^{2} (see figure). Assuming that the moduli of elasticity for the copper and the steel are $E_c = 16 \times 10^6$ psi and $E_s = 30 \times 10^6$ psi, respectively, calculate the magnitude of the tensile force P necessary to produce a total elongation $\delta = 0.05$ in.

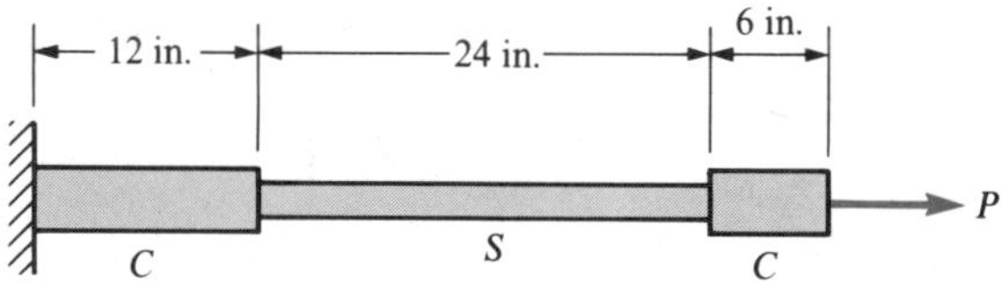

Prob. 2.2-17

2.2-18 A 4 m long steel pipe ($E = 210$ GPa) is loaded as shown in the figure. The cross-sectional area of the pipe is 1800 mm^2. (a) Calculate the displacement δ at the free end. (b) Find the distance x from the left-hand support to the point at which the displacement is zero.

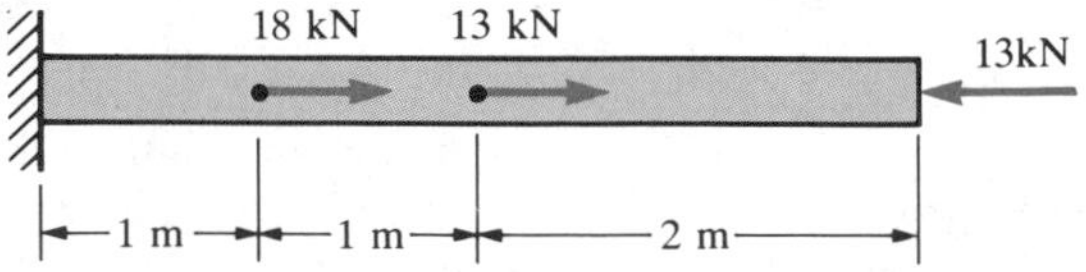

Prob. 2.2-18

2.2-19 Derive a formula for the elongation of a prismatic bar of length L and cross-sectional area A hanging vertically under its own weight. (Assume W = total weight of the bar and E = modulus of elasticity of the material.)

2.2-20 A uniform steel bar is 6 m long when lying on a horizontal surface. Determine its elongation when suspended vertically from one end. (Assume the modulus of elasticity $E = 200$ GPa and the specific weight $\gamma = 77.0\ \text{kN/m}^3$.)

2.2-21 A concrete pile, driven into the earth, supports a load P by friction along its sides (see figure). The friction force is assumed to be uniform and is denoted as f per unit length of pile. The pile has cross-sectional area A, modulus of elasticity E, and embedment length L. Derive a formula for the total shortening δ of the pile in terms of f, E, A, and L.

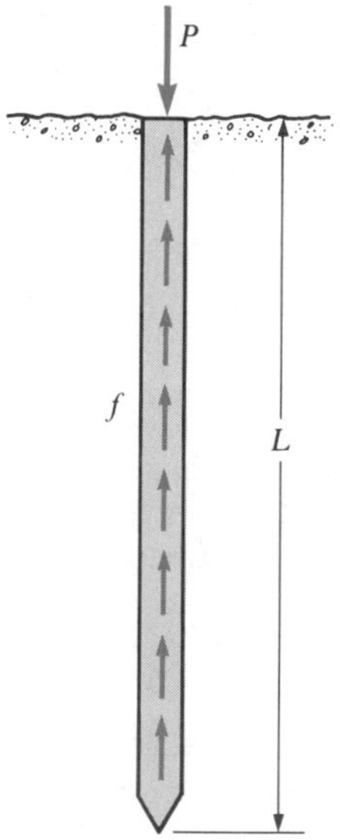

Prob. 2.2-21

2.2-22 A plastic bar ABC of total length $L = 1.2$ m and $E = 4.0$ GPa is made of two parts AB and BC of equal length (see figure). Part AB has diameter $d_1 = 100$ mm and part BC has diameter $d_2 = 60$ mm. A longitudinal hole is drilled through part AB for one-half of its length. The bar is compressed by forces $P = 110$ kN acting at the ends. If the shortening of the bar is limited to 8.0 mm, what is the maximum allowable hole diameter d?

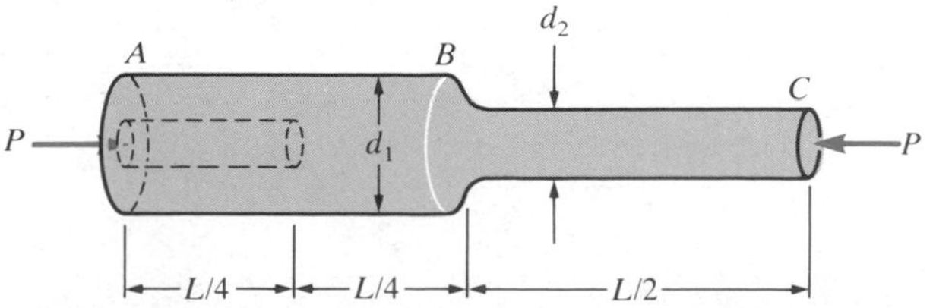

Prob. 2.2-22

2.2-23 The device shown in the figure consists of a horizontal bar ABC supported by two vertical bars BD and CE. The distance AC is 18 in., and the vertical bars are 6 in. apart. Bars BD and CE have lengths equal to 12 in. and 15 in., respectively, and cross-sectional areas equal to 0.4 in.^{2} They are constructed of aluminum with modulus of elasticity $E = 10{,}000$ ksi. Assuming that bar ABC is rigid, what is the maximum permissible load P if the displacement of end A is limited to 0.05 in.?

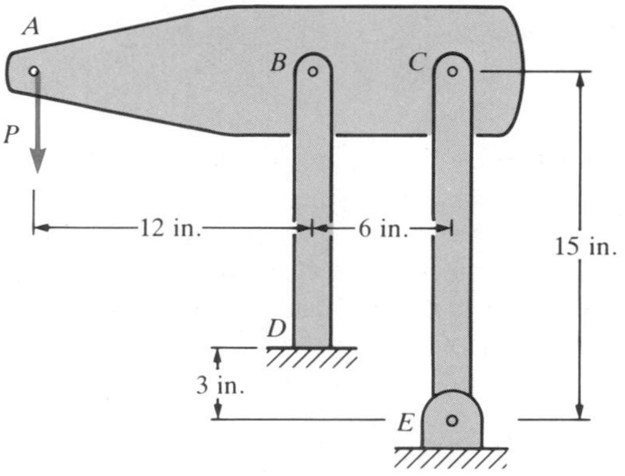

Prob. 2.2-23

2.2-24 A concrete pier of square cross section is 6 m high (see figure). The sides taper uniformly from a width of 0.5 m

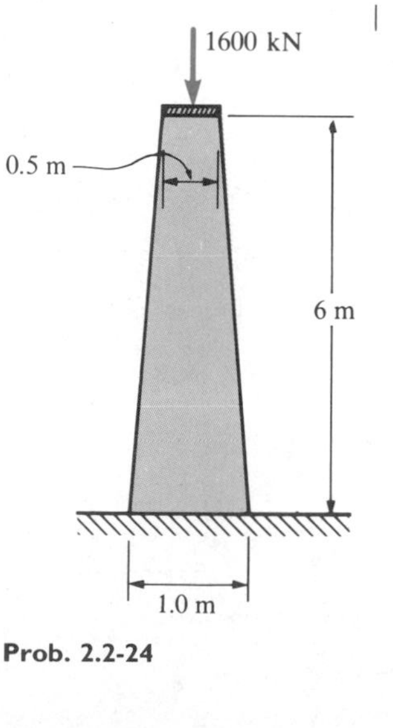

Prob. 2.2-24

at the top to 1.0 m at the bottom. Determine the shortening of the pier under a compressive load of 1600 kN (disregard the weight of the pier itself). Assume that the modulus of elasticity of concrete is 24 GPa.

2.2-25 A long, uniformly tapered bar AB of square cross section and length L is subjected to an axial load P (see figure). The cross-sectional dimensions vary from $d \times d$ at end A to $2d \times 2d$ at end B. Derive a formula for the elongation δ of the bar.

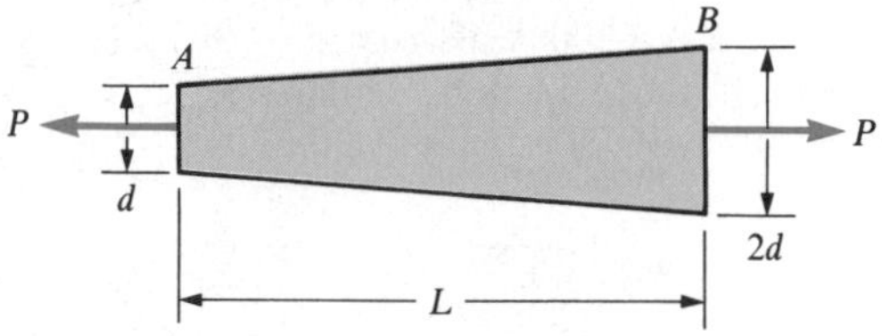

Prob. 2.2-25

2.2-26 A flat bar of rectangular cross section and constant thickness t is subjected to tension by forces P (see figure). The width of the bar varies linearly from b_1 at the left end to b_2 at the right end. (a) Derive a formula for the elongation δ of the bar. (b) Calculate the elongation assuming $b_1 = 100$ mm, $b_2 = 150$ mm, $L = 1.5$ m, $t = 25$ mm, $P = 35$ kN, and $E = 200$ GPa.

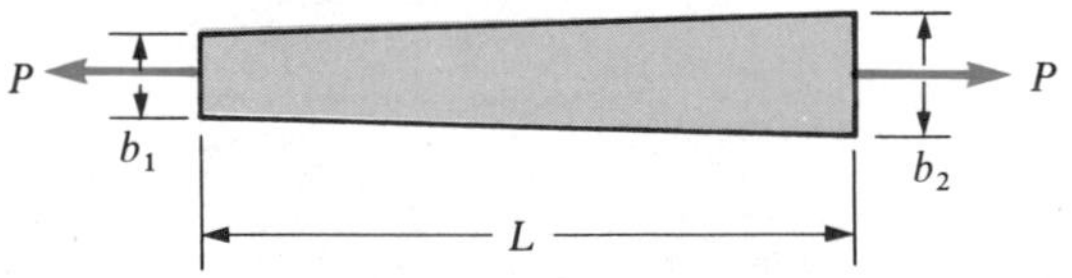

Prob. 2.2-26

2.2-27 Derive a formula for the elongation δ of a conical bar of circular cross section (see figure) under the action of its own weight if the length of the bar is L, the weight per unit volume is γ, and the modulus of elasticity is E.

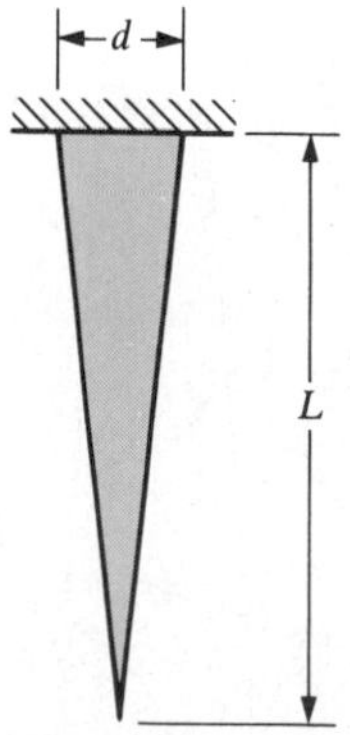

Prob. 2.2-27

***2.2-28** A rigid circular disc of radius R rotates with constant angular speed ω about an axis through its center C (see figure). Four identical prismatic blades, each of length L, project radially outward from the disc. The material of the blades has modulus of elasticity E and mass density ρ (mass per unit of volume). Determine the elongation δ of each blade due to the centrifugal effects.

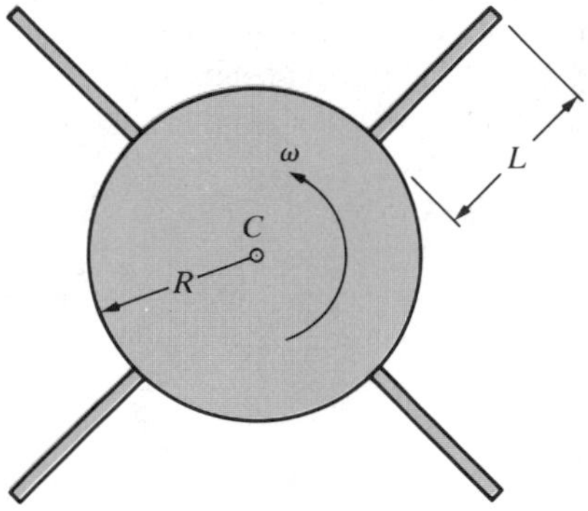

Prob. 2.2-28

****2.2-29** The main cables of a suspension bridge [see part (a) of the figure] follow a curve that is nearly parabolic because the primary load on the cables is the weight of the bridge deck, which is uniform in intensity along the horizontal. Therefore, let us represent the central span

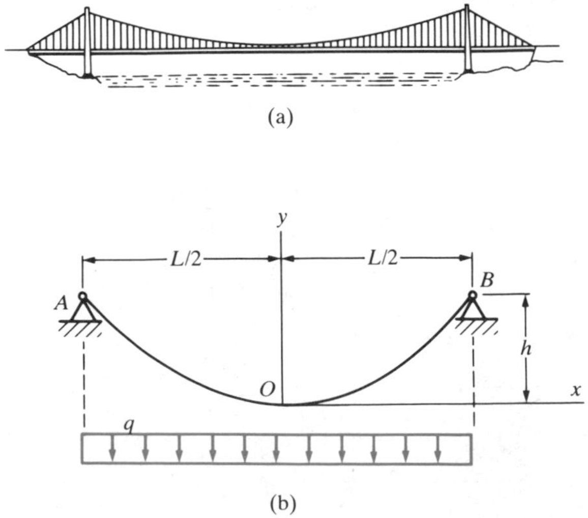

Prob. 2.2-29

AOB of one of the main cables as a parabolic cable supported at points A and B and carrying a uniform load of intensity q along the horizontal [see part (b) of the figure]. The span of the cable is L, the sag is h, the axial rigidity is EA, and the origin of coordinates is at midspan. (a) Derive a formula for the elongation δ of the cable. (b) Calculate the elongation δ for one of the main cables of the Golden Gate Bridge, for which the dimensions and properties are $L = 4200$ ft, $h = 470$ ft, $q = 12{,}700$ lb/ft, and $E = 28{,}800{,}000$ psi. The cable consists of 27,572 parallel wires of diameter 0.196 in. (Suggestions: Determine the tensile force T at any point in the cable from a free-body diagram of part of the cable; then determine the elongation of an element of the cable of length ds; finally, integrate along the curve of the cable to obtain the total elongation.)

2.3-1 Three bars, two of steel and one of brass, jointly support a load P acting through a rigid block (see figure). The steel bars are identical and are the same length as the brass bar. What fraction of the load P is carried by the brass bar if the total cross-sectional area of the steel bars is 50% larger than the cross-sectional area of the brass bar and the modulus of elasticity of the steel is twice that of the brass?

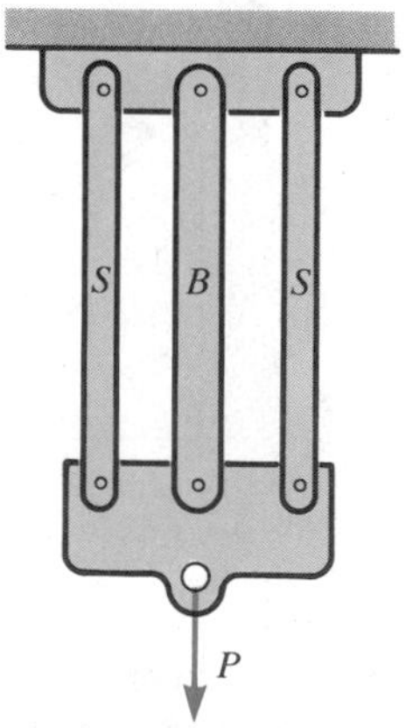

Prob. 2.3-1

2.3-2 A plastic rod AB of length $L = 0.5$ m has a diameter $d_1 = 30$ mm (see figure). A plastic sleeve CD of length $c = 0.3$ m and outer diameter $d_2 = 45$ mm is securely bonded to the rod so that no slippage can occur between the rod and the sleeve. The rod is made of an acrylic with modulus of elasticity $E_1 = 3.1$ GPa and the sleeve is made of a polyamide with $E_2 = 2.5$ GPa. Calculate the elongation δ of the rod when it is pulled by axial forces $P = 15$ kN.

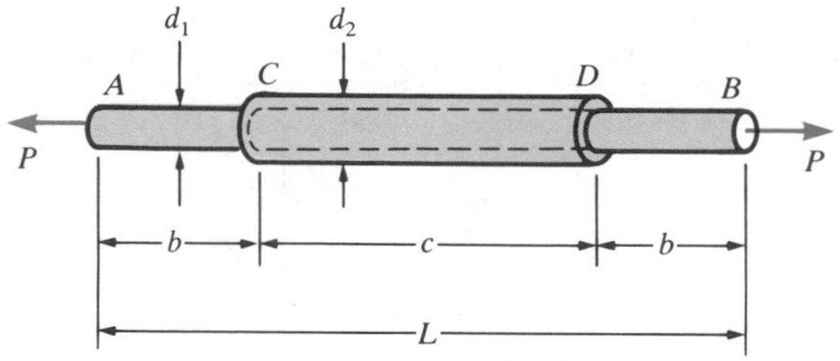

Prob. 2.3-2

2.3-3 A round steel pipe of inside diameter $d = 20$ in. and wall thickness $t = 0.5$ in. is filled with concrete and compressed between rigid plates (see figure). If the allowable stresses for the steel and concrete are 17,000 psi and 1,200 psi, respectively, find the maximum allowable load P_{max}. (Assume $E_s = 30 \times 10^3$ ksi and $E_c = 2 \times 10^3$ ksi.)

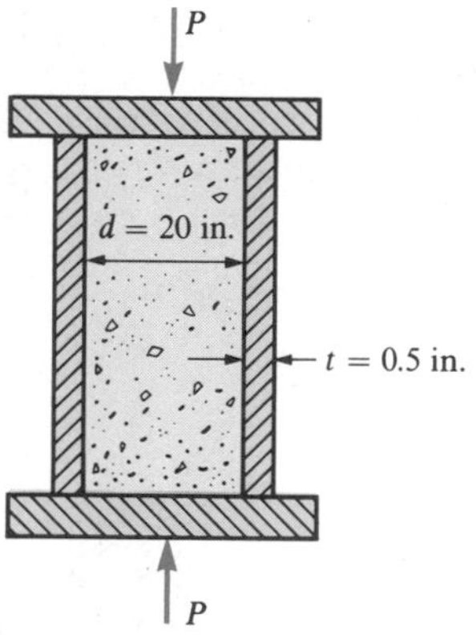

Prob. 2.3-3

2.3-4 A square column is formed of a 25 mm thick metal casing (outside dimensions 250 mm × 250 mm and inside dimensions 200 mm × 200 mm) that is filled with concrete (see figure on p. 132). The casing has modulus of elasticity $E_1 = 84$ GPa, and the concrete core has modulus of elasticity $E_2 = 14$ GPa. Find the maximum permissible load P_{max} on the column if the allowable stresses in the metal and the concrete are 36 MPa and 4.8 MPa, respectively. (Assume that the stresses in the metal and concrete are uniformly distributed over their respective areas.)

2.3-5 A reinforced concrete pedestal of square cross section (width $b = 20$ in. on each side) is constructed with 12 steel reinforcing bars, each having an effective diameter of 1.0 in. (see figure). The pedestal supports a compressive

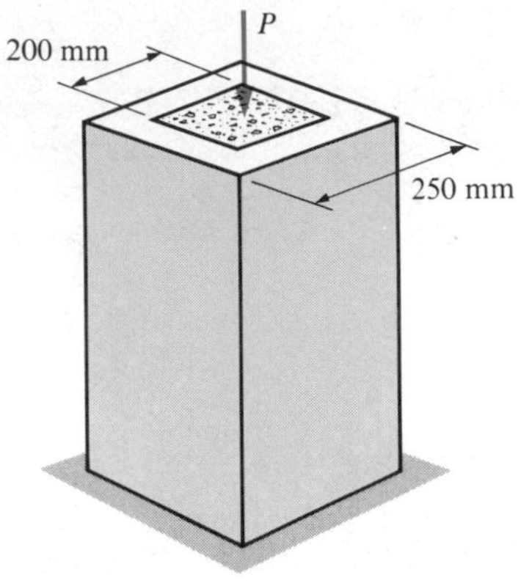

Prob. 2.3-4

load P applied through a rigid bearing plate. Calculate the maximum permissible load P_{max} if the allowable stresses in the steel and concrete are 12,000 psi and 1,200 psi, respectively. (Assume that the modulus of elasticity for steel is 29×10^6 psi and for concrete is 3.6×10^6 psi. Also, disregard the weight of the pedestal itself. Finally, note that the cross-sectional area of the concrete is *not* equal to b^2.)

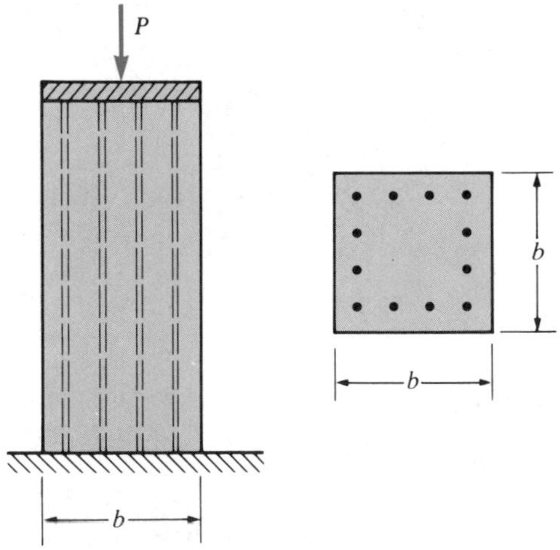

Prob. 2.3-5

2.3-6 A bar AB having length L and axial rigidity EA is supported at its upper end by an immovable support (see figure). At the lower end a small gap of size s exists between the end of the bar and a rigid surface. A load P acts on the bar at point C, which is one-third of the length from the lower end. If the support reactions produced by the load P are to be equal, what should be the size s of the gap?

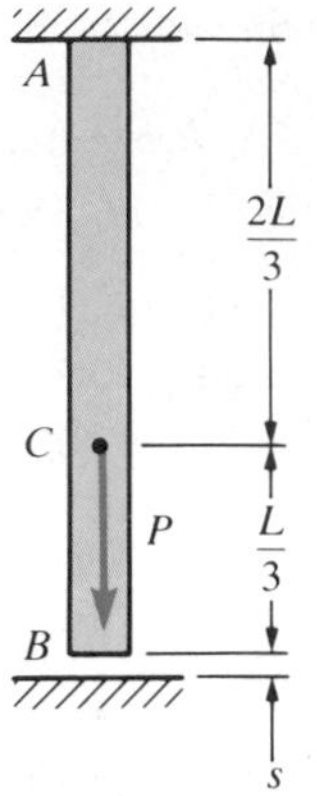

Prob. 2.3-6

2.3-7 A bar AB having two different cross-sectional areas A_1 and A_2 is held between rigid supports (see figure). A load P acts at point C. Determine the reactions R_a and R_b at supports A and B, respectively, due to the load P.

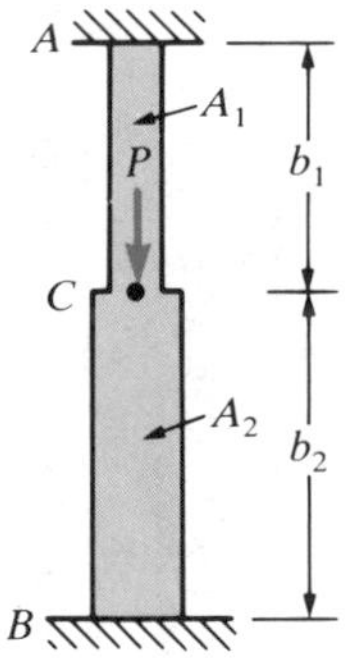

Prob. 2.3-7

2.3-8 The aluminum and steel pipes shown in the figure are fastened to rigid supports at ends A and B and to a rigid plate C where they are joined. Two equal loads P are symmetrically applied on the plate. Find the axial stresses σ_a and σ_s in the aluminum and steel pipes, respectively. (Use numerical data as follows: $P = 50$ kN, $A_a = 6000$ mm^2, $A_s = 600$ mm^2, $E_a = 70$ GPa, and $E_s = 200$ GPa.)

2.3-9 The axially loaded bar AB shown in the figure is held between rigid supports. The bar has cross-sectional area A_1 from A to C and $1.5A_1$ from C to B. What is the displacement δ_d of point D where the load P acts? What are the reactions R_a and R_b at supports A and B?

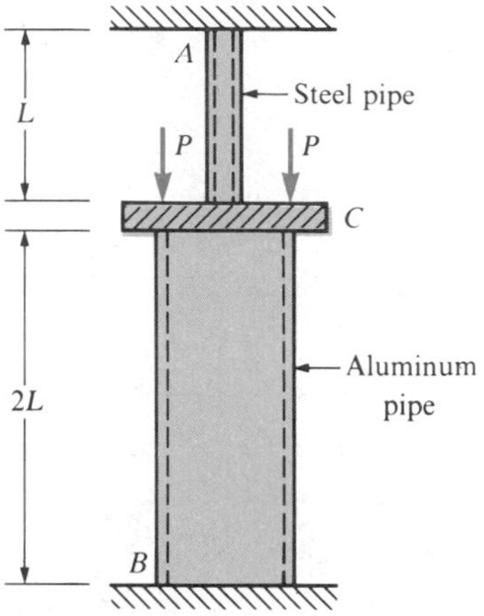

Prob. 2.3-8

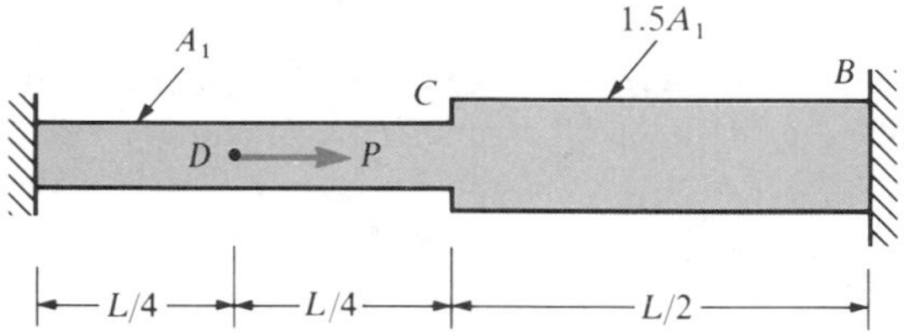

Prob. 2.3-9

2.3-10 A rigid block AB of weight W hangs on three equally spaced vertical wires, two of steel and one of aluminum (see figure). The wires also support a load P acting at the center of the block. The diameter of the steel wires is 4 mm, and the diameter of the aluminum wire is 6 mm. What load P can be supported if the allowable stress in the steel wires is 140 MPa and in the aluminum wire is 80 MPa? (Assume $W = 400$ N, $E_s = 210$ GPa, and $E_a = 70$ GPa.)

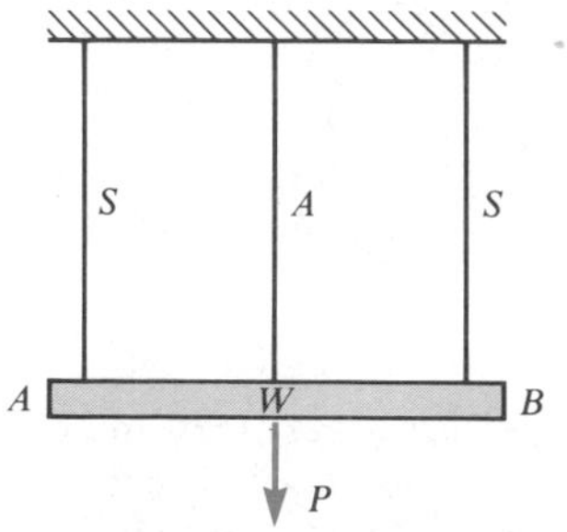

Prob. 2.3-10

2.3-11 A rod AB has two different cross-sectional areas as shown in the figure. The rod is rigidly attached to immovable supports at the ends and is loaded by equal and opposite forces P at the locations shown. Determine the axial stress σ at the middle of the bar, assuming A_1 is the cross-sectional area near the ends and A_2 is the cross-sectional area in the middle region. (Use numerical data as follows: $P = 5400$ lb, $A_1 = 0.6$ in.^{2}, $A_2 = 0.9$ in.^{2}, and $b = 1.5a$.)

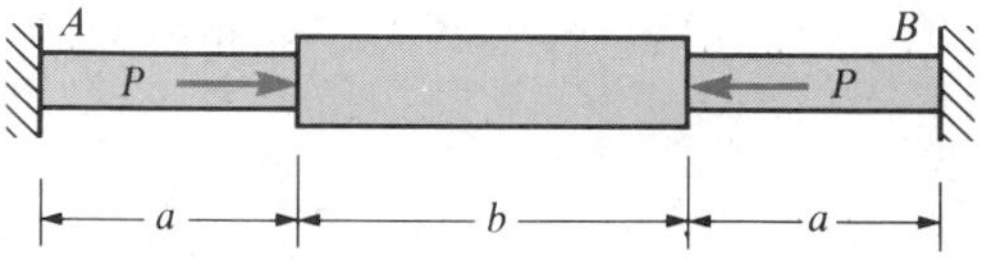

Prob. 2.3-11

2.3-12 A composite bar of square cross section with dimensions $2b \times 2b$ is constructed of two different materials having moduli of elasticity E_1 and E_2 (see figure). Both parts of the bar have the same cross-sectional dimensions. (a) Assuming that the end plates are rigid, derive a formula for the eccentricity e of the load P so that each part of the bar is stressed uniformly in compression. (b) Under these conditions, what axial forces P_1 and P_2 are carried by materials 1 and 2, respectively?

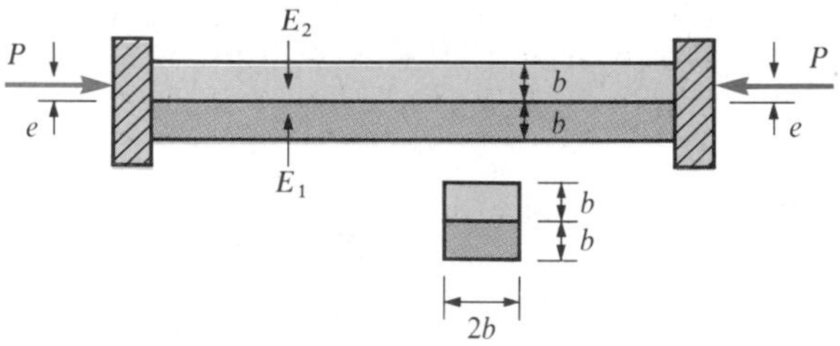

Prob. 2.3-12

2.3-13 Three wires A, B, and C having the same axial rigidity EA support a horizontal rigid beam (see figure).

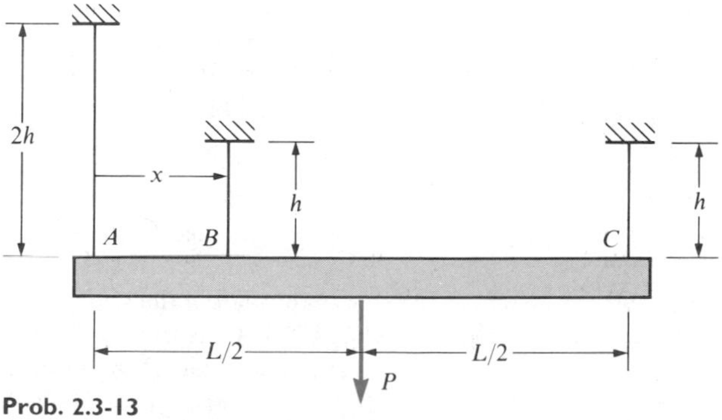

Prob. 2.3-13

Wires B and C have length h, and wire A has length $2h$. Determine the distance x between wires A and B in order that the rigid beam will remain horizontal when a load P is applied at its midpoint.

2.3-14 The apparatus shown in the figure consists of a rigid bar AB of weight $W = 20$ kN supported by three slender circular rods. The two outer rods are made of aluminum ($E_1 = 72$ GPa) with diameters $d_1 = 8$ mm, and the inner rod is magnesium ($E_2 = 45$ GPa) with diameter $d_2 = 12$ mm. The lengths of the rods are in the ratio $L_2/L_1 = 9/8$. What is the tensile stress σ_2 in the middle bar due to the weight of bar AB?

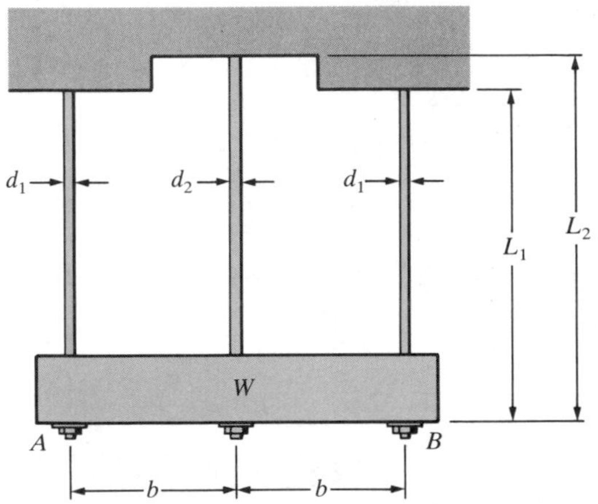

Probs. 2.3-14 and 2.3-15

2.3-15 A rigid bar AB weighing $W = 8500$ lb is supported by three slender circular rods (see figure). The two outer rods are made of aluminum ($E_1 = 10 \times 10^6$ psi) with diameters $d_1 = 0.4$ in. and lengths $L_1 = 50$ in. The inner rod is magnesium ($E_2 = 6.5 \times 10^6$ psi) with diameter d_2 and length L_2. The allowable stresses in the aluminum and magnesium are 24,000 psi and 13,000 psi, respectively. If it is desired to have all three rods loaded to their maximum allowable values, what should be the diameter d_2 and length L_2 of the middle rod?

2.3-16 A rigid bar AB of length L is hinged to a wall at A and supported by two vertical wires attached at points C and D (see figure). The wires have the same cross-sectional areas and are made of the same material, but the wire at D is twice as long as the wire at C. Find the tensile forces T_c and T_d in the wires due to the vertical load P acting at end B.

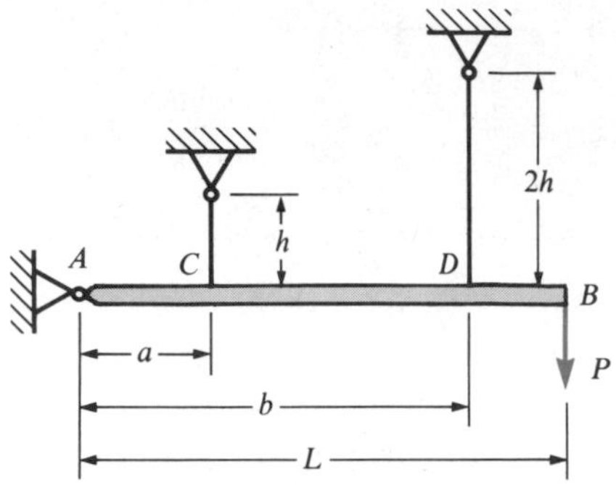

Prob. 2.3-16

2.3-17 A steel bar ABC ($E = 30 \times 10^6$ psi) has cross-sectional area A_1 from A to B and cross-sectional area A_2 from B to C (see figure). The bar is supported at end A and is subjected to a load P equal to 10,000 lb at end C. A steel collar BD having total cross-sectional area A_3 supports the bar at B. Determine the displacement δ_c at the lower end of the bar due to the load P, assuming that the collar fits snugly at B when there is no load. (Assume $L_1 = 2L_2 = 10$ in., $L_3 = 4$ in., $A_1 = 2A_3 = 1.6$ in.^{2}, and $A_2 = 0.5$ in.^{2})

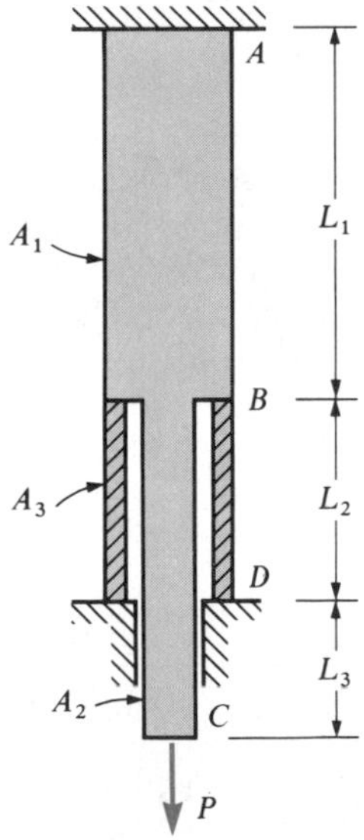

Prob. 2.3-17

2.3-18 Three identical wires A, B, and C support a rigid block (weight $= W$) to which a load $2W$ is applied at distance x from the middle as shown in the figure. Plot a graph showing how the forces F_a, F_b, and F_c in the three wires vary as a function of x. (Assume that x varies from zero to values greater than b, and note that the wires cannot resist compressive forces.)

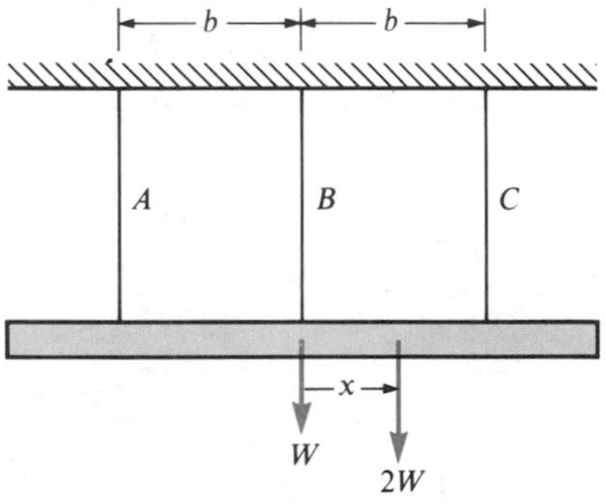

Prob. 2.3-18

2.3-19 The mechanical system shown in the figure consists of a vertical rigid bar AB that rotates about a pivot at C. Horizontal springs having stiffnesses k_1 and k_2 are attached to the bar at points A and D, respectively, and hold the bar in the vertical position when no loads are present. Determine the horizontal displacement δ_b at point B when the load P is applied at B.

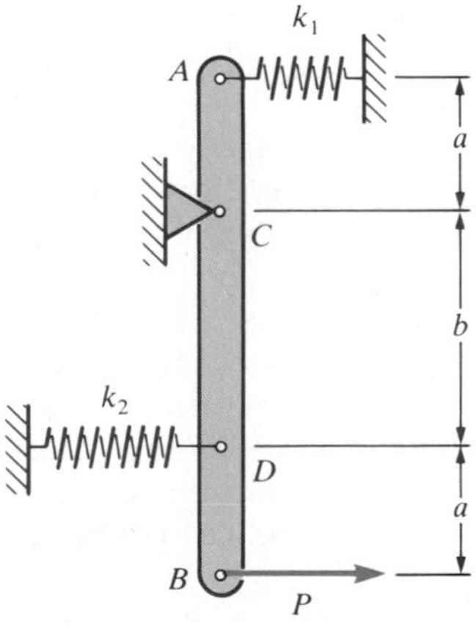

Prob. 2.3-19

2.3-20 A rigid bar AB is supported by two cables CE and DB that are identical except for their lengths (see figure). A load P acts on the bar at point F. Calculate the tensile force T in cable DB due to the force P if $L/h = \sqrt{12}$.

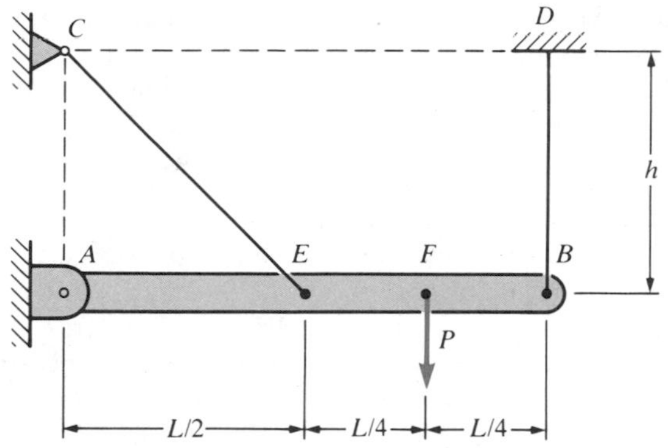

Prob. 2.3-20

2.3-21 A rigid bar AB of length $3b$ is supported by two inclined wires CD and CB (see figure). Both wires are the same except for their lengths. Determine the tensile forces T_1 and T_2 in wires CD and CB, respectively, due to the load P.

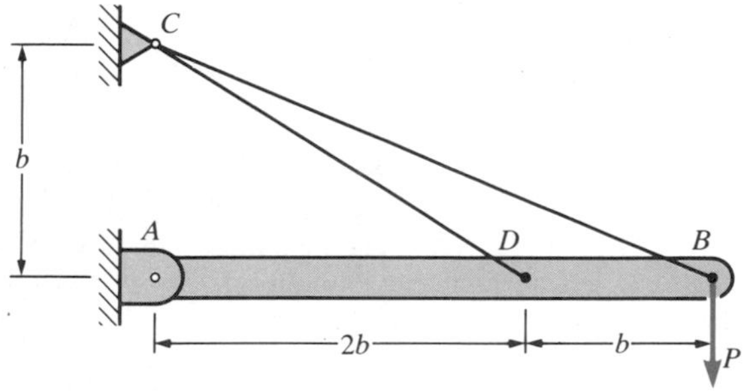

Prob. 2.3-21

2.3-22 A rectangular frame of width $7b$ and height $3b$ (see figure) is pivoted at C, supported at A and D by identical

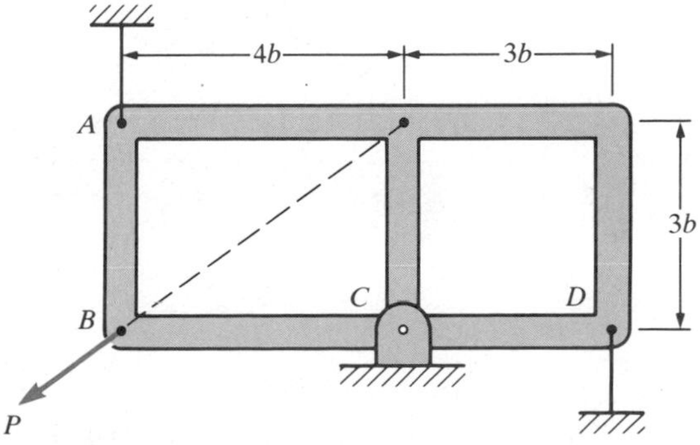

Prob. 2.3-22

vertical wires, and loaded by an inclined force P acting at B. The wires have cross-sectional area $A = 8\ \text{mm}^2$. If the stresses in the wires due to the load P are not to exceed a tensile stress $\sigma_t = 120$ MPa, what is the allowable load P_{allow}? (Disregard any deformations of the frame.)

***2.3-23** A horizontal rigid bar ACB of length $2b$ is supported by two inclined wires BD and CD (see figure). The wires are fastened to a support at D, which is distance h above point A. The wires have different diameters but are made of the same material with the same allowable stress σ_{allow}. Because the material of the wires is an expensive alloy, we wish to use as little of it as possible. Therefore, determine the height h (in terms of the distance b) so that the total volume of the wires will be a minimum. Disregard the weights of the bar ACB and the wires, and consider only the effects of the vertical load P acting on B. (Hint: Begin by finding the tensile forces in the wires in terms of h, b, and P. Next, determine the required cross-sectional areas by dividing the forces by the allowable stresses. Then obtain the total volume V of the wires by multiplying the areas by the lengths; this expression takes the form

$$V = \frac{2Pb}{\sigma_{allow}} f\left(\frac{h}{b}\right)$$

in which $f(h/b)$ is a function only of the nondimensional ratio h/b. Because P, b, and σ_{allow} are constants, the volume is a minimum when $f(h/b)$ is a minimum. The value of h/b that makes this function a minimum can be determined numerically.)

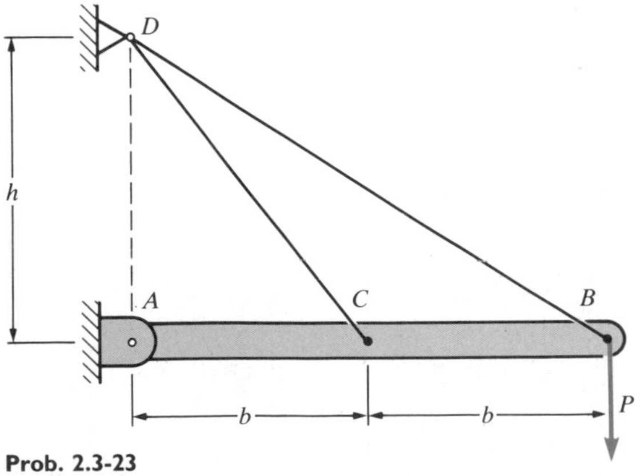

Prob. 2.3-23

***2.3-24** A trimetallic bar consisting of a circular steel core surrounded by brass and copper tubes [see part (a) of the figure] is uniformly compressed by an axial force P applied through rigid loading plates [see part (b) of the figure]. The steel core has diameter 10 mm, the brass tube has outer diameter 15 mm, and the copper tube has outer diameter 20 mm. The corresponding moduli of elasticity are $E_s = 210$ GPa, $E_b = 100$ GPa, and $E_c = 120$ GPa. Calculate the compressive stresses σ_s, σ_b, and σ_c in the steel, brass, and copper, respectively, when the force P equals 10 kN.

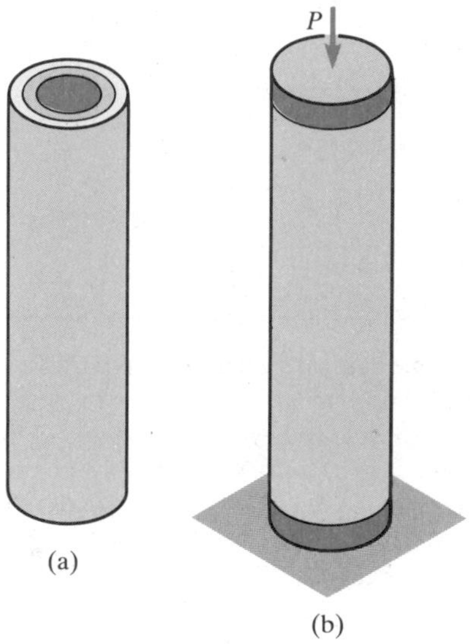

Prob. 2.3-24

2.4-1 A steel measuring tape lying on a flat pavement is used to measure the distance between two points A and B. The tape reads 95.49 ft when the temperature at the pavement surface is 115 °F and the tensile force in the tape is 20 lb. The cross-sectional dimensions of the tape are 0.3 in. × 0.014 in. The tape has modulus of elasticity $E = 30 \times 10^6$ psi and coefficient of thermal expansion $\alpha = 6.5 \times 10^{-6}$/°F. Calibration of the tape shows that it reads correctly on a flat surface at a temperature of 68 °F and a pull of 10 lb. What is the true distance d between the two points?

2.4-2 An aluminum pipe has a length of 60 m at a temperature of 18 °C. An adjacent steel pipe at the same temperature is 10 mm longer than the aluminum pipe. At what temperatures will the difference in lengths of the two pipes be 15 mm? (Assume that the coefficients of thermal expansion of aluminum and steel are $\alpha_a = 23 \times 10^{-6}$/°C and $\alpha_s = 12 \times 10^{-6}$/°C, respectively.)

2.4-3 The rails of a railroad track are welded together at their ends (to form a continuous rail and thus eliminate the clacking sound of the wheels) when the temperature is 50 °F. What compressive stress σ is produced in the rails when they are heated by the sun to 125 °F if $\alpha = 6.5 \times 10^{-6}/°F$ and $E = 30 \times 10^6$ psi?

2.4-4 The apparatus shown in the figure is made of a tungsten bar AC and a magnesium bar BD that are attached to a pointer CDP by pins at C and D. Let the coefficients of thermal expansion for magnesium and tungsten be denoted by α_m and α_t, respectively. Derive a formula for the vertical displacement δ (positive upward) of point P due to a uniform temperature increase ΔT. Could such a device be used as a thermometer?

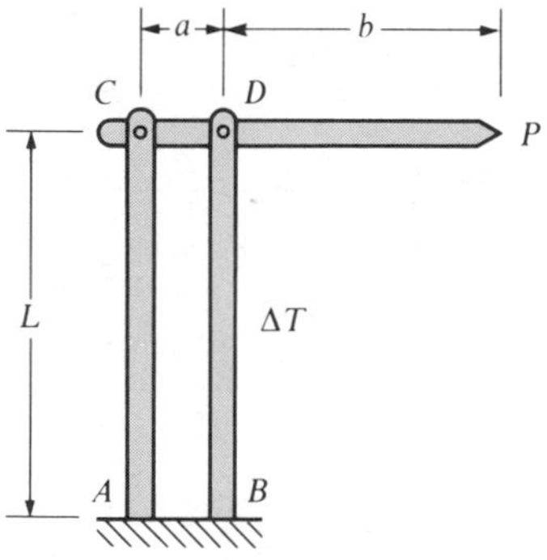

Prob. 2.4-4

2.4-5 A solid steel cylinder S is placed inside a copper tube C (see figure). The assembly is compressed between rigid plates by forces P. Obtain a formula for the increase in temperature ΔT that will cause all of the load to be carried by the copper tube. (Let α_s, α_c, E_s, E_c, A_s, and A_c represent the thermal expansion coefficients, moduli of elasticity, and cross-sectional areas, respectively, of the steel and copper.)

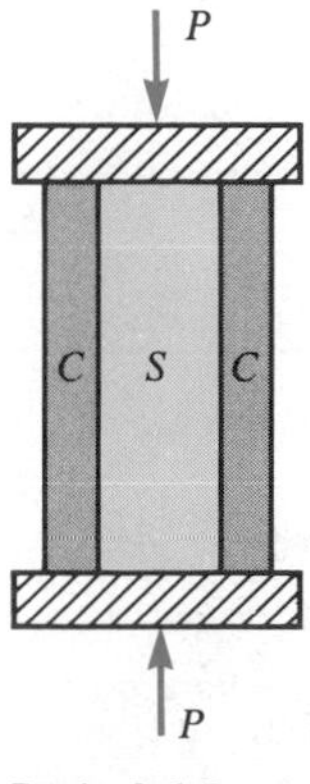

Prob. 2.4-5

2.4-6 A metal bar AB of length L is held between rigid supports and heated nonuniformly in such a manner that the temperature increase ΔT at distance x from end A is given by the expression $\Delta T = \Delta T_1 x^2/L^2$ (see figure). Find the compressive stress σ_c in the bar. (Assume that the material has modulus of elasticity E and coefficient of thermal expansion α.)

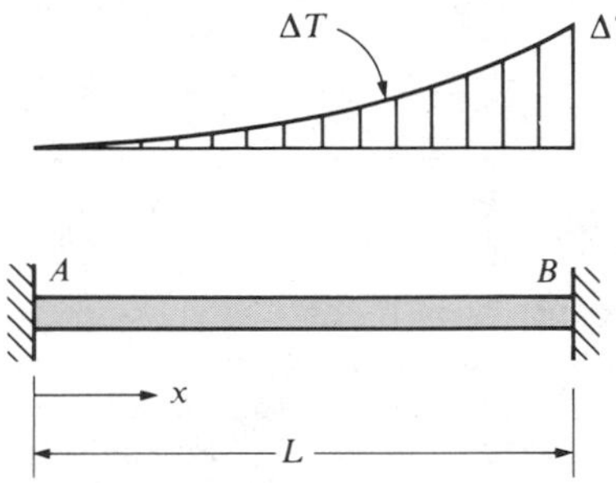

Prob. 2.4-6

2.4-7 A steel rod of diameter 0.6 in. is held snugly (but without any initial stresses) between rigid walls by the arrangement shown in the figure. Calculate the temperature drop ΔT (degrees Fahrenheit) at which the average shear stress in the 0.5 in. diameter bolt becomes 8000 psi. (For steel, use $\alpha_s = 6.5 \times 10^{-6}/°F$ and $E_s = 30 \times 10^6$ psi.)

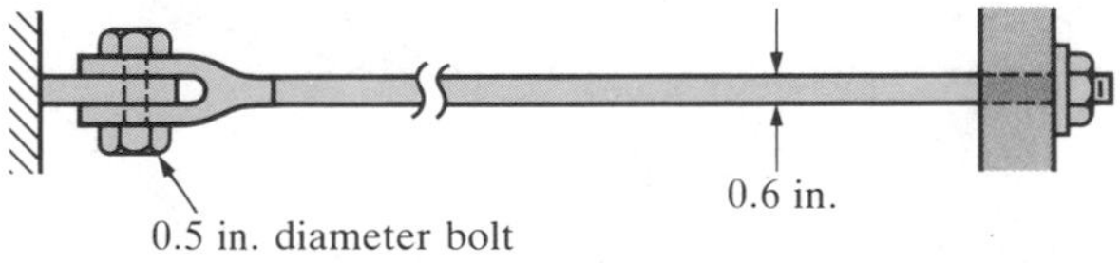

Prob. 2.4-7

2.4-8 A steel wire AB is stretched between rigid supports (see figure). The initial prestress in the wire is 40 MPa when the temperature is 20 °C. (a) What is the stress σ in the wire when the temperature drops to 0 °C? (b) At what temperature T will the stress in the wire become zero? (Assume $\alpha = 14 \times 10^{-6}/°C$ and $E = 210$ GPa.)

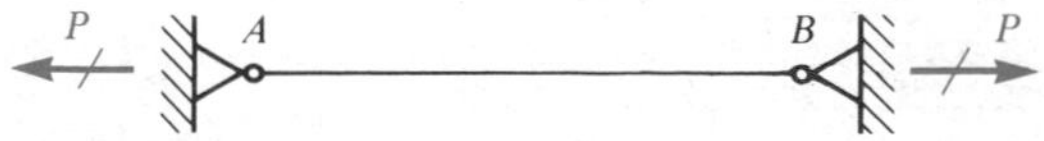

Prob. 2.4-8

2.4-9 A circular steel rod AB (diameter $d_1 = 1$ in., length $L_1 = 3$ ft) has a bronze sleeve (outer diameter $d_2 = 1.25$ in., length $L_2 = 1$ ft) shrunk onto it so that the two parts are securely bonded together (see figure). Calculate the total elongation δ of the steel bar due to a temperature rise $\Delta T = 600$ °F. (Material properties are as follows: for steel, $E_s = 30 \times 10^6$ psi and $\alpha_s = 6.5 \times 10^{-6}/°F$; for bronze, $E_b = 15 \times 10^6$ psi and $\alpha_b = 11 \times 10^{-6}/°F$.)

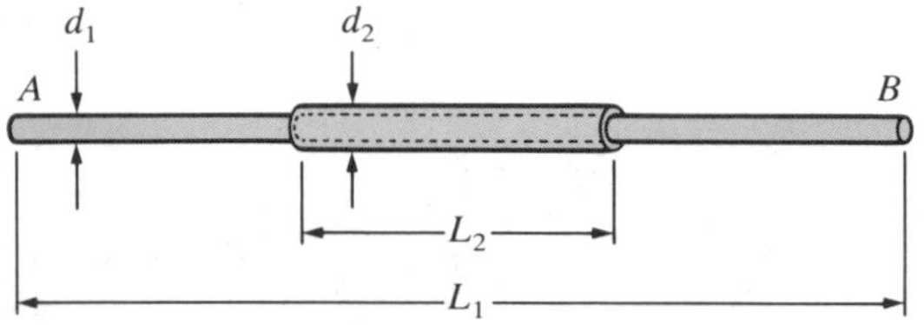

Prob. 2.4-9

2.4-10 A copper bar AB of length 1.0 m is placed in position at room temperature with a gap of 0.20 mm between end A and a rigid restraint (see figure). Calculate the axial compressive stress σ in the bar if the temperature rises 50 °C. (For copper, use $\alpha = 17 \times 10^{-6}/°C$ and $E = 110$ GPa.)

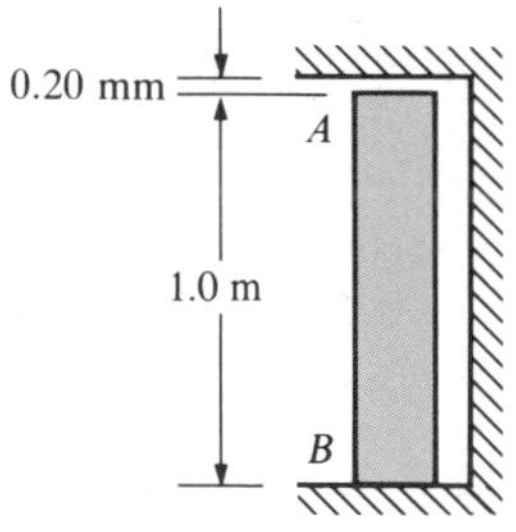

Prob. 2.4-10

2.4-11 Three parallel and adjacent steel bars ($E = 30 \times 10^6$ psi) jointly carry a tensile load $P = 300$ k (see figure). The cross-sectional area of each bar is 6 in.^{2}, and the length is 20 ft. If the middle bar is accidentally shorter than the other two by 0.03 in., what will be the final stress σ in the middle bar when the load P is applied? (Assume that the ends of the bars are pulled into alignment by the rigid end plate when the load is applied.)

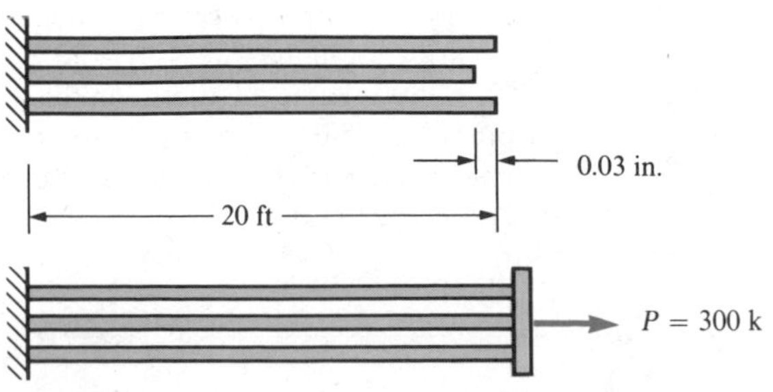

Prob. 2.4-11

2.4-12 A plastic bar ACB having two different cross sections is held between rigid supports as shown in the figure. The cross-sectional areas in the left- and right-hand parts are 1200 mm^2 and 1800 mm^2, respectively. The modulus of elasticity E is 6 GPa, and the coefficient of thermal expansion α is $100 \times 10^{-6}/°C$. The bar is subjected to a uniform temperature increase of 50 °C. Calculate the following quantities: (a) the axial force P in the bar; (b) the maximum axial stress σ; and (c) the displacement δ of point C.

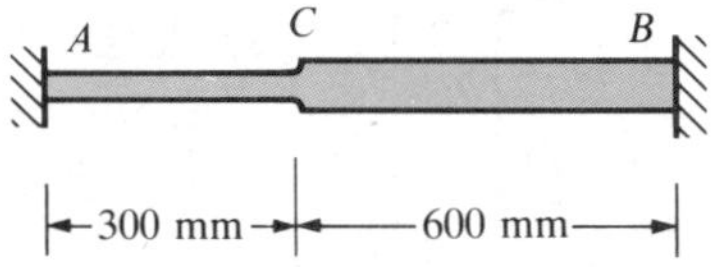

Prob. 2.4-12

2.4-13 A nonprismatic bar ABC of length L is held between rigid supports (see figure). The left half of the bar has cross-sectional area A_1 and the right half has cross-sectional area A_2. The modulus of elasticity is E and the coefficient of thermal expansion is α. Assuming that the bar is subjected to a uniform temperature change ΔT (positive ΔT means an increase in temperature) and that $A_2 > A_1$, derive expressions for: (a) the largest axial stress σ in the bar (positive means compressive stress), and (b) the displacement δ of point B (positive means the displacement is toward the left).

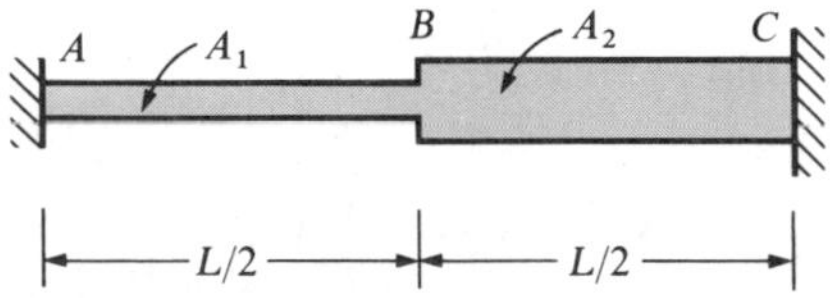

Prob. 2.4-13

2.4-14 A rigid steel plate is supported by three posts of high-strength concrete each having 200 mm × 200 mm square cross section and length $L = 2$ m (see figure). Before the load P is applied, the middle post is shorter than the others by an amount $s = 1.0$ mm. Determine the maximum permissible load P if the modulus of elasticity of the concrete is $E_c = 30$ GPa and the allowable stress in compression is $\sigma_{allow} = 20$ MPa.

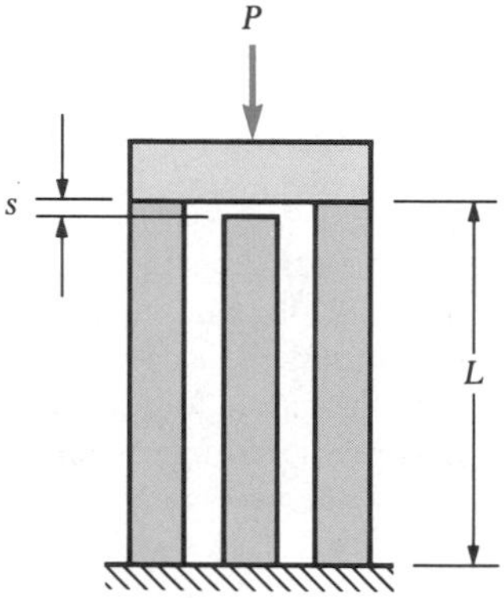

Prob. 2.4-14

2.4-15 The bimetallic thermal control shown in the figure is made of a brass bar (length $L_b = 0.75$ in. and cross-sectional area $A_b = 0.10$ in.^{2}) and a magnesium bar (length $L_m = 1.30$ in. and cross-sectional area $A_m = 0.20$ in.^{2}). The two bars are arranged so that the gap between their free ends is $\delta = 0.0050$ in. at room temperature. Calculate the following quantities: (a) the temperature increase ΔT (above room temperature) at which the two bars come into contact, and (b) the stress σ in the magnesium bar when the temperature increase ΔT is 300 °F. (Use the following material properties: $\alpha_b = 10 \times 10^{-6}$/°F, $\alpha_m = 14.5 \times 10^{-6}$/°F, $E_b = 15 \times 10^6$ psi, and $E_m = 6.5 \times 10^6$ psi.)

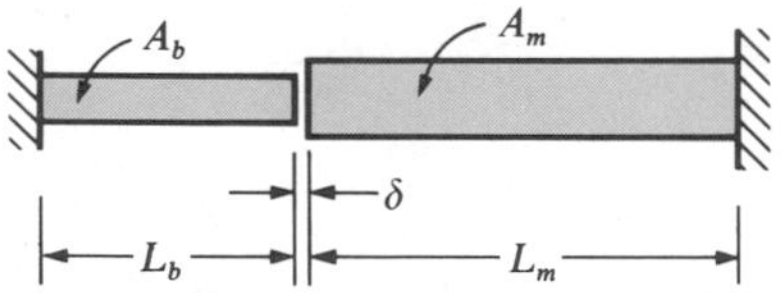

Prob. 2.4-15

2.4-16 A brass sleeve is fitted over a steel bolt (see figure), and the nut is tightened until it is just snug. The bolt has a diameter of 25 mm, and the sleeve has inside and outside diameters of 26 mm and 36 mm, respectively. Calculate the temperature rise ΔT that is required to produce a stress in the sleeve of 25 MPa compression. (Use material properties as follows: for brass, $\alpha_b = 20 \times 10^{-6}$/°C and $E_b = 100$ GPa; for steel, $\alpha_s = 12 \times 10^{-6}$/°C and $E_s = 200$ GPa.)

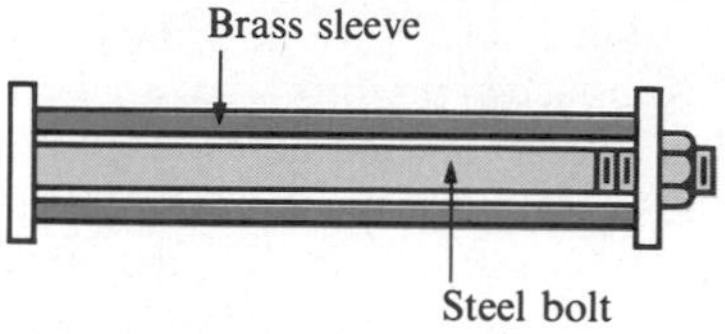

Prob. 2.4-16

2.4-17 A solid circular bar of aluminum fits inside a tube of copper having the same length (see figure). The outside diameter of the copper tube is 2 in., the inside diameter is 1.80 in., and the diameter of the aluminum bar is 1.75 in. At each end of the assembly, a metal pin of $\frac{1}{4}$ in. diameter goes through both bars at right angles to the axis. Find the average shear stress in the pins if the temperature is raised 30 °F. (For aluminum, $E_a = 10 \times 10^6$ psi and $\alpha_a = 13 \times 10^{-6}$/°F; for copper, $E_c = 17 \times 10^6$ psi and $\alpha_c = 9.3 \times 10^{-6}$/°F.)

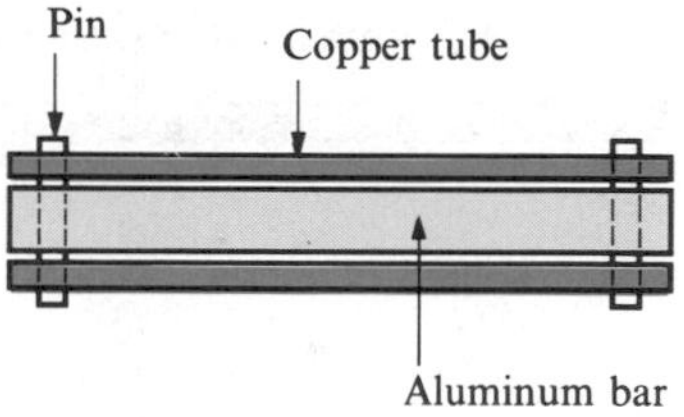

Prob. 2.4-17

2.4-18 A rigid triangular plate is pivoted at C and held by two identical horizontal wires at points A and B (see figure). Each wire has axial rigidity $EA = 540$ kN and coefficient of

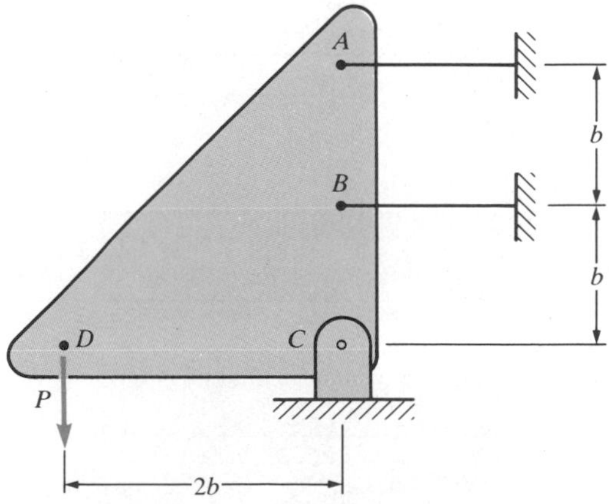

Prob. 2.4-18

thermal expansion $\alpha = 23 \times 10^{-6}/°C$. (a) If a vertical force $P = 2000$ N is applied at point D, what are the tensile forces T_a and T_b in the wires at A and B, respectively? (b) If, while the load P is acting, both wires have their temperatures raised uniformly by 130 °C, what are the forces T_a and T_b? (c) What increase ΔT in temperature will cause the wire at B to become slack?

2.4-19 What stresses will be produced in a steel bolt and a copper tube (see figure) by a quarter turn of the nut if the length of the tube is $L = 20$ in., the pitch of the threads is $p = \frac{1}{8}$ in., the area of the cross section of the bolt is $A_s = 1$ in.^{2}, and the area of the cross section of the tube is $A_c = 2$ in.^{2}? (Assume $E_s = 30 \times 10^6$ psi and $E_c = 16 \times 10^6$ psi.) (Note: The pitch of the threads is the same as the distance advanced by the nut in one complete turn.)

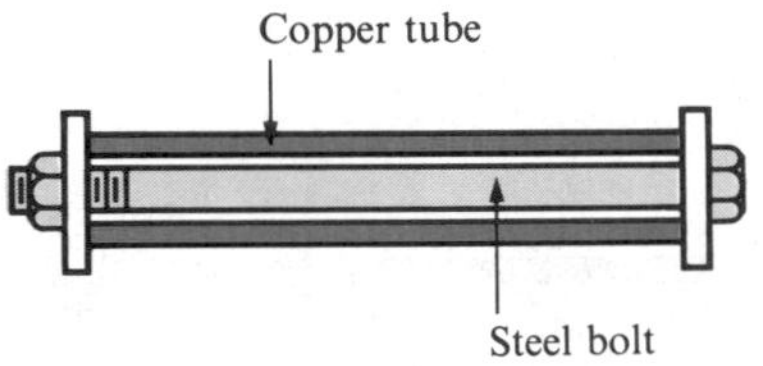

Prob. 2.4-19

2.4-20 Prestressed concrete beams are sometimes manufactured in the following manner. High-strength steel wires are stretched by a jacking mechanism that applies a force Q, as represented in part (a) of the figure. Concrete is then poured around the wires to form a beam, as shown in part (b). After the concrete sets properly, the jacks are released and the force Q is removed (see last part of figure). Thus, the beam is left in a prestressed condition, with the wires in tension and the concrete in compression. Let us assume that the prestressing force Q produces in the steel wires an initial stress $\sigma_0 = 760$ MPa. If the moduli of elasticity of the steel and concrete are in the ratio 8:1 and the cross-sectional areas are in the ratio 1:30, what are the final stresses σ_s and σ_c in the two materials?

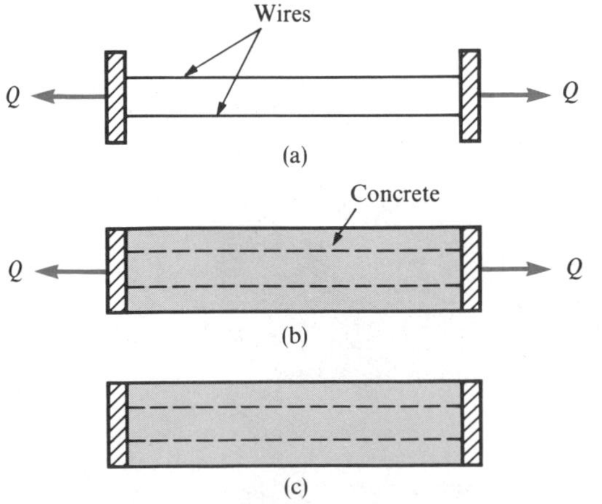

Prob. 2.4-20

2.4-21 A vertical cable AB (see figure) is prestressed to an initial tension of 1200 lb. Subsequently a weight $W = 1600$ lb is suspended from the cable at a height h above the base. Investigate the tensile forces P_a and P_b in the two parts of the cable as h varies from zero to L. (Note that the cable cannot sustain a compressive force.)

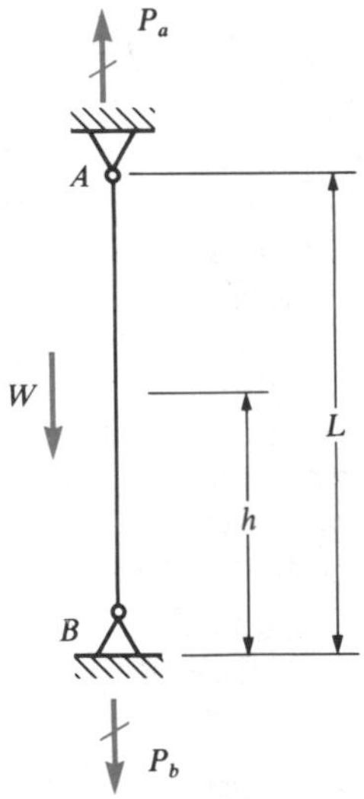

Prob. 2.4-21

2.4-22 A bimetallic bar consisting of a copper core securely bonded to two steel strips (see figure) is heated uniformly by an amount ΔT. Assuming that the width of the bar is b, the length is L, and the thickness of each layer is t, determine

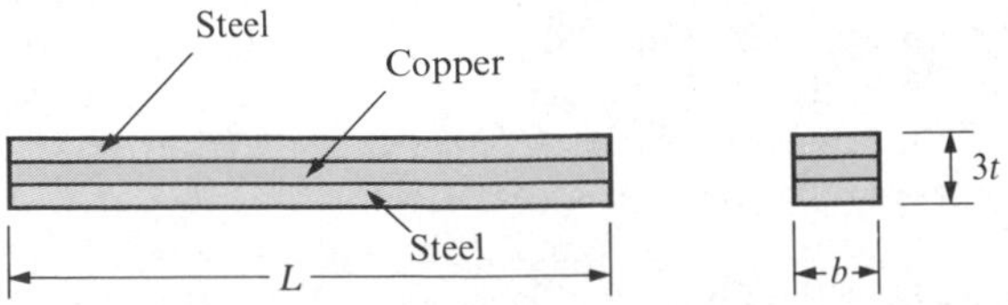

Prob. 2.4-22

the stresses σ_s and σ_c in the steel and copper, respectively. Also, draw free-body diagrams of each of the three strips. (Note: Coefficients of thermal expansion for steel and copper are α_s and α_c, respectively, and $\alpha_c > \alpha_s$. Moduli of elasticity are E_s and E_c.)

2.4-23 Two 20 ft long ropes are available for jointly supporting an object weighing $W = 1000$ lb (see figure). It is desired to have each rope support one-half of the load. The ropes are not the same size, and when tested individually with a 200 lb weight, rope A stretched 2 in. and rope B stretched 3 in. Which rope should be shorter, and by what amount, in order to support the load in the desired manner?

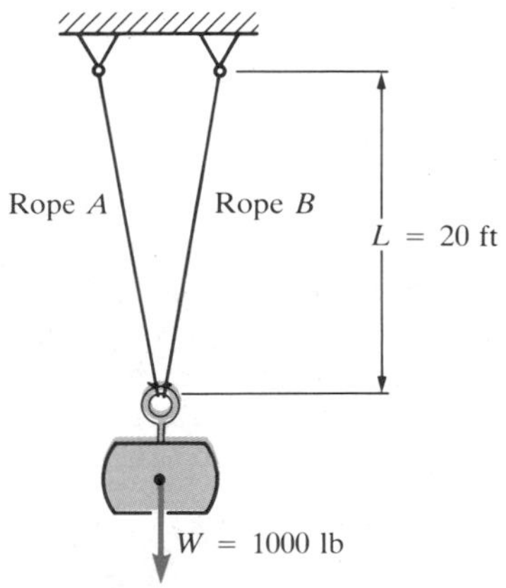

Prob. 2.4-23

2.4-24 A plastic cylinder P is held between a rigid plate and a foundation (see figure). Determine the compressive stress σ_p in the plastic when the nuts on the steel bolts S are tightened by one complete turn. Data for the assembly are as follows: length $L = 250$ mm, pitch of the bolt threads $p = 1.2$ mm, modulus of elasticity for steel $E_s = 200$ GPa, modulus of elasticity for the plastic $E_p = 7.5$ GPa, cross-sectional area of one bolt $A_s = 36.0\ \text{mm}^2$, and cross-sectional area of the plastic cylinder $A_p = 960\ \text{mm}^2$.

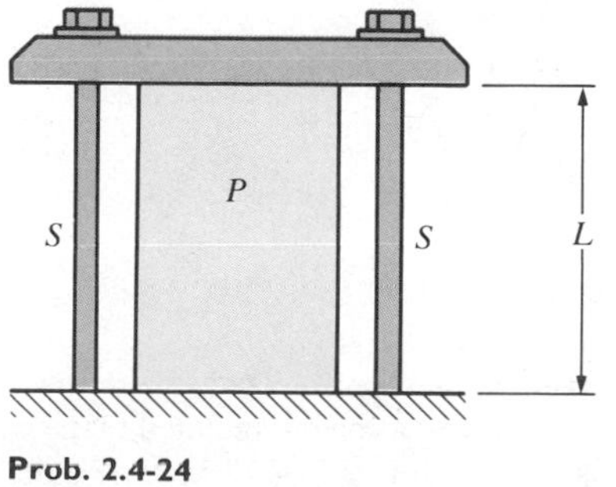

Prob. 2.4-24

2.5-1 A mild-steel tensile-test specimen shows an elongation reading of 0.00200 in. over a gage length of 2 in. Calculate the maximum shear stress τ_{max} in the material, assuming that $E = 29 \times 10^6$ psi.

2.5-2 A concrete test cylinder (see figure) having diameter $d = 150$ mm is subjected to axial compressive forces P in a testing machine. The ultimate shear stress for the concrete is 10 MPa and the ultimate normal stress in compression is 24 MPa. What axial load P is required to break the cylinder?

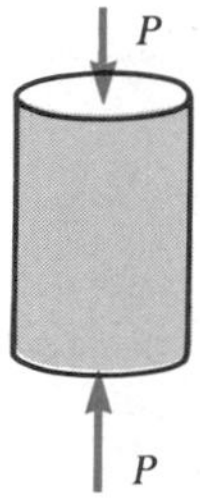

Prob. 2.5-2

2.5-3 Find the allowable tensile load P on a steel bar of 2 in. × 2 in. square cross section (see figure) if the allowable tensile stress is 18,000 psi and the allowable shear stress is 11,000 psi.

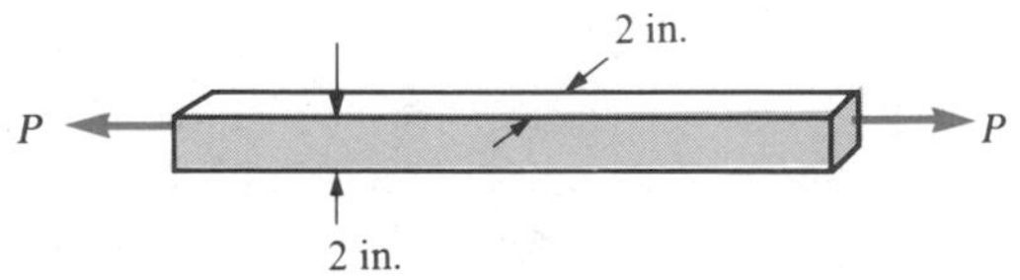

Prob. 2.5-3

2.5-4 A circular steel rod is to carry a tensile load $P = 140$ kN. The allowable stresses in tension and shear are $\sigma_{allow} = 120$ MPa and $\tau_{allow} = 55$ MPa, respectively. What is the minimum required diameter d of the rod?

2.5-5 A prismatic steel bar of 3 in. × 3 in. square cross section is subjected to a compressive load $P = 135$ kips (see figure on p. 142). Determine the normal and shear stresses on all faces of an element oriented at: (a) an angle $\theta = 45°$, and (b) an angle $\theta = 35°$.

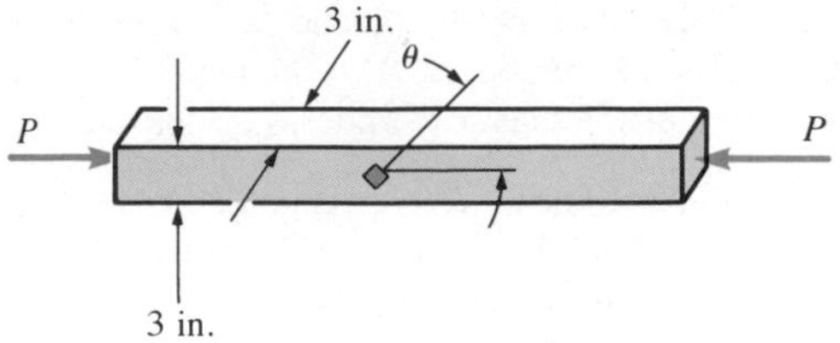

Prob. 2.5-5

2.5-6 A prismatic bar in tension has cross-sectional area $A = 1600\ \text{mm}^2$ and carries a tensile load $P = 160$ kN (see figure). Determine the stresses acting on all faces of an element oriented at: (a) an angle $\theta = 45°$, and (b) an angle $\theta = 65°$.

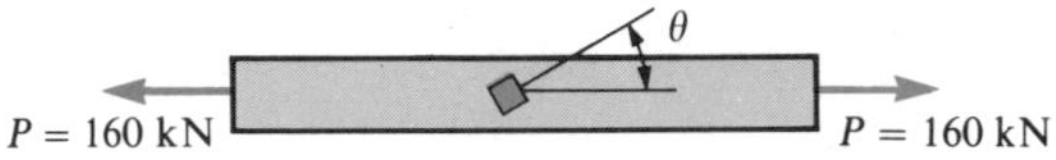

Prob. 2.5-6

2.5-7 A brass wire of diameter $d = \frac{1}{16}$ in. is tightly stretched between fixed points so that it is under a tensile force $T = 32$ lb (see figure). What is the maximum permissible temperature drop ΔT if the allowable shear stress in the wire is 10 ksi? (The coefficient of thermal expansion for the wire is $\alpha_b = 10.6 \times 10^{-6}/°\text{F}$ and the modulus of elasticity is $E_b = 15 \times 10^6$ psi.)

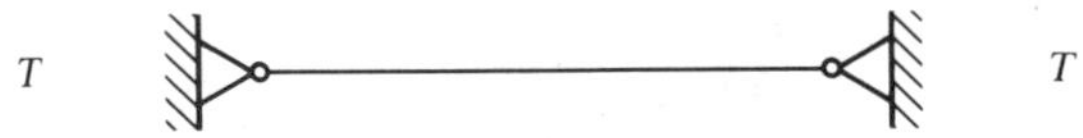

Prob. 2.5-7

2.5-8 Two boards are glued together along a scarf joint as shown in the figure. The angle between the plane of the joint and the faces of the boards is α. Under a tensile axial load, it is found that the average normal stress in the boards is $\sigma_1 = 18$ MPa. (a) What are the normal and shear stresses acting on the glued joint if $\alpha = 25°$? (b) If it is desired that the normal and shear stresses acting on the joint be in the ratio 6:5, what should be the angle α?

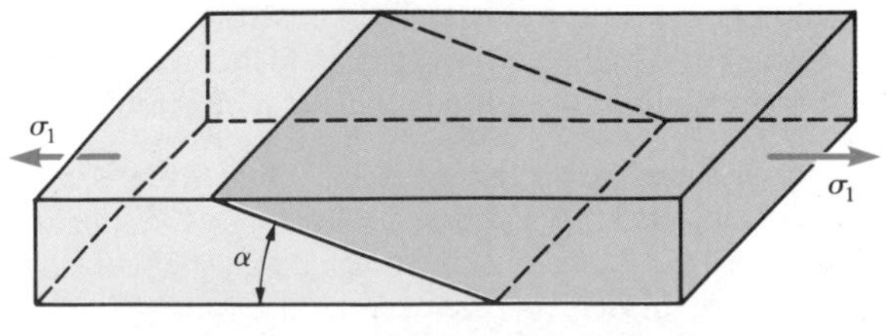

Prob. 2.5-8

2.5-9 A plastic bar fits between rigid supports at room temperature (68 °F) as shown in the figure. When the temperature of the bar is raised to 160 °F, the compressive normal stress on an inclined plane pq is found to be 1700 psi. What is the shear stress τ on plane pq? (Assume $\alpha = 60 \times 10^{-6}/°\text{F}$ and $E = 450 \times 10^3$ psi.)

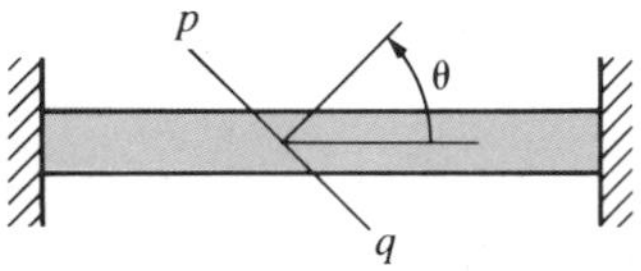

Probs. 2.5-9 and 2.5-10

2.5-10 A copper bar is held snugly between rigid supports (see figure). The allowable stresses on the inclined plane pq, for which $\theta = 55°$, are specified as 60 MPa in compression and 30 MPa in shear. What is the maximum permissible temperature rise ΔT if the allowable stresses are not to be exceeded? (Assume $\alpha = 17 \times 10^{-6}/°\text{C}$ and $E = 120$ GPa.)

2.5-11 A compression bar having a square cross section of width b is formed of two pieces of plastic glued along an inclined plane pq at an angle $\alpha = 40°$ (see figure). The allowable stresses on the glued joint are 1200 psi in compression and 1000 psi in shear, and the allowable stresses in the plastic are 2400 psi in compression and 1400 psi in shear. If the load $P = 32{,}000$ lb, what is the minimum required width $b_{\min}$ of the bar?

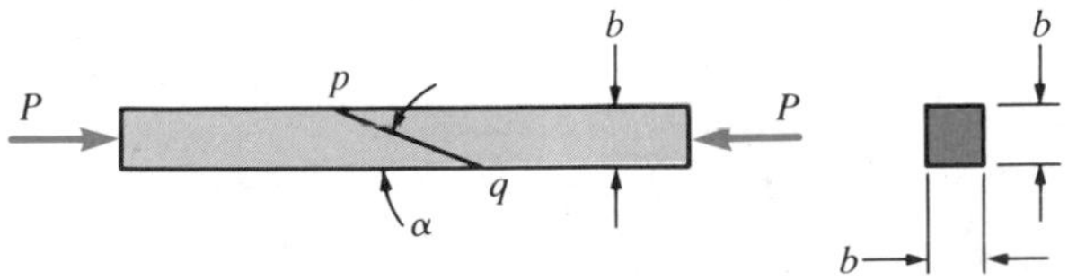

Prob. 2.5-11

2.5-12 A prismatic bar is subjected to a tensile force P (see figure). The stresses on a certain inclined plane pq are found to be $\sigma_\theta = 81$ MPa and $\tau_\theta = -27$ MPa. Find the normal stress σ_θ and shear stress τ_θ on an inclined plane for which $\theta = 30°$.

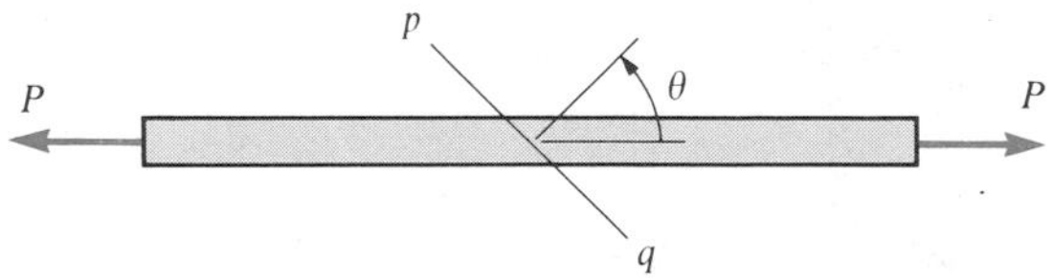

Prob. 2.5-12

2.5-13 Acting on the sides of an element cut from a bar in uniaxial stress are normal stresses of 12,000 psi and 6,000 psi (see figure). Determine the angle θ and the shear stress τ_θ. Also, determine the maximum normal stress σ_{max} and the maximum shear stress τ_{max} in the material.

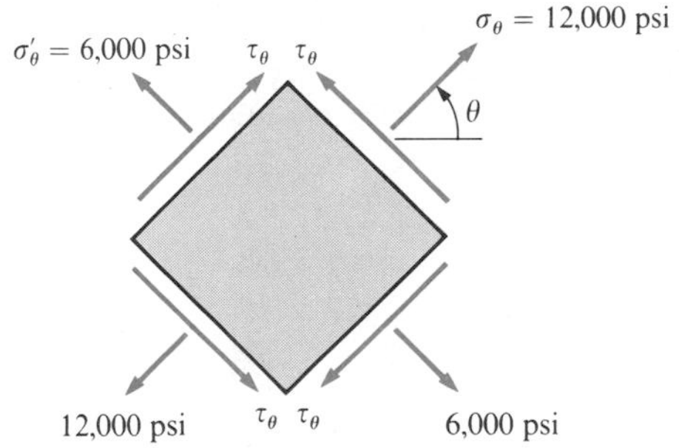

Prob. 2.5-13

2.5-14 A prismatic bar is subjected to an axial force, resulting in a compressive stress of 60 MPa on a plane at an angle $\theta = 30°$ (see figure). What are the stresses on an element rotated through an angle of 40°?

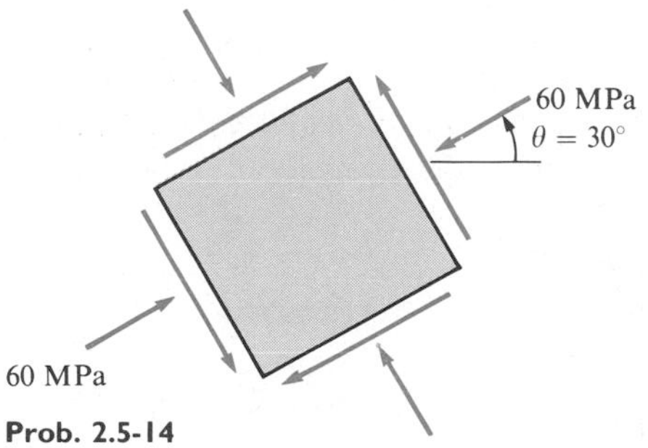

Prob. 2.5-14

***2.5-15** A tension member is to be made of two pieces of material glued together along the plane pq (see figure). For practical reasons, the angle θ is limited to the range of 0° to 70°. The allowable stress on the glued joint in shear is one-half of the allowable stress on the glued joint in tension. What should be the value of the angle θ in order that the bar will carry the greatest load P? (Assume that the strength of the glued joint controls the design.)

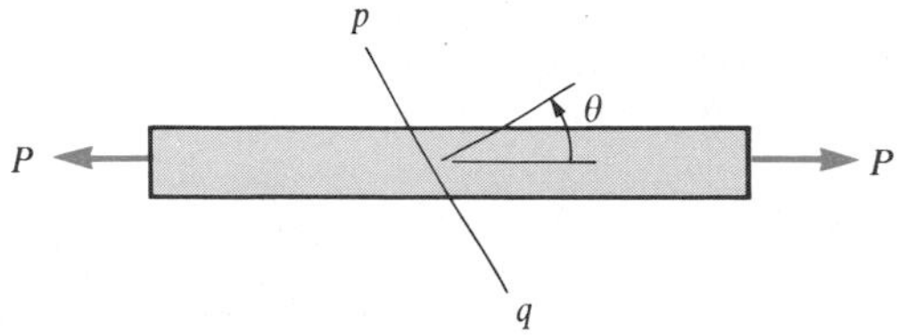

Probs. 2.5-15 and 2.5-16

***2.5-16** Solve the preceding problem if the allowable stresses on the glued joint in tension and shear are 15 MPa and 9 MPa, respectively, and the angle θ is limited to the range 0° to 50°. Also, determine the maximum allowable load P_{max} if the cross-sectional area of the bar is 900 mm^2.

2.6-1 A tensile force P acts on a bar having two different cross-sectional areas A and $4A$, as shown in the figure. (a) Derive a formula for the strain energy U stored in the bar, assuming E is the modulus of elasticity of the material. (b) Calculate the amount of strain energy for the following data: $P = 6000$ lb, $L = 20$ in., $E = 30 \times 10^6$ psi, and $A = 2$ in.^{2}

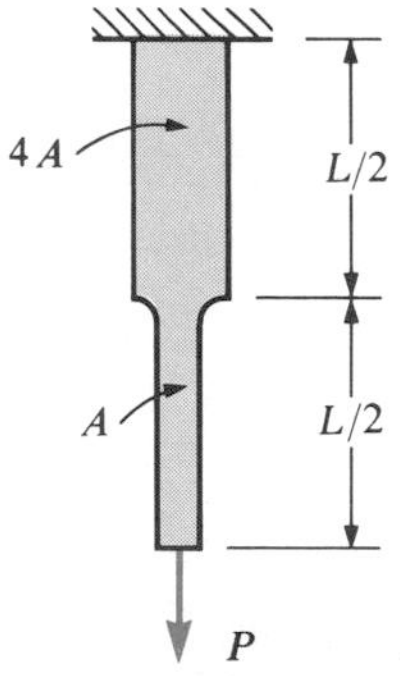

Prob. 2.6-1

2.6-2 A prismatic bar of length L and cross-sectional area A is subjected to loads P, $2P$, and $3P$, acting as shown in the figure on p. 144. (a) Derive a formula for the strain energy U stored in the bar. (b) Calculate the strain energy U for $P = 21$ kN, $L = 1.5$ m, $E = 200$ GPa, and $A = 882$ mm^2.

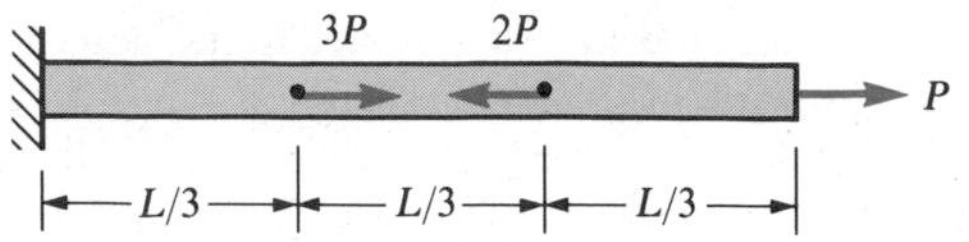

Prob. 2.6-2

2.6-3 Compute the strain energy per unit volume (psi) and the strain energy per unit weight (in.) that can be stored in each of the following materials without exceeding the proportional limit (see table).

Data for Problem 2.6-3

Material	Specific Weight γ (lb/in.^{3})	Modulus of Elasticity E (ksi)	Proportional Limit σ_{pl} (psi)
Mild steel	0.284	30,000	36,000
Tool steel	0.284	30,000	120,000
Aluminum	0.0984	10,500	50,000
Rubber (soft)	0.040	0.300	200

2.6-4 The bar ABC shown in the figure is loaded by a force P acting at end C and by a force Q acting at the midpoint B. The bar has constant axial rigidity EA. (a) Determine the strain energy U_1 of the bar when the force P acts alone ($Q = 0$). (b) Determine the strain energy U_2 when the force Q acts alone ($P = 0$). (c) Determine the strain energy U when the loads P and Q act simultaneously upon the bar.

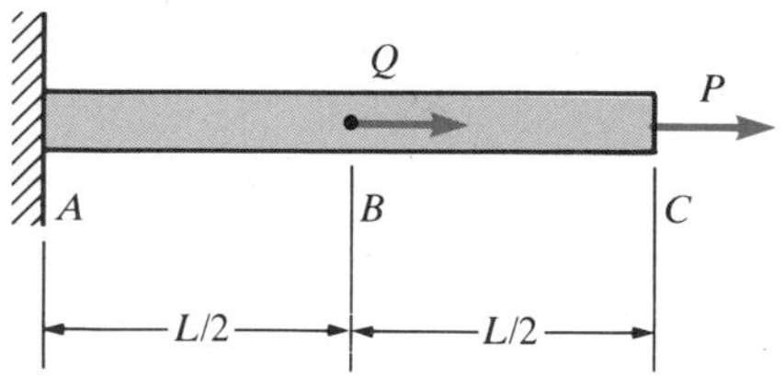

Prob. 2.6-4

2.6-5 The statically indeterminate structure shown in the figure consists of a massive bar AB, assumed to be rigid, that is supported by five equally spaced cables. The cables have various lengths (b, $2b$, and $3b$), but the same axial rigidity EA. (a) Determine the strain energy U of the cables in terms of the displacement δ of bar AB by using the equation $U = EA\delta^2/2L$ for the strain energy. (b) Determine the displacement δ by equating the strain energy of the structure to the work done by the load P.

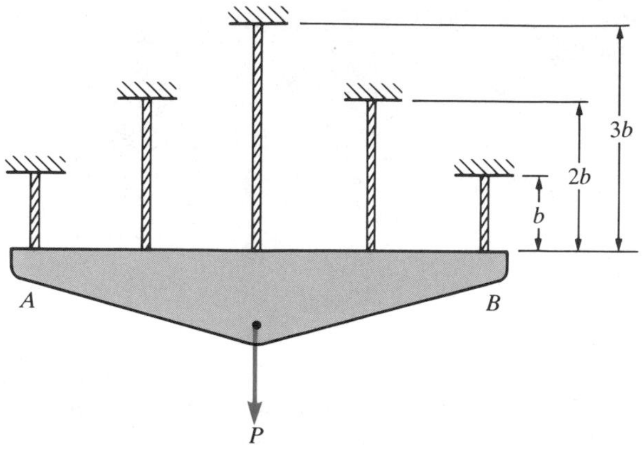

Prob. 2.6-5

2.6-6 A conical bar of diameter d at the support and length L hangs vertically under its own weight (see figure). Derive a formula for the strain energy U of the bar. (Let γ = specific weight and E = modulus of elasticity of the material.)

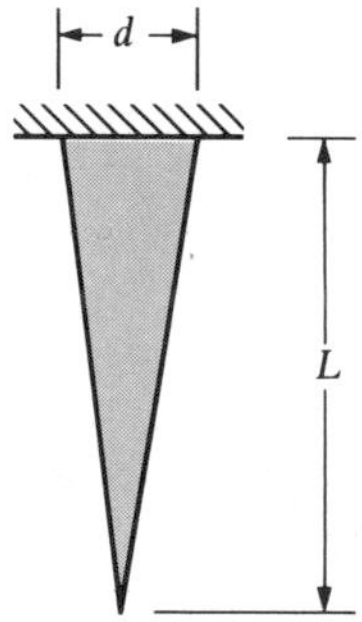

Prob. 2.6-6

2.6-7 A uniformly tapered bar AB of circular cross section is subjected to a load P at the free end, as shown in the figure. The diameters at the ends are d_1 and d_2, the length is L, and the modulus of elasticity is E. (a) Derive a formula

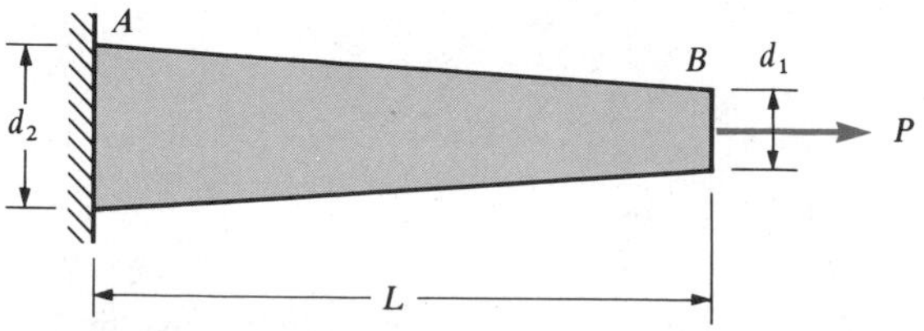

Prob. 2.6-7

for the strain energy U of the bar. (b) Determine the elongation δ of the bar due to the load P.

2.6-8 A flat, uniformly tapered bar AB of constant thickness t and length L is acted upon by a force P (see figure). The width of the bar at the support is b_2 and at the loaded end is b_1. (a) Determine the strain energy U in the bar. (b) Determine the elongation δ of the bar.

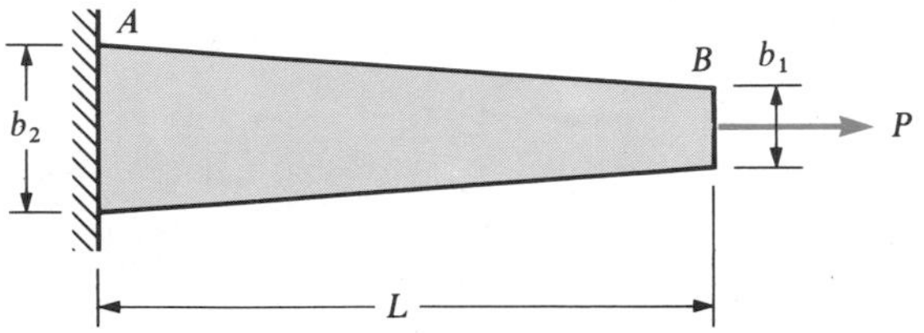

Prob. 2.6-8

2.6-9 A rotating bar ACB has constant angular speed ω about an axis through C (see figure). Determine the strain energy U stored in the bar due to the centrifugal effects. (Let L = length of each arm of the bar, A = cross-sectional area, E = modulus of elasticity, and ρ = mass density.)

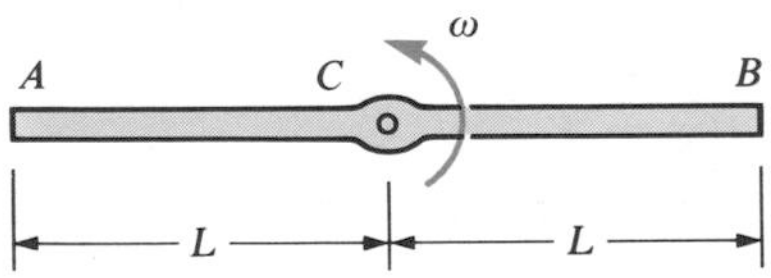

Prob. 2.6-9

2.6-10 The truss ABC shown in the figure supports a horizontal load P_1 and a vertical load P_2. The bars of the truss have the same axial rigidity EA. (a) What is the strain energy U_1 of the truss when $P_2 = 0$ and only P_1 is acting? (b) What is the strain energy U_2 when $P_1 = 0$ and only P_2 is acting? (c) What is the strain energy U when both P_1 and P_2 act simultaneously?

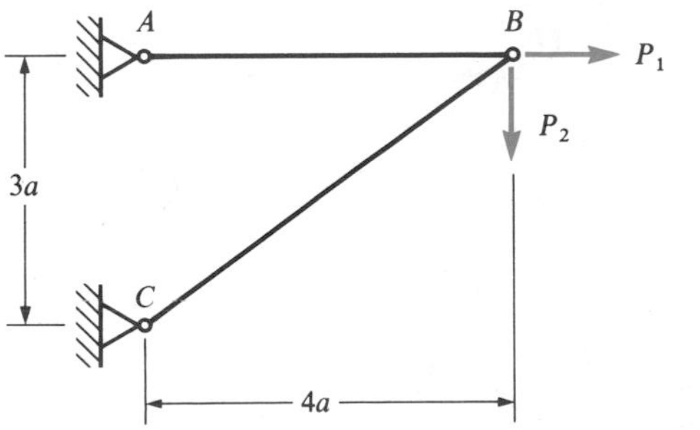

Prob. 2.6-10

2.6-11 A compressive load P is transmitted through a rigid plate to three bars that are identical except that initially the central bar is slightly shorter than the other two bars (see figure). The dimensions and properties of the assembly are as follows: length $L = 3$ ft, cross-sectional area of each bar $A = 5$ in.^{2}, modulus of elasticity $E = 6.5 \times 10^6$ psi, and the gap $s = 0.04$ in. (a) Calculate the load P required to close the gap. (b) Calculate the total downward displacement δ of the rigid plate when $P = 100$ k. (c) Determine the total strain energy U stored in the three bars when $P = 100$ k. (d) Explain why the strain energy U is *not* equal to $P\delta/2$. (Hint: Draw a load-displacement diagram.)

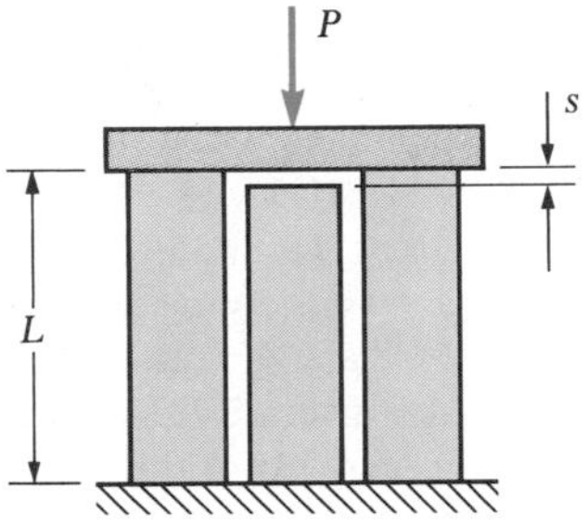

Prob. 2.6-11

2.6-12 Each bar of the plane truss ABC shown in the figure has cross-sectional area A and modulus of elasticity E. (a) Determine the total strain energy U in the truss due to the horizontal load P. (b) Using this expression for the strain energy, determine the horizontal displacement δ of joint B.

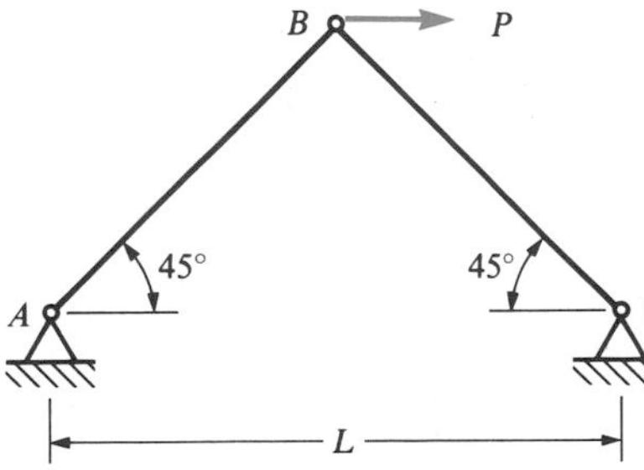

Prob. 2.6-12

2.6-13 All bars of the plane truss shown in the figure have the same axial rigidity EA. (a) Determine the total strain energy U stored in the truss due to the vertical load P. (b) Using this expression for the strain energy, determine the vertical displacement δ of joint D.

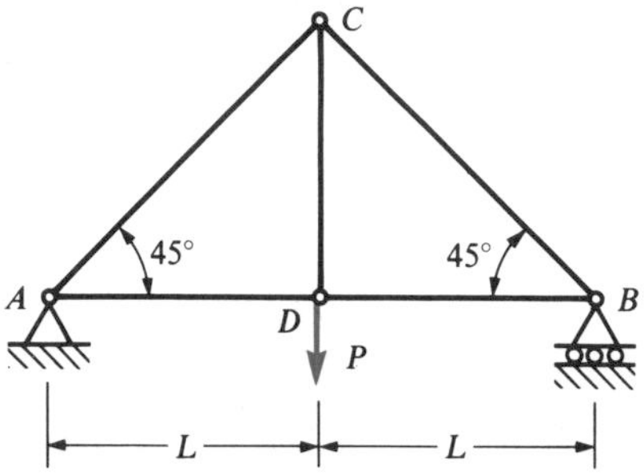

Prob. 2.6-13

2.6-14 A truss ABC supports a vertical load P as shown in the figure. The two bars AB and BC are identical with modulus of elasticity E and cross-sectional area A. The angle β can be varied by changing the lengths of the bars, but joint B must be at a distance L from the vertical wall. (a) Evaluate the strain energy U of the truss. (b) Determine the angle β in order that the strain energy will be a minimum. (c) Determine the corresponding vertical displacement δ of joint B.

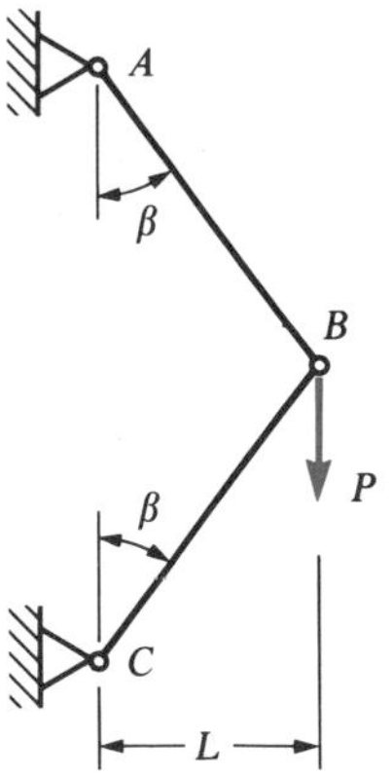

Prob. 2.6-14

***2.6-15** A rigid bar AB of weight W is supported on two springs, each of stiffness k (see figure). A third spring, also of stiffness k, hangs from a support above the bar. The distance between the hook on the upper spring and the eyebolt in the bar is s. A force P is applied to the bar, pushing it upward so that the hook can be inserted in the eyebolt. Then the force P is removed, and the system comes to rest in an equilibrium position. What is the total strain energy U of the springs? (Disregard the weights of the springs in the analysis.)

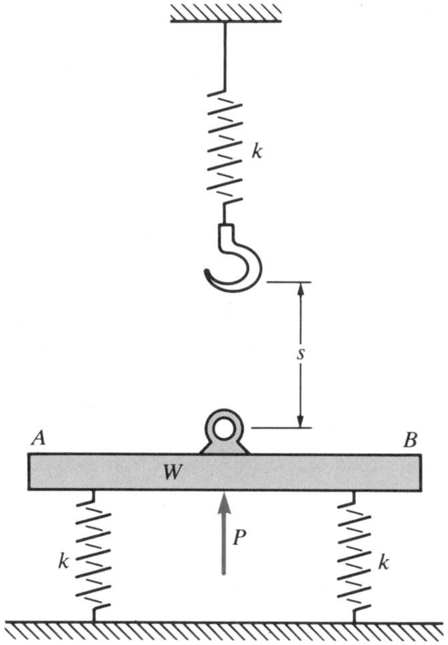

Prob. 2.6-15

****2.6-16** A bungee cord that behaves linearly elastically has an unstressed length $L_0 = 2$ ft and a stiffness $k = 1.0$ lb/in. The cord is attached to two pegs, distance $b = 16$ in. apart, and pulled at its midpoint by a force $P = 30$ lb (see figure). How much strain energy U is stored in the cord? (Note: The elongation of the cord is *not* small compared to its original length.)

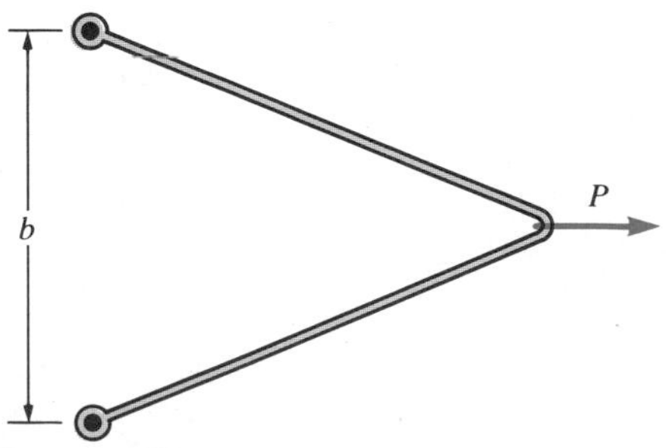

Prob. 2.6-16

The problems for Section 2.7 are to be solved on the basis of the idealizations described in the text. Also, disregard energy losses unless otherwise specified.

2.7-1 A small rubber ball (weight $W = 1$ oz) is attached by a rubber band to a wood paddle (see figure). The natural length of the rubber band is $L = 12$ in., its cross-sectional area is $A = 0.0025$ in.^{2}, and its modulus of elasticity is $E = 300$ psi. What was the velocity v of the ball when it left the paddle (after being struck by the paddle) if the rubber band stretched to a total length $3L = 36$ in. (Assume linearly elastic behavior of the band, and disregard the potential energy due to any change in elevation of the ball.)

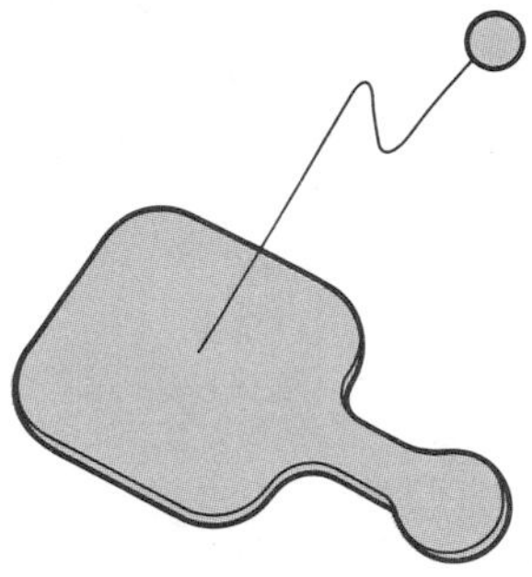

Prob. 2.7-1

2.7-2 A weight W rests on top of a wall and is attached to one end of a very flexible band having cross-sectional area A and modulus of elasticity E (see figure). The other end of the band is attached securely to the wall. The weight is then pushed off the wall and falls freely the full length of the band. (a) Obtain a formula for the impact factor, assuming that the band stretches elastically when it stops the falling weight. (b) Evaluate the impact factor if the weight, when hanging statically, elongates the band by 10% of its original length.

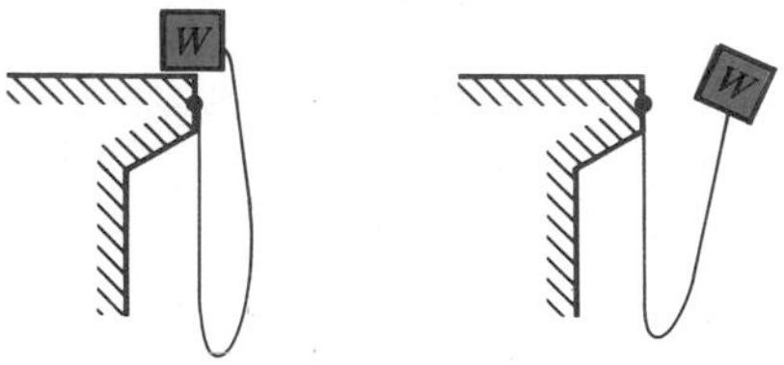

Prob. 2.7-2

2.7-3 A weight $W = 4000$ lb falls from a height $h = 1$ ft onto a vertical wood pole 20 ft long and 12 in. diameter (see figure). (a) Determine the maximum compressive stress σ_{max} in the pole assuming $E = 1.5 \times 10^6$ psi and disregarding energy losses. (b) Determine the maximum stress σ_{max} if the kinetic energy loss factor K_L equals 0.8.

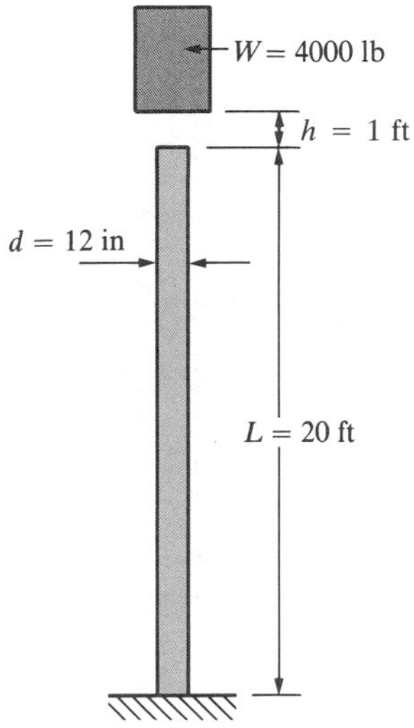

Prob. 2.7-3

2.7-4 A sliding collar of mass $M = 90$ kg falls onto a flange at the bottom of a slender vertical rod (see figure on p. 148). The rod has length $L = 3$ m, cross-sectional area $A = 350$ mm^2, and modulus of elasticity $E = 170$ GPa. Assuming an energy loss factor $K_L = 0.85$, determine the height h through which the mass M should drop to produce a maximum stress in the bar of $\sigma_{max} = 400$ MPa.

2.7-5 Solve the preceding problem if the collar weighs $W = 100$ lb, the length $L = 6$ ft, the cross-sectional area $A = 0.5$ in.^{2}, the modulus of elasticity $E = 30 \times 10^6$ psi, the energy loss factor $K_L = 0.85$, and the maximum stress $\sigma_{max} = 32{,}000$ psi.

2.7-6 The end plate of a cylindrical container is fastened to a flange by six bolts as shown in part (a) of the figure. The grip of the bolts is the distance b_1. Under a time varying load $F(t)$ that does a certain amount of work W in straining the bolts, the bolts develop a maximum tensile stress σ_1. Now suppose that the connection is redesigned with fittings for the bolts that increase the grip to $4b_1$, as shown in part (b) of the figure. What tensile stress σ_2 will now be developed in the bolts by the same time varying load $F(t)$ if

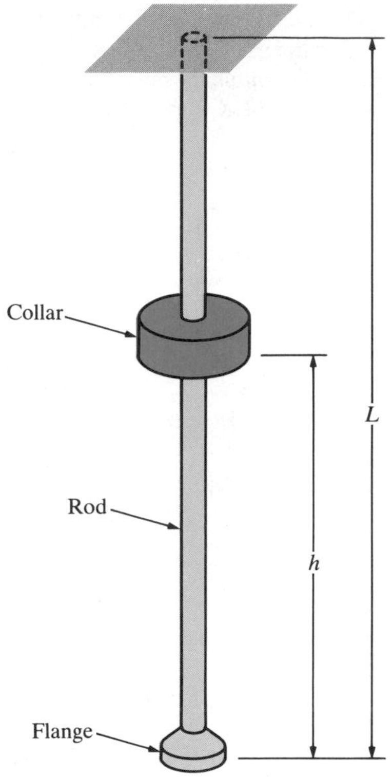

Probs. 2.7-4 and 2.7-5

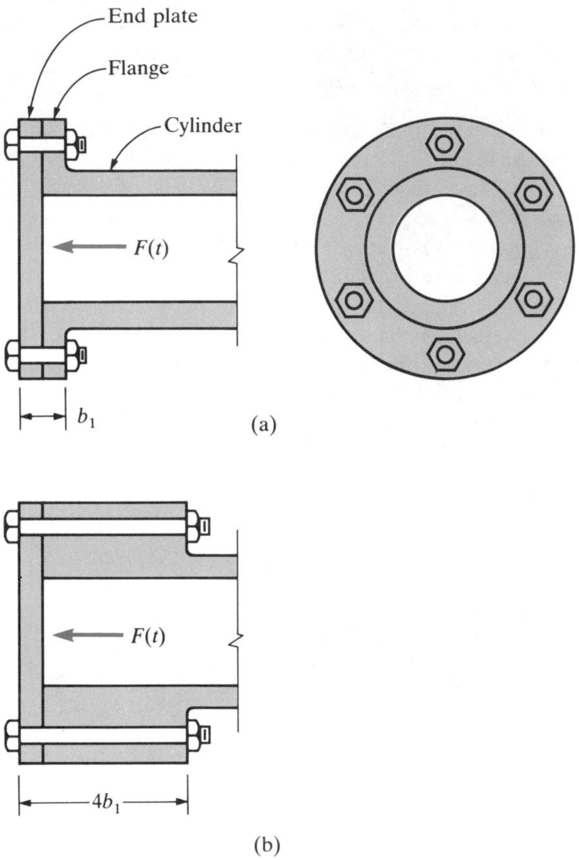

Prob. 2.7-6

the load does the same amount of work W in straining the bolts? (Note: Bolts subjected to dynamic loads often are designed with long grips to take advantage of the more favorable stress conditions.)

2.7-7 A cable with a restrainer at the bottom hangs vertically from its upper end (see figure). The cable has length $L = 50$ ft, effective cross-sectional area $A = 0.065$ in.^{2}, effective modulus of elasticity $E = 20 \times 10^6$ psi, and weighs $w = 0.24$ lb/ft. A slider of weight $W = 100$ lb drops from a height $h = 5$ ft onto the restrainer. Determine the maximum stress σ_{max} in the cable using Eq. (2-56) to estimate the energy losses. (Disregard the mass of the restrainer.)

2.7-8 Solve the preceding problem if the slider has mass $M = 40$ kg, $L = 12$ m, $A = 40$ mm^2, $E = 130$ GPa, $w = 3.5$ N/m, and $h = 2$ m.

2.7-9 A bumper for a mine car is constructed with a spring of stiffness $k = 1000$ lb/in. (see figure). If a car weighing 1500 lb is traveling at a speed $s = 4$ mph when it strikes the spring, what is the maximum shortening δ_{max} of the spring?

2.7-10 A "bungee jumper" having a mass of 50 kg leaps from a bridge, braking her fall with a long elastic shock cord having axial rigidity $EA = 2.1$ kN (see figure). If the jumpoff point is 60 m above the water, and if it is desired to maintain a clearance of 10 m between the jumper and the water, what length L of cord should be used?

2.7-11 A block weighing $W = 1$ lb drops inside a cylinder from a height $h = 12$ in. onto a spring having stiffness $k = 0.5$ lb/in. (see figure). The spring weighs $w = 0.06$ lb/in. and has a natural length $L = 12$ in. Using Eq. (2-56) for the energy loss factor, calculate the maximum shortening δ_{max} of the spring due to the falling weight.

2.7-12 Solve the preceding problem if the block weighs 5 N, $h = 200$ mm, $k = 90$ N/m, $w = 10$ N/m, and $L = 300$ mm.

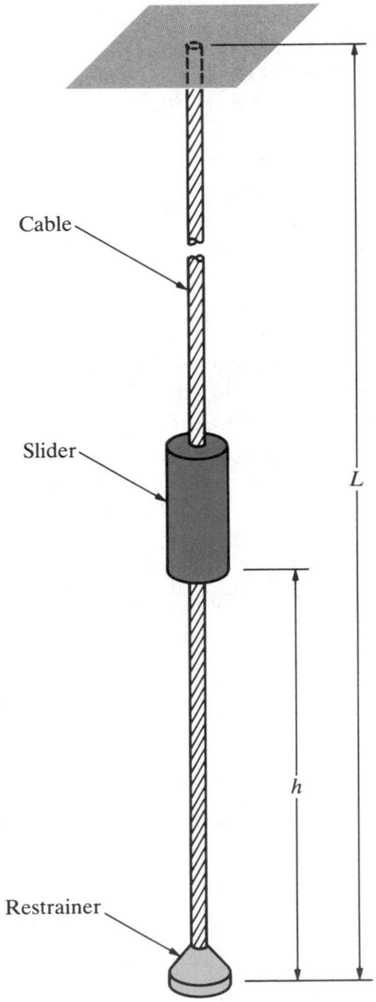

Probs. 2.7-7 and 2.7-8

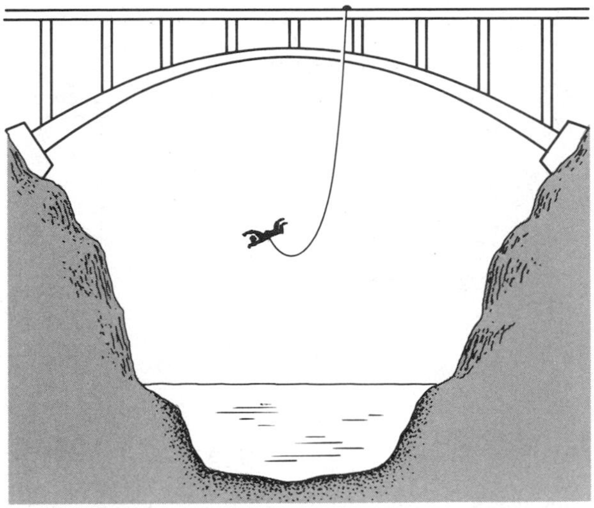

Prob. 2.7-10

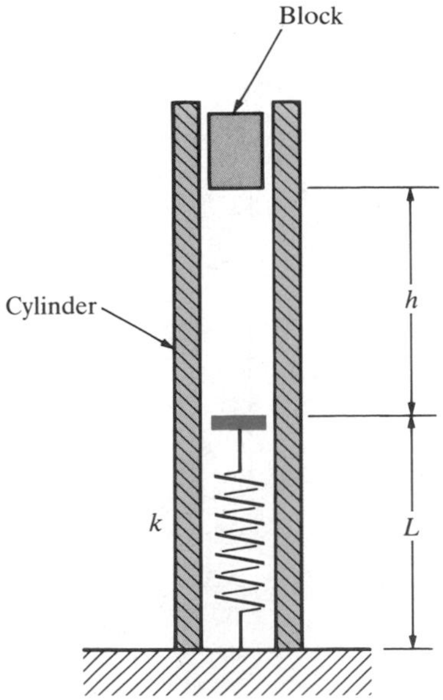

Probs. 2.7-11 and 2.7-12

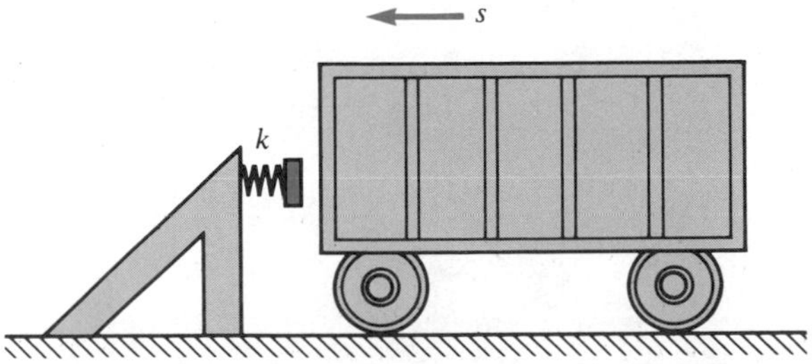

Prob. 2.7-9

2.7-13 A rigid bar AB weighing $W = 2$ lb and having length $L = 20$ in. is hinged at support A and supported at B by a nylon cord BC (see figure on p. 150). The cord has cross-sectional area $A = 0.04$ in.^{2}, length $b = 10$ in., and modulus of elasticity $E = 300$ ksi. If the bar is raised to its maximum height and then released, what is the maximum stress σ_{max} in the cord?

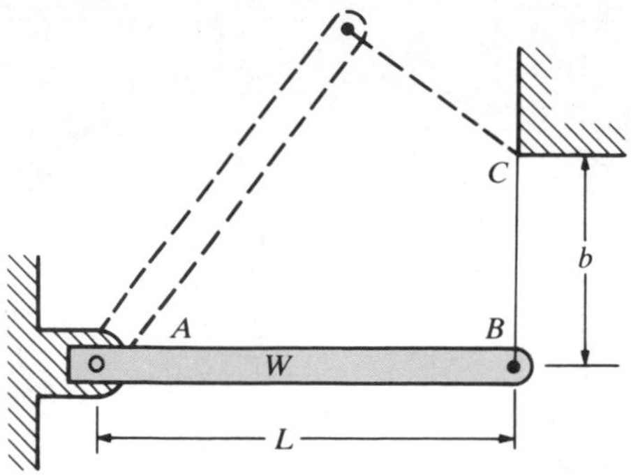

Prob. 2.7-13

2.7-14 A sliding collar of mass M drops from a height h onto the lower end of a nonprismatic bar (see figure). The lower part of the bar has length L_1 and diameter d_1, and the upper part has length L_2 and diameter d_2. The modulus of elasticity of the material is E. Determine the maximum displacement δ_{max} and the maximum tensile stress σ_{max} due to the falling mass if the dimensions and properties are as follows: $M = 5$ kg, $h = 10$ mm, $E = 200$ GPa, $d_1 = 5$ mm, $d_2 = 10$ mm, $L_1 = 300$ mm, and $L_2 = 100$ mm. Assume an energy loss factor K_L equal to 0.85.

2.7-15 Solve the preceding problem if the dimensions and properties are as follows: $M = 0.4$ slugs, $h = 0.5$ in., $E = 29 \times 10^6$ psi, $d_1 = 0.2$ in., $d_2 = 0.4$ in., $L_1 = 12$ in., and $L_2 = 4$ in. Assume an energy loss factor K_L equal to 0.8.

***2.7-16** The prismatic bar shown in the figure has length $L = 2.0$ m, diameter $d = 15$ mm, and modulus of elasticity $E = 200$ GPa. A spring of stiffness $k = 1.2$ MN/m is installed at the end of the bar. A sliding collar of mass $M = 20.0$ kg drops from a height $h = 50$ mm onto the spring. Determine the maximum elongation δ_{max} of the bar and the maximum tensile stress σ_{max} in the bar due to the impact, assuming no energy losses and disregarding the masses of the spring, flange, and bar. Compare your results with those obtained in part (a) of Example 1 of Section 2.7 for this same bar without the spring.

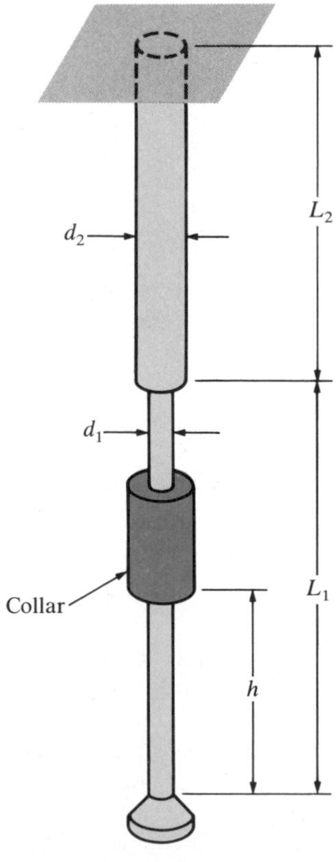

Probs. 2.7-14 and 2.7-15

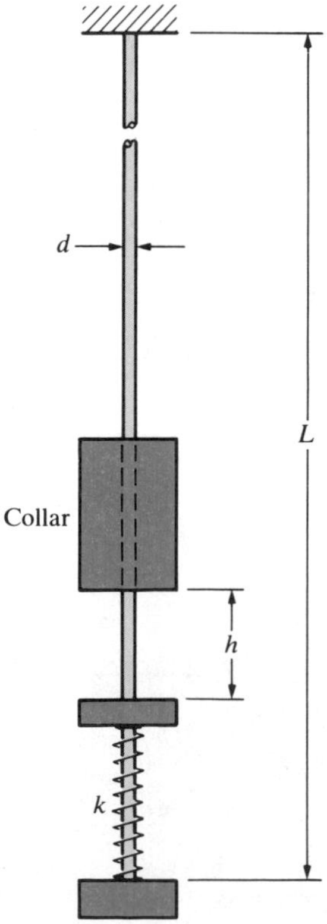

Prob. 2.7-16

2.8-1 A long bar AB of length L and specific weight γ hangs vertically under its own weight (see figure). The stress-strain relation for the material is given by the Ramberg-Osgood equation (Eq. 2-59):

$$\epsilon = \frac{\sigma}{E} + \frac{\sigma_0 \alpha}{E}\left(\frac{\sigma}{\sigma_0}\right)^m$$

Derive the following formula

$$\delta = \frac{\gamma L^2}{2E} + \frac{\sigma_0 \alpha L}{(m+1)E}\left(\frac{\gamma L}{\sigma_0}\right)^m$$

for the elongation of the bar.

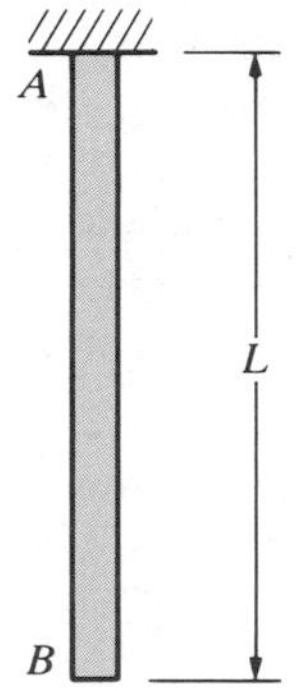

Prob. 2.8-1

2.8-2 A rigid bar AB, pinned at end A, is supported by a wire CD and loaded by a force P at end B (see figure). The wire is made of high-strength steel having modulus of elasticity $E = 210$ GPa and yield stress $\sigma_y = 820$ MPa. The length of the wire is $L = 1$ m and its diameter is $d = 3$ mm. The stress-strain diagram for the steel is defined by the modified power law, as follows:

$$\sigma = E\epsilon \qquad 0 \leq \sigma \leq \sigma_y$$

$$\sigma = \sigma_y\left(\frac{E\epsilon}{\sigma_y}\right)^n \qquad \sigma \geq \sigma_y$$

(a) Assuming $n = 0.2$, calculate the displacement δ_b at the end of the bar due to the load P. Take values of P from 2 to 6 kN in increments of 0.4 kN. (b) Plot a load-displacement diagram showing P versus δ_b.

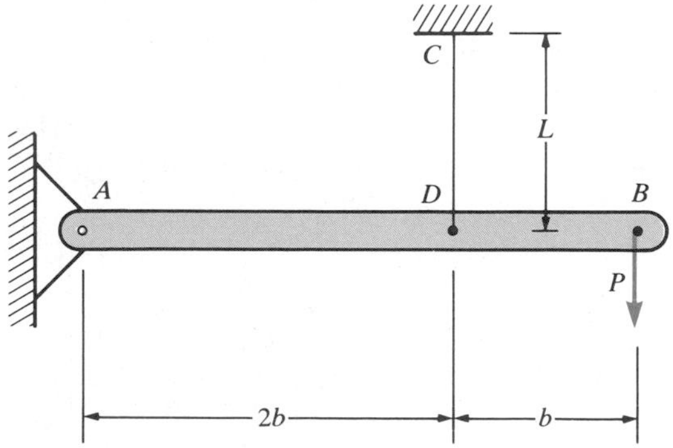

Prob. 2.8-2

2.8-3 A mass $M = 1.6$ slugs is attached to the end of a slender bar that rotates at high angular speed about an axis through point C (see figure). The bar has an effective length $L = 40$ in. and cross-sectional area $A = 0.10$ in.^{2} The material is a magnesium alloy having a stress-strain relation given by the Ramberg-Osgood equation

$$\epsilon = \frac{\sigma}{E} + \frac{\sigma_0 \alpha}{E}\left(\frac{\sigma}{\sigma_0}\right)^m$$

in which $E = 6.5 \times 10^6$ psi, $\sigma_0 = 25{,}000$ psi, $\alpha = 3/7$, and $m = 10$. What is the maximum permissible angular speed n (revolutions per minute) if the elongation of the bar is limited to 0.2 in.? (Disregard the mass of the bar itself.)

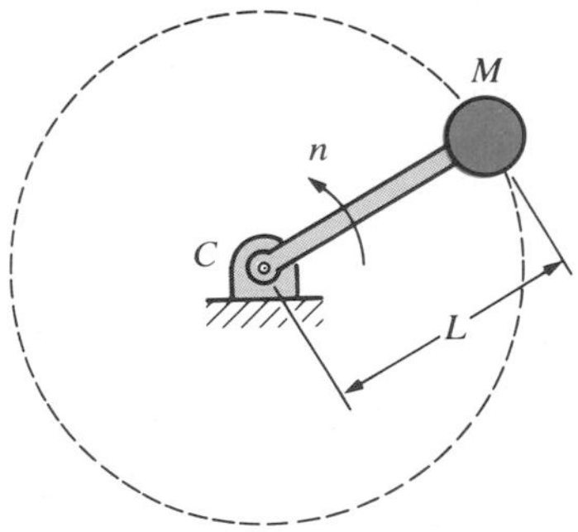

Prob. 2.8-3

2.8-4 A long rod hanging vertically in a well supports a load P at its lower end (see figure). The material has the bilinear stress-strain curve shown in the figure, in which E_1 and E_2 represent the slopes of the two parts of the diagram. Find the elongation δ of the bar due to its own weight and the force P if the unit weight $\gamma = 28\ \text{kN/m}^3$, the cross-sectional area $A = 960\ \text{mm}^2$, the length $L = 360$ m, and the load $P = 92$ kN.

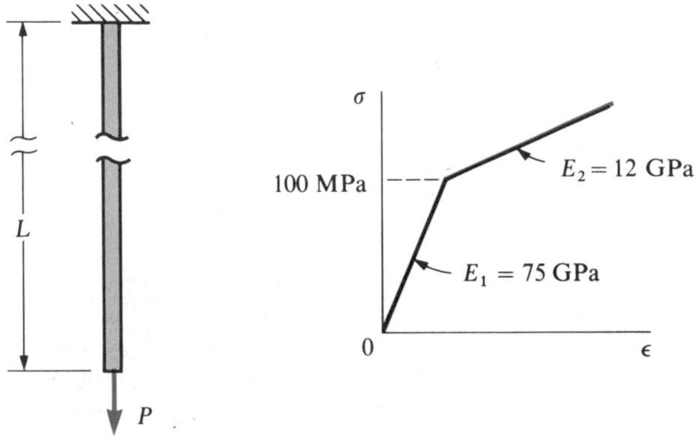

Prob. 2.8-4

***2.8-5** The stress-strain relation for the material of a bar AB hanging under its own weight (see figure) is given by a second-degree function:

$$\sigma = E\epsilon\left(1 - \frac{E\epsilon}{4\sigma_m}\right) \qquad 0 \le \epsilon \le \epsilon_m$$

$$\sigma = \sigma_m \qquad \epsilon \ge \epsilon_m$$

in which E is the initial slope of the stress-strain diagram, σ_m is the maximum stress, and $\epsilon_m = 2\sigma_m/E$ is the strain at which the diagram becomes horizontal. (a) Sketch the stress-strain relations, showing the shape of the diagram and labeling all special values. (b) Assuming that the bar has length L and specific weight γ, derive the following formula for the elongation of the bar:

$$\delta = \frac{2\sigma_m L}{E} - \frac{4\sigma_m^2}{3\gamma E}\left[1 - \left(1 - \frac{\gamma L}{\sigma_m}\right)^{3/2}\right]$$

This formula is valid provided $\sigma < \sigma_m$, which means that $L < \sigma_m/\gamma$. (c) Calculate the elongation δ for a bar of length $L = 2800$ ft and specific weight $\gamma = 170\ \text{lb/ft}^3$. The stress-strain diagram has initial slope $E = 10 \times 10^6$ psi and maximum stress $\sigma_m = 36{,}000$ psi.

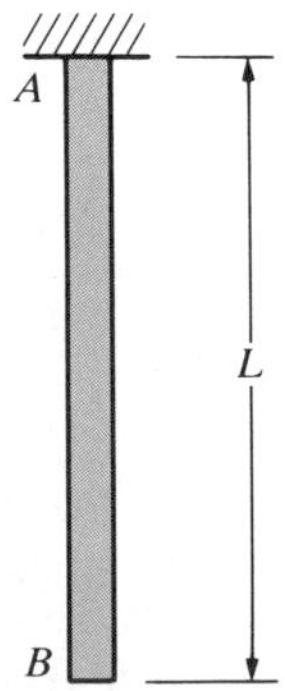

Prob. 2.8-5

Problems 2.8-6 through 2.8-16 are to be solved assuming that the material is steel with an elastic-plastic stress-strain diagram (see Fig. 2-48b). The yield stress is σ_y, the yield strain is ϵ_y, and the modulus of elasticity in the linearly elastic region is E.

2.8-6 A horizontal rigid bar AB carrying a load P is supported by five symmetrically placed wires, each of cross-sectional area A (see figure). (a) Determine the ultimate load P_u if the material of the wires is elastic-plastic with yield stress σ_y. (Disregard the weight of bar AB.) (b) How is P_u changed if bar AB is flexible instead of rigid?

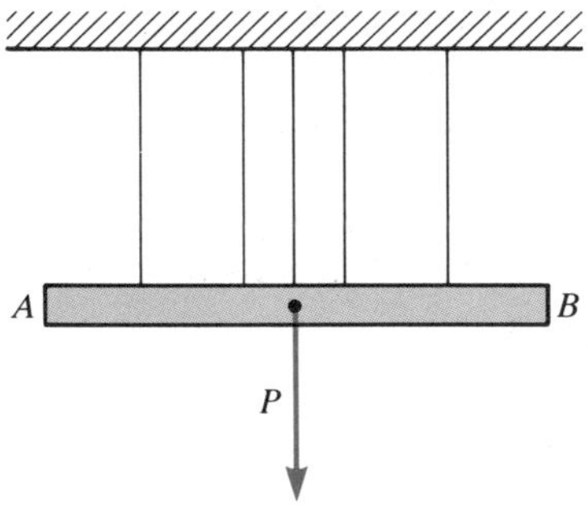

Prob. 2.8-6

2.8-7 A stepped bar having cross-sectional areas A_1 and A_2 is held between rigid supports (see figure). The material is elastic-plastic with yield stress σ_y. An axial load P acts at the change in cross section. What is the ultimate load P_u?

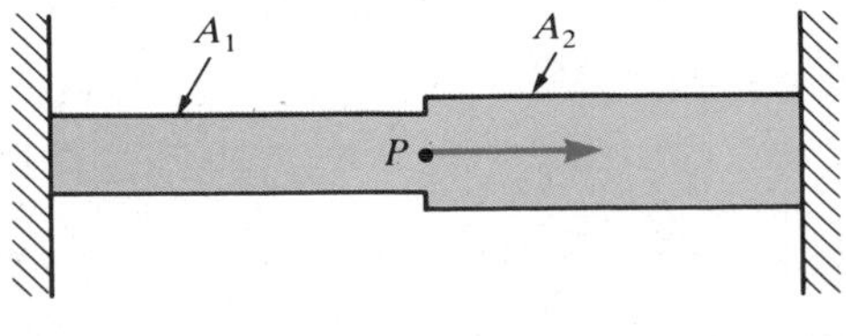

Prob. 2.8-7

2.8-8 A circular steel rod AB of diameter $d = 15$ mm is stretched tightly between two supports so that the tensile stress in the rod is 120 MPa (see figure). An axial force P is then applied to the rod at an intermediate location C. What is the ultimate load P_u if the material is elastic-plastic with yield stress $\sigma_y = 250$ MPa?

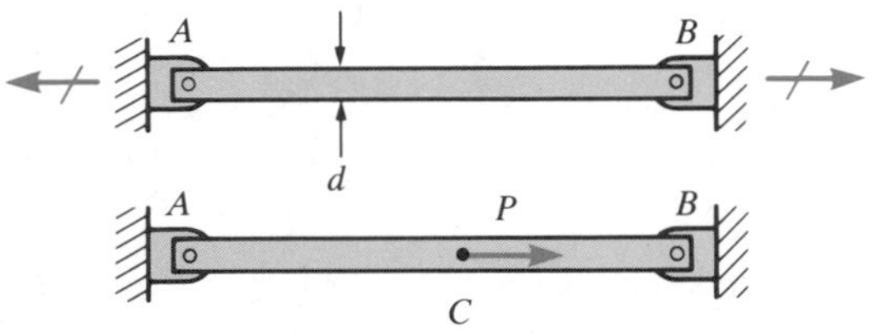

Prob. 2.8-8

2.8-9 Five bars, each having a diameter of 0.4 in., support a load P as shown in the figure. Determine the ultimate load P_u if the material is elastic-plastic with yield stress $\sigma_y = 42{,}000$ psi.

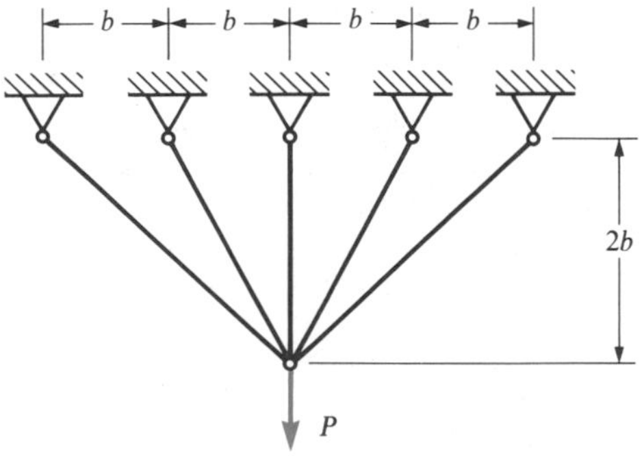

Prob. 2.8-9

2.8-10 A load P is carried by a horizontal beam that is supported by four rods arranged symmetrically as shown in the figure. Each rod has cross-sectional area A and is made of an elastic-plastic material having yield stress σ_y. Find the ultimate load P_u if the angles $\alpha = 62.5°$.

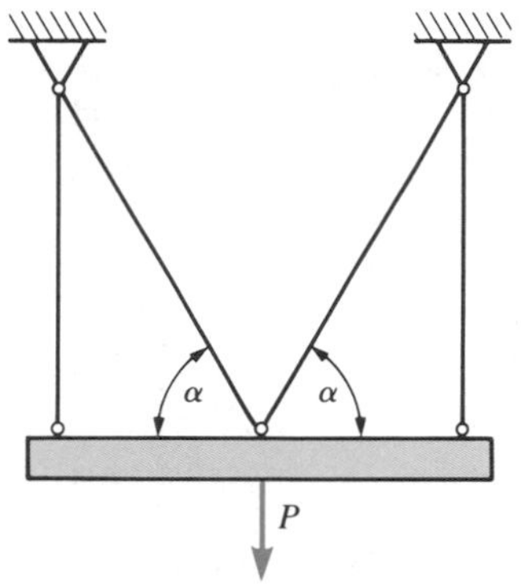

Prob. 2.8-10

2.8-11 A solid circular bar B, which is threaded into end fittings C and D, is surrounded by a circular sleeve A that fits snugly between, but is not attached to, the end fittings (see figure). Both sleeve A and bar B are made of an elastic-plastic material having yield stress σ_y. Sleeve A, which has a larger cross-sectional area than does bar B, is heated until the yield stress σ_y is reached in bar B. External tensile forces P are then applied to the end fittings. What is the ultimate value P_u of the loads P?

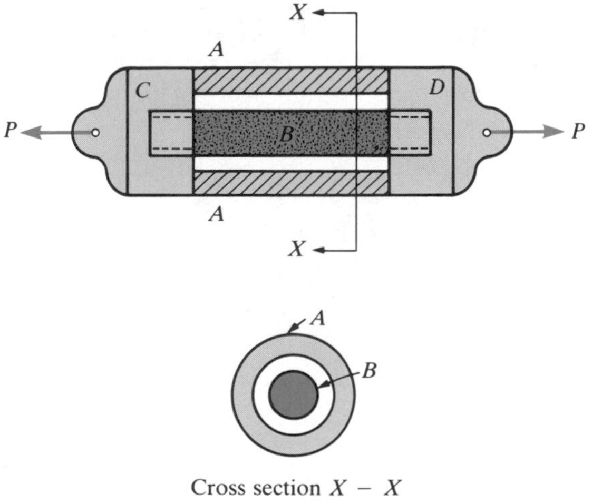

Prob. 2.8-11

2.8-12 The square truss $ABCD$ shown in the figure supports a horizontal load P acting at joint B. Each bar of the

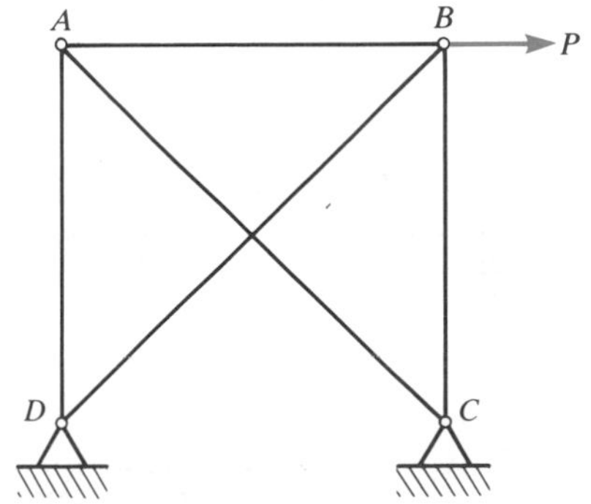

Prob. 2.8-12

truss has cross-sectional area A. The material is elastic-plastic with yield stress σ_y, and the properties are the same in both tension and compression. Buckling of the compression members is prevented by lateral restraints. Determine the ultimate load P_u. (Hint: The ultimate load is reached when the two most highly stressed bars become perfectly plastic. When that happens, the corresponding value of the load P can be calculated from static equilibrium.)

2.8-13 A rigid bar AB is supported on a fulcrum at C and loaded by a force P at end B (see figure). Three identical wires made of an elastic-plastic material (yield stress σ_y and modulus of elasticity E) support the rigid bar. Each wire has cross-sectional area A and length L. Determine the yield load P_y, the corresponding yield displacement δ_y at point B, the ultimate load P_u, and the displacement δ_u of point B when the load just reaches the value P_u. Also, draw the load-displacement diagram showing the force P versus the displacement δ_b of point B.

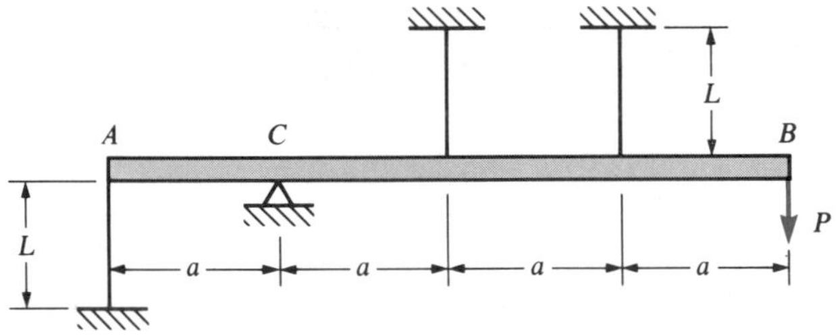

Prob. 2.8-13

2.8-14 The structure shown in the figure consists of a horizontal rigid bar AB supported by two steel wires, one having cross-sectional area A_1 and length L_1 and the other having properties A_2 and L_2. The material of the wires is elastic-plastic with yield stress σ_y and modulus of elasticity E. A vertical load P acts at end B of the bar. (a) Assuming that both wires have the same area ($A_1 = A_2 = A$) and that the lengths are in the ratio $L_2/L_1 = 3/4$ so that $L_1 = L$ and $L_2 = 3L/4$, determine the yield load P_y and the corresponding displacement δ_y of point B. (b) For the same conditions, determine the ultimate load P_u and the displacement δ_u of point B when the load just reaches the value P_u. (c) Construct a load-displacement diagram with the load P as ordinate and the displacement δ_b of point B as abscissa.

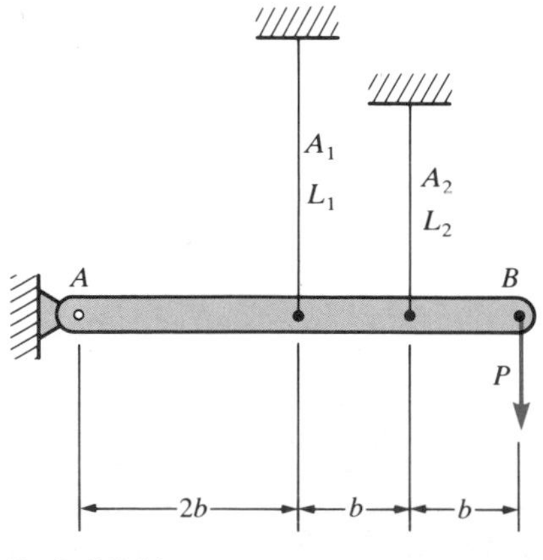

Prob. 2.8-14

***2.8-15** Two cables, each having a length L of approximately 125 ft, support a loaded container of weight W (see figure). The cables, which have effective cross-sectional area $A = 0.075$ in.^{2} and effective modulus of elasticity $E = 20 \times 10^6$ psi, are identical except that one cable is 4.0 in. longer than the other when they are hanging separately and unloaded. The cables are made of steel having an elastic-plastic stress-strain diagram with $\sigma_y = 80{,}000$ psi. Assume that the weight W is slowly increased by the addition of more material to the container. (a) Determine the weight W_y that first produces yielding of the shorter cable. Also, determine the corresponding elongation δ_y of the shorter cable. (b) Determine the weight W_u that produces yielding of both cables. Also, determine the elongation δ_u of the shorter cable when the weight W just reaches the value W_u. (c) Construct a load-displacement diagram showing the weight W as ordinate and the elongation δ of the shorter

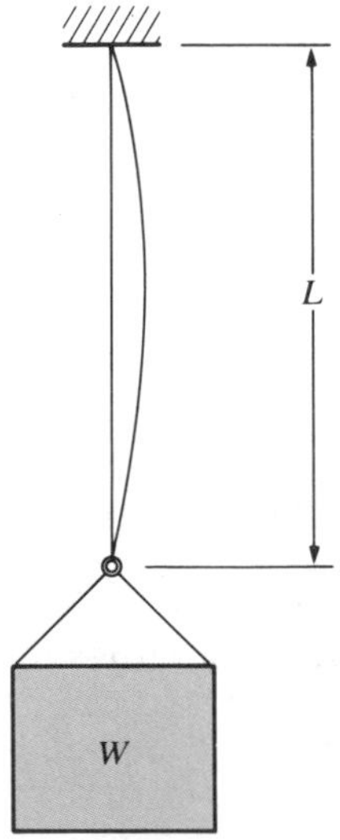

Prob. 2.8-15

cable as abscissa. (Hint: The load-displacement diagram is not a single straight line in the region $0 \leq W \leq W_y$.)

***2.8-16** A hollow circular tube T of length $L = 250$ mm is uniformly compressed by a force P (see figure). The outside and inside diameters of the tube are 80 and 70 mm, respectively. A concentric solid circular bar B of 40 mm diameter is mounted inside the tube. When no load is present, there is a clearance of 0.3 mm between the bar and the rigid top plate. Both bar and tube are made of steel having an elastic-plastic stress-strain diagram with $E = 200$ GPa and $\sigma_y = 380$ MPa. (a) Determine the yield load P_y, the shortening δ_y of the tube when $P = P_y$, the ultimate load P_u, and the shortening δ_u of the tube when P just reaches the value P_u. (b) Construct a load-displacement diagram showing the force P versus the shortening δ of the tube. (Hint: The load-displacement diagram is not a single straight line in the region $0 \leq P \leq P_y$.)

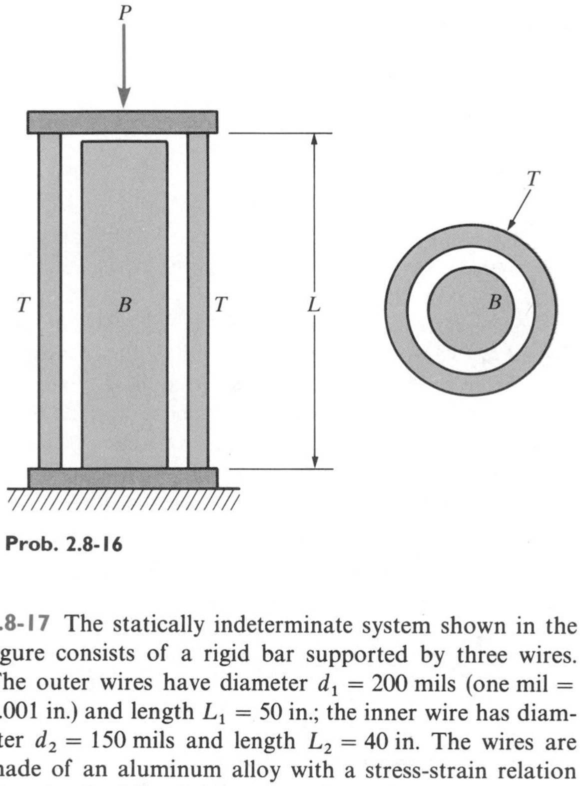

Prob. 2.8-16

***2.8-17** The statically indeterminate system shown in the figure consists of a rigid bar supported by three wires. The outer wires have diameter $d_1 = 200$ mils (one mil = 0.001 in.) and length $L_1 = 50$ in.; the inner wire has diameter $d_2 = 150$ mils and length $L_2 = 40$ in. The wires are made of an aluminum alloy with a stress-strain relation given by the following Ramberg-Osgood equation:

$$\epsilon = \frac{\sigma}{10 \times 10^6} + \frac{1}{614.0}\left(\frac{\sigma}{38{,}000}\right)^{10}$$

in which σ has units of psi. (See Fig. 2-51 for a graph of this equation.) Assuming that the total vertical load acting on the system is $P = 3000$ lb, calculate the tensile force F_1 in the outer wires, the tensile force F_2 in the inner wire, and the downward displacement δ of the rigid bar. (Hint: Use the alternative trial-and-error procedure described in the last paragraph of Section 2.8.)

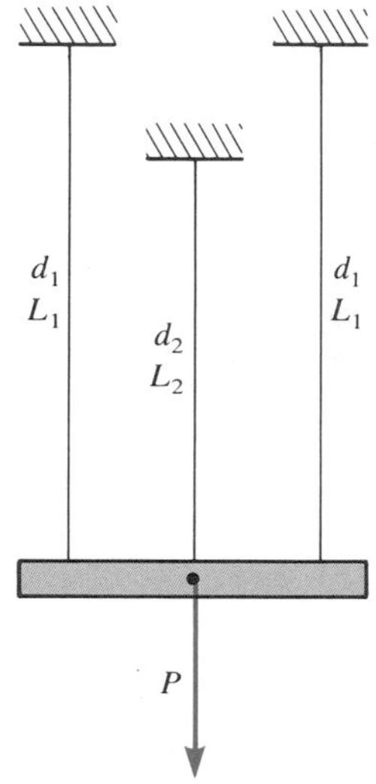

Prob. 2.8-17

The problems for Section 2.10 are to be solved by considering the stress-concentration factors.

2.10-1 The flat bars shown in the figure on p. 156 are subjected to tensile forces $P = 3.0$ k. Each bar has thickness $t = 0.25$ in. (a) For the bar with a hole, which has width $b = 6$ in., determine the maximum stresses for hole diameters $d = 1$ in. and $d = 2$ in. (b) For the stepped bar, which has widths $b = 4$ in. and $c = 2.5$ in., determine the maximum stresses for fillet radii $R = \frac{1}{4}$ in. and $R = \frac{1}{2}$ in.

2.10-2 The flat bars shown in the figure on p. 156 are subjected to tensile forces $P = 2.5$ kN. Each bar has thickness $t = 5$ mm. (a) For the bar with a hole, which has width $b = 60$ mm, determine the maximum stresses for hole diameters $d = 12$ mm and $d = 20$ mm. (b) For the stepped bar, which has widths $b = 60$ mm and $c = 40$ mm, determine the maximum stresses for fillet radii $R = 6$ mm and $R = 10$ mm.

2.10-3 A flat bar of thickness t has a hole drilled through it (see figure). The width of the bar is b, the diameter of the

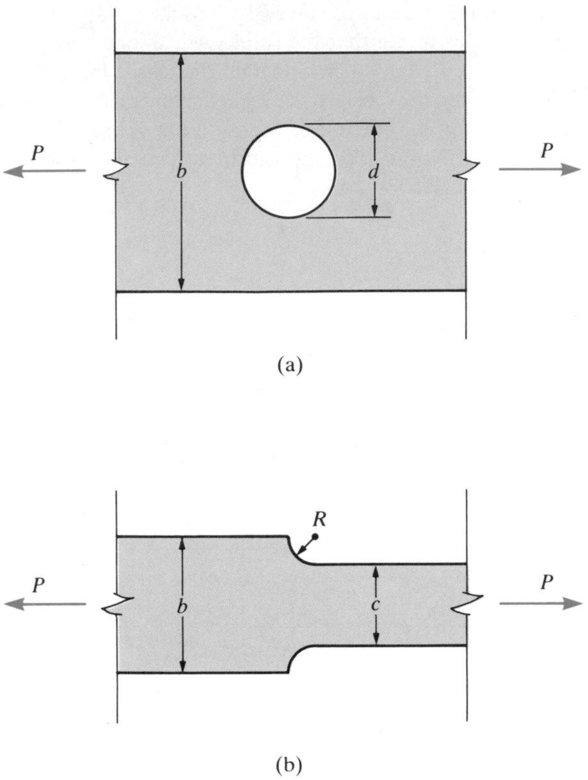

Probs. 2.10-1 and 2.10-2

hole is d, and the hole is equidistant from the sides of the bar. The hole may have any diameter that will fit within the bar. What is the maximum permissible tensile load P_{max} if the allowable tensile stress in the material is σ_t?

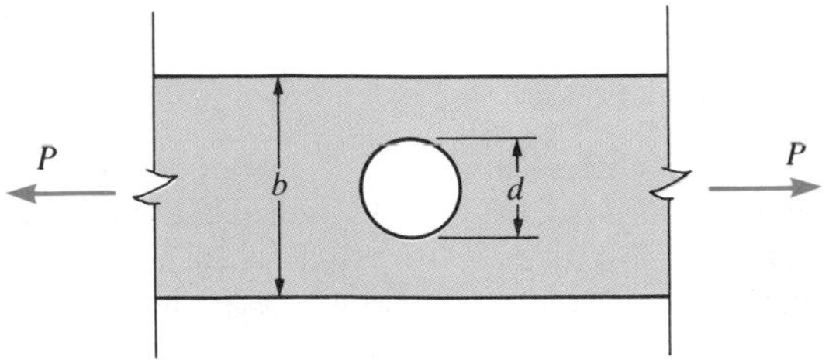

Prob. 2.10-3

2.10-4 A round prismatic bar of diameter $d_1 = 20$ mm is being compared with a bar of the same diameter that has an enlarged portion of diameter $d_2 = 25$ mm (see figure). The radius of the fillets is 2.0 mm. Does enlarging the bar in this manner make it stronger? Justify your answer by determining the maximum permissible load P_1 for the prismatic bar and the maximum permissible load P_2 for the enlarged bar, assuming that the allowable stress is 80 MPa.

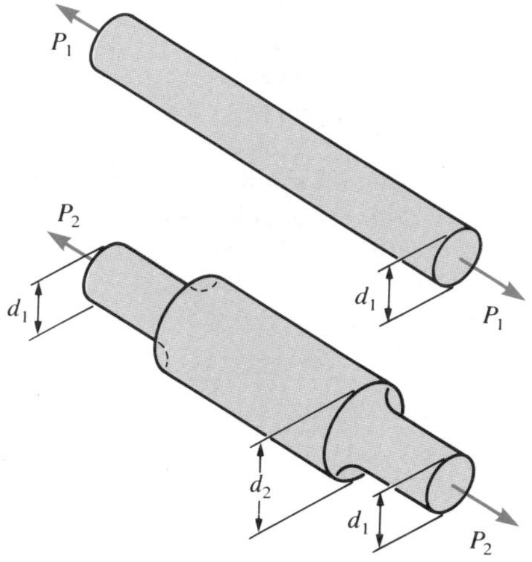

Prob. 2.10-4

2.10-5 A stepped flat bar with a hole (see figure) has widths $b_1 = 1.6$ in., $b_2 = 2.0$ in., and $b_3 = 1.4$ in. Also, the thickness $t = 0.25$ in., the hole diameter $d = 0.7$ in., and the fillets are quarter circles having the largest radii that will fit. (a) If the allowable stress is 6000 psi, what is the maximum permissible load P_{max}? (b) Assume that it is desired to increase the permissible load without changing the thickness t or the width b_2. If only one dimension can be changed (b_1, b_3, or d), which dimension would you change? Would you make it smaller or larger?

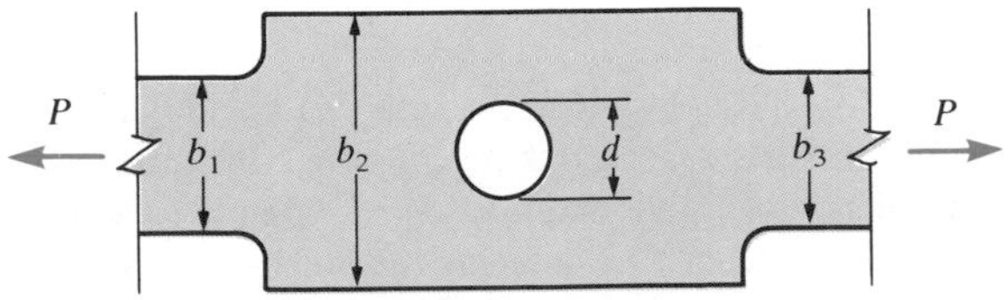

Probs. 2.10-5 and 2.10-6

2.10-6 Solve the preceding problem for the following data: $b_1 = 75$ mm, $b_2 = 100$ mm, $b_3 = 65$ mm, $t = 15$ mm, $d = 30$ mm, and the allowable stress is 40 MPa.

2.10-7 A flat bar with a hole (see figure) has widths $b = 2.4$ in. and $c = 1.6$ in. The fillets have radii equal to 0.2 in. What is the diameter d_{max} of the largest hole that can be drilled through the bar without reducing the load-carrying capacity?

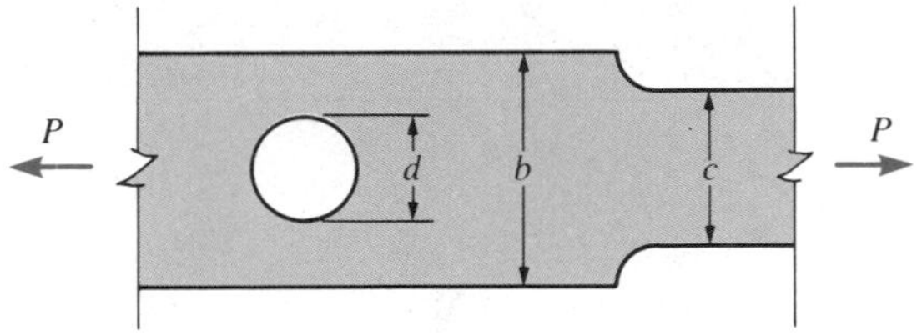

Prob. 2.10-7

CHAPTER 3

Torsion

3.1 INTRODUCTION

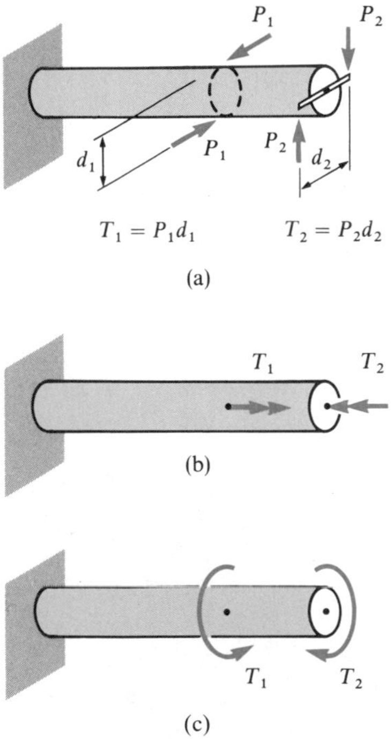

Fig. 3-1 Bar subjected to torsion by twisting couples T_1 and T_2

Torsion refers to the twisting of a structural member when it is loaded by couples that produce rotation about its longitudinal axis. Torsional loading is pictured in Fig. 3-1a, which shows a straight bar supported at one end and loaded by two pairs of forces. Each pair of forces forms a couple that tends to twist the bar about its longitudinal axis. The moment of a couple is equal to the product of one of the forces and the perpendicular distance between the lines of action of the forces; thus, the first couple has a moment $T_1 = P_1 d_1$ and the second has a moment $T_2 = P_2 d_2$.

For convenience in representing couples, we often will show the moment of a couple by a vector in the form of a double-headed arrow (Fig. 3-1b). The arrow is perpendicular to the plane containing the couple, and the direction, or sense, of the couple is indicated by the right-hand rule for moment vectors. An alternative representation is the use of a curved arrow acting in the direction of twist (Fig. 3-1c). Couples that produce twisting of a bar, such as couples T_1 and T_2 in Fig. 3-1, are called **torques, twisting couples**, or **twisting moments**.

In this chapter, we will develop formulas for the stresses and deformations produced in circular bars subjected to torsion. Examples of such bars are axles and drive shafts in machinery, propeller shafts, drill rods, torsional pendulums, screwdrivers, and steering rods. We will also consider thin-walled tubular members of the kind found in aerospace structures.

3.2 TORSION OF CIRCULAR BARS

Let us consider a bar or shaft of circular cross section twisted by couples T acting at the ends (Fig. 3-2a). A bar loaded in this manner is said to be in **pure torsion**. From considerations of symmetry, we can show that cross sections of the circular bar rotate as rigid bodies about the longitudinal axis, with radii remaining straight and the cross sections remaining plane and circular. Also, if the total angle of twist of the bar is small, neither the length of the bar nor its radius will change.

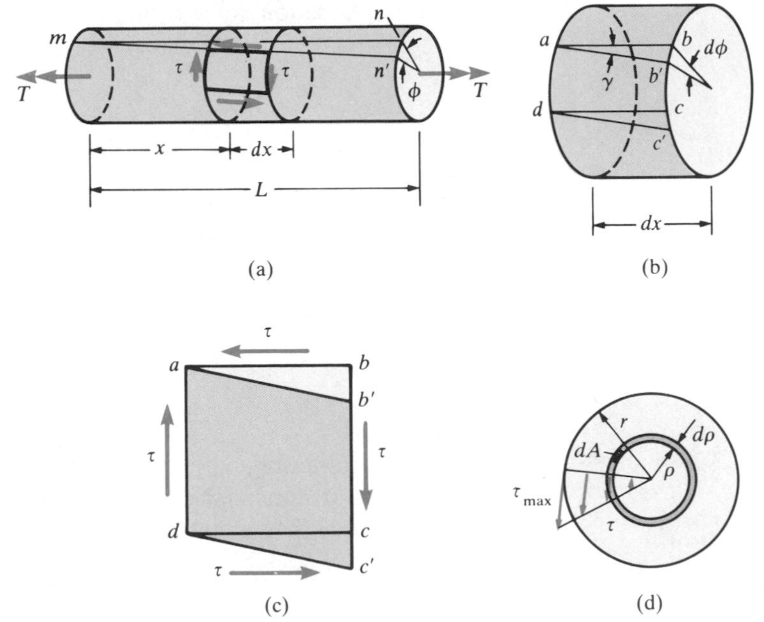

Fig. 3-2 Circular bar in pure torsion

During twisting, there will be a rotation about the longitudinal axis of one end of the bar with respect to the other. For instance, if the left-hand end of the bar is fixed, then the right-hand end will rotate through a small angle ϕ with respect to the left-hand end (Fig. 3-2a). The angle ϕ is known as the **angle of twist**. Because of this rotation, a longitudinal line mn on the surface of the bar will rotate through a small angle to position mn'.

Now consider an element of the bar between two cross sections that are distance dx apart (Fig. 3-2a). This volume element is isolated in Fig. 3-2b. On its outer surface we identify a small element $abcd$, with sides ab and cd that are parallel to the longitudinal axis. During twisting of the bar, the right-hand cross section rotates about the longitudinal axis with respect to the left-hand cross section, and points b and c move to b' and c', respectively. The lengths of the sides of the element do not change during this rotation, but the angles at the corners are no longer equal to 90°. The element is in a state of **pure shear** (see Section 1.6), and the magnitude of the shear strain γ is equal to the decrease in the angle at a. This decrease in angle is

$$\gamma = \frac{bb'}{ab}$$

The distance bb' is the length of a small arc of radius r subtended by the angle $d\phi$, which is the angle of rotation of one cross section with respect to the other. Thus, we obtain $bb' = r\,d\phi$. Also, the distance ab is equal to dx, the length of the element. Substituting these quantities into the preceding equation, we get

$$\gamma = \frac{r\,d\phi}{dx}$$

as the expression for the shear strain.

The quantity $d\phi/dx$ represents the rate of change of the **angle of twist** ϕ. We will denote $d\phi/dx$ by the symbol θ and refer to it as the **angle of twist per unit length**. Then we obtain

$$\gamma = \frac{r\,d\phi}{dx} = r\theta \tag{3-1}$$

In general, both ϕ and θ are functions of x. However, in the special case of pure torsion the angle of twist per unit length is constant along the length L of the bar, because every cross section is subjected to the same torque (as in Fig. 3-2a). Therefore, we obtain $\theta = \phi/L$ and Eq. (3-1) becomes

$$\gamma = r\theta = \frac{r\phi}{L} \tag{3-2}$$

for pure torsion. This equation can also be obtained directly from Fig. 3-2a by noting that the distance nn' equals $r\phi$ and that the angle γ is the angle between lines mn and mn'.

Finally, we observe that the preceding equations are based only upon geometric concepts, and therefore they are valid for any circular bar, regardless of whether the material behaves elastically or inelastically, linearly or nonlinearly.

The shear stresses τ in the circular bar act on an element in the directions shown in Fig. 3-2a. For clarity, these stresses are shown again in Fig. 3-2c. For a linearly elastic material, the shear stresses are related to the shear strains by Hooke's law in shear (Eq. 1-9); therefore, from

Eq. (3-1) we get

$$\tau = G\gamma = Gr\theta \tag{3-3}$$

in which G is the **shear modulus of elasticity**. Equations (3-1) and (3-3) relate the strains and stresses for an element at the surface of the bar to the angle of twist per unit length.

The strains and stresses within the interior of the bar can be determined in a manner similar to that used when analyzing an element at the surface. Because radii in the cross sections of a bar remain straight and undistorted during twisting, we see that the preceding discussion for an element *abcd* at the outer surface will also hold for a similar element situated at the surface of an interior cylinder of radius ρ (Fig. 3-2d). Such an interior element is also in pure shear with the corresponding shear strain and shear stress being given by the following equations:

$$\gamma = \rho\theta \qquad \tau = G\rho\theta \tag{3-4a,b}$$

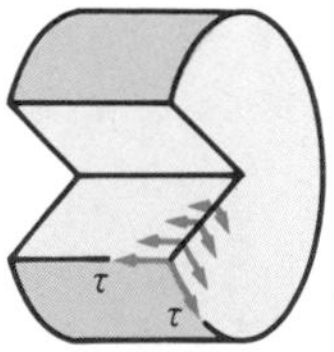

Fig. 3-3 Longitudinal shear stresses in a circular bar

These equations show that the shear strains and stresses in a circular bar vary linearly with the radial distance ρ from the center, with the maximum values occurring at the outer surface. This variation in the shear stress τ is illustrated by the triangular stress diagram in Fig. 3-2d.

The shear stresses acting on a cross-sectional plane are accompanied by shear stresses of the same magnitude acting on longitudinal planes (see Fig. 3-3). This conclusion follows from the fact that equal shear stresses always exist on mutually perpendicular planes, as explained in Section 1.6 and pictured in Fig. 3-2c. If a material is weaker in shear on longitudinal planes than on cross-sectional planes, as in the case of a circular bar made of wood, the first cracks due to twisting will appear on the surface in the longitudinal direction.

Fig. 3-4 Tensile and compressive stresses acting on an element oriented at 45° to the longitudinal axis

The state of pure shear stress at the surface of the shaft (Figs. 3-2a and c) is equivalent to equal tensile and compressive stresses acting on an element oriented at an angle of 45°, as explained later in Section 3.4. Therefore, a rectangular element with sides at 45° to the axis of the shaft will be submitted to the tensile and compressive stresses shown in Fig. 3-4. If a twisted bar is made of a material that is weaker in tension than in shear, failure will occur in tension along a helix inclined at 45° to the axis. This type of failure is easily demonstrated by twisting a piece of classroom chalk.

The relationship between the applied torque T and the angle of twist ϕ (Fig. 3-2a) may now be determined from the condition that the resultant couple of the shear stresses acting over the cross section (Fig. 3-2d) must be statically equivalent to the applied torque T. To determine the resultant couple, we consider a ring-shaped element of area having radius ρ and thickness $d\rho$ (Fig. 3-2d). Within this element we take another element (labeled dA in the figure) that extends over a small length of arc. The elemental shear force acting on this element of area is equal to $\tau\, dA$, where τ is the shear stress at radius ρ. The moment of this force about the axis of the bar is $\tau\rho\, dA$. Using Eq. (3-4b), we express this moment as $G\theta\rho^2\, dA$. The total torque T is the summation over the entire cross-sectional area of all such elemental moments:

$$T = \int G\theta\rho^2\, dA = G\theta \int \rho^2\, dA = G\theta I_p \qquad \text{(a)}$$

in which

$$I_p = \int \rho^2\, dA \qquad \text{(3-5)}$$

is the **polar moment of inertia** of the circular cross section. For a circle of radius r and diameter d, the polar moment of inertia (see Appendix D, Case 9) is

$$I_p = \frac{\pi r^4}{2} = \frac{\pi d^4}{32} \qquad \text{(3-6)}$$

Next, from Eq. (a) we obtain

$$\theta = \frac{T}{GI_p} \qquad \text{(3-7)}$$

which shows that θ, the angle of twist per unit length, is directly proportional to the torque T and inversely proportional to the product GI_p, known as the **torsional rigidity** of the shaft. The total angle of twist ϕ, equal to θL, is

$$\phi = \frac{TL}{GI_p} \qquad \text{(3-8)}$$

In this equation, the angle of twist ϕ is measured in radians. If SI units are used, the torque T may be expressed in newton meters (N·m), the length L in meters (m), the shear modulus G in pascals (Pa), and the polar moment of inertia I_p in meters to the fourth power (m^4). Similarly, if USCS units are used, T may be expressed in inch-pounds (in.-lb), L in inches (in.), G in pounds per square inch (psi), and I_p in inches to the fourth power ($in.^4$).

The quantity GI_p/L, called the **torsional stiffness**, is the torque required to produce a unit angle of rotation of one end of the bar with respect to the other. The **torsional flexibility** is defined as the reciprocal of the stiffness, or L/GI_p, which is the angle of rotation produced by a unit torque. The preceding expressions are analogous to those for axial stiffness EA/L and axial flexibility L/EA (see Eqs. 2-2 and 2-3).

Equation (3-8) may be used to determine the shear modulus of elasticity G for various materials. By conducting a torsion test on a circular specimen, we can measure the angle of twist ϕ produced by a known torque T. Then the magnitude of G can be calculated from Eq. (3-8).

The **maximum shear stress** τ_{max} in a circular bar may be found by substituting the expression for θ (Eq. 3-7) into the expression for τ (Eq. 3-3); thus,

$$\tau_{max} = \frac{Tr}{I_p} \qquad (3\text{-}9)$$

This equation, known as the **torsion formula**, shows that the maximum shear stress is proportional to the applied torque T and the radius r, and inversely proportional to the polar moment of inertia of the cross section. Substituting $r = d/2$ and $I_p = \pi d^4/32$, we get

$$\tau_{max} = \frac{16T}{\pi d^3} \qquad (3\text{-}10)$$

as the formula for the maximum shear stress in a solid circular bar. The shear stress at distance ρ from the center is

$$\tau = \frac{T\rho}{I_p} \qquad (3\text{-}11)$$

which is obtained from Eq. (3-4b). The units of shear stress are pascals (Pa) if SI units are used and either pounds per square inch (psi) or kips per square inch (ksi) if USCS units are used.*

HOLLOW CIRCULAR BARS

Hollow bars or tubes are much more efficient in resisting torsional loads than are solid bars. As explained in the preceding paragraphs, the shear stresses in a solid circular bar are maximum at the outer boundary of the

* The torsion theory for circular bars originated with the work of the famous French scientist C. A. de Coulomb (1736–1806); further developments were due to Thomas Young and A. Duleau (Ref. 3-1). The general theory of torsion (for bars of any shape) is due to the most famous elastician of all time, Barré de Saint-Venant (1797–1886); see Ref. 3-2.

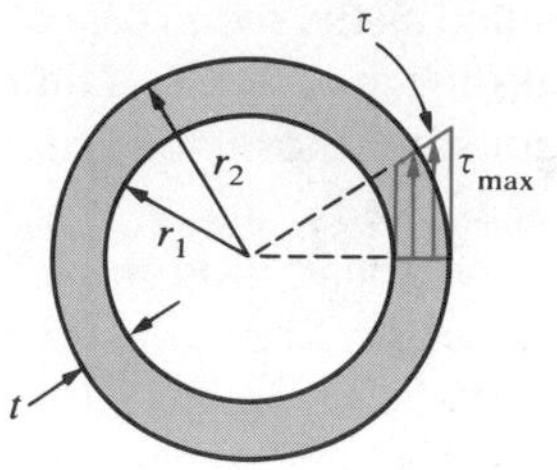

Fig. 3-5 Hollow circular bar

cross section and zero at the center. Therefore, most of the material in a solid shaft is stressed significantly below the maximum shear stress. If weight reduction and savings of material are important, it is advisable to use hollow shafts (Fig. 3-5).

The analysis of the torsion of a hollow circular bar is almost identical to that for a solid bar. The same basic expressions for shear strain γ and shear stress τ may still be used (see Eqs. 3-4a and b). Of course, the radial distance ρ appearing in those expressions is limited to the range r_1 to r_2, where r_1 is the inner radius and r_2 is the outer radius of the bar (Fig. 3-5).

The relationship between the applied torque T and the angle of twist θ per unit length is given by Eq. (a), except that the limits on the integral for the polar moment of inertia I_p (see Eq. 3-5) are $\rho = r_1$ and $\rho = r_2$. Thus, the polar moment of inertia I_p of the cross-sectional area shown in Fig. 3-5 is

$$I_p = \frac{\pi}{2}(r_2^4 - r_1^4) = \frac{\pi}{32}(d_2^4 - d_1^4) \tag{3-12}$$

If the tube is very thin and its thickness t (Fig. 3-5) is small compared to its radius, then the following approximate formulas (see Case 10 in Appendix D) may be used:

$$I_p \approx 2\pi r^3 t = \frac{\pi d^3 t}{4} \tag{3-13}$$

in which r and d are the average radius and diameter, respectively.

The equations for θ, ϕ, and τ derived for a solid bar (Eqs. 3-7 through 3-9) may be used for a hollow bar provided I_p is evaluated according to Eq. (3-12) or, if appropriate, Eq. (3-13). The shear stress distribution for a hollow bar is pictured in Fig. 3-5, and we see that the average stress in the material is closer to the maximum stress than it is in the case of a solid bar.

When designing a hollow shaft, the wall thickness must be large enough to prevent wrinkling or buckling of the shaft wall. Other design considerations include environmental and durability factors, which also may impose requirements for minimum wall thickness.

Finally, it is important to emphasize that the theory and equations for the torsion of circular bars cannot be used for bars of other shapes. Noncircular bars subjected to twisting moments generally behave quite differently from circular bars, and more advanced methods of analysis are required.

Example 1

A solid steel shaft of diameter $d = 60$ mm and length $L = 4$ m (Fig. 3-6) is to be designed using an allowable shear stress $\tau_{allow} = 40$ MPa and an allowable angle of twist per unit length $\theta = 1°$ per meter. (a) Determine the maximum permissible torque T_{max} that may be applied to the shaft, assuming $G = 80$ GPa. (b) Determine the angle of twist ϕ.

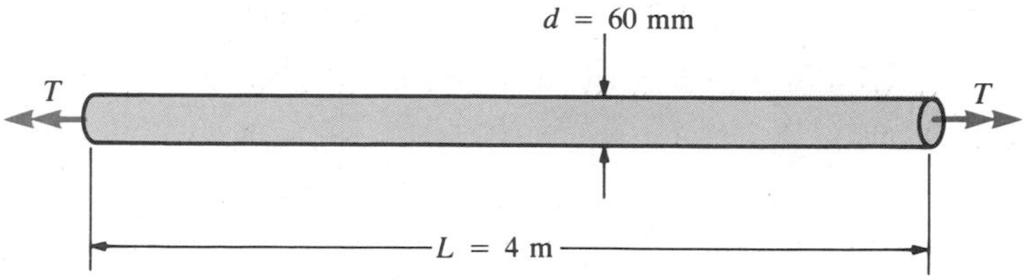

Fig. 3-6 Example 1

(a) The permissible torque T_1 based upon the allowable shear stress is obtained from the torsion formula $\tau_{max} = 16T/\pi d^3$ (see Eq. 3-10); thus,

$$T_1 = \frac{\pi d^3 \tau_{allow}}{16} = \frac{\pi}{16}(0.060 \text{ m})^3(40 \text{ MPa}) = 1700 \text{ N}\cdot\text{m}$$

The allowable angle of twist per unit length is 1° per meter or $\pi/180$ rad/m. The corresponding allowable torque T_2 is obtained from Eq. (3-7):

$$T_2 = GI_p\theta = (80 \text{ GPa})\left(\frac{\pi}{32}\right)(0.060 \text{ m})^4\left(\frac{\pi}{180} \text{ rad/m}\right) = 1780 \text{ N}\cdot\text{m}$$

The maximum permissible torque, equal to the smaller of the two preceding values, is $T_{max} = 1700$ N·m.

(b) The angle of twist ϕ may be calculated from Eq. (3-8) by substituting the maximum allowable torque for T:

$$\phi = \frac{TL}{GI_p} = \frac{(1700 \text{ N}\cdot\text{m})(4 \text{ m})}{(80 \text{ GPa})(\pi/32)(0.060 \text{ m})^4} = 0.0667 \text{ rad} = 3.8°$$

Alternatively, since the allowable shear stress (and not the allowable angle of twist per unit length) determined the allowable torque, we can eliminate T between Eqs. (3-8) and (3-9) and obtain

$$\phi = \frac{\tau_{max} L}{Gr} = \frac{(40 \text{ MPa})(4.0 \text{ m})}{(80 \text{ GPa})(0.030 \text{ m})} = 0.0667 \text{ rad} = 3.8°$$

If the allowable torque had been based upon the angle of twist per unit length, we would have found the angle of twist from the equation $\phi = \theta L = (1°/\text{m})(4 \text{ m}) = 4°$. Note that the actual angle of twist is equal to 4° times the ratio T_1/T_2.

Example 2

A hollow shaft and a solid shaft constructed of the same material have the same length and the same outside radius r (Fig. 3-7). The inside radius of the hollow shaft is $0.6r$. Assuming that both shafts are subjected to the same torque, compare the maximum shear stresses, the angles of rotation, and the weights of the two shafts.

The maximum shear stresses (see Eq. 3-9) are proportional to $1/I_p$, inasmuch as the torques T and radii r are the same. For the solid shaft, we get

$$I_p = \frac{\pi r^4}{2} = 0.5\pi r^4$$

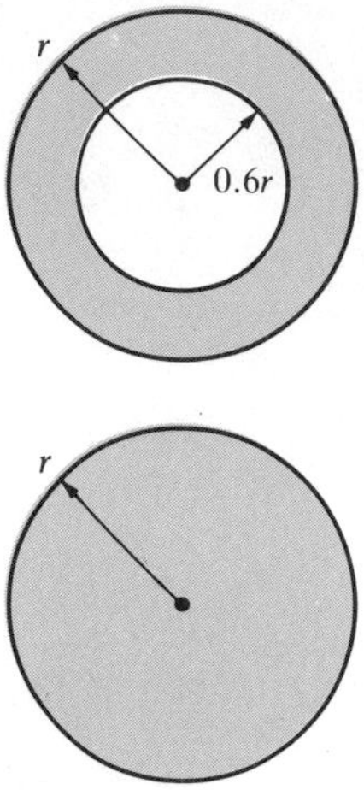

Fig. 3-7 Example 2

and for the hollow shaft

$$I_p = \frac{\pi r^4}{2} - \frac{\pi(0.6r)^4}{2} = 0.4352\pi r^4$$

Therefore, the ratio of the maximum shear stress in the hollow shaft to that in the solid shaft is 0.5/0.4352, or 1.15. Because the lengths are the same, the angles of rotation are in the same proportion as the stresses (see Eq. 3-8).

The weights of the shafts are proportional to the cross-sectional areas; hence the weight of the solid shaft is proportional to πr^2 and the weight of the hollow shaft is proportional to

$$\pi r^2 - \pi(0.6r)^2 = 0.64\pi r^2$$

Therefore, the weight of the hollow shaft is 64% of the weight of the solid shaft. These results show the inherent advantage of hollow shafts. In this example, the hollow bar has 15% greater stress and angle of rotation, but 36% less weight.

The relative efficiency of a structure is sometimes measured by the *strength-to-weight ratio*, which in this example may be defined as the allowable load divided by the weight. The allowable torque for the hollow shaft is

$$T_1 = \frac{\tau_{\max} I_p}{r} = \tau_{\max}(0.4352\pi r^3)$$

and for the solid shaft is

$$T_2 = \frac{\tau_{\max} I_p}{r} = \tau_{\max}(0.5\pi r^3)$$

The corresponding weights are

$$W_1 = (0.64\pi r^2)\gamma L \qquad W_2 = (\pi r^2)\gamma L$$

where γ is the specific weight of the material and L is the length of the shafts. Thus, the strength-to-weight ratios, denoted s_1 and s_2, are

$$s_1 = \frac{T_1}{W_1} = 0.68\frac{\tau_{\max} r}{\gamma L} \qquad s_2 = \frac{T_2}{W_2} = 0.5\frac{\tau_{\max} r}{\gamma L}$$

The hollow shaft in this example has a 36% greater strength-to-weight ratio than does the solid shaft. For a thinner shaft, this percentage will increase; for a thicker shaft, it will decrease.

3.3 NONUNIFORM TORSION

As explained in the preceding section, *pure torsion* refers to torsion of a prismatic bar subjected to torques acting only at the ends. **Nonuniform torsion** differs from pure torsion in that the bar need not be prismatic and the applied torques may vary along the length. In such cases, we usually can analyze the bar by applying the formulas of pure torsion in specialized ways, as illustrated in the following discussions.

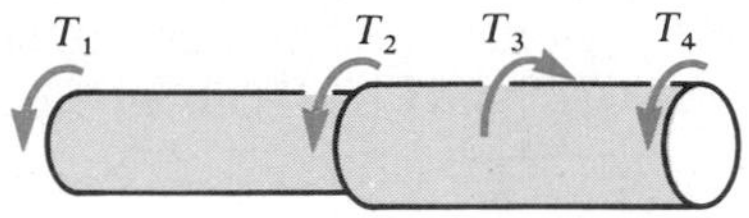

Fig. 3-8 Bar in nonuniform torsion

An example of nonuniform torsion is pictured in Fig. 3-8, which shows a bar made of two segments of different diameters and having torques applied at several cross sections. Each region of the bar between applied loads or between changes in cross section is in pure torsion, and therefore the formulas derived in the preceding section may be applied to each part separately. For this purpose, it is necessary to determine the magnitude and direction of the internal torque in each region. Then, knowing the internal torque, we can calculate the angle of twist and maximum shear stress for each part from Eqs. (3-8) and (3-9), respectively.

The total **angle of twist** of one end of the bar with respect to the other is obtained by summation, using the general formula

$$\phi = \sum_{i=1}^{n} \frac{T_i L_i}{G_i I_{pi}} \tag{3-14}$$

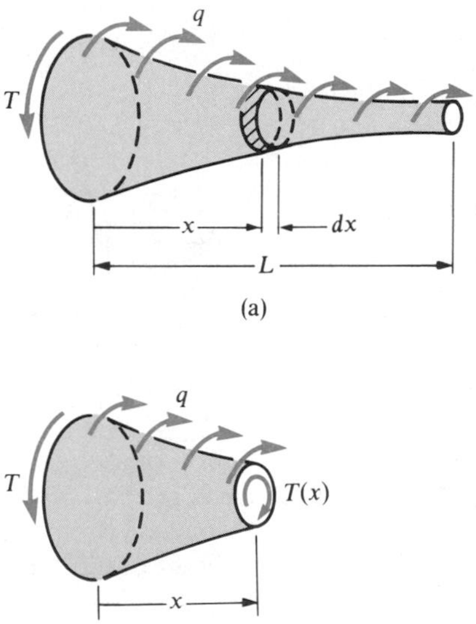

Fig. 3-9 Bar with varying cross section and varying torque

In this equation, the subscript i is a numbering index for the various parts of the bar, and n is the total number of parts. For part i of the bar, T_i is the torque (found from equilibrium), L_i is the length, G_i is the shear modulus, and I_{pi} is the polar moment of inertia. The expression $T_i L_i / G_i I_{pi}$ gives the angle of twist ϕ_i for part i. By summing algebraically all such angles of twist, we obtain the total angle of twist ϕ between the ends of the bar. High localized stresses occur at the sections where the diameter changes abruptly (see Section 3.10), but these stresses have relatively little effect on the total angle of twist ϕ, and so Eq. (3-14) gives satisfactory results.

Another case of nonuniform torsion occurs when either the torque or the cross section changes continuously along the axis of the bar. This situation is pictured in Fig. 3-9a, where the tapered bar is subjected to a torque of intensity q per unit distance along the axis of the bar. The torque $T(x)$ at a cross section located at distance x from the end of the bar may be found from equilibrium (Fig. 3-9b). Then, assuming that the taper is small and gradual, we can calculate the maximum shear stress at that cross section from the torsion formula (Eq. 3-9). The differential angle of rotation $d\phi$ for an element of length dx (Fig. 3-9a) is

$$d\phi = \frac{T(x)\,dx}{GI_p(x)}$$

where $I_p(x)$ is the polar moment of inertia of the cross section at distance x from the end. The total angle of twist between the ends of the bar is

$$\phi = \int_0^L d\phi = \int_0^L \frac{T(x)\,dx}{GI_p(x)} \tag{3-15}$$

This integral can be evaluated analytically in some cases, but usually it must be evaluated numerically. Equations (3-14) and (3-15) can be used for either solid or hollow bars having circular cross sections.

Example 1

A solid steel shaft $ABCD$ (Fig. 3-10) having diameter $d = 3$ in. turns freely in a bearing at D and is loaded at B and C by torques $T_1 = 20$ in.-k and $T_2 = 12$ in.-k. The shaft is connected in the gear box at A to gears that are temporarily locked in position. Determine the maximum shear stress in each part of the shaft and the angle of twist ϕ at end D. (Assume $L_1 = 20$ in., $L_2 = 30$ in., $L_3 = 20$ in., and $G = 11{,}500$ ksi.)

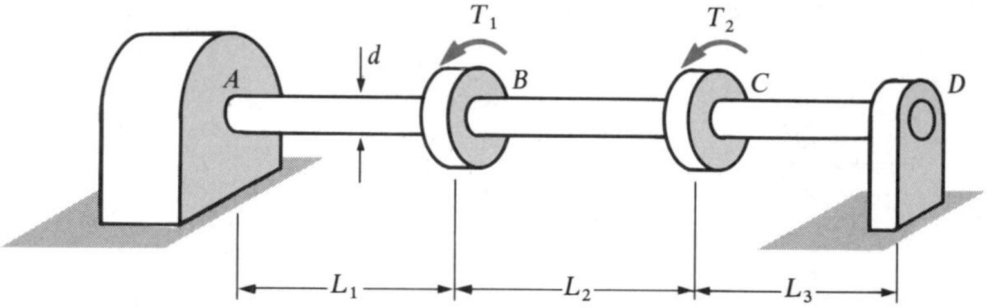

Fig. 3-10 Example 1

The maximum shear stresses are found from the torsion formula $\tau_{max} = 16T/\pi d^3$ (Eq. 3-10), in which T represents the internal torque in the part of the shaft under consideration. The torque is zero between C and D because no torque acts at D. From B to C the torque is equal to T_2, and from A to B the torque is the sum of T_1 and T_2. Thus, we obtain

$$T_{ab} = T_1 + T_2 = 32 \text{ in.-k} \qquad T_{bc} = T_2 = 12 \text{ in.-k} \qquad T_{cd} = 0$$

These torques are shown in the free-body diagrams of Fig. 3-11.

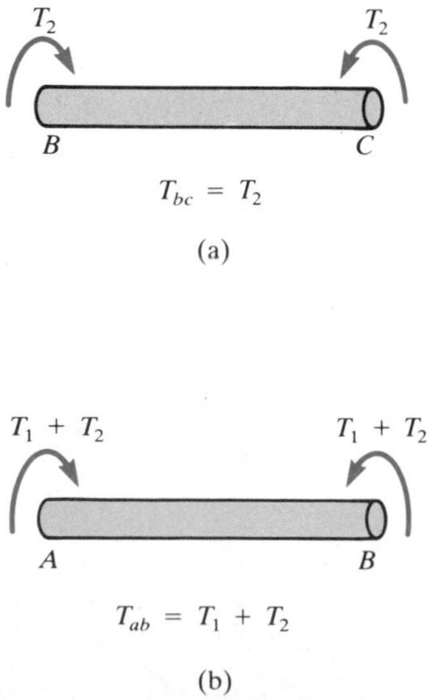

Fig. 3-11 Free-body diagrams of parts BC and AB of the shaft of Example 1

The maximum stresses in the shaft are calculated as follows:

$$\tau_{ab} = \frac{16T_{ab}}{\pi d^3} = \frac{16(32{,}000 \text{ in.-lb})}{\pi(3.0 \text{ in.})^3} = 6040 \text{ psi}$$

$$\tau_{bc} = \frac{16T_{bc}}{\pi d^3} = \frac{16(12{,}000 \text{ in.-lb})}{\pi(3.0 \text{ in.})^3} = 2260 \text{ psi}$$

$$\tau_{cd} = 0$$

The angle of twist ϕ at end D (with respect to end A) is determined by summing the angles of twist for each part of the bar, as follows:

$$\phi = \phi_{ab} + \phi_{bc} + \phi_{cd}$$

or, as expressed by Eq. (3-14),

$$\phi = \sum_{i=1}^{n} \frac{T_i L_i}{G_i I_{pi}} = \frac{T_{ab} L_1}{G I_p} + \frac{T_{bc} L_2}{G I_p} + \frac{T_{cd} L_3}{G I_p}$$

$$= \frac{(32 \text{ in.-k})(20 \text{ in.}) + (12 \text{ in.-k})(30 \text{ in.}) + 0}{(11{,}500 \text{ ksi})(\pi/32)(3 \text{ in.})^4}$$

$$= 0.006998 + 0.003937 + 0$$

$$= 0.0109 \text{ rad} = 0.627^\circ$$

In this example, the angles of twist ϕ_{ab} and ϕ_{bc} have the same sense and are added to calculate the total angle of twist ϕ from A to D. Depending upon the magnitudes and directions of the applied loads T_1 and T_2, we might find that these angles were of opposite sense. In that event, they would be subtracted instead of added. The relative angle of twist ϕ between the two ends of a system of shafts is sometimes called the *wind-up*.

The procedures illustrated in this example can be used for shafts having parts of different diameters or even of different materials.

Example 2

A tapered bar AB of solid circular cross section is twisted by torques T applied at the ends (Fig. 3-12). The diameter of the bar varies linearly from d_a at the left-hand end to d_b at the right-hand end. Derive a formula for the angle of twist ϕ of the bar.

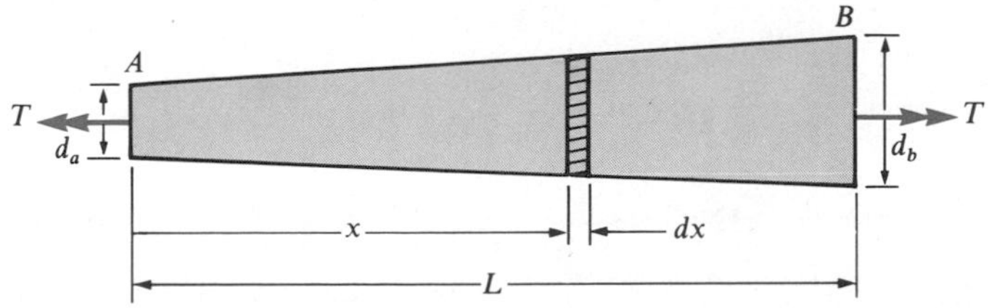

Fig. 3-12 Example 2

Because the polar moment of inertia of the cross section varies continuously with the distance x from end A, we may use Eq. (3-15) to determine ϕ. We note first that the diameter d at distance x from end A is

$$d = d_a + \frac{d_b - d_a}{L}x \tag{a}$$

Then the polar moment of inertia is

$$I_p(x) = \frac{\pi d^4}{32} = \frac{\pi}{32}\left(d_a + \frac{d_b - d_a}{L}x\right)^4 \tag{b}$$

Also, the torque $T(x)$ is constant and equal to the torque T applied at the ends. Therefore, the expression for the angle of twist (Eq. 3-15) becomes

$$\phi = \int_0^L \frac{T\,dx}{GI_p(x)} \tag{c}$$

in which L is the length of the bar, G is the shear modulus of the material, and $I_p(x)$ is given by Eq. (b). The integral in Eq. (c) is of the form

$$\int \frac{a\,dx}{(b + cx)^4}$$

in which

$$a = \frac{32T}{\pi G} \qquad b = d_a \qquad c = \frac{d_b - d_a}{L} \tag{d}$$

With the aid of a table of integrals, we find

$$\int \frac{a\,dx}{(b + cx)^4} = -\frac{a}{3c(b + cx)^3} \tag{e}$$

We can evaluate the integral by substituting for x the limits 0 and L and by substituting for a, b, and c the expressions in Eq. (d). The final result is

$$\phi = \frac{32TL}{3\pi G(d_b - d_a)}\left(\frac{1}{d_a^3} - \frac{1}{d_b^3}\right) \tag{3-16}$$

The preceding analysis could have been performed in a mathematically different way by taking the origin of the coordinate x at the apex of the cone formed by extending the surface of the bar in Fig. 3-12 to the left, as was done in the solution of Example 3 of Section 2.2 (see Fig. 2-9b).

A convenient form in which to write the solution for ϕ is

$$\phi = \frac{TL}{GI_{pa}}\left(\frac{\beta^2 + \beta + 1}{3\beta^3}\right) \tag{3-17}$$

in which $\beta = d_b/d_a$ and $I_{pa} = \pi d_a^4/32$ is the polar moment of inertia at end A. For example, if $\beta = 1$, we get $\phi = TL/GI_{pa}$, which is the angle of rotation of a prismatic bar of diameter d_a. If $\beta = 2$, we get $\phi = (7/24)(TL/GI_{pa})$, which is less than for the case of $\beta = 1$ because of the increased stiffness due to the larger diameter at end B.

3.4 PURE SHEAR

When a circular bar, either solid or hollow, is subjected to torsion, shear stresses τ act over the cross sections and on longitudinal planes, as explained in Section 3.2. A small, thin stress element *abcd* cut out between two cross sections and between two longitudinal planes (Fig. 3-13) is in a state of **pure shear**, because the only stresses acting on this element are the shear stresses on the four side faces. The directions of these stresses depend upon the directions of the applied torques *T*. In this discussion, we shall assume that the torques rotate the right-hand end of the bar clockwise when viewed from the right (Fig. 3-13a); hence the shear stresses acting on the element have the directions shown in Fig. 3-13b. The same state of stress exists for an element of similar shape cut out from the interior of the bar, except that the magnitude of the shear stresses is smaller because the radius to the element is smaller.

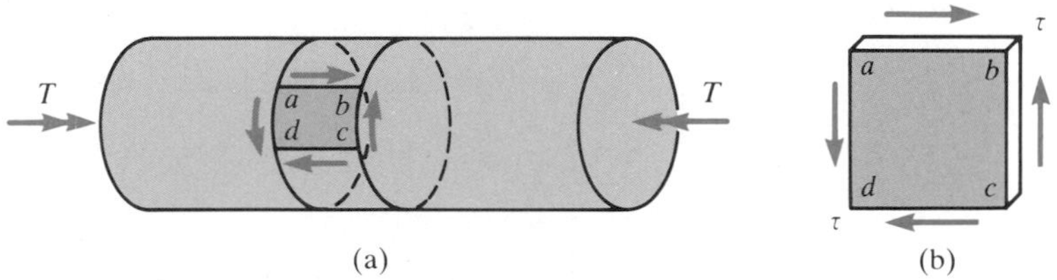

Fig. 3-13 Stresses acting on an element for a bar in torsion

A two-dimensional view of the stress element in pure shear is shown in Fig. 3-14a. A set of *xy* axes is oriented parallel to the sides of the element, with the *x* axis parallel to the longitudinal axis of the bar. The stresses acting on planes inclined to the axis of the bar can be determined by considering a wedge-shaped (or triangular) stress element having one face oriented at an angle θ to the *x* axis (Fig. 3-14b). The other side faces of this element are parallel to the sides of the element shown in Fig. 3-14a and have the shear stresses τ acting on them. Normal stresses σ_θ and shear stresses τ_θ act on the inclined face. These stresses are shown in their positive directions in Fig. 3-14b. The front and rear faces of the element are free of any stresses.

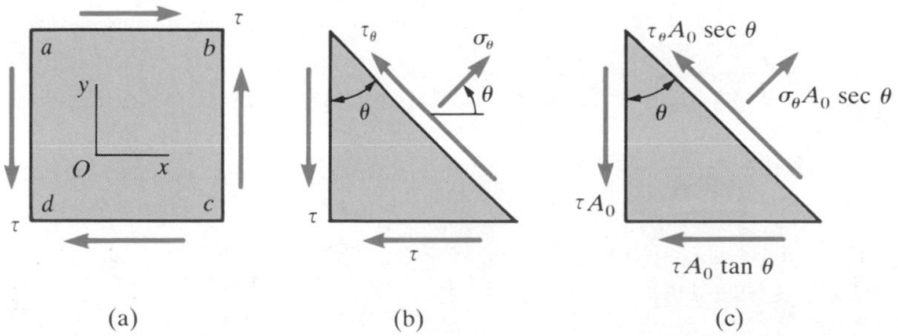

Fig. 3-14 Analysis of stresses on inclined planes for an element in pure shear

The stresses on the inclined face may be determined from the equilibrium of the triangular element (Fig. 3-14b), which is acted upon by forces on all three of its side faces. These forces can be obtained by multiplying the stresses by the areas of the faces on which they act. For instance, the force on the left-hand face is equal to τA_0, where A_0 is the area of the face. This force acts in the negative y direction (see Fig. 3-14c). Because the thickness of the element in the z direction is constant, we see that the area of the bottom face is $A_0 \tan\theta$ and the area of the inclined face is $A_0 \sec\theta$. Multiplying the stresses by the areas over which they act gives the forces acting on all faces of the element (Fig. 3-14c). The forces acting on the left-hand and bottom faces can be resolved into components acting perpendicular and parallel to the inclined face (that is, in the directions of σ_θ and τ_θ, respectively). Then we can write two equations of static equilibrium for the element, one in each of these directions. The first equation, obtained by summing forces in the direction of σ_θ, is

$$\sigma_\theta A_0 \sec\theta = \tau A_0 \sin\theta + \tau A_0 \tan\theta \cos\theta$$

or

$$\sigma_\theta = 2\tau \sin\theta \cos\theta \tag{3-18a}$$

The second equation is obtained by summing forces in the direction of τ_θ:

$$\tau_\theta A_0 \sec\theta = \tau A_0 \cos\theta - \tau A_0 \tan\theta \sin\theta$$

or

$$\tau_\theta = \tau(\cos^2\theta - \sin^2\theta) \tag{3-18b}$$

These equations can be expressed in alternative forms by introducing the following trigonometric identities:

$$2\sin\theta\cos\theta = \sin 2\theta \qquad \cos^2\theta - \sin^2\theta = \cos 2\theta$$

Then the equations for σ_θ and τ_θ become

$$\sigma_\theta = \tau \sin 2\theta \qquad \tau_\theta = \tau \cos 2\theta \tag{3-19a,b}$$

Equations (3-18) and (3-19) give the normal and shear stresses acting on any inclined plane in terms of the shear stresses τ acting on the x and y planes (Fig. 3-14a) and the angle θ defining the orientation of the inclined plane (Fig. 3-14b).

The manner in which the stresses σ_θ and τ_θ vary with the orientation of the inclined plane is shown by the graph in Fig. 3-15. We see that for $\theta = 0$, which is the right-hand face of the stress element in Fig. 3-14a, the graph gives $\sigma_\theta = 0$ and $\tau_\theta = \tau$, as expected. Also, for the top face of the element ($\theta = 90°$), we obtain $\sigma_\theta = 0$ and $\tau_\theta = -\tau$. The minus sign means that the shear stress acts in the negative τ_θ direction, that is, it acts to the right on face ab (Fig. 3-14a). These are the numerically largest shear stresses τ_θ acting on any planes.

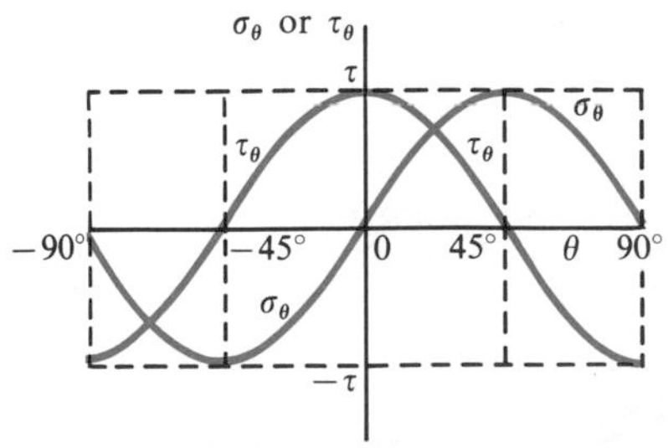

Fig. 3-15 Graph of normal stress σ_θ and shear stress τ_θ versus angle θ of the inclined plane

The normal stress σ_θ reaches a maximum value at $\theta = 45°$, where it is a tensile stress numerically equal to the shear stress τ. Similarly, σ_θ has its minimum value (which is compressive) at $\theta = -45°$. At these same angles, the shear stress τ_θ is equal to zero. Thus, a stress element oriented at an angle of 45° is acted upon by equal tensile and compressive stresses in perpendicular directions, but no shear stresses are present (Fig. 3-16b). The directions of the normal stresses shown in Fig. 3-16b are for an element subjected to shear stresses acting in the directions shown in Fig. 3-16a. If the shear stresses acting on the element of Fig. 3-16a have their directions reversed, the normal stresses acting on the 45° planes also will change directions.

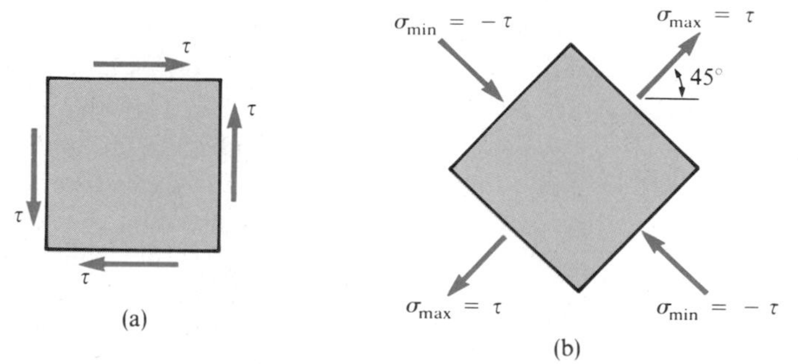

Fig. 3-16 Stress elements at $\theta = 0$ and $\theta = 45°$ for pure shear

If a stress element is oriented at an angle other than 45°, both normal and shear stresses will act on the side faces, as given by Eqs. (3-18) and (3-19). These more general stress conditions are discussed in detail in Chapter 6.

The equations derived in this section hold for a stress element in pure shear regardless of whether the element is cut from a bar in torsion or from some other structural element. Also, since Eqs. (3-18) and (3-19) were derived only from equilibrium, they are valid for any material, whether it is linearly elastic or not.

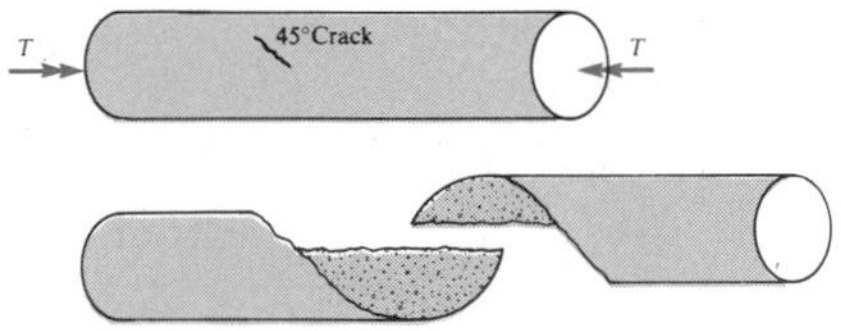

Fig. 3-17 Torsion failure of a brittle material by tension cracking along a 45° helical surface

The presence of maximum tensile stresses on planes at 45° to the x axis accounts for the fact that brittle materials that are weak in tension fail in torsion by cracking along a 45° helical surface (Fig. 3-17). As mentioned in Section 3.2, this type of failure is readily demonstrated by twisting a piece of classroom chalk.

Let us now consider the **strains** that occur when a material is subjected to pure shear. The stress element pictured in Fig. 3-16a will undergo shear strains γ, resulting in a deformation of the element, as shown in Fig. 3-18a and discussed previously in Section 1.6. The shear strain γ is measured as the change in angle between two planes that were originally perpendicular to each other. Thus, in Fig. 3-18a, the decrease in the angle at the lower left-hand corner of the element is the shear strain γ, measured in radians. This same change in angle occurs at the upper right-hand corner, where the angle decreases, and at the other two corners, where the angles increase. The lengths of the sides of the element,

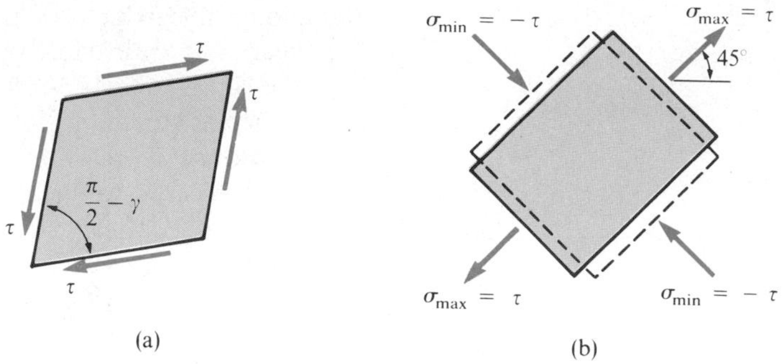

Fig. 3-18 Strains in pure shear: (a) element at $\theta = 0$, and (b) element at $\theta = 45°$

including the thickness perpendicular to the plane of the paper, do not change. Therefore, the element changes its shape from a rectangular parallelepiped to an oblique parallelepiped. This change in shape is referred to as **shear distortion**.

Next, consider the strains that occur in a stress element oriented at 45° (see Fig. 3-16b). The tensile stresses acting at 45° tend to elongate the element in that direction and, because of the Poisson effect, tend to shorten it in the perpendicular direction (the direction where $\theta = 135°$ or $-45°$). Similarly, the compressive stresses acting at 135° tend to shorten the element in the 135° direction and elongate it in the 45° direction. These changes in dimensions are shown in Fig. 3-18b. Note, however, that the element remains a rectangular parallelepiped, and there is no shear distortion.

If the material is linearly elastic, the shear strain for the element at $\theta = 0$ (Fig. 3-18a) is related to the shear stress by Hooke's law in shear:

$$\gamma = \frac{\tau}{G} \tag{3-20}$$

For the element at $\theta = 45°$ (Fig. 3-18b), we may use Poisson's ratio and Hooke's law for uniaxial stress to obtain the normal strains. The tensile stress $\sigma_{max} = \tau$ (at $\theta = 45°$) produces a positive strain equal to τ/E. It also produces a negative strain in the perpendicular direction, equal to $-\nu\tau/E$, where ν is Poisson's ratio. Similarly, the stress $\sigma_{min} = -\tau$ (at $\theta = 135°$) produces a negative strain equal to $-\tau/E$ and a positive strain in the perpendicular direction equal to $\nu\tau/E$. Therefore, the resultant normal strain in the 45° direction (that is, in the direction of the positive normal stress $\sigma_{max} = \tau$) is

$$\epsilon_{max} = \frac{\tau}{E} + \frac{\nu\tau}{E} = \frac{\tau}{E}(1 + \nu) \tag{3-21}$$

which is positive, representing elongation. The strain in the perpendicular direction is a negative strain of the same amount. We see that pure shear produces elongation in the 45° direction and shortening in the 135° direction, which is consistent with the shape of the deformed element of Fig. 3-18a.

We have already mentioned that the thickness of an element in pure shear does not change. Of course, the thickness of the 45° element (Fig. 3-18b) does not change either, because it represents the same state of stress. This observation is consistent with the discussion in the preceding paragraph, which pointed out that the stress $\sigma_{max} = \tau$ acting at 45° produces a lateral strain equal to $-\nu\tau/E$ and that the stress $\sigma_{min} = -\tau$ acting at 135° produces a lateral strain equal to $\nu\tau/E$. Therefore, the decrease in thickness of the element due to the tensile stress at 45° is exactly matched by the increase in thickness due to the compressive stress at 135°.

From the preceding discussions of the strains that exist in the stress elements of Fig. 3-18, we observe that no volume change occurs in pure shear as long as the strains are very small.

In the next section, we will use the geometry of the deformed element pictured in Fig. 3-18a to relate the shear strain γ and the normal strain ϵ_{max} (see Eqs. 3-20 and 3-21). Then we will be able to derive the relationship between the moduli of elasticity E and G.

Example

A hollow circular shaft (Fig. 3-19) with outside diameter equal to 100 mm and inside diameter equal to 80 mm is subjected to a torque $T = 12$ kN·m. What are the maximum tensile, compressive, and shear stresses in the shaft?

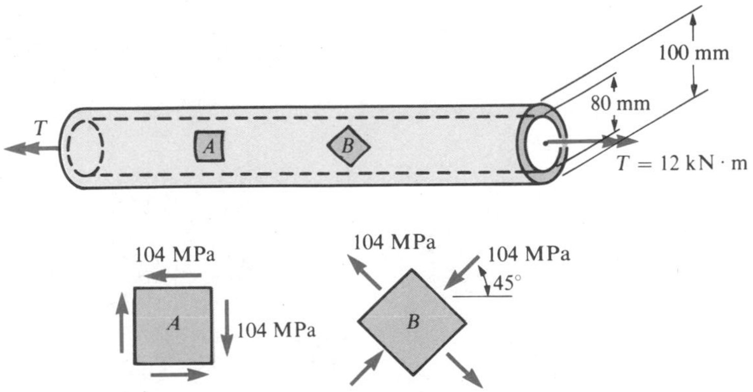

Fig. 3-19 Example

The maximum values of all three stresses are equal numerically, although they act on different planes. The maximum values are found from the torsion formula:

$$\tau_{max} = \frac{Tr}{I_p} = \frac{(12{,}000 \text{ N·m})(0.050 \text{ m})}{(\pi/32)[(0.100 \text{ m})^4 - (0.080 \text{ m})^4]} = 104 \text{ MPa}$$

The maximum shear stresses act on cross-sectional and longitudinal planes, as shown for element A in Fig. 3-19, and the maximum normal stresses act on planes at 45° to the axis (element B).

3.5 RELATIONSHIP BETWEEN MODULI OF ELASTICITY *E* AND *G*

The equations derived in the preceding section can be used to obtain an important relationship between the moduli of elasticity E and G. For this purpose, consider the stress element *abcd* shown in Fig. 3-20a. The front face of the element is assumed to be square, with the length of each side denoted as h. When this element is subjected to pure shear by stresses τ, the front face distorts into a rhombus (Fig. 3-20b) with sides of length h and with the shear strain $\gamma = \tau/G$, as explained in Section 3.4. Because of the distortion, diagonal *bd* is lengthened and diagonal *ac* is shortened. The length of diagonal *bd* can be obtained by multiplying its initial length $\sqrt{2}\,h$ by $1 + \epsilon_{max}$, where ϵ_{max} is the normal strain in the 45° direction:

$$L_{bd} = \sqrt{2}\,h(1 + \epsilon_{max}) \qquad \text{(a)}$$

This length can be related to the shear strain γ by considering the geometry of the distorted element (Fig. 3-20b).

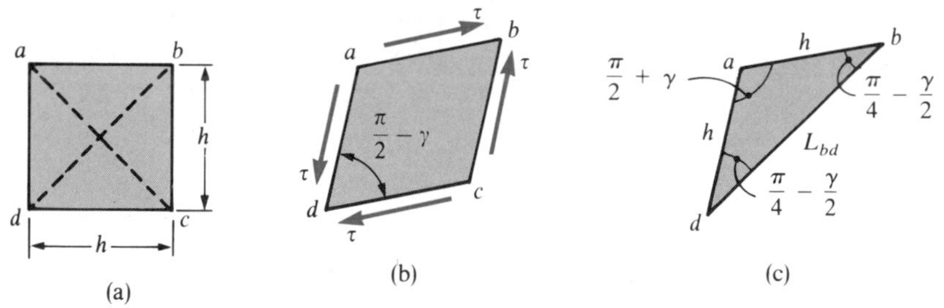

Fig. 3-20 Geometry of deformed element in pure shear

To obtain the geometric relationships, consider triangle *abd* (Fig. 3-20c), which represents one-half of the rhombus pictured in Fig. 3-20b. Side *bd* of this triangle has length L_{bd} (Eq. a), and the other sides have length h. Angle *adb* is equal to one-half of angle *adc* of the rhombus, or $\pi/4 - \gamma/2$, and angle *abd* is the same. Therefore, angle *dab* equals $\pi/2 + \gamma$. The law of cosines for triangle *abd* gives

$$L_{bd}^2 = h^2 + h^2 - 2h^2 \cos\left(\frac{\pi}{2} + \gamma\right)$$

Substituting for L_{bd} from Eq. (a) and simplifying, we get

$$(1 + \epsilon_{max})^2 = 1 - \cos\left(\frac{\pi}{2} + \gamma\right)$$

Expanding the term on the left-hand side, and also observing that $\cos(\pi/2 + \gamma) = -\sin\gamma$, we obtain

$$1 + 2\epsilon_{max} + \epsilon_{max}^2 = 1 + \sin\gamma$$

Because ϵ_{max} and γ are very small strains, we can disregard ϵ_{max}^2 in comparison with $2\epsilon_{max}$ and we can replace $\sin\gamma$ by γ. The resulting expression is

$$\epsilon_{max} = \frac{\gamma}{2} \qquad (3\text{-}22)$$

This equation shows that in pure shear the normal strain ϵ_{max} in the 45° direction is equal to one-half the shear strain γ.

The shear strain γ in Eq. (3-22) is equal to τ/G by Hooke's law (Eq. 3-20) and the normal strain ϵ_{max} is given by Eq. (3-21). Making these substitutions in Eq. (3-22) yields

$$G = \frac{E}{2(1 + v)} \qquad (3\text{-}23)$$

Thus, it is apparent that E, G, and v are not independent properties of a linearly elastic material. Instead, if two of them are known, the third can be calculated from the preceding equation. Some typical values of E, G, and v are listed in Table H-2, Appendix H.*

3.6 TRANSMISSION OF POWER BY CIRCULAR SHAFTS

The most important use of circular shafts is to transmit mechanical power from one device or machine to another, as in the drive shaft of an automobile, the propeller shaft of a ship, or the axle of a bicycle. The power is transmitted through the rotary motion of the shaft, and the amount of power transmitted depends upon the magnitude of the torque and the speed of rotation. A common design problem is the determination of the required size of a shaft so that it will transmit a specified amount of power at a specified rotational speed without exceeding the allowable stresses for the material.

Let us suppose that a motor-driven shaft (Fig. 3-21) is rotating at an angular speed ω, measured in radians per second (rad/s). The shaft is transmitting a torque T to a machine that is performing useful work. In general, the **work** W done by any torque of constant magnitude T is equal

* Equation (3-23) was derived originally by Poisson using the value $\frac{1}{4}$ for v; see Ref. 3-3.

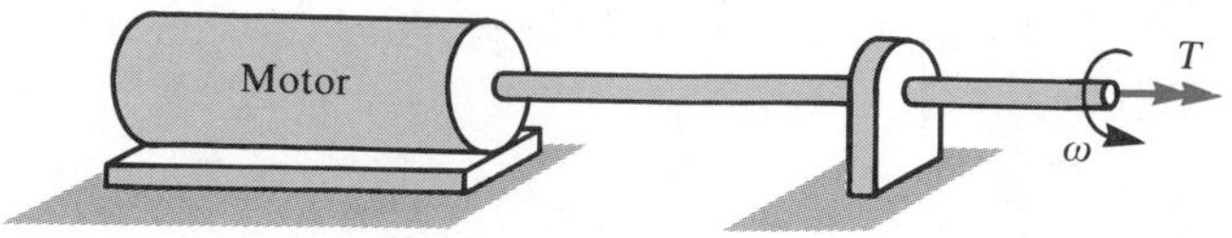

Fig. 3-21 Shaft transmitting a torque T at an angular speed ω

to the product of the torque and the angle through which it rotates; that is,

$$W = T\phi$$

where ϕ is the angular rotation in radians. **Power** is the time rate at which work is done, or

$$P = \frac{dW}{dt} = T\frac{d\phi}{dt}$$

in which P is the symbol for power and t represents time. The rate of change $d\phi/dt$ of the angular displacement ϕ is the angular speed ω, and therefore the preceding equation becomes

$$P = T\omega \qquad (\omega = \text{rad/s}) \tag{3-24}$$

This formula, which is familiar from elementary physics, gives the power transmitted by a rotating shaft. If T is expressed in newton meters, then the power is expressed in watts (W). One watt is equal to one newton meter per second (or one joule per second). If T is expressed in foot-pounds, then the power is expressed in foot-pounds per second.*

Angular speed is often expressed as the frequency f of rotation, or the number of revolutions per unit of time. The unit of frequency is the hertz (Hz), which is one revolution per second (s^{-1}). Inasmuch as one revolution equals 2π radians, we obtain

$$\omega = 2\pi f$$

The expression for power then becomes

$$P = 2\pi f T \qquad (f = \text{Hz} = \text{s}^{-1}) \tag{3-25}$$

Another commonly used unit is the number of revolutions per minute (rpm), denoted by the letter n. Therefore, we also have the following relationships:

$$n = 60f$$

and

$$P = \frac{2\pi n T}{60} \qquad (n = \text{rpm}) \tag{3-26}$$

* See Appendix A for units of power and conversion factors.

In Eqs. (3-25) and (3-26), the quantities P and T have the same units as in Eq. (3-24); that is, P has units of watts if T has units of newton meters, and P has units of foot-pounds per second if T has units of foot-pounds.

In U.S. engineering practice, power is often expressed in horsepower (hp), a unit equal to 550 ft-lb/s. The horsepower H being transmitted by a rotating shaft is

$$H = \frac{2\pi n T}{60(550)} = \frac{2\pi n T}{33{,}000} \qquad (n = \text{rpm},\ T = \text{ft-lb},\ H = \text{hp}) \qquad (3\text{-}27)$$

One horsepower is equal approximately to 746 watts.

The preceding equations relate transmitted power to the torque T in the shaft. The torque is related to the shear stresses, shear strains, and angles of twist by the formulas discussed in Sections 3.2 through 3.5.

Example 1

What is the minimum required diameter d for a solid circular shaft if it is to transmit 40 hp at 600 rpm without exceeding an allowable shear stress of 4000 psi?

The relationship between horsepower H and torque T is given by Eq. (3-27). Using that equation, we obtain

$$T = \frac{33{,}000\ H}{2\pi n} = \frac{33{,}000(40\ \text{hp})}{2\pi(600\ \text{rpm})} = 350.1\ \text{ft-lb}$$

as the torque that must be transmitted by the shaft.

The maximum shear stress produced by the torque T can be obtained from the torsion formula (Eq. 3-10). Solving that equation for the diameter of the shaft, and substituting τ_{allow} for τ_{max}, we get

$$d^3 = \frac{16T}{\pi\tau_{allow}} = \frac{16(350.1\ \text{ft-lb})(12\ \text{in./ft})}{\pi(4000\ \text{psi})} = 5.35\ \text{in.}^3$$

from which

$$d = 1.75\ \text{in.}$$

The diameter of the shaft must be at least this large if the allowable shear stress is not to be exceeded.

Example 2

A steel shaft ABC (Fig. 3-22a) of 50 mm diameter is driven at A by a motor that transmits 50 kW to the shaft at 10 Hz. The gears at B and C drive machinery requiring power equal to 30 kW and 20 kW, respectively. Compute the maximum shear stress τ in the shaft and the angle of twist ϕ between the motor at A and the gear at C.

The power transmitted by the shaft from A to B is the power supplied by the motor, or 50 kW. Therefore, the torque T_a acting in this part of the shaft (from

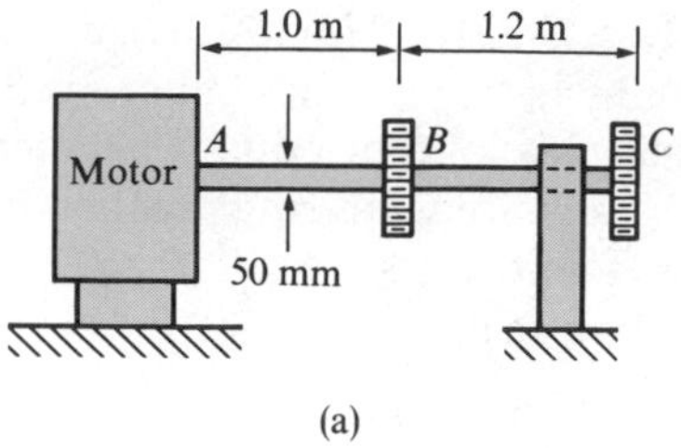

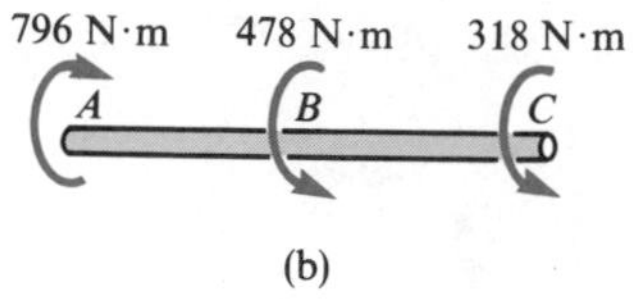

Fig. 3-22 Example 2

Eq. 3-25) is

$$T_a = \frac{P}{2\pi f} = \frac{50 \text{ kW}}{2\pi(10 \text{ Hz})} = 796 \text{ N·m}$$

This torque is applied to the shaft by the motor, as shown in the free-body diagram of the shaft (Fig. 3-22b).

At gear C, the power supplied by the shaft to the external machinery is 20 kW. It follows that the torque T_c in the shaft at C is

$$T_c = \frac{P}{2\pi f} = \frac{20 \text{ kW}}{2\pi(10 \text{ Hz})} = 318 \text{ N·m}$$

This torque is transmitted between the shaft and gear C, as shown in the free-body diagram (Fig. 3-22b). The direction of this torque must be opposite to the torque applied by the motor. (If we think of T_a as the "load" applied to the shaft, then the torques at gears B and C are the "reactions.") Finally, the torque T_b acting on the shaft at gear B can be obtained either by using Eq. (3-25) or by considering equilibrium of the shaft (Fig. 3-22b).

The shear stress and angle of twist in part AB of the shaft are

$$\tau_{ab} = \frac{16T}{\pi d^3} = \frac{16(796 \text{ N·m})}{\pi(50 \text{ mm})^3} = 32.4 \text{ MPa}$$

$$\phi_{ab} = \frac{TL}{GI_p} = \frac{(796 \text{ N·m})(1.0 \text{ m})}{(80 \text{ GPa})\left(\frac{\pi}{32}\right)(50 \text{ mm})^4} = 0.0162 \text{ rad}$$

in which we have assumed that $G = 80$ GPa. The corresponding quantities for part BC are

$$\tau_{bc} = \frac{16T}{\pi d^3} = \frac{16(318 \text{ N·m})}{\pi(50 \text{ mm})^3} = 13.0 \text{ MPa}$$

$$\phi_{bc} = \frac{TL}{GI_p} = \frac{(318 \text{ N·m})(1.2 \text{ m})}{(80 \text{ GPa})\left(\frac{\pi}{32}\right)(50 \text{ mm})^4} = 0.0078 \text{ rad}$$

Thus, the maximum shear stress in the shaft occurs in part AB and equals 32.4 MPa. Also, the total angle of twist is

$$\phi = \phi_{ab} + \phi_{bc} = 0.0162 \text{ rad} + 0.0078 \text{ rad} = 0.0240 \text{ rad}$$

inasmuch as both parts of the shaft twist in the same direction (see Fig. 3-22b).

3.7 STATICALLY INDETERMINATE TORSIONAL MEMBERS

The torsion examples discussed in preceding sections were statically determinate because the torques acting at all cross sections of the bars could be obtained from equations of equilibrium. By contrast, bars in torsion will be statically indeterminate if they are constrained by more supports than are required to hold them in stable equilibrium. Torsion members of this kind can be analyzed by supplementing the equilibrium equations with equations pertaining to the displacements (that is, by compatibility equations). The flexibility and stiffness methods, described in Section 2.3 for axially loaded members, may also be used for torsional members. However, the flexibility method is adequate for most torsion problems, and so we will limit our discussion to that method.

To illustrate the flexibility method as applied to bars in torsion, we will consider the shaft AB shown in Figs. 3-23a and b. This bar is fixed at both ends and is statically indeterminate. The bar has diameters d_a and d_b in parts AC and CB, respectively, and is loaded by a torque T_0 at C. The material of the bar is the same throughout both parts. The objectives of our analysis are to determine the reactive torques T_a and T_b at the ends, the maximum shear stresses, and the angle of rotation ϕ_c at the section where T_0 is applied.

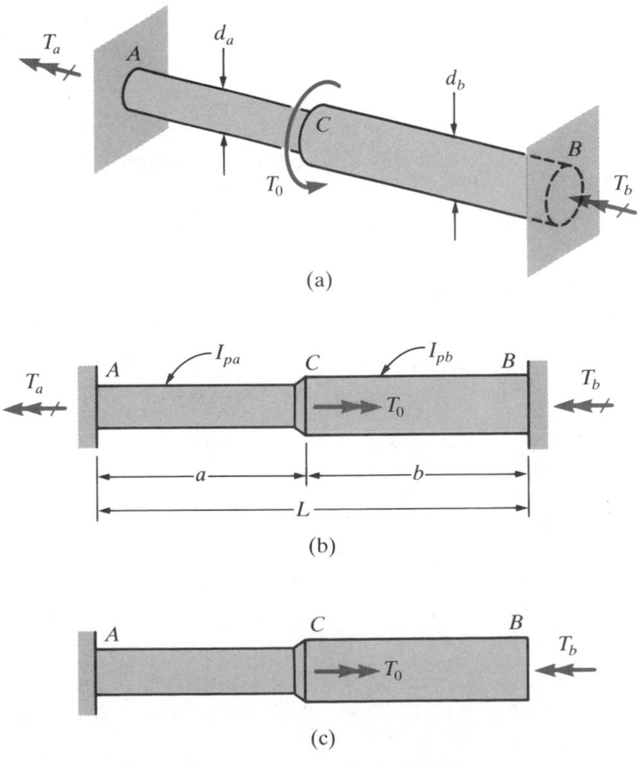

Fig. 3-23 Statically indeterminate bar in torsion

From *static equilibrium*, we obtain the following equation relating the torques (see Fig. 3-23b):

$$T_a + T_b = T_0 \tag{a}$$

To obtain a second equation relating T_a and T_b, we must select a redundant torque and then analyze the corresponding released structure. Let us select T_b as the redundant torque, which means that the released structure is obtained by removing support B (Fig. 3-23c). The torques T_0 and T_b act as loads on this released structure, and they produce an angle of twist ϕ_b at end B equal to the algebraic sum of the angles of twist of parts AC and CB. Thus, we get

$$\phi_b = \frac{(T_0 - T_b)a}{GI_{pa}} - \frac{T_b b}{GI_{pb}}$$

in which I_{pa} and I_{pb} are the polar moments of inertia of the left-hand and right-hand parts of the bar, respectively. Because the angle of rotation at end B in the original bar is equal to zero, the *equation of compatibility* is $\phi_b = 0$, or

$$\frac{T_0 a}{I_{pa}} - \frac{T_b a}{I_{pa}} - \frac{T_b b}{I_{pb}} = 0 \tag{b}$$

This equation can be solved for the redundant torque T_b, which then can be substituted into Eq. (a) to obtain T_a. The results are

$$T_a = T_0\left(\frac{bI_{pa}}{aI_{pb} + bI_{pa}}\right) \qquad T_b = T_0\left(\frac{aI_{pb}}{aI_{pb} + bI_{pa}}\right) \tag{3-28a,b}$$

If the bar has a constant cross section, so that $I_{pa} = I_{pb} = I_p$, then these equations simplify as follows:

$$T_a = \frac{T_0 b}{L} \qquad T_b = \frac{T_0 a}{L} \tag{3-29a,b}$$

These equations are analogous to those for the reactions of an axially loaded bar with fixed ends (see Eqs. 2-7 and 2-8).

The maximum shear stresses in each part of the bar are obtained directly from the torsion formula:

$$\tau_{ac} = \frac{T_a d_a}{2I_{pa}} \qquad \tau_{cb} = \frac{T_b d_b}{2I_{pb}}$$

Substituting from Eqs. (3-28a and b) gives

$$\tau_{ac} = \frac{T_0 b d_a}{2(aI_{pb} + bI_{pa})} \qquad \tau_{cb} = \frac{T_0 a d_b}{2(aI_{pb} + bI_{pa})} \tag{3-30a,b}$$

By comparing the product bd_a with the product ad_b, we can immediately determine which region of the bar has the larger stress.

Having found the torques T_a and T_b acting in the two parts of the bar, we can now obtain the angle of rotation ϕ_c at section C where the load is applied (Fig. 3-23b). This angle is equal to the angle of rotation of either part of the bar, since both parts must rotate through the same angle in order to have compatibility of rotations at C. Therefore, we obtain

$$\phi_c = \frac{T_a a}{GI_{pa}} = \frac{T_b b}{GI_{pb}} = \frac{abT_0}{G(aI_{pb} + bI_{pa})} \tag{3-31}$$

We note that if $a = b = L/2$ and $I_{pa} = I_{pb} = I_p$, the angle is

$$\phi_c = \frac{T_0 L}{4GI_p} \tag{c}$$

as anticipated from the symmetry of the bar and loading.

The preceding example illustrates the procedure that must be followed when analyzing a statically indeterminate torsional system by the flexibility method. The method is quite general and can be used in a great variety of situations, including the analysis of composite bars.

COMPOSITE BARS

An example of a composite bar is a bar made of concentric, circular torsional bars that are firmly bonded together to act as a single member, as shown in Fig. 3-24. If the two parts of the bar are constructed of the same material, the bar will behave as if it were a single bar, and all of the formulas derived in the preceding sections may be used directly. However, if the tube B and core A have different properties, the bar is statically indeterminate and a more detailed analysis is required.

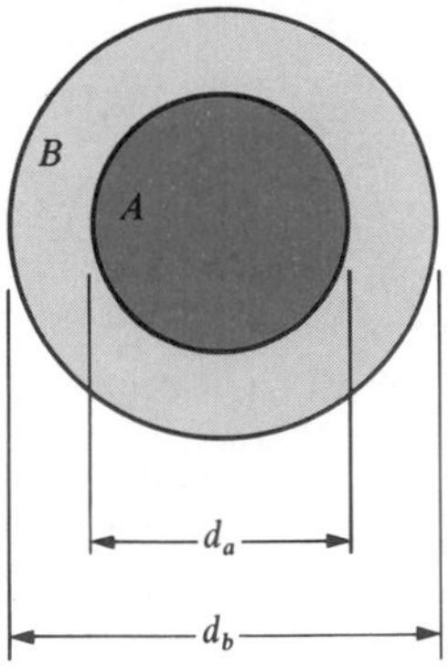

Fig. 3-24 Composite bar of two materials

For the purpose of making such an analysis, let us use the following notation:

G_a, G_b = shear moduli of elasticity for inner and outer parts, respectively

d_a, d_b = diameters of inner and outer parts, respectively

I_{pa}, I_{pb} = polar moments of inertia of inner and outer parts, respectively

Also, we will assume that the composite bar is acted upon by a total torque T, which is resisted by torques T_a and T_b developed in the core and tube, respectively.

The first equation is obtained from static equilibrium, as follows:

$$T = T_a + T_b \tag{d}$$

A second equation is obtained from the fact that the angles of twist ϕ are the same for both parts because they are bonded together. Therefore, from compatibility of rotations, we find the second equation:

$$\phi = \frac{T_a L}{G_a I_{pa}} = \frac{T_b L}{G_b I_{pb}} \tag{e}$$

in which L is the length of the bar. Solving Eqs. (d) and (e) gives the torques in the two parts of the bar:

$$T_a = T\left(\frac{G_a I_{pa}}{G_a I_{pa} + G_b I_{pb}}\right) \qquad T_b = T\left(\frac{G_b I_{pb}}{G_a I_{pa} + G_b I_{pb}}\right) \tag{3-32a,b}$$

The angle of rotation ϕ now becomes

$$\phi = \frac{TL}{G_a I_{pa} + G_b I_{pb}} \tag{3-33}$$

which is obtained by substituting from Eqs. (3-32) into Eq. (e).

The shear stresses in the bar can be obtained by applying the torsion formula to each part. For instance, the maximum shear stresses τ_a and τ_b in the core and tube, respectively, are

$$\tau_a = \frac{T_a(d_a/2)}{I_{pa}} \qquad \tau_b = \frac{T_b(d_b/2)}{I_{pb}} \tag{3-34a,b}$$

The ratio of the maximum stresses in the tube and core can be found by combining Eqs. (3-34) and (3-32); the result is

$$\frac{\tau_b}{\tau_a} = \frac{G_b d_b}{G_a d_a} \tag{3-35}$$

Note that it is possible for this ratio to be less than unity, which means that the stress in the core is larger than the stress in the tube.

The shear stress at the inner surface of the tube is not necessarily the same as the shear stress at the outer surface of the core. Although the shear strains in the two parts must be the same at the common boundary, the stresses are different whenever the materials have different shear moduli.

Example

A shaft AB of diameter d_1 has two circular flanges of diameter d_2 (Fig. 3-25a). Torques T_0 act at the ends of the shaft. While these loads are acting, a sleeve of the same material and with outer diameter d_3 and inner diameter d_2 is fitted over the flanges and welded securely to them (Fig. 3-25b). Then the torques T_0 are removed. Assuming that the flanges are rigid, determine the maximum shear stress τ_{max} in the sleeve.

When the shaft alone (Fig. 3-25a) is subjected to the torques T_0, the relative angle of rotation between the flanges is

$$\phi_0 = \frac{T_0 L}{G I_{p1}} \tag{f}$$

where L is the length of the shaft between the flanges, G is the shear modulus, and

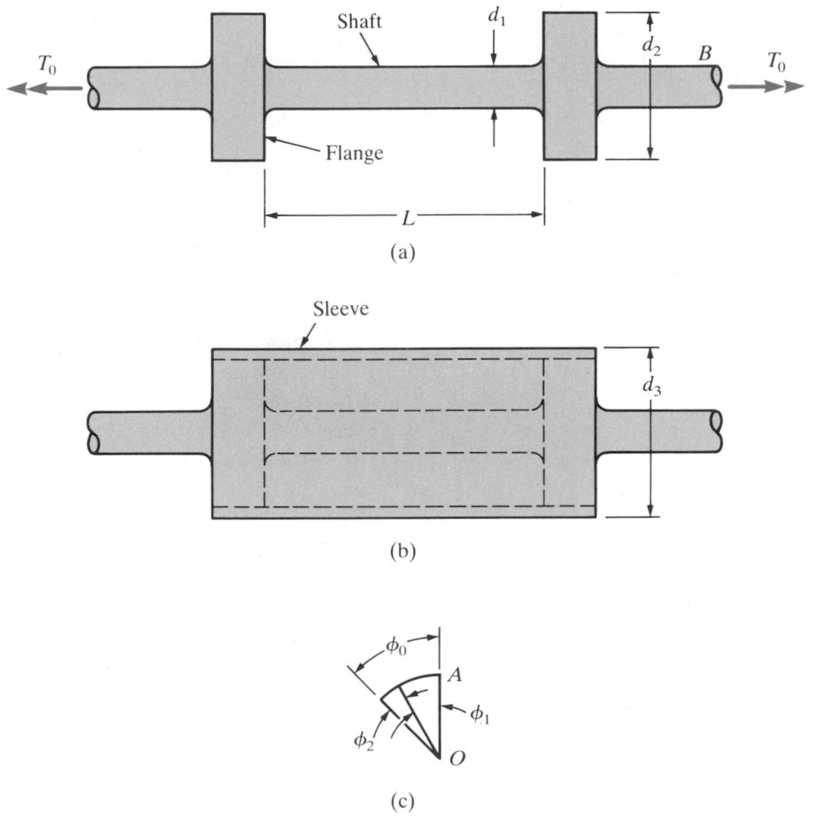

Fig. 3-25 Example

I_{p1} is the polar moment of inertia of the shaft:

$$I_{p1} = \frac{\pi d_1^4}{32} \tag{g}$$

After the sleeve is attached and the torques are removed, the shaft "unwinds" partially but is restrained by the twisting of the sleeve. We see that the system is statically indeterminate and that the angles of rotation of the two parts must be considered when finding the stresses.

Let us denote the final torques in the shaft and sleeve as T_1 and T_2, respectively. From static equilibrium it is evident that these two torques are equal, because the torque exerted by the shaft on the sleeve must be the same as the torque exerted by the sleeve on the shaft.

In its final configuration (Fig. 3-25b), the relative angle of rotation between the ends of the sleeve is

$$\phi_2 = \frac{T_2 L}{G I_{p2}} \tag{h}$$

in which I_{p2} is the polar moment of inertia of the sleeve:

$$I_{p2} = \frac{\pi}{32}(d_3^4 - d_2^4) \tag{i}$$

In a similar manner, we find the relative angle of rotation for the shaft between the flanges:

$$\phi_1 = \frac{T_1 L}{G I_{p1}} \tag{j}$$

The equation of compatibility expresses the relationship between the various angles of rotation. This relationship can be seen from the diagram in Fig. 3-25c, in which the vertical line OA represents a radius of one of the flanges before the torques T_0 are applied. When these torques are acting, the radius rotates through the counterclockwise angle ϕ_0 (see Eq. f), assuming that the other flange is held in position and does not rotate. (In other words, the diagram shows the relative angle of rotation of one flange with respect to the other.) Next, the sleeve is placed in position, the torques are removed, and the composite bar rotates back (clockwise) through the angle ϕ_2. In the final configuration, the shaft has a residual angle of rotation ϕ_1 and the sleeve has undergone a rotation through the angle ϕ_2 (Eq. h). In this manner we arrive at the equation of compatibility:

$$\phi_0 = \phi_1 + \phi_2$$

or

$$\frac{T_0 L}{G I_{p1}} = \frac{T_1 L}{G I_{p1}} + \frac{T_2 L}{G I_{p2}} \tag{k}$$

The two conditions, equilibrium and compatibility, enable us to find the unknown torques. From equilibrium we have $T_2 = T_1$. Substituting into Eq. (k) and solving yields

$$T_1 = T_2 = \frac{I_{p2} T_0}{I_{p1} + I_{p2}} \tag{l}$$

As a special case, suppose that the bar and sleeve have the same torsional stiffness ($I_{p1} = I_{p2}$). Then from Eq. (l) we see that each torque is equal to $T_0/2$. As another special case, suppose that the sleeve is rigid (I_{p2} becomes infinite); then each torque is equal to T_0 and there is no unwinding of the system. At the other extreme, suppose that the sleeve has no stiffness ($I_{p2} = 0$); then both torques equal zero. Thus, the torques T_1 and T_2 vary from zero to T_0 as the stiffness of the sleeve varies from zero to infinity.

The last step in this analysis is to find the maximum shear stress in the sleeve. Using the torsion formula and Eq. (l), we get

$$\tau_{max} = \frac{T_2(d_3/2)}{I_{p2}} = \frac{(d_3/2)T_0}{I_{p1} + I_{p2}} \tag{m}$$

If the dimensions and torque T_0 are known, we can calculate τ_{max} from this equation.

3.8 STRAIN ENERGY IN PURE SHEAR AND TORSION

When a structure is subjected to a statically applied load, work is done by the load and strain energy is absorbed by the structure, as discussed previously in Section 2.6. The evaluation of this strain energy is important in dynamic analyses and in many aspects of theory of structures. If no energy is lost in the form of heat, the work W will be equal to the strain energy U. This equality between work and energy may be used to obtain expressions for the strain energy stored in an element in pure shear. For this purpose, let us consider again a small element of material subjected to shear stresses τ on its side faces (Fig. 3-26a). For convenience, we will assume that the front face of the element is square, with each side having length h. The thickness of the element (perpendicular to the paper) is denoted by t. Under the action of the shear stresses τ, the element is distorted so that the front face becomes the rhombus shown in Fig. 3-26b, in which the shear strain is denoted as γ.

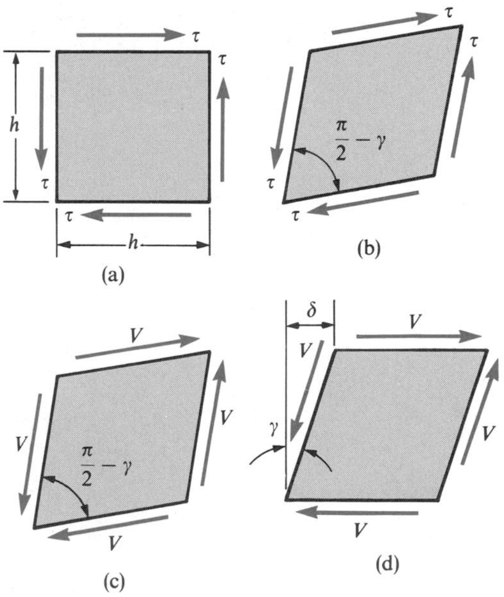

Fig. 3-26 Element in pure shear

The shear forces V acting on the side faces of the element (Fig. 3-26c) are found by multiplying the stresses by the areas on which they act:

$$V = \tau h t \tag{a}$$

These forces produce work as the element deforms from its initial shape (Fig. 3-26a) to its distorted shape (Fig. 3-26b). To calculate this work we need to determine the relative distances through which the shear forces move. This task is made easier if the element in Fig. 3-26c is rotated as a rigid body until two of its faces are horizontal, as in Fig. 3-26d. During the rigid-body rotation, the net work done by the forces V is zero because the forces occur in pairs that form two equal and opposite couples. As can be seen in Fig. 3-26d, the top face of the element is displaced horizontally through the distance δ (relative to the bottom face) as the shear force is gradually increased from zero to its final value V. The displacement δ is

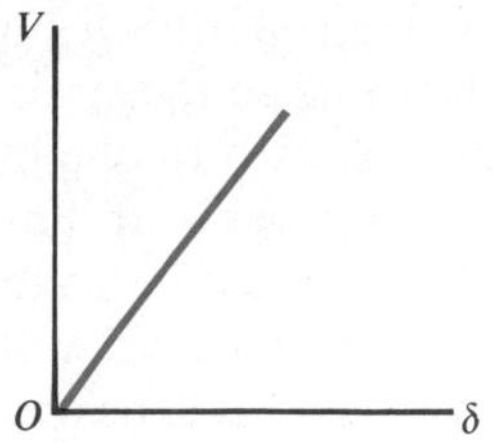

Fig. 3-27 Load-displacement diagram for a linearly elastic material

equal to the product of the shear strain γ (which is a small angle) and the vertical dimension of the element:

$$\delta = \gamma h \tag{b}$$

If we assume that the material is linearly elastic and follows Hooke's law, then the load-displacement diagram is linear and V is proportional to δ (Fig. 3-27). The strain energy U stored in the element is equal to the work W done by the shear force, which is equal to the area below the load-displacement curve:

$$U = W = \frac{V\delta}{2} \tag{c}$$

The forces on the side faces of the element (Fig. 3-26d) do not move along their lines of action; hence they do no work. Substituting from Eqs. (a) and (b) into Eq. (c), we get the total strain energy of the element:

$$U = \frac{\tau\gamma h^2 t}{2}$$

Because the volume of the element is $h^2 t$, the strain energy density (that is, the strain energy per unit volume) is

$$u = \frac{\tau\gamma}{2} \tag{3-36}$$

Finally, we introduce Hooke's law in shear ($\tau = G\gamma$) and obtain the following equations for the **strain energy density**:

$$u = \frac{\tau^2}{2G} \qquad u = \frac{G\gamma^2}{2} \tag{3-37a,b}$$

These equations are similar in form to those for uniaxial stress (Eqs. 2-37a and b). The strain energy density in pure shear may be visualized as the area below the shear stress-strain curve. The SI units for u are joules per cubic meter (J/m^3), and the USCS units are inch-pounds per cubic inch (or other similar units). Since these units are the same as those for stress, we may also express u in pascals or psi.

Now that we have obtained expressions for the strain energy density in pure shear, we can easily determine the amount of strain energy stored in a circular bar (either solid or hollow) subjected to pure torsion (Fig. 3-28a). Consider an elemental circular tube of material that extends for the length of the bar and has radius ρ and thickness $d\rho$. A small element from this tube is shown in Fig. 3-28b. It is subjected on its side faces to shear stresses τ given by the torsion formula $\tau = T\rho/I_p$. Hence,

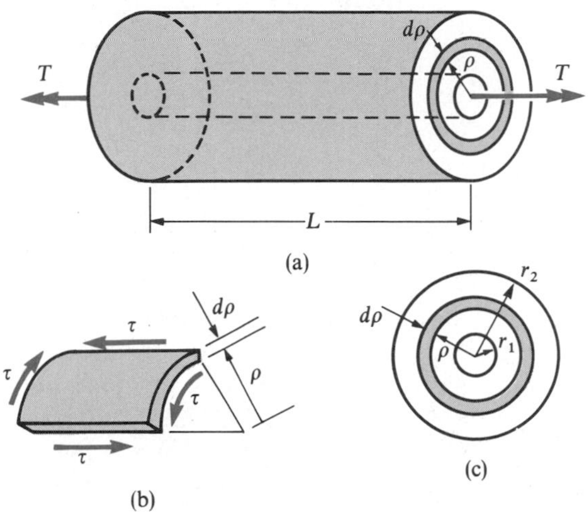

Fig. 3-28 Bar in pure torsion

the strain energy density at radius ρ is

$$u = \frac{\tau^2}{2G} = \frac{T^2\rho^2}{2GI_p^2}$$

The strain energy dU in the elemental tube is found by multiplying the strain energy density u by the volume of the tube:

$$dU = uL\,dA = \frac{T^2L\rho^2\,dA}{2GI_p^2}$$

in which $dA = 2\pi\rho\,d\rho$ is the ring-shaped area of the end face of the elemental tube. Next, the total strain energy of the bar may be obtained by integrating the preceding expression for dU between the limits $\rho = r_1$ and $\rho = r_2$ (Fig. 3-28c):

$$U = \int dU = \frac{T^2L}{2GI_p^2}\int_{\rho=r_1}^{\rho=r_2}\rho^2\,dA$$

We recognize the integral in this equation as the polar moment of inertia I_p. Hence, the elastic **strain energy** of a circular bar in pure torsion is

$$U = \frac{T^2L}{2GI_p} \qquad \text{(3-38a)}$$

This formula gives U in terms of the applied torque T. An alternative equation can be obtained by substituting from the formula for the angle of twist ($\phi = TL/GI_p$); thus,

$$U = \frac{GI_p\phi^2}{2L} \qquad \text{(3-38b)}$$

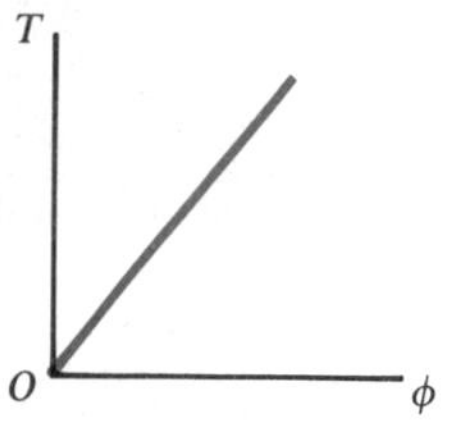

Fig. 3-29 Torque-rotation diagram for a bar in pure torsion

which gives U in terms of ϕ. The units for U are joules (J) in SI and inch-pounds (or other similar units) in USCS. Note the analogy between Eqs. (3-38a and b) for torsion and Eqs. (2-33a and b) for uniaxial loading.

A more direct way to obtain the preceding formulas for strain energy in pure torsion is to make use of a **torque-rotation diagram** for the bar (Fig. 3-29). This diagram is linear (with torque proportional to rotation) if the material follows Hooke's law and if the strains are small. During twisting of the bar, the torque T does work equal to the area below the line on the diagram; hence the corresponding elastic strain energy of the bar is

$$U = \frac{T\phi}{2} \qquad \text{(3-39)}$$

Combining this equation with the equation $\phi = TL/GI_p$ gives the same expressions for strain energy (Eqs. 3-38a and b).

NONUNIFORM TORSION

If the bar has a circular cross section with its radius varying along the axis of the bar, or if the torque changes along the axis of the bar (see Figs. 3-8 and 3-9), then we need more general formulas for the strain energy in torsion. To accomplish this result, consider an elemental disk of length dx at distance x from one end of the bar (Fig. 3-9). Assuming that the torque acting on this element is $T(x)$ and that the polar moment of inertia of its cross section is $I_p(x)$, we obtain from Eq. (3-38a) the following expression for the strain energy of the element:

$$dU = \frac{[T(x)]^2\,dx}{2GI_p(x)}$$

Therefore, the total strain energy of the bar is

$$U = \int_0^L \frac{[T(x)]^2\,dx}{2GI_p(x)} \qquad \text{(3-40a)}$$

An alternative expression for U is obtained by applying Eq. (3-38b) to the element of length dx:

$$dU = \frac{GI_p(x)(d\phi)^2}{2\,dx} = \frac{GI_p(x)}{2}\left(\frac{d\phi}{dx}\right)^2 dx$$

in which $d\phi$ is the angle of twist of the element and $d\phi/dx$ is the angle of twist θ per unit length. The total energy now becomes

$$U = \int_0^L \frac{GI_p(x)}{2}\left(\frac{d\phi}{dx}\right)^2 dx \qquad (3\text{-}40b)$$

Either Eq. (3-40a) or Eq. (3-40b) can be used to find the strain energy in nonuniform torsion, depending upon whether the torque or the angle of twist is known as a function of x.

Example 1

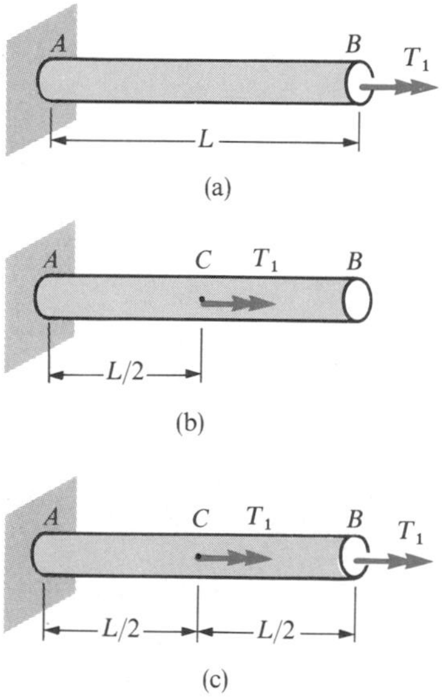

Fig. 3-30 Example 1

A circular bar AB of length L is fixed at end A and free at B (Fig. 3-30). Three different loading conditions are to be considered: (a) torque T_1 acting at end B; (b) torque T_1 acting at the midpoint C; and (c) torques T_1 acting simultaneously at B and C. For each case of loading, determine the strain energy U stored in the bar.

(a) For torque T_1 acting at end B, the strain energy is obtained directly from Eq. (3-38a):

$$U_a = \frac{T_1^2 L}{2GI_p}$$

(b) For torque T_1 acting at the midpoint C, we apply Eq. (3-38a) to part AC of the bar:

$$U_b = \frac{T_1^2(L/2)}{2GI_p} = \frac{T_1^2 L}{4GI_p}$$

(c) When both loads are acting, the torque in part CB is T_1, the torque in part AC is $2T_1$, and the strain energy is

$$U = \frac{T_1^2(L/2)}{2GI_p} + \frac{(2T_1)^2(L/2)}{2GI_p} = \frac{5T_1^2 L}{4GI_p}$$

It is important to observe that the strain energy produced by the two loads acting simultaneously is not equal to the sum of the strain energies due to the loads acting separately. As pointed out in Section 2.6, the explanation lies in the fact that strain energy is a quadratic function of the loads, not a linear function.

Example 2

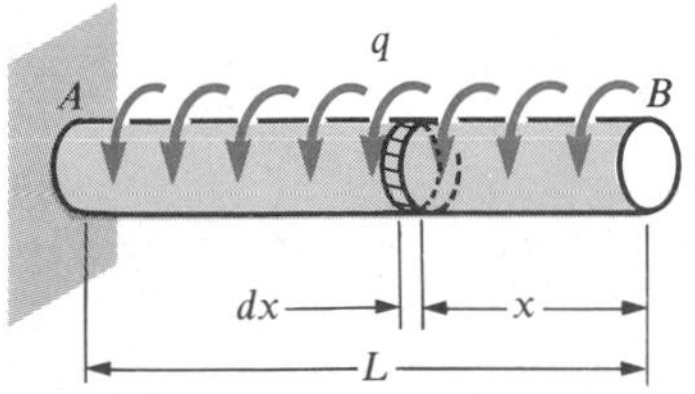

Fig. 3-31 Example 2

A circular bar AB, fixed at one end and free at the other, is loaded by a distributed torque of constant intensity q per unit distance along the axis of the bar (Fig. 3-31). Derive a formula for the amount of strain energy stored in the bar when the load is applied. Then evaluate the strain energy for the following numerical values: $L = 8$ m, $I_p = 120 \times 10^{-6}$ m^4, $q = 5$ kN·m/m, and $G = 78$ GPa.

The torque $T(x)$ acting at distance x from the free end of the bar is obtained from static equilibrium:

$$T(x) = qx$$

Substituting into Eq. (3-40a), we obtain

$$U = \int_0^L \frac{[T(x)]^2\,dx}{2GI_p} = \frac{1}{2GI_p}\int_0^L (qx)^2\,dx = \frac{q^2L^3}{6GI_p} \tag{3-41}$$

as the formula for the strain energy stored in the bar.

The strain energy can be evaluated numerically from Eq. (3-41) by substituting the given data:

$$U = \frac{(5\text{ kN}\cdot\text{m/m})^2(8\text{ m})^3}{6(78\text{ GPa})(120\times 10^{-6}\text{ m}^4)} = 228\text{ J}$$

Remember that one joule is equal to one newton meter (1 J = 1 N·m).

Example 3

A tapered bar AB of solid circular cross section is supported at one end and loaded by a torque T at the other end (Fig. 3-32). The diameter of the bar varies linearly from d_a at the left-hand end to d_b at the right-hand end. Determine the angle of twist ϕ of the bar.

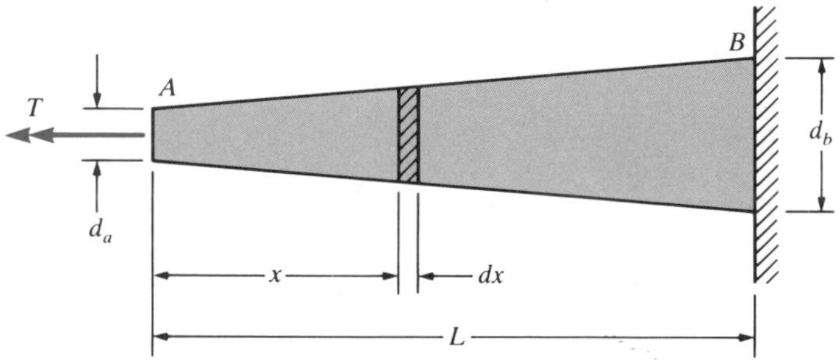

Fig. 3-32 Example 3

To find the angle of twist, we will use the principle of conservation of energy and observe that the work done by the torque T must equal the strain energy U stored in the bar. The work is

$$W = \frac{T\phi}{2} \tag{d}$$

and the strain energy U is given by Eq. (3-40a):

$$U = \int_0^L \frac{[T(x)]^2\,dx}{2GI_p(x)} \tag{e}$$

In this equation, the torque $T(x)$ is constant along the axis of the bar and equal to the load T. The polar moment of inertia is

$$I_p = \frac{\pi}{32}[d(x)]^4$$

in which $d(x)$ is the diameter of the bar at distance x from end A. From the geometry of the figure, we get

$$d(x) = d_a + (d_b - d_a)\frac{x}{L} \tag{f}$$

Substituting the preceding expressions into Eq. (e) gives

$$U = \int_0^L \frac{16T^2\,dx}{\pi G[d_a + (d_b - d_a)(x/L)]^4}$$

This expression can be integrated readily (see Eq. e in Example 2 of Section 3.3), and the result is

$$U = \frac{16T^2L}{3\pi G(d_b - d_a)}\left(\frac{1}{d_a^3} - \frac{1}{d_b^3}\right) \tag{g}$$

Equating this quantity to the work of the torque (Eq. d), we get

$$\phi = \frac{32TL}{3\pi G(d_b - d_a)}\left(\frac{1}{d_a^3} - \frac{1}{d_b^3}\right) \tag{h}$$

This equation, which is the same as Eq. (3-16) in Section 3.3, gives the angle of twist at end A of the tapered bar. Note that the method used in this example is suitable only when the bar is subjected to a single load, and then only when the desired displacement corresponds to that load.

3.9 THIN-WALLED TUBES

The torsion theory described in the preceding sections is applicable to solid or hollow bars of circular cross section. Such shapes are commonly used for torsional members, especially in machinery. However, in light-weight structures, such as aircraft and spacecraft, thin-walled tubular members of noncircular shapes are often required to resist torsion. In this section, we describe the analysis of structural members of this type.

To obtain formulas that are applicable to a variety of members, let us consider a thin-walled tube of arbitrary cross-sectional shape (Fig. 3-33a) subjected to pure torsion by torques T acting at the ends. The tube is cylindrical in shape; that is, all cross sections have the same dimensions and the longitudinal axis is a straight line. The thickness t of the wall of the tube may vary around the cross section, but t is assumed to be small in comparison with the total width of the tube. The shear stresses τ acting on the cross sections are pictured in Fig. 3-33b, which shows an element of the tube cut out between two cross sections a distance dx apart. The shear stresses are directed parallel to the edges of the cross section and "flow" around the tube. The intensity of the shear stresses varies so slightly across the thickness of the tube (because the tube is assumed to be thin) that for most purposes we may assume τ to be constant across the thickness.

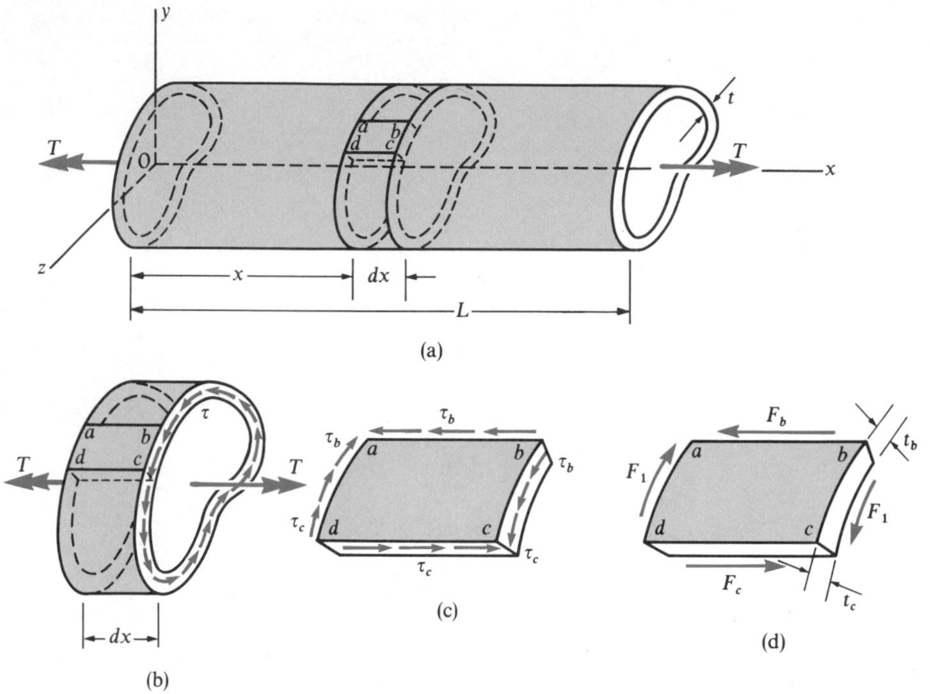

Fig. 3-33 Thin-walled tube of arbitrary cross-sectional shape

However, the manner in which τ varies around the cross section must be determined from equilibrium considerations.

To determine the magnitude of the shear stresses, consider a rectangular element obtained by making two longitudinal cuts *ab* and *cd* (Figs. 3-33a and b). This element is isolated as a free body in Fig. 3-33c. Acting on the cross-sectional face *bc* are the shear stresses τ shown in Fig. 3-33b. It is assumed that these stresses may vary in intensity as we move along the cross section from *b* to *c*; at *b* the shear stress is denoted τ_b, and at *c* it is denoted τ_c. As we know from equilibrium, identical shear stresses act in the opposite direction on the other cross-sectional face *ad*. On the longitudinal faces *ab* and *cd*, there will act shear stresses of the same magnitude as those on the cross sections, inasmuch as shear stresses on perpendicular planes are equal in magnitude (see Section 1.6). Thus, the constant shear stresses on faces *ab* and *cd* are equal to τ_b and τ_c, respectively.

The shear stresses acting on the longitudinal faces produce forces F_b and F_c (Fig. 3-33d) that can be obtained by multiplying the stresses by the areas on which they act:

$$F_b = \tau_b t_b \, dx \qquad F_c = \tau_c t_c \, dx$$

in which t_b and t_c represent the thicknesses of the tube at *b* and *c*, respectively. In addition, forces F_1 are produced by the stresses acting on faces *bc* and *ad*, but these forces do not enter into our discussion. From the equilibrium of the element in the x direction, we see that $F_b = F_c$, or

$$\tau_b t_b = \tau_c t_c$$

Because the locations of the longitudinal cuts *ab* and *cd* were selected arbitrarily, it follows from the preceding equation that the product of the shear stress τ and the thickness t of the tube is the same at every point in the cross section. This product is known as the **shear flow** and is denoted by the letter f:

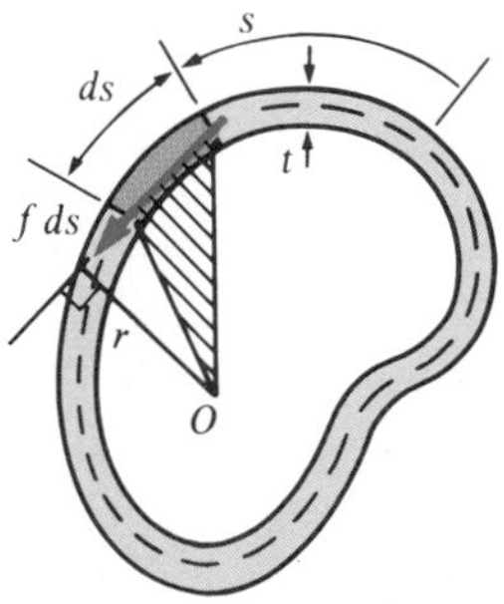

Fig. 3-34 Cross section of a thin-walled tube

$$f = \tau t = \text{constant} \tag{3-42}$$

The largest shear stress occurs where the thickness of the tube is smallest, and vice versa. Of course, in regions where the thickness is constant, the shear stress also is constant.

The next step in the analysis is to relate the shear flow f (and hence the shear stress τ) to the torque T acting on the tube. Consider an element of area of length ds in the cross section (Fig. 3-34). The distance s is measured along the **median line** (or the **centerline**) of the cross section (shown as a dashed line in the figure). The total shear force acting on the element of area is $f\,ds$, and the moment of this force about any point O is

$$dT = rf\,ds$$

in which r is the perpendicular distance from O to the line of action of the force. The latter is tangent to the median line of the cross section at the element ds. The torque T produced by the shear stresses is obtained by integrating along the entire length L_m of the median line of the cross section:

$$T = f\int_0^{L_m} r\,ds$$

The integral in this expression has a simple geometric interpretation. The quantity $r\,ds$ represents twice the area of the shaded triangle shown in Fig. 3-34. Note that the triangle has a base length ds and a height equal to r. Therefore, the integral represents double the area A_m enclosed by the median line of the cross section:

$$T = 2fA_m$$

From this equation, we get

$$f = \tau t = \frac{T}{2A_m} \qquad \tau = \frac{T}{2tA_m} \tag{3-43a,b}$$

The shear flow f and the shear stresses τ can be calculated from Eqs. (3-43) for any thin-walled tube subjected to a torque T.

The angle of twist ϕ can be determined by considering the strain energy of the tube. Because elements of the tube are in pure shear, the strain energy density is $\tau^2/2G$, as given by Eq. (3-37a). Then the strain energy of a small element of the tube having cross-sectional area $t\,ds$ (Fig. 3-34) and length dx (Fig. 3-33) is

$$dU = \frac{\tau^2}{2G} t\,ds\,dx = \frac{\tau^2 t^2}{2G}\frac{ds}{t}dx = \frac{f^2}{2G}\frac{ds}{t}dx$$

Therefore, the total strain energy of the tube is

$$U = \int dU = \frac{f^2}{2G}\int_0^{L_m}\left[\int_0^L dx\right]\frac{ds}{t}$$

in which we have utilized the fact that the shear flow f is a constant and may be placed outside the integral signs. Also, we note that t may vary with position around the median line and must remain with ds under the integral sign. Finally, we observe that the inner integral is equal to the length L of the tube. Hence the equation for U becomes

$$U = \frac{f^2 L}{2G}\int_0^{L_m}\frac{ds}{t}$$

Substituting for the shear flow from Eq. (3-43a), we obtain

$$U = \frac{T^2 L}{8GA_m^2}\int_0^{L_m}\frac{ds}{t} \tag{3-44}$$

as the equation for the strain energy of the tube in terms of the torque T.

This expression for strain energy can be written in simpler form by introducing a new property of the cross section, known in general as the **torsion constant** J. For a thin-walled tube, the torsion constant is

$$J = \frac{4A_m^2}{\displaystyle\int_0^{L_m}\frac{ds}{t}} \tag{3-45}$$

With this notation, the equation for strain energy (Eq. 3-44) becomes

$$U = \frac{T^2 L}{2GJ} \tag{3-46}$$

This equation has the same form as the one for strain energy in a circular bar (see Eq. 3-38a) except that the torsion constant J has replaced the polar moment of inertia I_p. In the special case of a cross section having constant thickness t, the expression for J (Eq. 3-45) simplifies to

$$J = \frac{4tA_m^2}{L_m} \tag{3-47}$$

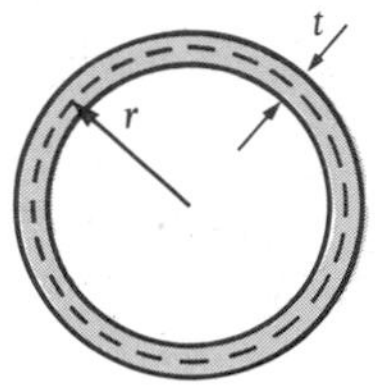

Fig. 3-35 Thin-walled circular tube

Note that J has units of length to the fourth power.

For each shape of cross section, we can evaluate J from one of the preceding equations (Eqs. 3-45 or 3-47). For example, consider a thin-walled circular tube (Fig. 3-35) of thickness t and radius r to the median line. The length of the median line and the area enclosed by it are

$$L_m = 2\pi r \qquad A_m = \pi r^2$$

Hence, the torsion constant is

$$J = 2\pi r^3 t \tag{3-48}$$

as obtained from Eq. (3-47).

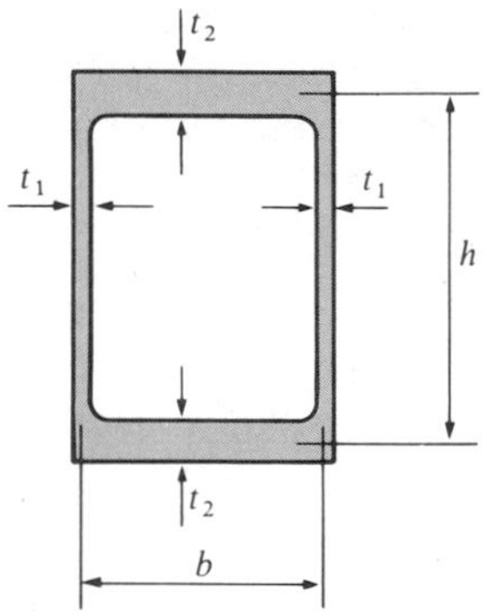

Fig. 3-36 Thin-walled rectangular tube

Figure 3-36 shows another example, a thin-walled rectangular tube. The tube has thickness t_1 on the sides and t_2 on the top and bottom. The height and width (to the median line of the cross section) are h and b, respectively. For this cross section, we have

$$L_m = 2(b + h) \qquad A_m = bh$$

and
$$\int_0^{L_m} \frac{ds}{t} = 2\int_0^h \frac{ds}{t_1} + 2\int_0^b \frac{ds}{t_2} = 2\left(\frac{h}{t_1} + \frac{b}{t_2}\right)$$

Thus, the torsion constant is

$$J = \frac{2b^2h^2t_1t_2}{bt_1 + ht_2} \tag{3-49}$$

as obtained from Eq. (3-45).

The angle of twist ϕ for a thin-walled tube may be determined by equating the work done by the applied torques T to the strain energy of the bar:

$$\frac{T\phi}{2} = \frac{T^2L}{2GJ}$$

from which

$$\phi = \frac{TL}{GJ} \tag{3-50}$$

Again we observe that the equation is of the same form as that for a circular bar (Eq. 3-8). If the angle of twist θ per unit length is needed, it can be obtained by dividing ϕ by L; thus, $\theta = T/GJ$ (compare with Eq. 3-7).* The quantity GJ is called the **torsional rigidity** of the bar.

THIN-WALLED CIRCULAR TUBES

Let us again consider the circular tube illustrated in Fig. 3-35. The shear flow and shear stresses in this tube are given by the formulas

$$f = \frac{T}{2\pi r^2} \qquad \tau = \frac{T}{2\pi r^2 t} \tag{3-51a,b}$$

which are obtained from Eqs. (3-43a and b) by substituting $A_m = \pi r^2$. Also, the angle of twist (see Eqs. 3-50 and 3-48) is

$$\phi = \frac{TL}{2\pi G r^3 t} \tag{3-52}$$

These results agree with those obtained from the previously derived equations for a hollow circular bar (Section 3.2) for the case in which the bar is thin walled. For a thin-walled tube of circular cross section, the polar moment of inertia (see Eq. 3-13) is approximately

$$I_p = 2\pi r^3 t$$

which agrees with Eq. (3-48) for J. Using this expression for I_p in the torsion formula (Eq. 3-9) leads to Eq. (3-51b) for the shear stress τ.

If a tube subjected to torsion has very thin walls, the possibility that the walls will buckle must be taken into account. For example, a long circular tube constructed of mild steel will buckle at normal working stresses when the ratio r/t (see Fig. 3-35) is about 60 (see Ref. 3-5). In all of our discussions, we assume that the wall thickness is great enough to prevent buckling by torsion.

Example 1

Compare the maximum shear stress in a circular tube (Fig. 3-35) as calculated from Eq. (3-51b) for a thin-walled tube with the stress calculated from the torsion formula (Eq. 3-9).

The formula for the shear stress in a thin-walled tube gives

$$\tau_1 = \frac{T}{2\pi r^2 t} = \frac{T}{2\pi t^3 \beta^2} \tag{a}$$

in which the relation $\beta = r/t$ is introduced. The exact maximum shear stress in the

* The torsion theory for thin-walled tubes was developed by R. Bredt, a German engineer who presented it in 1896 (Ref. 3-4).

tube is given by the torsion formula:

$$\tau_2 = \frac{T(r + t/2)}{I_p} \tag{b}$$

where $$I_p = \frac{\pi}{2}\left[\left(r + \frac{t}{2}\right)^4 - \left(r - \frac{t}{2}\right)^4\right]$$

as obtained from Eq. (3-12). After expansion, this expression for I_p simplifies to

$$I_p = \frac{\pi r t}{2}(4r^2 + t^2)$$

and the expression for τ_2 (Eq. b) becomes

$$\tau_2 = \frac{T(2r + t)}{\pi r t(4r^2 + t^2)} = \frac{T(2\beta + 1)}{\pi t^3 \beta(4\beta^2 + 1)} \tag{c}$$

The ratio τ_1/τ_2 is

$$\frac{\tau_1}{\tau_2} = \frac{4\beta^2 + 1}{2\beta(2\beta + 1)} \tag{3-53}$$

which depends only on the quantity β (that is, on the ratio r/t). For values of β equal to 5, 10, and 20, we obtain from Eq. (3-53) the ratios $\tau_1/\tau_2 = 0.92$, 0.95, and 0.98, respectively. Thus, Eq. (3-53) shows that the approximate formula for the shear stresses gives results that are only slightly less than those obtained from the exact formula and that the accuracy of the approximate formula increases as the tube becomes relatively thinner.

Example 2

A circular tube and a square tube (Fig. 3-37) are constructed of the same material. Both tubes have the same length, thickness, and cross-sectional area, and both are subjected to the same torque. What are the ratios of the shear stresses and of the angles of twist for the tubes? (Disregard the effects of stress concentrations at the corners of the square tube.)

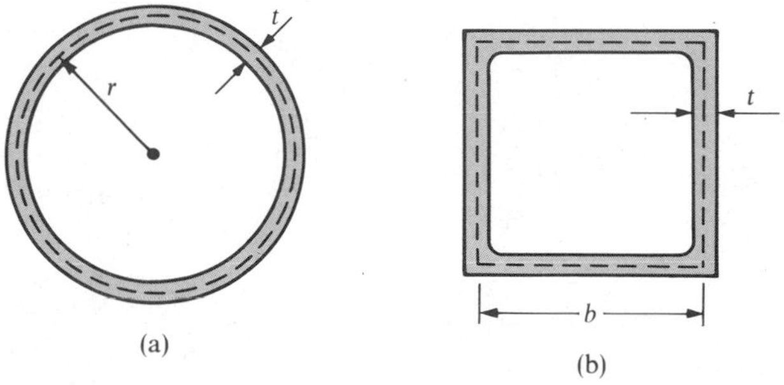

Fig. 3-37 Example 2

For the circular tube, the area A_{m1} enclosed by the median line of the cross section is $A_{m1} = \pi r^2$, where r is the radius to the median line. Also, the cross-sectional area of the circular tube is $A_1 = 2\pi rt$, and its torsion constant is $J_1 = 2\pi r^3 t$ (see Eq. 3-48).

For the square tube, the cross-sectional area is $A_2 = 4bt$, where b is the length of one side, measured along the median line. Inasmuch as the areas of both tubes are the same, we obtain $b = \pi r/2$. Also, the area enclosed by the median line of the square cross section is $A_{m2} = b^2$, and its torsion constant is $J_2 = \pi^3 r^3 t/8$ (obtained from Eq. 3-49).

The ratio τ_1/τ_2 of the shear stress in the circular tube to the shear stress in the square tube (see Eq. 3-43b) is

$$\frac{\tau_1}{\tau_2} = \frac{A_{m2}}{A_{m1}} = \frac{b^2}{\pi r^2} = \frac{\pi}{4} = 0.785 \tag{d}$$

The ratio of the angles of twist (see Eq. 3-50) is

$$\frac{\phi_1}{\phi_2} = \frac{J_2}{J_1} = \frac{\pi^2}{16} = 0.617 \tag{e}$$

These results show that the circular tube not only has a lower shear stress than does the square tube but also has greater stiffness against rotation.

*3.10 STRESS CONCENTRATIONS IN TORSION

In the previous sections we discussed stresses in members subjected to torsion assuming that the stress distribution varied in a smooth and continuous manner. This assumption is valid provided that there are no abrupt changes in the shape of the bar (no holes, grooves, notches, and the like) and provided that the loads are not concentrated over very small areas. If such disruptive conditions do exist, then high localized stresses will be created in the regions surrounding the discontinuities. In practical designs these **stress concentrations** are handled by means of **stress-concentration factors**, as explained previously in Section 2.10.

The effects of a stress concentration are confined to a small region around the discontinuity, in accord with Saint-Venant's principle (see Section 2.10). For instance, consider a stepped shaft consisting of two parts having different diameters (Fig. 3-38). The larger part has diameter D_2 and the smaller part diameter D_1. The junction between the parts forms a "shoulder" that is machined with a fillet of radius R. Without the fillet, the theoretical stress concentration factor would be infinitely large because of the sharp 90° reentrant "corner." Of course, infinite stresses cannot exist. Instead, the material at the reentrant corner will deform and partially relieve the high stress concentration. Such a situation is very dangerous under dynamic loads, and in good design a fillet is always used. The larger the radius of the fillet, the lower will be the stresses.

At a distance from the shoulder approximately equal to D_2 (see section A-A) the torsional stress is practically unaffected by the discontinuity. Therefore, the maximum stress τ_2 in the part of the bar to

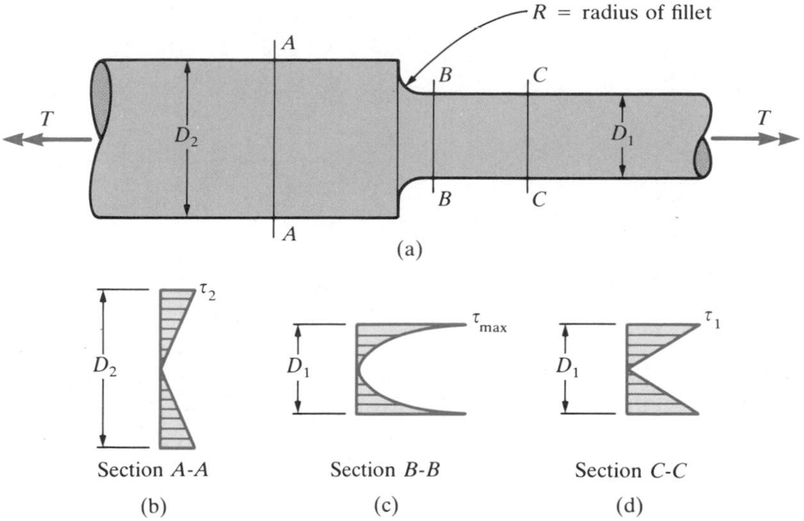

Fig. 3-38 Stepped shaft in torsion

the left of section A-A can be found from the torsion formula using D_2 as the diameter (see Fig. 3-38b). The same general comments apply at section C-C, which is distance D_1 (or greater) from the toe of the fillet. Because the diameter D_1 is less than the diameter D_2, the maximum stress τ_1 at section C-C (see Fig. 3-38d) is larger than τ_2.

The stress-concentration effect is greatest at section B-B, which cuts through the toe of the fillet. At this section the maximum stress is

$$\tau_{max} = K\tau_{nom} = K\frac{Tr}{I_p} = K\left(\frac{16T}{\pi D_1^3}\right) \tag{3-54}$$

In this equation, K is the stress-concentration factor and τ_{nom} (equal to τ_1) is the nominal shear stress that exists in the smaller part of the shaft (see Eq. 3-10). Values of the factor K are plotted in Fig. 3-39 as a function of the ratio R/D_1. Curves are plotted for various values of the ratio D_2/D_1. Note that when the fillet radius R becomes very small and the transition from one diameter to the other is abrupt, the value of K becomes quite large. Conversely, when R is large, the value of K approaches 1.0 and the effect of the stress concentration disappears. The dashed curve in Fig. 3-39 is for the special case of a full quarter-circular fillet ($D_2 = D_1 + 2R$).

Many other cases of stress concentrations for circular shafts, such as a shaft with a keyway or a shaft with a hole, are available in the engineering literature (see, for example, Ref. 2-14). As explained in Chapter 2, stress concentrations are important for static loads and brittle materials, and for all cases involving dynamic loads. As a case in point, fatigue failures are a major concern in the design of rotating shafts and axles (see Section 2.9 for a brief discussion of fatigue). The theoretical stress-concentration factors K given in this section are based upon linearly elastic behavior of the material. Experiments show that under fatigue loading, these factors are conservative and failures in ductile

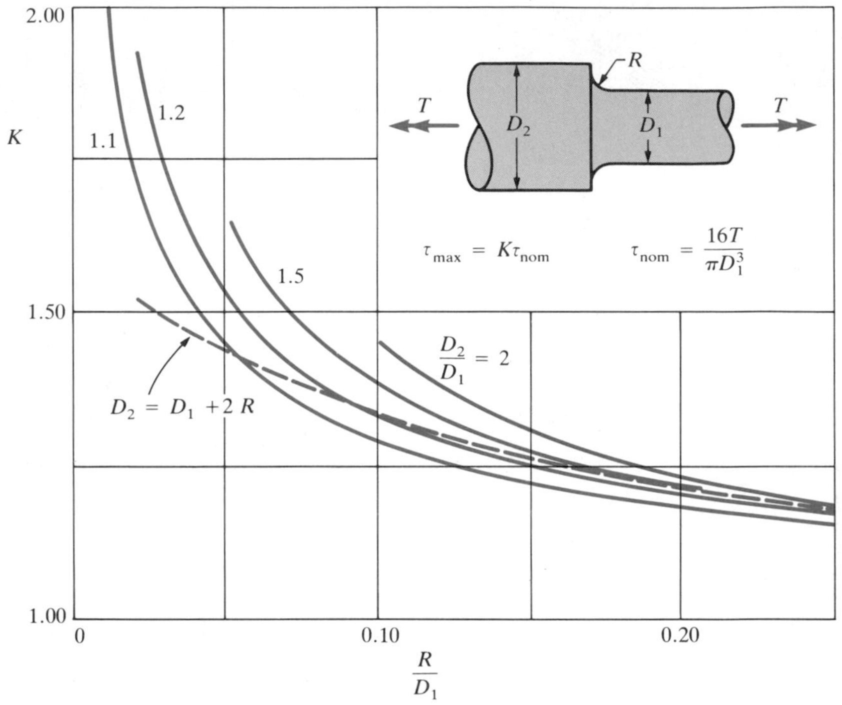

Fig. 3-39 Stress-concentration factor K for a stepped shaft in torsion. (The dashed line is for a full quarter-circular fillet.)

materials occur at larger loads than those predicted by the theoretical factors.

*3.11 NONLINEAR TORSION OF CIRCULAR BARS

The equations derived in the preceding sections for the torsional stresses in circular bars are valid only if the material follows Hooke's law. In this section we will consider the behavior of circular bars when the shear stresses exceed the proportional limit. Under such conditions, we may still assume from considerations of symmetry that cross sections remain plane and radii remain straight. Therefore, the shear strain γ at distance ρ from the axis of the bar (see Fig. 3-2d) is given by the same formula as in the case of linearly elastic torsion, namely,

$$\gamma = \rho\theta \tag{3-55}$$

in which θ is the angle of twist per unit length. Also, the maximum shear strain (at the outer edge of the cross section) is

$$\gamma_{max} = r\theta \tag{3-56}$$

in which r is the radius of the bar.

For any assumed value of θ, we can find the shear stress τ at any point in the bar if the shear stress-strain diagram for the material is known (Fig. 3-40a). The procedure consists of first obtaining the shear

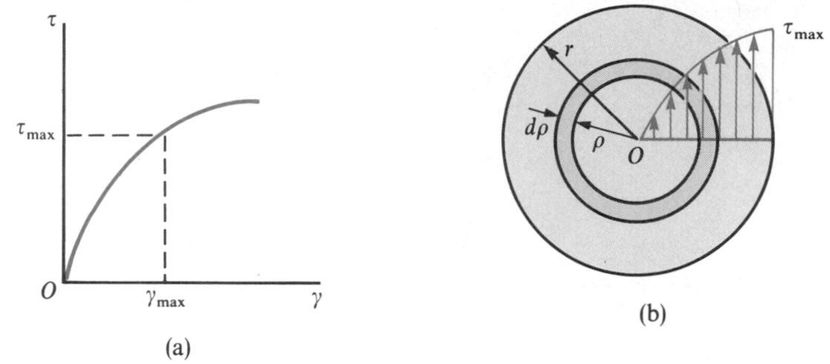

Fig. 3-40 Nonlinear torsion of a circular bar: (a) shear stress-strain diagram, and (b) distribution of shear stresses acting on the cross section

strain from Eqs. (3-55) and (3-56) and then obtaining the corresponding shear stress from the stress-strain diagram. For instance, at the outer edge of the cross section the strain is γ_{max} (calculated from Eq. 3-56), and the corresponding stress τ_{max} is obtained from the stress-strain diagram (Fig. 3-40a). At intermediate points in the cross section, the same procedure is used with Eq. (3-55). As a result, the distribution of shear stresses over the cross section (Fig. 3-40b) can be readily determined. The diagram of stresses has the same shape as the stress-strain diagram itself.

The torque T that produces the assumed angle of twist θ per unit length is the resultant of the shear stresses acting on the cross section (see Fig. 3-40b) and may be obtained from an equation of statics:

$$T = \int_0^r 2\pi\rho^2\tau \, d\rho \tag{3-57}$$

From Eq. (3-55) we obtain $\rho = \gamma/\theta$ and $d\rho = d\gamma/\theta$. Substituting these relations into Eq. (3-57), and also changing the upper limit of integration to γ_{max} to correspond with the new independent variable γ, we obtain

$$T = \frac{2\pi}{\theta^3}\int_0^{\gamma_{max}} \tau\gamma^2 \, d\gamma \tag{3-58}$$

This equation relates T and the assumed value of θ. To facilitate its evaluation, we note that the integral on the right-hand side has a simple geometric interpretation. It is the moment of inertia with respect to the vertical axis (that is, the τ axis) of the area below the stress-strain curve (Fig. 3-40a) between the origin O and the maximum strain γ_{max}. Thus, for any assumed value of θ, we can calculate γ_{max} from Eq. (3-56) and the corresponding moment of inertia from Fig. 3-40a. Then we can obtain the value of the torque T from Eq. (3-58). By repeating this procedure for various values of θ, we obtain a curve giving the relationship between T and θ. Having such a curve, we can determine θ for any given value of T.

If the material of the bar has a pronounced yield stress τ_y, the stress-strain diagram can be idealized as shown in Fig. 3-41a. The diagram consists of two straight lines, the first representing linearly elastic behavior and the second representing perfectly plastic behavior. If the maximum strain in the bar is less than the yield strain γ_y, then the bar behaves

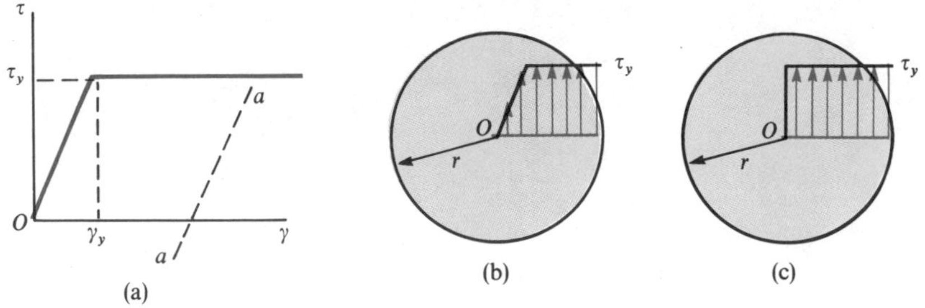

Fig. 3-41 Torsion of a circular bar of elastic-plastic material

elastically and the formulas derived in Section 3.2 can be used. When the strain at the outer edge of the cross section exceeds γ_y, the stress distribution on the cross section will have the form shown in Fig. 3-41b. Yielding begins at the outer edges of the bar and moves progressively inward as the strains increase. If the strains become very large, the region of yielding will approach the middle of the bar and the stress distribution will approach the uniform distribution shown in Fig. 3-41c. The corresponding torque T_u is the ultimate torque for the bar, and its value can be found from Eq. (3-57):

$$T_u = \int_0^r 2\pi\rho^2\tau_y\,d\rho = \frac{2\pi r^3\tau_y}{3} \tag{3-59}$$

When this value of torque is reached, further twisting of the shaft will occur without an increase in torque. Eventually, the effects of strain hardening will become noticeable and stresses greater than τ_y will develop.

The torque T_y at which yielding first begins in the bar is found from the torsion formula (Eq. 3-9) by substituting τ_y for τ_{max}:

$$T_y = \frac{\tau_y I_p}{r} = \frac{\pi r^3\tau_y}{2} \tag{3-60}$$

Comparing Eqs. (3-59) and (3-60), we see that the ratio of the ultimate torque to the yield torque is

$$\frac{T_u}{T_y} = \frac{4}{3} \tag{3-61}$$

From this result wc obscrvc that after yielding begins in the bar, it is only necessary to increase the torque by one-third to bring the bar to its ultimate load-carrying capacity.

RESIDUAL STRESSES

If a bar in torsion is loaded beyond the elastic limit and then the load is removed, some **residual stresses** will remain in the bar. To illustrate their calculation, assume that a solid circular bar is loaded to the ultimate torque T_u, thereby producing the stress distribution shown in Fig. 3-41c,

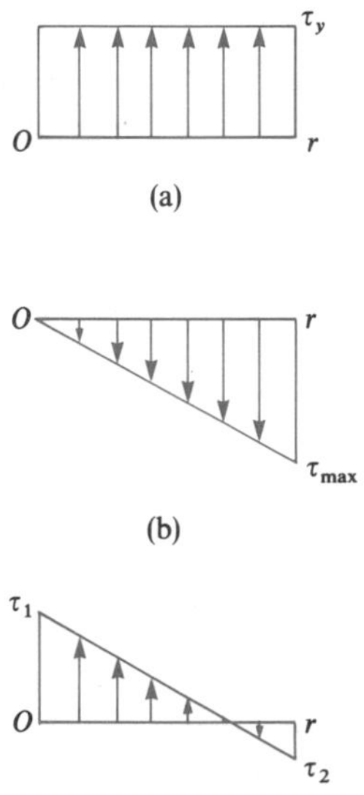

Fig. 3-42 Residual stresses in torsion

and that this torque is then removed. During the unloading process, the material follows a straight line on the stress-strain diagram (line *a-a* in Fig. 3-41a), parallel to the initial straight line representing Hooke's law. Thus, the stresses produced during unloading can be obtained from the equations for linearly elastic behavior (Eqs. 3-9 through 3-11).

The superposition of the stresses produced during loading and unloading is shown in Fig. 3-42. The stresses reached during loading are shown in part (a) of the figure, which is the same as the stress diagram of Fig. 3-41c. The corresponding torque is $T_u = 2\pi r^3 \tau_y/3$. This same torque is represented by the unloading stress diagram in Fig. 3-42b, except that now the torque acts in the opposite direction and the behavior is linearly elastic. The maximum stress is $\tau_{max} = T_u r/I_p$, which becomes $\tau_{max} = 4\tau_y/3$ when we substitute for T_u and I_p. The residual stresses in the bar, obtained by superimposing the stresses due to loading and unloading, are shown in Fig. 3-42c. At the center of the bar, the residual stress is

$$\tau_1 = \tau_y \tag{3-62}$$

and, at the outer edge of the cross section, the residual stress is

$$\tau_2 = \tau_{max} - \tau_y = \frac{\tau_y}{3} \tag{3-63}$$

This latter stress is opposite in direction to τ_1. This same procedure for calculating residual stresses can be used with other shapes of stress-strain diagrams.

PROBLEMS / CHAPTER 3

3.2-1 Determine the length L of a solid steel shaft of diameter $d = 1.5$ in. subjected to a torque T (see figure) if the maximum shear stress is 10,500 psi when the angle of twist ϕ is 3°. (Assume $G = 11.5 \times 10^6$ psi.)

3.2-2 A high-strength steel drill rod ($G = 80$ GPa) with a diameter of 12 mm is subjected to a torque T (see figure). What is the minimum required length L of the rod so that the cross section at one end can be rotated 180° with respect to the other end without exceeding an allowable shear stress of 300 MPa?

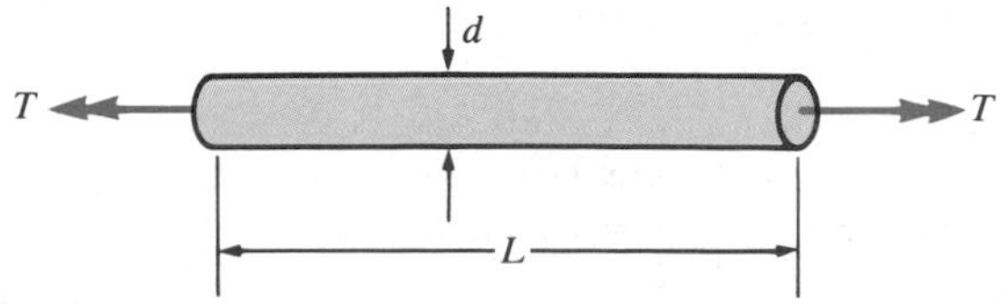

Probs. 3.2-1, 3.2-2, and 3.2-3

3.2-3 An aluminum bar of circular cross section is twisted by torques T acting at the ends (see figure). The dimensions and shear modulus are as follows: $L = 4.0$ ft, $d = 1.0$ in., and $G = 3.8 \times 10^6$ psi. If the angle of rotation at one end of the bar with respect to the other end is 5°, what are the maximum shear stress τ_{max} and the maximum shear strain γ_{max} in the bar?

3.2-4 A round axle (see figure) is made of oak having an allowable stress in shear, parallel to the grain, of 1800 kPa. If the axle has a diameter $d = 60$ mm, what is the maximum permissible torque T that can be applied?

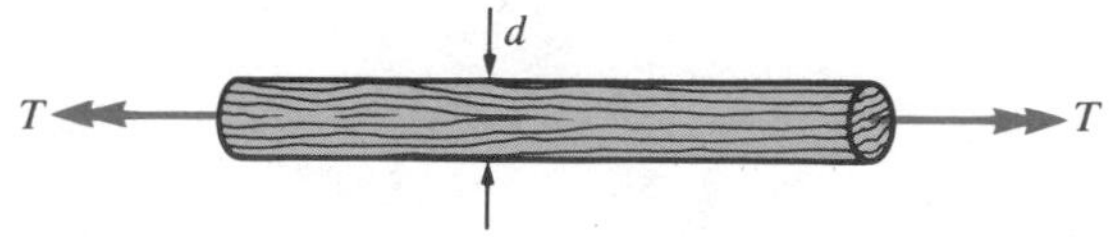

Prob. 3.2-4

3.2-5 The steel shaft of a socket wrench is 0.5 in. in diameter and 18 in. long (see figure). If the allowable stress in shear is 11,000 psi, what is the maximum permissible torque T that may be exerted with the wrench? Through what angle ϕ will the shaft twist under the action of the maximum torque? (Assume $G = 11.8 \times 10^6$ psi and disregard any bending of the shaft.)

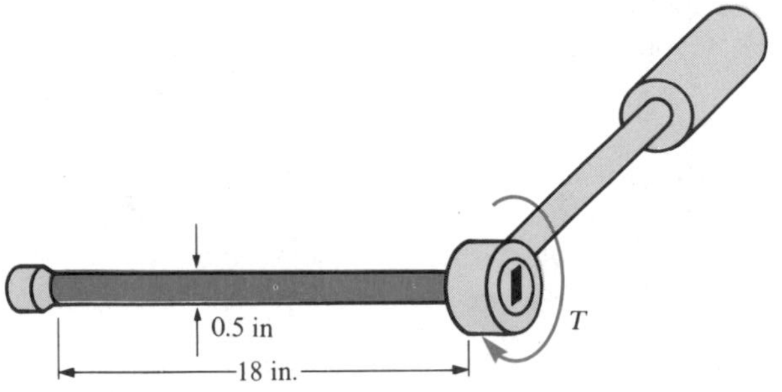

Prob. 3.2-5

3.2-6 Three identical circular disks A, B, and C are welded to the ends of three identical circular bars (see figure). The bars lie in a common plane and the disks are in planes perpendicular to the axes of the bars. The bars are welded at their intersection to form a rigid connection. Each bar has diameter $d_1 = 10$ mm and each disk has diameter $d_2 = 75$ mm. Forces P_1, P_2, and P_3 create couples acting on disks A, B, and C, respectively. If $P_1 = 100$ N, what is the maximum shear stress τ_{max} in any of the three bars?

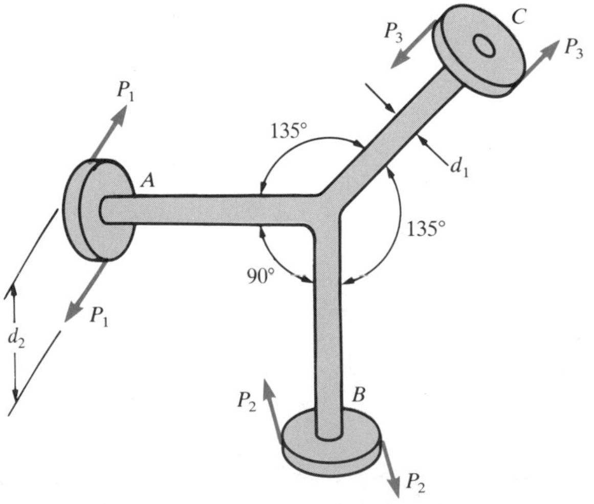

Prob. 3.2-6

3.2-7 A solid metal shaft of diameter 2 in. and length 6 ft is twisted in a testing machine until one end rotates through an angle $\phi = 5°$ with respect to the other end. For this angle of twist, the torque is measured as $T = 635$ ft-lb. Calculate the maximum shear stress τ_{max} in the shaft and the shear modulus of elasticity G.

3.2-8 A hollow circular tube of metal is subjected to twisting by torques T applied at the ends (see figure). The bar is 0.5 m long, and the inside and outside diameters are 30 mm and 40 mm, respectively. It is determined by measurement that the angle of rotation ϕ is 3.57° when the torque T is 600 N·m. Calculate the shear modulus of elasticity G for the material.

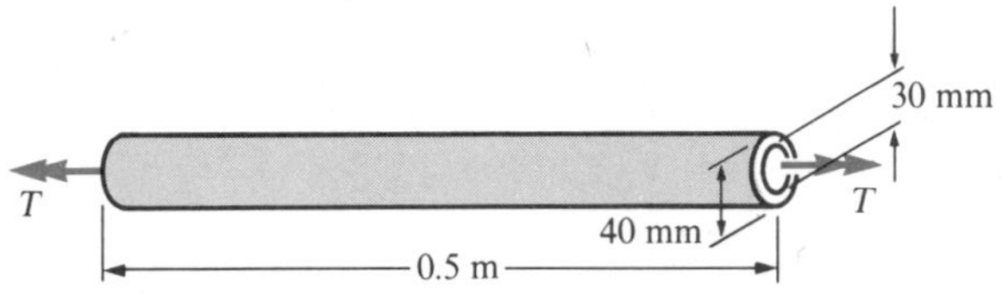

Prob. 3.2-8

3.2-9 A hollow shaft has outer diameter $d_2 = 3.6$ in. and inner diameter $d_1 = 2.5$ in. (see figure). Calculate the shear stresses τ_2 and τ_1 acting on elements at the outer and inner surfaces, respectively, due to the torque $T = 4000$ ft-lb. Draw a sketch showing how the shear stresses τ vary in magnitude along a radial line.

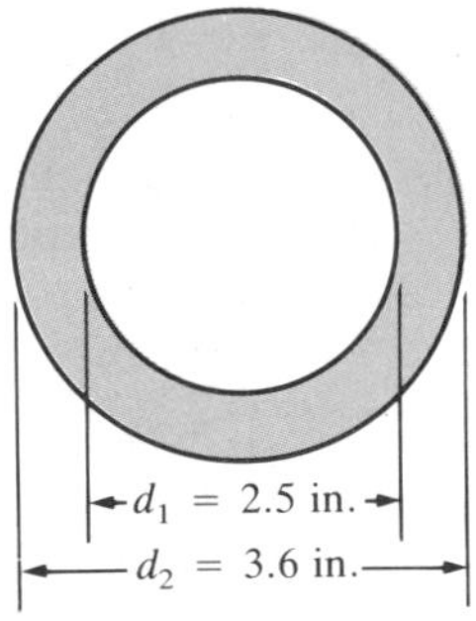

Prob. 3.2-9

3.2-10 A propeller shaft for a small ship is made of a solid steel bar 100 mm in diameter. The allowable stress in shear is 50 MPa, and the allowable angle of twist is 2° in 3 m. Assuming that $G = 80$ GPa, determine the maximum torque T that can be applied to the shaft.

3.2-11 What is the minimum required diameter d of a solid circular bar subjected to a torque $T = 30{,}000$ in.-lb if the allowable shear stress is 10,000 psi and the allowable angle of twist per unit length is 1° per 3 ft? (Assume $G = 11 \times 10^6$ psi.)

3.2-12 A hollow aluminum shaft ($G = 28$ GPa) with an outside diameter of 100 mm and an inside diameter of 86 mm is 2.4 m long. (a) If the shaft is twisted by torques at the ends, what is the total angle of twist ϕ when the maximum shear stress is 50 MPa? (b) What diameter d is required for a solid shaft to carry the same torque with the same maximum stress?

3.2-13 A hollow circular shaft and a solid circular shaft of the same material are to be designed to transmit the same torque T with the same maximum shear stress. If the inner radius of the hollow shaft is 0.85 times the outer radius, find: (a) the ratio of the outer diameter of the hollow shaft to the diameter of the solid shaft, and (b) the ratio of the weight of the hollow shaft to the weight of the solid shaft.

3.2-14 A solid circular bar initially has radius r (see figure). Then a hole of radius βr is bored longitudinally through the shaft. Derive formulas for: (a) the percent area removed, and (b) the percent reduction in the magnitude of the torque that may be applied to the bar. Plot a graph showing these percentages versus β.

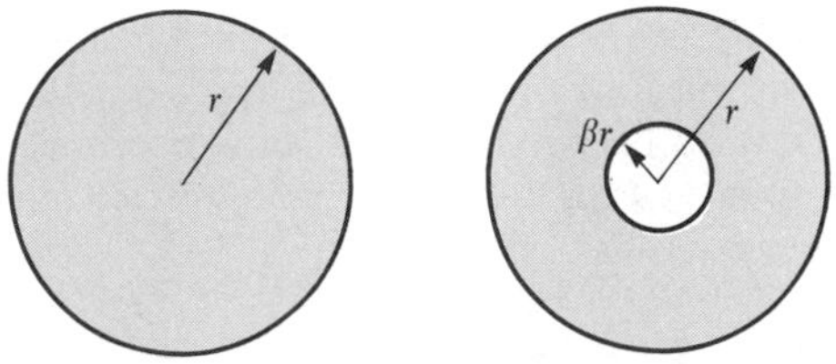

Prob. 3.2-14

3.3-1 A stepped shaft ABC consisting of two solid circular parts is subjected to torques T_1 and T_2 acting as shown in the figure. The larger part of the shaft has length $L_1 = 25$ in. and diameter $d_1 = 2.5$ in.; the smaller part has $L_2 = 18$ in. and $d_2 = 2.0$ in. The material is steel with shear modulus $G = 11 \times 10^6$ psi. If the torques are $T_1 = 9000$ in.-lb and $T_2 = 6000$ in.-lb, calculate: (a) the maximum shear stress τ_{max} in the shaft, and (b) the angle of rotation ϕ (degrees) at end C.

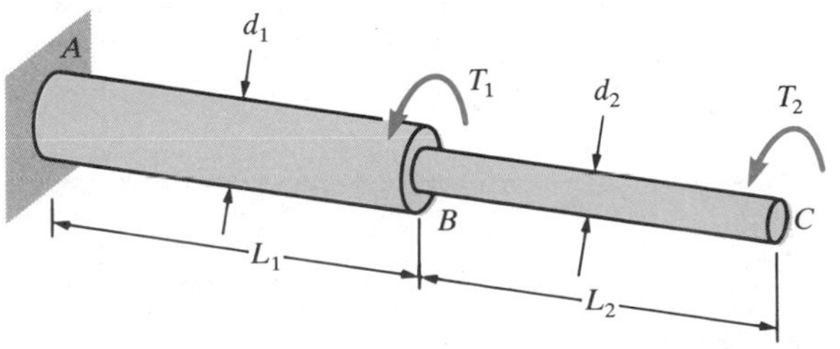

Prob. 3.3-1

3.3-2 A stepped shaft is subjected to three torques as shown in the figure. The length of each section is 0.5 m and the diameters are 80 mm, 60 mm, and 40 mm. The material is steel with shear modulus of elasticity $G = 80$ GPa. (a) Calculate the maximum shear stress τ_{max} in the shaft. (b) Calculate the angle of twist ϕ (degrees) at the free end.

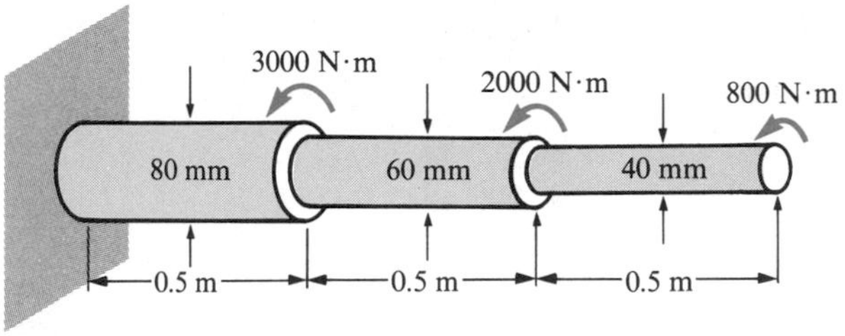

Prob. 3.3-2

3.3-3 A circular bar in torsion consists of two parts, as shown in the figure. One part has diameter $d_1 = 2.0$ in. and length $L_1 = 50$ in., and the other part has diameter $d_2 = 1.5$ in. and length $L_2 = 50$ in. What is the allowable torque T if the angle of twist between the ends of the bar is not to exceed 0.02 radians and the shear stress is not to exceed 4000 psi? (Assume $G = 12 \times 10^6$ psi.)

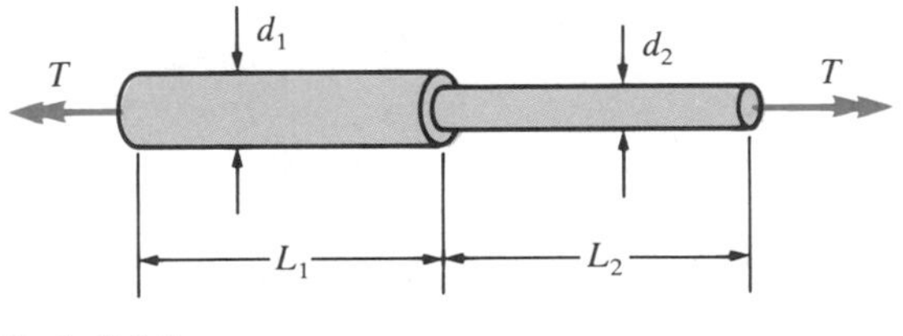

Prob. 3.3-3

3.3-4 A shaft ABC (see figure) consists of a solid bar of diameter d and length $L = 0.8$ m from A to B and a hollow bar of outside diameter $2d$, inside diameter d, and length $L = 0.8$ m from B to C. Torques are applied to the shaft through gears at A, B, and C. These torques are $T =$ 400 N·m, $7T = 2800$ N·m, and $8T = 3200$ N·m, respectively, acting in the directions shown. The shaft is made of steel having $G = 84$ GPa. What is the smallest permissible diameter d if the wind-up (the total angle of twist ϕ) between ends A and C is limited to 5°?

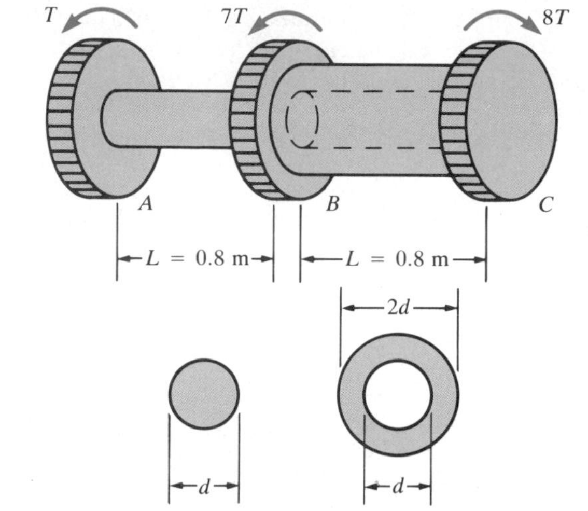

Prob. 3.3-4

3.3-5 Four gears are attached to a solid shaft and transmit the torques shown in the figure. Determine the required diameters d_{ab}, d_{bc}, and d_{cd} for each part of the shaft if the allowable stress in shear is 10,000 psi.

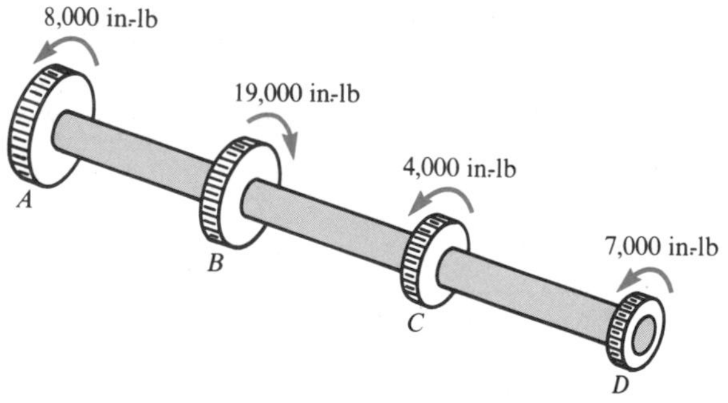

Probs. 3.3-5 and 3.3-7

3.3-6 A shaft of solid circular cross section with two different diameters is shown in the upper part of the figure. A prismatic hollow shaft made of the same material, having the same length, and having the same torsional stiffness is shown in the lower part of the figure. Determine the outside diameter d of the hollow shaft if the wall thickness t of the hollow shaft equals $d/10$.

3.3-7 Solve Prob. 3.3-5 assuming that the shaft is hollow with an inside diameter of 1 in. throughout the entire length of the shaft.

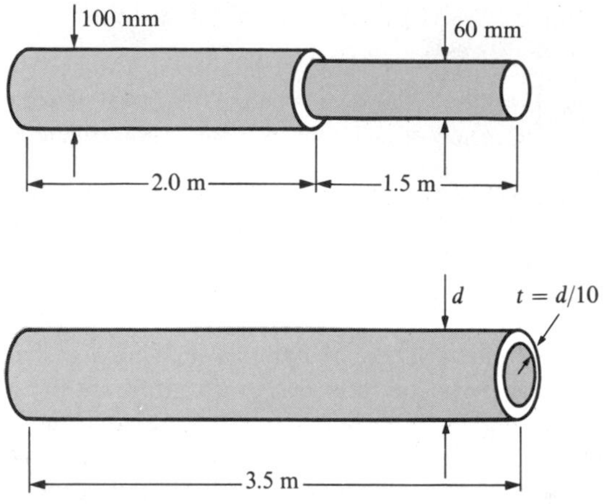

Prob. 3.3-6

3.3-8 Refer to the tapered bar shown in Fig. 3-12 and analyzed in Example 2 of Section 3.3. For what ratio d_b/d_a will the angle of twist ϕ be one-half the angle of twist for a prismatic bar of diameter d_a?

3.3-9 A uniformly tapered tube AB of hollow circular cross section is shown in the figure. The tube has constant wall thickness t and length L. The average diameters at the ends are d_a and $2d_a$. Because the wall of the tube is relatively thin, the polar moment of inertia may be approximated by the formula $I_p \approx \pi d^3 t/4$ (see Eq. 3-13). Derive a formula for the angle of twist ϕ of this tube when it is subjected to torques T acting at the ends.

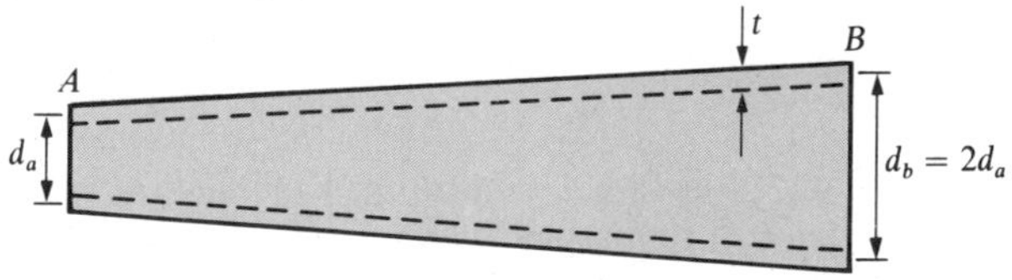

Prob. 3.3-9

3.3-10 A prismatic bar AB of solid circular cross section is fixed at the left-hand end and free at the right-hand end (see figure). The bar is subjected to a distributed torque of

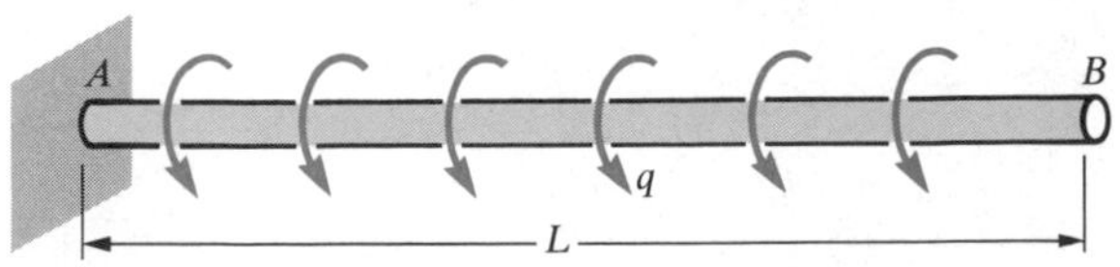

Prob. 3.3-10

constant intensity q per unit length. Derive a formula for the angle of rotation ϕ at end B of the bar.

3.3-11 Solve the preceding problem if the intensity q of the distributed torque varies linearly from a maximum value q_0 at end A to zero at end B.

***3.3-12** A horizontal circular bar of length L is supported at end A and free at end B (see figure). Horizontal vanes are attached to the side of the bar. These vanes have lengths that vary linearly from zero at end B to a maximum length c at end A. The total weight of the vanes is W. Determine the angle of rotation ϕ of end B of the bar. (Note: The vanes also produce bending of the bar, but this bending has no effect on the angle of twist.)

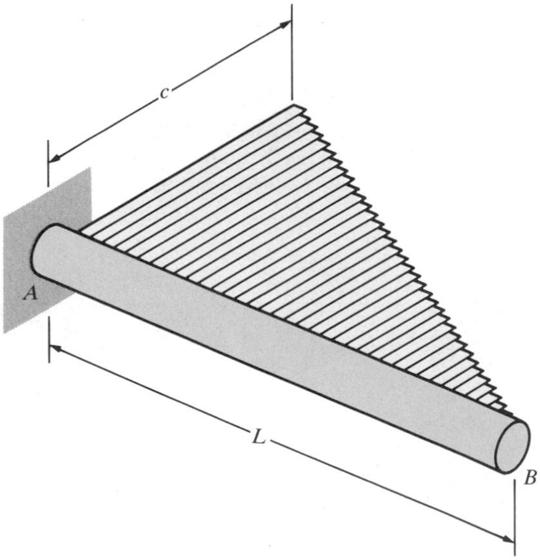

Prob. 3.3-12

3.4-1 A solid circular bar of diameter 4 in. is subjected to a torque of 85 in.-k (see figure). What are the maximum tensile, compressive, and shear stresses in the bar? (Show these stresses on sketches of stress elements.)

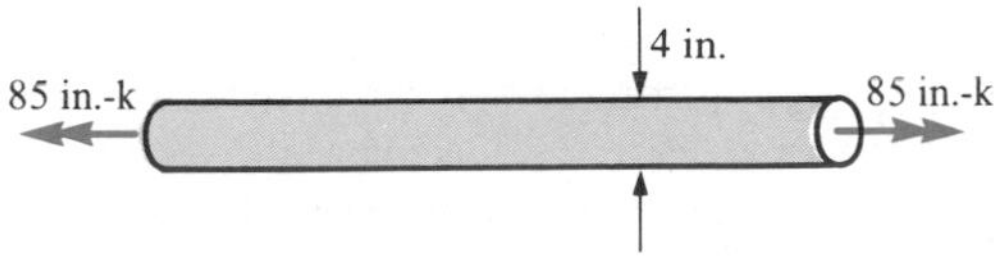

Prob. 3.4-1

3.4-2 A hollow circular shaft of aluminum ($G = 28$ GPa) has an outside diameter of 100 mm and an inside diameter of 50 mm. When twisted by a torque T, the bar has an angle of twist per unit length $\theta = 0.03$ rad/m. What is the maximum tensile stress σ_{max} in the shaft? What is the magnitude of the applied torque T?

3.4-3 A solid circular bar subjected to pure torsion is to be designed for allowable stresses in tension, compression, and shear of 9,000 psi, 20,000 psi, and 10,000 psi, respectively. What is the minimum required diameter d of the bar if it is to transmit a torque of 5,200 in.-lb?

3.4-4 A hollow circular steel bar ($G = 80$ GPa) is twisted by a torque T that produces a maximum shear strain $\gamma_{max} = 750 \times 10^{-6}$ rad. The bar has outside and inside radii of 75 and 60 mm, respectively. What is the maximum tensile stress σ_{max} in the bar? What is the magnitude of the applied torque T?

3.4-5 The normal strain in the 45° direction on the surface of a hollow circular tube (see figure) is 1860×10^{-6} when the torque $T = 1730$ in.-lb. The tube is made of beryllium copper (see Table H-2, Appendix H). If the outside diameter of the tube is 0.8 in., what is the inside diameter d?

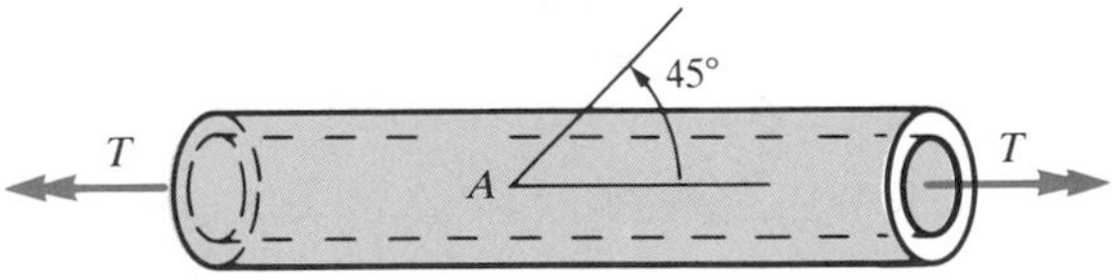

Prob. 3.4-5

3.5-1 A stress element in pure shear is subjected to shear stresses $\tau = 8000$ psi. Find the shear strain γ if $E = 30 \times 10^6$ psi and $\nu = 0.30$.

3.5-2 A hollow shaft of outside diameter 80 mm and inside diameter 50 mm is made of aluminum having shear modulus $G = 27$ GPa. When the shaft is subjected to a torque $T = 5.2$ kN·m, what are the values of the maximum shear strain γ_{max} and the maximum normal strain ϵ_{max}?

3.5-3 A solid circular bar of diameter $d = 2$ in. is twisted in a testing machine until the applied torque reaches the value $T = 11{,}000$ in.-lb (see figure). At this value of torque, a strain gage oriented at 45° to the axis of the bar gives a reading $\epsilon = 305 \times 10^{-6}$. Determine the shear modulus G of the material.

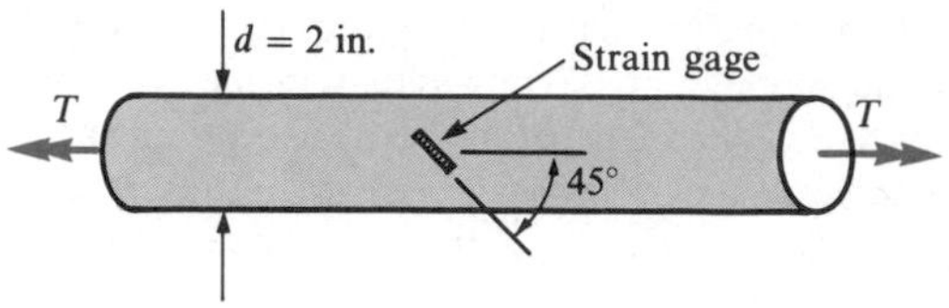

Prob. 3.5-3

3.5-4 A torque T is applied to a solid circular bar of diameter d, and the maximum normal strain ϵ on the surface of the bar (at 45° to the longitudinal axis) is measured. Obtain a formula for the shear modulus of elasticity G in terms of T, d, and ϵ.

3.5-5 A circular bar is subjected to a torque that produces a tensile stress $\sigma = 8000$ psi at 45° to the longitudinal axis. Determine the maximum normal strain ϵ and maximum shear strain γ in the bar, assuming $E = 11.5 \times 10^6$ psi and $\nu = 0.30$.

3.6-1 A solid circular shaft having a diameter of 3 in. turns at 75 rpm. What is the maximum power P (ft-lb/s) that this shaft may develop without exceeding an allowable stress in shear of 6000 psi?

3.6-2 How much power P (kW) may be transmitted by a solid circular shaft of diameter 80 mm turning at 0.75 Hz if the shear stress is not to exceed 40 MPa?

3.6-3 The propeller shaft of a certain ship is a hollow circular tube having outside diameter 18 in. and inside diameter 12 in. How much horsepower H can be transmitted by the shaft if it turns at 100 rpm and the maximum shear stress is limited to 4500 psi?

3.6-4 A solid circular shaft rotating at 4 Hz is required to transmit 150 kW. What is the minimum required shaft diameter d if the allowable shear stress is 40 MPa?

3.6-5 A solid circular shaft rotating at 200 rpm must transmit 150 hp. Calculate the minimum required diameter d of the shaft if the allowable shear stress is 8000 psi.

3.6-6 A hollow circular shaft is being designed to transmit 120 kW at 10 Hz. The inside diameter of the shaft is to be one-half of the outside diameter. Assuming that the allowable shear stress is 45 MPa, calculate the minimum required outside diameter d.

3.6-7 A hollow circular shaft is being designed with an inside diameter equal to three-fourths of the outside diameter. The shaft must transmit 400 hp at 400 rpm without exceeding the allowable shear stress of 6000 psi. Determine the minimum required outside diameter d.

3.6-8 What is the maximum power P (kW) that can be delivered by a hollow propeller shaft (outside diameter 50 mm, inside diameter 40 mm, and shear modulus of elasticity 80 GPa) turning at 600 rpm if the allowable tensile stress at the outside of the shaft is 200 MPa and the allowable angle of twist per unit length is 6°/m?

3.6-9 A hollow propeller shaft of outside diameter 6.0 in. is required to transmit 4000 hp while rotating at 1500 rpm. The material is steel with shear modulus of elasticity $G = 11.5 \times 10^6$ psi. The allowable stress in shear is 7500 psi and the allowable angle of twist per unit length is 0.01 degrees per inch. (a) Calculate the required inside diameter d of the propeller shaft. (b) Specify to the nearest $\frac{1}{4}$ in. the actual inside diameter to be used.

3.6-10 A propeller shaft of solid circular cross section and diameter d is spliced by a collar of the same material (see figure). The collar is securely bonded to both parts of the shaft. What should be the diameter d_1 of the collar in order that the splice can transmit the same power as the solid shaft?

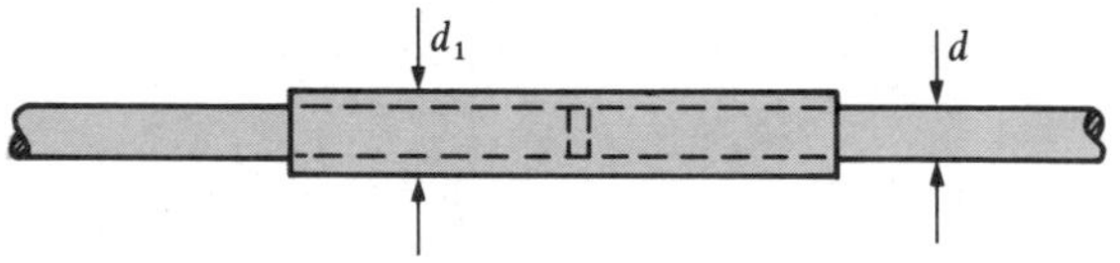

Prob. 3.6-10

3.6-11 Under certain conditions the drive shaft of a race car is required to transmit 500 hp at 2100 rpm. The shaft is hollow and has an outside diameter of 2.5 in. The material is high-strength steel and the modulus of elasticity in shear is 11.5×10^6 psi. Because of dynamic loads and severe environmental conditions, the maximum shear stress is limited to 12,000 psi. Also, the allowable angle of twist is 1° per foot. (a) Calculate the required inside diameter d of the drive shaft. (b) Specify to the nearest $\frac{1}{8}$ in. the actual inside diameter to be used.

3.6-12 The shaft ABC shown in the figure is driven by a motor at A, which delivers 300 kW at a rotational speed of 3.2 Hz. The gears at B and C take out 120 and 180 kW, respectively. The lengths of the two parts of the shaft are $L_1 = 1.5$ m and $L_2 = 0.9$ m. Calculate the required diameter d of the shaft if the allowable shear stress is 50 MPa, the allowable angle of twist in the shaft between points A and C is 0.02 radians, and $G = 75$ GPa.

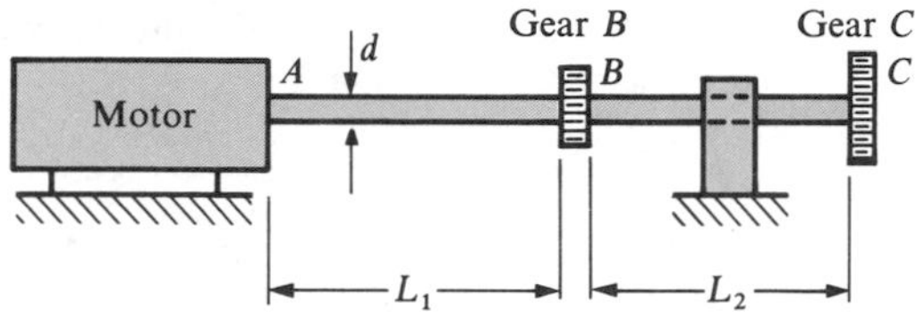

Probs. 3.6-12 and 3.6-13

3.6-13 A motor delivers 275 hp at 250 rpm to a shaft at A (see figure). The gears at B and C take out 125 and 150 hp, respectively. Determine the required diameter d of the shaft if the allowable shear stress is 7500 psi and the angle of twist between the motor and gear C is limited to 1.5°. (Assume $G = 11.5 \times 10^6$ psi, $L_1 = 6$ ft, and $L_2 = 4$ ft.)

3.7-1 A solid circular bar with fixed ends is acted upon by torques T_1 and T_2 at the locations shown in the figure. Obtain formulas for the reactive torques T_a and T_b.

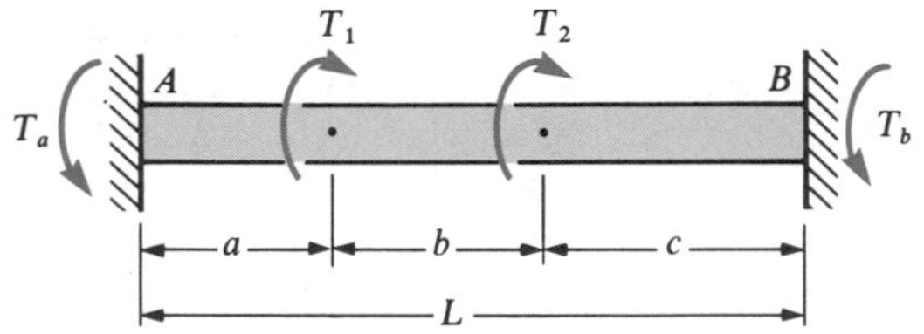

Prob. 3.7-1

3.7-2 A solid circular bar with fixed ends is acted upon by two oppositely directed torques T_0, as shown in the figure. Obtain formulas for the reactive torques T_a and T_d, the angle of twist ϕ_b at section B, and the angle of twist ϕ_m at the midsection of the bar.

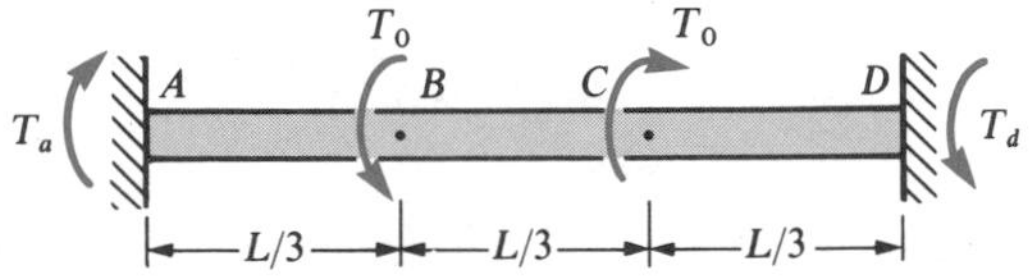

Prob. 3.7-2

3.7-3 A hollow steel shaft ACB of outside diameter 2 in. and inside diameter 1.5 in. is held against rotation at ends A and B (see figure). Horizontal forces P are applied at the ends of the vertical arm. Determine the allowable value of the forces P if the maximum permissible shear stress in the shaft is 8000 psi.

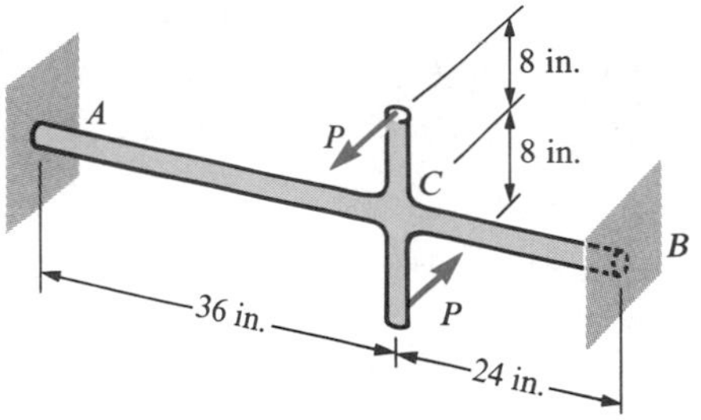

Prob. 3.7-3

3.7-4 A solid circular shaft AB of diameter d is fixed at both ends (see figure). A circular disk is attached to the shaft at the location shown. If the allowable shear stress in the shaft is τ_{allow}, and if $a > b$, what is the maximum permissible angle of rotation ϕ of the disk?

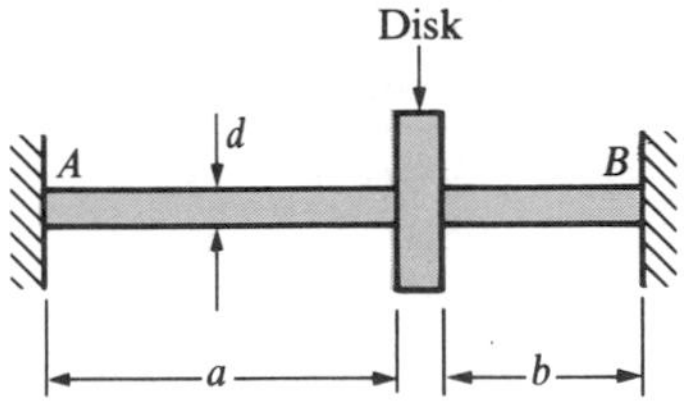

Prob. 3.7-4

3.7-5 A solid circular steel shaft AB, held rigidly at both ends, has two different diameters (see figure). Assuming that the maximum permissible shear stress is 12,000 psi, determine the allowable torque T that may be applied at the junction C.

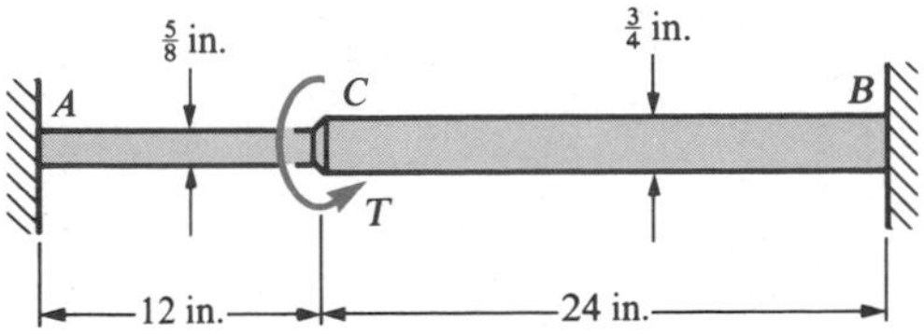

Prob. 3.7-5

3.7-6 A stepped shaft of solid circular cross section (see figure) is held against rotation at the ends. If the allowable stress in shear is 70 MPa, what is the allowable torque T that may be applied to the shaft at C?

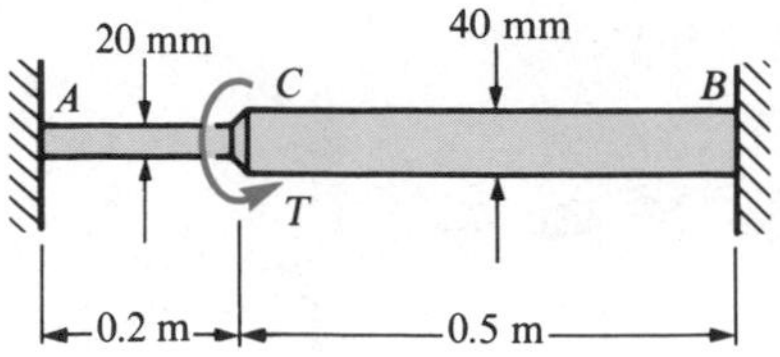

Prob. 3.7-6

3.7-7 A solid circular shaft AB of total length L is fixed against rotation at both ends (see figure). The shaft has diameters d_a and d_b in parts AC and CB, respectively. A torque T acts at section C. What should be the lengths a and b if the maximum shear stress is to be the same in both parts of the shaft?

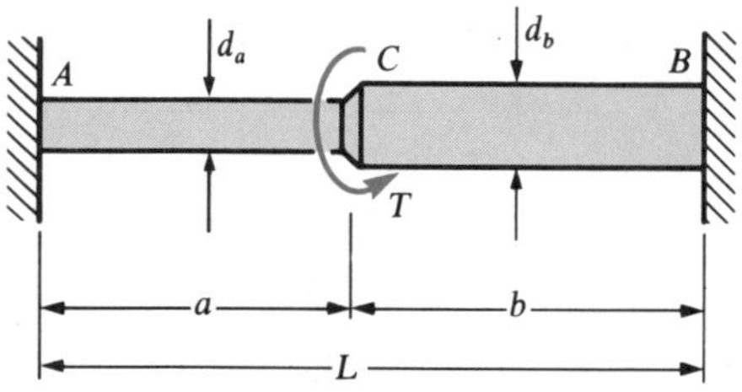

Prob. 3.7-7

3.7-8 A bar ABC that is fixed at both ends is subjected to a torque T at section B (see figure). The bar has a solid, circular cross section (diameter $= d_1$) from A to B and a hollow, circular cross section (outer diameter d_2, inner diameter d_1) from B to C. Derive an expression for the ratio a/L such that the reactive torques at A and C are equal numerically.

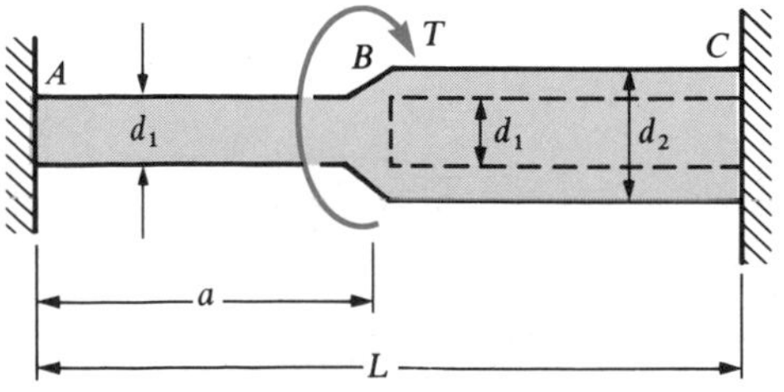

Prob. 3.7-8

3.7-9 A circular bar AB with fixed ends has a hole extending for half its length (see figure). The polar moments of inertia of the two parts of the bar are I_{pa} and I_{pb}. At what distance x from the left-hand end should a torque T be applied so that the reactive torques at the supports will be equal?

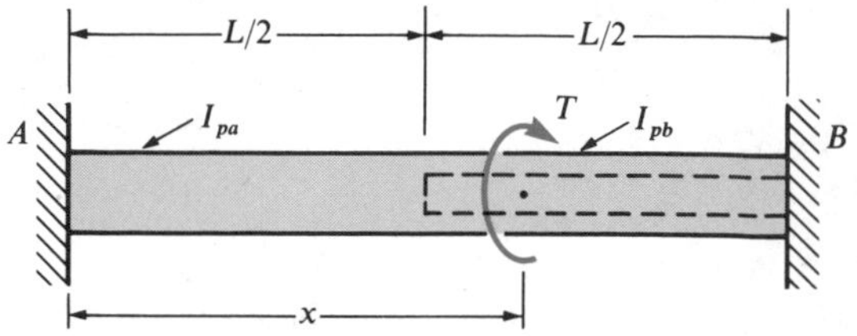

Prob. 3.7-9

3.7-10 A hollow tube AB of length L and torsional rigidity GI_p is rigidly supported at end A (see figure). At end B, a horizontal bar of length $2c$ is welded to the tube. Adjacent to the ends of the horizontal bar are two identical springs, each having a spring stiffness k (force per unit displacement). The length of the gap between the ends of the springs and the ends of the bar is b. The springs are stretched, attached to the ends of the bar, and then released, thus causing the bar and tube to rotate through a small angle β (see figure). Obtain a formula for the torque T in the tube.

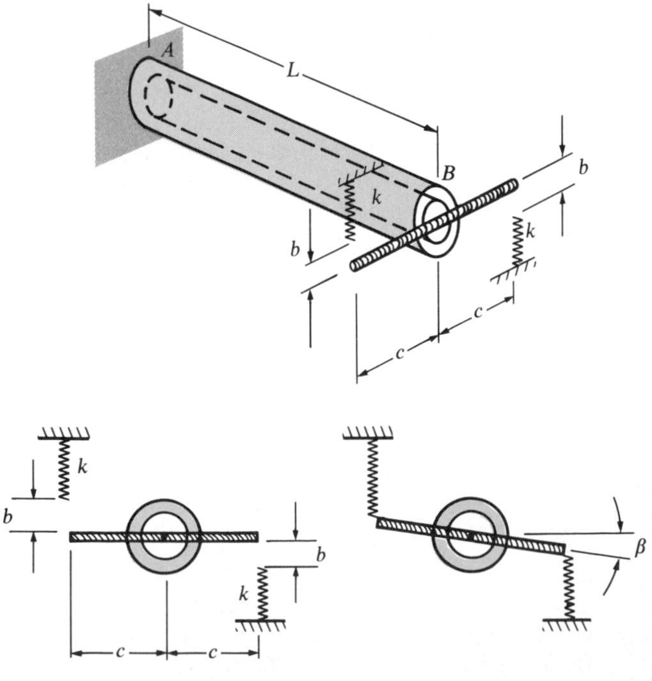

Prob. 3.7-10

3.7-11 A solid circular steel bar of diameter 2.0 in. is enclosed by a hollow steel tube having outer diameter 3.0 in. and inner diameter 2.5 in. (see figure). The two bars are held rigidly at end A and welded to a steel plate at B. (a) If a torque $T = 18$ in.-k is applied to the rigid plate, what are the maximum shear stresses τ_t and τ_b in the tube and bar, respectively? (b) What is the angle of rotation ϕ of the plate, assuming $G = 11.5 \times 10^6$ psi? (c) What is the torsional stiffness k of the device?

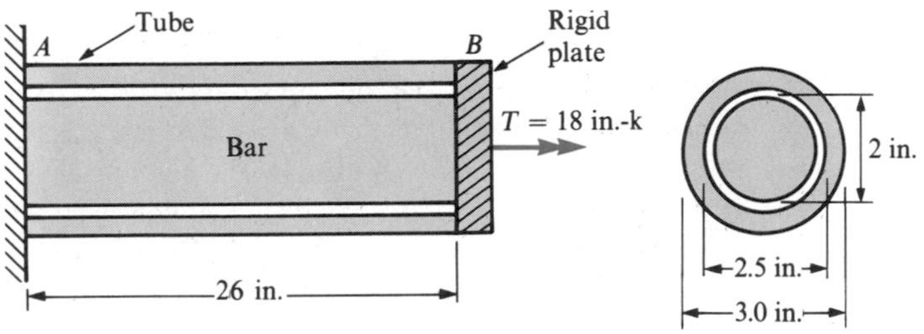

Prob. 3.7-11

3.7-12 A circular bar AB of polar moment of inertia I_p and length L is fixed at both ends (see figure). A distributed torque $q(x)$ acts along the length of the bar and varies linearly in intensity from zero at A to q_0 at B. Obtain formulas for the fixed-end torques T_a and T_b.

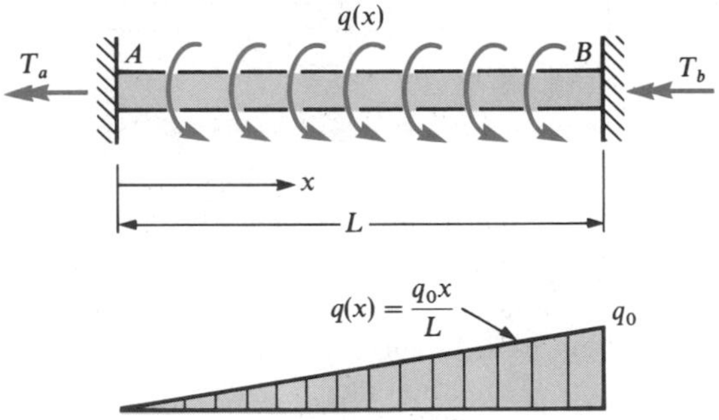

Prob. 3.7-12

3.7-13 Solve the preceding problem if the distributed torque $q(x)$ varies parabolically in intensity from zero at A to q_0 at B; that is, $q(x) = q_0x^2/L^2$.

3.7-14 A composite shaft is made by shrink fitting a steel tube over a brass core so that they act as a unit in torsion (see figure). The shear modulus of elasticity of the tube is $G_s = 75$ GPa and of the core is $G_b = 39$ GPa. The outside diameters are $d_1 = 25$ mm for the core and $d_2 = 40$ mm for the tube. Calculate the maximum shear stresses τ_s and τ_b in the steel and brass, respectively, due to a torque of 960 N · m.

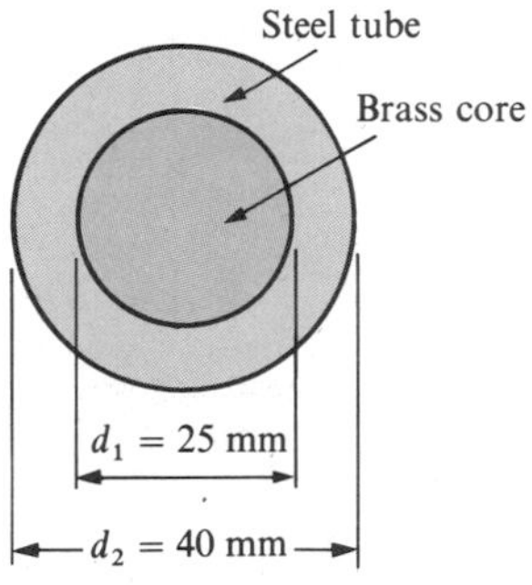

Prob. 3.7-14

3.7-15 A steel sleeve is fused onto a brass core to make a composite bar 30 in. long, as shown in the figure. The outside diameters of the sleeve and core are 2.4 in. and 2.0 in., respectively. The shear moduli of elasticity for steel and brass are $G_s = 11{,}500$ ksi and $G_b = 5{,}600$ ksi, respectively. The bar is subjected to a torque $T = 28{,}000$ in.-lb. (a) Calculate the maximum shear stresses τ_s and τ_b in the steel and brass parts, respectively. (b) Calculate the angle of twist ϕ.

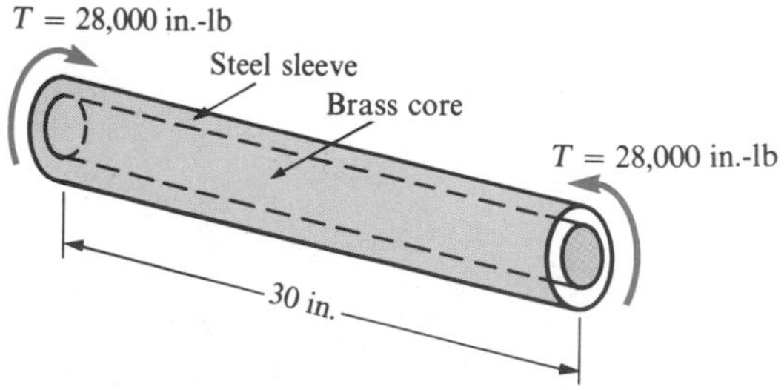

Prob. 3.7-15

3.7-16 A solid shaft is formed of two materials, an outer sleeve of steel ($G_s = 80$ GPa) and an inner rod of brass ($G_b = 36$ GPa), as shown in the figure. The outside diameters of the two parts are 75 mm and 60 mm. Assuming that the allowable shear stresses are $\tau_s = 80$ MPa and $\tau_b = 48$ MPa in the steel and brass, respectively, determine the maximum permissible torque T that may be applied to the shaft.

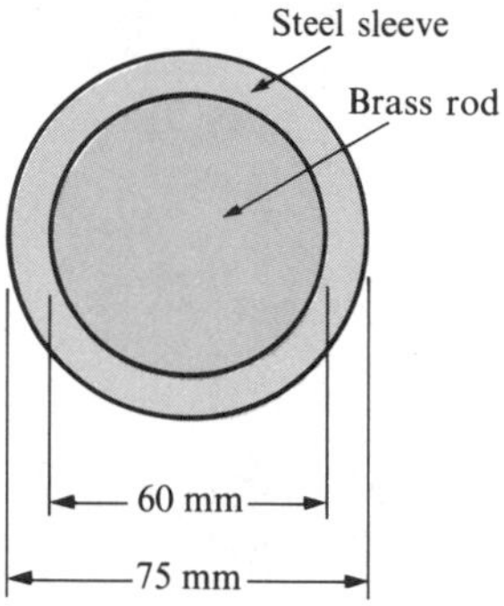

Prob. 3.7-16

3.7-17 A high-strength steel tube S is shrink fitted over an aluminum tube A to form a composite shaft as shown in the figure. The diameters are 3 in., 5 in., and 6 in. (see figure). The allowable shear stresses in the steel and aluminum are $\tau_s = 18$ ksi and $\tau_a = 8$ ksi, respectively. Determine the allowable torque T that may be applied to the shaft, assuming $G_s = 11{,}600$ ksi and $G_a = 4{,}000$ ksi.

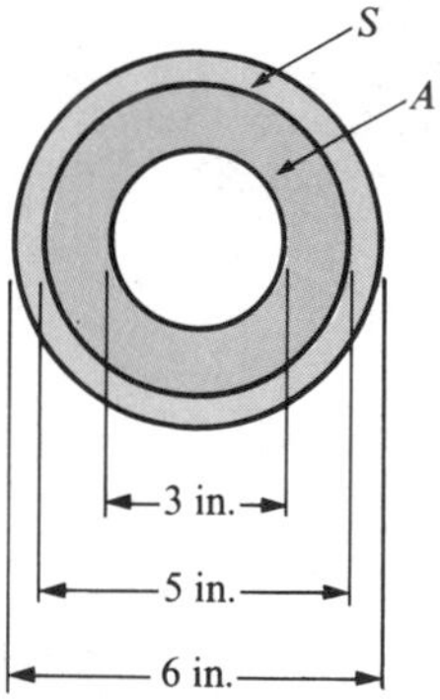

Prob. 3.7-17

3.7-18 A steel shaft ($G_s = 80$ GPa) of total length $L = 4.0$ m is encased for one-half of its length by a brass tube ($G_b = 40$ GPa) that is securely bonded to the steel (see figure). The diameters of the shaft and tube are $d_1 = 70$ mm and $d_2 = 90$ mm, respectively. (a) Determine the allowable torque T_1 if the angle of twist ϕ between ends A and C is limited to $\phi = 8°$. (b) Determine the allowable torque T_2 if the shear stress in the brass is limited to $\tau_b = 75$ MPa. (c) Determine the allowable torque T_3 if the shear stress in the steel is limited to $\tau_s = 70$ MPa. (d) What is the allowable torque T if all three of the preceding conditions must be satisfied?

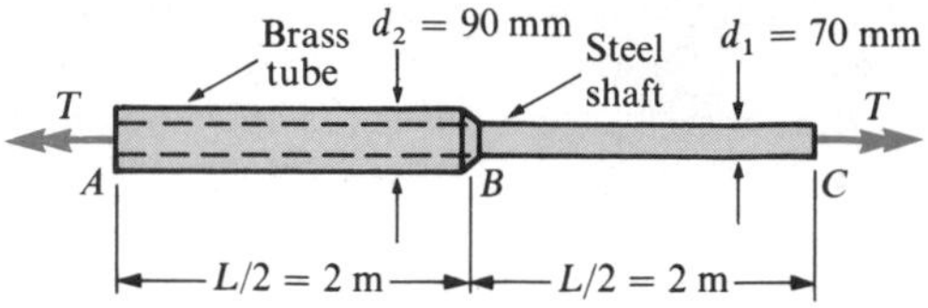

Prob. 3.7-18

3.8-1 A solid circular bar of copper ($G = 6.0 \times 10^6$ psi) with length $L = 4$ ft and diameter $d = 1.5$ in. is subjected to pure torsion by torques T. Calculate the amount of strain energy U stored in the bar when the maximum shear stress is 4000 psi.

3.8-2 A solid circular bar of steel ($G = 80$ GPa) with length $L = 3.5$ m and diameter $d = 120$ mm is subjected to pure torsion by torques T. How much strain energy U is stored in the bar when the maximum shear stress is 56 MPa?

3.8-3 How much strain energy U is stored in the stepped steel shaft shown in the figure when the angle of twist ϕ equals 0.015 radians? (Use $G = 11.8 \times 10^6$ psi.)

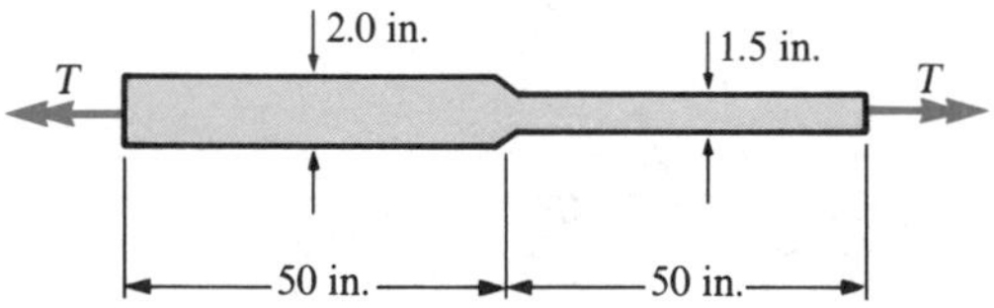

Prob. 3.8-3

3.8-4 Derive a formula for the strain energy U of the circular bar shown in the figure. The intensity q of distributed torque varies linearly from a maximum value of q_0 at the support to zero at the free end.

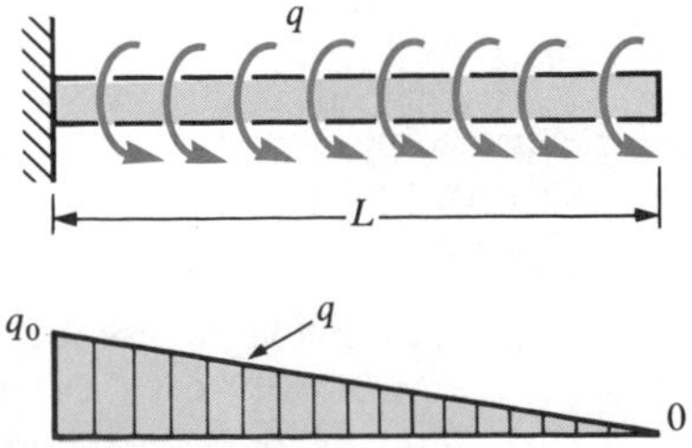

Prob. 3.8-4

3.8-5 A circular bar of length L is fixed at one end and free at the other (see figure). The bar is loaded by a torque T at the free end and by a continuously distributed torque of constant intensity q per unit distance along the axis of the bar. (a) What is the strain energy U_1 of the bar when the load T acts alone? (b) What is the strain energy U_2 when the load q acts alone? (c) What is the strain energy U_3 when both loads act simultaneously?

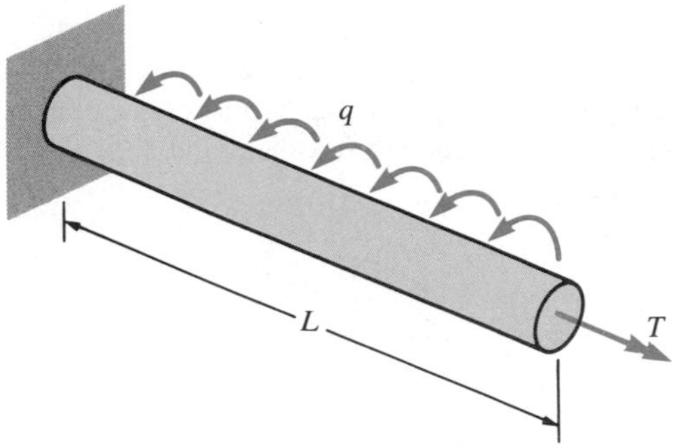

Prob. 3.8-5

3.8-6 A thin-walled hollow tube AB of conical shape has constant thickness t and average diameters d_a and d_b at the ends (see figure). (a) Derive a formula for the strain energy U of the tube when it is subjected to pure torsion by torques T. (Use Eq. 3-13 for the polar moment of inertia I_p.) (b) Obtain a formula for the angle of twist ϕ.

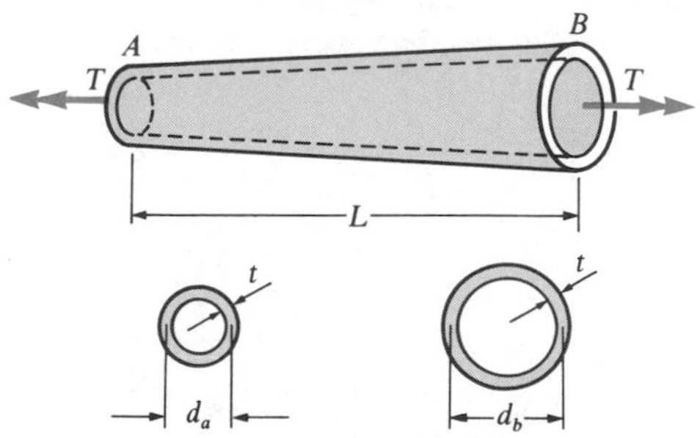

Prob. 3.8-6

3.8-7 Derive a formula for the strain energy U of the circular bar AB with fixed ends shown in the figure.

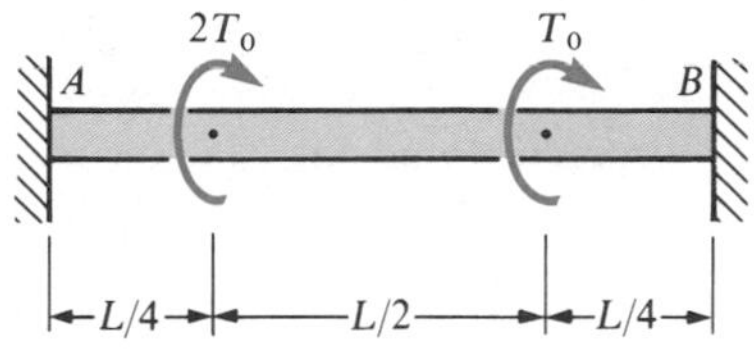

Prob. 3.8-7

3.8-8 A stepped shaft ABC is fixed at ends A and C and loaded by a torque T at B (see figure). The two parts of the bar are made of the same material, have lengths L_1 and L_2, and have polar moments of inertia I_{p1} and I_{p2}. Determine the angle of rotation ϕ at cross section B. (Hint: Use Eq. 3-38b to determine the strain energy U in terms of ϕ. Then equate the strain energy to the work done by the load T.)

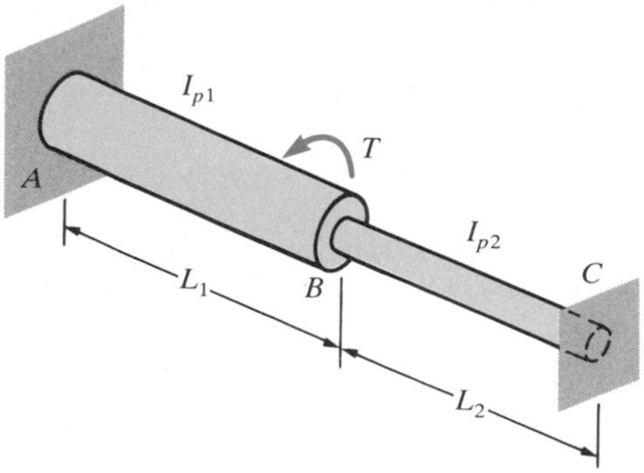

Prob. 3.8-8

***3.8-9** A hollow circular tube A fits over the end of a solid circular bar B, as shown in the figure. The far ends of both bars are fixed. A hole through bar B makes an angle β (in radians) with a line through two holes in tube A. Bar B is twisted until the holes are aligned, and then a pin is placed

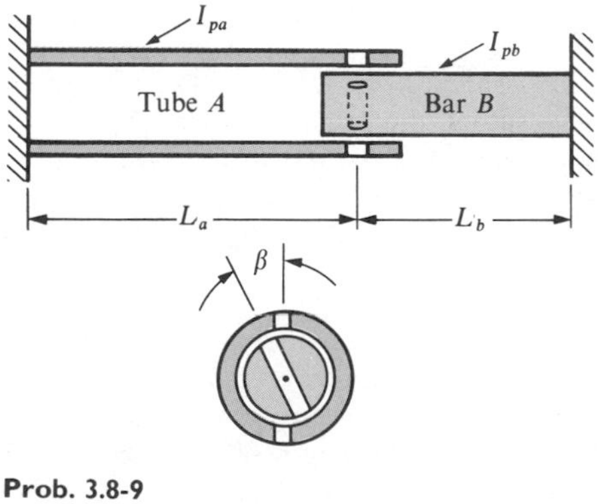

Prob. 3.8-9

through the holes. When the system returns to static equilibrium, what is the total strain energy of the two bars? (Let I_{pa} and I_{pb} represent the polar moments of inertia of bars A and B, respectively. The shear modulus of elasticity G is the same for both bars.)

3.9-1 A hollow circular tube having 9 in. inside diameter and 1 in. wall thickness (see figure) is subjected to a torque $T = 1200$ in.-k. Determine the maximum shear stresses τ_{max} in the tube using: (a) the approximate theory of thin-walled tubes (Eq. 3-51b), and (b) the exact torsion theory (Eq. 3-9).

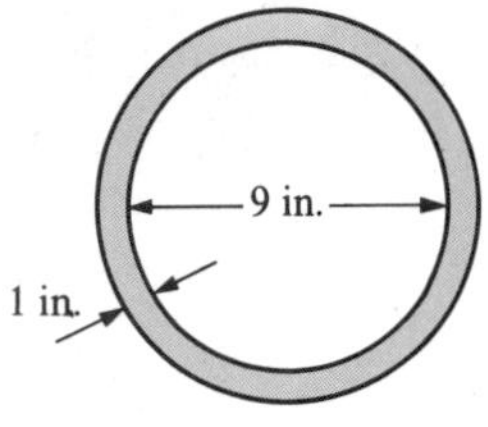

Prob. 3.9-1

3.9-2 Compare the angle of twist ϕ_1 for a thin-walled circular tube (see figure) as calculated from the approximate equation (Eq. 3-52) with the angle of twist ϕ_2 as calculated from the exact equation (Eq. 3-8). Express the ratio ϕ_1/ϕ_2 in terms of the nondimensional ratio $\beta = r/t$. Calculate the ratio of angles of twist for $\beta = 5$, 10, and 20.

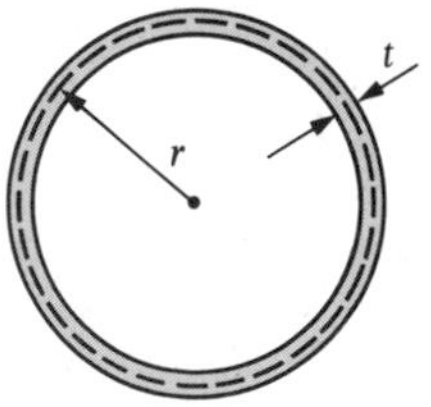

Prob. 3.9-2

3.9-3 A thin-walled circular tube and a solid circular bar of the same material, the same cross-sectional area, and the same length are subjected to torsion. What is the ratio of the strain energy U_1 in the tube to the strain energy U_2 in the solid bar if the maximum shear stresses are the same in both cases?

3.9-4 A thin-walled hollow tube of rectangular cross section is shown in the figure. (a) Calculate the shear stress τ in the tube due to a torque $T = 120$ N · m. (b) Calculate the angle of twist ϕ if the length of the tube is 0.25 m and $G = 26$ GPa.

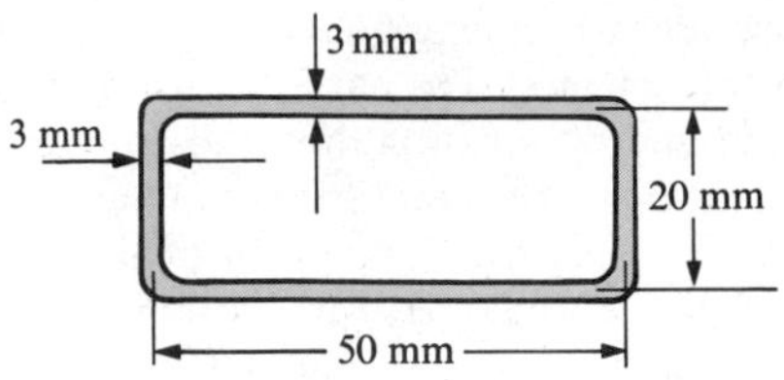

Prob. 3.9-4

3.9-5 A thin tubular shaft of circular cross section (see figure) with inside diameter 4 in. is subjected to a torque of 45,000 in.-lb. If the allowable shear stress is 12,000 psi, determine the required wall thickness t by using: (a) the approximate theory for a thin-walled tube, and (b) the exact torsion theory.

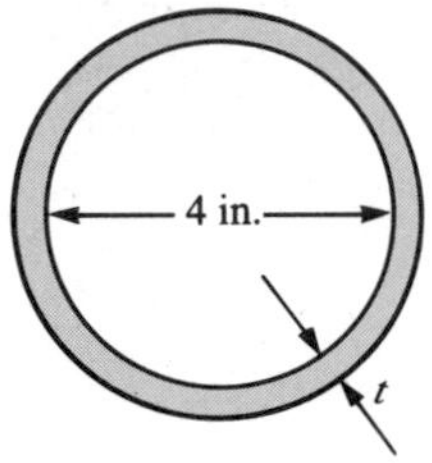

Prob. 3.9-5

3.9-6 The cross section of a stainless steel thin-walled tube ($G = 80$ GPa) is in the form of an equilateral triangle (see figure). The length along the median line of each side is $b = 150$ mm, and the wall thickness is $t = 8$ mm. If the allowable shear stress is 50 MPa, what is the maximum permissible torque T that may act on the tube? At this value of torque, what is the angle of twist θ per unit length?

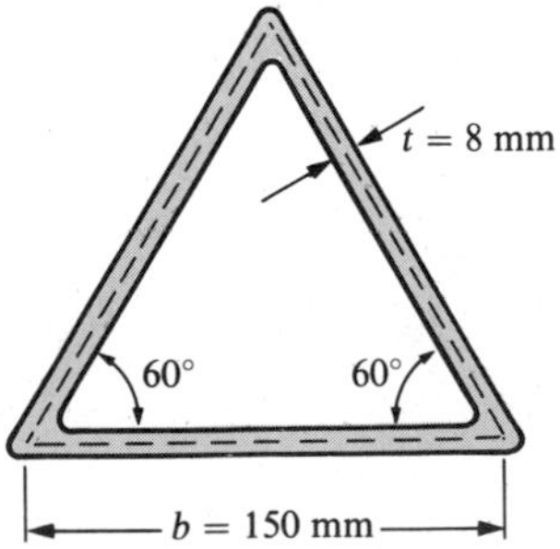

Prob. 3.9-6

3.9-7 A thin-walled tube having an elliptical cross section (see figure) is subjected to a torque $T = 48$ in.-k. Determine the shear stress τ and the angle of twist θ per unit length if $G = 12 \times 10^6$ psi, $t = 0.2$ in., $a = 3$ in., and $b = 2$ in. (Note: See Appendix D for the properties of an ellipse.)

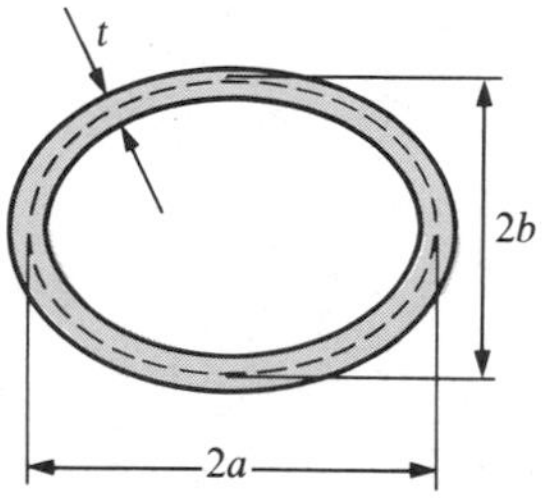

Prob. 3.9-7

3.9-8 Calculate the shear stress τ and the angle of twist ϕ for a steel tube ($G = 76$ GPa) having the cross section shown in the figure. The tube has length $L = 1.5$ m and is subjected to a torque $T = 15$ kN·m.

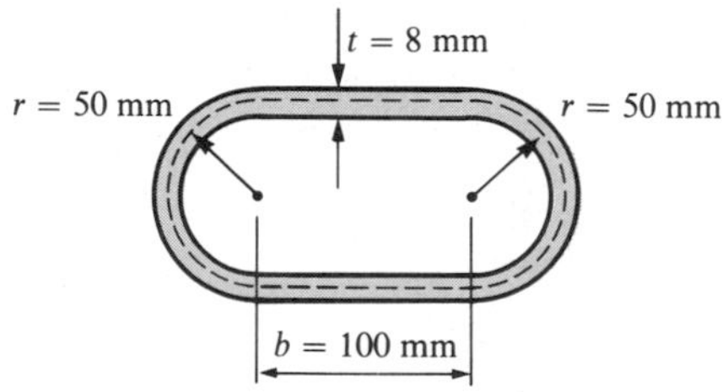

Prob. 3.9-8

3.9-9 A tubular aluminum bar ($G = 4 \times 10^6$ psi) of square cross section (see figure) with outside dimensions 2 in. × 2 in. must resist a torque $T = 3000$ in.-lb. Calculate the minimum required wall thickness t if the allowable shear stress is 4500 psi and the allowable angle of twist θ per unit length is 0.01 rad/ft.

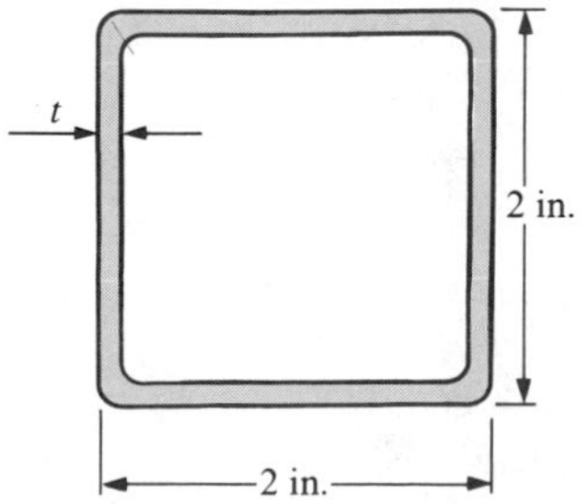

Prob. 3.9-9

3.9-10 A torque T is applied to a thin-walled tube having a cross section in the shape of a regular hexagon with constant wall thickness t and side length b (see figure). Obtain formulas for the shear stress τ and the angle of twist θ per unit length.

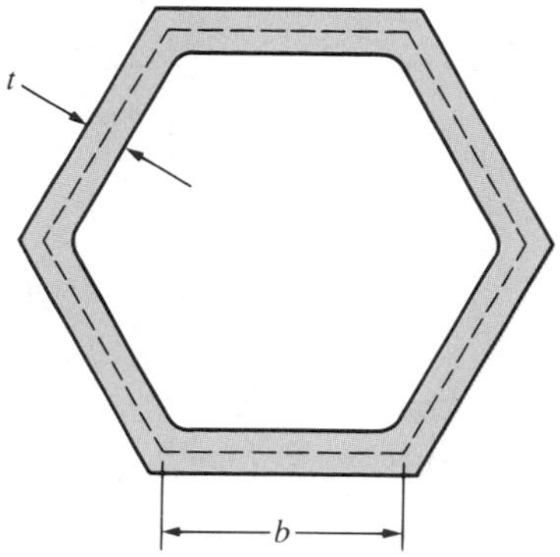

Prob. 3.9-10

3.9-11 A thin-walled rectangular tube has uniform thickness t and dimensions $a \times b$ to the median line of the cross section (see figure). How does the shear stress τ vary with the ratio $\beta = a/b$ if the total length L_m of the median line of the cross section and the torque T remain constant? From your results, show that the shear stress is smallest when the tube is square ($\beta = 1$).

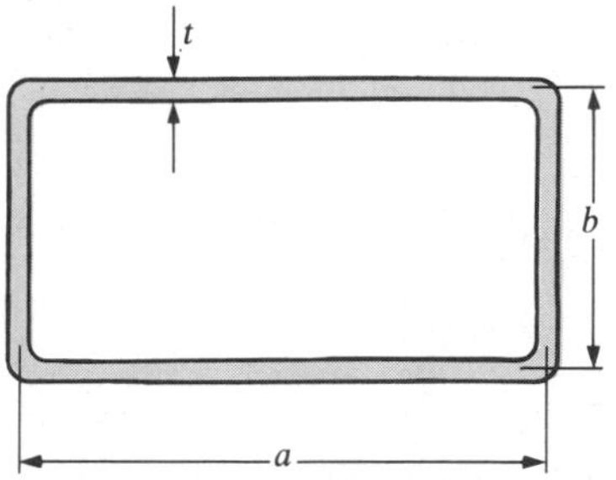

Probs. 3.9-11 and 3.9-12

3.9-12 Repeat the preceding problem for the angle of twist θ per unit length, and then show that the angle of twist is smallest when the tube is square ($\beta = 1$).

3.9-13 A long, thin-walled tapered tube AB of circular cross section (see figure) is subjected to a torque T. The tube has length L and constant wall thickness t. The diameters to the median lines of the cross sections at the ends A and B are d_a and d_b, respectively. Derive a formula for the angle of twist ϕ of the tube.

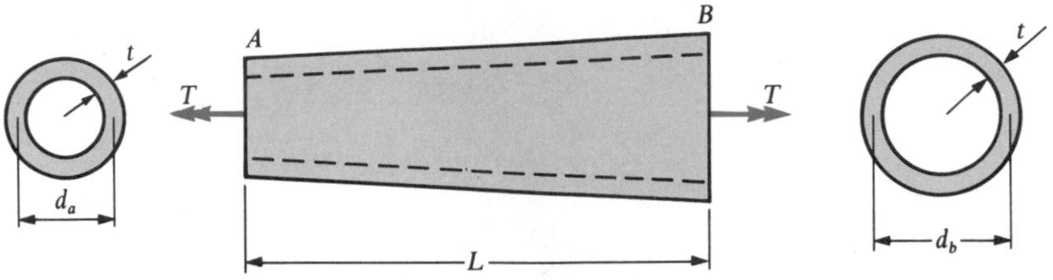

Prob. 3.9-13

The problems for Section 3.10 are to be solved by considering the stress-concentration factors.

3.10-1 A stepped shaft consisting of solid circular parts having diameters $D_1 = 2.0$ in. and $D_2 = 2.4$ in. (see figure) is subjected to a torque T. The radius of the fillet is $R = 0.1$ in. If the allowable maximum shear stress is 16,000 psi, what is the maximum permissible torque T_{max}?

3.10-2 A stepped shaft with diameters $D_1 = 40$ mm and $D_2 = 60$ mm carries a torque $T = 1100$ N·m (see figure). If the maximum allowable shear stress is 120 MPa, what is the smallest radius R_{min} that may be used for the fillet?

3.10-3 A full quarter-circular fillet is used at the shoulder of a stepped shaft having diameter $D_2 = 1.0$ in. (see figure). A torque $T = 1000$ in.-lb acts on the shaft. Determine the maximum shear stress τ_{max} in the shaft for values as follows: $D_1 = 0.7$, 0.8, 0.9, and 1.0 in. Plot a graph showing τ_{max} versus D_1.

3.10-4 The stepped shaft shown in the figure is required to transmit 600 kW of power at 400 rpm. The shaft has a full quarter-circular fillet, and the smaller diameter $D_1 =$ 100 mm. (a) If the maximum allowable shear stress is 100 MPa, what is the required diameter D_2? (b) Is the value determined in part (a) for diameter D_2 an upper or a lower limit on the value of D_2?

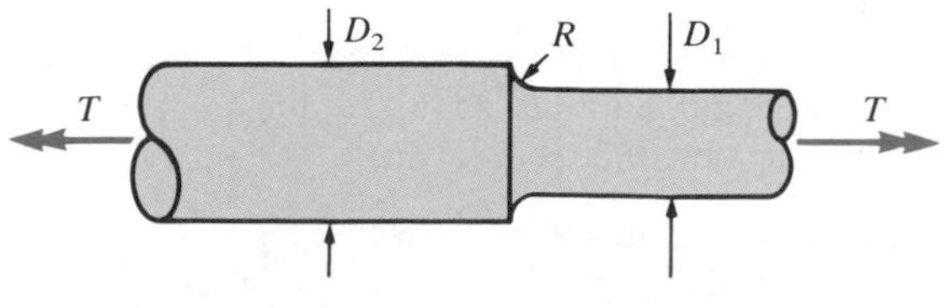

Probs. 3.10-1 through 3.10-5

3.10-5 A stepped shaft (see figure) has diameter $D_2 =$ 1.5 in. and a full quarter-circular fillet. The allowable shear stress is 15,000 psi, and the load $T = 4800$ in.-lb. What is the smallest permissible diameter D_1?

3.11-1 A bar of solid circular cross section is constructed of a material that follows Hooke's law ($\tau = G\gamma$). Derive the equation for the angle of twist per unit length $\theta = T/GI_p$ by using Eq. (3-58) for nonlinear torsion.

3.11-2 A solid circular bar (radius $= r$) subjected to a torque T is composed of a material for which the stress-strain diagram in shear is represented by the equation $\tau^n = B\gamma$, where n and B are constants. (a) Derive a formula for the shear stress τ_{max} at the outer edge of the cross section. (b) Verify that the formula reduces to the torsion formula, $\tau_{max} = Tr/I_p$, when $n = 1$.

3.11-3 Obtain a formula for the ratio of the ultimate torque T_u to the yield torque T_y for a bar of hollow, circular cross section (see figure) if the bar is made of an elastic-plastic material (see Fig. 3-41a for the stress-strain diagram). Plot a graph of the ratio T_u/T_y versus the ratio r_1/r_2.

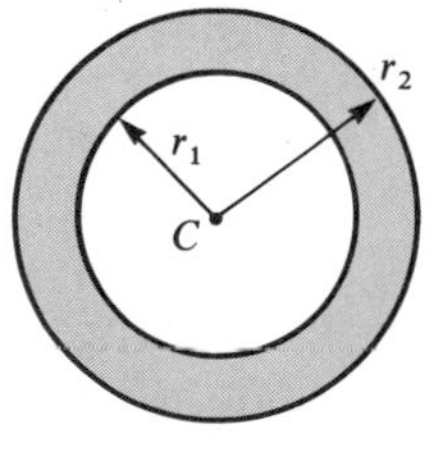

Prob. 3.11-3

3.11-4 A solid circular bar of elastic-plastic material (see the stress-strain diagram in Fig. 3-41a) is subjected to a torque T that is larger than the yield torque T_y and smaller than the ultimate torque T_u. Thus, the shear stresses acting on the cross section have the distribution shown in Fig. 3-41b. Let the shear strain at the outer edge of the cross section be γ_{max}. (a) Derive a formula for the torque T in terms of γ_{max}, γ_y, and T_y. (b) Verify that the formula reduces to $T = T_y$ when $\gamma_{max} = \gamma_y$ and that T approaches T_u as γ_{max} increases indefinitely. (c) Plot a graph of T/T_y versus γ_{max}/γ_y for values of γ_{max}/γ_y from 0 to 4.

***3.11-5** A hollow circular shaft has inside diameter $d_1 =$ 3 in. and outside diameter $d_2 = 6$ in. The shaft is made of steel having an elastic-plastic shear stress-strain diagram (see Fig. 3-41a) with $\tau_y = 24{,}000$ psi and $\gamma_y = 0.002$. (a) What torque T acting on the shaft will produce a maximum shear strain $\gamma_{max} = 0.003$ in the shaft? (b) What is the ultimate torque T_u? (c) What is the maximum shear strain γ_u when the torque just reaches the ultimate torque T_u?

***3.11-6** A solid circular bar of radius $r = 40$ mm is constructed of a material having the bilinear stress-strain diagram shown in the figure. Calculate the maximum shear stress τ_{max} in the bar when the torque $T = 6$ kN · m.

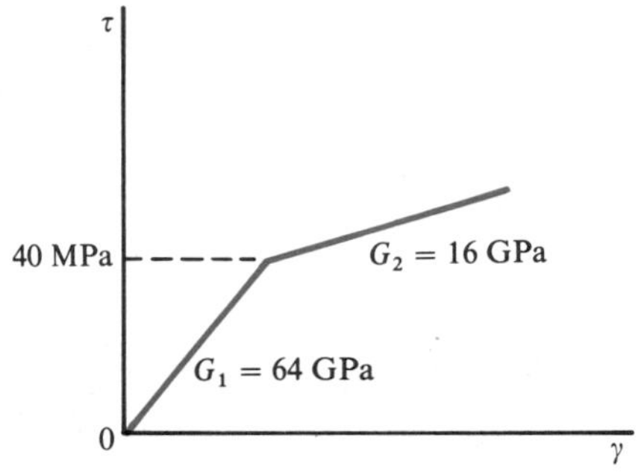

Prob. 3.11-6

CHAPTER 4

Shear Force and Bending Moment

4.1 TYPES OF BEAMS

A structural member that is designed to resist forces acting laterally to its axis is called a **beam**. Beams differ from bars in tension and torsion primarily because of the directions of the loads that act upon them. A bar in tension is subjected to loads directed along the axis, and a bar in torsion is subjected to torques having their vectors along the axis. By contrast, the loads on a beam are directed normal to the axis, as illustrated by the force P_1 acting on beam AB of Fig. 4-1a.

In this chapter we will consider only the simplest types of beams, such as those illustrated in Fig. 4-1. These beams are planar structures because all loads act in the plane of the figure and all deflections occur in that same plane, which is called the **plane of bending**. The beam of Fig. 4-1a, which has a pin support at one end and a roller support at the other, is called a **simply supported beam** or a **simple beam**. The essential feature of a pin support is that it restrains the beam from translating both horizontally and vertically but does not prevent rotation. Thus, end A of the beam does not translate, but the longitudinal axis of the beam may rotate in the plane of the figure. Therefore, a pin support is capable of developing a reactive force with both horizontal and vertical components, but there will be no moment reaction. At the roller support B, translation is prevented in the vertical direction but not in the horizontal direction; hence this support can resist a vertical force but not a horizontal force. Of course, the beam axis is free to rotate at B in the same manner as at A. The vertical reactions

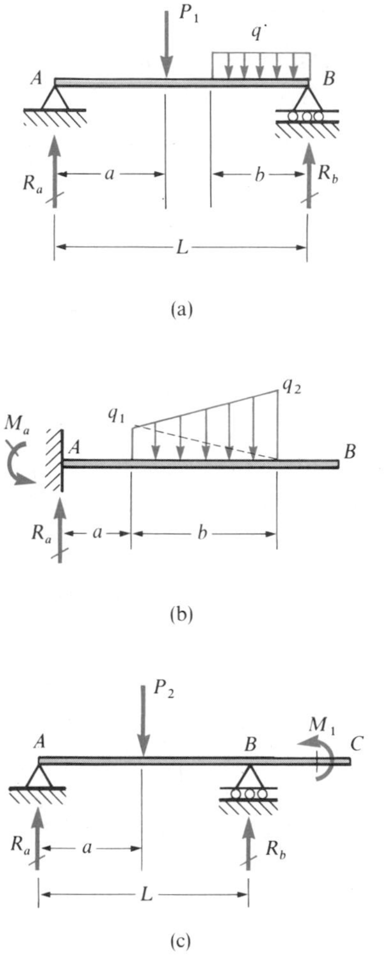

Fig. 4-1 Types of beams: (a) simple beam, (b) cantilever beam, and (c) beam with an overhang

at the supports of a simple beam may act either upward or downward, as required for equilibrium.

When making a sketch of a beam, we show the types of supports by conventional symbols that indicate the manner in which the beam is restrained. These symbols also indicate the nature of the reactive forces. However, the symbols are not intended to represent the actual physical construction. For instance, the end of a beam resting on a wall and bolted down against uplift is represented in a sketch as a pin connection, but in reality there is no pin at the end of the beam.

The beam shown in Fig. 4-1b, which is built-in or fixed at one end and free at the other, is called a **cantilever beam**. At the fixed (or clamped) support, the beam can neither translate nor rotate, whereas at the free end it may do both. Consequently, both force and moment reactions may exist at the fixed support. The third example in the figure is a beam with an **overhang** (Fig. 4-1c). This beam is simply supported at A and B, but it also projects beyond the support to point C, which is a free end.

Many other arrangements of supports for beams are possible, depending upon the particular applications. However, the examples given here and in later sections are sufficient to illustrate the basic concepts.

Loads acting on beams may be of several kinds, as illustrated in Fig. 4-1. **Concentrated loads** are forces such as P_1 and P_2. **Distributed loads** act over a distance, as shown by the load q in Fig. 4-1a. Such loads are measured by their **intensity**, which is expressed in units of force per unit distance along the axis of the beam (for example, newtons per meter or pounds per foot). A uniformly distributed load, or **uniform load**, has constant intensity q per unit distance. A varying load has an intensity that changes with distance along the axis; for instance, the **linearly varying load** of Fig. 4-1b has an intensity that varies from q_1 to q_2. Another kind of load is a **couple**, illustrated by the couple of moment M_1 acting on the overhanging beam (Fig. 4-1c).

As mentioned previously, we assume in this discussion that the loads on the beams act in the plane of the figure. This assumption requires that all forces have their vectors in the plane of the figure and that all couples have their moment vectors perpendicular to that plane. Furthermore, the beams must be symmetric about that plane; that is, the plane of bending must be a plane of symmetry of the beam itself, which means that the cross section of each beam must have a vertical axis of symmetry. Under these conditions, the beam will deflect only in the plane of bending. However, if these conditions are not met, the beam will bend out of its plane and a more general bending analysis is required.

The beams shown in Fig. 4-1 are **statically determinate** and their **reactions** can be determined from equilibrium equations. For instance, in the case of the simple beam AB of Fig. 4-1a, we note first that the only reactions are the vertical forces R_a and R_b at the ends. (If a horizontal load were to act on the beam, a horizontal reaction would also be induced at support A.) By summing moments about point B, we can calculate the reaction at A, and vice versa. The results are

$$R_a = \frac{P_1(L-a)}{L} + \frac{qb^2}{2L} \qquad R_b = \frac{P_1 a}{L} + \frac{qb(2L-b)}{2L}$$

An equation of equilibrium of forces in the vertical direction provides a check on these results.

The cantilever beam of Fig. 4-1b is subjected to a linearly varying distributed load; hence the diagram of load intensity is trapezoidal. This load is equilibrated by a vertical force R_a and a couple M_a acting at the fixed support. From the equilibrium of forces in the vertical direction, we get

$$R_a = \frac{(q_1 + q_2)b}{2}$$

in which we have utilized the fact that the resultant of the distributed load is equal to the area of the load-intensity diagram. The moment reaction M_a is found from equilibrium of moments. It is convenient in this example to sum moments about point A in order to eliminate R_a from the moment equation. For the purpose of obtaining the moment about A of the distributed load, we divide the trapezoidal diagram into two triangles, as shown by the dashed line in Fig. 4-1b. Then the moment about A of the lower triangular part of the load is

$$\frac{1}{2}(q_1 b)\left(a + \frac{b}{3}\right)$$

in which $q_1 b/2$ is the resultant force (equal to the area of the lower triangular load diagram) and $a + b/3$ is the moment arm of the resultant. A similar procedure may be used to obtain the moment of the upper triangular portion of the load, and the final result is

$$M_a = \frac{q_1 b}{2}\left(a + \frac{b}{3}\right) + \frac{q_2 b}{2}\left(a + \frac{2b}{3}\right)$$

This reactive moment acts counterclockwise, as shown in the figure.

The beam with an overhang (Fig. 4-1c) is subjected to a vertical force P_2 and a couple of moment M_1. Taking moments about points B and A gives the following equations of static equilibrium (counterclockwise moments are positive):

$$-R_a L + P_2(L-a) + M_1 = 0 \quad \text{or} \quad R_a = \frac{P_2(L-a)}{L} + \frac{M_1}{L}$$

$$-P_2 a + R_b L + M_1 = 0 \quad \text{or} \quad R_b = \frac{P_2 a}{L} - \frac{M_1}{L}$$

Again, summation of forces in the vertical direction provides a check on the results.

The preceding examples illustrate how the reactions (forces and couples) of statically determinate beams are calculated from equilibrium equations. Of course, the reactions of statically indeterminate beams cannot be found from equilibrium alone; their calculation requires consideration of the deflections caused by bending. These subjects are discussed in Chapters 7 and 8.

The support conditions shown in Fig. 4-1 are idealizations of actual conditions encountered in practice. Due to lack of perfect rigidity in the supporting structures or foundations, there may be a small amount of translation at a pin support or a small rotation at a fixed support. Also, it is rare to find complete lack of restraint against horizontal translation (as assumed for a roller support); instead, a small force may develop due to friction and other effects. Under most conditions, especially for statically determinate beams, these minor deviations from idealized conditions have little effect on the action of the beam and can be disregarded.

4.2 SHEAR FORCE AND BENDING MOMENT

When a beam is loaded by forces or couples, internal stresses and strains are created. To determine these stresses and strains, we first must find the **internal forces** and **internal couples** that act on cross sections of the beam. As an illustration, consider a cantilever beam acted upon by a vertical force P at its free end (Fig. 4-2a). Now imagine that we cut through the beam at a cross section mn located at distance x from the free end and isolate the left-hand part of the beam as a free body (Fig. 4-2b). The free body is held in equilibrium by the force P and by the stresses that act over the cut cross section mn. These stresses represent the action of the right-hand part of the beam on the left-hand part. Of course, at this stage of our discussion, we do not know the distribution of the stresses acting over the cross section; all we know is that the resultant of these stresses must be such as to maintain equilibrium of the free body we selected.

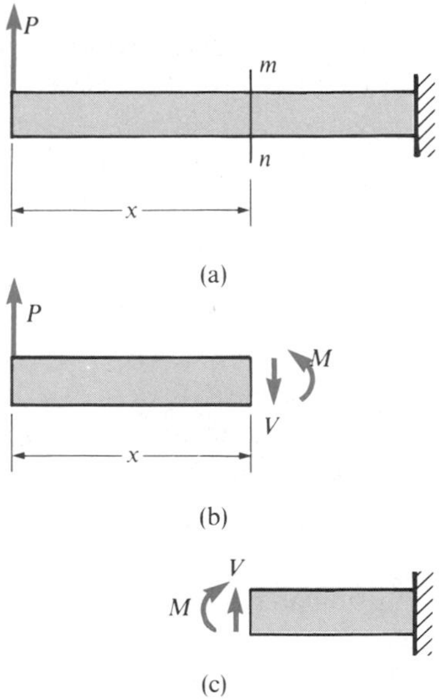

Fig. 4-2 Stress resultants V and M

It is convenient to reduce the resultant to a **shear force** V acting parallel to the cross section and a **bending couple** of moment M. Because the load P is transverse to the axis of the beam, no axial force exists at the cross section. Both the shear force and the bending couple act in the plane of the beam, which means that the moment vector for the couple is perpendicular to the plane of the figure. The moment of the bending couple is called the **bending moment** M. Because shear forces and bending moments, like axial forces in bars and twisting couples in shafts, are the resultants of stresses distributed over the cross section, they are known collectively as **stress resultants**.

The stress resultants in statically determinate beams can be calculated from equations of static equilibrium. As an example, consider again the cantilever beam of Fig. 4-2a. From the free-body diagram of

Fig. 4-2b, we obtain

$$V = P \qquad M = Px$$

where x is the distance from the free end to section mn. Thus, through the use of a free-body diagram and equations of equilibrium, we are able to calculate the stress resultants without difficulty. In the next chapter, we will see how to determine the internal stresses associated with V and M.

The shear force and bending moment are assumed to be positive when they act on the left-hand part of the beam in the directions shown in Fig. 4-2b. If we consider the right-hand part of the beam (Fig. 4-2c), then the directions of these same stress resultants are reversed. Therefore, we must recognize that the algebraic sign of a stress resultant does not depend upon its direction in space, such as upward or downward, or clockwise or counterclockwise, but rather the sign depends upon the direction of the stress resultant with respect to the material against which it acts. To make this point clear, the **sign conventions** for shear forces and bending moments are repeated in Fig. 4-3, where V and M are shown acting on an element of the beam cut out between two cross sections that are a small distance apart.

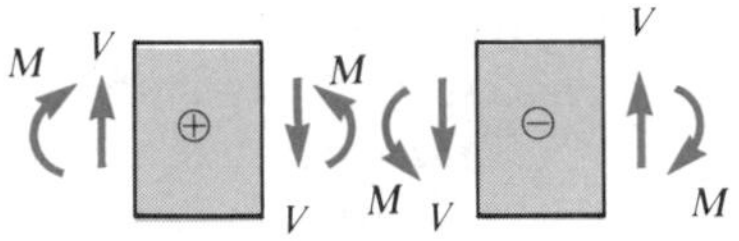

Fig. 4-3 Sign conventions for shear force V and bending moment M

The deformations of an element caused by both positive and negative shear forces and bending moments are sketched in Fig. 4-4. We see that a positive shear force tends to deform the element by causing the right-hand face to move downward with respect to the left-hand face, and a positive bending moment elongates the lower part of the beam and compresses the upper part. Because the signs for V and M are related to the deformations of the material, these sign conventions are called **deformation sign conventions**. We previously used a deformation sign convention for axial forces (tension is positive, compression is negative). A different kind of sign convention, called a **static sign convention**, is used in equations of static equilibrium. When using a static sign convention, forces are taken as positive when they act in the positive direction of a coordinate axis.

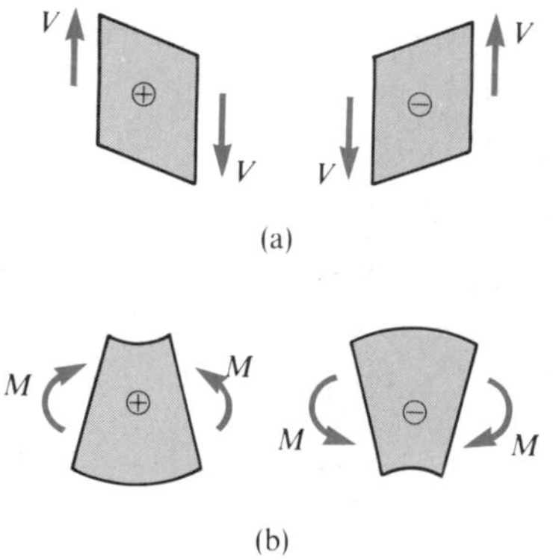

Fig. 4-4 Deformations (highly exaggerated) of an element caused by: (a) shear forces, and (b) bending moments

To illustrate the two types of sign conventions, let us write equations of equilibrium for the two parts of the beam shown in Fig. 4-2. Note that V and M are positive according to the deformation sign convention for stress resultants. However, if the y axis is positive upward, then the shear force V in Fig. 4-2b is given a negative sign in the equilibrium equation, which is written according to a static sign convention:

$$\Sigma F_y = 0 \quad \text{or} \quad P - V = 0$$

Of course, V would be given a positive sign in an equation of equilibrium for the right-hand part of the beam. Thus, a positive shear force may appear in a force equilibrium equation with either a positive or a negative sign, depending upon the free-body diagram that is considered. An analogous situation exists for bending moments when moment equilibrium equations are used. Difficulties with signs can be avoided by keeping in mind that two types of sign conventions are used in mechanics:

deformation sign conventions are used for stress resultants and static sign conventions are used in equations of equilibrium. The former are based upon how the material is deformed, and the latter are based upon directions in space.

Example 1

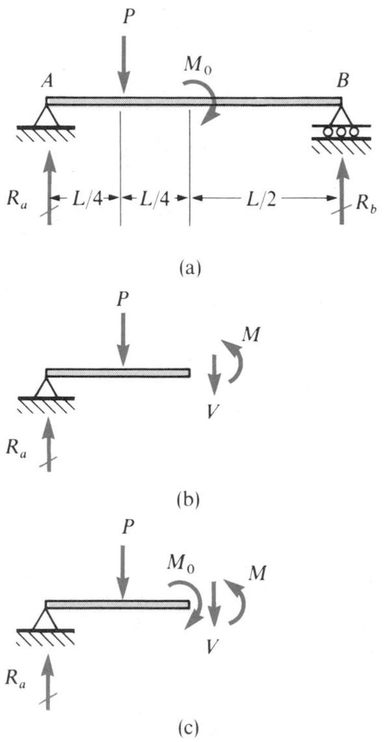

Fig. 4-5 Example 1

A simple beam AB supports two loads, a force P and a couple M_0, acting as shown in Fig. 4-5a. Find the shear force V and bending moment M in the beam at cross sections located as follows: (a) a small distance to the left of the middle of the beam and (b) a small distance to the right of the middle of the beam.

The first step in the analysis of this beam is to find the reactions R_a and R_b. Taking moments about ends A and B gives two equations of equilibrium, from which we find

$$R_a = \frac{3P}{4} - \frac{M_0}{L} \qquad R_b = \frac{P}{4} + \frac{M_0}{L}$$

Next, the beam is cut at a cross section just to the left of the middle, and a free-body diagram is drawn of either half of the beam. In this example, we choose the left-hand half of the beam, and the corresponding diagram is shown in Fig. 4-5b. The force P and the reaction R_a appear in this diagram, along with the unknown shear force V and bending moment M, both of which are shown in their positive directions. The couple M_0 does not appear in the figure because the beam is cut to the left of the point of application of M_0. A summation of forces in the vertical direction gives

$$R_a - P - V = 0 \quad \text{or} \quad V = -\frac{P}{4} - \frac{M_0}{L}$$

This result shows that when P and M_0 act in the directions shown in Fig. 4-5a, the shear force is negative and acts in the direction opposite to that assumed in Fig. 4-5b. Taking moments about an axis through the cross section where the beam is cut (Fig. 4-5b) gives

$$-R_a\left(\frac{L}{2}\right) + P\left(\frac{L}{4}\right) + M = 0 \quad \text{or} \quad M = \frac{PL}{8} - \frac{M_0}{2}$$

The bending moment M may be either positive or negative, depending upon the relative magnitudes of the terms in this equation.

To obtain the stress resultants at a cross section just to the right of the middle, we cut the beam at that section and again draw a free-body diagram (Fig. 4-5c). The difference between this diagram and the former one is that the couple M_0 now acts on the part of the beam to the left of the cut section. Again summing forces in the vertical direction, and also taking moments about an axis through the cut section, we obtain

$$V = -\frac{P}{4} - \frac{M_0}{L} \qquad M = \frac{PL}{8} + \frac{M_0}{2}$$

We see from these results that when the cut section is shifted from left to right of the couple M_0, the shear force does not change but the bending moment increases algebraically by an amount equal to M_0.

Example 2

A cantilever beam that is free at end A and fixed at end B is subjected to a distributed load of linearly varying intensity q (Fig. 4-6a). The maximum intensity of the load occurs at the fixed support and is denoted by q_0. Find the shear force V and bending moment M at distance x from the free end.

We begin by cutting through the beam at distance x from the left-hand end and isolating part of the beam as a free body (Fig. 4-6b). As in the preceding example, the unknown shear force V and bending moment M are assumed to be positive. The intensity of the distributed load is $q = q_0x/L$; therefore, the total downward load on the free body of Fig. 4-6b is equal to $q_0x^2/2L$. Equilibrium in the vertical direction gives

$$V = -\frac{q_0x^2}{2L} \qquad \text{(a)}$$

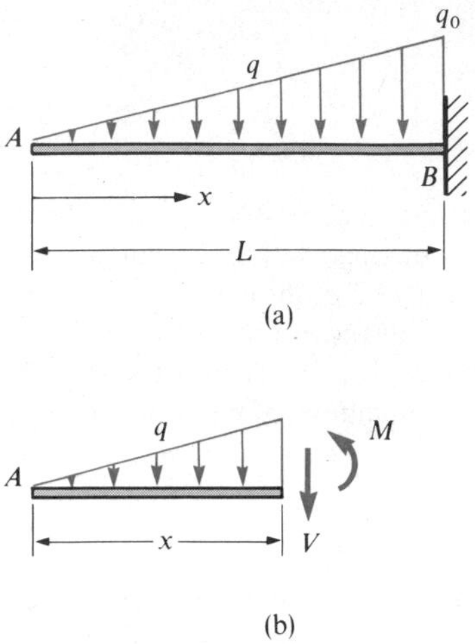

Fig. 4-6 Example 2

From this equation, we note that at the free end A ($x = 0$) the shear force is $V = 0$ and at the fixed end B ($x = L$) the shear force is $V = -q_0L/2$.

To find the bending moment in the beam, we write an equation of moment equilibrium about an axis through the cut section and solve for M:

$$M = -\frac{q_0x^3}{6L} \qquad \text{(b)}$$

Again considering the two ends of the beam, we see that the bending moment is equal to zero when $x = 0$ and equal to $-q_0L^2/6$ when $x = L$. Equations (a) and (b) can be used to obtain V and M at any point in the beam, and we see from these equations that both the shear force and the bending moment reach their numerically largest values at the fixed end of the beam.

Example 3

A beam ABC with an overhang supports a uniform load of intensity $q = 6$ kN/m and a concentrated load $P = 28$ kN (Fig. 4-7a). Calculate the shear force V and bending moment M at a cross section D located 5 m from the left-hand support.

We begin by calculating the reactions from equations of equilibrium for the entire beam. Thus, taking moments about the support at B, we get

$$-R_a(8\text{ m}) + (28\text{ kN})(5\text{ m}) + (6\text{ kN/m})(10\text{ m})(3\text{ m}) = 0$$

from which $R_a = 40$ kN. In a similar manner, equilibrium of moments about support A yields $R_b = 48$ kN. (We also observe that equilibrium of forces in the vertical direction is satisfied.)

Next, we make a cut at section D and construct a free-body diagram of the left-hand part of the beam (Fig. 4-7b). When drawing this diagram, we assume that the unknown stress resultants V and M are positive. The equations of equilibrium for the free body are as follows:

$$\Sigma F_y = 0 \qquad 40\text{ kN} - 28\text{ kN} - (6\text{ kN/m})(5\text{ m}) - V = 0$$

$$\Sigma M_d = 0 \qquad -(40\text{ kN})(5\text{ m}) + (28\text{ kN})(2\text{ m}) + (6\text{ kN/m})(5\text{ m})(2.5\text{ m}) + M = 0$$

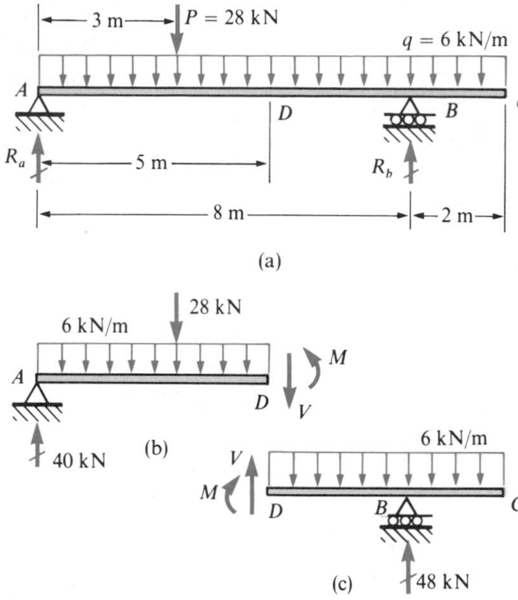

Fig. 4-7 Example 3

from which

$$V = -18 \text{ kN} \qquad M = 69 \text{ kN}\cdot\text{m}$$

The minus sign in the result for V means that the shear force acts in the negative direction (opposite to the direction shown in Fig. 4-7b).

An alternative method of solution is to obtain V and M from a free-body diagram of the right-hand part of the beam (Fig. 4-7c). When drawing this diagram, we again assume that the unknown shear force and unknown bending moment are positive. Then the two equations of equilibrium are

$$\Sigma F_y = 0 \qquad V - (6 \text{ kN/m})(5 \text{ m}) + 48 \text{ kN} = 0$$

$$\Sigma M_d = 0 \qquad -M - (6 \text{ kN/m})(5 \text{ m})(2.5 \text{ m}) + (48 \text{ kN})(3 \text{ m}) = 0$$

from which

$$V = -18 \text{ kN} \qquad M = 69 \text{ kN}\cdot\text{m}$$

as before.

4.3 RELATIONSHIPS BETWEEN LOAD, SHEAR FORCE, AND BENDING MOMENT

We will now obtain some important relationships between the loads on a beam, the shear force V, and the bending moment M. These relationships are quite useful when investigating the shear force and bending moment throughout the entire length of a beam, and they are especially helpful

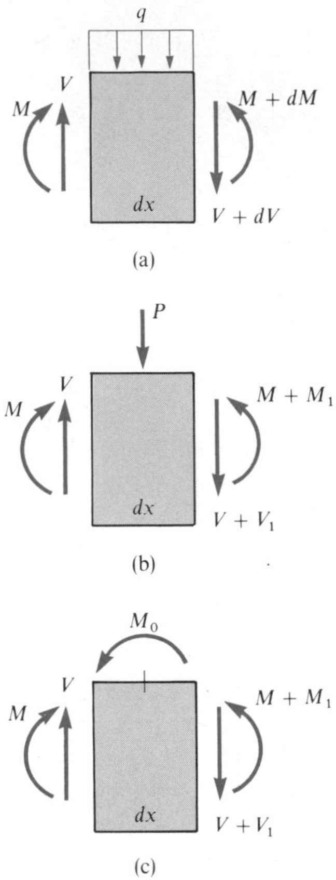

Fig. 4-8 Element of a beam used in deriving relationships between loads, shear forces, and bending moments

when constructing shear-force and bending-moment diagrams (see Section 4-4). As a means of obtaining the relationships, let us consider an element of a beam cut out between two cross sections that are distance dx apart (Fig. 4-8a). On the left-hand face of the element are shown the shear force V and bending moment M, acting in their positive directions. In general, V and M are functions of the distance x measured along the axis of the beam, which means that the shear force and bending moment have values on the right-hand face of the element that are slightly different from their values on the left-hand face. If we denote the increments in V and M by dV and dM, respectively, then the corresponding stress resultants on the right-hand face are $V + dV$ and $M + dM$. Again, they are shown acting in their positive directions.

The load acting on the top surface of the element may be a distributed load, a concentrated load, or a couple. Let us assume first that the load is distributed and of intensity q, as shown in Fig. 4-8a. Then, from equilibrium of forces in the vertical direction, we get

$$V - (V + dV) - q\,dx = 0$$

or

$$\frac{dV}{dx} = -q \qquad \text{(4-1)}$$

The significance of this equation is as follows. The shear force V changes with the distance x measured along the axis of the beam, and Eq. (4-1) shows that *the rate of change of V with respect to x is equal to $-q$.* For example, if there is no load on part of the beam ($q = 0$), then $dV/dx = 0$ and the shear force is constant in that part of the beam. If q is a constant over part of the beam (uniform load), then dV/dx is constant also and the shear force changes linearly in that part of the beam.

The **sign conventions** used with Eq. (4-1) are shown in Fig. 4-8a, where both V and q are positive. As explained in Section 4.2, a shear force is positive when it acts upward on the left face and downward on the right face of the element. We have now adopted a sign convention for distributed loads q, namely, *a distributed load is positive when it acts downward and negative when it acts upward.*

As an illustration of the use of Eq. (4-1), consider the cantilever beam with a linearly varying load that we discussed in Example 2 of the preceding section (see Fig. 4-6). The load on the beam is

$$q = \frac{q_0 x}{L}$$

which is positive because it acts downward. The shear force is

$$V = -\frac{q_0 x^2}{2L}$$

(see Eq. a, Section 4.2). Taking the derivative dV/dx gives

$$\frac{dV}{dx} = -\frac{q_0 x}{L} = -q$$

which is in accord with Eq. (4-1).

Equation (4-1) can be integrated along the axis of the beam to obtain a useful equation pertaining to the shear forces acting at two different cross sections. To obtain this relationship, we multiply both sides of Eq. (4-1) by dx and then integrate between two points A and B on the axis of the beam. The result is

$$\int_A^B dV = -\int_A^B q\,dx \tag{a}$$

The left-hand side of this equation equals the difference $V_b - V_a$ of the shear forces at sections B and A. The integral on the right-hand side represents the area of the load-intensity diagram between A and B, which is equal in magnitude to the resultant of the distributed load between A and B. Thus,

$$\begin{aligned} V_b - V_a &= -\int_A^B q\,dx \\ &= -(\text{area of load-intensity diagram between } A \text{ and } B) \end{aligned} \tag{4-2}$$

The area of the load-intensity diagram may be treated as either positive or negative, depending upon whether q acts downward or upward, respectively.

Because Eq. (4-1) was derived for an element of the beam subjected only to a continuously distributed load (or to no load), we cannot use Eq. (4-1) at a point where a concentrated load is applied. In a similar manner, we cannot use Eq. (4-2) if a concentrated load P acts on the beam between points A and B, because the intensity of load q is undefined for a concentrated load.

Returning to the beam element shown in Fig. 4-8a, let us now consider equilibrium by summing moments about an axis through the left-hand face of the element and perpendicular to the plane of the figure. Taking moments as positive when counterclockwise, we obtain

$$-M - q\,dx\left(\frac{dx}{2}\right) - (V + dV)\,dx + M + dM = 0$$

Discarding products of differentials because they are negligible compared to the other terms, we obtain the following relationship:

$$\frac{dM}{dx} = V \tag{4-3}$$

This equation shows that *the rate of change of the bending moment M with respect to x is equal to the shear force*. For instance, if the shear force is zero in a region of the beam, then the bending moment is constant in that same region. Equation (4-3) applies only in regions where distributed loads act on the beam. At a point where a concentrated load acts, a sudden change (or discontinuity) in the shear force occurs and the derivative dM/dx is undefined.

Again using the cantilever beam in Fig. 4-6a as an example, we recall that the bending moment (see Eq. b, Section 4.2) is

$$M = -\frac{q_0 x^3}{6L}$$

The derivative dM/dx is

$$\frac{dM}{dx} = -\frac{q_0 x^2}{2L}$$

which is the same as the shear force V in the beam (see Eq. a, Section 4.2); thus, Eq. (4-3) is satisfied.

Integrating Eq. (4-3) between two points A and B on the beam axis gives

$$\int_A^B dM = \int_A^B V\,dx \tag{b}$$

The integral on the left-hand side of this equation is equal to the difference $M_b - M_a$ of the bending moments at points A and B. To interpret the integral on the right-hand side, we need to consider V as a function of x and to visualize a diagram showing the manner in which V varies with x. Then we see that the integral on the right represents the area below the diagram between A and B. Therefore, we can express Eq. (b) in the following manner:

$$M_b - M_a = \int_A^B V\,dx$$

$$= \text{area of shear-force diagram between } A \text{ and } B \tag{4-4}$$

This equation can be used even when concentrated loads are acting on the beam between points A and B. However, it is not valid if a couple acts between A and B, because a couple produces a sudden change in the bending moment and the left-hand side of Eq. (b) cannot be integrated across such a discontinuity.

Now let us consider a concentrated load P acting on the beam element (Fig. 4-8b). In order to have a **sign convention** for concentrated loads, we assume that *a downward load is positive*. As before, the stress resultants on the left-hand face are denoted by V and M. On the right-hand face, they are denoted by $V + V_1$ and $M + M_1$, where V_1 and M_1 represent the possible increments in the shear force and bending moment. From equilibrium of forces in the vertical direction, we get

$$V_1 = -P \tag{4-5}$$

which means that an abrupt change in the shear force occurs at any point where a concentrated load acts. As we pass from left to right through the point of load application, the shear force decreases by an amount equal to the magnitude of the downward load. From equilibrium of moments (Fig. 4-8b), we get

$$-M - P\left(\frac{dx}{2}\right) - (V + V_1)\,dx + M + M_1 = 0$$

or

$$M_1 = P\left(\frac{dx}{2}\right) + V\,dx + V_1\,dx$$

Since the length dx of the element is infinitesimally small, we see from this equation that the increment M_1 in the bending moment is also infinitesimally small. Thus, we conclude that the bending moment does not change as we pass through the point of application of a concentrated load.

Even though the bending moment M does not change at a concentrated load, its rate of change dM/dx undergoes an abrupt change. At the left-hand side of the element (Fig. 4-8b), the rate of change of the bending moment (see Eq. 4-3) is $dM/dx = V$. At the right-hand side, the rate of change is $dM/dx = V + V_1$. Therefore, we conclude that at the point of application of a concentrated load P, the rate of change dM/dx decreases abruptly by an amount equal to P.

The last case is a load in the form of a couple M_0 (Fig. 4-8c). The **sign convention** for a couple is that *a couple acting as a load on a beam is positive when it is counterclockwise*. From equilibrium of the element in the vertical direction, we obtain $V_1 = 0$, which shows that the shear force does not change at the point of application of a couple. Equilibrium of moments for the element gives

$$-M + M_0 - (V + V_1)\,dx + M + M_1 = 0$$

or, disregarding terms that contain differentials,

$$M_1 = -M_0 \tag{4-6}$$

This equation shows that there is an abrupt decrease in the bending moment in the beam due to the applied couple M_0 as we move from left to right through the point of load application.

Equations (4-1) through (4-6) are useful when making a complete investigation of the shear forces and bending moments in a beam, as illustrated in the next section.

4.4 SHEAR-FORCE AND BENDING-MOMENT DIAGRAMS

The shear forces V and bending moments M in a beam are functions of the distance x measured along the longitudinal axis. When designing a beam, it is desirable to know the values of V and M at all cross sections.

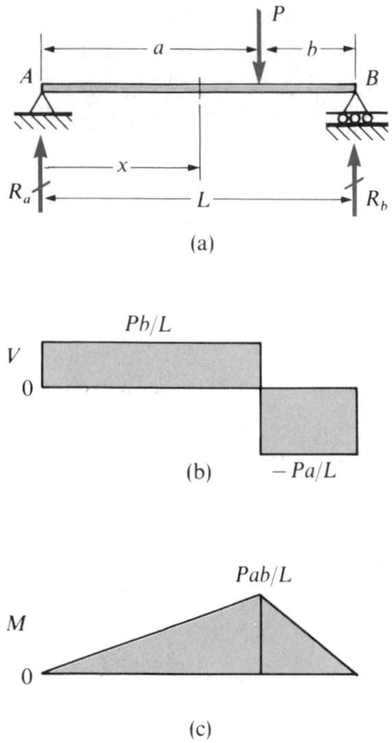

Fig. 4-9 Shear-force and bending-moment diagrams for a simple beam with a concentrated load

A convenient way to provide this information is to draw graphs showing how V and M vary with x. To plot such a graph, we take the abscissa as the position of the cross section (that is, the distance x), and we take the ordinate as the corresponding value of either the shear force or the bending moment. These graphs are called **shear-force and bending-moment diagrams**.

To illustrate the construction of the diagrams, let us consider a simple beam AB carrying a concentrated load P (Fig. 4-9a). The reactions for this beam are

$$R_a = \frac{Pb}{L} \qquad R_b = \frac{Pa}{L} \tag{a}$$

as found from equilibrium of the entire beam. Next, we cut through the beam to the left of the load P and at distance x from support A. Then we construct a free-body diagram of the left-hand part of the beam, and we find from equilibrium that

$$V = R_a = \frac{Pb}{L} \qquad M = R_a x = \frac{Pbx}{L} \tag{b}$$

These equations show that the shear force is constant from support A to the point of application of the load P and that the bending moment varies linearly with x. The expressions for V and M are plotted directly beneath the sketch of the beam (Fig. 4-9). In the case of the shear force, the diagram begins at the end of the beam with an abrupt jump in the shear force equal to the reaction R_a. Then the shear force remains constant up to $x = a$. In this same region, the bending-moment diagram is a straight line increasing from $M = 0$ at the support to $M = Pab/L$ at $x = a$.

Next we cut through the beam to the right of the load P (that is, in the region $a < x < L$), and, from equilibrium of the left-hand part of the beam, we obtain the following expressions:

$$V = \frac{Pb}{L} - P = -\frac{Pa}{L} \tag{c}$$

$$M = \frac{Pbx}{L} - P(x - a) = Pa\left(1 - \frac{x}{L}\right) \tag{d}$$

Again we see that the shear force is constant and that the bending moment is a linear function of x. At $x = a$, the bending moment is equal to Pab/L; at $x = L$, it is zero. Equations (c) and (d) for V and M are plotted in Fig. 4-9.

In deriving Eqs. (c) and (d) for the shear force and bending moment to the right of the load P, we considered the equilibrium of the left-hand part of the beam, which is acted upon by the two forces R_a and P. It would have been slightly simpler in this example to consider the right-hand portion of the beam as a free body. From equilibrium of the right-hand part, we obtain the equations

$$V = -R_b = -\frac{Pa}{L} \qquad M = R_b(L - x) = \frac{Pa}{L}(L - x)$$

which are the same results as obtained from Eqs. (c) and (d).

Let us now observe certain characteristics of the diagrams in Fig. 4-9. We note first that the slope dV/dx of the shear-force diagram is zero in the regions $0 < x < a$ and $a < x < L$, which is in accord with the equation $dV/dx = -q$ (Eq. 4-1). Also, in these same regions, the slope dM/dx of the bending moment is equal to V (Eq. 4-3). At the point of application of the load P, there is an abrupt change in the shear-force diagram (equal to P) and a corresponding change in the slope of the bending-moment diagram. To the left of the load P, the slope of the moment diagram is positive and equal to Pb/L; to the right, it is negative and equal to $-Pa/L$.

Consider next the area of the shear-force diagram (see Eq. 4-4). As we move from $x = 0$ to $x = a$, the area of the shear-force diagram is $(Pb/L)a$, or Pab/L. This quantity represents the increase in bending moment between these same two points. From $x = a$ to $x = L$, the area of the shear-force diagram is $-Pab/L$, which means that in this region the bending moment decreases by that amount. Thus, the bending moment is zero at end B of the beam, as expected. If we consider the entire shear-force diagram and if we note that $M = 0$ at both ends of the beam, then Eq. (4-4) requires that the area of the diagram between the ends of the beam be zero. This conclusion would not apply if the beam were subjected to a load in the form of a couple.

The maximum or minimum values of shear forces and bending moments are needed when designing beams. For a simple beam with a single concentrated load, the maximum shear force occurs at the end of the beam nearest to the concentrated load, and the maximum bending moment occurs under the load itself.

To further illustrate the construction of shear-force and bending-moment diagrams, let us consider a simple beam with a uniformly distributed load (Fig. 4-10a). In this case, each of the reactions R_a and R_b is equal to $qL/2$; hence at a cross section at distance x from the left-hand end A, we obtain

$$V = \frac{qL}{2} - qx \qquad M = \frac{qLx}{2} - \frac{qx^2}{2} \tag{e}$$

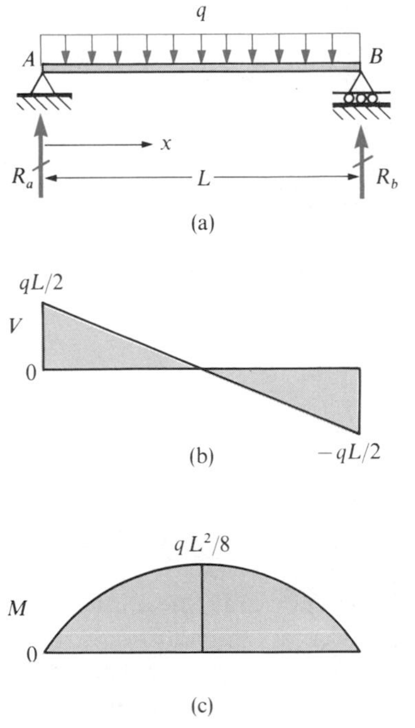

Fig. 4-10 Shear-force and bending-moment diagrams for a simple beam with a uniform load

The first of these equations shows that the shear-force diagram consists of an inclined straight line having ordinates $qL/2$ and $-qL/2$ at $x = 0$ and $x = L$, respectively (Fig. 4-10b). The slope of this line is $-q$, as expected from the equation $dV/dx = -q$ (Eq. 4-1). The bending-moment diagram is a parabolic curve that is symmetric about the middle of the beam (Fig. 4-10c). At each cross section, the slope of the bending-moment

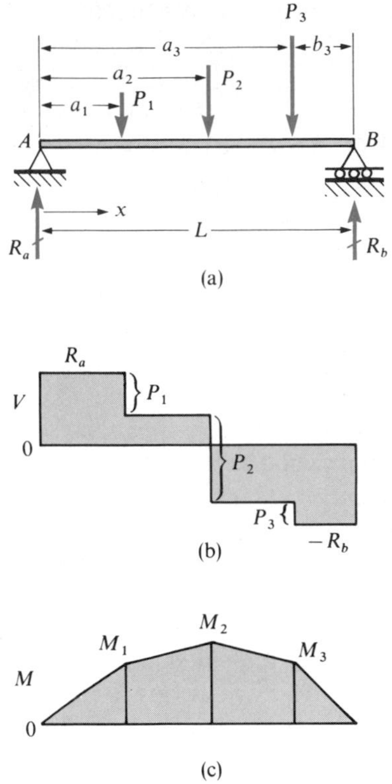

Fig. 4-11 Shear-force and bending-moment diagrams for a simple beam with several concentrated loads

diagram is equal to the shear force (see Eq. 4-3):

$$\frac{dM}{dx} = \frac{d}{dx}\left(\frac{qLx}{2} - \frac{qx^2}{2}\right) = \frac{qL}{2} - qx = V$$

The maximum value of the bending moment occurs at the point where $dM/dx = 0$ (that is, at the cross section where the shear force is zero). This section is at the middle of the beam in our example; hence we substitute $x = L/2$ into the expression for M to obtain

$$M_{\max} = \frac{qL^2}{8} \tag{f}$$

as shown in the bending-moment diagram.

The diagram of load intensity (Fig. 4-10a) has area qL. In accord with Eq. (4-2), the shear force V decreases by this amount as we move along the beam from A to B. The area of the shear-force diagram between $x = 0$ and $x = L/2$ is $qL^2/8$, and this area represents the increase in the bending moment between those same two points. In a similar manner, the bending moment decreases by $qL^2/8$ as we proceed from $x = L/2$ to $x = L$.

If several concentrated loads act on a simple beam (Fig. 4-11a), expressions for V and M may be determined for each region of the beam between the points of load application. Again measuring distance x from end A of the beam, we obtain for the first region of the beam ($0 < x < a_1$) the following equations:

$$V = R_a \qquad M = R_a x \tag{g}$$

For the second region ($a_1 < x < a_2$), we get

$$V = R_a - P_1 \qquad M = R_a x - P_1(x - a_1) \tag{h}$$

For the third region of the beam ($a_2 < x < a_3$), it is advantageous to consider the right-hand part of the beam rather than the left, because fewer loads act on the corresponding free body. Hence, we obtain

$$V = -R_b + P_3 \tag{i}$$

and

$$M = R_b(L - x) - P_3(L - b_3 - x) \tag{j}$$

Finally, for the fourth portion of the beam, we obtain

$$V = -R_b \qquad M = R_b(L - x) \tag{k}$$

From Eqs. (g) through (k) we see that in each region of the beam the shear force remains constant; hence the shear-force diagram has the shape shown in Fig. 4-11b. Also, the bending moment in each part of the beam is a linear function of x; therefore, the corresponding part of the diagram is represented by an inclined straight line. To assist in drawing these lines, we obtain the bending moments under the concentrated loads by substituting $x = a_1$, $x = a_2$, and $x = a_3$ into Eqs. (g), (h), and (k), respectively. In this manner we obtain for the bending moments the values

$$M_1 = R_a a_1 \qquad M_2 = R_a a_2 - P_1(a_2 - a_1) \qquad M_3 = R_b b_3 \qquad (l)$$

From these values, we can readily construct the bending-moment diagram (Fig. 4-11c) because the diagram consists of straight lines between points of loading.

Note that, at each point where a concentrated load acts, the shear-force diagram changes abruptly by an amount equal to the load. Furthermore, at each such discontinuity in the shear force, there is a corresponding change in the slope dM/dx of the bending-moment diagram. Also, the change in bending moment between two load points equals the area of the shear-force diagram between those same two points (see Eq. 4-4). For example, the change in bending moment between loads P_1 and P_2 is $M_2 - M_1$. Substituting from Eqs. (l), we get

$$M_2 - M_1 = (R_a - P_1)(a_2 - a_1)$$

which is the area of the rectangular shear-force diagram between $x = a_1$ and $x = a_2$.

The maximum bending moment in a beam with only concentrated loads must occur under one of the loads or at a reaction. From the equation $dM/dx = V$, we know that the slope of the bending-moment diagram is equal to the shear force. It follows that the bending moment has a maximum or minimum value at cross sections where the shear force changes sign, which occurs only under a load. If, as we proceed along the x axis, the shear force changes from a positive to a negative value (as in Fig. 4-11b), then the slope in the bending-moment diagram also changes from positive to negative. Therefore, we must have a maximum bending moment at this cross section. Conversely, a change in shear force from a negative to a positive value indicates a minimum bending moment. The shear-force diagram can (but usually does not) intersect the horizontal axis at several points. Corresponding to each such intersection point, there is a local maximum or minimum in the bending-moment diagram. The values of all local maximums and minimums must be calculated in order to find the absolute maximum positive and negative bending moments in the beam for use in design.

In general, the maximum positive or negative bending moments in a beam may occur at a concentrated load (provided the shear force changes sign at the load), at a reaction, at a cross section where the shear force equals zero (see Fig. 4-10), or at a section where a couple is applied. The discussions and examples in this section illustrate all of these possibilities.

When several loads act on a beam, the shear-force and bending-moment diagrams can be obtained by superposition (or summation) of the diagrams obtained from each of the loads acting separately. For instance, the shear-force diagram of Fig. 4-11b is actually the sum of three separate diagrams, each of the type shown in Fig. 4-9b for a single concentrated load. We can make an analogous comment for the bending-moment diagram of Fig. 4-11c. These conclusions concerning superposition of shear-force and bending-moment diagrams follow from the fact that shear forces and bending moments are linear functions of the applied loads.

Example 1

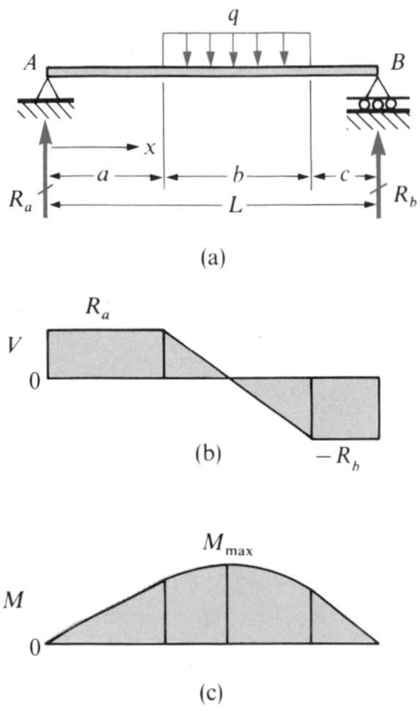

Fig. 4-12 Example 1

Construct the shear-force and bending-moment diagrams for a simple beam with a uniform load of constant intensity q acting over a part of the span (Fig. 4-12a).

We begin the analysis by determining the reactions for the beam:

$$R_a = \frac{qb}{L}\left(c + \frac{b}{2}\right) \qquad R_b = \frac{qb}{L}\left(a + \frac{b}{2}\right) \tag{m}$$

To obtain the shear forces and bending moments, we consider separately the three regions of the beam. For the left-hand part of the beam ($0 < x < a$), we find

$$V = R_a \qquad M = R_a x \tag{n}$$

For a cross section within the loaded portion of the beam, the shear force is obtained by subtracting from the reaction R_a the load $q(x - a)$ acting on the beam to the left of the cross section. The bending moment in this same region is obtained by subtracting the moment of the load to the left of the cross section from the moment of the reaction R_a. In this manner, we find

$$V = R_a - q(x - a) \tag{o}$$

$$M = R_a x - \frac{q(x - a)^2}{2} \tag{p}$$

For the unloaded portion of the beam at the right-hand end, we get

$$V = -R_b \qquad M = R_b(L - x) \tag{q}$$

Using Eqs. (n) through (q), we can readily construct the shear-force and bending-moment diagrams.

The shear-force diagram (Fig. 4-12b) consists of horizontal straight lines in the unloaded regions of the beam and an inclined straight line in the loaded region, as expected from the equation $dV/dx = -q$. The bending-moment diagram (Fig. 4-12c) consists of two inclined straight lines in the unloaded portions of the beam and a parabolic curve in the loaded portion. The inclined lines have slopes equal to R_a and $-R_b$, respectively (see Eq. 4-3). Each of these lines is tangent to the parabolic curve at the point where it meets the curve. This conclusion follows from the fact that there are no abrupt changes in the magnitude of the shear force at these points. Hence, from Eq. (4-3), we see that the slope of the bending-moment diagram cannot change abruptly.

The maximum bending moment M_{max} occurs where the shear force equals zero. Its value can be found by setting the shear force V (from Eq. o) equal to zero, solving for x, and then substituting this value of x into the expression for the bending moment (Eq. p). The result is

$$M_{\max} = \frac{qb}{8L^2}(b + 2c)(4aL + 2bc + b^2) \qquad \text{(r)}$$

This moment always occurs at a cross section within the region of the uniform load and is always positive. As a special case, consider a beam with a uniform load extending over the entire span (see Fig. 4-10). Then we substitute $a = c = 0$ and $b = L$ into Eq. (r) and obtain $M_{\max} = qL^2/8$ (see Eq. f).

Example 2

Construct the shear-force and bending-moment diagrams for the cantilever beam with two concentrated loads shown in Fig. 4-13a.

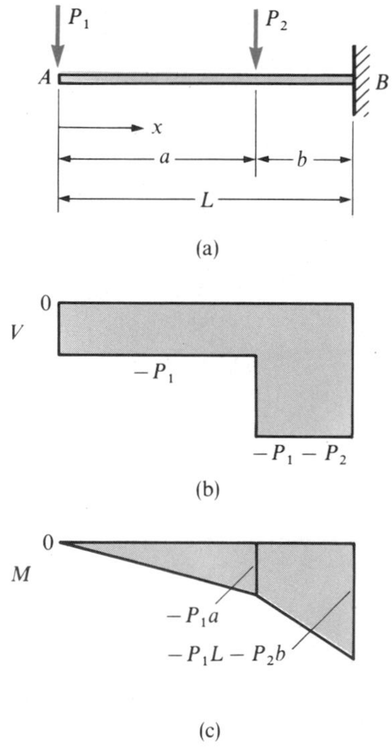

Fig. 4-13 Example 2

Again measuring distance x from the left-hand end of the beam, and considering first the region $0 < x < a$, we obtain

$$V = -P_1 \qquad M = -P_1 x \qquad \text{(s)}$$

For the right-hand portion of the beam ($a < x < L$), we obtain

$$V = -P_1 - P_2 \qquad M = -P_1 x - P_2(x - a) \qquad \text{(t)}$$

The corresponding shear-force and bending-moment diagrams are shown in Figs. 4-13b and c. The shear force is constant between the loads, and it reaches its maximum numerical value at the support. The bending-moment diagram consists of two inclined straight lines, the slopes of which are equal to the shear forces in the corresponding portions of the cantilever beam. The maximum bending moment occurs at the support ($x = L$), and it is equal to the area of the shear-force diagram, as expected from Eq. (4-4).

Example 3

Construct shear-force and bending-moment diagrams for the beam with an overhang shown in Fig. 4-14a. The beam is subjected to a uniform load of constant intensity $q = 1.0$ k/ft on the overhang and a counterclockwise couple $M_0 = 12.0$ ft-k acting midway between the supports.

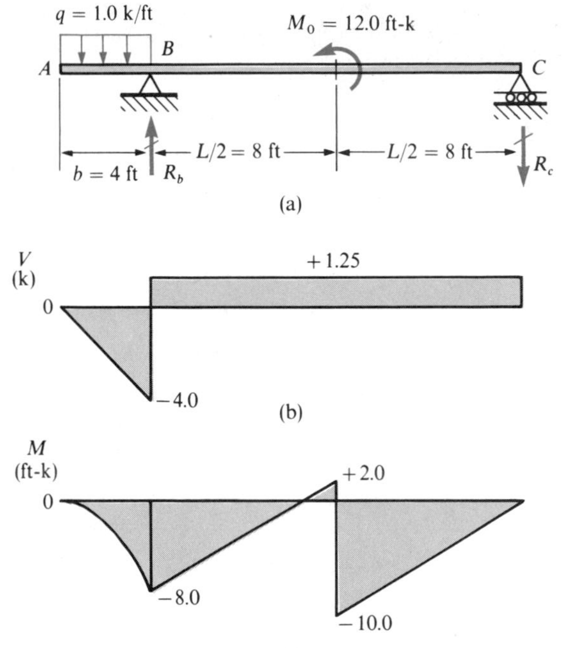

Fig. 4-14 Example 3

We can readily calculate the reactions R_b and R_c, and we find that R_b is upward and R_c is downward, as shown in the figure. Their numerical values are as follows:

$$R_b = 5.25 \text{ k} \qquad R_c = 1.25 \text{ k}$$

Using the techniques already described, we now draw the shear-force diagram (Fig. 4-14b). Note that the shear force does not change at the point of application of the couple M_0. The bending-moment diagram has the shape shown in Fig. 4-14c. At B, the moment is

$$M_b = -\frac{qb^2}{2} = -\frac{1}{2}(1.0 \text{ k/ft})(4 \text{ ft})^2 = -8.0 \text{ ft-k}$$

which is also equal to the area of the shear-force diagram between A and B. The slope of the bending-moment diagram from B to C is 1.25 k (that is, the slope equals the shear force), but the bending moment changes abruptly due to the couple M_0. This change is equal to M_0 (see Eq. 4-6). Maximum or minimum values of the bending moment occur where the shear force changes sign and where the couple is applied.

Example 4

A cantilever beam AB supporting a uniform load of constant intensity q is shown in Fig. 4-15a. Draw the shear-force and bending-moment diagrams for this beam.

The reactions R_b and M_b, obtained by considering the equilibrium of the entire beam, are

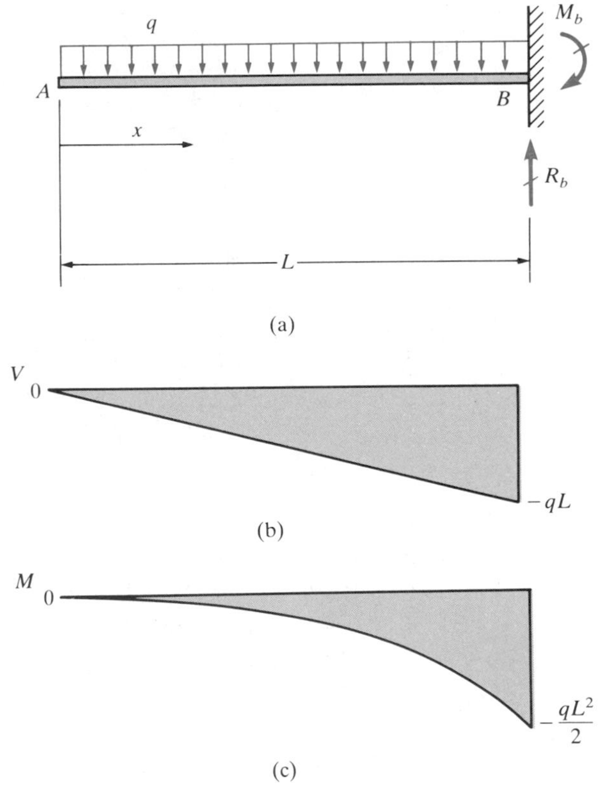

Fig. 4-15 Example 4

$$R_b = qL \qquad M_b = qL^2/2 \tag{u}$$

The shear force V and bending moment M at any cross section located at distance x from end A can easily be found with the aid of a free-body diagram similar to the one shown in Example 2 of Section 4.2 (see Fig. 4-6b). By this means we obtain

$$V = -qx \qquad M = -qx^2/2 \tag{v}$$

Now we can plot these equations to obtain the shear-force and bending-moment diagrams (Figs. 4-15b and c). The values of V and M at support B are consistent with the values of R_b and M_b previously determined (see Eqs. u). We also observe that the slope of the shear force diagram is equal to $-q$ and that the slope of the bending-moment diagram at any point is equal to the shear force V.

In this example it is also feasible to determine V and M by integrating the differential relationships between load, shear force, and bending moment. The shear force is obtained from Eq. (4-2), which is repeated here:

$$V_b - V_a = -\int_A^B q\,dx \tag{w}$$

For our purposes, we take A at the left-hand end of the beam where the shear force is zero ($V_a = 0$), and we take B at distance x from A. Then, from Eq. (w), we get

$$V = V_b = -\int_0^x q\,dx = -qx \tag{x}$$

which agrees with the previous result. The bending moment is obtained from Eq. (4-4):

$$M_b - M_a = \int_A^B V\,dx \tag{y}$$

Since the bending moment at end A is zero, we obtain

$$M = M_b = \int_0^x -qx\,dx = -\frac{qx^2}{2} \tag{z}$$

Integrating the differential relationships is quite simple in this example because the loading pattern is continuous and there are no concentrated loads or couples. If concentrated loads or couples are present, discontinuities in the V and M diagrams will exist, and we cannot integrate Eqs. (w) and (y) across such discontinuities, as explained in Section 4.3.

PROBLEMS / CHAPTER 4

4.2-1 Determine the shear force V and bending moment M at the middle of the simple beam AB shown in the figure.

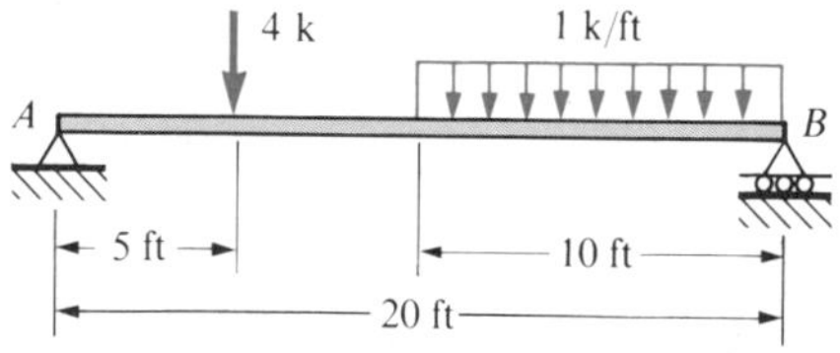

Prob. 4.2-1

4.2-2 Calculate the shear force V and bending moment M at a section located 0.5 m from the fixed support A of the cantilever beam shown in the figure.

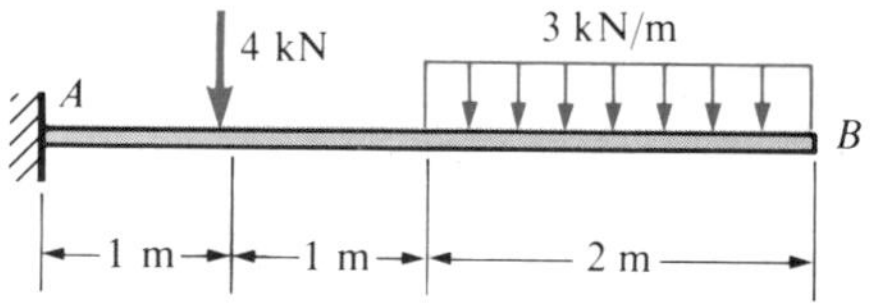

Prob. 4.2-2

4.2-3 What are the shear force V and bending moment M at the midpoint of the overhanging beam shown in the figure?

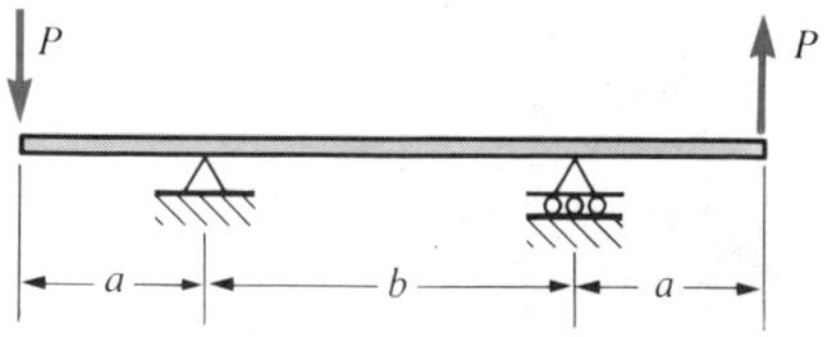

Prob. 4.2-3

4.2-4 Find the shear force V and bending moment M at the midpoint of the simple beam AB shown in the figure.

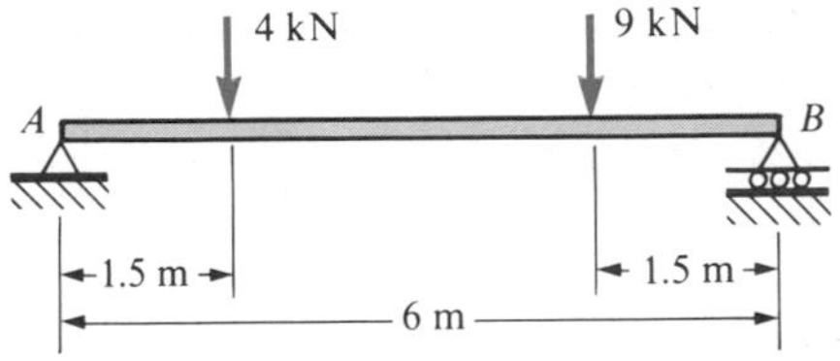

Prob. 4.2-4

4.2-5 Find the shear force V and bending moment M at a section located 16 ft from the left-hand end A of the beam with an overhang shown in the figure.

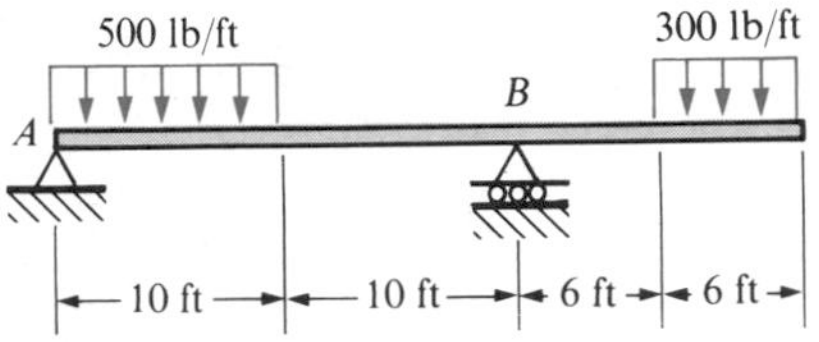

Prob. 4.2-5

4.2-6 The beam shown in the figure is simply supported at A and B. The loads consist of a couple $M_0 = 4$ kN·m at A and a concentrated load $P = 9$ kN at the end of the overhang. Determine the shear force V and bending moment M at a section 2 m from the left-hand support.

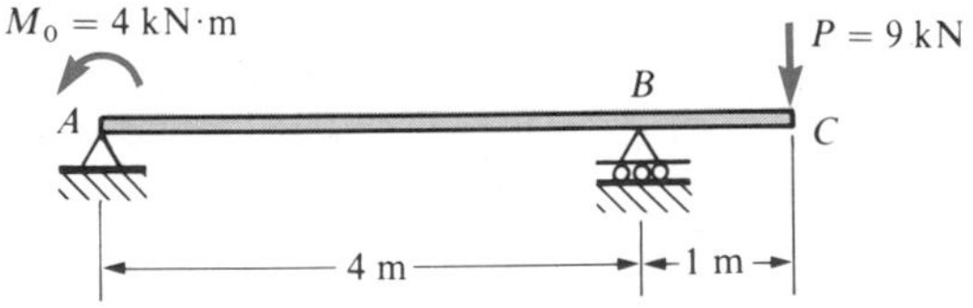

Prob. 4.2-6

4.2-7 A lumberjack weighing 240 lb stands at the midpoint of a floating log that is 16 ft long. What is the maximum bending moment in the log?

4.2-8 A simply supported beam AB supports a trapezoidal load (see figure). The intensity of the distributed load varies linearly from 50 kN/m at A to 30 kN/m at B. Calculate the shear force V and bending moment M at the midpoint of the beam.

4.2-9 A curved bar ABC is subjected to loads in the form of two equal and opposite forces P, as shown in the figure. The axis of the bar forms a semicircle of radius r. Determine the axial force N, shear force V, and bending moment M acting at a cross section defined by the angle θ (see figure).

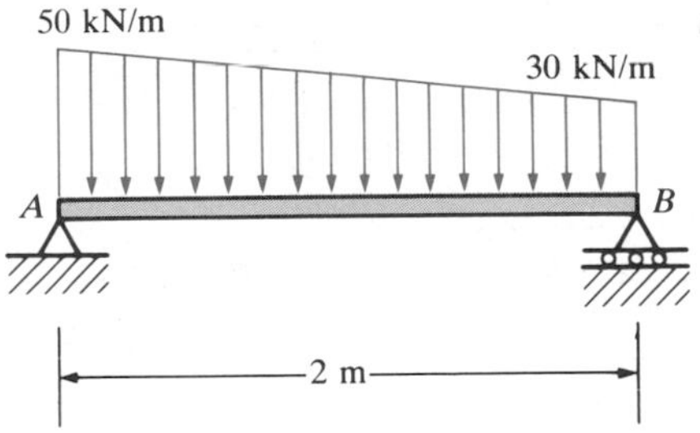

Prob. 4.2-8

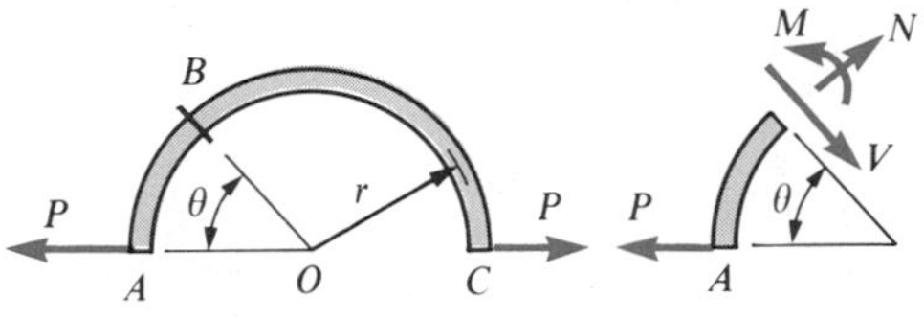

Prob. 4.2-9

4.2-10 The beam ABC shown in the figure is attached to a pin support at C and a roller support at A. A uniform load of intensity q acts on part AB and a triangular load of maximum intensity $2q$ acts on part BC. (a) Obtain an expression for the bending moment M in part AB at distance x from support A. (b) From this expression, determine the maximum bending moment M_{max} in part AB.

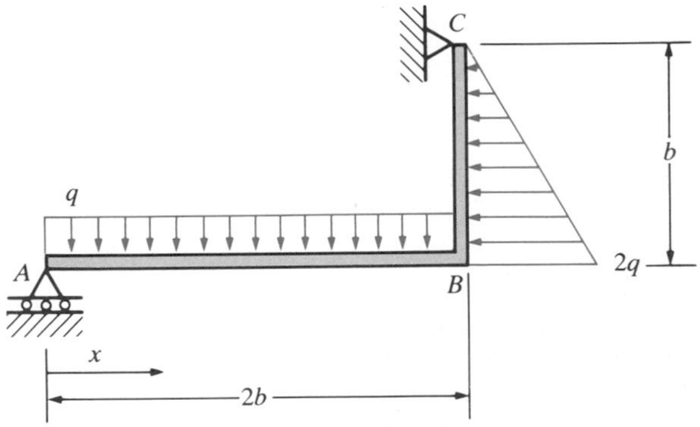

Prob. 4.2-10

4.2-11 A beam ABC with a vertical arm BD is supported as a simple beam at A and C (see figure). A cable passes over a small pulley that is attached to the arm at D. One end of the cable is attached to the beam at E. What is the force P in the cable if the bending moment in the beam just to the left of B is equal numerically to 64 in.-k?

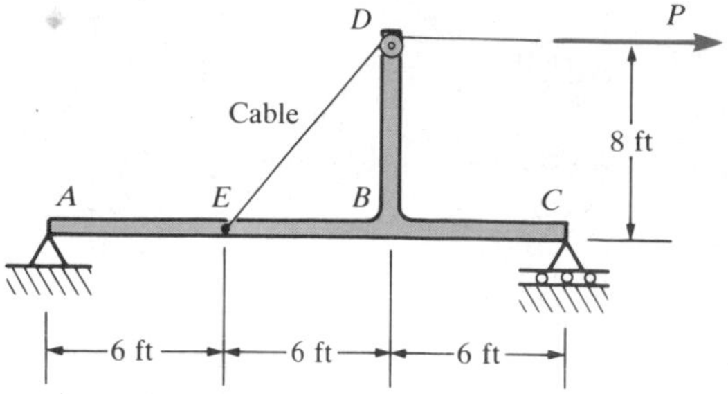

Prob. 4.2-11

4.2-12 The beam *ABCD* is loaded by a force $W = 30$ kN by the arrangement shown in the figure. The cable passes over a small frictionless pulley at *B* and is attached at *E* to the vertical arm. Calculate the axial force *N*, shear force *V*, and bending moment *M* at section *C*, which is just to the left of the vertical arm.

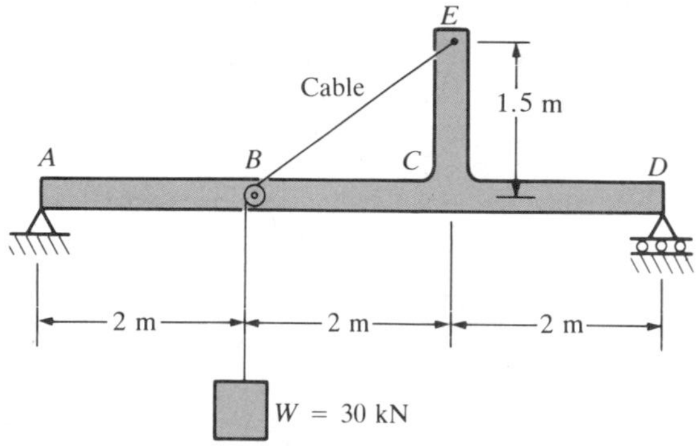

Prob. 4.2-12

4.2-13 The beam *ABCD* is held in equilibrium by uniformly distributed loads of intensities q_1 and q_2 as shown in the figure. Find the shear force *V* and bending moment *M* at the following cross sections of the beam: (a) cross section at point *B*, and (b) cross section at midpoint of the beam. (Assume $a = 3$ ft, $b = 8$ ft, and $q_1 = 2800$ lb/ft.)

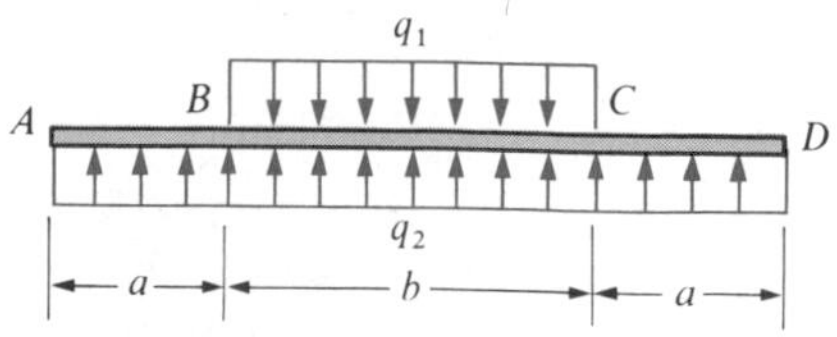

Prob. 4.2-13

4.2-14 The beam *ABCD* shown in the figure has overhanging ends and carries a distributed load of linearly varying intensity. For what ratio a/L will the shear force *V* always be zero at the midpoint of the beam?

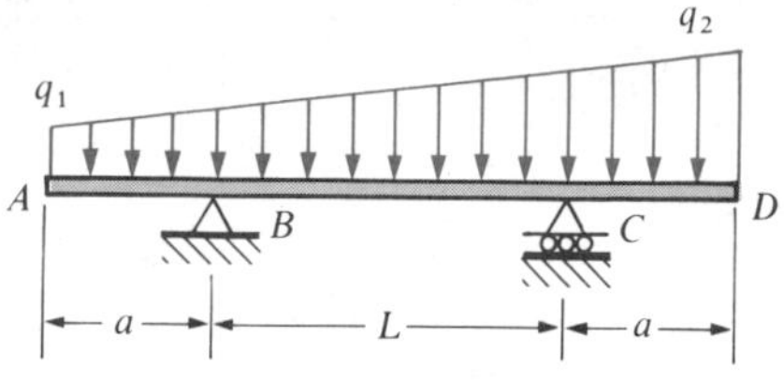

Prob. 4.2-14

***4.2-15** A beam *ABC* with an overhang (see figure) supports a concentrated load $P = 8$ k at the left-hand end and a uniform load of intensity *q* throughout its entire length. It is known that the shear force and bending moment at section *X-X* are equal to -16 k and -96 ft-k, respectively. Find the shear force *V* and bending moment *M* at section *Y-Y*.

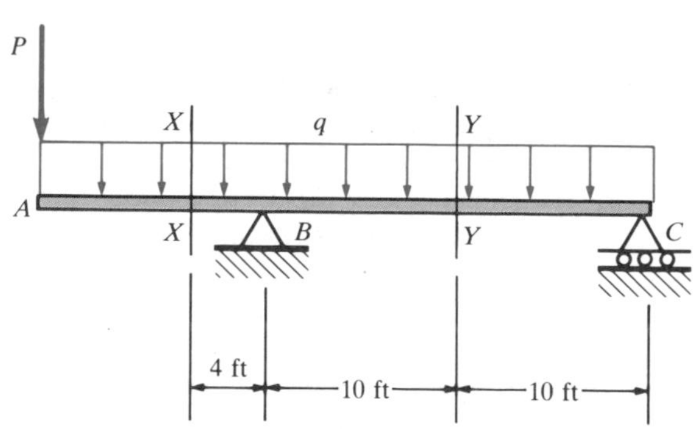

Prob. 4.2-15

***4.2-16** The centrifuge shown in the figure rotates in a horizontal plane (the *xy* plane) on a smooth surface about

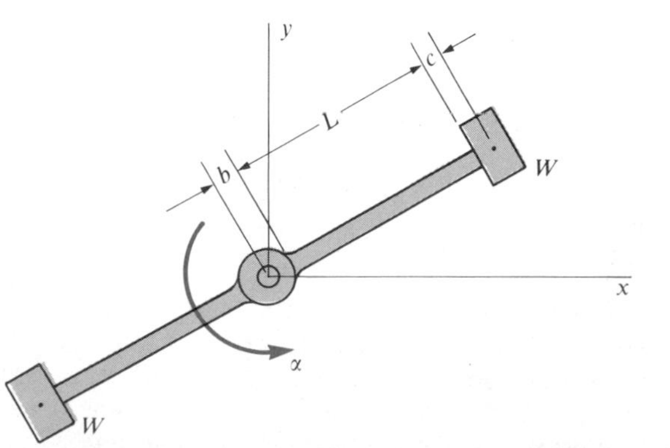

Prob. 4.2-16

the z axis (which is vertical) with an angular acceleration α. Each of the two arms has weight w per unit length and supports weight $W = 10wL$ at its end. Derive formulas for the maximum shear force and maximum bending moment in the arms, assuming $b = L/9$ and $c = L/10$.

When solving Problems 4.4-1 through 4.4-38, draw approximately to scale the shear-force and bending-moment diagrams and label all critical ordinates, including the maximum and minimum values.

4.4-1 Construct the shear-force and bending-moment diagrams for a simple beam supporting two equal concentrated loads (see figure).

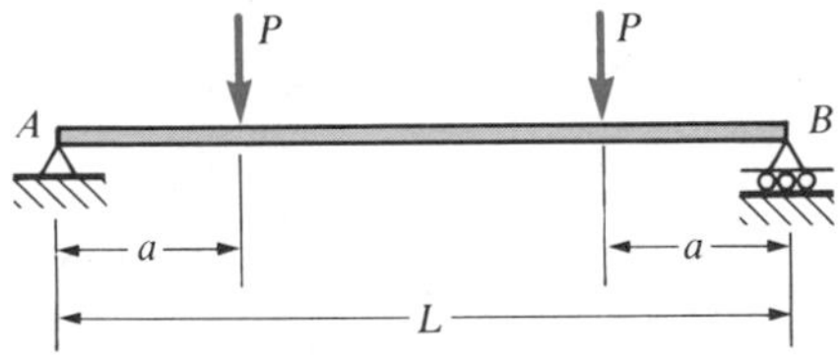

Prob. 4.4-1

4.4-2 Construct the shear-force and bending-moment diagrams for a cantilever beam carrying a uniform load of intensity q over one-half of the length (see figure).

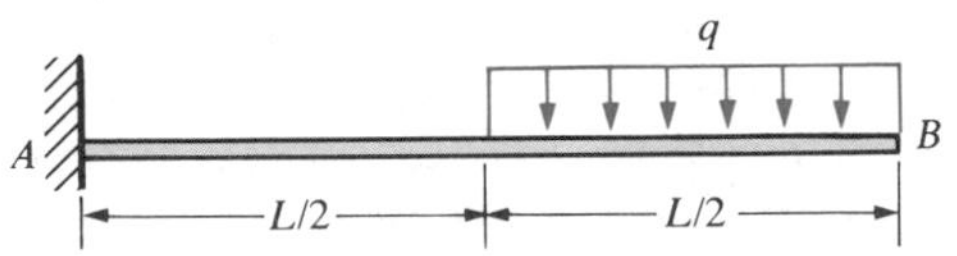

Prob. 4.4-2

4.4-3 A simple beam AB is subjected to a couple of moment M_0 acting at distance a from the left-hand support (see figure). Draw the shear-force and bending-moment diagrams for this beam.

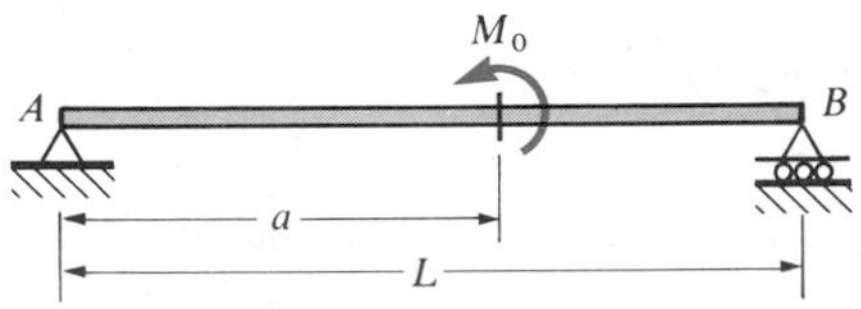

Prob. 4.4-3

4.4-4 The simple beam AB shown in the figure is subjected to a concentrated load P and a couple $M_1 = PL/4$ acting at the positions indicated. Draw the shear-force and bending-moment diagrams for this beam.

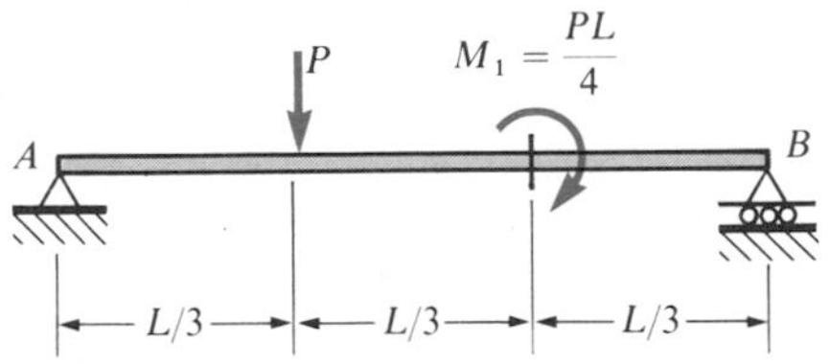

Prob. 4.4-4

4.4-5 A simply supported beam ABC supports a vertical load P by means of a bracket BDE (see figure). Draw the shear-force and bending-moment diagrams for the beam.

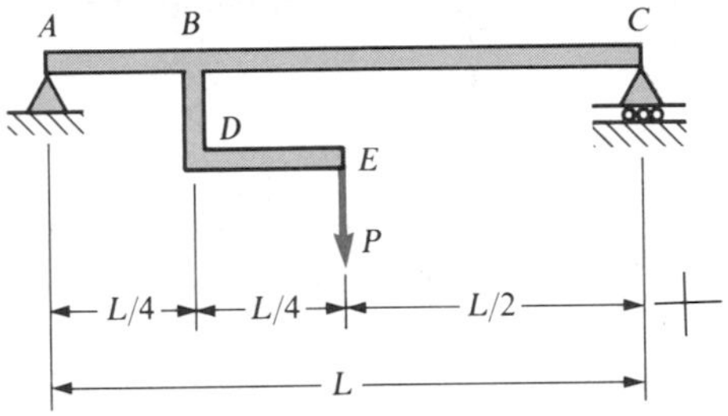

Prob. 4.4-5

4.4-6 A simple beam AB supports a uniform load of intensity $q = 6.0$ kN/m over a portion of the span (see figure). Assuming that $L = 10$ m, $a = 4$ m, and $b = 2$ m, draw the shear-force and bending-moment diagrams for this beam.

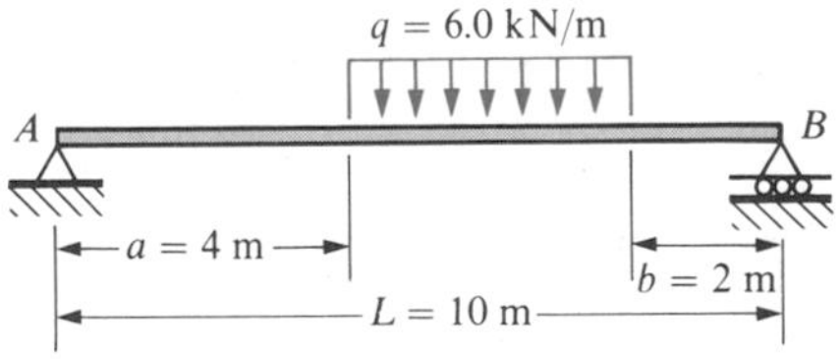

Prob. 4.4-6

4.4-7 A simple beam AB subjected to two bending couples is shown in the figure. Construct the shear-force and bending-moment diagrams for this beam.

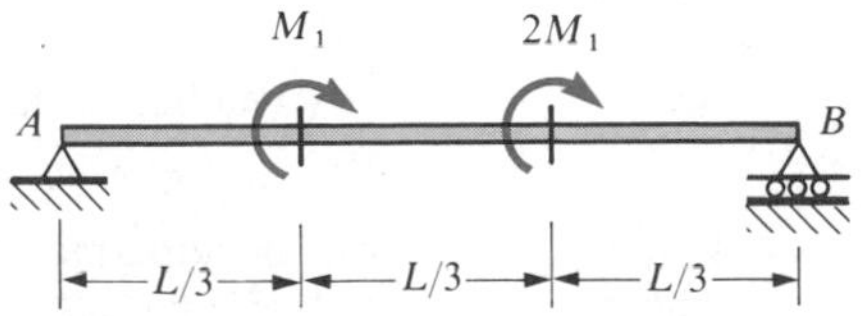

Prob. 4.4-7

4.4-8 Draw the shear-force and bending-moment diagrams for the beam *ABC* loaded as shown in the figure.

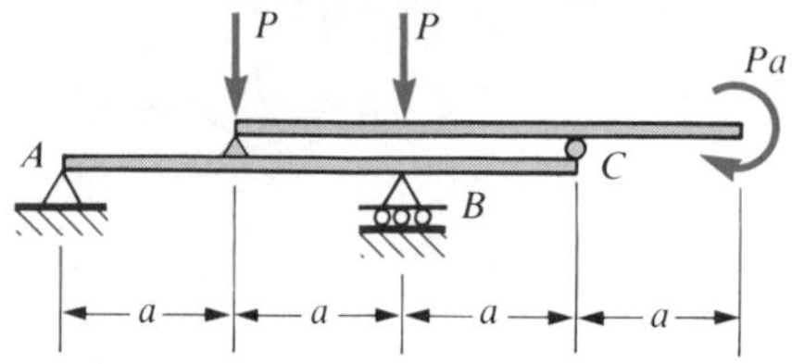

Prob. 4.4-8

4.4-9 The cantilever beam *AB* supports a concentrated load and a couple as shown in the figure. Construct shear-force and bending-moment diagrams for the beam.

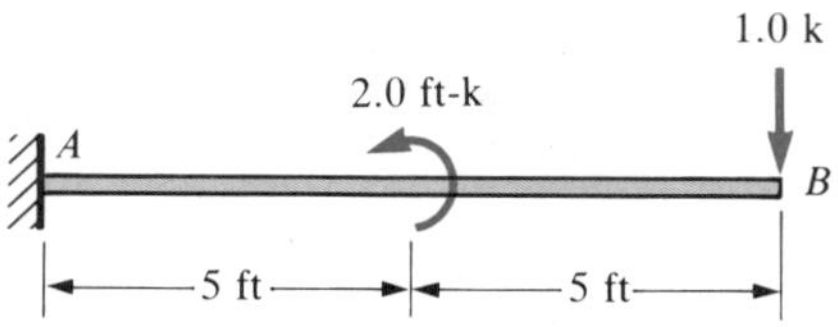

Prob. 4.4-9

4.4-10 Construct shear-force and bending-moment diagrams for the beam *ABC* loaded as shown in the figure. The cable passes over a small frictionless pulley at *C* and supports a weight $W = 5.0$ kN.

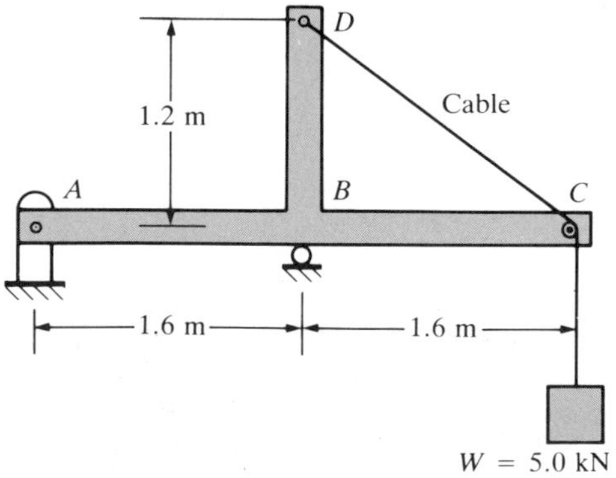

Prob. 4.4-10

4.4-11 Draw the shear-force and bending-moment diagrams for the simple beam of Problem 4.2-1.

4.4-12 Draw the shear-force and bending-moment diagrams for the cantilever beam of Problem 4.2-2.

4.4-13 Draw the shear-force and bending-moment diagrams for the overhanging beam of Problem 4.2-3.

4.4-14 Draw the shear-force and bending-moment diagrams for the simple beam shown in Problem 4.2-4.

4.4-15 Draw the shear-force and bending-moment diagrams for the beam with an overhang shown in Problem 4.2-5.

4.4-16 Draw the shear-force and bending-moment diagrams for the beam with an overhang shown in Problem 4.2-6.

4.4-17 Draw the shear-force and bending-moment diagrams for a cantilever beam with a linearly varying load of maximum intensity q_0 (see figure).

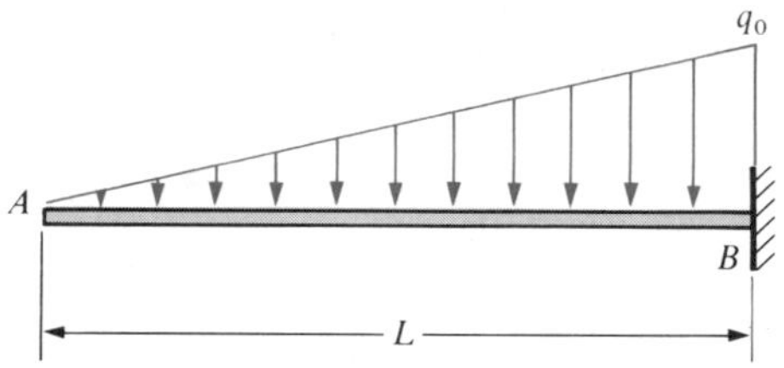

Prob. 4.4-17

4.4-18 A beam with two equal overhangs (see figure), supporting a uniform load of intensity *q*, has total length *L*. Find the distance *a* between supports *A* and *B* so that the maximum bending moment in the beam has the smallest possible numerical value. Draw the shear-force and bending-moment diagrams for this condition.

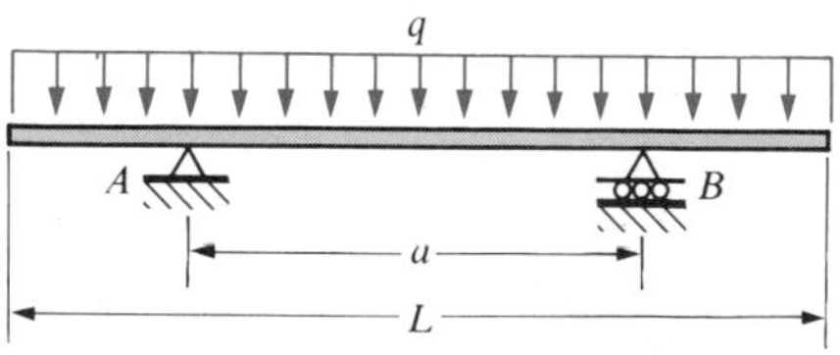

Prob. 4.4-18

4.4-19 Draw the shear-force and bending-moment diagrams for the beam with an overhang shown in the figure.

4.4-20 through **4.4-30** Construct the shear-force and bending-moment diagrams for the beams shown in the figures.

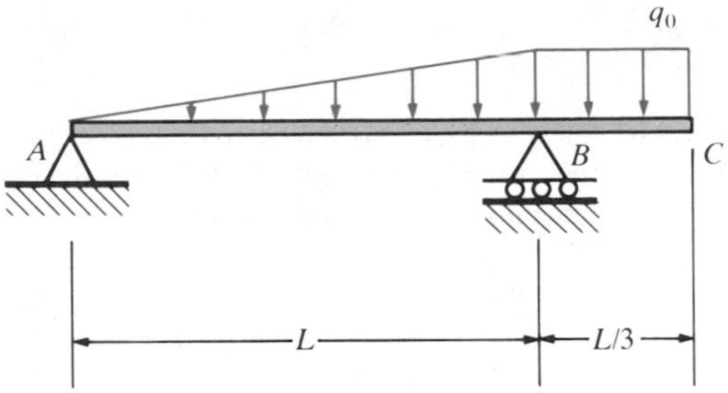

Prob. 4.4-19

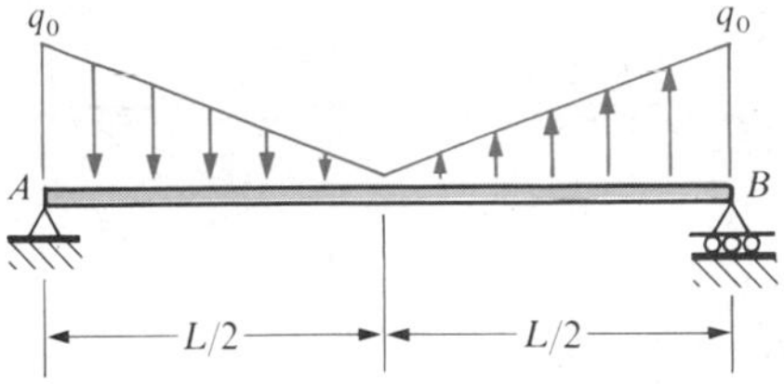

Prob. 4.4-24

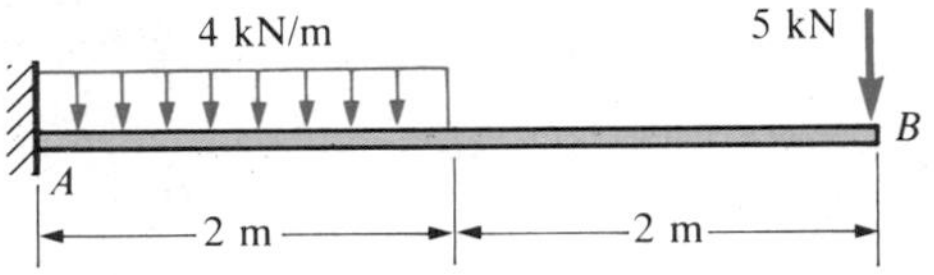

Prob. 4.4-20

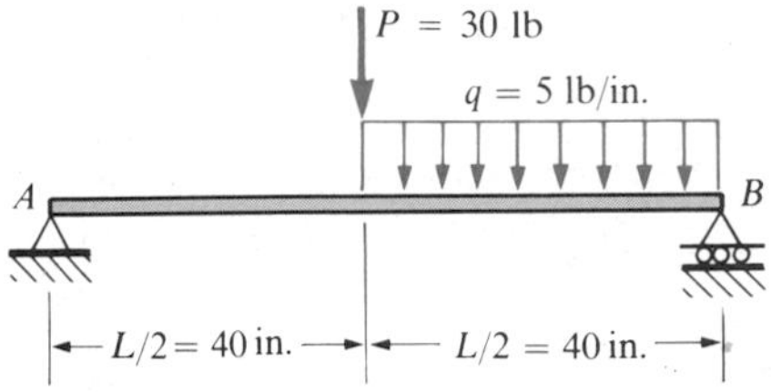

Prob. 4.4-25

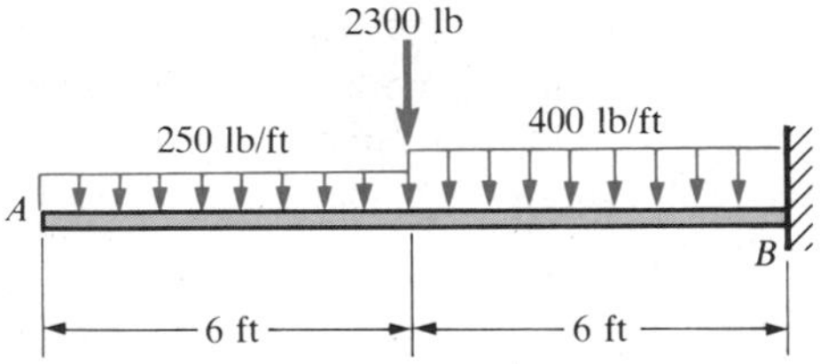

Prob. 4.4-21

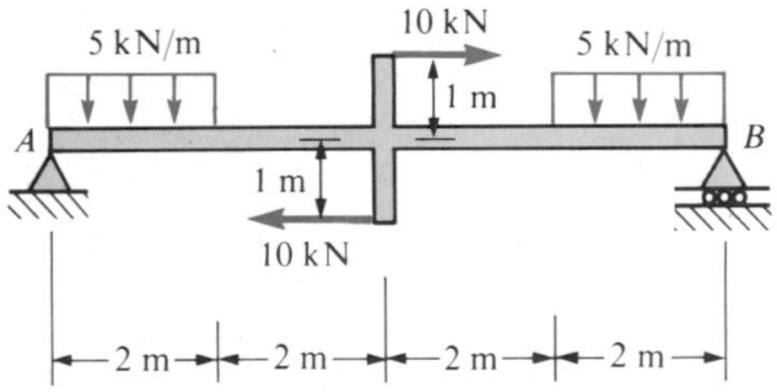

Prob. 4.4-26

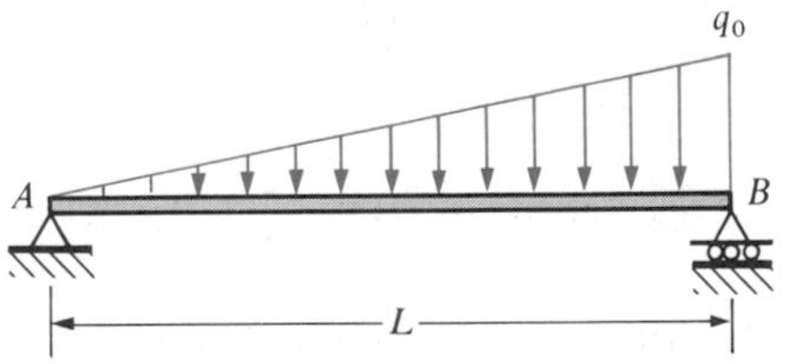

Prob. 4.4-22

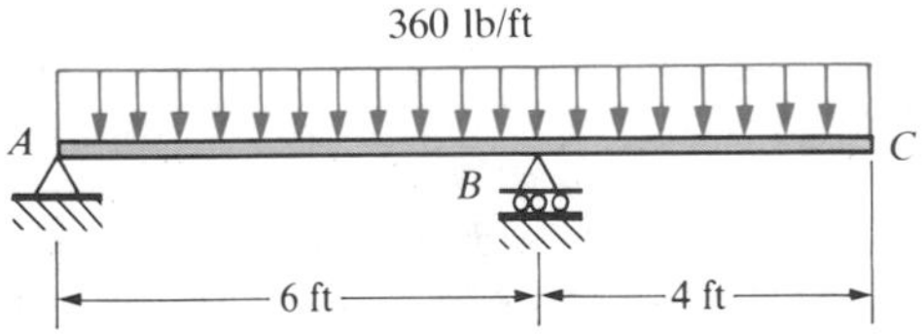

Prob. 4.4-27

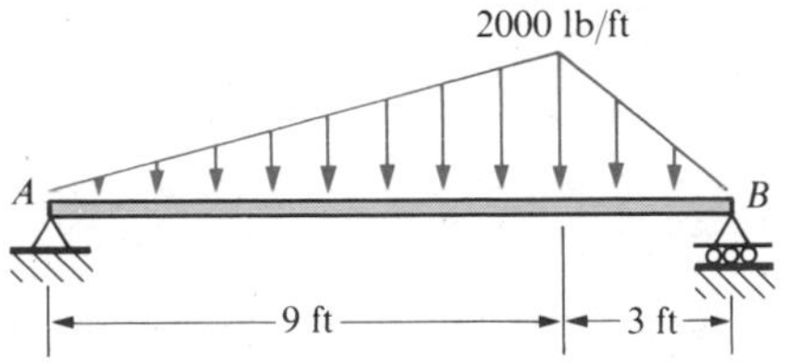

Prob. 4.4-23

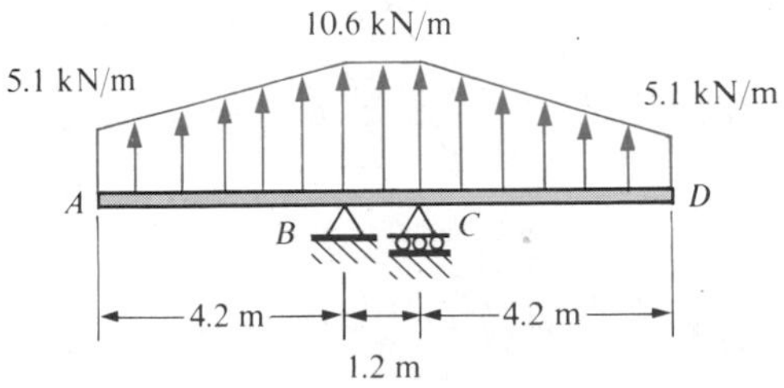

Prob. 4.4-28

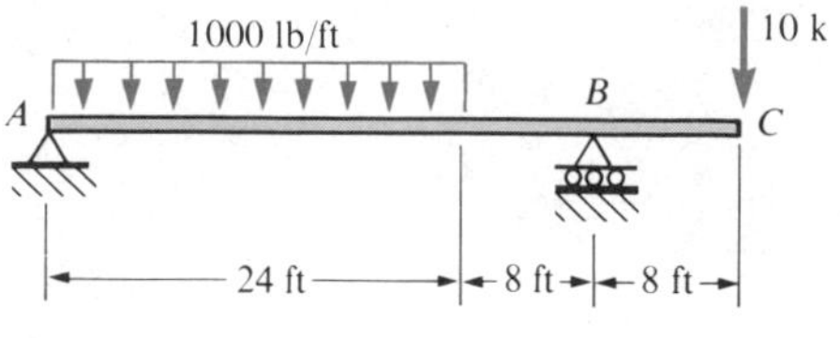

Prob. 4.4-29

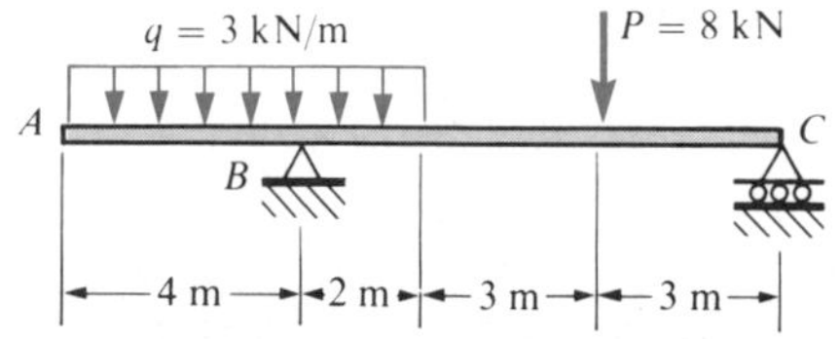

Prob. 4.4-30

4.4-31 Beam $ABCD$ has two equal overhangs and supports a symmetric triangular load (see figure). Determine the distance b between the supports so that the bending moment at the midpoint of the beam is zero. Draw the shear-force and bending-moment diagrams for this condition.

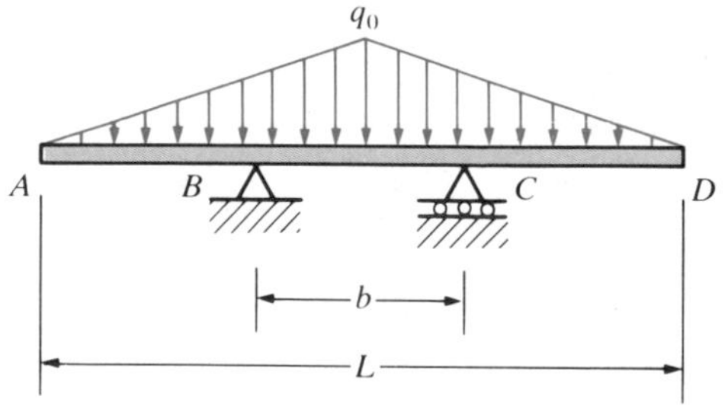

Prob. 4.4-31

4.4-32 The beam ABC shown in the figure consists of a cantilever part AB attached to a simple span BC by a pin at B. The pin can transmit a shear force but not a bending moment. Draw the shear-force and bending-moment diagrams for the beam.

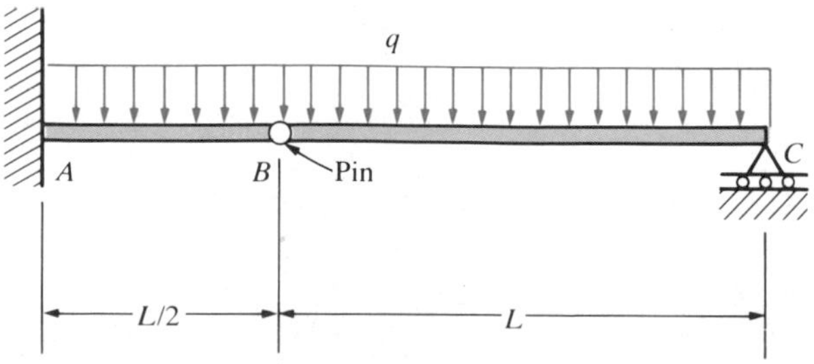

Prob. 4.4-32

4.4-33 The beam $ABCD$ supports a distributed load of linearly varying intensity (see figure). Draw the shear-force and bending-moment diagrams for this beam.

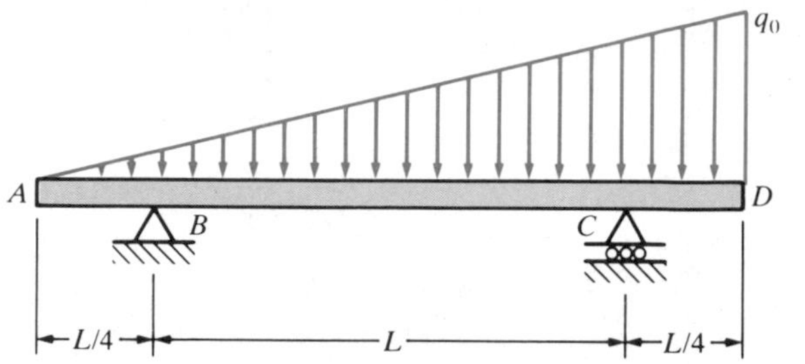

Prob. 4.4-33

4.4-34 The beam $ABCDE$ shown in the figure has simple supports at A, C, and E and a hinge (or pin) at D. A load of 4 kN acts at the end of the bracket that extends from the beam at B, and a load of 2 kN acts at the midpoint of part DE. Draw the shear-force and bending-moment diagrams for the beam. (Note that the pin at D can transmit a shear force but not a bending moment.)

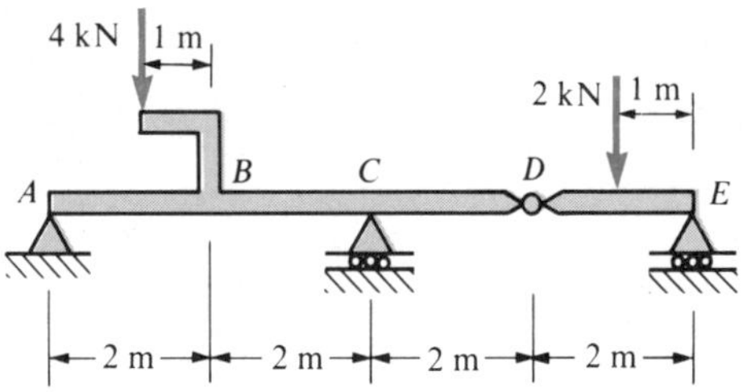

Prob. 4.4-34

4.4-35 The shear-force diagram for a simple beam is shown in the figure. Determine the loading on the beam and draw

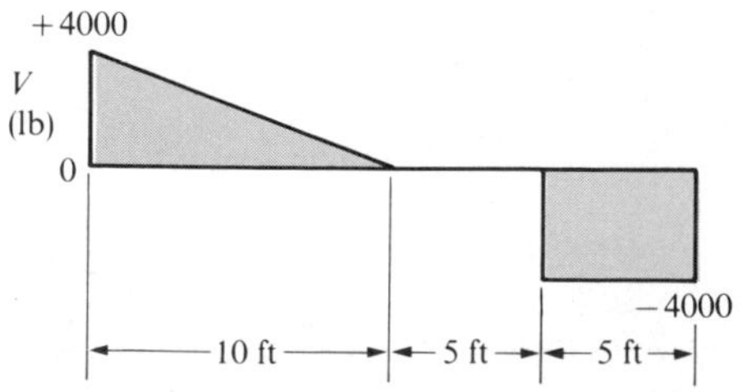

Prob. 4.4-35

the bending-moment diagram, assuming that no couples act as loads on the beam. (Note that the shear force has units of pounds.)

4.4-36 The shear-force diagram for a beam is shown in the figure. Assuming that no couples act as loads on the beam, draw the bending-moment diagram. (Note that the shear force has units of kilonewtons.)

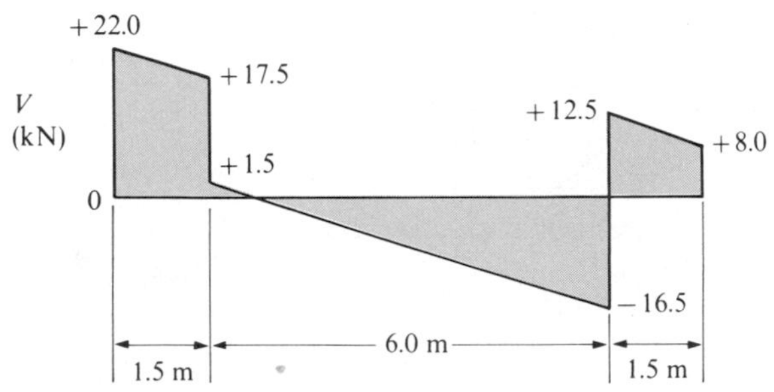

Prob. 4.4-36

4.4-37 Construct the shear-force and bending-moment diagrams for a simple beam with a triangular load (maximum intensity q_0) that extends for three-fourths of the span length L (see figure).

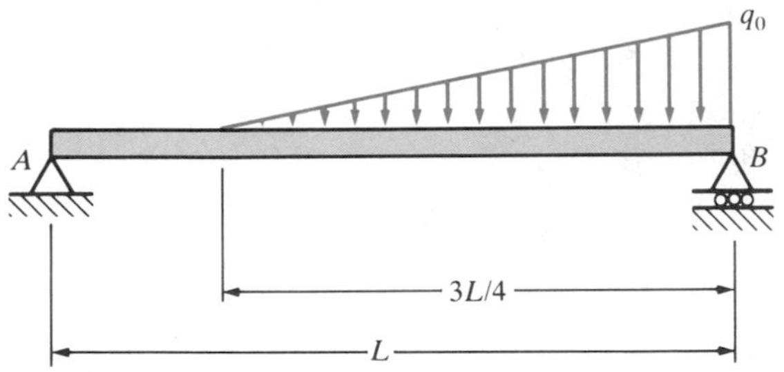

Prob. 4.4-37

4.4-38 A beam AB is subjected to concentrated loads P and $2P$ as shown in the figure. The beam rests on a foundation

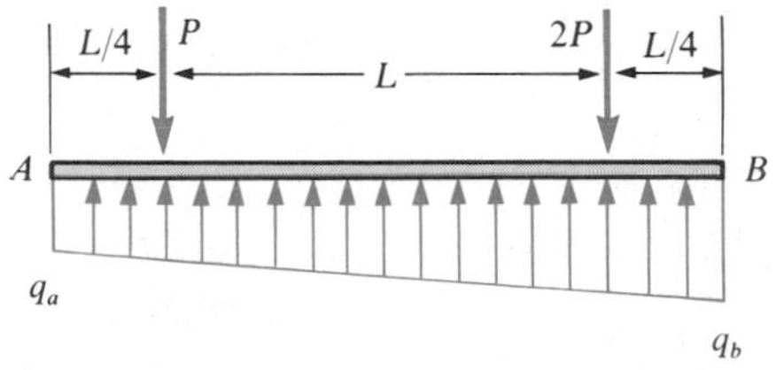

Prob. 4.4-38

that exerts a continuously distributed reaction against the beam. Assuming that the distributed reaction varies linearly in intensity from A to B, determine the intensities q_a and q_b of the reaction at the ends A and B, respectively. Also, construct the shear-force and bending-moment diagrams for the beam.

4.4-39 Two equal loads P are separated by a fixed distance d (see figure). This load combination may be located at any distance x from the left-hand support of the simple beam AB. (a) For what distances x will the shear force in the beam be a maximum? Also, derive a formula for the maximum shear force V_{max}. (b) Derive a formula for the distance x so as to produce the maximum bending moment M_{max} in the beam; also, obtain an expression for M_{max}.

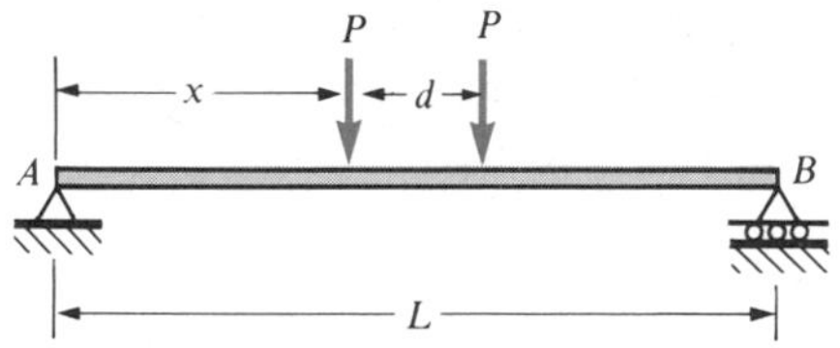

Prob. 4.4-39

4.4-40 A simple beam AB carries two connected wheel loads P and $2P$ that are distance d apart (see figure). The loads may be placed at any distance x from the left-hand support of the beam. Determine the distance x for (a) maximum shear force in the beam and (b) maximum bending moment in the beam if $P = 6$ kN, $d = 1.6$ m, and $L = 8$ m. Also, determine the maximum shear force V_{max} and maximum bending moment M_{max}.

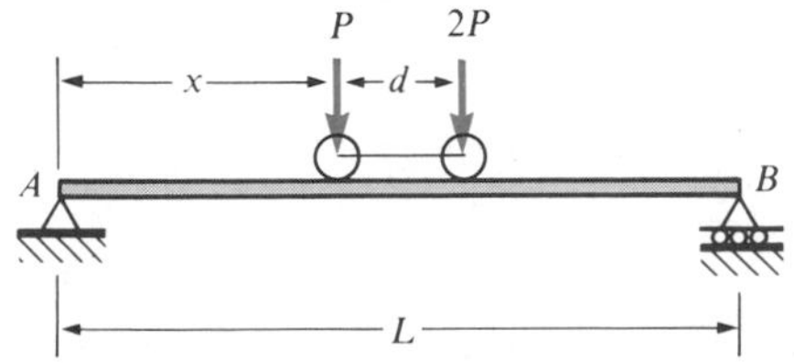

Prob. 4.4-40

4.4-41 Three wheel loads W_1, W_2, and W_3 move across a simply supported beam as shown in the figure. Determine the position of the wheels, as defined by the distance x from end A, so as to produce the maximum bending moment in the beam, assuming that $W_1 = 4$ k and $W_2 = W_3 = 16$ k. Also, determine the maximum bending moment M_{max}.

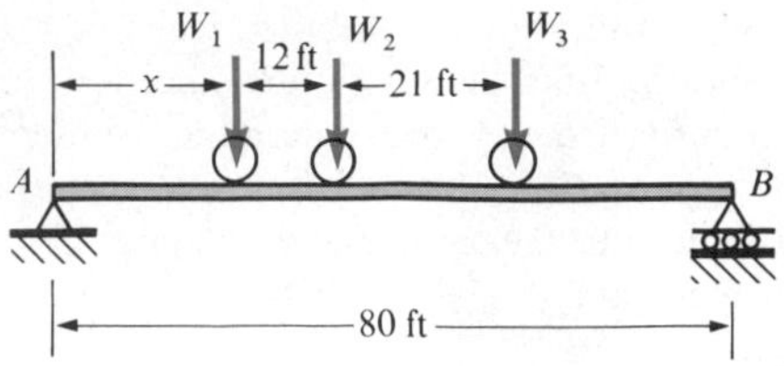

Prob. 4.4-41

4.4-42 A small retaining wall 2.4 m high is constructed of vertical wood planks ABC, as shown in the figure. The planks are 125 mm wide and are simply supported at B and C by a horizontal steel beam and a concrete foundation wall, respectively. The steel beam is 0.6 m from the top of the wall. The soil pressure against the wall is assumed to vary linearly from zero at the top (point A) to $p = 18$ kPa at the bottom (point C). Calculate the maximum bending moment M_{max} in the planks.

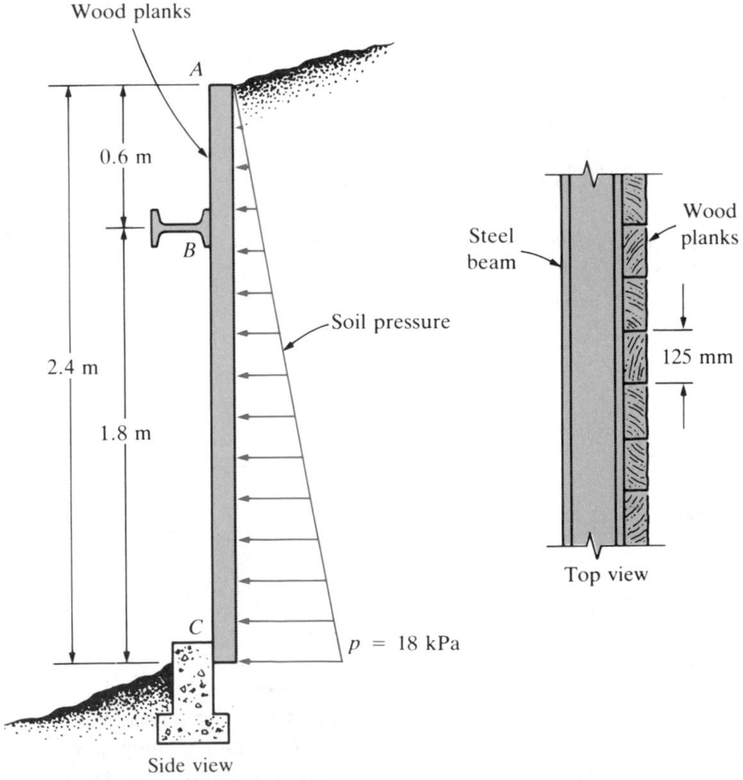

Prob. 4.4-42

***4.4-43** The loads on a simple beam consist of n equally spaced forces (see figure). The total applied load is P; therefore, each force is equal to P/n. The length of the beam is L; hence the spacing between loads is $L/(n + 1)$. (a) Derive general formulas for the maximum bending moment in the beam. (b) From these formulas, determine the maximum bending moment for several successive values of n ($n = 1, 2, 3, 4, \ldots$). (c) Compare these results with the maximum bending moment in the beam due to a uniformly distributed load of intensity q such that $qL = P$.

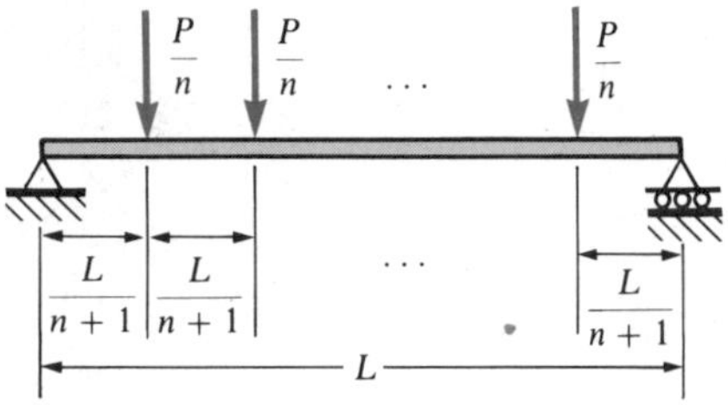

Prob. 4.4-43

*4.4-44 A beam ABC of length L supports a triangularly distributed load of maximum intensity q_0 as shown in the figure. The beam is simply supported at points B and C, with C at the right-hand end and B at any desired distance a from the left-hand end. (a) Determine the ratio a/L so that the numerically largest bending moment in the beam is a minimum. (b) What is the maximum bending moment in the beam under this condition?

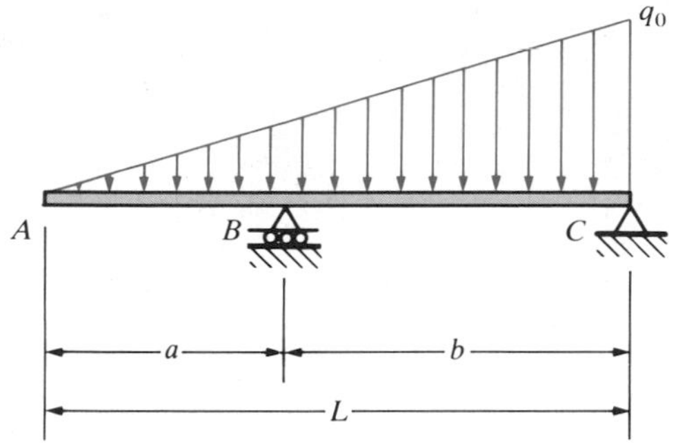

Prob. 4.4-44

CHAPTER 5

Stresses in Beams

5.1 INTRODUCTION

A beam is a structural member subjected to loads that act laterally to the longitudinal axis, as explained in the preceding chapter. The loads create internal actions, or stress resultants, in the form of shear forces and bending moments. In this chapter, we discuss the stresses and strains associated with the shear forces and bending moments, and we also examine numerous topics of practical importance in the analysis and design of beams. We will consider only beams having initially straight longitudinal axes.

Lateral loads acting on a beam cause the beam to bend, or flex, thereby deforming the axis of the beam into a curve. An illustration is given in Fig. 5-1, which shows a cantilever beam AB subjected to a load P at the free end. Before the load is applied, the longitudinal axis of the beam is a straight line. After loading, the axis is bent into a curve (Fig. 5-1b), called the **deflection curve** of the beam.

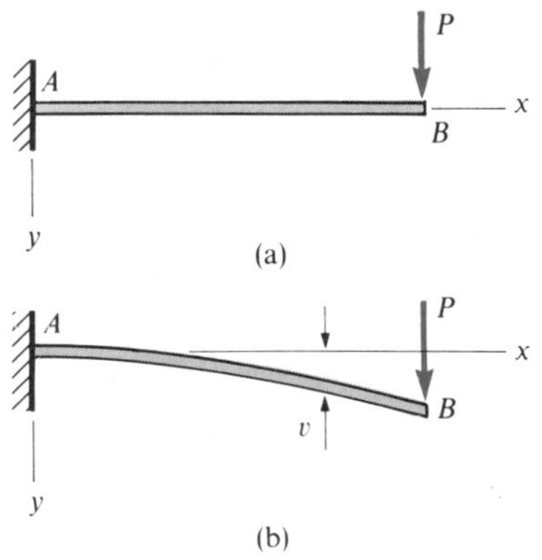

Fig. 5-1 Bending of a cantilever beam

For reference purposes, we construct a system of coordinate axes with its origin at the support. The positive x axis is directed to the right along the longitudinal axis of the beam, and the positive y axis is downward. The z axis, which is not shown in this figure, is directed inward (that is, away from the viewer), so that the axes form a right-handed coordinate system.

The beams considered in this chapter are assumed to be symmetric about the xy plane, which means that the y axis is an axis of symmetry of the cross sections. In addition, all loads are assumed to act in the xy plane. As a consequence, the bending deflections occur in this same plane, which is known as the **plane of bending**. Thus, the deflection curve AB of the

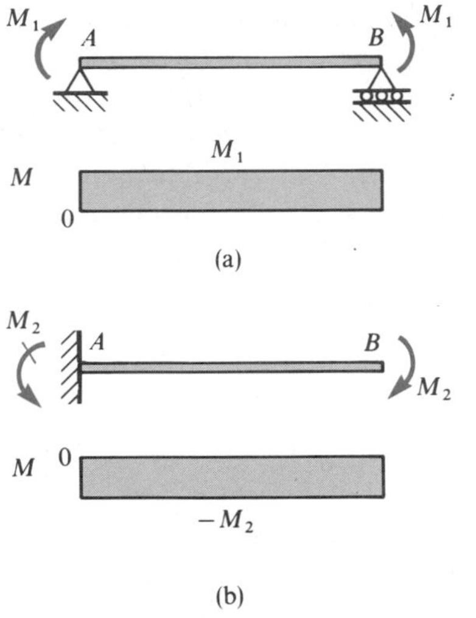

Fig. 5-2 Beams in pure bending: (a) simple beam subjected to positive bending moments $M = M_1$, and (b) cantilever beam subjected to negative bending moments $M = -M_2$

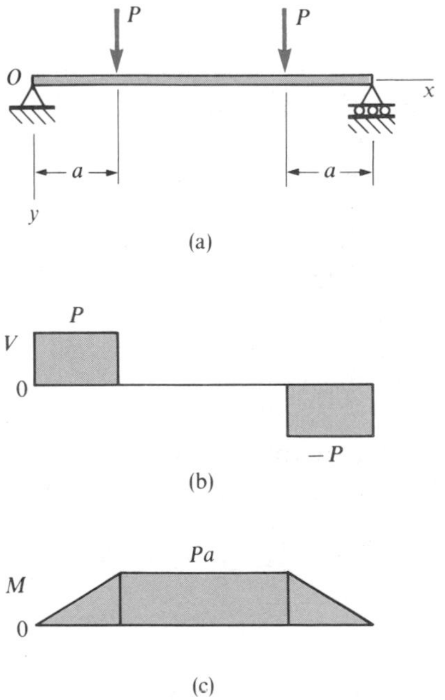

Fig. 5-3 Beam with central region in pure bending

beam shown in Fig. 5-1b is a plane curve lying in the plane of bending. We will denote the deflection in the y direction by the letter v.

When considering the bending of beams, it is sometimes necessary to distinguish between pure bending and nonuniform bending. **Pure bending** refers to flexure of a beam under a constant bending moment, which means that the shear force is zero (because $V = dM/dx$; see Eq. 4-3). In contrast, **nonuniform bending** refers to flexure in the presence of shear forces, which means that the bending moment changes as we move along the axis of the beam. To illustrate these definitions, consider a simple beam loaded by couples M_1 acting at the ends (Fig. 5-2a). These loads produce a constant bending moment $M = M_1$ acting throughout the length of the beam, as shown by the bending moment diagram drawn below the sketch of the beam. Furthermore, the shear force V in the beam is zero at all cross sections. Therefore, this beam is in pure bending.

Another illustration of pure bending is given in Fig. 5-2b, where the cantilever beam AB is subjected to a clockwise couple M_2 at the free end. In this case, the bending moment in the beam is negative ($M = -M_2$), but the beam is in pure bending because M is constant and there are no shear forces.

As yet another example, consider the symmetrically loaded simple beam of Fig. 5-3. The two forces P produce the shear-force and bending-moment diagrams shown in Figs. 5-3b and c. The region between the loads P has no shear force and is subjected only to a constant bending moment $M = Pa$; hence the central region of this beam is in pure bending. The regions of length a near the ends of the beam are in nonuniform bending because the bending moment M varies and shear forces V are present. In the remainder of this chapter, we will investigate the strains and stresses in a great variety of beams subjected to both pure and nonuniform bending.

5.2 NORMAL STRAINS IN BEAMS

When a beam is loaded and bent, its longitudinal axis is deformed into a curve, as explained in the preceding section (see Fig. 5-1). The strains and stresses in the beam are directly related to the shape of the deflection curve, which is determined by the curvature. Therefore, in this section we begin by discussing the curvature of a bent beam, and then we consider the normal strains, both longitudinal and transverse.

CURVATURE

To illustrate what is meant by curvature, consider again a cantilever beam subjected to a lateral load P acting at the free end (Fig. 5-4). We identify two points m_1 and m_2 on the deflection curve; point m_1 is located at distance x from the y axis, and point m_2 is situated a small distance ds

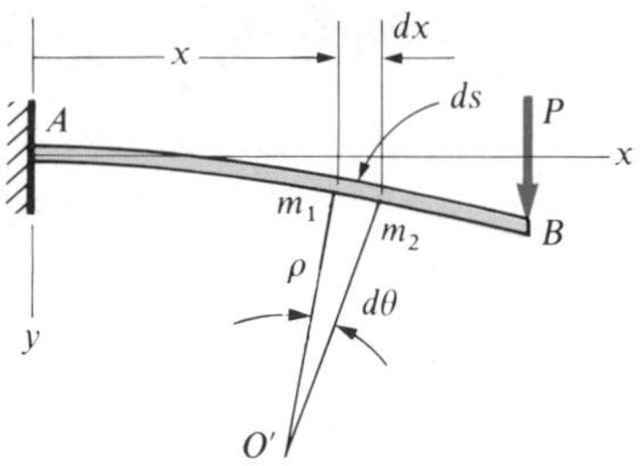

Fig. 5-4 Curvature of a bent beam

further along the curve. At each of these points we draw a line normal to the tangent to the deflection curve. These normals intersect at point O', which is the **center of curvature** for the deflection curve at distance x from the support. The length of a normal (that is, the distance from the center of curvature to the curve itself) is called the **radius of curvature** ρ (Greek letter rho). As defined in calculus and analytic geometry, the **curvature** κ (Greek letter kappa) is the reciprocal of the radius of curvature:

$$\kappa = \frac{1}{\rho} \tag{a}$$

Also, from the geometry of the figure, we obtain

$$\rho \, d\theta = ds \tag{b}$$

in which $d\theta$ is the small angle between the normals and ds is the distance along the curve between the normals. If the deflections of the beam are small, which is the most common situation, then the deflection curve is very flat and the distance ds along the curve may be set equal to its horizontal projection dx (Fig. 5-4). We then obtain

$$\kappa = \frac{1}{\rho} = \frac{d\theta}{dx} \tag{5-1}$$

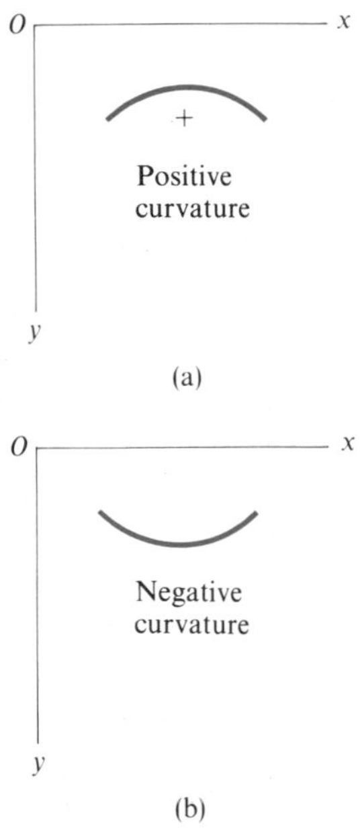

Fig. 5-5 Sign convention for curvature

The curvature κ is a function of the distance x measured along the axis of the beam. We will see later that for a beam in pure bending, κ is constant, and for a beam in nonuniform bending, κ changes with x.

The **sign convention for curvature** is related to the orientation of the coordinate axes. If the x axis is positive to the right and the y axis is positive downward, as shown in Fig. 5-4, then the curvature of the deflection curve is positive when the beam is bent concave downward and negative when it is bent concave upward. This sign convention is portrayed in Fig. 5-5. We have not adopted this sign convention arbitrarily; rather, it is established mathematically from the directions of the coordinate axes. Also, we note that positive curvature corresponds to a positive value of $d\theta/dx$ (see Eq. 5-1); the latter is positive when the beam deforms as in Fig. 5-5a and negative when it deforms as in Fig. 5-5b. These sign conventions are explained later in the more complete discussion of curvature given in Section 7.2.

NORMAL STRAINS

To find the internal strains in a beam, we must consider the curvature of the beam and the associated deformations. For this purpose, let us take a portion ab of a beam in pure bending produced by couples M_0 (Fig. 5-6a). These couples act in directions that produce positive curvature of the beam (see Fig. 5-5a). The bending moment M in the beam due to the

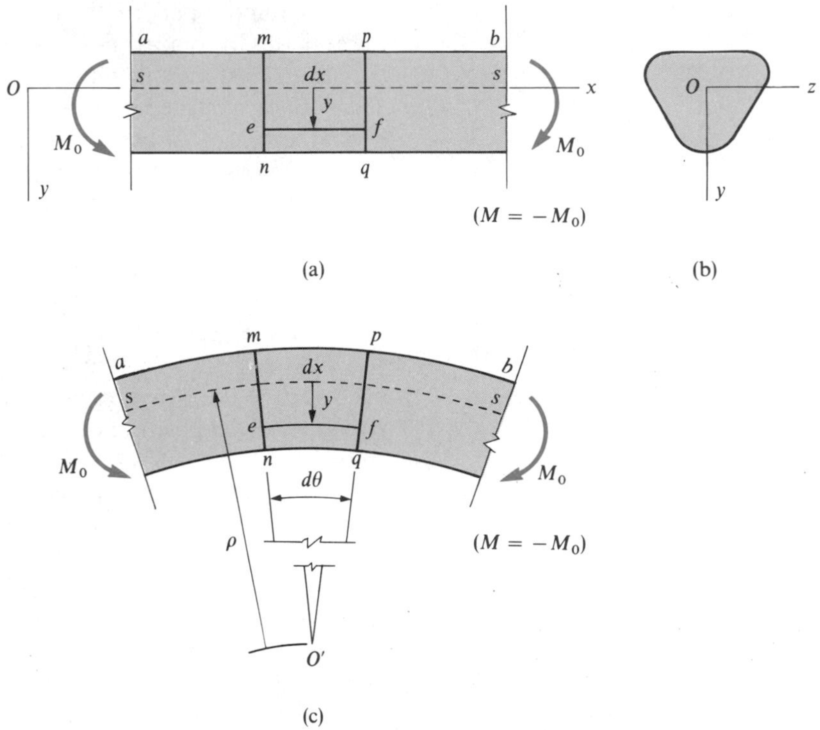

Fig. 5-6 Deformations of a beam in pure bending produced by couples M_0. (Note: The bending moment M is negative and equals $-M_0$.)

couples M_0 is negative ($M = -M_0$), as can be seen from Figs. 4-3 and 4.4. We assume that the beam initially has a straight longitudinal axis (the x axis in Fig. 5-6a) and that the cross sections are symmetric about the y axis (Fig. 5-6b).

Under the action of the couples M_0, the beam deflects in the xy plane (the plane of bending) and its longitudinal axis *ss* is bent into a circular curve (Fig. 5-6c). Cross sections of the beam, such as *mn* and *pq*, remain plane and normal to longitudinal lines within the beam. The fact that cross sections of a beam in pure bending remain plane can be established experimentally by making precise strain measurements, or it can be proven theoretically using arguments of symmetry. The symmetry of the beam and its loading (Fig. 5-6a) requires that all elements of the beam (such as element *mnqp*) deform in an identical manner, which is possible only if the deflection curve is circular (Fig. 5-6c) and if cross sections remain plane during bending. These conclusions are valid for a beam of any material (elastic or inelastic, linear or nonlinear); of course, the material properties, like the dimensions, must be symmetric about the y axis (Fig. 5-6b).

As a result of the bending deformations shown in Fig. 5-6c, cross sections *mn* and *pq* rotate with respect to each other about axes perpendicular to the xy plane. Longitudinal lines on the convex side of the beam are elongated, whereas those on the concave side are shortened.

Thus, the upper part of the beam is in tension longitudinally, and the lower part is in compression. Somewhere between the top and bottom of the beam is a surface in which longitudinal lines do not change in length. This surface, indicated by the dashed line *ss* in Figs. 5-6a and c, is called the **neutral surface of the beam**. Its intersection with any cross-sectional plane is called the **neutral axis of the cross section**; for instance, the z axis is the neutral axis for the cross section shown in Fig. 5-6b.

The planes of cross sections *mn* and *pq* of the deformed beam intersect in a line through the center of curvature O' (Fig. 5-6c). The angle between these planes is denoted $d\theta$, and the distance from O' to the neutral surface is the radius of curvature ρ. The initial distance dx between the two planes (Fig. 5-6a) is unchanged at the neutral surface (Fig. 5-6c); hence $\rho\, d\theta = dx$. However, all other longitudinal lines between the two planes either lengthen or shorten, thereby creating longitudinal **normal strains** ϵ_x.

To evaluate the longitudinal normal strains, consider a typical longitudinal line *ef* located within the beam between planes *mn* and *pq* (Fig. 5-6a). We locate line *ef* by its distance y from the neutral surface in the initially straight beam; thus, we are assuming that the x axis lies along the neutral surface of the undeformed beam. Of course, when the beam deflects, the neutral axis of any cross section is displaced from the x axis by an amount equal to the deflection v (see Fig. 5-1b). Nevertheless, for the purpose of locating a longitudinal line such as *ef* (Fig. 5-6), it is sufficient to give its distance y from the neutral surface in the undeflected beam (Figs. 5-1a and 5-6a). Thus, the length L_1 of line *ef* after bending takes place is

$$L_1 = (\rho - y)\, d\theta = dx - \frac{y}{\rho} dx$$

Inasmuch as the original length of line *ef* is dx, it follows that its elongation is $L_1 - dx$, or $-y\, dx/\rho$. The corresponding strain is equal to the elongation divided by the initial length dx; therefore,

$$\epsilon_x = -\frac{y}{\rho} = -\kappa y \qquad (5\text{-}2)$$

where κ is the curvature. This equation shows that the longitudinal normal strains in the beam are proportional to the curvature and vary linearly with the distance y from the neutral surface. When the point under consideration is below the neutral surface, the distance y is positive; if the curvature also is positive (as in Fig. 5-6c), then ϵ_x will be a negative strain, representing a shortening. When the point is above the neutral surface, the distance y is negative; then, for positive curvature, ϵ_x will be positive, which represents an elongation. Note that the sign convention for ϵ_x in Eq. (5-2) is the same as that used for normal strains in earlier chapters.

We derived Eq. (5-2) solely from the geometry of the deformed beam; the properties of the material did not enter into the discussion. Therefore, the equation is valid irrespective of the shape of the stress-strain diagram of the material.

TRANSVERSE STRAINS

The axial strains ϵ_x given by Eq. (5-2) are accompanied by transverse strains ϵ_z due to the effects of Poisson's ratio, as discussed in Section 1.5. Positive strains ϵ_x above the neutral surface *ss* (Fig. 5-7a) are accompanied by negative transverse strains, whereas below the neutral surface the transverse strains are positive. Thus, the transverse strains are given by the equation

$$\epsilon_z = -\nu\epsilon_x = \nu\kappa y \tag{5-3}$$

in which ν is Poisson's ratio. As a result of these strains, the shape of the cross section changes. For instance, consider a beam of rectangular cross section (Fig. 5-7b). The strains ϵ_x cause the width of the cross section to increase below the z axis and to decrease above it. Since these changes in width are directly proportional to y (Eq. 5-3), the sides of the rectangular cross section become inclined to each other (Fig. 5-7c). Also, all straight lines in the cross section that were originally parallel to the z axis become slightly curved so as to remain normal to the sides of the section. The center of curvature O'' of these lines is above the beam, and the corresponding transverse radius of curvature ρ_1 is larger than the longitudinal radius of curvature ρ in the same proportion that ϵ_x is numerically

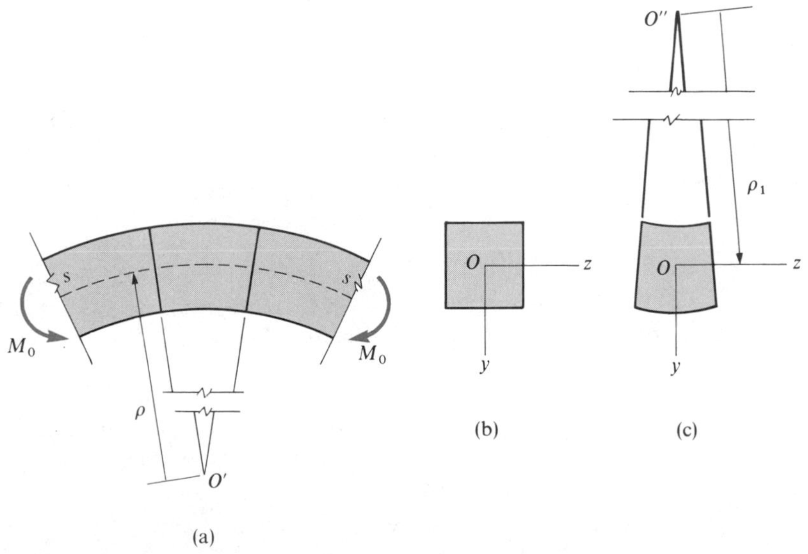

Fig. 5-7 Transverse deformations of a beam in pure bending

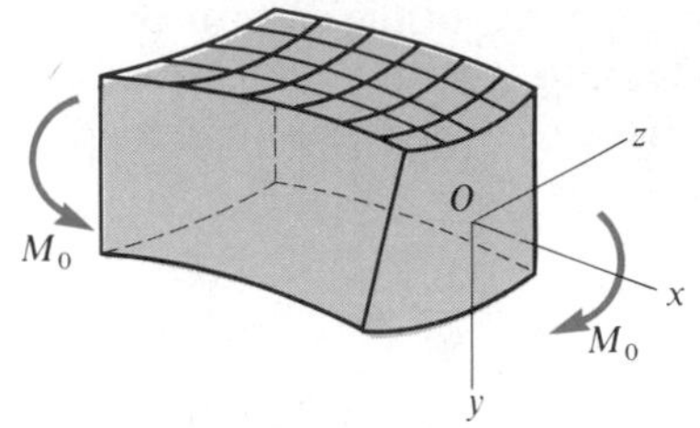

Fig. 5-8 Deformed shape of a beam of rectangular cross section in pure bending showing anticlastic curvature

larger than ϵ_z (see Eq. 5-3); hence we obtain

$$\rho_1 = \frac{\rho}{\nu} \qquad \kappa_1 = \nu\kappa \tag{5-4a,b}$$

in which $\kappa_1 = 1/\rho_1$ is the transverse curvature.

The deformed shape of a rectangular beam in pure bending by couples M_0 is shown highly exaggerated in Fig. 5-8. The longitudinal curvature in the xy plane is positive, whereas the transverse curvature in the yz plane is negative. As a result, the top surface of the beam becomes saddle shaped. When a surface has curvatures of opposite signs in two perpendicular planes, as in the case of a saddle, it is said to have **anticlastic curvature**. By contrast, if the curvatures are of the same sign, as when the surface has the shape of a dome, the curvature is **synclastic**. Anticlastic curvature can be seen clearly if you taken an ordinary rubber eraser (the kind that is rectangular in shape, not too thick, and usually pink in color), hold it in your hand, and bend it between your thumb and fingers.

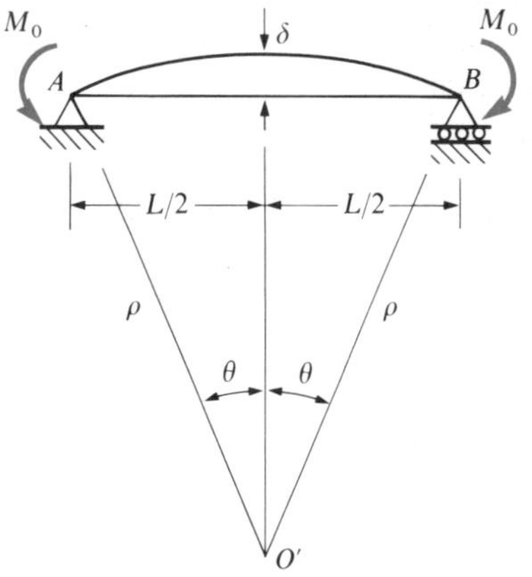

Fig. 5-9 Example

Example

A simply supported steel beam AB (Fig. 5-9) of length $L = 12$ ft is bent by couples M_0 that produce a longitudinal normal strain at the top surface of the beam equal to the yield strain of the steel. The distance from the top surface of the beam to the neutral surface is 6 in. Calculate the radius of curvature ρ, the curvature κ, and the vertical deflection δ at the midpoint of the beam, assuming that the yield strain is equal to 0.0014.

Equation (5-2) relates ρ and κ to the normal strain ϵ_x. Substituting into that equation and using only absolute values, we obtain

$$\rho = \frac{y}{\epsilon_x} = \frac{6 \text{ in.}}{0.0014} = 4286 \text{ in.} = 357 \text{ ft}$$

$$\kappa = \frac{1}{\rho} = 0.0028 \text{ ft}^{-1}$$

These results show that even at relatively large strains in the material, the radius of curvature is extremely large and the deflection curve of the beam is nearly flat.

To further emphasize this point, let us now calculate the deflection δ at the midpoint of the beam. Because the deflection curve is a circular arc, we see that the deflection δ is

$$\delta = \rho(1 - \cos\theta) \tag{c}$$

where θ is the angle shown in the figure. For a nearly flat curve, we can assume that the distance between supports is equal to the length L of the beam. Therefore,

$$\theta = \arcsin\frac{L}{2\rho} \tag{d}$$

Substitution of numerical values produces the following results:

$$\frac{L}{2\rho} = 0.0168 \qquad \theta = \arcsin\frac{L}{2\rho} = 0.0168 \text{ rad}$$

$$\delta = \rho(1 - \cos\theta) = (4286 \text{ in.})(1 - 0.9998589) = 0.605 \text{ in.}$$

Thus, the ratio of the span length of the beam to the deflection at the midpoint is

$$\frac{L}{\delta} = \frac{12 \text{ ft}}{0.605 \text{ in.}} = 238$$

which confirms that the deflection curve is almost flat. Of course, the deflection δ shown in Fig. 5-9 is highly exaggerated for clarity.

5.3 NORMAL STRESSES IN BEAMS

The stresses σ_x acting normal to the cross section of a beam can be obtained from the normal strains ϵ_x. Longitudinal elements of the beam are subjected only to tension or compression (that is, they are in uniaxial stress); therefore, the stress-strain diagram for the material will provide the relationship between σ_x and ϵ_x. If the material is elastic with a linear stress-strain diagram, we can use Hooke's law for uniaxial stress ($\sigma = E\epsilon$) and obtain

$$\sigma_x = E\epsilon_x = -E\kappa y \tag{5-5}$$

(see Eq. 5-2). Thus, the normal stresses acting on the cross section vary linearly with the distance y from the neutral surface. This type of stress distribution is pictured in Fig. 5-10a, where the stresses are negative (compression) below the neutral surface and positive (tension) above the

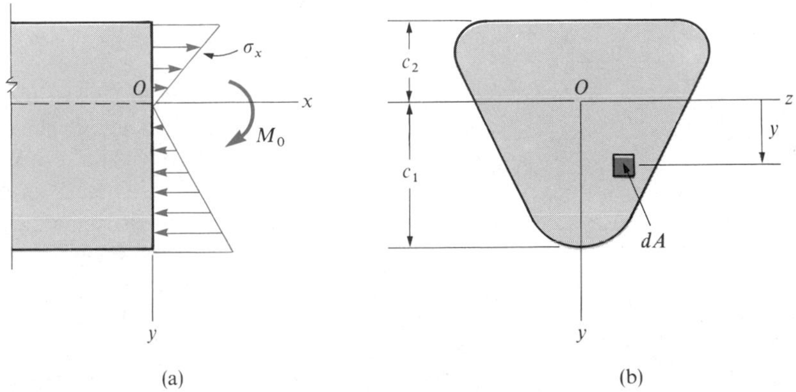

Fig. 5-10 Distribution of normal stresses σ_x in a beam of linearly elastic material

surface when the applied couple M_0 acts in the direction shown. As explained in the preceding section, this couple produces a positive curvature κ in the beam, although it represents a negative bending moment M.

Let us now consider the resultant of the normal stresses σ_x acting over the cross section. In general, the resultant of such stresses may consist of a horizontal force in the x direction and a couple acting about the z axis. However, in our situation no axial force acts on the cross section and the only resultant is the couple M_0. Therefore, we obtain two equations of statics, the first expressing that the resultant force in the x direction is zero and the second expressing that the resultant moment is equal to M_0. To evaluate these resultants, consider an element of area dA in the cross section at distance y from the neutral axis (Fig. 5-10b). The force acting on this element is normal to the cross section and has a magnitude $\sigma_x\,dA$. Because no resultant normal force acts on the cross section, the integral of $\sigma_x\,dA$ over the entire area of the cross section must vanish; thus,

$$\int \sigma_x\,dA = -\int E\kappa y\,dA = 0$$

Because the curvature κ and modulus of elasticity E are constants at the cross section, we conclude that

$$\int y\,dA = 0 \tag{5-6}$$

for a beam in pure bending. This equation shows that the first moment of the area of the cross section with respect to the z axis is zero, which means that the z axis must pass through the centroid of the cross section. Since the z axis is also the neutral axis, we conclude that *the neutral axis passes through the centroid of the cross section* when the beam is in pure bending and the material follows Hooke's law. This property can be used to quickly determine the position of the neutral axis. Of course, our discussion is limited to beams for which the y axis is an axis of symmetry, as explained previously. As a consequence, the y axis also must pass through the centroid, and we see that *the origin of coordinates O is located at the centroid of the cross section.* Furthermore, the symmetry of the cross section about the y axis means that the y axis is a principal axis (see Section C.8, Appendix C, for a discussion of principal axes). The z axis is also a principal axis since it is perpendicular to the y axis. Therefore, when a beam of linearly elastic material is subjected to pure bending, the y and z axes are **principal centroidal axes**.

Let us consider next the moment resultant of the stresses σ_x acting over the cross section (Fig. 5-10a). The element of force $\sigma_x\,dA$ acting on the element dA goes in the positive direction of the x axis when σ_x is positive and in the negative direction when σ_x is negative. We see from Fig. 5-10 that a positive stress σ_x acting on the element dA produces a counterclockwise moment about the z axis, which is a negative contribution to the moment M_0. Thus, we obtain

$$dM_0 = -\sigma_x y\, dA$$

The integral of all such elemental moments over the entire cross-sectional area must result in the total moment M_0:

$$M_0 = -\int \sigma_x y\, dA$$

Noting again that the bending moment M is equal to $-M_0$, and then substituting for σ_x from Eq. (5-5), we get the following expressions for the bending moment:

$$M = \int \sigma_x y\, dA = -\kappa E \int y^2\, dA$$

This equation can be expressed in the simpler form

$$M = -\kappa EI \tag{5-7}$$

in which

$$I = \int y^2\, dA \tag{5-8}$$

is the **moment of inertia** of the cross-sectional area with respect to the z axis (that is, with respect to the neutral axis). Moments of inertia have dimensions of length to the fourth power, and typical units are in.^{4}, m^4, and mm^4 for beam calculations.* Equation (5-7) can be rearranged as follows:

$$\kappa = \frac{1}{\rho} = -\frac{M}{EI} \tag{5-9}$$

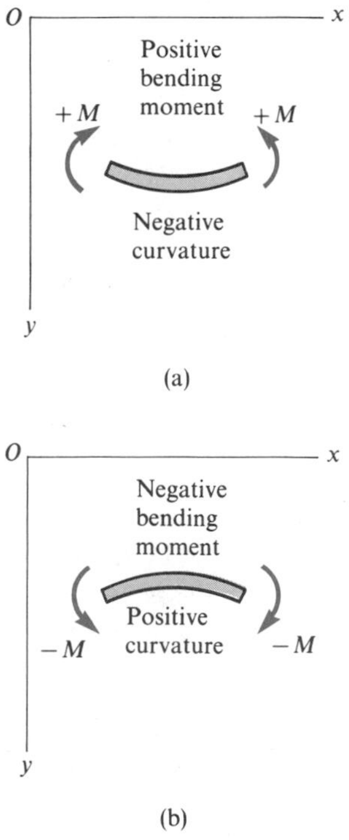

Fig. 5-11 Relationships between signs of bending moments and signs of curvatures (Eq. 5-9)

This equation shows that the curvature of the longitudinal axis of a beam is proportional to the bending moment M and inversely proportional to the quantity EI, which is the **flexural rigidity** of the beam.

The minus sign in the moment-curvature equation (Eq. 5-9) is a consequence of the sign convention we have adopted for bending moments. Comparing the sign convention for moments (Fig. 4-3) with that for curvatures (Fig. 5-5), we see that *a positive bending moment produces negative curvature and a negative bending moment produces positive curvature*, as shown in Fig. 5-11. (If the opposite sign convention

* Moments of inertia of areas are discussed in Appendix C.

for bending moments is used, or if the y axis is positive upward, then the minus sign is omitted in Eq. 5-9.)

The normal stresses in the beam can be related to the bending moment by substituting the expression for curvature (Eq. 5-9) into the expression for σ_x (Eq. 5-5), yielding

$$\sigma_x = \frac{My}{I} \tag{5-10}$$

This equation shows that the stresses are proportional to the bending moment M and inversely proportional to the moment of inertia I of the cross section. Also, the stresses vary linearly with the distance y from the neutral axis. If a positive bending moment acts on the beam, the stresses are positive (tension) over the part of the cross section where y is positive. If a negative moment acts, negative stresses (compression) are produced where y is positive. These relationships are shown in Fig. 5-12. Equation (5-10) for the normal stresses is usually called the **flexure formula**, and stresses calculated from this formula are called **bending stresses**.

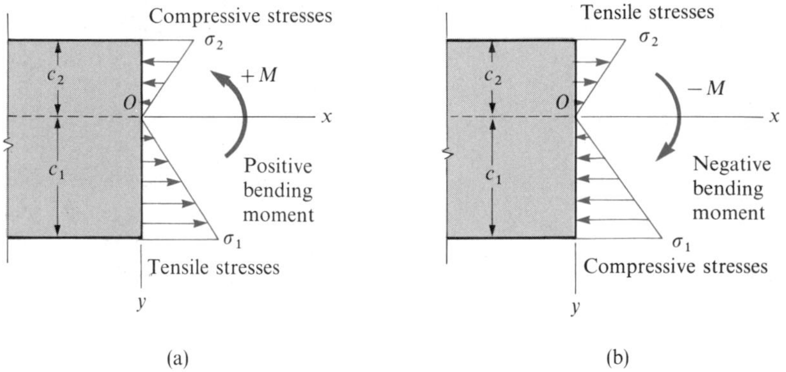

Fig. 5-12 Relationships between signs of bending moments and signs of normal stresses (Eq. 5-10)

The maximum tensile and compressive stresses in the beam occur at points located farthest from the neutral axis. Let us denote by c_1 and c_2 the distances from the neutral axis to the extreme elements in the positive and negative y directions, respectively (see Figs. 5-10 and 5-12). Then the maximum normal stresses (from Eq. 5-10) are

$$\sigma_1 = \frac{Mc_1}{I} = \frac{M}{S_1} \qquad \sigma_2 = -\frac{Mc_2}{I} = -\frac{M}{S_2} \tag{5-11a,b}$$

in which

$$S_1 = \frac{I}{c_1} \qquad S_2 = \frac{I}{c_2} \tag{5-12a,b}$$

The quantities S_1 and S_2 are known as the **section moduli** of the cross-sectional area. From Eqs. (5-12) we see that a section modulus has dimensions of length to the third power (for example, in.^{3}, m^3, and mm^3).

If the cross section is symmetric with respect to the z axis (doubly symmetric cross section), then $c_1 = c_2 = c$, and the maximum tensile and compressive stresses are equal numerically:

$$\sigma_1 = -\sigma_2 = \frac{Mc}{I} = \frac{M}{S} \qquad \text{(5-13)}$$

in which

$$S = \frac{I}{c} \qquad \text{(5-14)}$$

is the section modulus.

For a beam of **rectangular cross section** with width b and height h (Fig. 5-13a), the moment of inertia and section modulus are

$$I = \frac{bh^3}{12} \qquad S = \frac{bh^2}{6} \qquad \text{(5-15a,b)}$$

For a **circular cross section** of diameter d (Fig. 5-13b), these properties are

$$I = \frac{\pi d^4}{64} \qquad S = \frac{\pi d^3}{32} \qquad \text{(5-16a,b)}$$

Properties of many plane figures are listed in Appendix D. For cross-sectional shapes not listed, we must obtain the location of the neutral axis, the moment of inertia, and the section moduli by direct calculation, using the techniques described in Appendix C. The dimensions and properties of beams that are commercially available are listed in Appendices E and F and in various engineering handbooks, as explained in more detail in the next section.

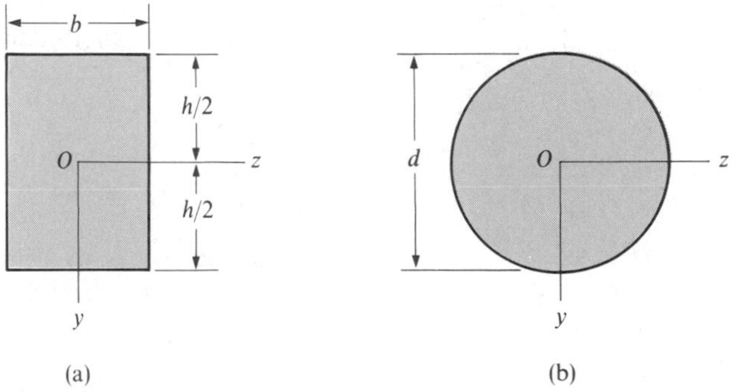

Fig. 5-13 Doubly symmetric cross-sectional shapes

The preceding analysis of normal stresses in beams is for pure bending, which means that no shear forces act on the cross sections. In the case of nonuniform bending, the presence of shear forces produces warping, or out-of-plane distortion, of the cross sections, and a cross section that is plane before bending is no longer plane after bending. Warping due to shear greatly complicates the behavior of the beam, but more elaborate analyses show that the normal stresses σ_x calculated from the flexure formula are not significantly altered by the presence of the shear stresses and the associated warping (Ref. 5-1). Thus, we may justifiably use the theory of pure bending for calculating normal stresses even when we have nonuniform bending. The calculation of the shear stresses is considered in Section 5.5.*

The flexure formula gives results that are accurate only in the regions of the beam where the stress distribution is not disrupted by irregularities in the shape of the beam or by discontinuities in loading. Such irregularities produce localized stresses, or *stress concentrations*, that are much greater than the stresses obtained from the flexure formula (see Section 5.10).

Example 1

A steel wire of diameter d is bent around a drum of radius r (Fig. 5-14). Calculate the maximum bending stress σ_{max} and the bending moment M in the wire, assuming $E = 200$ GPa, $d = 4$ mm, and $r = 0.5$ m.

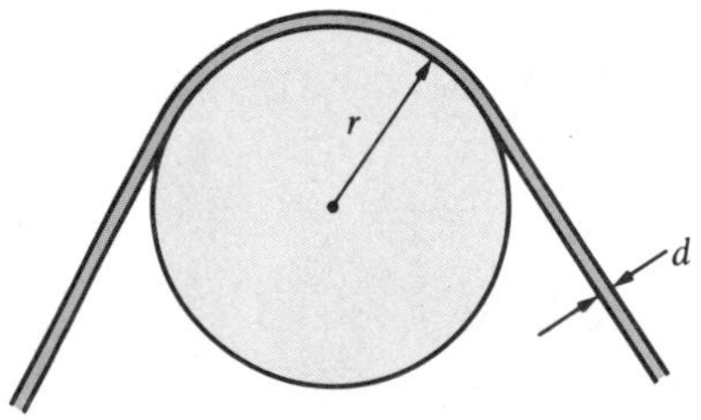

Fig. 5-14 Example 1

The radius of curvature of the bent wire, measured to the neutral axis of the cross section, is

$$\rho = r + \frac{d}{2} \tag{a}$$

The maximum tensile and compressive stresses, which are numerically equal, are obtained from Eq. (5-5) by substituting $\kappa = 1/\rho$ and $y = d/2$:

$$\sigma_{max} = \frac{Ed}{2r + d} \tag{5-17}$$

If the radius of the drum is large compared to the diameter of the wire, the second term in the denominator may be disregarded.

To calculate the maximum stress, we now substitute numerical values for E, d, and r:

$$\sigma_{max} = \frac{(200 \text{ GPa})(4 \text{ mm})}{2(500 \text{ mm}) + 4 \text{ mm}} = 797 \text{ MPa}$$

If d is disregarded in the denominator, the result is $\sigma_{max} = 800$ MPa, which differs from the preceding result by less than 1%.

* Beam theory began with Galileo Galilei (1564–1642), who investigated the behavior of various types of beams. His work in mechanics of materials is described in his famous book *Two New Sciences*, first published in 1638 (Ref. 5-2). Although Galileo made many important discoveries regarding beams, he did not correctly obtain the distribution of stresses in a beam. Further progress was made by Mariotte, Jacob Bernoulli, Euler, Parent, Saint-Venant, and others (Ref. 5-3).

The maximum bending moment in the wire may be found from the equation $M = \sigma S$ (see Eq. 5-13), in which σ is obtained from Eq. (5-17) and $S = \pi d^3/32$; thus,

$$M_{max} = \sigma_{max} S = \frac{\pi E d^4}{32(2r + d)} \tag{5-18}$$

Again substituting numerical values, we obtain $M_{max} = 5.01$ N·m.

Example 2

A simple beam AB of span length $L = 22$ ft (Fig. 5-15) supports a uniform load of intensity $q = 1.5$ k/ft and a concentrated load $P = 12$ k. The beam is constructed of glued laminated wood with width $b = 8.75$ in. and depth $d = 27$ in. The uniform load q includes the weight of the beam. Determine the maximum tensile and compressive stresses in the beam due to bending.

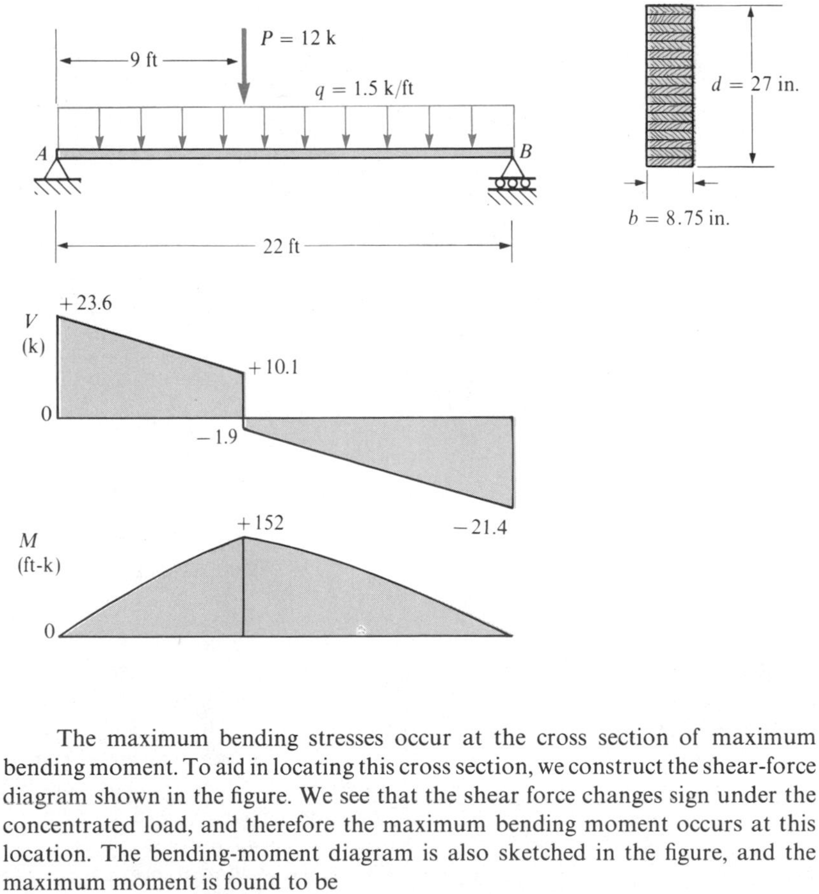

Fig. 5-15 Example 2

The maximum bending stresses occur at the cross section of maximum bending moment. To aid in locating this cross section, we construct the shear-force diagram shown in the figure. We see that the shear force changes sign under the concentrated load, and therefore the maximum bending moment occurs at this location. The bending-moment diagram is also sketched in the figure, and the maximum moment is found to be

$$M_{max} = 152 \text{ ft-k}$$

Next, we calculate the section modulus of the cross-sectional area (Eq. 5-15b):

$$S = \frac{bh^2}{6} = \frac{1}{6}(8.75 \text{ in.})(27 \text{ in.})^2 = 1063 \text{ in.}^3$$

Finally, we determine the maximum tensile and compressive stresses σ_t and σ_c from Eq. (5-13):

$$\sigma_t = \sigma_1 = \frac{M}{S} = \frac{152 \text{ ft-k}}{1063 \text{ in.}^3} = 1720 \text{ psi}$$

$$\sigma_c = \sigma_2 = -\frac{M}{S} = -1720 \text{ psi}$$

In this example, the bending moment is positive; therefore, the maximum tensile stress occurs at the bottom of the beam (σ_1), and the maximum compressive stress occurs at the top (σ_2).

Example 3

The beam ABC shown in Fig. 5-16 has simple supports at A and B and an overhang from B to C. A uniform load of intensity $q = 3.0$ kN/m acts throughout the length of the beam. The beam is constructed of steel plates (12 mm thick) welded to form a channel section, the dimensions of which are shown in Fig. 5-17a. Calculate the maximum tensile and compressive stresses in the beam due to the load q.

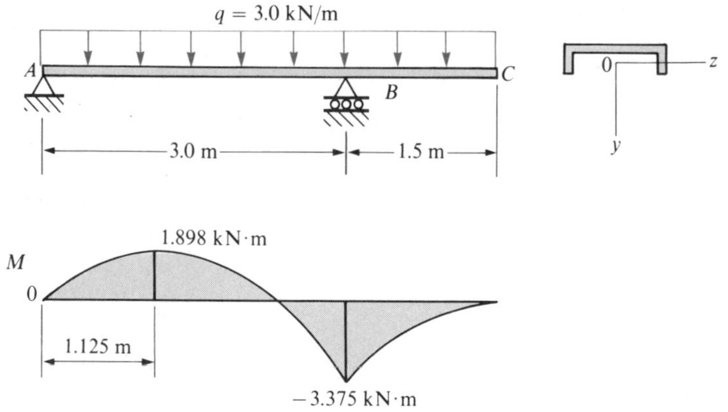

Fig. 5-16 Example 3

The bending-moment diagram for this beam can be obtained by the methods described in Chapter 4. The resulting diagram is shown in Fig. 5-16, and we see that the maximum positive and negative bending moments are 1.898 kN·m and −3.375 kN·m, respectively.

The position of the neutral axis is found by locating the centroid of the cross-sectional area shown in Fig. 5-17a. Using the techniques described in Appendix C, Section C.2, we divide the area into three parts (A_1, A_2, and A_3), as shown in

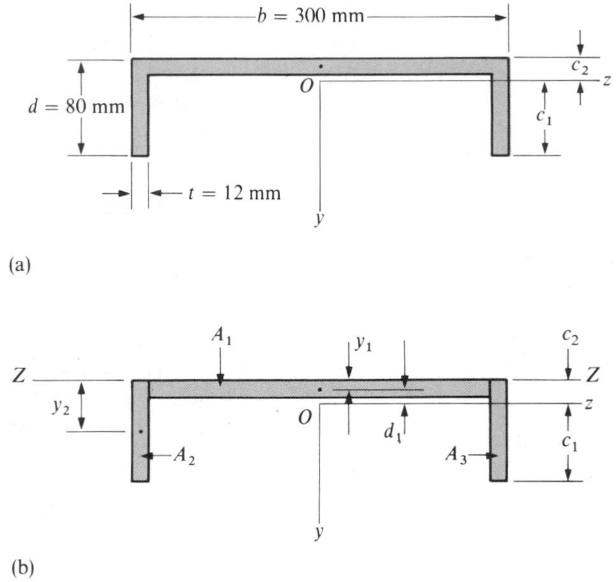

Fig. 5-17 Cross section of beam discussed in Example 3

Fig. 5-17b, and we take an axis Z-Z at the top of the cross section as a reference axis. Then the calculations are as follows:

$$y_1 = 6 \text{ mm} \qquad A_1 = (276 \text{ mm})(12 \text{ mm}) = 3312 \text{ mm}^2$$

$$y_2 = 40 \text{ mm} \qquad A_2 = (12 \text{ mm})(80 \text{ mm}) = 960 \text{ mm}^2$$

$$y_3 = y_2 \qquad A_3 = A_2$$

$$c_2 = \frac{\Sigma y_i A_i}{\Sigma A_i} = 18.48 \text{ mm}$$

$$c_1 = 80 \text{ mm} - c_2 = 61.52 \text{ mm}$$

Thus, the position of the neutral axis (the z axis) is determined.

To calculate the moment of inertia of the cross-sectional area about the neutral axis, we make use of the parallel-axis theorem (see Section C.5). Beginning with area A_1, we obtain its moment of inertia about the z axis from the equation

$$I_{z_1} = I_{z_c} + A_1 d_1^2 \tag{b}$$

The quantity I_{z_c} represents the moment of inertia of area A_1 about its own centroidal axis:

$$I_{z_c} = \frac{1}{12}(276 \text{ mm})(12 \text{ mm})^3 = 39{,}744 \text{ mm}^4$$

The distance d_1 extends from the centroidal axis of area A_1 to the z axis:

$$d_1 = c_2 - 6 \text{ mm} = 12.48 \text{ mm}$$

Therefore, the moment of inertia of area A_1 about the z axis (from Eq. b) is

$$I_{z_1} = 39{,}744 \text{ mm}^4 + (3312 \text{ mm}^2)(12.48 \text{ mm})^2 = 555{,}600 \text{ mm}^4$$

Proceeding in the same manner for areas A_2 and A_3, we get

$$I_{z_2} = I_{z_3} = 956{,}600 \text{ mm}^4$$

The centroidal moment of inertia of the entire cross-sectional area is

$$I = I_{z_1} + I_{z_2} + I_{z_3} = 2.469 \times 10^6 \text{ mm}^4$$

The section moduli for the bottom and top of the beam (see Eqs. 5-12a and b) are

$$S_1 = \frac{I}{c_1} = 40{,}100 \text{ mm}^3 \qquad S_2 = \frac{I}{c_2} = 133{,}600 \text{ mm}^3$$

With these properties determined, we can now proceed to calculate the stresses from Eqs. (5-11a and b).

At the cross section of maximum positive bending moment, the largest tensile stress occurs at the bottom of the beam (σ_1) and the largest compressive stress occurs at the top (σ_2):

$$\sigma_t = \sigma_1 = \frac{M}{S_1} = \frac{1.898 \text{ kN·m}}{40{,}100 \text{ mm}^3} = 47.3 \text{ MPa}$$

$$\sigma_c = \sigma_2 = -\frac{M}{S_2} = -\frac{1.898 \text{ kN·m}}{133{,}600 \text{ mm}^3} = -14.2 \text{ MPa}$$

Similarly, the largest stresses at the section of maximum negative moment are

$$\sigma_t = \sigma_2 = -\frac{M}{S_2} = -\frac{-3.375 \text{ kN·m}}{133{,}600 \text{ mm}^3} = 25.3 \text{ MPa}$$

$$\sigma_c = \sigma_1 = \frac{M}{S_1} = \frac{-3.375 \text{ kN·m}}{40{,}100 \text{ mm}^3} = -84.2 \text{ MPa}$$

A comparison of the preceding four stresses shows that the maximum tensile stress in the beam due to the load q is 47.3 MPa and the maximum compressive stress is -84.2 MPa.

5.4 DESIGN OF BEAMS

The overall process of designing a beam requires consideration of numerous factors, such as the materials, loads, environmental conditions, and type of structure. However, in many cases the task eventually reduces to the selection of a particular beam shape and size, subject to the constraint that the actual stresses in the beam must not exceed the allowable stresses. In the following discussion, we will consider only the bending stresses (that is, the stresses obtained from the flexure formula, Eq. 5-10). A more thorough design also requires that the shear stresses be kept below their allowable values (see Section 5.5) and that the effects of buckling and stress concentrations be considered.

For the purpose of selecting a beam, it is convenient to determine the required section modulus S by dividing the maximum bending moment by the allowable stress in the material (see Eqs. 5-11):

$$S = \frac{M_{\text{max}}}{\sigma_{\text{allow}}} \tag{5-19}$$

In this equation, σ_{allow} is the maximum permissible normal stress, which is based upon the properties of the material and the magnitude of the desired factor of safety. To ensure that the allowable stresses are not exceeded, the selected beam must have a cross-sectional area that provides a section modulus at least as large as that obtained from Eq. (5-19).

If the allowable stresses are the same for both tension and compression, it is logical to choose a cross-sectional shape that is doubly symmetric and that has its centroid (and neutral axis) at the midheight of the beam. If the allowable stresses are different for tension and compression, it may be desirable to use an unsymmetric cross section such that the distances to the extreme points in tension and compression are in nearly the same ratio as the respective allowable stresses. Of course, to minimize the weight of a beam and to save material, it is common practice to select a beam that has not only the required section modulus but also the smallest cross-sectional area.

Beams of *steel*, *aluminum*, and *wood* are manufactured in standard sizes. The dimensions and properties of these beams are listed in engineering handbooks such as those published by the American Institute of Steel Construction (AISC), the Aluminum Association, and the National Forest Products Association (see Refs. 5-4 through 5-8). However, for use in solving problems in this book, some abridged tables of structural steel and wood sections are given in Appendices E and F. These tables give the dimensions of the cross section and important properties such as weight per foot of length, cross-sectional area, moment of inertia, and section modulus.

Structural steel sections are given a designation such as W 30 × 211, which means that the section is of W shape (also called a wide-flange shape) with a nominal depth of 30 in. and a weight of 211 lb per ft of length. Analogous designations are used for S shapes (also called I-beams) and C shapes (also called channels), as shown in the tables in Appendix E. Angle sections, or L shapes, are designated by giving the lengths of the two legs and the thickness; for example, L 8 × 6 × 1 denotes an angle of thickness 1 in. and with unequal legs, one of length 8 in. and the other of length 6 in. All of these sections, as well as the others listed in the AISC Manual (Ref. 5-4), are manufactured by the rolling process. In this process, a hot billet of steel is passed back and forth between rolls until it is formed into the desired shape.

Aluminum structural sections are made by the extrusion process, in which a hot billet is pushed, or extruded, through a shaped die. Since dies are relatively easy to make and the material is workable, aluminum beams can be extruded in almost any desired shape. The Aluminum Construction Manual (Ref. 5-6) lists many standard shapes of wide-flange beams, I-beams, channels, angles, and other sections, but custom-made shapes also can be ordered.

Wood beams have rectangular cross sections that are denoted by nominal dimensions, such as 4 × 8 (inches). These dimensions represent the rough-cut size of the lumber. The net dimensions of a beam are smaller if the surfaces of the rough lumber are planed, or *surfaced*, to make them smooth. Thus, a 4 × 8 section is actually 3.5 × 7.25 in. in size after it is surfaced. The net dimensions should be used in all computations for surfaced lumber.

Let us now consider the relative efficiency in bending of various cross-sectional shapes. In general, a beam is more efficient if the material is located farther from the neutral axis, where it is more highly stressed and provides a larger section modulus. Consider first a rectangle of width b and height h (Fig. 5-18a). The section modulus (from Eq. 5-15b) is

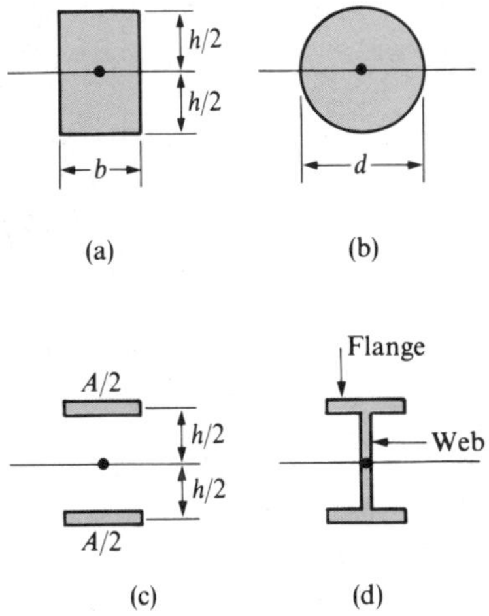

Fig. 5-18 Various cross-sectional shapes of beams

$$S = \frac{bh^2}{6} = \frac{Ah}{6} = 0.167Ah \tag{a}$$

where A denotes the cross-sectional area. This equation shows that a rectangular cross section of given area becomes more efficient as the height h is increased (and the width b is decreased to keep the area constant). However, there is a limit to the increase in height, because the beam becomes laterally unstable when the ratio of height to width becomes too large. Thus, a beam of very narrow rectangular section may fail because of lateral (sideways) buckling rather than insufficient strength of the material (Ref. 5-9).

Next, let us compare a solid circular cross section of diameter d (Fig. 5-18b) with a square cross section of the same area. If the diameter of the circle is d, then the side h_1 of the square of equal area is $h_1 = (d/2)\sqrt{\pi}$. The corresponding section moduli are

$$S_{\text{circle}} = \frac{\pi d^3}{32} = 0.0982d^3$$

$$S_{\text{square}} = \frac{h_1^3}{6} = \frac{\pi\sqrt{\pi}\,d^3}{48} = 0.1160d^3$$

or

$$\frac{S_{\text{square}}}{S_{\text{circle}}} = 1.18$$

which shows that a beam of square cross section is more efficient in bending than a circular one of the same area.

The ideal cross-sectional shape for a given cross-sectional area A and height h would be to distribute each half of the area at a distance $h/2$ from the neutral axis, as shown in Fig. 5-18c. Then we obtain

$$I = 2\left(\frac{A}{2}\right)\left(\frac{h}{2}\right)^2 = \frac{Ah^2}{4} \qquad S = 0.5Ah \tag{b}$$

This ideal limit may be approximated in practice by the use of a wide-flange section or an I-section with most of the material in the flanges (Fig. 5-18d). Because of the necessity of putting part of the material in the web of the beam, the limiting condition (Eq. b) can never be realized. For standard wide-flange beams, the section modulus is approximately

$$S \approx 0.35Ah \tag{c}$$

Comparison of Eq. (c) with Eq. (a) shows that a wide-flange section is more efficient than a rectangular section of the same area and same height. The reason is that more of the material in a rectangular beam is located near the neutral axis where it is understressed. By contrast, in a wide-flange beam, most of the material is located in the flanges, at the greatest available distance from the neutral axis. An additional consideration is the fact that a wide-flange beam is wider and therefore more stable with respect to sideways buckling than a beam of rectangular section of the same height and section modulus. Of course, if the web of a wide-flange beam is too thin, it will be susceptible to localized buckling or it may be overstressed in shear, as discussed in the next section.

Example 1

A temporary wood dam is constructed of horizontal planks A supported by vertical wood posts B that are built into the ground so that they act as cantilever beams (Fig. 5-19). The vertical posts are of square cross section (dimensions $b \times b$) and spaced at distance $s = 0.8$ m. The water level is at the full height $h = 2$ m of the

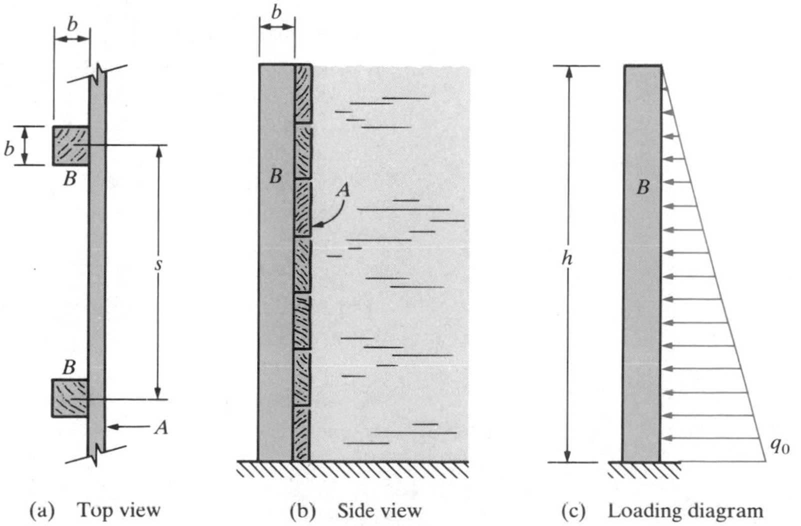

Fig. 5-19 Example 1

dam. Determine the minimum required dimension b of the posts if the allowable bending stress in the wood is $\sigma_{allow} = 8$ MPa.

The loading diagram for one of the posts is triangular in shape (Fig. 5-19c), and the maximum intensity of the load is

$$q_0 = \gamma h s \tag{d}$$

where γ is the specific weight of water. Therefore, the maximum bending moment in a post is

$$M_{max} = \frac{q_0 h}{2}\left(\frac{h}{3}\right) = \frac{\gamma h^3 s}{6}$$

and the required section modulus is

$$S = \frac{M_{max}}{\sigma_{allow}} = \frac{\gamma h^3 s}{6\sigma_{allow}} \tag{e}$$

For a beam of square cross section, the section modulus is $S = b^3/6$. Substituting this expression for S into Eq. (e) yields

$$b^3 = \frac{\gamma h^3 s}{\sigma_{allow}}$$

Substituting numerical values, we obtain

$$b^3 = \frac{(9.81 \text{ kN/m}^3)(2 \text{ m})^3(0.8 \text{ m})}{8 \text{ MPa}} = 0.007848 \text{ m}^3 \qquad b = 199 \text{ mm}$$

and the minimum required dimension b of the square posts is 199 mm. Any larger dimension, such as 200 mm, will ensure that the actual bending stress is less than the allowable stress.

Example 2

A simple beam AB of span length 21 ft must support a uniform load $q = 2000$ lb/ft distributed as shown in Fig. 5-20. Determine the required section modulus S if the allowable bending stress is $\sigma_{allow} = 18{,}000$ psi, and then select a wide-flange beam from Table E-1 in Appendix E. Next, calculate a new value of S taking into account the weight of the beam, and select a new beam size if necessary.

To locate the section of maximum bending moment, it is helpful to construct the shear-force diagram, as shown in the figure. The reactions at the supports are

$$R_a = 18{,}860 \text{ lb} \qquad R_b = 17{,}140 \text{ lb}$$

The distance x to the section of zero shear force is obtained from the equation

$$R_a - qx = 0$$

Solving this equation, we obtain $x = R_a/q = 9.429$ ft. The maximum bending moment occurs at this same section:

$$M_{max} = R_a x - \frac{qx^2}{2} = 88{,}900 \text{ ft-lb}$$

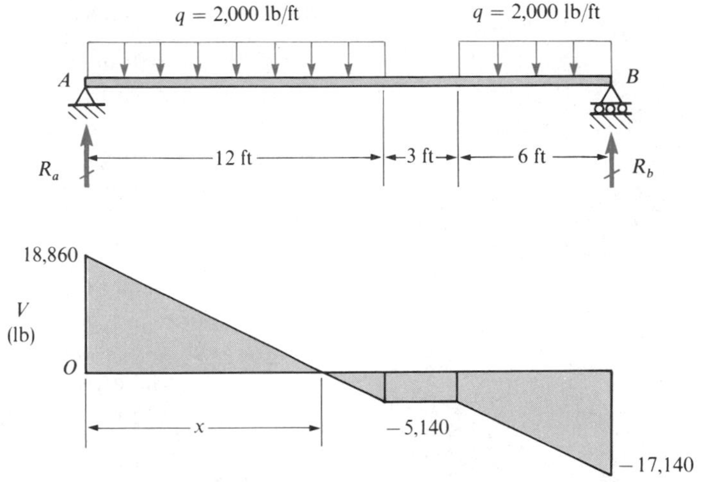

Fig. 5-20 Example 2

The required section modulus (disregarding the weight of the beam) is

$$S = \frac{M_{\text{max}}}{\sigma_{\text{allow}}} = \frac{88{,}900 \text{ ft-lb}}{18{,}000 \text{ psi}} = 59.3 \text{ in.}^3$$

We now enter Table E-1 and select a wide-flange beam having a section modulus greater than 59.3 in.^{3} The lightest beam that provides the required section modulus is W 12 × 50 with $S = 64.7$ in.^{3} Of course, the table in Appendix E is abridged, and lighter sections may be available commercially.

When the weight of this particular beam (50 lb/ft) is taken into account, the reaction R_a becomes 19,380 lb, the distance x to the cross section of zero shear becomes 9.455 ft, and the maximum bending moment increases to 91,600 ft-lb. The required section modulus is

$$S = \frac{M_{\text{max}}}{\sigma_{\text{allow}}} = \frac{91{,}600 \text{ ft-lb}}{18{,}000 \text{ psi}} = 61.1 \text{ in.}^3$$

and the W 12 × 50 beam is still satisfactory. If it were not, a new beam with a larger section modulus would be selected and the process repeated.

5.5 SHEAR STRESSES IN BEAMS

When a beam is subjected to lateral loads, both bending moments M and shear forces V act on the cross sections. The normal stresses σ_x associated with the bending moments are obtained from the flexure formula, as described in Section 5.3. In this section and in the following section, we will investigate the distribution of shear stresses τ associated with the shear force V.

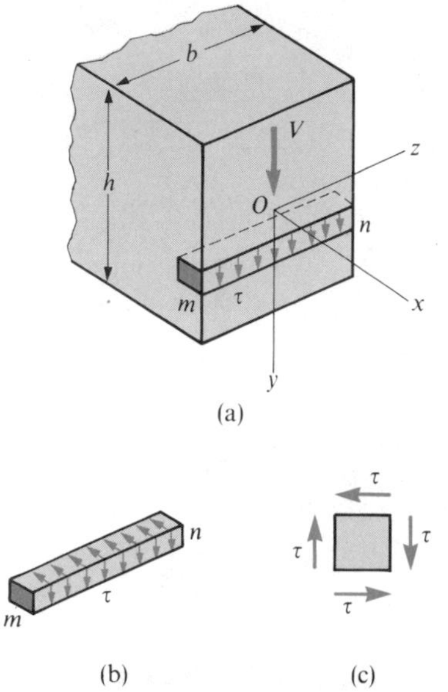

Fig. 5-21 Shear stresses in a beam of rectangular cross section

BEAMS OF RECTANGULAR CROSS SECTION

Let us begin with the simplest case, which is a beam of rectangular cross section having width b and height h (Fig. 5-21a). We can reasonably assume that the shear stresses τ act parallel to the shear force V (that is, parallel to the vertical sides of the cross section). Let us also assume that the distribution of shear stresses is uniform across the width of the beam. The use of these two assumptions will enable us to determine completely the distribution of the shear stresses acting on the cross section.

A small element of the beam may be cut out between two adjacent cross sections and between two planes that are parallel to the neutral surface, as shown by element *mn* in Fig. 5-21a. In accord with the foregoing assumptions, the vertical shear stresses τ are uniformly distributed on the vertical faces of this element. We also know from the discussion of shear stresses in Section 1.6 that shear stresses acting on one side of an element are accompanied by shear stresses of equal magnitude acting on perpendicular faces of the element (Figs. 5-21b and c). Thus, there will be horizontal shear stresses between horizontal layers of the beam as well as transverse shear stresses on the vertical cross sections. At any point within the beam, these complementary shear stresses are equal in magnitude.

This observation about the equality of the horizontal and vertical shear stresses leads to an interesting conclusion regarding the shear stresses at the top and bottom of the beam. If we consider the element *mn* shown in Fig. 5-21 to be at either the top or the bottom, we will see that the horizontal shear stresses must vanish because there are no stresses on the outer surfaces of the beam. Therefore, the vertical shear stress τ also must vanish at the top and bottom of the beam (that is, $\tau = 0$ where $y = \pm h/2$).

The existence of horizontal shear stresses in a beam can be demonstrated by a simple experiment. Take two equal rectangular bars of height h and place them on simple supports, as shown in Fig. 5-22a, and then load them by a force P. If no friction exists between the bars, bending of the two bars will occur independently. Each bar will be in compression above its neutral axis and in tension below its neutral axis, and each will deform as indicated in Fig. 5-22b. The bottom surface of the upper bar will slide with respect to the top surface of the lower bar. If, instead of two bars, we have a solid bar of depth $2h$, shear stresses must exist along the neutral plane and must be of such magnitude as to prevent the sliding shown in Fig. 5-22b. Because of the presence of shear stresses that prevent sliding, the single bar of depth $2h$ is much stiffer and stronger than two separate bars each of depth h.

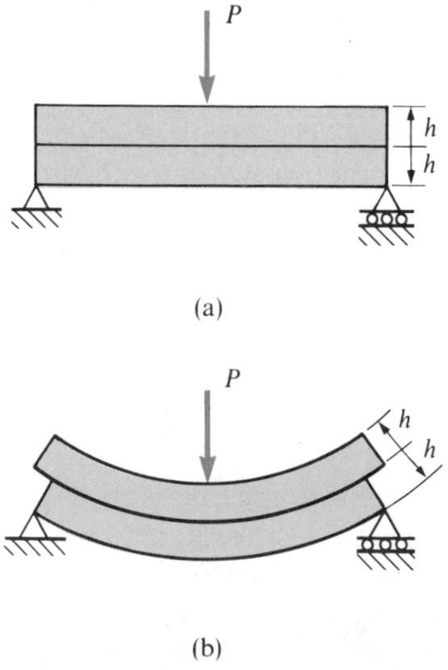

Fig. 5-22 Bending of two separate beams

To evaluate these shear stresses, let us consider the equilibrium of an element pp_1n_1n (Fig. 5-23a) cut out from a beam between two adjacent cross sections mn and m_1n_1 separated by a distance dx. The bottom face of this element is the lower surface of the beam and is free from stress; its top face is parallel to the neutral surface and at an arbitrary distance y_1 from that surface. The top face is acted upon by the horizontal shear stresses τ

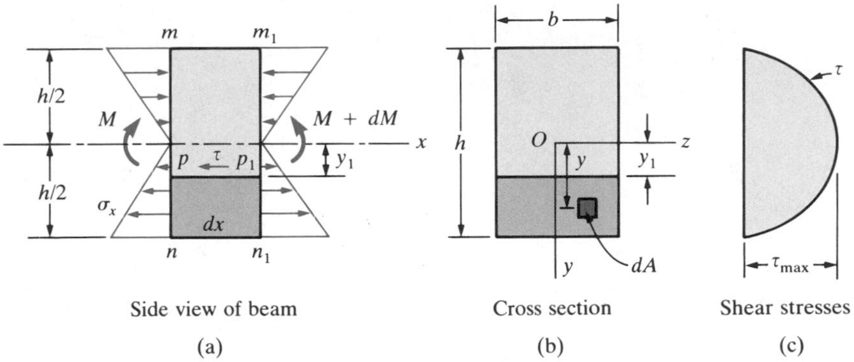

Fig. 5-23 Shear stresses in a beam of rectangular cross section

existing at this level in the beam. The end faces of the element are acted upon by the normal bending stresses σ_x produced by the bending moments. In addition, there are vertical shear stresses on the end faces, but these stresses will not enter into the equation of equilibrium of the element in the horizontal direction (the x direction), so they are not shown in Fig. 5-23a.

If the bending moments at cross sections mn and m_1n_1 are equal (that is, if the beam is in pure bending), the normal stresses σ_x acting over the sides np and n_1p_1 also will be equal. Under these conditions, the element will be in equilibrium under the action of the normal stresses alone, and hence the shear stress τ on the top face of the element will be equal to zero. This conclusion is obvious inasmuch as no shear force V exists if the beam is in pure bending.

In the more general case of a varying bending moment (nonuniform bending), let us denote by M and $M + dM$ the bending moments acting at cross sections mn and m_1n_1, respectively. Considering an element of area dA at distance y from the neutral axis (Fig. 5-23b), we see that the normal force acting on this element is $\sigma_x\,dA$, in which σ_x is the normal stress obtained from the flexure formula (Eq. 5-10). If the element of area is located on the left-hand face pn of the element, the normal force is

$$\sigma_x\,dA = \frac{My}{I}\,dA$$

Summing these elemental forces over the area of the face pn of the solid element pp_1n_1n gives the total horizontal force F_1 acting on that face:

$$F_1 = \int \frac{My}{I}\,dA$$

This integration is performed over the area of the cross section shown shaded in Fig. 5-23b (that is, over the area of the cross section from $y = y_1$ to $y = h/2$).

In the same manner, we find that the total force F_2 acting on the right-hand face p_1n_1 of the solid element pp_1n_1n is

$$F_2 = \int \frac{(M + dM)y}{I} dA$$

Finally, the horizontal force F_3 acting on the top face pp_1 of the element is

$$F_3 = \tau b\, dx \qquad \text{(a)}$$

in which $b\,dx$ is the area of the top face.

The forces F_1, F_2, and F_3 must be in static equilibrium in the horizontal or x direction; thus,

$$F_3 = F_2 - F_1 \qquad \text{(b)}$$

or

$$\tau b\, dx = \int \frac{(M + dM)y}{I} dA - \int \frac{My}{I} dA$$

from which

$$\tau = \frac{dM}{dx}\left(\frac{1}{Ib}\right) \int y\, dA$$

Substituting $V = dM/dx$ (see Eq. 4-3), we obtain the following formula for the shear stress τ:

$$\tau = \frac{V}{Ib} \int y\, dA \qquad \text{(5-20)}$$

The integral in this equation represents the first moment of the shaded portion of the cross section (Fig. 5-23b) with respect to the neutral axis (the z axis). In different terms, the integral is the first moment of the cross-sectional area below the level y_1 at which the shear stress τ acts. When y_1 is measured above the neutral axis, the integral is the first moment of the area above the level at which the shear stress acts. Denoting the first moment of the shaded area by Q, we can write Eq. (5-20) in the form

$$\tau = \frac{VQ}{Ib} \qquad \text{(5-21)}$$

This equation, known as the **shear formula**, can be used to determine the shear stress τ at any point in the cross section. To determine how the stress varies, we must examine how Q varies, because V, I, and b are constants for a given rectangular cross section.*

* In the derivation, Q is the first moment of the cross-sectional area shown shaded in Fig. 5-23b, and this area is usually used for calculating Q. However, we could also take the first moment of the remaining cross-sectional area, because it too is equal to Q (except for sign). The reason is that the first moment of the entire cross-sectional area (about the neutral axis) is equal to zero; hence the value of Q for the area below the level y_1 is the negative of Q for the area above that level.

The first moment Q for the shaded area of Fig. 5-23b is obtained by multiplying the area by the distance from the centroid of the area to the neutral axis:

$$Q = b\left(\frac{h}{2} - y_1\right)\left(y_1 + \frac{h/2 - y_1}{2}\right) = \frac{b}{2}\left(\frac{h^2}{4} - y_1^2\right) \tag{c}$$

Of course, the first moment can also be determined by integration:

$$Q = \int y\,dA = \int_{-b/2}^{b/2}\int_{y_1}^{h/2} y\,dy\,dz = \int_{y_1}^{h/2} yb\,dy$$

$$= b\left[\frac{y^2}{2}\right]_{y_1}^{h/2} = \frac{b}{2}\left(\frac{h^2}{4} - y_1^2\right)$$

as obtained before. Now substituting for Q (Eq. c) into the shear formula (Eq. 5-21), we get

$$\tau = \frac{V}{2I}\left(\frac{h^2}{4} - y_1^2\right) \tag{5-22}$$

This equation shows that the shear stress in a rectangular beam varies quadratically with the distance y_1 from the neutral axis; thus, when plotted along the height of the beam, τ varies as shown in Fig. 5-23c. Note that the stress is zero when $y_1 = \pm h/2$.

The maximum value of the shear stress occurs at the neutral axis, where $y_1 = 0$; substituting this value into Eq. (5-22), we get

$$\tau_{max} = \frac{Vh^2}{8I} = \frac{3V}{2A} \tag{5-23}$$

in which $A = bh$ is the cross-sectional area. Thus, the maximum shear stress is 50% larger than the average shear stress (equal to V/A). Note that the preceding equations for τ (Eqs. 5-20 through 5-23) can be used to calculate either vertical shear stresses acting on cross sections or horizontal shear stresses acting between horizontal layers of the beam.*

The preceding formulas for shear stresses were derived without regard to any particular sign conventions for V and τ. It is sufficient to observe that the shear force is the resultant of the shear stresses; therefore, the stresses act in the same direction as the force. For most purposes, only absolute values are used in the shear formulas, and the directions of the stresses are determined by inspection, as in Fig. 5-21. If we wish to use the sign convention for shear stresses described in Section 1.6 (see Fig. 1-24), then a positive shear force (Fig. 5-21a) produces a negative shear stress on

* The shear-stress analysis presented in this section was developed by the Russian engineer D. J. Jourawski; see Refs. 5-10 and 5-11.

the right-hand face of the element (Fig. 5-21c). To reconcile these sign conventions, we must insert a minus sign in the shear formula.

The formulas for shear stresses in rectangular beams are valid for beams of ordinary proportions and are subject to the same restrictions as the flexure formula from which they are derived. Thus, the formulas are valid only for beams of linearly elastic materials with small deflections. The formulas may be considered to be exact for narrow beams (b much less than h), but they become less accurate as b increases relative to h. For instance, when $b = h$, the true maximum shear stress is about 13% larger than the value given by Eq. (5-23).*

A common error is to apply the shear formula to cross-sectional shapes, such as a triangle, for which it is not applicable. The shear formula is limited to cross-sectional shapes that have sides parallel to the y axis (because we assumed that the shear stresses τ acted parallel to the y axis) and for which the shear stress distribution is uniform across the width b of the cross section.

EFFECTS OF SHEAR STRAINS

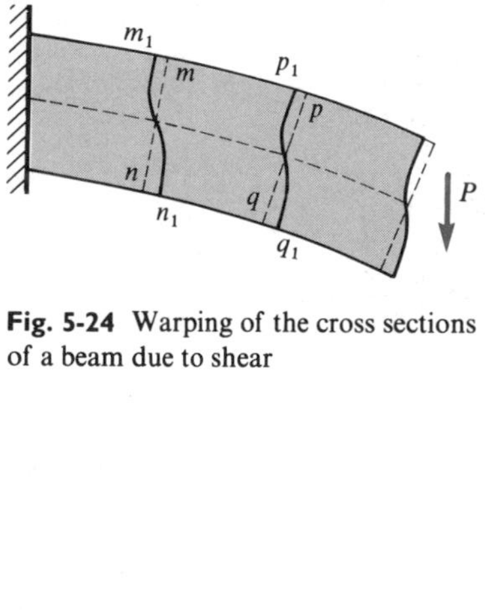

Fig. 5-24 Warping of the cross sections of a beam due to shear

Because the shear stress τ varies parabolically from top to bottom of the beam, it follows that the shear strain $\gamma = \tau/G$ varies in the same manner. As a result, cross sections of the beam that were originally plane surfaces become warped. This warping can be demonstrated by bending a beam on which vertical lines, such as lines mn and pq in Fig. 5-24, have been drawn. The lines will not remain straight, but will become curves, with the maximum shear strain occurring at the neutral surface. At the points m_1, p_1, n_1, and q_1, the shear strain is zero and the curves m_1n_1 and p_1q_1 remain normal to the upper and lower surfaces of the bar after bending. At the neutral surface, the angles between the tangents to the curves m_1n_1 and p_1q_1 and the normal sections mn and pq are equal to the shear strain $\gamma = \tau_{max}/G$. If the shear force V remains constant along the beam, the warping of all cross sections is the same, so that $m_1m = p_1p$ and $nn_1 = qq_1$. Therefore, the stretching or the shortening of longitudinal elements due to the bending moments is unaffected by the shear strains, and the distribution of the normal stresses σ is the same as in pure bending.

A more elaborate investigation of the strains shows that warping of cross sections due to shear strains does not substantially affect the longitudinal strains even when a distributed load acts on the beam and the shear force varies continuously along the length. For concentrated loads, the stress distribution near the loads is more complicated and stress concentrations are present (see Sections 2.10 and 5.10), but these irregularities are very localized and do not appreciably affect the overall stress distribution in the beam. For these reasons, it is quite justifiable to

* For a discussion of the limitations of the shear formula, see Ref. 5-12.

use the flexure formula (Eq. 5-10), which was derived for pure bending, in the case of nonuniform bending.

Example 1

Calculate the normal and shear stresses acting at point C in the steel beam AB shown in Fig. 5-25a. The beam is simply supported and has a span length $L = 3$ ft. The cross section is rectangular with width 1 in. and height 4 in. (Fig. 5-25b). The total uniform load on the beam (including its own weight) is $q = 160$ lb/in. (The beam is adequately supported against sideways buckling.)

The bending moment M and shear force V at the cross section through point C are found (from static equilibrium) to have the following values:

$$M = 17{,}920 \text{ in.-lb} \qquad V = -1{,}600 \text{ lb}$$

The signs in these terms are based upon the sign conventions for M and V shown in Fig. 4-3. The moment of inertia of the cross-sectional area is

$$I = \frac{bh^3}{12} = \frac{1}{12}(1.0 \text{ in.})(4.0 \text{ in.})^3 = 5.333 \text{ in.}^4$$

Therefore, the bending stress at point C, which is located at $y = -1.0$ in. (see Fig. 5-10b for the sign convention for y), is

$$\sigma_x = \frac{My}{I} = \frac{(17{,}920 \text{ in.-lb})(-1.0 \text{ in.})}{5.333 \text{ in.}^4} = -3{,}360 \text{ psi}$$

The minus sign indicates that the stress is compressive, which is apparent by inspection of the beam and its loading.

To obtain the shear stress, we need to evaluate the first moment Q of the cross-sectional area between point C and the outer edge of the cross section; this area is shown shaded in Fig. 5-25b. The first moment of this area about the z axis is equal to the product of the area and its centroidal distance from the z axis; thus, using only numerical values, we get

$$Q = (1 \text{ in.})(1 \text{ in.})(1.5 \text{ in.}) = 1.5 \text{ in.}^3$$

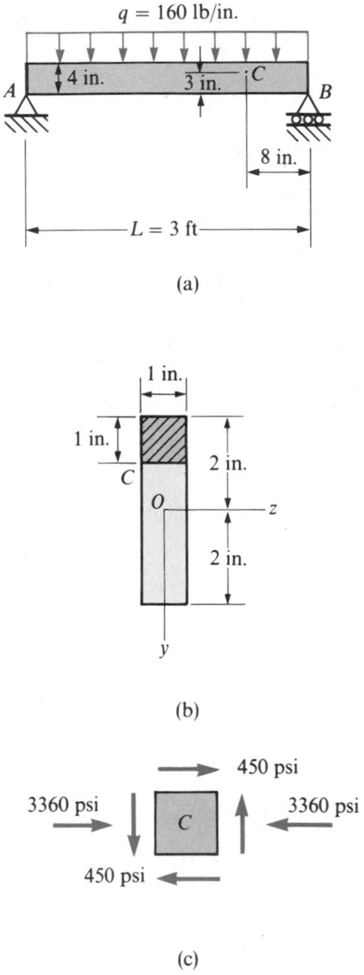

Fig. 5-25 Example 1

Now we can substitute into the shear formula, again using only numerical values, and obtain the shear stress at C:

$$\tau = \frac{VQ}{Ib} = \frac{(1{,}600 \text{ lb})(1.5 \text{ in.}^3)}{(5.333 \text{ in.}^4)(1.0 \text{ in.})} = 450 \text{ psi}$$

Because the shear force is negative, it acts upward on the left-hand part of the beam when the beam is cut by a section through C; therefore, the shear stress τ acts in this same direction. A convenient way to show the directions of the stresses is to draw a stress element at point C (Fig. 5-25c).

Example 2

A simple beam AB carrying two concentrated loads P (Fig. 5-26) has a rectangular cross section of width $b = 100$ mm and height $h = 150$ mm. The distance a from

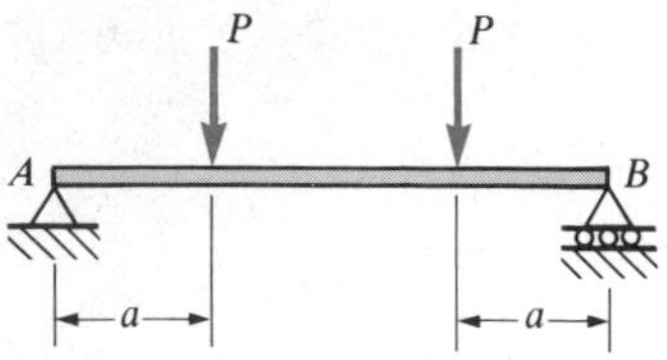

Fig. 5-26 Example 2

the end of the beam to one of the loads is 0.5 m. Determine the allowable value of P if the beam is constructed of wood having an allowable stress in bending $\sigma_{\text{allow}} = 11$ MPa and an allowable stress in horizontal shear $\tau_{\text{allow}} = 1.2$ MPa. Disregard the weight of the beam itself.

The maximum bending moment M and the maximum shear force V in the beam are

$$M = Pa \qquad V = P$$

Also, the section modulus S and cross-sectional area A are

$$S = \frac{bh^2}{6} \qquad A = bh$$

The maximum normal and shear stresses in the beam (from Eqs. 5-13 and 5-23) are

$$\sigma = \frac{M}{S} = \frac{6Pa}{bh^2} \qquad \tau = \frac{3V}{2A} = \frac{3P}{2bh}$$

Therefore, the maximum permissible values of the load P are

$$P = \frac{\sigma_{\text{allow}} bh^2}{6a} \quad \text{and} \quad P = \frac{2\tau_{\text{allow}} bh}{3}$$

Substituting numerical values into these formulas, we get

$$P = 8.25 \text{ kN} \quad \text{and} \quad P = 12.0 \text{ kN}$$

Thus, the bending stress governs the design, and the allowable load is $P = 8.25$ kN.

BEAMS OF CIRCULAR CROSS SECTION

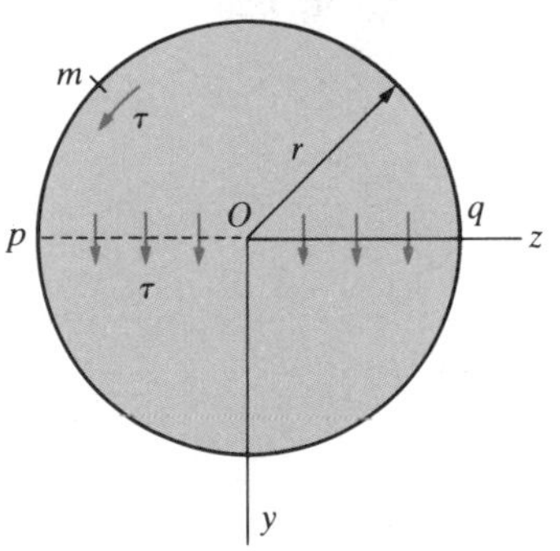

Fig. 5-27 Shear stresses in a beam of circular cross section

When a beam has a circular cross section (Fig. 5-27), we can no longer assume that all of the shear stresses act parallel to the y axis. For instance, we can easily prove that at a point such as m on the boundary of the cross section, the shear stress τ must act tangent to the boundary. This follows from the fact that the outer surface of the beam is free of stress, and therefore the shear stress τ can have no component in the radial direction (because shear stresses acting on perpendicular planes must be equal in magnitude).

The largest shear stresses occur at the neutral axis (the z axis), where the shear stresses act parallel to the y axis. Furthermore, we may assume with good accuracy that these stresses are constant across the width of the beam (from point p to point q in Fig. 5-27). Inasmuch as these assumptions are the same as those used in deriving the shear formula $\tau = VQ/Ib$ (Eq. 5-21), we may use that formula to calculate the shear stresses at the neutral axis. For a circular cross section of radius r, we obtain

$$I = \frac{\pi r^4}{4} \qquad b = 2r \qquad Q = \frac{\pi r^2}{2}\left(\frac{4r}{3\pi}\right) = \frac{2r^3}{3} \tag{d}$$

(see Cases 9 and 11, Appendix D). Substituting these expressions for I, b, and Q into the shear formula, we obtain

$$\tau_{max} = \frac{VQ}{Ib} = \frac{4V}{3\pi r^2} = \frac{4V}{3A} \tag{5-24}$$

in which A is the area of the cross section. This equation shows that the maximum shear stress in a circular beam is $\frac{4}{3}$ times the average shear stress V/A.

Although the preceding theory for a circular beam is approximate, it gives results that differ from the exact theory of elasticity by only a few percent (see Refs. 5-12 and 5-13).

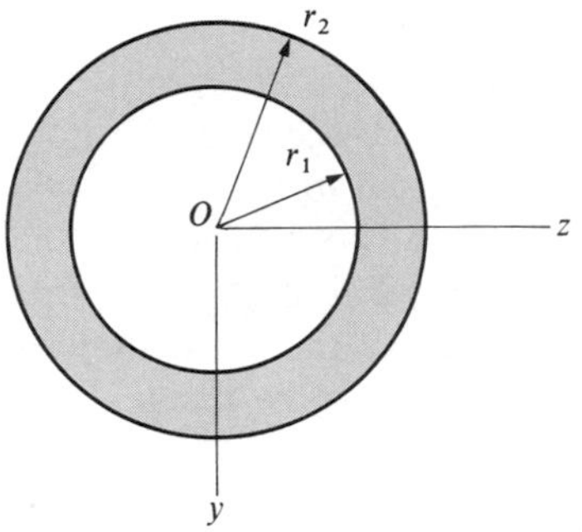

Fig. 5-28 Hollow circular cross section

If a beam has a **hollow circular cross section** (Fig. 5-28), we may again assume with good accuracy that the shear stresses along the neutral axis are parallel to the y axis and uniformly distributed. Under these conditions, we may obtain the maximum shear stress from the shear formula (Eq. 5-21). The properties of the cross section are

$$I = \frac{\pi}{4}(r_2^4 - r_1^4) \qquad b = 2(r_2 - r_1) \qquad Q = \frac{2}{3}(r_2^3 - r_1^3)$$

and the maximum stress is

$$\tau_{max} = \frac{VQ}{Ib} = \frac{4V}{3A}\left(\frac{r_2^2 + r_2 r_1 + r_1^2}{r_2^2 + r_1^2}\right) \tag{5-25}$$

in which $A = \pi(r_2^2 - r_1^2)$ is the area of the hollow section. If $r_1 = 0$, this equation reduces to Eq. (5-24) for a solid circular beam.

5.6 SHEAR STRESSES IN THE WEBS OF BEAMS WITH FLANGES

When a beam of wide-flange shape (Fig. 5-29a) is subjected to a shear force V, shear stresses are developed throughout the cross section. Because of the shape, the distribution of these stresses is much more complicated than in the case of a rectangular beam. For instance, in the flanges of the beam, shear stresses act on the cross sections in both the horizontal and vertical directions. However, most of the vertical shear force V is carried by shear stresses in the web, and we can determine those stresses (which include the maximum shear stress) using the same techniques we used for rectangular beams.

Let us begin by considering the shear stresses at location ef in the web of the beam (Fig. 5-29a). We will make the same assumptions as in the case of a rectangular beam; namely, the shear stresses act parallel to the y axis and are uniformly distributed across the thickness t of the web. Then the derivation made in the preceding section will still be valid, and the

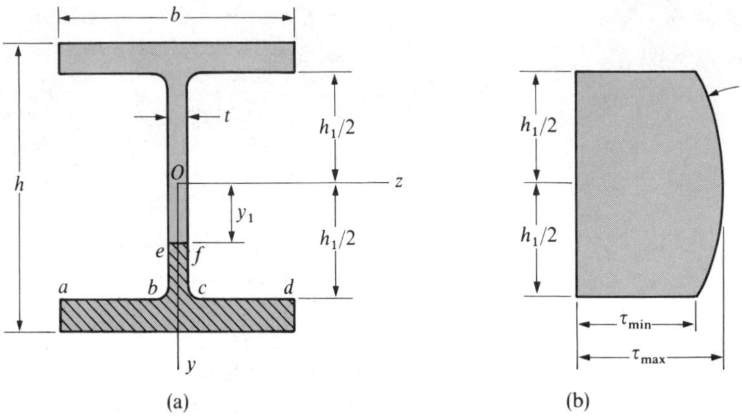

Fig. 5-29 Shear stresses in the web of a wide-flange beam

shear formula $\tau = VQ/Ib$ will still apply. However, the width b now becomes the thickness t of the web, and the area used in calculating the first moment Q becomes the area between ef and the bottom edge of the cross section (that is, the shaded area of Fig. 5-29a). This area consists of two rectangles, because we will disregard the effects of the small fillets at the juncture of the web and flange. The first rectangle is the flange itself, which has area

$$A_f = b\left(\frac{h}{2} - \frac{h_1}{2}\right)$$

The second rectangle is the part of the web between ef and the flange, that is, rectangle $efcb$, which has area

$$A_w = t\left(\frac{h_1}{2} - y_1\right)$$

The first moments of these areas about the neutral axis are obtained by multiplying the areas by the distances from the z axis to the centroids of the areas:

$$Q = b\left(\frac{h}{2} - \frac{h_1}{2}\right)\left(\frac{h_1}{2} + \frac{h/2 - h_1/2}{2}\right) + t\left(\frac{h_1}{2} - y_1\right)\left(y_1 + \frac{h_1/2 - y_1}{2}\right)$$

Upon simplifying, we get

$$Q = \frac{b}{8}(h^2 - h_1^2) + \frac{t}{8}(h_1^2 - 4y_1^2)$$

Therefore, the shear stress τ in the web of the beam is

$$\tau = \frac{VQ}{It} = \frac{V}{8It}[b(h^2 - h_1^2) + t(h_1^2 - 4y_1^2)] \qquad (5\text{-}26)$$

From this equation, we see that τ varies quadratically throughout the height of the web, as shown by the graph in Fig. 5-29b. Upon introducing the following expression for the moment of inertia I,

$$I = \frac{bh^3}{12} - \frac{(b-t)h_1^3}{12} = \frac{1}{12}(bh^3 - bh_1^3 + th_1^3) \tag{5-27}$$

we can rewrite Eq. (5-26) in the form

$$\tau = \frac{3V(bh^2 - bh_1^2 + th_1^2 - 4ty_1^2)}{2t(bh^3 - bh_1^3 + th_1^3)} \tag{5-28}$$

which expresses τ in terms of the dimensions of the cross section.

The maximum shear stress occurs at the neutral axis ($y_1 = 0$), and the minimum shear stress in the web occurs at the juncture with the flange ($y_1 = \pm h_1/2$). Thus, we find

$$\tau_{max} = \frac{V}{8It}(bh^2 - bh_1^2 + th_1^2) = \frac{3V(bh^2 - bh_1^2 + th_1^2)}{2t(bh^3 - bh_1^3 + th_1^3)} \tag{5-29}$$

and

$$\tau_{min} = \frac{Vb}{8It}(h^2 - h_1^2) = \frac{3Vb(h^2 - h_1^2)}{2t(bh^3 - bh_1^3 + th_1^3)} \tag{5-30}$$

Depending upon the beam dimensions, the maximum shear stress typically is from 10% to 60% greater than the minimum stress in the web.

The shear force carried by the web alone may be determined by multiplying the area of the stress diagram (Fig. 5-29b) by the width t of the web itself. The area of the stress diagram consists of two parts, a rectangle of area $h_1\tau_{min}$ and a parabolic segment of area

$$\frac{2}{3}(h_1)(\tau_{max} - \tau_{min})$$

Thus, the shear force in the web is

$$V_{web} = h_1\tau_{min}t + \frac{2}{3}(h_1)(\tau_{max} - \tau_{min})t$$

$$= \frac{th_1}{3}(2\tau_{max} + \tau_{min}) \tag{5-31}$$

For beams of typical proportions, the shear stresses in the web account for 90 to 98% of the total shear force; the remainder is carried by shear in the flanges.

In design work, it is common to calculate an approximation to the maximum shear stress by dividing the total shear force by the area of the web. This stress represents an average shear stress in the web:

$$\tau_{aver} = \frac{V}{th_1} \tag{5-32}$$

For typical wide-flange beams, the average shear stress is within 10% (plus or minus) of the actual maximum shear stress.

The elementary theory presented in this section is quite accurate when used for calculating shear stresses in the web. However, when considering the distribution of shear stresses in the flanges, the assumption of constant shear stress across the width b of the flanges cannot be made. For example, we see immediately that for $y_1 = h_1/2$ the shear stress over the free surfaces ab and cd (Fig. 5-29a) must be zero, whereas across the junction bc the stress is $\tau_{\min}$. This observation indicates that at the junction of the web and flange, the distribution of shear stresses follows a more complicated law, one that cannot be investigated by an elementary analysis. The stresses would become very large at the junction if the internal corners were square, so fillets are used to reduce the stresses, as shown in the figure. Because of the localized nature of the stress distribution near the intersection of the web and flange, the shear formula does not give accurate results for the vertical shear stresses in the flange. However, the formula does give good results for the *horizontal* shear stresses in the flange, as discussed later in Section 5.14.

Example

Determine the maximum shear stress in the web of a beam having the T-shaped cross section shown in Fig. 5-30 if $b = 4$ in., $t = 1$ in., $h = 8$ in., $h_1 = 7$ in., and the shear force $V = 10{,}000$ lb.

The distance c to the centroid of the cross section is determined as follows:

$$c = \frac{(3\text{ in.})(1\text{ in.})(0.5\text{ in.}) + (8\text{ in.})(1\text{ in.})(4\text{ in.})}{(3\text{ in.})(1\text{ in.}) + (8\text{ in.})(1\text{ in.})}$$

$$= \frac{33.5\text{ in.}^3}{11.0\text{ in.}^2} = 3.045\text{ in.}$$

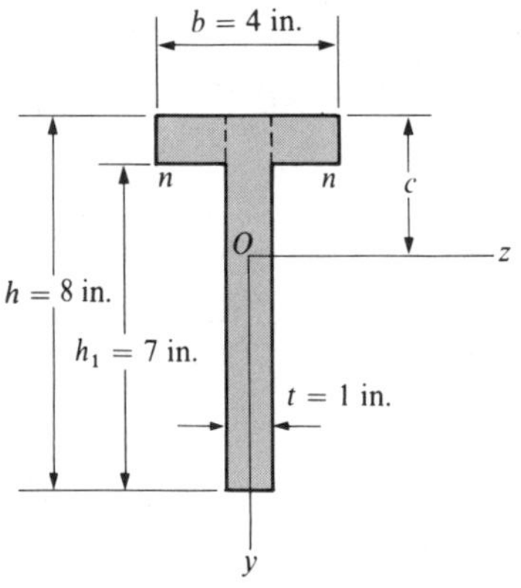

Fig. 5-30 Example

The moment of inertia I of the cross section about the neutral axis can be found by first obtaining the moment of inertia about axis nn and then using the parallel-axis theorem (see Appendix C). The calculations are as follows:

$$I = \frac{1}{3}(4\text{ in.})(1\text{ in.})^3 + \frac{1}{3}(1\text{ in.})(7\text{ in.})^3 - (11.0\text{ in.}^2)(2.045\text{ in.})^2 = 69.66\text{ in.}^4$$

The maximum shear stress occurs at the neutral axis, which is in the web. The first moment Q of the area below the neutral axis is

$$Q = (1\text{ in.})(4.955\text{ in.})^2\left(\frac{1}{2}\right) = 12.28\text{ in.}^3$$

Substituting into the shear formula, we get

$$\tau = \frac{VQ}{It} = \frac{(10{,}000\text{ lb})(12.28\text{ in.}^3)}{(69.66\text{ in.}^4)(1\text{ in.})} = 1760\text{ psi}$$

which is the maximum shear stress in the beam.

5.7 BUILT-UP BEAMS

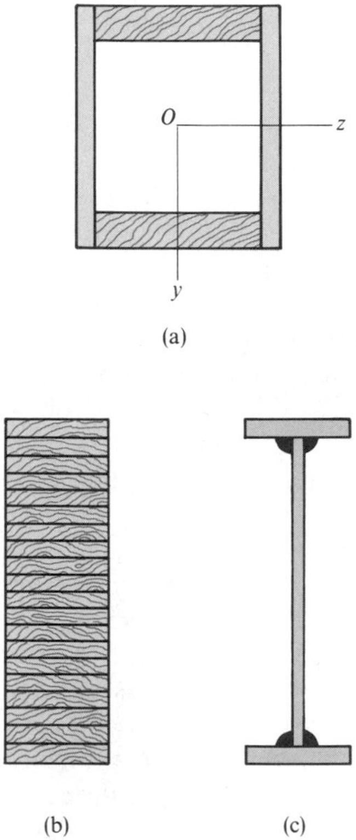

Fig. 5-31 Examples of built-up beams: (a) wood box beam, (b) glulam beam, and (c) steel plate girder

A **built-up beam** is fabricated of two or more pieces of material joined together to form a single, solid beam. Such beams can be constructed in a great variety of shapes to meet special needs or to provide larger cross sections than are ordinarily available. Figure 5-31 shows some typical cross sections of built-up beams. Part (a) of the figure shows a wood **box beam** constructed of two planks, which serve as flanges, connected by plywood webs. The pieces are joined together with nails, screws, or glue; the design must ensure that the entire cross section acts as a single unit. The second example is a glued laminated beam (called a **glulam beam**) made of boards glued together to form a much larger beam than could be cut from a tree as a single member. The last example is a welded steel **plate girder**, commonly used in bridge construction and fabricated from three steel plates by fillet welding at the junctions.

A built-up beam is usually designed under the assumption that the parts will be adequately connected so that the beam behaves as a single member. The design calculations involve two phases. In the first phase, the beam is proportioned as a solid beam, taking into account both bending and shear stresses, as explained in the preceding sections. In the second phase, the connecting elements (such as nails, bolts, welds, and glue) are designed to ensure that the beam behaves as a solid beam. The loads carried by the connecting elements are the horizontal shear forces transmitted between the parts of the beam.

A formula for the horizontal shear forces acting between the parts can be obtained by referring to the derivation of the shear formula (see Fig. 5-23, Section 5.5). In that derivation, we assumed that the shear stress τ acting on the top face of the element pp_1n_1n was uniformly distributed across the width b of the beam. However, in a more general case, this assumption may not be valid. Therefore, instead of calculating the shear stresses, we can calculate the total horizontal force F_3 acting on the top face of the element (see Eq. a of Section 5.5). This force may be expressed as $F_3 = f\,dx$, in which f is the shear force per unit distance along the axis of the beam. The quantity f is called the **shear flow**.

To evaluate the shear flow, we proceed as in the derivation of the shear formula. Beginning with Eq. (b) of Section 5.5, we have

$$\begin{aligned} F_3 &= f\,dx \\ &= F_2 - F_1 \end{aligned}$$

or, upon substituting for F_2 and F_1 and solving,

$$f = \frac{dM}{dx}\left(\frac{1}{I}\right)\int y\,dA$$

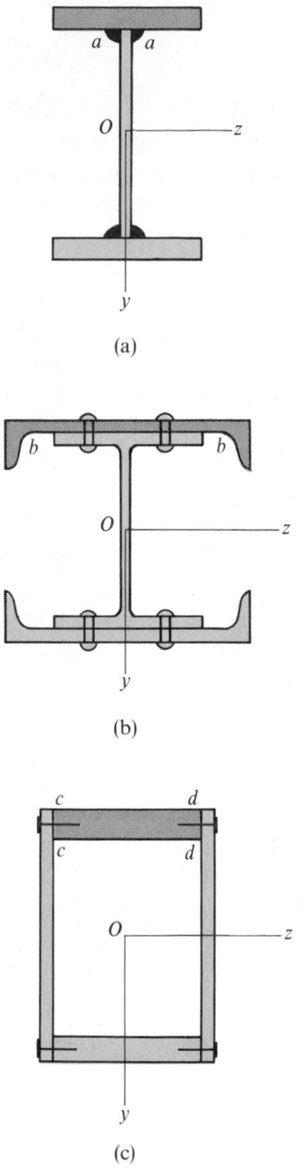

Fig. 5-32 Areas used when calculating the shear flow f

Again replacing dM/dx with V and the integral with Q, we obtain

$$f = \frac{VQ}{I} \tag{5-33}$$

This equation gives the shear flow f acting between the shaded element shown in Fig. 5-23 and the remainder of the beam. In deriving Eq. (5-33), we made no assumption about the distribution of shear stresses across the width of the section. (When the shear stress is uniformly distributed across the width of the beam, as we assumed for a beam of rectangular cross section, then $f = \tau b$.)

The formula for shear flow f is not limited to beams of rectangular cross section; it is valid for any beam having a cross section that is symmetric about the y axis. In practice, the cross-sectional face p_1n_1 of the shaded element of Fig. 5-23 may have any of a variety of shapes. In such cases, the shear flow f is the force per unit distance acting along the line in the cross section that divides the shaded area from the rest of the beam. To clarify this point, let us consider the cross sections shown in Fig. 5-32. In the case of a welded steel plate girder (Fig. 5-32a), the welds must transmit the horizontal shear force between the flanges and the web. This force (per unit distance) is the shear flow along the contact surface *aa*. It may be calculated from the shear-flow formula $f = VQ/I$ with Q as the first moment of the cross-sectional area above the contact surface *aa*. In other words, Q is the first moment of the flange area, shown shaded in the figure. Of course, the first moment is always calculated with respect to the z axis. Having calculated f in this manner, we then may select a weld size that is adequate to resist the force f per unit distance (acting longitudinally).

A second example is the beam of Fig. 5-32b, which consists of a wide-flange beam that is strengthened by riveting a channel section to each flange. The horizontal shear force acting between a channel and the main beam must be transmitted by the rivets. This force is calculated from the shear-flow formula using Q as the first moment of the entire channel (shown shaded in the figure). The resulting shear flow f is the longitudinal force per unit distance acting along the contact surface *bb*, and the rivets must be of adequate size and longitudinal spacing to resist this force. The last example (Fig. 5-32c) is similar, except that the shear flow is the force per unit distance along both contact surfaces *cc* and *dd*, and Q is calculated for the area shown shaded. In this case, the shear flow f is resisted by the combined action of the nails on both sides of the beam. These ideas will now be illustrated by an example.

Example

A wood box beam (Fig. 5-33a) is constructed of two boards (each 40 mm × 180 mm in cross section) that serve as flanges and two webs of plywood, each 15 mm thick. The total height of the beam is 280 mm. The plywood is fastened

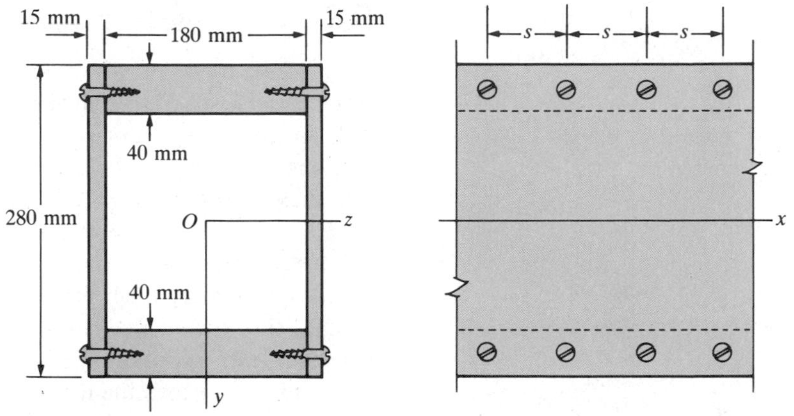

Fig. 5-33 Example

to the flanges by screws having an allowable load in shear of $F = 1100$ N per screw. If the shear force V acting on the cross section is 10.5 kN, determine the maximum permissible longitudinal spacing s of the screws (Fig. 5-33b).

The horizontal shear force transmitted between one of the flanges and the two webs can be found from the shear-flow formula $f = VQ/I$. The quantity Q is the first moment of the cross-sectional area between the top surface of the beam and the contact surfaces where the shear flow is being calculated. In this case, the section is cut along lines *cc* and *dd* of Fig. 5-32c; hence the area is that of one flange (shown shaded in Fig. 5-32c). Using the dimensions shown in Fig. 5-33a, we calculate Q for one flange as follows:

$$Q = \bar{y}A = (120\text{ mm})(180\text{ mm})(40\text{ mm}) = 864 \times 10^3\text{ mm}^3$$

Also, the moment of inertia of the entire cross-sectional area about the neutral axis is

$$I = \frac{1}{12}(210\text{ mm})(280\text{ mm})^3 - \frac{1}{12}(180\text{ mm})(200\text{ mm})^3$$

$$= 264.2 \times 10^6\text{ mm}^4$$

Substituting into the shear-flow formula, we obtain

$$f = \frac{VQ}{I} = \frac{(10{,}500\text{ N})(864 \times 10^3\text{ mm}^3)}{264.2 \times 10^6\text{ mm}^4} = 34.3\text{ N/mm}$$

which is the shear force per millimeter of length that must be carried by the screws.

The load capacity of the screws per unit length is $2F/s$, because there are two lines of screws (one on each side of the flange). Equating $2F/s$ to the shear flow and solving for s, we obtain

$$s = \frac{2F}{f} = \frac{2(1100\text{ N})}{34.3\text{ N/mm}} = 64.1\text{ mm}$$

This value of s represents the maximum permissible spacing of the screws. For convenience in fabrication, a spacing such as $s = 60$ mm would be selected.

5.8 NONPRISMATIC BEAMS

The beam theories presented in the preceding sections apply rigorously only to prismatic beams, that is, straight beams having the same cross sections throughout their lengths. For such beams, we derived the flexure and shear formulas ($\sigma = My/I$, $\tau = VQ/Ib$, and $f = VQ/I$). However, in practice we often encounter nonprismatic beams with cross-sectional dimensions that vary along their lengths. Familiar examples are parts of vehicles, airplanes, machinery, and buildings. As in the case of prismatic beams, the stress analysis of nonprismatic beams requires that we calculate both normal and shear stresses. The former can be calculated from the flexure formula if the changes in cross-sectional dimensions are gradual, but the calculation of the shear stresses requires a more complicated theory, as explained in the following discussions.

NORMAL STRESSES

If the cross-sectional dimensions change gradually as we move from one end of the beam to the other, we may calculate the normal stresses σ from the flexure formula $\sigma = My/I$ and obtain good accuracy. For instance, consider the tapered cantilever beam of solid circular cross section shown in Fig. 5-34. If the angle between the top edge of the beam and the horizontal is less than 20° (which means that the angle between the top and bottom surfaces is less than 40°), the error in calculating the normal stresses from the flexure formula is less than 10%. Of course, for smaller angles, the error is less.

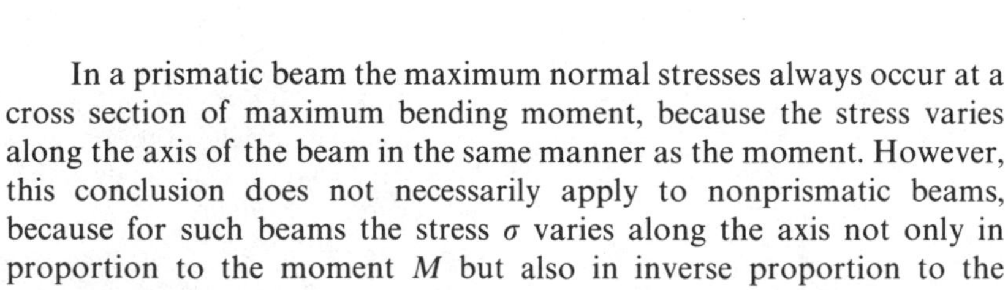

Fig. 5-34 Tapered cantilever beam of circular cross section

In a prismatic beam the maximum normal stresses always occur at a cross section of maximum bending moment, because the stress varies along the axis of the beam in the same manner as the moment. However, this conclusion does not necessarily apply to nonprismatic beams, because for such beams the stress σ varies along the axis not only in proportion to the moment M but also in inverse proportion to the moment of inertia I.

An illustration of this situation is the tapered cantilever beam of Fig. 5-34. To facilitate the analysis of this beam, let us assume that the fixed end B has a diameter twice that of the free end A:

$$\frac{d_b}{d_a} = 2 \tag{a}$$

Then the diameter d of the beam at distance x from the left-hand end is

$$d = d_a + (d_b - d_a)\frac{x}{L} = d_a\left(1 + \frac{x}{L}\right) \tag{b}$$

and the corresponding section modulus is

$$S = \frac{\pi d^3}{32} = \frac{\pi}{32} d_a^3 \left(1 + \frac{x}{L}\right)^3 \tag{c}$$

If the length L is much larger than the diameter d, we can use the flexure formula to calculate (with good accuracy) the maximum normal stress σ at any cross section due to the load P:

$$\sigma = \frac{M}{S} = \frac{Px}{S} = \frac{32Px}{\pi d_a^3(1 + x/L)^3} \tag{d}$$

Taking the derivative $d\sigma/dx$ and equating it to zero, we can find the value of x for which σ is a maximum; the result is $x = L/2$. The corresponding maximum stress is

$$\sigma_{\max} = \frac{128PL}{27\pi d_a^3} = 4.741\frac{PL}{\pi d_a^3} \tag{5-34}$$

At the section of maximum bending moment (support B), the largest stress is

$$\sigma_b = \frac{4PL}{\pi d_a^3} \tag{e}$$

In this particular example, the maximum stress occurs at the midpoint of the beam and is 19% greater than the stress at the built-in end. If the taper of the beam is reduced, the section of maximum normal stress will move from the midpoint toward the fixed support. For very small tapers of the beam, the maximum stress occurs at end B, just as it does in a prismatic cantilever.

The preceding example illustrates the use of the flexure formula for a nonprismatic beam; the main considerations are that the degree of taper must be small and that the maximum stress may occur elsewhere than at the section of maximum bending moment.

SHEAR STRESSES

The distribution of shear stresses in a nonprismatic beam is quite different from that in a prismatic beam. Instead of using the shear formula, we must derive a new relationship that incorporates the effect of the changing cross sections of the beam. For this purpose, let us consider an element of small

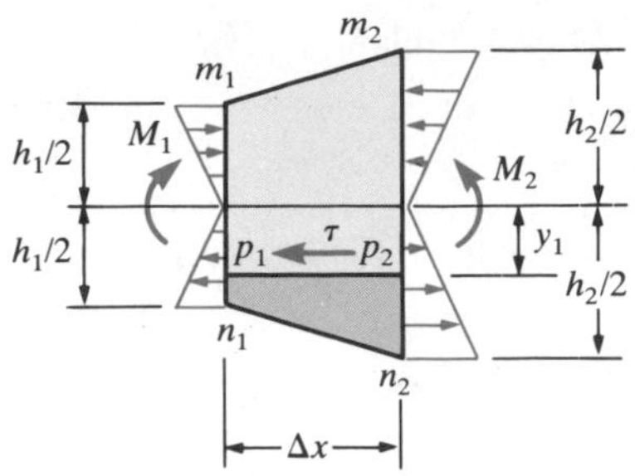

Fig. 5-35 Element of length Δx from a nonprismatic beam

length Δx cut out from a nonprismatic beam, as shown in Fig. 5-35. On the left-hand face m_1n_1 of this element, the bending moment is M_1; on the right-hand face m_2n_2, it is M_2. The corresponding heights of the beam are denoted by h_1 and h_2, respectively, and y_1 is the distance from the neutral axis to the point where the shear stress τ acts (compare with Fig. 5-23a for a prismatic beam).

The shear flow f acting along the top face p_1p_2 of the shaded element $n_1p_1p_2n_2$ is obtained from an equation of static equilibrium for the element. The forces due to the normal bending stresses acting on the left-hand and right-hand faces of the element are M_1Q_1/I_1 and M_2Q_2/I_2, respectively (see the equations for F_1 and F_2 in Section 5.5). In these expressions, I_1 and I_2 are the moments of inertia of the two cross sections, and Q_1 and Q_2 are the first moments about the neutral axis of the areas of the sides p_1n_1 and p_2n_2, respectively. The total force on the top face p_1p_2 of the element is $f\Delta x$. Thus, the equation of static equilibrium becomes

$$f\Delta x = \frac{M_2Q_2}{I_2} - \frac{M_1Q_1}{I_1} \tag{5-35}$$

If the shear stress τ is uniform across the width of the beam, then $f = \tau b$, where b is the width of the beam at that location.

For any particular beam, Eq. (5-35) can be used to find the average shear flow between the cross sections by substituting numerical values for M, Q, I, and Δx. The procedure is as follows. Select the cross section of the beam where the shear flow f is to be calculated. Then take an adjoining section a small distance Δx away; for instance, take $\Delta x = L/100$. At each of these sections, determine the following: (1) bending moment M, (2) first moment Q of the part of the cross-sectional area located between the outer edge of the beam and the line in the cross section where f is to be determined, and (3) moment of inertia I of the entire cross section. Then substitute these values into Eq. (5-35) and solve for f. The smaller the distance Δx, the more accurate the results will be (in theory). However, if Δx is too small, the numerical accuracy may be inadequate because Eq. (5-35) requires taking the difference between two terms that are nearly equal. Equation (5-35) is quite general and is useful in many practical situations to obtain the shear flow for beams of varying cross section when exact formulas are not available.

SHEAR STRESSES IN RECTANGULAR BEAMS

In the special case of a beam of rectangular cross section with constant width b and varying height h, we can convert Eq. (5-35) to a more exact equation by letting Δx approach zero. Also, in this case it is reasonable to assume that the shear stresses τ are uniformly distributed across the width b of the beam; hence $f = \tau b$ in Eq. (5-35). To derive the desired formula for the shear stresses, we begin by evaluating Q and I at cross sections 1 and 2, which are a distance Δx apart:

$$Q_1 = \frac{b}{2}\left(\frac{h_1^2}{4} - y_1^2\right) \qquad Q_2 = \frac{b}{2}\left(\frac{h_2^2}{4} - y_1^2\right)$$

$$I_1 = \frac{bh_1^3}{12} \qquad I_2 = \frac{bh_2^3}{12}$$

Now, letting $h_2 = h_1 + \Delta h$, where Δh is the increment in h as we go from section 1 to section 2, we obtain from the preceding equations

$$Q_2 = Q_1 + \frac{bh_1\,\Delta h}{4} \qquad I_2 = I_1 + \frac{bh_1^2\,\Delta h}{4}$$

In deriving these equations, terms containing the square and cube of Δh have been dropped because they are small in comparison to the terms retained. Noting also that $M_2 = M_1 + \Delta M$, we now substitute the expressions for Q_2, I_2, and M_2 into Eq. (5-35):

$$\tau b\,\Delta x = \frac{(M_1 + \Delta M)\left(Q_1 + \dfrac{bh_1\,\Delta h}{4}\right)}{I_1 + \dfrac{bh_1^2\,\Delta h}{4}} - \frac{M_1 Q_1}{I_1}$$

To simplify this equation, we may multiply all terms by the denominator of the second term, expand the remaining products, and drop all terms containing the products of two small quantities. The result is

$$\tau b I_1\,\Delta x = \frac{M_1 b h_1}{4}\Delta h + Q_1\,\Delta M - \frac{M_1 Q_1}{4I_1} b h_1^2\,\Delta h$$

This equation may now be divided by Δx. Then, in the limit as Δx becomes smaller and smaller, the term $\Delta h/\Delta x$ becomes dh/dx and the term $\Delta M/\Delta x$ becomes dM/dx, which is equal to the shear force V. With these changes, the preceding equation becomes

$$\tau b I_1 = \frac{M_1 b h_1}{4}\frac{dh}{dx} + Q_1 V - \frac{M_1 Q_1 b h_1^2}{4I_1}\frac{dh}{dx}$$

As a final step, we divide all terms by bI_1 and drop the numerical subscripts, which are no longer needed. Thus, the final form of the expression for the shear stress τ in a nonprismatic beam of rectangular cross section is

$$\tau = \frac{VQ}{Ib} + \frac{Mh}{4I}\left(1 - \frac{Qh}{I}\right)\frac{dh}{dx} \tag{5-36}$$

This equation is valid for a beam of constant width b and varying height h. We observe that the shear stress τ at any cross section is dependent not only upon the shear force V but also upon the bending moment M and the rate of change of h with respect to x. Equation (5-36) gives values that are in good agreement with more exact theories, provided the angle of taper is small.

As a specific example, let us investigate the distribution of shear stresses in the cantilever beam of rectangular cross section shown in Fig. 5-36a. The beam has heights h_a and $h_b = 2h_a$ at its ends and a uniform taper. Therefore, the quantity dh/dx is constant:

$$\frac{dh}{dx} = \frac{h_b - h_a}{L} = \frac{h_a}{L}$$

At the left end A, the bending moment is zero and Eq. (5-36) gives the same parabolic distribution of shear stresses as in a prismatic beam. This distribution is plotted in Fig. 5-36b. The maximum shear stress occurs at the neutral axis and is equal to $1.5\ P/bh_a$.

At the midpoint of the beam ($x = L/2$), the following values are obtained:

$$V = P \qquad h = 1.5h_a \qquad Q = \frac{b}{2}\left(\frac{h^2}{4} - y_1^2\right) \qquad M = \frac{PL}{2} \qquad I = \frac{bh^3}{12}$$

Substituting into Eq. (5-36) yields

$$\tau = \frac{2P}{3bh_a}$$

which is plotted in Fig. 5-36c. In this particular instance, we obtain the interesting result that the shear stress is uniformly distributed over the

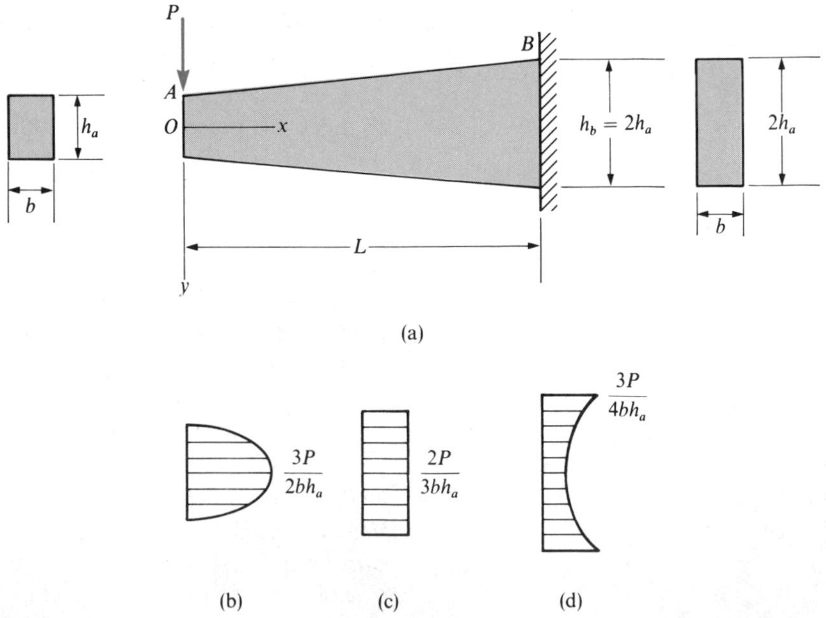

Fig. 5-36 Shear stresses in a tapered cantilever beam of rectangular cross section: (a) dimensions of beam, (b) distribution of shear stresses τ at $x = 0$, (c) shear stresses at $x = L/2$, and (d) shear stresses at $x = L$

height of the beam. Between the left end A of the beam and the midpoint, there is a gradual change from the shear-stress distribution pictured in Fig. 5-36b to the uniform distribution shown in Fig. 5-36c.

At the right-hand end B of the beam ($x = L$), we have

$$V = P \qquad h = 2h_a \qquad Q = \frac{b}{2}\left(\frac{h^2}{4} - y_1^2\right) \qquad M = PL \qquad I = \frac{bh^3}{12}$$

Substituting into Eq. (5-36) gives

$$\tau = \frac{3P}{8bh_a}\left(1 + \frac{y_1^2}{h_a^2}\right)$$

which is plotted in Fig. 5-36d. Note that the maximum shear stress on this cross section, equal to $3P/4bh_a$, occurs at the outer edges of the beam. The minimum stress, equal to one-half the maximum, occurs at the neutral axis, where $y_1 = 0$.

Shear stresses in rectangular beams with the top surface horizontal and the bottom surface inclined can be found by methods similar to those described here for a beam with both surfaces inclined. A discussion of such cases is given in Ref. 5-14.

FULLY STRESSED BEAMS

To minimize the amount of material in a beam, we may vary the dimensions of the cross sections in an attempt to maintain the maximum allowable stress at every section. A beam in this condition is called a **fully stressed beam**. Of course, ideal conditions are seldom attained because of practical problems in constructing the beam and the possibility of the loads being different from those assumed in design. Leaf springs in automobiles and bridge girders with cover plates are familiar examples of structures with varying dimensions that are designed to maintain constant maximum stresses (as nearly as practicable).

A cantilever beam with a concentrated load P at the end (Fig. 5-37) will serve as an example of a fully stressed beam. The cross section of the beam is assumed to be rectangular with constant width b. It is planned to vary the height h in order to maintain a constant maximum normal stress σ_{allow} due to the load P. Hence, at every cross section, the following

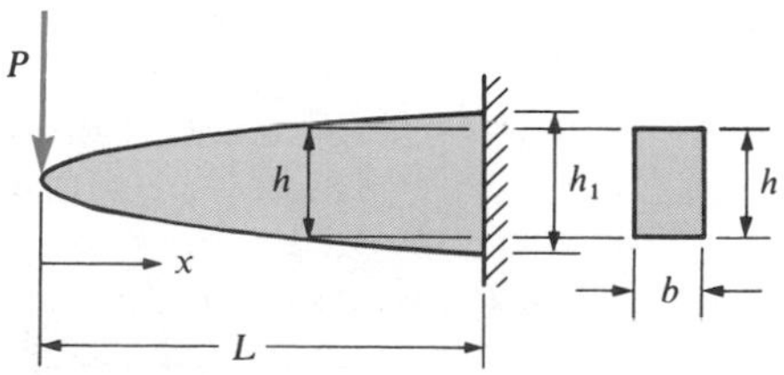

Fig. 5-37 Fully stressed beam having constant maximum normal stress (shear stresses disregarded)

equation must hold:

$$\sigma_{\text{allow}} = \frac{M}{S} = \frac{6M}{bh^2} = \frac{6Px}{bh^2}$$

Therefore, the height h_1 of the beam at the fixed support is

$$h_1 = \sqrt{\frac{6PL}{b\sigma_{\text{allow}}}}$$

and the height h at any other cross section is

$$h = \sqrt{\frac{6Px}{b\sigma_{\text{allow}}}} = h_1\sqrt{\frac{x}{L}} \tag{f}$$

This last equation shows that the height h of the beam varies quadratically with x, and thus the beam has the shape shown in Fig. 5-37. At the loaded end, the cross-sectional area is calculated to be zero, because we considered only the normal stresses due to bending. Of course, shear stresses are also present, and the cross sections of the beam (especially near the free end) must be designed to resist the shear force.

5.9 BEAMS WITH AXIAL LOADS

Structural members often are subjected to the simultaneous action of bending loads and axial forces. A simple example is shown in Fig. 5-38a, which portrays a cantilever beam subjected to an inclined force P acting through the centroid of the end cross section. The load P can be resolved into two components, a lateral load Q and an axial load S. These loads produce stress resultants in the form of bending moments M, shear forces V, and axial forces N. For a cross section at distance x from the support (Fig. 5-38b), these stress resultants are

$$M = Q(L - x) \qquad V = -Q \qquad N = S \tag{a}$$

The stresses associated with these stress resultants can be determined at any point in the cross section by means of the appropriate formulas ($\sigma = My/I$, $\tau = VQ/Ib$, and $\sigma = N/A$).

Let us begin with the axial force N, which produces a uniform stress distribution $\sigma = N/A$ (Fig. 5-38c). In this example the stress σ is tensile, or positive, as shown by the plus signs enclosed in circles. Next, the bending moment M produces a linearly varying stress $\sigma = My/I$, as sketched in Fig. 5-38d, with tension on the lower part of the beam and compression on the upper part. The final distribution of the normal stresses acting on the cross section is obtained by combining (or superimposing) the stresses produced by the force N and the moment M. Thus, the total stresses are

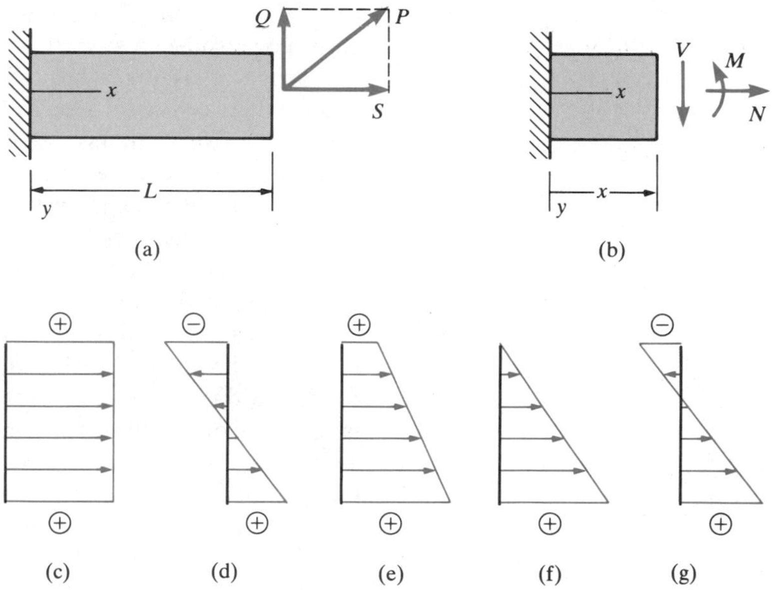

Fig. 5-38 Stress distributions in a cantilever beam subjected simultaneously to lateral and axial loads: (a) beam and loads, (b) stress resultants at a cross section, (c) tensile stresses due to the axial force N, (d) tensile and compressive stresses due to the bending moment M, and (e), (f), (g) possible final stress distributions

obtained from the following equation:

$$\sigma = \frac{N}{A} + \frac{My}{I} \tag{5-37}$$

Note that N is positive when it produces tension and M is positive according to the bending-moment sign convention (positive bending moment M produces compression in the upper part of the beam; see Fig. 4-3). With these sign conventions, the sign of the normal stress σ in Eq. (5-37) is positive for tension and negative for compression, as expected.

The final stress distribution in the beam depends upon the relative algebraic values of the terms in Eq. (5-37). For our example, the three possibilities are shown in Figs. 5-38e, f, and g. If the maximum bending stress (Fig. 5-38d) is less than the axial stress due to N (Fig. 5-38c), the entire cross section will be in tension, as shown in Fig. 5-38e. If the maximum bending stress equals the axial stress, the distribution will be triangular (Fig. 5-38f) and, finally, if the bending stress is larger than the axial stress, the cross section will be partially in tension and partially in compression (Fig. 5-38g). Of course, if the axial force is a compressive force, or if the bending moment is reversed, the stress distributions will change accordingly.

Whenever bending and axial loads act simultaneously, the neutral axis (that is, the line in the cross section where the normal stress is zero) is no longer through the centroid of the cross section. As shown in

Figs. 5-38e, f, and g, the neutral axis may be outside the cross section, at the edge of the cross section, or anywhere within the section.

When analyzing a beam with axial loads, it is important to distinguish between two cases, as follows: (1) The beam is relatively short and stiff, or stocky. In this case the lateral deflections of the beam are very small compared to the length, and the presence of the deflections produces only an insignificant change in the line of action of the axial forces. As a consequence, the bending moment M does not depend upon the deflection. (2) The beam is relatively long and flexible, or slender. Then the bending deflections (even though small in magnitude) may be large enough to alter significantly the line of action of the axial forces. Then an additional bending moment, equal to the product of the axial force and the deflection, is created at every cross section. In other words, there is an interaction, or coupling, between the axial effects and the bending effects. This second type of behavior is discussed in Chapter 9 on columns. In this section, we are considering only beams that are relatively stiff and that meet the first set of conditions.

The distinction between a stocky beam and a slender beam is obviously not a precise one. In general, the only way to know whether interaction effects are important is to analyze the beam with and without the interaction effects and then notice whether the results differ significantly. However, this procedure may require considerable calculating effort. As a guideline for practical use, we may usually consider a beam with a length-to-depth ratio of 10 or less to be a stocky beam.

Example

A simple beam AB having a rectangular cross section (width b and height h) and span length L is loaded by a force P acting at the end of an arm of length a (see Fig. 5-39a). Determine the maximum tensile and compressive stresses in the beam.

We begin by determining the reactions of the beam, which are shown in the free-body diagram of Fig. 5-39b. Next, we construct axial-force and bending-moment diagrams (Figs. 5-39c and d). The axial-force diagram shows that the axial force in the left-hand half of the beam is compressive and equal to P. The bending-moment diagram shows that the maximum positive moment occurs just to the left of midspan and the maximum negative moment just to the right. At a cross section located a small distance to the left of the midpoint of the beam, the axial force and the bending moment are, respectively,

$$N = -P \qquad M = \frac{Pa}{2}$$

Next, we find from Eq. (5-37) that the stresses at the bottom and top edges of the beam ($y = h/2$ and $y = -h/2$, respectively) are as follows:

$$\sigma_b = -\frac{P}{bh} + \frac{3Pa}{bh^2} \qquad \sigma_t = -\frac{P}{bh} - \frac{3Pa}{bh^2}$$

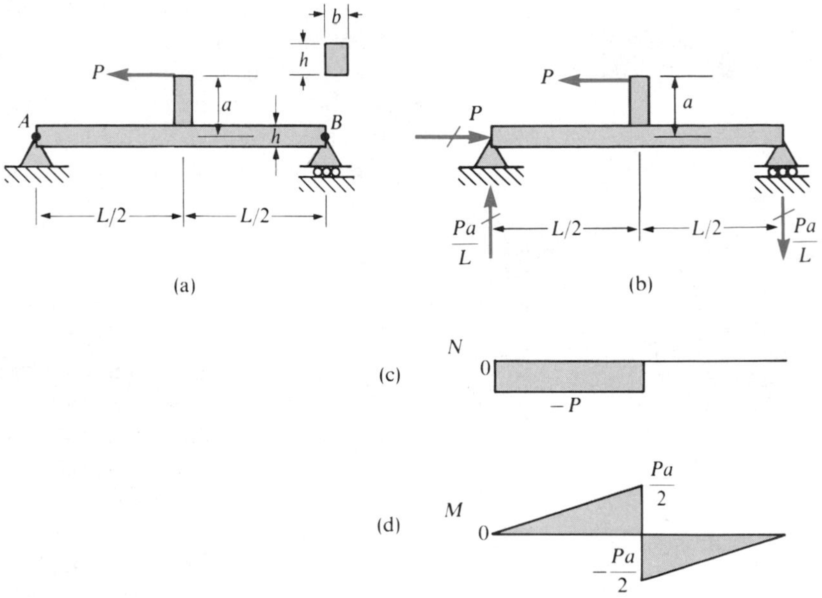

Fig. 5-39 Example

Similarly, the stresses at a cross section just to the right of the center of the beam (where $M = -Pa/2$ and $N = 0$) are

$$\sigma_b = -\frac{3Pa}{bh^2} \qquad \sigma_t = \frac{3Pa}{bh^2}$$

Upon comparing these four stresses, we see that the maximum tensile stress in the beam occurs at the top to the right of the midpoint and the maximum compressive stress occurs at the top to the left of the midpoint. These stresses are, respectively,

$$\sigma_{\text{tens}} = \frac{3Pa}{bh^2} \qquad \sigma_{\text{comp}} = -\frac{P}{bh} - \frac{3Pa}{bh^2}$$

In this example, the compressive stress is numerically larger than the tensile stress.

ECCENTRIC AXIAL LOADS

An important case of practical interest occurs when a bar is subjected to an axial load applied eccentrically, as illustrated in Fig. 5-40. In this example, the tensile load P acts normal to the end cross section at a distance e from the z axis, which is a principal axis through the centroid C (Fig. 5-40b). As in previous discussions, the y axis is an axis of symmetry.

The eccentric load P is statically equivalent to a force P applied at the centroid plus a couple Pe. Therefore, the normal stress at any point in a cross section (from Eq. 5-37) is

$$\sigma = \frac{P}{A} + \frac{Pey}{I} \qquad (5\text{-}38)$$

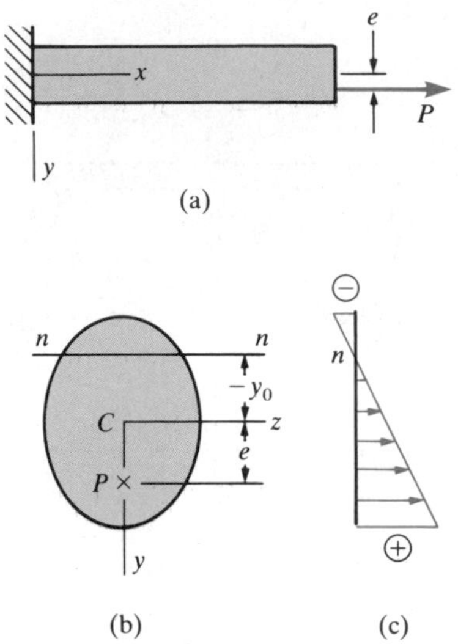

Fig. 5-40 Bar subjected to an eccentric axial force *P*

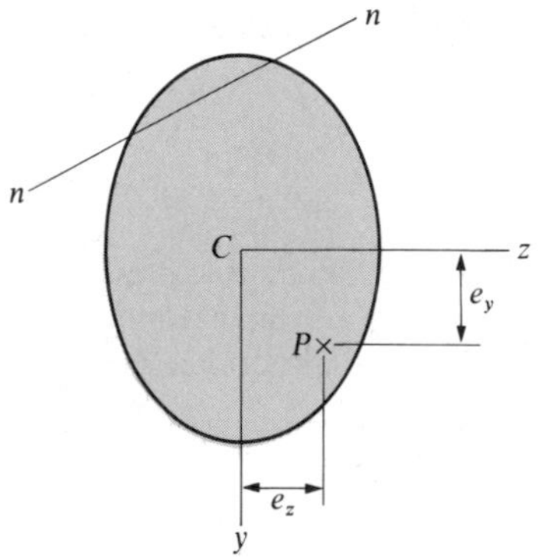

Fig. 5-41 Eccentric axial force *P* producing bending about both centroidal principal axes

This stress distribution is shown in Fig. 5-40c. If the axial load is compressive, the value of P in Eq. (5-38) is negative.

The line in the cross section where the stress is equal to zero (line *nn* in Fig. 5-40b) is the neutral axis of the beam. The position of this line can be obtained from Eq. (5-38) by setting the normal stress σ equal to zero and then solving for the coordinate y, which we now denote as y_0. The result is

$$y_0 = -\frac{I}{Ae} \tag{5-39}$$

The coordinate y_0 is measured from the z axis (which is the neutral axis under pure bending) to the line *nn* of zero stress (the neutral axis under combined bending and axial load). Because y and y_0 are positive downward in Fig. 5-40, the minus sign in Eq. (5-39) means that the neutral axis lies above the z axis. (Note that the load P is tensile and that the eccentricity e is assumed to be positive when P acts below the z axis.) If the eccentricity e is increased, the neutral axis will move toward the centroid; if e is reduced, the neutral axis will move away from the centroid.

When the point of application of the eccentric load P is not on one of the principal axes of the cross section, there will be simultaneous bending about both centroidal principal axes. Denoting the coordinates of the point of application of P by e_y and e_z (Fig. 5-41), we see that the bending moments about the y and z axes are numerically equal to Pe_z and Pe_y, respectively. The resultant normal stress σ at any point in the cross section (a point defined by coordinates y and z) then becomes

$$\sigma = \frac{P}{A} + \frac{Pe_z z}{I_y} + \frac{Pe_y y}{I_z} \tag{5-40}$$

where I_y and I_z are the moments of inertia about the y and z axes, respectively. In Eq. (5-40), the axial force P is positive if it is tensile, and e_y and e_z are positive in the coordinate directions shown in Fig. 5-41. Equation (5-40) reduces to Eq. (5-38) when P lies on the y axis and e_z equals zero.

The equation of the neutral axis can be found by setting the normal stress σ equal to zero in Eq. (5-40). With the corresponding coordinates denoted as y_0 and z_0, that equation becomes

$$\frac{Ae_y}{I_z} y_0 + \frac{Ae_z}{I_y} z_0 + 1 = 0 \tag{5-41}$$

This equation is linear in y_0 and z_0, and therefore the neutral axis is a straight line, such as line *nn* in Fig. 5-41. The intercepts of line *nn* with the y and z axes can be found by setting z_0 and y_0, respectively, equal to zero in Eq. (5-41) and solving for the intercepts. Then the neutral axis can be drawn as a straight line through those two points. The neutral axis may or may not intersect the cross section, depending upon the shape of the cross section and the position of the load P.

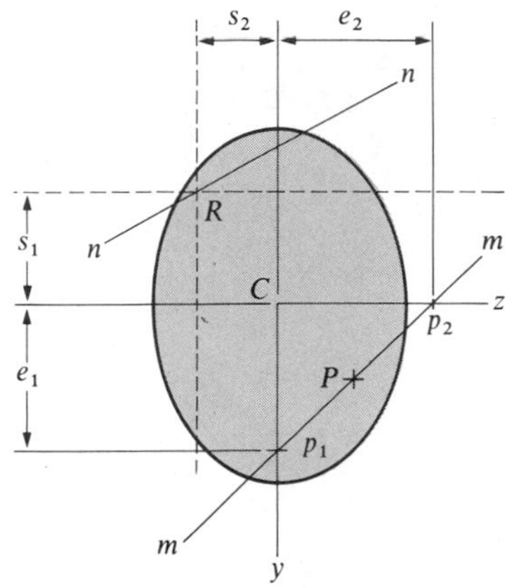

Fig. 5-42 Relationship between position of load P and neutral axis nn

An interesting relationship exists between the point of application of the eccentric axial force P and the position of the neutral axis nn, as follows. If the force P moves along any straight line mm (Fig. 5-42), the neutral axis rotates about a fixed point R. To demonstrate this fact, we observe first that the force P can be resolved into two parallel components, one acting at p_1 and the other at p_2. The component at p_1 acts in a principal plane of bending; hence the corresponding line of zero stress is parallel to the z axis and is located at distance $s_1 = I_z/Ae_1$ from the z axis (see Fig. 5-42 and Eq. 5-39). Similarly, the component at p_2 produces bending about the y axis, and the line of zero stress is located at distance $s_2 = I_y/Ae_2$ from the y axis. Point R, at the intersection of the two dashed lines in the figure, will always be on the neutral axis nn when both components of load act simultaneously. As the load P moves along the line mm, point R remains fixed in position, and the neutral axis always passes through it. We will now use this relationship to establish the core of a rectangular section.

THE CORE OF A CROSS SECTION

When the eccentricity e of the axial load P (Fig. 5-40) is small, the neutral axis will lie outside the cross section and the normal stresses will have the same sign throughout the cross section. A condition of this kind is important, for instance, when a compressive load acts on a material that is very weak in tension, such as glass, concrete, stone, and ceramic materials. For such materials, it may be necessary to ensure that the load produces no tension at any point of the cross section. This condition exists if the load remains within a certain small region surrounding the centroid. A compressive force acting within that region produces compression over the entire cross section, and a tensile force acting within that region produces tension over the entire cross section. This region is called the **core** (or the **kern**) of the section.*

The core of a rectangular cross section (Fig. 5-43a) can be found in the following manner. If the load lies along the positive y axis, the neutral axis nn will coincide with the upper edge of the section when the load is at point p, a distance e_1 from the centroid. The distance e_1 can be found from Eq. (5-39) by substituting $y_0 = -h/2$, $I = bh^3/12$, and $A = bh$; thus, $e_1 = h/6$. Similarly, the neutral axis coincides with the left-hand edge of the section when the load P acts on the positive z axis at point q, a distance $e_2 = b/6$ from the centroid. As the load moves along a straight line between points p and q, the neutral axis will rotate about point R at the corner of the rectangular cross section. Hence, line pq is one of the sides of the core; the other three sides can be located by symmetry. We see that the core is a rhombus with diagonals of lengths $b/3$ and $h/3$ (Fig. 5-43b).

* The concept of the core of a cross section was introduced by the French engineer J. A. C. Bresse in 1854; see Ref. 5-15.

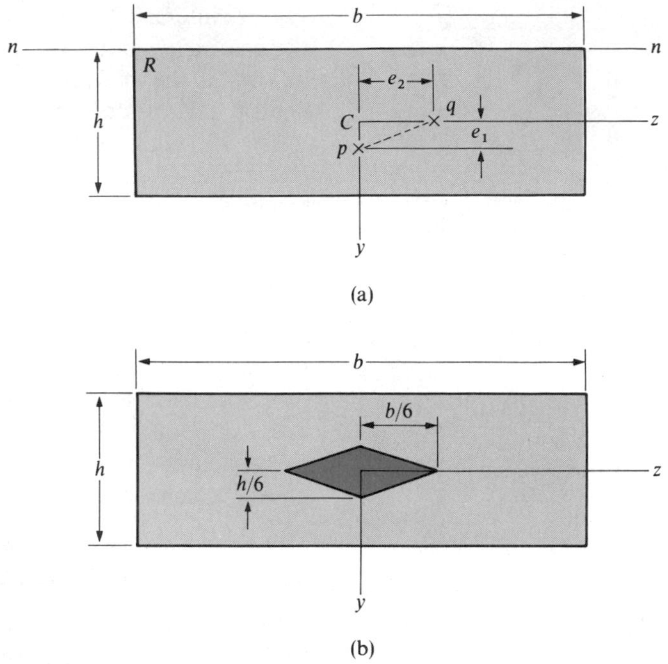

Fig. 5-43 The core of a rectangular cross section

If the point of application of a compressive load P is within this rhombus, the neutral axis will not intersect the cross section and the entire section will be in compression. The core of other cross-sectional shapes can be found by similar techniques.

*5.10 STRESS CONCENTRATIONS IN BENDING

The flexure and shear formulas discussed in earlier sections of this chapter are valid for beams without holes, notches, or other abrupt changes in dimensions. Whenever such discontinuities exist, high localized stresses are produced. These *stress concentrations* can be extremely important when the member is subjected to dynamic loads or when the material is brittle, as discussed previously in Sections 2.9, 2.10, and 3.10.

For illustrative purposes, two cases of stress concentrations in beams are described in this section. The first case is a beam of rectangular cross section with a hole through it at the neutral axis (Fig. 5-44). The beam has height h and width b (perpendicular to the plane of the figure) and is in pure bending under the action of bending moments M. When the diameter d of the hole is small compared to the height h, the stress distribution on the cross section through the hole is approximately as shown by the diagram in Fig. 5-44a. At point B on the edge of the hole, the actual stress is much larger than the stress that would exist at that point if

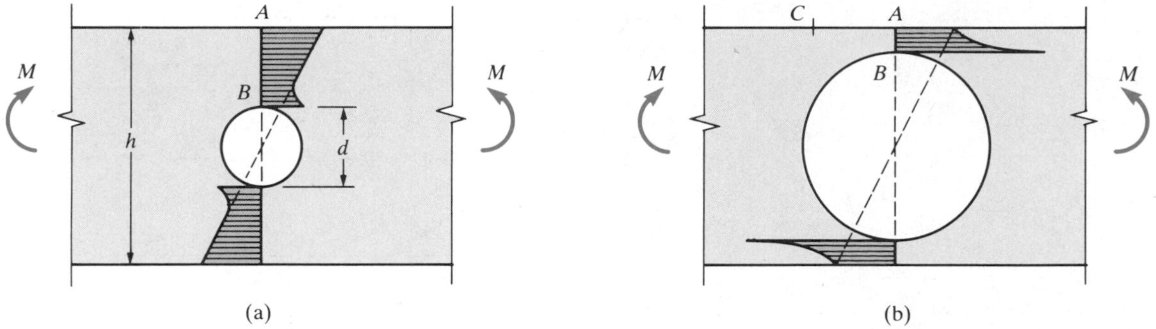

Fig. 5-44 Stress distributions in a beam in pure bending (M = bending moment) with a circular hole (d = diameter) through the beam at the neutral axis. The beam has a rectangular cross section with height h and width b (perpendicular to the plane of the figure).

the hole were not present. (The dashed line in the figure shows the stress distribution with no hole.) However, near the outer edges of the beam (point A), the stress distribution varies linearly with distance from the neutral axis and is only slightly affected by the presence of the hole.

When the hole is relatively large, the stress pattern is approximately as shown in Fig. 5-44b. There is a large increase in stress at point B and a small change in stress at point A as compared to the stress distribution in the beam without a hole (again shown by the dashed line). The stress at point C is larger than the stress at A, although it is smaller than the stress at B.

Extensive investigations have shown that the stress at the edge of the hole (point B) is approximately twice the *nominal stress* at that point. The nominal stress is the stress calculated from the flexure formula $\sigma = My/I$, in which y is the distance $d/2$ from the neutral axis to point B and I is the moment of inertia of the net cross section (the cross section through the hole). Thus, we have the following approximate formula for the stress at point B:

$$\sigma_b \approx 2\frac{My}{I} = \frac{12Md}{b(h^3 - d^3)} \tag{5-42}$$

At the outer edge of the beam (point C), the stress is approximately equal to the nominal stress at point A (where $y = h/2$):

$$\sigma_c \approx \frac{My}{I} = \frac{6Mh}{b(h^3 - d^3)} \tag{5-43}$$

From Eqs. (5-42) and (5-43) we see that when the ratio d/h of hole

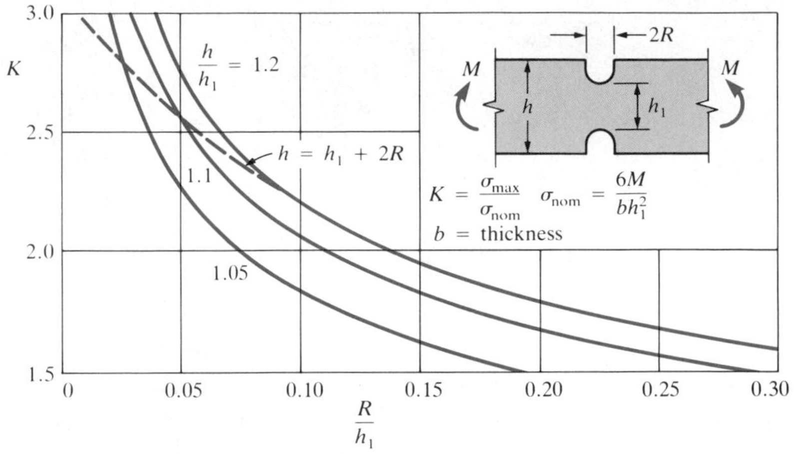

Fig. 5-45 Stress-concentration factor K for a notched beam of rectangular cross section in pure bending (h = height of beam; b = width of beam, perpendicular to the plane of the figure). The dashed line is for semicircular notches ($h = h_1 + 2R$).

diameter to height of beam exceeds 1/2, the largest stress is at B, and when d/h is less than 1/2, the largest stress is at C.

Another situation of practical interest is a rectangular beam with notches (Fig. 5-45). The beam shown in the figure is subjected to pure bending and has height h and width b (perpendicular to the plane of the figure). Also, the net height of the beam (that is, the distance between the bases of the notches) is h_1 and the radius at the base of each notch is R. The maximum stress in this beam occurs at the base of the notches and may be much larger than the nominal stress at that same point. The nominal stress is calculated from the flexure formula with $y = h_1/2$ and $I = bh_1^3/12$; thus,

$$\sigma_{nom} = \frac{My}{I} = \frac{6M}{bh_1^2} \tag{5-44}$$

The maximum stress is equal to the stress-concentration factor K times the nominal stress:

$$\sigma_{max} = K\sigma_{nom} \tag{5-45}$$

The stress-concentration factor K is plotted in Fig. 5-45 for a few values of the ratio h/h_1. Note that when the notch becomes "sharper," that is, thc ratio R/h_1 becomes smaller, the stress-concentration factor increases.*

The effects of the stress concentrations are confined to small regions around the holes and notches, as explained in the discussion of Saint-Venant's principle in Section 2.10. At a distance equal to h or greater from the hole or notch, the stress-concentration effect is negligible and the ordinary formulas for stresses may be used.

* Figure 5-45 is plotted from the formulas given in Ref. 2-14. This reference also contains formulas for stress-concentration factors for many other cases of bending.

*5.11 COMPOSITE BEAMS

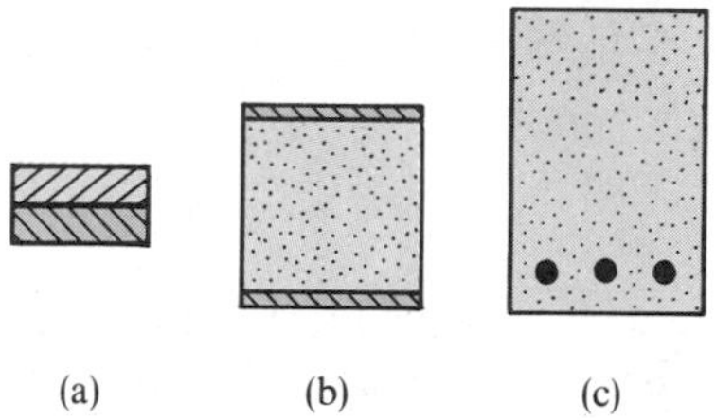

Fig. 5-46 Cross sections of composite beams: (a) bimetallic beam, (b) sandwich beam, and (c) reinforced concrete beam

Beams that are fabricated of more than one material are called **composite beams**. Examples are bimetallic beams, sandwich beams, and reinforced concrete beams, as shown in Fig. 5-46. Composite beams may be analyzed by the same bending theory we used for ordinary beams (see Sections 5.2 and 5.3), because the assumption that plane cross sections before bending remain plane after bending (see Fig. 5-6) is valid in pure bending regardless of the material. It follows from this assumption that the longitudinal strains ϵ_x vary linearly from top to bottom of the beam (Eq. 5-2). This strain distribution is shown in Fig. 5-47b for a composite beam made of two different materials, labeled materials 1 and 2 in the cross-sectional sketch (Fig. 5-47a). The position of the neutral axis for this beam is *not* at the centroid of the cross-sectional area, as explained later.

The normal stresses σ_x acting on the cross section can be obtained from the strains ϵ_x by using the stress-strain relationships for the materials. Let us assume that the materials behave in a linearly elastic manner so that Hooke's law for uniaxial stress is valid. Then the stresses in each material are obtained by multiplying the strains by the appropriate modulus of elasticity. Denoting the moduli of elasticity for materials 1 and 2 as E_1 and E_2, respectively, and also assuming that $E_2 > E_1$, we obtain the stress diagram shown in Fig. 5-47c. The normal stresses σ_x at distance y from the neutral axis are given by the following equations (compare with Eq. 5-5):

$$\sigma_{x_1} = -E_1 \kappa y \qquad \sigma_{x_2} = -E_2 \kappa y \tag{5-46a,b}$$

in which σ_{x_1} is the stress in material 1 and σ_{x_2} is the stress in material 2.

The position of the **neutral axis** can be found by using the condition that the resultant axial force acting on the cross section is zero; therefore,

$$\int_1 \sigma_{x_1}\, dA + \int_2 \sigma_{x_2}\, dA = 0$$

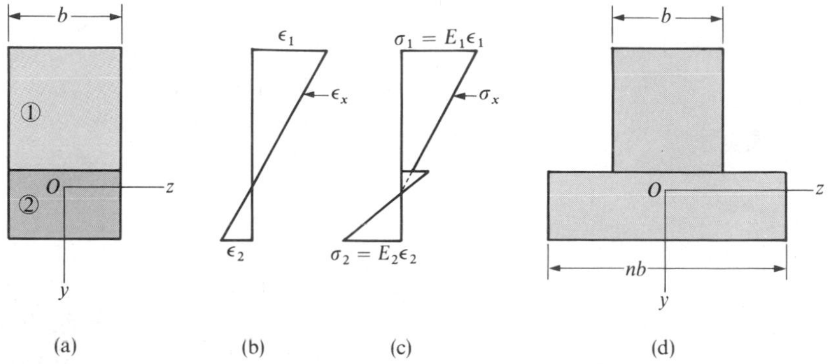

Fig. 5-47 Composite beam of two materials: (a) cross section, (b) strain distribution, (c) stress distribution, and (d) transformed section

where it is understood that the first integral is evaluated over the cross-sectional area of material 1 and the second integral is evaluated over the cross-sectional area of material 2. Replacing σ_{x_1} and σ_{x_2} in the preceding equation by their expressions from Eqs. (5-46a and b), we get

$$E_1 \int_1 y\,dA + E_2 \int_2 y\,dA = 0 \tag{5-47}$$

This equation, which may be considered as a generalized form of Eq. (5-6), can be used to locate the neutral axis for a beam of two materials. (If there are more than two materials, additional terms are required in Eq. 5-47.) The integrals in the equation represent the first moments of the two parts of the cross-sectional area with respect to the neutral axis. The use of the equation is illustrated later in a numerical example. Of course, if the cross section is doubly symmetric (as in the case of the sandwich beam of Fig. 5-46b), the neutral axis is located at the midheight of the cross section.

The relationships between the bending moment M and the stresses in the beam may be found by the same procedures used to obtain the flexure formula (see Eqs. 5-7 through 5-10). The derivation is as follows:

$$\begin{aligned} M &= \int \sigma_x y\,dA = \int_1 \sigma_{x_1} y\,dA + \int_2 \sigma_{x_2} y\,dA \\ &= -\kappa E_1 \int_1 y^2\,dA - \kappa E_2 \int_2 y^2\,dA \\ &= -\kappa(E_1 I_1 + E_2 I_2) \end{aligned} \tag{5-48}$$

where I_1 and I_2 are the moments of inertia about the neutral axis of cross-sectional areas 1 and 2, respectively. Note that $I = I_1 + I_2$, where I is the moment of inertia of the entire cross-sectional area about the neutral axis. Equation (5-48) can be solved for the curvature:

$$\kappa = \frac{1}{\rho} = -\frac{M}{E_1 I_1 + E_2 I_2} \tag{5-49}$$

The denominator on the right-hand side may be considered as the flexural rigidity of the composite beam.

The stresses in the beam are now obtained by substituting the expression for curvature (Eq. 5-49) into the expressions for σ_{x_1} and σ_{x_2} (Eqs. 5-46a and b); thus, we find

$$\sigma_{x_1} = \frac{MyE_1}{E_1 I_1 + E_2 I_2} \qquad \sigma_{x_2} = \frac{MyE_2}{E_1 I_1 + E_2 I_2} \tag{5-50a,b}$$

These expressions, known as the **flexure formulas for a composite beam**, give the normal stresses in materials 1 and 2, respectively. Of course, if $E_1 = E_2 = E$, both equations reduce to the flexure formula for a beam of one material.

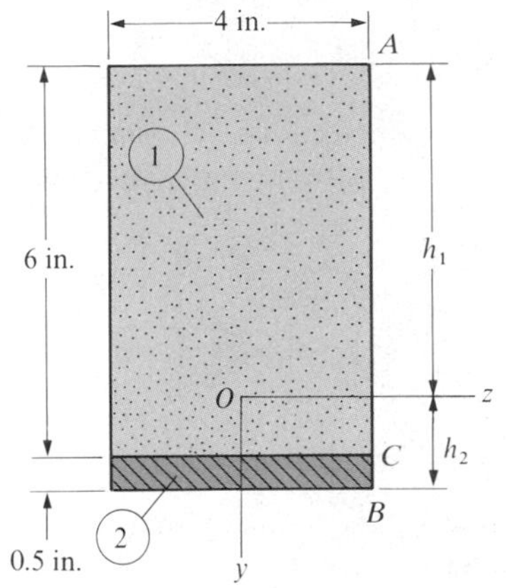

Fig. 5-48 Example 1. Composite beam

Example 1

A composite beam having the cross-sectional dimensions shown in Fig. 5-48 is subjected to a positive bending moment $M = 30{,}000$ in.-lb. Calculate the maximum and minimum stresses in both materials of the beam assuming $E_1 = 1{,}000{,}000$ psi and $E_2 = 20{,}000{,}000$ psi.

The first step in the analysis is to locate the neutral axis of the cross section. Let us assume that the neutral axis lies within material 1, as shown in the figure, and let us denote the distances from this axis to the top and bottom of the beam as h_1 and h_2, respectively. To obtain the distances h_1 and h_2, we use Eq. (5-47). The integrals in that equation may be evaluated by taking the first moments of cross-sectional areas 1 and 2 about the z axis, as follows:

$$\int_1 y\,dA = -\frac{h_1}{2}(h_1)(4\text{ in.}) + \frac{6\text{ in.} - h_1}{2}(6\text{ in.} - h_1)(4\text{ in.})$$

$$= (24\text{ in.}^2)(3\text{ in.} - h_1)$$

$$\int_2 y\,dA = (6.25\text{ in.} - h_1)(0.5\text{ in.})(4\text{ in.}) = (2\text{ in.}^2)(6.25\text{ in.} - h_1)$$

in which h_1 has units of inches. Substituting these expressions into Eq. (5-47) gives

$$(1 \times 10^6\text{ psi})(24\text{ in.}^2)(3\text{ in.} - h_1) + (20 \times 10^6\text{ psi})(2\text{ in.}^2)(6.25\text{ in.} - h_1) = 0$$

Solving this equation, we obtain

$$h_1 = 5.031\text{ in.} \qquad h_2 = 6.5\text{ in.} - h_1 = 1.469\text{ in.}$$

and the position of the neutral axis is established.

The moments of inertia I_1 and I_2 about the neutral axis can be found by using the parallel-axis theorem (see Section C.4, Appendix C). Beginning with area 1 (Fig. 5-48), we write

$$I_1 = \frac{1}{12}(4\text{ in.})(6\text{ in.})^3 + (4\text{ in.})(6\text{ in.})(h_1 - 3\text{ in.})^2 = 171.0\text{ in.}^4$$

Similarly, we also obtain

$$I_2 = \frac{1}{12}(4\text{ in.})(0.5\text{ in.})^3 + (4\text{ in.})(0.5\text{ in.})(h_2 - 0.25\text{ in.})^2 = 3.014\text{ in.}^4$$

To check these calculations for moments of inertia, we can determine the moment of inertia I of the entire cross-sectional area about the z axis as follows:

$$I = \frac{1}{3}(4\text{ in.})\,h_1^3 + \frac{1}{3}(4\text{ in.})\,h_2^3 = 169.8 + 4.2 = 174.0\text{ in.}^4$$

which is equal to the sum of I_1 and I_2.

Now we can calculate the bending stresses in materials 1 and 2 from the flexure formulas for composite beams (Eqs. 5-50a and b). The maximum compressive stress in material 1 occurs at the top of the beam (point A) where $y = -h_1 = -5.031$ in. Substituting this value into Eq. (5-50a) for the normal stress in material 1, we get

$$\sigma_{1A} = -653 \text{ psi}$$

The largest tensile stress in material 1 occurs at the contact plane between the two materials (point C) where $y = h_2 - 0.5$ in. $= 0.969$ in.; hence from Eq. (5-50a),

$$\sigma_{1C} = 126 \text{ psi}$$

Material 2 is in tension throughout. The maximum tensile stress σ_{2B} occurs at point B at the bottom of the beam ($y = h_2 = 1.469$ in.); hence, from Eq. (5-50b) we get

$$\sigma_{2B} = 3810 \text{ psi}$$

The minimum tensile stress in material 2 is obtained at the contact plane (point C); this stress equals 2514 psi. Note that the ratio of stresses at C is $2514/126 = 20$, which is equal to the ratio E_2/E_1 of the moduli.

TRANSFORMED-SECTION METHOD

The transformed-section method, which is based upon Eqs. (5-46) to (5-50), provides a convenient way to analyze a composite beam. The method is to transform the cross section, consisting of more than one material, into an equivalent cross section composed of only one material. Then the latter, called the **transformed section**, is analyzed in the usual manner for a beam of one material.

The transformed section must have the same location for the neutral axis and the same moment-resisting capacity as the original beam if it is to be equivalent. In order to see how this equivalence is accomplished, let us refer to the equation used for locating the neutral axis (Eq. 5-47). Introducing the notation

$$n = \frac{E_2}{E_1} \tag{5-51}$$

where n is the **modular ratio**, we can rewrite Eq. (5-47) in the form

$$\int_1 y\,dA + \int_2 yn\,dA = 0 \tag{5-52}$$

This equation shows that the neutral axis will have the same position in the transformed beam as in the original beam if each element of area dA in material 2 is multiplied by the factor n, provided that the distance y for each such element of area is unchanged. Therefore, we can consider the cross section as consisting of two parts: (1) area 1 with its dimensions unchanged, and (2) area 2 with its width multiplied by n. Thus, we have a new cross section consisting of only one material, namely, material 1.

The transformed section for the composite beam of Fig. 5-47a is shown in Fig. 5-47d. As explained in the preceding paragraph, material 1 remains unchanged, but the width of material 2 is multiplied by n. (We assume in this illustration that $n > 1$, but this assumption is not necessary.) The transformed section consists entirely of material 1, and its neutral axis is in the same location as the neutral axis of the original beam (Fig. 5-47a).

The bending-moment capacity of the transformed section must be the same as for the original cross section. To show that this is the case, we note that the stresses in the transformed beam, which consists of only one material, are given by Eq. (5-5):

$$\sigma_x = -E_1 \kappa y$$

Therefore, the bending moment M can be obtained as follows (see Fig. 5-47d):

$$\begin{aligned} M &= \int \sigma_x y \, dA = \int_1 \sigma_x y \, dA + \int_2 \sigma_x y \, dA \\ &= -\kappa E_1 \int_1 y^2 \, dA - \kappa E_1 \int_2 y^2 \, dA \\ &= -\kappa (E_1 I_1 + E_1 n I_2) = -\kappa (E_1 I_1 + E_2 I_2) \end{aligned}$$

This result is the same as Eq. (5-48), which demonstrates that the bending moment is unchanged between the original beam and the transformed beam.

The stresses in the transformed beam can be found by the flexure formula for a beam of one material. Thus, the stresses in the beam transformed to material 1 are

$$\sigma_{x_1} = \frac{My}{I_t} \qquad (5\text{-}53)$$

where I_t is the moment of inertia about the neutral axis of the transformed section; that is,

$$I_t = I_1 + nI_2 = I_1 + \frac{E_2}{E_1} I_2 \qquad (5\text{-}54)$$

Substituting the expression for I_t from Eq. (5-54) into Eq. (5-53) gives

$$\sigma_{x_1} = \frac{MyE_1}{E_1 I_1 + E_2 I_2} \qquad (5\text{-}55)$$

which is the same as Eq. (5-50a), thus showing that the stresses in material 1 in the original beam are the same as the stresses in the transformed beam. This conclusion results from the fact that we transformed the beam to material 1. However, the stresses in material 2 in the original beam are not the same as the stresses in the corresponding part of the transformed beam. Instead, the stresses in the transformed beam (Eq. 5-55) must be multiplied by the modular ratio n to obtain the stresses in material 2 of the original beam (see Eq. 5-50b).

We may also transform the original beam to a beam consisting entirely of material 2. In that case, the stresses in the original beam in material 2 are the same as the stresses in the corresponding parts of the transformed beam. However, the stresses in material 1 are obtained by multiplying the stresses in the transformed beam by the modular ratio n, which in this case is $n = E_1/E_2$.

The transformed-section method may be extended to composite beams of more than two materials. Also, it is possible to transform the beam into a material having any arbitrary value of E, in which case all parts of the beam must be transformed to the fictitious material. Of course, it is usually simpler to transform to one of the original materials, the choice among them being arbitrary.

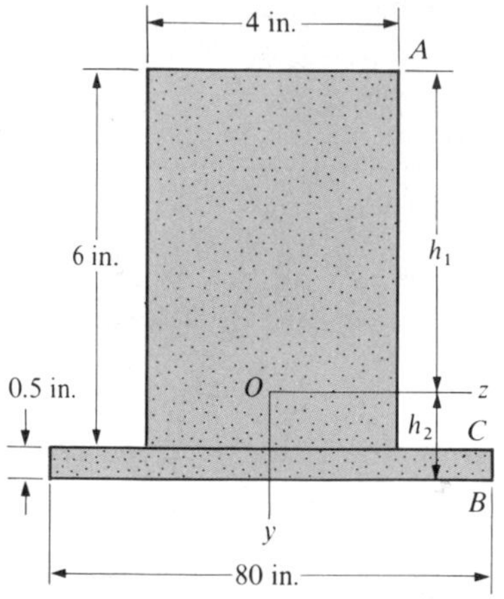

Fig. 5-49 Example 2. Composite beam of Example 1 as transformed to material 1

Example 2

The composite beam described in Example 1 and pictured in Fig. 5-48 is to be analyzed by the transformed-section method. Let us transform the original beam into a beam of material 1 (Fig. 5-49). Therefore, the upper part of the beam is not altered, but the lower part has its width multiplied by the modular ratio, which is

$$n = \frac{E_2}{E_1} = \frac{20{,}000{,}000 \text{ psi}}{1{,}000{,}000 \text{ psi}} = 20$$

Thus, the width of part 2 becomes 80 in. in the transformed section.

Because the transformed beam is of one material, the neutral axis must pass through the centroid of the cross-sectional area. Taking the top edge of the cross section as a reference line, we calculate the centroidal distance h_1 as follows:

$$h_1 = \frac{\sum y_i A_i}{\sum A_i} = \frac{(3 \text{ in.})(4 \text{ in.})(6 \text{ in.}) + (6.25 \text{ in.})(80 \text{ in.})(0.5 \text{ in.})}{(4 \text{ in.})(6 \text{ in.}) + (80 \text{ in.})(0.5 \text{ in.})}$$

$$= \frac{322 \text{ in.}^3}{64 \text{ in.}^2} = 5.031 \text{ in.}$$

Also, the distance h_2 is

$$h_2 = 6.5 \text{ in.} - h_1 = 1.469 \text{ in.}$$

The position of the neutral axis is now determined.

Utilizing the parallel-axis theorem, we calculate the moment of inertia I_t of the entire cross-sectional area about the neutral axis as follows:

$$I_t = \frac{1}{12}(4 \text{ in.})(6 \text{ in.})^3 + (4 \text{ in.})(6 \text{ in.})(h_1 - 3 \text{ in.})^2$$

$$+ \frac{1}{12}(80 \text{ in.})(0.5 \text{ in.})^3 + (80 \text{ in.})(0.5 \text{ in.})(h_2 - 0.25 \text{ in.})^2$$

$$= 231.3 \text{ in.}^4$$

The flexure formula is now used to calculate the stresses in the transformed beam at the top (point A), at the contact of the two parts (point C), and at the bottom (point B), respectively:

$$\sigma_a = \frac{My}{I_t} = \frac{(30{,}000 \text{ in.-lb})(-5.031 \text{ in.})}{231.3 \text{ in.}^4} = -652.5 \text{ psi}$$

$$\sigma_c = \frac{My}{I_t} = \frac{(30{,}000 \text{ in.-lb})(0.969 \text{ in.})}{231.3 \text{ in.}^4} = 125.7 \text{ psi}$$

$$\sigma_b = \frac{My}{I_t} = \frac{(30{,}000 \text{ in.-lb})(1.469 \text{ in.})}{231.3 \text{ in.}^4} = 190.5 \text{ psi}$$

The stresses in the original beam are the same as in the transformed beam for material 1; therefore, the maximum compressive stress in material 1 is

$$\sigma_{1A} = -653 \text{ psi}$$

at the top of the beam. Also, the largest tensile stress in material 1 (at the contact surface) is

$$\sigma_{1C} = 126 \text{ psi}$$

For material 2, we multiply the stresses in the transformed beam by n. The maximum tension occurs at the bottom of the beam:

$$\sigma_{2B} = n(190.5 \text{ psi}) = 20(190.5 \text{ psi}) = 3810 \text{ psi}$$

Also, the tension in material 2 at the contact plane is 20(125.7 psi) or 2514 psi. All of these results agree with those found in Example 1.

SANDWICH BEAMS

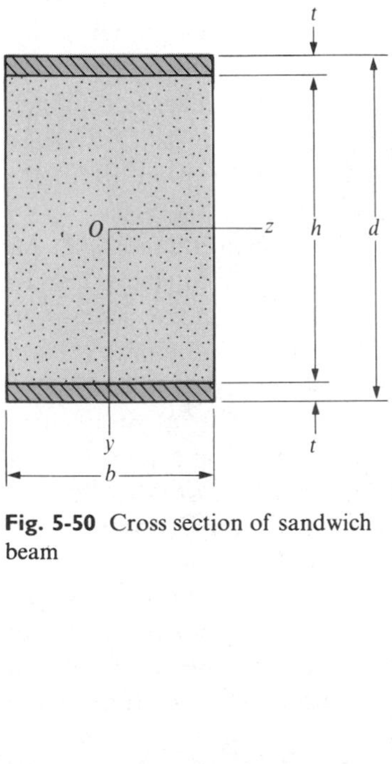

Fig. 5-50 Cross section of sandwich beam

A sandwich beam consists of two thin layers of material, called the **faces**, on either side of a thick **core** (see Fig. 5-50). The core is usually of lightweight, low-strength material serving primarily as a filler or spacer, whereas the faces are high-strength material. Sandwich construction is used where light weight combined with high strength and high stiffness are needed; examples are skis and aircraft parts.

Sandwich beams can be analyzed for flexure in the manner just described for beams of two materials. However, an approximate theory for bending of sandwich beams has been developed on the assumption that the faces carry all of the longitudinal bending stresses. This assumption is reasonable when the core has a low modulus of elasticity in the longitudinal direction relative to that of the faces. Under these conditions, the normal stress at the outermost edges of the beam (see Fig. 5-50) are

$$\sigma_x = \frac{Md}{2I_f} \tag{5-56}$$

where d is the depth of the beam and I_f is the moment of inertia of the faces about the z axis, as follows:

$$I_f = \frac{b}{12}(d^3 - h^3) \tag{5-57}$$

If the faces are thin, we can assume that the core carries all of the shear stresses. Therefore, the average shear stress and shear strain in the core are, respectively,

$$\tau = \frac{V}{bh} \qquad \gamma = \frac{V}{bhG_c} \tag{5-58a,b}$$

where h is the height of the core, V is the shear force, and G_c is the shear modulus for the core material.

More accurate theories of bending of sandwich beams have been developed; for instance, bending of the faces as individual beams can be taken into account. For detailed treatments of sandwich beams and other forms of composite construction, consult textbooks devoted entirely to those subjects.

*5.12 BEAMS WITH SKEW LOADS

The bending theories developed in preceding sections apply to beams having a plane of symmetry (the xy plane) through the longitudinal axis. In addition, we assumed that all lateral loads acted in the same plane, which then became the plane of bending. Sometimes the loads on a beam act in directions that are inclined to the plane of symmetry, as illustrated in Fig. 5-51a by a cantilever beam subjected to a skew load P at the free end. The simplest case of this kind arises when the beam is doubly symmetric, that is, both the xy and xz planes are planes of symmetry. Then we can analyze the beam by resolving the load into two components, one in each plane of symmetry. A bending analysis can be performed for each load component acting separately, and the final stresses can be obtained by superimposing the results. Of course, the load P must act through the centroid of the cross section if there is to be no twisting about the longitudinal axis (see Section 5.13).

Beams supporting skew loads are subjected to bending moments M_y and M_z acting about both principal axes of the cross section (axes y and z in Fig. 5-51), and the analysis is facilitated if we establish a consistent sign convention for these moments, as follows. The bending moments M_y and M_z will be represented vectorially by double-headed arrows (Fig. 5-51b), and the moments are positive when their vectors point in the positive directions of the y and z axes, respectively. The right-hand rule of vectors give the sense of the moment (that is, the direction of rotation of the moment). This means that a positive moment M_y produces compression on the left-hand side of the beam in Fig. 5-51 and tension on the right-hand side. Similarly, a positive moment M_z produces compression on the bottom and tension on the top. Note that the bending moments shown in Fig. 5-51b act on the positive x face of a portion of the beam, that is, on a cross section with its outward normal in the positive direction of the x axis.

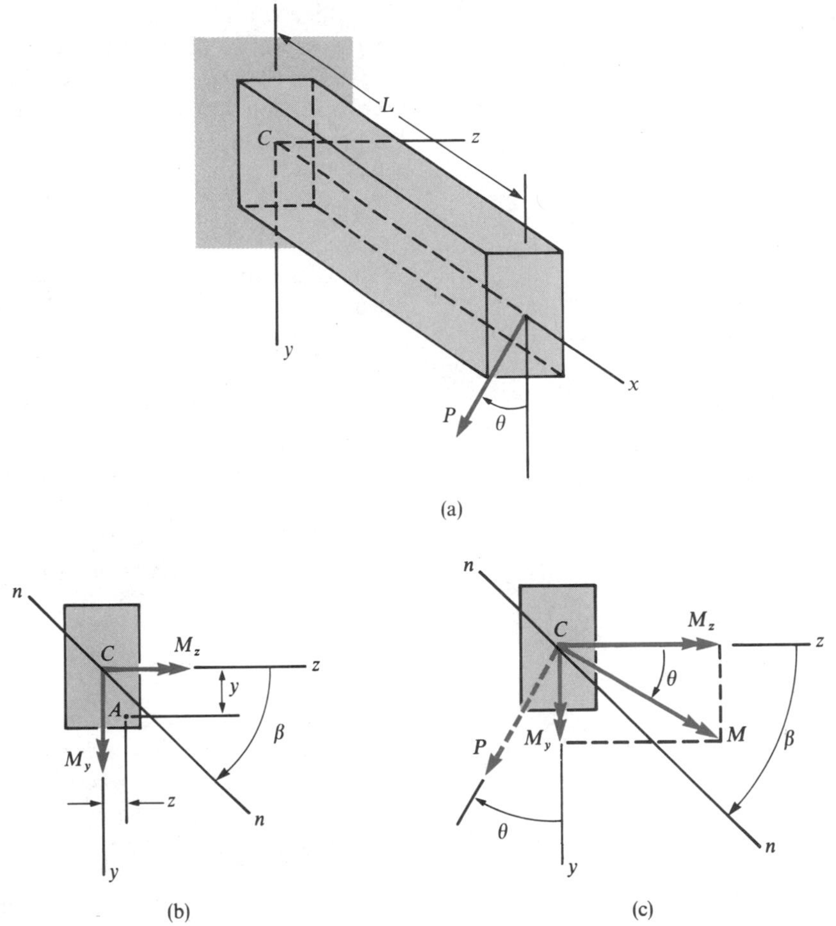

Fig. 5-51 Doubly symmetric beam with a skew load

We will now derive formulas for the stresses in the **doubly symmetric cantilever beam** of Fig. 5-51. The components of the load P are $P\cos\theta$ in the positive y direction and $P\sin\theta$ in the negative z direction. Therefore, the bending moments acting on a cross section (Fig. 5-51b) located at distance x from the fixed support are

$$M_z = (P\cos\theta)(L - x) \qquad M_y = (P\sin\theta)(L - x) \tag{a}$$

in which L is the length of the beam.

Inasmuch as the bending moments M_y and M_z act in the planes of symmetry of the beam, the bending stresses can be obtained from the flexure formula. Let us consider any point A in the cross section with coordinates y and z (Fig. 5-51b). The **normal stress** at this point is

$$\sigma_x = \frac{M_y z}{I_y} - \frac{M_z y}{I_z} \tag{5-59}$$

in which I_y and I_z are the moments of inertia of the cross-sectional area with respect to the y and z axes, respectively.

The equation of the line of zero stress (that is, the **neutral axis** of the cross section when the beam is bent by the skew load) can be determined by equating the stress σ_x (Eq. 5-59) to zero:

$$\frac{M_y}{I_y}z - \frac{M_z}{I_z}y = 0 \tag{5-60}$$

This equation shows that the neutral axis nn (Fig. 5-51b) is an inclined line through the centroid C. The angle β between this line and the z axis is determined as follows:

$$\tan\beta = \frac{y}{z} = \frac{M_y I_z}{M_z I_y} \tag{5-61}$$

Depending upon the magnitudes and directions of the bending moments, the angle β may vary from $-90°$ to $+90°$. The maximum normal stresses occur at points located farthest from the neutral axis.

For the cantilever beam of Fig. 5-51, the bending moments are given by Eqs. (a) and have the ratio

$$\frac{M_y}{M_z} = \tan\theta \tag{b}$$

which shows that the resultant moment vector M is at the angle θ with the z axis (Fig. 5-51c). This vector is perpendicular to the longitudinal plane containing the force P, which is shown by a dashed arrow in Fig. 5-51c. Substituting from Eq. (b) into Eq. (5-61) gives the following equation for the angle β locating the neutral axis:

$$\tan\beta = \frac{M_y I_z}{M_z I_y} = \frac{I_z}{I_y}\tan\theta \tag{5-62}$$

From this equation we see that the angle β is generally not equal to θ, which means that the neutral axis is not perpendicular to the plane of the load. Exceptions occur in three special cases: (1) If $\theta = 0$, the load lies in the xy plane, and the z axis ($\beta = 0$) is the neutral axis. (2) If $\theta = 90°$, the load lies in the xz plane, and the y axis ($\beta = 90°$) is the neutral axis. (3) If $I_z = I_y$, the principal moments of inertia are equal (as for a square or circular cross section) and all axes through the centroid are principal axes with the same moment of inertia. Under these conditions, the plane of loading, no matter what its direction, is always a principal plane, and the neutral axis is always perpendicular to it ($\beta = \theta$).

The fact that the neutral axis is not necessarily perpendicular to the plane of the load can greatly affect the stresses in the beam, especially if the ratio of the principal moments of inertia is very large. Then the behavior of the beam may be very sensitive to slight changes in the direction of the load or in the alignment of the beam itself. This characteristic is illustrated in the following example.

Example

A certain cantilever beam of I-shape is very narrow compared to its height (Fig. 5-52). The beam is an S 24 × 80 section (see Appendix E) with length L = 12 feet. A load P = 10 k acts in the vertical direction at the free end of the beam.

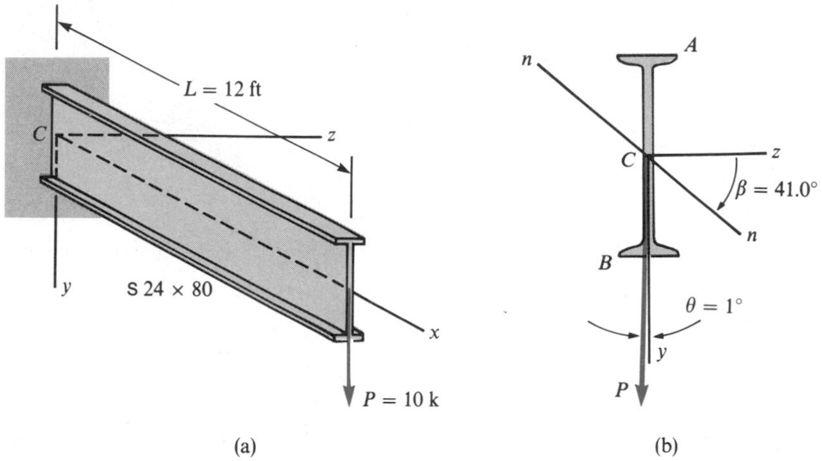

Fig. 5-52 Example. Cantilever beam with moment of inertia I_z much larger than I_y

If this beam is constructed in perfect alignment, the load P will act in the direction of the y axis and the z axis will be the neutral axis. Then the maximum stress in the beam (at the fixed end) is obtained from the flexure formula:

$$\sigma_{max} = \frac{Mc}{I_z} = \frac{PLc}{I_z} = \frac{(10\text{ k})(12\text{ ft})(12.00\text{ in.})}{2100\text{ in.}^4} = 8230\text{ psi}$$

This stress is tensile at the top of the beam and compressive at the bottom of the beam.

Now let us suppose that the beam is constructed with a very slight inclination (Fig. 5-52b), so that the angle between the load P and the y axis is $\theta = 1°$. This inclination might be caused by defects in manufacture, errors in construction, or a small movement of the supporting structure. The angle β giving the orientation of the neutral axis nn is obtained from Eq. (5-62):

$$\tan\beta = \frac{I_z}{I_y}\tan\theta = \frac{2100\text{ in.}^4}{42.2\text{ in.}^4}\tan 1° = 0.8686$$

and $\beta = 41.0°$. This calculation shows that the neutral axis is inclined 41° from the z axis even though the plane of the load is inclined only 1° from the y axis. The great sensitivity of the position of the neutral axis to the angle of the load is a consequence of the large I_z/I_y ratio.

The maximum stresses in the beam occur at the support at points A and B (Fig. 5-52b), located at the farthest distances from the neutral axis. At the support, the bending moments (from Eqs. a) are

$$M_z = (P\cos\theta)L = (10\text{ k})(\cos 1^\circ)(12\text{ ft}) = 1440\text{ in.-k}$$

$$M_y = (P\sin\theta)L = (10\text{ k})(\sin 1^\circ)(12\text{ ft}) = 25.13\text{ in.-k}$$

The coordinates of point A are

$$z = 3.50\text{ in.} \qquad y = -12.0\text{ in.}$$

Therefore, the tensile stress at A(see Eq. 5-59) is

$$\sigma_a = \frac{M_y z}{I_y} - \frac{M_z y}{I_z}$$

$$= \frac{(25.13\text{ in.-k})(3.50\text{ in.})}{42.2\text{ in.}^4} - \frac{(1440\text{ in.-k})(-12.0\text{ in.})}{2100\text{ in.}^4}$$

$$= 2084\text{ psi} + 8229\text{ psi} = 10{,}310\text{ psi}$$

The stress at B has the same magnitude but is a compressive stress:

$$\sigma_b = -10{,}310\text{ psi}$$

These stresses are 25% larger than the maximum stress $\sigma_{max} = 8230$ psi obtained for the beam with $\theta = 0$. In addition to having larger stresses, the beam now undergoes a lateral deflection in the z direction, whereas the original beam does not.

This example shows that beams with I_z much larger than I_y may develop large stresses if the beam or its loads deviate even a small amount from their planned alignment. Therefore, such beams should be used with caution, because they are highly susceptible to lateral or sideways buckling. One remedy is to provide adequate lateral support for the beam, thereby preventing lateral bending.

*5.13 SHEAR CENTER

Lateral loads acting on a beam produce both bending moments and shear forces. If the loads act in a plane of symmetry, we can calculate the stresses associated with the bending moments by using the flexure formula, because the plane of symmetry is the plane of bending. However, if the loads act in a plane that is not a plane of symmetry, the beam may be subjected to torsion as well as bending.

To demonstrate the effects of lateral loads under these more general circumstances, consider a cantilever beam of singly symmetric cross section supporting a load P at the free end (Fig. 5-53a). The force P is perpendicular to the xz plane, which is the plane of symmetry of the beam. Note that the z axis is an axis of symmetry of the cross section (Fig. 5-53b) and that the origin of coordinates is taken at the centroid C of the cross section. Therefore, the y and z axes are principal centroidal axes.

Let us make the assumption that under the action of the load P, the beam bends with the z axis as the neutral axis. Then two stress resultants exist on every cross section of the beam—the bending moment M_z acting about the z axis and the shear force V_y (equal to P) acting in the y direction

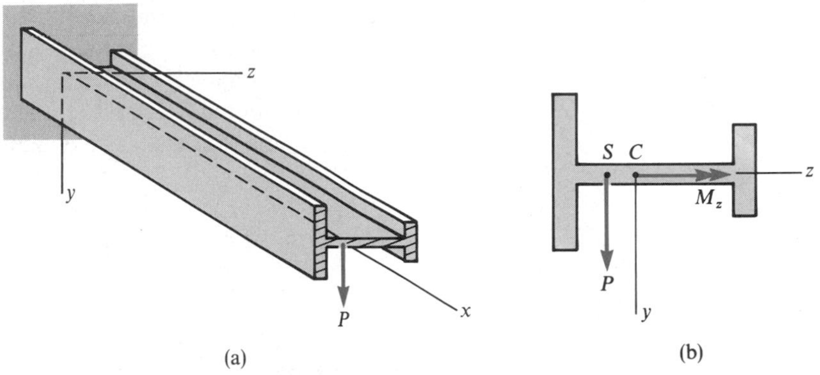

Fig. 5-53 Cantilever beam with singly symmetric cross section: (a) beam with load, and (b) cross section of beam showing centroid C and shear center S

(Fig. 5-53b). The moment M_z is the resultant of the normal stresses acting on the cross section, and the force P is the resultant of the shear stresses. If the material follows Hooke's law, the normal stresses will vary linearly with the distance from the neutral axis. Since we obtain the shear stresses from the normal stresses by means of static equilibrium, the distribution of normal stresses also determines the distribution of the shear stresses. Thus, the shear stresses have a specific resultant force P that has its line of action through a point S lying on the z axis (Fig. 5-53b). This point is known as the **shear center** (or the **center of flexure**) of the cross section. It does not coincide with the centroid C except in special cases that are mentioned later.

From this discussion, we see that the load P must act through the shear center S if the only resultant of the shear stresses is to be the force P itself. Under these conditions, bending occurs about the z axis as the neutral axis with the beam deflecting in the xy plane as the plane of bending. If P does not act through S, it can be replaced by a force P through S plus a twisting couple (that is, a couple having its vector in the x direction). The effect of the force is to produce bending about the z axis as just described, and the effect of the couple is to produce torsion of the beam. Thus, we observe that *a lateral load acting on a beam will produce bending without twisting only if it acts through the shear center.* As a consequence, locating the shear center S is an important aspect of beam design whenever the plane of bending is not a plane of symmetry.

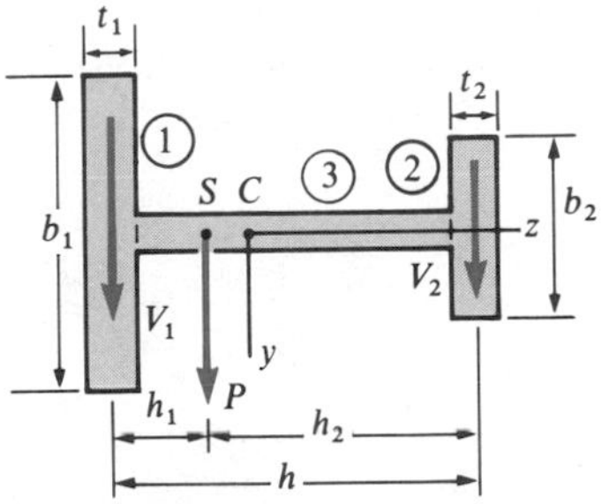

Fig. 5-54 Cross section of singly symmetric I-beam showing location of shear center S

The shear center of the singly symmetric I-beam shown in Fig. 5-53 can be determined as follows. We consider the cross section to consist of three rectangular parts, namely, the two flanges and the web (Fig. 5-54). All three parts are subject to the same curvature when bending takes place, because they are integral parts of the same cross section. Therefore, the bending moment carried by each part is in proportion to its moment of inertia about the z axis:

$$\kappa = \frac{M_1}{EI_1} = \frac{M_2}{EI_2} = \frac{M_3}{EI_3} \tag{a}$$

in which M_1, M_2, and M_3 are the moments resisted by parts 1, 2, and 3, respectively, and I_1, I_2, and I_3 are their respective moments of inertia about the z axis. If the web is thin, its moment of inertia I_3 will be very small compared to I_1 and I_2. Then we may disregard the effect of the web and assume that all of the moment is resisted by the flanges:

$$M_z = M_1 + M_2$$

Also, from Eq. (a) we get

$$\frac{M_1}{I_1} = \frac{M_2}{I_2}$$

Combining the preceding two equations, we obtain

$$M_1 = \frac{M_z I_1}{I_1 + I_2} \qquad M_2 = \frac{M_z I_2}{I_1 + I_2} \tag{b}$$

The shear forces V_1 and V_2 in the flanges are in the same proportions as the bending moments (because $V = dM/dx$); hence

$$\frac{V_1}{V_2} = \frac{M_1}{M_2}$$

Also, the total shear force V_y (equal to P) is

$$V_y = V_1 + V_2$$

By comparing this equation with Eqs. (b), we see that the shear forces are

$$V_1 = \frac{V_y I_1}{I_1 + I_2} \qquad V_2 = \frac{V_y I_2}{I_1 + I_2} \tag{c}$$

The line of action of the resultant of these two shear forces determines the location of the shear center S.

To locate the shear center S in terms of the dimensions of the cross section, let us denote the distance between the centerlines of the flanges by h, the widths of the flanges by b_1 and b_2, and their thicknesses by t_1 and t_2 (Fig. 5-54). We wish to determine the distances h_1 and h_2 from the centerlines of the flanges to the shear center S. Inasmuch as P is the resultant of V_1 and V_2, the forces V_1 and V_2 must produce no moment resultant about point S; therefore,

$$V_1 h_1 = V_2 h_2$$

or, using Eqs. (c),

$$\frac{h_1}{h_2} = \frac{I_2}{I_1} \tag{d}$$

The moments of inertia I_1 and I_2 are

$$I_1 = \frac{t_1 b_1^3}{12} \qquad I_2 = \frac{t_2 b_2^3}{12}$$

Substituting these expressions into Eq. (d), and also noting that $h = h_1 + h_2$, we find

$$h_1 = \frac{t_2 b_2^3 h}{t_1 b_1^3 + t_2 b_2^3} \qquad h_2 = \frac{t_1 b_1^3 h}{t_1 b_1^3 + t_2 b_2^3} \tag{5-63a,b}$$

These expressions can be used to locate the shear center S of a singly symmetric I-beam. Note particularly that h, h_1, and h_2 are measured to the centerlines of the flanges. If $b_1 > b_2$ and $t_1 \geq t_2$, or if $t_1 > t_2$ and $b_1 \geq b_2$, the shear center is located to the left of the centroid C in Fig. 5-54; that is, it is located between C and the larger flange.

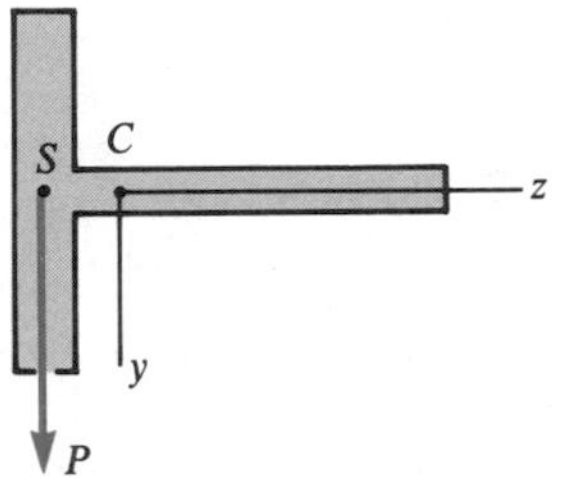

Fig. 5-55 Location of shear center S for a T-beam

A special case occurs when the beam is a T-beam and has only one flange (Fig. 5-55). For a beam of this shape, we substitute $b_2 = t_2 = 0$ into Eqs. (5-63) and obtain

$$h_1 = 0 \qquad h_2 = h$$

This result shows that the shear center is located at the intersection of the centerlines of the flange and web. We could have anticipated this result because in the derivation we assumed that the web was very thin so that the shear force was carried entirely by the flanges.

If the beam of Fig. 5-54 has two identical flanges ($t_1 = t_2$ and $b_1 = b_2$), we have a doubly symmetric I-beam for which the y and z axes are axes of symmetry. Then Eqs. (5-63) yield

$$h_1 = h_2 = \frac{h}{2}$$

This result shows that the shear center coincides with the centroid. In general, the shear center (like the centroid) lies on any axes of symmetry, and the two points always coincide for doubly symmetric cross sections. Thus, any lateral load that acts through the centroid of a doubly symmetric cross section will produce bending without torsion. The method for finding stresses in such beams was discussed in the preceding section.

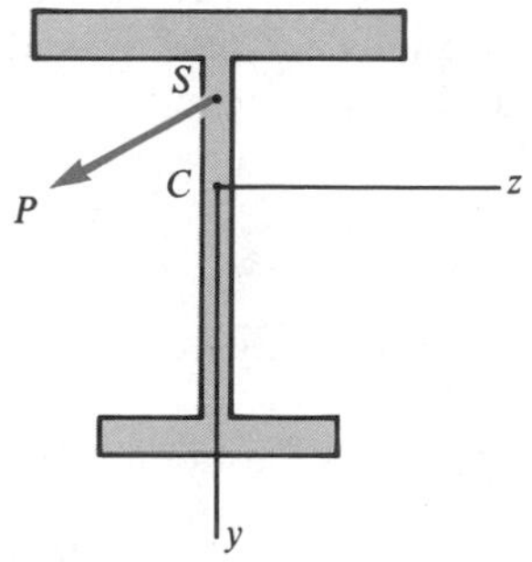

Fig. 5-56 Singly symmetric beam with lateral load P acting through the shear center S

If a beam is singly symmetric, both the centroid and the shear center lie on the axis of symmetry, as already mentioned. Then, if a lateral load acts through the shear center in any direction (Fig. 5-56), it can be resolved into two components, one in the direction of the y axis and the other parallel to the z axis. The first component will produce bending in the xy plane with the z axis as the neutral axis. The second component will produce bending (without torsion) in the xz plane with the y axis as the neutral axis. The stresses (and deflections) produced by these loads acting separately can be superimposed to obtain the stresses (and deflections) caused by the original inclined load P.

If the lateral load P does not act through the shear center (Fig. 5-57a), it can be replaced by a statically equivalent system consisting of a parallel force P acting through the shear center and a twisting couple T

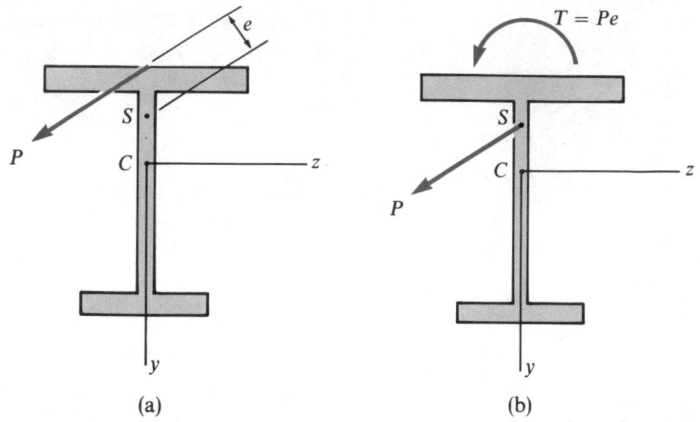

Fig. 5-57 Singly symmetric beam with lateral load P that does not act through the shear center S

(Fig. 5-57b). The couple T has a magnitude equal to the product of the force P and the distance e between its old and new lines of action. The effects of the load P acting through the shear center can be determined as described in the preceding paragraph, and the effects of the twisting couple can be handled by an appropriate torsion analysis.

Locating the shear center for singly symmetric or unsymmetric cross sections is not always an easy task. For solid sections and closed hollow sections, it is usually located near the centroid. Such sections have high torsional rigidities, and usually we can ignore the effects of twisting if the load is applied at or near the centroid. Beams with thin-walled open cross sections (such as channels, angles, T-beams, and Z-sections) are torsionally very weak; therefore, it is important to locate the shear centers and to take into account the effects of twisting. Cross sections of this type are considered in the following two sections.

*5.14 BEAMS OF THIN-WALLED OPEN CROSS SECTIONS

The distribution of shear stresses in rectangular beams and in the webs of beams with flanges was described previously in Sections 5.5 and 5.6, and we derived the following **shear formula** for calculating the stresses:

$$\tau = \frac{VQ}{Ib} \tag{5-64}$$

This formula is valid if the stresses are distributed uniformly across the width b of the beam, a requirement that is satisfied for beams of thin rectangular cross section, for the webs of wide-flange beams, and for a few other shapes.

We will now consider a special class of beams, known as beams of **thin-walled open cross section**, for which we can determine the shear stresses by the same method that we used when deriving the shear formula.

Beams of this kind are distinguished by two features: (1) the cross section has a wall thickness that is small compared to its overall height or width, and (2) the cross section is open, as in the case of an I-beam or channel beam, rather than closed, as in the case of a hollow box beam. Beams of thin-walled open sections are widely used in engineering work and are also called **structural sections** or **profile sections**.

We begin our study of shear stresses in structural sections by examining a beam having its cross-sectional centerline *mm* of arbitrary shape (Fig. 5-58a). The y and z axes are principal centroidal axes of the cross section, and the load P acts parallel to the y axis (Fig. 5-58b). If the load P acts through the shear center S, there will be no twisting of the beam; bending will occur in the xy plane and the z axis will be the neutral axis. Then the normal stresses at any point in the beam are given by

$$\sigma_x = -\frac{M_z y}{I_z} \tag{5-65}$$

where M_z is the bending moment about the z axis (positive as defined in Section 5.12 and Fig. 5-51) and y is the ordinate to the point under consideration.

Now consider an element *abcd* cut out between two cross sections, distance dx apart, and having length s measured along the centerline of the cross section (Fig. 5-58a). The resultant of the normal stresses acting on face *ad* is denoted F_1 (Fig. 5-58c) and the resultant on face *bc* is denoted F_2. Since the bending moment at face *ad* is larger than the moment at *bc*, the force F_1 will be larger than F_2; hence shear stresses τ must act along the face *cd* in order to have equilibrium of the element. These shear stresses must be parallel to the top and bottom surfaces of the element, which are free of stress, and must be accompanied by complementary shear stresses

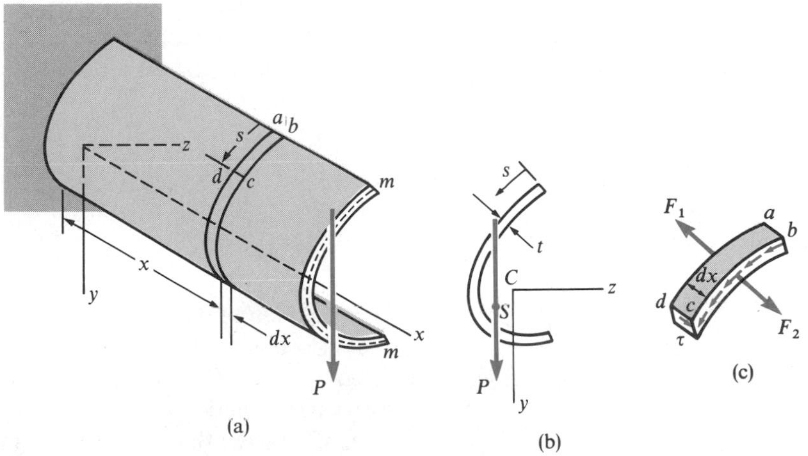

Fig. 5-58 Shear stresses in a beam of thin-walled open cross section. (Note: The y and z axes are principal centroidal axes.)

acting on cross sections *ad* and *bc*. Summing forces in the x direction for the element *abcd* (Fig. 5-58c), we get

$$\tau t\,dx = F_1 - F_2 \tag{a}$$

where t is the thickness of the cross section at *cd*; that is, t is the thickness at distance s from the free edge of the cross section (Fig. 5-58b). Using Eq. (5-65), we conclude that

$$F_1 = \int_0^s \sigma_x\,dA = -\frac{M_{z1}}{I_z}\int_0^s y\,dA$$

where dA is an element of area on the side *ad* of the element, y is the coordinate of the element dA, and M_{z1} is the bending moment at this cross section. An analogous expression is obtained for the force F_2:

$$F_2 = \int_0^s \sigma_x\,dA = -\frac{M_{z2}}{I_z}\int_0^s y\,dA$$

Substituting the expressions for F_1 and F_2 into Eq. (a), we get

$$\tau = \frac{M_{z2} - M_{z1}}{dx}\,\frac{1}{I_z t}\int_0^s y\,dA$$

The quantity $(M_{z2} - M_{z1})/dx$ is the rate of change of the bending moment and is equal to $-V_y$, where V_y is the shear force in the y direction (equal to P in Fig. 5-58). Therefore, the equation for the shear stresses is

$$\tau = -\frac{V_y}{I_z t}\int_0^s y\,dA$$

This equation gives the shear stresses at any point in the cross section at distance s from the free edge. The integral on the right-hand side represents the first moment with respect to the neutral axis (the z axis) of the area of the cross section from $s = 0$ to $s = s$. Denoting this first moment by Q_z, and also using only the absolute value of the shear stress because its direction can be determined by inspection, we write the equation for the **shear stresses** τ in the simpler form

$$\tau = \frac{V_y Q_z}{I_z t} \tag{5-66}$$

which is analogous to the shear formula (Eq. 5-64). The shear stresses τ are directed along the centerline of the cross section, parallel to the edges of the section, and are assumed to be of constant intensity across the thickness t of the wall. The wall thickness need not be constant but may vary as a function of s.

The **shear flow** at any point in the cross section, equal to the product of the shear stress and the thickness at that point, is

$$f = \tau t = \frac{V_y Q_z}{I_z} \tag{5-67}$$

Because V_y and I_z are constants, the shear flow is directly proportional to Q_z. At the top and bottom edges of the cross section, Q_z is zero and hence the shear flow is zero. The shear flow varies continuously between these end points and reaches its maximum value when Q_z is a maximum, which occurs at the neutral axis.

If the beam shown in Fig. 5-58 is bent by loads that act through the shear center and that are parallel to the z axis, then the y axis will be the neutral axis for bending. In such a case, we can repeat the same type of analysis and arrive at the following equations in place of Eqs. (5-66) and (5-67):

$$\tau = \frac{V_z Q_y}{I_y t} \qquad f = \tau t = \frac{V_z Q_y}{I_y} \tag{5-68a,b}$$

In these equations, V_z is the shear force parallel to the z axis and Q_y is the first moment with respect to the y axis.

In summary, we have derived expressions for the shear stresses in beams of thin-walled open cross sections provided that the shear force acts through the shear center S and is parallel to one of the principal centroidal axes. If the shear force is inclined to the y and z axes (but still acts through the shear center), it can be resolved into components parallel to the principal axes. Then two separate analyses can be made, and the results may be superimposed.

Example

Investigate the shear stresses in a wide-flange beam loaded by a force P acting in the vertical direction in the plane of the web (Fig. 5-59a).

Let us begin by taking a cross section of the beam (Fig. 5-59b) and considering the stresses in the right-hand part of the top flange. The distance s for this part of the beam will be measured from point a, where the shear stress is zero, leftward to section bb. The cross-sectional area between point a and section bb is st_f, where t_f is the thickness of the flange. The distance from the centroid of this area to the neutral axis is $h/2$ (note that h is the height of the beam between the centerlines of the flanges). Thus, for section bb we have $Q_z = st_f h/2$, and the shear stress is

$$\tau = \frac{shP}{2I_z} \tag{5-69}$$

as obtained from Eq. (5-66). The direction of this stress can be found by considering the forces acting on an element cut out of the flange between point a

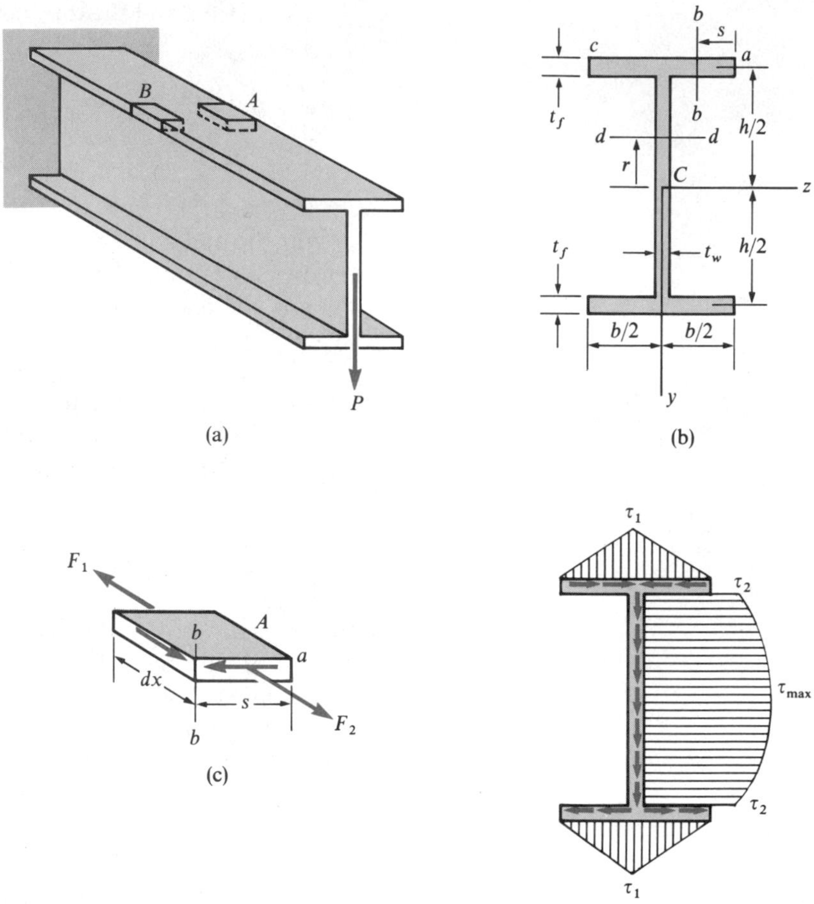

Fig. 5-59 Example. Shear stresses in a wide-flange beam

and section *bb* (see element *A* in Fig. 5-59a). This element is drawn to a larger scale in Fig. 5-59c, in order to show clearly the forces acting on it. We see immediately that the tensile force F_1 is larger than the force F_2, because the bending moment is larger on the rear face of the element than it is on the front face. It follows that, for equilibrium, the shear stress τ on the left-hand face of element *A* must act toward the reader. This conclusion then dictates the direction of the shear stresses on the cross section; namely, they must act toward the left. Returning now to Fig. 5-59b, we see that we have completely determined the magnitude and direction of the shear stress at section *bb*. This section may be taken anywhere between point *a* and the junction of the flange and web; hence throughout this region the shear stress is horizontal and to the left, and its magnitude is given by Eq. (5-69). We see also from Eq. (5-69) that the stress increases linearly with the distance *s*, as shown graphically in Fig. 5-59d. The maximum value τ_1 is reached when $s = b/2$, where *b* is the flange width; thus,

$$\tau_1 = \frac{bhP}{4I_z} \tag{5-70}$$

and the corresponding shear flow is

$$f_1 = \tau_1 t_f = \frac{bht_f P}{4I_z} \tag{5-71}$$

Note that we have calculated the shear stress at the junction of the centerlines of the flange and web, utilizing only centerline dimensions of the cross section. This approximate procedure is satisfactory for thin-walled sections.

By beginning at point c on the left-hand part of the top flange (Fig. 5-59b) and measuring s toward the right, we can repeat the same kind of analysis. We will find that the magnitude of the shear stresses is again given by Eqs. (5-69) and (5-70). However, by cutting out an element B (Fig. 5-59a) and considering its equilibrium, we find that the shear stresses on the cross section now act toward the right, as shown in Fig. 5-59d.

The next step is to determine the shear stresses in the web. Considering a horizontal cut at the top of the web, just below the flange, we find the first moment to be

$$Q_z = \frac{bt_f h}{2}$$

and the corresponding shear stress is

$$\tau_2 = \frac{bht_f P}{2I_z t_w} \tag{5-72}$$

where t_w is the thickness of the web. The associated shear flow is

$$f_2 = \tau_2 t_w = \frac{bht_f P}{2I_z} \tag{5-73}$$

which is equal to twice the shear flow f_1, as expected. The shear stresses in the web act vertically downward and increase in magnitude until the neutral axis is reached. At section dd, located at distance r from the neutral axis, the shear stress is calculated as follows:

$$Q_z = \frac{bt_f h}{2} + \left(\frac{h}{2} - r\right)(t_w)\left(\frac{h/2 + r}{2}\right) = \frac{bt_f h}{2} + \frac{t_w}{2}\left(\frac{h^2}{4} - r^2\right)$$

and

$$\tau = \left(\frac{bt_f h}{t_w} + \frac{h^2}{4} - r^2\right)\frac{P}{2I_z} \tag{5-74}$$

When $r = h/2$, this equation reduces to Eq. (5-72), and when $r = 0$ it gives the maximum shear stress:

$$\tau_{\max} = \left(\frac{bt_f}{t_w} + \frac{h}{4}\right)\frac{Ph}{2I_z} \tag{5-75}$$

Again it should be pointed out that we have made all calculations on the basis of the centerline dimensions of the cross section, which gives reasonably accurate results for thin sections. For this reason, however, the shear stresses calculated for the web of a wide-flange beam by using Eq. (5-74) may be slightly different from those obtained in our previous analysis (Eq. 5-26).

The shear stresses in the web vary parabolically, as shown in Fig. 5-59d, although the variation is not great. This fact can be seen from the ratio of τ_{max} to τ_2, which is

$$\frac{\tau_{max}}{\tau_2} = 1 + \frac{ht_w}{4bt_f} \tag{5-76}$$

The second term on the right-hand side is usually small; for example, if we take typical values such as $h = 2b$ and $t_f = 2t_w$, the ratio is $\tau_{max}/\tau_2 = 1.25$.

Finally, we may investigate the shear stresses in the lower flange by the same methods used for the top flange. We find that the magnitudes of the stresses are the same as in the top flange, but the directions are as shown in Fig. 5-59d.

From Fig. 5-59d, we see that the shear stresses on the cross section "flow" inward from the outermost edges of the top flange, then down through the web, and finally outward to the edges of the bottom flange. Because this flow is always continuous in any structural section, it serves as a convenient way to determine the directions of the stresses. For instance, if the shear force acts downward on the beam of Fig. 5-59, we know immediately that the shear flow in the web must be downward. Then, knowing the direction of the shear flow in the web, we also know the directions in the flanges, because of the required continuity in the shear flow. Using this simple technique to get the directions of the shear stresses is easier than visualizing the directions of the forces acting on elements such as A (Fig. 5-59c) cut out from the beam.

The resultant of all shear stresses on the cross section is clearly a vertical force, because the horizontal stresses in the flanges produce no resultant. The shear stresses in the web have a resultant R, which can be found by integrating the shear stresses over the height of the web, as follows:

$$R = \int \tau \, dA = 2\int_0^{h/2} \tau t_w \, dr$$

Substituting from Eq. (5-74), we get

$$R = 2t_w \int_0^{h/2} \left(\frac{bt_f h}{t_w} + \frac{h^2}{4} - r^2\right)\left(\frac{P}{2I_z}\right) dr$$

$$= \left(\frac{bt_f}{t_w} + \frac{h}{6}\right)\frac{h^2 t_w P}{2I_z} \tag{b}$$

The term I_z can be calculated as follows:

$$I_z = \frac{t_w h^3}{12} + \frac{bt_f h^2}{2} \tag{5-77}$$

where the first term is the moment of inertia of the web and the second term is the moment of inertia of the flanges, again calculated by using centerline dimensions. When this expression for I_z is substituted into Eq. (b), we get $R = P$, which establishes the fact that the resultant of the shear stresses acting on the cross section is equal to the vertical force P. The resultant passes through the centroid C, which is also the shear center for a doubly symmetric beam.

*5.15 SHEAR CENTERS OF THIN-WALLED OPEN SECTIONS

In the preceding section we developed techniques for determining the shear stresses in beams of thin-walled open cross sections. Now we will use those techniques to locate the shear centers. Inasmuch as the shear center is the point through which the resultant shear force acts, we can locate the shear center by evaluating the shear stresses and then locating the line of action of the resultant of those stresses. To illustrate the method and also to obtain some useful formulas, we will now locate the shear centers of several thin-walled sections.

The first beam we consider is a **channel section** (Fig. 5-60a) bent about the z axis and subjected to a vertical shear force V_y acting parallel to the y axis. The distribution of shear stresses for the channel is shown in Fig. 5-60b. To find the maximum stress τ_1 in the flange, we use Eq. (5-66) with Q_z equal to the first moment of the flange area about the z axis:

$$Q_z = \frac{bt_f h}{2}$$

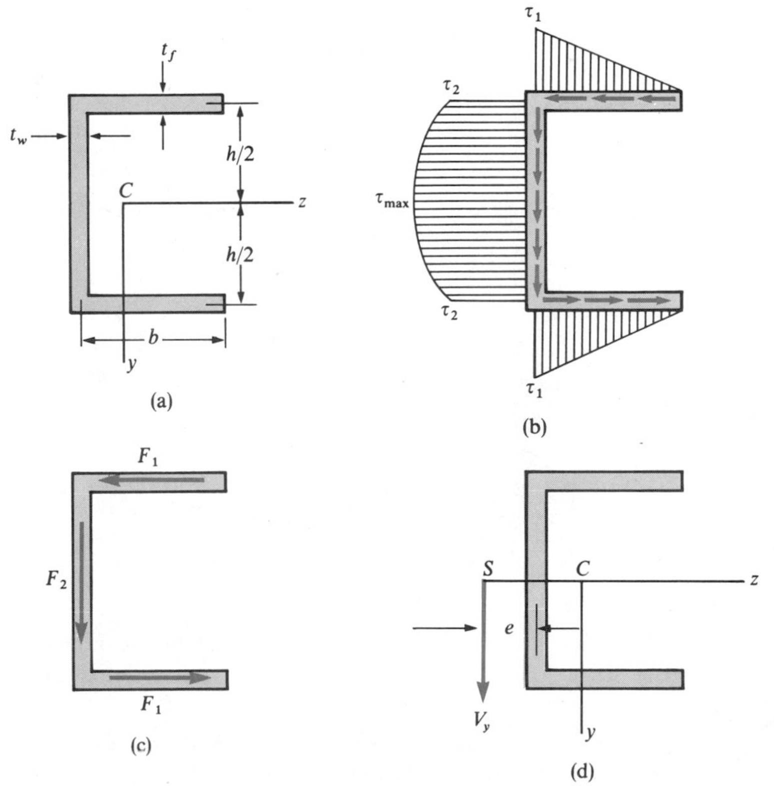

Fig. 5-60 Shear center of a channel section

Thus, the stress τ_1 in the flange is

$$\tau_1 = \frac{bhV_y}{2I_z} \tag{5-78}$$

In the same manner, we find the stress τ_2 at the top of the web:

$$\tau_2 = \frac{bt_f hV_y}{2t_w I_z} \tag{5-79}$$

Also, at the neutral axis the stress is

$$\tau_{max} = \left(\frac{bt_f}{t_w} + \frac{h}{4}\right)\frac{hV_y}{2I_z} \tag{5-80}$$

These stresses are shown in Fig. 5-60b.

The total shear force F_1 in either flange (Fig. 5-60c) can be found from the triangular diagrams of shear stress. Each force is equal to the area of the stress triangle multiplied by the thickness of the flange:

$$F_1 = \left(\frac{\tau_1 b}{2}\right)(t_f) = \frac{hb^2 t_f V_y}{4I_z} \tag{a}$$

The vertical force F_2 in the web must be equal to the shear force V_y, since the forces in the flanges have no vertical components. We can easily verify that $F_2 = V_y$ by considering the parabolic stress diagram of Fig. 5-60b. We note that the diagram is made up of two parts: a rectangle of dimensions τ_2 and h, and a parabolic area equal to

$$\frac{2}{3}(\tau_{max} - \tau_2)h$$

Thus, the total shear force, equal to the area of the stress diagram times the web thickness, is

$$F_2 = \tau_2 h t_w + \frac{2}{3}(\tau_{max} - \tau_2)ht_w$$

Substituting the expressions for τ_2 and τ_{max}, we obtain

$$F_2 = \left(\frac{t_w h^3}{12} + \frac{bh^2 t_f}{2}\right)\frac{V_y}{I_z}$$

Finally, we note that the expression for the moment of inertia is

$$I_z = \frac{t_w h^3}{12} + \frac{bh^2 t_f}{2} \tag{5-81}$$

in which we again base the calculations upon centerline dimensions only. Substituting this expression for I_z into the equation for F_2, we get

$$F_2 = V_y \tag{b}$$

as expected.

The three forces acting on the cross section (Fig. 5-60c) must be statically equivalent to the resultant force V_y acting through the shear center S (Fig. 5-60d). Hence, the moment of the force V_y about any point in the cross section is equal to the moment about that same point of the three forces. This moment relationship provides an equation from which the distance e to the shear center may be found. It is usually convenient to select the center of moments at the shear center S. Then, for the channel being analyzed, the equation of moments is

$$F_1 h - F_2 e = 0$$

Substituting the expressions for F_1 and F_2 (Eqs. a and b) and solving for e, we find

$$e = \frac{b^2 h^2 t_f}{4I_z} \tag{5-82}$$

When the expression for I_z (Eq. 5-81) is substituted, Eq. (5-82) becomes

$$e = \frac{3b^2 t_f}{ht_w + 6bt_f} \tag{5-83}$$

Note that e is the distance from the centerline of the web to the shear center.

Instead of taking moments about the shear center, we can also use a point on the centerline of the web as the center of moments. Then the moment equation for the channel beam is

$$F_1 h = V_y e$$

from which

$$e = \frac{b^2 h^2 t_f}{4I_z}$$

as before. This alternative choice of the center of moments is sometimes convenient, although it offers no significant reduction in calculating effort.

As explained in Section 5.13, a channel beam will undergo simple bending whenever it is loaded by forces acting through the shear center S. If the loads act parallel to the y axis but through some point other than the shear center (for example, the loads may act in the plane of the web), they can be replaced by a statically equivalent force system consisting of loads through the shear center and twisting couples. We then have a combination of bending and torsion of the beam. If the loads act in the direction of the z axis, through points S and C, we will have simple bending about the y axis. If the loads act in skew directions, they can be replaced by statically equivalent loads parallel to the y and z axes.

The next shape to be considered is an **equal-leg angle section** (Fig. 5-61a), which we assume is subjected to a shear force V_y. Each leg of the angle has length b and thickness t. At distance s from one end, the

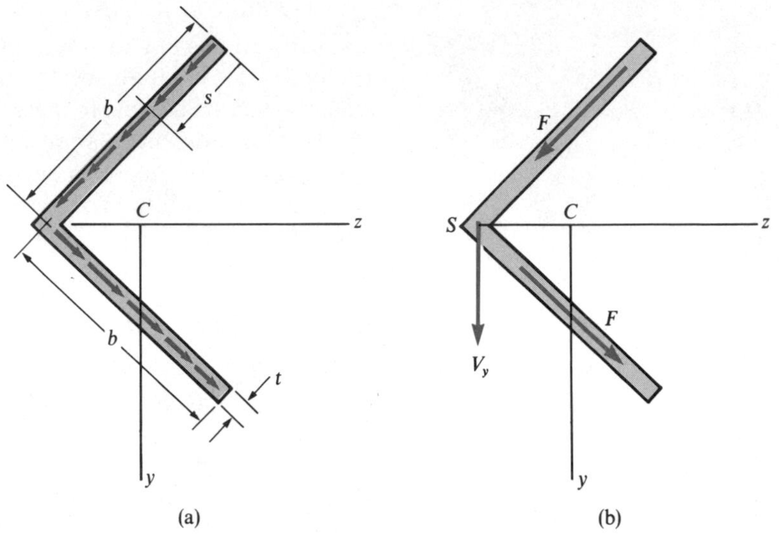

Fig. 5-61 Shear center of an equal-leg angle section

shear stress is

$$\tau = \frac{V_y Q_z}{I_z t} = \frac{3sV_y}{\sqrt{2}b^3 t}\left(b - \frac{s}{2}\right) \tag{c}$$

in which we have used the relations

$$Q_z = \frac{st}{\sqrt{2}}\left(b - \frac{s}{2}\right)$$

and
$$I_z = \frac{b^3 t}{3} \tag{5-84}$$

Equation (c) shows that τ varies quadratically with s; it reaches its maximum value when $s = b$:

$$\tau_{\max} = \frac{3V_y}{2\sqrt{2}bt} \tag{d}$$

The total shear force F in each leg (Fig. 5-61b) is

$$F = \int_0^b \tau t\, ds = \frac{3V_y}{\sqrt{2}b^3}\int_0^b \left(bs - \frac{s^2}{2}\right) ds = \frac{V_y}{\sqrt{2}} \tag{e}$$

Taking the vertical components of the forces F, we see that the resultant of these forces is a vertical force equal to V_y, as expected. Furthermore, we see that this resultant force must pass through the point where the lines of action of the two forces F intersect, which shows that the shear center S of the angle section is at the junction of the two legs.

For all cross sections consisting of two narrow intersecting rectangles, as in the examples of Fig. 5-62, the shear stresses have resultant

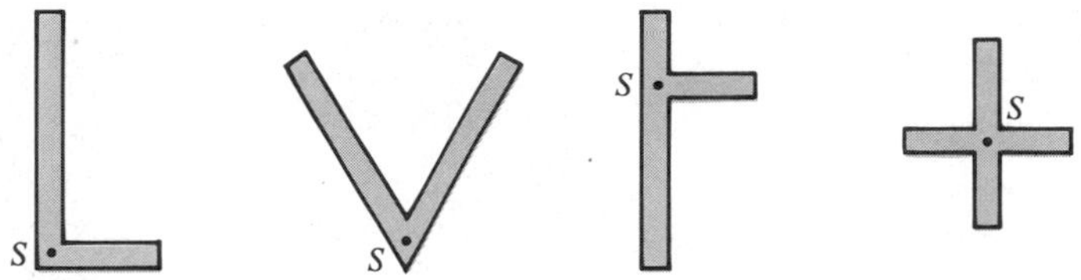

Fig. 5-62 Shear centers S of sections consisting of two intersecting narrow rectangles

forces that intersect at the junction of the rectangles. Therefore, the shear center S is located at this point, as shown in the figure.

The locations of the shear centers for most structural shapes of practical interest are given in this section, either in the preceding discussions or in the examples and problems that follow.*

Example 1

Locate the shear center S of the thin-walled semicircular cross section shown in Fig. 5-63.

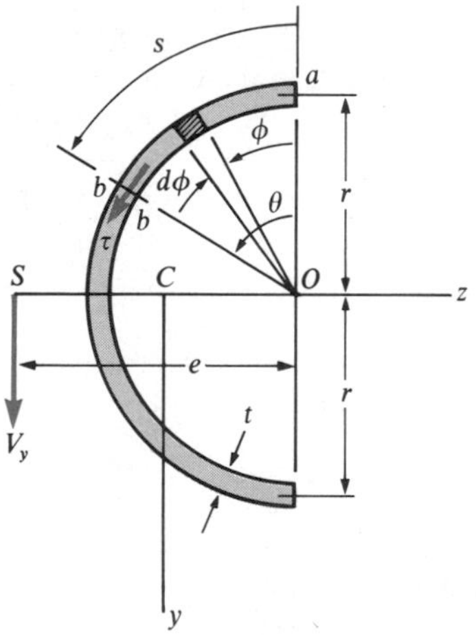

Fig. 5-63 Example 1. Shear center of a thin-walled semicircular section

Let us consider a section bb defined by the distance s measured along the centerline of the cross section. The central angle subtended between point a, which is at the edge of the section, and section bb is denoted by θ. Therefore, we have $s = r\theta$, where r is the radius of the centerline. The first moment of the area between a and section bb is

$$Q_z = \int y\,dA = \int_0^\theta (r\cos\phi)(rt)\,d\phi = r^2 t \sin\theta$$

where t is the thickness of the section. Thus, the shear stress τ at section bb is

$$\tau = \frac{V_y Q_z}{I_z t} = \frac{V_y r^2 \sin\theta}{I_z}$$

Substituting $I_z = \pi r^3 t/2$, we get

$$\tau = \frac{2V_y \sin\theta}{\pi r t} \tag{5-85}$$

When $\theta = 0$ or $\theta = \pi$, this expression gives $\tau = 0$, and when $\theta = \pi/2$ it gives the maximum shear stress.

The moment about the center O due to the shear stresses τ is

$$T = \int \tau r\,dA = \int_0^\pi \frac{2V_y r \sin\theta\,d\theta}{\pi} = \frac{4rV_y}{\pi}$$

which must be equal to the moment of the force V_y acting at the shear center; hence

$$T = V_y e = \frac{4rV_y}{\pi}$$

* The first determination of a shear center was made by Timoshenko in 1913 (Ref. 5-16). For additional information and for the historical development of the shear-center concept, see Refs. 5-16 through 5-29.

Thus, the distance e from point O to the shear center is

$$e = \frac{4r}{\pi} \approx 1.27r \tag{5-86}$$

The shear center of a more general thin-walled circular section is given in Problem 5.15-10.

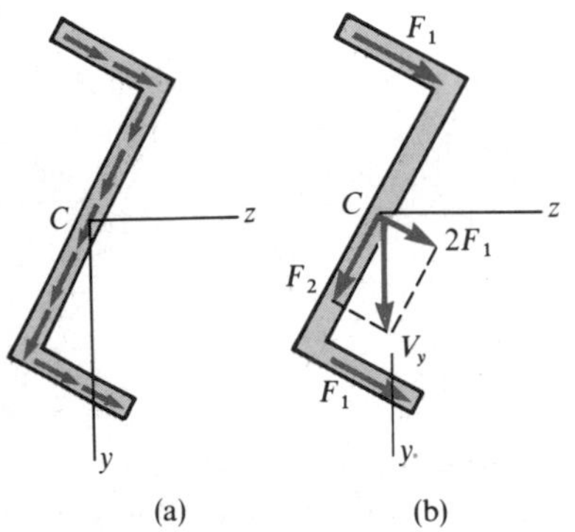

Fig. 5-64 Example 2. Shear center of a Z-section

Example 2

Locate the shear center of the Z-section shown in Fig. 5-64a. (The y and z axes shown in the figure are principal axes through the centroid C.)

Let us assume that a shear force V_y acts parallel to the y axis. Then the shear stresses in the flanges and web will be directed as shown in Fig. 5-64a. From symmetry considerations, we see that the forces F_1 in the two flanges must be equal to each other (Fig. 5-64b). The resultant of all three forces acting on the cross section (F_1 in the flanges and F_2 in the web) must be equal to the shear force V_y. The forces F_1 have a resultant $2F_1$ acting through the centroid and parallel to the flanges. This force intersects the force F_2 at the centroid C, and therefore this point must be the shear center. The composition of the forces $2F_1$ and F_2 into the resultant shear force V_y is shown in Fig. 5-64b.

If the beam is subjected to a shear force V_z parallel to the z axis, we arrive at a similar conclusion, namely, that the shear center coincides with the centroid.

*5.16 INELASTIC BENDING

Inelastic bending refers to the bending of beams when the material does not follow Hooke's law, and therefore it occurs whenever the beam is loaded in such a manner that the stresses exceed the proportional limit of the material. Naturally, the behavior of a beam in elastic bending depends upon the shape of the stress-strain diagram. The material might have a stress-strain diagram that is curved beyond the proportional limit, as shown for aluminum in Fig. 1-10, or the material might be steel, which exhibits pronounced yielding and has a stress-strain diagram as shown in Fig. 1-7. In any case, if the stress-strain diagram is known, it is always possible to determine the stresses, strains, and deflections of the beam.

The ultimate load-carrying capacity of a beam can also be determined from an inelastic analysis. The ultimate load is usually much greater than the load at the proportional limit (that is, the largest load the beam can carry without exceeding the proportional limit at any point). For design purposes, we often need the ultimate load in order to ascertain the true factor of safety against failure. Of course, this factor is much larger than the factor of safety with respect to the proportional-limit load. The concepts of ultimate-load design (see Section 1.7) are often used in the design of aluminum, steel, and reinforced concrete structures.

The analysis of an inelastic beam is based on the fact that plane cross sections of a beam remain plane under pure bending, a condition that is valid for both nonlinearly inelastic materials and linearly elastic materials (see Section 5.2). Therefore, the strains in an inelastic beam vary linearly over the height of the beam. Then, with the aid of the stress-strain diagram and equations of statics, we can determine the magnitudes of the stresses.

Let us begin with a beam in **pure bending** subjected to a positive bending moment M (Fig. 5-65a). The bending couples act in the xy plane, which is assumed to be a plane of symmetry of the cross section (Fig. 5-65b). Thus, the beam will deflect in the same plane, which is the plane of bending. The z axis is the neutral axis of the cross section, but its location remains to be determined.

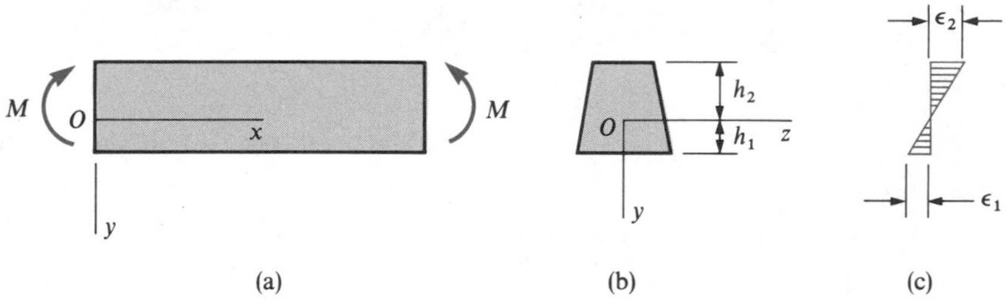

Fig. 5-65 Inelastic bending of a beam

The strains in the beam vary from top to bottom in the manner pictured in Fig. 5-65c, with the strains at the lower and upper surfaces being denoted by ϵ_1 and ϵ_2, respectively. Denoting by ρ the radius of curvature of the deflection curve, we see that the strain at distance y from the neutral surface (see Eq. 5-2) is

$$\epsilon = -\frac{y}{\rho} = -\kappa y \tag{5-87}$$

in which $\kappa = 1/\rho$ is the curvature. The strains at the outermost surfaces are

$$\epsilon_1 = -\kappa h_1 \qquad \epsilon_2 = \kappa h_2 \tag{5-88a,b}$$

where h_1 and h_2 are the distances from the neutral axis to the lower and upper surfaces of the beam, respectively. From these equations we see that the strains can be determined provided we know the curvature and the position of the neutral axis.

The location of the **neutral axis** can be found by making use of the stress-strain diagram for the material and an equation of statics expressing the fact that the resultant horizontal force due to the normal stresses σ acting on any cross section of the beam vanishes; thus,

$$\int \sigma \, dA = 0 \tag{5-89}$$

in which dA is an element of area in the cross section and the integration is performed over the entire cross section.

The curvature can be obtained by making use of a second equation of statics, namely, an equation stating that the resultant of the stresses acting on the cross section is equal to the bending moment M:

$$\int \sigma y \, dA = M \tag{5-90}$$

Of course, Eqs. (5-89) and (5-90) are the same basic equations used previously in analyzing beams of linearly elastic material (see Section 5.3). In the next two sections we will use these equations to analyze inelastic beams.

*5.17 PLASTIC BENDING

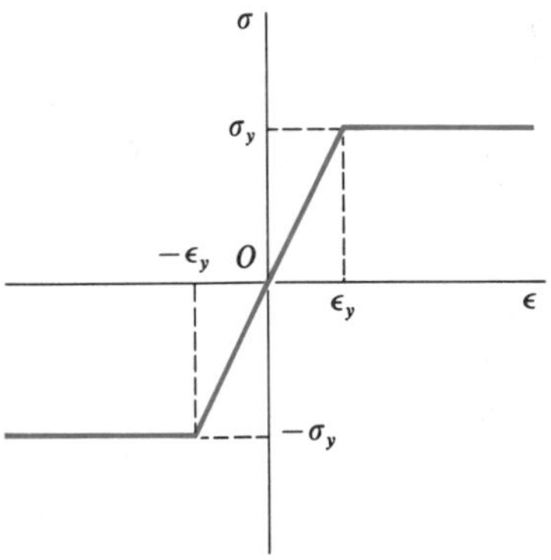

Fig. 5-66 Stress-strain diagram for an elastic-plastic material

The simplest case of inelastic bending is **plastic bending**, which occurs when the material of the beam is elastic-plastic. Such a material follows Hooke's law up to the yield stress and then yields plastically under constant stress. The stress-strain diagram for an **elastic-plastic material** having the same yield stress σ_y and the same modulus of elasticity E in both tension and compression is shown in Fig. 5-66. We see that an elastic-plastic material has a region of linear elasticity between regions of perfect plasticity. For this reason, the term *perfectly plastic material* is often applied to an elastic-plastic material.

Structural steels may be idealized as elastic-plastic materials because they have sharply defined yield points and they undergo large strains during yielding. The assumption of perfect plasticity after the yield stress is reached means that the effects of strain hardening are disregarded, but, since strain hardening provides an increase in the strength of the steel, it is generally safe to disregard it.

Now let us consider a beam of elastic-plastic material subjected to pure bending (Fig. 5-65). When the applied bending couples M are small, the maximum stress in the beam is less than the yield stress σ_y, and the beam is in the condition of ordinary elastic bending with a linear stress distribution, as shown in Fig. 5-67a. Under these conditions we learn from Eqs. (5-87) through (5-90) that the neutral axis passes through the centroid of the cross section, that the normal stress is $\sigma = My/I$, and that the curvature is $-M/EI$. These results are valid until the stress in the beam at the point farthest from the neutral axis reaches the yield stress (Fig. 5-67b). The corresponding moment acting on the beam is called the **yield moment** M_y:

$$M_y = \frac{\sigma_y I}{c} = \sigma_y S \tag{5-91}$$

in which c is the distance to the point farthest from the neutral axis and S is

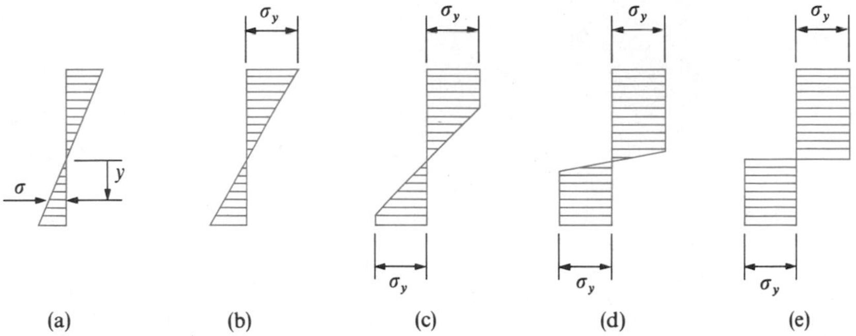

Fig. 5-67 Stress distributions in a beam of elastic-plastic material

the corresponding section modulus. For a beam of rectangular cross section, the yield moment is

$$M_y = \frac{\sigma_y bh^2}{6} \qquad \text{(5-92)}$$

in which b is the width and h is the height of the cross section.

If we now increase the bending moment above the yield moment M_y, the strains at the extreme points of the cross section will continue to increase and the maximum strain will exceed the yield strain ϵ_y. However, because of perfectly plastic yielding, the maximum stresses will remain constant and equal to σ_y. The stress condition then will be as pictured in Fig. 5-67c; the outer regions of the beam will have become plastic while a central core remains elastic. Also, the position of the neutral axis changes, unless the cross section is doubly symmetric. However, this shift in the neutral axis is too small to be shown in Fig. 5-67.

As the bending moment increases, the plastic region extends farther inward toward the neutral axis until the condition shown in Fig. 5-67d is reached. At this stage the strains in the extreme fibers are perhaps 10 or 15 times the yield strain ϵ_y, and the elastic core has almost disappeared. Thus, for all practical purposes the beam has reached its ultimate moment-resisting capacity, and we can idealize the ultimate stress distribution as consisting of two rectangular parts (Fig. 5-67e). The bending moment corresponding to this idealized stress distribution is called the **plastic moment** M_p, and it represents the maximum moment that can be sustained by a beam of elastic-plastic material.

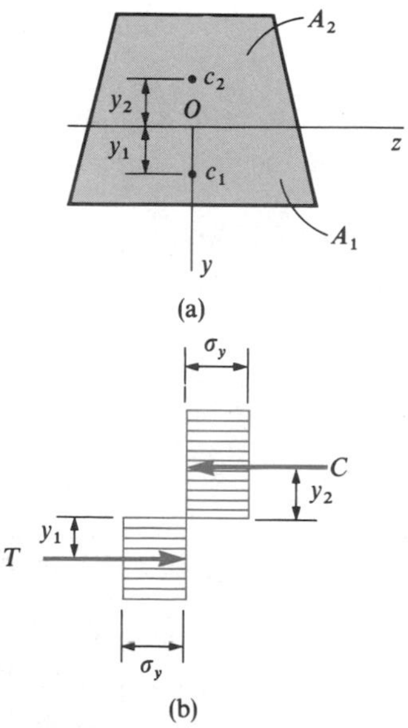

Fig. 5-68 Determination of the plastic moment M_p

The determination of the plastic moment is obviously of great importance because it is the limiting or maximum moment for the beam. In order to find M_p, we begin by locating the **neutral axis** of the cross section (Fig. 5-68a). Above the neutral axis, every element in the cross section has a compressive stress equal to σ_y (Fig. 5-68b); below the neutral axis, the stress is tension and also equal to σ_y. The total tensile force T is equal to $\sigma_y A_1$, where A_1 is the area of the cross section below the neutral axis. Similarly, the compressive force C is $\sigma_y A_2$, where A_2 is the cross-sectional area above the neutral axis. As shown by Eq. (5-89), the resultant force on the cross section must be zero, so that

$$T - C = 0 \quad \text{or} \quad A_1 = A_2 \tag{a}$$

Since the total area of the cross section is $A = A_1 + A_2$, it is apparent from the preceding equation that

$$A_1 = A_2 = \frac{A}{2} \tag{5-93}$$

and we conclude that *the neutral axis divides the cross section into two equal areas.* Thus, the neutral axis for the plastic moment M_p generally is different from the neutral axis for linearly elastic bending. For instance, for a trapezoidal cross section that is narrower at the top than at the bottom, as shown in Fig. 5-68a, the neutral axis is slightly lower for the plastic case than for the elastic case. Of course, if the cross section is doubly symmetric, as for a rectangular beam or a wide-flange beam, the neutral axis will have the same position for both plastic and elastic bending.

The plastic moment M_p can now be found from Eq. (5-90) by integrating over the cross section or by the equivalent procedure of taking moments about the neutral axis of the forces T and C shown in Fig. 5-68b; thus, we obtain

$$M_p = Ty_1 + Cy_2$$

where y_1 and y_2 are the distances from the neutral axis to the centroids c_1 and c_2 of the areas A_1 and A_2, respectively. Replacing T and C by $\sigma_y A/2$, we get

$$M_p = \frac{\sigma_y A(y_1 + y_2)}{2} \tag{5-94}$$

The procedure for each particular beam is to divide the cross section into two equal areas, locate the centroid of each half, and then use Eq. (5-94) to calculate M_p.

The expression for the plastic moment can be written in a form analogous to that for the yield moment (see Eq. 5-91), as follows:

$$M_p = \sigma_y Z \tag{5-95}$$

in which

$$Z = \frac{A(y_1 + y_2)}{2} \tag{5-96}$$

is the **plastic modulus** for the cross section. The plastic modulus may be interpreted geometrically as the first moment (taken about the neutral axis) of the area of the cross section above the neutral axis plus the first moment of the area below the neutral axis.

The ratio of the plastic moment to the yield moment is solely a function of the shape of the cross section and is called the **shape factor** f:

$$f = \frac{M_p}{M_y} = \frac{Z}{S} \tag{5-97}$$

If we take a beam of rectangular cross section (b = width, h = height), the plastic modulus (see Eq. 5-96) becomes

$$Z = \frac{bh}{2}\left(\frac{h}{4} + \frac{h}{4}\right) = \frac{bh^2}{4} \tag{5-98}$$

Recalling that the section modulus is $S = bh^2/6$, we see that the shape factor for a rectangular beam is

$$f = \frac{3}{2} \tag{5-99}$$

Thus, the plastic moment for a rectangular beam is 50% greater than the yield moment.

For a wide-flange beam (see Fig. 5-69), the plastic modulus is calculated by taking the first moment of one flange and the part of the web above the neutral axis and then multiplying by 2. The result is

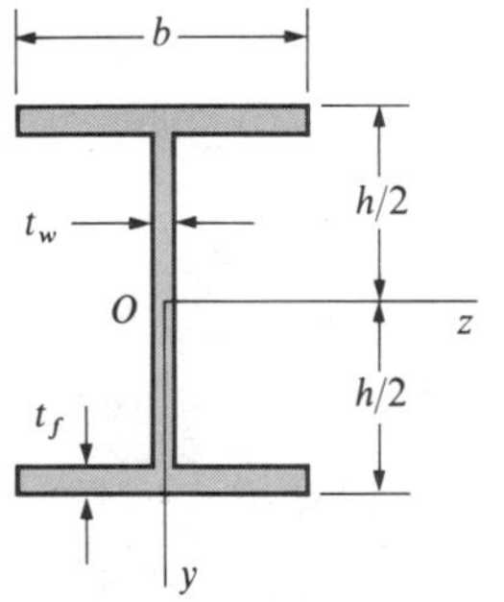

Fig. 5-69 Wide-flange beam

$$Z = bt_f(h - t_f) + t_w\left(\frac{h}{2} - t_f\right)^2 \tag{5-100}$$

in which the first term on the right-hand side comes from the flanges and the second term from the web. Values of Z for standard shapes of wide-flange beams are tabulated in the AISC manual (Ref. 5-4). The shape factor f for wide-flange beams is typically in the range 1.1 to 1.2, depending upon the proportions of the cross section.

MOMENT-CURVATURE RELATION

It has already been observed that for values of bending moment less than the yield moment M_y, the curvature κ is $-M/EI$. If we denote by κ_y the **yield curvature** (that is, the curvature when M equals M_y), we have

$$\kappa_y = -\frac{M_y}{EI} \tag{5-101}$$

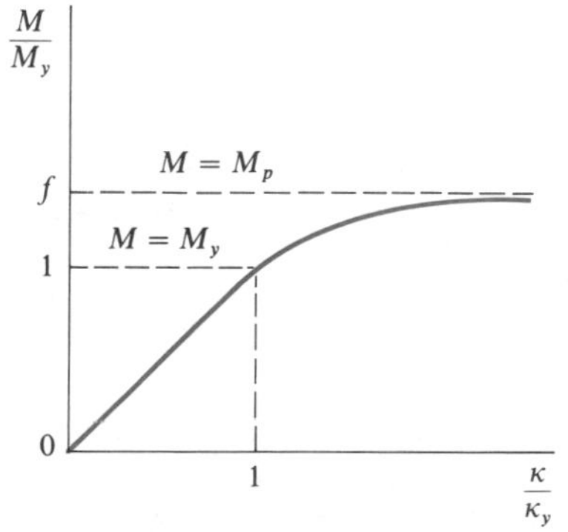

Fig. 5-70 Moment-curvature diagram for a beam of elastic-plastic material

Now we can express the moment-curvature relation for a beam in the linearly elastic range in nondimensional form as follows:

$$\frac{M}{M_y} = \frac{\kappa}{\kappa_y} \qquad (0 \le M \le M_y) \tag{5-102}$$

This equation is represented by the straight-line portion of the moment-curvature diagram shown in Fig. 5-70.

When the moment becomes greater than M_y, part of the beam will become fully plastic as discussed earlier in connection with Fig. 5-67c. The moment-curvature relationship then becomes nonlinear, as shown in Fig. 5-70. As the plastic zone penetrates farther toward the neutral axis of the beam, the curve in Fig. 5-70 becomes flatter and approaches a horizontal line as an asymptote. This asymptote represents the plastic moment M_p; hence the ordinate to the asymptote is the shape factor f. The equation of the curved line depends upon the shape of the cross section.

Let us take again the case of a beam of **rectangular cross section** (Fig. 5-71a), and let e represent the distance from the neutral axis to the edge of the elastic core. The fully plastic zone is shown darker in part (a) of the figure and the stress distribution in the beam is shown in part (b). The bending moment represented by this stress distribution is

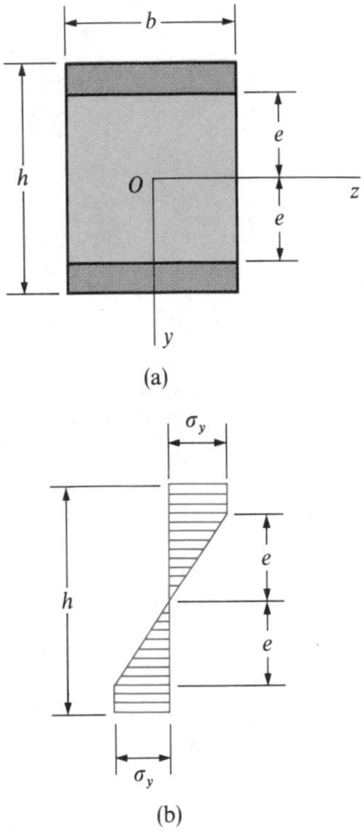

Fig. 5-71 Stress distribution in a beam of rectangular cross section

$$M = \sigma_y b\left(\frac{h}{2} - e\right)\left(\frac{h}{2} + e\right) + \sigma_y b\left(\frac{2e^2}{3}\right) \tag{b}$$

in which the first term on the right-hand side is the moment due to the stresses in the fully plastic zone and the second term is due to the stresses in the elastic core. Simplifying Eq. (b), we get

$$\begin{aligned} M &= \frac{\sigma_y bh^2}{6}\left(\frac{3}{2} - \frac{2e^2}{h^2}\right) \\ &= M_y\left(\frac{3}{2} - \frac{2e^2}{h^2}\right) \end{aligned} \tag{c}$$

Note that when $e = h/2$, we get $M = M_y$ from Eq. (c), and when $e = 0$, we get $M = 3M_y/2$, which is equal to the plastic moment M_p for a rectangular cross section.

The curvature of the rectangular beam shown in Fig. 5-71 can easily be found by applying the equation $\kappa = -\epsilon/y$ (see Eq. 5-87) to a point in the beam at the outer edge of the elastic core. At such a point we have $\epsilon = \sigma_y/E$ and $y = e$, so that

$$\kappa = -\frac{\sigma_y}{Ee} \tag{d}$$

This equation can be expressed in nondimensional form by introducing the following expression for the yield curvature κ_y for a rectangular beam:

$$\kappa_y = -\frac{M_y}{EI} = -\frac{2\sigma_y}{Eh} \tag{e}$$

Combining Eqs. (d) and (e), we get

$$\frac{\kappa}{\kappa_y} = \frac{h}{2e} \tag{f}$$

We can now eliminate e/h between Eqs. (c) and (f) and obtain the moment-curvature equation in the following nondimensional form:

$$\frac{M}{M_y} = \frac{3}{2} - \frac{\kappa_y^2}{2\kappa^2} \qquad (M_y \le M \le M_p) \tag{5-103}$$

Solving this equation for the curvature in terms of the moment, we get

$$\frac{\kappa}{\kappa_y} = \frac{1}{\sqrt{3 - 2M/M_y}} \qquad (M_y \le M \le M_p) \tag{5-104}$$

A graph of the moment-curvature relationship for a rectangular beam (from either Eq. 5-103 or Eq. 5-104) is given in Fig. 5-72.

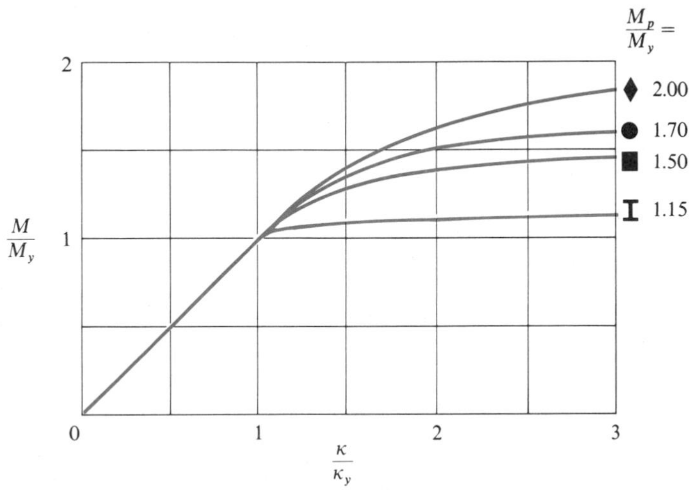

Fig. 5-72 Moment-curvature diagrams for beams of elastic-plastic material

For **other cross sections**, a similar procedure can be followed to obtain moment-curvature equations. Graphs of these equations for a rhombus, a circle, and a typical wide-flange beam are also plotted in Fig. 5-72. In each instance the diagram consists of a straight line representing the linearly elastic region followed by a curved line representing the region where the beam is partially plastic and partially elastic. In the latter region of the diagram, additional yielding may occur in the plastic zone of the beam without an increase in stress, whereas in the central elastic zone of the beam, additional deformation occurs simultaneously with an increase in stress. Thus, the deformation of the beam is controlled by the elastic zone; this type of behavior is sometimes called *contained plastic flow*. When the curvature becomes very large, each curve in Fig. 5-72 approaches a horizontal asymptote. At this stage the beam can continue to deform without any increase in the applied bending moment. Then we have the condition of *unrestricted plastic flow*, and the corresponding bending moment is the plastic moment M_p. The presence of unrestricted plastic flow leads to the concept of a *plastic hinge*, which is used in a method of structural analysis called *plastic analysis* (see, for example, Refs. 5-30, 5-31, and 5-32).

*5.18 ANALYSIS OF INELASTIC BEAMS

In the preceding section we dealt with the special case of an elastic-plastic material. Because this idealization represents very closely the behavior of structural steel, it is especially important in engineering work. Now, however, we shall consider more general kinds of inelastic bending in which the material may have a stress-strain diagram of any shape, typified by curve *AOB* in Fig. 5-73. Let us also consider a beam of rectangular cross section (Fig. 5-74), and let us denote by h_1 and h_2 the distances from the neutral axis to the lower and upper surfaces of the beam, respectively. These distances are equal only if the stress-strain diagram has the same shape in both tension and compression.

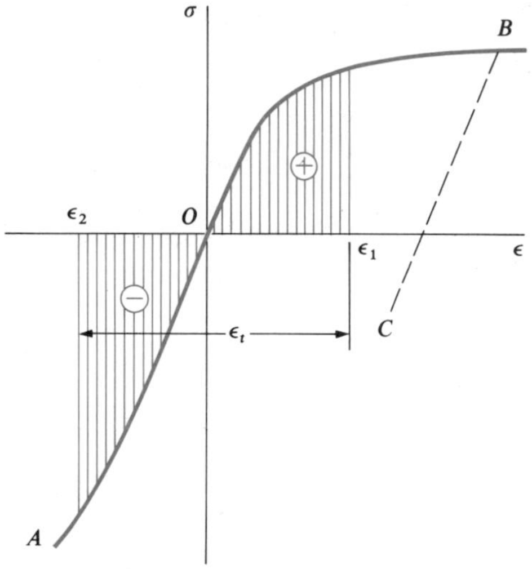

Fig. 5-73 Stress-strain diagram for an inelastic material

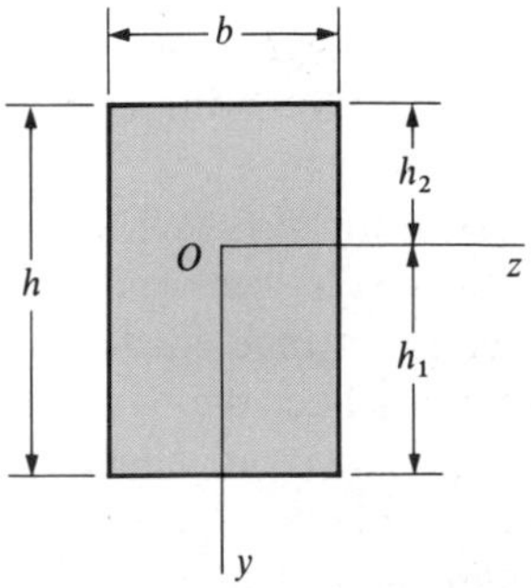

Fig. 5-74 Inelastic beam of rectangular cross section

To locate the neutral axis of the rectangular beam, we use Eqs. (5-87) and (5-89). From Eq. (5-87) we get

$$y = -\rho\epsilon \qquad dy = -\rho\, d\epsilon \tag{a}$$

Substituting into Eq. (5-89), we obtain

$$\int \sigma\, dA = \int_{-h_2}^{h_1} \sigma b\, dy = -\rho b \int_{\epsilon_2}^{\epsilon_1} \sigma\, d\epsilon = 0$$

in which ϵ_1 and ϵ_2 are the strains at the bottom and top of the beam, respectively (see Fig. 5-65). Note that these strains are also indicated on the stress-strain curve of Fig. 5-73, assuming that the curvature is negative and that the bottom of the beam is in tension. The preceding equation shows that the position of the neutral axis is such that

$$\int_{\epsilon_2}^{\epsilon_1} \sigma \, d\epsilon = 0 \tag{5-105}$$

Let us now denote by ϵ_t the sum of the absolute values of the maximum positive and negative strains in the beam:

$$\epsilon_t = \epsilon_1 - \epsilon_2 = -\kappa h_1 - \kappa h_2 = -\kappa h \tag{5-106}$$

To satisfy Eq. (5-105) and thereby locate the neutral axis, we begin by assuming a value of ϵ_t. Then we place the length ϵ_t on the horizontal axis in Fig. 5-73 in such a way as to make the tension area of the diagram equal to the compression area. In this manner we obtain the strains ϵ_1 and ϵ_2 at the extreme points; these strains correspond to the assumed value of the total strain ϵ_t. The corresponding position of the neutral axis is then obtained from Eqs. (5-88), as follows:

$$\frac{h_1}{h_2} = \frac{\epsilon_1}{-\epsilon_2} = \left|\frac{\epsilon_1}{\epsilon_2}\right| \tag{5-107}$$

Because the strains ϵ are linear with distance from the neutral axis (see Fig. 5-65c), we conclude that the stress-strain diagram AOB in Fig. 5-73 represents the distribution of bending stresses over the height of the beam if the distance h is substituted for ϵ_t. Thus, for an assumed value of ϵ_t, we now know the position of the neutral axis and the stresses and strains over the height of the beam. From Eq. (5-106) we also know the curvature of the beam.

The next step is to find the bending moment M through the use of Eq. (5-90). Substituting from Eqs. (a), we can put Eq. (5-90) into the form

$$\int \sigma y \, dA = \int_{-h_2}^{h_1} \sigma y b \, dy = \rho^2 b \int_{\epsilon_2}^{\epsilon_1} \sigma \epsilon \, d\epsilon = M \tag{b}$$

By observing from Eq. (5-106) that $\rho = 1/\kappa = -h/\epsilon_t$, we can write Eq. (b) in the form

$$M = \frac{bh^2}{\epsilon_t^2} \int_{\epsilon_2}^{\epsilon_1} \sigma \epsilon \, d\epsilon \tag{5-108}$$

The integral in this expression represents the first moment with respect to the vertical axis of the area below the stress-strain curve (the shaded area in Fig. 5-73). Note that the first moments of both areas are positive, because both σ and ϵ are positive to the right of the vertical axis and negative to the left. By evaluating these first moments and adding them, we obtain the value of the integral in Eq. (5-108). Then we can calculate the bending moment M, thereby completing the analysis of the beam in pure bending for the assumed value of ϵ_t.

The entire process can be repeated for other values of ϵ_t, each such calculation giving a value of curvature and a corresponding

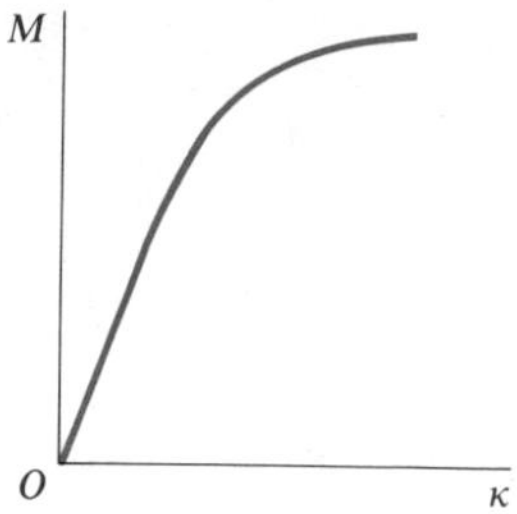

Fig. 5-75 Moment-curvature diagram for inelastic bending

moment. From these data the moment-curvature diagram can be plotted (Fig. 5-75). Such a diagram is for a particular stress-strain curve and for a particular beam of rectangular cross section.

The calculations are simplified if the tension and compression parts of the stress-strain diagram are the same, because then we know immediately that the neutral axis passes through the centroid of the rectangular cross section. The following equations apply to this case;

$$h_1 = h_2 = \frac{h}{2} \tag{5-109}$$

$$\epsilon_1 = -\epsilon_2 = \frac{\epsilon_t}{2} \tag{5-110}$$

$$M = \frac{2bh^2}{\epsilon_t^2}\int_0^{\epsilon_1} \sigma\epsilon\, d\epsilon \tag{5-111}$$

From the last of these equations, we can calculate M for any assumed value of ϵ_t, and then the corresponding curvature can be found from Eq. (5-106).

In the special case of a rectangular beam of linearly elastic material, we have $\sigma = E\epsilon$, and Eq. (5-111) gives

$$M = \frac{2bh^2}{\epsilon_t^2}\int_0^{\epsilon_1} E\epsilon^2\, d\epsilon = \frac{2bh^2E\epsilon_1^3}{3\epsilon_t^2} \tag{c}$$

Now substituting $\epsilon_t = 2\epsilon_1$ and $\sigma_{max} = E\epsilon_1$ into Eq. (c), we get

$$M = \frac{\sigma_{max}bh^2}{6} = \sigma_{max}S \tag{d}$$

where σ_{max} is the stress at the bottom of the beam. Also, if we substitute $\epsilon_t = 2\epsilon_1$ and $\epsilon_1 = -\kappa h/2$ into Eq. (c), we get

$$\kappa = -\frac{M}{Ebh^3/12} = -\frac{M}{EI} \tag{e}$$

Equations (d) and (e) are the familiar equations of linearly elastic bending.

If instead of a rectangle we have some other shape of cross section for which the width b is variable (see Fig. 5-65b, for instance), then instead of Eqs. (5-105) and (5-108) we get

$$\int_{\epsilon_2}^{\epsilon_1} \sigma b\, d\epsilon = 0 \tag{5-112}$$

$$M = \frac{h^2}{\epsilon_t^2}\int_{\epsilon_2}^{\epsilon_1} \sigma b\epsilon\, d\epsilon \tag{5-113}$$

Note that the width b has been retained under the integral sign.

As a specific example, let us consider a T-section (Fig. 5-76). Denoting by ϵ_3 the strain at the junction of the web and flange, we can write the

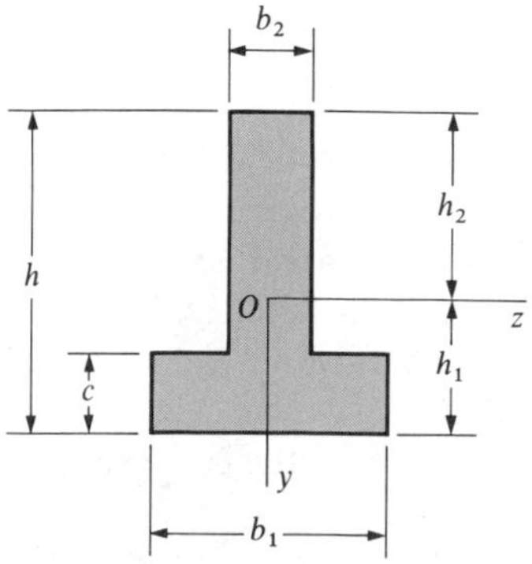

Fig. 5-76 Inelastic beam of T-section

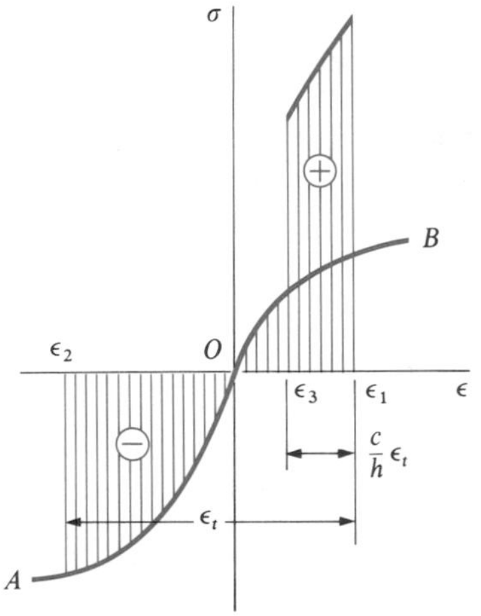

Fig. 5-77 Modified stress-strain diagram for a T-section

preceding equations in the following form:

$$\int_{\epsilon_2}^{\epsilon_3} \sigma\, d\epsilon + \int_{\epsilon_3}^{\epsilon_1} \sigma \frac{b_1}{b_2}\, d\epsilon = 0 \qquad \text{(f)}$$

$$M = \frac{b_2 h^2}{\epsilon_t^2}\left[\int_{\epsilon_2}^{\epsilon_3} \sigma\epsilon\, d\epsilon + \int_{\epsilon_3}^{\epsilon_1} \sigma \frac{b_1}{b_2}\epsilon\, d\epsilon\right] \qquad \text{(g)}$$

We see from these equations that the ordinates of the stress-strain diagram in the region corresponding to the flange of the cross section must be magnified in the ratio b_1/b_2 (see Fig. 5-77). In determining the position of the neutral axis, we proceed as in the earlier case and mark on the horizontal axis of Fig. 5-77 the position of the assumed strain ϵ_t, such that the two shaded areas become numerically equal. The position must be found, of course, by trial and error. For each trial position, the strain ϵ_3 at the junction of the web and flange is obtained from the equation

$$\frac{\epsilon_1 - \epsilon_3}{\epsilon_t} = \frac{c}{h} \qquad \text{(h)}$$

By proceeding in this manner, we can obtain the strains ϵ_1 and ϵ_2 at the bottom and top of the beam.

Having located the neutral axis of the T-section from Eq. (f), we can now find the bending moment from Eq. (g). We note that the two integrals within the brackets represent the first moments of the shaded areas in Fig. 5-77 with respect to the vertical axis through the origin *O*. By calculating this first moment and then substituting into Eq. (g), we get the bending moment *M* corresponding to the assumed value of ϵ_t. We can also find the corresponding curvature κ from Eq. (5-106), and then we can construct the moment-curvature diagram for the beam of T-section.

The preceding description of the analysis of a T-beam in inelastic bending is quite general and can be used for any stress-strain curve and for other shapes of cross sections. Whenever the stress-strain curve can be approximated by a mathematical expression, it is theoretically possible to determine the stresses, strains, and curvature by direct analysis. However, this procedure is feasible only in relatively simple cases, as illustrated in the following example for a beam of rectangular cross section.

Example

A beam of rectangular cross section (see Fig. 5-74) is constructed of material having a stress-strain diagram consisting of two straight lines, as shown in Fig. 5-78. The modulus of elasticity in tension is E_1 and in compression is E_2, so that

$$\sigma = E_1\epsilon \qquad (\epsilon \geq 0) \qquad \text{(i)}$$

$$\sigma = E_2\epsilon \qquad (\epsilon \leq 0) \qquad \text{(j)}$$

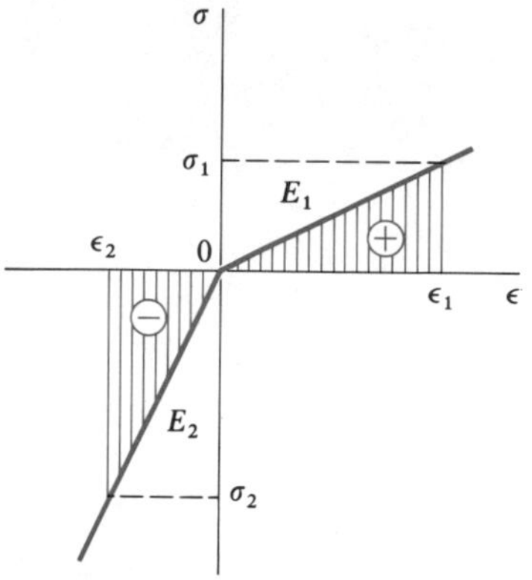

Fig. 5-78 Example. Stress-strain diagram consisting of two straight lines

The beam is assumed to be subjected to a positive bending moment M. The objectives of the analysis are to locate the neutral axis, obtain the moment-curvature expression, and find the maximum stresses and strains in the beam.

Let us denote the strains at the bottom and top of the beam as ϵ_1 and ϵ_2, respectively. The corresponding maximum stresses are σ_1 and σ_2. These stresses and strains are indicated in Fig. 5-78. To locate the neutral axis, we observe that the two shaded areas under the stress-strain curve must be equal, so that

$$\frac{\sigma_1 \epsilon_1}{2} = \frac{\sigma_2 \epsilon_2}{2} \tag{k}$$

From Eqs. (i) and (j) we know that

$$\sigma_1 = E_1 \epsilon_1 \qquad \sigma_2 = E_2 \epsilon_2 \tag{l}$$

Also, we have the relations

$$\epsilon_1 = -\kappa h_1 \qquad \epsilon_2 = \kappa h_2 \tag{m}$$

Substituting from Eqs. (l) and (m) into Eq. (k), we get

$$E_1 h_1^2 = E_2 h_2^2 \tag{n}$$

as the first equation relating h_1 and h_2. In addition, we have the equation

$$h = h_1 + h_2 \tag{o}$$

The distances h_1 and h_2 can now be found by simultaneous solution of Eqs. (n) and (o), yielding

$$h_1 = \frac{h\sqrt{E_2}}{\sqrt{E_1} + \sqrt{E_2}} \qquad h_2 = \frac{h\sqrt{E_1}}{\sqrt{E_1} + \sqrt{E_2}} \tag{5-114a,b}$$

Thus, the position of the neutral axis is determined.

Next we calculate the bending moment by substituting from Eqs. (i) and (j) into Eq. (5-108); thus,

$$M = \frac{bh^2}{\epsilon_t^2}\left[\int_{\epsilon_2}^{0} E_2 \epsilon^2 \, d\epsilon + \int_0^{\epsilon_1} E_1 \epsilon^2 \, d\epsilon\right]$$

which becomes

$$M = \frac{bh^2}{3\epsilon_t^2}(-E_2\epsilon_2^3 + E_1\epsilon_1^3) \tag{p}$$

The strains ϵ_1 and ϵ_2 are related to the curvature by Eqs. (m). Substituting for h_1 and h_2(Eqs. 5-114), we get

$$\epsilon_1 = -\kappa h \frac{\sqrt{E_2}}{\sqrt{E_1} + \sqrt{E_2}} \qquad \epsilon_2 = \kappa h \frac{\sqrt{E_1}}{\sqrt{E_1} + \sqrt{E_2}}$$

Also, the strain ϵ_t is equal to $-\kappa h$ (see Eq. 5-106). We can now substitute these expressions for ϵ_1, ϵ_2, and ϵ_t into Eq. (p), obtaining

$$M = -\frac{4E_1 E_2 I\kappa}{(\sqrt{E_1} + \sqrt{E_2})^2} \tag{q}$$

in which $I = bh^3/12$. Finally, let us introduce the notation

$$E_r = \frac{4E_1E_2}{(\sqrt{E_1} + \sqrt{E_2})^2} \tag{5-115}$$

so that Eq. (q) becomes $M = -E_r I\kappa$ and the curvature becomes

$$\kappa = -\frac{M}{E_r I} \tag{5-116}$$

The quantity E_r, called the *reduced modulus of elasticity*, has a value between E_1 and E_2. In the special case when both moduli are equal to E, the reduced modulus E_r also equals E.

The stresses and strains at the extreme points of the cross section are now readily found in terms of the bending moment. We take Eqs. (m) for ϵ_1 and ϵ_2 and substitute Eq. (5-116) for the curvature, obtaining

$$\epsilon_1 = \frac{Mh_1}{E_r I} \qquad \epsilon_2 = -\frac{Mh_2}{E_r I} \tag{5-117a,b}$$

Next, the stress-strain equations (Eqs. 1) yield

$$\sigma_1 = \frac{Mh_1}{I}\frac{E_1}{E_r} \qquad \sigma_2 = -\frac{Mh_2}{I}\frac{E_2}{E_r} \tag{5-118a,b}$$

and the analysis of this beam is completed. Equations (5-118) are flexure formulas for a rectangular beam constructed of a material having different moduli in tension and compression.

PROBLEMS / CHAPTER 5

5.2-1 Determine the maximum normal strain ϵ_{max} produced in a steel wire of diameter $d = \frac{1}{16}$ in. when it is bent around a cylinder of radius $r = 20$ in., as shown in the figure. (Assume that the neutral surface passes through the center of the cross section of the wire.)

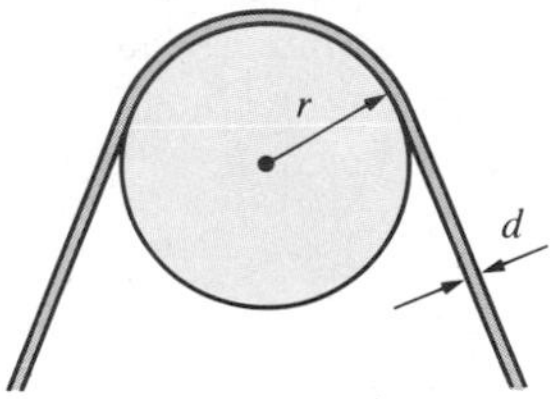

Prob. 5.2-1

5.2-2 A copper wire having diameter $d = 3$ mm is bent into a circle and held with the ends just touching (see figure). If the maximum permissible strain in the copper is $\epsilon_{max} = 2.4 \times 10^{-3}$, what is the shortest length L of wire that can be used? (Assume that the neutral surface passes through the center of the cross section of the wire.)

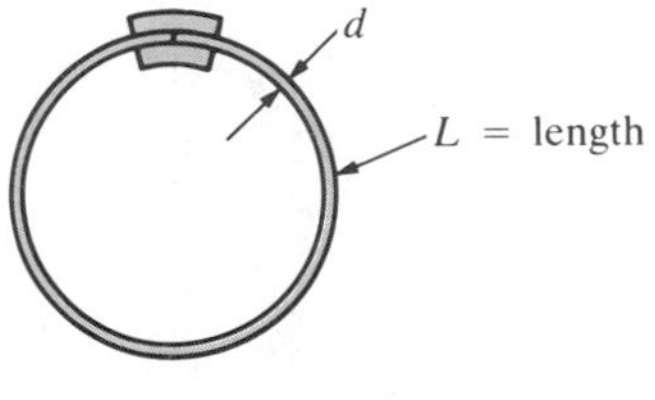

Prob. 5.2-2

5.2-3 A cantilever beam AB is subjected to a couple M_0 acting at its free end (see figure). The length of the beam is $L = 4$ ft and the longitudinal normal strain at the top surface is 0.0008. The distance from the top surface of the beam to the neutral surface is 2 in. Calculate the radius of curvature ρ, the curvature κ, and the vertical deflection δ at the end of the beam.

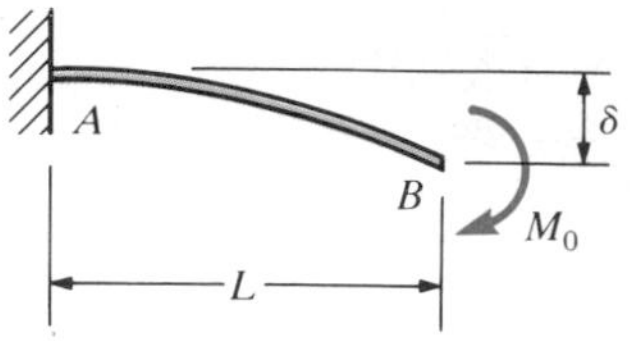

Prob. 5.2-3

5.2-4 A thin strip of steel of length $L = 0.8$ m and thickness $t = 2$ mm is bent by couples M_0 (see figure). The deflection δ at the midpoint of the strip (measured from a line joining its end points) is found to be 143 mm. (a) Calculate the radius of curvature ρ of the strip. (b) Assuming that the neutral surface is equidistant from the top and bottom surfaces of the strip, calculate the longitudinal normal strain ϵ at the top surface of the strip.

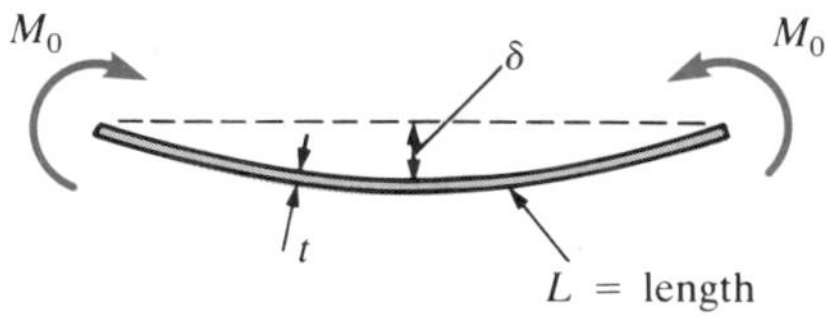

Prob. 5.2-4

5.3-1 A steel wire ($E = 30 \times 10^6$ psi) of diameter $d = \frac{1}{32}$ in. is bent around a pulley of radius $r = 8$ in. (see figure). (a) What is the maximum stress σ_{max} in the wire? (b) Does the stress increase or decrease if the radius of the pulley is increased?

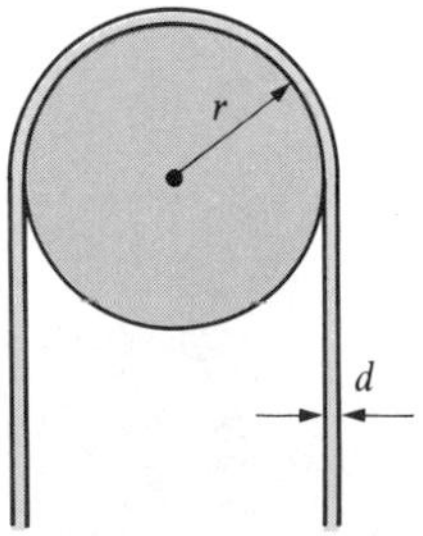

Prob. 5.3-1

5.3-2 A thin strip of copper ($E = 120$ GPa) having length $L = 2$ m and thickness $t = 1$ mm is bent into a circle and held with the ends just touching (see figure). (a) Calculate the maximum bending stress σ_{max} in the strip. (b) Does the stress increase or decrease if the thickness of the strip is increased?

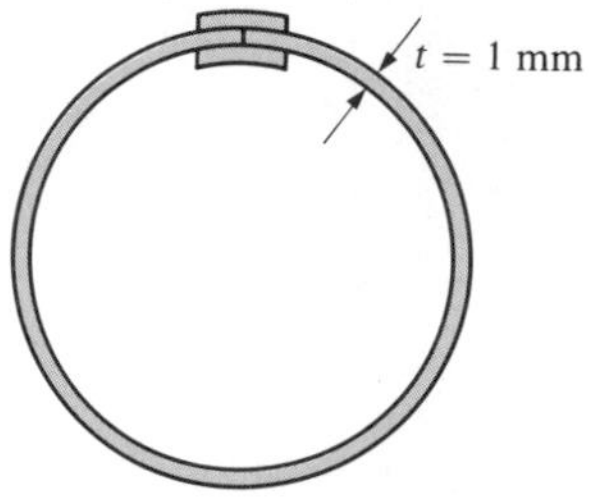

Prob. 5.3-2

5.3-3 A thin steel rule ($E = 30 \times 10^6$ psi) having thickness $t = \frac{1}{32}$ in. and length $L = 10$ in. is bent by couples M_0 into a circular arc subtending a central angle $\alpha = 52.5°$ (see figure). (a) What is the maximum bending stress σ_{max} in the rule? (b) Does the stress increase or decrease if the central angle is increased?

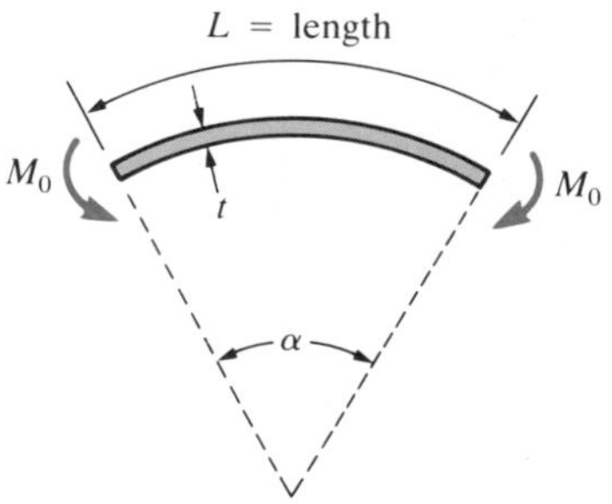

Prob. 5.3-3

5.3-4 A simple beam AB with span length $L = 3.5$ m carries a uniform load of intensity $q = 6.4$ kN/m (see figure). Calculate the maximum bending stress σ_{max} due to the load q if the beam has a rectangular cross section with width $b = 140$ mm and height $h = 240$ mm.

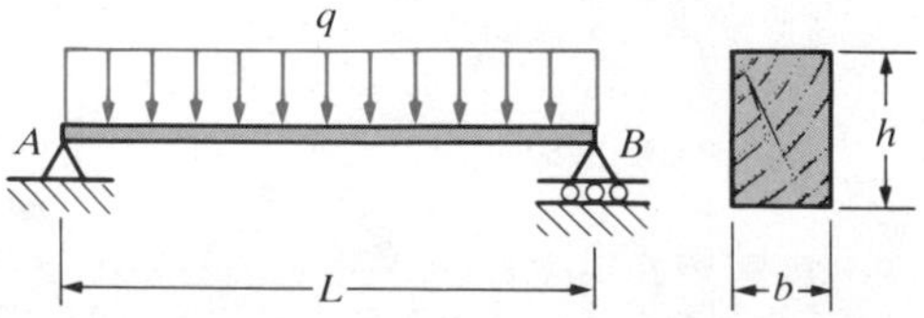

Probs. 5.3-4 and 5.3-5

5.3-5 A simple beam AB with span length $L = 16$ ft carries a uniform load of intensity $q = 250$ lb/ft (see figure). Calculate the maximum bending stress σ_{max} due to the load q if the beam has a rectangular cross section with width $b = 5\frac{5}{8}$ in. and height $h = 7\frac{5}{8}$ in.

5.3-6 A freight-car axle AB is loaded as shown in the figure, with the forces P representing the car loads (transmitted through the axle boxes) and the forces R representing the rail loads (transmitted through the wheels). The diameter of the axle is $d = 80$ mm, the wheel gauge is $L = 1.45$ m, and the distance $b = 200$ mm. Calculate the maximum bending stress σ_{max} in the axle if $P = 8.9$ kN.

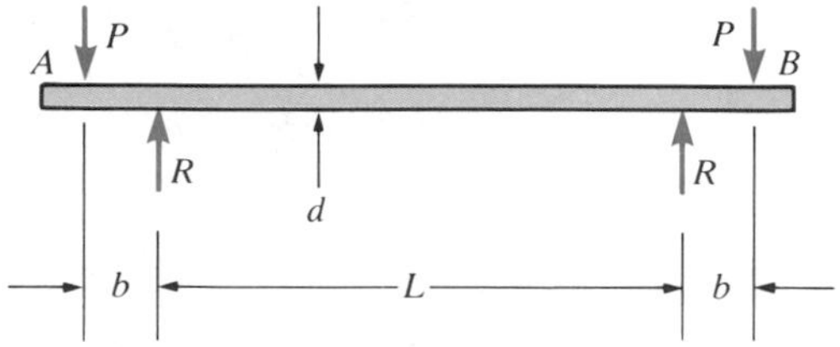

Prob. 5.3-6

5.3-7 A steel beam with overhanging ends and wide-flange cross section is supported as shown in the figure. The beam carries a uniform load of intensity $q = 8$ k/ft on each overhang. Assuming that the section modulus of the cross-sectional area is $S = 414$ in.^{3}, determine the maximum bending stress σ_{max} in the beam due to the load q.

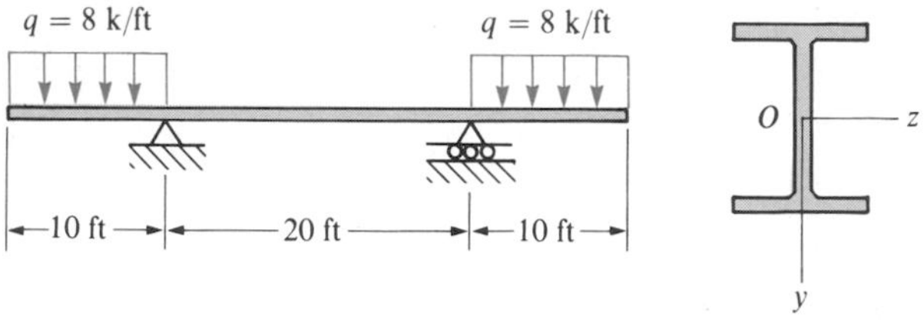

Prob. 5.3-7

5.3-8 A simple beam with span length $L = 2$ m is subjected to a uniform load of intensity $q = 60$ kN/m (see figure). The beam has a rectangular cross section with width $b = 50$ mm and height $h = 150$ mm. (a) Determine the maximum normal stress σ_{max} due to the load q. (b) Also determine the normal stress σ_c at point C located by the distances $c = 500$ mm and $d = 25$ mm.

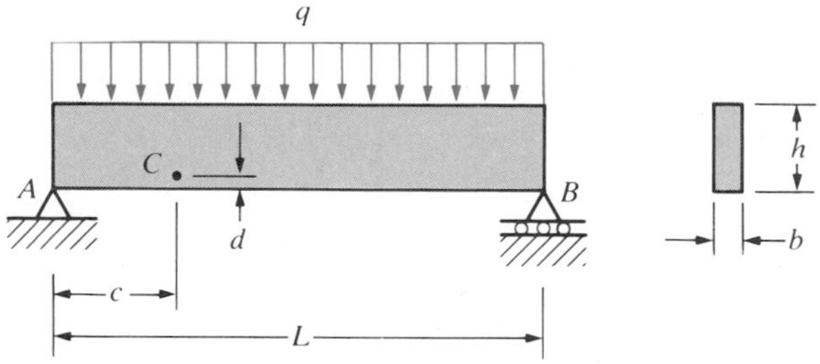

Prob. 5.3-8

5.3-9 A bar AB functioning as a simple beam is loaded by two forces $P = 8$ k (see figure). The span length is 48 in. and the loads act at the quarter points of the beam. The cross section is rectangular with width 2 in. and height 4 in., and the material is steel with modulus of elasticity $E = 30 \times 10^6$ psi and Poisson's ratio $\nu = 0.30$. Disregard the weight of the beam itself. (a) Calculate the maximum bending stress σ_{max} in the beam. (b) Calculate the change Δb in the width of the top of the beam at the midpoint of the span, and state whether the change is an increase or a decrease.

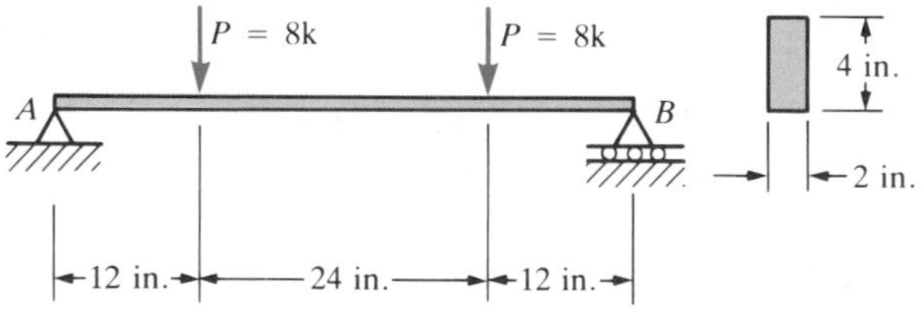

Prob. 5.3-9

5.3-10 A small dam of height $h = 2.4$ m is constructed of vertical wood beams AB of thickness $t = 160$ mm, as shown in the figure. Consider the beams to be simply supported at the top and bottom. Determine the maximum bending stress σ_{max} in the beams, assuming that the specific weight of water is $\gamma = 9.81$ kN/m^3.

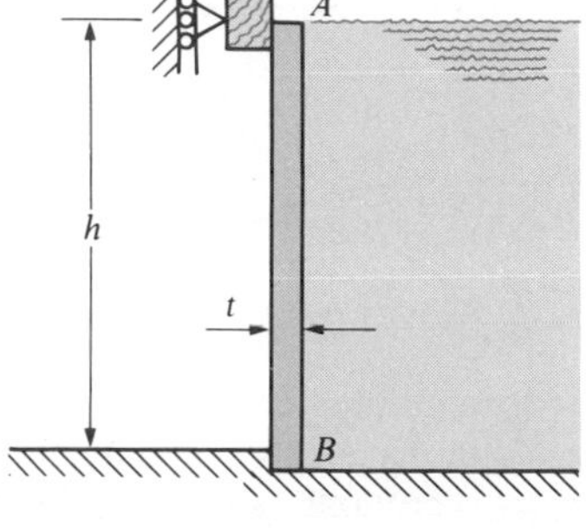

Prob. 5.3-10

5.3-11 A railroad tie (or *sleeper*) is subjected to two concentrated loads $P = 45$ k, acting as shown in the figure. The reaction q of the ballast may be assumed to be uniformly distributed over the length of the tie, which has cross-sectional dimensions $b = 12$ in. and $h = 10$ in. Calculate the maximum bending stress σ_{max} in the tie due to the loads P, assuming $L = 57$ in. and $a = 19.5$ in.

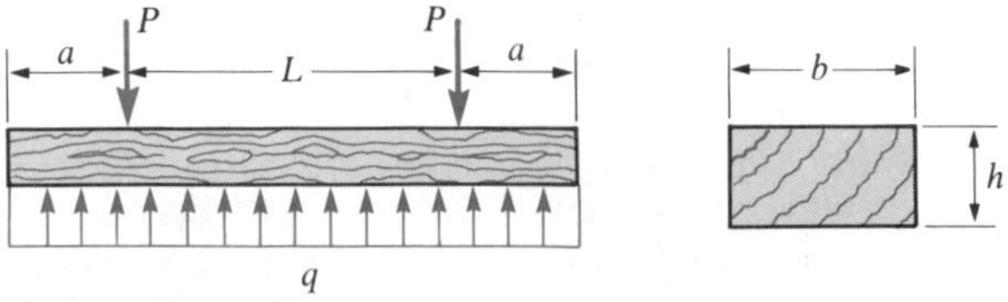

Prob. 5.3-11

5.3-12 Obtain formulas for the maximum normal stress σ_{max} (due to pure bending by moments M) for beams having cross sections as follows (see figure): (a) a hollow circle with $d_1 = d$ and $d_2 = 1.5d$; (b) the core of a circle of diameter d with the angle $\beta = 45°$; and (c) a regular hexagon with sides of length b.

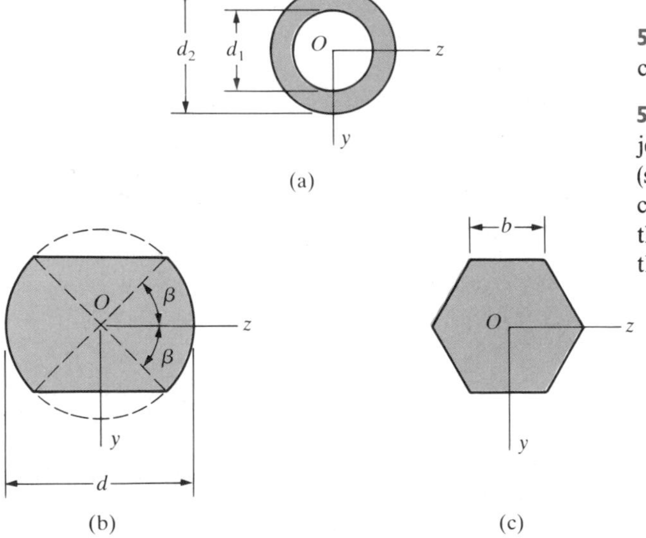

Prob. 5.3-12

5.3-13 Obtain formulas for the maximum tensile stress σ_t (due to pure bending by positive bending moments M) for beams having cross sections as follows: (a) an equilateral triangle with sides of length b; (b) a semicircle of diameter d; and (c) an isosceles trapezoid with bases $b_1 = b$, $b_2 = 3b/2$, and altitude h.

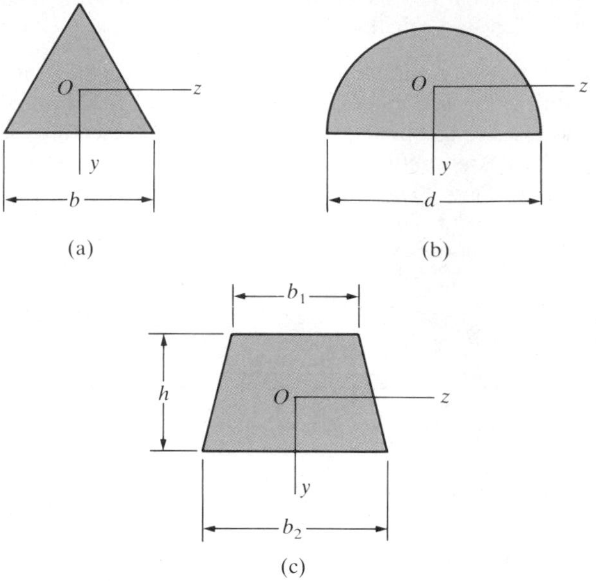

Probs. 5.3-13 and 5.3-14

5.3-14 Solve the preceding problem for the maximum compressive stress σ_c.

5.3-15 A simple beam AB of span length $L = 24$ ft is subjected to two wheel loads acting at distance $d = 5$ ft apart (see figure). Each wheel transmits a load $P = 3$ k, and the carriage may occupy any position on the beam. Determine the maximum bending stress σ_{max} due to the wheel loads if the beam is an I-beam having section modulus $S = 16.2$ in.^{3}

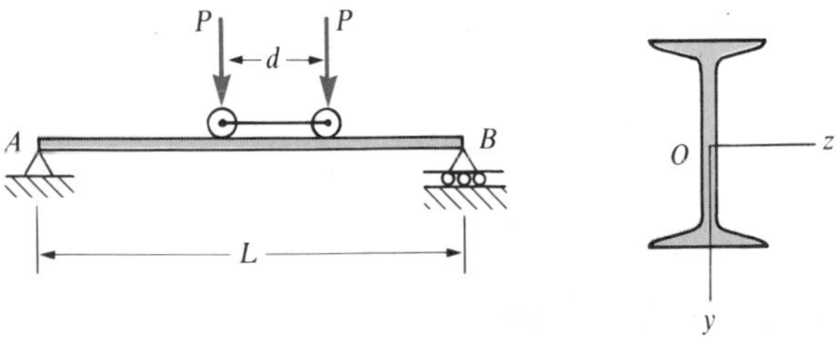

Prob. 5.3-15

5.3-16 A wood beam $ABCD$ of rectangular cross section 100 mm × 250 mm is supported at points B and C (see figure). Forces $P = 16$ kN act at the ends of the overhangs.

The span length $L = 2.5$ m and the overhangs are of length $a = 0.3$ m. Determine the maximum bending stress σ_{max} due to the following loads: (a) the forces P acting alone; (b) the weight of the beam acting alone; and (c) the forces P and the weight of the beam acting simultaneously. (Assume the specific weight of the wood $\gamma = 5.5$ kN/m^3.)

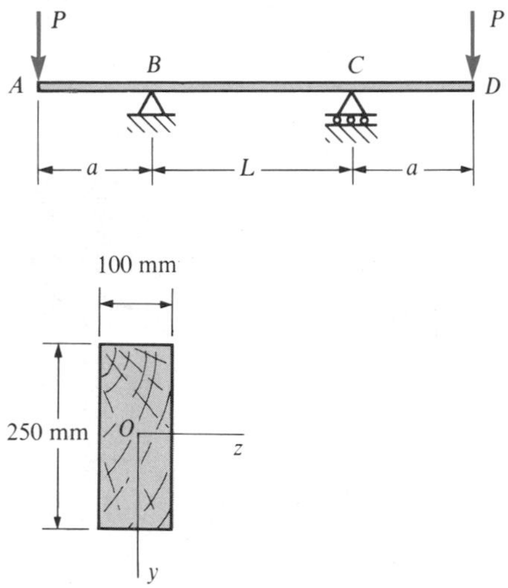

Prob. 5.3-16

5.3-17 A cantilever beam AB, loaded as shown in the figure, is constructed of a channel section. Find the maximum tensile stress σ_t and maximum compressive stress σ_c if the cross section has the dimensions indicated and the moment of inertia about the neutral axis is $I = 2.81$ in.^{4} (Note: The uniform load represents the weight of the beam.)

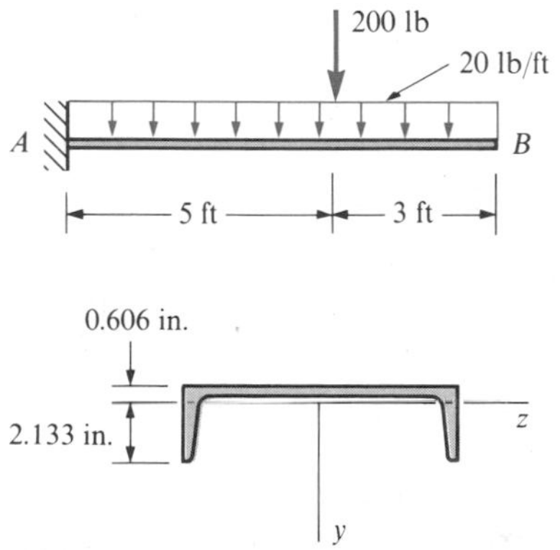

Prob. 5.3-17

5.3-18 A frame ABC travels horizontally with an acceleration a_0 (see figure). Obtain a formula for the maximum stress σ_{max} in the vertical arm AB, which has length L, thickness t, and mass density ρ.

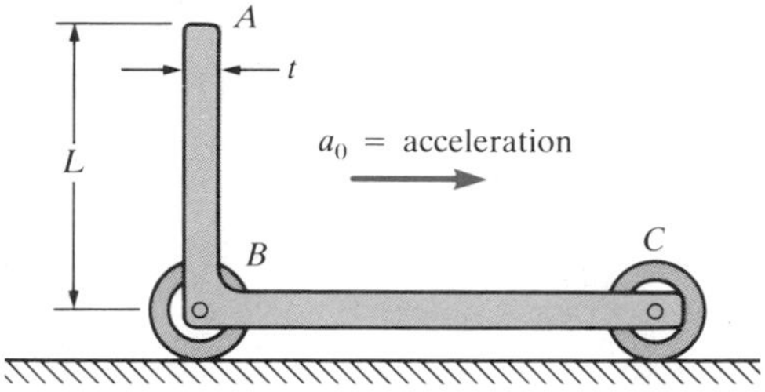

Prob. 5.3-18

5.3-19 Determine the maximum tensile stress σ_t and maximum compressive stress σ_c due to the concentrated load P acting on the simple beam AB (see figure). Data are as follows: $P = 1.8$ k, $L = 9$ ft, $c = 3.5$ ft, $b = 3$ in., $b_1 = 1$ in., $h = 4$ in., and $h_1 = 3$ in.

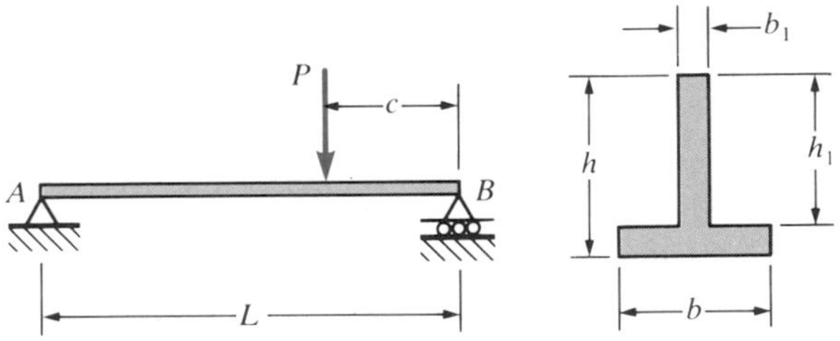

Prob. 5.3-19

5.3-20 A beam ABC supports a concentrated load $P = 4.0$ kN at the end of the overhang (see figure). The cross section of the beam is T-shaped with dimensions as shown. Calculate the maximum tensile stress σ_t and maximum compressive stress σ_c due to the load P.

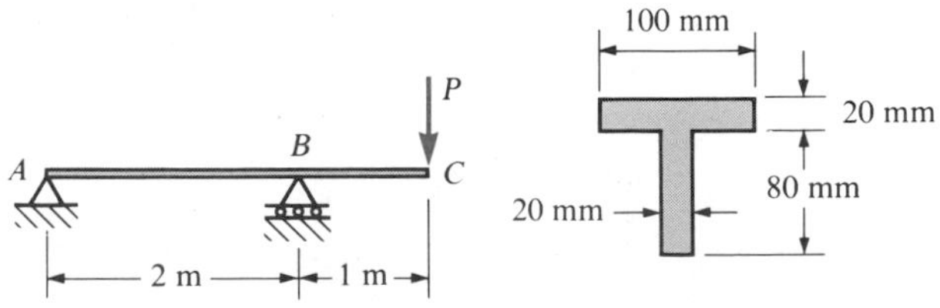

Prob. 5.3-20

5.3-21 A beam with an overhang supports a uniform load of 200 lb/ft throughout its length (see figure). The cross section of the beam has the shape of a channel with dimensions as shown in the figure. The moment of inertia about the z axis equals 5.14 in.^{4} Calculate the maximum tensile stress σ_t and maximum compressive stress σ_c in the beam due to the uniform load.

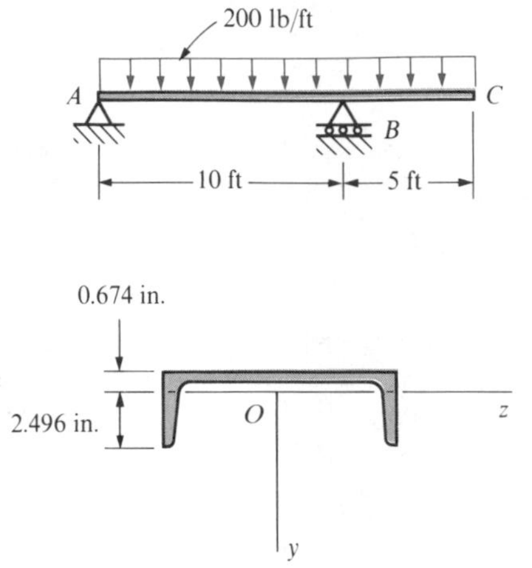

Prob. 5.3-21

5.3-22 A channel beam having the cross-sectional shape shown in the figure is simply supported at the ends (span length $L = 3$ m) and carries a concentrated load $P = 90$ kN at the midpoint. Determine the maximum tensile stress σ_t and maximum compressive stress σ_c due to the load P, assuming $b = 450$ mm, $h = 180$ mm, and $t = 30$ mm.

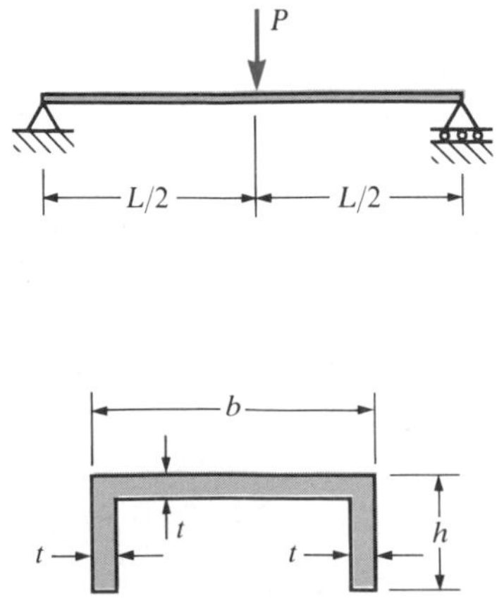

Prob. 5.3-22

***5.3-23** A small dam of height $h = 6$ ft is constructed of vertical wood beams AB, as shown in the figure. The wood beams, which have thickness $t = 2.5$ in., are simply supported by horizontal steel beams at A and B. Construct a graph showing the maximum normal stress σ_{max} in the wood beams versus the depth d of the water above the lower support at B. Plot σ_{max} (psi) as the ordinate of the graph and d (ft) as the abscissa. (Specific weight of water $\gamma = 62.4$ lb/ft^3.)

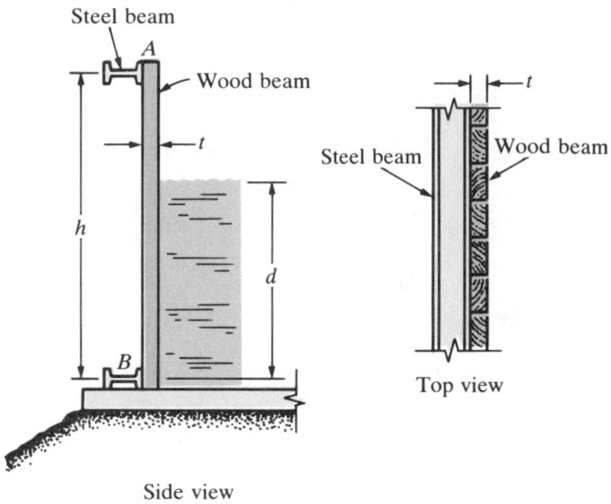

Prob. 5.3-23

5.4-1 A cantilever beam of length $L = 6$ ft supports a uniform load of intensity $q = 200$ lb/ft and a concentrated load $P = 2500$ lb acting at the free end (see figure). Calculate the required section modulus S if $\sigma_{allow} = 16{,}000$ psi. Select a suitable wide-flange beam (W shape) from Table E-1, Appendix E, and recalculate S taking into account the weight of the beam. Select a new beam size if necessary.

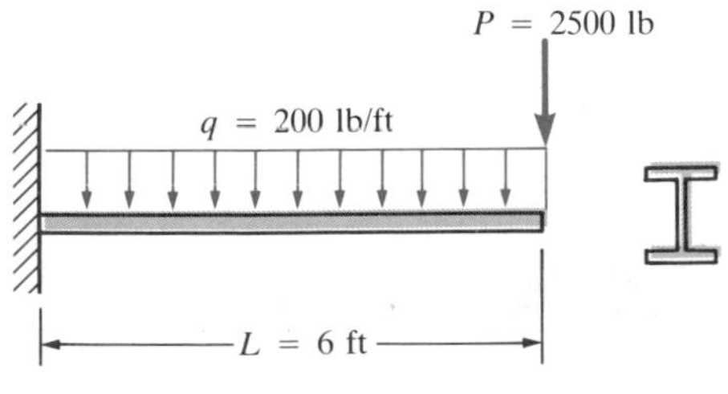

Prob. 5.4-1

5.4-2 A simple beam of length $L = 15$ ft carries a uniform load of intensity $q = 400$ lb/ft acting throughout the length of the beam and a concentrated load $P = 4000$ lb acting at the midpoint (see figure). Assuming $\sigma_{allow} = 18{,}000$ psi, calculate the required section modulus S. Select a suitable wide-flange beam (W shape) from Table E-1, Appendix E, and recalculate S taking into account the weight of the beam. Select a new beam size if necessary.

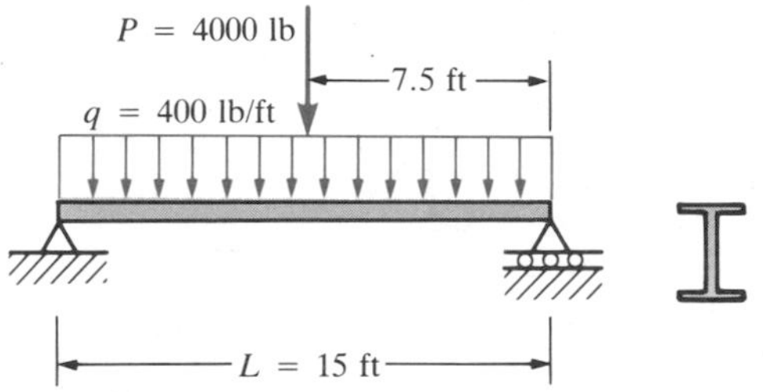

Prob. 5.4-2

5.4-3 A simple beam AB is loaded as shown in the figure. Calculate the required section modulus S if $\sigma_{allow} = 16{,}000$ psi, $L = 24$ ft, $P = 2000$ lb, and $q = 300$ lb/ft. Select a suitable I-beam (S shape) from Table E-2, Appendix E, and recalculate S taking into account the weight of the beam. Select a new beam size if necessary.

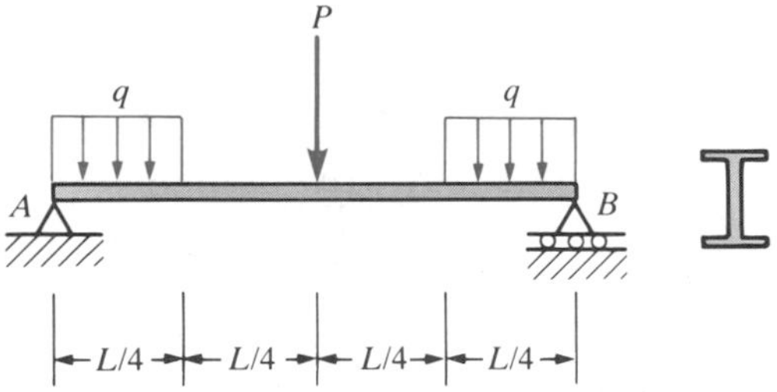

Prob. 5.4-3

5.4-4 A fiberglass bracket $ABCD$ of solid circular cross section has the shape shown in the figure. A vertical load $P = 36$ N acts at the free end D. Determine the minimum permissible diameter d_{min} of the bracket if the allowable bending stress in the material is 30 MPa and $b = 100$ mm. (Disregard the weight of the bracket itself.)

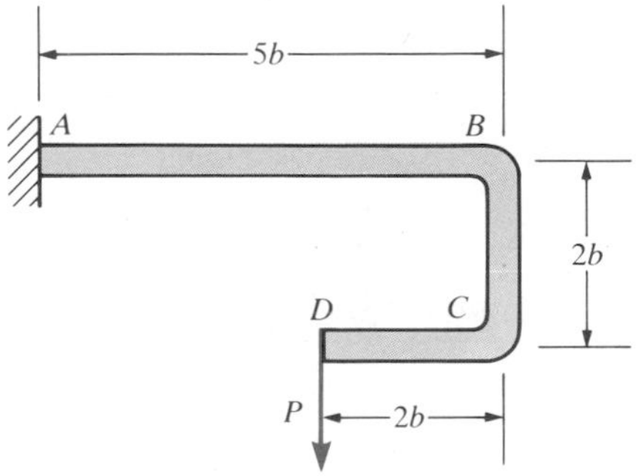

Prob. 5.4-4

5.4-5 A cantilever beam AB (see figure) is made of wood and has a circular cross section. The length $L = 7.2$ ft and the allowable bending stress is 1200 psi. A concentrated load $P = 750$ lb acts at the free end. Calculate the required diameter d_{min} of the beam, disregarding the weight of the beam itself.

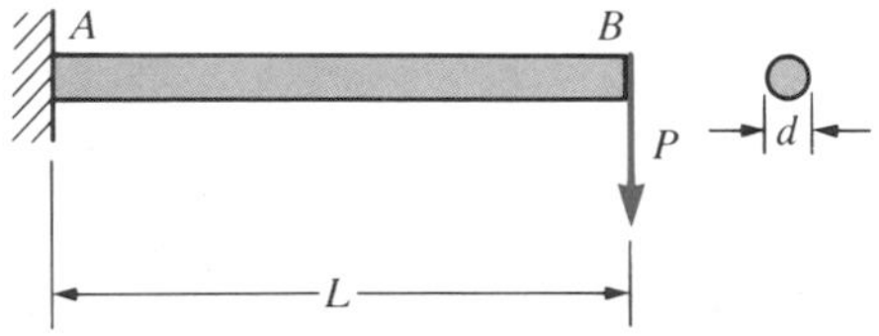

Probs. 5.4-5 and 5.4-6

5.4-6 A cantilever beam of circular cross section and length $L = 1.0$ m supports a load $P = 50$ N acting at the free end (see figure). The beam is made of steel with an allowable bending stress of 80 MPa. Determine the required diameter d_{min} of the beam, including the effect of the beam's own weight.

5.4-7 A two-axle carriage that is part of a crane in a testing laboratory moves slowly across a simple beam AB (see figure). The load transmitted to the beam from the front axle is 2000 lb and from the rear axle is 4000 lb. The weight of the beam itself may be disregarded. (a) Determine the minimum required section modulus S for the beam if the allowable bending stress is 15,000 psi, the length of the beam is 16 ft, and the wheelbase of the carriage is 5 ft. (b) Select a suitable I-beam (S shape) from Table E-2, Appendix E.

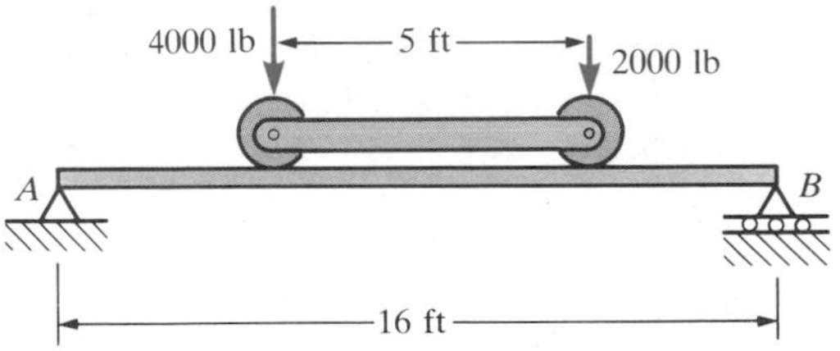

Prob. 5.4-7

5.4-8 A wood beam ABC of square cross section is supported at A and B and has an overhang BC (see figure). A uniform load of intensity $q = 1.5$ kN/m acts on the beam.

Calculate the required side dimension b of the square cross section assuming $L = 2.5$ m and $\sigma_{allow} = 12$ MPa. Include the effect of the weight of the beam and assume the specific weight of the wood is $\gamma = 5.5\ \mathrm{kN/m^3}$.

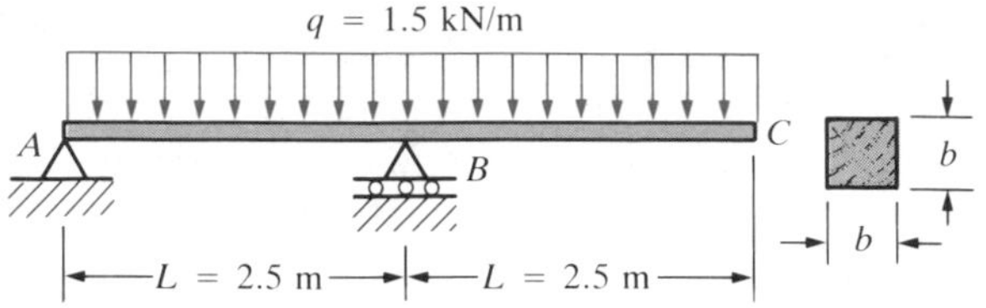

Prob. 5.4-8

5.4-9 A compound beam *ABCD* is supported at points *A*, *B*, and *D* and has a splice (represented by the pin connection) at point *C*, as shown in the figure. The distance $a = 8$ ft and the beam is a W 10 × 60 wide-flange section with an allowable stress of 16,000 psi. Find the allowable uniform load q that may be placed on top of the beam, taking into account the weight of the beam.

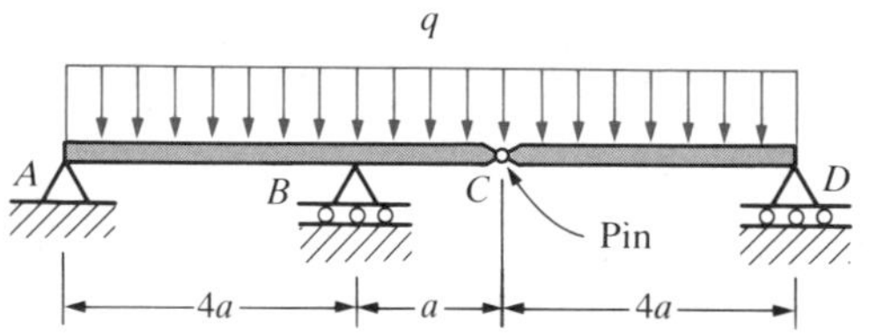

Prob. 5.4-9

5.4-10 A metal beam *ABC* is simply supported at *A* and *B* and has an overhang *BC* of length $L = 2$ m (see figure). The beam supports a uniform load of intensity $q = 100$ kN/m over its entire length of 6 m. The cross section of the beam is rectangular with its height equal to twice the width b. The allowable bending stress in the metal is $\sigma_{allow} = 90$ MPa and its specific weight is $\gamma = 77.0\ \mathrm{kN/m^3}$. (a) Disregarding the weight of the beam, calculate the required width b of the rectangular cross section. (b) Taking into account the weight of the beam, calculate the required width b.

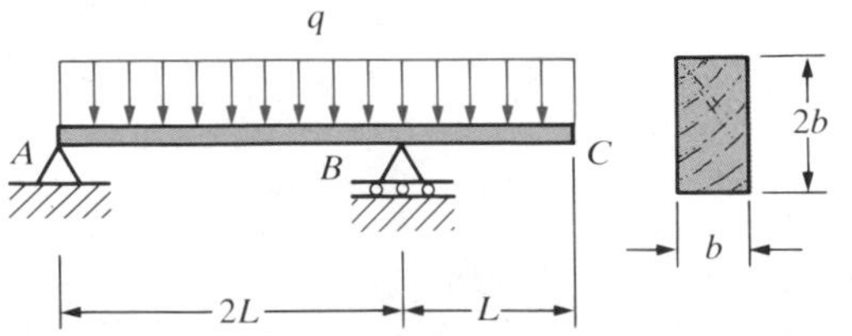

Prob. 5.4-10

5.4-11 A rectangular wood beam is to be cut from a circular log of diameter d (see figure). What should be the dimensions b and h in order that the beam will resist the largest possible bending moment?

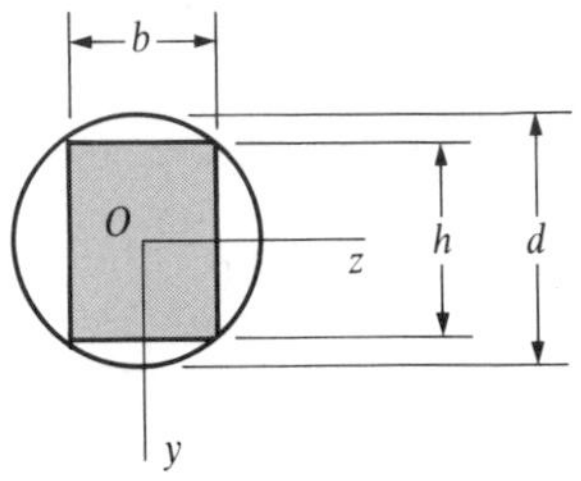

Prob. 5.4-11

5.4-12 A beam having a cross section in the form of a channel (see figure) is subjected to a bending moment acting about the z axis. Calculate the thickness t of the channel in order that the bending stresses at the top and bottom of the beam will be in the ratio 7:3, respectively.

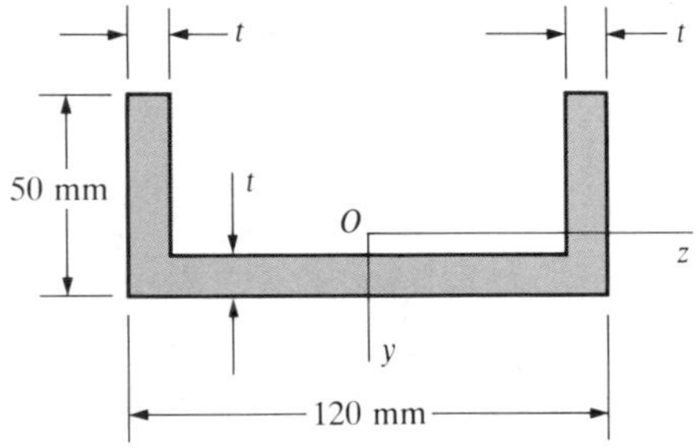

Prob. 5.4-12

5.4-13 Determine the width b of the flange of the T-beam shown in the figure so that the normal stresses at the top and bottom of the beam will be in the ratio 3:1, respectively, when the beam is bent about the z axis.

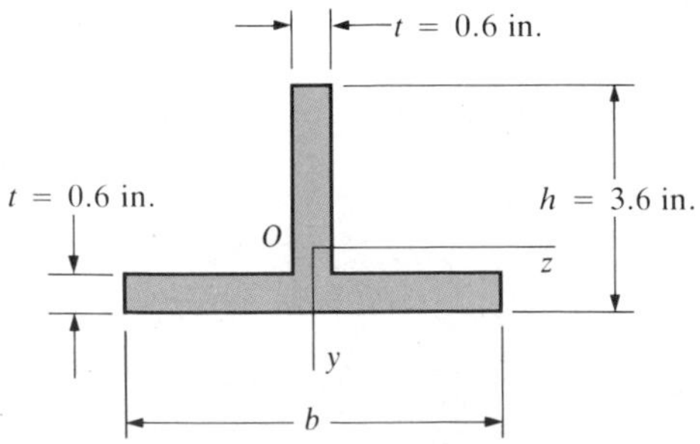

Prob. 5.4-13

5.4-14 A beam having a cross section in the form of an unsymmetric I section (see figure) is subjected to a bending moment acting about the z axis. Determine the width b of the top flange in order that the stresses at the top and bottom of the beam will be in the ratio 4:3, respectively.

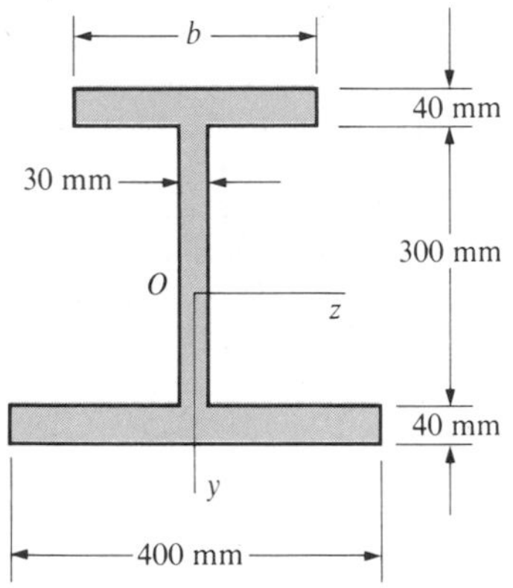

Prob. 5.4-14

5.4-15 Determine the ratios of the weights of three beams that have the same length, are made of the same material, are subjected to the same maximum bending moment, and have the same maximum normal stress if their cross sections are: (1) a rectangle with height equal to twice the width, (2) a square, and (3) a circle (see figure).

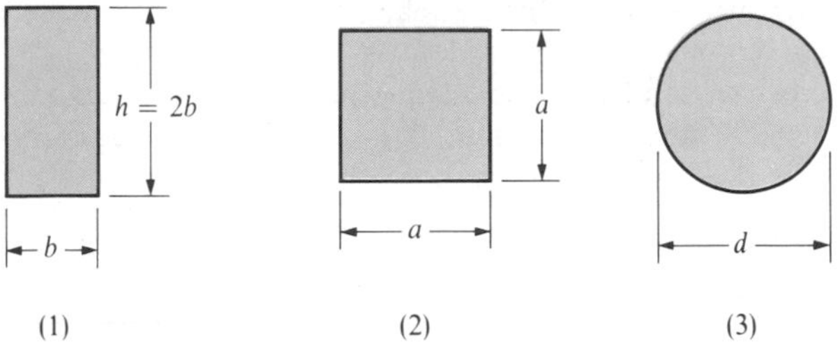

Prob. 5.4-15

5.4-16 A steel plate (called a *cover plate*) having cross-sectional dimensions 4.0 in. × 0.5 in. is welded along the full length of the top flange of a W 12 × 35 wide-flange beam (see figure, which shows the beam cross section). What is the ratio r of the smaller section modulus of the cover-plated beam to the section modulus of the wide-flange beam alone?

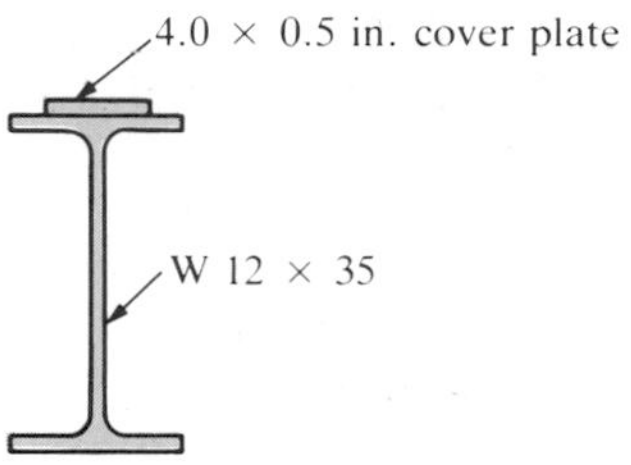

Prob. 5.4-16

5.4-17 A floor system in a small building consists of wood planking supported by 2 in. (nominal width) joists spaced at distance s, measured from center to center (see figure). The span length L of each joist is 10.5 ft, the spacing is $s = 16$ in., and the allowable bending stress is 1200 psi. The uniform floor load is 70 lb/ft^2, which includes an allowance for the weight of the floor system itself. Calculate the required section modulus S for the joists, and then select a suitable joist size (surfaced lumber), assuming that each joist may be represented as a simple beam carrying a uniform load.

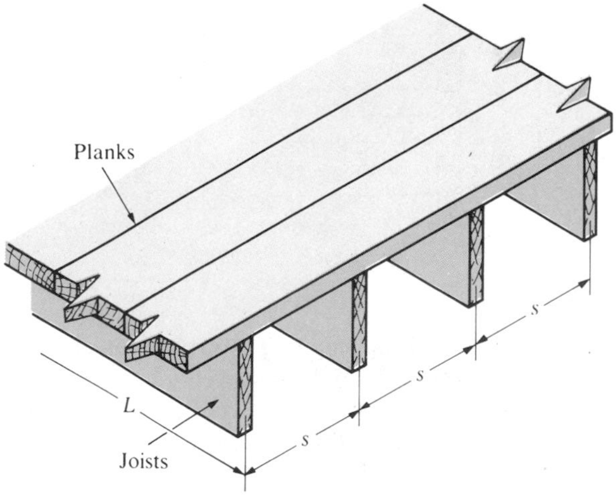

Probs. 5.4-17 and 5.4-18

5.4-18 The wood joists supporting a floor (see figure) are 2 in. × 8 in. (nominal dimensions of surfaced lumber) and have a span length $L = 13$ ft. The total floor load is 50 lb/ft^2, which includes the weight of the joists and the floor. Calculate the maximum permissible spacing s of the joists if the allowable bending stress is 1250 psi. Assume that each joist may be represented as a simple beam carrying a uniform load.

5.4-19 A retaining wall 5 ft high is constructed of horizontal wood planks 3 in. thick (actual dimension) that are supported by vertical wood piles of 12 in. diameter (actual dimension), as shown in the figure. The lateral earth pressure is $p_1 = 100\ \text{lb/ft}^2$ at the top of the wall and $p_2 = 400\ \text{lb/ft}^2$ at the bottom. Assuming that the allowable stress in the wood is 1200 psi, calculate the maximum permissible spacing s of the piles. (Hint: Observe that the spacing of the piles may be governed by the load-carrying capacity of either the planks or the piles. Consider the piles to act as cantilever beams subjected to a trapezoidal distribution of load, and consider the planks to act as simple beams between the piles. To be on the safe side, assume that the pressure on the bottom plank is uniform and equal to the maximum pressure.)

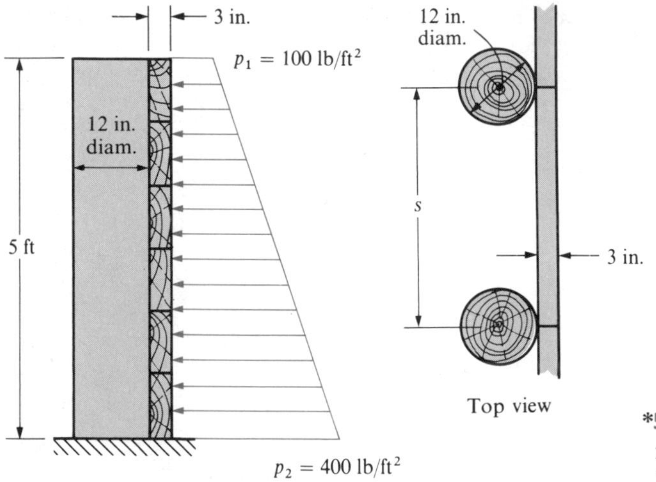

Prob. 5.4-19

5.4-20 A beam AB of length $L = 0.5$ m is supported at points C and D (see figure). The supports may be moved to any desired positions between the ends of the beam. The beam carries a uniform load of intensity q, which includes the weight of the beam itself. The cross section of the beam is rectangular with width $b = 30$ mm and height $h = 6$ mm. Determine the maximum permissible value of the load q if the allowable bending stress is $\sigma_{\text{allow}} = 400$ MPa.

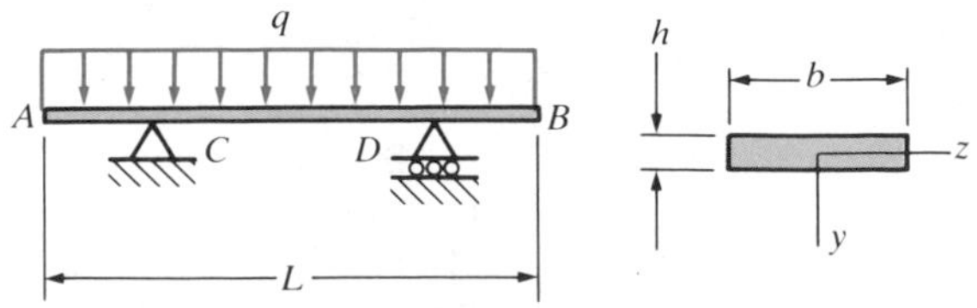

Prob. 5.4-20

***5.4-21** Water pressure acts against an inclined panel ABC that serves as a barrier (see figure). The panel is pivoted at point B, which is height h above the base, and presses against the base at A when the water level is not too high (note that the panel will rotate about the pin at B when the depth d of the water reaches a certain maximum depth d_{max}). The panel has thickness t and is inclined at an angle α to the horizontal. Derive a formula for the minimum required thickness t_{min} of the panel if the maximum bending stress in the panel is not to exceed an allowable stress σ_{allow}. (Note that the maximum stress in the panel occurs when the depth of the water reaches the maximum possible depth d_{max}. When deriving the formula, consider only the effects of bending in the panel, disregard the weight of the panel itself, and let γ be the specific weight of water.)

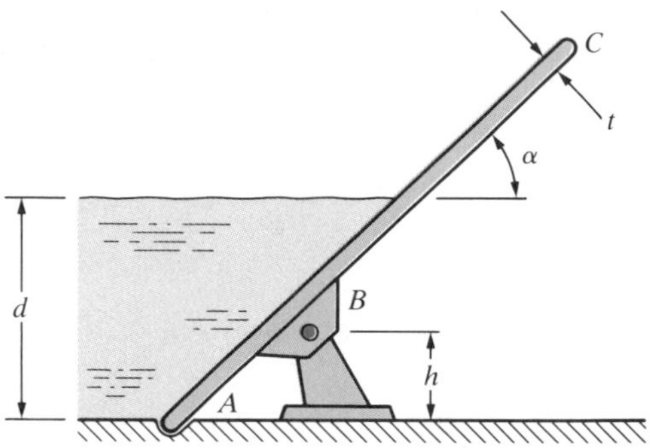

Prob. 5.4-21

***5.4-22** An extruded aluminum beam has the doubly symmetric cross-sectional shape shown in part (a) of the figure. For purposes of a bending analysis, the fillets and rounded corners may be disregarded, as shown in part (b). The cross section has the following dimensions: $b = 220$ mm, $h = 200$ mm, $h_1 = 80$ mm, $t_1 = t_2 = 10$ mm, and $t_3 =$

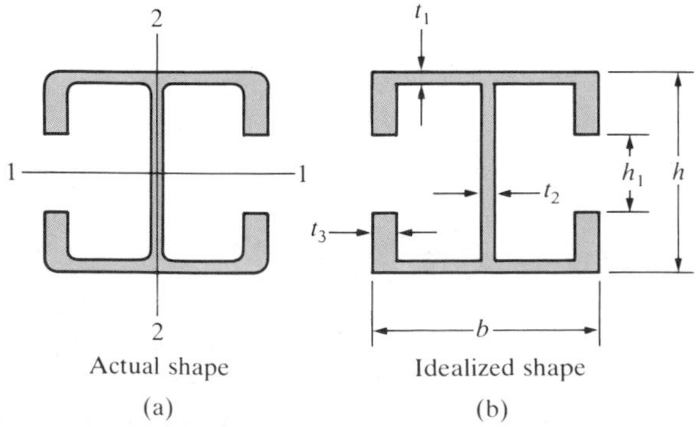

Prob. 5.4-22

20 mm. (a) Is the beam stronger when bent about axis 1-1 (that is, when axis 1-1 is the neutral axis) or when bent about axis 2-2? (b) What is the ratio s of the larger section modulus to the smaller section modulus?

5.4-23 A beam of square cross section (a = length of each side) is bent in the plane of a diagonal (see figure). By removing a small amount of material at the top and bottom corners, as shown by the shaded triangles in the figure, we can increase the section modulus and obtain a stronger beam, even though the area of the cross section is reduced. (a) Determine the ratio β defining the areas that should be removed in order to obtain the strongest cross section in bending. (b) By what percent is the section modulus increased when the areas are removed?

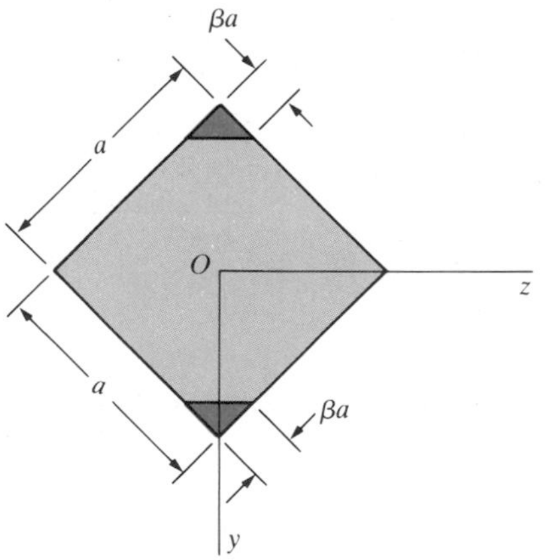

Prob. 5.4-23

5.5-1 The shear stresses τ in a rectangular beam are given by Eq. (5-22):

$$\tau = \frac{V}{2I}\left(\frac{h^2}{4} - y_1^2\right)$$

(see Fig. 5-23). By integrating over the cross-sectional area, show that the resultant of these shear stresses is the shear force V.

5.5-2 (a) A simple beam of length L and rectangular cross section of height h is subjected to a uniform load (see figure). Derive a formula for the maximum shear stress τ_{max} in the beam in terms of the maximum bending stress σ_{max}. (b) Repeat part (a) for a cantilever beam (see figure).

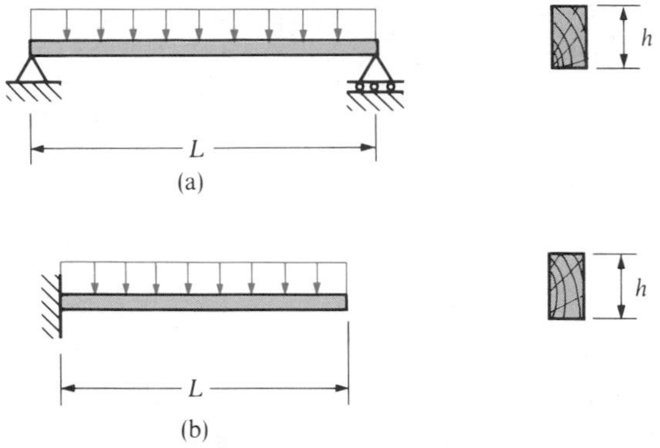

Prob. 5.5-2

5.5-3 Calculate the maximum shear stress τ_{max} and the maximum bending stress σ_{max} in a simply supported wood beam (see figure) carrying a uniform load of 1800 lb/ft (which includes the weight of the beam) if the length is 6 ft and the cross section is rectangular with width 8 in. and height 10 in. (actual dimensions).

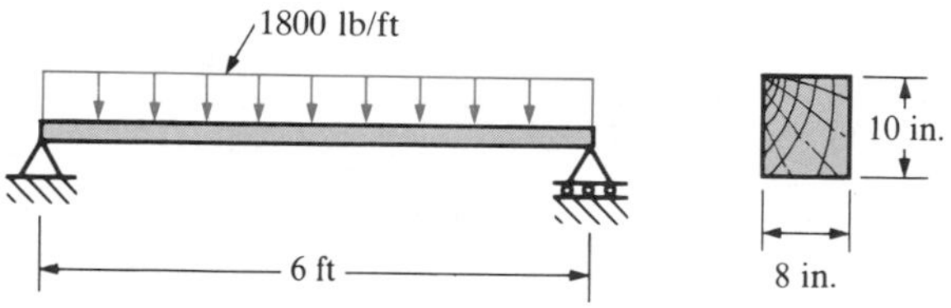

Prob. 5.5-3

5.5-4 Calculate the maximum shear stress τ_{max} and the maximum bending stress σ_{max} in a circular pole (see figure) of length $L = 2$ m and diameter $d = 150$ mm subjected to a concentrated load $P = 2.5$ kN.

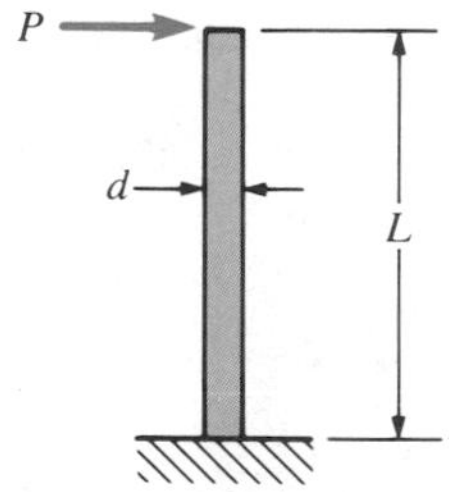

Prob. 5.5-4

5.5-5 A simple beam of length $L = 16$ in. and cross-sectional dimensions $b = 0.5$ in. and $h = 2$ in. (see figure) supports a uniform load of intensity $q = 250$ lb/in., which includes the weight of the beam. Calculate the shear stresses in the beam (at the cross section of maximum shear force V) at points located $\frac{1}{4}$ in., $\frac{1}{2}$ in., $\frac{3}{4}$ in., and 1 in. from the top surface of the beam. From these calculations, plot a graph showing the distribution of shear stresses from top to bottom of the beam.

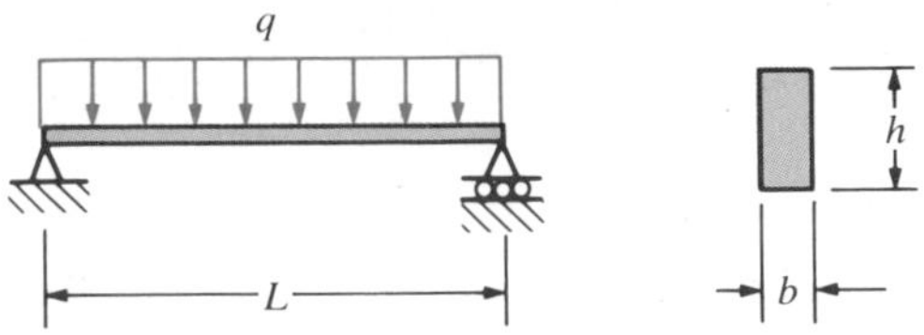

Prob. 5.5-5

5.5-6 A cantilever beam of length $L = 2$ m supports a load $P = 15$ kN (see figure). The beam is made of wood with cross-sectional dimensions 150 mm × 200 mm. Calculate the shear stresses due to the load P at points located 25 mm, 50 mm, 75 mm, and 100 mm from the top surface of the beam. From these results, plot a graph showing the distribution of shear stresses from top to bottom of the beam.

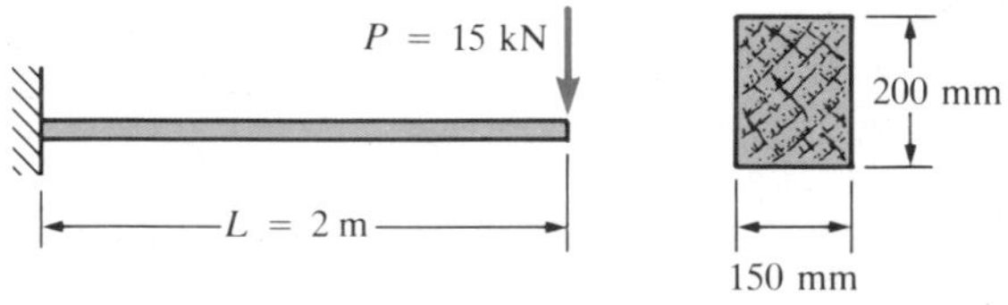

Prob. 5.5-6

5.5-7 A laminated wood beam is built up by gluing together three 2 in. × 4 in. boards (actual dimensions) to form a solid beam 4 in. × 6 in. in cross section, as shown in the figure. The allowable shear stress in the glued joints is 50 psi and the allowable bending stress is 1600 psi. If the beam is a 3-ft long cantilever, what is the allowable load P at the free end? (Disregard the weight of the beam.)

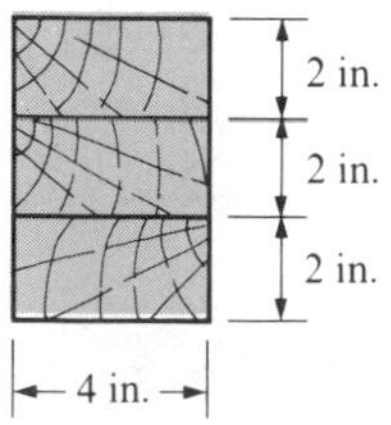

Prob. 5.5-7

5.5-8 Refer to the vertical wood posts B analyzed in Example 1 of Section 5.4 (see Fig. 5-19). Determine the minimum required dimension b of the posts based upon shear if the allowable shear stress in the wood is $\tau_{allow} =$ 0.8 MPa.

5.5-9 A cantilever beam of rectangular cross section (h = height of the beam) carries a uniform load. The allowable stresses in bending and shear are σ_{allow} and τ_{allow}, respectively. Derive a formula for the length L_0 of the beam below which the shear stress governs the permissible load and above which the bending stress governs.

5.5-10 A wood beam of rectangular cross section is simply supported and uniformly loaded. The height of the beam is 300 mm, and the allowable stresses in bending and shear are $\sigma_{allow} = 8.2$ MPa and $\tau_{allow} = 1.0$ MPa, respectively. Determine the span length L_0 below which the shear stress governs the permissible load and above which the bending stress governs.

5.5-11 A simply supported wood beam of rectangular cross section and span length 4 ft carries a concentrated load P at midspan in addition to its own weight (see figure). The cross section has nominal dimensions 6 in. × 10 in. (see Appendix F for actual dimensions and weight of the beam). Calculate the maximum permissible value of the load P if $\sigma_{allow} = 1000$ psi and $\tau_{allow} = 140$ psi.

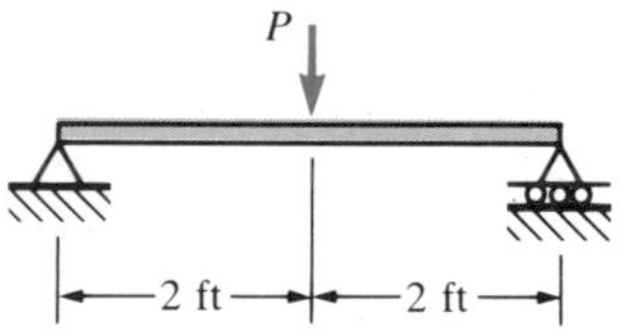

Prob. 5.5-11

5.5-12 A laminated plastic beam of square cross section is built up by gluing together three strips, each 10 mm × 30 mm in cross section (see figure). The beam has a total weight of 3.2 N and is simply supported with a span length of 320 mm. Considering the weight of the beam, calculate the maximum permissible load P that may be placed at the

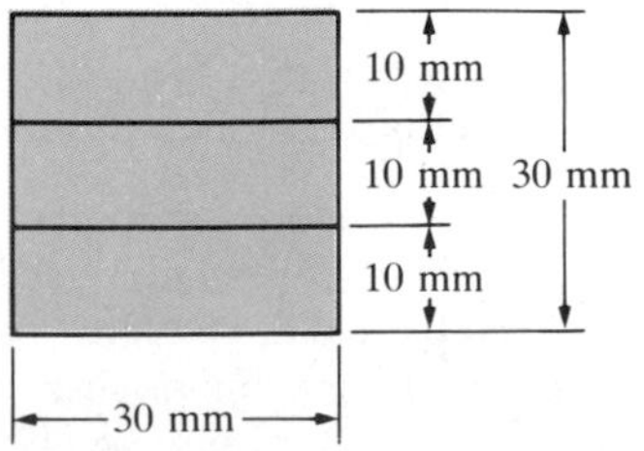

Prob. 5.5-12

midpoint of the beam if: (a) the allowable shear stress in the glued joints is 0.3 MPa, and (b) the allowable bending stress is 8 MPa.

5.5-13 A wood beam ABC with simple supports at A and B and an overhang BC is subjected to loads $3P$ and P acting as shown in the figure. The beam has a nominal 4 in. × 12 in. cross section with actual dimensions 3.5 in. × 11.25 in. (see Appendix F). (a) Determine the maximum permissible load P based upon an allowable bending stress in the wood of 1200 psi. (b) Determine the maximum permissible load P based upon an allowable shear stress in the wood of 100 psi. (Disregard the weight of the beam.)

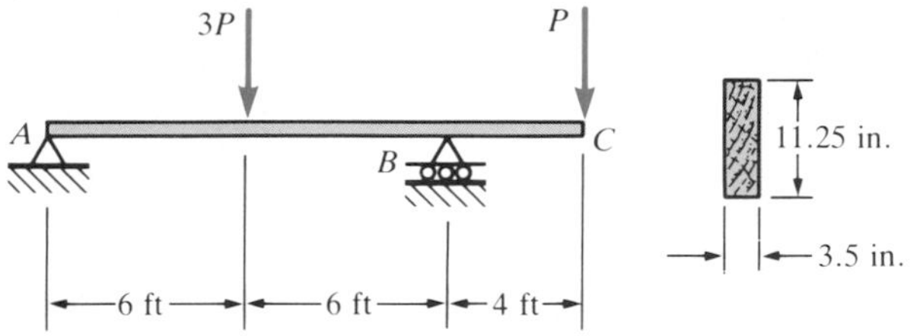

Prob. 5.5-13

***5.5-14** A simple log bridge in the woods consists of two parallel logs with cross planks between them. Each log is Douglas fir, is 300 mm in diameter, and spans 2.5 m (see figure). A truck moves slowly across the bridge, with its weight equally distributed between the two logs. Because the wheelbase of the truck is greater than 2.5 m, only one set of wheels is on the bridge at a time. Thus, the wheel load on one log is equivalent to a concentrated load W acting at any position along the span. In addition, the weight of one log and the planks it supports is equivalent to a uniform load of 850 N/m acting on the log. Determine the maximum permissible wheel load W based upon: (a) an allowable bending stress of 7.0 MPa, and (b) an allowable shear stress of 0.8 MPa.

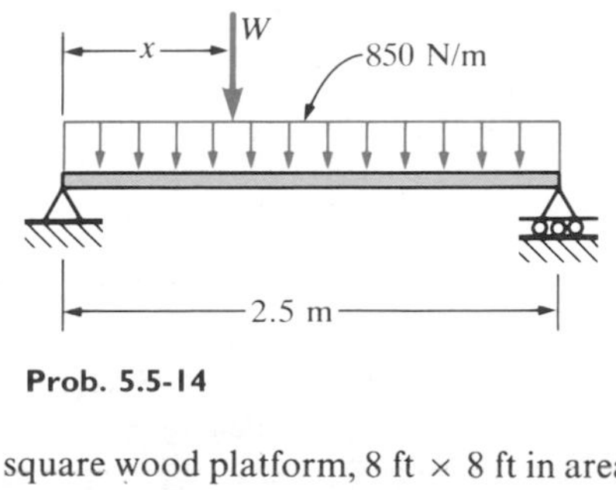

Prob. 5.5-14

****5.5-15** A square wood platform, 8 ft × 8 ft in area, rests on 6-in. wide masonry walls (see figure). The deck of the platform is constructed of 2 in. nominal thickness tongue-and-groove planks (actual thickness 1.5 in.; see Appendix F) supported on two 8-ft long beams. The beams have 4 in. × 6 in. nominal dimensions (actual dimensions 3.5 in. × 5.5 in.) and are supported at their ends by the walls. The structure is designed to support a uniformly distributed load w (lb/ft^2) acting over the entire top surface of the platform. The allowable normal stress in bending for both planks and beams is 2400 psi and the allowable shear stress is 100 psi. When analyzing the planks, it is assumed that their reactions are uniformly distributed over the top surfaces of the supporting beams, and when analyzing the beams it is assumed that their reactions are uniformly distributed over the area of contact with the supporting walls. The weight of the planks and beams is to be taken into account in the analysis, and the wood is assumed to have a specific weight $\gamma = 30$ lb/ft^3. (a) Determine the allowable loads w_1, w_2, w_3, and w_4 (lb/ft^2) based upon the following parameters, respectively: (1) normal stress in the planks, (2) shear stress in the planks, (3) normal stress in the beams, and (4) shear stress in the beams. (b) Which of the preceding values becomes the allowable load w_{allow} on the platform? (Hints: Use care in constructing the loading diagrams, noting especially that the reactions are distributed loads instead of concentrated loads. Also, note that the maximum shear forces occur at the inside faces of the supports.)

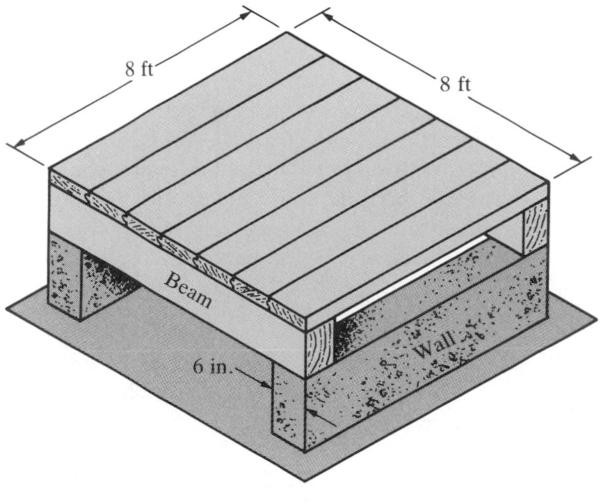

Prob. 5.5-15

5.6-1 (a) Calculate the maximum shear stress τ_{max} in the web of a wide-flange beam (see figure) if $b = 6$ in., $t = 0.5$ in., $h = 12$ in., $h_1 = 10.5$ in., and $V = 30{,}000$ lb. (b) Calculate the average shear stress τ_{aver} by dividing V by the area of the web, and then obtain the ratio τ_{max}/τ_{aver}.

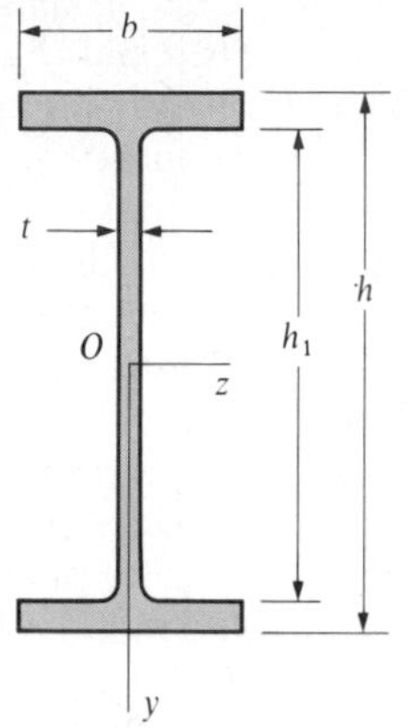

Probs. 5.6-1, 5.6-2, and 5.6-4

5.6-2 (a) Calculate the maximum shear stress τ_{max} in the web of a wide-flange beam (see figure) if $b = 180$ mm, $t = 12$ mm, $h = 420$ mm, $h_1 = 380$ mm, and $V = 125$ kN. (b) Calculate the average shear stress τ_{aver} by dividing V by the area of the web, and then obtain the ratio τ_{max}/τ_{aver}.

5.6-3 (a) Calculate the maximum shear stress τ_{max} in the web of a W 18 × 71 steel wide-flange beam (see Table E-1, Appendix E) due to a shear force $V = 21$ k. (b) Calculate the average shear stress τ_{aver} by dividing V by the area of the web, and then obtain the ratio τ_{max}/τ_{aver}.

5.6-4 A wide-flange beam (see figure) having cross-sectional dimensions $b = 180$ mm, $t = 12$ mm, $h = 620$ mm, and $h_1 = 570$ mm is subjected to a shear force $V = 270$ kN. (a) Calculate the maximum and minimum shear stresses τ_{max} and τ_{min} in the web. (b) Calculate the average shear stress τ_{aver} by dividing V by the area of the web, and then obtain the ratio τ_{max}/τ_{aver}. (c) Calculate the shear force V_{web} carried in the web and obtain the ratio V_{web}/V.

5.6-5 (a) Calculate the maximum and minimum shear stresses τ_{max} and τ_{min} in the web of a W 24 × 94 steel wide-flange beam (see Table E-1, Appendix E) due to a shear force $V = 32$ k. (b) Calculate the average shear stress τ_{aver} by dividing V by the area of the web, and then obtain the ratio τ_{max}/τ_{aver}. (c) Calculate the shear force V_{web} carried in the web and obtain the ratio V_{web}/V.

5.6-6 A simple beam AB with span length $L = 14$ m supports a uniform load of intensity q that includes the weight of the beam (see figure). The beam is constructed of three plates welded together to form the cross section shown. Determine the maximum permissible load q based upon: (a) an allowable bending stress $\sigma_{allow} = 110$ MPa, and (b) an allowable shear stress $\tau_{allow} = 50$ MPa.

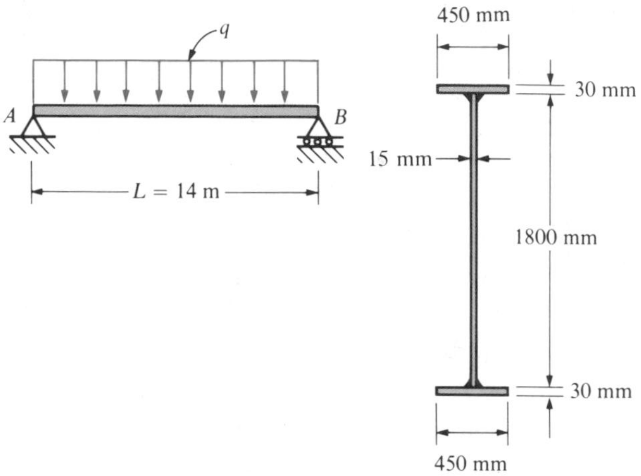

Prob. 5.6-6

5.6-7 A cantilever beam of length $L = 5$ ft supports a uniform load of intensity q that includes the weight of the beam. The beam is a steel W 10 × 12 wide-flange section (see Table E-1, Appendix E). Calculate the maximum permissible load q based upon: (a) an allowable bending stress $\sigma_{allow} = 16$ ksi, and (b) an allowable shear stress $\tau_{allow} = 8.5$ ksi.

5.6-8 A hollow square aluminum box beam has the cross section shown in the figure. Calculate the maximum and minimum shear stresses τ_{max} and τ_{min} in the webs of the beam due to a shear force $V = 60$ kN.

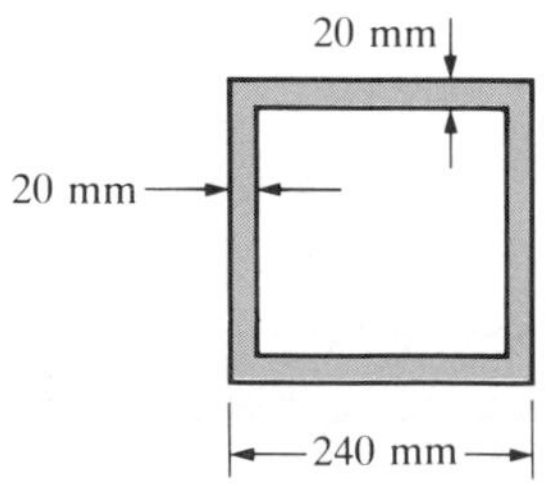

Prob. 5.6-8

5.6-9 A hollow steel box beam has the rectangular cross section shown in the figure. Calculate the maximum and minimum shear stresses τ_{max} and τ_{min} in the webs of the beam due to a shear force $V = 35$ k.

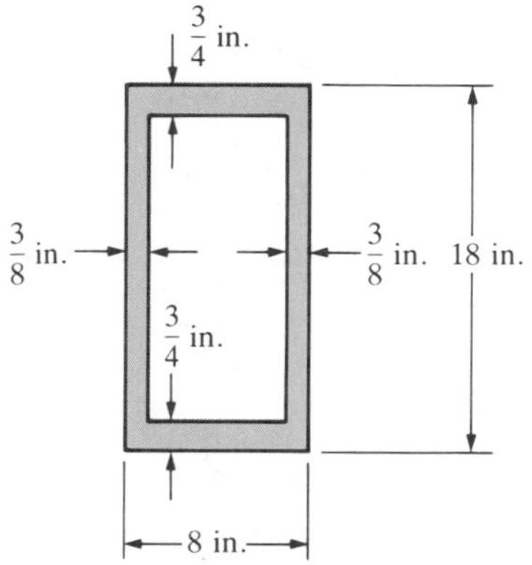

Prob. 5.6-9

5.6-10 The T-beam shown in the figure has cross-sectional dimensions as follows: $b = 220$ mm, $t = 15$ mm, $h = 300$ mm, and $h_1 = 275$ mm. The beam is subjected to a shear force $V = 62$ kN. Calculate the maximum shear stress τ_{max} in the web of the beam.

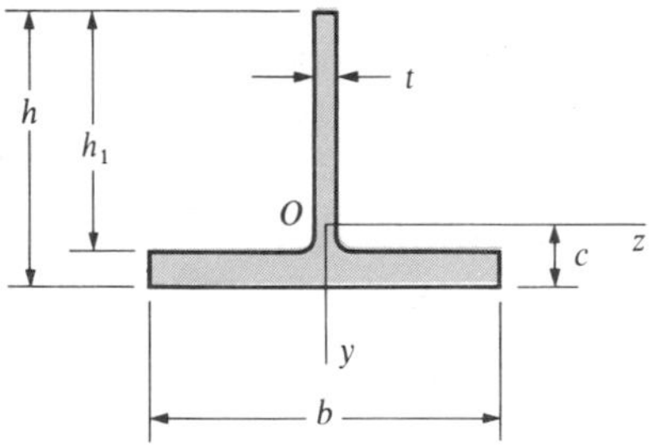

Probs. 5.6-10 and 5.6-11

5.6-11 Calculate the maximum shear stress τ_{max} in the web of the T-beam shown in the figure if $b = 10$ in., $t = 0.6$ in., $h = 8$ in., $h_1 = 7$ in., and the shear force $V = 5000$ lb.

5.6-12 A projecting arm in a mechanical device is modeled as a cantilever beam AB (see figure). The arm is loaded at the free end by a force $P = 900$ N. The cross section has the dimensions shown in the figure. If the length L is 0.15 m, what are the maximum normal stress σ_{max} and the maximum shear stress τ_{max} due to the load P? (Hint: The maximum shear stress does not necessarily occur at the neutral axis.)

***5.6-13** An unsymmetric I-beam (see figure) has cross-sectional dimensions as follows: $b_1 = 10$ in., $b_2 = 6$ in., $t = 0.4$ in., $h = 12$ in., and $t_1 = 1$ in. The beam is subjected to a shear force $V = 16$ k. (a) Calculate the maximum and minimum shear stresses τ_{max} and τ_{min} in the web of the beam. (b) Calculate the average shear stress τ_{aver} by dividing V by the area of the web, and then obtain the ratio τ_{max}/τ_{aver}. (c) Calculate the shear force V_{web} carried in the web and obtain the ratio V_{web}/V.

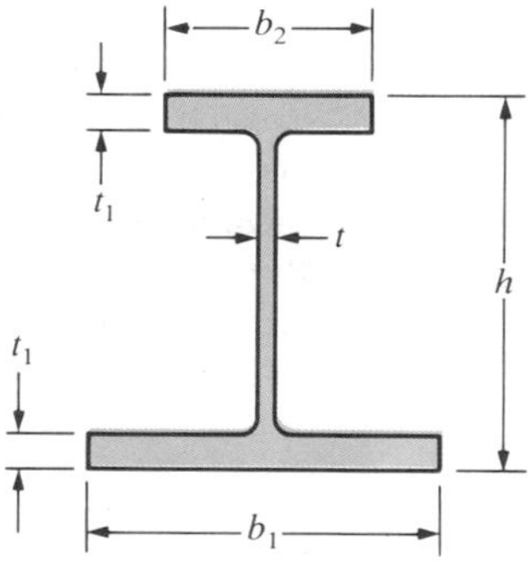

Prob. 5.6-13

***5.6-14** Calculate the maximum shear stress τ_{max} in the beam of tubular cross section shown in the figure on p. 356 if it is subjected to a shear force $V = 40$ kN. The cross section has outside height $h = 250$ mm, outside width $b = 125$ mm, constant thickness $t = 20$ mm, and inside radius $R = 30$ mm at the "corners."

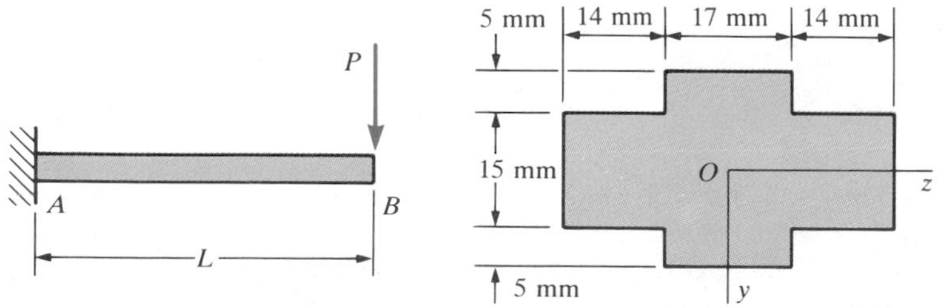

Prob. 5.6-12

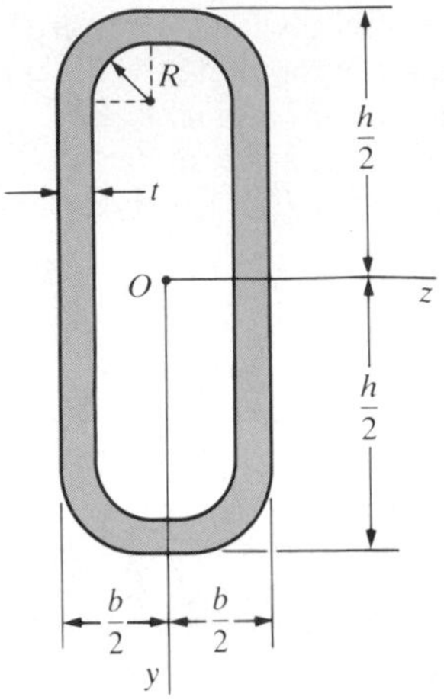

Prob. 5.6-14

5.7-1 A welded steel girder having the cross section shown in the figure is fabricated of two 18 in. × 1 in. flange plates and a web plate that is $\frac{3}{8}$ in. thick and 64 in. deep. The plates are joined by four longitudinal fillet welds that run continuously throughout the length of the girder. If the girder is subjected to a shear force of 300 kips, what force F (per unit length of weld) must be resisted by each weld?

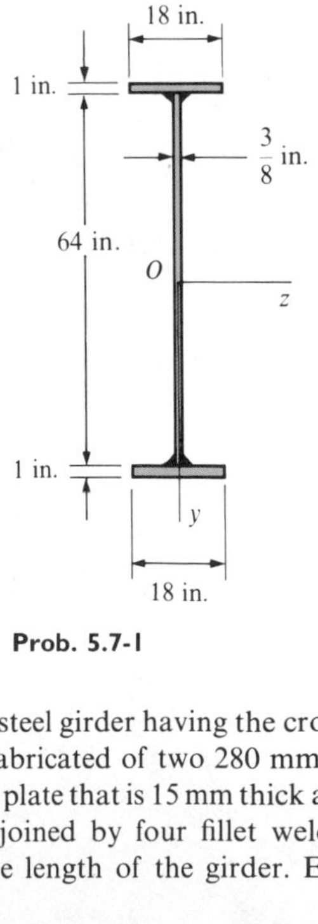

Prob. 5.7-1

5.7-2 A welded steel girder having the cross section shown in the figure is fabricated of two 280 mm × 25 mm flange plates and a web plate that is 15 mm thick and 600 mm deep. The plates are joined by four fillet welds that run continuously for the length of the girder. Each weld has an allowable load in shear of 900 kN/m. Calculate the maximum allowable shear force V for the girder.

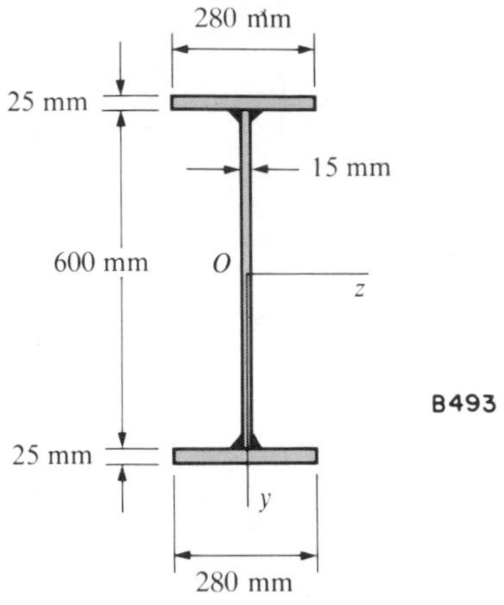

Prob. 5.7-2

5.7-3 A box beam constructed of wood boards of size 1 in. × 6 in. (actual dimensions) is shown in the figure. The boards are joined by screws for which the allowable load in shear is $F = 250$ lb per screw. Calculate the maximum permissible longitudinal spacing s of the screws if the shear force V is 1200 lb.

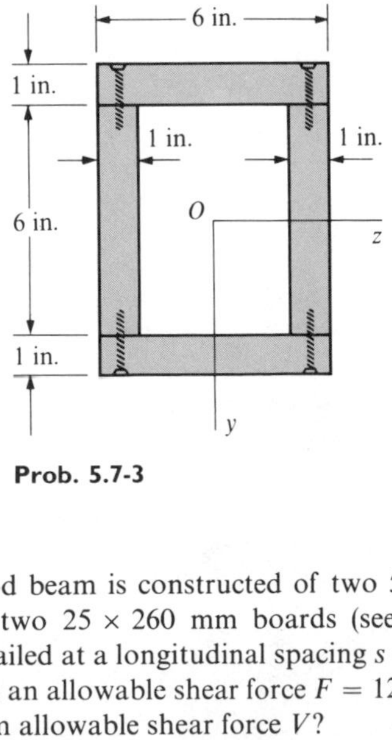

Prob. 5.7-3

5.7-4 A wood beam is constructed of two 50 × 260 mm boards and two 25 × 260 mm boards (see figure). The boards are nailed at a longitudinal spacing $s = 100$ mm. If each nail has an allowable shear force $F = 1200$ N, what is the maximum allowable shear force V?

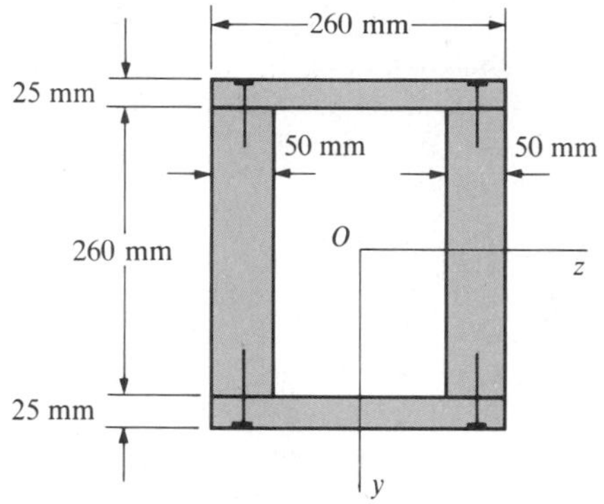

Prob. 5.7-4

5.7-5 A wood box beam (see figure) is built up of four 2 in. × 8 in. members (nominal dimensions). If the longitudinal spacing of the nails is $s = 4$ in. and the allowable load per nail is $F = 300$ lb, calculate the maximum allowable shear force V. (Note: See Appendix F for actual dimensions of the members.)

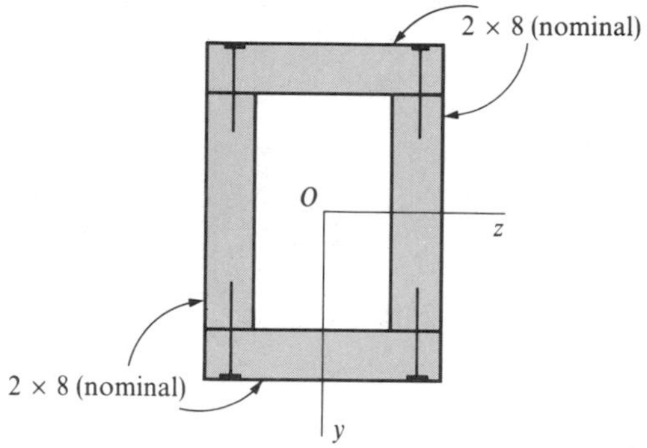

Prob. 5.7-5

5.7-6 A beam of T cross section is formed by nailing together two boards having the dimensions shown in the figure. If the total shear force V acting on the cross section is 1600 N and each nail may carry 750 N in shear, what is the maximum allowable nail spacing s?

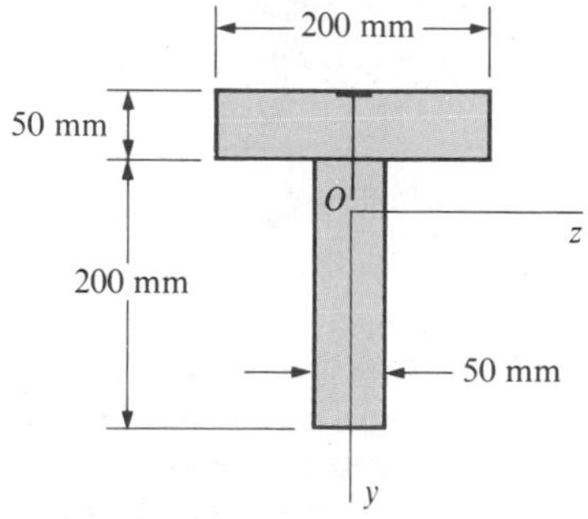

Prob. 5.7-6

5.7-7 A hollow wood beam with plywood webs has the cross-sectional dimensions shown in the figure. The plywood is attached to the flanges by means of small nails having an allowable load in shear of 30 lb. Find the maximum allowable spacing s of the nails at cross sections where the shear force V is equal to (a) 200 lb and (b) 300 lb.

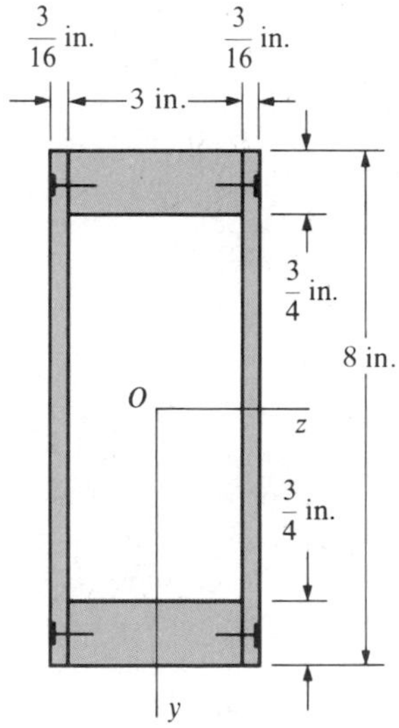

Prob. 5.7-7

5.7-8 Two W 10 × 45 steel wide-flange beams (see Table E-1, Appendix E) are bolted together to form a built-up beam as shown in the figure. What is the maximum permissible bolt spacing s if the shear force $V = 20$ kips and the allowable load in shear on each bolt is $F = 3.1$ kips?

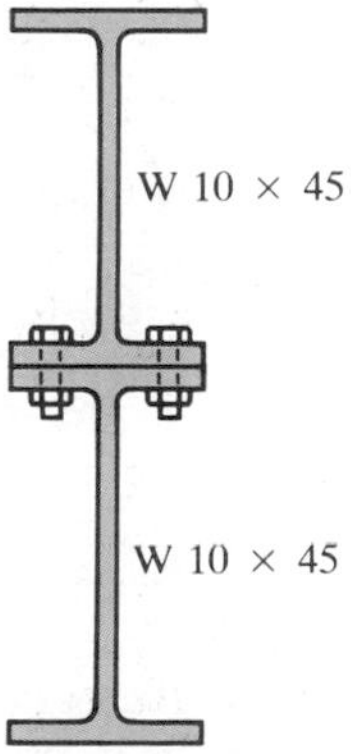

Prob. 5.7-8

5.7-9 A built-up beam is fabricated from a W 16 × 77 wide-flange beam and two 10 in. × $\frac{1}{2}$ in. cover plates (see figure). The allowable load in shear on each bolt is 2.8 kips. What is the required bolt spacing s if the shear force $V = 40$ kips?

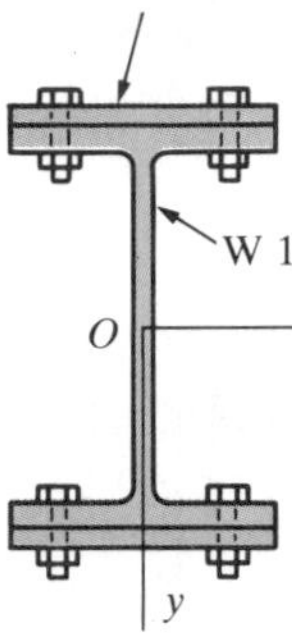

Prob. 5.7-9

5.7-10 A built-up beam is fabricated from a W 12 × 50 wide-flange section and two C 12 × 30 channel sections (see figure). The sections are joined by bolts spaced at 8 in. longitudinally. What is the maximum allowable shear force V if each bolt may carry 2.4 kips in shear?

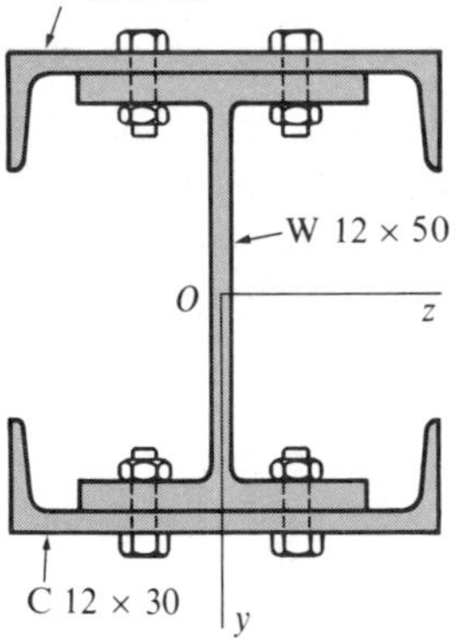

Prob. 5.7-10

5.8-1 A tapered cantilever beam AB of length L has circular cross sections and supports a concentrated load P at the free end (see Fig. 5-34). The diameter of the beam varies linearly from d_a at the free end to d_b at the fixed end. At what distance x from the free end does the maximum normal stress due to the load P occur if $d_b/d_a = 2.5$? What is the magnitude of the maximum normal stress σ_{max}? What is the ratio of this stress to the largest stress σ_b at the support?

5.8-2 Refer again to the tapered cantilever beam of circular cross section shown in Fig. 5-34. Considering only the effects of bending due to the load P, determine the range of values of the ratio d_b/d_a for which the maximum normal stress occurs at the support.

5.8-3 A tapered cantilever beam AB of length L has square cross sections and supports a concentrated load P at the free end (see figure). The width and height of the beam vary linearly from h_a at the free end to h_b at the fixed end. Considering only the effects of bending due to the load P, determine the distance x from the free end to the cross section of maximum normal stress if $h_b = 3h_a$. What is the magnitude of the maximum normal stress σ_{max}? What is the ratio of this stress to the largest stress σ_b at the support?

5.8-4 Refer to the tapered cantilever beam of rectangular cross section shown in Fig. 5-36. Considering only the effects of the load P, determine the distribution of shear stresses τ at cross sections where $x = L/4$ and $x = 3L/4$. Draw graphs showing how the shear stresses vary from top to bottom of the beam at each of those sections.

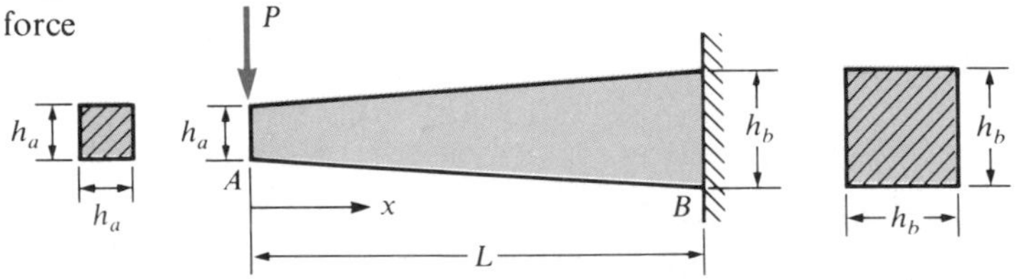

Prob. 5.8-3

5.8-5 A tapered cantilever beam AB of rectangular cross section is subjected to a concentrated load and a couple at the free end as shown in the figure. The width of the beam is constant and equal to 1 in., but the height varies linearly from 2 in. at the loaded end to 3 in. at the support. At what distance x from the free end does the maximum normal stress σ_{max} due to bending occur? What is the magnitude of the maximum normal stress σ_{max}? What is the ratio of this stress to the largest stress σ_b at the support? (Disregard the weight of the beam itself.)

***5.8-6** A spoke in a large flywheel is modeled as a beam fixed at one end and loaded by a force P and a couple M_0 at the other (see figure). The cross section of the spoke is elliptical in shape with major and minor axes (height and width, respectively) having the lengths shown in the figure. The cross-sectional dimensions vary linearly from end A to end B. Considering only the effects of bending due to the loads

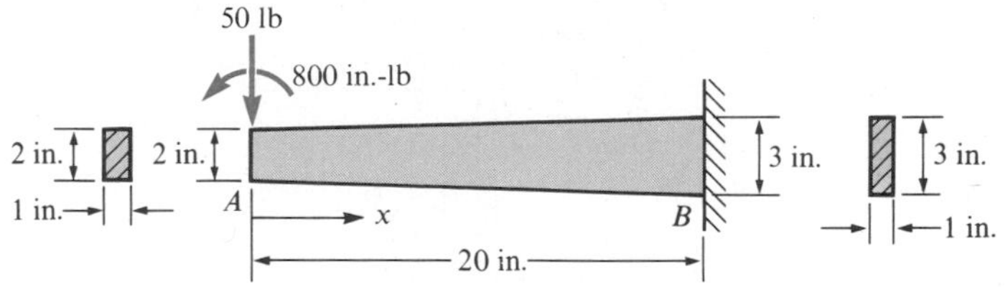

Probs. 5.8-5 and 5.8-7

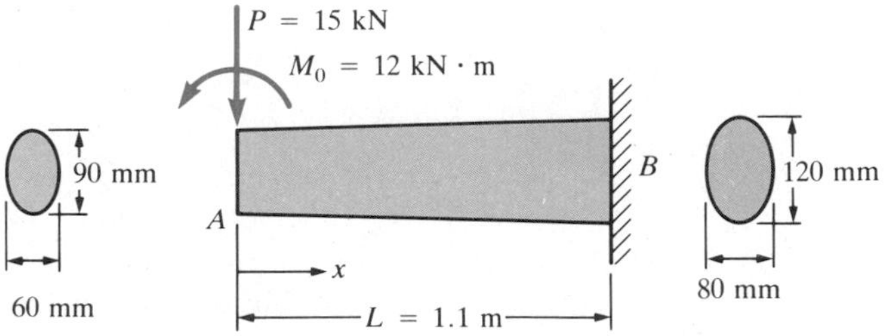

Prob. 5.8-6

P and M_0, determine the distance x from the free end to the cross section of maximum normal stress. What is the magnitude of the maximum normal stress σ_{max}?

***5.8-7** The tapered cantilever beam AB shown in the figure has a rectangular cross section of constant width equal to 1 in. and height varying linearly from 2 in. to 3 in. A concentrated load and a couple act at the free end. Investigate the shear stresses τ in the beam by determining the stresses at the cross sections where $x = 0$, 10 in., and 20 in. For each of these sections, draw a graph showing how the shear stress varies from top to bottom of the beam and label all critical values.

****5.8-8** Refer again to the tapered cantilever beam of circular cross section shown in Fig. 5-34. (a) Considering only the effects of bending due to the load P, determine the ratio d_b/d_a for which the volume (and hence the weight) of the beam is a minimum if the allowable stress in bending is σ_{allow}. (b) Obtain a formula for the minimum volume V_{min}. (c) What is the ratio of the minimum volume V_{min} to the volume of a prismatic beam with $d_b = d_a$?

Problems 5.8-9 to 5.8-12 pertain to fully stressed beams. Consider only the loads shown in the figures and disregard the weights of the beams.

5.8-9 A cantilever beam AB of rectangular cross sections with constant height h and varying width b is subjected to a concentrated load P at the free end (see figure). How should the width b vary as a function of x (measured from the free end of the beam) in order to have a fully stressed beam? (Consider only the normal stresses due to bending, and assume that the allowable stress is σ_{allow}.)

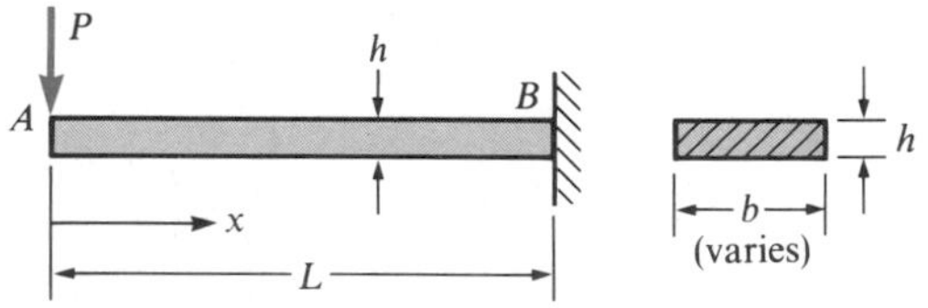

Prob. 5.8-9

5.8-10 A cantilever beam AB of rectangular cross sections with constant width b and varying height h is subjected to a uniform load of intensity q (see figure). How should the height h vary as a function of x (measured from the free end of the beam) in order to have a fully stressed beam? (Consider only the normal stresses due to bending, and assume that the allowable stress is σ_{allow}.)

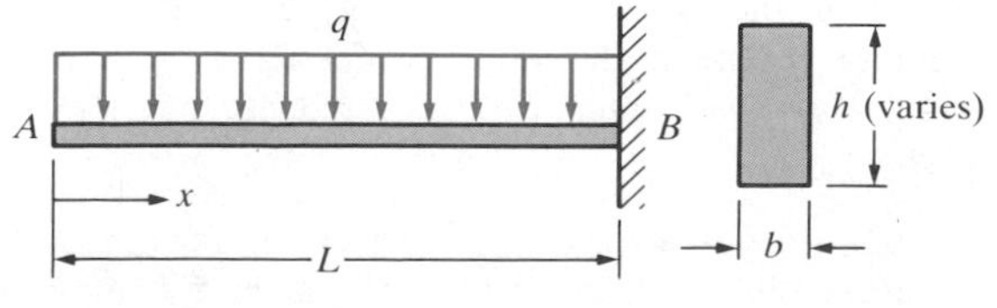

Prob. 5.8-10

5.8-11 A simple beam AB of rectangular cross sections with constant width b and varying height h is subjected to a triangular loading system as shown in the figure. How should the height h vary as a function of x (measured from the midpoint of the beam) in order to have a fully stressed beam? (Consider only the normal stresses due to bending, and assume that the allowable stress is σ_{allow}.)

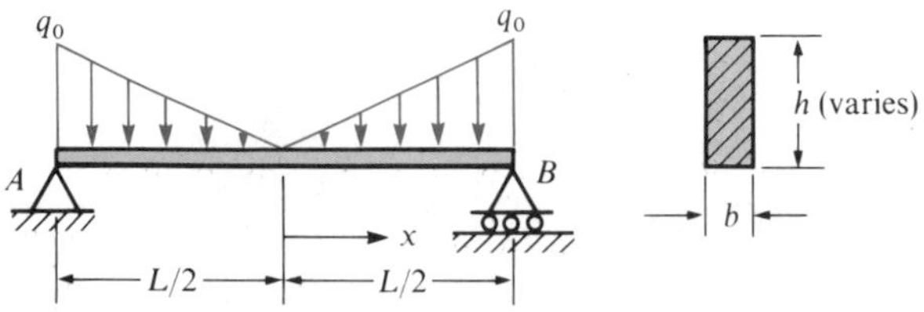

Prob. 5.8-11

5.8-12 A simple beam AB supporting a concentrated load P (see figure) has rectangular cross sections with constant width b and varying height h. The concentrated load may act anywhere along the span of the beam. How should the height h vary as a function of x (measured from the midpoint of the beam) in order to minimize the weight of the beam? (Consider only the normal stresses due to bending, and assume that the allowable stress is σ_{allow}.)

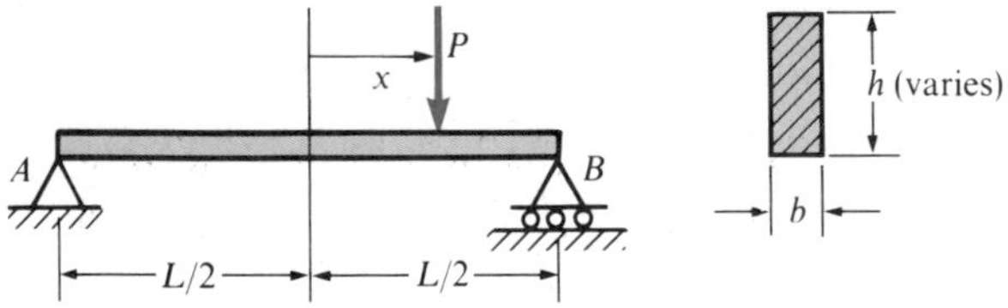

Prob. 5.8-12

5.9-1 The end regions of bar AB (see figure) have square cross sections with sides of length b. In the middle region of the bar the cross-sectional area is reduced by one-half by cutting a notch of depth $b/2$. An axial load P acts at the centroid of the cross section at end B. Determine the maximum tensile and compressive stresses σ_t and σ_c, respectively, at section mn within the reduced region of the bar.

5.9-2 Solve the preceding problem if the end regions of the bar have circular cross sections of diameter b and the middle region has a semicircular cross section.

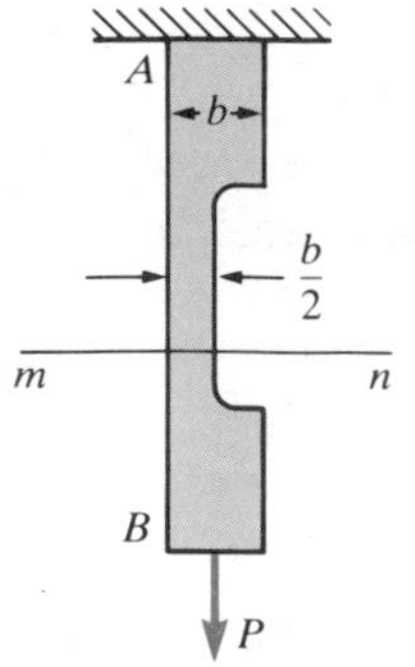

Probs. 5.9-1 and 5.9-2

5.9-3 A curved bar ABC having a circular axis (radius $r = 12$ in.) is loaded by forces $P = 400$ lb, as shown in the figure. The cross section of the bar is rectangular with height h and thickness t (perpendicular to the plane of the figure). If the allowable stress in the bar is 12,000 psi and the height $h = 1.25$ in., what is the minimum required thickness t_{min}?

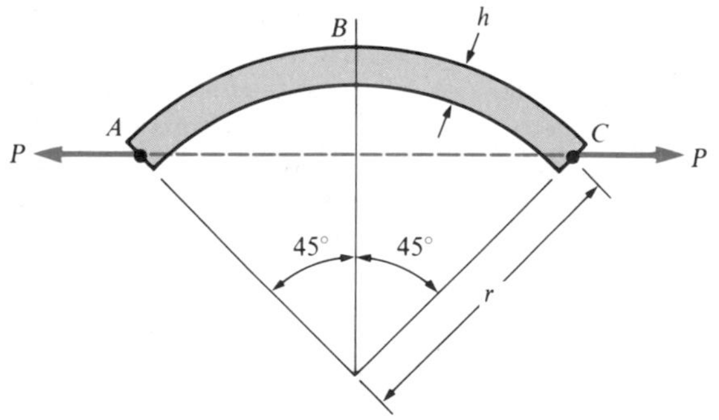

Prob. 5.9-3

5.9-4 A vertical aluminum pole AB is fixed at the base and pulled at the top by a cable having a tensile force T acting at an angle $\alpha = 25°$ to the pole (see figure). The cable is attached at the outer surface of the pole. The pole has length $L = 2.0$ m and a hollow circular cross section with outer diameter $d_2 = 260$ mm and inner diameter $d_1 = 200$ mm. Determine the allowable force T_{allow} in the cable if the allowable compressive stress in the aluminum pole is 90 MPa.

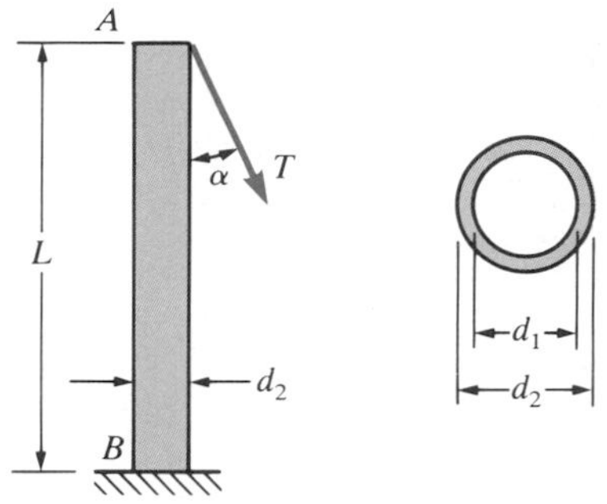

Prob. 5.9-4

5.9-5 A frame ABC is formed by welding two steel pipes at B (see figure). Each pipe has cross-sectional area $A =$ 16.1 in.^{2}, moment of inertia $I = 212$ in.^{4}, and outside diameter $d = 10.75$ in. Find the maximum tensile and compressive stresses σ_t and σ_c, respectively, in the frame due to a load $P = 3200$ lb if $L = 8$ ft and $H = 6$ ft.

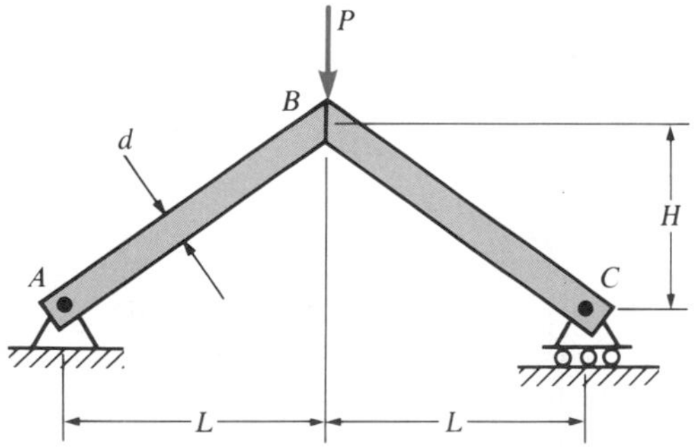

Probs. 5.9-5 and 5.9-6

5.9-6 Two steel pipes are welded together to form a frame ABC, as shown in the figure. Each pipe has outside diameter 240 mm and inside diameter 200 mm. Assuming $L = H =$ 1.5 m and $P = 8$ kN, calculate the maximum tensile and compressive stresses σ_t and σ_c, respectively, in the frame due to the load P.

5.9-7 A circular, cylindrical tower having height h, inside diameter d_1, and outside diameter d_2 begins to lean slightly (see figure). What is the maximum permissible angle of inclination α from the vertical if there is to be no tension in the tower? (Consider that the only load is the uniformly distributed weight of the tower itself, and assume that the angle α is very small.)

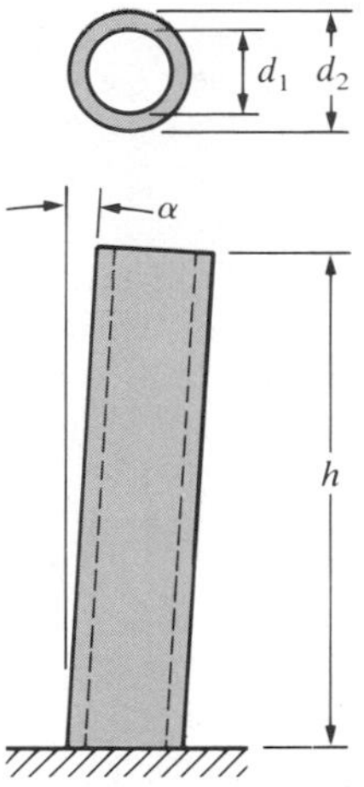

Prob. 5.9-7

5.9-8 A solid bar of circular cross section is subjected to an axial tensile force $T = 26$ kN and a bending moment $M =$ 3.2 kN · m (see figure). Based upon an allowable stress in tension of 120 MPa, determine the required diameter d of the bar. (Disregard the weight of the bar itself.)

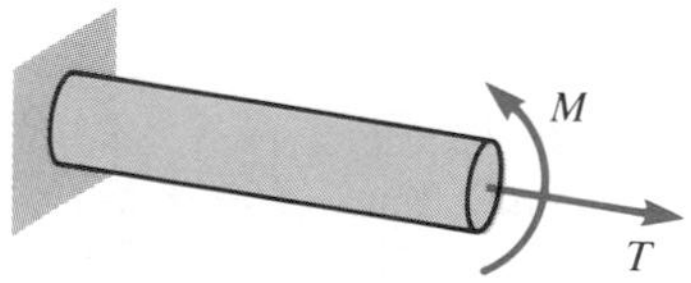

Prob. 5.9-8

5.9-9 A square pillar is subjected to a compressive force $P = 750$ k and a bending moment $M = 60$ ft-k (see figure). What is the required side dimension b of the pillar if the allowable stresses are 2600 psi in compression and 900 psi in tension? (Disregard the weight of the pillar itself.)

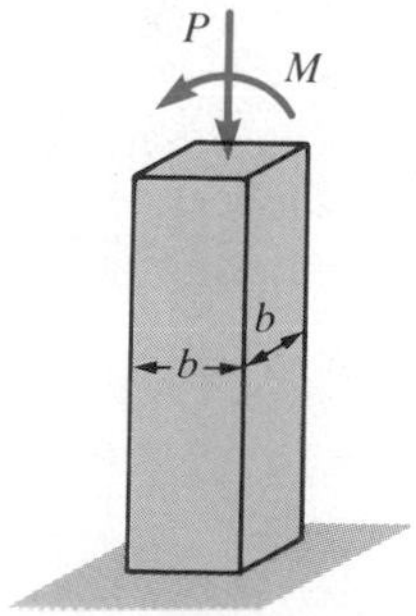

Prob. 5.9-9

5.9-10 A plain concrete wall (i.e., a wall with no steel reinforcement) rests on a secure foundation and serves as a small dam (see figure). The height of the wall is $h = 2$ m and the thickness of the wall is $t = 0.3$ m. (a) Determine the maximum tensile and compressive stresses σ_t and σ_c, respectively, at the base of the wall when the water level reaches the top ($d = h$). Assume plain concrete has specific weight $\gamma_c = 23$ kN/m^3. (b) Determine the maximum permissible depth d_{max} of the water if there is to be no tension in the concrete.

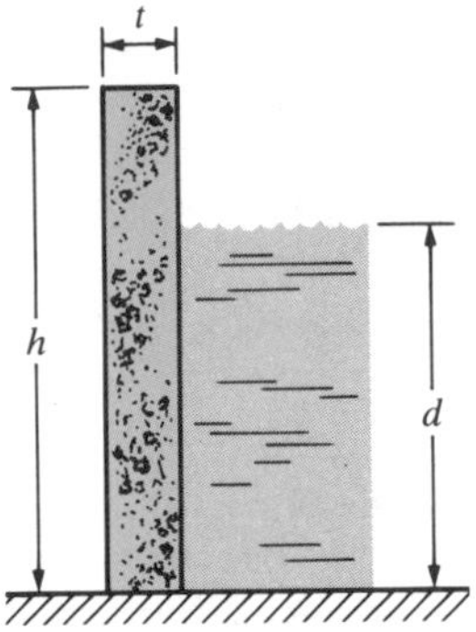

Probs. 5.9-10, 5.9-11, and 5.9-12

5.9-11 Solve the preceding problem for the following data: $h = 6$ ft, $t = 12$ in., and $\gamma_c = 145$ lb/ft^3.

5.9-12 A plain concrete wall having height h and thickness t rests on a secure foundation and serves as a small dam (see figure). Derive a formula for the maximum permissible depth d_{max} of the water if there is to be no tension in the concrete. (Let γ_c and γ_w represent the specific weights of concrete and water, respectively.)

5.9-13 A flying buttress transmits a load $P = 5400$ lb, acting at an angle of 60° to the horizontal, to the top of a vertical buttress AB (see figure). The vertical buttress has height $h = 16$ ft and rectangular cross section of thickness $t = 5$ ft and width $b = 3$ ft (perpendicular to the plane of the figure). The stone used in the construction weighs $\gamma = 165$ lb/ft.^{3} What is the required weight W of the pedestal and statue above the vertical buttress (that is, above section A) to avoid any tensile stresses in the vertical buttress?

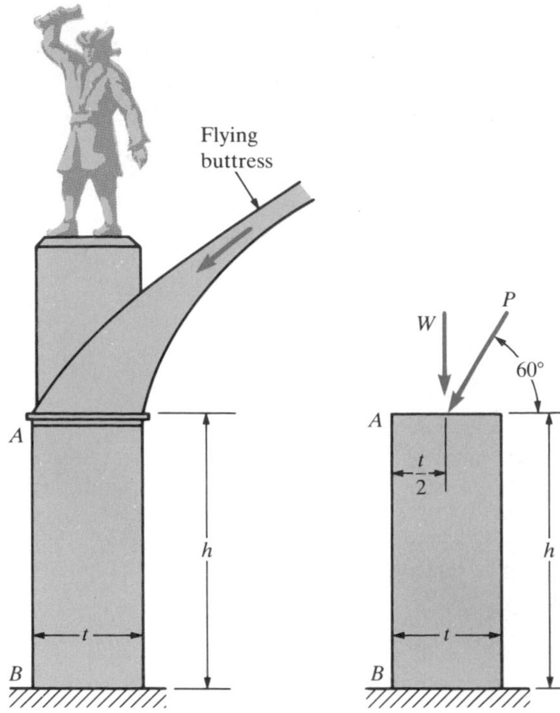

Prob. 5.9-13

5.9-14 The cantilever beam shown in the figure has a rectangular cross section of height h. When the loads P have the directions shown in the figure, the neutral axis nn is always located above the centroid of the cross section. Let s represent the distance from the z axis to the neutral axis. (a) Obtain an expression for s as a function of the distance x from the fixed support, disregarding the weight of the beam. (b) Plot a graph showing s versus x for the case where $L = 1$ m and $h = 120$ mm.

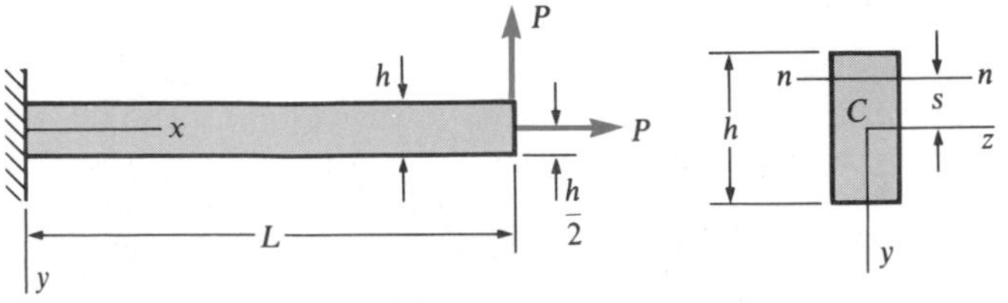

Prob. 5.9-14

Problems 5.9-15 through 5.9-18 pertain to bars with eccentric axial loads.

5.9-15 A short length of a C 6 × 13 channel section (see Appendix E) is subjected to an axial compressive force P that has its line of action through the midpoint of the web of the channel (see figure). (a) Determine the equation of the neutral axis under these loading conditions. (b) If the allowable stresses in tension and compression are 8,000 psi and 12,000 psi, respectively, find the maximum permissible load P_{max}.

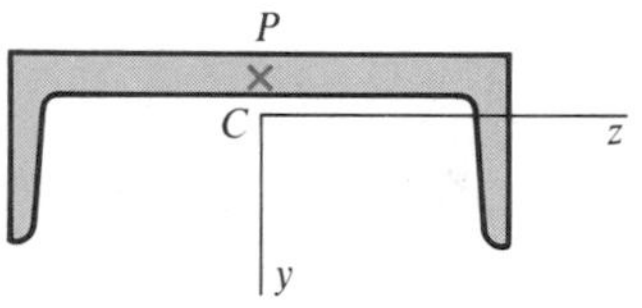

Probs. 5.9-15 and 5.9-16

5.9-16 A short length of a C 8 × 11.5 channel section (see Appendix E) is subjected to an axial compressive force P that has its line of action through the midpoint of the web of the channel (see figure). (a) Determine the equation of the neutral axis under these loading conditions. (b) If the allowable stresses in tension and compression are 10,000 psi and 16,000 psi, respectively, find the maximum permissible load P_{max}.

5.9-17 A short column constructed of a W 10 × 60 wide-flange section (see Appendix E) is subjected to a compressive load $P = 30$ kips that has its line of action at one of the outside corners of the cross section (see figure). (a) Determine the equation of the neutral axis under these loading conditions. (b) Find the maximum tensile and compressive stresses σ_t and σ_c, respectively, in the column.

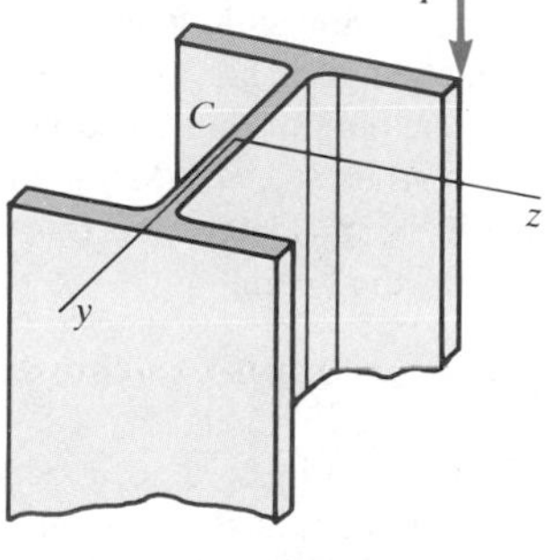

Probs. 5.9-17 and 5.9-18

5.9-18 A short column constructed of a W 12 × 87 wide-flange section (see Appendix E) is subjected to a compressive load $P = 40$ kips that has its line of action at one of the outside corners of the cross section (see figure). (a) Determine the equation of the neutral axis under these loading conditions. (b) Find the maximum tensile and compressive stresses σ_t and σ_c, respectively, in the column.

Problems 5.9-19 through 5.9-24 pertain to the core of a cross section.

5.9-19 A hollow circular cross section with inner radius r_1 and outer radius r_2 is shown in the figure. (a) Determine the radius r_c of the core of the section. (b) From the result for part (a), obtain the radius r_c of the core of a solid circular cross section. (c) From the result for part (a), obtain the core radius r_c for the case in which r_1 approaches r_2 and the cross-sectional area becomes a thin ring.

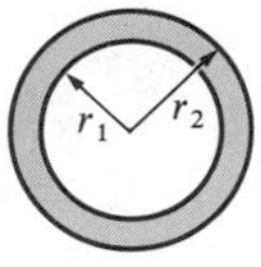

Prob. 5.9-19

5.9-20 Determine the core of a W 8 × 28 structural steel section (see Appendix E).

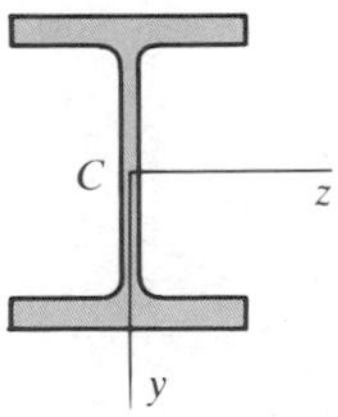

Probs. 5.9-20 and 5.9-21

5.9-21 Determine the core of a W 12 × 50 structural steel section (see Appendix E).

5.9-22 The cross section of a square chimney (b = length of each side) with a circular flue of diameter d is shown in the figure on p. 364. Determine the core of the section if $d = 2b/3$.

5.9-23 Determine the core of a cross section in the form of an equilateral triangle with sides of length b (see figure on p. 364).

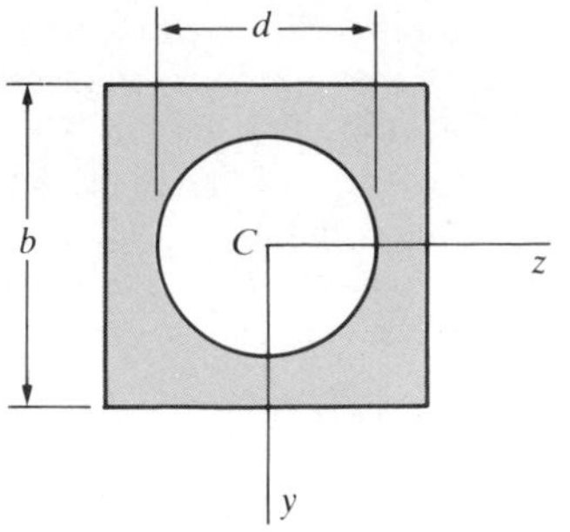

Prob. 5.9-22

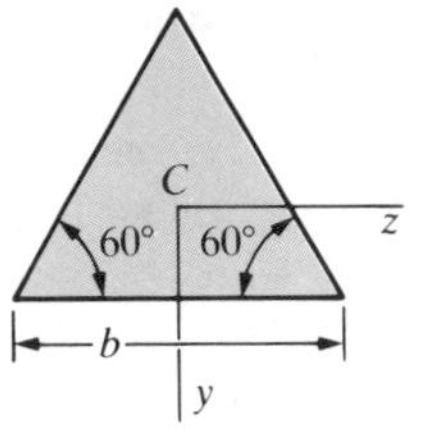

Prob. 5.9-23

5.9-24 The doubly symmetric, hollow square cross section shown in the figure has outer dimension b and inner dimension c. What is the largest inner dimension c_{max} such that the square hole lies entirely within the core of the cross section?

Prob. 5.9-24

The problems for Section 5.10 are to be solved considering the stress-concentration factors.

5.10-1 The beams shown in the figure are subjected to bending moments $M = 2100$ in.-lb. Each beam has a rectangular cross section with height $h = 1.5$ in. and width $b = 0.375$ in. (a) For the beam with a hole at midheight, determine the maximum stresses for hole diameters $d =$ 0.25, 0.50, 0.75, and 1.00 in. (b) For the beam with two identical notches (inside height $h_1 = 1.25$ in.), determine the maximum stresses for notch radii $R = 0.05$, 0.10, 0.15, and 0.20 in.

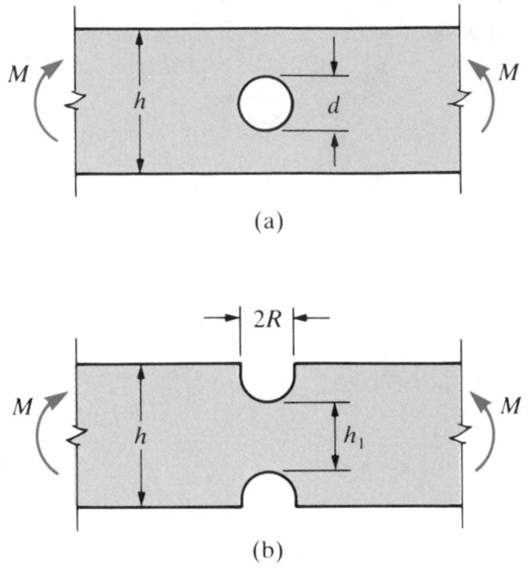

Probs. 5.10-1 through 5.10-4

5.10-2 The beams shown in the figure are subjected to bending moments $M = 250$ N·m. Each beam has a rectangular cross section with height $h = 44$ mm and width $b = 10$ mm. (a) For the beam with a hole at midheight, determine the maximum stresses for hole diameters $d = 10$, 16, 22, and 28 mm. (b) For the beam with two identical notches (inside height $h_1 = 40$ mm), determine the maximum stresses for notch radii $R = 2$, 4, 6, and 8 mm.

5.10-3 A rectangular beam with semicircular notches, part (b) of the figure, has dimensions $h = 0.88$ in. and $h_1 =$ 0.80 in. The maximum allowable bending stress in the material (metal) is $\sigma_{max} = 60$ ksi, and the bending moment is $M = 600$ in.-lb. Determine the minimum permissible width b_{min} of the beam.

5.10-4 A rectangular beam with semicircular notches, as shown in part (b) of the figure, has dimensions $h = 120$ mm and $h_1 = 100$ mm. The maximum allowable bending stress in the material (plastic) is $\sigma_{max} = 6$ MPa, and the bending moment is $M = 150$ N·m. Determine the minimum permissible width b_{min} of the beam.

5.10-5 A rectangular beam with notches and a hole (see figure) has dimensions $h = 5.5$ in., $h_1 = 5$ in., and width $b = 1.6$ in. The beam is subjected to a bending moment $M = 130$ in.-k, and the maximum allowable bending stress in the material is $\sigma_{max} = 42{,}000$ psi. (a) What is the smallest radius R_{min} that should be used in the notches? (b) What is the diameter d_{max} of the largest hole that should be drilled at the midheight of the beam?

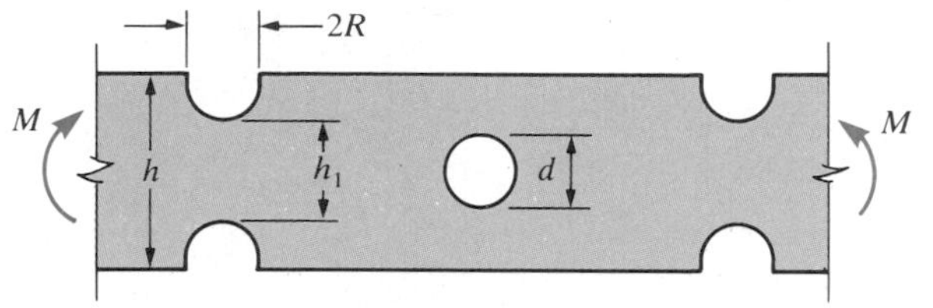

Prob. 5.10-5

5.11-1 A simple beam of length 12 ft supports a total uniform load of intensity 1000 lb/ft (see figure). The beam consists of a wood member of 4 in. × 11.5 in. (actual dimensions) that is reinforced by 0.25 in. thick steel plates on top and bottom. The moduli of elasticity for the steel and wood are $E_s = 30 \times 10^6$ psi and $E_w = 1.5 \times 10^6$ psi, respectively. Calculate the maximum stress σ_s in the steel plates and the maximum stress σ_w in the wood member.

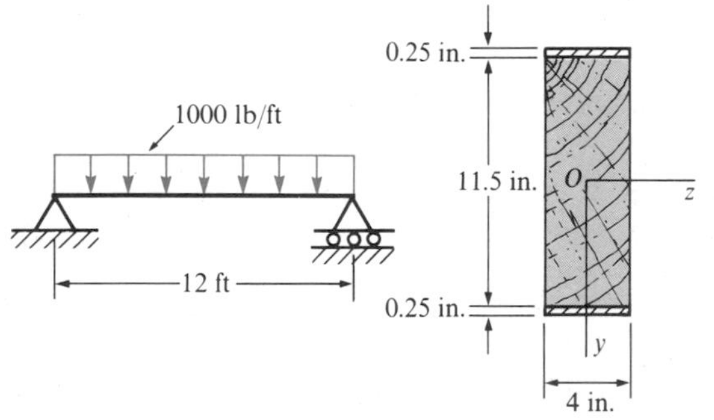

Prob. 5.11-1

5.11-2 The composite beam shown in the figure is simply supported and carries a total uniform load of 40 kN/m on a span length of 5 m. The beam is built of a wood member having cross-sectional dimensions 150 mm × 250 mm and two steel plates of dimensions 50 mm × 150 mm. Determine the maximum stresses σ_s and σ_w in the steel and wood, respectively, if the moduli of elasticity are $E_s = 209$ GPa and $E_w = 11$ GPa

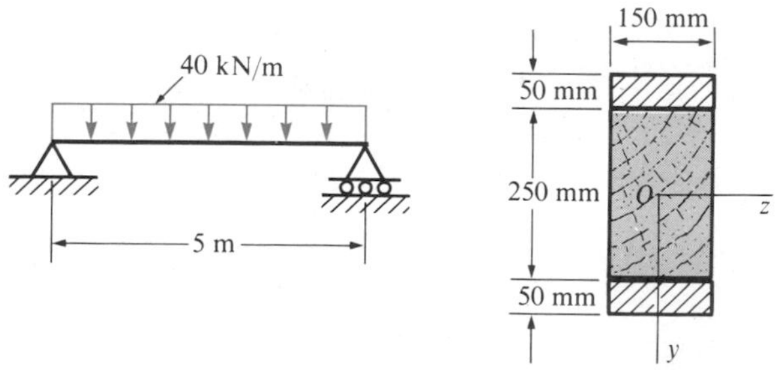

Prob. 5.11-2

5.11-3 A wood beam 8 in. wide × 12 in. deep (actual dimensions) is reinforced on both top and bottom by steel plates 0.5 in. thick (see figure). Find the allowable bending moment M_{max} about the z axis if the allowable stress in the wood is 1,050 psi and in the steel is 18,000 psi. (Assume that the ratio of the moduli of elasticity of steel and wood is 20.)

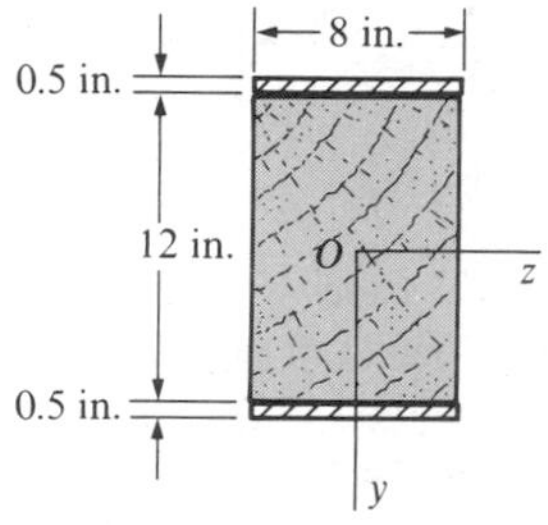

Prob. 5.11-3

5.11-4 A wood beam of dimensions 200 mm × 300 mm is reinforced on its sides by steel plates 12 mm thick (see figure). The moduli of elasticity for the steel and wood are $E_s = 204$ GPa and $E_w = 8.5$ GPa, respectively. Also, the corresponding allowable stresses are $\sigma_s = 140$ MPa and $\sigma_w = 10$ MPa. Calculate the maximum permissible bending moment M_{max} about the z axis.

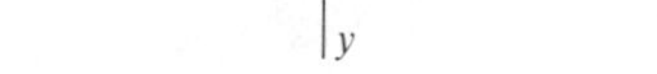

Prob. 5.11-4

5.11-5 A hollow box beam is constructed with webs of Douglas-fir plywood and flanges of pine as shown in the figure. The plywood is 1 in. thick × 12 in. wide; the flanges are 2 in. × 4 in. (actual size). The modulus of elasticity for the plywood is 1,600,000 psi and for the pine is 1,200,000 psi.

If the allowable stresses are 2100 psi for the plywood and 1800 psi for the pine, find the allowable bending moment M_{max} for the beam.

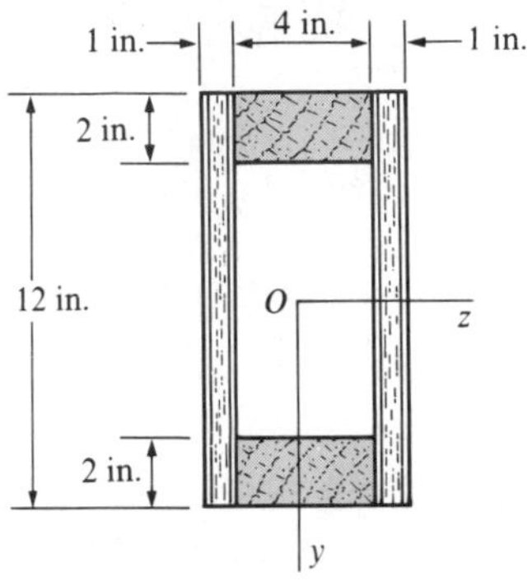

Prob. 5.11-5

5.11-6 A round steel tube of outside diameter d and an aluminum core of diameter $d/2$ are bonded to form a composite beam as shown in the figure. Derive a formula for the allowable bending moment M that can be carried by the beam based upon an allowable stress in the steel of σ_s. (Assume that the moduli of elasticity for the steel and aluminum are E_s and E_a, respectively.)

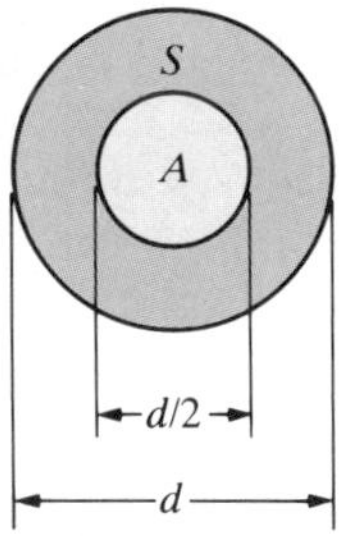

Prob. 5.11-6

5.11-7 A simple beam of span length 10.5 ft carries a uniform load of 3200 lb/ft. The cross section of the beam is a hollow box with wood flanges and steel webs, as shown in the figure. What is the required thickness t of the steel plates if the allowable stresses are 18,000 psi for steel and 1,000 psi for the wood? (Assume that the moduli of elasticity for the steel and the wood are 30×10^6 psi and 1.5×10^6 psi, respectively.)

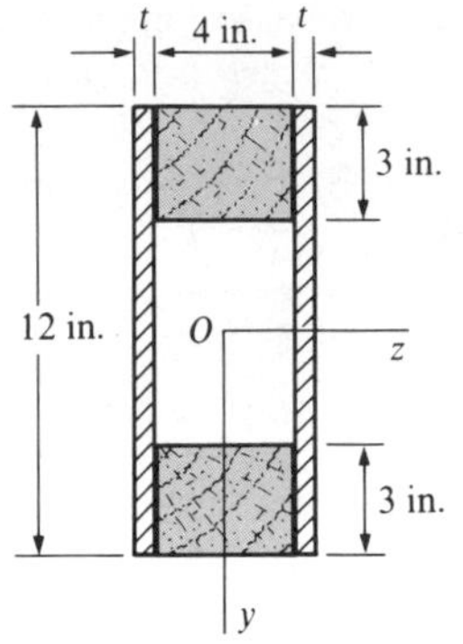

Prob. 5.11-7

5.11-8 A simply supported composite beam 3 m long carries a load $P = 5$ kN at the middle of the span (see figure). The beam is constructed of a wood member 100 mm wide × 150 mm deep reinforced on its lower side by a steel bar 8 mm thick × 100 mm wide. Find the maximum bending stresses σ_s and σ_w in the steel and wood, respectively, due to the load P if the modulus of elasticity is $E_w = 10$ GPa for wood and $E_s = 210$ GPa for steel.

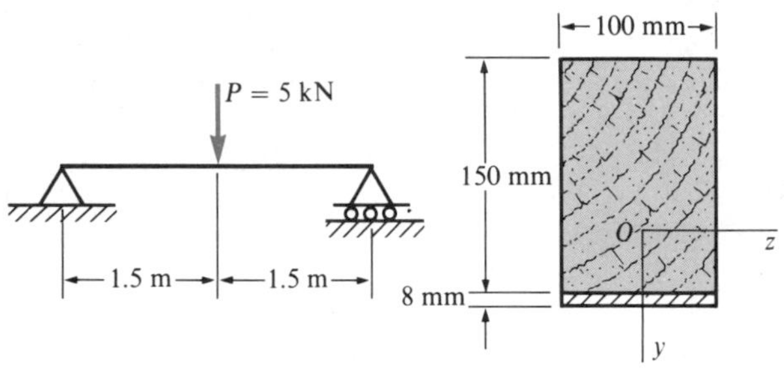

Prob. 5.11-8

5.11-9 A composite beam made of wood and steel has the cross-sectional dimensions shown in the figure. The beam is simply supported with a span length of 11 ft. A concentrated load $P = 1800$ lb acts at the midpoint. Calculate the maximum bending stresses σ_s and σ_w in the steel and wood, respectively, due to the load P if $E_s/E_w = 20$.

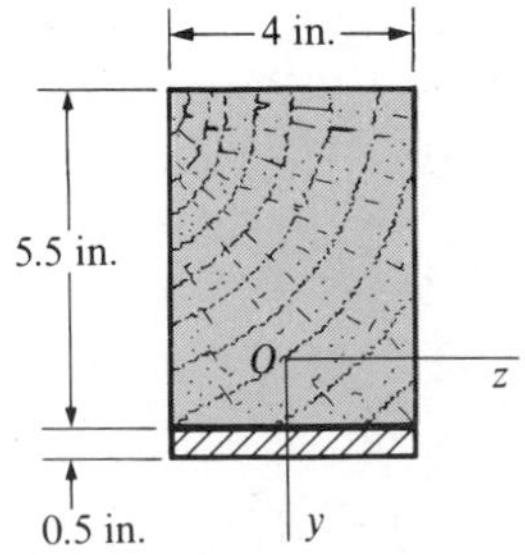

Prob. 5.11-9

5.11-10 The cross section of a composite beam made of aluminum and steel is shown in the figure. The moduli of elasticity are $E_a = 75$ GPa and $E_s = 200$ GPa. Under the action of a bending moment that produces a maximum stress of 50 MPa in the aluminum, what is the maximum stress σ_s in the steel?

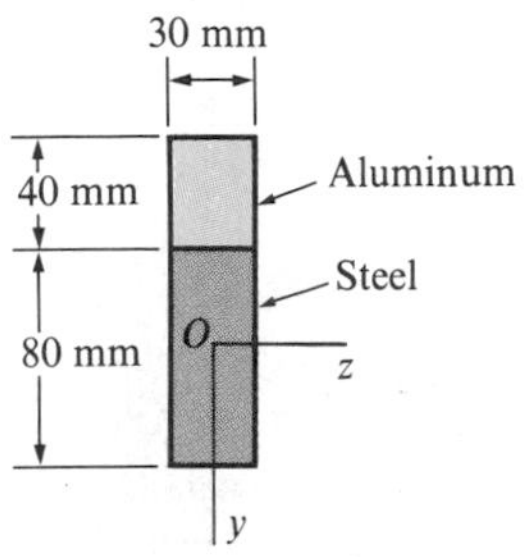

Prob. 5.11-10

5.11-11 A bimetallic beam used in a temperature-control switch consists of strips of aluminum and copper bonded together (see figure). The width of the beam is 1 in., and each layer has a thickness of $\frac{1}{16}$ in. Under the action of a bending moment $M = 12$ in.-lb, what are the maximum stresses σ_a and σ_c in the aluminum and copper, respectively? (Assume $E_a = 10{,}500{,}000$ psi and $E_c = 16{,}800{,}000$ psi.)

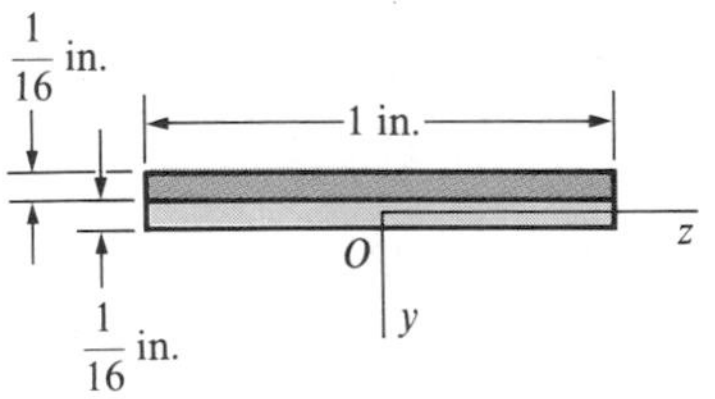

Prob. 5.11-11

5.11-12 The cross section of a bimetallic strip is shown in the figure. Assuming that the moduli of elasticity for metals A and B are $E_a = 280$ GPa and $E_b = 150$ GPa, respectively, determine the smaller of the two section moduli for the beam (that is, the ratio of bending moment to maximum bending stress). In which material does the maximum stress occur?

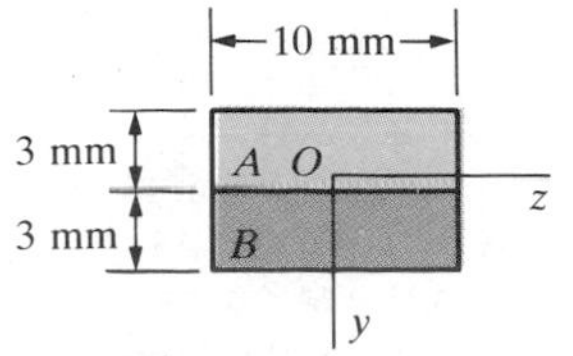

Prob. 5.11-12

5.11-13 A composite beam is constructed of a wood beam 6 in. wide × 8 in. deep (actual dimensions) reinforced on the lower side by a 0.5 in. × 6 in. steel plate (see figure). The modulus of elasticity for the wood is $E_w = 1.2 \times 10^6$ psi and for the steel is $E_s = 30 \times 10^6$ psi. Find the allowable bending moment M_{max} for the beam if the allowable stress in the wood is $\sigma_w = 1{,}500$ psi and in the steel is $\sigma_s = 16{,}000$ psi.

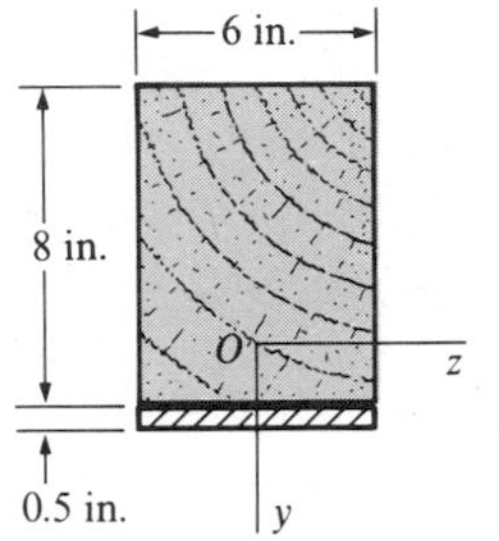

Prob. 5.11-13

***5.11-14** The cross section of a reinforced concrete beam is shown in the figure. The diameter of each of the three steel reinforcing bars is 25 mm, and the modular ratio $n = 10$. The allowable compressive stress in the concrete is $\sigma_c = 12$ MPa, and the allowable tensile stress in the steel is 120 MPa. Calculate the maximum allowable bending moment M_{max} for this beam. (Assume that the concrete resists only compression, and consider that the area of the steel bars is concentrated in a horizontal line at distance 360 mm below the top of the beam.)

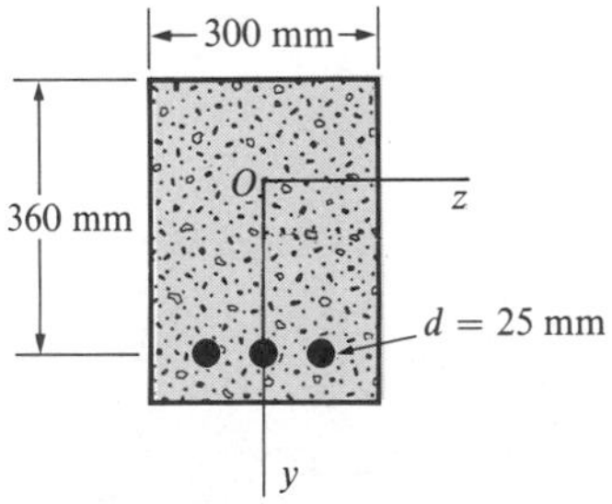

Prob. 5.11-14

5.12-1 A cantilever beam of rectangular cross section (see figure) supports an inclined load P at its free end. Demonstrate that if the load P has its line of action along a diagonal of the cross section, the neutral axis will lie along the other diagonal.

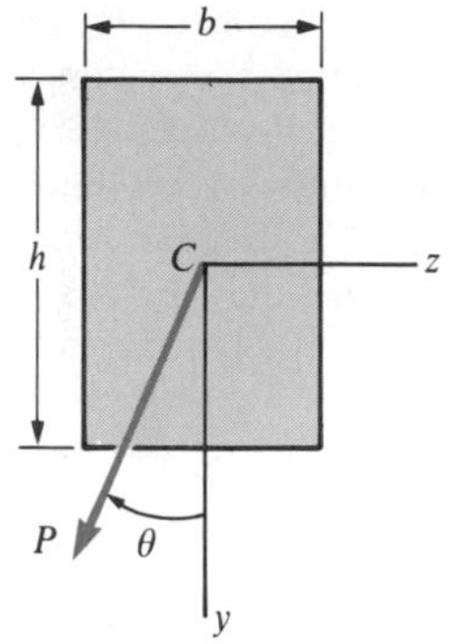

Probs. 5.12-1, 5.12-2, and 5.12-3

5.12-2 A wood cantilever beam of rectangular cross section (see figure) supports an inclined load P at its free end. Calculate the maximum tensile stress σ_{max} due to the load P. Data for the beam are as follows: $b = 75$ mm, $h = 150$ mm, $L = 1.4$ m, $P = 800$ N, and $\theta = 30°$.

5.12-3 Solve the preceding problem for a cantilever beam with data as follows: $b = 4$ in., $h = 8$ in., $L = 6$ ft, $P = 500$ lb, and $\theta = 45°$.

5.12-4 A steel beam of I-section (see figure) is simply supported at the ends. Two equal and oppositely directed bending couples M_0 act at the ends of the beam, so that the beam is in pure bending. The couples act in plane *mm*, which is oriented at an angle α to the xy plane. Determine the maximum tensile stress σ_{max} due to the couples M_0. Data for the beam are as follows: S 8 × 18.4 section, $M_0 = 45$ in.-k, and $\alpha = 30°$.

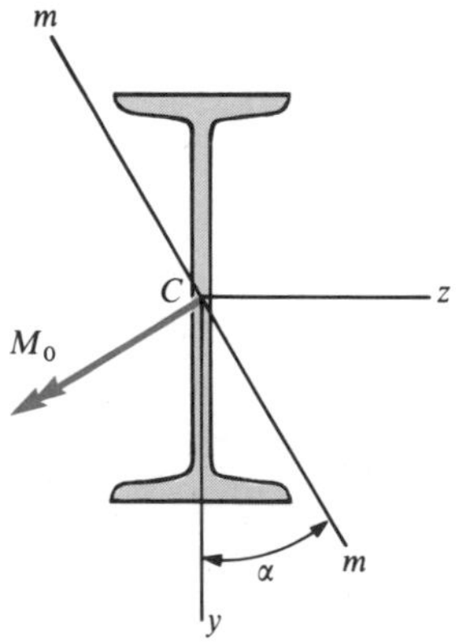

Prob. 5.12-4

5.12-5 A cantilever beam of wide-flange cross section (see figure) supports an inclined load P at its free end. Calculate the maximum tensile stress σ_{max} due to the load P. Data for the beam are as follows: W 10 × 45 section, $L = 6$ ft, $P = 3.5$ k, and $\theta = 35°$.

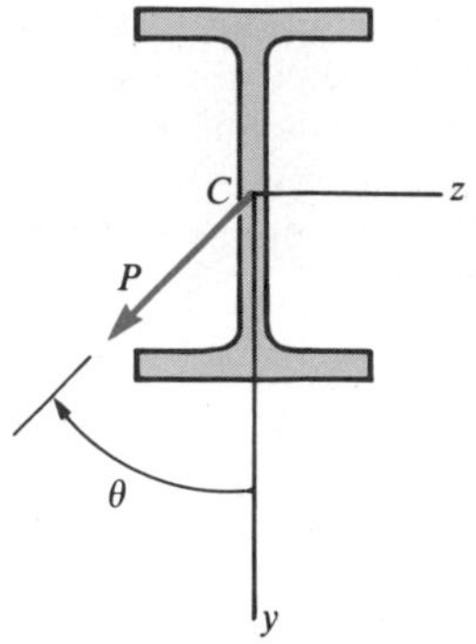

Probs. 5.12-5 and 5.12-6

5.12-6 Solve the preceding problem using the following data: W 8 × 35 section, $L = 6$ ft, $P = 2.0$ k, and $\theta = 30°$.

5.12-7 A simply supported wide-flange beam of span length L carries a vertical concentrated load P acting through the centroid C at the midpoint of the span. The beam is attached to supports inclined at an angle θ to the horizontal, as shown in the figure. Calculate the stresses at the four outside corners of the cross section (points A, B, D, and E) due to the load P. Data for the beam are as follows: W 10 × 60 section, $L = 10$ ft, $P = 5$ k, and $\theta = 26.57°$.

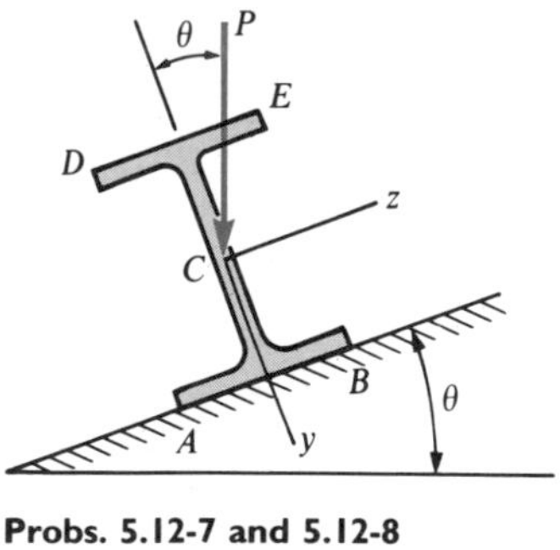

Probs. 5.12-7 and 5.12-8

5.12-8 Solve the preceding problem using the following data: W 8 × 28 section, $L = 8$ ft, $P = 3.8$ k, and $\theta = 20°$.

5.12-9 A wood beam of rectangular cross section (see figure) is simply supported at the ends. The longitudinal axis of the beam is horizontal, but the cross section is inclined at an angle θ. The load on the beam is a vertical uniform load of intensity q that acts through the centroid C. Calculate the maximum tensile stress σ_{max} if $b = 4$ in., $h = 8$ in., $L = 8$ ft, $\tan\theta = \frac{1}{3}$, and $q = 200$ lb/ft.

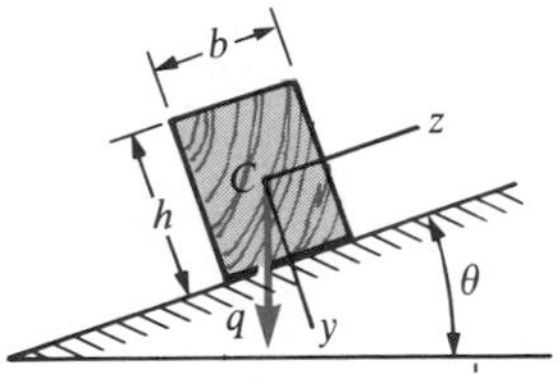

Probs. 5.12-9 and 5.12-10

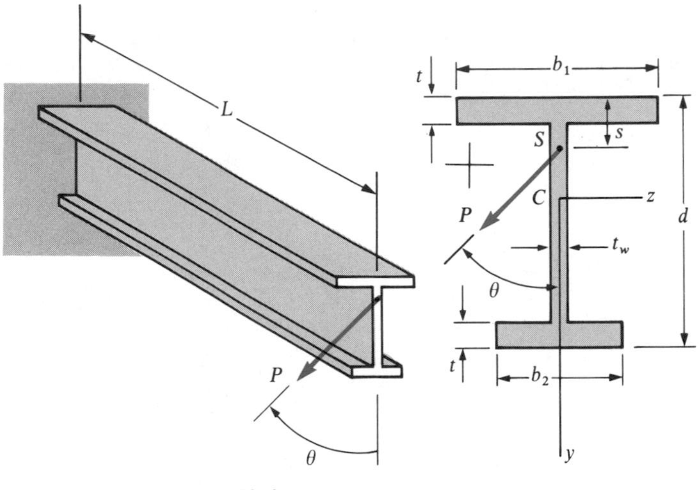

Probs. 5.13-1 through 5.13-4

5.12-10 Solve the preceding problem using the following data: $b = 200$ mm, $h = 50$ mm, $L = 2$ m, $\theta = 30°$, and $q = 1.0$ kN/m.

5.12-11 A cantilever beam of W 12 × 14 section and length $L = 8$ ft supports a slightly inclined load $P = 600$ lb at the free end (see figure). (a) Plot a graph of the stress σ_a at point A as a function of the angle θ. (b) Plot a graph of the angle β, which locates the neutral axis nn, as a function of the angle θ. Let θ vary from 0 to 10°.

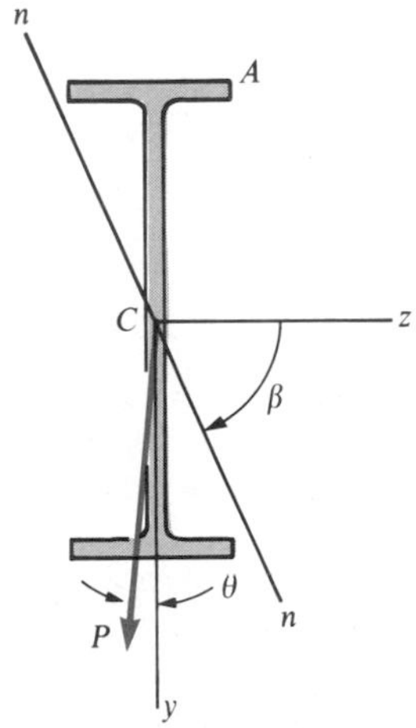

Prob. 5.12-11

5.13-1 A singly symmetric I-beam of length L is supported as a cantilever and subjected to a lateral load P at its free end (see figure). The load acts at an angle θ to the y axis. (a) At what distance s from the top surface of the beam (measured along the centerline of the web) should the load P be applied in order that the beam will bend without twisting? (b) Assuming that no twisting occurs, calculate the maximum tensile stress σ_t and the maximum compressive stress σ_c due to bending. Use the following data: $d = 20$ in., $b_1 = 18$ in., $b_2 = 8$ in., $t = 2$ in., $t_w = 1$ in., $P = 10$ k, $\theta = 25°$, and $L = 12$ ft.

5.13-2 Solve the preceding problem for the following beam and load: $d = 600$ mm, $b_1 = 300$ mm, $b_2 = 200$ mm, $t =$ 30 mm, $t_w = 20$ mm, $P = 10$ kN, $\theta = 20°$, and $L = 5$ m.

5.13-3 Solve Problem 5.13-1 for the following beam and load: $d = 15$ in., $b_1 = 12$ in., $b_2 = 9$ in., $t = 1.25$ in., $t_w =$ 1.0 in., $P = 1.5$ k, $\theta = 35°$, and $L = 10$ ft.

5.13-4 Solve Problem 5.13-1 for the following beam and load: $d = 360$ mm, $b_1 = 320$ mm, $b_2 = 240$ mm, $t = 30$ mm, $t_w = 25$ mm, $P = 10$ kN, $\theta = 45°$, and $L = 3$ m.

Unless otherwise stated, all calculations and derivations for the problems in Section 5.14 are to be made using centerline dimensions.

5.14-1 A simple beam of I-section and span length $L =$ 12 ft supports a uniform load of intensity $q = 3.2$ k/ft (see figure). The dimensions of the cross section are $h = 11.5$ in., $b = 8$ in., and $t_f = t_w = 0.5$ in. (a) Calculate the maximum shear stress τ_{max} on cross section A-A located at distance

$d = 3$ ft from the end of beam. (b) Calculate the shear stress τ_b at point B on cross section A-A. Point B is located at a distance $a = 1$ in. from the edge of the lower flange.

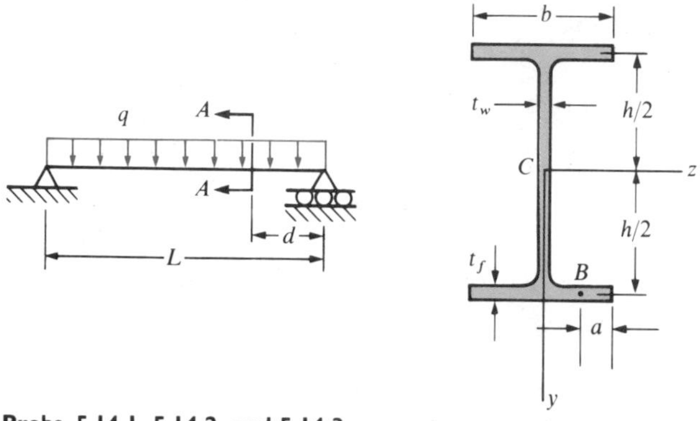

Probs. 5.14-1, 5.14-2, and 5.14-3

Section A-A

5.14-2 Solve the preceding problem for the following data: $L = 4$ m, $q = 48$ kN/m, $h = 300$ mm, $b = 200$ mm, $t_f = 15$ mm, $t_w = 10$ mm, $d = 1$ m, and $a = 25$ mm.

5.14-3 Solve Problem 5.14-1 by making all calculations based upon the actual dimensions of the cross section instead of the centerline dimensions. Compare the results obtained from the two solutions.

5.14-4 A singly symmetric I-beam is subjected to a shear force P acting through the shear center S parallel to the y axis (see figure). Obtain a formula for the maximum shear stress τ_{max} in the beam, assuming $b_1 > b_2$.

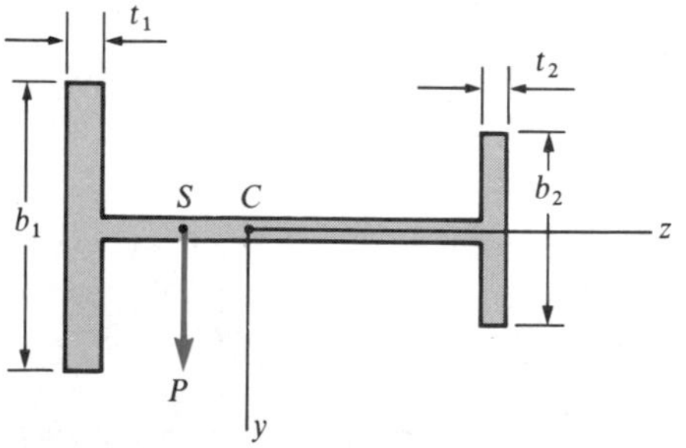

Prob. 5.14-4

5.14-5 A T-beam is subjected to a shear force P acting through the shear center S parallel to the y axis (see figure). The flange has width b and thickness t. Investigate the shear stresses acting on the cross section, as follows: (a) Obtain a formula for the maximum shear stress τ_{max}. (b) Plot a graph showing how the shear stress τ varies over the height of the flange. (c) Demonstrate that the resultant of the shear stresses is equal to P.

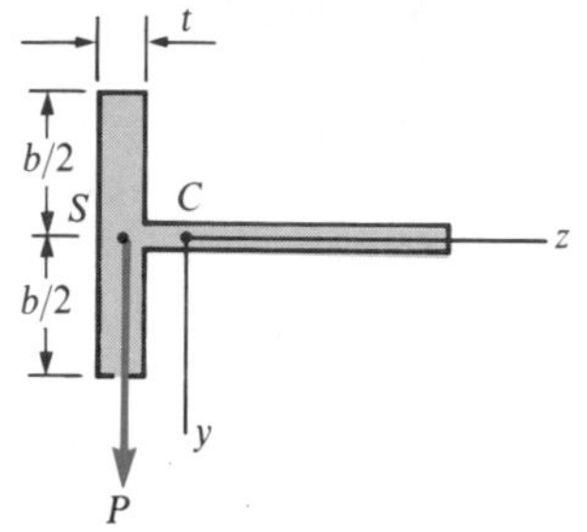

Prob. 5.14-5

5.14-6 An equal-leg angle section is subjected to a shear force P acting in the direction of the y axis, which is an axis of symmetry (see figure). Each leg of the angle has centerline length b and thickness t. Investigate the shear stresses acting on the cross section, as follows: (a) Derive a formula for the maximum shear stress τ_{max} in the angle. (b) Plot a graph showing how the shear stress τ varies along the centerline of the cross section. (c) Demonstrate that the resultant of the shear stresses is equal to P.

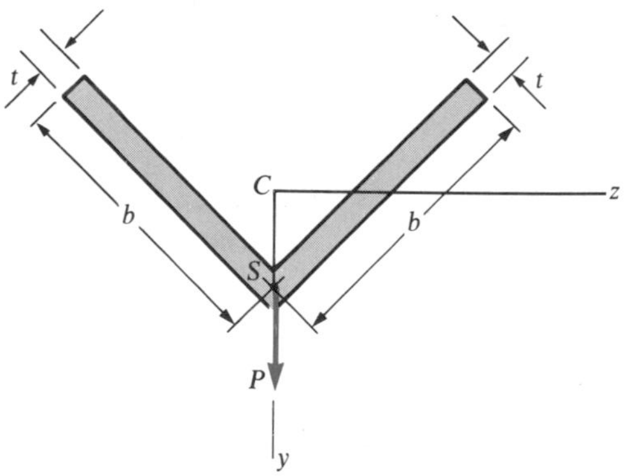

Prob. 5.14-6

5.14-7 A beam of channel cross section is subjected to a shear force $P = 12$ k acting along the y axis, which is an axis of symmetry (see figure). Investigate the shear stresses acting on the cross section, as follows: (a) Calculate the maximum shear stress τ_{max}. (b) Plot graphs showing how the shear stress τ varies over the height b of the flanges and along the length h of the web. (c) Demonstrate that the resultant of the shear stresses is equal to P. The dimensions of the cross section are $h = 18$ in., $b = 6$ in., and $t = 1$ in.

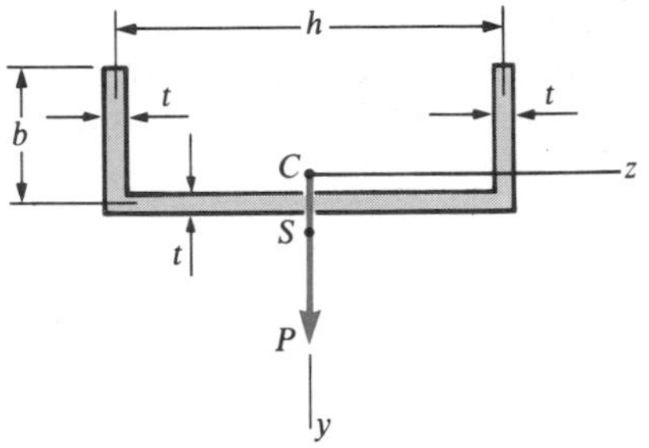

Probs. 5.14-7 and 5.14-8

5.14-8 Solve the preceding problem if $P = 20$ kN and the dimensions of the cross section are $h = 225$ mm, $b = 90$ mm, and $t = 20$ mm.

***5.14-9** A cross section in the form of an unbalanced I-beam is turned sideways and subjected to a shear force $P = 20$ k acting in the direction of the y axis, which is an axis of symmetry (see figure). Investigate the shear stresses acting on the cross section, as follows: (a) Calculate the maximum shear stress τ_{max}. (b) Plot a graph showing how the shear stress τ varies over the height b of the flanges and the length h of the web. (c) Demonstrate that the resultant of the shear stresses is equal to P. The dimensions of the cross section are $h = 16$ in., $b = 9$ in., $b_1 = 6$ in., $b_2 = 3$ in., $t_f = 1$ in., and $t_w = 0.75$ in.

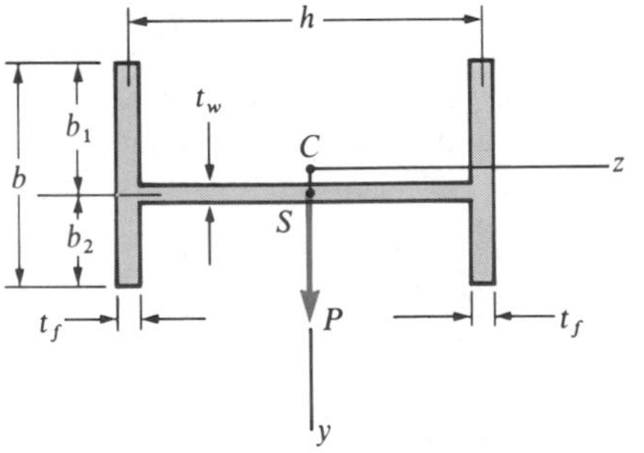

Probs. 5.14-9 and 5.14-10

***5.14-10** Solve the preceding problem if $P = 120$ kN and the dimensions of the cross section are $h = 300$ mm, $b = 160$ mm, $b_1 = 120$ mm, $b_2 = 40$ mm, $t_f = 20$ mm, and $t_w = 15$ mm.

When locating the shear centers in the problems for Section 5.15, assume that the cross sections are thin-walled and use centerline dimensions for all calculations and derivations.

5.15-1 Calculate the distance e_0 from the back of a C 10 × 25 channel section to the shear center S (see figure).

5.15-2 Calculate the distance e_0 from the back of a C 8 × 13.75 channel section to the shear center S (see figure).

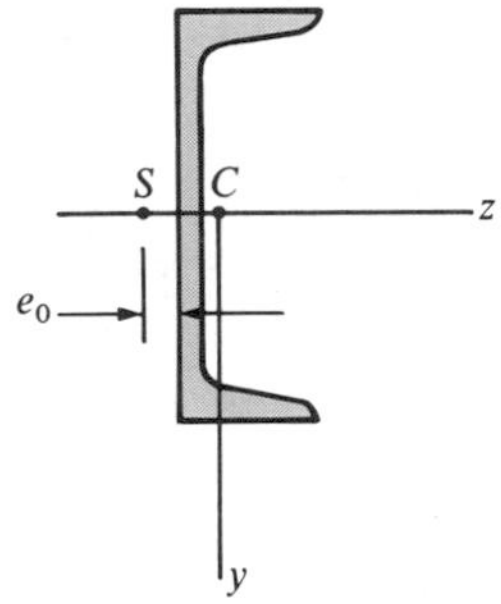

Probs. 5.15-1 and 5.15-2

5.15-3 A cross section in the shape of an unbalanced I-beam is shown in the figure. Derive the following formula for the distance e from the centerline of the web to the shear center S:

$$e = \frac{3t_f(b_2^2 - b_1^2)}{ht_w + 6t_f(b_1 + b_2)}$$

Also, check the formula for the special cases in which $b_1 = 0$ and $b_2 = b$ (channel section) and $b_1 = b_2 = b/2$ (doubly symmetric I-beam).

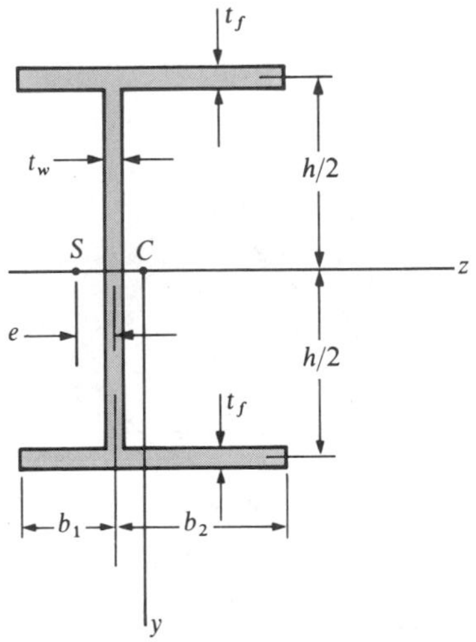

Prob. 5.15-3

5.15-4 A cross section of a channel beam with double flanges and constant thickness throughout the section is shown in the figure. Derive the following formula for the distance e from the centerline of the web to the shear center S:

$$e = \frac{3b^2(h_1^2 + h_2^2)}{h_2^3 + 6b(h_1^2 + h_2^2)}$$

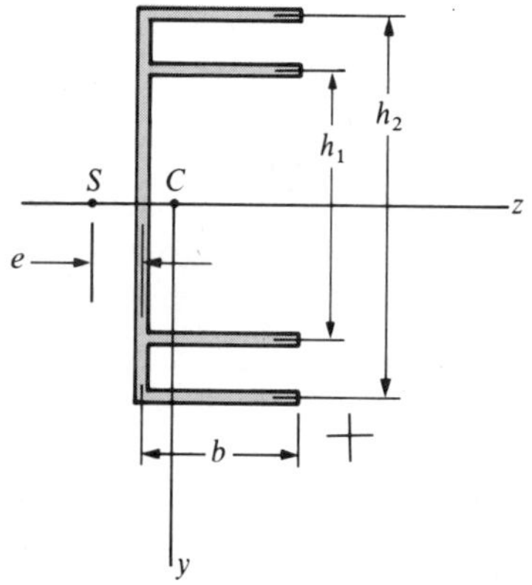

Prob. 5.15-4

5.15-5 A cross section of a slit square tube of constant thickness is shown in the figure. Derive the following formula for the distance e from the corner of the cross section to the shear center S:

$$e = \frac{b}{2\sqrt{2}}$$

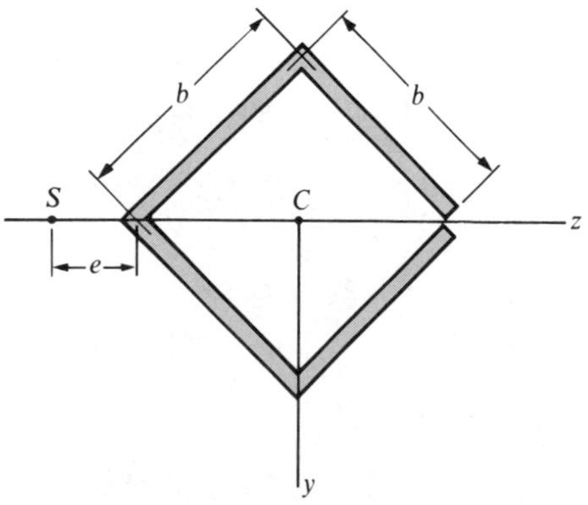

Prob. 5.15-5

5.15-6 A cross section of a slit rectangular tube of constant thickness is shown in the figure. Derive the following formula for the distance e from the centerline of the wall of the tube to the shear center S:

$$e = \frac{b(2h + 3b)}{2(h + 3b)}$$

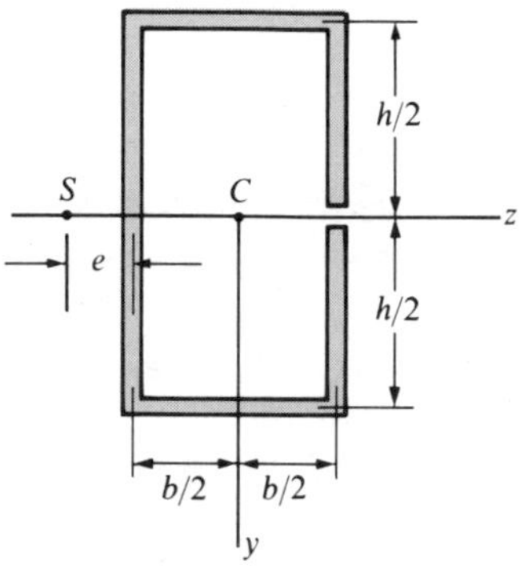

Prob. 5.15-6

***5.15-7** Derive the following formula for the distance e from the centerline of the wall to the shear center S for the C-section of constant thickness shown in the figure:

$$e = \frac{3bh^2(b + 2a) - 8ba^3}{h^2(h + 6b + 6a) + 4a^2(2a - 3h)}$$

Also, check the formula for the special cases in which $a = 0$ (channel section) and $a = h/2$ (slit rectangular tube).

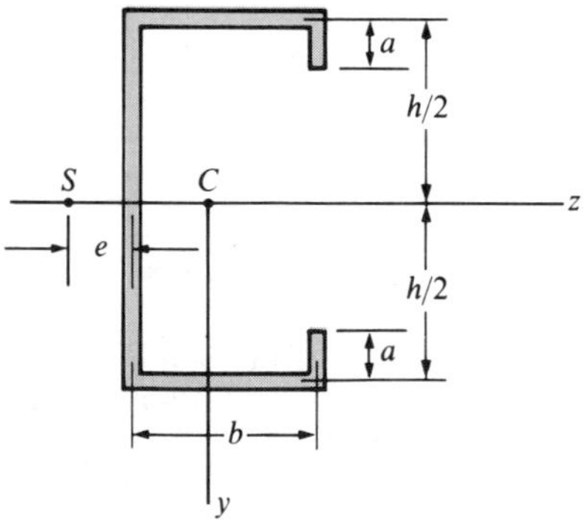

Prob. 5.15-7

***5.15-8** Derive the following formula for the distance e from the centerline of the wall to the shear center S for the hat section of constant thickness shown in the figure:

$$e = \frac{3bh^2(b + 2a) - 8ba^3}{h^2(h + 6b + 6a) + 4a^2(2a + 3h)}$$

Also, check the formula for the special case in which $a = 0$ (channel section).

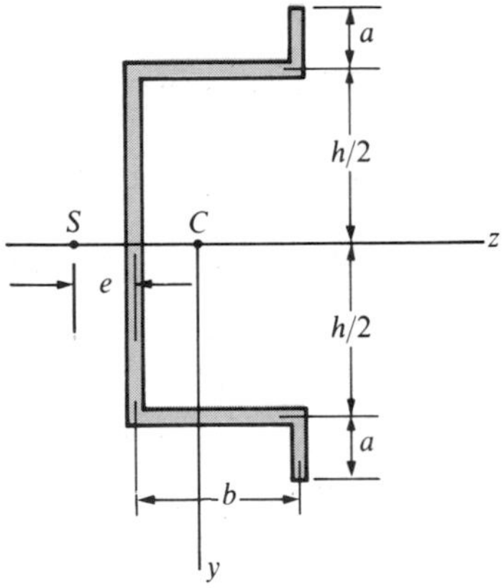

Prob. 5.15-8

5.15-9 A cross section of a slit circular tube of constant thickness is shown in the figure. Show that the distance e from the center of the circle to the shear center S is given by the formula $e = 2r$.

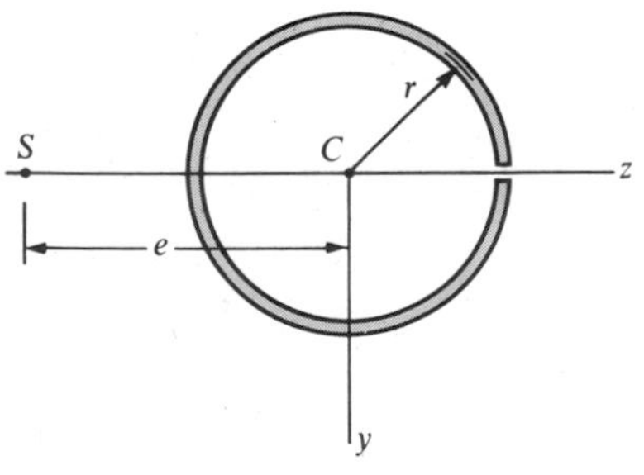

Prob. 5.15-9

*5.15-10 A cross section in the shape of a circular arc of constant thickness is shown in the figure. Derive the following formula for the distance e from the center of the circle to the shear center S:

$$e = \frac{2r(\sin\beta - \beta\cos\beta)}{\beta - \sin\beta\cos\beta}$$

Also, plot a graph showing how the distance e varies as β goes from 0 to π.

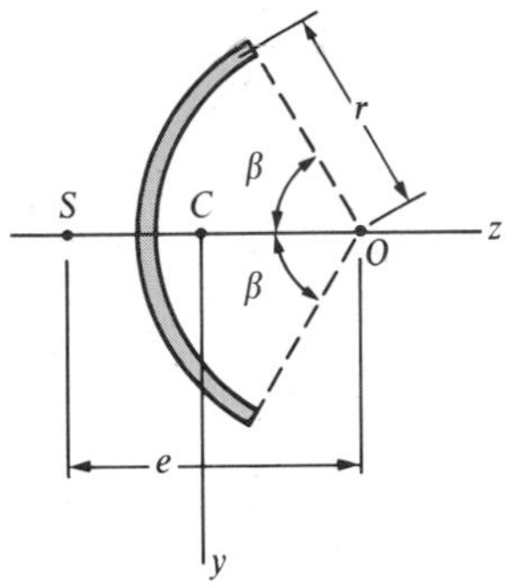

Prob. 5.15-10

5.15-11 A U-shaped cross section of constant thickness is shown in the figure. Derive the following formula for the distance e from the center of the semicircle to the shear center S:

$$e = \frac{2(2r^2 + b^2 + \pi br)}{4b + \pi r}$$

Also, plot a graph showing how the distance e (expressed as the nondimensional ratio e/r) varies as a function of the ratio b/r.

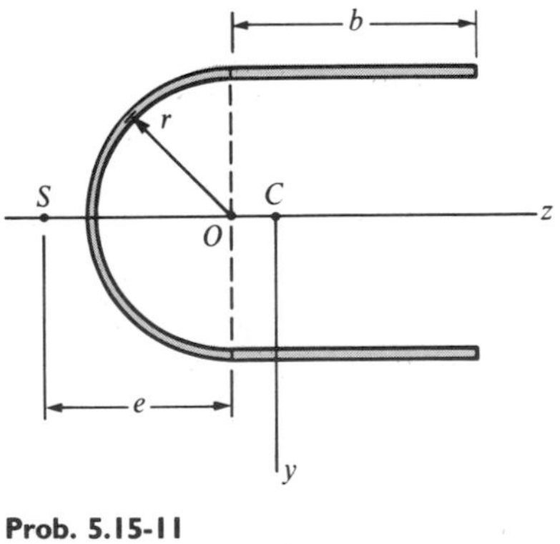

Prob. 5.15-11

5.15-12 Calculate the distance e from the centerline of the wall to the shear center S for the singly symmetric cross section of constant thickness shown in the figure. (Procedure: Determine the shear stresses acting on the cross section and find the line of action of their resultant. Use the formula of Problem 5.15-13 only as a check.)

*5.15-13 Derive the following formula for the distance e from the centerline of the web to the shear center S for the singly symmetric cross section of constant thickness shown in the figure:

$$e = \frac{3bh^2(b + 2a) - 8ba^3}{h^2(h + 6b) + 4a(3h^2 + 4a^2 + 6ah)}$$

Also, check the formula for the special case in which $a = 0$ (channel section).

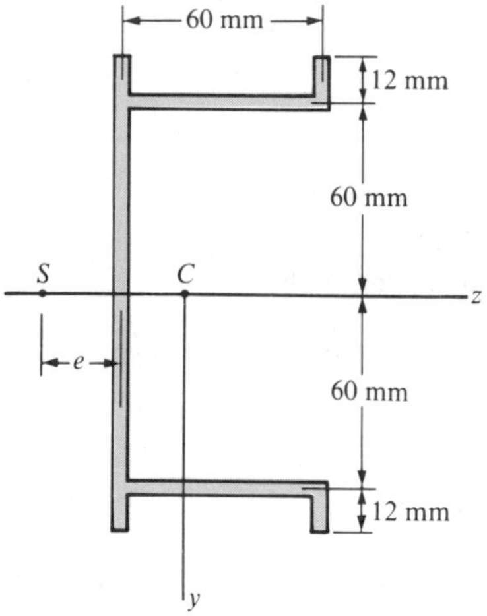

Prob. 5.15-12

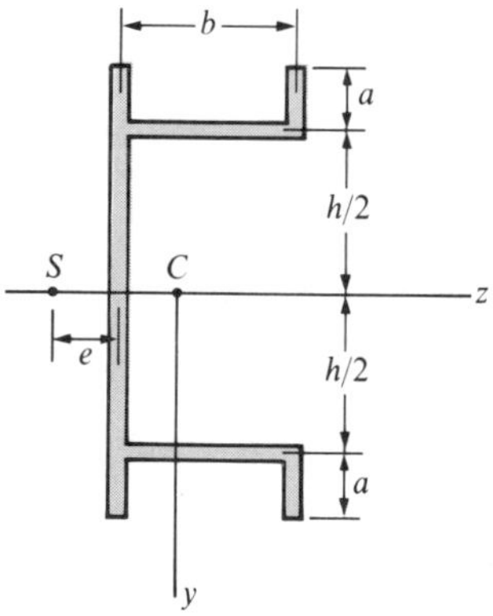

Prob. 5.15-13

The problems for Section 5.17 are to be solved using the assumption that the material is elastic-plastic.

5.17-1 Determine the shape factor f for a solid circular cross section (see figure).

5.17-2 Determine the shape factor f for a cross section in the form of a rhombus (see figure).

5.17-3 (a) Determine the shape factor f for a hollow circular cross section (see figure). (b) In the special case of a very thin section, what is the shape factor f?

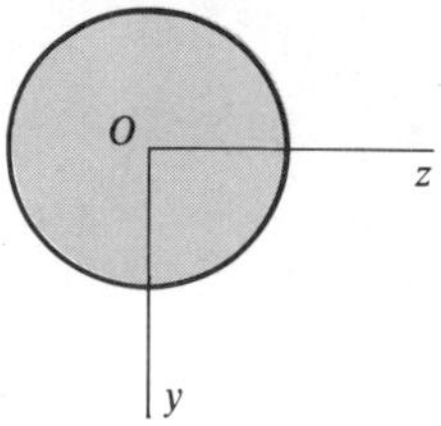

Prob. 5.17-1

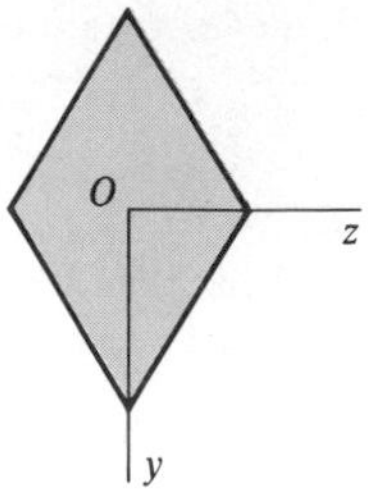

Prob. 5.17-2

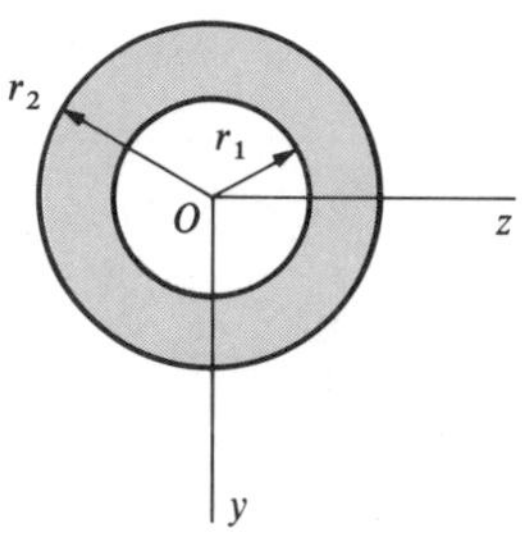

Prob. 5.17-3

5.17-4 A beam of rectangular cross section (see figure) with height $h = 120$ mm and width $b = 40$ mm is constructed of steel with yield stress $\sigma_y = 250$ MPa and $E = 210$ GPa. Obtain numerical values for the yield moment M_y, the plastic moment M_p, and the plastic modulus Z for this beam. Also, plot to scale (in numerical form) the moment-curvature diagram for this particular beam.

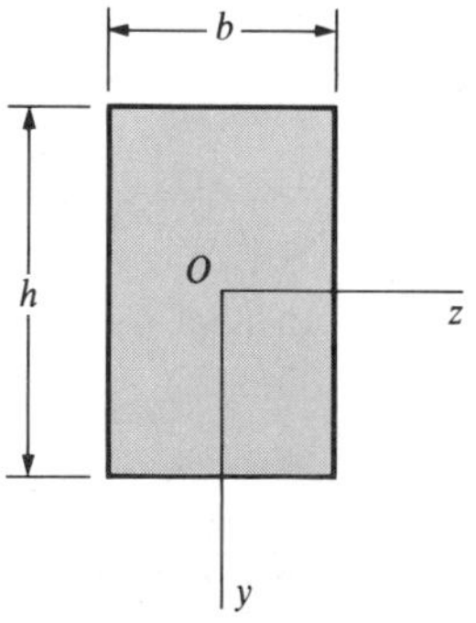

Probs. 5.17-4 and 5.17-5

5.17-5 Solve the preceding problem for a rectangular beam with $h = 4.0$ in., $b = 2.0$ in., $\sigma_y = 36$ ksi, and $E = 30 \times 10^3$ ksi.

5.17-6 Calculate the plastic moment M_p, the plastic modulus Z, and the shape factor f for the wide-flange beam shown in the figure. Assume $h = 400$ mm, $b = 160$ mm, $t_f = 12$ mm, $t_w = 8$ mm, and $\sigma_y = 290$ MPa.

5.17-7 Solve the preceding problem for a wide-flange beam with $h = 12.0$ in., $b = 6.0$ in., $t_f = 0.6$ in., $t_w = 0.4$ in., and $\sigma_y = 36$ ksi.

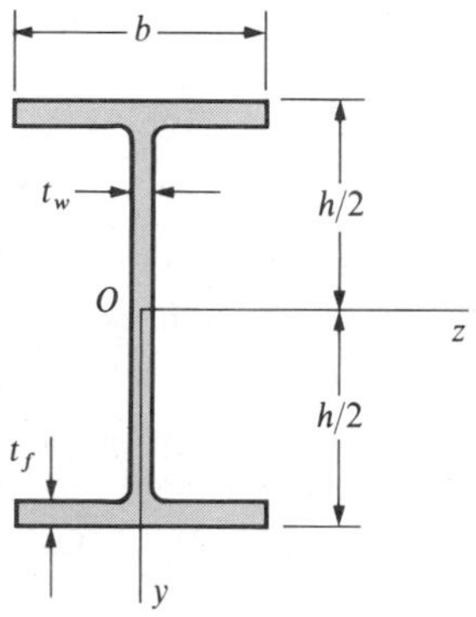

Probs. 5.17-6 and 5.17-7

5.17-8 Determine the plastic modulus Z and the shape factor f for a W 12 × 87 section (see Appendix E). (Note: Obtain these properties by calculation using the dimensions of the cross section.)

5.17-9 Solve the preceding problem for a W 10 × 45 section.

5.17-10 Solve Problem 5.17-8 for a W 8 × 28 section.

5.17-11 A doubly symmetric hollow box section with height $h = 16$ in., width $b = 8$ in., and constant wall thickness $t = 0.75$ in. is shown in the figure. The beam is constructed of steel with yield stress $\sigma_y = 36{,}000$ psi. Calculate the plastic moment M_p, the plastic modulus Z, and the shape factor f.

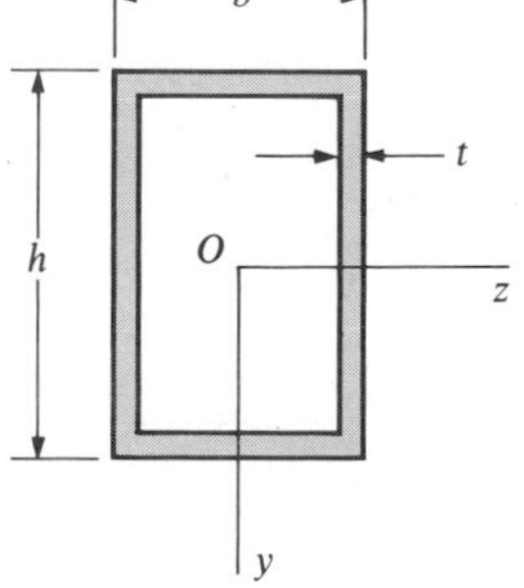

Probs. 5.17-11 and 5.17-12

5.17-12 Solve the preceding problem for a box beam with dimensions $h = 0.4$ m, $b = 0.2$ m, and $t = 20$ mm. (Assume $\sigma_y = 230$ MPa.)

5.17-13 A singly symmetric beam of T-section (see figure) has dimensions $b = 6$ in., $a = 8$ in., $t_w = 1.5$ in., and $t_f = 1.5$ in. Calculate the plastic modulus Z and the shape factor f.

5.17-14 Solve the preceding problem for a T-beam with dimensions $b = 140$ mm, $a = 200$ mm, $t_w = 20$ mm, and $t_f = 25$ mm.

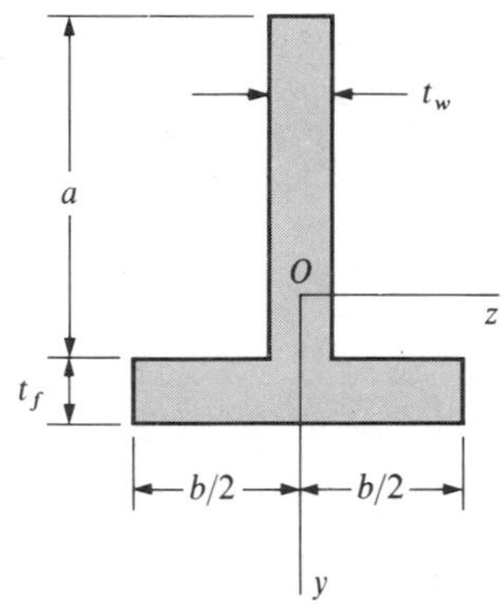

Probs. 5.17-13 and 5.17-14

5.17-15 A singly symmetric wide-flange beam of steel ($\sigma_y = 42{,}000$ psi) has the cross section shown in the figure. Determine the plastic moment M_p.

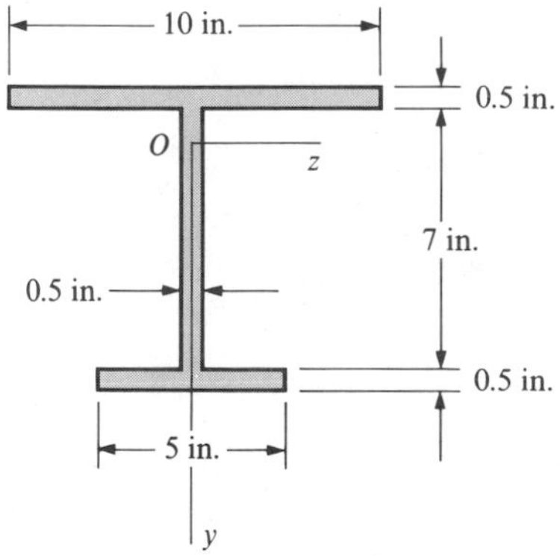

Prob. 5.17-15

5.17-16 A W 14 × 53 steel wide-flange beam (see Appendix E) is subjected to a bending moment M that produces plastic yielding throughout the flanges but the web remains elastic. Calculate the moment M and the corresponding curvature κ, assuming $\sigma_y = 36{,}000$ psi and $E = 30 \times 10^6$ psi.

5.17-17 Solve the preceding problem for a W 12 × 50 section.

5.17-18 Solve Problem 5.17-16 for a W 8 × 28 section.

5.17-19 The doubly symmetric hollow box beam shown in the figure is subjected to a bending moment M of such magnitude that the flanges yield but the webs remain elastic. Calculate the ratios M/M_y and κ/κ_y, assuming $h = 14$ in., $h_1 = 12.5$ in., $b = 8$ in., and $b_1 = 7$ in. Also, calculate the shape factor $f = M_p/M_y$. (Note that M/M_y for this beam must be between 1 and f.)

5.17-20 Solve the preceding problem for a box beam with dimensions $h = 400$ mm, $h_1 = 360$ mm, $b = 200$ mm, and $b_1 = 160$ mm.

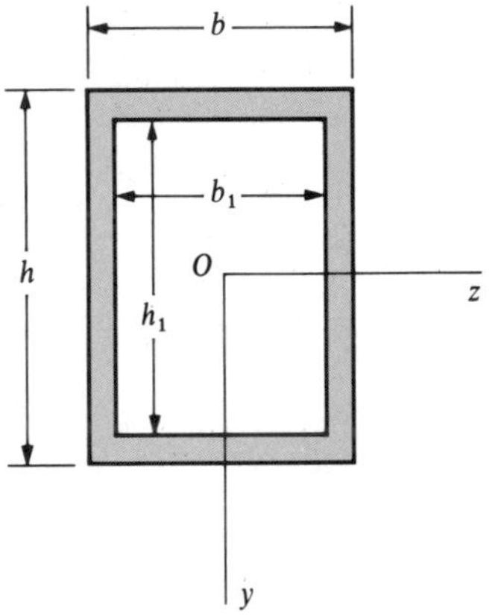

Probs. 5.17-19 and 5.17-20

***5.17-21** Derive the moment-curvature relationship in nondimensional form (that is, the equation relating M/M_y to κ/κ_y) for an elastic-plastic beam having a cross section in the shape of a rhombus (see figure for Problem 5.17-2).

****5.17-22** Derive the moment-curvature relationship in nondimensional form (that is, the equation relating M/M_y to κ/κ_y) for an elastic-plastic beam of solid circular cross section.

5.18-1 A cantilever beam of rectangular cross section and length L supports a concentrated load P at its free end (see figure). The material of the beam has a modulus of elasticity E_1 in tension and E_2 in compression. Numerical values are as follows: $L = 4$ ft, $P = 1$ k, $b = 2.5$ in., $h = 6$ in., $E_1 = 10 \times 10^6$ psi, and $E_2 = 40 \times 10^6$ psi. (a) Determine the distances h_1 and h_2 from the neutral axis to the tension and compression surfaces, respectively. (b) Calculate the maximum tensile stress σ_t and maximum compressive stress σ_c due to bending.

5.18-2 Repeat the preceding problem for the following conditions: $L = 1.0$ m, $P = 1.0$ kN, $b = 40$ mm, $h = 100$ mm, $E_1 = 70$ GPa, and $E_2 = 250$ GPa.

5.18-3 A beam of rectangular cross section is made of a material represented by the bilinear stress-strain diagram shown in the figure. The moduli of elasticity are E_1 and E_2, and the properties are the same in both tension and compression. The following numerical values apply to this beam: $b = 2.0$ in., $h = 6.0$ in., $E_1 = 10 \times 10^6$ psi, $E_2 = 4 \times 10^6$ psi, and $\sigma_y = 24{,}000$ psi. (a) If the maximum stress σ_{max} in the beam is 28,000 psi, what is the bending moment M? (b) What is the radius of curvature ρ?

5.18-4 Solve the preceding problem using the following data: $b = 40$ mm, $h = 120$ mm, $E_1 = 80$ GPa, $E_2 = 30$ GPa, $\sigma_y = 160$ MPa, and $\sigma_{max} = 200$ MPa.

***5.18-5** Determine the maximum stress σ_{max} in the beam of Problem 5.18-3 if the bending moment $M = 330$ in.-k.

***5.18-6** Determine the maximum stress σ_{max} in the beam of Problem 5.18-4 if the bending moment $M = 18$ kN·m.

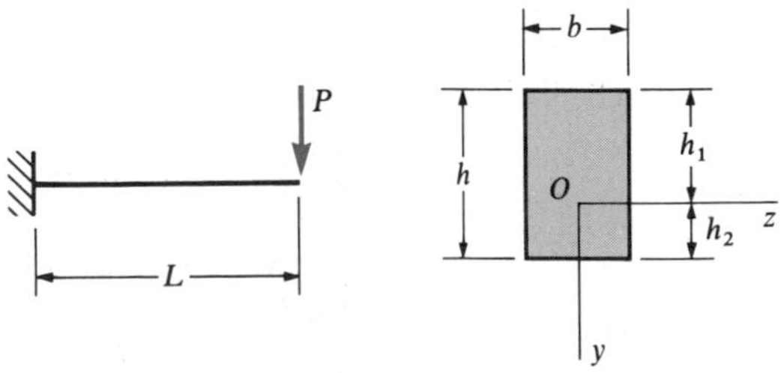

Probs. 5.18-1 and 5.18-2

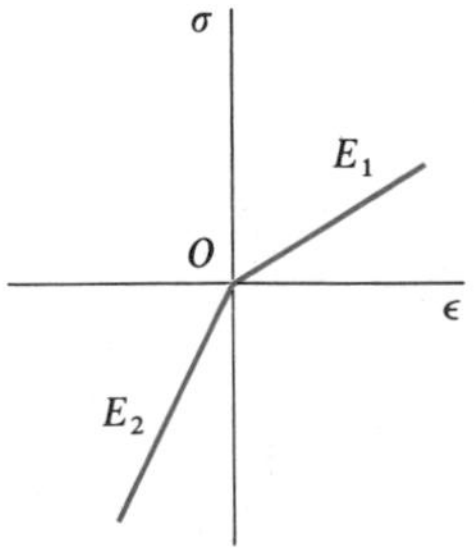

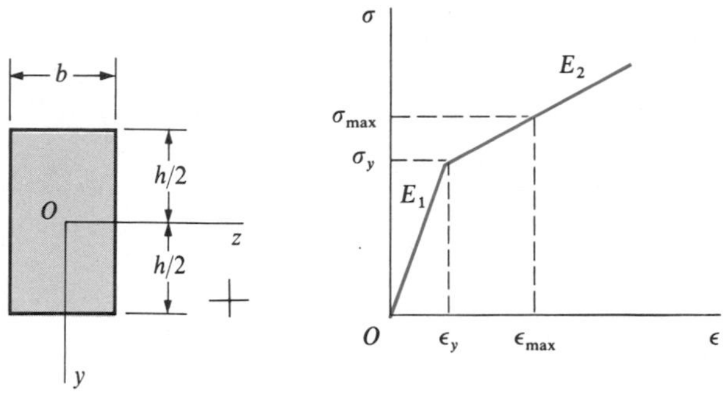

Probs. 5.18-3 through 5.18-7

***5.18-7** Derive the following moment-curvature equation for a beam of rectangular cross section having a bilinear stress-strain diagram (see figure):

$$\frac{M}{M_y} = \frac{1}{2}\left[\left(1 - \frac{E_2}{E_1}\right)\left(3 - \frac{\kappa_y^2}{\kappa^2}\right) + \frac{2E_2\kappa}{E_1\kappa_y}\right] \quad \left(\frac{M}{M_y} \geq 1\right)$$

in which $M_y = \sigma_y S = \sigma_y bh^2/6$ and $\kappa_y = 2\epsilon_y/h = 2\sigma_y/E_1 h$. Also, plot a graph of M/M_y versus κ/κ_y.

5.18-8 A beam of rectangular cross section (width b, height h) is made of material having a stress-strain diagram in tension given by the equation

$$\sigma = B_1\epsilon - B_2\epsilon^2$$

in which B_1 and B_2 are constants. The diagram in compression is the same as in tension. Derive a formula for the resisting moment M of the beam if the maximum strain is ϵ_1.

5.18-9 A beam of rectangular cross section with width $b = 3.0$ in. and height $h = 6.0$ in. is constructed of high-strength steel having a stress-strain diagram defined by the numerical data listed in the accompanying table. The diagram is the same for both tension and compression. Calculate the value of the bending moment M if the maximum stress in the beam is 105 ksi.

Stress-Strain Data for Problem 5.18-9

Stress (ksi)	Strain
0	0
70	0.0024
75	0.0030
80	0.0038
90	0.0063
100	0.0105
110	0.0170
120	0.0274

5.18-10 The tension stress-strain law for the material of a beam is assumed to be of the form $\sigma = B\epsilon^n$ in which B and n are constants ($0 \leq n \leq 1$). The stress-strain diagram for compression is the same as for tension. The cross section of the beam is rectangular with width b and height h. (a) Derive the following formula for the moment-curvature relationship for this beam:

$$M = \frac{bh^{n+2}B\kappa^n}{2^{n+1}(n+2)}$$

(b) Derive the following formula for the maximum stress in the beam:

$$\sigma_1 = \frac{Mc}{I}\frac{n+2}{3}$$

in which $c = h/2$ and $I = bh^3/12$. (c) Derive the following formula for the stress σ in the beam at distance y from the neutral axis:

$$\frac{\sigma}{\sigma_1} = \left(\frac{2y}{h}\right)^n$$

Plot a graph showing the stress distribution in the beam for various values of n. (For convenience, plot σ/σ_1 versus $2y/h$ and choose $n = 1, \frac{1}{2}, \frac{1}{4}, 0$.) (d) For all of the preceding results, investigate and interpret the special cases in which $n = 1$ and $n = 0$.

5.18-11 The stress distribution from the neutral axis to the lower surface of a rectangular beam (width b, height h) is given by the equation

$$\frac{\sigma}{\sigma_1} = 1 - \left(1 - \frac{2y}{h}\right)^m$$

in which σ_1 is the maximum stress at the lower surface, y is measured from the neutral axis downward, and m is a constant ($m \geq 1$). The neutral axis is at the midheight of the beam, and the stress distribution for the upper half of the beam is the same (except for sign) as for the lower half. (a) Find the bending moment M for this beam. (b) Plot a graph showing the stress distribution in the beam for various values of m. (For convenience, plot σ/σ_1 versus $2y/h$ and choose $m = 1, 2, 4, 10$.)

CHAPTER 6

Analysis of Stress and Strain

6.1 INTRODUCTION

Normal and shear stresses in beams, shafts, bars, and other structural elements can be obtained from the basic formulas discussed in the preceding chapters. For instance, the stresses in a beam may be calculated from the flexure and shear formulas ($\sigma = My/I$ and $\tau = VQ/Ib$). The values obtained from the basic formulas are for stresses that act on cross sections of the members (recall that a cross section is a plane section cut perpendicular to the longitudinal axis). Now, in this chapter, we discuss methods for finding the normal and shear stresses acting on inclined sections cut through the members.

For the special cases of uniaxial stress and pure shear, we already derived expressions for the stresses on inclined planes in Sections 2.5 and 3.4, respectively. We found that the maximum shear stresses for an axially loaded bar occur on planes inclined at 45°, and the maximum tensile and compressive stresses for a bar in torsion also occur on 45° planes. In a similar manner, inclined sections cut through a beam may be subjected to both normal and shear stresses, and these stresses may be larger than the stresses acting on a cross section. To handle analyses of this type, we need a more general approach for determining the stresses on inclined planes.

Our approach uses **stress elements** to represent the state of stress at a point in a body. Stress elements were discussed previously in a specialized context (see Sections 2.5 and 3.4), but now we will use them in a more formalized manner by associating a set of coordinate axes with each position of a stress element. Our objective is to derive the transformation relationships that give the stress components for any orientation of these axes. In other words, we wish to determine the stresses acting on the sides of a stress element oriented in any desired direction. This process is called a **transformation of axes** or a **stress transformation**, although it must be emphasized that only one intrinsic state of stress exists at a point, regardless of the orientation of the stress element being used to portray that state of stress. When we have two elements with different orientations at the same point in the body, the stresses acting on the *faces* of the two elements are different, but they still represent the same state of stress, namely, the stress at the point under consideration. This situation is analogous to the representation of a force vector by its components; although the force is represented by different components when the coordinate axes are rotated to a new orientation, the force itself is not changed. However, the state of stress at a point in a body is a much more complex quantity than is a force, and the transformation relationships for stresses are more complicated than those for vectors. In mathematical terms, stress is known as a *tensor*; other tensor quantities in mechanics are strains and moments of inertia. The similarities among the transformation equations for stresses, strains, and moments of inertia are pointed out later in this chapter.

6.2 PLANE STRESS

The stress conditions encountered in axially loaded bars, shafts in torsion, and beams in bending are examples of a state of stress called **plane stress**. To explain plane stress, let us consider the infinitesimal element shown in Fig. 6-1a. This element is a rectangular parallelepiped with its edges parallel to the x, y, and z axes. The faces of the element are designated by

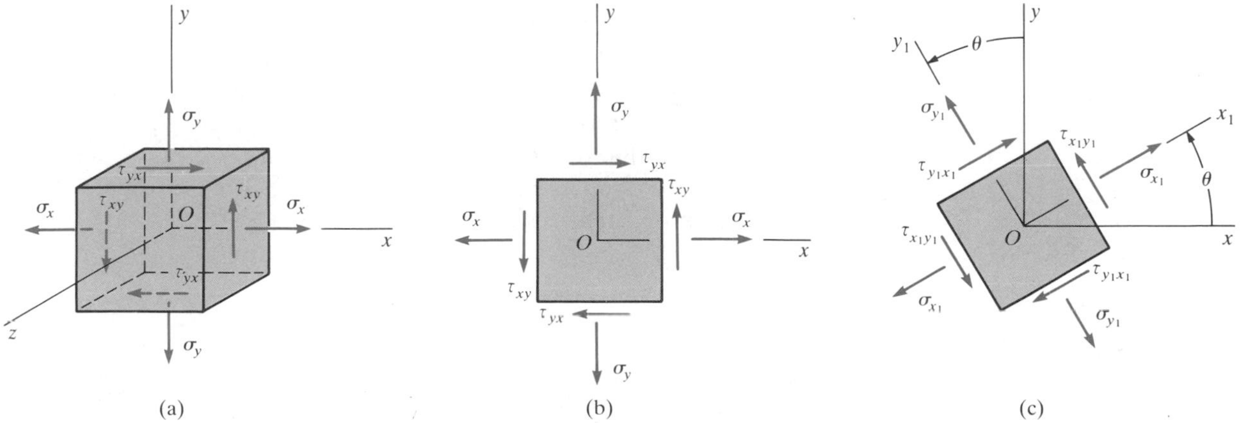

Fig. 6-1 Elements in plane stress

the directions of their outward normals, as explained previously in Section 1.6. Thus, the right-hand face of the element in this figure is referred to as the positive x face, and the left-hand face (hidden from the viewer) is referred to as the negative x face. Similarly, the top face is the positive y face, and the front face is the positive z face.

In plane stress, only the x and y faces of the element are subjected to stresses, and all stresses act parallel to the x and y axes (Fig. 6-1a). The symbols for the stresses have the following meanings. A **normal stress** σ has a subscript that identifies the face on which the stress acts. For instance, the stress σ_x acts on the x face of the element. Of course, equal normal stresses act on opposite faces of the element. The sign convention for normal stresses is that tension is positive. A **shear stress** τ has two subscripts; the first denotes the face on which the stress acts, and the second gives the direction on that face. Thus, the stress τ_{xy} acts on the x face in the direction of the y axis, and the stress τ_{yx} acts on the y face in the direction of the x axis. The sign convention for shear stresses is as follows. A shear stress is positive when it acts on a positive face of the element in the positive direction of an axis, and it is negative when it acts on a positive face in the negative direction of an axis. Therefore, the stresses τ_{xy} and τ_{yx} shown on the positive x and y faces (Fig. 6-1a) are positive shear stresses. Similarly, on a negative face of the element, the shear stress is positive in the negative direction of an axis. Hence, the stresses τ_{xy} and τ_{yx} shown on the negative x and y faces of the element in the figure are also positive. This sign convention for shear stresses is easily remembered by the rule that when the directions associated with the subscripts are plus-plus or minus-minus, the stress is positive; when the directions are plus-minus or minus-plus, the stress is negative.

The preceding sign convention for shear stresses is consistent with static equilibrium of the element, because we know that shear stresses on opposite faces of an element must be equal in magnitude and opposite in direction. Thus, according to our sign convention, a positive stress τ_{xy} acts upward on the positive face (Fig. 6-1a) and downward on the negative face. In a similar manner, the stresses τ_{yx} acting on the top and bottom faces of the element are positive and have opposite directions. Finally, we recall that shear stresses on perpendicular planes are equal in magnitude and have directions such that both stresses point toward, or away from, the line of intersection of the faces. Inasmuch as τ_{xy} and τ_{yx} are positive in the directions shown in the figure, they are consistent with this observation. Therefore, we note that

$$\tau_{xy} = \tau_{yx} \tag{6-1}$$

This relationship was derived previously from static equilibrium of the element (see Section 1.6).

For convenience in sketching plane-stress elements, we usually draw only a two-dimensional view of the element, as shown in Fig. 6-1b.

Although a figure of this kind shows clearly all of the stresses acting on a plane-stress element, we must always keep in mind that the element is a solid body with a constant thickness perpendicular to the plane of the figure.

We are now ready to consider the stresses acting on **inclined sections**, our assumption being that the stresses σ_x, σ_y, and τ_{xy} are known (for example, from a bending or torsion analysis). To portray the stresses acting on an inclined section, we consider another stress element whose faces are parallel and perpendicular to the inclined section (Fig. 6-1c). Associated with this new element are axes x_1, y_1, and z_1, such that the z_1 axis coincides with the z axis and the x_1y_1 axes are rotated counterclockwise through an angle θ with respect to the xy axes. The normal and shear stresses acting on this inclined element are denoted σ_{x_1}, σ_{y_1}, $\tau_{x_1y_1}$, and $\tau_{y_1x_1}$, using the same subscript designations and sign conventions as for the stresses acting on the xy element. The previous conclusions regarding the shear stresses still apply, and therefore

$$\tau_{x_1y_1} = \tau_{y_1x_1} \tag{6-2}$$

An important observation is that the shear stresses acting on all four side faces of the element are known if we know the shear stress acting on any one of the faces.

The stresses acting on the inclined x_1y_1 element can be expressed in terms of the stresses on the xy element by using equations of static equilibrium. For this purpose, we choose a wedge-shaped element whose inclined face is the x_1 face of the inclined element and whose other two side faces are parallel to the x and y axes (Fig. 6-2a). In order to write equations of equilibrium, we need to obtain the forces acting on these faces and construct a free-body diagram.

Let us denote the area of the left-hand side face (that is, the negative x face) as A_0. Then the normal and shear forces acting on this face are $\sigma_x A_0$ and $\tau_{xy}A_0$, as shown in the free-body diagram of Fig. 6-2b. The area of the bottom face (or negative y face) is $A_0 \tan\theta$, and the area of the inclined face (or positive x_1 face) is $A_0 \sec\theta$. Thus, the normal and shear forces acting on these faces have the magnitudes and directions shown. The four forces acting on the left-hand and bottom faces can now be resolved into orthogonal components acting in the x_1 and y_1 directions. Then we can sum forces in those directions and obtain two equations of equilibrium for the element. The first equation, obtained by summing forces in the x_1 direction, is

$$\sigma_{x_1}A_0\sec\theta - \sigma_x A_0\cos\theta - \tau_{xy}A_0\sin\theta$$
$$-\sigma_y A_0\tan\theta\sin\theta - \tau_{yx}A_0\tan\theta\cos\theta = 0$$

In the same manner, summation of forces in the y_1 direction gives

$$\tau_{x_1y_1}A_0\sec\theta + \sigma_x A_0\sin\theta - \tau_{xy}A_0\cos\theta$$
$$-\sigma_y A_0\tan\theta\cos\theta + \tau_{yx}A_0\tan\theta\sin\theta = 0$$

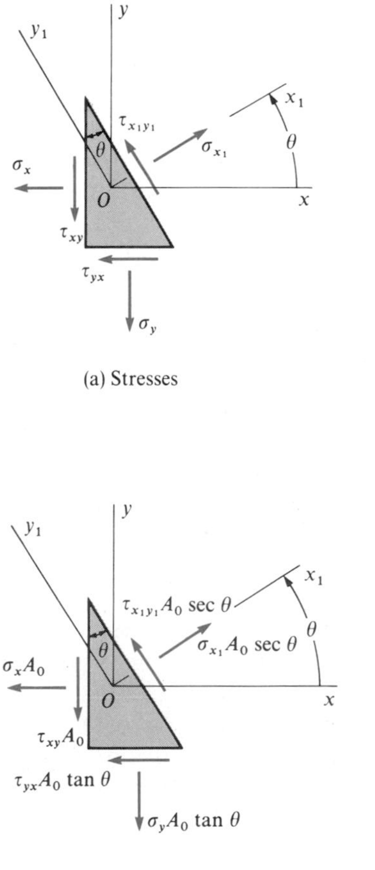

(a) Stresses

(b) Forces

Fig. 6-2 Wedge-shaped stress element in plane stress: (a) stresses acting on the element, and (b) forces acting on the element

Using the relationship $\tau_{xy} = \tau_{yx}$, and also simplifying and rearranging, we obtain the following equations:

$$\sigma_{x_1} = \sigma_x \cos^2\theta + \sigma_y \sin^2\theta + 2\tau_{xy}\sin\theta\cos\theta \qquad \text{(6-3a)}$$

$$\tau_{x_1y_1} = -(\sigma_x - \sigma_y)\sin\theta\cos\theta + \tau_{xy}(\cos^2\theta - \sin^2\theta) \qquad \text{(6-3b)}$$

Equations (6-3) give the normal and shear stresses acting on the x_1 plane in terms of the angle of rotation θ and the stresses σ_x, σ_y, and τ_{xy} acting on the x and y planes.

For the special value $\theta = 0$, we note that Eqs. (6-3) give $\sigma_{x_1} = \sigma_x$ and $\tau_{x_1y_1} = \tau_{xy}$, as expected. Also, when $\theta = 90°$, the equations give $\sigma_{x_1} = \sigma_y$ and $\tau_{x_1y_1} = -\tau_{xy}$. In the latter case, the x_1 axis is vertical; hence the stress $\tau_{x_1y_1}$ is positive to the left, which is opposite to the positive direction of τ_{yx}.

Equations (6-3) can be expressed in a useful alternative form by introducing the following trigonometric identities:

$$\cos^2\theta = \frac{1}{2}(1 + \cos 2\theta) \qquad \sin^2\theta = \frac{1}{2}(1 - \cos 2\theta)$$

$$\sin\theta\cos\theta = \frac{1}{2}\sin 2\theta$$

Then the equations become

$$\sigma_{x_1} = \frac{\sigma_x + \sigma_y}{2} + \frac{\sigma_x - \sigma_y}{2}\cos 2\theta + \tau_{xy}\sin 2\theta \qquad \text{(6-4a)}$$

$$\tau_{x_1y_1} = -\frac{\sigma_x - \sigma_y}{2}\sin 2\theta + \tau_{xy}\cos 2\theta \qquad \text{(6-4b)}$$

These equations for σ_{x_1} and $\tau_{x_1y_1}$ are known as the **transformation equations for plane stress** because they transform the stress components from one set of axes to another. However, as explained previously, the intrinsic state of stress at the point under consideration is the same whether represented by the stresses on the xy element or on the x_1y_1 element (Fig. 6-1). The transformation equations were derived solely from equilibrium considerations, so they are applicable to stresses in any kind of material.

An important observation concerning the normal stresses can be obtained from the transformation equations. As a preliminary matter, we note that the normal stress σ_{y_1} acting on the y_1 face of an inclined element (Fig. 6-1c) can be obtained from Eq. (6-4a) by substituting $\theta + 90°$ for θ, which produces the following equation for σ_{y_1}:

$$\sigma_{y_1} = \frac{\sigma_x + \sigma_y}{2} - \frac{\sigma_x - \sigma_y}{2}\cos 2\theta - \tau_{xy}\sin 2\theta \tag{6-5}$$

Summing Eqs. (6-4a) and (6-5) gives

$$\sigma_{x_1} + \sigma_{y_1} = \sigma_x + \sigma_y \tag{6-6}$$

This equation shows that the sum of the normal stresses acting on perpendicular faces of plane-stress elements at a point is constant and independent of the angle θ.

The general manner in which the normal and shear stresses vary is shown in Fig. 6-3, which is a graph of σ_{x_1} and $\tau_{x_1y_1}$ versus the angle θ (from Eqs. 6-3 or 6-4). The graph is plotted for the particular case of $\sigma_y = 0.2\sigma_x$ and $\tau_{xy} = 0.8\sigma_x$. We see from the plot that the stresses vary continuously as the orientation of the element is changed. At certain angles, the normal

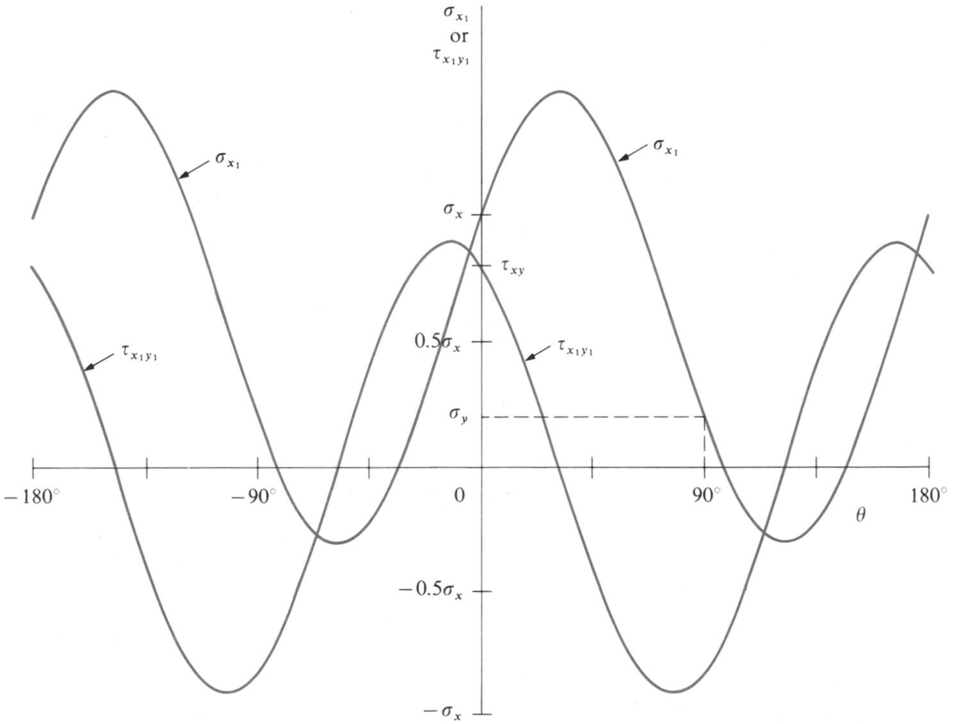

Fig. 6-3 Graph of normal stress σ_{x_1} and shear stress $\tau_{x_1y_1}$ versus angle θ (for $\sigma_y = 0.2\sigma_x$ and $\tau_{xy} = 0.8\sigma_x$)

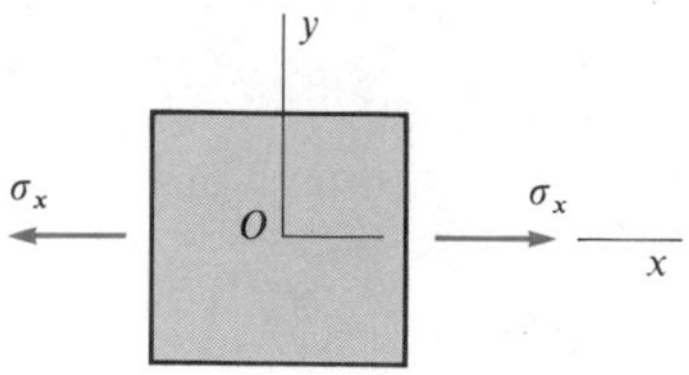

Fig. 6-4 Uniaxial stress

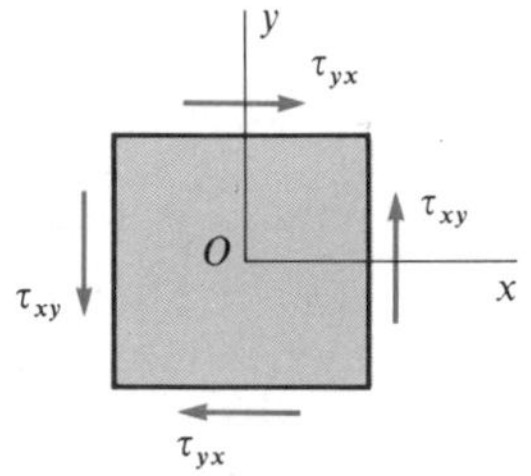

Fig. 6-5 Pure shear

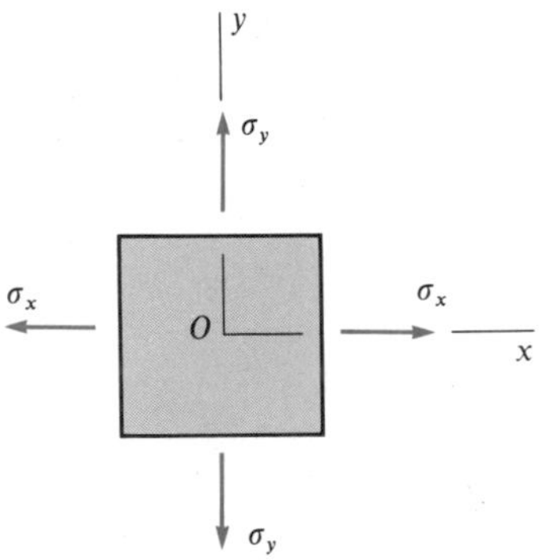

Fig. 6-6 Biaxial stress

stress reaches a maximum or minimum value; at other angles, it becomes zero. Similarly, the shear stress has maximum, minimum, and zero values at certain angles. A more detailed investigation of these special values of the stresses is made in the next section.

The general case of plane stress reduces to simpler states of stress under special conditions. For instance, if all stresses acting on the xy element (Fig. 6-1b) are zero except for the normal stress σ_x, then the element is in **uniaxial stress** (Fig. 6-4). The corresponding transformation equations, obtained by setting σ_y and τ_{xy} equal to zero in Eqs. (6-3), are

$$\sigma_{x_1} = \sigma_x \cos^2 \theta \tag{6-7a}$$

$$\tau_{x_1y_1} = -\sigma_x \sin\theta \cos\theta \tag{6-7b}$$

These equations agree with the equations derived previously in Section 2.5 (see Eqs. 2-27), except that now we are using a more generalized notation for the stresses acting on an inclined element.

Another special case is **pure shear** (Fig. 6-5), for which the transformation equations are obtained by substituting $\sigma_x = 0$ and $\sigma_y = 0$ into Eqs. (6-3):

$$\sigma_{x_1} = 2\tau_{xy} \sin\theta \cos\theta \tag{6-8a}$$

$$\tau_{x_1y_1} = \tau_{xy}(\cos^2\theta - \sin^2\theta) \tag{6-8b}$$

Again, the equations correspond to those derived earlier (see Eqs. 3-18 in Section 3.4).

The last special case is **biaxial stress**, a stress condition in which the xy element is subjected to normal stresses in both the x and y directions without shear stresses (Fig. 6-6). The equations for biaxial stress are obtained from Eqs. (6-3) or (6-4) by dropping the terms containing τ_{xy}. Biaxial stress, which occurs in thin-walled pressure vessels, is discussed in more detail in Section 6.6.

Example 1

An element in plane stress is subjected to stresses $\sigma_x = 16{,}000$ psi, $\sigma_y = 6{,}000$ psi, and $\tau_{xy} = \tau_{yx} = 4{,}000$ psi as shown in Fig. 6-7a. Determine the stresses acting on an element inclined at an angle $\theta = 45°$.

Although the stresses acting on the inclined element can be obtained from either Eqs. (6-3) or (6-4), we usually find the latter more convenient for practical use. Using the given numerical values, we obtain

$$\frac{\sigma_x + \sigma_y}{2} = 11{,}000 \text{ psi} \qquad \frac{\sigma_x - \sigma_y}{2} = 5{,}000 \text{ psi}$$

$$\sin 2\theta = \sin 90° = 1 \qquad \cos 2\theta = \cos 90° = 0$$

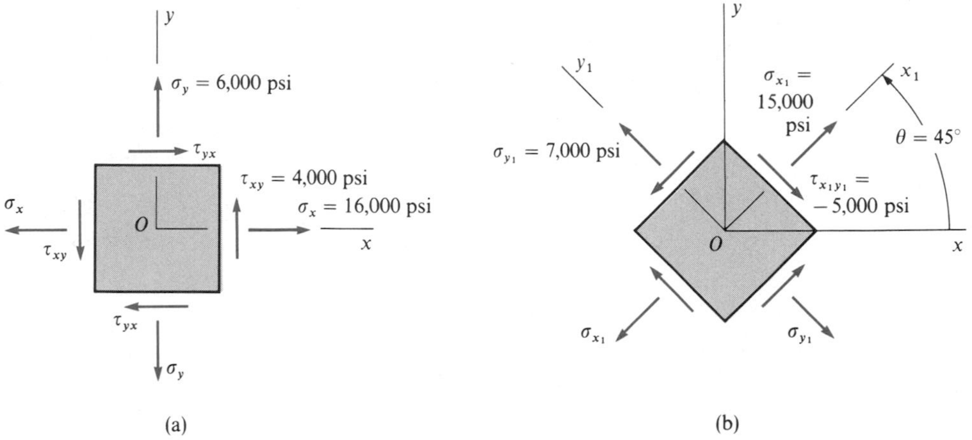

Fig. 6-7 Example 1

Substituting these values and $\tau_{xy} = 4{,}000$ psi into Eqs. (6-4), we get

$$\sigma_{x_1} = \frac{\sigma_x + \sigma_y}{2} + \frac{\sigma_x - \sigma_y}{2}\cos 2\theta + \tau_{xy}\sin 2\theta$$

$$= 11{,}000 \text{ psi} + (5{,}000 \text{ psi})(0) + (4{,}000 \text{ psi})(1) = 15{,}000 \text{ psi}$$

$$\tau_{x_1y_1} = -\frac{\sigma_x - \sigma_y}{2}\sin 2\theta + \tau_{xy}\cos 2\theta$$

$$= -(5{,}000 \text{ psi})(1) + (4{,}000 \text{ psi})(0) = -5{,}000 \text{ psi}$$

In addition, the stress σ_{y_1} may be obtained from Eq. (6-5):

$$\sigma_{y_1} = \frac{\sigma_x + \sigma_y}{2} - \frac{\sigma_x - \sigma_y}{2}\cos 2\theta - \tau_{xy}\sin 2\theta$$

$$= 11{,}000 \text{ psi} - (5{,}000 \text{ psi})(0) - (4{,}000 \text{ psi})(1) = 7{,}000 \text{ psi}$$

From these results, we can obtain the stresses acting on the sides of an element at $\theta = 45°$, as shown in Fig. 6-7b. The arrows show the true directions in which the stresses act. Note especially the directions of the shear stresses, all of which have the same magnitude. Also, observe that the sum of the normal stresses remains constant and equal to 22,000 psi (Eq. 6-6). The stress components of Fig. 6-7b represent the same state of stress as that shown in Fig. 6-7a; the difference lies in the orientations of the planes on which the stress components act.

Example 2

A plane-stress condition exists at a point in a loaded structure. The stresses have the magnitudes and directions shown on the stress element of Fig. 6-8a. Calculate the stresses acting on the planes obtained by rotating the axes clockwise through an angle of 15°.

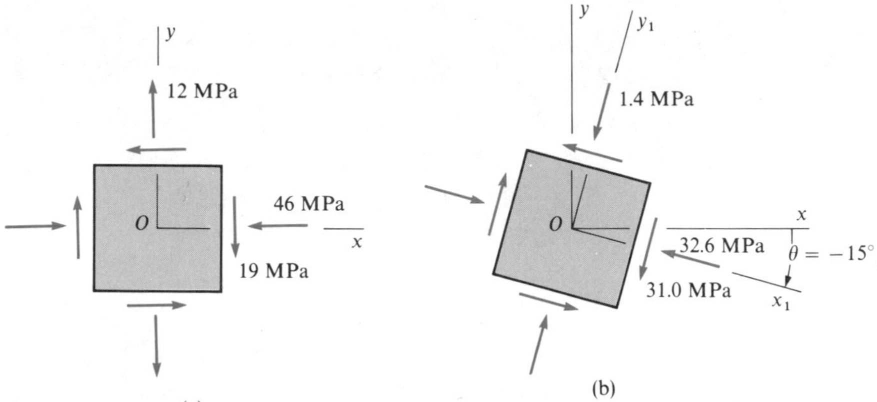

Fig. 6-8 Example 2

The stresses acting on the element shown in Fig. 6-8a have the following values:

$$\sigma_x = -46 \text{ MPa} \qquad \sigma_y = 12 \text{ MPa} \qquad \tau_{xy} = -19 \text{ MPa}$$

A clockwise rotation of the axes through an angle of 15° puts the element in the position shown in Fig. 6-8b with the x_1 axis at an angle $\theta = -15°$ to the x axis. (Alternatively, the x_1 axis could be placed at an angle $\theta = 75°$.)

We can readily calculate the stresses on the x_1 face of the element at $-15°$ by using Eqs. (6-4). The calculations proceed as follows:

$$\frac{\sigma_x + \sigma_y}{2} = -17 \text{ MPa} \qquad \frac{\sigma_x - \sigma_y}{2} = -29 \text{ MPa}$$

$$\sin 2\theta = \sin(-30°) \qquad \cos 2\theta = \cos(-30°)$$

$$\sigma_{x_1} = \frac{\sigma_x + \sigma_y}{2} + \frac{\sigma_x - \sigma_y}{2}\cos 2\theta + \tau_{xy}\sin 2\theta$$

$$= -17 \text{ MPa} - (29 \text{ MPa})\cos(-30°) - (19 \text{ MPa})\sin(-30°)$$

$$= -32.6 \text{ MPa}$$

$$\tau_{x_1y_1} = -\frac{\sigma_x - \sigma_y}{2}\sin 2\theta + \tau_{xy}\cos 2\theta$$

$$= (29 \text{ MPa})\sin(-30°) - (19 \text{ MPa})\cos(-30°)$$

$$= -31.0 \text{ MPa}$$

Also, the normal stress acting on the y_1 face (from Eq. 6-5) is

$$\sigma_{y_1} = \frac{\sigma_x + \sigma_y}{2} - \frac{\sigma_x - \sigma_y}{2}\cos 2\theta - \tau_{xy}\sin 2\theta$$

$$= -17 \text{ MPa} + (29 \text{ MPa})\cos(-30°) + (19 \text{ MPa})\sin(-30°)$$

$$= -1.4 \text{ MPa}$$

This stress can also be calculated by substituting $\theta = 75°$ into Eq. (6-4a). Again we note that $\sigma_{x_1} + \sigma_{y_1} = \sigma_x + \sigma_y$. The stresses acting on the inclined planes are shown on the stress element of Fig. 6-8b, with the arrows indicating the true directions of the stresses.

6.3 PRINCIPAL STRESSES AND MAXIMUM SHEAR STRESSES

The transformation equations for plane stress show that the normal stress σ_{x_1} and shear stress $\tau_{x_1y_1}$ vary continuously as the axes are rotated through the angle θ. This variation is pictured in Fig. 6-3 for a selected combination of stresses. For design purposes, the largest stresses (both normal and shear stresses) are usually needed.

To determine the maximum and minimum normal stresses, known as the **principal stresses**, we begin with the expression for σ_{x_1} (Eq. 6-4a). By taking the derivative of σ_{x_1} with respect to θ and setting it equal to zero, we obtain an equation that can be solved for the values of θ at which σ_{x_1} is a maximum or minimum. The equation for the derivative is

$$\frac{d\sigma_{x_1}}{d\theta} = -(\sigma_x - \sigma_y)\sin 2\theta + 2\tau_{xy}\cos 2\theta = 0 \tag{a}$$

from which we get

$$\tan 2\theta_p = \frac{2\tau_{xy}}{\sigma_x - \sigma_y} \tag{6-9}$$

The subscript p indicates that the angle θ_p defines the orientation of the **principal planes**, which are the planes on which the principal stresses act. Two values of $2\theta_p$ in the range from 0 to 360° can be obtained from Eq. (6-9). These values differ by 180°, with the lower value being between 0 and 180° and the higher between 180° and 360°. Therefore, the angle θ_p has two values that differ by 90°, one between 0 and 90° and the other between 90° and 180°. For one of these angles, the normal stress σ_{x_1} is a maximum principal stress; for the other, it is a minimum principal stress. Because the two values of θ_p differ by 90°, we conclude that *the principal stresses occur on mutually perpendicular planes.*

The values of the principal stresses can be calculated by substituting each of the two values of θ_p into the stress transformation equation (Eq. 6-4a) and solving for σ_{x_1}. By this procedure we also know which of the two principal stresses is associated with each of the two principal angles θ_p.

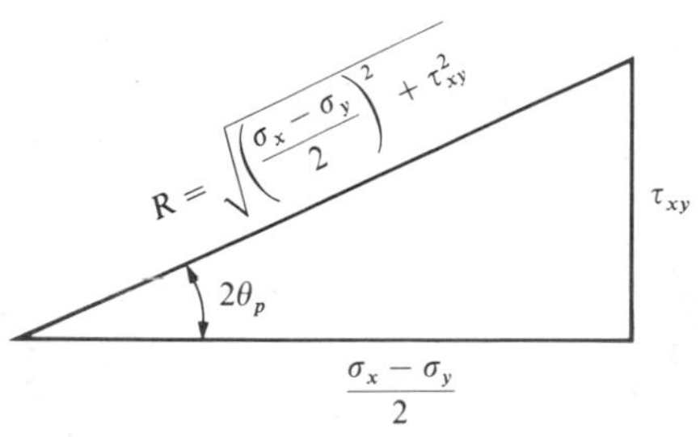

Fig. 6-9

General formulas for the principal stresses can also be obtained. To do so, refer to Fig. 6-9, which is constructed from Eq. (6-9). From the figure we see that

$$\cos 2\theta_p = \frac{\sigma_x - \sigma_y}{2R} \qquad \sin 2\theta_p = \frac{\tau_{xy}}{R} \tag{6-10a,b}$$

in which

$$R = \sqrt{\left(\frac{\sigma_x - \sigma_y}{2}\right)^2 + \tau_{xy}^2} \tag{6-11}$$

The quantity R is always a positive number having units of stress. Now substitute the expressions for $\cos 2\theta_p$ and $\sin 2\theta_p$ into Eq. (6-4a) and obtain the algebraically larger of the two principal stresses, denoted by σ_1:

$$\sigma_1 = \frac{\sigma_x + \sigma_y}{2} + \sqrt{\left(\frac{\sigma_x - \sigma_y}{2}\right)^2 + \tau_{xy}^2}$$

The smaller of the principal stresses, denoted by σ_2, may be found from the condition that

$$\sigma_1 + \sigma_2 = \sigma_x + \sigma_y \tag{6-12}$$

inasmuch as σ_1 and σ_2 act on perpendicular planes. Substituting the expression for σ_1 into Eq. (6-12) and solving for σ_2, we get

$$\sigma_2 = \frac{\sigma_x + \sigma_y}{2} - \sqrt{\left(\frac{\sigma_x - \sigma_y}{2}\right)^2 + \tau_{xy}^2}$$

This expression has the same form as the expression for σ_1 but differs by the presence of the minus sign before the square root. Thus, the preceding formulas can be combined into a single formula for the principal stresses:

$$\sigma_{1,2} = \frac{\sigma_x + \sigma_y}{2} \pm \sqrt{\left(\frac{\sigma_x - \sigma_y}{2}\right)^2 + \tau_{xy}^2} \tag{6-13}$$

The plus sign gives the algebraically larger principal stress σ_1 and the minus sign gives the smaller principal stress σ_2.

In summary, the two angles defining the principal planes are denoted θ_{p_1} and θ_{p_2}, corresponding to the principal stresses σ_1 and σ_2, respectively. Both angles can be determined from the equation for $\tan 2\theta_p$ (Eq. 6-9), but we cannot tell from that equation which angle is θ_{p_1} and which is θ_{p_2}. A simple procedure for making this determination is to take one of the values and substitute it into the equation for σ_{x_1} (Eq. 6-4a). The resulting value of σ_{x_1} will be recognized as either σ_1 or σ_2, thus correlating the principal angles with the principal stresses.

Another method for correlating the principal angles and stresses is to use Eqs. (6-10) for finding θ_p, since the only angle that satisfies both of those equations is θ_{p_1}. Thus, we can rewrite those equations as follows:

$$\cos 2\theta_{p_1} = \frac{\sigma_x - \sigma_y}{2R} \qquad \sin 2\theta_{p_1} = \frac{\tau_{xy}}{R} \tag{6-14a,b}$$

Only one angle exists between 0 and 360° that satisfies both of these equations. Thus, the value of θ_{p_1}, corresponding to the maximum principal stress σ_1, can be determined uniquely from Eqs. (6-14). The angle θ_{p_2}, corresponding to σ_2, defines a plane that is perpendicular to the plane defined by θ_{p_1}. Therefore, θ_{p_2} can be taken as 90° larger or smaller than θ_{p_1}.

An important characteristic concerning the principal planes can be observed from the equation for the shear stresses (Eq. 6-4b). If we set the shear stress $\tau_{x_1y_1}$ equal to zero (compare with Eq. a) and solve for the angle 2θ, we obtain for $\tan 2\theta$ the same expression as before (Eq. 6-9). Thus, we observe that *the shear stresses are zero on the principal planes.*

The principal planes for elements in uniaxial and biaxial stress are the x and y planes themselves (Fig. 6-10), because $\tan 2\theta_p = 0$ (see Eq. 6-9) and the two values of θ_p are 0 and 90°.

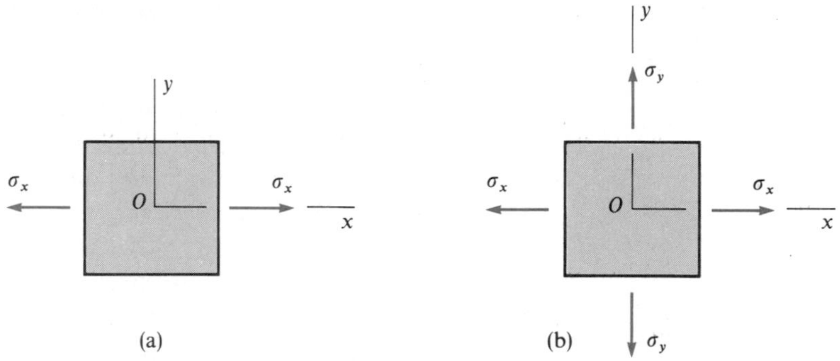

Fig. 6-10 Elements in uniaxial and biaxial stress

For an element in pure shear (Fig. 6-11a), the principal planes are oriented at 45° to the x axis (Fig. 6-11b), because $\tan 2\theta_p$ is infinite and the two values of θ_p are 45° and 135°. If τ_{xy} is positive, the principal stresses are $\sigma_1 = \tau_{xy}$ and $\sigma_2 = -\tau_{xy}$ (see Section 3.4 for a discussion of pure shear).

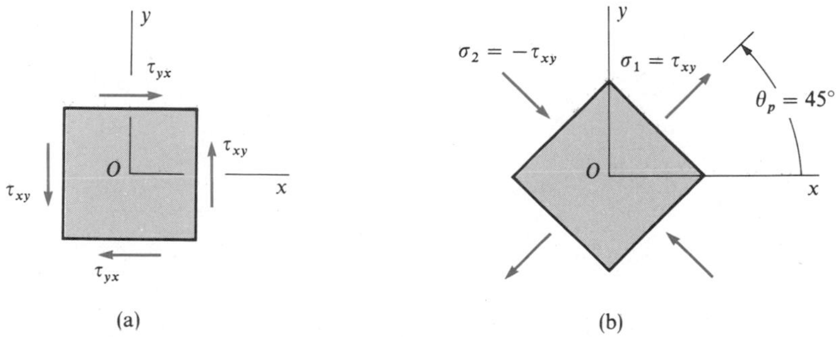

Fig. 6-11 Element in pure shear

The preceding discussion of principal stresses refers only to rotation of axes in the xy plane, that is, rotation about the z axis (Fig. 6-12a). Therefore, the two principal stresses determined from Eq. (6-13) are called

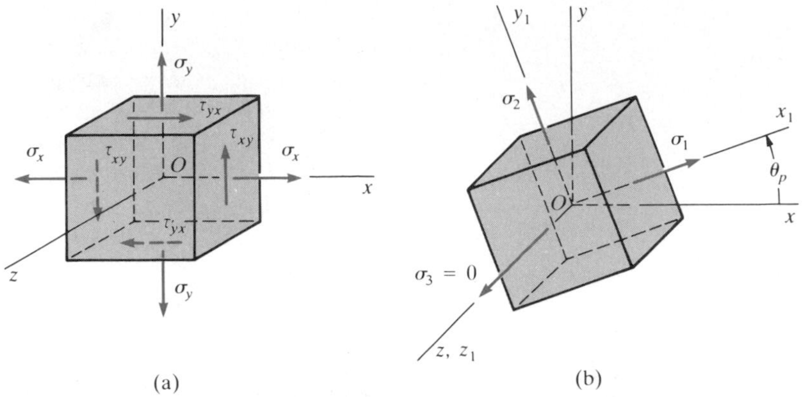

Fig. 6-12 Principal planes for an element in plane stress

the **in-plane principal stresses**. We must not overlook the fact that the stress element is actually three-dimensional and has three (not two) principal stresses acting on three mutually perpendicular planes. By making a more complete **three-dimensional analysis**, it can be shown that the three principal planes for a plane stress element are the two principal planes we have already described plus the *z* face of the element. These principal planes are shown in Fig. 6-12b, where a stress element has been rotated about the *z* axis through the principal angle θ_p, which is one of the two angles found from Eq. (6-9). The principal stresses are σ_1, σ_2, and σ_3, where σ_1 and σ_2 are given by Eq. (6-13) and σ_3 equals zero. By definition, σ_1 is algebraically larger than σ_2, but σ_3 may be algebraically larger or smaller than either or both of σ_1 and σ_3. Note again that there are no shear stresses on the principal planes.*

MAXIMUM SHEAR STRESSES

Having found the maximum normal stresses acting on an element in plane stress, we now consider the determination of the maximum shear stresses and the planes on which they act. The shear stresses $\tau_{x_1y_1}$ acting on inclined elements are given by Eq. (6-4b). Taking the derivative of $\tau_{x_1y_1}$ with respect to θ and setting it equal to zero, we obtain

$$\frac{d\tau_{x_1y_1}}{d\theta} = -(\sigma_x - \sigma_y)\cos 2\theta - 2\tau_{xy}\sin 2\theta = 0 \tag{b}$$

from which

$$\tan 2\theta_s = -\frac{\sigma_x - \sigma_y}{2\tau_{xy}} \tag{6-15}$$

* The determination of principal stresses is an example of a type of mathematical analysis known as *eigenvalue analysis*, which is described in books on matrix algebra (such as Ref. 6-1). The stress-transformation equations and the concept of principal stresses are due to the French mathematicians A. L. Cauchy (1789–1857) and Barré de Saint-Venant (1797–1886) and the Scottish scientist and engineer W. J. M. Rankine (1820–1872); see Refs. 6-2, 6-3, and 6-4, respectively.

The subscript s indicates that the angle θ_s defines the orientation of the planes of maximum shear stresses. Equation (6-15) yields one value of θ_s between 0 and 90° and another between 90° and 180°; these two values differ by 90°. Hence, the maximum and minimum values of $\tau_{x_1y_1}$ occur on perpendicular planes. Because shear stresses on perpendicular planes are equal in absolute value, the maximum and minimum shear stresses differ only in sign. Furthermore, comparing Eq. (6-15) with Eq. (6-9) shows that

$$\tan 2\theta_s = -\frac{1}{\tan 2\theta_p} = -\cot 2\theta_p \qquad (6\text{-}16)$$

From trigonometry, we know that

$$\tan(\alpha \pm 90°) = -\cot \alpha$$

Hence, we see that $2\theta_s = 2\theta_p \pm 90°$, or

$$\theta_s = \theta_p \pm 45° \qquad (6\text{-}17)$$

and we conclude that *the planes of maximum shear stress occur at 45° to the principal planes.*

The plane of the algebraically maximum shear stress τ_{max} is defined by the angle θ_{s_1}, for which we can obtain the following equations:

$$\cos 2\theta_{s_1} = \frac{\tau_{xy}}{R} \qquad \sin 2\theta_{s_1} = -\frac{\sigma_x - \sigma_y}{2R} \qquad (6\text{-}18a,b)$$

in which R is given by Eq. (6-11). Also, the angle θ_{s_1} is related to the angle θ_{p_1} (see Eqs. 6-14) as follows:

$$\theta_{s_1} = \theta_{p_1} - 45° \qquad (6\text{-}19)$$

The corresponding maximum shear stress is obtained by substituting the expressions for $\cos 2\theta_{s_1}$ and $\sin 2\theta_{s_1}$ into Eq. (6-4b), yielding

$$\tau_{max} = \sqrt{\left(\frac{\sigma_x - \sigma_y}{2}\right)^2 + \tau_{xy}^2} \qquad (6\text{-}20)$$

The algebraically minimum shear stress τ_{min} has the same magnitude but opposite sign.

A useful expression for the maximum shear stress can be obtained from the principal stresses σ_1 and σ_2, which are given by Eq. (6-13). Subtracting the expression for σ_2 from that for σ_1, and then comparing with Eq. (6-20), we see that

$$\tau_{max} = \frac{\sigma_1 - \sigma_2}{2} \qquad (6\text{-}21)$$

Thus, *the maximum shear stress is equal to one-half the difference of the principal stresses.*

Normal stresses also act on the planes of maximum shear stresses. The normal stress on the plane of the maximum shear stress can be evaluated by substituting the expressions for the angle θ_{s_1} (Eqs. 6-18) into the equation for σ_{x_1} (Eq. 6-4a). The resulting stress, denoted σ_{aver}, is equal to the average of the normal stresses on the x and y planes:

$$\sigma_{\text{aver}} = \frac{\sigma_x + \sigma_y}{2} \tag{6-22}$$

The stress σ_{aver} acts on both the plane of maximum shear stress and the plane of minimum shear stress.

In the particular cases of uniaxial stress and biaxial stress (Fig. 6-10), the planes of maximum shear stress occur at 45° to the x and y axes. In the case of pure shear (Fig. 6-11), the maximum shear stresses occur on the x and y planes.

The preceding analysis of shear stresses has dealt only with **in-plane shear stresses**. If we make a **three-dimensional analysis**, we can establish that there are three possible positions of the element for maximum shear stress. With reference to the plane stress element of Fig. 6-12b, which is oriented in the principal directions, we can obtain the three positions by making 45° rotations (not shown in the figure) about the x_1, y_1, and z_1 axes. The corresponding maximum and minimum shear stresses (see Eq. 6-21) are

$$(\tau_{\max})_{x_1} = \pm\frac{\sigma_2}{2} \qquad (\tau_{\max})_{y_1} = \pm\frac{\sigma_1}{2} \qquad (\tau_{\max})_z = \pm\frac{\sigma_1 - \sigma_2}{2} \tag{6-23a,b,c}$$

The algebraic values of σ_1 and σ_2 determine which of these expressions gives the numerically largest shear stress. If σ_1 and σ_2 are of the same sign, then one of the first two expressions is numerically largest; if they have opposite signs, then the last expression gives the largest value.

Example

An element in plane stress is subjected to stresses $\sigma_x = 12{,}300$ psi, $\sigma_y = -4{,}200$ psi, and $\tau_{xy} = -4{,}700$ psi, as shown in Fig. 6-13a. (a) Determine the principal stresses and show them on a sketch of a properly oriented element. (b) Determine the maximum shear stresses and show them on a sketch of a properly oriented element. (Consider only the in-plane stresses.)

(a) *Principal stresses.* The principal angles θ_p that locate the principal planes can be obtained by solving Eq. (6-9):

$$\tan 2\theta_p = \frac{2\tau_{xy}}{\sigma_x - \sigma_y} = \frac{2(-4{,}700\text{ psi})}{12{,}300\text{ psi} - (-4{,}200\text{ psi})} = -0.5697$$

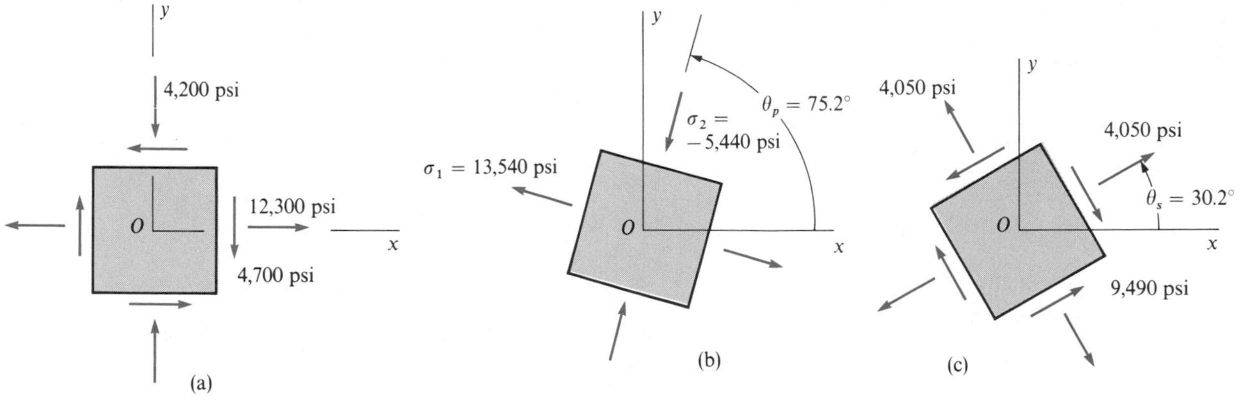

Fig. 6-13 Example. (a) Element in plane stress, (b) principal stresses, and (c) maximum shear stresses

from which

$$2\theta_p = 150.3° \quad \text{and} \quad \theta_p = 75.2°$$

or

$$2\theta_p = 330.3° \quad \text{and} \quad \theta_p = 165.2°$$

The principal stresses are obtained by substituting the two values of $2\theta_p$ into the equation for σ_{x_1} (Eq. 6-4a). As a preliminary calculation, we determine the following quantities:

$$\frac{\sigma_x + \sigma_y}{2} = \frac{12{,}300 \text{ psi} - 4{,}200 \text{ psi}}{2} = 4{,}050 \text{ psi}$$

$$\frac{\sigma_x - \sigma_y}{2} = \frac{12{,}300 \text{ psi} + 4{,}200 \text{ psi}}{2} = 8{,}250 \text{ psi}$$

Now we substitute the first value of $2\theta_p$ into Eq. (6-4a):

$$\sigma_{x_1} = \frac{\sigma_x + \sigma_y}{2} + \frac{\sigma_x - \sigma_y}{2}\cos 2\theta + \tau_{xy}\sin 2\theta$$

$$= 4{,}050 \text{ psi} + (8{,}250 \text{ psi})(\cos 150.3°) - (4{,}700 \text{ psi})(\sin 150.3°)$$

$$= -5{,}440 \text{ psi}$$

In a similar manner, we substitute the second value of $2\theta_p$ and obtain $\sigma_{x_1} =$ 13,540 psi. Thus, the principal stresses and their corresponding principal angles are

$$\sigma_1 = 13{,}540 \text{ psi} \quad \text{and} \quad \theta_{p_1} = 165.2°$$

$$\sigma_2 = -5{,}440 \text{ psi} \quad \text{and} \quad \theta_{p_2} = 75.2°$$

Note that θ_{p_1} and θ_{p_2} differ by 90° and that $\sigma_1 + \sigma_2 = \sigma_x + \sigma_y$. The principal stresses are shown on the element of Fig. 6-13b. Of course, no shear stresses act on the principal planes.

Alternative solution. The principal stresses may be calculated directly from Eq. (6-13):

$$\sigma_{1,2} = \frac{\sigma_x + \sigma_y}{2} \pm \sqrt{\left(\frac{\sigma_x - \sigma_y}{2}\right)^2 + \tau_{xy}^2}$$

$$= 4{,}050 \text{ psi} \pm \sqrt{(8{,}250 \text{ psi})^2 + (-4{,}700 \text{ psi})^2}$$

$$= 4{,}050 \text{ psi} \pm 9{,}490 \text{ psi}$$

Therefore, $\sigma_1 = 13{,}540 \text{ psi} \qquad \sigma_2 = -5{,}440 \text{ psi}$

The angle θ_{p_1} to the plane on which σ_1 acts is obtained from Eqs. (6-14):

$$\cos 2\theta_{p_1} = \frac{\sigma_x - \sigma_y}{2R} = \frac{8{,}250 \text{ psi}}{9{,}490 \text{ psi}} = 0.8689$$

$$\sin 2\theta_{p_1} = \frac{\tau_{xy}}{R} = \frac{-4{,}700 \text{ psi}}{9{,}490 \text{ psi}} = -0.4950$$

in which we have noted that R is the square-root term in the preceding calculation for σ_1 and σ_2. The only angle between 0 and 360° satisfying both of these conditions is $2\theta_{p1} = 330.3°$; hence, $\theta_{p_1} = 165.2°$. This angle is associated with the algebraically larger principal stress $\sigma_1 = 13{,}540$ psi. The other angle is 90° larger or smaller than θ_{p_1}; hence $\theta_{p_2} = 75.2°$, with $\sigma_2 = -5{,}440$ psi. These results agree with those found previously.

(b) *Maximum shear stresses.* The maximum shear stresses are given by Eq. (6-20):

$$\tau_{max} = \sqrt{\left(\frac{\sigma_x - \sigma_y}{2}\right)^2 + \tau_{xy}^2}$$

$$= 9{,}490 \text{ psi}$$

which has already been calculated in part (a). The angle θ_{s_1} to the plane having the positive maximum shear stress is calculated from Eq. (6-19):

$$\theta_{s_1} = \theta_{p_1} - 45° = 165.2° - 45° = 120.2°$$

It follows that the negative shear stress acts on the plane for which $\theta_{s_2} = 120.2° - 90° = 30.2°$. The normal stresses acting on the planes of maximum shear stresses are calculated from Eq. (6-22):

$$\sigma_{aver} = \frac{\sigma_x + \sigma_y}{2} = 4{,}050 \text{ psi}$$

The stress element subjected to the maximum shear stresses is shown in Fig. 6-13c.

As an alternative approach for finding the maximum shear stresses, we can use Eq. (6-15) to determine the angles θ_{s_1} and θ_{s_2}, and then we can use Eq. (6-4b) to obtain the corresponding shear stresses.

6.4 MOHR'S CIRCLE FOR PLANE STRESS

The transformation equations for plane stress (Eqs. 6-4) can be represented in a graphical form known as **Mohr's circle**. This representation is extremely useful in visualizing the relationships between normal and shear stresses acting on various inclined planes at a point in a stressed body. To establish Mohr's circle, we rearrange Eqs. (6-4) as follows:

$$\sigma_{x_1} - \frac{\sigma_x + \sigma_y}{2} = \frac{\sigma_x - \sigma_y}{2}\cos 2\theta + \tau_{xy}\sin 2\theta \qquad \text{(6-4a) repeated}$$

$$\tau_{x_1y_1} = -\frac{\sigma_x - \sigma_y}{2}\sin 2\theta + \tau_{xy}\cos 2\theta \qquad \text{(6-4b) repeated}$$

These equations are the equations of a circle in parametric form, with the angle 2θ as the parameter. Squaring both sides of each equation and then adding eliminates the parameter; the resulting equation is

$$\left(\sigma_{x_1} - \frac{\sigma_x + \sigma_y}{2}\right)^2 + \tau_{x_1y_1}^2 = \left(\frac{\sigma_x - \sigma_y}{2}\right)^2 + \tau_{xy}^2 \qquad \text{(a)}$$

This equation can be written in simpler form by using the following notation from Section 6.3:

$$\sigma_{\text{aver}} = \frac{\sigma_x + \sigma_y}{2} \qquad R = \sqrt{\left(\frac{\sigma_x - \sigma_y}{2}\right)^2 + \tau_{xy}^2} \qquad \text{(6-24a,b)}$$

Equation (a) now becomes

$$(\sigma_{x_1} - \sigma_{\text{aver}})^2 + \tau_{x_1y_1}^2 = R^2 \qquad \text{(6-25)}$$

which is the equation of a circle with σ_{x_1} and $\tau_{x_1y_1}$ as the coordinates. The circle has radius R, and its center has coordinates $\sigma_{x_1} = \sigma_{\text{aver}}$ and $\tau_{x_1y_1} = 0$.

Our next task is to construct Mohr's circle from Eqs. (6-4) and (6-25). The normal stress σ_{x_1} is plotted as the abscissa and the shear stress $\tau_{x_1y_1}$ is plotted as the ordinate. However, there are two ways the circle can be drawn. In the first form of Mohr's circle, we plot σ_{x_1} positive to the right and $\tau_{x_1y_1}$ positive downward, as shown in Fig. 6-14a. The advantage of this choice of axes is that the angle 2θ on Mohr's circle is positive when counterclockwise, which is the same as the positive direction of 2θ in the transformation equations (see Fig. 6-1). In the second form of Mohr's circle, $\tau_{x_1y_1}$ is plotted positive upward, but the angle 2θ is now positive clockwise (Fig. 6-14b).

Both forms of Mohr's circle are mathematically correct, and either one can be used. However, it is easier to visualize the orientation of the stress element in space if the angle 2θ has the same positive direction in

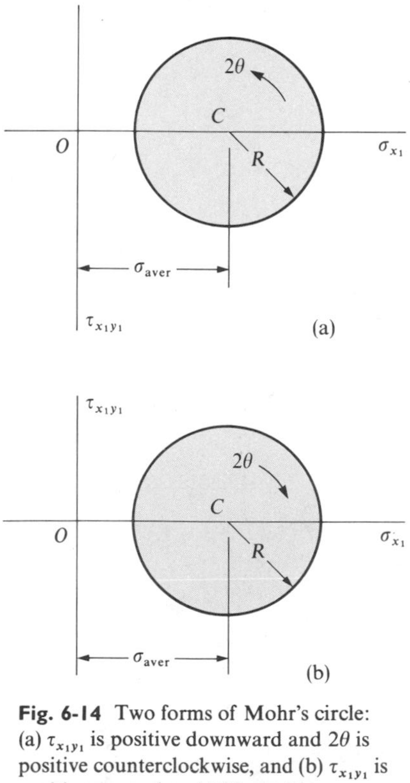

Fig. 6-14 Two forms of Mohr's circle: (a) $\tau_{x_1y_1}$ is positive downward and 2θ is positive counterclockwise, and (b) $\tau_{x_1y_1}$ is positive upward and 2θ is positive clockwise. (Note: The first form is used in this book.)

Mohr's circle as it has for the element itself. Therefore, we will opt for the first form of Mohr's circle (Fig. 6-14a) in which *the angle 2θ is always positive when counterclockwise.*

Let us now construct Mohr's circle for an element in plane stress (Figs. 6-15a and b). The procedure is as follows: (1) Locate the center C of the circle at the point having coordinates $\sigma_{x_1} = \sigma_{\text{aver}}$ and $\tau_{x_1 y_1} = 0$ (Fig. 6-15c). (2) Locate point A, which is the point on the circle representing the stress conditions on the x face of the element ($\theta = 0$); for this point we have $\sigma_{x_1} = \sigma_x$ and $\tau_{x_1 y_1} = \tau_{xy}$. (3) Locate point B, representing the stress conditions on the y face of the element ($\theta = 90°$). The coordi-

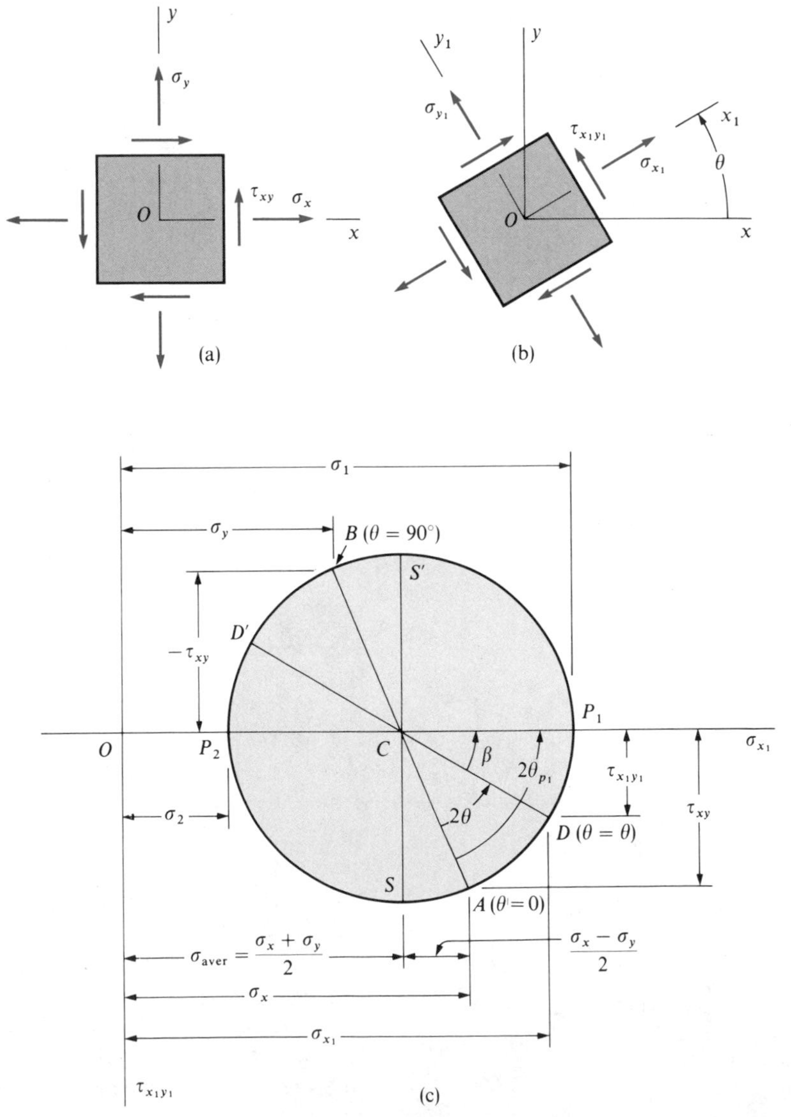

Fig. 6-15 Mohr's circle for plane stress

nates of this point are $\sigma_{x_1} = \sigma_y$ and $\tau_{x_1y_1} = -\tau_{xy}$, because, when the element is oriented at an angle $\theta = 90°$, the normal stress σ_{x_1} becomes σ_y and the shear stress $\tau_{x_1y_1}$ becomes the negative of τ_{xy}. Note that a line from A to B passes through the center C. Hence, points A and B, representing the stresses on planes at 90° to each other, are at opposite ends of a diameter (180° apart on the circle). (4) Draw the circle through points A and B with center at C.

The radius R of the circle is the length of line CA. To calculate this length, we observe that the abscissas of points C and A are $(\sigma_x + \sigma_y)/2$ and σ_x, respectively. The difference in these abscissas is $(\sigma_x - \sigma_y)/2$, as shown in Fig. 6-15c. Also, the ordinate to point A is τ_{xy}. Therefore, line CA is the hypotenuse of a right triangle having one side of length $(\sigma_x - \sigma_y)/2$ and the other side of length τ_{xy}. Taking the square root of the sum of the squares of these two sides gives R (see Eq. 6-24b).

Let us now determine the stresses acting on an inclined face of the element oriented at an angle θ from the x axis (Fig. 6-15b). On Mohr's circle, we take an angle 2θ counterclockwise from the radius CA, because A is the point for which $\theta = 0$. The angle 2θ locates point D on the circle; this point has coordinates σ_{x_1} and $\tau_{x_1y_1}$, representing the stresses on the x_1 face of the stress element. To show that the coordinates of point D are given by the stress transformation equations (Eqs. 6-4), we let β represent the angle between the radial line CD and the σ_{x_1} axis. Then, from the geometry of the figure, we obtain the following four relationships:

$$\sigma_{x_1} = \frac{\sigma_x + \sigma_y}{2} + R\cos\beta \qquad \tau_{x_1y_1} = R\sin\beta \tag{b}$$

$$\cos(2\theta + \beta) = \frac{\sigma_x - \sigma_y}{2R} \qquad \sin(2\theta + \beta) = \frac{\tau_{xy}}{R}$$

Expanding the cosine and sine expressions gives

$$\cos 2\theta\cos\beta - \sin 2\theta\sin\beta = \frac{\sigma_x - \sigma_y}{2R}$$

$$\sin 2\theta\cos\beta + \cos 2\theta\sin\beta = \frac{\tau_{xy}}{R}$$

Multiplying the first equation by $\cos 2\theta$ and the second by $\sin 2\theta$, and then adding the two equations, we obtain

$$\cos\beta = \frac{1}{R}\left(\frac{\sigma_x - \sigma_y}{2}\cos 2\theta + \tau_{xy}\sin 2\theta\right)$$

Also, multiplying the first equation by $\sin 2\theta$ and the second by $\cos 2\theta$ and then subtracting, we get

$$\sin\beta = \frac{1}{R}\left(-\frac{\sigma_x - \sigma_y}{2}\sin 2\theta + \tau_{xy}\cos 2\theta\right)$$

When these expressions for $\cos\beta$ and $\sin\beta$ are substituted into Eqs. (b), we obtain the stress transformation equations (Eqs. 6-4). Thus, we have shown that point D on Mohr's circle, defined by the angle 2θ, represents the stress conditions on the x_1 face of the stress element, defined by the angle θ.

Point D', diametrically opposite point D, is located by an angle that is 180° greater than the angle 2θ to point D (see Fig. 6-15c). Therefore, point D' represents the stresses on a face of the stress element 90° from the face represented by point D; hence point D' gives the stresses σ_{y_1} and $-\tau_{x_1y_1}$ on the y_1 face.

As we rotate the axes counterclockwise through an angle θ (Fig. 6-15b), the point on Mohr's circle corresponding to the x_1 face moves counterclockwise through an angle 2θ. Similarly, if we rotate the axes clockwise, the point on the circle moves clockwise. At point P_1 on the circle, the normal stress reaches its algebraically largest value and the shear stress is zero. Hence, P_1 represents a **principal plane**. The other principal plane, associated with the algebraically smallest normal stress, is represented by point P_2. From the geometry of the circle, we see that the larger principal stress is

$$\sigma_1 = OC + CP_1 = \frac{\sigma_x + \sigma_y}{2} + R$$

which, upon substitution of the expression for R (Eq. 6-24b), agrees with Eq. (6-13). In a similar manner, we can verify the expression for σ_2.

The principal angle θ_{p_1} between the x axis and the plane of the algebraically larger principal stress for the rotated stress element (Fig. 6-15b) is one-half the angle $2\theta_{p_1}$ between radii CA and CP_1 on Mohr's circle. The cosine and sine of the angle $2\theta_{p_1}$ can be obtained by inspection from the circle:

$$\cos 2\theta_{p_1} = \frac{\sigma_x - \sigma_y}{2R} \qquad \sin 2\theta_{p_1} = \frac{\tau_{xy}}{R}$$

These expressions agree with Eqs. (6-14). The angle $2\theta_{p_2}$ to the other principal point is 180° larger than $2\theta_{p_1}$; hence $\theta_{p_2} = \theta_{p_1} + 90°$.

Points S and S', representing the planes of **maximum and minimum shear stresses**, are located on the circle at 90° angles from points P_1 and P_2. Therefore, the planes of maximum shear stress are at 45° to the principal planes, as discussed in Section 6.3. The maximum shear stress is numerically equal to the radius of the circle (compare Eq. 6-24b for R with Eq. 6-20 for τ_{max}). Also, the normal stresses on the planes of maximum shear stress are equal to the abscissa of point C, which is the average normal stress (Eq. 6-22).

From the preceding discussions, it is apparent that we can find the stresses on any inclined planes, as well as the principal stresses and maximum shear stresses, from Mohr's circle. The diagram of Fig. 6-15 was drawn with both σ_x and σ_y as positive stresses, but the same procedures are followed if either or both stresses are negative. In such cases, part or all of

Mohr's circle will be located to the left of the origin. We should also note that point A, representing the stresses on the plane $\theta = 0$, may be situated anywhere around the circle, depending upon the relative values of the stresses σ_x, σ_y, and τ_{xy}. However, the angle 2θ is always measured counterclockwise from the radius CA, regardless of where point A is located.

It is also possible to use Mohr's circle inversely. If we know the stresses σ_{x_1}, σ_{y_1}, and $\tau_{x_1y_1}$ acting on a rotated element, as well as the angle θ itself, then we can construct the circle and determine the stresses σ_x, σ_y, and τ_{xy} for $\theta = 0$. The procedure is to locate points D and D' from the known stresses and then draw the circle using line DD' as a diameter. By measuring the angle 2θ in a negative sense (that is, clockwise) from radius CD, we can locate point A, corresponding to the x face of the element. Next, we locate point B by constructing a diameter from A. Finally, we use the coordinates of A and B to obtain the stresses acting on all faces of the element for which $\theta = 0$.

An alternative convention for the directions of the shear stresses is sometimes used. In this convention the direction of a shear stress acting on an element of the material is indicated by the sense of the rotation that it tends to produce. If the shear stress tends to rotate the stress element clockwise, it is called a clockwise shear stress, and if it tends to rotate it counterclockwise, it is a counterclockwise stress. Then, when constructing Mohr's circle, clockwise shear stresses are plotted upward and counterclockwise shear stresses are plotted downward. This method of constructing the circle is identical to the method already described in this section, because a positive shear stress $\tau_{x_1y_1}$ is also a counterclockwise shear stress, and both are plotted positive downward. Thus, the alternative approach represents only a change in our point of view. Instead of thinking of the vertical axis as having positive shear stresses downward and negative upward, we think of the vertical axis as having clockwise shear stresses upward and counterclockwise downward.

If desired, we can construct Mohr's circle to scale and measure values of stress from the drawing. However, it is usually preferable to perform numerical calculations for the stresses, using trigonometry and the geometry of the circle. Mohr's circle makes it possible to visualize the relationships between stresses acting on planes at various angles, and it also serves as a simple memory device for obtaining the stress transformation equations. Although most graphical techniques are no longer used in engineering work, Mohr's circle remains valuable because it provides a simple and clear picture of an otherwise complicated analysis. The circle is also applicable to transformations involving two-dimensional strains and moments of inertia of plane areas, because these quantities follow the same transformation laws as do stresses (see Section 6.11 and Appendix C, Section C.7).*

* Mohr's circle is named after the famous German civil engineer Otto Christian Mohr (1835–1918), who developed the circle in 1882 (Ref. 6-5).

Example 1

An element in plane stress is subjected to stresses $\sigma_x = 15{,}000$ psi, $\sigma_y = 5{,}000$ psi, and $\tau_{xy} = 4{,}000$ psi, as shown in Fig. 6-16a. Using Mohr's circle, determine (a) the stresses acting on an element oriented at an angle $\theta = 40°$, (b) the principal stresses, and (c) the maximum shear stresses. Show all results on sketches of properly oriented elements.

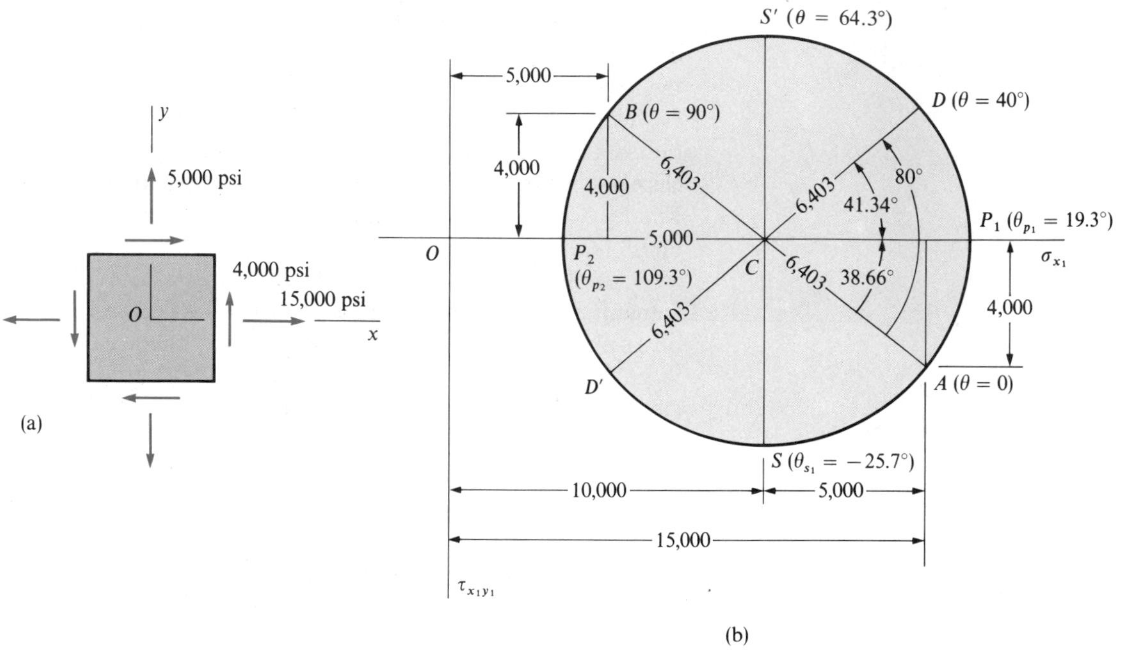

Fig. 6-16 Example 1. (Note: All stresses on Mohr's circle have units of psi.)

The center C of the circle (Fig. 6-16b) is located on the σ_{x_1} axis at the point where σ_{x_1} equals σ_{aver}:

$$\sigma_{\text{aver}} = \frac{\sigma_x + \sigma_y}{2} = \frac{15{,}000 \text{ psi} + 5{,}000 \text{ psi}}{2} = 10{,}000 \text{ psi}$$

Point A, representing the stresses on the x face of the element, has coordinates

$$\sigma_{x_1} = 15{,}000 \text{ psi} \qquad \tau_{x_1y_1} = 4{,}000 \text{ psi}$$

Similarly, the coordinates of point B, representing the stresses on the y face, are

$$\sigma_{x_1} = 5{,}000 \text{ psi} \qquad \tau_{x_1y_1} = -4{,}000 \text{ psi}$$

The circle is now drawn through points A and B, with center at C and radius

$$R = \sqrt{(5{,}000 \text{ psi})^2 + (4{,}000 \text{ psi})^2} = 6{,}403 \text{ psi}$$

The angle ACP_1 is the angle $2\theta_{p_1}$ from point A to point P_1, representing the principal plane having the larger principal stress σ_1; this angle is obtained as follows:

$$\tan 2\theta_{p_1} = \frac{4{,}000 \text{ psi}}{5{,}000 \text{ psi}} = 0.8$$

from which

$$2\theta_{p_1} = 38.66^\circ \qquad \theta_{p_1} = 19.33^\circ$$

Thus, we now have calculated the important angles and stresses that determine the geometry of Mohr's circle.

(a) The stresses acting on a plane at $\theta = 40^\circ$ are given by point D, which is at an angle $2\theta = 80^\circ$ from point A. Therefore, angle DCP_1 is

$$\text{Angle } DCP_1 = 80^\circ - 2\theta_{p_1} = 80^\circ - 38.66^\circ = 41.34^\circ$$

This angle is between line CD and the σ_{x_1} axis. Therefore, by inspection of the figure, we see that the coordinates of point D are

$$\sigma_{x_1} = 10{,}000 \text{ psi} + (6{,}403 \text{ psi})(\cos 41.34^\circ) = 14{,}810 \text{ psi}$$

$$\tau_{x_1y_1} = -(6{,}403 \text{ psi})(\sin 41.34^\circ) = -4{,}230 \text{ psi}$$

In an analogous manner, we can find the stresses represented by point D':

$$\sigma_{x_1} = 10{,}000 \text{ psi} - (6{,}403 \text{ psi})(\cos 41.34^\circ) = 5{,}190 \text{ psi}$$

$$\tau_{x_1y_1} = (6{,}403 \text{ psi})(\sin 41.34^\circ) = 4{,}230 \text{ psi}$$

These results are shown in Fig. 6-17a on a sketch of an element oriented at an angle $\theta = 40^\circ$.

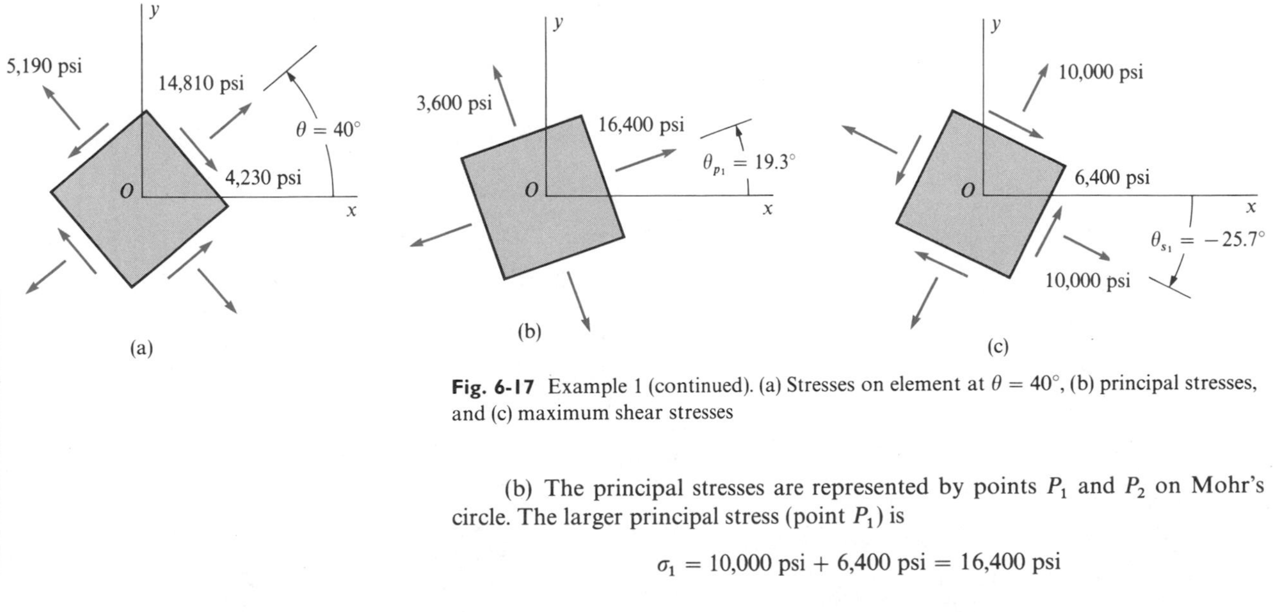

Fig. 6-17 Example 1 (continued). (a) Stresses on element at $\theta = 40^\circ$, (b) principal stresses, and (c) maximum shear stresses

(b) The principal stresses are represented by points P_1 and P_2 on Mohr's circle. The larger principal stress (point P_1) is

$$\sigma_1 = 10{,}000 \text{ psi} + 6{,}400 \text{ psi} = 16{,}400 \text{ psi}$$

as seen by inspection of the circle. This stress acts on a plane defined by the angle $\theta_{p_1} = 19.3°$. Similarly, the smaller principal stress is

$$\sigma_2 = 10{,}000 \text{ psi} - 6{,}400 \text{ psi} = 3{,}600 \text{ psi}$$

The angle $2\theta_{p_2}$ is $38.66° + 180° = 218.66°$; thus, the principal plane is defined by the angle $\theta_{p_2} = 109.3°$. The principal stresses and principal planes are shown in Fig. 6-17b.

(c) The maximum and minimum shear stresses are represented by points S and S'; therefore,

$$\tau_{\text{max}} = 6{,}400 \text{ psi}$$

which is the radius of the circle. The angle ACS is $90° - 38.66° = 51.34°$; hence the angle $2\theta_{s_1}$ to point S on the circle is

$$2\theta_{s_1} = -51.34°$$

because it is measured clockwise. The corresponding angle θ_{s_1} to the plane of the maximum shear stress is one-half that value, or $\theta_{s_1} = -25.7°$. The maximum and minimum shear stresses and the associated normal stresses σ_{aver} are shown in Fig. 6-17c.

Example 2

An element in plane stress is subjected to stresses $\sigma_x = -50$ MPa, $\sigma_y = 10$ MPa, and $\tau_{xy} = -40$ MPa, as shown in Fig. 6-18a. Using Mohr's circle, determine (a) the stresses acting on an element oriented at an angle $\theta = 45°$, (b) the principal stresses, and (c) the maximum shear stresses. Show all results on sketches of properly oriented elements.

The center of the circle is on the σ_{x_1} axis at point C where σ_{x_1} equals σ_{aver}, which is

$$\sigma_{\text{aver}} = \frac{\sigma_x + \sigma_y}{2} = -20 \text{ MPa}$$

The stresses on the x face of the element determine the coordinates of point A:

$$\sigma_{x_1} = -50 \text{ MPa} \qquad \tau_{x_1y_1} = -40 \text{ MPa}$$

The coordinates of point B represent the stresses on the y face of the element:

$$\sigma_{x_1} = 10 \text{ MPa} \qquad \tau_{x_1y_1} = 40 \text{ MPa}$$

These points define the circle, which has radius

$$R = \sqrt{(30 \text{ MPa})^2 + (40 \text{ MPa})^2} = 50 \text{ MPa}$$

The angle ACP_2 is the angle $2\theta_{p_2}$ from point A to point P_2, representing the principal plane having the algebraically smaller principal stress σ_2. This angle is found by noting that

$$\tan 2\theta_{p2} = \frac{40 \text{ MPa}}{30 \text{ MPa}} = \frac{4}{3}$$

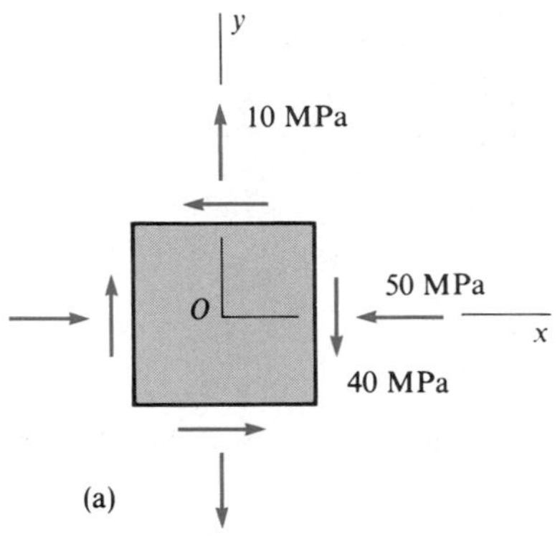

(a)

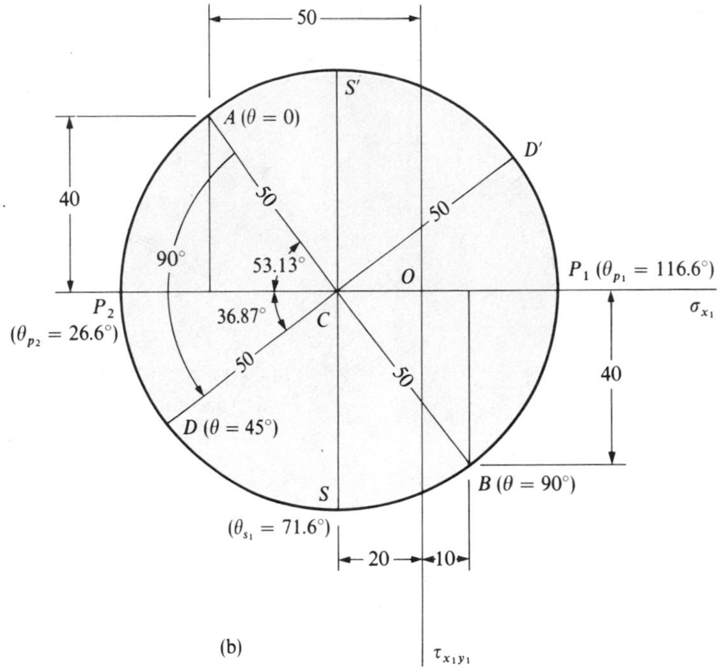

(b)

Fig. 6-18 Example 2. (Note: All stresses on Mohr's circle have units of MPa.)

Hence,

$$2\theta_{p_2} = 53.13^\circ \qquad \theta_{p_2} = 26.57^\circ$$

Thus, all of the required angles and stresses have been obtained, as shown on the circle.

(a) The stresses acting on a plane at $\theta = 45^\circ$ are represented by point D, which is at an angle $2\theta = 90^\circ$ from point A. The angle DCP_2 is

$$\text{Angle } DCP_2 = 90^\circ - 2\theta_{p_2} = 90^\circ - 53.13^\circ = 36.87^\circ$$

This angle is between line CD and the negative σ_{x_1} axis; therefore, by inspection we obtain the coordinates of point D:

$$\sigma_{x_1} = -20\text{ MPa} - (50\text{ MPa})(\cos 36.87^\circ) = -60\text{ MPa}$$

$$\tau_{x_1y_1} = (50\text{ MPa})(\sin 36.87^\circ) = 30\text{ MPa}$$

Similarly, the coordinates of point D' are

$$\sigma_{x_1} = -20\text{ MPa} + (50\text{ MPa})(\cos 36.87^\circ) = 20\text{ MPa}$$

$$\tau_{x_1y_1} = (-50\text{ MPa})(\sin 36.87^\circ) = -30\text{ MPa}$$

These stresses, which act on an element at $\theta = 45^\circ$, are shown in Fig. 6-19a.

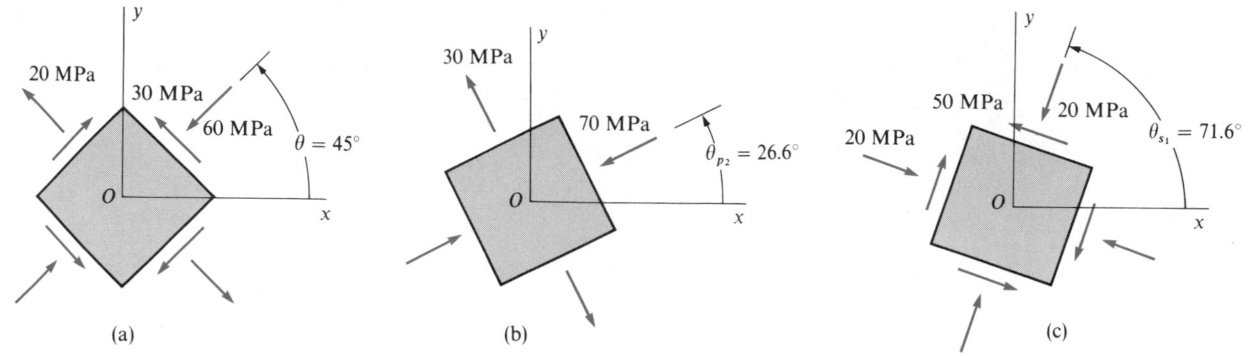

Fig. 6-19 Example 2 (continued). (a) Stresses on element at $\theta = 45°$, (b) principal stresses, and (c) maximum shear stresses

(b) The principal stresses are represented by points P_1 and P_2 on the circle. Their values are

$$\sigma_1 = -20 \text{ MPa} + 50 \text{ MPa} = 30 \text{ MPa}$$

$$\sigma_2 = -20 \text{ MPa} - 50 \text{ MPa} = -70 \text{ MPa}$$

as obtained by inspection of the circle. The angle $2\theta_{p_1}$ on the circle (measured counterclockwise from A to P_1) is $53.1° + 180° = 233.1°$; hence $\theta_{p_1} = 116.6°$. The angle to point P_2 is $2\theta_{p_2} = 53.1°$, or $\theta_{p_2} = 26.6°$. The principal planes and principal stresses are shown in Fig. 6-19b.

(c) The maximum and minimum shear stresses, represented by points S and S', are $\tau_{max} = 50$ MPa and $\tau_{min} = -50$ MPa. The angle ACS (equal to $2\theta_{s_1}$) is $53.13° + 90° = 143.13°$; hence the angle $\theta_{s_1} = 71.6°$. The maximum shear stresses and the associated normal stresses σ_{aver} are shown in Fig. 6-19c.

6.5 HOOKE'S LAW FOR PLANE STRESS

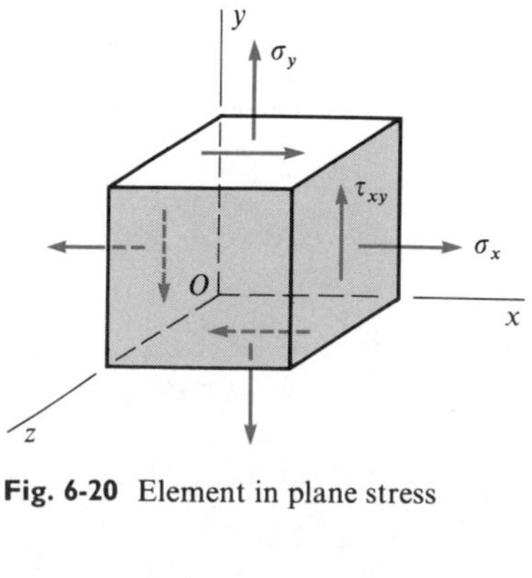

Fig. 6-20 Element in plane stress

The stresses acting on inclined planes for an element in plane stress (Fig. 6-20) were analyzed in the preceding sections. In those discussions, we derived the equations solely from static equilibrium, and the properties of the materials were not needed. Now we will investigate the strains in the material, which requires that we consider the material properties. We will assume that the material is uniform, or homogeneous, throughout the body and that it has the same properties in all directions (isotropic material). Furthermore, we will assume that Hooke's law holds, which means that the material behaves in a linearly elastic manner. Under these conditions, we can readily obtain the relationships between the stresses and strains in the body.

Let us begin by considering the normal strains ϵ_x, ϵ_y, and ϵ_z, which are shown in Fig. 6-21 as the changes in dimensions of an infinitesimally

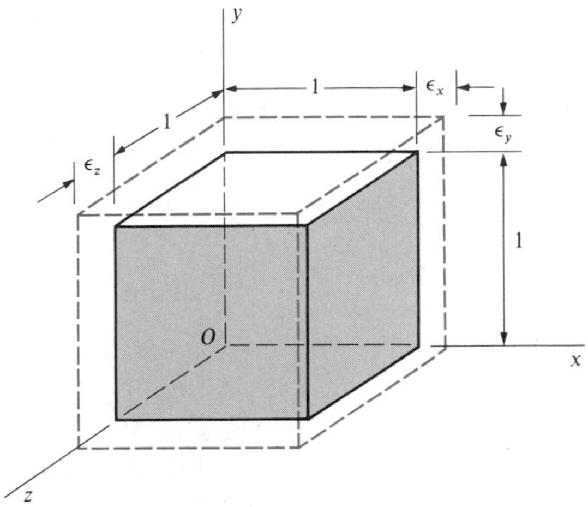

Fig. 6-21 Normal strains ϵ_x, ϵ_y, and ϵ_z

small cube having edges of unit length. All three strains are shown positive (elongation) in the figure. These strains can be expressed in terms of the stresses (Fig. 6-20) by superimposing the effects of the individual stresses. For instance, the stress σ_x produces a strain ϵ_x equal to σ_x/E, and the stress σ_y produces a strain ϵ_x equal to $-\nu\sigma_y/E$. Of course, the shear stress τ_{xy} produces no normal strain in the x direction. Thus, the resultant strain ϵ_x is

$$\epsilon_x = \frac{1}{E}(\sigma_x - \nu\sigma_y) \tag{6-26a}$$

In a similar way, we obtain the strains in the y and z directions:

$$\epsilon_y = \frac{1}{E}(\sigma_y - \nu\sigma_x) \qquad \epsilon_z = -\frac{\nu}{E}(\sigma_x + \sigma_y) \tag{6-26b,c}$$

These equations may be used to find the normal strains when the stresses are known.

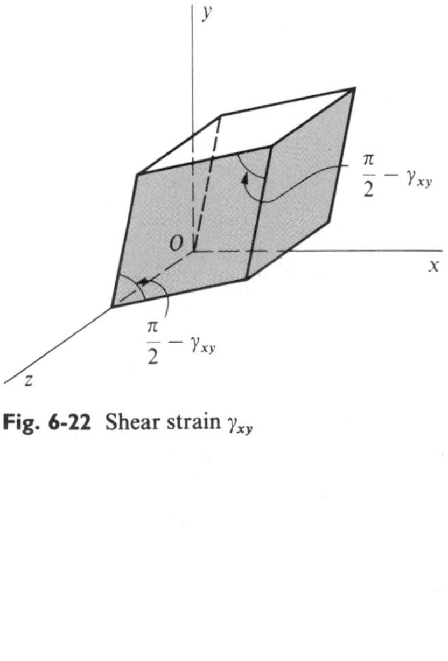

Fig. 6-22 Shear strain γ_{xy}

The shear stress τ_{xy} causes a distortion of the element such that each z face becomes a rhombus (Fig. 6-22). The shear strain γ_{xy} represents the decrease in angle between the positive (or negative) x and y faces of the element. Because no other shear stresses act on the sides of the plane stress element (Fig. 6-20), the x and y faces are not distorted and remain square. The shear strain is related to the shear stress by Hooke's law in shear, as follows:

$$\gamma_{xy} = \frac{\tau_{xy}}{G} \tag{6-27}$$

Of course, the normal stresses σ_x and σ_y have no effect on the shear strain γ_{xy}. Hence, Eqs. (6-26) and (6-27) give the strains due to all stresses (σ_x, σ_y, and τ_{xy}) acting simultaneously.

The first two equations for the normal strains (Eqs. 6-26a and b) can be solved simultaneously for the stresses in terms of the strains:

$$\sigma_x = \frac{E}{1 - \nu^2}(\epsilon_x + \nu\epsilon_y) \tag{6-28a}$$

$$\sigma_y = \frac{E}{1 - \nu^2}(\epsilon_y + \nu\epsilon_x) \tag{6-28b}$$

In addition, we have

$$\tau_{xy} = G\gamma_{xy} \tag{6-29}$$

These equations may be used to find the stresses when the strains are known.

Equations (6-26) through (6-29) are known collectively as **Hooke's law for plane stress**. They contain three elastic constants (E, G, and ν), but only two are independent because of the relationship $G = E/2(1 + \nu)$.

UNIT VOLUME CHANGE

The unit volume change at a point in a body subjected to plane stress can be found by considering the element of Fig. 6-21. The original volume of this element is $V_o = (1)(1)(1) = 1$, and its final volume is

$$V_f = (1 + \epsilon_x)(1 + \epsilon_y)(1 + \epsilon_z) \tag{a}$$

or, disregarding products of small quantities,

$$V_f = 1 + \epsilon_x + \epsilon_y + \epsilon_z \tag{b}$$

Therefore, the change in volume is

$$\Delta V = V_f - V_o = \epsilon_x + \epsilon_y + \epsilon_z \tag{c}$$

and the unit volume change (or dilatation) becomes

$$e = \frac{\Delta V}{V_o} = \epsilon_x + \epsilon_y + \epsilon_z \tag{6-30}$$

This equation is valid for any material, whether or not it follows Hooke's law. Note that the shear strains produce no change in volume.

When the material follows Hooke's law, we can substitute Eqs. (6-26) into Eq. (6-30) and obtain the following expression for the unit volume change in plane stress:

$$e = \frac{\Delta V}{V_o} = \frac{1 - 2\nu}{E}(\sigma_x + \sigma_y) \tag{6-31}$$

If $\sigma_y = 0$, this equation reduces to Eq. (1-7) for the unit volume change in uniaxial stress. Knowing the value of e from Eq. (6-31), we can find by integration the total volume change for any body subjected to plane stress.

STRAIN ENERGY DENSITY

The strain energy density u is the strain energy stored in a unit volume of the material (see previous discussions in Sections 2.6 and 3.8). For an element in plane stress, we can use the elements pictured in Figs. 6-21 and 6-22. Inasmuch as the normal and shear strains occur independently, we can add their strain energies to obtain the total energy.

Beginning with the normal strains (Fig. 6-21), we observe that the total force acting on the x face of the element is equal algebraically to σ_x, inasmuch as the area of the face is unity. This force moves through a distance equal to ϵ_x as the stresses are applied to the element. Assuming that Hooke's law holds for the material, we see that the work done by this force is $\sigma_x \epsilon_x / 2$. Similarly, the force σ_y on the y face does work equal to $\sigma_y \epsilon_y / 2$. The sum of these work terms, which is the same as the strain energy density, is

$$\frac{1}{2}(\sigma_x \epsilon_x + \sigma_y \epsilon_y)$$

The strain energy density associated with the shear strains (Fig. 6-22) was evaluated in Section 3.8 (see Eq. 3-36) and equals $\tau_{xy}\gamma_{xy}/2$. By adding the strain energy densities for normal and shear strains, we obtain the following formula for plane stress:

$$u = \frac{1}{2}(\sigma_x \epsilon_x + \sigma_y \epsilon_y + \tau_{xy}\gamma_{xy}) \qquad (6\text{-}32)$$

Substituting for the strains from Eqs. (6-26) and (6-27), we obtain the strain energy density in terms of stresses alone:

$$u = \frac{1}{2E}(\sigma_x^2 + \sigma_y^2 - 2\nu\sigma_x\sigma_y) + \frac{\tau_{xy}^2}{2G} \qquad (6\text{-}33)$$

In the same manner, we can substitute for the stresses from Eqs. (6-28) and (6-29) to obtain the strain energy density in terms of strains:

$$u = \frac{E}{2(1-\nu^2)}(\epsilon_x^2 + \epsilon_y^2 + 2\nu\epsilon_x\epsilon_y) + \frac{G\gamma_{xy}^2}{2} \qquad (6\text{-}34)$$

For the special case of uniaxial stress, we substitute the values

$$\sigma_y = 0 \qquad \tau_{xy} = 0 \qquad \epsilon_y = -\nu\epsilon_x \qquad \gamma_{xy} = 0$$

into Eqs. (6-33) and (6-34) to obtain

$$u = \frac{\sigma_x^2}{2E} \quad \text{and} \quad u = \frac{E\epsilon_x^2}{2}$$

These equations agree with Eqs. (2-37) of Section 2.6.

6.6 SPHERICAL AND CYLINDRICAL PRESSURE VESSELS (BIAXIAL STRESS)

Pressure vessels are closed structures that contain liquids or gases under pressure. Familiar examples are spherical water-storage tanks, cylindrical tanks for compressed air, pressurized pipes, and inflated balloons. The curved walls of pressure vessels are often very thin in relation to the radius and the length, and in such cases they are in the general class of structural forms known as **shells**. Other examples of shell structures are thin curved roofs, domes, and fuselages.

In this section, we consider only thin-walled pressure vessels of spherical and circular cylindrical shape (Fig. 6-23). The term *thin-walled* is not precise, but a general rule is that the ratio of radius r to wall thickness t should be greater than 10 in order for us to determine the stresses in the walls with reasonable accuracy using statics alone. A second limitation is that the internal pressure must exceed the external pressure; otherwise, the shell may collapse due to buckling of the walls.

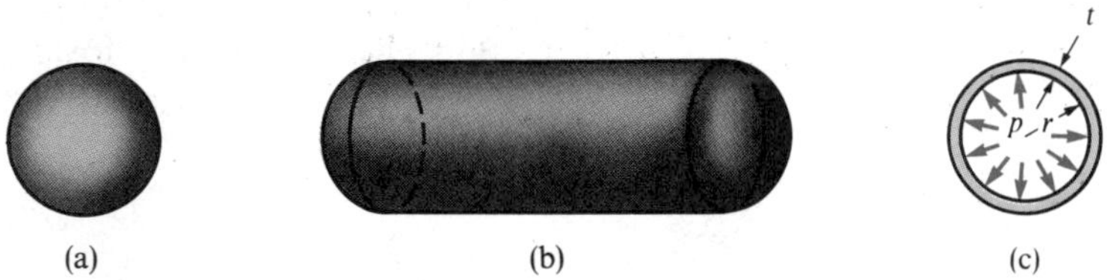

Fig. 6-23 Thin-walled pressure vessels: (a) spherical vessel, (b) cylindrical vessel, and (c) cross section showing internal pressure p

SPHERICAL PRESSURE VESSELS

A tank of spherical shape is the ideal container for resisting internal pressure. We need only contemplate the familiar soap bubble to recognize that a sphere is the "natural" shape for this purpose. To obtain the stresses in the wall, let us cut through the sphere on a vertical diametral plane and isolate half of the shell and its contents as a free body (Fig. 6-24a). Acting on this free body are the stresses σ in the wall and the internal pressure p. The weight of the tank and its contents is disregarded in this analysis. The pressure acts horizontally against the plane circular area formed by the

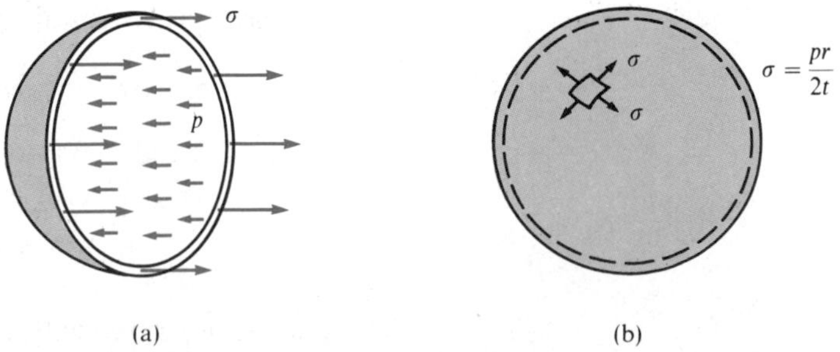

Fig. 6-24 Stresses in a spherical pressure vessel

cut, and the resultant force equals $p(\pi r^2)$, where r is the inside radius of the sphere. Note that the pressure p is the net internal pressure, or gage pressure (that is, the pressure above the atomospheric or external pressure).

The tensile stress σ in the wall of the sphere is uniform around the circumference of the tank, because of the symmetry of the tank and its loading. Furthermore, because the wall is very thin, we can assume with good accuracy that the stress is uniform across the thickness t. The accuracy of this approximation increases as the shell becomes thinner and decreases as it becomes thicker. The resultant force obtained from the normal stress is $\sigma(2\pi r_m t)$, where t is the thickness and r_m is the mean radius of the shell ($r_m = r + t/2$). However, since our analysis is valid only for very thin shells, we can assume that $r_m \approx r$; then the resultant force becomes $\sigma(2\pi rt)$. Equilibrium of forces in the horizontal direction gives

$$\sigma(2\pi rt) - p(\pi r^2) = 0$$

from which we obtain

$$\sigma = \frac{pr}{2t} \tag{6-35}$$

As is evident from the symmetry of a spherical shell, this same equation for the stress σ will be obtained if we cut a plane through the center of the sphere in any direction whatsoever. Therefore, we conclude that a pressurized sphere is subjected to uniform tensile stresses σ in all directions. This stress condition is represented in Fig. 6-24b by the small stress element with stresses σ acting in mutually perpendicular directions. Stresses of this kind, which act tangential (rather than perpendicular) to the curved surface are known as **membrane stresses**. The name arises from the fact that stresses of this type exist in true membranes, such as soap films or thin sheets of rubber.

At the **outer surface** of a spherical pressure vessel, there are no stresses acting normal to the surface. Hence, the stress condition is a special case of biaxial stress in which σ_x and σ_y are the same (Fig. 6-25a). Because no shear stresses act on this element, we obtain exactly the same normal stresses

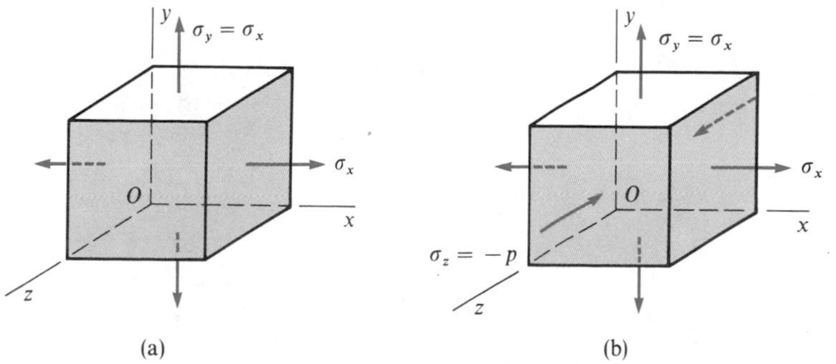

Fig. 6-25 Stresses in a spherical pressure vessel at: (a) the outer surface, and (b) the inner surface

when the element is rotated about the z axis through any angle. Thus, Mohr's circle for this stress condition reduces to a point, and every inclined plane is a principal plane. The principal stresses are

$$\sigma_1 = \sigma_2 = \frac{pr}{2t} \tag{6-36}$$

Because every direction is a principal direction (for rotations about the z axis), there are no in-plane shear stresses.

However, as discussed in Section 6.3, we must not overlook the fact that a plane-stress element is actually three dimensional. In the present case, shear stresses exist in the material even though there are no in-plane shear stresses. To obtain the maximum shear stress, we note that the third principal stress (in the z direction) is zero. Therefore, the absolute maximum shear stress, obtained by a 45° rotation about either the x or the y axis, is

$$\tau_{max} = \frac{\sigma_1}{2} = \frac{pr}{4t} \tag{6-37}$$

as obtained from Eqs. (6-23).

At the **inner surface** of the wall of the spherical vessel, the stress element has the same membrane stresses (Eq. 6-35) but, in addition, a compressive stress equal to p acts in the z direction (Fig. 6-25b). These three normal stresses are the principal stresses:

$$\sigma_1 = \sigma_2 = \frac{pr}{2t} \qquad \sigma_3 = -p \tag{6-38}$$

The maximum in-plane shear stress is zero, but the maximum out-of-plane shear stress (obtained by a 45° rotation about either the x or y axis) is

$$\tau_{max} = \frac{\sigma_1 + p}{2} = \frac{pr}{4t} + \frac{p}{2} \tag{6-39}$$

If the ratio r/t is sufficiently large, the last term in this equation can be disregarded. Then the equation becomes the same as Eq. (6-37), and we can assume that the maximum shear stress is constant across the thickness of the shell.

Every spherical tank used as a pressure vessel will have at least one opening in the wall as well as various attachments and supports. These features result in nonuniformities in the stress distribution that cannot be analyzed by elementary methods. High localized stresses are produced near the discontinuities in the shell, which must be reinforced in these regions. Therefore, the equations we have derived for the membrane stresses are valid everywhere in the wall of the vessel except close to the discontinuities. Other considerations entering into the design of tanks include the effects of corrosion, accidental impacts, and temperature effects.

CYLINDRICAL PRESSURE VESSELS

Next we shall examine the stresses in a thin-walled circular cylindrical tank with closed ends and internal pressure p (Fig. 6-26a). A stress element with faces parallel and perpendicular to the axis of the tank is shown in the figure. The normal stresses σ_1 and σ_2 acting on the side faces of this element represent the membrane stresses in the wall. No shear stresses act on the faces of this element because of the symmetry of the vessel. Therefore, the stresses σ_1 and σ_2 are principal stresses. Because of its direction, the stress σ_1 is called the **circumferential stress** or the **hoop stress**; similarly, the stress σ_2 is the **longitudinal stress** or the **axial stress**. Each of these stresses can be calculated from equilibrium by using appropriate free-body diagrams.

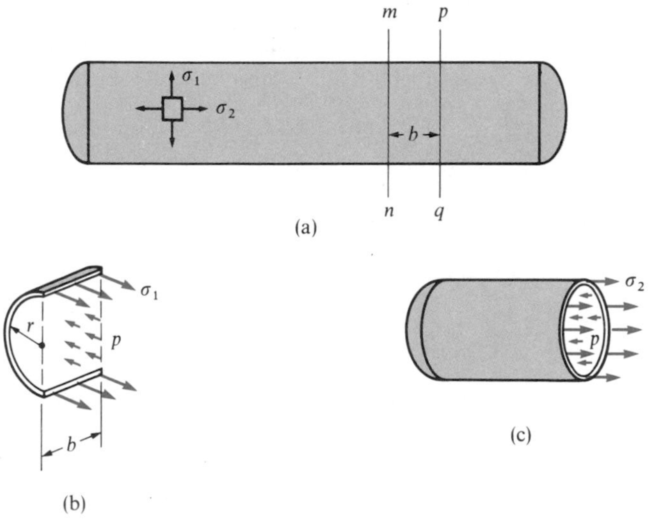

Fig. 6-26 Stresses in a circular cylindrical pressure vessel

To calculate the circumferential stress σ_1, we isolate a free body by making two cuts (mn and pq) a distance b apart and perpendicular to the longitudinal axis (Fig. 6-26a). Also, we make a third cut in a vertical plane through the axis itself; the resulting free-body is shown in Fig. 6-26b. Acting on the longitudinal vertical face of this free body are the stresses σ_1 in the wall and the internal pressure p. Stresses and pressures also act on the transverse faces of this free body, but they are not shown in the figure because they do not enter the equation of equilibrium that we will use. Also, we again disregard the weight of the vessel and its contents. The horizontal forces due to the stress σ_1 and the pressure p act in opposite directions; hence we have the following equation of equilibrium:

$$\sigma_1(2bt) - p(2br) = 0$$

in which t is the thickness of the wall and r is the inside radius of the cylinder. From the preceding equation, we obtain

$$\sigma_1 = \frac{pr}{t} \tag{6-40}$$

as the formula for the circumferential stress. As discussed previously, this stress is uniformly distributed over the thickness of the wall provided the wall is very thin.

The longitudinal stress σ_2 is obtained from a free body of the part of the tank to the left of a section (such as *mn*) that is perpendicular to the longitudinal axis (Fig. 6-26c). In this case, the equation of equilibrium is

$$\sigma_2(2\pi rt) - p(\pi r^2) = 0$$

in which, as explained previously, we have used the inside radius of the shell in place of the mean radius when calculating the force due to the stress σ_2. Solving the preceding equation for σ_2, we obtain

$$\sigma_2 = \frac{pr}{2t} \tag{6-41}$$

which is the same as the membrane stress in a spherical shell. Comparing Eqs. (6-40) and (6-41), we see that

$$\sigma_1 = 2\sigma_2 \tag{6-42}$$

and the circumferential stress in a cylindrical shell is twice the longitudinal stress.

The principal stresses σ_1 and σ_2 at the **outer surface** of the shell are shown acting on a stress element in Fig. 6-27a. The third principal stress, acting in the *z* direction, is zero. Thus, we again have biaxial stress. The in-plane maximum shear stresses occur when the element is rotated 45°

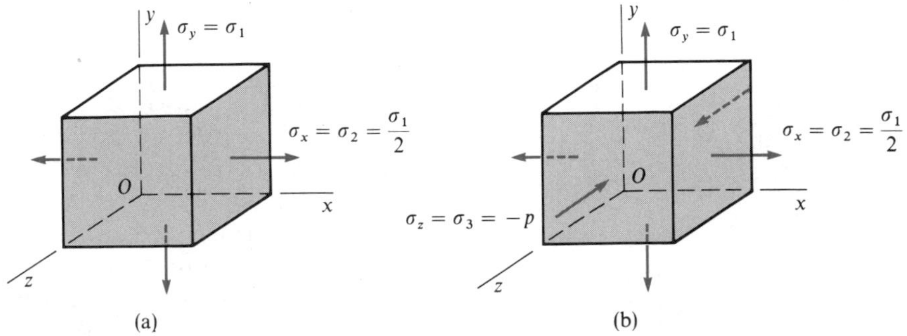

Fig. 6-27 Stresses in a circular cylindrical pressure vessel at: (a) the outer surface, and (b) the inner surface

about the z axis; this stress is

$$(\tau_{max})_z = \frac{\sigma_1 - \sigma_2}{2} = \frac{\sigma_1}{4} = \frac{pr}{4t} \tag{6-43}$$

(see Eq. 6-21). The maximum out-of-plane shear stresses are obtained by 45° rotations about the x and y axes, respectively:

$$(\tau_{max})_x = \frac{\sigma_1}{2} = \frac{pr}{2t} \qquad (\tau_{max})_y = \frac{\sigma_2}{2} = \frac{pr}{4t}$$

Thus, the absolute maximum shear stress is

$$\tau_{max} = \frac{\sigma_1}{2} = \frac{pr}{2t} \tag{6-44}$$

and it occurs when the element is rotated 45° about the x axis.

The stress conditions at the **inner surface** of the shell are shown in Fig. 6-27b. The principal normal stresses are

$$\sigma_1 = \frac{pr}{t} \qquad \sigma_2 = \frac{pr}{2t} \qquad \sigma_3 = -p \tag{6-45}$$

The three maximum shear stresses, obtained by 45° rotations about the x, y, and z axes, are

$$(\tau_{max})_x = \frac{\sigma_1 + p}{2} = \frac{pr}{2t} + \frac{p}{2} \qquad (\tau_{max})_y = \frac{\sigma_2 + p}{2} = \frac{pr}{4t} + \frac{p}{2}$$
$$(\tau_{max})_z = \frac{\sigma_1 - \sigma_2}{2} = \frac{pr}{4t} \tag{6-46}$$

The first of these stresses is the largest. However, as explained in the discussion of shear stresses in a spherical shell, we often may disregard the additional term $p/2$ in these expressions and assume that the maximum shear stress is constant across the thickness and given by Eq. (6-44).

The preceding stress formulas are valid in parts of the cylinder away from any discontinuities. An obvious discontinuity exists at the end of the cylinder where the head is attached. Others occur at openings in the cylinder or where objects are attached to the cylinder.

Example 1

A spherical pressure vessel having 18 in. inside diameter and $\frac{1}{4}$ in. wall thickness is to be constructed by welding two aluminum hemispheres (Fig. 6-28). From tests, it is found that the ultimate and yield stresses in tension at the weld are 24 ksi and 16 ksi, respectively. The tank must have a factor of safety of 2.1 with respect to the ultimate stress and 1.5 with respect to the yield stress. What is the maximum permissible pressure in the tank?

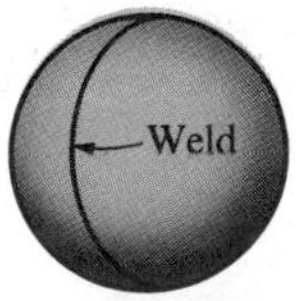

Fig. 6-28 Example 1. Spherical pressure vessel

The allowable stress based upon the ultimate stress σ_u is

$$\sigma_{\text{allow}} = \frac{\sigma_u}{n} = \frac{24 \text{ ksi}}{2.1} = 11.43 \text{ ksi}$$

(see Eq. 1-14). Based upon the yield stress σ_y, we find that the allowable stress (Eq. 1-13) is

$$\sigma_{\text{allow}} = \frac{\sigma_y}{n} = \frac{16 \text{ ksi}}{1.5} = 10.67 \text{ ksi}$$

The latter stress is lower; hence it governs the design.

The maximum tensile stress in the tank is given by the formula $\sigma = pr/2t$ (see Eq. 6-36). Solving this equation for the pressure, we get

$$p = \frac{2t\sigma_{\text{allow}}}{r} = \frac{2(0.25 \text{ in.})(10.67 \text{ ksi})}{9 \text{ in.}} = 592.8 \text{ psi}$$

Thus, the maximum allowable pressure is $p_{\text{max}} = 592$ psi. (Note that in a calculation of this kind, we round downward, not upward.)

Example 2

A cylindrical pressure vessel is constructed with a helical weld that makes an angle of 55° with the longitudinal axis (Fig. 6-29a). The tank has inside radius $r = 1.8$ m and wall thickness $t = 8$ mm. The maximum internal pressure is 600 kPa. Calculate the following quantities for the cylindrical part of the tank: (a) the circumferential and longitudinal stresses; (b) the absolute maximum shear stress; and (c) the normal and shear stresses acting perpendicular and parallel to the weld, respectively.

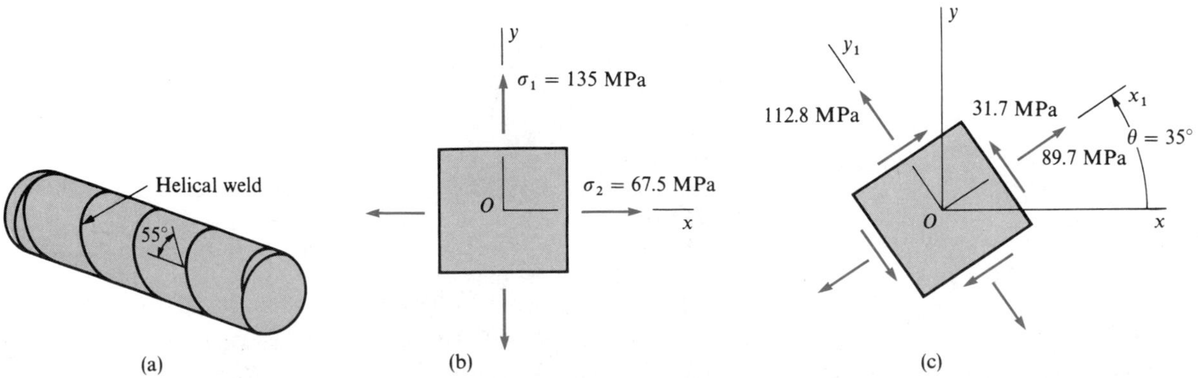

Fig. 6-29 Example 2. Cylindrical pressure vessel

(a) The circumferential and longitudinal stresses are obtained from Eqs. (6-40) and (6-42), respectively:

$$\sigma_1 = \frac{pr}{t} = \frac{(600\text{ kPa})(1.8\text{ m})}{8\text{ mm}} = 135\text{ MPa} \qquad \sigma_2 = \frac{\sigma_1}{2} = 67.5\text{ MPa}$$

These principal stresses are shown on the biaxial stress element of Fig. 6-29b.

(b) The largest in-plane shear stress is obtained from Eq. (6-43):

$$\tau = \frac{\sigma_1 - \sigma_2}{2} = \frac{\sigma_1}{4} = 33.8\text{ MPa}$$

However, the absolute maximum shear stress in the cylinder wall is

$$\tau_{\text{max}} = \frac{\sigma_1}{2} = 67.5\text{ MPa}$$

as obtained from Eq. (6-44).

(c) An element oriented at an angle $\theta = 35°$ so that its sides are parallel and perpendicular to the weld is shown in Fig. 6-29c. Either the stress transformation equations (Eqs. 6-4) or Mohr's circle may be used to obtain the normal and shear stresses acting on the side faces of this element. The Mohr's circle construction is shown in Fig. 6-30. Point A represents the stress σ_2 on the x face ($\theta = 0$), and point B represents the stress σ_1 on the y face ($\theta = 90°$). A counterclockwise angle $2\theta = 70°$ on the circle locates point D, which corresponds to the stresses on the x_1 face ($\theta = 35°$). Since the radius of the circle is

$$R = \frac{135\text{ MPa} - 67.5\text{ MPa}}{2} = 33.75\text{ MPa}$$

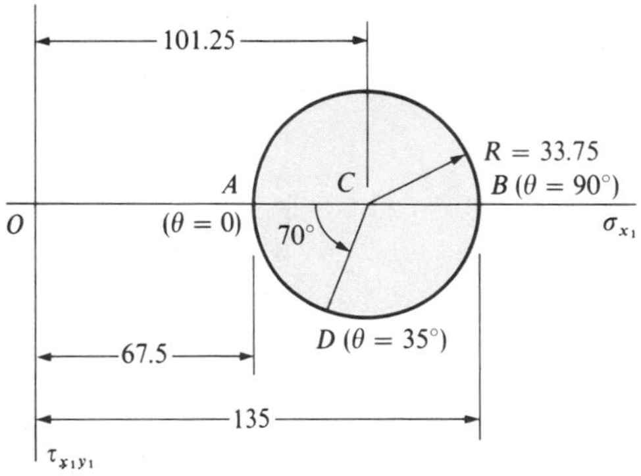

Fig. 6-30 Mohr's circle for the biaxial stress element of Fig. 6-29. (Note: All stresses have units of MPa.)

we see that the coordinates of point D are

$$\sigma_{x_1} = 101.25\text{ MPa} - (33.75\text{ MPa})\cos 70° = 89.7\text{ MPa}$$

$$\tau_{x_1y_1} = (33.75\text{ MPa})\sin 70° = 31.7\text{ MPa}$$

These stresses act on the x_1 face of the element (Fig. 6-29c). The normal stress on the y_1 face is found from the equality of the sum of the normal stresses on perpendicular planes:

$$\sigma_1 + \sigma_2 = \sigma_{x_1} + \sigma_{y_1}$$

from which

$$\sigma_{y_1} = \sigma_1 + \sigma_2 - \sigma_{x_1} = 135 \text{ MPa} + 67.5 \text{ MPa} - 89.7 \text{ MPa}$$
$$= 112.8 \text{ MPa}$$

Thus, the normal and shear stresses acting on planes parallel and perpendicular to the weld are as shown on the inclined stress element of Fig. 6-29c. We see that the tensile stress acting across the weld is 89.7 MPa and the shear stress acting along the weld is 31.7 MPa.

6.7 COMBINED LOADINGS (PLANE STRESS)

Structural members often are required to resist more than one type of loading. For example, a shaft in torsion may also be subjected to bending, or a beam may be subjected to the simultaneous action of bending moments and axial forces. The stress analysis of a member subjected to such **combined loadings** can usually be performed by superimposing the stresses due to each load acting separately. Superposition is permissible if the stresses are linear functions of the loads and if there is no interaction effect between the various loads (that is, if the stresses due to one load are not affected by the presence of any other loads). The latter requirement is usually met if the displacements and rotations of the structure are small.

The analysis begins with the determination of the stresses due to the axial forces, torques, shear forces, and bending moments, using the stress formulas derived in previous chapters. Then these stresses are combined at any particular point in the structure to obtain the resultant stresses at that point. Thus, the stresses σ_x, σ_y, and τ_{xy} acting on a stress element at the point are known. Then the stresses acting in inclined directions can be investigated by using either the stress-transformation equations or Mohr's circle. In particular, the principal stresses and maximum shear stresses can be calculated.

Any number of critical locations in the member can be analyzed in this manner, thereby either confirming the adequacy of the design or, if the stresses are too large or too small, showing that design changes are needed.

To illustrate the general method, consider the solid circular cantilever bar shown in Fig. 6-31a. The bar is loaded at the free end by a twisting couple T and a lateral bending force P. These loads produce at every cross section a bending moment M, a shear force V, and a twisting couple T, each of which produces stresses acting over the cross sections. If we isolate a stress element A at the top of the bar, we see that it is subjected

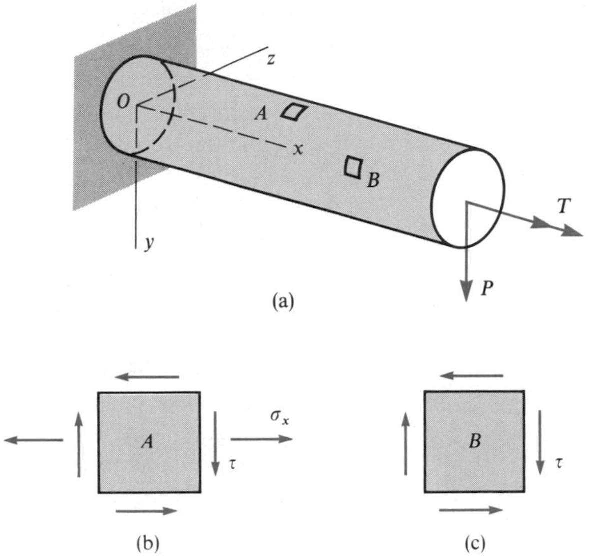

Fig. 6-31 Combined bending and torsion

to bending stresses $\sigma_x = Mr/I$ and torsional shear stresses $\tau = Tr/I_p$. In these expressions, r is the radius of the bar, I is the moment of inertia about the z axis (the neutral axis), and I_p is the polar moment of inertia. At the top of the bar, there are no shear stresses associated with the shear force V. Thus, the element at A is subjected to plane stress, as shown in Fig. 6-31b.

If the stresses σ_x and τ have been calculated at point A, we can proceed to determine the stresses on an element at A oriented at any desired angle. However, the principal stresses and the maximum shear stresses are often of the greatest interest, and they can be calculated either from the equations of Section 6.3 or from Mohr's circle. Then these maximum stresses can be compared with the allowable normal and shear stresses to verify the adequacy of the bar. Of course, in our example, the stresses are largest when element A is located at the fixed end of the beam where the bending moment M has its maximum value. Hence, the top of the beam at the support is one of the critical points where the stresses must be investigated.

Another critical point is on the side of the bar at the neutral axis (point B in Fig. 6-31a). At this location, the bending stress σ_x is zero but the shear stress produced by the shear force V has its largest value. The element at B is in a state of pure shear (Fig. 6-31c), with the resultant shear stress τ consisting of two parts: first, the shear stress τ_1 due to the torque T and obtained from the formula $\tau_1 = Tr/I_p$; and, second, the shear stress τ_2 due to the shear force V(equal to the load P) and obtained from the formula $\tau_2 = 4V/3A$ for a solid circular bar. Thus, the total shear stress acting on element B is $\tau = \tau_1 + \tau_2$. The principal stresses at this point occur on planes at 45° to the axis and have the same magnitudes as the shear stress itself. Of course, the maximum shear stress at B is the stress τ itself.

The maximum normal and shear stresses at B should be compared with those obtained for elements (such as A) at the top and bottom of the bar in order to ascertain the absolute maximum stresses for use in design.

The preceding discussion illustrates the general approach to problems involving combined loadings. The variety of practical situations is seemingly endless, so it is not worthwhile to derive specific formulas for design use. Instead, the best procedure is to analyze each structure at various critical points, and then compare the results. When selecting the points to be investigated, it is natural to choose those locations where either the normal or the shear stresses are a maximum. By using good judgment in the selection of the critical points, we can be reasonably certain of obtaining the absolute maximum stresses without the necessity of analyzing a large number of stress elements.

Example

A bracket consisting of two parts, a vertical bar of length $L_1 = 0.5$ m and a horizontal bar of length $L_2 = 0.3$ m, is subjected to a load $P = 1.0$ kN at its free end (Fig. 6-32a). The line of action of the load P is parallel to the yz plane and at an angle $\alpha = 30°$ from the vertical. The bracket has a solid circular cross section of diameter $d = 30$ mm. Considering only the action of the load P, calculate the maximum tensile stress σ_t, the maximum compressive stress σ_c, and the maximum in-plane shear stress τ_{max} in the bracket.

To analyze this bracket, we resolve the force P into two components (Fig. 6-32a). The component P_y acts vertically downward, and the component P_z acts horizontally in the negative z direction. With $P = 1.0$ kN and $\alpha = 30°$, these forces are

$$P_y = P\cos\alpha = 0.8660 \text{ kN} \qquad P_z = P\sin\alpha = 0.5 \text{ kN}$$

Our approach is to determine the stresses caused by P_y and P_z acting separately and then to combine those stresses to obtain the total stresses.

From an inspection of the bracket in Fig. 6-32a, we see that the maximum stresses occur at the base, which is the part of the bracket that is farthest from the line of action of the force P. Furthermore, the forces P_y and P_z produce bending about the z and x axes, respectively, as explained later in more detail, which means that the maximum bending stresses occur at points A, B, C, and D (Figs. 6-32a and b). Therefore, we expect the maximum principal stresses and maximum shear stresses to occur at those same points. Thus, points A, B, C, and D are the critical points to be investigated in this analysis.

The force P_y produces both an axial force and a bending moment in the vertical member (Fig. 6-32c). The axial force is compressive and equal to P_y, and the bending moment is equal to P_yL_2. The latter causes bending in the xy plane with the z axis as the neutral axis; this bending produces a compressive stress at point B, an equal tensile stress at point D, and no stresses at points A and C. Thus,

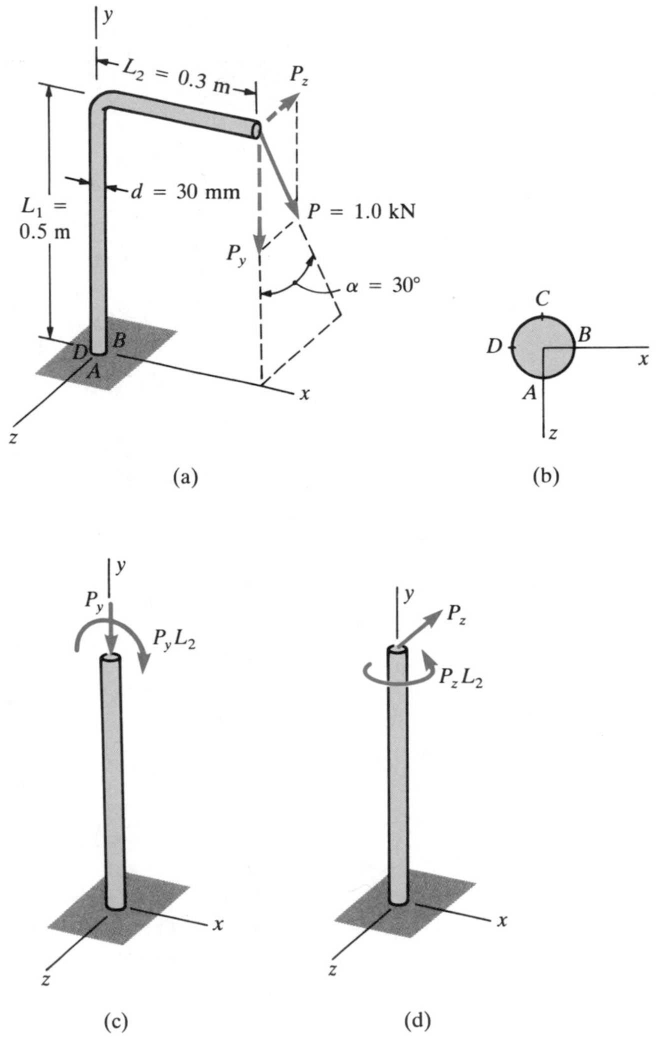

Fig. 6-32 Example. (a) Bracket of solid circular cross section (d = diameter) subjected to an inclined load P having components P_y and P_z; (b) cross section at base of bracket showing critical points A, B, C, and D; (c) axial force and bending moment at the top of the vertical member due to the force P_y; and (d) lateral force and twisting couple at the top of the vertical member due to the force P_z.

the stresses at A, B, C, and D due to the force P_y acting alone are

$$\sigma_{ay} = \sigma_{cy} = -\frac{P_y}{A} \qquad \sigma_{by} = -\frac{P_y}{A} - \frac{P_yL_2}{S} \qquad \sigma_{dy} = -\frac{P_y}{A} + \frac{P_yL_2}{S}$$

in which $A = \pi d^2/4 = 706.9 \times 10^{-6}\ \text{m}^2$ is the cross-sectional area and $S = \pi d^3/32 = 2.651 \times 10^{-6}\ \text{m}^3$ is the section modulus. Upon substituting numerical values into the preceding equations, we obtain

$$\sigma_{ay} = \sigma_{cy} = -1.23\ \text{MPa}$$

$$\sigma_{by} = -1.23\ \text{MPa} - 98.01\ \text{MPa} = -99.24\ \text{MPa}$$

$$\sigma_{dy} = -1.23\ \text{MPa} + 98.01\ \text{MPa} = 96.79\ \text{MPa}$$

These stresses are shown in Fig. 6-33a.

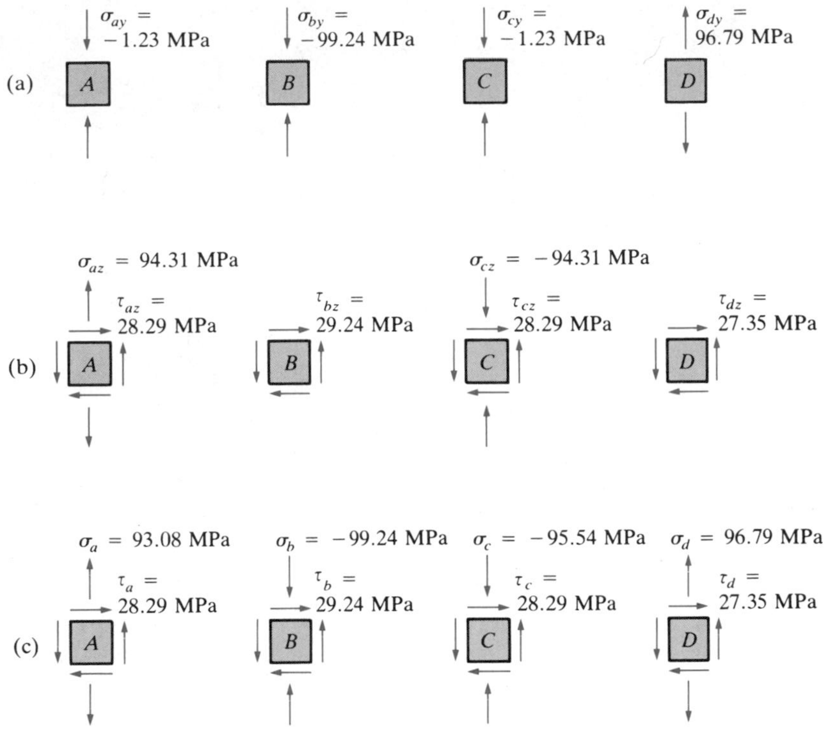

Fig. 6-33 Example (continued). Stress elements at points A, B, C, and D of the bracket shown in Fig. 6-32. (a) Stresses produced by the force P_y acting alone; (b) stresses produced by the force P_z acting alone; and (c) combined stresses.

Next we consider the force P_z, which produces a lateral load equal to P_z and a twisting couple equal to P_zL_2 at the top of the vertical member (Fig. 6-32d). The lateral load creates a bending moment at the base equal to P_zL_1 and a shear force equal to P_z. The bending moment results in tension at A, compression at C, and no bending stresses at B and D. The shear force results in maximum shear stresses at B and D (given by the shear formula for circular bars, $\tau_{max} = 4V/3A$, Eq. 5-24) and no shear stresses at A and C. The twisting couple P_zL_2 results in torsional shear stresses (given by the torsion formula, $\tau_{max} = 16T/\pi d^3$, Eq. 3-10) at all four points. Thus, the stresses due to the force P_z are as follows:

$$\sigma_{az} = \frac{P_zL_1}{S} = 94.31 \text{ MPa} \qquad \sigma_{cz} = -\frac{P_zL_1}{S} = -94.31 \text{ MPa} \qquad \sigma_{bz} = \sigma_{dz} = 0$$

$$\tau_{az} = \tau_{cz} = \frac{16P_zL_2}{\pi d^3} = 28.29 \text{ MPa}$$

$$\tau_{bz} = \frac{4P_z}{3A} + \frac{16P_zL_2}{\pi d^3} = 0.94 \text{ MPa} + 28.29 \text{ MPa} = 29.24 \text{ MPa}$$

$$\tau_{dz} = -\frac{4P_z}{3A} + \frac{16P_zL_2}{\pi d^3} = -0.94 \text{ MPa} + 28.29 \text{ MPa} = 27.35 \text{ MPa}$$

as shown in Fig. 6-33b.

The combined normal stresses at points A, B, C, and D are obtained by adding the normal stresses due to P_y and P_z:

$$\sigma_a = \sigma_{ay} + \sigma_{az} = -1.23 \text{ MPa} + 94.31 \text{ MPa} = 93.08 \text{ MPa}$$

$$\sigma_b = \sigma_{by} + \sigma_{bz} = -99.24 \text{ MPa} + 0 = -99.24 \text{ MPa}$$

$$\sigma_c = \sigma_{cy} + \sigma_{cz} = -1.23 \text{ MPa} - 94.31 \text{ MPa} = -95.54 \text{ MPa}$$

$$\sigma_d = \sigma_{dy} + \sigma_{dz} = 96.79 \text{ MPa} + 0 = 96.79 \text{ MPa}$$

These stresses are pictured in Fig. 6-33c.

The final shear stresses at A, B, C, and D (see Fig. 6-33c) are the same as those due to P_z alone (Fig. 6-33b) because P_y produces no shear stresses.

Now that we have determined all stresses acting at the critical points, we can investigate the principal stresses and maximum shear stresses. Beginning with point A, we obtain the principal stresses from Eq. (6-13), which in this case reduces to

$$\sigma_{1,2} = \frac{\sigma_a}{2} \pm \sqrt{\left(\frac{\sigma_a}{2}\right)^2 + \tau_a^2} = 46.54 \text{ MPa} \pm 54.47 \text{ MPa}$$

from which we get

$$\sigma_1 = 101.0 \text{ MPa} \quad \text{and} \quad \sigma_2 = -7.9 \text{ MPa}$$

Also, the maximum in-plane shear stress is obtained from Eq. (6-20), which reduces to

$$\tau = \sqrt{\left(\frac{\sigma_a}{2}\right)^2 + \tau_a^2} = 54.47 \text{ MPa}$$

The maximum stresses for the remaining three points are found in the same manner; the results are

Point B: $\sigma_1 = 8.0$ MPa	$\sigma_2 = -107.2$ MPa	$\tau = 57.6$ MPa
Point C: $\sigma_1 = 7.8$ MPa	$\sigma_2 = -103.3$ MPa	$\tau = 55.5$ MPa
Point D: $\sigma_1 = 104.0$ MPa	$\sigma_2 = -7.2$ MPa	$\tau = 55.6$ MPa

Comparing the maximum stresses at the four critical points leads to the following results:

1. The maximum tensile stress occurs at point D: $\sigma_t = 104.0$ MPa
2. The maximum compressive stress occurs at point B: $\sigma_c = -107.2$ MPa
3. The maximum shear stress occurs at point B: $\tau_{max} = 57.6$ MPa

Thus, the maximum stresses in the bracket have been calculated.

If desired, the principal stresses and maximum shear stresses can be shown acting on stress elements oriented in the proper directions, as was done in earlier sections. Also, with foresight and good judgment, some of the critical points may be eliminated from consideration, thereby reducing the amount of calculating effort required.

6.8 PRINCIPAL STRESSES IN BEAMS

The normal and shear stresses acting at any point in the cross section of a beam can be obtained from the flexure and shear formulas ($\sigma = My/I$ and $\tau = VQ/Ib$). The normal stress is a maximum at the outer edges of the beam and equals zero at the neutral axis, whereas the shear stress is zero at the outer edges and usually is a maximum at the neutral axis. In most circumstances, only these stresses are needed in order to design the beam. However, a more detailed study requires that we calculate the principal stresses and maximum shear stresses at various locations.

To visualize how the principal stresses vary in a beam, let us examine the stresses in a beam of rectangular cross section (Fig. 6-34a). Five points, denoted in the figure by A, B, C, D, and E, are selected at the cross section. Points A and E are at the upper and lower surfaces, point C is at

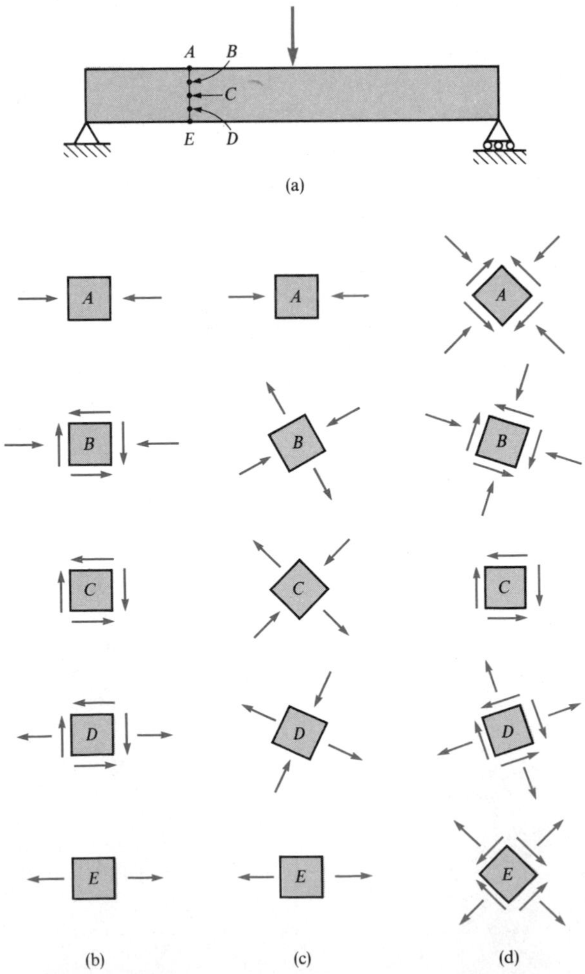

Fig. 6-34 Stresses in a beam of rectangular cross section: (a) points A, B, C, D, and E in the cross section, (b) normal and shear stresses acting on horizontal and vertical planes, (c) principal stresses, and (d) maximum shear stresses

the midheight of the beam, and points B and D are in between. The cross-sectional stresses at each of these points can be readily calculated if the bending moment and shear force are known. These stresses can be pictured as acting on plane stress elements having horizontal and vertical faces (Fig. 6-34b). The stress condition of the beam is uniaxial at the top and bottom of the beam and pure shear at the neutral axis. At other locations, both normal and shear stresses act on the stress element. To find the principal stresses and maximum shear stresses at such locations, we may use either the equations of plane stress (Section 6.3) or Mohr's circle (Section 6.4). The directions of the principal stresses at each point are shown schematically in Fig. 6-34c, and the maximum shear stresses are shown in Fig. 6-34d.

From the sketches in Fig. 6-34c, we can observe how the principal stresses change. At the top of the beam, the compressive principal stress acts in the horizontal direction. As we move toward the neutral axis, this principal stress becomes inclined to the horizontal, and at the neutral axis (point C) it acts at 45°. As we approach the bottom of the beam, the direction of the compressive principal stress approaches the vertical direction. The magnitude of this stress varies continuously between the top of the beam and the bottom (where it becomes zero). The maximum numerical value of this stress (in a rectangular beam) usually occurs at point A, although it is theoretically possible (for deep beams with high shear) for the maximum value to occur at a point such as B. Analogous comments apply to the tensile principal stress, which also varies in both magnitude and direction as we move from A to E.

By investigating the stresses at many cross sections of the beam, we can determine how the principal stresses vary throughout the beam. Then it is possible to construct two systems of orthogonal curves, called *stress trajectories*, that give the directions of the principal stresses. Two examples of stress trajectories for rectangular beams are shown in Fig. 6-35; part (a) shows a cantilever beam with a force acting at the free end, and part (b) shows a simple beam with a uniform load. In the figure, solid lines are used for tensile principal stresses and dashed lines for compressive principal stresses. Only the stresses obtained from the flexure and shear formulas are considered in these figures; the direct compressive stresses caused by the uniform load bearing on the top of the beam are disregarded. The two sets of curves always intersect at right angles, and every trajectory crosses the neutral surface at 45°. At the top and bottom surfaces of the beam, where the shear stress is zero, the trajectories become either horizontal or vertical.*

Another type of curve that may be plotted from the principal stresses is a *stress contour*, which is a curve that connects points of equal principal stress. Stress contours for a cantilever beam of rectangular cross section are shown in Fig. 6-36 (for tensile principal stresses only).

* Stress trajectories were originated by the German engineer Karl Culmann (1821–1881); see Refs. 6-8 and 6-9.

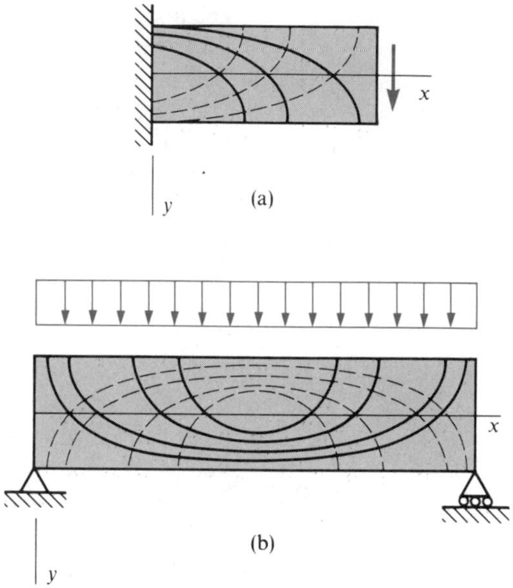

Fig. 6-35 Principal stress trajectories for beams of rectangular cross section: (a) cantilever beam, and (b) simple beam. (Solid lines represent tensile principal stresses, and dashed lines represent compressive principal stresses.)

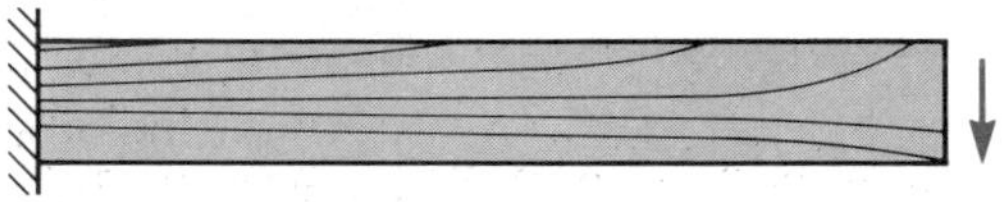

Fig. 6-36 Typical stress contours (tensile principal stresses only) for a cantilever beam

Beams having other shapes of cross sections can be analyzed for the principal stresses in a manner similar to that described for rectangular beams. The maximum principal stress in a wide-flange or I-beam usually occurs at the top or bottom but may sometimes occur in the web where it meets the flange. Similarly, the maximum shear stress usually occurs at the neutral axis, but under extraordinary loading conditions it may occur away from the neutral axis.

When analyzing a beam for the maximum stresses, remember that high stresses (or stress concentrations) exist near supports, points of load application, fillets, and holes. Such stresses are confined to the region very close to the discontinuity, and they cannot be calculated by the elementary beam formulas used in this chapter.

6.9 TRIAXIAL STRESS

An element of material subjected to normal stresses σ_x, σ_y, and σ_z acting in mutually perpendicular directions (Fig. 6-37a) is said to be in a state of **triaxial stress**. Note that no shear stresses act on the x, y, and z faces of the element; hence this stress condition is *not* the most general case of three-

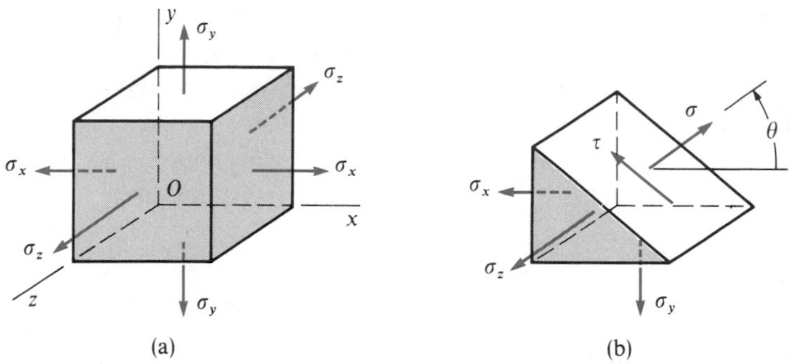

Fig. 6-37 Element in triaxial stress

dimensional stress (which is discussed in the next section). The absence of shear stresses shows that the stresses σ_x, σ_y, and σ_z are the **principal stresses** for the element.

If an inclined plane parallel to the z axis is cut through the element (Fig. 6-37b), the only stresses on the inclined face are the normal stress σ and shear stress τ acting in the xy plane. These stresses are the same as the stresses σ_{x_1} and $\tau_{x_1y_1}$ encountered previously in our discussions of plane stress. Because these stresses are found from equations of force equilibrium in the xy plane, they are independent of the stress σ_z. Thus, we conclude that we can use the transformation equations of plane stress, as well as Mohr's circle, when determining the stresses σ and τ. The same conclusion holds for the normal and shear stresses acting on inclined planes cut through the element parallel to the x and y axes.

From the previous discussions of plane stress, we know that the maximum shear stresses occur on planes oriented at 45° to the principal planes. To obtain these planes for an element in triaxial stress, we orient the elements at angles of 45° about the x, y, and z axes (Fig. 6-37a). For example, let us consider a 45° rotation about the z axis; then the **maximum shear stresses** acting on an element at this orientation are

$$(\tau_{\max})_z = \pm\frac{\sigma_x - \sigma_y}{2} \tag{6-47a}$$

Similarly, if we rotate about the x axis through an angle of 45°, we obtain the following maximum shear stresses:

$$(\tau_{\max})_x = \pm\frac{\sigma_y - \sigma_z}{2} \tag{6-47b}$$

Finally, rotating about the y axis through a 45° angle gives the stresses

$$(\tau_{\max})_y = \pm\frac{\sigma_x - \sigma_z}{2} \tag{6-47c}$$

The **absolute maximum shear stress** is the algebraically largest of the stresses determined from Eqs. (6-47). It is equal to one-half the difference between the algebraically largest and algebraically smallest of the three principal stresses.

The stresses acting on elements oriented at various angles to the x, y, and z axes can be visualized with the aid of Mohr's circles. For elements obtained by rotation about the z axis, the corresponding circle is labeled A in Fig. 6-38; this circle is drawn for the case in which $\sigma_x > \sigma_y$ and both σ_x and σ_y are tension. In a similar manner, we can construct circles B and C for elements obtained by rotations about the x and y axes, respectively. The radii of the circles represent the maximum shear stresses given by Eqs. (6-47), and the absolute maximum shear stress is equal to the radius of the largest circle. The normal stresses acting on the planes of maximum shear stresses have magnitudes given by the abscissas of the centers of the circles.

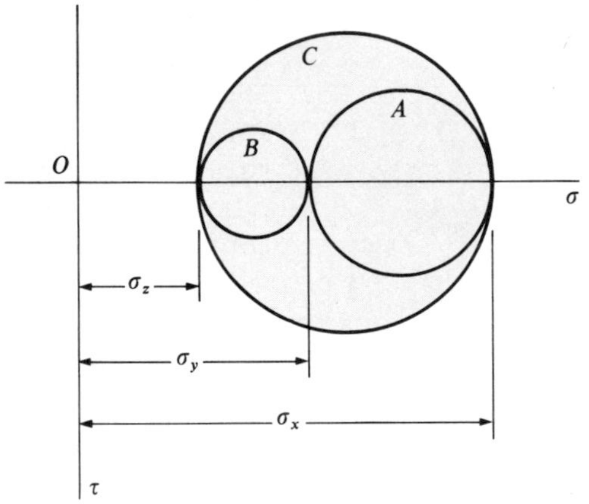

Fig. 6-38 Mohr's circles for an element in triaxial stress

In the preceding discussion, we considered only the stresses acting on planes obtained by rotating about the x, y, and z axes. Thus, every such plane is parallel to one of the axes. For instance, the inclined plane of Fig. 6-37b is parallel to the z axis, and its normal is parallel to the xy plane. Of course, we can also cut through the element in skew directions, so that the resulting inclined planes have normals that are skew to all three coordinate axes. The normal and shear stresses acting on such planes can be obtained by a more complicated three-dimensional analysis (see Section 6.10). However, those normal stresses are intermediate in value between the algebraically maximum and minimum principal stresses, and the shear stresses are less than the absolute maximum shear stress obtained from Eqs. (6-47).

HOOKE'S LAW FOR TRIAXIAL STRESS

The relationships between normal stresses and normal strains in the x, y, and z directions for triaxial stress can be obtained for a material that follows Hooke's law by using the same procedure as for plane stress (see Section 6.5). The strains produced by the stresses σ_x, σ_y, and σ_z acting independently are superimposed to obtain the resultant strains. Thus, we readily arrive at the following equations for the strains:

$$\epsilon_x = \frac{\sigma_x}{E} - \frac{\nu}{E}(\sigma_y + \sigma_z) \tag{6-48a}$$

$$\epsilon_y = \frac{\sigma_y}{E} - \frac{\nu}{E}(\sigma_z + \sigma_x) \tag{6-48b}$$

$$\epsilon_z = \frac{\sigma_z}{E} - \frac{\nu}{E}(\sigma_x + \sigma_y) \tag{6-48c}$$

In these equations, the standard sign conventions for σ and ϵ are used; that is, tensile stress σ and extensional strain ϵ are positive.

The preceding equations can be solved simultaneously for the stresses in terms of the strains:

$$\sigma_x = \frac{E}{(1+\nu)(1-2\nu)}[(1-\nu)\epsilon_x + \nu(\epsilon_y + \epsilon_z)] \tag{6-49a}$$

$$\sigma_y = \frac{E}{(1+\nu)(1-2\nu)}[(1-\nu)\epsilon_y + \nu(\epsilon_z + \epsilon_x)] \tag{6-49b}$$

$$\sigma_z = \frac{E}{(1+\nu)(1-2\nu)}[(1-\nu)\epsilon_z + \nu(\epsilon_x + \epsilon_y)] \tag{6-49c}$$

Equations (6-48) and (6-49) represent Hooke's law for triaxial stress.

UNIT VOLUME CHANGE

The unit volume change for an element in triaxial stress is obtained in the same manner as for plane stress (see Section 6.5). Beginning with a cube of unit dimensions (see Fig. 6-21), we see that its initial volume is $V_o = 1$ and its final volume is

$$V_f = (1 + \epsilon_x)(1 + \epsilon_y)(1 + \epsilon_z) \tag{a}$$

The **unit volume change** is defined as

$$e = \frac{\Delta V}{V_o} = \frac{V_f - V_o}{V_o} = \frac{V_f}{V_o} - 1 \tag{6-50}$$

which, upon substitution from Eq. (a), becomes

$$\begin{aligned} e &= (1 + \epsilon_x)(1 + \epsilon_y)(1 + \epsilon_z) - 1 \\ &= \epsilon_x + \epsilon_y + \epsilon_z + \epsilon_x\epsilon_y + \epsilon_x\epsilon_z + \epsilon_y\epsilon_z + \epsilon_x\epsilon_y\epsilon_z \end{aligned} \tag{6-51}$$

When the strains are small quantities, we may disregard the terms containing their products and obtain the following expression for the unit volume change:

$$e = \epsilon_x + \epsilon_y + \epsilon_z \tag{6-52}$$

Substituting for the strains from Eqs. (6-48), we get

$$e = \frac{1 - 2\nu}{E}(\sigma_x + \sigma_y + \sigma_z) \tag{6-53}$$

as the expression for the unit volume change in the general case of triaxial stress. (The quantity e is also called the **dilatation** or the **volumetric strain.**)

STRAIN ENERGY DENSITY

Let us assume for convenience that the triaxial stress element of Fig. 6-37 has unit dimensions. Then the forces acting on its faces are equal algebraically to the respective stresses. Each force moves through a distance equal to the corresponding strain as the stresses are applied to the element. The work done by these forces is the same as the strain energy density u of the element, inasmuch as the element has unit volume. Therefore, assuming that Hooke's law holds for the material, we obtain the following expression for the strain energy density:

$$u = \frac{1}{2}(\sigma_x\epsilon_x + \sigma_y\epsilon_y + \sigma_z\epsilon_z) \tag{6-54}$$

Substituting for the strains from Eqs. (6-48), we obtain the strain energy density in terms of the stresses:

$$u = \frac{1}{2E}(\sigma_x^2 + \sigma_y^2 + \sigma_z^2) - \frac{\nu}{E}(\sigma_x\sigma_y + \sigma_x\sigma_z + \sigma_y\sigma_z) \tag{6-55a}$$

In a similar manner, we can express the strain energy density in terms of the strains:

$$u = \frac{E}{2(1 + \nu)(1 - 2\nu)}[(1 - \nu)(\epsilon_x^2 + \epsilon_y^2 + \epsilon_z^2) + 2\nu(\epsilon_x\epsilon_y + \epsilon_x\epsilon_z + \epsilon_y\epsilon_z)] \tag{6-55b}$$

When calculating from these expressions, we must substitute the stresses and strains with their proper algebraic signs.

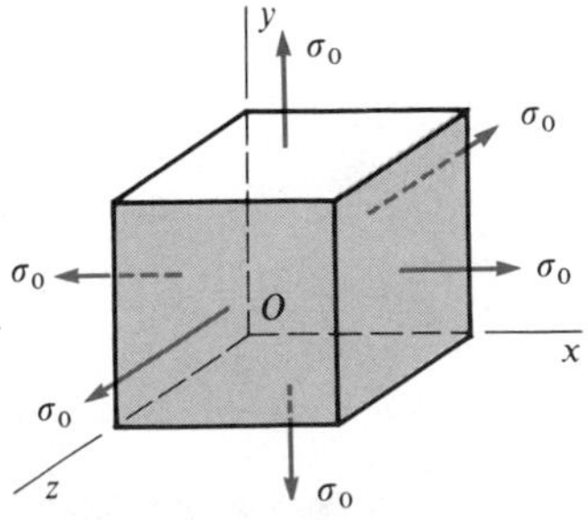

Fig. 6-39 Spherical stress

SPHERICAL STRESS

A special state of triaxial stress, called **spherical stress**, exists when all three normal stresses are equal (see Fig. 6-39):

$$\sigma_x = \sigma_y = \sigma_z = \sigma_0 \tag{6-56}$$

Under these stress conditions, any plane cut through the element is subjected to the same normal stress σ_0. Thus, we have equal normal stresses in every direction and no shear stresses. Every plane is a principal plane, and the three Mohr's circles shown in Fig. 6-38 reduce to a single point.

The normal strains ϵ_0 in spherical stress are also the same in all directions:

$$\epsilon_0 = \frac{\sigma_0}{E}(1 - 2\nu) \tag{6-57}$$

as obtained from Eqs. (6-48). Since there are no shear strains, a cube changes in size but remains a cube. In general, any body subjected to spherical stress will maintain its relative proportions but will either expand or contract in volume depending upon whether σ_0 is tension or compression.

The expression for the unit volume change can be obtained from Eq. (6-53) by substituting σ_0 for the stresses; the result is

$$e = \frac{\Delta V}{V_o} = \frac{3(1 - 2\nu)\sigma_0}{E} \tag{6-58a}$$

or

$$e = 3\epsilon_0 \tag{6-58b}$$

Equation (6-58a) is usually simplified by introducing a new quantity K called the **volume modulus of elasticity** or **bulk modulus of elasticity**:

$$K = \frac{E}{3(1 - 2\nu)} \tag{6-59}$$

With this notation, the expression for the volumetric strain becomes

$$e = \frac{\sigma_0}{K} \tag{6-60}$$

and, therefore,

$$K = \frac{\sigma_0}{e} \tag{6-61}$$

The volume modulus K can be defined as the ratio of the spherical stress to the volumetric strain, which is analogous to the definition of the modulus E. Note that the preceding formulas for e and K are based upon the assumption that the strains are small.

From Eq. (6-59) for K, we see that if Poisson's ratio equals 1/3, the moduli K and E are equal. If $\nu = 0$, then K has the value $E/3$. When $\nu = 0.5$, K becomes infinite, which corresponds to a rigid material having no change in volume. Thus, the theoretical maximum value of Poisson's ratio is 0.5.

If the spherical stress σ_0 is a pressure p, as in the case of an object submerged in a fluid, or rock deep within the earth, the stress state is known as **hydrostatic stress**.

*6.10 THREE-DIMENSIONAL STRESS

In the most general case of three-dimensional stress, a stress element will be subjected to normal and shear stresses on all faces (see Fig. 6-40). As described in Section 6.2 for plane stress, the shear stresses have two subscripts, the first denoting the plane on which the stress acts and the second identifying the direction in that plane. All stresses are shown acting in their positive directions in Fig. 6-40.

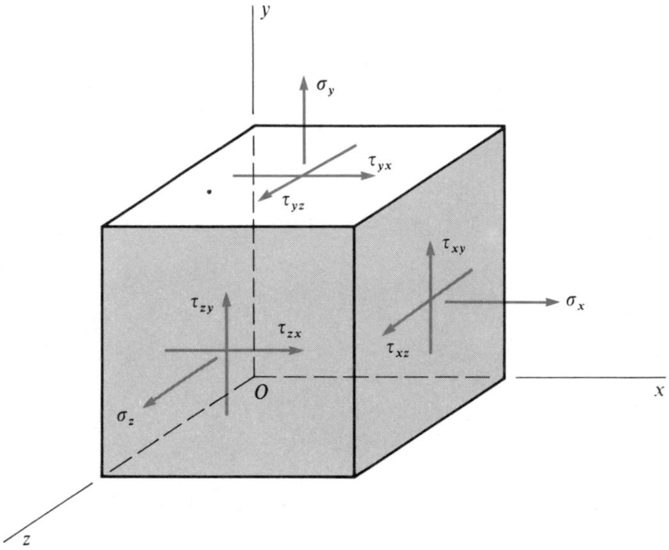

Fig. 6-40 Element subjected to three-dimensional stress. (Only stresses acting on the positive faces are shown. Oppositely directed stresses act on the negative faces.)

From equilibrium of the element, it can be proven that shear stresses acting on perpendicular planes and directed perpendicular to the line of intersection of the planes are equal in magnitude. Therefore, the following relationships hold:

$$\tau_{xy} = \tau_{yx} \qquad \tau_{xz} = \tau_{zx} \qquad \tau_{yz} = \tau_{zy} \tag{6-62}$$

This concept of equality of shear stresses has been discussed previously, and the first of Eqs. (6-62) was used in our discussion of plane stress (see Eq. 6-1).

Inclined planes cut through the element are subjected to normal and shear stresses, analogous to the stresses σ_{x_1} and $\tau_{x_1y_1}$ acting on inclined planes in plane stress (see Fig. 6-1). However, in the three-dimensional case, the normal to the inclined plane is not necessarily parallel to one of the coordinate planes; that is, it may be skew to all three axes. Nevertheless, the normal and shear stresses acting on any inclined plane can be determined by static equilibrium. The formulas for these stresses are rather long and complicated; they can be found in textbooks on theory of elasticity (such as Ref. 2-1).

Of particular importance are the three **principal stresses**, which are obtained as the three real roots of the following cubic equation:

$$\sigma^3 - A\sigma^2 + B\sigma - C = 0 \qquad \text{(6-63)}$$

in which

$$A = \sigma_x + \sigma_y + \sigma_z$$
$$B = \sigma_x\sigma_y + \sigma_x\sigma_z + \sigma_y\sigma_z - \tau_{xy}^2 - \tau_{xz}^2 - \tau_{yz}^2$$
$$C = \sigma_x\sigma_y\sigma_z + 2\tau_{xy}\tau_{xz}\tau_{yz} - \sigma_x\tau_{yz}^2 - \sigma_y\tau_{xz}^2 - \sigma_z\tau_{xy}^2$$

The quantities A, B, and C are known as **stress invariants**, because they do not change in value when the axes are rotated to new positions.

The procedure for evaluating the principal stresses for the three-dimensional stress element of Fig. 6-40 is as follows. After establishing the normal and shear stresses acting on the faces of the element, calculate the stress invariants A, B, and C. Then solve the cubic equation (6-63) for its three roots; these roots are the principal stresses σ_1, σ_2, and σ_3.

After finding the principal stresses, it is relatively easy to obtain the **maximum shear stresses**. Because no shear stresses act on the principal planes, it follows that an element oriented to the principal directions is in a state of triaxial stress. Therefore, the three maximum shear stresses (see Eqs. 6-47) are

$$(\tau_{max})_3 = \pm\frac{\sigma_1 - \sigma_2}{2} \qquad (\tau_{max})_2 = \pm\frac{\sigma_1 - \sigma_3}{2}$$
$$(\tau_{max})_1 = \pm\frac{\sigma_2 - \sigma_3}{2} \qquad \text{(6-64)}$$

The absolute maximum shear stress is the numerically largest of the stresses determined from these three equations.

Hooke's law for three-dimensional stress consists of the triaxial equations relating normal stresses and strains (Eqs. 6-48 and 6-49) and the equations relating shear stresses and strains. The latter have the same form as Eqs. (6-27) and (6-29) for plane stress, but now there are three sets of shear equations instead of one. Thus, the equations for three-dimensional stress, giving the strains in terms of the stresses, are as follows:

$$\epsilon_x = \frac{\sigma_x}{E} - \frac{v}{E}(\sigma_y + \sigma_z)$$
$$\epsilon_y = \frac{\sigma_y}{E} - \frac{v}{E}(\sigma_z + \sigma_x)$$
$$\epsilon_z = \frac{\sigma_z}{E} - \frac{v}{E}(\sigma_x + \sigma_y) \qquad \text{(6-65)}$$
$$\gamma_{xy} = \frac{\tau_{xy}}{G} \qquad \gamma_{xz} = \frac{\tau_{xz}}{G} \qquad \gamma_{yz} = \frac{\tau_{yz}}{G}$$

The inverse equations, giving the stresses in terms of the strains, are

$$\sigma_x = \frac{E}{(1+\nu)(1-2\nu)}[(1-\nu)\epsilon_x + \nu(\epsilon_y + \epsilon_z)]$$

$$\sigma_y = \frac{E}{(1+\nu)(1-2\nu)}[(1-\nu)\epsilon_y + \nu(\epsilon_z + \epsilon_x)] \qquad (6\text{-}66)$$

$$\sigma_z = \frac{E}{(1+\nu)(1-2\nu)}[(1-\nu)\epsilon_z + \nu(\epsilon_x + \epsilon_y)]$$

$$\tau_{xy} = G\gamma_{xy} \qquad \tau_{xz} = G\gamma_{xz} \qquad \tau_{yz} = G\gamma_{yz}$$

Equations (6-65) and (6-66) are often referred to as the **generalized Hooke's law**.

Since the shear strains produce no change in volume, the expression for the unit volume change e is the same as for triaxial stress:

$$e = \frac{\Delta V}{V_o} = (1+\epsilon_x)(1+\epsilon_y)(1+\epsilon_z) - 1 \qquad (6\text{-}67a)$$

or, for small strains,

$$e = \epsilon_x + \epsilon_y + \epsilon_z \qquad (6\text{-}67b)$$

Finally, the expression for strain energy density (see Eqs. 6-32 and 6-54) is

$$u = \frac{1}{2}(\sigma_x\epsilon_x + \sigma_y\epsilon_y + \sigma_z\epsilon_z + \tau_{xy}\gamma_{xy} + \tau_{xz}\gamma_{xz} + \tau_{yz}\gamma_{yz}) \qquad (6\text{-}68)$$

By substituting from Hooke's law, we can express u in terms of stresses or strains only; thus, in terms of stresses,

$$u = \frac{1}{2E}(\sigma_x^2 + \sigma_y^2 + \sigma_z^2) - \frac{\nu}{E}(\sigma_x\sigma_y + \sigma_x\sigma_z + \sigma_y\sigma_z)$$

$$+ \frac{1}{2G}(\tau_{xy}^2 + \tau_{xz}^2 + \tau_{yz}^2) \qquad (6\text{-}69a)$$

and, in terms of strains,

$$u = \frac{E}{2(1+\nu)(1-2\nu)}[(1-\nu)(\epsilon_x^2 + \epsilon_y^2 + \epsilon_z^2)$$

$$+ 2\nu(\epsilon_x\epsilon_y + \epsilon_x\epsilon_z + \epsilon_y\epsilon_z)] + \frac{G}{2}(\gamma_{xy}^2 + \gamma_{xz}^2 + \gamma_{yz}^2) \qquad (6\text{-}69b)$$

The equations of three-dimensional stress include all of the simpler stress states, such as triaxial stress and plane stress, as special cases.

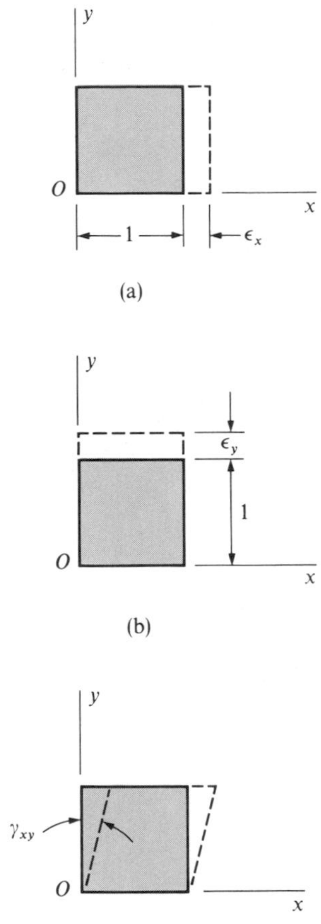

Fig. 6-41 Strain components ϵ_x, ϵ_y, and γ_{xy} in the xy plane

*6.11 PLANE STRAIN

The normal and shear strains at a point in a body vary with direction, in a manner analogous to that for stresses. In this section, we will derive the equations that give the strains in inclined directions in terms of the strains in the coordinate directions. These relationships are especially important in experimental investigations, where strains are measured with strain gages. The gages are oriented in specific directions, and it is usually necessary to calculate the strains in other directions.

Recall that in the xy plane three strain components may exist, as shown in the three parts of Fig. 6-41. These strains are the normal strain ϵ_x in the x direction, the normal strain ϵ_y in the y direction, and the shear strain γ_{xy}. An element of material subjected only to these strains is said to be in a state of **plane strain**. It follows that an element in plane strain has no normal strain ϵ_z and no shear strains γ_{xz} and γ_{yz} in the xz and yz planes, respectively. Thus, plane strain is defined by the following conditions:

$$\epsilon_z = 0 \qquad \gamma_{xz} = 0 \qquad \gamma_{yz} = 0 \tag{6-70}$$

The remaining strains (ϵ_x, ϵ_y, and γ_{xy}) may have nonzero values.

The preceding definition of plane strain is analogous to that for plane stress. In plane stress, the following stresses must be zero:

$$\sigma_z = 0 \qquad \tau_{xz} = 0 \qquad \tau_{yz} = 0 \tag{6-71}$$

whereas the remaining stresses (σ_x, σ_y, and τ_{xy}) may have nonzero values. A comparison of plane stress and plane strain is given in Fig. 6-42.

It should not be inferred from the similarities in the definitions of plane stress and plane strain that both occur simultaneously. In general, an element in plane stress undergoes a strain in the z direction (see Fig. 6-42); hence it clearly is not in plane strain. Also, most elements subjected to plane strain will have stresses σ_z acting on them because of the requirement that $\epsilon_z = 0$; again, we see that plane strain and plane stress do not occur simultaneously. An exception is when an element in plane stress is subjected to equal and opposite normal stresses (that is, when $\sigma_x = -\sigma_y$). In this special case, there is no normal strain in the z direction ($\epsilon_z = 0$; see Eq. 6-26c) and the element is in a state of plane s rain as well as plane stress. Another special case, albeit a hypothetical one, is when a material has $v = 0$; then every plane stress element is also in plane strain because $\epsilon_z = 0$ (see Eq. 6-26c).*

The stress transformation equations derived for plane stress in the xy plane (Eqs. 6-4) may also be used if a normal stress σ_z is present. The reason is that the stress σ_z does not enter the equations of equilibrium used in determining the stresses σ_{x_1} and $\tau_{x_1y_1}$ acting on inclined planes. An

* In this discussion, we are not considering the effects of temperature changes and prestrains, both of which produce additional strains that would alter some of our comments.

	Plane stress	Plane strain
	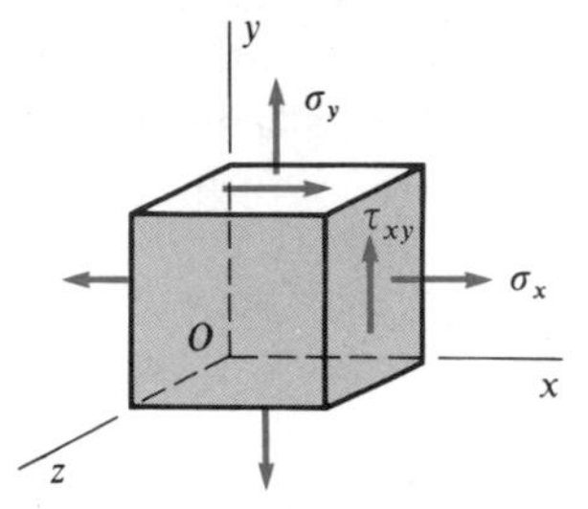	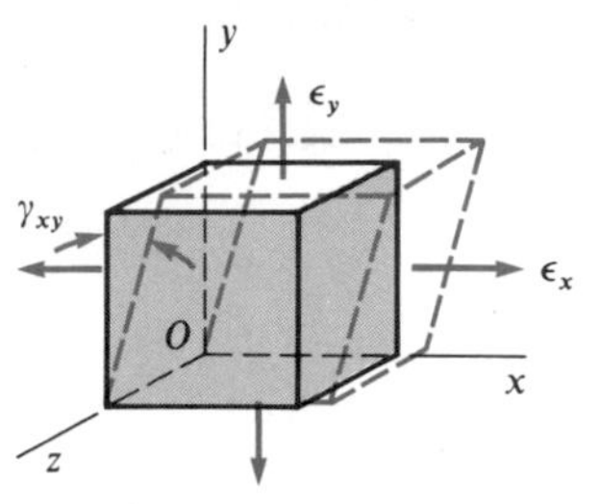
Stresses	$\sigma_z = 0$ $\tau_{xz} = 0$ $\tau_{yz} = 0$ σ_x, σ_y, and τ_{xy} may have nonzero values	$\tau_{xz} = 0$ $\tau_{yz} = 0$ σ_x, σ_y, σ_z, and τ_{xy} may have nonzero values
Strains	$\gamma_{xz} = 0$ $\gamma_{yz} = 0$ ϵ_x, ϵ_y, ϵ_z, and γ_{xy} may have nonzero values	$\epsilon_z = 0$ $\gamma_{xz} = 0$ $\gamma_{yz} = 0$ ϵ_x, ϵ_y, and γ_{xy} may have nonzero values

Fig. 6-42 Comparison of plane stress and plane strain

analogous situation exists for plane strain. Although we will derive the strain transformation equations for the case of plane strain, the equations actually are valid even when a strain ϵ_z exists. Therefore, the transformation equations for plane stress can be used for the stresses in the xy plane that occur in the case of plane strain, and the transformation equations for plane strain can be used for the strains in the xy plane that occur in the case of plane stress.

In the derivation of the transformation equations for plane strain, we will use the coordinate axes shown in Fig. 6-43. We will assume that the normal strains ϵ_x and ϵ_y and the shear strain γ_{xy} associated with the xy axes are known (see Fig. 6-41). The objective of the analysis is to determine the normal strain ϵ_{x_1} and shear strain $\gamma_{x_1y_1}$ associated with the x_1y_1 axes, which are rotated counterclockwise through an angle θ from the xy axes. (We need not derive a separate equation for the normal strain ϵ_{y_1}, because it can be obtained from the equation for ϵ_{x_1} by substituting $\theta + 90°$ for θ.)

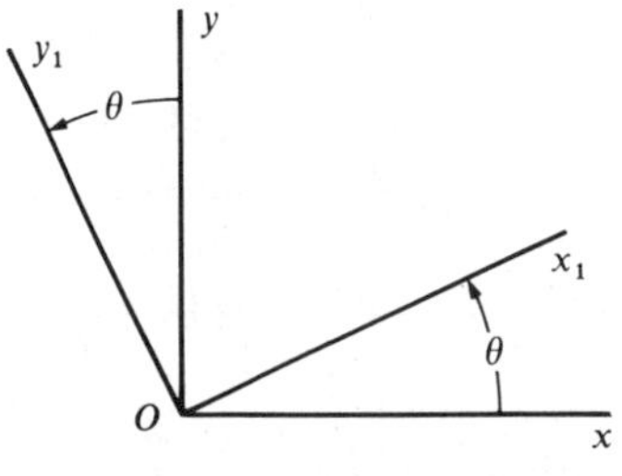

Fig. 6-43 Rotated axes x_1 and y_1

The positive z face of a plane-strain element having rectangular faces is shown in Fig. 6-44. The diagonal of the rectangle is in the direction of the x_1 axis, and the sides have lengths dx and dy. The strains ϵ_x, ϵ_y, and γ_{xy} in the xy plane produce an elongation of the element in the x direction equal to $\epsilon_x\,dx$ (Fig. 6-44a), an elongation in the y direction equal to $\epsilon_y\,dy$

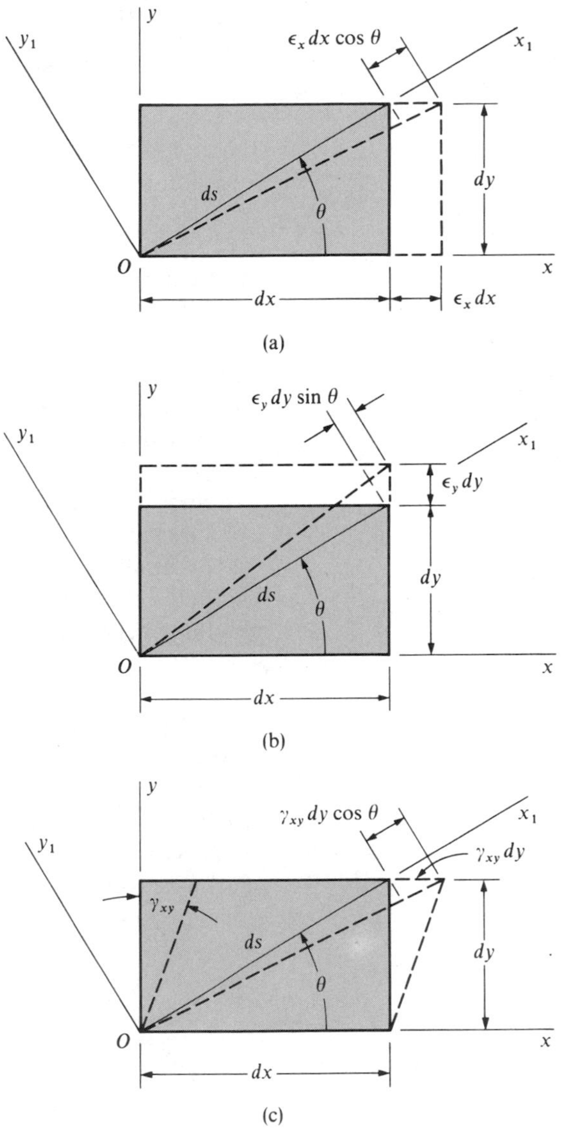

Fig. 6-44 Deformations of an element in plane strain due to: (a) normal strain ϵ_x, (b) normal strain ϵ_y, and (c) shear strain γ_{xy}

(Fig. 6-44b), and a decrease in the angle between the x and y faces equal to γ_{xy} (Fig. 6-44c). These deformations cause the diagonal to increase in length by amounts equal to $\epsilon_x dx \cos\theta$, $\epsilon_y dy \sin\theta$, and $\gamma_{xy} dy \cos\theta$, respectively. The total increase Δd in the length of the diagonal is the sum of these three expressions:

$$\Delta d = \epsilon_x dx \cos\theta + \epsilon_y dy \sin\theta + \gamma_{xy} dy \cos\theta$$

The normal strain ϵ_{x_1} in the x_1 direction is equal to this increase in length divided by the initial length ds of the diagonal:

$$\epsilon_{x_1} = \frac{\Delta d}{ds} = \epsilon_x \frac{dx}{ds} \cos\theta + \epsilon_y \frac{dy}{ds} \sin\theta + \gamma_{xy} \frac{dy}{ds} \cos\theta$$

Observing that $dx/ds = \cos\theta$ and $dy/ds = \sin\theta$, we obtain the following equation for the normal strain:

$$\epsilon_{x_1} = \epsilon_x \cos^2\theta + \epsilon_y \sin^2\theta + \gamma_{xy} \sin\theta\cos\theta \tag{6-72a}$$

As mentioned previously, the normal strain ϵ_{y_1} in the y_1 direction is obtained from this equation by substituting $\theta + 90°$ for θ.

Next, consider the shear strain $\gamma_{x_1y_1}$ associated with the rotated axes. This strain is equal to the decrease in angle between lines in the material that were initially along the lines of the x_1 and y_1 axes. To clarify this idea, let line Oa in Fig. 6-45 represent a line in the material that initially was along the x_1 axis (that is, along the diagonal of the element). The deformations pictured in Fig. 6-44 cause this line to rotate through a counterclockwise angle α from the x_1 axis (Fig. 6-45). Similarly, line Ob was originally along the y_1 axis, but because of the deformations it rotates through a clockwise angle β. The shear strain $\gamma_{x_1y_1}$ is the total decrease in the angle between the two lines that originally were at right angles; thus,

$$\gamma_{x_1y_1} = \alpha + \beta$$

The angle α can be found from Fig. 6-44 as follows. The strain ϵ_x (Fig. 6-44a) produces a clockwise rotation of line Oa equal to the distance $\epsilon_x d_x \sin\theta$ divided by ds. Therefore, the contribution of ϵ_x to the angle α is $-\epsilon_x d_x \sin\theta$. Similarly, the strain ϵ_y produces a counterclockwise rotation of line Oa equal to $\epsilon_y d_y \cos\theta$ divided by ds, and the strain γ_{xy} produces a clockwise rotation equal to $\gamma_{xy} dy \sin\theta$ divided by ds. Thus, the resultant counterclockwise rotation of line Oa is

$$\alpha = -\epsilon_x \frac{dx}{ds} \sin\theta + \epsilon_y \frac{dy}{ds} \cos\theta - \gamma_{xy} \frac{dy}{ds} \sin\theta$$

or

$$\begin{aligned} \alpha &= -\epsilon_x \sin\theta\cos\theta + \epsilon_y \sin\theta\cos\theta - \gamma_{xy} \sin^2\theta \\ &= -(\epsilon_x - \epsilon_y)\sin\theta\cos\theta - \gamma_{xy}\sin^2\theta \end{aligned}$$

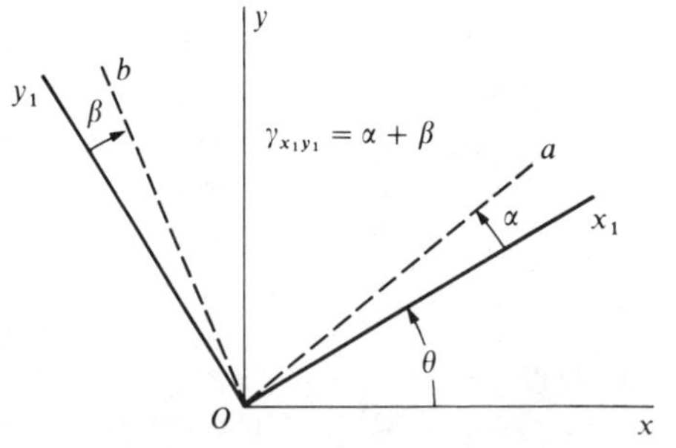

Fig. 6-45 Shear strain $\gamma_{x_1y_1}$ associated with x_1y_1 axes

The rotation of line Ob, which initially was at 90° to line Oa, can be found by substituting $\theta + 90°$ for θ in the expression for α. The resulting expression is counterclockwise when positive; hence it is equal to the negative of the angle β, since β is a clockwise rotation. Thus, we get

$$\beta = (\epsilon_x - \epsilon_y)\sin(\theta + 90°)\cos(\theta + 90°) + \gamma_{xy}\sin^2(\theta + 90°)$$
$$= -(\epsilon_x - \epsilon_y)\sin\theta\cos\theta + \gamma_{xy}\cos^2\theta$$

Adding α and β gives the shear strain $\gamma_{x_1y_1}$:

$$\gamma_{x_1y_1} = -2(\epsilon_x - \epsilon_y)\sin\theta\cos\theta + \gamma_{xy}(\cos^2\theta - \sin^2\theta)$$

To put the equation in a more useful form, we divide each term by 2 and write the equation as follows:

$$\frac{\gamma_{x_1y_1}}{2} = -(\epsilon_x - \epsilon_y)\sin\theta\cos\theta + \frac{\gamma_{xy}}{2}(\cos^2\theta - \sin^2\theta) \qquad \text{(6-72b)}$$

Equations (6-72) give the normal and shear strains with respect to rotated axes in terms of the strains oriented to the x and y axes. These equations are similar in form to Eqs. (6-3) for plane stress, with ϵ_{x_1} corresponding to σ_{x_1}, $\gamma_{x_1y_1}/2$ corresponding to $\tau_{x_1y_1}$, ϵ_x corresponding to σ_x, ϵ_y corresponding to σ_y, and $\gamma_{xy}/2$ corresponding to τ_{xy}.

The equations for plane strain can be expressed in terms of the angle 2θ by substituting the following trigonometric identities:

$$\cos^2\theta = \frac{1}{2}(1 + \cos 2\theta) \qquad \sin^2\theta = \frac{1}{2}(1 - \cos 2\theta)$$

$$\sin\theta\cos\theta = \frac{1}{2}\sin 2\theta$$

The **transformation equations for plane strain** now become

$$\epsilon_{x_1} = \frac{\epsilon_x + \epsilon_y}{2} + \frac{\epsilon_x - \epsilon_y}{2}\cos 2\theta + \frac{\gamma_{xy}}{2}\sin 2\theta \qquad \text{(6-73a)}$$

$$\frac{\gamma_{x_1y_1}}{2} = -\frac{\epsilon_x - \epsilon_y}{2}\sin 2\theta + \frac{\gamma_{xy}}{2}\cos 2\theta \qquad \text{(6-73b)}$$

These equations are the counterparts of Eqs. (6-4) for plane stress. The corresponding variables in the two sets of equations are listed in Table 6-1.

The analogy between the transformation equations for plane stress and plane strain shows that all of the observations madc in Sections 6.2, 6.3, and 6.4 concerning plane stress have their counterparts in plane strain.

Table 6-1 Corresponding variables in the transformation equations for plane stress (Eqs. 6-3 and 6-4) and plane strain (Eqs. 6-72 and 6-73)

Stresses	Strains
σ_x	ϵ_x
σ_y	ϵ_y
τ_{xy}	$\gamma_{xy}/2$
σ_{x_1}	ϵ_{x_1}
$\tau_{x_1y_1}$	$\gamma_{x_1y_1}/2$

For instance, the sum of the normal strains in perpendicular directions is a constant:

$$\epsilon_{x_1} + \epsilon_{y_1} = \epsilon_x + \epsilon_y \tag{6-74}$$

This equality can be verified easily by substituting the expressions for ϵ_{x_1} (Eq. 6-73a) and ϵ_{y_1} (obtained from Eq. 6-73a with θ replaced by $\theta + 90°$) into Eq. (6-74).

Principal strains exist in perpendicular directions calculated from the following equation (compare with Eq. 6-9):

$$\tan 2\theta_p = \frac{\gamma_{xy}}{\epsilon_x - \epsilon_y} \tag{6-75}$$

The principal strains can be calculated from the equation

$$\epsilon_{1,2} = \frac{\epsilon_x + \epsilon_y}{2} \pm \sqrt{\left(\frac{\epsilon_x - \epsilon_y}{2}\right)^2 + \left(\frac{\gamma_{xy}}{2}\right)^2} \tag{6-76}$$

which corresponds to Eq. (6-13) for the principal stresses. In the directions of the principal strains, the shear strains are zero. The two principal strains can be correlated with the two principal directions by the techniques described in Section 6.3 for stresses. (Note that the third principal strain is $\epsilon_z = 0$.)

The **maximum shear strains** in the xy plane are associated with axes at 45° to the directions of the principal strains. The algebraically maximum shear strain (in the xy plane) is given by the following equation:

$$\frac{\gamma_{max}}{2} = \sqrt{\left(\frac{\epsilon_x - \epsilon_y}{2}\right)^2 + \left(\frac{\gamma_{xy}}{2}\right)^2} \tag{6-77}$$

The minimum shear strain has the same magnitude but is negative. In the directions of maximum shear strain, the normal strains are equal to $(\epsilon_x + \epsilon_y)/2$.

An element in plane stress that is oriented to the principal directions (see Fig. 6-12b) has no shear stresses acting on its faces. Therefore, the shear strain $\gamma_{x_1y_1}$ for this element also is zero. It follows that the normal strains in this element are the principal strains. Thus, the principal strains occur in the same directions as the principal stresses.

Mohr's circle for plane strain is constructed in the same general manner as for plane stress, as illustrated in Fig. 6-46. Normal strains ϵ_{x_1} are plotted as the abscissas, and half the shear strains $(\gamma_{x_1y_1}/2)$ are plotted downward as the ordinates. The center C of the circle has an abscissa equal to $(\epsilon_x + \epsilon_y)/2$. Point A, representing the strains associated with the x direction ($\theta = 0$), has coordinates ϵ_x and $\gamma_{xy}/2$. Point B, at the opposite end of a diameter from A, has coordinates ϵ_y and $-\gamma_{xy}/2$, representing the

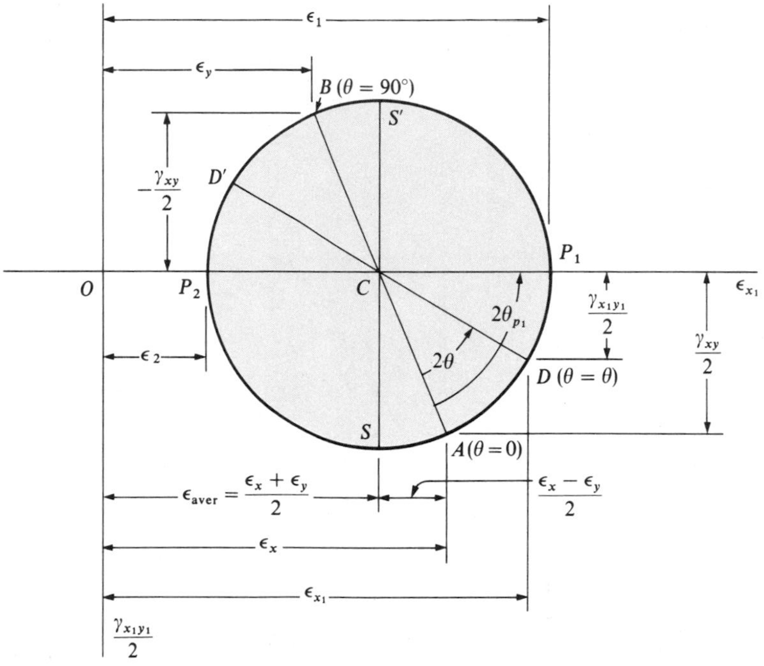

Fig. 6-46 Mohr's circle for plane strain

strains associated with a pair of axes rotated through an angle $\theta = 90°$. The strains associated with axes at an angle θ are given by point D, which is located by measuring an angle 2θ from radius CA. The principal strains are represented by points P_1 and P_2 and the maximum shear strains by points S and S'. All of these strains can be determined directly from the circle or from the equations given previously.

An important use of the transformation equations for strain and Mohr's circle is the interpretation of strain-gage measurements, which are discussed further in Example 2. However, references on experimental stress analysis should be consulted for detailed information about experimental techniques (see, for example, Refs. 6-10 and 6-11).

Example 1

An element of material subjected to plane strain has strains as follows: $\epsilon_x = 340 \times 10^{-6}$, $\epsilon_y = 110 \times 10^{-6}$, and $\gamma_{xy} = 180 \times 10^{-6}$. These strains are shown highly exaggerated in Fig. 6-47a, which portrays an element of unit dimensions aligned with the x and y axes. Since the edges of the element have unit lengths, the changes in dimensions are equal to the normal strains. For convenience, the shear strain is shown as the change in angle at the corner of the element located at the origin.

Calculate the following quantities: (a) the strains for an element oriented at an angle $\theta = 30°$, (b) the principal strains, and (c) the maximum shear strains. (Consider only the in-plane strains.)

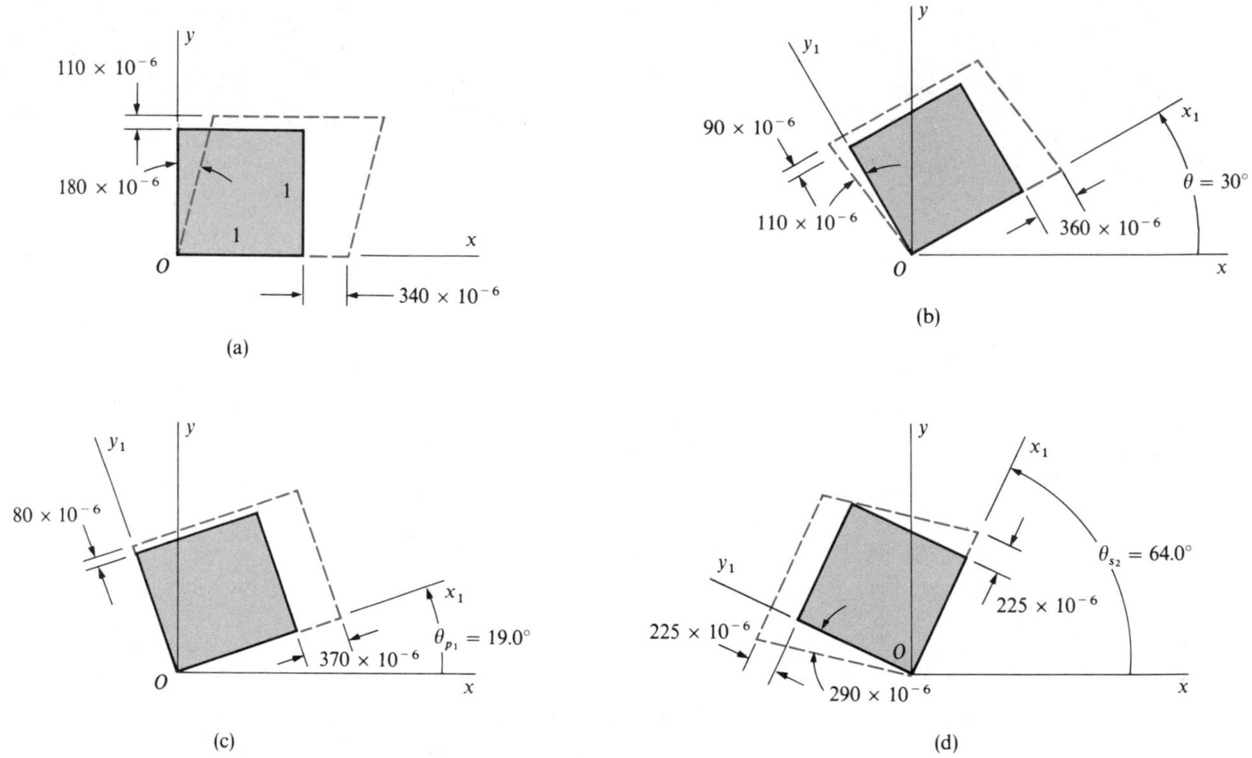

Fig. 6-47 Example 1. (a) Element in plane strain, (b) element at $\theta = 30°$, (c) principal strains, and (d) maximum shear strains. (Note: The edges of the elements have unit lengths.)

(a) The strains for an element oriented at an angle of 30° are found from the transformation equations (Eqs. 6-73). However, before substituting into those equations, we make the following preliminary calculations:

$$\frac{\epsilon_x + \epsilon_y}{2} = \frac{(340 + 110)10^{-6}}{2} = 225 \times 10^{-6}$$

$$\frac{\epsilon_x - \epsilon_y}{2} = \frac{(340 - 110)10^{-6}}{2} = 115 \times 10^{-6}$$

$$\frac{\gamma_{xy}}{2} = 90 \times 10^{-6}$$

Now we substitute into Eqs. (6-73) and get

$$\epsilon_{x_1} = \frac{\epsilon_x + \epsilon_y}{2} + \frac{\epsilon_x - \epsilon_y}{2}\cos 2\theta + \frac{\gamma_{xy}}{2}\sin 2\theta$$

$$= (225 \times 10^{-6}) + (115 \times 10^{-6})(\cos 60°) + (90 \times 10^{-6})(\sin 60°)$$

$$= 360 \times 10^{-6}$$

$$\frac{\gamma_{x_1y_1}}{2} = -\frac{\epsilon_x - \epsilon_y}{2}\sin 2\theta + \frac{\gamma_{xy}}{2}\cos 2\theta$$

$$= -(115 \times 10^{-6})(\sin 60^\circ) + (90 \times 10^{-6})(\cos 60^\circ)$$

$$= -55 \times 10^{-6}$$

or

$$\gamma_{x_1y_1} = -110 \times 10^{-6}$$

The strain ϵ_{y_1} can be obtained from Eq. (6-74):

$$\epsilon_{x_1} + \epsilon_{y_1} = \epsilon_x + \epsilon_y$$

Thus,

$$\epsilon_{y_1} = \epsilon_x + \epsilon_y - \epsilon_{x_1}$$

$$= (340 + 110 - 360)10^{-6} = 90 \times 10^{-6}$$

These strains are shown in Fig. 6-47b for an element at $\theta = 30^\circ$. Note that the angle at the corner of the element at the origin increases because $\gamma_{x_1y_1}$ is negative.

(b) The principal strains are calculated from Eq. (6-76), as follows:

$$\epsilon_{1,2} = \frac{\epsilon_x + \epsilon_y}{2} \pm \sqrt{\left(\frac{\epsilon_x - \epsilon_y}{2}\right)^2 + \left(\frac{\gamma_{xy}}{2}\right)^2}$$

$$= 225 \times 10^{-6} \pm \sqrt{(115 \times 10^{-6})^2 + (90 \times 10^{-6})^2}$$

$$= 225 \times 10^{-6} \pm 146 \times 10^{-6}$$

Therefore,

$$\epsilon_1 = 370 \times 10^{-6} \qquad \epsilon_2 = 80 \times 10^{-6}$$

The angles to the principal directions can be obtained from Eq. (6-75):

$$\tan 2\theta_p = \frac{\gamma_{xy}}{\epsilon_x - \epsilon_y} = \frac{180}{340 - 110} = 0.783$$

Hence, the two values of $2\theta_p$ between 0 and 360° are 38.0° and 218°, and the angles to the principal directions are

$$\theta_p = 19.0^\circ \text{ and } 109.0^\circ$$

To determine which value of θ_p is associated with each principal strain, we substitute θ_p into Eq. (6-73a) and solve for the strain. Thus, using $\theta_p = 19.0^\circ$, we get

$$\epsilon_{x_1} = \frac{\epsilon_x + \epsilon_y}{2} + \frac{\epsilon_x - \epsilon_y}{2}\cos 2\theta + \frac{\gamma_{xy}}{2}\sin 2\theta$$

$$= (225 \times 10^{-6}) + (115 \times 10^{-6})(\cos 38.0^\circ) + (90 \times 10^{-6})(\sin 38.0^\circ)$$

$$= 370 \times 10^{-6}$$

This result shows that the larger principal strain ϵ_1 is at the angle $\theta_{p_1} = 19.0^\circ$. Then the smaller strain ϵ_2 is 90° from that direction ($\theta_{p_2} = 109.0^\circ$). The principal strains are portrayed in Fig. 6-47c.

(c) The maximum shear strain is calculated from Eq. (6-77):

$$\frac{\gamma_{max}}{2} = \sqrt{\left(\frac{\epsilon_x - \epsilon_y}{2}\right)^2 + \left(\frac{\gamma_{xy}}{2}\right)^2} = 146 \times 10^{-6}$$

or

$$\gamma_{max} = 290 \times 10^{-6}$$

to two significant digits. The orientation of the element for maximum shear strains is at 45° to the principal directions; therefore, $\theta_s = 19.0° + 45° = 64.0°$ and $2\theta_s = 128.0°$. By substituting into Eq. (6-73b) we can determine the sign of the shear strain associated with this direction. The calculations are as follows:

$$\frac{\gamma_{x_1y_1}}{2} = -\frac{\epsilon_x - \epsilon_y}{2}\sin 2\theta + \frac{\gamma_{xy}}{2}\cos 2\theta$$

$$= -(115 \times 10^{-6})(\sin 128.0°) + (90 \times 10^{-6})(\cos 128.0°)$$

$$= -146 \times 10^{-6}$$

This result shows that an element oriented at an angle $\theta_{s_2} = 64.0°$ has the maximum negative shear strain.

We can arrive at the same result by observing that the angle θ_{s_1} to the direction of maximum positive shear strain is always 45° less than θ_{p_1}. Hence,

$$\theta_{s_1} = \theta_{p_1} - 45° = 19.0° - 45° = -26.0°$$

and

$$\theta_{s_2} = \theta_{s_1} + 90° = 64.0°$$

The corresponding shear strains are $\gamma_{max} = 290 \times 10^{-6}$ and $\gamma_{min} = -290 \times 10^{-6}$, respectively.

The normal strains on an element having the maximum and minimum shear strains are

$$\epsilon_{aver} = \frac{\epsilon_x + \epsilon_y}{2} = 225 \times 10^{-6}$$

A sketch of this element is given in Fig. 6-47d. Note that $\gamma_{max}/2$ is equal to one-half the difference of the principal strains.

In this example, we chose to solve for the strains by using the transformation equations. However, all of the results can be obtained just as easily from Mohr's circle.

Example 2

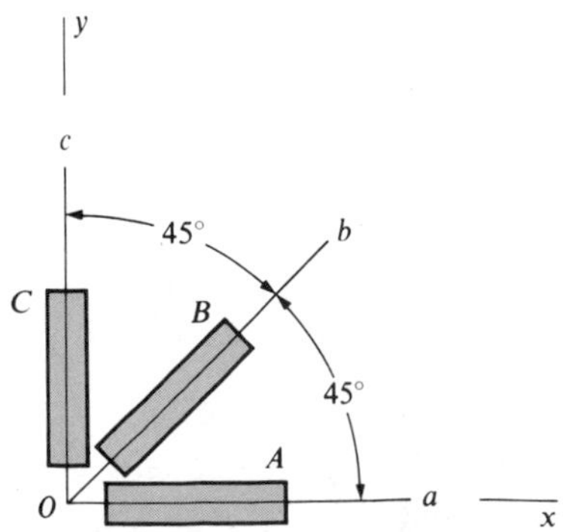

Fig. 6-48 Example 2. Strain rosette

An electrical-resistance **strain gage** is a small device that is glued to the surface of an object. The gage contains wires that are stretched or shortened when the object is strained at that point. The electrical resistance of the wires is altered when the wires change in length. This change in resistance is measured and converted into a strain measurement. The gages are extremely sensitive and can measure strains as small as 1×10^{-6}. Since each gage measures the normal strain in only one direction, it is often necessary to use three gages in combination, with each gage measuring the strain in a different direction. From three such measurements, it is possible to calculate the strains in any direction. A group of three gages arranged in a particular pattern is called a **strain rosette**. Because the rosette is mounted on the surface of the body, where the material is in plane stress, we can use the transformation equations for plane strain to calculate the strains in various directions at the surface.

A 45° strain rosette consists of three electrical-resistance strain gages arranged as shown in Fig. 6-48. Gages A, B, and C measure the normal strains ϵ_a,

ϵ_b, and ϵ_c in the directions of lines *Oa*, *Ob*, and *Oc*, respectively. Show how to obtain the strains ϵ_x, ϵ_y, and γ_{xy} associated with the *xy* axes.

Because gages *A* and *C* are aligned with the *x* and *y* axes, they give the strains ϵ_x and ϵ_y directly:

$$\epsilon_x = \epsilon_a \qquad \epsilon_y = \epsilon_c$$

To obtain the shear strain γ_{xy}, we may use the transformation Eq. (6-73a) for the strain ϵ_{x_1}:

$$\epsilon_{x_1} = \frac{\epsilon_x + \epsilon_y}{2} + \frac{\epsilon_x - \epsilon_y}{2}\cos 2\theta + \frac{\gamma_{xy}}{2}\sin 2\theta$$

For the angle $\theta = 45°$, we see that $\epsilon_{x_1} = \epsilon_b$; therefore, the preceding equation gives

$$\epsilon_b = \frac{\epsilon_a + \epsilon_c}{2} + \frac{\epsilon_a - \epsilon_c}{2}(\cos 90°) + \frac{\gamma_{xy}}{2}(\sin 90°)$$

Solving for γ_{xy}, we get

$$\gamma_{xy} = 2\epsilon_b - \epsilon_a - \epsilon_c$$

Thus, the strains ϵ_x, ϵ_y, and γ_{xy} are easily determined from the strain-gage readings. Knowing these strains, we can calculate the strains in any other directions by means of Mohr's circle or the transformation equations, as illustrated in the preceding example. Also, we can calculate the principal strains and maximum shear strains in the material.

PROBLEMS / CHAPTER 6

6.2-1 An element in plane stress is subjected to stresses $\sigma_x = 7500$ psi, $\sigma_y = 4500$ psi, and $\tau_{xy} = 3000$ psi, as shown in the figure. Determine the stresses acting on an element oriented at an angle $\theta = 50°$ from the x axis, where the angle θ is positive when counterclockwise. Also, show these stresses on a sketch of an element oriented at the angle θ.

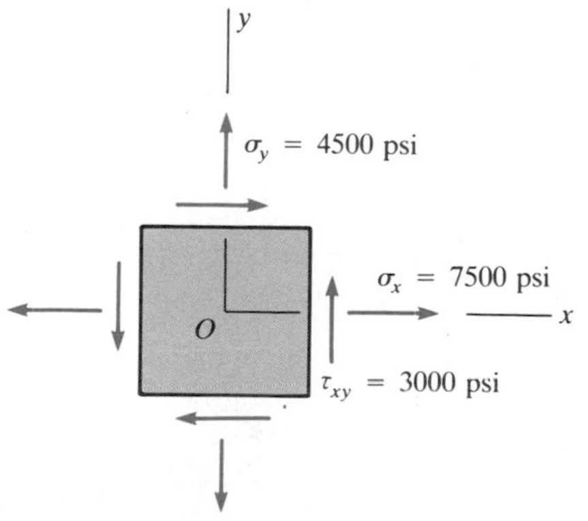

Prob. 6.2-1

6.2-2 Solve the preceding problem for $\sigma_x = 110$ MPa, $\sigma_y = 40$ MPa, $\tau_{xy} = 30$ MPa, and $\theta = 30°$ (see figure).

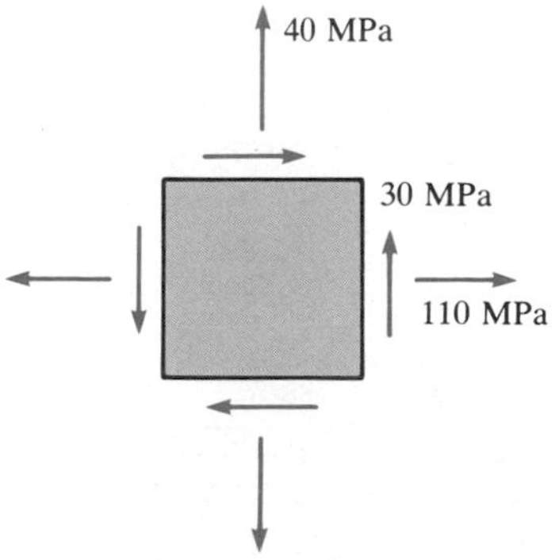

Prob. 6.2-2

6.2-3 Solve Problem 6.2-1 for $\sigma_x = 9{,}300$ psi, $\sigma_y = -22{,}100$ psi, $\tau_{xy} = -4{,}800$ psi, and $\theta = 60°$ (see figure).

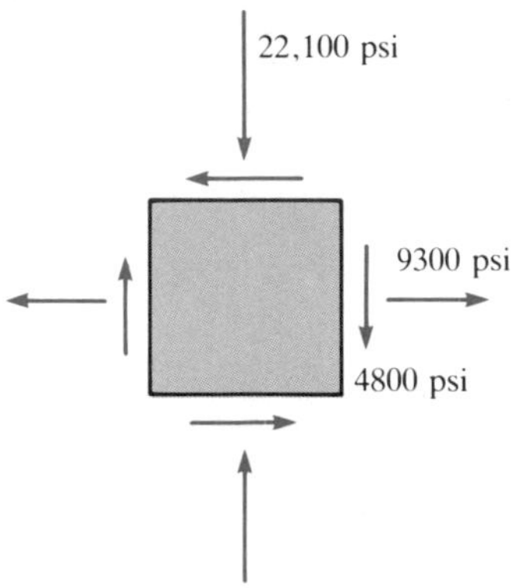

Prob. 6.2-3

6.2-4 Solve Problem 6.2-1 for $\sigma_x = 52$ MPa, $\sigma_y = -16$ MPa, $\tau_{xy} = -31$ MPa, and $\theta = 55°$ (see figure).

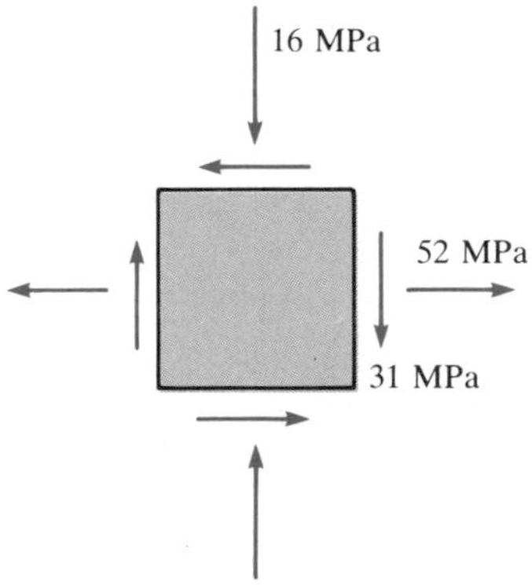

Prob. 6.2-4

6.2-5 A thin plate of dimensions 6 in. by 3 in. is formed by welding two triangular plates (see figure). The plate is subjected to a tensile stress of 2000 psi in the long direction and a compressive stress of 400 psi in the short direction. Determine the normal stress σ_w acting perpendicular to the line of the weld and the shear stress τ_w acting parallel to the weld.

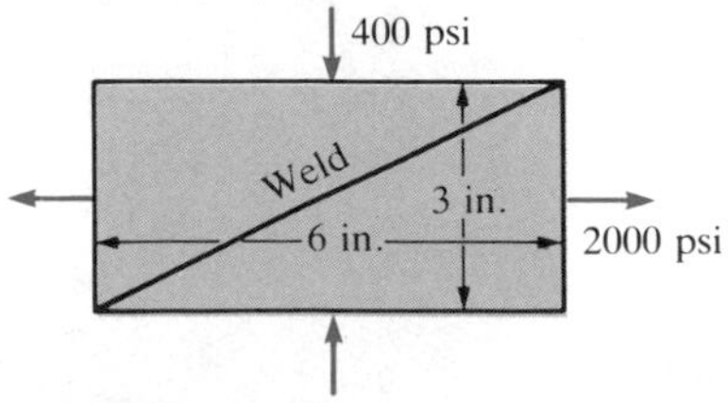

Prob. 6.2-5

6.2-6 Solve the preceding problem for a plate of dimensions 200 mm by 75 mm subjected to a compressive stress of 3 MPa in the long direction and a tensile stress of 10 MPa in the short direction (see figure).

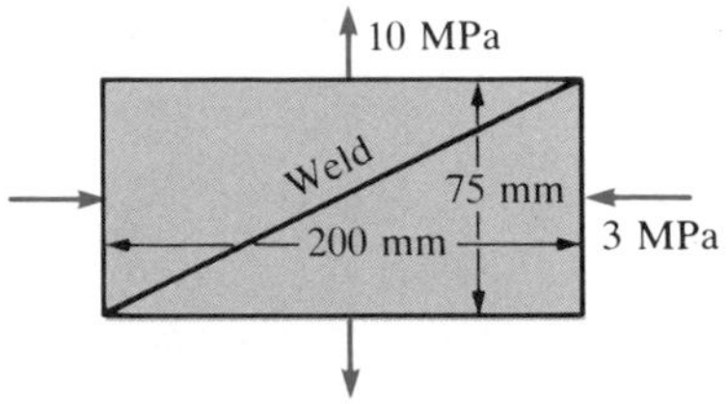

Prob. 6.2-6

6.2-7 At a particular point in a machine the material is in biaxial stress with $\sigma_x = 2000$ psi and $\sigma_y = -8000$ psi (see figure). What is the orientation of an inclined plane through the point such that the normal stress on the plane is zero? Sketch a stress element having that plane as one of its sides and show all stresses acting on the element.

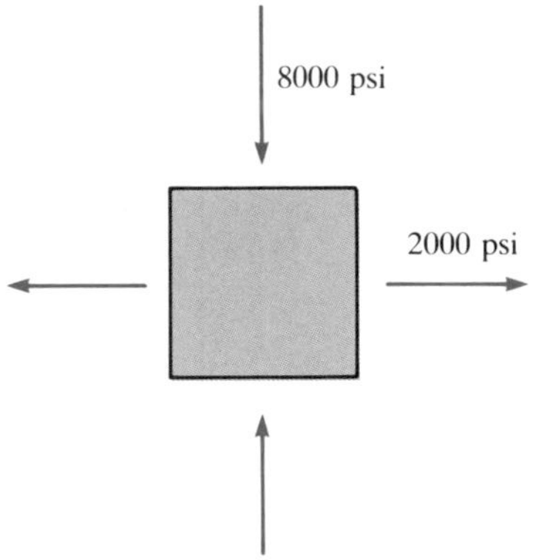

Prob. 6.2-7

6.2-8 Solve the preceding problem for $\sigma_x = 120$ MPa and $\sigma_y = -40$ MPa (see figure).

6.2-9 and 6.2-10 An element in plane stress is oriented at a known angle θ (see figure). On the inclined element, the normal and shear stresses have the magnitudes and directions shown in the figure. Determine the normal and shear stresses on an element whose sides are parallel to the xy axes; that is, determine σ_x, σ_y, and τ_{xy}.

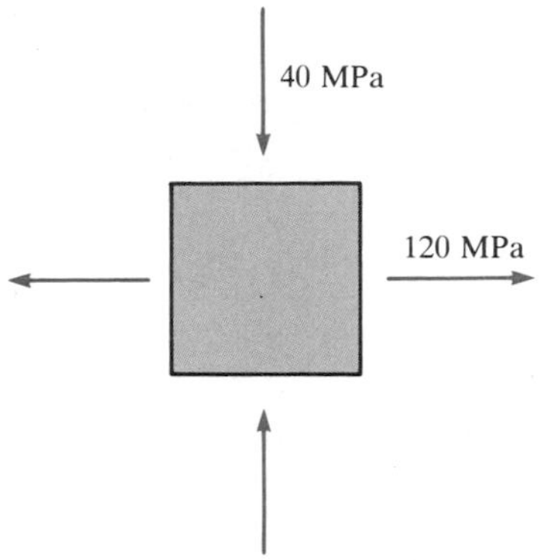

Prob. 6.2-8

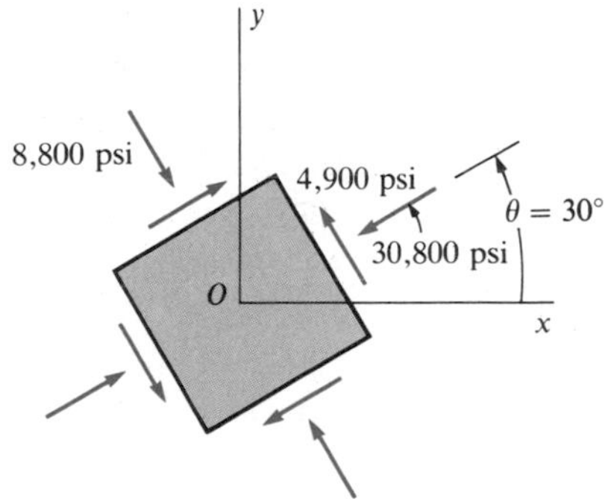

Prob. 6.2-9

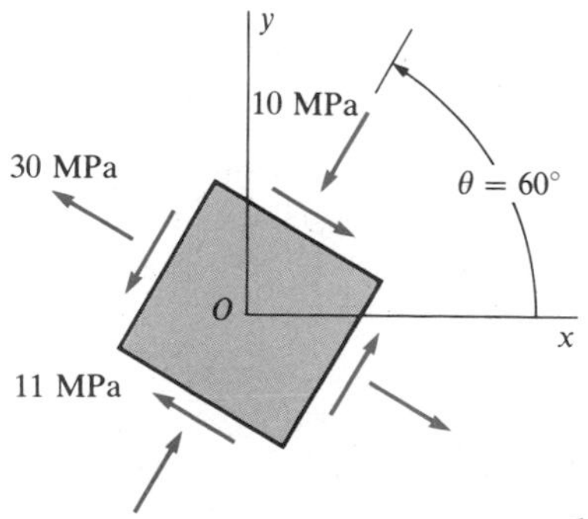

Prob. 6.2-10

6.2-11 A thin plate in plane stress is subjected to stresses σ_x, σ_y, and τ_{xy} (see figure). The stress σ_x equals 1500 psi (tension). It is also known that on planes oriented at $\theta = 40°$ and 80° the normal stresses are 4000 psi (tension). Determine the stresses σ_y and τ_{xy}.

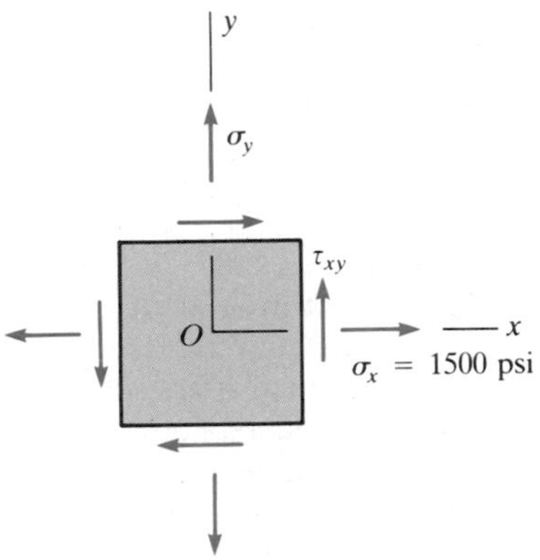

Prob. 6.2-11

6.2-12 Solve the preceding problem if $\sigma_x = 20$ MPa and it is known that on planes oriented at $\theta = 40°$ and 80° the normal stresses are 50 MPa (tension).

6.2-13 and 6.2-14 At a point in a structure subjected to plane stress, the stresses have the magnitudes and directions shown acting on element *A* in the first part of the figure. Element *B*, located at the same point in the structure, is oriented at an angle θ_1 (between zero and 90°) such that the stresses are as shown in the second part of the figure. Calculate the normal stress σ_b, the shear stress τ_b, and the angle θ_1.

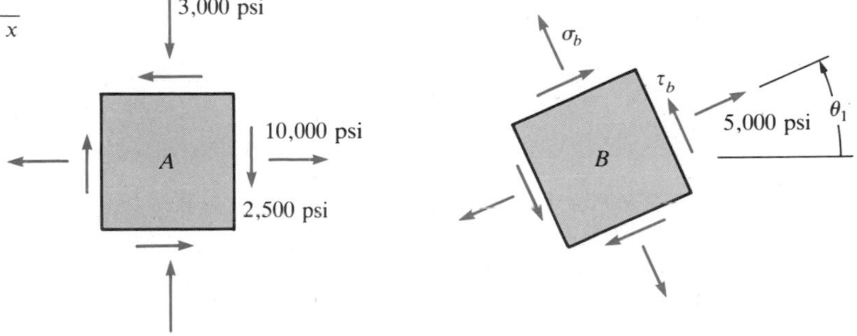

Prob. 6.2-13

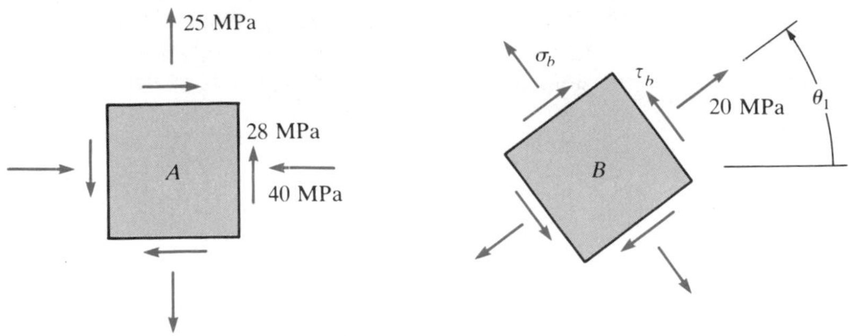

Prob. 6.2-14

***6.2-15** At a certain point in a structure the material is subjected to plane stress such that on a vertical plane through the point, the normal stress is 14 ksi tension and the shear stress is 12 ksi counterclockwise [see part (a) of the figure], and on a plane oriented at an angle θ_1, the normal stress is 21 ksi tension and the shear stress is 5 ksi counterclockwise [part (b) of the figure]. What is the angle θ_1?

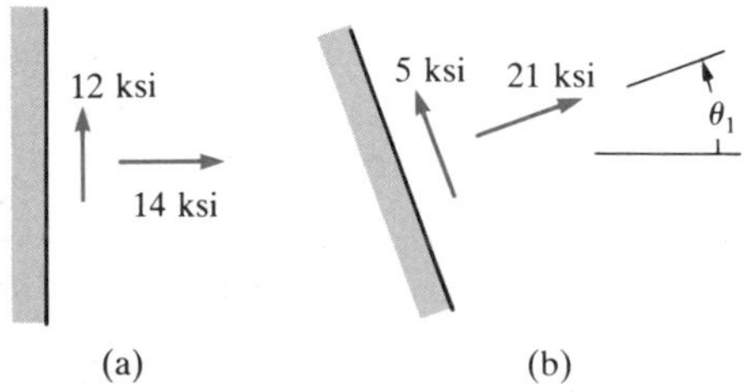

Prob. 6.2-15

***6.2-16** At a certain point in a structure the material is subjected to plane stress such that on a vertical plane through the point, the normal stress is 35 MPa tension and the shear stress is 30 MPa counterclockwise [see part (a) of the figure], and on a plane oriented at an angle θ_1, the normal stress is 42 MPa tension and the shear stress is 26 MPa clockwise [part (b) of the figure]. What is the angle θ_1?

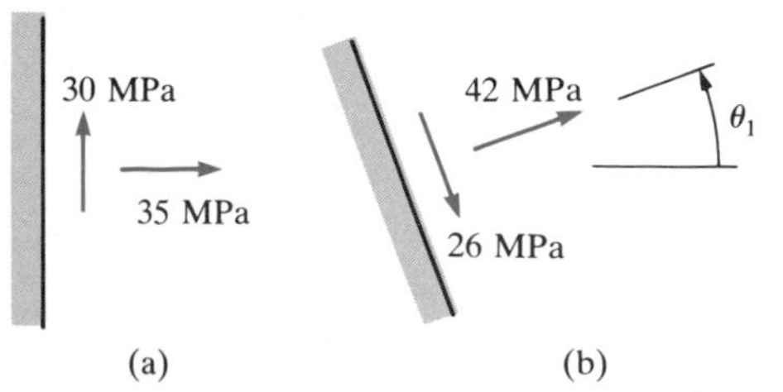

Prob. 6.2-16

When solving the problems for Section 6.3, consider only the in-plane stresses.

6.3-1 An element in plane stress is subjected to stresses $\sigma_x = 7500$ psi, $\sigma_y = 4500$ psi, and $\tau_{xy} = 3000$ psi (see figure for Problem 6.2-1). (a) Determine the principal stresses and show them on a sketch of a properly oriented element. (b) Determine the maximum shear stresses and associated normal stresses and show them on a sketch of a properly oriented element.

6.3-2 An element in plane stress is subjected to stresses $\sigma_x = 110$ MPa, $\sigma_y = 40$ MPa, and $\tau_{xy} = 30$ MPa (see figure for Problem 6.2-2). (a) Determine the principal stresses and show them on a sketch of a properly oriented element. (b) Determine the maximum shear stresses and associated normal stresses and show them on a sketch of a properly oriented element.

6.3-3 An element in plane stress is subjected to stresses $\sigma_x = 9{,}300$ psi, $\sigma_y = -22{,}100$ psi, and $\tau_{xy} = -4{,}800$ psi (see figure for Problem 6.2-3). (a) Determine the principal stresses and show them on a sketch of a properly oriented element. (b) Determine the maximum shear stresses and associated normal stresses and show them on a sketch of a properly oriented element.

6.3-4 An element in plane stress is subjected to stresses $\sigma_x = 52$ MPa, $\sigma_y = -16$ MPa, and $\tau_{xy} = -31$ MPa (see figure for Problem 6.2-4). (a) Determine the principal stresses and show them on a sketch of a properly oriented element. (b) Determine the maximum shear stresses and associated normal stresses and show them on a sketch of a properly oriented element.

6.3-5 to 6.3-12 An element in plane stress (see figure) is subjected to stresses σ_x, σ_y, and τ_{xy}. (a) Determine the principal stresses and show them on a sketch of a properly

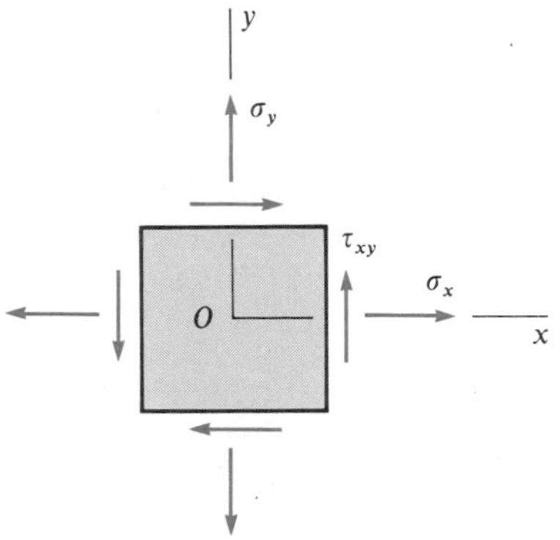

Probs. 6.3-5 to 6.3-12 and Probs. 6.3-15 to 6.3-20

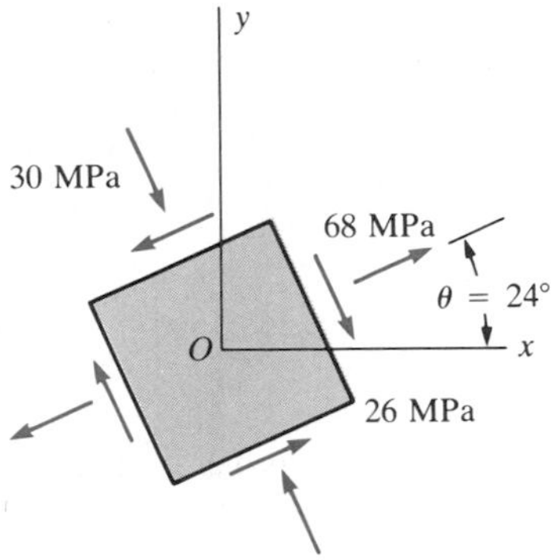

Prob. 6.3-14

oriented element. (b) Determine the maximum shear stresses and associated normal stresses and show them on a sketch of a properly oriented element.

6.3-5 $\sigma_x = 16{,}000$ psi, $\sigma_y = 2{,}000$ psi, $\tau_{xy} = 1{,}800$ psi
6.3-6 $\sigma_x = 72$ MPa, $\sigma_y = 30$ MPa, $\tau_{xy} = 28$ MPa
6.3-7 $\sigma_x = 6{,}500$ psi, $\sigma_y = -27{,}000$ psi, $\tau_{xy} = -8{,}000$ psi
6.3-8 $\sigma_x = 16$ MPa, $\sigma_y = -96$ MPa, $\tau_{xy} = -49$ MPa
6.3-9 $\sigma_x = -11{,}000$ psi, $\sigma_y = 31{,}000$ psi, $\tau_{xy} = -29{,}000$ psi
6.3-10 $\sigma_x = -81$ MPa, $\sigma_y = 129$ MPa, $\tau_{xy} = -36$ MPa
6.3-11 $\sigma_x = -21{,}200$ psi, $\sigma_y = -4{,}200$ psi, $\tau_{xy} = 3{,}900$ psi
6.3-12 $\sigma_x = -71$ MPa, $\sigma_y = -33$ MPa, $\tau_{xy} = 29$ MPa

6.3-13 and 6.3-14 At a certain point on the surface of a machine, the stresses acting on an element oriented at an angle θ to the x axis have the magnitudes and directions shown in the figure. (a) Determine the principal stresses and show them on a sketch of a properly oriented element. (b) Determine the maximum shear stresses and associated normal stresses and show them on a sketch of a properly oriented element.

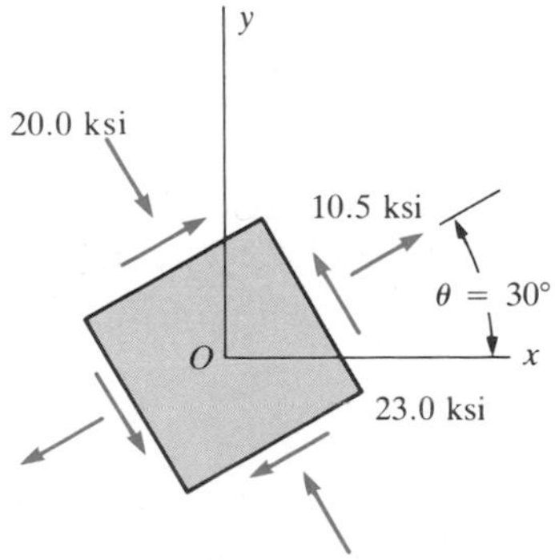

Prob. 6.3-13

6.3-15 For an element in plane stress (see figure) the normal stresses are $\sigma_x = -2{,}400$ psi and $\sigma_y = 10{,}400$ psi. Calculate the maximum permissible value τ_{max} of the shear stress τ_{xy} if the allowable shear stress in the material is 11,000 psi.

6.3-16 For an element in plane stress (see figure) the normal stresses are $\sigma_x = 66.2$ MPa and $\sigma_y = -114.1$ MPa. Calculate the maximum permissible value τ_{max} of the shear stress τ_{xy} if the allowable shear stress in the material is 150 MPa.

6.3-17 An element in plane stress (see figure) is subjected to stresses $\sigma_x = 7620$ psi and $\tau_{xy} = -2910$ psi. It is also known that one of the principal stresses equals 8400 psi in tension. (a) Determine the stress σ_y. (b) Determine the other principal stress and the orientation of the principal planes; then show the principal stresses on a sketch of a properly oriented element.

6.3-18 An element in plane stress (see figure) is subjected to stresses $\sigma_x = -68.5$ MPa and $\tau_{xy} = 39.2$ MPa. It is also known that one of the principal stresses equals 26.1 MPa in tension. (a) Determine the stress σ_y. (b) Determine the other principal stress and the orientation of the principal planes; then show the principal stresses on a sketch of a properly oriented element.

***6.3-19** At a certain point on the surface of a machine, an element in plane stress (see figure) is subjected to shear stresses $\tau_{xy} = -3000$ psi. It is also known that σ_y is a compressive stress. One of the principal stresses equals 7600 psi in tension and the maximum shear stress equals 5600 psi. (a) Determine the stresses σ_x and σ_y. (b) Determine the other principal stress and the orientation of the principal planes; then show the principal stresses on a sketch of a properly oriented element.

***6.3-20** At a certain point on the surface of a structure, an element in plane stress (see figure) is subjected to shear stresses $\tau_{xy} = -21$ MPa. It is also known that σ_x is a compressive stress. One of the principal stresses equals 52 MPa in tension and the maximum shear stress equals 38 MPa. (a) Determine the stresses σ_x and σ_y. (b) Determine the other principal stress and the orientation of the principal planes; then show the principal stresses on a sketch of a properly oriented element.

The problems for Section 6.4 are to be solved using Mohr's circle. Consider only the in-plane stresses.

6.4-1 Construct Mohr's circle for an element in uniaxial stress (see figure). (a) From the circle, derive the following stress transformation equations:

$$\sigma_{x_1} = \frac{\sigma_x}{2}(1 + \cos 2\theta) \qquad \tau_{x_1} = -\frac{\sigma_x}{2}\sin 2\theta$$

(b) Show from the circle that the principal stresses are $\sigma_1 = \sigma_x$ and $\sigma_2 = 0$. (c) Obtain from the circle the maximum shear stresses and associated normal stresses and show them on a sketch of a properly oriented element.

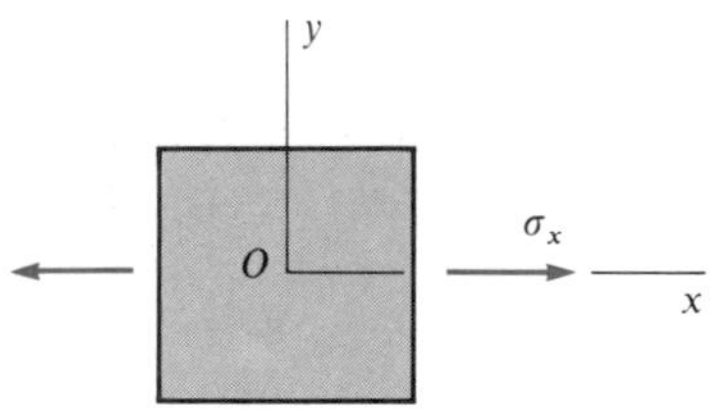

Prob. 6.4-1

6.4-2 Construct Mohr's circle for an element in pure shear (see figure). (a) From the circle, derive the following stress transformation equations:

$$\sigma_{x_1} = \tau_{xy}\sin 2\theta \qquad \tau_{x_1y_1} = \tau_{xy}\cos 2\theta$$

(b) Obtain from the circle the principal stresses and show them on a sketch of a properly oriented element. (c) Show from the circle that the maximum and minimum shear stresses are $\pm\tau_{xy}$.

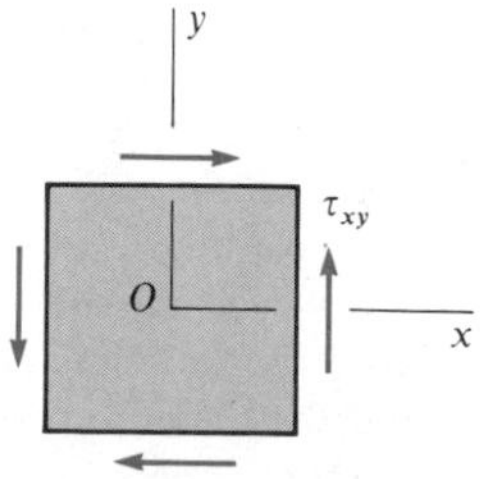

Prob. 6.4-2

6.4-3 Construct Mohr's circle for an element in biaxial stress (see figure), assuming $\sigma_x > \sigma_y$. (a) From the circle, derive the following stress transformation equations:

$$\sigma_{x_1} = \frac{\sigma_x + \sigma_y}{2} + \frac{\sigma_x - \sigma_y}{2}\cos 2\theta$$

$$\tau_{x_1y_1} = -\frac{\sigma_x - \sigma_y}{2}\sin 2\theta$$

(b) Show that the principal stresses are $\sigma_1 = \sigma_x$ and $\sigma_2 = \sigma_y$. (c) Obtain the maximum shear stresses and associated normal stresses and show them on a sketch of a properly oriented element.

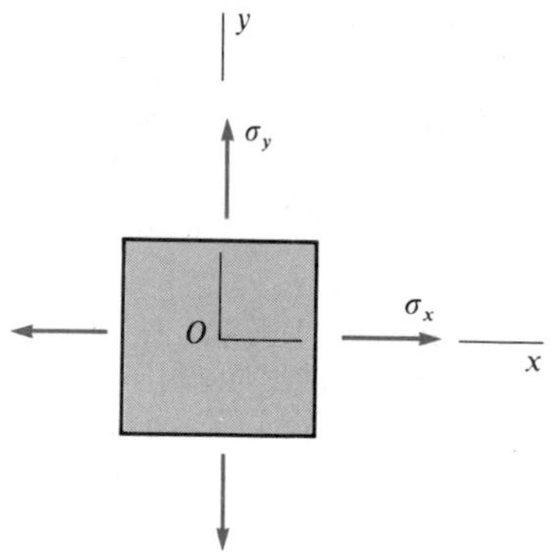

Prob. 6.4-3

6.4-4 Construct Mohr's circle for an element in biaxial stress subjected to two equal stresses ($\sigma_x = \sigma_y = \sigma_0$) as shown in the figure. Determine the normal and shear stresses on inclined planes, the principal stresses and angles to the principal planes, and the maximum shear stresses.

6.4-5 and 6.4-6 An element in uniaxial stress is subjected to stresses σ_x as shown in the figure. Using Mohr's circle, determine: (a) the stresses acting on an element oriented at an angle $\theta = 30°$ from the x axis, and (b) the maximum shear stresses and associated normal stresses. Show the stresses on sketches of properly oriented elements.

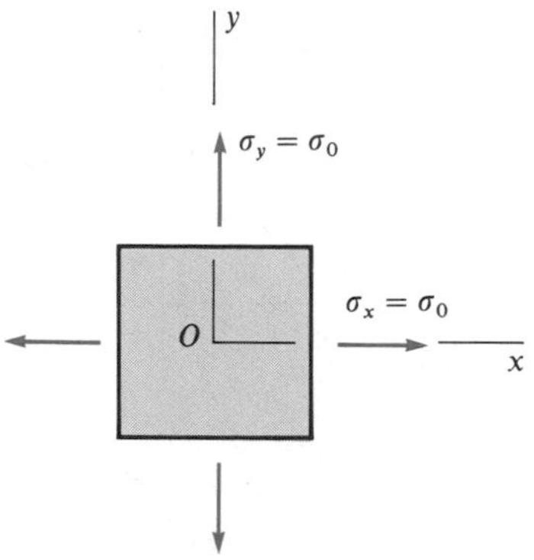

Prob. 6.4-4

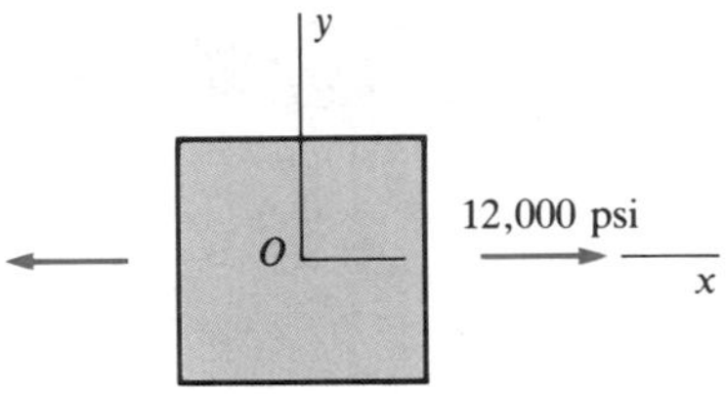

Prob. 6.4-5

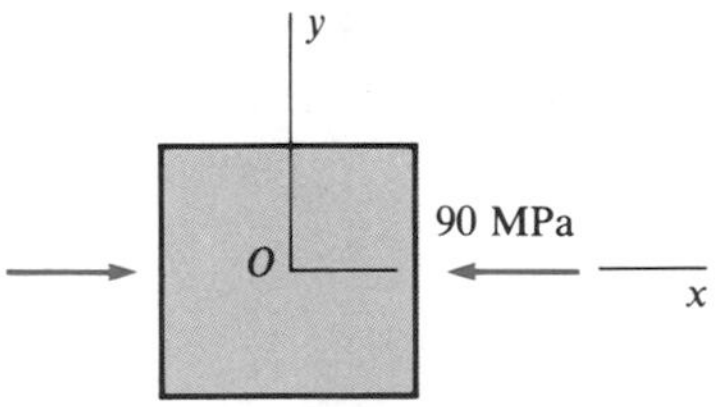

Prob. 6.4-6

6.4-7 and 6.4-8 An element in pure shear is subjected to stresses τ_{xy} as shown in the figure. Using Mohr's circle, determine: (a) the stresses acting on an element oriented at an angle $\theta = 75°$ from the x axis, and (b) the principal stresses. Show the stresses on sketches of properly oriented elements.

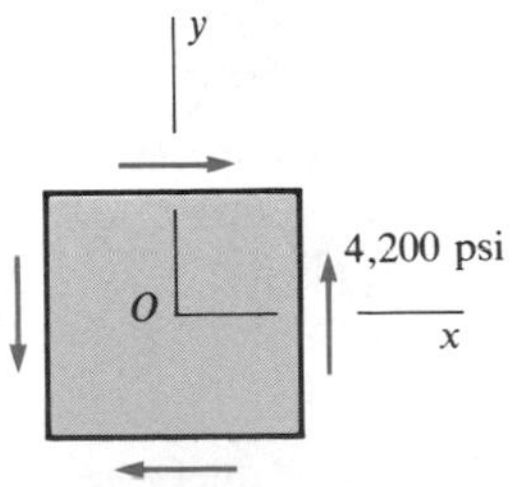

Prob. 6.4-7

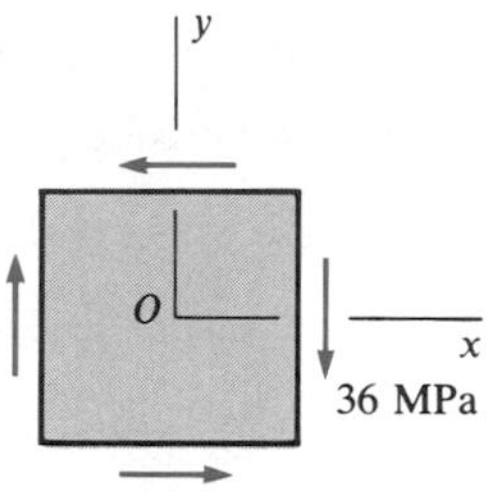

Prob. 6.4-8

6.4-9 and 6.4-10 An element in biaxial stress is subjected to stresses σ_x and σ_y as shown in the figure. Using Mohr's circle, determine: (a) the stresses acting on an element rotated through an angle $\theta = 22.5°$ from the x axis, and (b) the maximum shear stresses and associated normal stresses. Show the stresses on sketches of properly oriented elements.

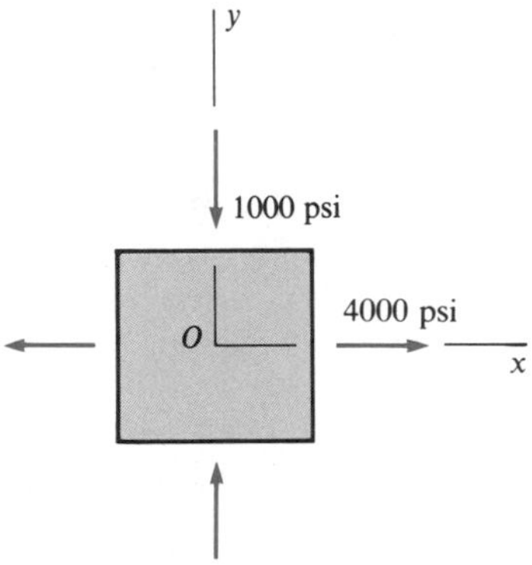

Prob. 6.4-9

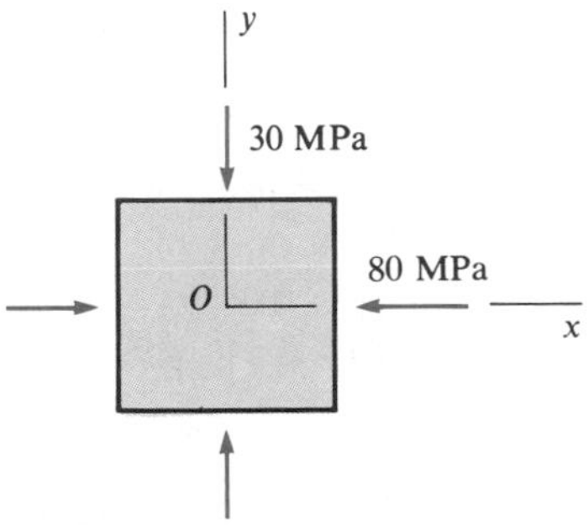

Prob. 6.4-10

6.4-11 Solve Problem 6.2-1 by using Mohr's circle.

6.4-12 Solve Problem 6.2-2 by using Mohr's circle.

6.4-13 Solve Problem 6.2-3 by using Mohr's circle.

6.4-14 Solve Problem 6.2-4 by using Mohr's circle.

6.4-15 Solve Problem 6.2-5 by using Mohr's circle.

6.4-16 Solve Problem 6.2-6 by using Mohr's circle.

6.4-17 Solve Problem 6.2-7 by using Mohr's circle.

6.4-18 Solve Problem 6.2-8 by using Mohr's circle.

6.4-19 Solve Problem 6.2-9 by using Mohr's circle.

6.4-20 Solve Problem 6.2-10 by using Mohr's circle.

6.4-21 Solve Problem 6.3-1 by using Mohr's circle.

6.4-22 Solve Problem 6.3-2 by using Mohr's circle.

6.4-23 Solve Problem 6.3-3 by using Mohr's circle.

6.4-24 Solve Problem 6.3-4 by using Mohr's circle.

6.4-25 Solve Problem 6.3-5 by using Mohr's circle.

6.4-26 Solve Problem 6.3-6 by using Mohr's circle.

6.4-27 Solve Problem 6.3-7 by using Mohr's circle.

6.4-28 Solve Problem 6.3-8 by using Mohr's circle.

6.4-29 Solve Problem 6.3-9 by using Mohr's circle.

6.4-30 Solve Problem 6.3-10 by using Mohr's circle.

6.4-31 Solve Problem 6.3-11 by using Mohr's circle.

6.4-32 Solve Problem 6.3-12 by using Mohr's circle.

6.4-33 Solve Problem 6.3-13 by using Mohr's circle.

6.4-34 Solve Problem 6.3-14 by using Mohr's circle.

6.4-35 Solve Problem 6.3-15 by using Mohr's circle.

6.4-36 Solve Problem 6.3-16 by using Mohr's circle.

6.4-37 Solve Problem 6.3-17 by using Mohr's circle.

6.4-38 Solve Problem 6.3-18 by using Mohr's circle.

6.4-39 Solve Problem 6.3-19 by using Mohr's circle.

6.4-40 Solve Problem 6.3-20 by using Mohr's circle.

6.5-1 A thin rectangular steel plate is subjected to uniform normal stresses σ_x and σ_y as shown in the figure. Strain gages A and B, oriented in the x and y directions, respectively, are attached to the plate. The gage readings give normal strains $\epsilon_x = 0.0012$ and $\epsilon_y = -0.0008$. Knowing that $E = 30 \times 10^6$ psi and $\nu = 0.3$, determine the stresses σ_x and σ_y.

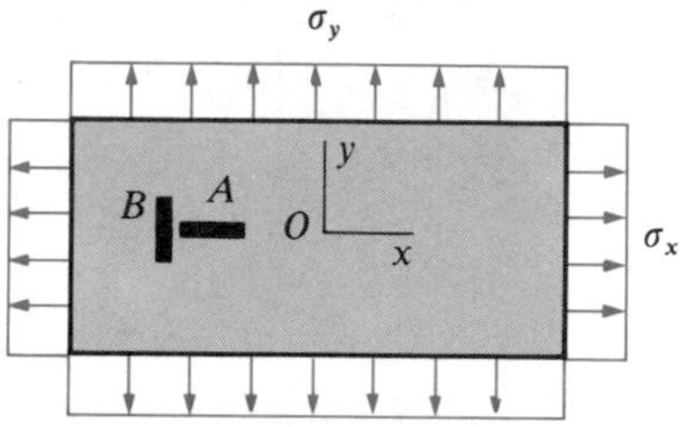

Probs. 6.5-1 and 6.5-2

6.5-2 Strain gages A and B, oriented in the x and y directions, respectively, are attached to a thin rectangular steel plate as shown in the figure. The plate is subjected to uniform normal stresses σ_x and σ_y. The strain gages give readings $\epsilon_x = 480 \times 10^{-6}$ and $\epsilon_y = 130 \times 10^{-6}$. Assuming that $E = 200$ GPa and $\nu = 0.30$, calculate the stresses σ_x and σ_y.

6.5-3 The normal strains ϵ_x and ϵ_y for an element in plane stress (see figure) are measured with strain gages. (a) Obtain a formula for the normal strain ϵ_z in the z direction in terms of ϵ_x, ϵ_y, and Poisson's ratio ν. (b) If $\epsilon_x = 180 \times 10^{-6}$, $\epsilon_y = -50 \times 10^{-6}$, and $\nu = 0.3$, what is the strain ϵ_z?

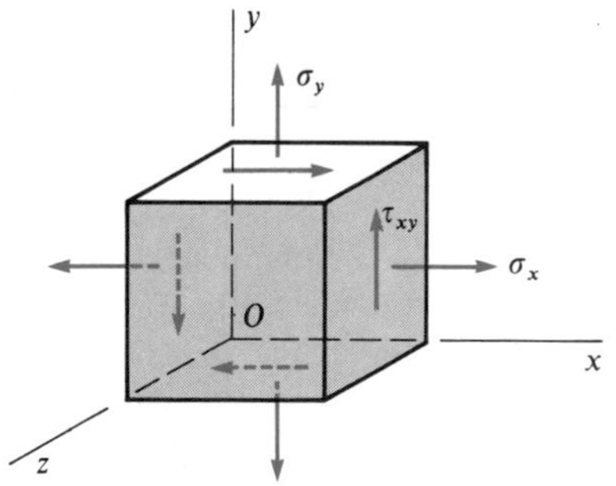

Prob. 6.5-3

6.5-4 A cube of metal is compressed on two opposite faces by uniformly distributed compressive forces of magnitude P (see figure). What compressive force F, also uniformly distributed, must be applied to one of the other pairs of faces in order that those faces remain the same distance apart?

6.5-5 A rectangular plate of thickness t, width b, and height h is subjected to normal stresses $\sigma_x = 12$ ksi and $\sigma_y = -3$ ksi,

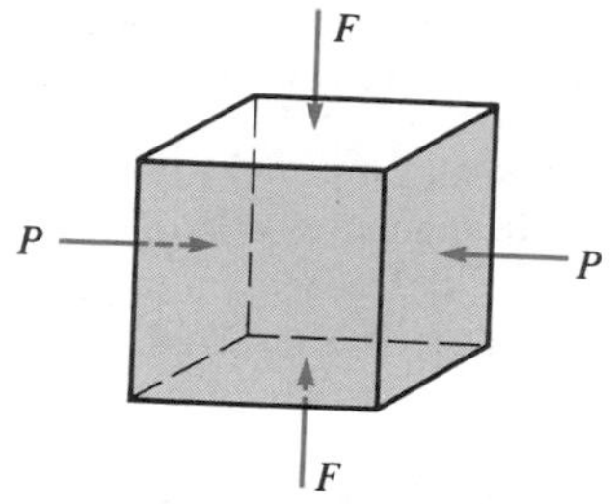

Prob. 6.5-4

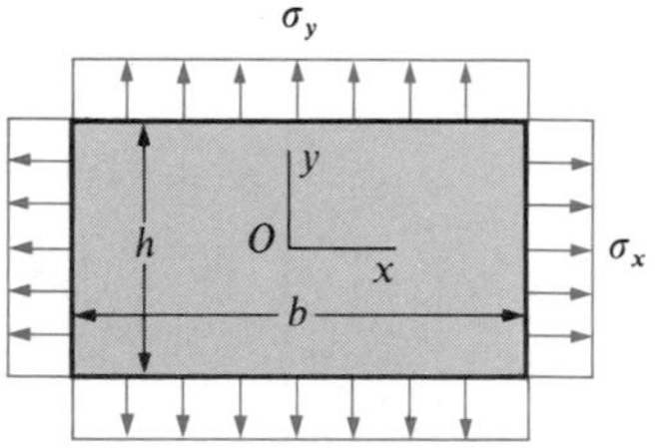

Probs. 6.5-5 and 6.5-6

as shown in the figure. (a) Calculate the maximum in-plane shear strain γ_{max} in the plate. (b) Calculate the change Δt in thickness and the change ΔV in volume of the plate if the dimensions are as follows: $t = 0.5$ in., $b = 30$ in., and $h = 20$ in. (The material is aluminum with $E = 10{,}500$ ksi and $\nu = 0.33$.)

6.5-6 A rectangular plate of thickness t, width b, and height h is subjected to normal stresses $\sigma_x = 90$ MPa and $\sigma_y = -20$ MPa, as shown in the figure. (a) Calculate the maximum in-plane shear strain γ_{max} in the plate. (b) Calculate the change Δt in thickness and the change ΔV in volume of the plate if the dimensions are as follows: $t = 20$ mm, $b = 800$ mm, and $h = 400$ mm. (The material is steel with $E = 200$ GPa and $\nu = 0.3$.)

6.5-7 A cube of concrete 4 in. on each edge is compressed in two perpendicular directions by forces $P = 15{,}000$ lb. Determine the change ΔV in volume of the cube and the total strain energy U stored in the cube, assuming $E = 3.8 \times 10^6$ psi and $\nu = 0.1$.

6.5-8 A brass cube 40 mm on each edge is compressed in two perpendicular directions by forces $P = 120$ kN. Calculate the change ΔV in volume of the cube and the total strain energy U stored in the cube, assuming $E = 100$ GPa and $\nu = 0.34$.

6.5-9 A square plate of width b and thickness t is loaded by normal forces P_x and P_y and shear forces V as shown in the figure. These forces produce uniformly distributed stresses acting on the side faces of the plate. Calculate the change ΔV in the volume of the plate and the total strain energy U stored in the plate if the dimensions are $b = 10$ in. and $t = 1$ in., the plate is made of aluminum with $E = 10{,}600$ ksi and $\nu = 0.33$, and the forces are $P_x = 120$ kips, $P_y = 30$ kips, and $V = 20$ kips.

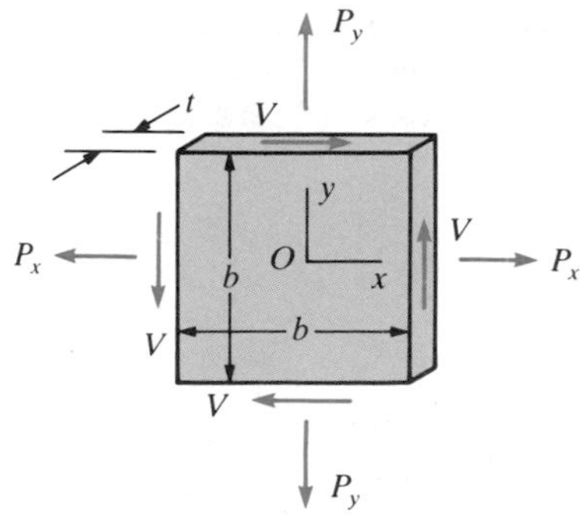

Probs. 6.5-9 and 6.5-10

6.5-10 Solve the preceding problem for a magnesium plate with $b = 500$ mm, $t = 40$ mm, $E = 45$ GPa, $\nu = 0.35$, $P_x = 450$ kN, $P_y = 150$ kN, and $V = 110$ kN.

6.5-11 An element in plane stress is subjected to stresses σ_x, σ_y, and τ_{xy} (see figure) such that $\sigma_y = -0.5\sigma_x$ and $\tau_{xy} = 0.5\sigma_x$. The strain energy density of the element is $u = 33$ psi. Assuming that the material is steel with $E = 30 \times 10^6$ psi and $\nu = 0.3$, determine the stresses σ_x, σ_y, and τ_{xy}.

6.5-12 An element in plane stress is subjected to stresses σ_x, σ_y, and τ_{xy} (see figure). The stresses are related as follows: $\sigma_y = -0.6\sigma_x$ and $\tau_{xy} = \sigma_x$. The strain energy density of the

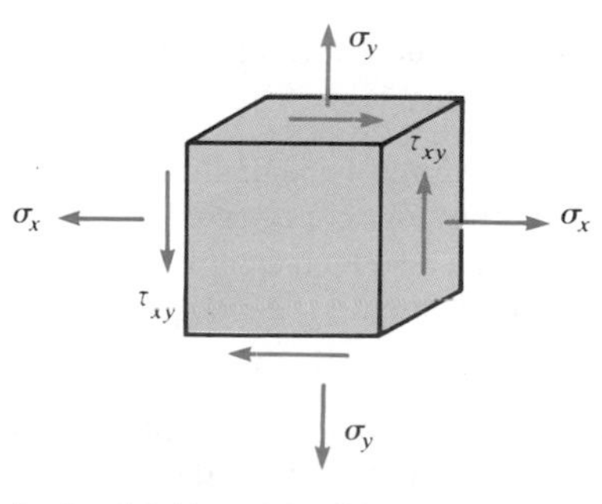

Probs. 6.5-11 and 6.5-12

element is $u = 280$ kPa, and the material is magnesium with $E = 45$ GPa and $\nu = 0.35$. Determine the stresses σ_x, σ_y, and τ_{xy}.

6.5-13 A circle of diameter $d = 8$ in. is etched on an unstressed brass plate (see figure). The plate is square with dimensions 16 in. × 16 in. × 0.5 in. Later, forces are applied that produce uniformly distributed normal stresses $\sigma_x = 6000$ psi and $\sigma_y = 2000$ psi along the edges of the plate. Calculate the following quantities: (a) the change in length Δac of diameter ac; (b) the change in length Δbd of diameter bd; (c) the change Δt in thickness of the plate; (d) the dilatation e; and (e) the total strain energy U stored in the plate. (Assume $E = 15 \times 10^6$ psi and $\nu = 0.34$.)

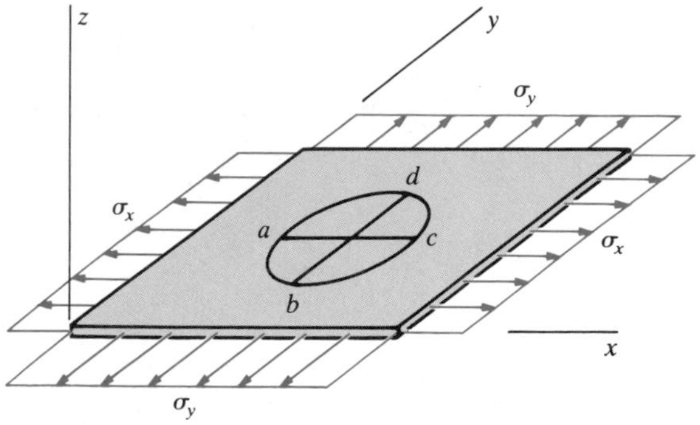

Prob. 6.5-13

6.6-1 A spherical stainless steel tank having an inside diameter of 16 in. is used as a pressurized fuel tank. The thickness of the shell is 0.093 in. and the allowable stress in tension is 120,000 psi. Determine the maximum permissible pressure p inside the tank.

6.6-2 A spherical steel pressure vessel with an inside diameter of 600 mm is being designed for a pressure of 8 MPa. The yield stress of the steel is 400 MPa. What is the minimum required thickness t for a factor of safety against yielding of 2.5?

6.6-3 A spherical tank of inside diameter 48 in. and wall thickness 2 in. contains compressed air at a pressure of 2400 psi. The tank is constructed of two hemispheres joined by welding. What is the tensile load f (lb per in. of length) carried by the weld?

6.6-4 The internal pressure in a spherical tank is $p = 3.6$ MPa. The inside diameter of the tank is 220 mm and the wall thickness is 5 mm. (a) Determine the maximum in-plane shear stress τ in the wall of the tank. (b) Determine the absolute maximum shear stress τ_{max}.

6.6-5 A spherical shell subjected to internal pressure $p = 500$ psi has an inside diameter of 36 in. and a wall thickness of 0.5 in. (a) What is the maximum in-plane shear stress τ in the shell? (b) What is the absolute maximum shear stress τ_{max}?

6.6-6 The vertical standpipe shown in the figure has inside diameter $d = 2$ m and wall thickness $t = 10$ mm. What height h of water will produce a circumferential stress of 16 MPa in the wall of the pipe?

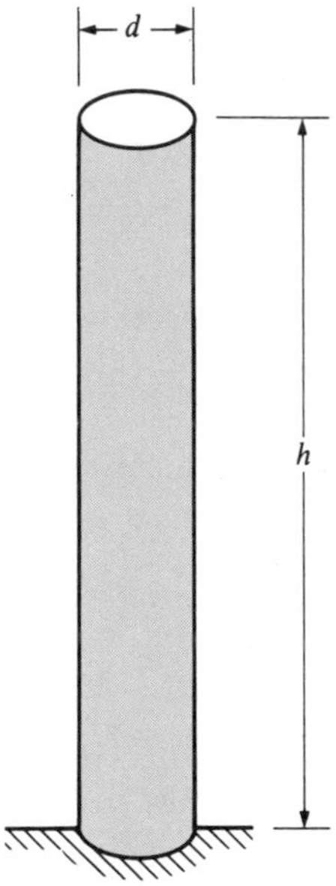

Probs. 6.6-6 and 6.6-7

6.6-7 A vertical steel standpipe of height $h = 50$ ft and inside diameter $d = 8$ ft is filled with water (see figure). Considering only the circumferential stress, find the minimum required wall thickness t_{min} if the allowable tensile stress in the steel is 12 ksi.

6.6-8 A cylindrical tank with an inside diameter of 300 mm is subjected to internal gas pressure $p = 2.0$ MPa. The tank is constructed of aluminum sections that are welded circumferentially (see figure). The heads of the tank are hemispherical. The allowable tensile stress in the wall of the tank is 60 MPa, and the allowable tensile stress perpendicular to a weld is 40 MPa. Considering only the membrane stresses in the tank, determine the minimum required thickness t_{min} of: (a) the cylindrical part of the tank, and (b) the hemispherical heads.

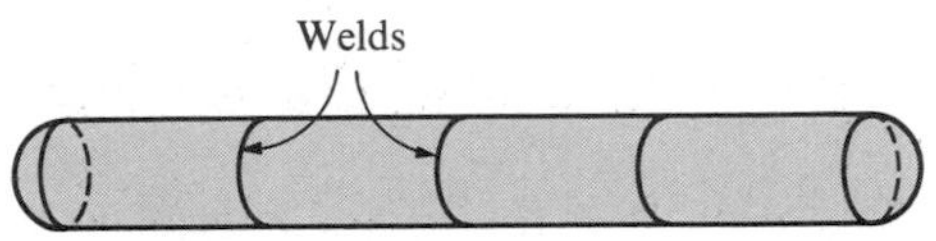

Probs. 6.6-8 and 6.6-9

6.6-9 A cylindrical tank with hemispherical heads is constructed of steel sections that are welded circumferentially (see figure). The tank diameter is 4 ft, the wall thickness is 0.75 in., and the maximum internal pressure is 250 psi. (a) Determine the maximum tensile stress σ in the heads of the tank. (b) Determine the maximum circumferential stress σ_c in the cylindrical part of the tank. (c) Determine the maximum tensile stress σ_w acting perpendicular to the welded joints.

6.6-10 A cylindrical tank with closed ends (see figure) contains compressed air at a maximum pressure of 850 kPa. The inside diameter of the tank is 1.2 m, and the wall thickness is 10 mm. (a) Calculate the principal membrane stresses σ_1 and σ_2 in the wall of the cylinder, and show these stresses on a sketch of a properly oriented element. (b) Determine the maximum in-plane shear stresses τ_{max}, and show them on a sketch of a properly oriented element. (c) Calculate the absolute maximum shear stress τ_{max} in the cylinder.

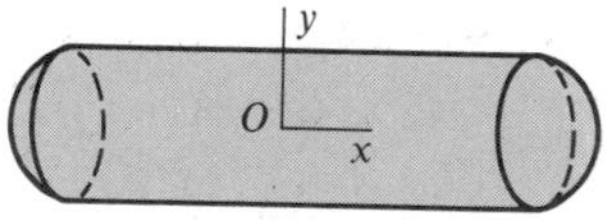

Probs. 6.6-10 and 6.6-11

6.6-11 Solve the preceding problem for a tank of inside diameter $d = 10$ in., wall thickness $t = 0.5$ in., and internal pressure $p = 1200$ psi.

6.6-12 A thin-walled cylindrical tank of inside radius r is subjected simultaneously to internal gas pressure p and a compressive force F at the ends (see figure). What should be the magnitude of the force F in order to produce pure shear in the wall of the cylinder?

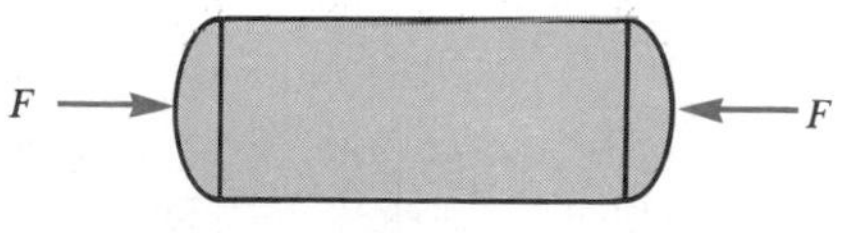

Prob. 6.6-12

6.6-13 A cylindrical pressure tank is constructed with a helical weld that makes an angle $\alpha = 75°$ with the longitudinal axis (see figure). The tank has inside radius $r =$ 18 in., wall thickness $t = 0.6$ in., and internal pressure $p =$ 200 psi. Determine the following quantities for the cylindrical part of the tank: (a) the circumferential and longitudinal stresses, (b) the maximum in-plane shear stress, (c) the absolute maximum shear stress, and (d) the normal and shear stresses acting on planes perpendicular and parallel to the weld.

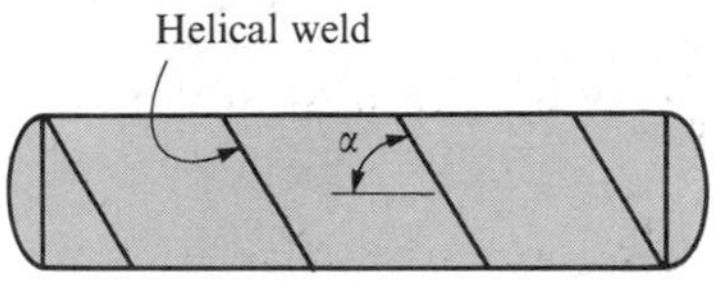

Probs. 6.6-13 and 6.6-14

6.6-14 Solve the preceding problem for a tank with $\alpha = 60°$, $r = 0.5$ m, $t = 15$ mm, and $p = 2.4$ MPa.

6.6-15 A cylindrical tank containing compressed air has wall thickness $t = 0.5$ in. and inside radius $r = 10$ in. (see figure). The stresses in the wall of the tank acting on an inclined element have the values shown in the figure. What is the air pressure p in the tank?

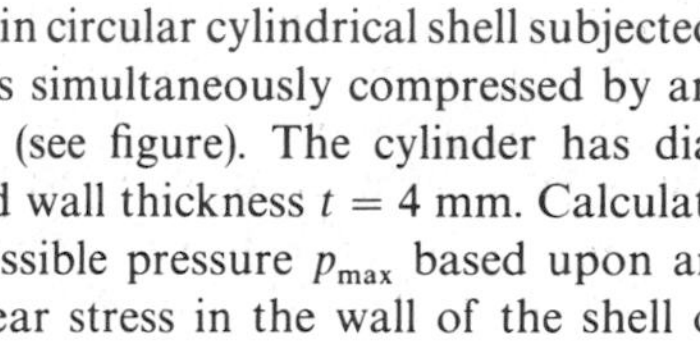

Prob. 6.6-15

6.6-16 A thin circular cylindrical shell subjected to internal pressure p is simultaneously compressed by an axial load $P = 72$ kN (see figure). The cylinder has diameter $d =$ 100 mm and wall thickness $t = 4$ mm. Calculate the maximum permissible pressure p_{max} based upon an allowable in-plane shear stress in the wall of the shell of 60 MPa.

6.7-1 A bar of solid circular cross section (diameter $d =$ 3 in.) is subjected simultaneously to an axial tensile load $P = 45$ k and a torque $T = 25$ in.-k (see figure). Calculate the maximum tensile stress σ_t, maximum compressive stress σ_c, and maximum in-plane shear stress τ_{max} in the bar.

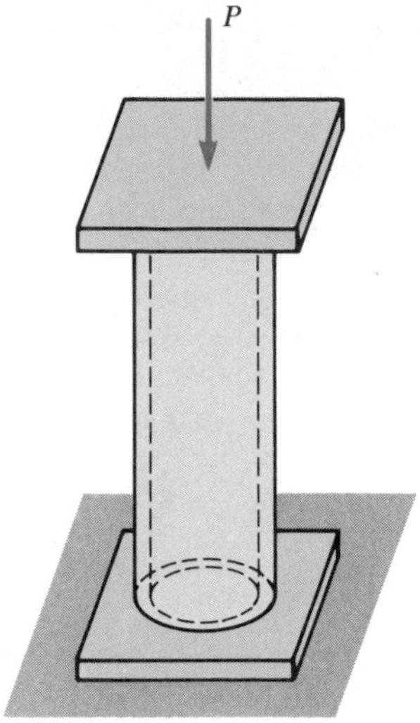

Prob. 6.6-16

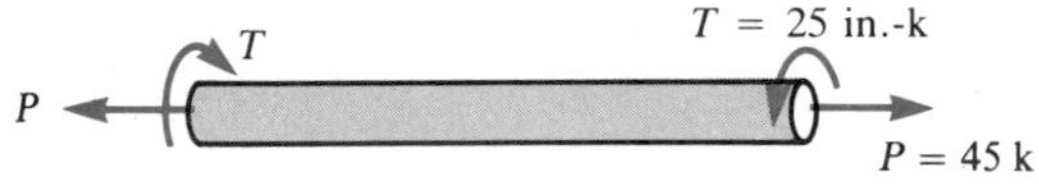

Prob. 6.7-1

6.7-2 A generator shaft of hollow circular cross section (outside diameter 200 mm and inside diameter 160 mm) is subjected simultaneously to a torque $T = 22.2$ kN·m and an axial compressive load $P = 724$ kN (see figure). Determine the maximum tensile stress σ_t, maximum compressive stress σ_c, and maximum in-plane shear stress τ_{max} in the shaft.

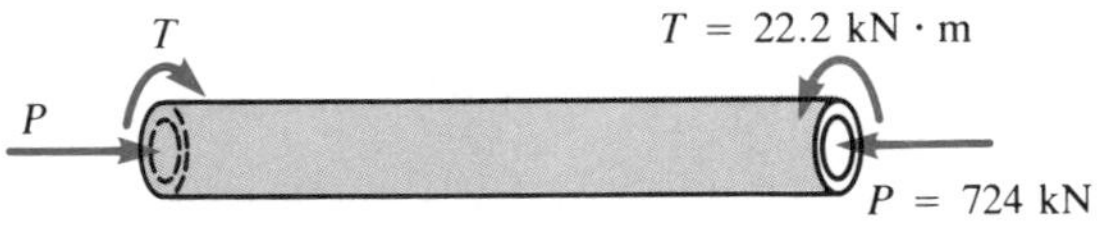

Prob. 6.7-2

6.7-3 A post having a hollow circular cross section supports a horizontal load $P = 360$ lb acting at the end of an arm that is 4 ft long (see figure). The height of the post is 25 ft, and its section modulus is $S = 10$ in.^{3} (a) Calculate the maximum tensile stress σ_{max} and maximum in-plane shear stress τ_{max} at point A due to the load P. Point A is located where the normal stress due to bending alone is a maximum. (b) If the maximum tensile stress and the maximum in-plane shear stress at point A are limited to 16,000 psi and 6,000 psi, respectively, what is the largest permissible value of the load P?

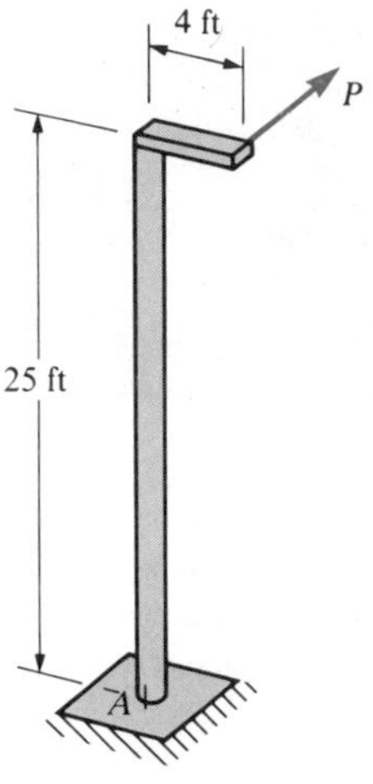

Prob. 6.7-3

6.7-4 A sign is supported by a pipe (see figure) having outside diameter 100 mm and inside diameter 80 mm. The dimensions of the sign are 2.0 m × 0.75 m, and its lower edge is 3.2 m above the support. The wind pressure against the sign is 1.8 kPa. Determine the maximum in-plane shear stresses due to the wind pressure on the sign at points A, B, and C, located at the base of the pipe.

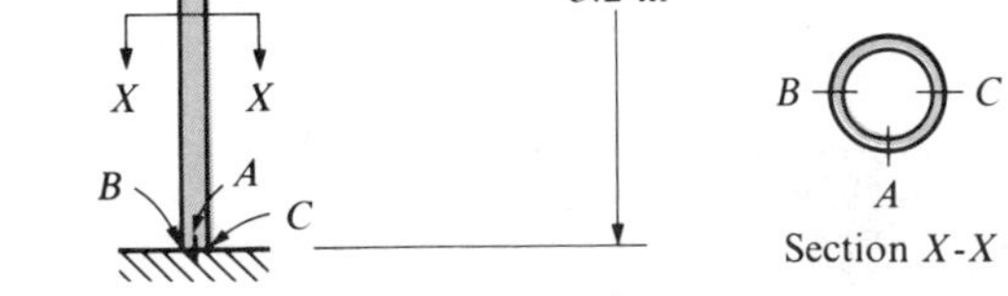

Prob. 6.7-4

6.7-5 An L-shaped bracket ABC lying in a horizontal plane supports a load $P = 150$ lb (see figure). The bracket has a hollow rectangular cross section with outside dimensions 2 in. × 3.5 in. and wall thickness 0.125 in. The centerline length of arm AB is 20 in. and of arm BC is 30 in. Considering only the force P, calculate the maximum tensile stress σ_t, maximum compressive stress σ_c, and maximum in-plane shear stress τ_{max} at point A, which is located on the top of the bracket at the support.

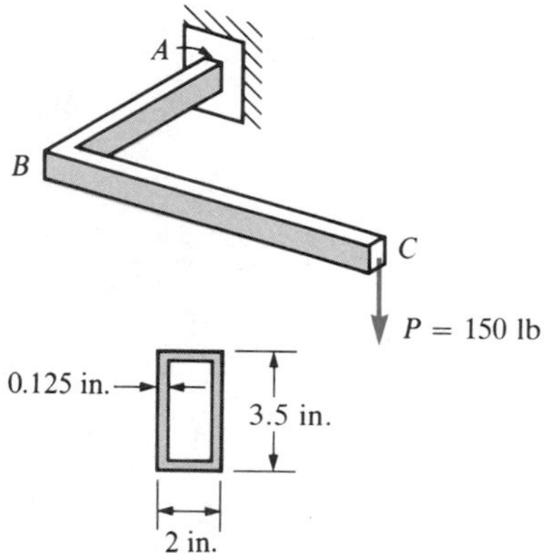

Prob. 6.7-5

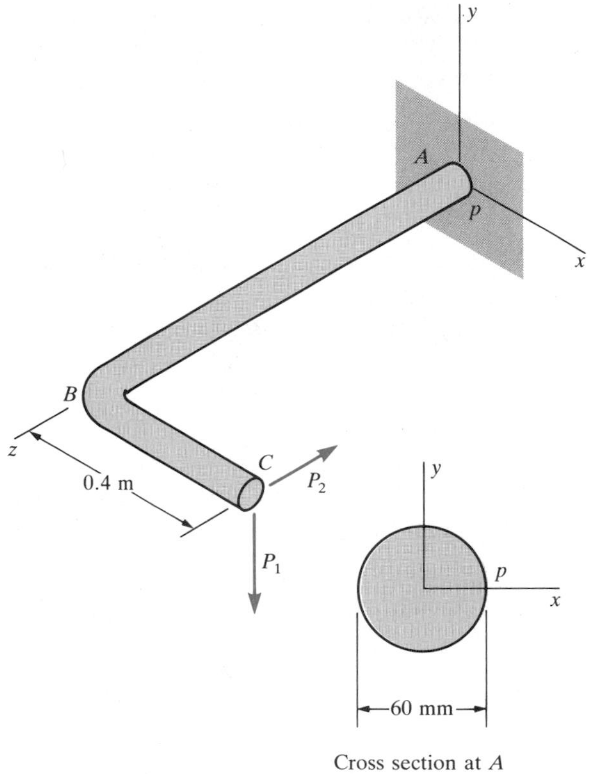

Prob. 6.7-6

6.7-6 A horizontal bracket ABC (see figure) consists of two perpendicular arms AB and BC, the latter having a length of 0.4 m. Arm AB has a solid circular cross section with diameter equal to 60 mm. At point C a load $P_1 = 2.02$ kN acts vertically and a load $P_2 = 3.07$ kN acts horizontally parallel to arm AB. Considering only the forces P_1 and P_2, calculate the maximum tensile stress σ_t, the maximum compressive stress σ_c, and the maximum in-plane shear stress τ_{max} at point p, which is located at support A on the side of the bracket at midheight.

6.7-7 A shaft of diameter $d = 2.5$ in. supports a 30 in. diameter pulley weighing 450 lb (see figure). The belt tensions (horizontal forces) are 1200 lb and 200 lb. Determine the maximum tensile stress σ_{max} and the maximum shear stress τ_{max} in the shaft at the first bearing, which is located 6 in. from the pulley. (Hint: Combine the vertical and horizontal forces acting on the pulley into a single resultant force, acting at the center, plus a couple. Then the projecting part of the shaft is a circular cantilever beam loaded at its end by a lateral bending load and a twisting couple.)

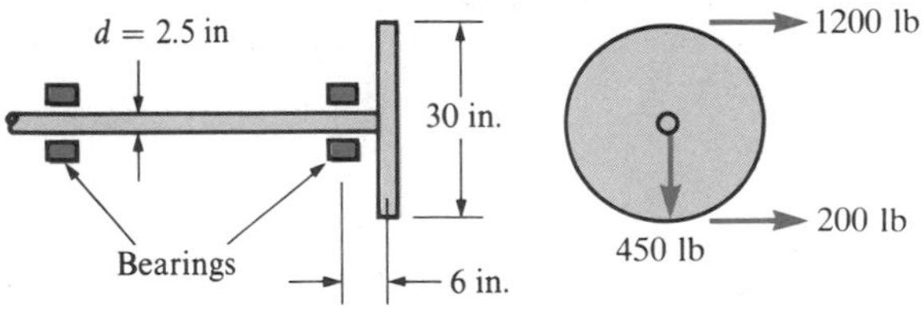

Prob. 6.7-7

6.7-8 The torsional pendulum shown in the figure consists of a circular disk of mass $M = 50$ kg suspended by a steel wire ($G = 80$ GPa) of length $L = 1$ m and diameter $d = 4$ mm. Calculate the maximum angle of rotation ϕ_{max} (that is, the maximum amplitude of torsional vibrations) that

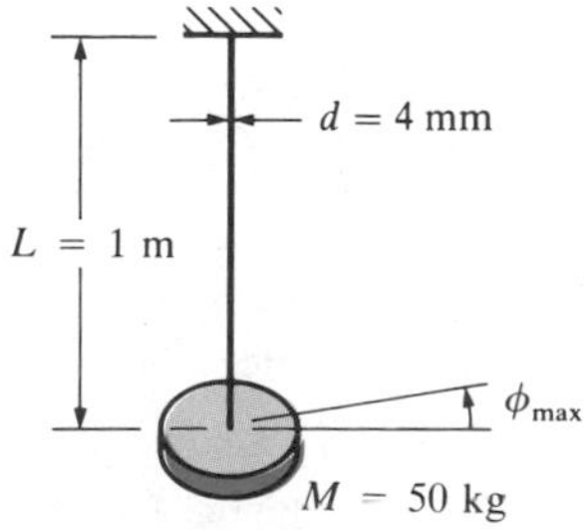

Prob. 6.7-8

the disk can have without exceeding an allowable stress in tension of 80 MPa and an allowable stress in shear of 50 MPa in the wire. (Consider only the in-plane shear stresses.)

6.7-9 A cylindrical pressure vessel having radius $r = 12$ in. and wall thickness $t = 0.6$ in. is subjected to internal pressure $p = 360$ psi. In addition, a torque $T = 90$ ft-k is applied to the closed ends of the cylinder (see figure). (a) Determine the stresses σ_x, σ_y, and τ_{xy} acting on a stress element at point A in the wall of the cylinder. (b) Determine the maximum tensile stress σ_{max} and the maximum in-plane shear stress τ_{max} in the wall of the cylinder.

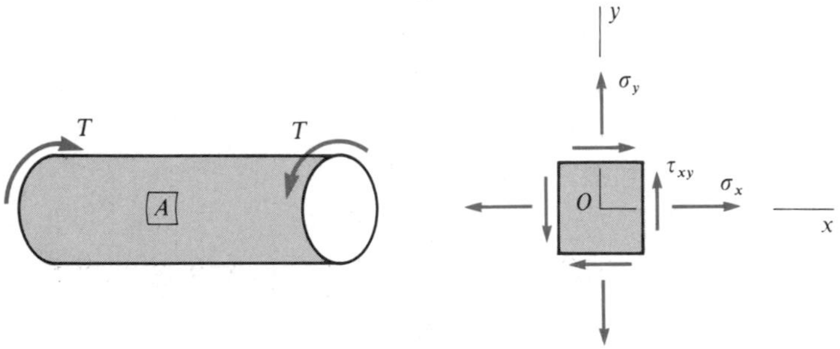

Prob. 6.7-9

6.7-10 A pressurized cylindrical tank is loaded by torques T and tensile forces P (see figure). The tank has radius $r = 50$ mm and wall thickness $t = 3$ mm. The internal pressure $p = 3.5$ MPa and the torque $T = 450$ N·m. What is the maximum permissible value of the forces P if the allowable tensile stress in the wall of the cylinder is 72 MPa?

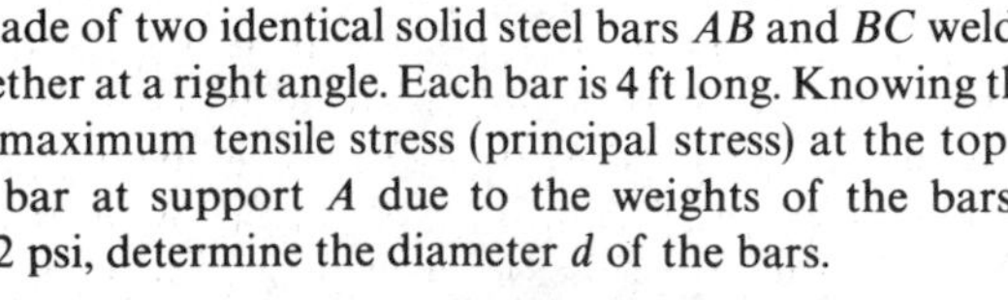

Prob. 6.7-10

6.7-11 An arm ABC lying in a horizontal plane (see figure) is made of two identical solid steel bars AB and BC welded together at a right angle. Each bar is 4 ft long. Knowing that the maximum tensile stress (principal stress) at the top of the bar at support A due to the weights of the bars is 8052 psi, determine the diameter d of the bars.

6.7-12 A semicircular bar AB lying in a horizontal plane is supported at B (see figure). The bar has centerline radius R and weight q per unit of length (total weight of the bar

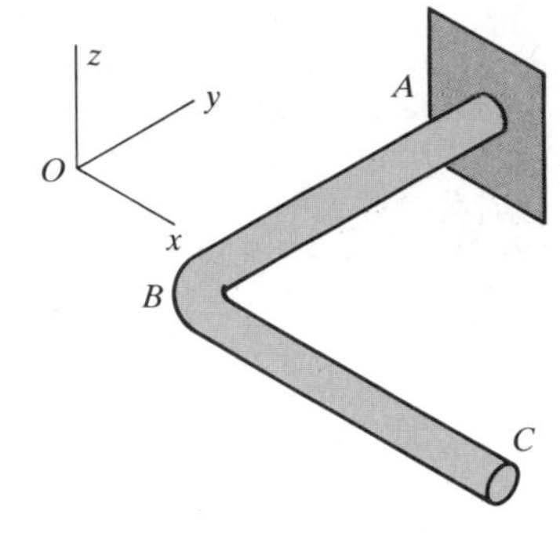

Prob. 6.7-11

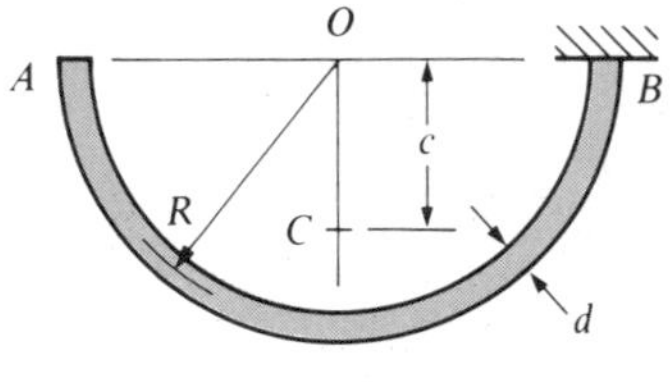

Prob. 6.7-12

equals πqR). The cross section of the bar is circular with diameter d. Obtain formulas for the maximum tensile stress σ_t, maximum compressive stress σ_c, and maximum in-plane shear stress τ_{max} at the top of the bar at the support due to the weight of the bar. (Note: The center of gravity of the bar is at point C, a distance $c = 2R/\pi$ from the center O.)

When solving the problems for Section 6.8, disregard the weights of the beams.

6.8-1 A cantilever beam of rectangular cross section is subjected to a concentrated load P at the free end (see figure). Calculate the principal stresses and maximum in-plane shear stresses at point A, and show these stresses on sketches of properly oriented elements. Use the following values: $P = 30$ k, $b = 4$ in., $h = 10$ in., $c = 2$ ft, and $d = 3$ in.

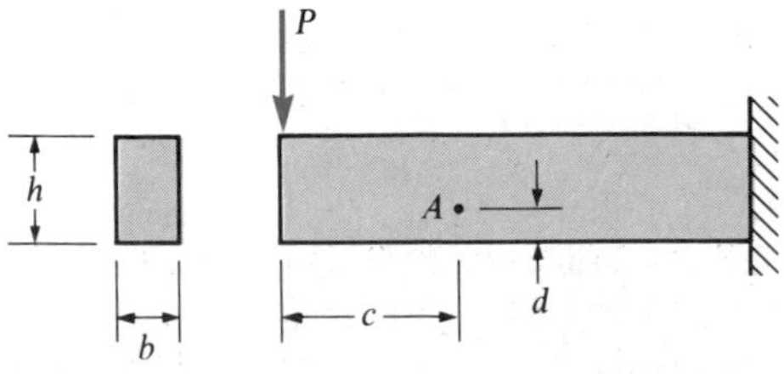

Probs. 6.8-1 and 6.8-2

6.8-2 Solve the preceding problem for the following values: $P = 180$ kN, $b = 100$ mm, $h = 200$ mm, $c = 0.5$ m, and $d = 150$ mm.

6.8-3 A W 10 × 30 wide-flange beam (see Table E-1, Appendix E) is simply supported with a span length of 8 ft. The beam supports a concentrated load of 30 kips at midspan. At a cross section located 2 ft from the left-hand support, determine the principal stresses σ_1 and σ_2 and the maximum in-plane shear stress τ_{max} at each of the following points: (a) the top of the beam, (b) the top of the web, and (c) the neutral axis.

6.8-4 A beam of I-section (see figure) has the following dimensions: $b = 120$ mm, $t = 10$ mm, $h = 300$ mm, and $h_1 = 260$ mm. The beam is simply supported with span length $L = 3.0$ m. A concentrated load $P = 120$ kN acts at the midpoint. At a cross section located 1.0 m from the left-hand support, calculate the principal stresses σ_1 and σ_2 and the maximum in-plane shear stress τ_{max} at each of the following points: (a) the top of the beam, (b) the top of the web, and (c) the neutral axis.

6.8-5 An overhanging beam *ABC* of rectangular cross section supports a concentrated load *P* at the free end (see figure). The span length from *A* to *B* is *L*, and the length of the overhang is $L/2$. The cross section has width *b* and height *h*. Point *D* is located midway between the supports at a distance *d* from the top face of the beam. Knowing that the maximum tensile stress (principal stress) at point *D* is $\sigma_1 = 3620$ psi, determine the magnitude of the load *P*. Data for the beam are $L = 50$ in., $b = 1$ in., $h = 6$ in., and $d = 1$ in.

6.8-6 A cantilever beam of T-section is loaded by an inclined force of magnitude 10 kN applied at the point shown in the figure. Obtain the principal stresses σ_1 and σ_2 and the maximum in-plane shear stress τ_{max} at points *A* and *B* in the web of the beam.

6.8-7 A simple beam of rectangular cross section has span length $L = 60$ in. and supports a concentrated load $P = 18$ k at the midpoint (see figure). The height of the beam is $h = 6$ in. and the width is $b = 2$ in. Cross section *mm* is located 20 in. from the left-hand support. Plot graphs showing how the principal stresses σ_1 and σ_2 and the maximum in-plane shear stress τ_{max} vary over the height of the beam.

6.8-8 Solve the preceding problem for a cross section *mm* located 0.15 m from the support if $L = 0.7$ m, $P = 144$ kN, $h = 120$ mm, and $b = 20$ mm.

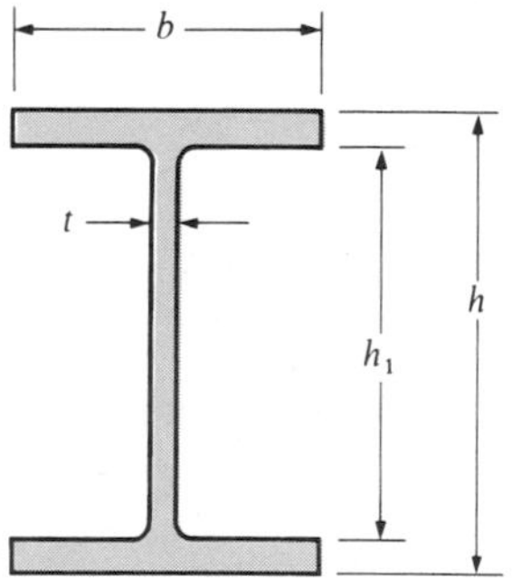

Prob. 6.8-4

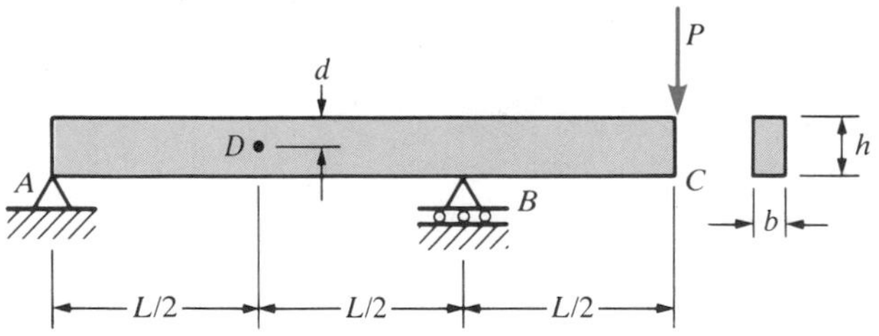

Prob. 6.8-5

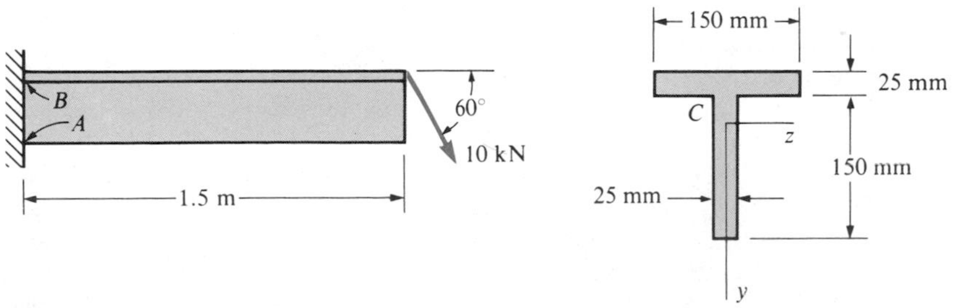

Prob. 6.8-6

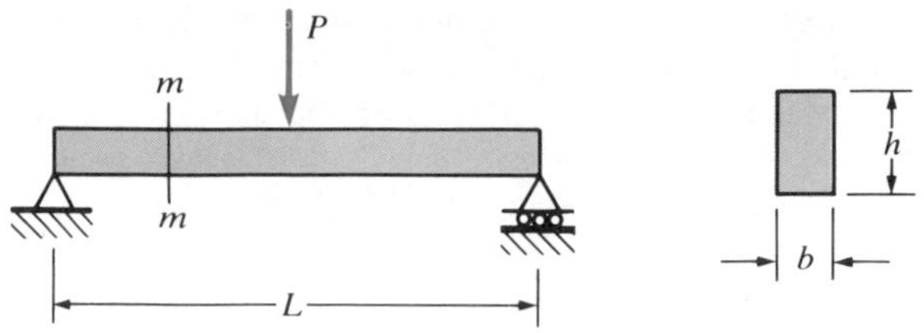

Probs. 6.8-7 and 6.8-8

6.9-1 A block of aluminum in the form of a rectangular parallelepiped (see figure) of dimensions $a = 5$ in., $b = 4$ in., and $c = 3$ in. is subjected to triaxial stresses $\sigma_x = 11{,}000$ psi, $\sigma_y = -5{,}000$ psi, and $\sigma_z = -1{,}500$ psi acting on the x, y, and z faces, respectively. Calculate the following quantities: (a) the maximum shear stress τ_{max} in the material; (b) the changes Δa, Δb, and Δc in the dimensions of the block; (c) the change ΔV in the volume; and (d) the total strain energy U stored in the block. (Assume $E = 10{,}400$ ksi and $\nu = 0.33$.)

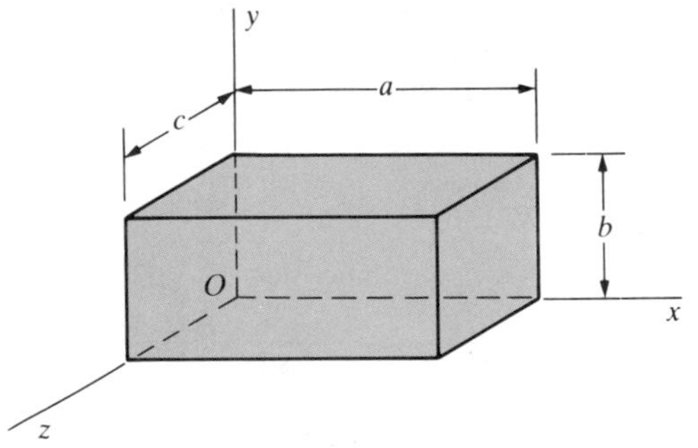

Probs. 6.9-1 and 6.9-2

6.9-2 Solve the preceding problem if the block is steel ($E = 200$ GPa, $\nu = 0.30$) with dimensions $a = 250$ mm, $b = 150$ mm, and $c = 150$ mm and the stresses are $\sigma_x = -50$ MPa, $\sigma_y = -40$ MPa, and $\sigma_z = -36$ MPa.

6.9-3 A cube of cast iron with sides of length $a = 3$ in. (see figure) is tested in a laboratory under triaxial stress. Strain gages mounted on the faces of the block record the following strains: $\epsilon_x = -350 \times 10^{-6}$ and $\epsilon_y = \epsilon_z = -65 \times 10^{-6}$. Calculate the following quantities: (a) the normal stresses σ_x, σ_y, and σ_z acting on the x, y, and z faces of the element; (b) the maximum shear stress τ_{max} in the material; (c) the change ΔV in the volume of the block; and (d) the total strain energy U stored in the block. (Assume $E = 14{,}000$ ksi and $\nu = 0.25$.)

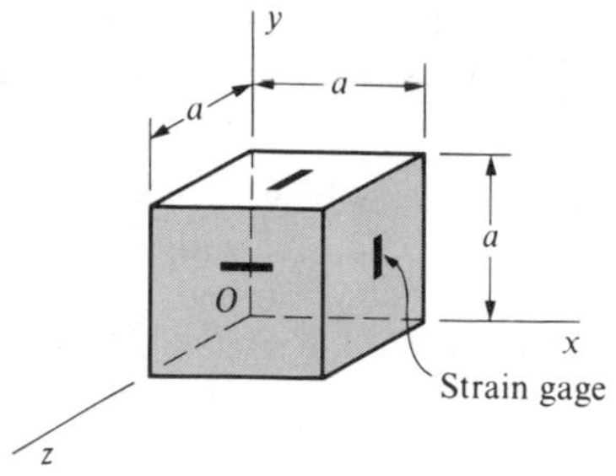

Probs. 6.9-3 and 6.9-4

6.9-4 Solve the preceding problem if the cube is granite ($E = 60$ GPa, $\nu = 0.25$) with dimensions $a = 50$ mm and the measured strains are $\epsilon_x = -620 \times 10^{-6}$ and $\epsilon_y = \epsilon_z = -250 \times 10^{-6}$.

6.9-5 A rubber cylinder A of diameter d is compressed inside a steel cylinder B by a force F (see figure). (a) Obtain a formula for the lateral pressure p between the rubber and the steel. Express p in terms of F, d, and Poisson's ratio ν for the rubber. Disregard friction between the rubber and the steel, and assume that the steel cylinder is rigid. (b) Calculate the pressure p if $F = 800$ lb, $d = 2$ in., and $\nu = 0.45$.

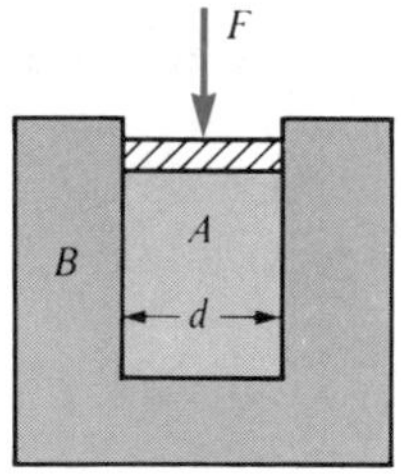

Prob. 6.9-5

6.9-6 A block A of hard rubber is confined between plane parallel rigid walls B (see figure). The rubber is not confined in the direction perpendicular to the plane of the paper. A uniformly distributed pressure p_0 is applied to the top of the rubber block. (a) Obtain a formula for the lateral pressure p between the rubber and the rigid walls. Express p in terms of p_0 and Poisson's ratio ν for the rubber, and disregard friction between the rubber and the walls. (b) Obtain a formula for the unit volume change e in terms of p_0, ν, and the modulus E. Assume that the strains are small quantities. (c) Obtain a formula for the unit volume change e without assuming that the strains are small. (d) Calculate the lateral pressure p and the unit volume change e if the applied pressure $p_0 = 2.75$ MPa and the rubber has the following properties: $E = 4.1$ MPa, $\nu = 0.48$. Explain why the formula for the unit volume change based upon small strains is not valid in this case.

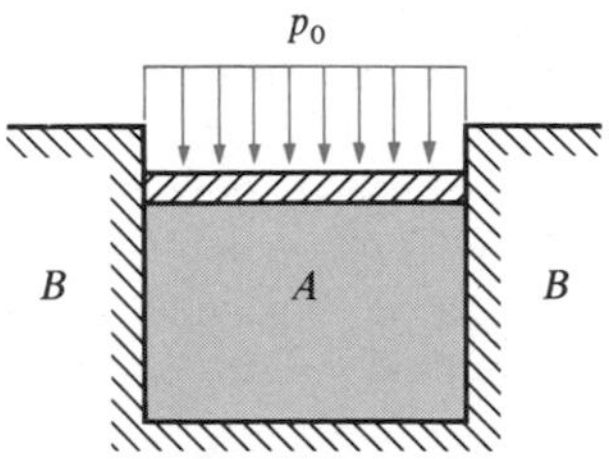

Prob. 6.9-6

6.9-7 A solid steel sphere ($E = 30 \times 10^6$ psi, $\nu = 0.3$) is subjected to hydrostatic pressure p such that its volume is reduced by 0.4%. (a) Calculate the pressure p. (b) Calculate the volume modulus of elasticity K for the steel. (c) Calculate the strain energy U stored in the sphere if its diameter is $d = 6$ in.

6.9-8 A cube of magnesium 100 mm on each side is lowered into the ocean to a depth such that the length of each side shortens by 0.018 mm. Assuming that $E = 45$ GPa and $\nu = 0.35$, calculate the following quantities: (a) the depth d to which the cube is lowered, and (b) the percent increase in the density of the magnesium.

6.9-9 A solid bronze sphere (volume modulus of elasticity $K = 14.5 \times 10^6$ psi) is suddenly heated at its outer surface. The tendency of the heated outer part of the sphere to expand produces uniform tension in all directions at the center of the sphere. If the stress at the center is 12,000 psi, what is the strain? Also, calculate the unit volume change e and the strain energy density u at the center.

Consider only the in-plane strains when solving the problems for Section 6.11.

6.11-1 An element of material subjected to plane strain (see figure) has strains as follows: $\epsilon_x = 220 \times 10^{-6}$, $\epsilon_y = 480 \times 10^{-6}$, and $\gamma_{xy} = 180 \times 10^{-6}$. Calculate the strains for an element oriented at an angle $\theta = 50°$.

6.11-2 Solve the preceding problem for the following data: $\epsilon_x = 420 \times 10^{-6}$, $\epsilon_y = -170 \times 10^{-6}$, $\gamma_{xy} = 310 \times 10^{-6}$, and $\theta = 37.5°$.

6.11-3 The strains for an element of material in plane strain (see figure) are as follows: $\epsilon_x = 480 \times 10^{-6}$, $\epsilon_y = 140 \times 10^{-6}$, and $\gamma_{xy} = -350 \times 10^{-6}$. Determine the principal strains and maximum shear strains.

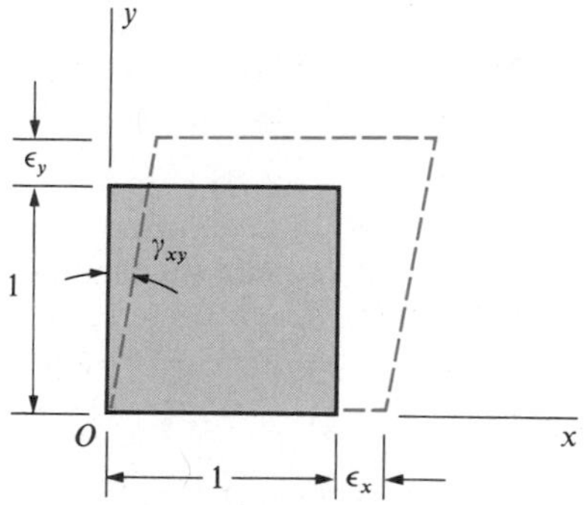

Probs. 6.11-1 to 6.11-6

6.11-4 Solve the preceding problem for the following strains: $\epsilon_x = 120 \times 10^{-6}$, $\epsilon_y = -450 \times 10^{-6}$, and $\gamma_{xy} = -360 \times 10^{-6}$.

6.11-5 An element of material in plane strain (see figure) is subjected to strains $\epsilon_x = 480 \times 10^{-6}$, $\epsilon_y = 70 \times 10^{-6}$, and $\gamma_{xy} = 420 \times 10^{-6}$. Determine the following quantities: (a) the strains for an element oriented at an angle $\theta = 75°$, (b) the principal strains, and (c) the maximum shear strains.

6.11-6 Solve the preceding problem for the following data: $\epsilon_x = -1120 \times 10^{-6}$, $\epsilon_y = -430 \times 10^{-6}$, $\gamma_{xy} = 780 \times 10^{-6}$, and $\theta = 45°$.

6.11-7 An element in plane stress is subjected to stresses $\sigma_x = -8400$ psi, $\sigma_y = 1100$ psi, and $\tau_{xy} = -1700$ psi as shown in the figure. The material is aluminum with modulus of elasticity $E = 10{,}000$ ksi and Poisson's ratio $\nu = 0.33$. Determine the following quantities: (a) the strains for an element oriented at an angle $\theta = 30°$, (b) the principal strains, and (c) the maximum shear strains.

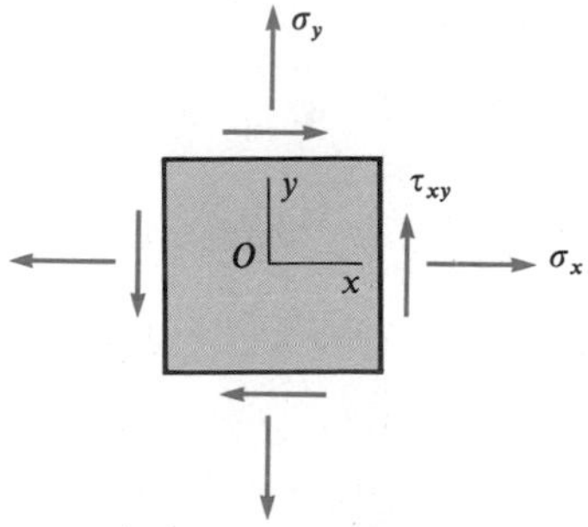

Probs. 6.11-7 and 6.11-8

6.11-8 Solve the preceding problem for the following data: $\sigma_x = -150$ MPa, $\sigma_y = -210$ MPa, $\tau_{xy} = -16$ MPa, and $\theta = 50°$. The material is brass with $E = 100$ GPa and $\nu = 0.34$.

6.11-9 During a static test of an airplane wing, the strain gage readings from a 45° rosette (see figure) are as follows: gage A, 520×10^{-6}; gage B, 360×10^{-6}, and gage C, -80×10^{-6}. Determine the principal strains and maximum shear strains.

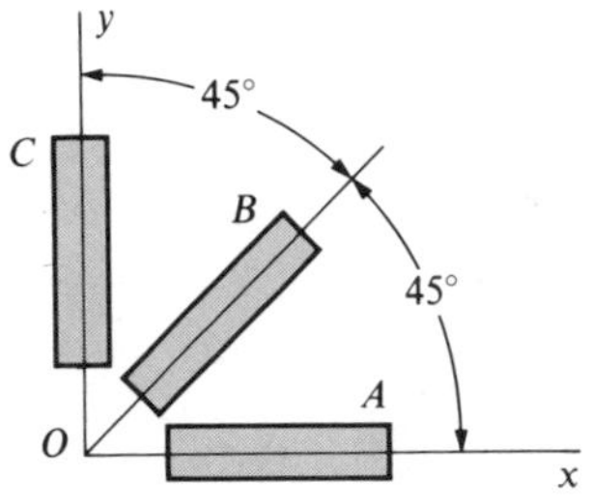

Probs. 6.11-9 and 6.11-10

6.11-10 A 45° strain rosette (see figure) mounted on the surface of an automobile frame that is being tested gives the following readings: gage A, 310×10^{-6}; gage B, 180×10^{-6}; and gage C, -160×10^{-6}. Determine the principal strains and maximum shear strains.

6.11-11 A cantilever beam AB of rectangular cross section (1 in. × 4 in.) is loaded by a force P that acts at the midheight of the beam and is inclined at an angle α to the vertical (see figure). Two strain gages are placed at point C, which is at the midheight of the beam. Gage 1 measures the strain in the horizontal direction and gage 2 measures the strain at an angle of 60° to the horizontal [see part (c) of the figure]. The measured strains are $\epsilon_1 = 125 \times 10^{-6}$ and $\epsilon_2 = -375 \times 10^{-6}$. Determine the force P and the angle α, assuming $E = 30 \times 10^6$ psi and $\nu = 1/3$.

6.11-12 A 60° strain rosette, or delta rosette, consists of three electrical-resistance strain gages arranged as shown in the figure. Gage A measures the normal strain ϵ_a in the direction of the x axis. Gages B and C measure the strains ϵ_b and ϵ_c in the inclined directions shown. Obtain the equations for the strains ϵ_x, ϵ_y, and γ_{xy} associated with the xy axes.

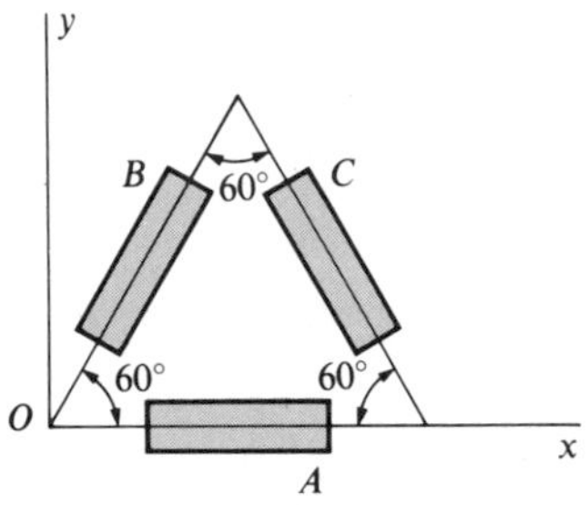

Prob. 6.11-12

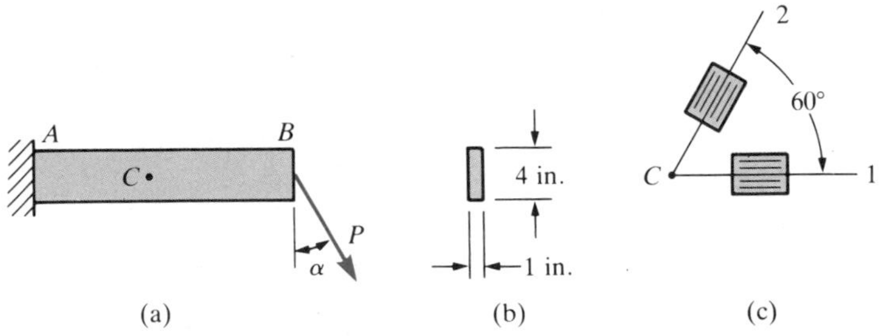

Prob. 6.11-11

CHAPTER 7

Deflections of Beams

7.1 INTRODUCTION

When a beam is loaded, the initially straight longitudinal axis is deformed into a curve, called the **deflection curve** of the beam. In this chapter we describe methods for determining the equation of the deflection curve and for finding deflections at specific points along the axis of the beam. The calculation of deflections is essential to the analysis of statically indeterminate beams, as explained in the next chapter. In addition, deflections often must be calculated in order to verify that they do not exceed the maximum permissible values. This situation arises in building design, where there is usually an upper limit on deflections because large deflections are associated with poor appearance and with too much flexibility in the structure. In machine design and aircraft design, the control of deflections is important in maintaining clearances and preventing undesirable vibrations.

7.2 DIFFERENTIAL EQUATIONS OF THE DEFLECTION CURVE

To obtain the general equations for the deflection curve of a beam, let us consider the cantilever beam AB shown in Fig. 7-1a. We take the origin of coordinates at the fixed end, with the x axis directed to the right and the y axis directed downward. As in previous discussions, we assume that the xy plane is a plane of symmetry and that all loads act in this plane; thus, the xy plane is the plane of bending. The **deflection** v of the beam at any point m_1 at distance x from the origin (Fig. 7-1a) is the translation (or displacement) of that point in the y direction, measured from the x axis to the deflection curve. For the axes we have selected, a downward deflection is positive and an upward deflection is negative. When v is expressed as a function of x, we have the equation of the deflection curve.

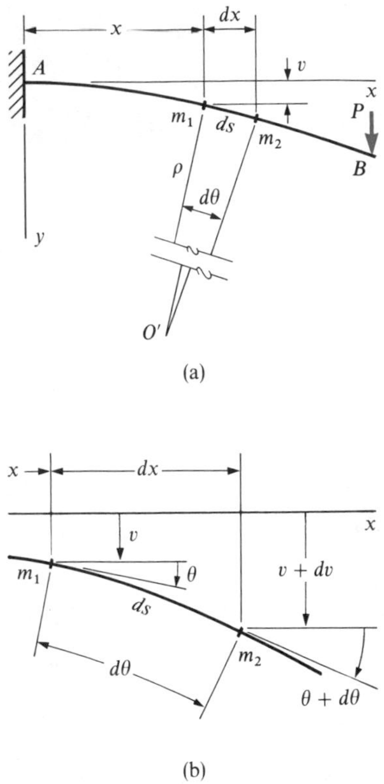

Fig. 7-1 Deflection curve of a beam

The **angle of rotation** θ of the axis of the beam at any point m_1 is the angle between the x axis and the tangent to the deflection curve (Fig. 7-lb). This angle is positive when clockwise, provided the x and y axes have the directions shown.

Now consider a second point m_2, located on the deflection curve at a small distance ds further along the curve and at distance $x + dx$ (measured parallel to the x axis) from the origin. The deflection at this point is $v + dv$, where dv is the increment in deflection as we move from m_1 to m_2. Also, the angle of rotation at m_2 is $\theta + d\theta$, where $d\theta$ is the increment in angle of rotation. At points m_1 and m_2, we can construct lines normal to the tangents to the deflection curve. The intersection of these normals locates the **center of curvature** O', and the distance from O' to the curve is the **radius of curvature** ρ. From the figure, we see that $\rho\, d\theta = ds$; hence the **curvature** κ (equal to the reciprocal of the radius of curvature) is given by the equation:

$$\kappa = \frac{1}{\rho} = \frac{d\theta}{ds} \qquad (7\text{-}1)$$

The **sign convention** for curvature is pictured in Fig. 5-5. Note that positive curvature corresponds to a positive value of $d\theta/ds$, which means that the angle θ increases as we move along the beam in the positive x direction.

The **slope** of the deflection curve is the first derivative dv/dx, as we know from calculus. From Fig. 7-1b, we see that the slope is equal to the tangent of the angle of rotation θ, because dx is infinitesimally small; thus

$$\frac{dv}{dx} = \tan\theta \qquad \text{or} \qquad \theta = \arctan\frac{dv}{dx} \qquad (7\text{-}2a,b)$$

Equations (7-1) and (7-2) are based only upon geometric considerations and apply to a beam of any material. Furthermore, there is no restriction on the magnitudes of the slopes and deflections.

Most beams undergo only very small rotations when they are loaded, resulting in very flat deflection curves with extremely small curvatures. Under these conditions, the angle θ is a very small quantity, and we can make some approximations that simplify the analysis. From Fig. 7-1b we see that

$$ds = \frac{dx}{\cos\theta}$$

Since $\cos\theta \approx 1$ when θ is small, we obtain

$$ds \approx dx \qquad (a)$$

Therefore, Eq. (7-1) becomes

$$\kappa = \frac{1}{\rho} = \frac{d\theta}{dx} \qquad (7\text{-}3)$$

Also, since $\tan\theta \approx \theta$ when θ is a small quantity, we can approximate Eq. (7-2a) as follows:

$$\theta \approx \tan\theta = \frac{dv}{dx} \tag{b}$$

Thus, for small rotations of a beam, the angle of rotation and the slope may be assumed to be equal. (Note that the angle of rotation is measured in radians.) Taking the derivative of θ with respect to x, we get

$$\frac{d\theta}{dx} = \frac{d^2v}{dx^2} \tag{c}$$

Now combining this equation with Eq. (7-3), we obtain

$$\kappa = \frac{1}{\rho} = \frac{d\theta}{dx} = \frac{d^2v}{dx^2} \tag{7-4}$$

This equation relates the curvature to the deflection v of the beam. It is valid for a beam of any material, provided the rotations are small.

If the material of the beam is linearly elastic and follows Hooke's law, the curvature (see Eq. 5-9) is:

$$\kappa = \frac{1}{\rho} = -\frac{M}{EI} \tag{7-5}$$

in which M is the bending moment and EI is the flexural rigidity of the beam. Note that Eq. (7-5) is valid for large as well as small rotations. The relationships between the signs for bending moments and for curvatures are portrayed in Fig. 5-11. Combining Eq. (7-4), which is limited to small rotations, with Eq. (7-5) yields

$$\frac{d\theta}{dx} = \frac{d^2v}{dx^2} = -\frac{M}{EI} \tag{7-6}$$

which is the basic **differential equation of the deflection curve** of a beam. This equation can be integrated in each particular case to find the angle of rotation θ or the deflection v, provided the bending moment M is known.

In summary, the **sign conventions** to be used with Eq. (7-6) are as follows: (1) the x and y axes are positive to the right and downward, respectively; (2) the angle of rotation θ is positive when clockwise from the x axis; (3) the deflection v is positive downward; (4) the bending moment M is positive when it produces compression in the upper part of the beam; and (5) the curvature is positive when the beam is bent concave downward. If the sign convention for M is reversed, or if the y axis (and hence v) is taken positive upward, then the minus sign in Eq. (7-6) should be changed to a plus sign. If both M and y are reversed in sign, the equation is unchanged.

By differentiating Eq. (7-6) with respect to x and then substituting the equations $q = -dV/dx$ and $V = dM/dx$ (see Eqs. 4-1 and 4-3), we

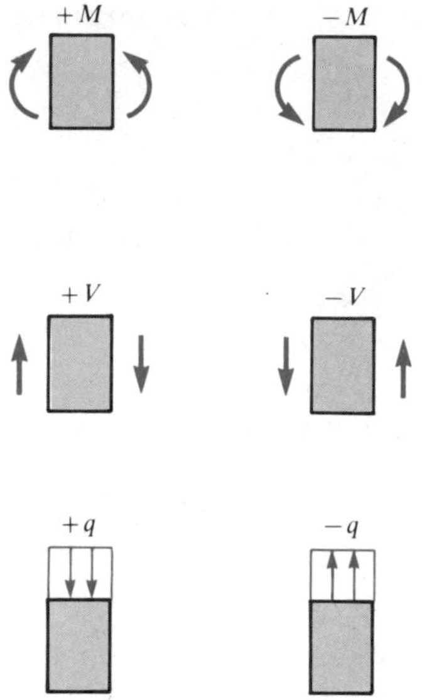

Fig. 7-2 Sign conventions for bending moment M, shear force V, and intensity of distributed load q

obtain:

$$\frac{d^3v}{dx^3} = -\frac{V}{EI} \tag{7-7}$$

$$\frac{d^4v}{dx^4} = \frac{q}{EI} \tag{7-8}$$

where V is the shear force and q is the intensity of distributed load. The deflection v can be found by solving any one of Eqs. (7-6) through (7-8), depending on mathematical convenience and personal preference. The sign conventions for M, V, and q are repeated in Fig. 7-2.

For simplicity in succeeding discussions, we often will use **primes** to denote differentiation:

$$v' \equiv \frac{dv}{dx} \qquad v'' \equiv \frac{d^2v}{dx^2} \qquad v''' \equiv \frac{d^3v}{dx^3} \qquad v'''' \equiv \frac{d^4v}{dx^4} \tag{7-9}$$

Using this notation, we can express the **differential equations** given above in the following form:

$$EIv'' = -M \qquad EIv''' = -V \qquad EIv'''' = q \tag{7-10a,b,c}$$

In the next two sections, we will use these equations to find deflections of beams. The procedure consists of successive integrations of the equations, with the resulting constants of integration being evaluated from the boundary conditions of the beam.

From the derivation of Eqs. (7-10), we see that they are valid only when Hooke's law applies for the material and when the slopes of the deflection curve are very small. Also, it should be realized that the equations were derived by considering the deformations due to pure bending and disregarding the shear deformations. These limitations are satisfactory for most practical purposes, although on rare occasions it may be necessary to consider the additional deflections due to shear effects (see Section 10.9).

EXACT EXPRESSION FOR CURVATURE

If the deflection curve of a beam has large slopes, we cannot use the approximations given by Eqs. (a) and (b) Instead, we must resort to the exact expressions for curvature (Eq. 7-1) and angle of rotation (Eq. 7-2b). Combining those expressions, we get

$$\kappa = \frac{1}{\rho} = \frac{d\theta}{ds} = \frac{d(\arctan v')}{dx}\,\frac{dx}{ds}$$

Noting from Fig. 7-1b that $ds^2 = dx^2 + dv^2$, we obtain

$$\frac{ds}{dx} = \left[1 + \left(\frac{dv}{dx}\right)^2\right]^{1/2} = [1 + (v')^2]^{1/2}$$

Also, by differentiation, we get

$$\frac{d}{dx}(\arctan v') = \frac{v''}{1 + (v')^2}$$

Substitution of these expressions into the equation for curvature yields

$$\kappa = \frac{1}{\rho} = \frac{v''}{[1 + (v')^2]^{3/2}} \tag{7-11}$$

Comparing this equation with Eq. (7-4), we see that the assumption of small slopes is equivalent to disregarding $(v')^2$ in comparison to unity, thus making the denominator in Eq. (7-11) equal to one. Equation (7-11) must be used for the curvature when solving problems that involve large deflections of beams (see Section 7.12).*

7.3 DEFLECTIONS BY INTEGRATION OF THE BENDING-MOMENT EQUATION

The equation of the deflection curve in terms of the bending moment (Eq. 7-10a) may be integrated to obtain the deflection v as a function of x. Since the differential equation is of second order, two integrations are required. The first step in the solution is to write the equations for the bending moment, using free-body diagrams and static equilibrium to obtain the equations.** If the loading on the beam changes abruptly as we move along the axis of the beam, there will be separate moment expressions for each region of the beam between the points at which such changes occur. For each of these regions, we substitute the expression for M into the differential equation. Then the equation is integrated to obtain the slope v', and a constant of integration is introduced by this process. A second integration gives the deflection v, and another constant of integration is introduced. Thus, there are two constants of integration for each region of the beam. These constants can be evaluated from boundary conditions pertaining to v and v' at the supports of the beam and from continuity conditions on v and v' at the points where the regions of integration meet. Each such condition gives an equation containing one or more of the constants of integration. Since the number of conditions always matches the number of constants, we can solve these equations for the constants. Then the evaluated constants can be substituted back into

* The basic relationship stating that the curvature of a beam is proportional to the bending moment (Eq. 7-5) was first obtained by Jacob Bernoulli, although he obtained an incorrect value for the constant of proportionality. The relationship was used later by Euler, who solved the differential equation of the deflection curve both for large deflections (using Eq. 7-11 for the curvature) and for small deflections (using Eq. 7-6). For the history of elastic curves, see Ref. 7-1.

** Only statically determinate beams are considered in this chapter. The analysis of statically indeterminate beams is described in Chapter 8.

the expressions for v, thus yielding the final equations of the deflection curve. This method for finding deflections is sometimes called the **method of successive integrations**. The following examples illustrate the procedure for both simple and cantilever beams.

Example 1

Determine the equation of the deflection curve for a simple beam AB supporting a uniform load of intensity q (Fig. 7-3). Also, determine the maximum deflection δ at the middle of the beam and the angles of rotation θ_a and θ_b at the supports.

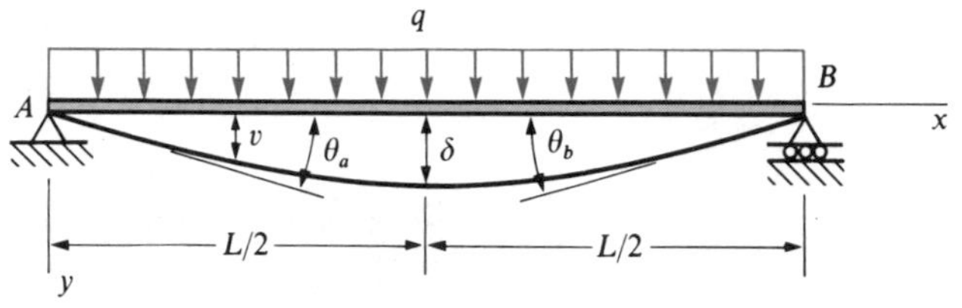

Fig. 7-3 Example 1. Deflections of a simple beam with a uniform load

With the origin of coordinates taken at the left-hand support, the equation for the bending moment in the beam is

$$M = \frac{qLx}{2} - \frac{qx^2}{2}$$

This expression for the moment is obtained by constructing a free-body diagram and writing an equation of static equilibrium. The second-order differential equation (Eq. 7-10a) becomes

$$EIv'' = -\frac{qLx}{2} + \frac{qx^2}{2}$$

Multiplying both sides of this equation by dx and integrating, we obtain

$$EIv' = -\frac{qLx^2}{4} + \frac{qx^3}{6} + C_1 \qquad \text{(a)}$$

where C_1 is a constant of integration. To evaluate this constant, we observe from symmetry that the slope v' of the deflection curve at midspan is zero. Thus, we have the condition

$$v' = 0 \qquad \text{when } x = \frac{L}{2}$$

which may be expressed more succinctly as

$$v'\left(\frac{L}{2}\right) = 0$$

Applying this condition to Eq. (a) gives

$$C_1 = \frac{qL^3}{24}$$

Equation (a) then becomes

$$EIv' = -\frac{qLx^2}{4} + \frac{qx^3}{6} + \frac{qL^3}{24} \tag{7-12}$$

Again multiplying both sides of the equation by dx and integrating, we get

$$EIv = -\frac{qLx^3}{12} + \frac{qx^4}{24} + \frac{qL^3x}{24} + C_2 \tag{b}$$

The constant of integration C_2 may be evaluated from the condition that $v = 0$ when $x = 0$, or

$$v(0) = 0$$

Applying this condition to Eq. (b) yields $C_2 = 0$; hence the equation for the deflection curve is

$$v = \frac{qx}{24EI}(L^3 - 2Lx^2 + x^3) \tag{7-13}$$

This equation gives the deflection at any point along the axis of the beam.

The maximum deflection δ occurs at the midpoint of the span and is obtained by setting x equal to $L/2$ in Eq. (7-13). The result is

$$\delta = v_{\max} = \frac{5qL^4}{384EI} \tag{7-14}$$

The maximum angles of rotation occur at the supports of the beam. At the left-hand end, the angle θ_a is equal to the slope v'; thus, by substituting $x = 0$ into Eq. (7-12), we obtain

$$\theta_a = v'(0) = \frac{qL^3}{24EI} \tag{7-15}$$

In a similar manner, we obtain the angle θ_b at the other end:

$$\theta_b = -v'(L) = \frac{qL^3}{24EI} \tag{7-16}$$

Since the beam and loading are symmetric about the midpoint, the angles of rotation at the ends are equal. Note that the angles of rotation are assumed to be positive when the ends of the beam rotate as shown in Fig. 7-3. Of course, the positive sense of the slope v' is determined by the directions of the coordinate axes. For the beam in this example, v' is positive at support A, negative at B, and zero at the midpoint.

Example 2

Determine the equation of the deflection curve for a cantilever beam AB subjected to a uniform load of intensity q (Fig. 7-4). Also, determine the deflection δ_b and angle of rotation θ_b at the free end.

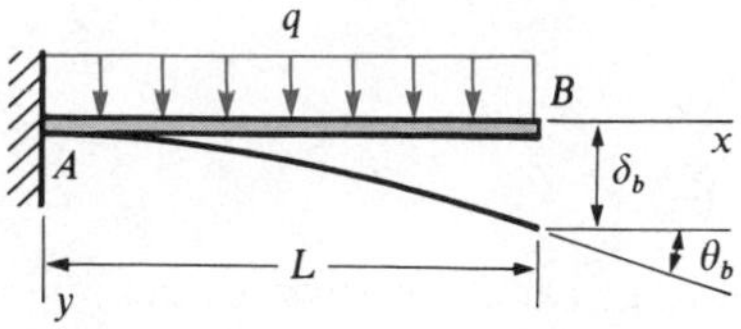

Fig. 7-4 Example 2. Deflections of a cantilever beam with a uniform load

We again take the origin of coordinates at the left-hand support. Then the expression for the bending moment (obtained from a free-body diagram and an equation of static equilibrium) is

$$M = -\frac{q(L-x)^2}{2}$$

Then the differential equation (Eq. 7-10a) becomes

$$EIv'' = \frac{q(L-x)^2}{2}$$

The first integration of this equation gives

$$EIv' = -\frac{q(L-x)^3}{6} + C_1$$

The constant of integration C_1 can be found from the condition that the slope of the beam is zero at the support; thus, we have $v'(0) = 0$, which gives $C_1 = qL^3/6$. Therefore, the slope is

$$v' = \frac{qx}{6EI}(3L^2 - 3Lx + x^2) \qquad \text{(7-17)}$$

Integration of this equation yields

$$v = \frac{qx^2}{24EI}(6L^2 - 4Lx + x^2) + C_2$$

The boundary condition on the deflection at the support is $v(0) = 0$, which shows that $C_2 = 0$. Thus, the equation of the deflection curve is

$$v = \frac{qx^2}{24EI}(6L^2 - 4Lx + x^2) \qquad \text{(7-18)}$$

The angle of rotation θ_b and the deflection δ_b at the free end of the beam (Fig. 7-4) are readily found by substituting $x = L$ into Eqs. (7-17) and (7-18), as follows:

$$\theta_b = v'(L) = \frac{qL^3}{6EI} \qquad \delta_b = v(L) = \frac{qL^4}{8EI} \qquad \text{(7-19a,b)}$$

These quantities are the maximum angle of rotation and maximum deflection, respectively.

Example 3

Determine the equations of the deflection curve for a simple beam AB supporting a concentrated load P (Fig. 7-5). Also, determine the angles of rotation θ_a and θ_b at the supports, the maximum deflection v_{max}, and the deflection δ_c at the midpoint C of the beam. (Note that the position of the load P is defined by the distances a and b from the supports A and B respectively.)

To obtain the deflections, we first must determine the expression for the bending moment in each part of the beam. Then we must write Eq. (7-10a) twice,

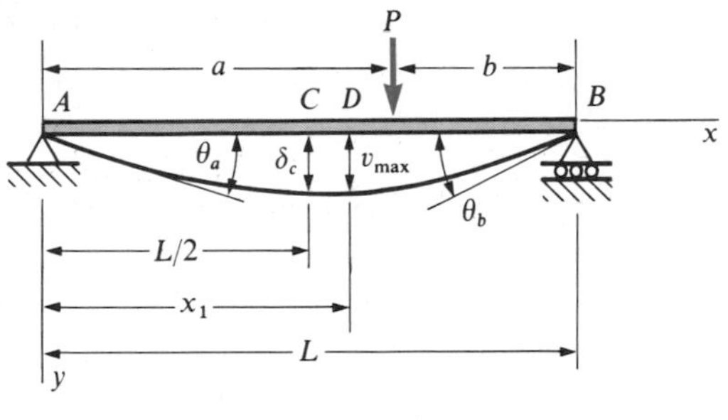

Fig. 7-5 Example 3. Deflections of a simple beam with a concentrated load

once for each part of the beam. The results are

$$EIv'' = -\frac{Pbx}{L} \qquad (0 \le x \le a)$$

$$EIv'' = -\frac{Pbx}{L} + P(x - a) \qquad (a \le x \le L)$$

The first integrations of these equations give

$$EIv' = -\frac{Pbx^2}{2L} + C_1 \qquad (0 \le x \le a) \tag{c}$$

$$EIv' = -\frac{Pbx^2}{2L} + \frac{P(x-a)^2}{2} + C_2 \qquad (a \le x \le L) \tag{d}$$

Performing second integrations, we obtain

$$EIv = -\frac{Pbx^3}{6L} + C_1 x + C_3 \qquad (0 \le x \le a) \tag{e}$$

$$EIv = -\frac{Pbx^3}{6L} + \frac{P(x-a)^3}{6} + C_2 x + C_4 \qquad (a \le x \le L) \tag{f}$$

The four constants of integration appearing in the preceding equations can be found from the following conditions: (1) at $x = a$, the slopes v' for the two parts of the beam must be equal; (2) at $x = a$, the deflections v for the two parts of the beam must be equal; (3) at $x = 0$, the deflection is zero; and (4) at $x = L$, the deflection is zero. From the first condition we see that the slopes determined from Eqs. (c) and (d) must be equal when $x = a$; thus,

$$-\frac{Pba^2}{2L} + C_1 = -\frac{Pba^2}{2L} + C_2$$

from which $C_1 = C_2$. The second condition means that the deflections found from Eqs. (e) and (f) must be equal when $x = a$:

$$-\frac{Pba^3}{6L} + C_1 a + C_3 = -\frac{Pba^3}{6L} + C_2 a + C_4$$

which gives $C_3 = C_4$. Finally, when conditions (3) and (4) are applied to Eqs. (e) and (f), respectively, we obtain

$$C_3 = 0 \quad \text{and} \quad -\frac{PbL^2}{6} + \frac{Pb^3}{6} + C_2 L = 0$$

From the preceding results, we see that

$$C_1 = C_2 = \frac{Pb(L^2 - b^2)}{6L} \qquad C_3 = C_4 = 0$$

Substitution of these values into Eqs. (e) and (f) gives the equations for the deflection curve:

$$EIv = \frac{Pbx}{6L}(L^2 - b^2 - x^2) \qquad (0 \leq x \leq a) \tag{7-20a}$$

$$EIv = \frac{Pbx}{6L}(L^2 - b^2 - x^2) + \frac{P(x-a)^3}{6} \qquad (a \leq x \leq L) \tag{7-20b}$$

The first of these equations gives the deflection curve for the part of the beam to the left of the load P, and the second gives the deflection to the right of the load.

The slopes for the two parts of the beam are found by substituting the values of C_1 and C_2 into Eqs. (c) and (d), which gives

$$EIv' = \frac{Pb}{6L}(L^2 - b^2 - 3x^2) \qquad (0 \leq x \leq a) \tag{7-21a}$$

$$EIv' = \frac{Pb}{6L}(L^2 - b^2 - 3x^2) + \frac{P(x-a)^2}{2} \qquad (a \leq x \leq L) \tag{7-21b}$$

From these equations the slope at any point of the deflection curve can be calculated.

To obtain the angles of rotation θ_a and θ_b at the ends of the beam, we substitute $x = 0$ into Eq. (7-21a) and $x = L$ into Eq. (7-21b):

$$\theta_a = v'(0) = \frac{Pb(L^2 - b^2)}{6LEI} = \frac{Pab(L+b)}{6LEI} \tag{7-22a}$$

$$\theta_b = -v'(L) = \frac{Pab(L+a)}{6LEI} \tag{7-22b}$$

The angle θ_a has its maximum value when $b = L/\sqrt{3}$.

The maximum deflection of the beam occurs at point D (Fig. 7-5) where the deflection curve has a horizontal tangent. If $a > b$, this point is in the part of the beam to the left of the load. We can locate this point by equating the slope v' from Eq. (7-21a) to zero. Denoting the distance from end A of the beam to the point of maximum deflection by x_1, we find from Eq. (7-21a) the following formula for x_1:

$$x_1 = \sqrt{\frac{L^2 - b^2}{3}} \qquad (a \geq b) \tag{7-23}$$

From this equation we see that as the load P moves from the middle of the beam ($b = L/2$) to the right-hand end (b approaches 0), the distance x_1 varies from $L/2$ to $L/\sqrt{3} = 0.577L$. Thus, the maximum deflection occurs at a point very close to the midpoint of the beam and always between the midpoint of the beam and the load. The maximum deflection is found by substituting x_1 (from Eq. 7-23) into Eq. (7-20a), yielding

$$v_{max} = \frac{Pb(L^2 - b^2)^{3/2}}{9\sqrt{3}\,LEI} \qquad (a \geq b) \qquad (7\text{-}24)$$

The deflection at the middle of the beam is obtained by substituting $x = L/2$ into Eq. (7-20a):

$$\delta_c = v\left(\frac{L}{2}\right) = \frac{Pb(3L^2 - 4b^2)}{48EI} \qquad (a \geq b) \qquad (7\text{-}25)$$

Because the maximum deflection always occurs near the midpoint of the beam, Eq. (7-25) gives a good approximation of the maximum deflection. In the most unfavorable case (when b approaches zero), the difference between the maximum deflection and the deflection at the midpoint is less than 3% of the maximum deflection.

An important special case occurs when the load P acts at the midpoint of the beam. Then $a = b = L/2$, and the preceding results simplify to the following:

$$v' = \frac{P}{16EI}(L^2 - 4x^2) \qquad \left(0 \leq x \leq \frac{L}{2}\right) \qquad (7\text{-}26)$$

$$v = \frac{Px}{48EI}(3L^2 - 4x^2) \qquad \left(0 \leq x \leq \frac{L}{2}\right) \qquad (7\text{-}27)$$

$$\theta_a = \theta_b = \frac{PL^2}{16EI} \qquad (7\text{-}28)$$

$$\delta_c = v_{max} = \frac{PL^3}{48EI} \qquad (7\text{-}29)$$

In this case the deflection curve is symmetric about the midpoint of the beam; however, the equations for v' and v (Eqs. 7-26 and 7-27) are valid only for the left-hand half of the beam.

7.4 DEFLECTIONS BY INTEGRATION OF THE SHEAR-FORCE AND LOAD EQUATIONS

The equations of the deflection curve in terms of the shear force V and the load q (Eqs. 7-10b and c, respectively) may also be used to obtain beam deflections. The procedure is similar to that for the bending-moment equation, except that more integrations are required. For instance, if we begin with the load equation, which is of fourth order, four integrations are needed in order to arrive at the deflection equation. These additional steps introduce additional constants of integration, which can be obtained from the boundary and continuity conditions. The conditions now include conditions on the shear forces and bending moments, as well as on the slopes and deflections.

The choice among the three differential equations depends upon mathematical convenience and personal preference. For instance, if the expression for the load q is easy to write but the expression for the bending moment M is not easily obtained by static equilibrium from free-body diagrams, then the load equation should be used. The following examples illustrate the techniques for making the analyses.

Example 1

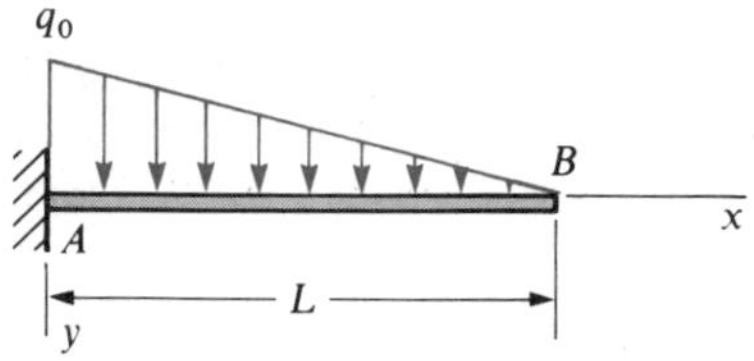

Fig. 7-6 Example 1. Deflections of a cantilever beam with a triangular load

Determine the equation of the deflection curve for a cantilever beam AB supporting a triangularly distributed load of maximum intensity q_0 (Fig. 7-6). Also, determine the deflection δ_b and angle of rotation θ_b at the free end.

The intensity of the distributed load is given by the following equation:

$$q = \frac{q_0(L - x)}{L}$$

Hence, the fourth-order differential equation (Eq. 7-10c) becomes

$$EIv'''' = \frac{q_0(L - x)}{L}$$

The first integration gives

$$EIv''' = \frac{q_0 x}{2L}(2L - x) + C_1 \tag{a}$$

The right-hand side of this equation represents the negative of the shear force V (see Eq. 7-10b). Because the shear force is zero at $x = L$, we obtain the following boundary condition:

$$v'''(L) = 0$$

Using this condition with Eq. (a), we get $C_1 = -q_0L/2$. When this expression for C_1 is substituted into Eq. (a), it becomes

$$EIv''' = -\frac{q_0}{2L}(L^2 - 2Lx + x^2) = -\frac{q_0}{2L}(L - x)^2$$

Integrating a second time, we obtain

$$EIv'' = \frac{q_0}{6L}(L - x)^3 + C_2 \tag{b}$$

which is the bending-moment equation (Eq. 7-10a). A second boundary condition is obtained at the free end of the beam, where the bending moment is zero:

$$v''(L) = 0$$

Applying this condition to Eq. (b), we obtain $C_2 = 0$; therefore,

$$EIv'' = \frac{q_0}{6L}(L - x)^3$$

The third and fourth integrations yield

$$EIv' = -\frac{q_0}{24L}(L - x)^4 + C_3 \tag{c}$$

$$EIv = \frac{q_0}{120L}(L - x)^5 + C_3x + C_4 \tag{d}$$

From the boundary conditions at the fixed support, which are

$$v'(0) = 0 \qquad v(0) = 0$$

we obtain

$$C_3 = \frac{q_0L^3}{24} \qquad C_4 = -\frac{q_0L^4}{120}$$

Substituting these expressions for the constants into Eqs. (c) and (d), we obtain the following equations for the slope and deflection of the beam:

$$v' = \frac{q_0x}{24LEI}(4L^3 - 6L^2x + 4Lx^2 - x^3) \tag{7-30}$$

$$v = \frac{q_0x^2}{120LEI}(10L^3 - 10L^2x + 5Lx^2 - x^3) \tag{7-31}$$

The angle of rotation θ_b and deflection δ_b at the free end ($x = L$) are now easily obtained from Eqs. (7-30) and (7-31); the results are

$$\theta_b = \frac{q_0L^3}{24EI} \qquad \delta_b = \frac{q_0L^4}{30EI} \tag{7-32a,b}$$

Thus, we have determined the required deflections and slopes of the beam by solving the fourth-order differential equation of the deflection curve.

Example 2

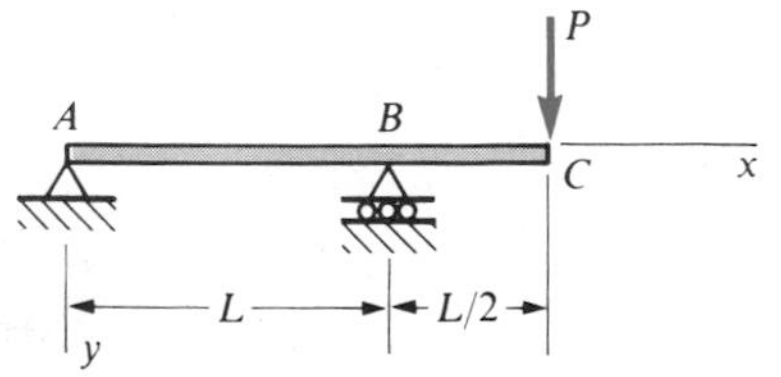

Fig. 7-7 Example 2. Deflections of a beam with an overhang

Determine the equations of the deflection curve for a simple beam with an overhang subjected to a concentrated load P (Fig. 7-7). The main span has length L and the overhang has length $L/2$.

Because reactive forces act at supports A and B, we must write separate differential equations for parts AB and BC of the beam. The shear forces in those parts are as follows:

$$V = -\frac{P}{2} \qquad (0 < x < L)$$

$$V = P \qquad \left(L < x < \frac{3L}{2}\right)$$

as found from static equilibrium. Thus, the third-order differential equations (see

Eq. 7-10b) are

$$EIv''' = \frac{P}{2} \qquad (0 < x < L)$$

$$EIv''' = -P \qquad \left(L < x < \frac{3L}{2}\right)$$

Integration of these equations yields the following bending-moment equations:

$$EIv'' = \frac{Px}{2} + C_1 \qquad (0 \le x \le L) \tag{e}$$

$$EIv'' = -Px + C_2 \qquad \left(L \le x \le \frac{3L}{2}\right) \tag{f}$$

The bending moments at points A and C are zero; hence we have the following boundary conditions:

$$v''(0) = 0 \qquad v''\left(\frac{3L}{2}\right) = 0$$

Using these conditions with Eqs. (e) and (f), we get

$$C_1 = 0 \qquad C_2 = \frac{3PL}{2}$$

from which

$$EIv'' = \frac{Px}{2} \qquad (0 \le x \le L)$$

$$EIv'' = \frac{P(3L - 2x)}{2} \qquad \left(L \le x \le \frac{3L}{2}\right)$$

The next integrations yield

$$EIv' = \frac{Px^2}{4} + C_3 \qquad (0 \le x \le L)$$

$$EIv' = \frac{Px(3L - x)}{2} + C_4 \qquad \left(L \le x \le \frac{3L}{2}\right)$$

The only condition on the slope is the continuity condition at support B:

$$v'(L) \text{ for } AB = v'(L) \text{ for } BC$$

or

$$\frac{PL^2}{4} + C_3 = PL^2 + C_4$$

This equation eliminates one constant of integration because we can express C_4 in terms of C_3:

$$C_4 = C_3 - \frac{3PL^2}{4} \tag{g}$$

The third and last integration gives

$$EIv = \frac{Px^3}{12} + C_3x + C_5 \qquad (0 \le x \le L)$$

$$EIv = \frac{Px^2(9L - 2x)}{12} + C_4x + C_6 \qquad \left(L \le x \le \frac{3L}{2}\right)$$

Since the deflections at points A and B are zero, we obtain three more boundary conditions:

$$v(0) = 0 \qquad v(L) \text{ for } AB = 0 \qquad v(L) \text{ for } BC = 0$$

The resulting equations for the constants are

$$C_5 = 0 \qquad C_3 = -\frac{PL^2}{12} \qquad C_4L + C_6 = -\frac{7PL^3}{12}$$

Combining these equations with Eq. (g) we get

$$C_4 = -\frac{5PL^2}{6} \qquad C_6 = \frac{PL^3}{4}$$

Thus, the deflection equations for the beam are

$$v = -\frac{Px}{12EI}(L^2 - x^2) \qquad (0 \le x \le L) \tag{7-33a}$$

$$v = \frac{P}{12EI}(3L^3 - 10L^2x + 9Lx^2 - 2x^3)$$

$$= \frac{P}{12EI}(3L - x)(L - x)(L - 2x) \qquad \left(L \le x \le \frac{3L}{2}\right) \tag{7-33b}$$

From Eq. (7-33b) we find the deflection at the end of the overhang ($x = 3L/2$):

$$\delta_c = \frac{PL^3}{8EI} \tag{7-34}$$

Note that the deflection of the beam is downward between B and C and upward between A and B.

7.5 MOMENT-AREA METHOD

In this section we consider another method for finding deflections of beams. Known as the **moment-area method**, it utilizes properties of the area of the bending-moment diagram. The method is suitable when the deflection or angle of rotation at only one point of the beam is desired, because it is possible to find such quantities without first finding the complete equation of the deflection curve.

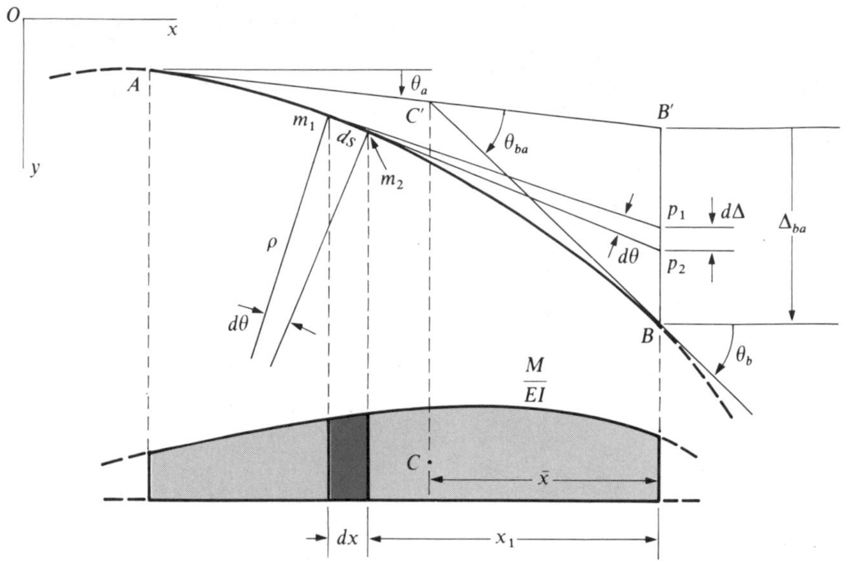

Fig. 7-8 Moment-area method

To explain the method, let us consider a segment AB of the deflection curve of a beam in a region where the curvature is positive (Fig. 7-8). At point A the tangent AB' to the deflection curve has a positive angle of rotation θ_a from the x axis, and at point B the tangent $C'B$ is at an angle θ_b. The angle between the tangents, denoted θ_{ba}, is equal to the difference between θ_b and θ_a:

$$\theta_{ba} = \theta_b - \theta_a \tag{7-35}$$

Thus, θ_{ba} represents the relative angle of rotation of the tangent at B with respect to the tangent at A. The relative angle θ_{ba} is positive when θ_b is larger than θ_a, as shown in the figure.

Next, consider two points m_1 and m_2 on the axis of the beam, a distance ds apart. The tangents to the deflection curve at these points are shown in the figure as lines m_1p_1 and m_2p_2. The normals to these tangents intersect at the center of curvature at an angle $d\theta$, which is equal to ds/ρ, where ρ is the radius of curvature. It follows that the angle between the two tangents also is equal to $d\theta$. The angle $d\theta$ can be obtained from Eq. (7-6):

$$d\theta = -\frac{M\,dx}{EI} \tag{a}$$

in which M is the bending moment in the beam and EI is the flexural rigidity.

The quantity $M\,dx/EI$ has a simple geometric interpretation. Directly below the beam in Fig. 7-8, we draw the M/EI diagram (that is, a diagram in which the ordinate at any point is equal to the bending moment M at that point divided by the flexural rigidity EI at that point). Thus, the M/EI diagram has the same shape as the bending-moment diagram *only* when EI is constant. The term $M\,dx/EI$ is the area of the shaded strip (Fig. 7-8) within the M/EI diagram. (Note that Fig. 7-8 is drawn with both the curvature and the bending moment shown as positive quantities. This is done so that the derivation of the moment-area theorems will be in terms of positive quantities.)

Let us now integrate Eq. (a) between points A and B:

$$\int_A^B d\theta = -\int_A^B \frac{M\,dx}{EI} \tag{b}$$

The integral on the left is equal to $\theta_b - \theta_a$, which is the relative angle θ_{ba} between the tangents at B and A. The integral on the right is equal to the area of the M/EI diagram between points A and B. Note that the area of the M/EI diagram is an algebraic quantity and that it may be positive or negative, depending upon whether the bending moment is positive or negative.

Now we can write Eq. (b) as follows:

$$\begin{aligned}\theta_{ba} &= -\int_A^B \frac{M\,dx}{EI}\\ &= -[\text{area of } M/EI \text{ diagram between } A \text{ and } B]\end{aligned} \tag{7-36}$$

This equation may be stated as a theorem:

> **First moment-area theorem:** The angle θ_{ba} between the tangents to the deflection curve at two points A and B is equal to the negative of the area of the M/EI diagram between those points.

The sign conventions used with this theorem are summarized as follows: (1) The relative angle θ_{ba} between the tangents is positive when the angle θ_b is algebraically larger than the angle θ_a, as illustrated in Fig. 7-8. Note that point B must be to the right of point A; that is, it must be further along the axis of the beam as we move in the x direction. (2) The bending moment M is positive according to our usual sign convention; that is, M is positive when it produces compression in the upper part of the beam. (3) The area of the M/EI diagram is given a positive or negative sign according to whether the bending moment is positive or negative. If part of the bending-moment diagram is positive and part is negative, then the corresponding parts of the M/EI diagram are given those same signs. Thus, the areas of the M/EI diagram are treated as algebraic quantities.

The first moment-area theorem is used in deflection calculations to relate the angles of rotation between selected points along the axis of the beam, as illustrated later.

As the next step in the analysis, let us consider the vertical offset Δ_{ba} between point B on the deflection curve and point B' on the tangent at A

(Fig. 7-8). Recalling that the angles of rotation θ_a and θ_b are very small quantities (and hence the tangents at A and B are nearly horizontal lines), we observe from the figure that the vertical distance $d\Delta$ (equal to p_1p_2) is $x_1\, d\theta$, where x_1 is the horizontal distance from the element m_1m_2 to point B. Since $d\theta = -M\,dx/EI$, we can write

$$d\Delta = x_1\, d\theta = -x_1 \frac{M\,dx}{EI} \tag{c}$$

The distance $d\Delta$ represents the contribution made by the bending of element m_1m_2 to the total offset Δ_{ba}. The expression $x_1 M\,dx/EI$ may be interpreted geometrically as the first moment of the shaded element of area ($M\,dx/EI$) of the M/EI diagram taken about a vertical line through B. Integrating Eq. (c) between points A and B, we get

$$\int_A^B d\Delta = -\int_A^B x_1 \frac{M\,dx}{EI} \tag{d}$$

The integral on the left is equal to Δ_{ba}, which is the vertical offset of point B from the tangent at A. The integral on the right represents the first moment with respect to point B of the area of the M/EI diagram between A and B. Therefore, we can write Eq. (d) as follows:

$$\begin{aligned}\Delta_{ba} &= -\int_A^B x_1 \frac{M\,dx}{EI}\\ &= -[\text{first moment of the area of the } M/EI \text{ diagram}\\ &\qquad \text{between } A \text{ and } B\text{, taken with respect to } B]\end{aligned} \tag{7-37}$$

This equation represents the second theorem:

> **Second moment-area theorem:** The offset Δ_{ba} of point B from the tangent at A is equal to the negative of the first moment of the area of the M/EI diagram between A and B, taken with respect to B.

Note that the offset Δ_{ba} is positive in the y direction. If, as we move from A to B in the x direction, the area of the M/EI diagram is negative, then the first moment is also negative and the offset is positive, which means that point B is below the tangent at A. This situation is illustrated in Fig. 7-8. If the area is positive, then the first moment is positive, the offset is negative, and point B is above the tangent at A.

The first moment of the area of the M/EI diagram can be obtained by taking the product of the area of the diagram and the distance $\bar{x}$ from point B to the centroid C of the area (Fig. 7-8). This procedure is usually more convenient than integrating, because the diagram usually consists of familiar geometric figures such as rectangles, triangles, and parabolic segments. The areas and centroidal distances of such figures are tabulated in Appendix D.

The second moment-area theorem is useful for finding deflections because it relates the position of a point on the beam to the tangent at some other point. The techniques to be used are illustrated in the examples that follow.

In many beams, it is obvious whether the beam deflects upward or downward and whether the angle of rotation is clockwise or counterclockwise. In such cases, it is not necessary to follow the complicated sign conventions that have been described for the moment-area method. Instead, we can calculate with absolute values and determine the directions by inspection.

The moment-area theorems apply only to linearly elastic beams, because they are based upon Eq. (7-6). Later, in Section 7.13, we will use more general curvature-area theorems that apply to inelastic beams.*

Example 1

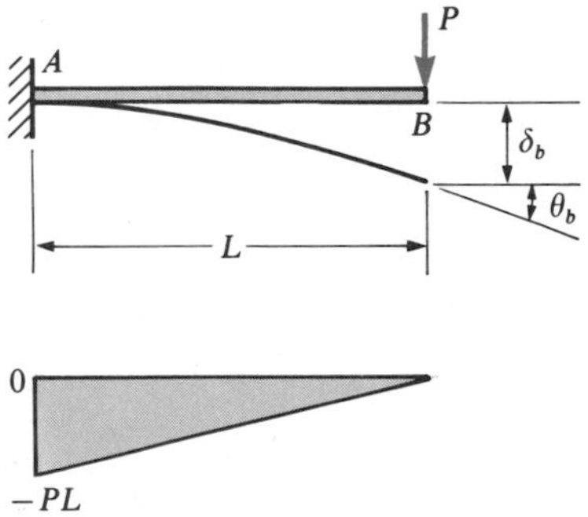

Fig. 7-9 Example 1. Cantilever beam with a concentrated load

Determine the angle of rotation θ_b and deflection δ_b at the free end B of a cantilever beam AB supporting a concentrated load P (Fig. 7-9).

The bending-moment diagram for this beam is triangular in shape, as shown in the lower part of the figure. Since the flexural rigidity EI is constant, the M/EI diagram has the same shape as the bending-moment diagram. From the first moment-area theorem, we know that the relative angle of rotation θ_{ba} between the tangents at B and A is equal to the negative of the area of the M/EI diagram. The area of the diagram is

$$A_1 = \frac{1}{2}(L)(-PL)\left(\frac{1}{EI}\right) = -\frac{PL^2}{2EI}$$

and, therefore,

$$\theta_{ba} = \theta_b - \theta_a = -A_1 = \frac{PL^2}{2EI}$$

The tangent to the deflection curve at A is horizontal ($\theta_a = 0$); hence

$$\theta_b = \frac{PL^2}{2EI} \tag{7-38}$$

The end of the beam rotates clockwise, as shown in the figure.

The deflection δ_b at the free end can be obtained from the second theorem. The offset Δ_{ba} of point B from the tangent at A is the same as the deflection δ_b in this case. The first moment of the area of the M/EI diagram, taken with respect to point B, is

$$Q_1 = A_1\left(\frac{2L}{3}\right) = -\frac{PL^2}{2EI}\left(\frac{2L}{3}\right)$$

$$= -\frac{PL^3}{3EI}$$

* The moment-area method was introduced by Saint-Venant (Refs. 7-6 and 7-7). The method was developed more fully by Mohr (Refs. 7-8 and 7-9) and by Greene (Ref. 7-10).

From the second moment-area theorem, we obtain $\delta_b = -Q_1$, or

$$\delta_b = \frac{PL^3}{3EI} \tag{7-39}$$

A positive value means that both the offset and the deflection are downward.

Example 2

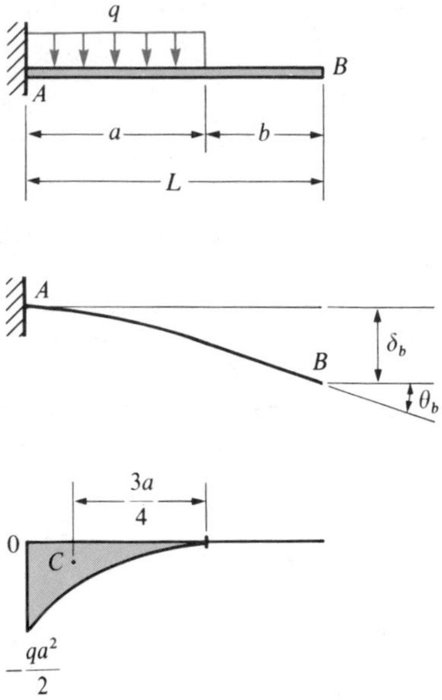

Fig. 7-10 Example 2. Cantilever beam with a uniform load over part of the length

Find the angle of rotation θ_b and deflection δ_b at the free end B of a cantilever beam AB with a uniform load of intensity q acting over part of the length (Fig. 7-10).

The bending-moment diagram consists of a parabolic spandrel of second degree (see Case 19, Appendix D, for properties of this area). The M/EI diagram has the same shape, inasmuch as EI is constant. Note that the deflected axis of the beam is curved in the region under the uniform load and is straight in the unloaded part.

The angle of rotation θ_b is equal to the relative angle θ_{ba}, inasmuch as the tangent at A is horizontal. The area of the M/EI diagram is

$$A_1 = \frac{1}{3}(a)\left(-\frac{qa^2}{2}\right)\left(\frac{1}{EI}\right) = -\frac{qa^3}{6EI}$$

From the first moment-area theorem, we obtain $\theta_b = -A_1$, or

$$\theta_b = \frac{qa^3}{6EI} \tag{7-40}$$

This angle is equal to the slope of the beam throughout the unloaded region.

The deflection δ_b is equal to the offset Δ_{ba} in this case. Hence, δ_b equals the negative of the first moment of the area of the M/EI diagram, taken with respect to B. The centroid C of the diagram is at distance $3a/4$ from the edge of the load, or distance $b + 3a/4$ from B. Thus, the first moment is

$$Q_1 = A_1\left(b + \frac{3a}{4}\right) = \left(-\frac{qa^3}{6EI}\right)\left(b + \frac{3a}{4}\right) = -\frac{qa^3}{24EI}(4L - a)$$

since $b = L - a$. The deflection at the end is $\delta_b = -Q_1$, or

$$\delta_b = \frac{qa^3}{24EI}(4L - a) \tag{7-41}$$

If $a = L$, this equation becomes $\delta_b = qL^4/8EI$, which is the deflection of a fully loaded cantilever beam (Eq. 7-19b).

Example 3

Find the angle of rotation θ_b and deflection δ_b at the free end B of a cantilever beam AB with a uniform load of intensity q acting over the right-hand half of the length (Fig. 7-11).

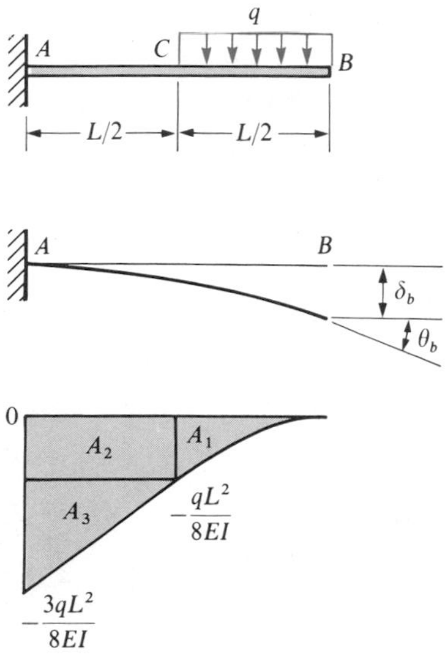

Fig. 7-11 Example 3. Cantilever beam with a uniform load over one-half of the length

The bending-moment diagram consists of a parabolic curve from B to C and a straight line from C to A. The M/EI diagram has the same shape, since EI is constant. For the purpose of evaluating the area and first moment of the M/EI diagram, it is convenient to divide this diagram into three parts having areas A_1, A_2, and A_3. These parts are a parabolic spandrel, a rectangle, and a triangle, respectively, with areas as follows:

$$A_1 = \frac{1}{3}\left(\frac{L}{2}\right)\left(-\frac{qL^2}{8EI}\right) = -\frac{qL^3}{48EI}$$

$$A_2 = \frac{L}{2}\left(-\frac{qL^2}{8EI}\right) = -\frac{qL^3}{16EI}$$

$$A_3 = \frac{1}{2}\left(\frac{L}{2}\right)\left(-\frac{qL^2}{4EI}\right) = -\frac{qL^3}{16EI}$$

The angle of rotation θ_b is equal to the negative of the area of the M/EI diagram:

$$\theta_b = -(A_1 + A_2 + A_3) = \frac{7qL^3}{48EI} \qquad (7\text{-}42)$$

The deflection δ_b is equal to the negative of the first moment of the M/EI diagram, taken about B:

$$\delta_b = -(A_1\bar{x}_1 + A_2\bar{x}_2 + A_3\bar{x}_3)$$

in which $\bar{x}_1$, $\bar{x}_2$, $\bar{x}_3$ are the distances from B to the centroids of the respective areas. Thus,

$$\delta_b = \frac{qL^3}{48EI}\left(\frac{3L}{8}\right) + \frac{qL^3}{16EI}\left(\frac{3L}{4}\right) + \frac{qL^3}{16EI}\left(\frac{5L}{6}\right) = \frac{41qL^4}{384EI} \qquad (7\text{-}43)$$

This example illustrates how the area and first moment of a complicated M/EI diagram can be determined by dividing the diagram into parts having known properties.

Example 4

A simple beam AB supports a concentrated load P as shown in Fig. 7-12. Determine the angle of rotation θ_a at support A, the deflection δ under the load P, and the maximum deflection δ_{max}.

We begin by constructing the tangent AB' at support A. Then we note that the distance BB' is the offset Δ_{ba} of point B from the tangent. Hence, we can calculate the distance BB' by evaluating the first moment of the M/EI diagram with respect to B and using the second moment-area theorem. The bending-moment diagram is triangular, with maximum ordinate Pab/L, as shown in the figure, and the M/EI diagram has the same shape, since EI is constant. Thus, the area of the M/EI diagram is

$$A_1 = \frac{1}{2}(L)\left(\frac{Pab}{L}\right)\left(\frac{1}{EI}\right) = \frac{Pab}{2EI}$$

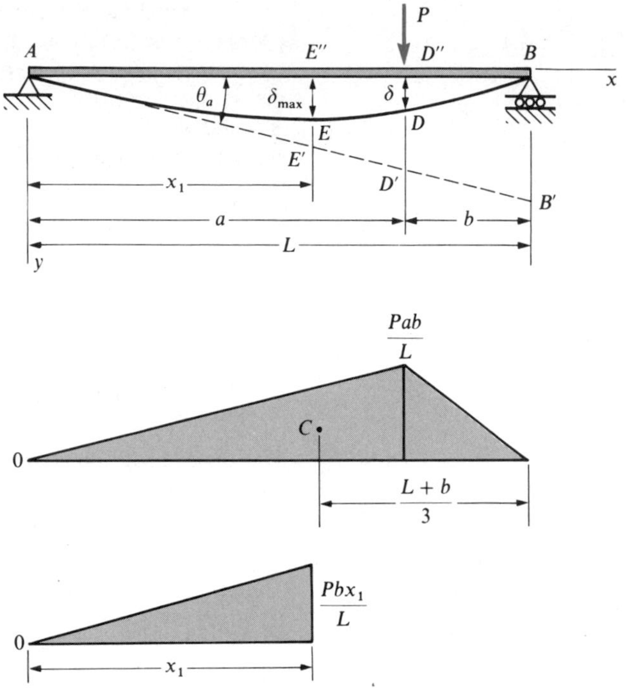

Fig. 7-12 Example 4. Simple beam with a concentrated load

The centroid C of this area is at distance $(L + b)/3$ from B (see Case 3, Appendix D). Therefore, the offset Δ_{ba} is

$$\Delta_{ba} = -A_1\left(\frac{L+b}{3}\right) = -\frac{Pab}{6EI}(L+b)$$

The minus sign shows that the offset is in the negative y direction, or upward; that is, point B is above the tangent. The distance BB' is numerically equal to Δ_{ba}, as follows:

$$BB' = \frac{Pab}{6EI}(L+b)$$

Also, we see from the figure that the angle θ_a is equal to the distance BB' divided by the length of the beam:

$$\theta_a = \frac{BB'}{L} = \frac{Pab}{6LEI}(L+b) \tag{7-44}$$

Thus, the angle of rotation at support A has been found.

As can be seen from Fig. 7-12, the deflection δ under the load P is equal to the distance $D'D''$ minus the distance $D'D$. The distance $D'D''$ is equal to $a\theta_a$, and the distance $D'D$ can be found from the second moment-area theorem. This latter distance is numerically equal to the offset of point D from the tangent at A. Hence, we take the first moment of the area of the M/EI diagram between A and D with respect to D. The desired area of the diagram is the area of the left-hand triangle divided by EI:

$$A_2 = \frac{1}{2}(a)\left(\frac{Pab}{L}\right)\left(\frac{1}{EI}\right) = \frac{Pa^2b}{2LEI}$$

Its first moment about D is

$$Q_1 = A_2\left(\frac{a}{3}\right) = \frac{Pa^3b}{6LEI}$$

The offset Δ_{da} is the negative of this expression; however, we need the actual distance $D'D$, which is equal to Q_1 itself:

$$D'D = \frac{Pa^3b}{6LEI}$$

Thus, for the deflection δ at point D, we obtain

$$\delta = D'D'' - D'D = a\theta_a - \frac{Pa^3b}{6LEI} = \frac{Pa^2b^2}{3LEI} \qquad \text{(7-45)}$$

The preceding formulas for θ_a and δ are valid for any position of the load P.

To determine the maximum deflection, let us assume that $a \geq b$, so that the maximum deflection occurs to the left of the load (or under the load in the special case when $a = b$). The maximum deflection occurs at point E (distance x_1 from support A), where the deflection curve has a horizontal tangent. The relative angle θ_{ea} between the tangents at E and A is equal to the negative of the area of the M/EI diagram between A and E, according to the first moment-area theorem. The portion of the bending-moment diagram between A and E is shown in the lowest part of Fig. 7-12. The area of the corresponding M/EI diagram is

$$A_3 = \frac{1}{2}(x_1)\left(\frac{Pbx_1}{L}\right)\left(\frac{1}{EI}\right) = \frac{Pbx_1^2}{2LEI}$$

From the first theorem, we now obtain

$$\theta_{ea} = \theta_e - \theta_a = -A_3 = -\frac{Pbx_1^2}{2LEI}$$

The angle θ_e is zero, and the angle θ_a is given by Eq. (7-44); substituting these values, we get an equation that can be solved for the distance x_1:

$$-\frac{Pab}{6LEI}(L + b) = -\frac{Pbx_1^2}{2LEI}$$

from which

$$x_1 = \sqrt{\frac{a(2L - a)}{3}} = \sqrt{\frac{L^2 - b^2}{3}} \qquad \text{(7-46)}$$

Thus, the distance to point E is determined.

The maximum deflection δ_{max} is equal to $E'E''$ minus $E'E$; the former is equal to $x_1\theta_a$, and the latter can be obtained from the second moment-area theorem in the same way that we found the distance $D'D$. The resulting calculation is as follows:

$$\delta_{max} = x_1\theta_a - A_3\left(\frac{x_1}{3}\right) = \frac{Pb}{9\sqrt{3}\,LEI}(L^2 - b^2)^{3/2} \tag{7-47}$$

An alternate and slightly simpler scheme for finding δ_{max} is to note that δ_{max} is numerically equal to the offset of point A from the tangent at E. Hence, we can find δ_{max} by determining the first moment of the area of the M/EI diagram between A and E with respect to A:

$$\delta_{max} = A_3\left(\frac{2x_1}{3}\right) = \frac{Pbx_1^3}{3LEI} \tag{7-48}$$

Substituting for x_1 from Eq. (7-46), we get the same expression as before (Eq. 7-47). Equations (7-46) through (7-48) are valid when $a \geq b$.

7.6 METHOD OF SUPERPOSITION

The differential equations of the deflection curve of a beam (Eqs. 7-10) are linear differential equations, which means that all terms containing the deflection v and its derivatives are raised to the first power only. Therefore, solutions of the equations for various loading conditions may be superimposed. Thus, the deflection of the beam caused by several different loads acting simultaneously can be found by superimposing the deflections caused by the loads acting separately. For instance, if v_1 represents the deflection due to a load q_1 and if v_2 represents the deflection due to a load q_2, the total deflection produced by q_1 and q_2 acting simultaneously is $v_1 + v_2$.

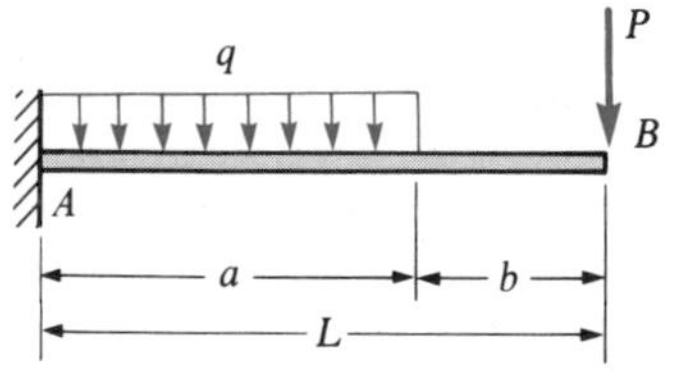

Fig. 7-13 Cantilever beam with two loads

To illustrate this idea, consider the cantilever beam shown in Fig. 7-13. This beam supports a uniform load of intensity q over part of the span and a concentrated load P acting at the end. Suppose we wish to find the deflection δ_b at the free end. When the load P acts alone, the deflection at B is $PL^3/3EI$, as shown in Example 1 of the preceding section (Eq. 7-39). When the uniform load acts alone, the deflection is $qa^3(4L - a)/24EI$, as obtained in Example 2 of the preceding section (Eq. 7-41). Hence, the deflection δ_b due to the combined loading is

$$\delta_b = \frac{PL^3}{3EI} + \frac{qa^3(4L - a)}{24EI} \tag{7-49}$$

The deflection and angle of rotation at any point of the beam can be found by this procedure.

The method of superposition is most useful when the loading system on the beam can be subdivided into loading conditions that produce deflections that are already known, as illustrated in the example just given. For convenient use in cases of this kind, **tables of beam deflections** are given in Appendix G. Using these tables and the method of superposition,

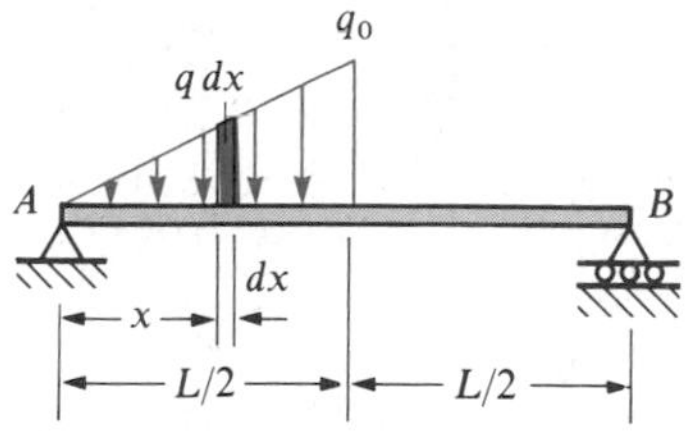

Fig. 7-14 Simple beam with a triangular load

we can find deflections and angles of rotation for many different loading conditions for beams. Examples of this type are given at the end of this section.

Superposition may also be used for distributed loadings that are not already given in the tables by considering an element of the distributed load as if it were a concentrated load and then integrating throughout the region of the load. This procedure can be easily understood from the example shown in Fig. 7-14. The load on the simple beam AB is triangularly distributed over the left-hand half of the beam, and we wish to obtain the deflection δ at the midpoint. An element $q\,dx$ of the distributed load can be visualized as a concentrated load. The deflection at the midpoint produced by a concentrated load P acting at distance x from the left end is

$$\frac{Px}{48EI}(3L^2 - 4x^2)$$

which is obtained from Case 5 of Table G-2 in Appendix G. Substituting $q\,dx$ for P in this expression, and noting that $q = 2q_0x/L$, we obtain for the deflection

$$\delta = \int_0^{L/2} \frac{qx\,dx}{48EI}(3L^2 - 4x^2)$$

$$= \frac{q_0}{24LEI}\int_0^{L/2}(3L^2 - 4x^2)x^2\,dx = \frac{q_0L^4}{240EI} \qquad (7\text{-}50)$$

By this same procedure of superimposing elements of the distributed load, we can calculate the angle of rotation θ_a at the left end of the beam. The expression for this angle due to a concentrated load P (see Case 5 of Table G-2) is

$$\frac{Pab(L + b)}{6LEI}$$

In this expression, we must replace P with $2q_0x\,dx/L$, a with x, and b with $L - x$; thus;

$$\theta_a = \int_0^{L/2} \frac{q_0x\,dx}{3L^2EI}(x)(L - x)(2L - x) = \frac{41q_0L^3}{2880EI} \qquad (7\text{-}51)$$

Another illustration of this superposition technique is given in Example 2.

In each of the preceding illustrations, we have used the **principle of superposition** to obtain deflections of beams. This concept is widely used in mechanics and is valid whenever the quantity to be determined is a linear function of the applied loads. Under such conditions, the desired quantity may be found due to each load acting separately, and then the results may be superimposed to obtain the total value due to all loads acting simultaneously. In the case of deflections of beams, the principle of superposition is valid if Hooke's law holds for the material and if the deflections and rotations of the beam are small. The requirement of small rotations ensures that the differential equation of the deflection curve is linear, and the requirement of small deflections ensures that the lines of action of the loads and reactions are not changed significantly from their original positions.

The following examples further illustrate the use of the principle of superposition for calculating deflections and rotations of beams.

Example 1

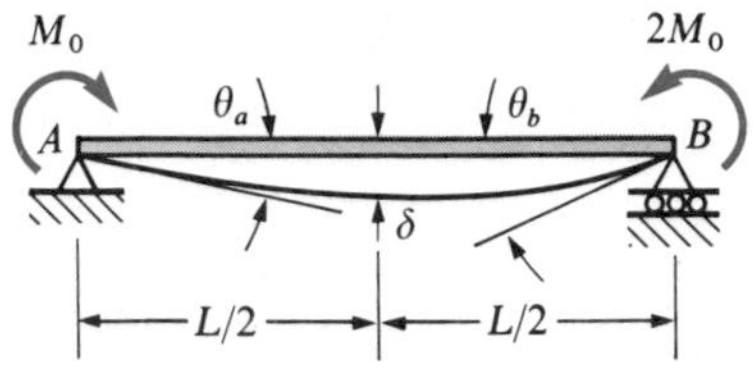

Fig. 7-15 Example 1. Simple beam with couples acting at the ends

A simple beam AB is acted upon by couples M_0 and $2M_0$ at the ends (see Fig. 7-15). Obtain expressions for the angles of rotation θ_a and θ_b at the ends of the beam and the deflection δ at the middle.

Using Case 7 of Table G-2, we obtain by superposition

$$\theta_a = \frac{M_0L}{3EI} + \frac{(2M_0)L}{6EI} = \frac{2M_0L}{3EI}$$

$$\theta_b = \frac{M_0L}{6EI} + \frac{(2M_0)L}{3EI} = \frac{5M_0L}{6EI}$$

$$\delta = \frac{M_0L^2}{16EI} + \frac{(2M_0)L^2}{16EI} = \frac{3M_0L^2}{16EI}$$

Thus, the required quantities have been found.

Example 2

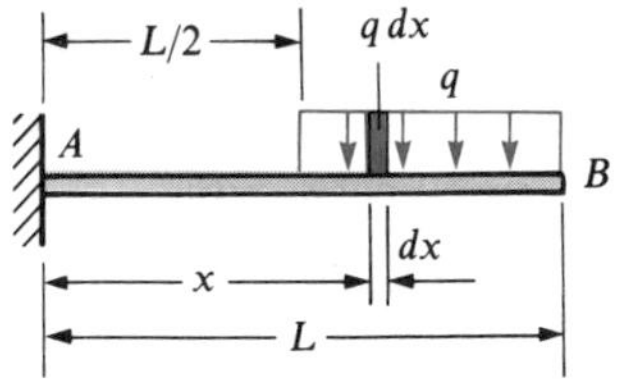

Fig. 7-16 Example 2. Cantilever beam with a uniform load over one-half of the length

A cantilever beam AB carries a uniform load of intensity q over the right-hand half of its length, as shown in Fig. 7-16. Find the deflection δ_b and the angle of rotation θ_b at the free end.

We begin by considering an element $q\,dx$ of the load located at distance x from the support. This element of load produces a deflection $d\delta$ and an angle $d\theta$ at the free end equal to

$$d\delta = \frac{(q\,dx)(x^2)(3L - x)}{6EI} \qquad d\theta = \frac{(q\,dx)(x^2)}{2EI}$$

as found from Case 5 of Table G-1. By integrating over the loaded region we get

$$\delta_b = \frac{q}{6EI}\int_{L/2}^{L} x^2(3L - x)dx = \frac{41qL^4}{384EI} \tag{7-52}$$

$$\theta_b = \frac{q}{2EI}\int_{L/2}^{L} x^2\,dx = \frac{7qL^3}{48EI} \tag{7-53}$$

These same results can be obtained more simply by using the formulas in Case 3 of Table G-1 and substituting $a = b = L/2$.

Example 3

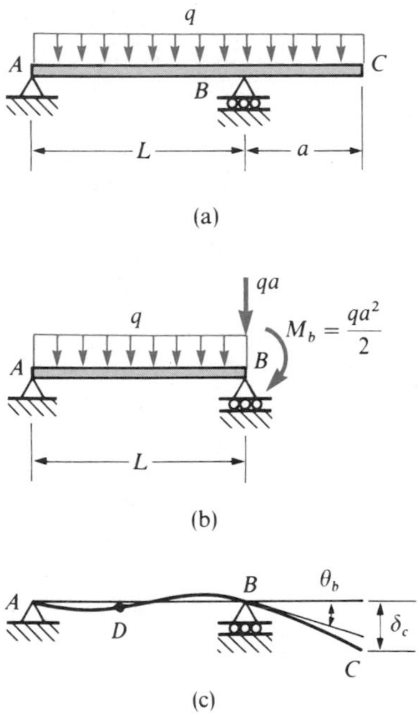

Fig. 7-17 Example 3. Simple beam with an overhang

A simple beam with an overhang is loaded as shown in Fig. 7-17a. Find the deflection δ_c at the end of the overhang.

The deflection of point C is made up of two parts: (1) a deflection δ_1 caused by the rotation of the beam axis at support B, and (2) a deflection δ_2 caused by the bending of part BC acting as a cantilever beam. To obtain the first part of the deflection, we observe that portion AB of the beam is in the same condition as a simple beam carrying a uniform load and subjected to a couple M_b (equal to $qa^2/2$) and a vertical load (equal to qa) acting at the right-hand end, as shown in Fig. 7-17b. The angle θ_b at end B (see Cases 1 and 7 of Table G-2) is:

$$\theta_b = -\frac{qL^3}{24EI} + \frac{M_b L}{3EI} = \frac{qL(4a^2 - L^2)}{24EI}$$

in which clockwise rotation is positive. The deflection δ_1 of point C, due only to the rotation at B, is equal to $a\theta_b$, or

$$\delta_1 = \frac{qaL(4a^2 - L^2)}{24EI}$$

This deflection is positive when downward.

The bending of the overhang itself produces an additional downward deflection δ_2 at C. This deflection is equal to the deflection of a cantilever beam of length a (see Case 1 of Table G-1):

$$\delta_2 = \frac{qa^4}{8EI}$$

The total deflection of point C, assumed to be positive when downward, is

$$\delta_c = \delta_1 + \delta_2 = \frac{qa}{24EI}(3a^3 + 4a^2L - L^3) \qquad \text{(7-54)}$$

From this result, we can show that when a is less than $L(\sqrt{13} - 1)/6$, or $0.434\,L$, the deflection δ_c is negative and point C deflects upward.

The shape of the deflection curve for the beam in this example is shown in Fig. 7-17c for the case where a is large enough ($a > 0.434L$) to produce a downward deflection at C and small enough ($a < L$) to ensure that the reaction at A is upward. Under these conditions the beam has a positive bending moment from A to a point D, and the deflection curve is convex downward in this part of the beam. From D to C, the bending moment is negative, and the deflection curve is convex upward.

At point D, the curvature of the beam axis equals zero because the bending moment is zero. A point such as D where the curvature and bending moment change sign is called a **point of inflection** or **point of contraflexure**. The bending moment M and the second derivative d^2v/dx^2 of the deflection v with respect to x always vanish at an inflection point (see Eq. 7-10a). (However, a point where M and d^2v/dx^2 equal zero is not *necessarily* an inflection point because it is possible for those quantities to be zero without changing signs at that point; for example, they could have maximum or minimum values.)

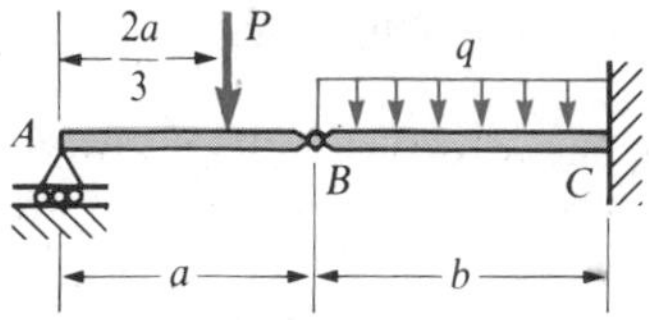

Fig. 7-18 Example 4. Compound beam with a hinge

Example 4

Determine the deflection δ_b at the hinge B for the compound beam shown in Fig. 7-18. Note that the beam is composed of two parts: (1) a beam AB, simply supported at A, and (2) a cantilever beam BC, fixed at C. The two beams are linked together by a pin connection at B.

Considering beam AB as a free body, we see that it has vertical reactions $P/3$ and $2P/3$ at ends A and B, respectively. Therefore, beam BC is in the condition of a cantilever beam subjected to a uniform load of intensity q and a concentrated load at the end equal to $2P/3$. The deflection of the end of this cantilever, which is the same as the deflection of the hinge, is

$$\delta_b = \frac{qb^4}{8EI} + \frac{2Pb^3}{9EI}$$

as found from Cases 1 and 4 of Table G-1.

7.7 NONPRISMATIC BEAMS

The methods presented in the preceding sections for calculating deflections of prismatic beams (that is, beams with constant cross sections throughout their lengths) can also be used to find deflections of nonprismatic beams. Such beams include those with different cross-sectional areas in various parts of the beam (see Fig. 7-19a for an example) and tapered beams (see Fig. 7-20). When a beam has abrupt changes in cross-sectional dimensions, there are local stress concentrations at the points where changes occur; however, these local stresses have no noticeable effect on the calculation of deflections. For a tapered beam, the bending theory derived previously for a prismatic beam gives satisfactory results provided that the angle of taper is small.

The first method for finding deflections is integration of the **differential equation** of the deflection curve. To illustrate this method, consider the example pictured in Fig. 7-19a. This beam is assumed to be reinforced along its central region so that the moment of inertia in that part of the beam is twice the moment of inertia in the end regions. The differential equation of the deflection curve in terms of the bending moment (Eq. 7-10a) for the left-hand half of the beam may be written in two parts, as follows:

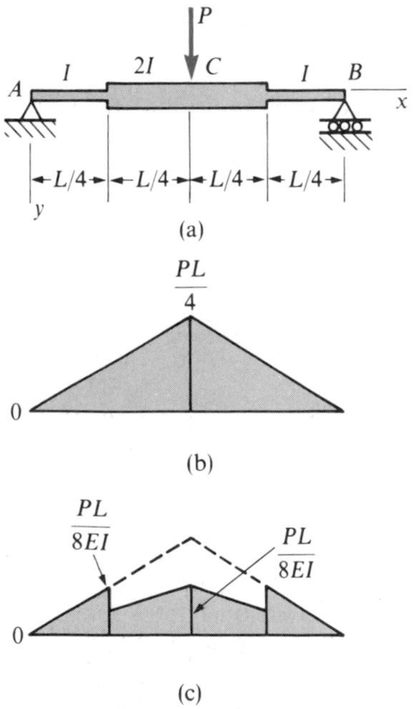

Fig. 7-19 Simple beam with two different moments of inertia

$$EIv'' = -\frac{Px}{2} \qquad \left(0 \le x \le \frac{L}{4}\right) \tag{a}$$

$$E(2I)v'' = -\frac{Px}{2} \qquad \left(\frac{L}{4} \le x \le \frac{L}{2}\right) \tag{b}$$

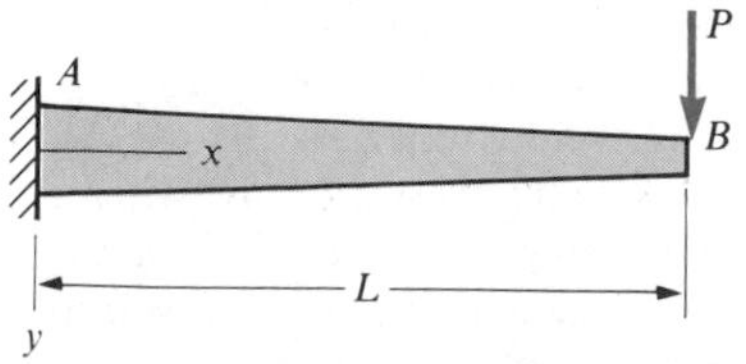

Fig. 7-20 Tapered cantilever beam

Each of these equations can be integrated twice to obtain expressions for the slopes and deflections. The four constants of integration can be found from the following conditions: (1) at $x = 0$, $v = 0$; (2) at $x = L/2$, $v' = 0$; (3) at $x = L/4$, the slope of the beam obtained from Eq. (a) is equal to the slope obtained from Eq. (b); and (4) at $x = L/4$, the deflection obtained from Eq. (a) is equal to the deflection obtained from Eq. (b). Having found the constants of integration from these conditions, we then will know the deflection curve of the beam for each of the two regions under consideration.

The use of the differential equation for finding deflections is practical only if the number of equations to be solved is limited to one or two and only if the integrations are easily performed, as in the preceding illustration. In the case of a tapered beam (Fig. 7-20), it may be difficult to solve the differential equation mathematically. The reason is that the expression for the moment of inertia I as a function of x is often complicated and produces a differential equation with a variable coefficient instead of constant coefficients.

The method of integration for nonprismatic beams should be performed in most cases with the moment equation (Eq. 7-10a) rather than with the shear and load equations (Eqs. 7-10b and c). The reason is that the moment equation can be written in the following simple form:

$$v'' = -\frac{M}{EI_x} \tag{7-55}$$

in which I_x is the moment of inertia of the cross section at distance x from the origin of coordinates. If the right-hand side of Eq. (7-55) can be integrated, then the method of integration is feasible. However, the shear and load equations are obtained by differentiating Eq. (7-55), which leads to much more complicated differential equations whenever I_x is also a variable and must be differentiated.

The second method we will consider is the **moment-area method**. The use of this method is illustrated in Fig. 7-19 for the beam discussed previously. The bending-moment diagram for the beam is given in Fig. 7-19b and the M/EI diagram in Fig. 7-19c. The areas and first moments of the various parts of the M/EI diagram are used to find the angles of rotation and the deflections. For example, let us find the angle of rotation at the left-hand support and the deflection at the midpoint. From the symmetry of the beam, we know that the tangent to the deflection curve at the midpoint C is horizontal. It follows from the first moment-area theorem that the angle of rotation θ_a at the left support is equal to the area of the M/EI diagram between points A and C. Therefore, this angle is obtained as follows:

$$\theta_a = \text{(area of triangle)} + \text{(area of trapezoid)}$$

$$= \frac{1}{2}\left(\frac{L}{4}\right)\left(\frac{PL}{8EI}\right) + \frac{1}{2}\left(\frac{PL}{16EI} + \frac{PL}{8EI}\right)\left(\frac{L}{4}\right) = \frac{5PL^2}{128EI} \tag{c}$$

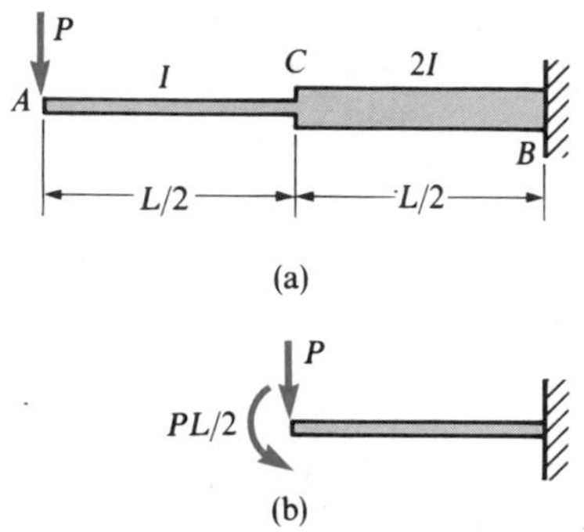

Fig. 7-21 Cantilever beam with two different moments of inertia

The offset of point A from a tangent to the deflection curve at point C, which is equal to the deflection δ_c at the middle of the beam, is obtained by taking the first moment of the area of the M/EI diagram between A and C about point A, in accordance with the second moment-area theorem. Thus,

$$\delta_c = \text{(first moment of triangle)} + \text{(first moment of trapezoid)}$$

$$= \left(\frac{2}{3}\right)\left(\frac{L}{4}\right)\left(\frac{PL^2}{64EI}\right) + \left(\frac{L}{4} + \frac{5L}{36}\right)\left(\frac{3PL^2}{128EI}\right) = \frac{3PL^3}{256EI} \tag{d}$$

This example shows that the use of the moment-area theorem for nonprismatic beams is similar to that for prismatic beams.

Another method for finding deflections is the **method of superposition**. To illustrate this method for a nonprismatic beam, consider the cantilever beam of Fig. 7-21 and assume that we wish to calculate the deflection δ_a at the free end. This deflection can be obtained in two steps. First, we imagine that the beam is held rigidly at point C, so that it neither deflects nor rotates at that point. Then we can calculate the deflection of A due to bending of AC as a cantilever beam. Since this beam has length $L/2$ and moment of inertia I, its deflection δ_1 is

$$\delta_1 = \frac{P(L/2)^3}{3EI} = \frac{PL^3}{24EI}$$

In addition, part CB of the beam also behaves like a cantilever (Fig. 7-21b) and contributes to the deflection of point A. The deflection δ_c and angle of rotation θ_c of the free end of this cantilever are

$$\delta_c = \frac{P(L/2)^3}{3(2EI)} + \frac{(PL/2)(L/2)^2}{2(2EI)} = \frac{5PL^3}{96EI}$$

$$\theta_c = \frac{P(L/2)^2}{2(2EI)} + \frac{(PL/2)(L/2)}{2EI} = \frac{3PL^2}{16EI}$$

The deflection δ_c and angle of rotation θ_c make an additional contribution δ_2 to the deflection under the load P, as follows:

$$\delta_2 = \delta_c + \theta_c\left(\frac{L}{2}\right) = \frac{7PL^3}{48EI}$$

Therefore, the total deflection of the free end A is

$$\delta_a = \delta_1 + \delta_2 = \frac{3PL^3}{16EI} \tag{e}$$

Thus, the techniques of the method of superposition may be readily adapted to certain kinds of nonprismatic beams, as this example shows. The choice of method to be used for finding deflections of nonprismatic beams and the details of the procedure depend upon the problem to be solved and the preferences of the analyst.

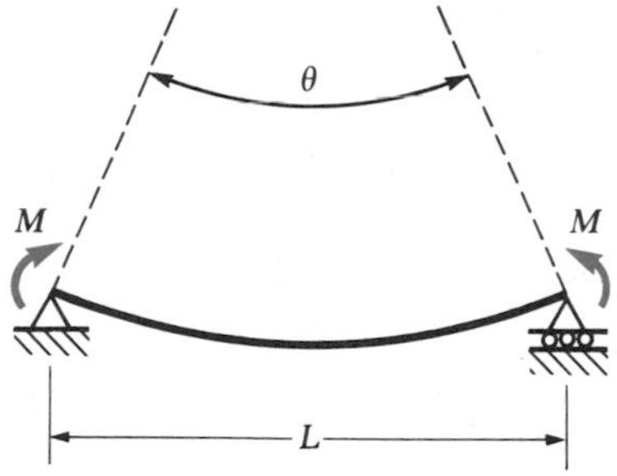

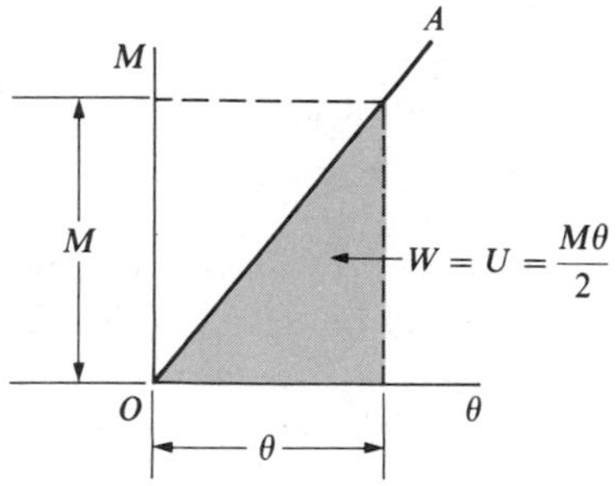

Fig. 7-22 (a) Beam in pure bending by couples M, and (b) diagram showing linear relationship between bending moment M and angle of rotation θ

7.8 STRAIN ENERGY OF BENDING

The concept of strain energy was explained previously in our discussions of axially loaded members and bars in torsion (see Sections 2.6 and 3.8, respectively). Now we will apply those concepts to bending of beams. We will only consider beams that behave in a linearly elastic manner, which means that the material must follow Hooke's law and the deflections and rotations must be small.

Let us begin with a beam in pure bending subjected to couples M (Fig. 7-22a). The deflection curve is a circular arc of constant curvature $\kappa = -M/EI$ (see Eq. 7-5). The angle θ subtended by this arc equals L/ρ, where L is the length of the beam and ρ is the radius of curvature; thus, considering only absolute values, we obtain

$$\theta = \frac{ML}{EI} \tag{7-56}$$

This linear relationship between the couples M and the angle θ is shown graphically in Fig. 7-22b by line OA on the moment-rotation diagram. As the bending couples are gradually increased in magnitude from zero to their maximum values M, they perform work W represented by the area below line OA (the shaded area in Fig. 7-22b). This work, which is equal to the strain energy U stored in the beam, is

$$U = W = \frac{M\theta}{2} \tag{7-57}$$

This equation is analogous to Eq. (2-32) for the strain energy in an axially loaded bar.

By combining Eqs. (7-56) and (7-57), we can express the strain energy stored in a beam in pure bending in either of the following forms:

$$U = \frac{M^2L}{2EI} \qquad U = \frac{EI\theta^2}{2L} \tag{7-58a,b}$$

The first of these equations expresses the strain energy in terms of the load M and the second expresses it in terms of the angle θ. The equations are similar in form to those for strain energy in an axially loaded bar (see Eqs. 2-33).

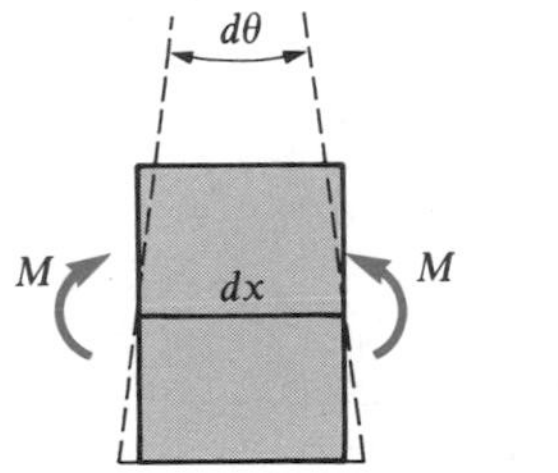

Fig. 7-23 Element of a beam

If the bending moment M varies along the length of the beam (nonuniform bending), then we may obtain the strain energy by applying Eqs. (7-58) to an element of the beam and integrating along the length. Considering an element of length dx subjected to a moment M (Fig. 7-23), we note that the angle $d\theta$ between the sides of the element (see Eq. 7-6) is

$$d\theta = \frac{d^2v}{dx^2}dx = \frac{M\,dx}{EI}$$

in which only absolute values are considered. Thus, the strain energy dU stored in the element (from Eqs. 7-58) is

$$dU = \frac{M^2\,dx}{2EI} \quad \text{or} \quad dU = \frac{EI}{2}\left(\frac{d^2v}{dx^2}\right)^2 dx$$

By integrating the preceding equations, we can express the total strain energy stored in the beam in either of the following forms:

$$U = \int \frac{M^2\,dx}{2EI} \qquad U = \int \frac{EI}{2}\left(\frac{d^2v}{dx^2}\right)^2 dx \qquad \text{(7-59a,b)}$$

in which the integrations are carried out over the length of the beam. The first equation is used when the bending moment is known, and the second is used when the equation of the deflection curve is known.

Equations (7-59) give the strain energy in a beam when only the effects of bending moments are taken into account. In addition to the strain energy of bending, strain energy of shear will be stored in the elements of the beam; this strain energy is discussed later in Section 10.9. However, for beams of usual proportions in which the lengths are much greater than the depths (say, $L/d > 6$), the strain energy of shear is relatively small in comparison with the strain energy of bending and may be disregarded.

Principles of mechanics based upon strain energy have an important role in structural analysis and in the design of structures to resist dynamic loads. Some of these principles are discussed in Chapter 10. In this section, however, we are concerned primarily with evaluating strain energy in beams and then using the strain energy to solve simple deflection and impact problems. For instance, if a beam is subjected to only one concentrated load P or one couple M_0, we can find the deflection δ or the angle of rotation θ at the point where the load acts, in the direction of the load, by equating the work done by the load ($P\delta/2$ or $M\theta/2$) to the strain energy of the beam. This same technique was discussed in Section 2.6. However, the method is limited in its usefulness because there can be only one load on the structure, and the only deflection that can be found is the one corresponding to that load.

Example 1

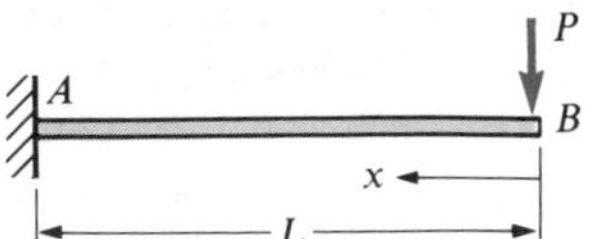

(a)

Fig. 7-24 (a) Example 1. Strain energy of a beam

Determine the strain energy U stored in a cantilever beam of length L carrying a concentrated load P at the free end (Fig. 7-24a). Also, find the vertical deflection δ_b at the free end.

The bending moment at a cross section located at distance x from the free end is $M = -Px$. Substituting this expression for M into Eq. (7-59a) gives

$$U = \int \frac{M^2\,dx}{2EI} = \int_0^L \frac{(-Px)^2\,dx}{2EI} = \frac{P^2L^3}{6EI} \qquad \text{(a)}$$

as the strain energy of the beam. Note that the strain energy is always a positive quantity and the load appears to the second power.

To obtain the deflection under the load P, we equate the work done by the load with the strain energy:

$$\frac{P\delta_b}{2} = \frac{P^2L^3}{6EI}$$

from which

$$\delta_b = \frac{PL^3}{3EI} \tag{b}$$

This expression agrees with the result obtained in Example 1, Section 7.5 (see Eq. 7-39). Again, it is important to observe that the only deflection we can find in this example is the deflection corresponding to P (that is, the vertical deflection at the free end).

Example 2

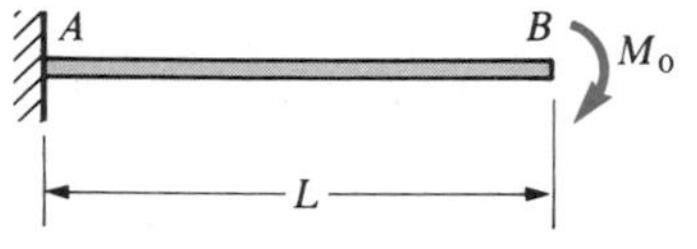

Fig. 7-24(b) Example 2. Strain energy of a beam

The cantilever beam shown in Fig. 7-24b is subjected to a couple M_0 acting at the free end. Determine the strain energy U of the beam and the angle of rotation θ_b at the free end.

In this case the bending moment is constant and equal to $-M_0$; hence from Eq. (7-59a) we obtain

$$U = \int \frac{M^2\,dx}{2EI} = \int_0^L \frac{(-M_0)^2\,dx}{2EI} = \frac{M_0^2L}{2EI} \tag{c}$$

The work done by the couple M_0 during loading of the beam is $M_0\theta_b/2$; therefore,

$$\frac{M_0\theta_b}{2} = \frac{M_0^2L}{2EI}$$

and

$$\theta_b = \frac{M_0L}{EI} \tag{d}$$

The angle of rotation has the same sense as the moment, which is clockwise in this example.

Example 3

In this example the cantilever beam is subjected simultaneously to both a concentrated load P and a couple M_0 (Fig. 7-24c). Determine the strain energy U of the beam.

The bending moment in the beam is given by the expression

$$M = -Px - M_0$$

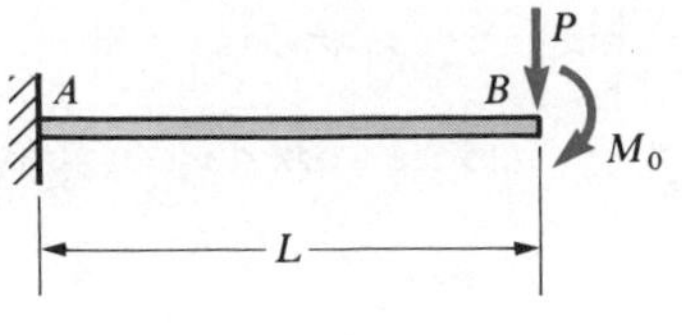

Fig. 7-24 (c) Example 3. Strain energy of a beam

where x is measured from the free end. Therefore, the strain energy is

$$U = \int \frac{M^2\,dx}{2EI} = \frac{1}{2EI}\int_0^L (-Px - M_0)^2\,dx$$

$$= \frac{P^2L^3}{6EI} + \frac{PM_0L^2}{2EI} + \frac{M_0^2L}{2EI} \tag{e}$$

The first term in the last expression gives the strain energy due to P acting alone ($M_0 = 0$) and the last term gives the strain energy due to M_0 alone ($P = 0$). However, when both loads act simultaneously, the term in the middle also appears in the expression for strain energy. This result shows that *the strain energy in a structure due to two or more loads cannot be obtained by adding the strain energies due to the loads acting separately.* The explanation is that strain energy is a quadratic function of the load, not a linear function, and the principle of superposition does not apply.

We also observe that we cannot calculate a deflection for the beam of Fig. 7-24c by equating work and strain energy because there are too many unknowns in the expression for work:

$$W = \frac{P\delta_b}{2} + \frac{M_0\theta_b}{2}$$

When this expression is equated with U, we have one equation with two unknowns. Although the equation is quite correct, we cannot obtain any useful information from it.

Example 4

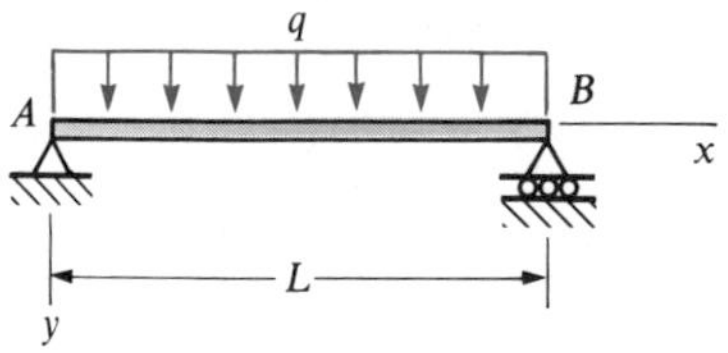

Fig. 7-25 Example 4. Strain energy of a beam

The equation of the deflection curve for a simple beam with a uniform load of intensity q (Fig. 7-25) is

$$v = \frac{qx}{24EI}(L^3 - 2Lx^2 + x^3)$$

as given in Case 1 of Table G-2. Using this expression, determine the strain energy U stored in the beam.

To obtain the strain energy, we will use Eq. (7-59b), which contains d^2v/dx^2. Taking derivatives of the expression for v, we get

$$\frac{dv}{dx} = \frac{q}{24EI}(L^3 - 6Lx^2 + 4x^3)$$

$$\frac{d^2v}{dx^2} = -\frac{qx}{2EI}(L - x)$$

Now substituting into Eq. (7-59b) and integrating, we obtain

$$U = \int \frac{EI}{2}\left(\frac{d^2v}{dx^2}\right)^2 dx = \frac{EI}{2}\int_0^L \left[-\frac{qx}{2EI}(L-x)\right]^2 dx = \frac{q^2L^5}{240EI} \tag{f}$$

as the amount of strain energy stored in the beam. Note again that the load appears to the second power.

DEFLECTION PRODUCED BY IMPACT

The dynamic deflection of a beam subjected to an impact load may be determined under certain simplified conditions by equating the work done by the load with the strain energy stored in the beam. This approach was described in Section 2.7 for impact loads on an axially loaded bar. The assumptions described in that discussion apply also to beams; namely, the falling weight sticks to the beam and moves with it, no energy losses occur, the beam is linearly elastic, the deflected shape of the beam is the same under a dynamic load as under a static load, and the potential energy of the beam due to its change in position is disregarded. In general, these assumptions are reasonable if the mass of the falling object is very large compared to the mass of the beam. Otherwise, this approximate analysis is not valid and a more advanced analysis is required.

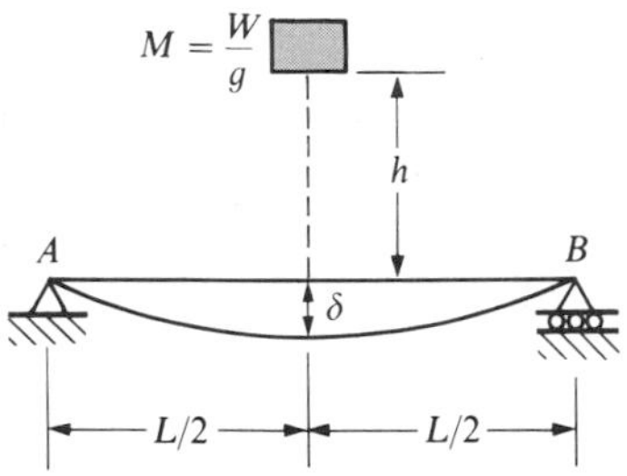

Fig. 7-26 Deflection of a beam produced by a falling body

As an illustration, consider a simple beam AB that is struck at the middle by a falling body of weight W (Fig. 7-26). In accord with the preceding idealizations, we may assume that all of the work done by the body during its fall is transformed into elastic strain energy of the beam. Since the distance through which it falls is $h + \delta$, where h is the initial height above the beam and δ is the maximum dynamic deflection of the beam, the work done is

$$W(h + \delta) \tag{g}$$

If we let P denote the force exerted on the beam when its deflection is a maximum, then the relationship between P and δ is

$$\delta = \frac{PL^3}{48EI} \quad \text{or} \quad P = \frac{48EI\delta}{L^3}$$

because we have assumed that the deflection shape under a dynamic load is the same as under a static load. Therefore, the strain energy of the beam, equal to the work of the force P, is

$$U = \frac{P\delta}{2} = \frac{24EI\delta^2}{L^3}$$

Equating the work done by the falling mass (Eq. g) with the strain energy gives

$$W(h + \delta) = \frac{24EI\delta^2}{L^3}$$

This equation is quadratic in δ and can be solved for its positive root:

$$\delta = \frac{WL^3}{48EI} + \left[\left(\frac{WL^3}{48EI}\right)^2 + 2h\left(\frac{WL^3}{48EI}\right)\right]^{1/2} \tag{7-60}$$

Let us denote the static deflection of the beam due to the weight W as δ_{st}:

$$\delta_{st} = \frac{WL^3}{48EI} \tag{7-61}$$

Then Eq. (7-60) for the maximum dynamic deflection can be written in simpler form as follows:

$$\delta = \delta_{st} + (\delta_{st}^2 + 2h\delta_{st})^{1/2} \tag{7-62}$$

From this equation we observe that the dynamic deflection is always larger than the static deflection. If the height $h = 0$, which means that the load is applied suddenly but without any free fall, the dynamic deflection is twice the static deflection. If h is very large compared to the deflection, then the term containing h in Eq. (7-62) predominates, and the equation can be simplified to

$$\delta = \sqrt{2h\delta_{st}} \tag{7-63}$$

These and other observations are analogous to those encountered previously for impact on a bar in tension (Section 2.7).

The deflection δ calculated from Eq. (7-62) generally represents an upper limit, because we assumed there were no losses of energy during impact. A more realistic analysis can be made by introducing a kinetic energy loss factor, as explained in Section 2.7. Local deformation of the contact surfaces, the tendency of the falling mass to bounce upward, and the mass of the beam itself tend to reduce the deflection.

*7.9 DISCONTINUITY FUNCTIONS

Discontinuity functions are utilized in a variety of engineering applications, including beam analysis, electrical circuits, and heat transfer. These functions probably are the easiest to use and understand when applied to beams, and therefore the study of mechanics of materials offers an excellent opportunity to become familiar with them. The mathematics of the functions are described in this section, and we also will see how the functions are used to represent loads on beams. Then, in the next section, we will use them for finding slopes and deflections of beams.

The unique feature of discontinuity functions is that they permit the writing of a discontinuous function by a single expression, whereas the more conventional approach requires that a discontinuous function be described by a series of expressions, one for each region in which the function is distinct. For instance, if the loading on a beam consists of a mixture of concentrated and distributed loads, we can write with discontinuity functions a single equation that applies throughout the entire length, whereas ordinarily we must write separate equations for each segment of the beam between changes in loading. In a similar manner, we can express shear forces, bending moments, slopes, and deflections of a beam by one equation each, even though there may be several changes in loads along the axis of the beam.

These results are achievable because the functions themselves are discontinuous; that is, they have different values in different regions of the independent variable. In effect, these functions can pass through discontinuities in a manner that is not possible with ordinary continuous functions. However, because they differ significantly from the functions we are accustomed to, discontinuity functions must be used with care and caution.

Two kinds of functions, called **Macaulay functions** and **singularity functions**, will be discussed in this section. Although these functions have different definitions and properties, together they form a family of **discontinuity functions.***

MACAULAY FUNCTIONS

Macaulay functions are used to represent quantities that "begin" at some particular point on the x axis (such as point $x = a$) and that have the value zero to the left of that point. For instance, one of the Macaulay functions, denoted F_1, is defined as follows:

$$F_1(x) = \langle x - a \rangle^1 = \begin{cases} 0 & \text{when } x \leq a \\ x - a & \text{when } x \geq a \end{cases} \tag{7-64}$$

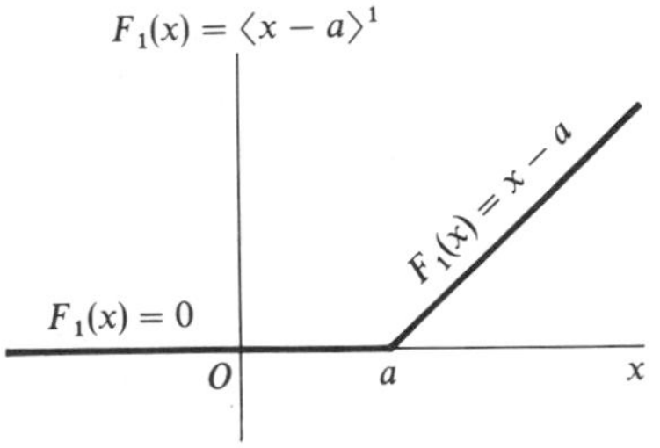

Fig. 7-27 Graph of the Macaulay function F_1 (the unit ramp function)

In this equation, x is the independent variable and a is the value of x at which the function "begins." The *pointed brackets* (or *angle brackets*) are the mathematical symbol for a discontinuity function. In the case of the function F_1, the pointed brackets tell us that the function has the value zero when x is less than or equal to a (that is, when the expression within the brackets is negative or zero), and a value equal to $x - a$ when x is greater than or equal to a. A graph of this function, called the **unit ramp function**, is given in Fig. 7-27.

In general terms, the Macaulay functions are defined by the following expressions:

$$F_n(x) = \langle x - a \rangle^n = \begin{cases} 0 & \text{when } x \leq a \\ (x - a)^n & \text{when } x \geq a \end{cases} \tag{7-65}$$

$$n = 0, 1, 2, 3, \ldots$$

We see from this definition that the Macaulay functions have the value zero to the left of the point $x = a$ and the value $(x - a)^n$ to the right of that point. Except for the case $n = 0$, which is discussed later, the function equals zero at $x = a$. Another way to express this definition is the following: If the quantity $x - a$ within the pointed brackets is negative or zero, the Macaulay function has the value zero; if the function $x - a$ is positive or zero, the Macaulay function has the value obtained by replacing the pointed brackets with parentheses.

The preceding definition of the Macaulay functions holds for values of n equal to positive integers and zero. When $n = 0$, note that the function

* Sometimes both kinds of discontinuity functions are called *singularity functions*, but this usage obscures the distinctions between the two functions, which obey different mathematical laws. Furthermore, Macaulay functions do not have singularities.

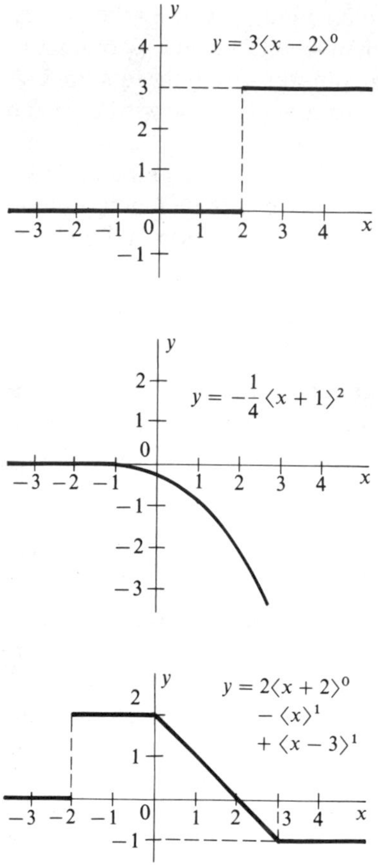

Fig. 7-28 Graphs of expressions involving Macaulay functions

takes on the following special values:

$$F_0(x) = \langle x - a \rangle^0 = \begin{cases} 0 & \text{when } x \leq a \\ 1 & \text{when } x \geq a \end{cases} \qquad (7\text{-}66)$$

This function has a vertical "step" at the point of discontinuity $x = a$; thus, at $x = a$ it has two values, zero and one. The function F_0, called the **unit step function**, is pictured in Table 7-1 along with other Macaulay functions.*

The Macaulay functions of higher degree can be expressed in terms of the unit step function, as follows:

$$F_n(x) = \langle x - a \rangle^n = (x - a)^n \langle x - a \rangle^0 \qquad (7\text{-}67)$$

This equation is easily confirmed by comparing Eqs. (7-65) and (7-66).

Some of the basic algebraic operations, such as addition, subtraction, and multiplication by a constant, can be performed on the Macaulay functions. Illustrations of these elementary operations are given in Fig. 7-28. The important thing to note from these examples is that a function y having different algebraic expressions for different regions along the x axis can be written as a single equation through the use of Macaulay functions. The reader should verify each of the graphs in Fig. 7-28 in order to become familiar with the functions.

The Macaulay functions can be integrated and differentiated according to the formulas given in the last column of Table 7-1. These formulas can be verified by ordinary differentiation and integration of the functions in the two regions $x \leq a$ and $x \geq a$.

The **units** of the Macaulay functions are the same as the units of x^n; that is, F_0 is dimensionless, F_1 has units of x, F_2 has units of x^2, and so forth.**

SINGULARITY FUNCTIONS

The second kind of discontinuity functions are the singularity functions, defined by the following expressions:

$$F_n(x) = \langle x - a \rangle^n = \begin{cases} 0 & \text{when } x \neq a \\ \pm\infty & \text{when } x = a \end{cases} \qquad (7\text{-}68)$$

$$n = -1, -2, -3, \ldots$$

Note that singularity functions are defined for negative integer values of n, whereas Macaulay functions are defined for positive integers and for zero.

* The unit step function is also known as the *Heaviside step function*, denoted by $H(x - a)$, after Oliver Heaviside (1850–1925), an English physicist and electrical engineer.

** The use of special brackets for discontinuity functions was introduced by the English mathematician W. H. Macaulay in 1919 (Ref. 7-11), and the brackets are often called *Macaulay's* brackets. However, Macaulay used braces { } to identify the functions; the pointed brackets ⟨ ⟩ in common use today were introduced at a later time. The general concept of combining two or more expressions into a single function by using a special symbol predates Macaulay's work (see Refs. 7-12 through 7-15).

Table 7-1 Discontinuity functions

	Name	Definition	Graph	Derivative and integral
Singularity functions	Unit doublet function	$F_{-2} = \langle x - a \rangle^{-2} = \begin{cases} 0 & x \neq a \\ \pm\infty & x = a \end{cases}$	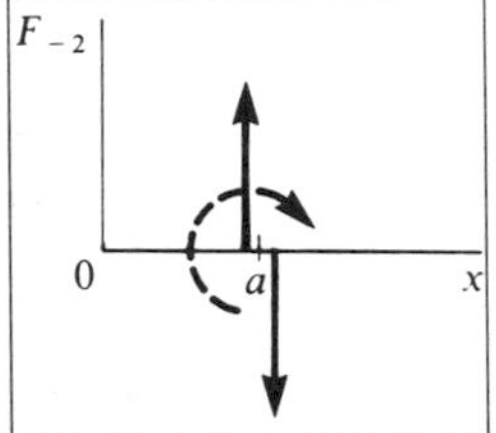	$\int_{-\infty}^{x} F_{-2}\,dx = F_{-1}$
	Unit impulse function	$F_{-1} = \langle x - a \rangle^{-1} = \begin{cases} 0 & x \neq a \\ +\infty & x = a \end{cases}$	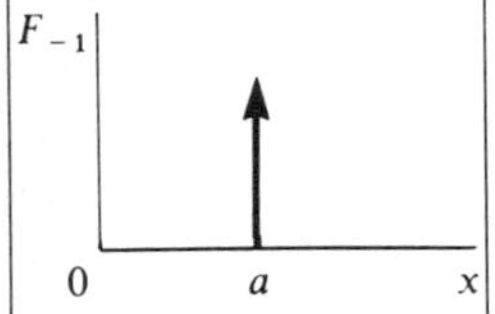	$\int_{-\infty}^{x} F_{-1}\,dx = F_0$
Macaulay functions	Unit step function	$F_0 = \langle x - a \rangle^{0} = \begin{cases} 0 & x \leq a \\ 1 & x \geq a \end{cases}$	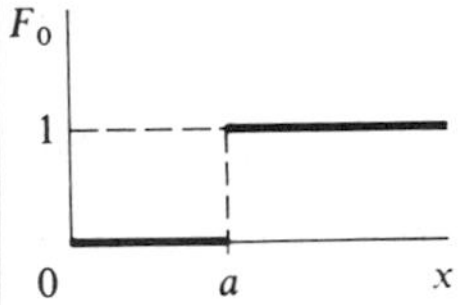	$\int_{-\infty}^{x} F_0\,dx = F_1$
	Unit ramp function	$F_1 = \langle x - a \rangle^{1} = \begin{cases} 0 & x \leq a \\ x - a & x \geq a \end{cases}$	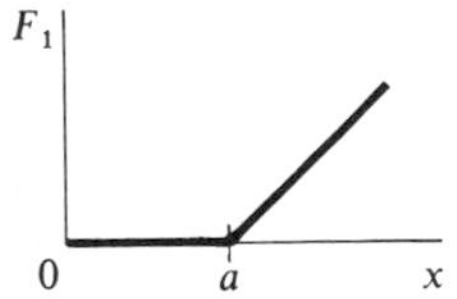	$\frac{d}{dx} F_1 = F_0$ $\int_{-\infty}^{x} F_1\,dx = \frac{F_2}{2}$
	Unit second-degree function	$F_2 = \langle x - a \rangle^{2} = \begin{cases} 0 & x \leq a \\ (x - a)^2 & x \geq a \end{cases}$	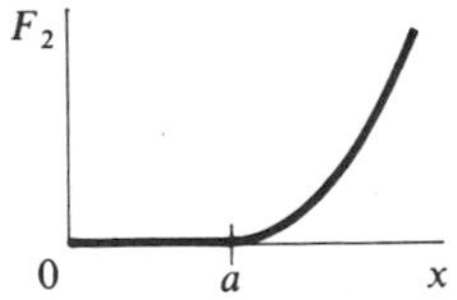	$\frac{d}{dx} F_2 = 2F_1$ $\int_{-\infty}^{x} F_2\,dx = \frac{F_3}{3}$
	General Macaulay function	$F_n = \langle x - a \rangle^{n} = \begin{cases} 0 & x \leq a \\ (x - a)^n & x \geq a \end{cases}$ $n = 0, 1, 2, 3, \ldots$	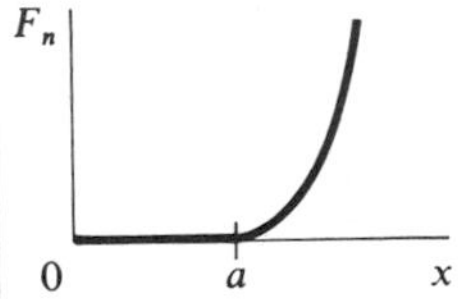	$\frac{d}{dx} F_n = nF_{n-1}$ $n = 1, 2, 3, \ldots$ $\int_{-\infty}^{x} F_n\,dx = \frac{F_{n+1}}{n+1}$ $n = 0, 1, 2, 3, \ldots$

The pointed brackets identify both kinds of functions, but the brackets have different meanings in the two cases (compare Eqs. 7-65 and 7-68).

The singularity functions have the value zero everywhere except at the singular point $x = a$. The singularities arise because, when n is a negative integer, the function $(x - a)^n$ can be written as a fraction with the expression $x - a$ in the denominator; thus, when $x = a$ the function becomes infinite.

The two most important singularity functions are pictured in Table 7-1, where we see that the kind of singularity depends upon the value of n. The **unit doublet function** ($n = -2$) has a singularity that can be pictured by two arrows of infinite extent, one directed upward and the other downward, with the arrows being infinitesimally close to each other. For convenience, these arrows can be visualized as forces, and then the doublet can be represented by a curved arrow that is the moment of the two forces. This moment is equal to the product of an infinite force and a vanishingly small distance; the moment turns out to be finite and equal to unity. For this reason, the doublet is also known as the **unit moment**. (A third name in common usage is **dipole**.)

The **unit impulse function** ($n = -1$) is also infinite at $x = a$, but in a different way. It can be pictured by a single arrow, as shown in Table 7-1. If the arrow is visualized as a force, then the force has infinite intensity and acts over an infinitesimally small distance along the x axis. The force is equal to the intensity times the distance over which it acts; this product also turns out to be finite and equal to unity. Thus, the name **unit force** is also used for this function.*

Singularity functions are sometimes categorized as *pathological functions* or *improper functions*, because they are neither continuous nor differentiable at $x = a$. However, singularity functions can be integrated through the singularities; the integration formula is as follows:

$$\int_{-\infty}^{x} F_n\,dx = \int_{-\infty}^{x} \langle x - a\rangle^n dx = \langle x - a\rangle^{n+1} = F_{n+1} \qquad (7\text{-}69)$$

$$n = -1, -2, -3, \ldots$$

Note that this formula is not the same as the integration formula for the Macaulay functions, which is given in the last line of Table 7-1. Equation (7-69) shows that the integral of the unit doublet function is the unit impulse function, and the integral of the unit impulse function is the unit step function (see the last column of Table 7-1).

The **units** of the singularity functions, like those of the Macaulay functions, are the same as the units of x^n. Thus, the doublet function has units of $1/x^2$ and the impulse function has units of $1/x$.

* The terminology *unit impulse function* comes from the use of this function in dynamics, where the x axis is the time axis. In physics and mathematics, this function is denoted by $\delta(x - a)$ and is called the *Dirac delta function*, named for the English theoretical physicist Paul A. M. Dirac (1902–1984) who developed it.

REPRESENTATION OF LOADS ON BEAMS BY DISCONTINUITY FUNCTIONS

The discontinuity functions listed in Table 7-1 are ideally suited for representing loads on beams, such as couples, forces, uniform loads, and varying loads. The shapes of the various loading diagrams exactly match the shapes of the corresponding functions F_{-2}, F_{-1}, F_0, F_1, and so on. It is necessary only to multiply the functions given in the table (which are the unit functions) by the appropriate load intensities in order to obtain mathematical representations of the loads. Several standard load cases, explained in detail in the following paragraphs, are listed in Table 7-2 for ready reference. More complicated cases of loading can be handled by superposition of these elementary ones.

To explain how the expressions in Table 7-2 are obtained, let us consider first the uniform load of Case 3. This load can be expressed in terms of the unit step function F_0 (see Table 7-1), which is given by the formula

$$F_0(x) = \langle x - a \rangle^0$$

This function has the value 0 for $x \leq a$ and the value $+1$ for $x \geq a$. If the function is multiplied by the constant q_0, representing intensity of uniform load, it becomes an expression for the uniformly distributed load on a beam:

$$q(x) = q_0 \langle x - a \rangle^0 \tag{a}$$

The load $q(x)$ defined by this expression has the value 0 for $x \leq a$ and the value q_0 for $x \geq a$. Thus, at $x = a$, the function equals zero if we approach from the left, and it equals q_0 if we approach from the right. Equation (a) is listed for Case 3 in the last column of Table 7-2. The direction of the load that is represented by Eq. (a) can be either upward or downward, depending upon the sign convention that we adopt for distributed loads. Since we assumed previously that downward uniform load is positive, the expression for $q(x)$ represents the load pictured in the middle column of the table. With our sign convention, the quantity q_0 is positive for downward load and negative for upward load.* Note that the uniform load q_0 continues indefinitely to the right along the x axis.

Cases 4 and 5 in Table 7-2 can be explained in a manner similar to that for Case 3, using the ramp and second-degree functions. Both loading functions continue indefinitely to the right. To define the functions, we must specify a particular point on each graph. A convenient way to do so is to give the ordinate q_0 at some arbitrarily selected point located at distance b from the point $x = a$, as shown in the table.

* If we had assumed positive load to be upward, the equation for $q(x)$ would still be the same, but q_0 would be positive for upward load and negative for downward load, because $\langle x - a \rangle^0$ equals 1 for $x \geq a$.

Table 7-2 Load intensities represented by discontinuity functions

Case	Load on beam (shown positive)	Intensity $q(x)$ of equivalent distributed load (positive downward)
1	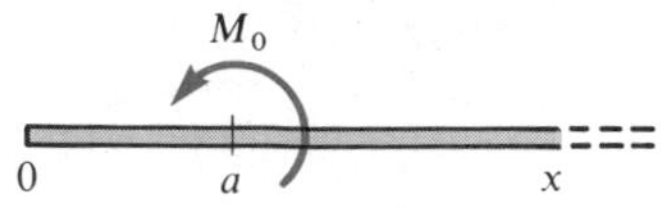	$q(x) = M_0\langle x - a\rangle^{-2}$
2	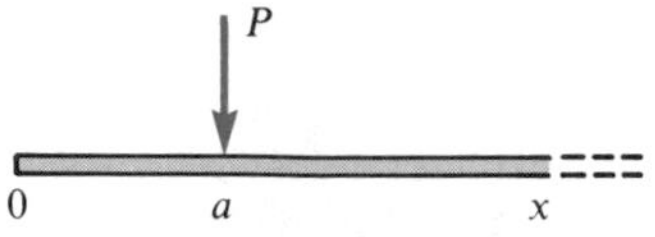	$q(x) = P\langle x - a\rangle^{-1}$
3	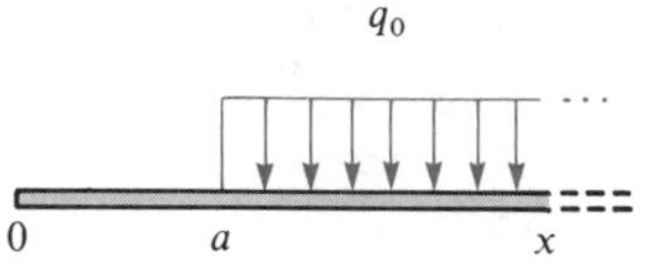	$q(x) = q_0\langle x - a\rangle^{0}$
4	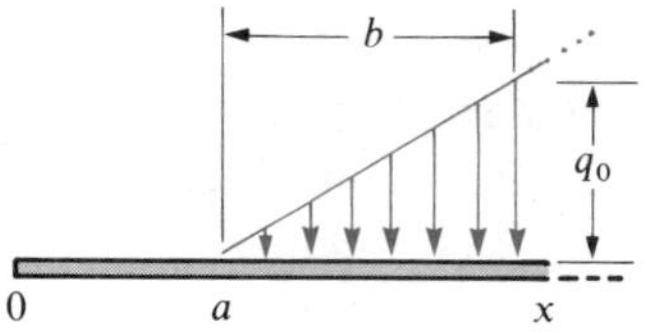	$q(x) = \frac{q_0}{b}\langle x - a\rangle^{1}$
5	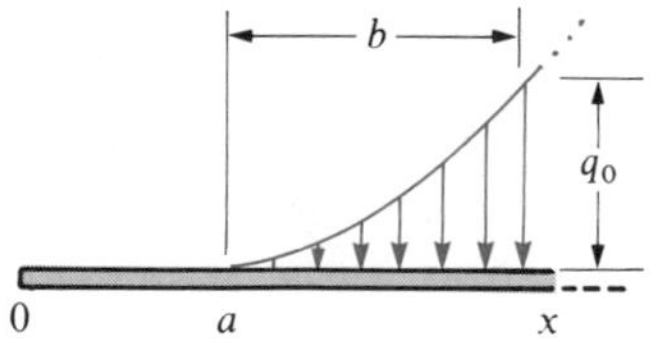	$q(x) = \frac{q_0}{b^2}\langle x - a\rangle^{2}$
6	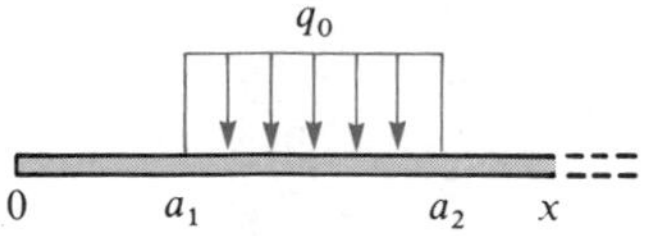	$q(x) = q_0\langle x - a_1\rangle^{0} - q_0\langle x - a_2\rangle^{0}$
7	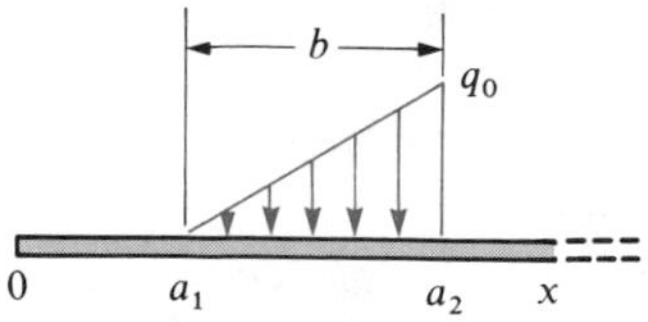	$q(x) = \frac{q_0}{b}\langle x - a_1\rangle^{1} - \frac{q_0}{b}\langle x - a_2\rangle^{1} - q_0\langle x - a_2\rangle^{0}$
8	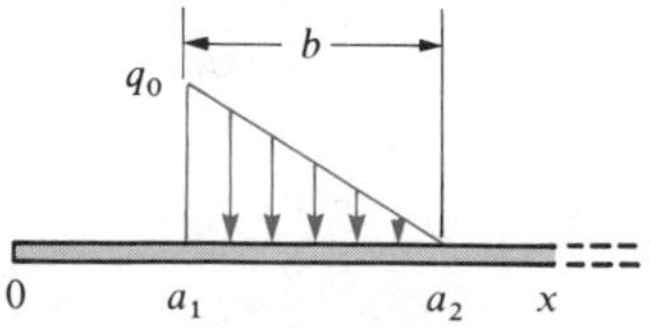	$q(x) = q_0\langle x - a_1\rangle^{0} - \frac{q_0}{b}\langle x - a_1\rangle^{1} + \frac{q_0}{b}\langle x - a_2\rangle^{1}$

The loading in Case 6 is a segment of uniform load that begins at $x = a_1$ and ends at $x = a_2$. This loading can be expressed as the superposition of two loads. The first load is a uniform load of intensity q_0 that begins at $x = a_1$ and continues indefinitely to the right (see Case 3); the second load has intensity $-q_0$, begins at $x = a_2$, and also continues indefinitely to the right. Thus, the second load cancels the first load in the region to the right of the point $x = a_2$.

The loads in Cases 7 and 8 also consist of segments of distributed loads and are obtained by combining more elementary load patterns. The reader can easily verify the expressions given in the last column of the table for these two cases. More complicated loading patterns that involve distributed loads are obtained by similar techniques of superposition using the Macaulay functions.

Loads in the form of couples or forces (Cases 1 and 2) are handled by the singularity functions, with the unit doublet function representing a unit couple and the unit impulse function representing a unit force. When the unit doublet function is multiplied by M_0, it represents a couple as an equivalent distributed load of intensity $q(x)$. The units of M_0 are force times length, and the units of the unit doublet function are length to the power -2. Thus, their product has units of force divided by length, which are the correct units for intensity $q(x)$ of distributed load. The situation is similar for a concentrated load; the product of force P and the unit impulse function has units of load intensity. The equations for $q(x)$ given in Cases 1 and 2 are mathematical expressions that define the equivalent load intensities for a moment and a force. Their signs have physical meaning only when we adopt sign conventions for the loads themselves. We will assume that loads in the form of couples and forces are positive as shown in the table (that is, positive when counterclockwise and downward, respectively). As before, the equivalent loads $q(x)$ are positive when downward.

Example 1

The simple beam AB shown in Fig. 7-29a supports a concentrated load P and a couple M_0. (a) Write the expression for the intensity $q(x)$ of the equivalent distributed loads acting on the beam in the region between the supports ($0 < x < L$). (b) Write the expression for the intensity $q(x)$ including the reactions ($0 \le x \le L$).

(a) The loads on the beam between the supports are the concentrated load P and the couple M_0. Their equivalent distributed loads are obtained from Cases 2 and 1, respectively, of Table 7-2. Thus, with the origin of coordinates at support A, we can write by inspection the following expression:

$$q(x) = P\left\langle x - \frac{L}{3}\right\rangle^{-1} + M_0\left\langle x - \frac{2L}{3}\right\rangle^{-2} \quad \text{(b)}$$

This single equation gives the equivalent distributed load at any point along the

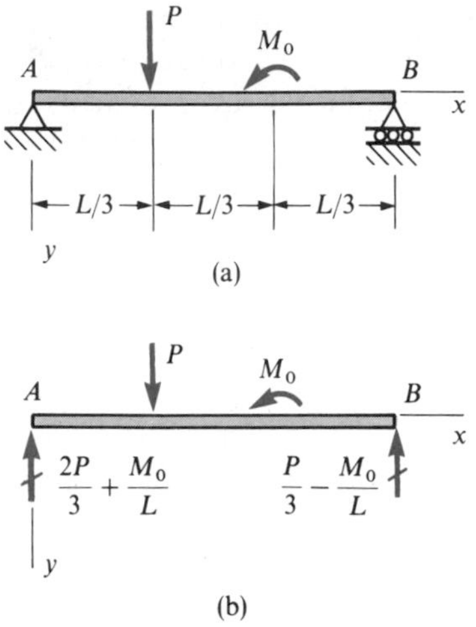

Fig. 7-29 Example 1. Representation of loads by discontinuity functions

axis of the beam (except at the supports). Note that each term on the right-hand side is zero everywhere except at the point where the load acts.

(b) It is sometimes desirable to include the reactions in the expression for the load $q(x)$; this expression then is valid at all points along the axis of the beam, including the end points. To obtain the expression for $q(x)$, we begin by determining the reactions from static equilibrium and showing them on a free-body diagram of the beam (Fig. 7-29b). Then we can write the expression for $q(x)$ by inspection, using Table 7-2:

$$q(x) = -\left(\frac{2P}{3} + \frac{M_0}{L}\right)\langle x\rangle^{-1} + P\left\langle x - \frac{L}{3}\right\rangle^{-1}$$
$$+ M_0\left\langle x - \frac{2L}{3}\right\rangle^{-2} - \left(\frac{P}{3} - \frac{M_0}{L}\right)\langle x - L\rangle^{-1} \quad \text{(c)}$$

The last term in this equation, containing $\langle x - L\rangle^{-1}$, is zero at every point along the beam except at the right support. Hence, it has no role in calculating the deflections and may be omitted.

In the next section, we will integrate equations for the load $q(x)$, such as Eqs. (b) and (c), to obtain successively the shear forces, bending moments, slopes, and deflections of beams in terms of discontinuity functions.

Example 2

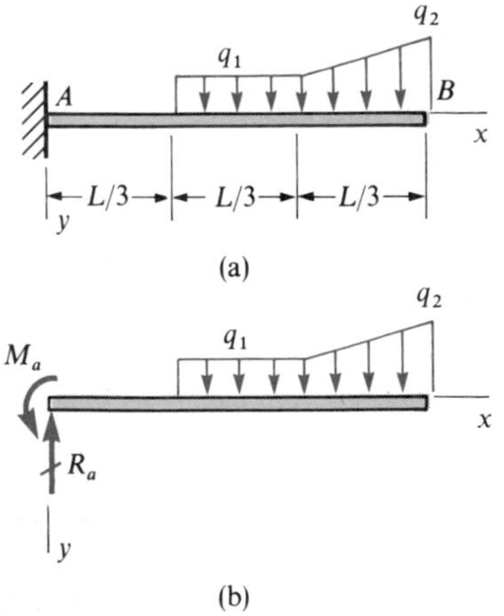

Fig. 7-30 Example 2. Representation of loads by discontinuity functions

A cantilever beam AB supports a distributed load of varying intensity as shown in Fig. 7-30a. (a) Write the expression for the intensity $q(x)$ of the distributed load acting on the beam to the right of the support. (b) Write the expression for the intensity $q(x)$ including the reactions.

(a) The loads on the beam can be divided into two parts, a uniform load acting from $x = L/3$ to $x = L$ and a triangular load acting from $x = 2L/3$ to $x = L$. Thus, using Cases 6 and 7 of Table 7-2, we get

$$q(x) = q_1\left\langle x - \frac{L}{3}\right\rangle^0 - q_1\langle x - L\rangle^0 + \frac{3(q_2 - q_1)}{L}\left\langle x - \frac{2L}{3}\right\rangle^1$$
$$- \frac{3(q_2 - q_1)}{L}\langle x - L\rangle^1 - (q_2 - q_1)\langle x - L\rangle^0$$

The three terms containing $\langle x - L\rangle$ are zero at every point along the axis of the beam; hence they have no physical meaning and will not affect the calculation of the deflections. Thus, those terms may be omitted, and the expression for $q(x)$ becomes

$$q(x) = q_1\left\langle x - \frac{L}{3}\right\rangle^0 + \frac{3(q_2 - q_1)}{L}\left\langle x - \frac{2L}{3}\right\rangle^1$$

As a check on this expression, we can substitute specific values of x and determine the load. For instance, if $x = L/6$, we get $q(x) = 0$; if $x = L/2$, we get $q(x) = q_1$; if $x = L$, we get $q(x) = q_1 + (3/L)(q_2 - q_1)(L - 2L/3) = q_2$; and so on.

(b) To obtain an expression for $q(x)$ that includes the reactions, we draw the free-body diagram of Fig. 7-30b. The vertical reaction R_a and the moment reaction

M_a at support A are

$$R_a = \frac{(3q_1 + q_2)L}{6} \qquad M_a = \frac{4(2q_1 + q_2)L^2}{27} \tag{d}$$

Using Table 7-2, we obtain the following expression for $q(x)$:

$$q(x) = -R_a\langle x\rangle^{-1} + M_a\langle x\rangle^{-2} + q_1\left\langle x - \frac{L}{3}\right\rangle^0 + \frac{3(q_2 - q_1)}{L}\left\langle x - \frac{2L}{3}\right\rangle^1$$

To put this equation in final form, we merely substitute the expressions for R_a and M_a from Eqs. (d).

Example 3

A beam ABC with simple supports at A and B and an overhang from B to C (Fig. 7-31a) supports a uniform load $q = 12$ kN/m over part of the span and a concentrated load $P = 15$ kN at the free end. Write the expression for the intensity $q(x)$ of the equivalent distributed loads acting on the beam.

We begin by calculating the reactions of the beam, which are shown on the free-body diagram of Fig. 7-31b. Also, we take the origin of coordinates at support A. Then we can write the expression for $q(x)$ with the aid of Table 7-2, as follows:

$$q(x) = -(9 \text{ kN})\langle x\rangle^{-1} + (12 \text{ kN/m})(\langle x - 5 \text{ m}\rangle^0 - \langle x - 10 \text{ m}\rangle^0)$$
$$-(66 \text{ kN})\langle x - 10 \text{ m}\rangle^{-1} + (15 \text{ kN})\langle x - 14 \text{ m}\rangle^{-1}$$

in which x has units of meters. The last term equals zero at all points along the beam except at the right end; hence it may be omitted. Also, it is cumbersome to write the units with every term. Hence in practical calculations we would write $q(x)$ as

$$q(x) = -9\langle x\rangle^{-1} + 12\langle x - 5\rangle^0 - 12\langle x - 10\rangle^0 - 66\langle x - 10\rangle^{-1}$$

and we would accompany this equation with the following note:

x has units of meters
$q(x)$ has units of kN/m

Without this note, the equation could easily be misinterpreted and erroneous results could be obtained.

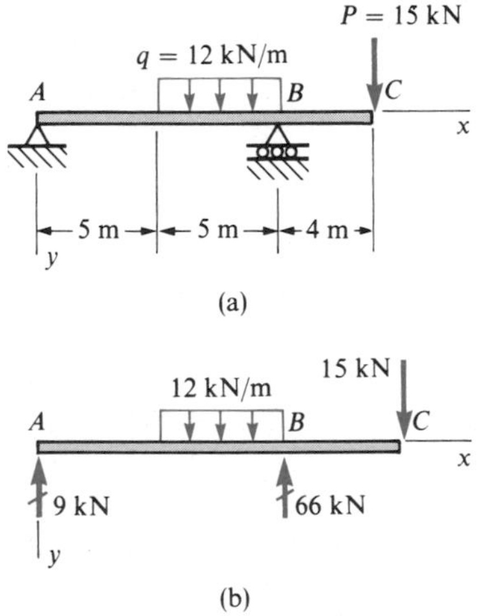

Fig. 7-31 Example 3. Representation of loads by discontinuity functions

*7.10 USE OF DISCONTINUITY FUNCTIONS TO OBTAIN BEAM DEFLECTIONS

In the conventional integration method for finding beam deflections (see Eqs. 7-10 and Sections 7.3 and 7.4), we write an expression for the load q, shear force V, or bending moment M for each segment of the beam between points where the load changes. These expressions are then integrated separately for each segment of the beam. Both boundary and continuity conditions are needed in order to evaluate the resulting constants of integration. This method of integration is quite satisfactory for very simple loadings but is likely to become unwieldy if the number of segments exceeds two. The use of discontinuity functions makes it possible to write one expression that is valid throughout the entire length of the beam. When this expression is integrated, only one constant of integration appears. Thus, in certain types of problems, the use of discontinuity functions can be very helpful.

The procedure for the use of **discontinuity functions** is quite straightforward. First, we write the expression for the equivalent distributed load $q(x)$, using the techniques described in the preceding section. This expression is then substituted into the differential equation of the deflection curve (Eq. 7-10c). Next, the differential equation is integrated successively to obtain the shear force V, the bending moment M, the slope v', and the deflection v. Each integration produces one constant of integration, which can be evaluated from known boundary conditions. Thus, we ultimately arrive at a single expression that gives the deflection v at every point of the beam. We need not integrate a separate differential equation for each segment of the beam, because the use of discontinuity functions makes it possible to integrate across discontinuities and singularities without introducing conditions of continuity of slope and deflection. Continuity of slopes and deflections is ensured automatically by the integration process.

The discontinuity method for obtaining beam deflections is also called **Clebsch's method**, after the German engineer who developed it in 1883 (Refs. 7-12 through 7-15). There are many variations in the details of the procedure, depending upon the particular beam being analyzed as well as the personal preferences of the analyst. Some of these variations are brought out in the examples that follow; others will be encountered in the problems. The analysis may begin with the load, shear-force, or bending-moment equation (Eqs. 7-10a, b, or c); however, in our examples we will start with the load equation because the solution then automatically includes the other two equations.

Example 1

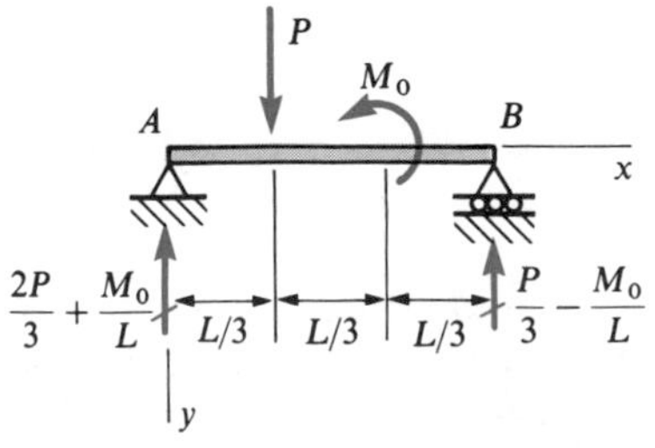

Fig. 7-32 Example 1. Discontinuity functions

Obtain the equation of the deflection curve for a simple beam AB supporting a concentrated load P and a couple M_0 (Fig. 7-32).

We will begin by setting up the differential equation of the deflection curve in terms of the distributed load q (Eq. 7-10c). With the origin of coordinates at end A, the loads P and M_0 can be expressed in terms of an equivalent distributed load $q(x)$ as follows:

$$q(x) = P\left\langle x - \frac{L}{3}\right\rangle^{-1} + M_0\left\langle x - \frac{2L}{3}\right\rangle^{-2} \tag{a}$$

(see Eq. b of Example 1, Section 7.9). Therefore, the differential equation becomes

$$EIv'''' = q = P\left\langle x - \frac{L}{3}\right\rangle^{-1} + M_0\left\langle x - \frac{2L}{3}\right\rangle^{-2} \qquad \text{(b)}$$

Integration of this equation yields

$$EIv''' = -V = P\left\langle x - \frac{L}{3}\right\rangle^{0} + M_0\left\langle x - \frac{2L}{3}\right\rangle^{-1} + C_1 \qquad \text{(c)}$$

The constant of integration is evaluated from a condition on the shear force. Let us use the shear condition at the left end of the beam, namely, when $x = 0$ (or, more precisely, when x is slightly greater than zero), the shear force equals the reaction R_a; thus,

$$V(0) = R_a = \frac{2P}{3} + \frac{M_0}{L}$$

Applying this condition to Eq. (c), we obtain

$$C_1 = -R_a = -\left(\frac{2P}{3} + \frac{M_0}{L}\right)$$

Hence, we get

$$EIv''' = -V = -\left(\frac{2P}{3} + \frac{M_0}{L}\right) + P\left\langle x - \frac{L}{3}\right\rangle^{0} + M_0\left\langle x - \frac{2L}{3}\right\rangle^{-1} \qquad \text{(d)}$$

as the differential equation in terms of the shear force.

Instead of beginning with Eq. (a), which represents only the applied loads between the supports, we could have started the analysis with an expression for $q(x)$ that contains the reactions in addition to the loads. For the beam of Fig. 7-32, this expression was obtained in the preceding section (see Eq. c in Example 1, Section 7.9). Using that expression for $q(x)$, we get the following differential equation:

$$\begin{aligned} EIv'''' = q = &-\left(\frac{2P}{3} + \frac{M_0}{L}\right)\langle x\rangle^{-1} + P\left\langle x - \frac{L}{3}\right\rangle^{-1} \\ &+ M_0\left\langle x - \frac{2L}{3}\right\rangle^{-2} - \left(\frac{P}{3} - \frac{M_0}{L}\right)\langle x - L\rangle^{-1} \qquad \text{(e)} \end{aligned}$$

The last term in this equation, representing the reaction at support B, is zero at every point along the axis of the beam up to the support. Because this term will have no effect on the calculation of shear forces, bending moments, slopes, and deflections of the beam, it can be omitted. Then we proceed to integrate Eq. (e) to obtain the equation for the shear force:

$$\begin{aligned} EIv''' = -V = &-\left(\frac{2P}{3} + \frac{M_0}{L}\right)\langle x\rangle^{0} + P\left\langle x - \frac{L}{3}\right\rangle^{0} \\ &+ M_0\left\langle x - \frac{2L}{3}\right\rangle^{-1} + C_1' \qquad \text{(f)} \end{aligned}$$

The constant of integration can be evaluated from the known shear force at any point along the beam; for variety, let us evaluate it at a point just to the left of support B:

$$V(L) = R_a - P = \frac{2P}{3} + \frac{M_0}{L} - P$$

Then from Eq. (f) we find

$$-\left(\frac{2P}{3} + \frac{M_0}{L} - P\right) = -\left(\frac{2P}{3} + \frac{M_0}{L}\right) + P(1) + M_0(0) + C_1'$$

from which $C_1' = 0$. We could have anticipated this result, because, when the expression for $q(x)$ includes all forces (both loads and reactions), no constant of integration is required. Also, we note that the term $\langle x \rangle^0$ in Eq. (f) is equal to unity at every point along the axis of the beam. Therefore, substituting 1 for $\langle x \rangle^0$ and 0 for C_1' into Eq. (f), we get

$$EIv''' = -V = -\left(\frac{2P}{3} + \frac{M_0}{L}\right) + P\left\langle x - \frac{L}{3}\right\rangle^0 + M_0\left\langle x - \frac{2L}{3}\right\rangle^{-1} \quad \text{(g)}$$

which is identical to Eq. (d).

The two preceding methods lead to the same equation for the shear force. One method begins with an equation for the loads only, after which the shear force is evaluated from a boundary condition; the other method begins with an equation that includes the reactions but requires no boundary condition. The two methods differ only in the details, and the choice between them is a matter of individual preference.

Now let us examine Eq. (g) for the shear force. The first term on the right-hand side is the shear force due to the reaction R_a, and the second term is the shear force due to the load P. In an ordinary analysis using free-body diagrams, these two terms would be the only ones to appear in an expression for V. The third term, involving the couple M_0, represents a singularity in the shear-force diagram. This term is present because the action of the couple M_0, applied at a point, must produce concentrations of shear force. However, this term has no effect on the construction of the shear-force diagram; for all practical purposes, it can be disregarded when considering the shear force. Nevertheless, the term must be retained in the equation, because it affects the next integration, which gives the bending moment.

A second integration of the differential equation produces the equation for the bending moment M:

$$EIv'' = -M = -\left(\frac{2P}{3} + \frac{M_0}{L}\right)x + P\left\langle x - \frac{L}{3}\right\rangle^1 + M_0\left\langle x - \frac{2L}{3}\right\rangle^0 + C_2 \quad \text{(h)}$$

We can evaluate the constant C_2 from the known bending moment at either end of the beam. At the left end we have $M(0) = 0$, from which we get $C_2 = 0$. Again we could have anticipated this result, because, when the expression for the shear force includes the effects of all loads and reactions (both forces and couples), no constant of integration is required.

Note that we could have obtained Eq. (h) for the bending moment directly from free-body diagrams and static equilibrium; then we could have started the deflection analysis with the second-order differential equation (Eq. 7-10a).

The slopes and deflections of the beam are obtained by two more integrations, each of which requires the introduction of a constant of integration:

$$EIv' = -\left(\frac{2P}{3} + \frac{M_0}{L}\right)\frac{x^2}{2} + \frac{P}{2}\left\langle x - \frac{L}{3}\right\rangle^2 + M_0\left\langle x - \frac{2L}{3}\right\rangle^1 + C_3 \qquad \text{(i)}$$

$$EIv = -\left(\frac{2P}{3} + \frac{M_0}{L}\right)\frac{x^3}{6} + \frac{P}{6}\left\langle x - \frac{L}{3}\right\rangle^3 + \frac{M_0}{2}\left\langle x - \frac{2L}{3}\right\rangle^2 + C_3 x + C_4 \qquad \text{(j)}$$

The two additional boundary conditions are

$$v(0) = 0 \quad \text{and} \quad v(L) = 0$$

from which we find

$$C_4 = 0 \qquad C_3 = \frac{5PL^2}{81} + \frac{M_0 L}{9}$$

Hence the final equations for the slope and deflection are

$$EIv' = -\left(\frac{2P}{3} + \frac{M_0}{L}\right)\frac{x^2}{2} + \frac{P}{2}\left\langle x - \frac{L}{3}\right\rangle^2 + M_0\left\langle x - \frac{2L}{3}\right\rangle^1 + \frac{5PL^2}{81} + \frac{M_0 L}{9} \qquad \text{(k)}$$

$$EIv = -\left(\frac{2P}{3} + \frac{M_0}{L}\right)\frac{x^3}{6} + \frac{P}{6}\left\langle x - \frac{L}{3}\right\rangle^3 + \frac{M_0}{2}\left\langle x - \frac{2L}{3}\right\rangle^2 + \left(\frac{5PL^2}{81} + \frac{M_0 L}{9}\right)x \qquad \text{(l)}$$

From these equations we can find the slope and deflection at any point of the beam. For instance, at the left-hand support ($x = 0$), Eq. (k) yields

$$\theta_a = v'(0) = \frac{5PL^2}{81EI} + \frac{M_0 L}{9EI} \qquad \text{(m)}$$

At the center of the beam ($x = L/2$), the angle of rotation and deflection are, respectively,

$$\theta_c = v'\left(\frac{L}{2}\right) = -\frac{5PL^2}{648EI} - \frac{M_0 L}{72EI} \qquad \text{(n)}$$

$$\delta_c = v\left(\frac{L}{2}\right) = \frac{23PL^3}{1296EI} + \frac{5M_0 L^2}{144EI} \qquad \text{(o)}$$

In a similar manner, we can find any other desired deflections and angles of rotation. Thus, the analysis of the beam by the use of discontinuity functions is complete.

Example 2

Obtain the equation of the deflection curve for the cantilever beam AB shown in Fig. 7-33.

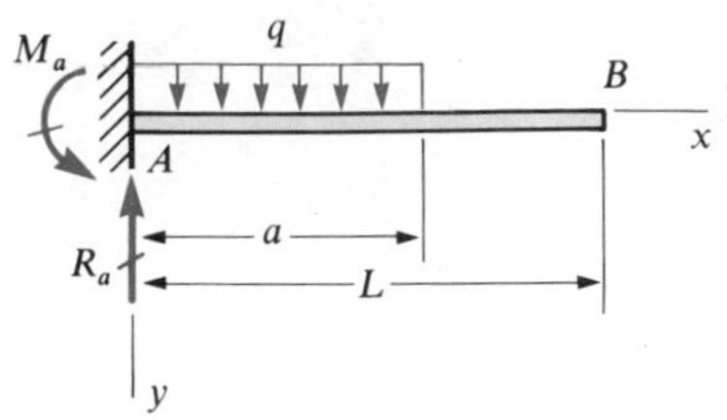

Fig. 7-33 Example 2. Discontinuity functions

Let us begin by determining the reactions at the fixed support:

$$R_a = qa \qquad M_a = \frac{qa^2}{2} \tag{p}$$

Then we can express the intensity $q(x)$ of the equivalent distributed load as follows:

$$q(x) = -R_a\langle x\rangle^{-1} + M_a\langle x\rangle^{-2} + q\langle x\rangle^0 - q\langle x - a\rangle^0 \tag{q}$$

Substituting from Eqs. (p) into Eq. (q), and then using the differential equation in terms of the load (Eq. 7-10c), we obtain

$$EIv'''' = q = -qa\langle x\rangle^{-1} + \frac{qa^2}{2}\langle x\rangle^{-2} + q\langle x\rangle^0 - q\langle x - a\rangle^0 \tag{r}$$

Replacing $\langle x\rangle^0$ with unity and then integrating, we get

$$EIv''' = -V = -qa\langle x\rangle^0 + \frac{qa^2}{2}\langle x\rangle^{-1} + qx - q\langle x - a\rangle^1 \tag{s}$$

Because we began with the complete expression for q, including the reactions, no constant of integration is required. Also, note that the second term on the right comes from the moment reaction; it has no effect on the shear-force diagram, but it must be retained because it affects the next step of integration. Also, a disadvantage of using discontinuity functions is evident from Eq. (s): for the right-hand part of the beam (where $a < x < L$), the shear force V equals zero, but this fact may not be recognized from the equation.

The next step is to integrate Eq. (s) to obtain the bending moment. Again we replace $\langle x\rangle^0$ with unity before integrating:

$$EIv'' = -M = -qax + \frac{qa^2}{2}\langle x\rangle^0 + \frac{qx^2}{2} - \frac{q}{2}\langle x - a\rangle^2 \tag{t}$$

Since we retained all terms in the equation for V, including the term based upon the couple M_a, no constant of integration is needed in the preceding equation. Now we can replace $\langle x\rangle^0$ with unity and perform two more integrations:

$$EIv' = \frac{q}{6}(x^3 - 3ax^2 + 3a^2x) - \frac{q}{6}\langle x - a\rangle^3 + C_1 \tag{u}$$

$$EIv = \frac{q}{24}(x^4 - 4ax^3 + 6a^2x^2) - \frac{q}{24}\langle x - a\rangle^4 + C_1x + C_2 \tag{v}$$

The constants C_1 and C_2 are evaluated from the conditions at the fixed support:

$$v'(0) = 0 \qquad v(0) = 0$$

These conditions give $C_1 = 0$ and $C_2 = 0$; hence the final equations for v' and v are

$$EIv' = \frac{qx}{6}(3a^2 - 3ax + x^2) - \frac{q}{6}\langle x - a\rangle^3 \tag{w}$$

$$EIv = \frac{qx^2}{24}(6a^2 - 4ax + x^2) - \frac{q}{24}\langle x - a\rangle^4 \tag{x}$$

Thus, we have obtained the equation of the entire deflection curve in terms of discontinuity functions.

The slope and deflection at any specific point may be obtained by substituting the appropriate value of x into Eqs. (w) and (x). For example, at the point $x = a$ we get

$$v'(a) = \frac{qa^3}{6EI} \qquad v(a) = \frac{qa^4}{8EI} \tag{y}$$

and at $x = L$ we get

$$\theta_b = v'(L) = \frac{qa^3}{6EI} \qquad \delta_b = v(L) = \frac{qa^3}{24EI}(4L - a) \tag{z}$$

Note that the deflection curve is a straight line in the unloaded part of the beam.

Example 3

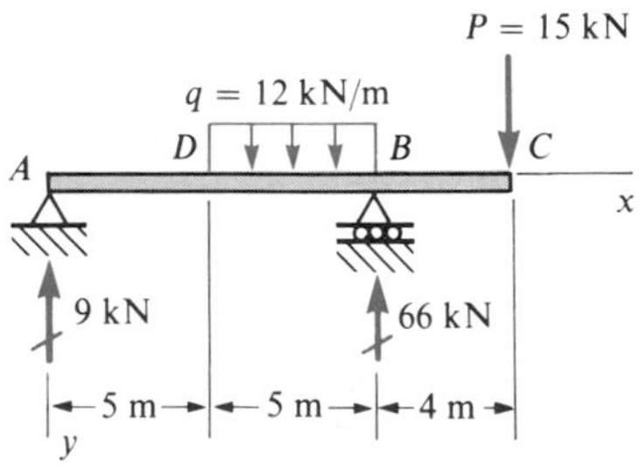

Fig. 7-34 Example 3. Discontinuity functions

The beam ABC shown in Fig. 7-34 consists of a simple span AB with an overhang BC. Determine the vertical deflections δ_c and δ_d at points C and D, respectively. The beam has modulus of elasticity $E = 200$ GPa and moment of inertia $I = 118 \times 10^6$ mm^4.

The loads and reactions of this beam were expressed in terms of discontinuity functions in Example 3 of the preceding section. Therefore, we can take the expression for $q(x)$ from that example:

$$q(x) = -9\langle x\rangle^{-1} + 12\langle x - 5\rangle^0 - 12\langle x - 10\rangle^0 - 66\langle x - 10\rangle^{-1}$$

in which x has units of meters (m) and $q(x)$ has units of kilonewtons per meter (kN/m). Substituting into Eq. (7-10c), we get the differential equation of the deflection curve:

$$EIv'''' = q = -9\langle x\rangle^{-1} + 12\langle x - 5\rangle^0 - 12\langle x - 10\rangle^0 - 66\langle x - 10\rangle^{-1}$$

In this equation, v has units of meters (m), E has units of kilonewtons per square meter (kPa), and I has units of meters to the fourth power (m^4). (Note that v'''' has units of m^{-3}.)

The first and second integrations give the equations for the shear force and bending moment:

$$EIv''' = -V = -9\langle x\rangle^0 + 12\langle x - 5\rangle^1 - 12\langle x - 10\rangle^1 - 66\langle x - 10\rangle^0$$

$$EIv'' = -M = -9x + 6\langle x - 5\rangle^2 - 6\langle x - 10\rangle^2 - 66\langle x - 10\rangle^1$$

As explained previously, no integration constants for V and M are required when we use the complete expression for $q(x)$, that is, the expression that includes the reactions and the loads.

The next two integrations give the slope and deflection:

$$EIv' = -\frac{9x^2}{2} + 2\langle x - 5\rangle^3 - 2\langle x - 10\rangle^3 - 33\langle x - 10\rangle^2 + C_1$$

$$EIv = -\frac{3x^3}{2} + \frac{1}{2}\langle x - 5\rangle^4 - \frac{1}{2}\langle x - 10\rangle^4 - 11\langle x - 10\rangle^3 + C_1 x + C_2$$

The boundary conditions on the deflection are

$$v(0) = 0 \qquad v(10) = 0$$

from which we get

$$C_1 = \frac{475}{4} \qquad C_2 = 0$$

Thus, the final expressions for v' and v are

$$EIv' = -\frac{9x^2}{2} + \frac{475}{4} + 2\langle x-5\rangle^3 - 2\langle x-10\rangle^3 - 33\langle x-10\rangle^2$$

$$EIv = -\frac{3x^3}{2} + \frac{475x}{4} + \frac{1}{2}\langle x-5\rangle^4 - \frac{1}{2}\langle x-10\rangle^4 - 11\langle x-10\rangle^3$$

in which the variables have the following units:

x has units of meters (m)
v has units of meters (m)
E has units of kilonewtons per square meter (kPa)
I has units of meters to fourth power (m^4)

The deflection δ_d at point D is found by substituting the following numerical values into the preceding equation:

$$x = 5 \text{ m} \qquad E = 200 \text{ GPa} = 200 \times 10^6 \text{ kPa}$$

$$I = 118 \times 10^6 \text{ mm}^4 = 118 \times 10^{-6} \text{ m}^4$$

The calculations proceed as follows:

$$EI\delta_d = EIv(5) = -\frac{3}{2}(5)^3 + \frac{475(5)}{4} = 406.25 \text{ kN}\cdot\text{m}^3$$

$$\delta_d = \frac{406.25 \text{ kN}\cdot\text{m}^3}{(200 \times 10^6 \text{ kPa})(118 \times 10^{-6} \text{ m}^4)} = 0.0172 \text{ m} = 17.2 \text{ mm}$$

In a similar manner, we calculate the deflection δ_c at the end of the overhang:

$$x = 14 \text{ m} \qquad E = 200 \times 10^6 \text{ kPa} \qquad I = 118 \times 10^{-6} \text{ m}^4$$

$$EI\delta_c = EIv(14) = -\frac{3}{2}(14)^3 + \frac{475(14)}{4} + \frac{1}{2}(14-5)^4$$

$$-\frac{1}{2}(14-10)^4 - 11(14-10)^3$$

$$= -5.00 \text{ kN}\cdot\text{m}^3$$

$$\delta_c = -\frac{5.00 \text{ kN}\cdot\text{m}^3}{(200 \times 10^6 \text{ kPa})(118 \times 10^{-6} \text{ m}^4)} = -0.00021 \text{ m} = -0.21 \text{ mm}$$

The minus sign means that the deflection is upward.

*7.11 TEMPERATURE EFFECTS

A uniform temperature increase causes an unconstrained bar or beam to have its length increased by the amount

$$\delta_t = \alpha(\Delta T)L \tag{7-70}$$

in which α is the coefficient of thermal expansion,* ΔT is the increase in temperature, and L is the length of the bar (see Fig. 2-17 and Eq. 2-19). If a beam is supported in such a manner that longitudinal expansion is free to occur, as is the case for all beams considered in this chapter, then a uniform temperature change will not produce any stresses in the beam. Neither will there be any lateral deflections of such a beam because there is no tendency for the beam to bend.

The behavior of a beam is quite different if the temperature is not constant across its height. For example, assume that a simple beam, initially straight and at a uniform temperature T_0, has its temperature changed to T_1 on its upper surface and T_2 on its lower surface, as pictured in Fig. 7-35a. If we assume that the variation in temperature is linear between the top and bottom of the beam, then the average temperature of the beam is $T_{aver} = (T_1 + T_2)/2$ and it occurs at the midheight. Any difference between this average temperature and the initial temperature T_0 results in a change in the length of the beam, as described in the preceding paragraph. The **temperature differential** $T_2 - T_1$ between the bottom and top of the beam results in a **curvature** of the axis of the beam, which means that lateral deflections are produced.

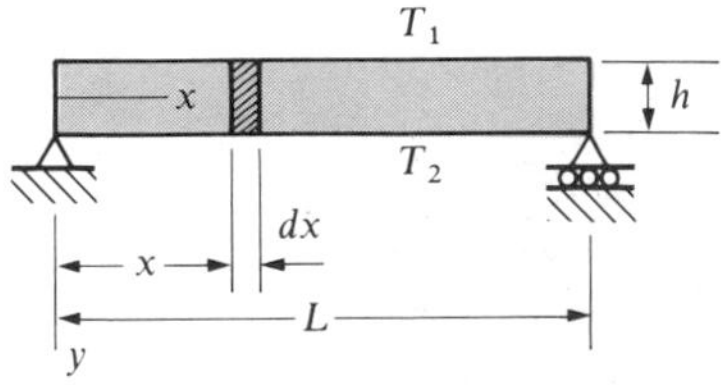

(a)

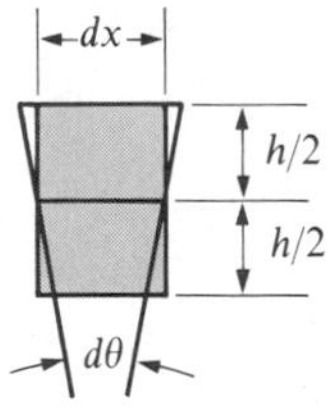

(b)

Fig. 7-35 Temperature effects in a beam

To investigate the deflections, we consider the deformation of an element having length dx (Fig. 7-35b). The changes in length of the element at the lower and upper surfaces are $\alpha(T_2 - T_0)\,dx$ and $\alpha(T_1 - T_0)\,dx$, respectively. If T_1 is greater than T_2, the sides of the element will rotate with respect to each other through an angle $d\theta$, as shown in Fig. 7-35b. The angle $d\theta$ is related to the changes in dimension by the following equation, obtained from the geometry of the figure:

$$h\,d\theta = \alpha(T_1 - T_0)\,dx - \alpha(T_2 - T_0)\,dx$$

or

$$\frac{d\theta}{dx} = -\frac{\alpha(T_2 - T_1)}{h} \tag{7-71}$$

in which h is the height of the beam. We have already seen that the quantity $d\theta/dx$ represents the curvature of the deflection curve of the beam (see Eq. 7-4); therefore, we are led to the following **differential**

* Values of the thermal coefficient α are given in Appendix H, Table H-4.

equation of the deflection curve:

$$\frac{d^2v}{dx^2} = -\frac{\alpha(T_2 - T_1)}{h} \tag{7-72}$$

Note that when T_2 is greater than T_1, the curvature is negative and the beam is bent concave upward. The quantity $\alpha(T_2 - T_1)/h$ is the counterpart of the quantity M/EI, which appears in the differential equation for the deflection curve that we used previously (Eq. 7-6).

Now that we have established Eq. (7-72) as the basic differential equation for a beam subjected to a temperature change between top and bottom of the beam, we can proceed to solve this equation using the same techniques as already described for the effects of bending moments. We can integrate the equation successively to obtain dv/dx and v, and then we can use boundary conditions to evaluate the constants of integration. Alternatively, theorems of the moment-area method may be used; all that is required is to replace M/EI by $\alpha(T_2 - T_1)/h$ in those theorems.

*7.12 LARGE DEFLECTIONS OF BEAMS

The beam deflections discussed previously in this chapter were obtained by solving the differential equations for a beam with small angles of rotation (Eqs. 7-10). When the slopes and deflections become large, the exact differential equation of the deflection curve must be used. This equation, which is also based upon the assumption that the material of the beam remains linearly elastic, is:

$$\kappa = \frac{d\theta}{ds} = -\frac{M}{EI} \tag{7-73}$$

(see Eqs. 7-1 and 7-5). The quantity $d\theta/ds$, which is the curvature of the beam, is the rate of change of θ (the angle of rotation of the deflection curve) with respect to s (the distance measured along the curve itself). When the rotations are very small, the distance s becomes the same as the distance x, and the angle of rotation θ becomes the same as the slope dv/dx; hence $d\theta/ds$ is approximated by d^2v/dx^2. However, for large deflections these simplifications are not valid, and Eq. (7-73) must be solved using the exact expression for curvature (Eq. 7-11). Thus, the differential equation of the deflection curve becomes

$$\frac{\dfrac{d^2v}{dx^2}}{\left[1 + \left(\dfrac{dv}{dx}\right)^2\right]^{3/2}} = -\frac{M}{EI} \tag{7-74}$$

The exact shape of the elastic deflection curve, as found from this equation, is called the **elastica**.

The mathematical solution of the problem of the elastica has been obtained for many different types of beams and loading conditions. Because the solutions are lengthy, they will not be worked out here. Instead, we will give the final results for one beam of practical interest together with references showing where other solutions can be found.*

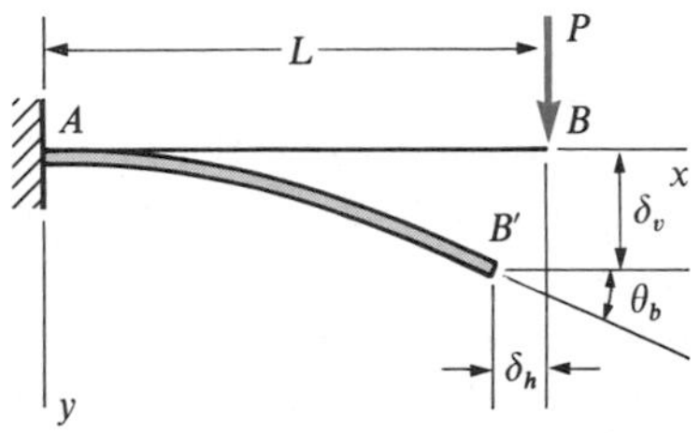

Fig. 7-36 Large deflections of a cantilever beam

Let us consider the cantilever beam AB shown in Fig. 7-36. The load P is assumed to produce large deflections of the beam, resulting in the end of the beam moving from point B to B'. The angle of rotation of the end of the beam is denoted θ_b, and the horizontal and vertical displacements are δ_h and δ_v, respectively. The length AB' of the deflection curve is equal to the initial length L because axial changes in length due to direct tension are disregarded. Because the beam is statically determinate, the expression for the bending moment M can be obtained easily and substituted into the differential equation (Eq. 7-74). Then, after considerable manipulation of the equation, including a change of the dependent variable, and also after applying the appropriate boundary conditions, the solution of the equation can be obtained in terms of elliptic functions. From this solution, expressions for θ_b, δ_v, and δ_h are obtained in terms of elliptic integrals. For instance, the equation for finding the angle θ_b is as follows:

$$F(k) - F(k, \phi) = \sqrt{\frac{PL^2}{EI}} \tag{7-75}$$

The terms in this equation have the following definitions:

$$k = \sqrt{\frac{1 + \sin\theta_b}{2}} \tag{a}$$

$$\phi = \arcsin\frac{1}{k\sqrt{2}} \tag{b}$$

$$F(k) = \text{complete elliptic integral of the first kind}$$

$$= \int_0^{\pi/2} \frac{dt}{\sqrt{1 - k^2\sin^2 t}} \tag{c}$$

$$F(k, \phi) = \text{incomplete elliptic integral of the first kind}$$

$$= \int_0^{\phi} \frac{dt}{\sqrt{1 - k^2\sin^2 t}} \tag{d}$$

Elliptic integrals can be evaluated by numerical integration using a programmable calculator or a computer program. Also, mathematical

* The problem of the elastica was first investigated by Jacob Bernoulli, Euler, Lagrange, and Plana (Refs. 7-18 through 7-23). Solutions to the problem of the elastica can be found in Refs. 7-24 through 7-27. A complete reference on large deflections of beams is the book by R. Frisch-Fay (Ref. 7-27), which contains many references. Another bibliography can be found in Ref. 7-28.

handbooks contain tables that give numerical values of the elliptic integrals for various values of the modulus k and the amplitude ϕ. Note that the difference of the two elliptic integrals in Eq. (7-75) can be expressed as a single integral with limits from ϕ to $\pi/2$.

Because of the transcendental nature of Eq. (7-75), it must be solved numerically to obtain θ_b. The procedure is as follows: (1) Assume a value of θ_b between 0 and 90°; (2) calculate k from Eq. (a); (3) obtain the corresponding value of $F(k)$ from Eq. (c); (4) calculate ϕ from Eq. (b); (5) knowing k and ϕ, obtain the value of $F(k, \phi)$ from Eq. (d); and (6) calculate the load P from Eq. (7-75). This process gives the load P that corresponds to the particular assumed value of θ_b. By repeating the calculations for other values of θ_b, we can determine as many corresponding values of P and θ_b as we desire. A listing of values determined in this manner is presented in Table 7-3, which is adapted from Ref. 7-29.

Table 7-3 Angle of rotation and deflections for a cantilever beam with a concentrated load (Fig. 7-36)

$\frac{PL^2}{EI}$	$\frac{\theta_b}{\pi/2}$	$\frac{\delta_v}{L}$	$\frac{\delta_h}{L}$
0	0	0	0
0.2	0.063	0.066	0.003
0.4	0.126	0.131	0.010
0.6	0.185	0.192	0.022
0.8	0.241	0.249	0.038
1.0	0.294	0.302	0.056
1.5	0.407	0.411	0.108
2	0.498	0.493	0.161
3	0.628	0.603	0.254
4	0.714	0.670	0.329
5	0.774	0.714	0.388
6	0.817	0.745	0.435
7	0.850	0.767	0.473
8	0.875	0.785	0.505
9	0.895	0.799	0.532
10	0.911	0.811	0.555
15	0.956	0.848	0.635
∞	1	1	1

The equation for the vertical deflection at the end of the cantilever beam is

$$\frac{\delta_v}{L} = 1 - \sqrt{\frac{4EI}{PL^2}}\,[E(k) - E(k, \phi)] \qquad (7\text{-}76)$$

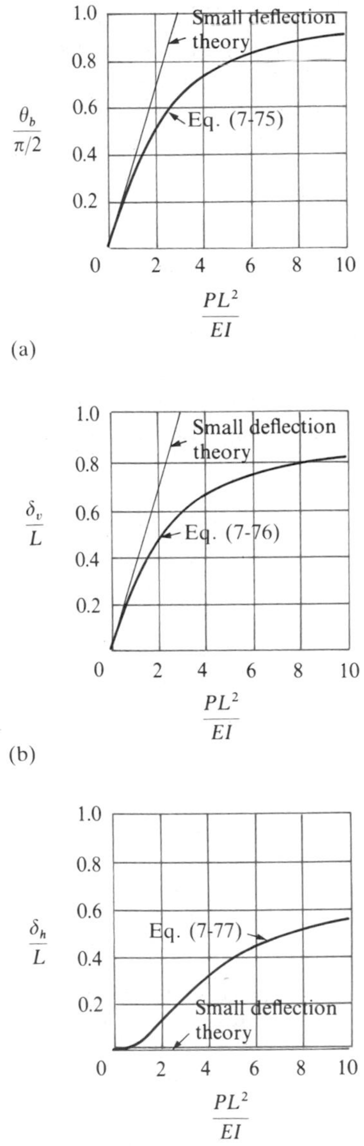

Fig. 7-37 Large deflections of a cantilever beam with a concentrated load P at the free end (Fig. 7-36): (a) angle of rotation θ_b, (b) vertical deflection δ_v, and (c) horizontal deflection δ_h

in which

$$E(k) = \text{complete elliptic integral of the second kind}$$

$$= \int_0^{\pi/2} \sqrt{1 - k^2 \sin^2 t}\, dt \tag{e}$$

$$E(k, \phi) = \text{incomplete elliptic integral of the second kind}$$

$$= \int_0^{\phi} \sqrt{1 - k^2 \sin^2 t}\, dt \tag{f}$$

Equation (7-76) can be solved by using the values of θ_b, k, ϕ, and PL^2/EI that were obtained in the solution of Eq. (7-75). The only additional step is to determine the values of $E(k)$ and $E(k, \phi)$ and substitute them into Eq. (7-76). Again we note that for purposes of calculation the difference between the elliptic integrals in Eq. (7-76) is equal to the value of a single integral having limits from ϕ to $\pi/2$. Some numerical results are given in Table 7-3.

Finally, we can obtain the horizontal deflection from the equation

$$\frac{\delta_h}{L} = 1 - \sqrt{\frac{2EI \sin \theta_b}{PL^2}} \tag{7-77}$$

The results for this case are shown in the last column of Table 7-3. The angle of rotation θ_b and the deflections δ_v and δ_h are also shown graphically in Figs. 7-37a, b, and c, respectively. In each instance, the large deflection results merge with those obtained by small-deflection theory as the load approaches zero.

The complete solution for a cantilever beam with a vertical load at the end (Fig. 7-36) is given in Refs. 7-27, 7-30, and 7-31, and the deflection of a cantilever beam with a uniform load is given in Ref. 7-32. The solutions for cantilever beams can be adapted to symmetrically loaded, simply supported beams by considering one-half of the simple beam to be the same as a cantilever beam.

*7.13 DEFLECTIONS OF INELASTIC BEAMS

The deflections of inelastic beams may be determined by methods similar to those used for elastic beams. However, the calculations are usually much more complicated because of the nonlinear relationships between bending moments and curvatures. To begin the analysis, recall the basic relationship between curvature and deflection (Eq. 7-4):

$$\kappa = \frac{d^2 v}{dx^2} \tag{7-78}$$

Because this equation was derived solely from geometric considerations, it is valid for a beam of any material. The principal limitations of the equation are that it is valid only for small rotations and the effects of shear are disregarded. However, the equation is sufficiently accurate for most purposes.

To use Eq. (7-78) for finding deflections, we must know the curvature κ, which is equal to $-M/EI$ for a beam of linearly elastic material (see Eq. 7-5). In the case of an inelastic beam, the appropriate expression for curvature (such as Eq. 5-104 of Section 5.17) must be used instead of $-M/EI$. Once the curvature is known, we may solve Eq. (7-78) for the deflection v using some of the same methods described earlier for elastic beams. For instance, we may use the successive-integration method and integrate Eq. (7-78) twice; however, there are practical difficulties in performing the integrations in all but the most elementary beam problems.

Another method, which we shall call the **curvature-area method**, is a generalized version of the moment-area method (Section 7.5). The curvature-area method, which is based upon the properties of the curvature diagram instead of the M/EI diagram, uses the following **curvature-area theorems**:

1. The angle θ_{ba} between the tangents to the deflection curve at two points A and B is equal to the negative of the area of the curvature diagram between those points.
2. The offset Δ_{ba} of point B from the tangent at A is equal to the negative of the first moment of the area of the curvature diagram between A and B, taken with respect to B.

These theorems can be used to find angles of rotation and deflections for inelastic beams in the same manner that the moment-area theorems were used for elastic beams.

Because the curvature diagram usually cannot be expressed in terms of simple functions, numerical techniques are generally needed for finding the deflections. For instance, the curvature can be calculated at a set of discrete points along the axis of the beam, and ordinates of the curvature diagram can be plotted at each of these points. These ordinates can be connected by straight lines to give an approximation of the exact diagram. Then the areas and first moments of the approximate diagram can be found numerically, after which the deflections and slopes can be determined from the curvature-area theorems. Only very simple problems can be handled by these methods, which are valid only for beams with doubly-symmetric cross sections and for materials having the same stress-strain diagrams in both tension and compression.

When calculating deflections, we must keep in mind that the principle of superposition does not apply to any beam for which the proportional limit has been exceeded. Therefore, we cannot use the method of superposition when determining deflections of inelastic beams.

Example

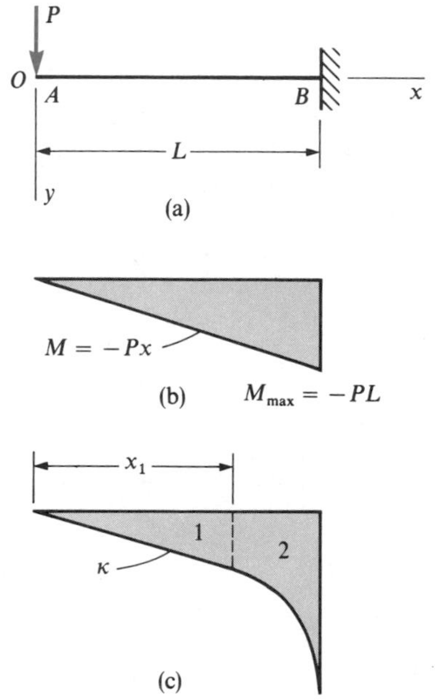

Fig. 7-38 Example. Deflections of an elastic-plastic beam of rectangular cross section

A cantilever beam AB supporting a concentrated load P at the free end (see Fig. 7-38a) is constructed of an elastic-plastic material (see Fig. 5-66). Determine the angle of rotation θ and the deflection δ at the free end of the beam from the onset of loading up to failure, assuming that the beam is of rectangular cross section.

Let us begin by sketching the bending moment diagram for the beam (Fig. 7-38b). The maximum moment is equal to PL, and if this value is less than the yield moment M_y, the beam is fully elastic. For the elastic range we have

$$\theta = \frac{PL^2}{2EI} \qquad \delta = \frac{PL^3}{3EI} \tag{a,b}$$

The yield load P_y that first produces yielding of the beam is given by the equation

$$P_y = \frac{M_y}{L} \tag{c}$$

The angle θ_y and the deflection δ_y caused by this load are

$$\theta_y = \frac{P_y L^2}{2EI} \qquad \delta_y = \frac{P_y L^3}{3EI} \tag{d,e}$$

We can express the angle of rotation and the deflection for the entire elastic range in nondimensional form by the following equations:

$$\frac{\theta}{\theta_y} = \frac{P}{P_y} \qquad \frac{\delta}{\delta_y} = \frac{P}{P_y} \qquad \left(0 \le \frac{P}{P_y} \le 1\right) \tag{7-79a,b}$$

When the maximum moment in the beam exceeds M_y, the beam will have two regions: (1) a region of fully elastic behavior, and (2) a region of elastic-plastic behavior, as shown in the curvature diagram of Fig. 7-38c. In region 1 the curvature is

$$\kappa = \frac{Px}{EI} \tag{f}$$

and in region 2 (see Eq. 5-104) it is

$$\kappa = \frac{\kappa_y}{\sqrt{3 - 2Px/M_y}} \tag{g}$$

where $\kappa_y = M_y/EI$. The length x_1 of the elastic region is found from the equation $Px_1 = M_y$, so that

$$x_1 = \frac{M_y}{P} \tag{h}$$

The angle at the end of the beam, from the first curvature-area theorem, is equal to the total area of the curvature diagram:

$$\theta = \int_0^{x_1} \frac{Px\,dx}{EI} + \int_{x_1}^{L} \frac{\kappa_y\,dx}{\sqrt{3 - 2Px/M_y}} = \frac{Px_1^2}{2EI} + \frac{\kappa_y M_y}{P}\left[\sqrt{3 - 2Px_1/M_y} - \sqrt{3 - 2PL/M_y}\right]$$

If we substitute M_y/EI for κ_y, M_y/P for x_1, and P_yL for M_y, the preceding equation becomes

$$\frac{\theta}{\theta_y} = \frac{P_y}{P}\left[3 - 2\sqrt{3 - 2P/P_y}\right] \qquad \left(1 \le \frac{P}{P_y} \le \frac{3}{2}\right) \tag{7-80}$$

in which θ_y is given by Eq. (d). This equation is valid until the maximum moment in the beam becomes equal to the plastic moment M_p, which corresponds to $P/P_y = 3/2$ (see Eq. 5-99). At the instant when P/P_y reaches this value, the angle of rotation is $\theta/\theta_y = 2$. Subsequently, this angle increases indefinitely. A graph of load versus angle of rotation, expressed as P/P_y versus θ/θ_y, is shown in Fig. 7-39.

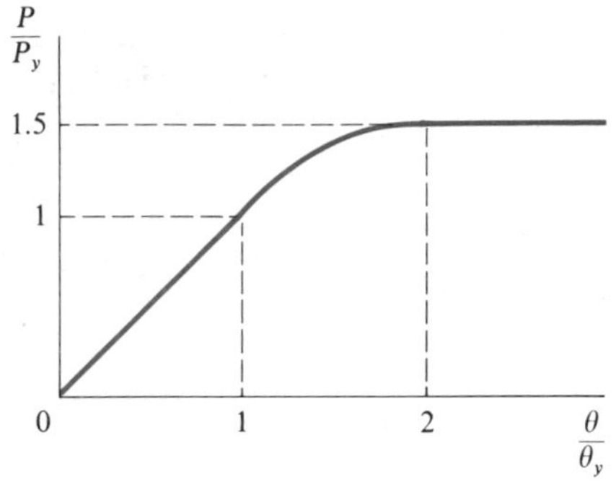

Fig. 7-39 Load versus angle of rotation for cantilever beam shown in Fig. 7-38

The deflection at the end of the beam is calculated from the second curvature-area theorem as follows:

$$\delta = \int_0^{x_1} \frac{Px^2\,dx}{EI} + \int_{x_1}^{L} \frac{\kappa_y x\,dx}{\sqrt{3 - 2Px/M_y}}$$

By evaluating these integrals and making the same substitutions as for Eq. (7-80), we get

$$\frac{\delta}{\delta_y} = \left(\frac{P_y}{P}\right)^2\left[5 - \left(3 + \frac{P}{P_y}\right)\sqrt{3 - \frac{2P}{P_y}}\right] \qquad \left(1 \le \frac{P}{P_y} \le \frac{3}{2}\right) \tag{7-81}$$

in which δ_y is given by Eq. (e). When P/P_y equals 3/2, the deflecion is $\delta/\delta_y = 20/9$. The load-deflection diagram is plotted in Fig. 7-40.

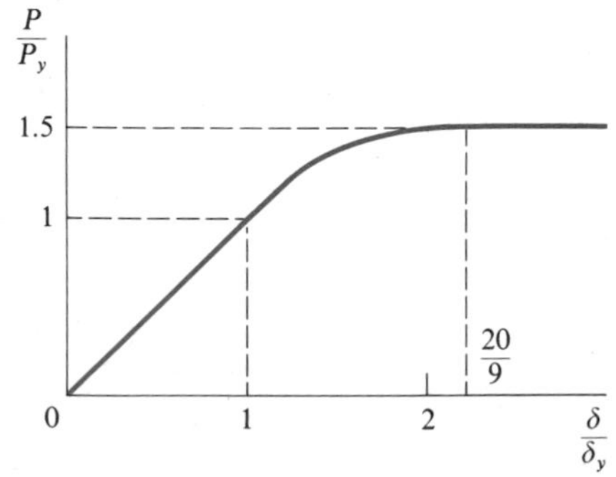

Fig. 7-40 Load versus deflection for cantilever beam shown in Fig. 7-38

PROBLEMS / CHAPTER 7

The beams described in Problems 7.3-1 through 7.3-11 have constant flexural rigidity EI.

7.3-1 By successive differentiations of the deflection curve for a simple beam with a uniform load (see Eq. 7-13 of Example 1), obtain expressions for the bending moment M and shear force V in the beam. Verify these results by using static equilibrium.

7.3-2 The equation of the deflection curve for a simple beam AB (see figure) is

$$v = \frac{q_0 x}{360LEI}(7L^4 - 10L^2x^2 + 3x^4)$$

Describe the load acting on the beam.

7.3-3 The equation of the deflection curve for a simple beam AB (see figure) is

$$v = \frac{q_0 L^4}{\pi^4 EI}\sin\frac{\pi x}{L}$$

Describe the load acting on the beam.

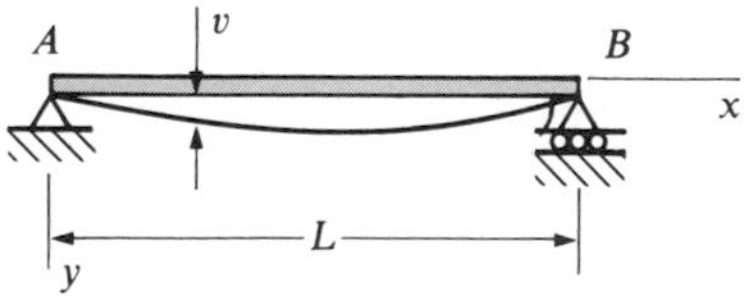

Probs. 7.3-2 and 7.3-3

7.3-4 A simple beam of rectangular cross section supports a load that deflects the beam 27 mm at the midpoint. The beam is to be replaced by a new beam of the same material and same span length. The new beam also has a rectangular cross section, but its width is only one-half the width of the original beam. What must be the height h_2 of the new beam compared to the height h_1 of the original beam if the new beam is to deflect only 16 mm under the same load?

7.3-5 A wide-flange beam (W 12 × 35) supporting a uniform load q (see Fig. 7-3 and Example 1) has a span length $L = 14$ ft. Calculate the maximum deflection δ and the angle of rotation θ at the supports if $q = 1.8$ k/ft and $E = 30 \times 10^6$ psi.

7.3-6 A cantilever beam with a uniform load (see Fig. 7-4 and Example 2) has a height h equal to $\frac{1}{12}$ of the length L. The beam is a steel wide-flange section with $E = 210$ GPa and an allowable stress of 140 MPa in both tension and compression. Calculate the ratio δ/L of the deflection at the free end to the length, assuming that the beam carries the maximum allowable load.

7.3-7 What is the span length L of a uniformly loaded simple beam (see Fig. 7-3 and Example 1) of wide-flange cross section if the maximum bending stress is 12,000 psi, the maximum deflection is 0.1 in., the height of the beam is 12 in., and the modulus of elasticity is 30×10^6 psi?

7.3-8 A uniformly loaded steel beam with simple supports (see Figure 7-3 and Example 1) has a deflection of 10 mm at the midpoint and a slope of 0.01 radians at the ends. Calculate the height h of the beam if the maximum bending stress σ is 90 MPa and E equals 200 GPa.

7.3-9 Calculate the maximum deflection δ of a uniformly loaded simple beam (see Fig. 7-3 and Example 1) if the span length $L = 7$ ft, $q = 150$ lb/ft, and the maximum bending stress $\sigma = 9000$ psi. The beam has a square cross section and is made of aluminum having modulus of elasticity $E = 10 \times 10^6$ psi.

7.3-10 A simple beam AB supports a concentrated load P as shown in Fig. 7-5 and discussed in Example 3. Plot a graph showing how the ratio θ_a/θ_b of the angles of rotation at the supports varies with the ratio a/L that defines the position of the load P $(0 < a/L < 1)$.

7.3-11 Obtain a formula for the ratio δ_c/δ_{max} of the deflection at the midpoint to the maximum deflection for a simple beam supporting a concentrated load P (see Fig. 7-5 and Example 3). From the formula, plot a graph of δ_c/δ_{max} versus the ratio a/L that defines the position of the load $(0.5 < a/L < 1)$.

Problems 7.3-12 through 7.3-19 are to be solved by integrating the second-order differential equation of the deflection curve (the bending-moment equation). All beams have constant flexural rigidity EI.

7.3-12 Determine the equation of the deflection curve for a cantilever beam AB supporting a load P at the free end (see figure). Also, determine the deflection δ_b and angle of rotation θ_b at the free end.

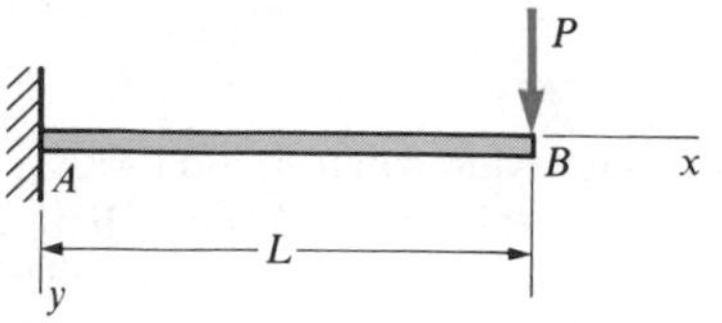

Prob. 7.3-12

7.3-13 Determine the equation of the deflection curve for a simple beam AB loaded by a couple M_0 at the left-hand support (see figure). Also, determine the maximum deflection $v_{\max}$.

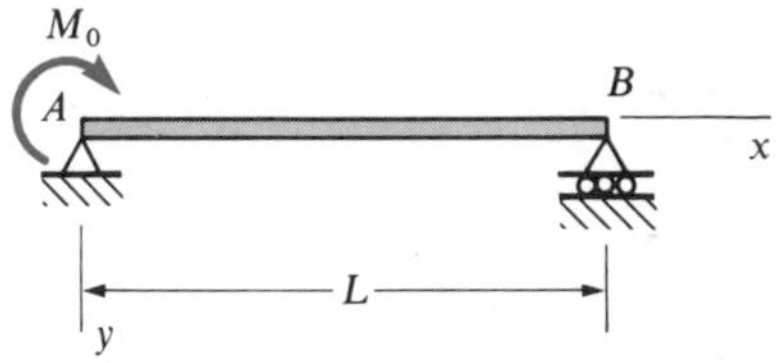

Prob. 7.3-13

7.3-14 A cantilever beam AB supporting a triangularly distributed load of maximum intensity q_0 is shown in the figure. Obtain formulas for the deflection δ_b and angle of rotation θ_b at the free end.

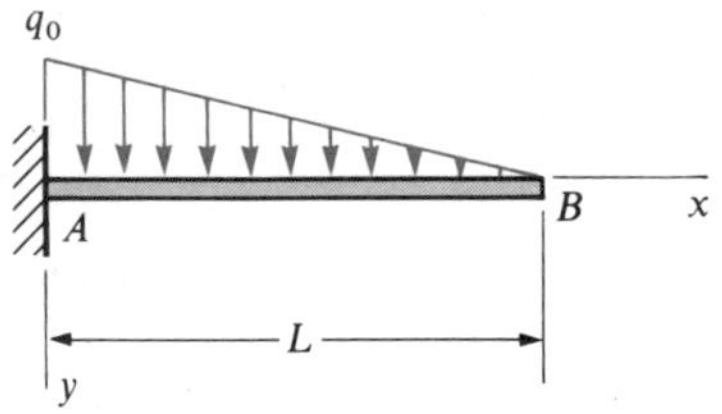

Prob. 7.3-14

7.3-15 A cantilever beam AB is acted upon by a uniformly distributed moment (bending moment, not torque) of intensity m per unit distance along the axis of the beam (see figure). Derive the equation of the deflection curve and obtain formulas for the deflection δ_b and angle of rotation θ_b at the free end.

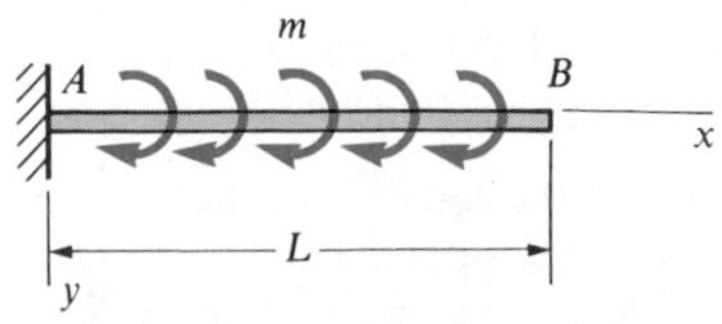

Prob. 7.3-15

7.3-16 Derive the equations of the deflection curve for a simple beam AB loaded by a couple M_0 acting at distance a from the left-hand support (see figure).

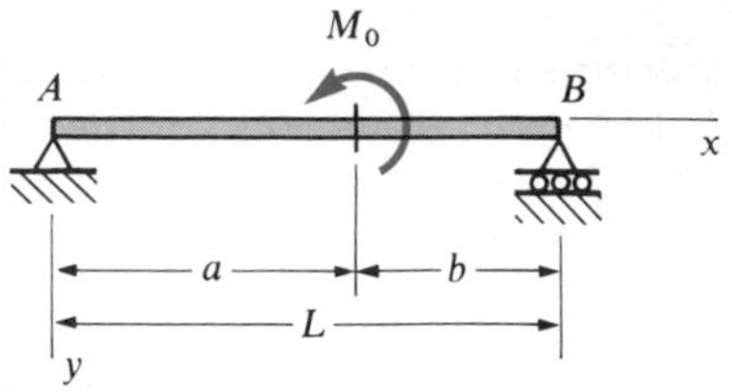

Prob. 7.3-16

7.3-17 Determine the equations of the deflection curve for a cantilever beam AB carrying a uniform load of intensity q over part of the span (see figure).

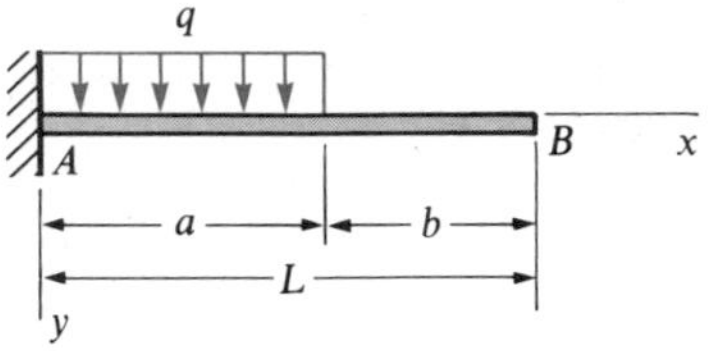

Prob. 7.3-17

7.3-18 Obtain formulas for the deflections δ_b and δ_c at points B and C, respectively, for the cantilever beam AB supporting a uniform load of intensity q acting over one-half of the length (see figure).

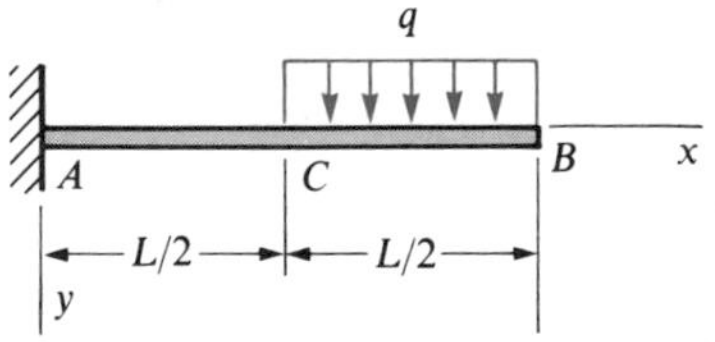

Prob. 7.3-18

7.3-19 Derive the equations of the deflection curve for a simple beam AB with a uniform load of intensity q acting over the left-hand half of the span (see figure).

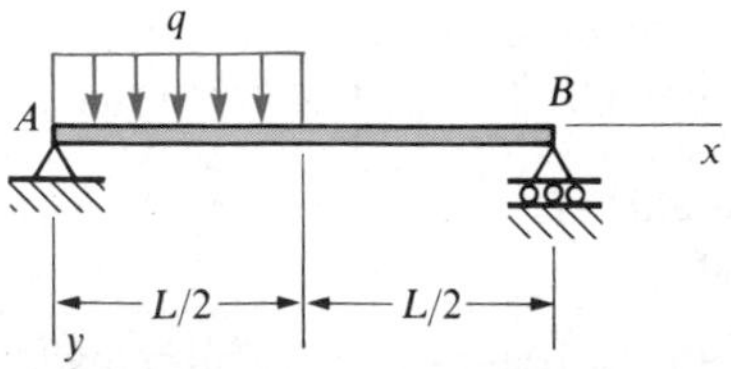

Prob. 7.3-19

The beams described in the problems for Section 7.4 have constant flexural rigidity EI.

7.4-1 Determine the equation of the deflection curve for a simple beam supporting a uniform load of intensity q (see Fig. 7-3 and Example 1 of Section 7.3). Use the fourth-order differential equation of the deflection curve (the load equation).

7.4-2 Solve Problem 7.3-12 (a cantilever beam with a concentrated load at the free end) using the third-order differential equation of the deflection curve (the shear-force equation).

7.4-3 Determine the equation of the deflection curve for a cantilever beam AB when a couple M_0 acts clockwise at the free end (see figure). Also, determine the deflection δ_b and slope θ_b at the free end. Use the third-order differential equation of the deflection curve (the shear-force equation).

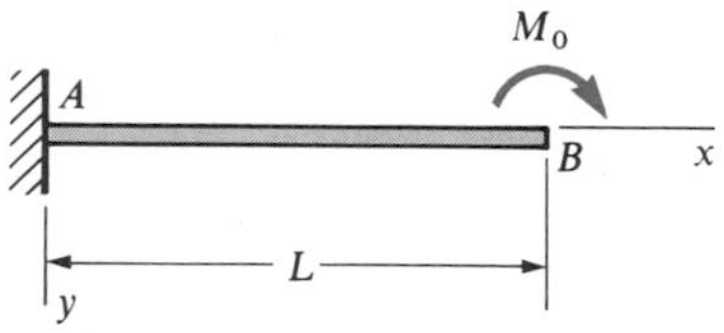

Prob. 7.4-3

7.4-4 The distributed load acting on a cantilever beam AB has an intensity q given by the expression $q_0 \cos \pi x/2L$, where q_0 is the maximum intensity of the load (see figure). Determine the equation of the deflection curve and the deflection δ_b at the free end. Use the fourth-order differential equation of the deflection curve (the load equation).

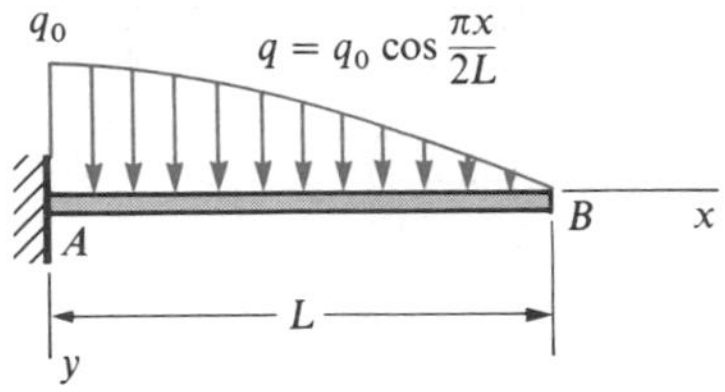

Prob. 7.4-4

7.4-5 A simple beam AB is subjected to a distributed load of intensity $q = q_0 \sin \pi x/L$ (see figure). What is the deflection δ at the midpoint of the beam? Use the fourth-order differential equation of the deflection curve (the load equation).

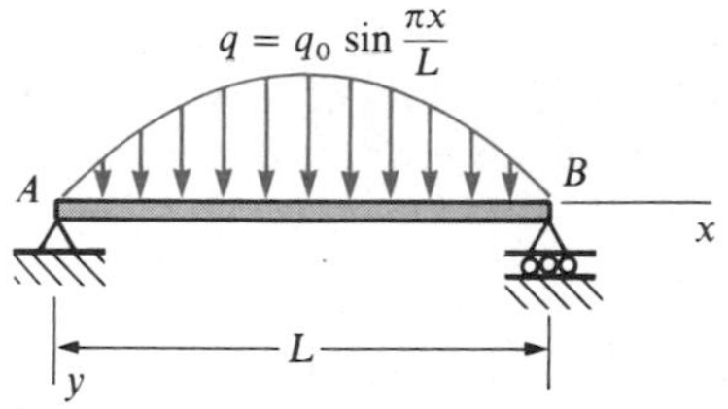

Prob. 7.4-5

7.4-6 A cantilever beam AB is subjected to a parabolically varying load of intensity $q = q_0(L^2 - x^2)/L^2$ as shown in the figure. Determine the deflection δ_b and angle of rotation θ_b at the free end. Use the fourth-order differential equation of the deflection curve (the load equation).

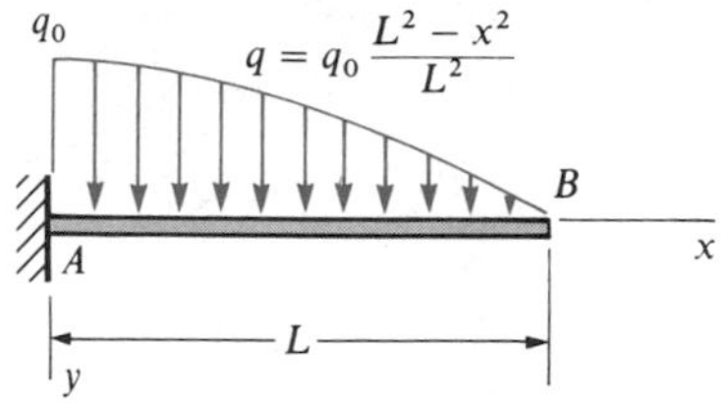

Prob. 7.4-6

7.4-7 Derive the equation of the deflection curve for a simple beam AB carrying a triangularly distributed load of maximum intensity q_0 (see figure). Also, determine the maximum deflection δ_{max} of the beam. Use the fourth-order differential equation of the deflection curve (the load equation).

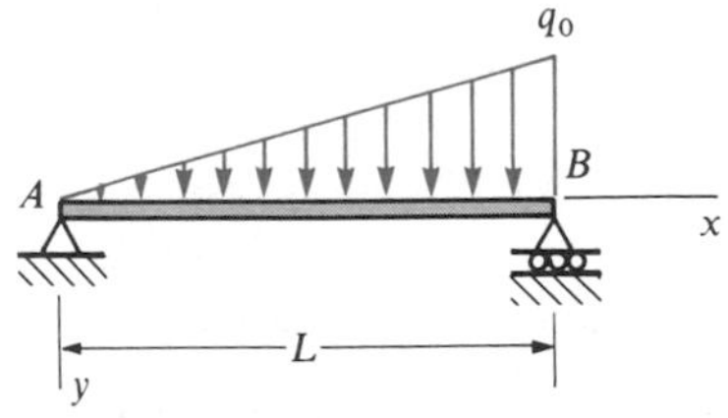

Prob. 7.4-7

7.4-8 Determine the equations of the deflection curve for a simple beam AB supporting a concentrated load P (see Fig. 7-5 and Example 3 of Section 7.3). Use the third-order differential equation of the deflection curve (the shear-force equation).

7.4-9 Solve Problem 7.3-19 (a simple beam with a uniform load over the left-hand half of the span) using the fourth-order differential equation of the deflection curve (the load equation).

7.4-10 Solve Problem 7.3-18 (a cantilever beam with a uniform load over one-half the length) using the fourth-order differential equation of the deflection curve (the load equation).

7.4-11 Determine the equations of the deflection curve for an overhanging beam *ABC* subjected to a uniform load of intensity *q* acting on the overhang (see figure). Also, obtain formulas for the deflection δ_c and angle of rotation θ_c at the end of the overhang. Use the fourth-order differential equation of the deflection curve (the load equation).

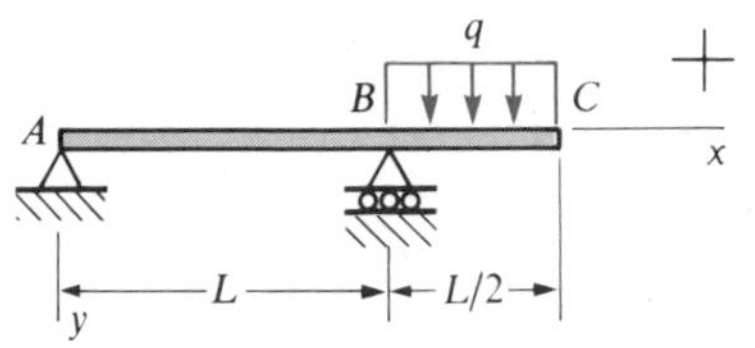

Prob. 7.4-11

The problems for Section 7.5 are to be solved by the moment-area method. All beams have constant flexural rigidity EI.

7.5-1 Referring to Fig. 7-8, which was used in deriving the moment-area theorems, prove that point *C*′, the intersection of the tangents at *A* and *B*, is directly above the centroid *C* of the area of the *M*/*EI* diagram.

7.5-2 A cantilever beam *AB* is subjected to a uniform load of intensity *q* acting throughout its length (see Fig. 7-4 of Example 2 in Section 7.3). Determine the angle of rotation θ_b and the deflection δ_b at the free end.

7.5-3 The load on a cantilever beam *AB* has a triangular distribution with maximum intensity q_0 (see Fig. 7-6 of Example 1 in Section 7.4). Determine the angle of rotation θ_b and the deflection δ_b at the free end.

7.5-4 A cantilever beam *AB* is subjected to a concentrated load *P* and a couple M_0 acting at the free end (see figure). Obtain formulas for the angle of rotation θ_b and the deflection δ_b at end *B*.

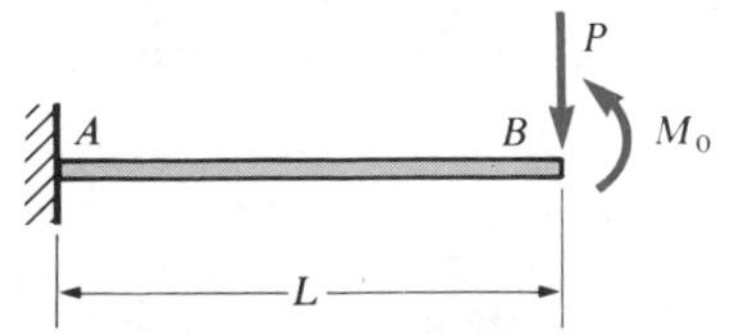

Prob. 7.5-4

7.5-5 Determine the angle of rotation θ_b and the deflection δ_b at the free end of a cantilever beam *AB* with a uniform load of intensity *q* acting over the middle third of the length (see figure).

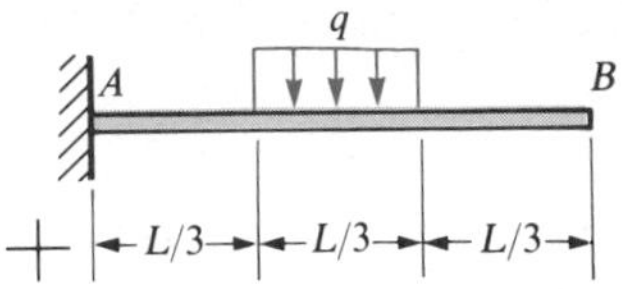

Prob. 7.5-5

7.5-6 A cantilever beam *ACB* supports two concentrated loads P_1 and P_2 as shown in the figure. Calculate the deflections δ_b and δ_c at points *B* and *C*, respectively. Assume $P_1 = 8$ kN, $P_2 = 6$ kN, $L = 2.5$ m, $E = 200$ GPa, and $I = 18.5 \times 10^6$ mm^4.

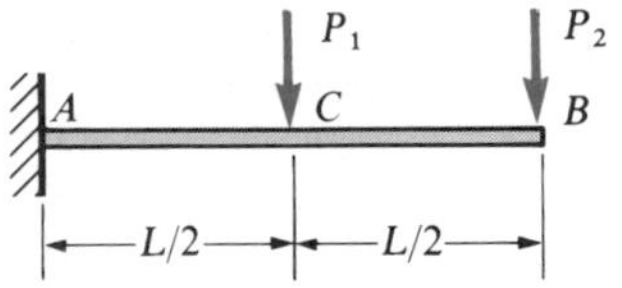

Prob. 7.5-6

7.5-7 Calculate the deflections δ_b and δ_c at points *B* and *C*, respectively, of the cantilever beam *ACB* shown in the figure. Assume $M_0 = 35$ in.-k, $P = 3.6$ k, $L = 8$ ft, and $EI = 2.1 \times 10^9$ lb-in.^{2}

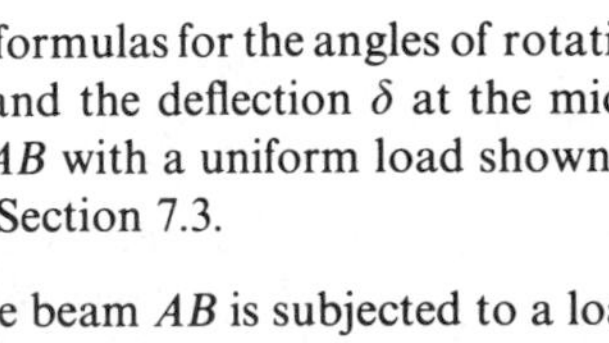

Prob. 7.5-7

7.5-8 Obtain formulas for the angles of rotation θ_a and θ_b at the supports and the deflection δ at the midpoint for the simple beam *AB* with a uniform load shown in Fig. 7-3 of Example 1 in Section 7.3.

7.5-9 A simple beam *AB* is subjected to a load in the form of a couple M_0 acting at end *B* (see figure). Determine the angles of rotation θ_a and θ_b and the maximum deflection δ_{max}.

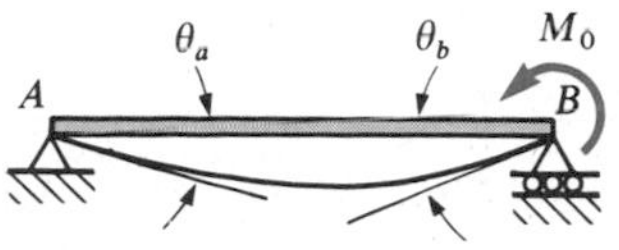

Prob. 7.5-9

7.5-10 A simple beam AB supports two concentrated loads P at the positions shown in the figure. A support C at the midpoint of the beam is positioned at distance d below the beam before the loads are applied. Assuming that $d = 10$ mm, $L = 6$ m, $E = 200$ GPa, and $I = 164 \times 10^6$ mm^4, calculate the magnitude of the loads P so that the beam just touches the support at C.

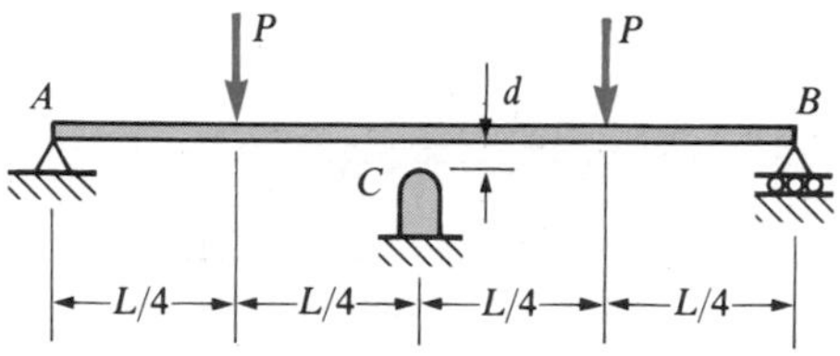

Prob. 7.5-10

7.5-11 The simple beam AB shown in the figure supports two equal concentrated loads P, one acting downward and the other upward. Determine the angle of rotation θ_a at the left-hand end, the deflection δ_1 under the downward load, and the deflection δ_2 at the midpoint of the beam.

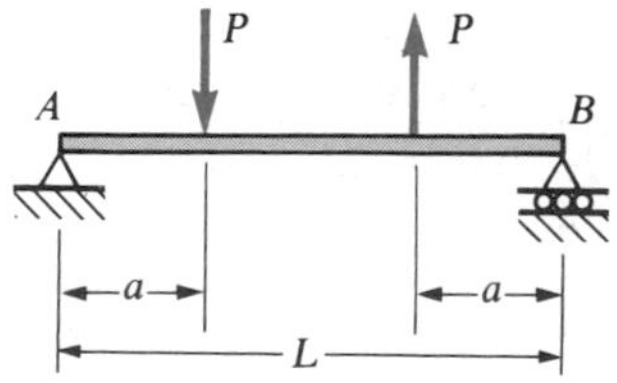

Prob. 7.5-11

7.5-12 A simple beam AB is subjected to couples M_0 and $2M_0$ acting as shown in the figure. Determine the angles of rotation θ_a and θ_b at the ends of the beam and the maximum deflection δ_{max}.

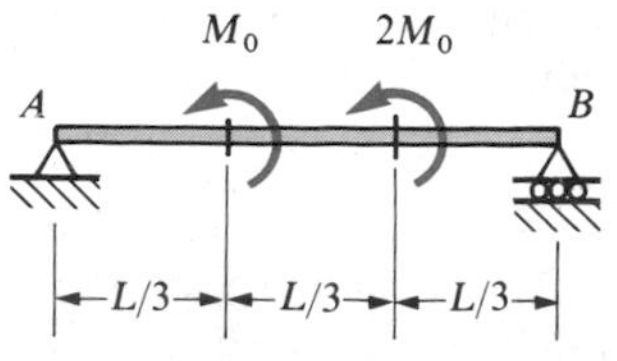

Prob. 7.5-12

7.5-13 A simple beam AB supports two concentrated loads P_1 and P_2 acting as shown in the figure. Calculate the maximum deflection δ_{max} of the beam, assuming $P_1 = 20$ k, $P_2 = 40$ k, $L = 30$ ft, $E = 30 \times 10^6$ psi, and $I = 1800$ in.^{4}

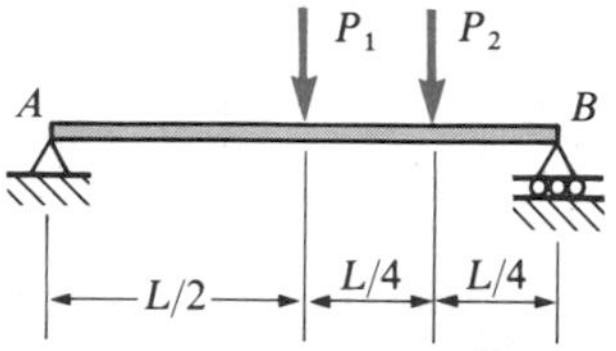

Prob. 7.5-13

7.5-14 Determine the deflection δ_c at the end of the overhang for the beam ABC shown in the figure.

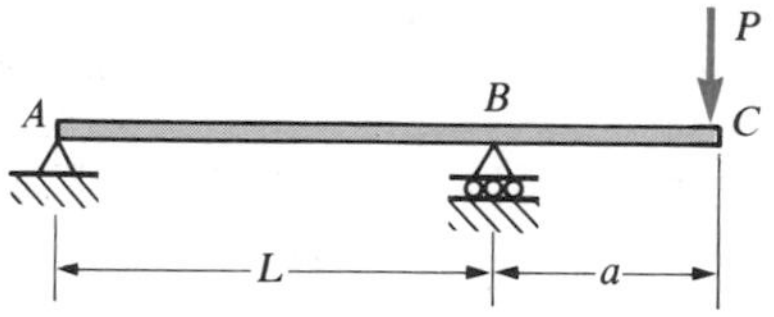

Prob. 7.5-14

7.5-15 A beam with an overhang supports loads P and Q as shown in the figure. Determine the ratio P/Q that will make the deflection at C equal to zero.

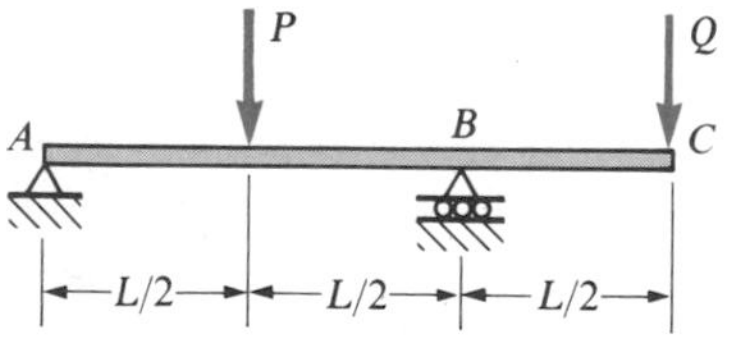

Prob. 7.5-15

7.5-16 A beam ABC with an overhang BC supports a uniform load of intensity q throughout its length, as shown in the figure. Determine the deflection δ_c at the free end C.

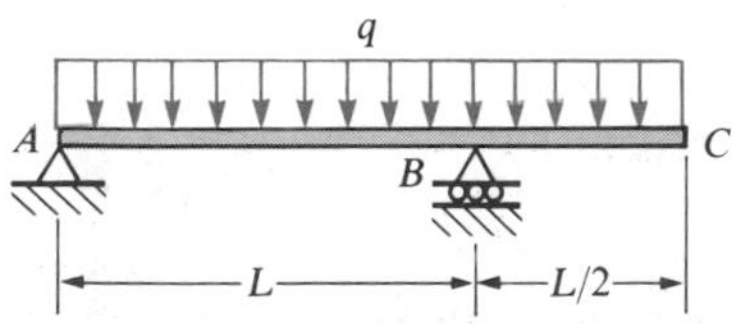

Prob. 7.5-16

7.5-17 A beam $ABECD$ is simply supported at B and C and loaded by an upward force P at A and a downward force P at D (see figure). (a) Determine the angles of rotation θ_b and θ_c at supports B and C, respectively. (b) Determine the deflections δ_a and δ_d at the ends A and D, respectively. (c) Determine the deflection δ_e at the midpoint of the beam.

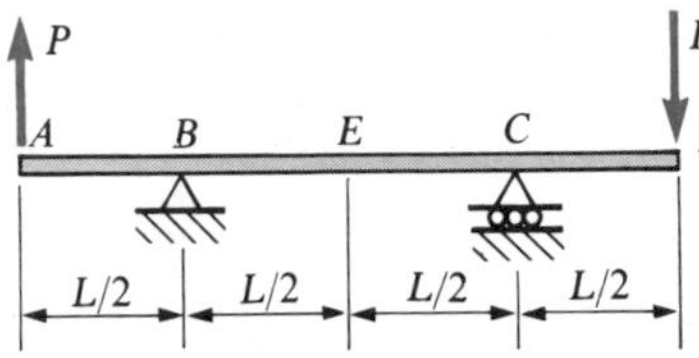

Prob. 7.5-17

7.5-18 The beam shown in the figure has overhangs at both ends and supports a uniform load of intensity q on each overhang. A concentrated load P acts at the midpoint of the beam. What should be the magnitude of the load P (in terms of qL) in order that no deflection will occur at the midpoint of the beam?

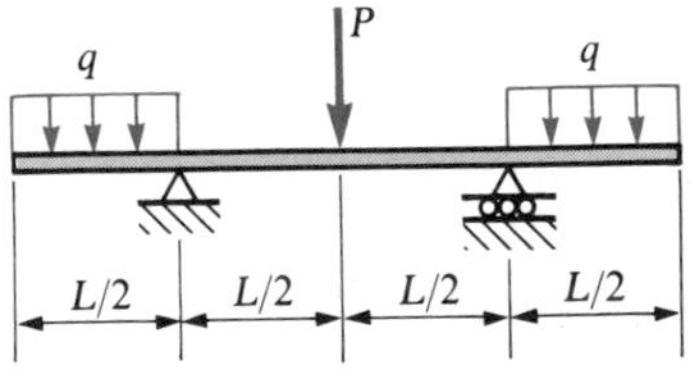

Prob. 7.5-18

The problems for Section 7.6 are to be solved by the method of superposition. All beams have constant flexural rigidity EI.

7.6-1 A cantilever beam AB carries two concentrated loads P acting at the third points, as shown in the figure. Find the deflection δ_b at the free end of the beam.

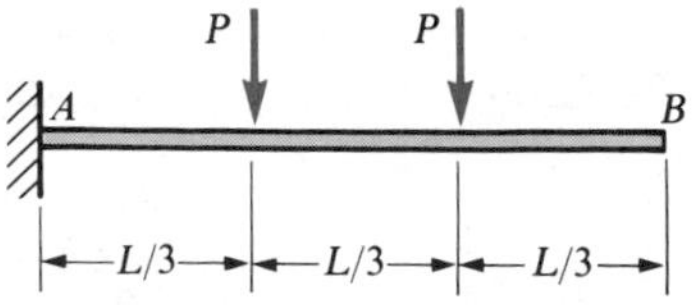

Prob. 7.6-1

7.6-2 What is the deflection δ at the midpoint of a simple beam (see figure) supporting three equal and equally spaced concentrated loads P?

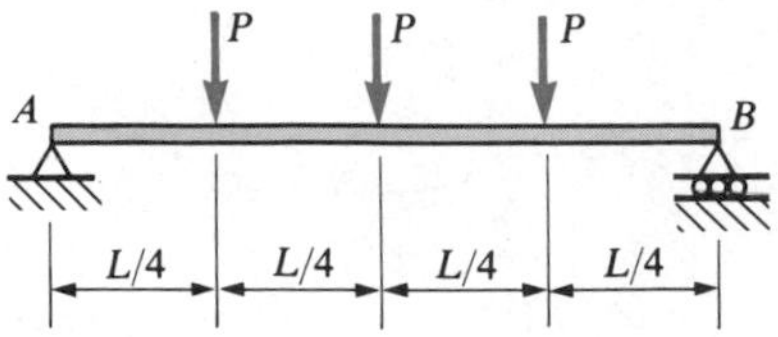

Prob. 7.6-2

7.6-3 The cantilever beam AB shown in the figure has a bracket BCD attached to its free end. A force P acts at the end of the bracket. (a) Find the ratio a/L so that the vertical deflection of point B will be zero. (b) Find the ratio a/L so that the angle of rotation of the beam at point B will be zero.

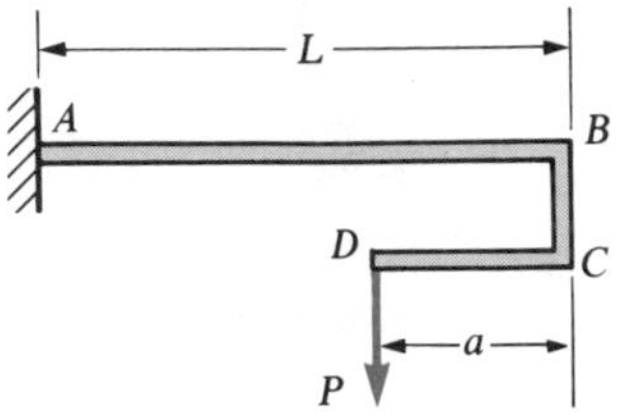

Prob. 7.6-3

7.6-4 What must be the equation $y = f(x)$ of the axis of the curved beam AB (see figure) before the load is applied in order that the load P, moving along the bar, always stays at the same level?

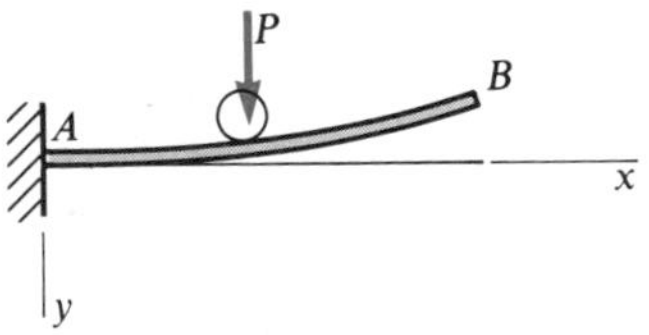

Prob. 7.6-4

7.6-5 Solve Problem 7.5-5 (a cantilever beam with a uniform load acting over the middle third of its length).

7.6-6 Solve Problem 7.5-6 (a cantilever beam with two concentrated loads).

7.6-7 Solve Problem 7.5-7 (a cantilever beam with a couple M_0 at the midpoint and a concentrated load P at the free end).

7.6-8 Determine the angle of rotation θ_b and deflection δ_b at the free end of a cantilever beam AB supporting a parabolic load defined by the equation $q = q_0x^2/L^2$ (see figure).

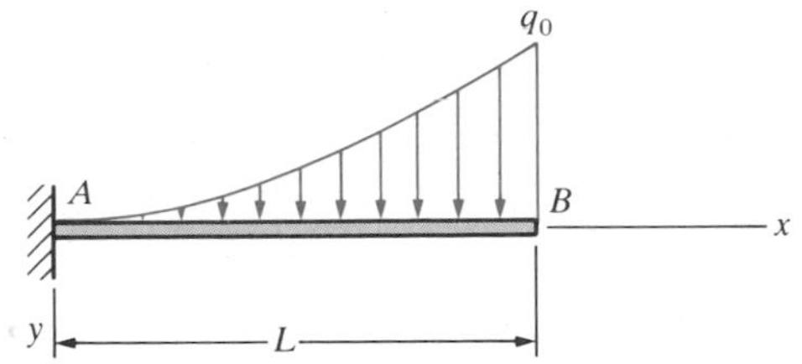

Prob. 7.6-8

7.6-9 Two prismatic beams of the same material are geometrically similar to one another; that is, each dimension of the second beam is n times the corresponding dimension of the first beam. Both beams are supported in the same manner, and the only loads are the weights of the beams themselves. Find the ratio δ_2/δ_1 of the deflection of the second beam to the corresponding deflection of the first beam.

7.6-10 A simple beam AB supports a uniform load of intensity q acting over the middle region of the span (see figure). Determine the angle of rotation θ_a at the left-hand support and the deflection δ at the midpoint.

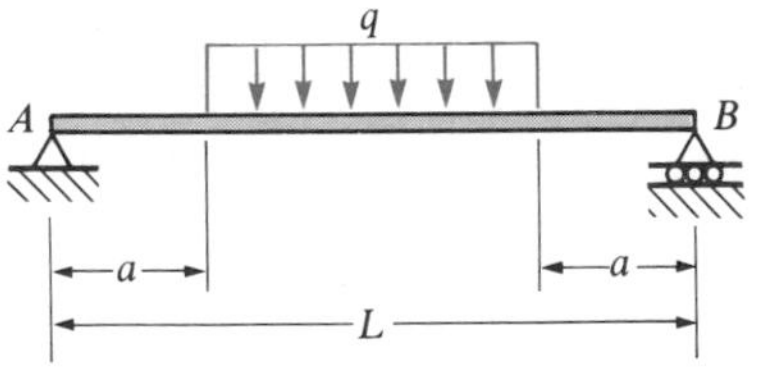

Prob. 7.6-10

7.6-11 Solve Problem 7.5-11 (a simple beam with two equal concentrated loads, one acting downward and the other upward).

7.6-12 Solve Problem 7.5-12 (a simple beam subjected to couples M_0 and $2M_0$ acting at the third points).

7.6-13 The simple beam AB shown in the figure supports a triangular load of maximum intensity q_0 on the left-hand half of the span. Find the deflection δ at the midpoint of the beam.

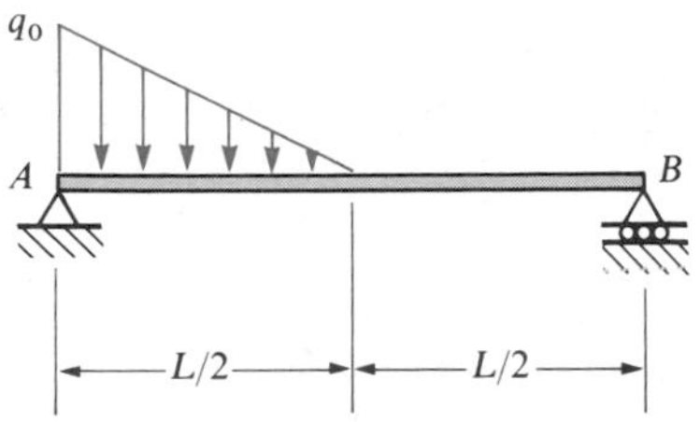

Prob. 7.6-13

7.6-14 A beam $ABCD$ with an overhang CD supports two concentrated loads P and Q, acting as shown in the figure. (a) Determine the deflection δ_b at point B. (b) For what ratio P/Q will the deflection at B equal zero?

7.6-15 The overhanging beam shown in the figure supports two concentrated loads P and Q. (a) Determine the deflection δ_d at point D. (b) For what ratio P/Q will the deflection at D equal zero?

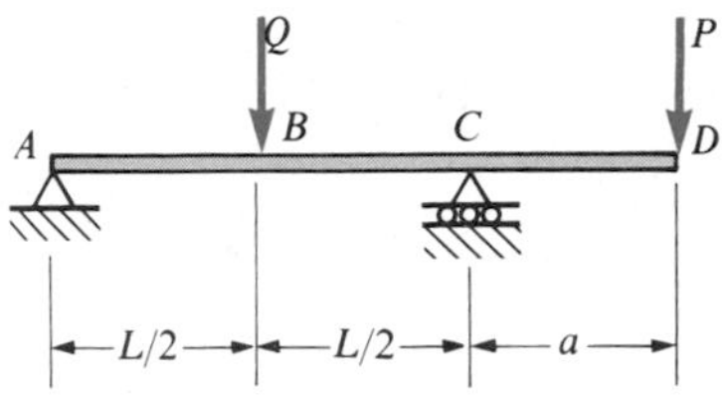

Probs. 7.6-14 and 7.6-15

7.6-16 Determine the angle of rotation θ_a at support A of the overhanging beam with a uniform load discussed in Example 3 (see Fig. 7-17).

7.6-17 Solve Problem 7.5-17 (a beam with overhangs at both ends subjected to concentrated loads).

7.6-18 Solve Problem 7.5-18 (a beam with overhangs at both ends subjected to uniform loads and a concentrated load).

7.6-19 A thin metal strip of total weight W and length L is placed across the top of a flat table of width $L/3$ as shown in the figure. What is the clearance δ between the strip and the middle of the table? (The strip of metal has flexural rigidity EI.)

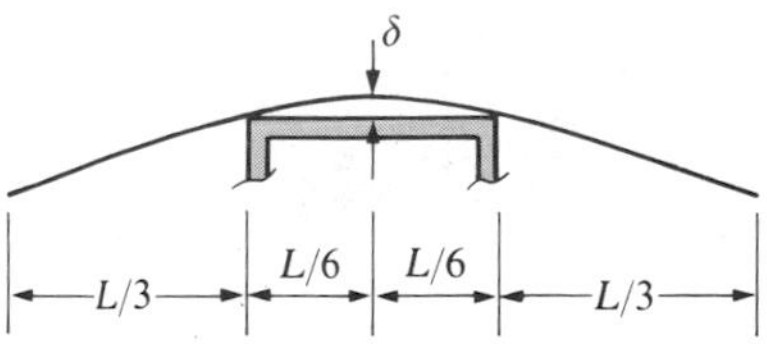

Prob. 7.6-19

7.6-20 Determine the angle of rotation θ_a at support A of the compound beam with a pin discussed in Example 4 (see Fig. 7-18) if $a = L$, $b = L/2$, and $P = 2qb$.

7.6-21 The compound beam $ABCD$ shown in the figure has fixed supports at ends A and D and consists of three members joined by pin connections at B and C. Find the deflection δ under the load P.

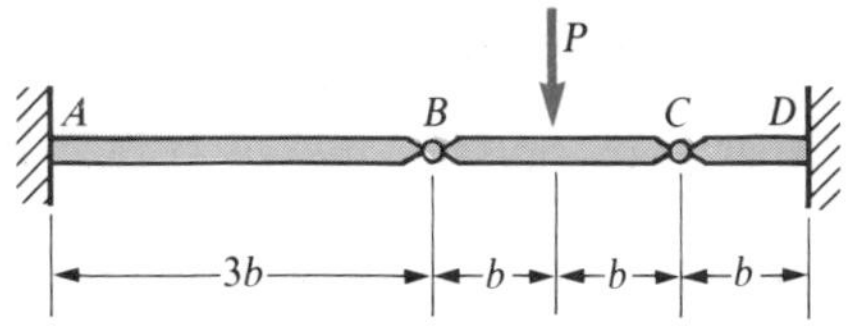

Prob. 7.6-21

7.6-22 A compound beam $ABCDE$ (see figure) consists of two parts (ABC and CDE) connected by a hinge at C. Determine the deflection δ_e at the free end E due to the load P acting at E.

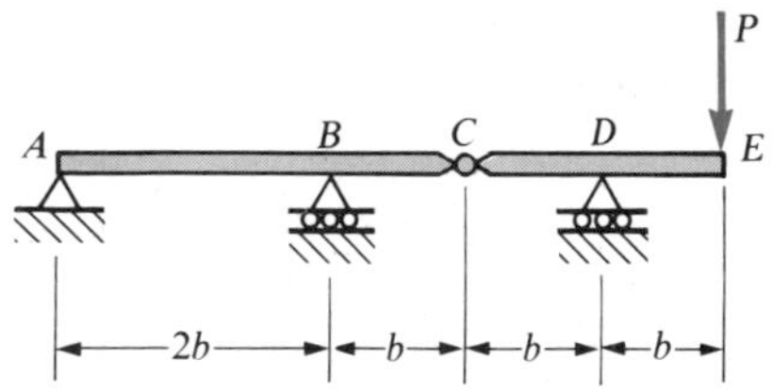

Prob. 7.6-22

7.6-23 A steel beam ABC is simply supported at A and held by a high-strength steel wire at B (see figure). A load $P = 220$ lb acts at the free end C. The wire has axial rigidity $EA = 300 \times 10^3$ lb, and the beam has flexural rigidity $EI = 30 \times 10^6$ lb-in.^{2} If $b = 12$ in., what is the deflection δ_c of point C due to the load P?

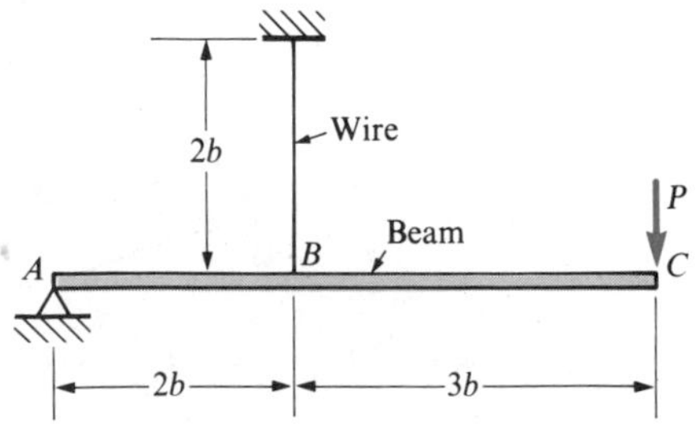

Prob. 7.6-23

7.6-24 An overhanging beam ABC is supported at end A by a pin support and at B by a spring of stiffness k (see figure). Span AB has length L and carries a uniform load of intensity q. The overhang BC has length b. For what stiffness of the spring will the uniform load produce no deflection at the free end C?

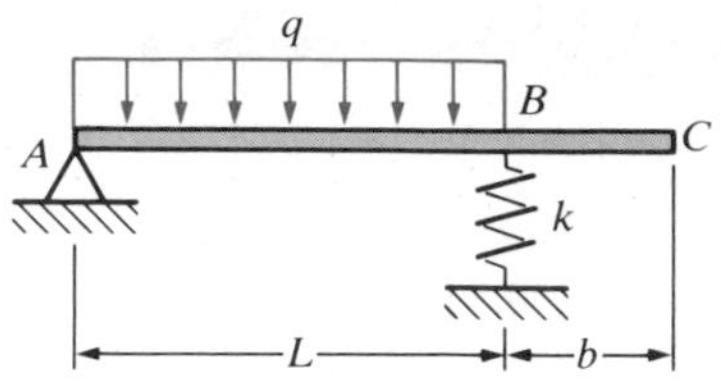

Prob. 7.6-24

***7.6-25** Two equal wheel loads P, acting at distance $L/4$ apart, move slowly across a simple beam AB of span length L (see figure). Determine the maximum value of the deflection δ_c at the midpoint of the beam.

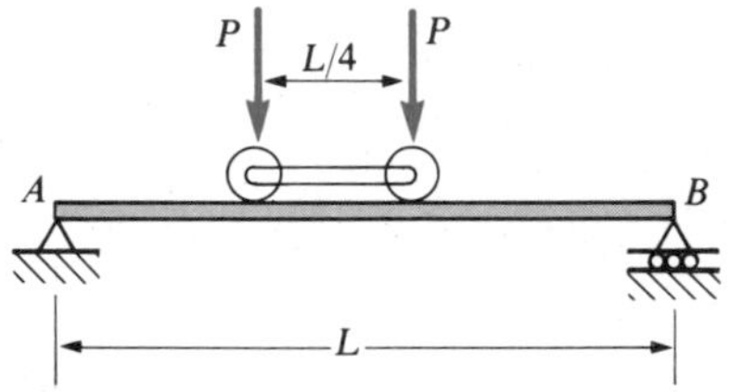

Prob. 7.6-25

***7.6-26** Find the horizontal deflection δ_h and vertical deflection δ_v at the free end C of the frame ABC shown in the figure. (Assume the flexural rigidity EI is constant throughout the frame. Also, disregard the effects of axial deformations and consider only the effects of bending due to the load P.)

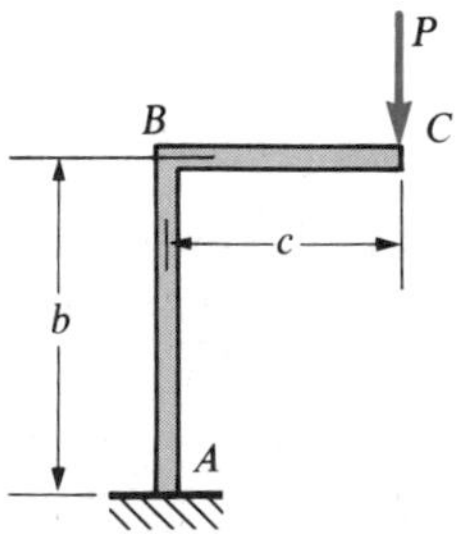

Prob. 7.6-26

***7.6-27** The frame $ABCD$ shown in the figure is squeezed by two collinear forces P acting at A and D. What is the decrease δ in the distance between points A and D when the loads P are applied? (Assume the flexural rigidity EI is

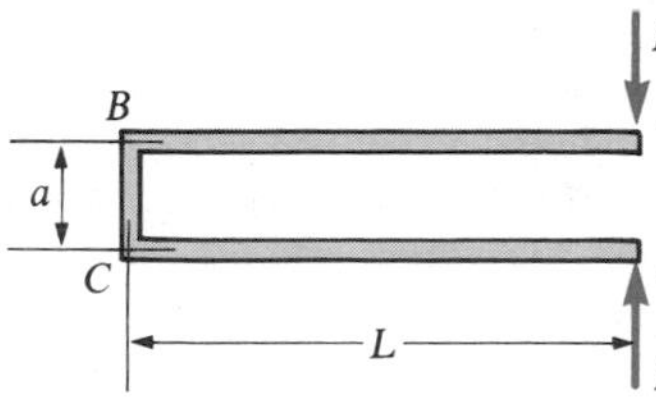

Prob. 7.6-27

constant throughout the frame and disregard the effects of axial deformations.)

7.7-1 The cantilever beam AB shown in the figure has moments of inertia I_2 and I_1 in parts AC and CB, respectively. (a) Determine the deflection δ_b at the free end B due to the load P. (b) Determine the ratio r of the deflection δ_b to the deflection at B for a prismatic cantilever with moment of inertia I_1.

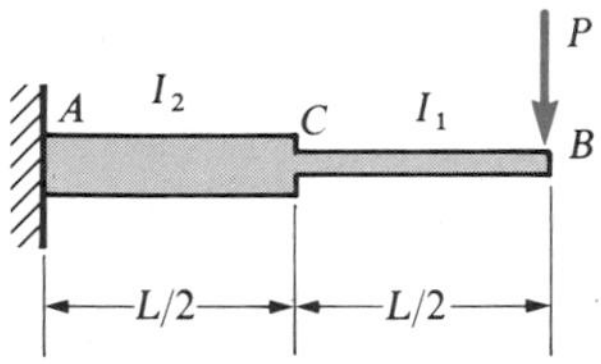

Prob. 7.7-1

7.7-2 A cantilever beam ACB with two different moments of inertia I_1 and I_2 (see figure) supports a uniform load of intensity q throughout its length. Determine the deflection δ_b at the free end B due to the load q assuming $I_1 = I$ and $I_2 = 3I$.

7.7-3 Determine the deflection δ_b at the free end B of the cantilever beam ACB shown in the figure. The beam supports a uniform load of intensity q throughout its length and has moments of inertia I_2 and I_1 in parts AC and CB, respectively.

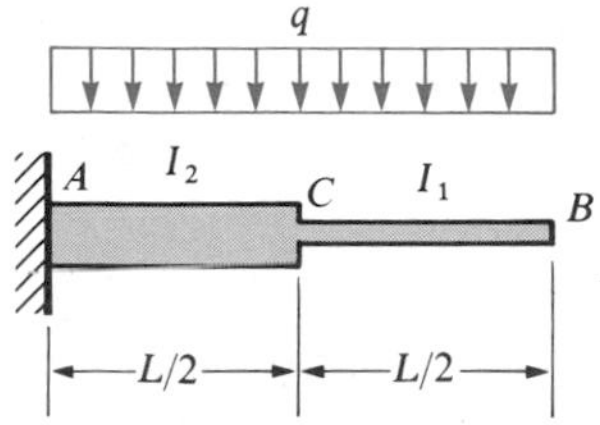

Probs. 7.7-2 and 7.7-3

7.7-4 The nonprismatic simple beam ACB shown in the figure has two different moments of inertia I_1 and I_2. Determine the angle of rotation θ_a at support A and the deflection δ_c at the midpoint due to the load P.

7.7-5 A simple beam ACB has moment of inertia I_1 near the supports and moment of inertia I_2 in the middle region, as shown in the figure. Determine the angle of rotation θ_a at support A and the deflection δ_c at the midpoint due to a uniform load of intensity q acting over the entire length of the beam. Assume $I_1 = I$ and $I_2 = 3I$.

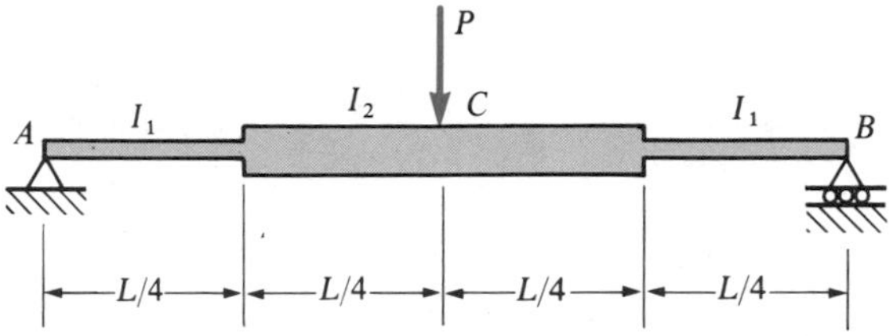

Probs. 7.7-4 and 7.7-5

7.7-6 A tapered cantilever beam AB supports a concentrated load P at the free end B (see figure). Assume that the cross sections of the beam are rectangular with constant width b, depth d_a at support A, and depth $d_b = 2d_a/3$ at the free end B. Determine the deflection δ_b at end B.

7.7-7 A tapered cantilever beam AB (see figure) supports a concentrated load P at the free end B. The beam has rectangular cross sections of constant width b. The depths of the beam at ends A and B are d_a and d_b, respectively. Determine the deflection δ_b at end B due to the load P.

7.7-8 Assume that the tapered cantilever beam AB shown in the figure has thin-walled, hollow circular cross sections of constant thickness t. The diameters at ends A and B are d_a and $d_b = d_a/2$, respectively. Find the deflection δ_b at the free end of the beam due to the load P acting at B.

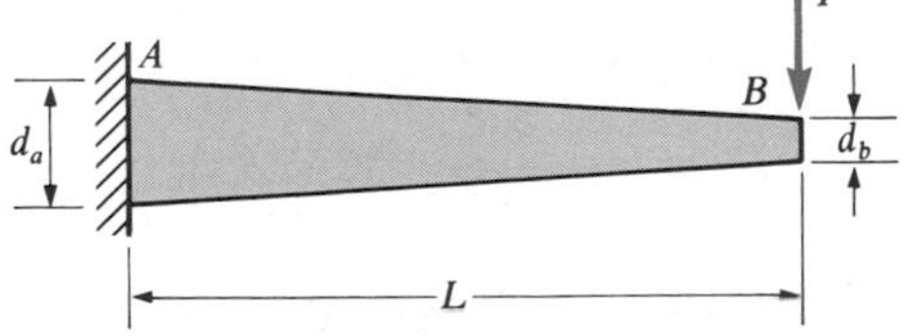

Probs. 7.7-6, 7.7-7, and 7.7-8

7.7-9 Determine the deflection δ_c at the midpoint of the simple beam AB shown in the figure. The beam has constant height h, but the width varies as shown in the lower part of the figure. (Disregard the weight of the beam and consider only the effects of the concentrated load P acting at the midpoint.)

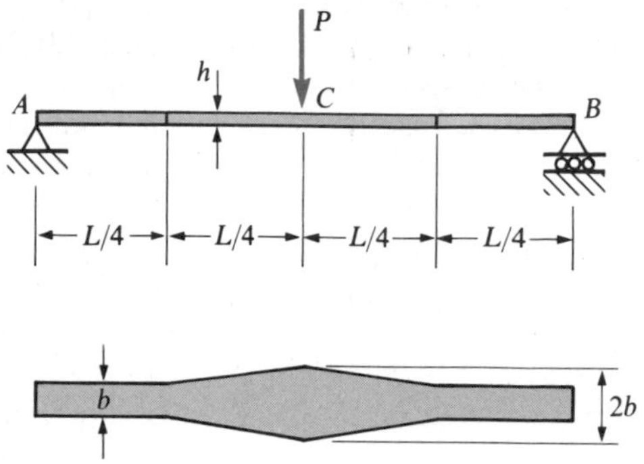

Prob. 7.7-9

The beams described in the problems for Section 7.8 have constant flexural rigidity EI.

7.8-1 A simple beam AB of length L supports a concentrated load P acting at the midpoint (see figure). (a) Determine the strain energy U stored in the beam due to the load P. (b) From this result, find the deflection δ under the load P.

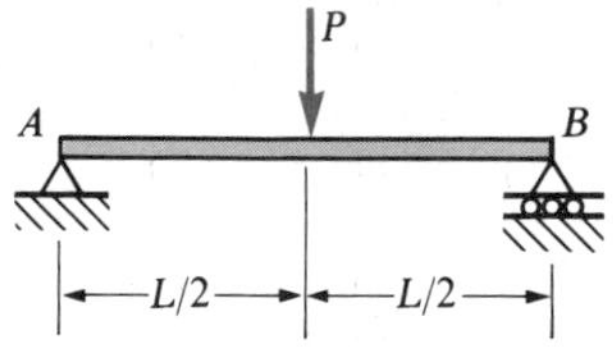

Prob. 7.8-1

7.8-2 A beam ABC with simple supports at A and B and an overhang BC supports a concentrated load P at the free end C (see figure). (a) Determine the strain energy U stored in the beam due to the load P. (b) From this result, find the deflection δ_c under the load P.

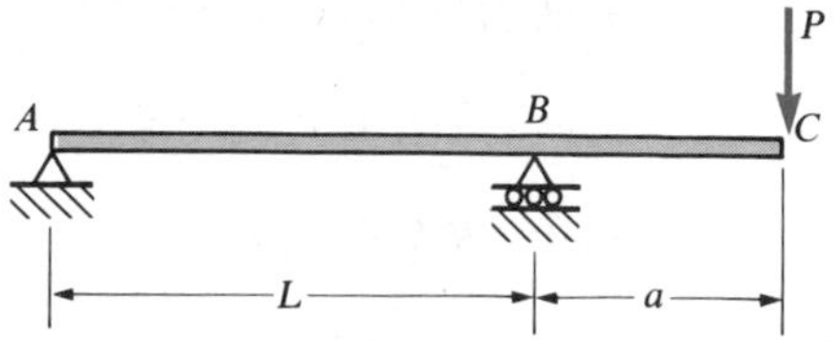

Prob. 7.8-2

7.8-3 (a) How much strain energy U is stored in a cantilever beam of length L due to a uniform load of intensity q acting along the entire length of the beam? (b) Calculate the value of U if the beam is a W 8 × 21 steel wide-flange section 6 ft long loaded to a maximum bending stress of 15,000 psi. (Assume $E = 30 \times 10^6$ psi.)

7.8-4 Two parallel beams of the same material support only their own weights. The beams have the same types of supports, but every dimension of the second beam (including its longitudinal dimensions) is n times that of the first beam. What is the ratio U_2/U_1 of their strain energies?

7.8-5 A uniformly loaded simple beam of span length L and rectangular cross section (b = width, h = height) has a maximum bending stress σ_{max}. Determine the strain energy U stored in the beam.

7.8-6 A simple beam AB of length L is loaded in such a manner that its deflection curve is a parabola (symmetric about the midpoint) with midpoint deflection equal to δ, as shown in the figure. How much strain energy U is stored in the beam?

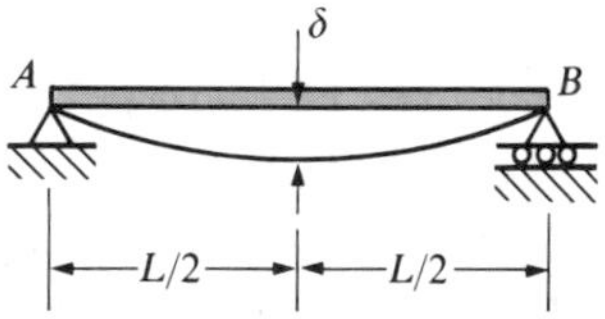

Probs. 7.8-6 and 7.8-7

7.8-7 Solve the preceding problem if the deflection curve is a half-wave of a sine curve.

7.8-8 A simple beam AB supporting a concentrated load P at the midpoint and a couple M_0 at one end is shown in the figure. Determine the strain energy U stored in the beam due to the force P and the couple M_0 acting simultaneously.

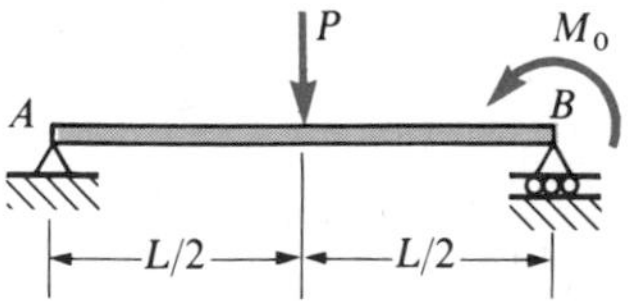

Prob. 7.8-8

7.8-9 An object of weight W is dropped onto the midpoint of a simple beam from a height h (see figure). The beam has a rectangular cross section of area A. Assuming that h is large compared to the static deflection of the beam due to

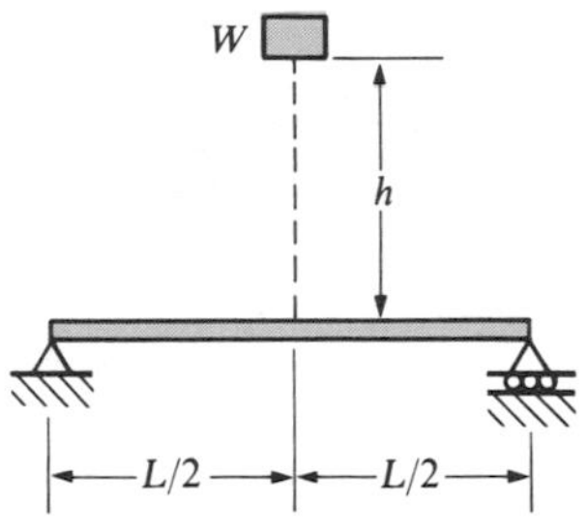

Probs. 7.8-9 and 7.8-10

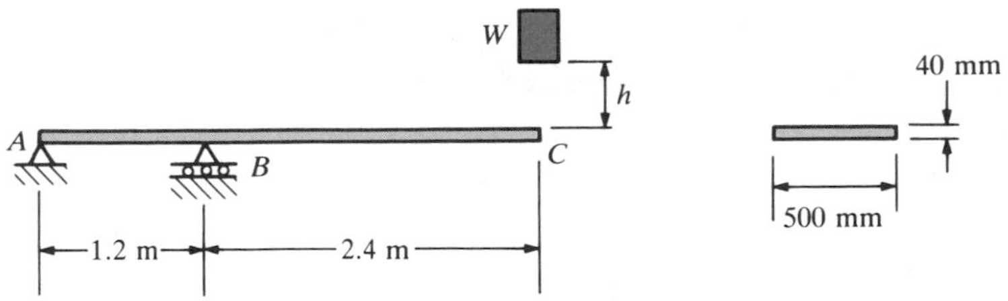

Prob. 7.8-14

W, and also assuming that the mass of the falling object is large compared to the mass of the beam, obtain a formula for the maximum bending stress σ_{max} in the beam due to the falling weight.

7.8-10 A very heavy object of weight W is dropped onto the midpoint of a simple beam from a height h (see figure). Obtain a formula for the maximum bending stress σ_{max} due to the falling weight in terms of h, σ_{st}, and δ_{st}, where σ_{st} is the maximum bending stress and δ_{st} is the deflection when the weight W acts on the beam as a statically applied load. Plot a graph of the ratio σ_{max}/σ_{st} (that is, the ratio of the dynamic stress to the static stress) versus the ratio h/δ_{st}. (Let h/δ_{st} vary from 0 to 10.)

7.8-11 A cantilever beam of length $L = 8$ ft is constructed of a W 8 × 21 wide-flange section ($E = 30 \times 10^6$ psi). A weight $W = 1600$ lb falls through a height $h = 0.25$ in. onto the end of the beam. Calculate the maximum deflection δ_{max} of the end of the beam and the maximum bending stress σ_{max} due to the falling weight.

7.8-12 A weight $W = 20$ kN falls through a height $h = 1.0$ mm onto the middle of a simple beam of length $L = 3$ m. The beam is made of wood with square cross section (dimension d on each side) and $E = 12$ GPa. If the allowable bending stress in the wood is $\sigma_{allow} = 10$ MPa, what is the minimum required dimension d? (Disregard the weight of the beam itself.)

7.8-13 A weight $W = 4000$ lb falls through a height $h = 0.5$ in. onto the middle of a simple beam of length $L = 10$ ft. Assuming that the allowable bending stress in the beam is $\sigma_{allow} = 18{,}000$ psi and $E = 30 \times 10^6$ psi, select the lightest wide-flange beam listed in Table E-1 in Appendix E that will be satisfactory. (Disregard the weight of the beam itself.)

7.8-14 An overhanging beam ABC of rectangular cross section has the dimensions shown in the figure. A weight $W = 750$ N drops onto end C of the beam. If the allowable normal stress in bending is 45 MPa, what is the maximum height h from which the weight may be dropped? (Assume $E = 12$ GPa and disregard the weight of the beam itself.)

***7.8-15** A heavy flywheel in the form of a solid circular disk of radius r is attached by a bearing to the end of a simply supported beam of flexural rigidity EI and length L (see figure). The flywheel has weight W and rotates at angular velocity ω. If the bearing suddenly freezes, what will be the reaction R at support A of the beam? (Disregard the mass of the beam itself.)

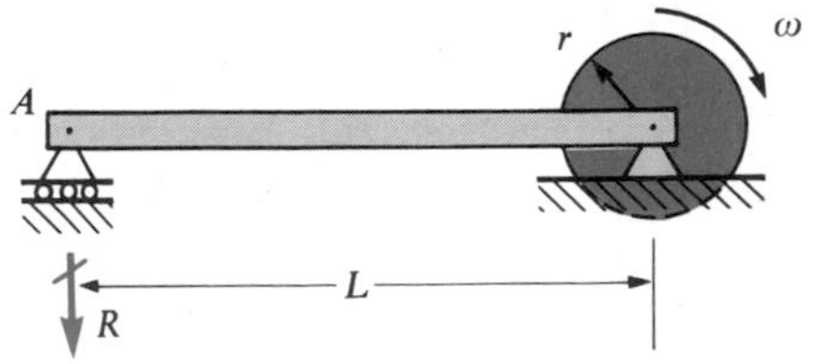

Prob. 7.8-15

7.9-1 through **7.9-12** A beam and its loading are shown in the figure. Using discontinuity functions, write the expression for the intensity $q(x)$ of the equivalent distributed load acting on the beam (include the reactions in the expression for the equivalent load).

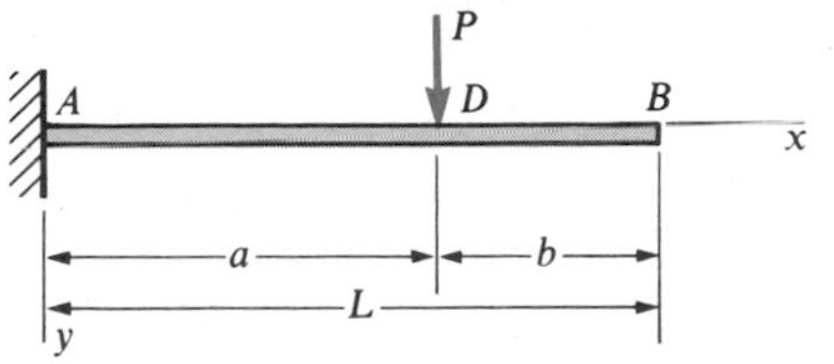

Probs. 7.9-1 and 7.10-1

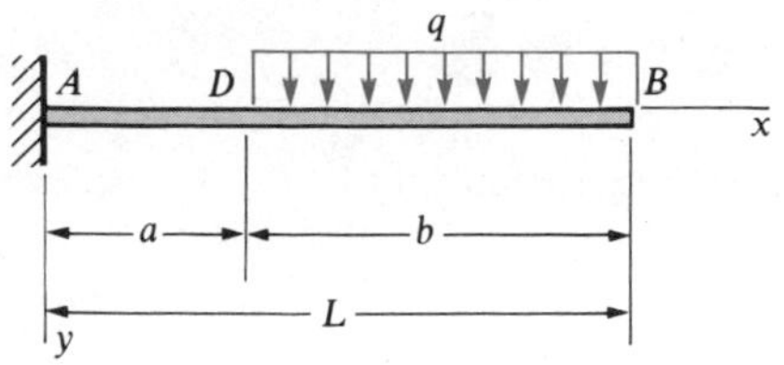

Probs. 7.9-2 and 7.10-2

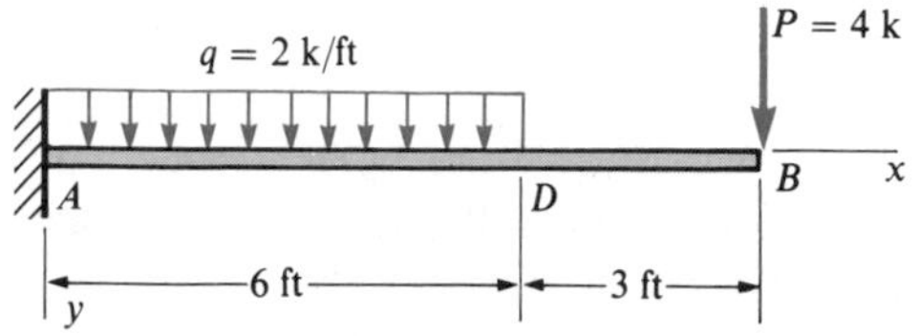

Probs. 7.9-3 and 7.10-3

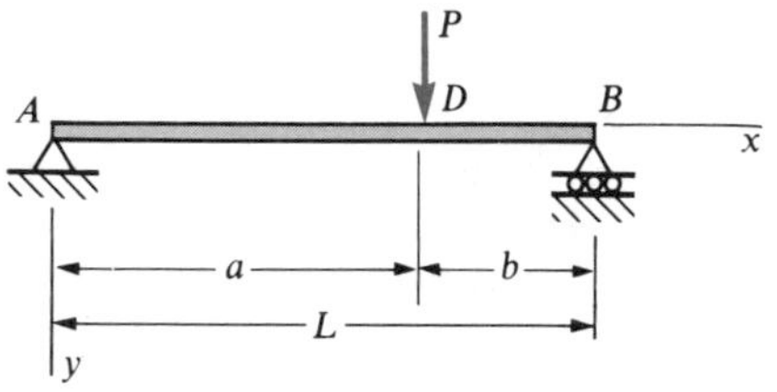

Probs. 7.9-4 and 7.10-4

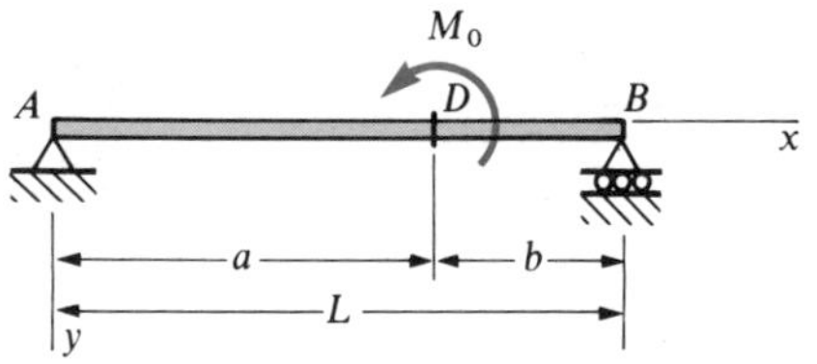

Probs. 7.9-5 and 7.10-5

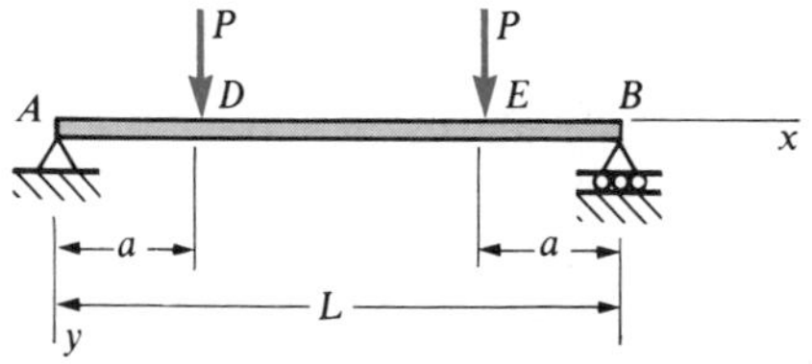

Probs. 7.9-6 and 7.10-6

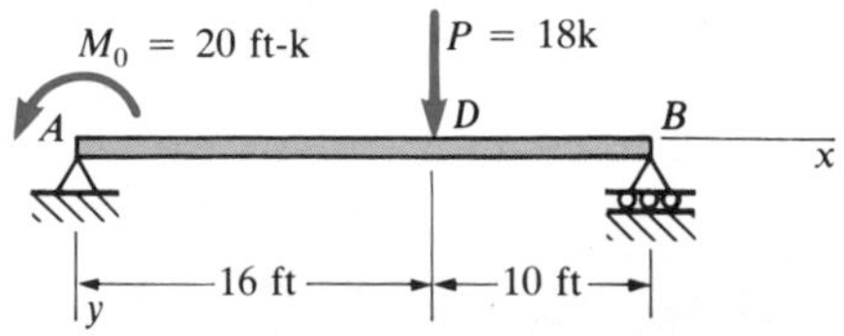

Probs. 7.9-7 and 7.10-7

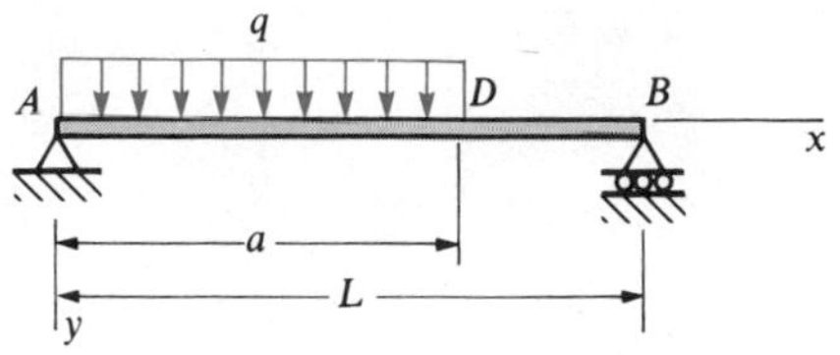

Probs. 7.9-8 and 7.10-8

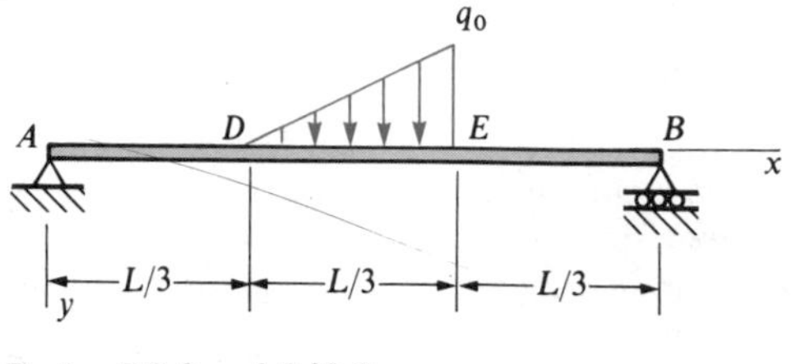

Probs. 7.9-9 and 7.10-9

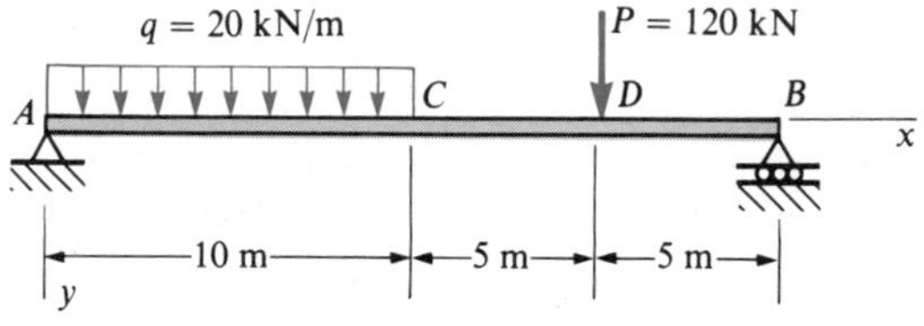

Probs. 7.9-10 and 7.10-10

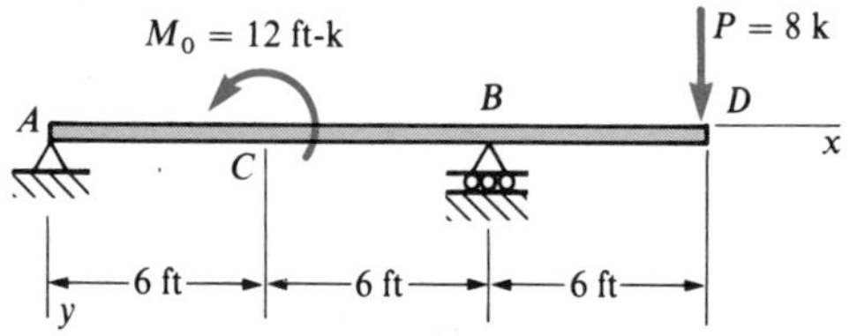

Probs. 7.9-11 and 7.10-11

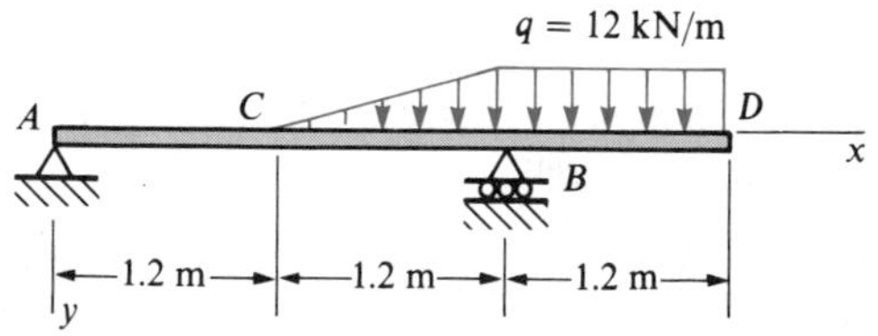

Probs. 7.9-12 and 7.10-12

The problems for Section 7.10 are to be solved by using discontinuity functions. All beams have constant flexural rigidity EI.

7.10-1, **7.10-2**, and **7.10-3** Determine the equation of the deflection curve for the cantilever beam *AB* shown in the figure. Also, obtain the angle of rotation θ_b and deflection δ_b at the free end. (For the beam of Problem 7.10-3, assume $E = 10 \times 10^3$ ksi and $I = 450$ in.^{4})

7.10-4, 7.10-5, and **7.10-6** Determine the equation of the deflection curve for the simple beam AB shown in the figure. Also, obtain the angle of rotation θ_a at the left-hand support and the deflection δ_d at point D.

7.10-7 Determine the equation of the deflection curve for the simple beam AB shown in the figure. Also, obtain the angle of rotation θ_a at the left-hand support and the deflection δ_d at point D. Assume $E = 30 \times 10^6$ psi and $I = 720$ in.^{4}

7.10-8, 7.10-9, and **7.10-10** Obtain the equation of the deflection curve for the simple beam AB (see figure). Also, determine the angle of rotation θ_b at the right-hand support (positive when counterclockwise) and the deflection δ_d at point D. (For the beam of Problem 7.10-10, assume $E = 200$ GPa and $I = 2.60 \times 10^9$ mm^4.)

7.10-11 A beam ABD with simple supports at A and B and an overhang BD is shown in the figure. (a) Obtain the equation of the deflection curve for the beam. (b) Calculate the deflections δ_c and δ_d at points C and D, respectively. (Assume $E = 30 \times 10^6$ psi and $I = 280$ in.^{4})

7.10-12 The overhanging beam ABD shown in the figure is simply supported at A and B. Obtain the equation of the deflection curve and the deflections δ_c and δ_d at points C and D, respectively. (Assume $E = 200$ GPa and $I = 15 \times 10^6$ mm^4.)

The beams described in the problems for Section 7.11 have constant flexural rigidity EI.

7.11-1 A simple beam of length L and height h undergoes a temperature change such that the bottom of the beam is at temperature T_2 and the top of the beam is at temperature T_1 (see Fig. 7-35a). Determine the equation of the deflection curve of the beam, the angle of rotation θ at the supports, and the deflection δ at the midpoint.

7.11-2 A cantilever beam of length L and height h (see figure) is subjected to a temperature change such that the temperature on the upper surface is T_1 and on the lower surface is T_2. Obtain expressions for the angle of rotation θ (positive when clockwise) and the deflection δ (positive when downward) at the free end.

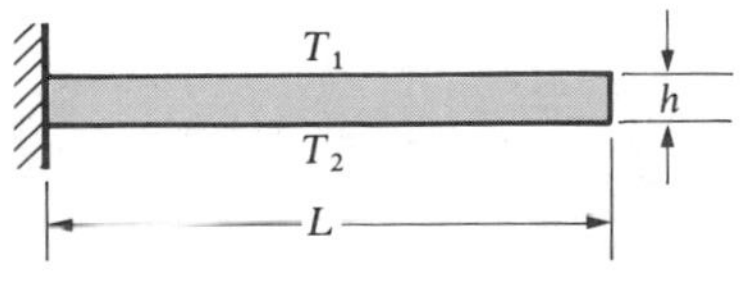

Prob. 7.11-2

7.11-3 An overhanging beam ABC of height h is heated to a temperature T_1 on the top and T_2 on the bottom (see figure). Determine the deflection δ_c (positive when downward) at the free end of the overhang.

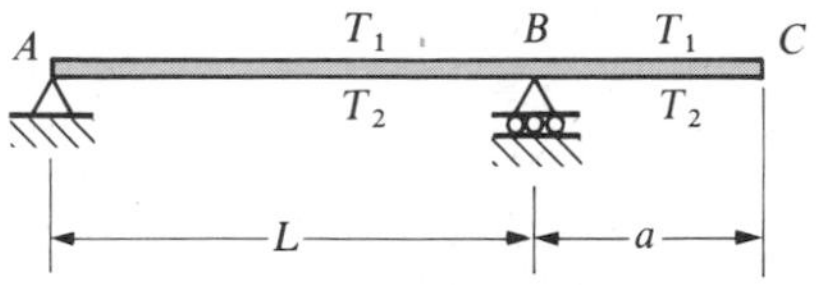

Prob. 7.11-3

7.11-4 A simple beam of length L and height h (see Fig. 7-35a) is heated nonuniformly so that the temperature difference $T_2 - T_1$ between the bottom and top of the beam varies linearly along the length of the beam; that is, $T_2 - T_1 = T_0 x$, where T_0 is a constant. Determine the maximum deflection δ_{max} of the beam.

The problems for Section 7.13 are to be solved using the assumption that the material is elastic-plastic. All beams are prismatic.

7.13-1 A simple beam AB of length L and rectangular cross section supports a concentrated load P at the midpoint (see figure). Obtain the following formulas for the angle of rotation θ at the supports and the deflection δ at the midpoint:

$$\frac{\theta}{\theta_y} = \frac{P_y}{P}\left[3 - 2\sqrt{3 - \frac{2P}{P_y}}\right] \qquad \left(1 \le \frac{P}{P_y} \le \frac{3}{2}\right)$$

$$\frac{\delta}{\delta_y} = \left(\frac{P_y}{P}\right)^2\left[5 - \left(3 + \frac{P}{P_y}\right)\sqrt{3 - \frac{2P}{P_y}}\right] \qquad \left(1 \le \frac{P}{P_y} \le \frac{3}{2}\right)$$

in which $\theta_y = P_y L^2/16EI$, $\delta_y = P_y L^3/48EI$, and $P_y = 4M_y/L$. (Hint: Use the formulas from the Example of Section 7.13 and apply them to one-half of the beam.)

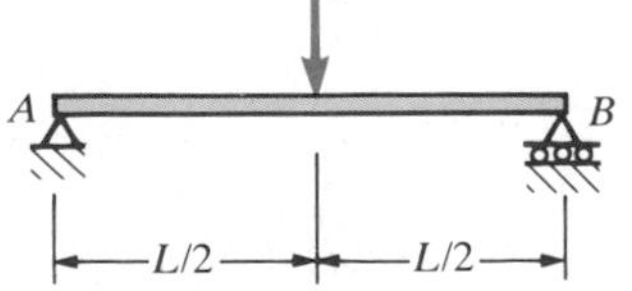

Prob. 7.13-1

7.13-2 A rectangular beam $ABCD$ with overhangs at both ends is loaded by two forces P as shown in the figure. Derive the following formula for the deflection δ at the midpoint of the beam:

$$\frac{\delta}{\delta_y} = \left(3 - \frac{2P}{P_y}\right)^{-1/2} \qquad \left(1 \leq \frac{P}{P_y} \leq \frac{3}{2}\right)$$

in which $\delta_y = P_y cL^2/8EI$ and $P_y = M_y/c$. Also, draw to scale a graph of P/P_y versus δ/δ_y.

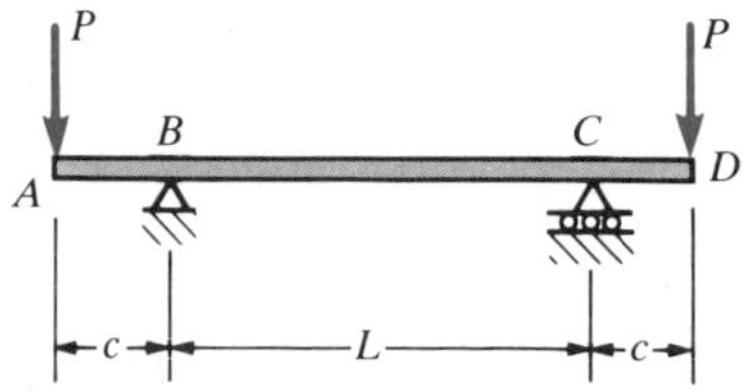

Prob. 7.13-2

***7.13-3** Derive the following expressions for the angle of rotation θ and the deflection δ at the free end of a cantilever beam of rectangular cross section and length L if the beam supports a uniform load of intensity q (see figure):

$$\frac{\theta}{\theta_y} = \sqrt{\frac{9q_y}{2q}}\left[\arcsin\sqrt{\frac{2q}{3q_y}} - 0.4839\right] \qquad \left(1 \leq \frac{q}{q_y} \leq \frac{3}{2}\right)$$

$$\frac{\delta}{\delta_y} = \frac{q_y}{q}\left[3 - 2\sqrt{3 - \frac{2q}{q_y}}\right] \qquad \left(1 \leq \frac{q}{q_y} \leq \frac{3}{2}\right)$$

in which $\theta_y = q_y L^3/6EI$, $\delta_y = q_y L^4/8EI$, and $q_y = 2M_y/L^2$. Also, draw to scale the graphs of q/q_y versus θ/θ_y and q/q_y versus δ/δ_y.

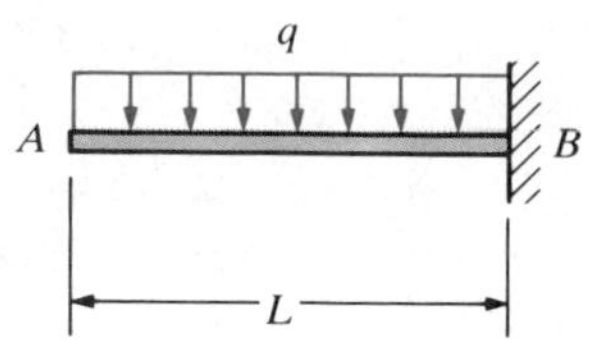

Prob. 7.13-3

***7.13-4** A simple beam AB of length L and rectangular cross section is subjected to a couple M_0 acting at one end (see figure). Derive the following formulas for the angles of rotation θ_a and θ_b at the supports:

$$\frac{\theta_a}{\theta_{ay}} = 2\left(\frac{M_y}{M_0}\right)^2\left[\frac{9M_0}{2M_y} - 5 + \sqrt{\left(3 - \frac{2M_0}{M_y}\right)^3}\,\right] \qquad \left(1 \leq \frac{M_0}{M_y} \leq \frac{3}{2}\right)$$

$$\frac{\theta_b}{\theta_{by}} = \left(\frac{M_y}{M_0}\right)^2\left[5 - \left(3 + \frac{M_0}{M_y}\right)\sqrt{3 - \frac{2M_0}{M_y}}\,\right] \qquad \left(1 \leq \frac{M_0}{M_y} \leq \frac{3}{2}\right)$$

in which $\theta_{ay} = M_y L/6EI$ and $\theta_{by} = M_y L/3EI$. Also, draw to scale the graphs of M_0/M_y versus θ_a/θ_{ay} and M_0/M_y versus θ_b/θ_{by}.

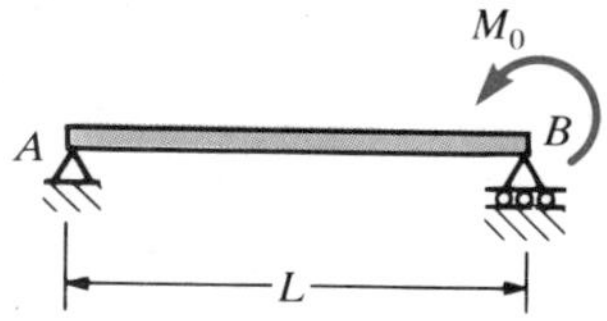

Prob. 7.13-4

CHAPTER 8

Statically Indeterminate Beams

8.1 STATICALLY INDETERMINATE BEAMS

In this chapter we consider the analysis of beams that have a larger number of reactions than can be found from equations of static equilibrium. Such beams are said to be **statically indeterminate**. Only statically determinate beams were considered in the previous chapters, and in each instance we could immediately obtain the reactions of the beam by solving equations of static equilibrium. Knowing the reactions, we then could obtain the bending moments and shear forces, which in turn made it possible to find the stresses and deflections. However, when the beam is statically indeterminate, we cannot solve for the forces on the basis of equilibrium alone. Instead, we must take into account the deflections of the beam and obtain equations of compatibility to supplement the equations of equilibrium. This same procedure was discussed in Chapter 2 for statically indeterminate structures involving members in tension and compression.

Several types of statically indeterminate beams are illustrated in Fig. 8-1. The beam in part (a) of the figure is fixed (or clamped) at support A and is simply supported at B; such a beam is called either a **propped cantilever beam** or a **fixed-simple beam**. The reactions of the beam consist of horizontal and vertical forces at A, a couple at A, and a vertical force at B. Because there are only three independent equations of static equilibrium for the beam, it is not possible to calculate all four of these reactions by statics. The number of reactions in excess of the number of equilibrium equations is called the **degree of statical indeterminacy**. Thus,

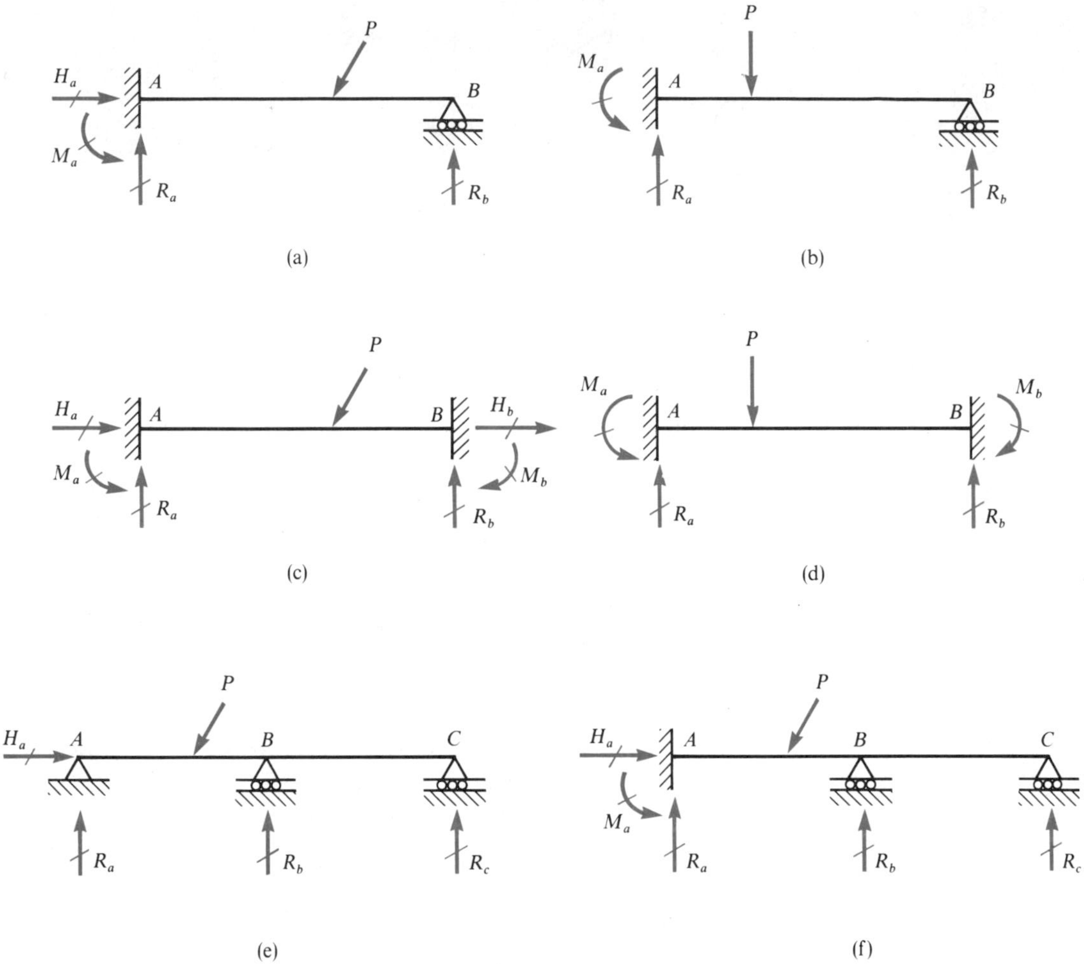

Fig. 8-1 Statically indeterminate beams: (a) and (b) propped cantilever beams; (c) and (d) fixed-end beams; and (e) and (f) continuous beams

the beam pictured in Fig. 8-1a is said to be statically indeterminate to the first degree.

Any reactions in excess of the number needed to support a structure in a statically determinate manner are called **statical redundants**, and the number of such redundants necessarily is the same as the degree of statical indeterminacy. For example, the reaction R_b shown in Fig. 8-1a may be considered a redundant reaction. Note that the structure becomes a cantilever beam when the support at B is removed. The statically determinate structure that remains when the redundant is released is called the **released structure** or the **primary structure**.

Another approach to the beam in Fig. 8-1a is to consider the reactive moment M_a as the redundant; if the moment restraint is removed, the released structure is a simple beam with a pin support at A and a roller support at B.

A special case arises if all loads on the beam are vertical (Fig. 8-1b) because then the horizontal reaction vanishes. However, the beam is still statically indeterminate to the first degree inasmuch as there are now two independent equations of static equilibrium and three reactions.

A **fixed-end beam**, sometimes called a *fixed-fixed beam* or a *clamped beam*, is shown in Fig. 8-1c. At each support there are three reactive quantities, giving a total of six unknown reactions. Because there are three equations of equilibrium, the beam is statically indeterminate to the third degree. If we take the reactions at one end as the three redundants, and if we remove them from the structure, a cantilever beam will remain as the released structure. If we remove the two fixed-end moments and one horizontal reaction, the released structure is a simple beam.

Again considering the special case of vertical loads only (Fig. 8-1d), we find that there are only four reactions to be determined. The number of static equilibrium equations is two, and the beam is statically indeterminate to the second degree.

The remaining two beams shown in Fig. 8-1 are examples of **continuous beams**, so called because they have more than one span and are continuous over a support. The beam shown in Fig. 8-1e is statically indeterminate to the first degree because there are four reactive forces and only three equations of static equilibrium. If R_b is selected as the redundant, and if we remove support B from the beam, there will remain a statically determinate simple beam AC. If R_c is selected as the redundant, the released structure will be a simple beam AB with an overhang BC. The last beam shown in Fig. 8-1 is statically indeterminate to the second degree. If we select R_b and R_c as the redundant reactions, the released structure is a cantilever beam.

In the following sections, we will discuss various methods for analyzing statically indeterminate beams. The objective in each case is to determine the redundant reactions because, after they are known, the remaining force quantities can be found from equilibrium. With the forces known, we then can determine the stresses and deflections at any point using the methods of beam analysis described in preceding chapters.

8.2 ANALYSIS BY THE DIFFERENTIAL EQUATIONS OF THE DEFLECTION CURVE

Statically indeterminate beams may be analyzed by solving one of the differential equations of the deflection curve. The procedure is essentially the same as that for a determinate beam (see Sections 7.2, 7.3, and 7.4) and consists of writing the differential equation, integrating to obtain its general solution, and then applying boundary conditions to evaluate the constants of integration. There always are enough boundary conditions not only to ascertain the constants of integration but also to find the redundant reactions that appear in the solution. Either the second-order equation in terms of the bending moment (Eq. 7-10a), the third-order equation in terms of the shear force (Eq. 7-10b), or the fourth-order equation in terms of the intensity of distributed load (Eq. 7-10c) may be used.

Because of the computational difficulties that arise when a large number of constants is to be evaluated, the method is practical only for relatively simple cases of loading and for beams of only one span. The following examples illustrate the procedure.*

Example 1

A propped cantilever beam AB with a uniform load of intensity q is shown in Fig. 8-2. Determine the reactions R_a, R_b, and M_a for this beam.

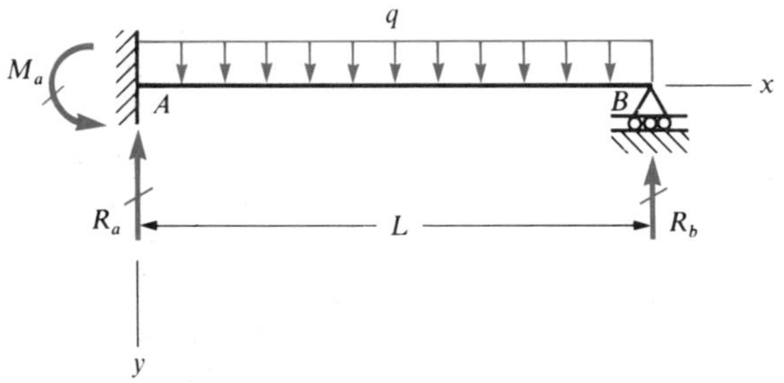

Fig. 8-2 Example 1. Propped cantilever beam

Let us begin with the second-order differential equation of the deflection curve (the bending-moment equation). The expression for the bending moment M will be in terms not only of the load but also of the redundant reaction. Therefore, we need to select a redundant reaction and express the other reactions in terms of it. Let us choose the reaction R_b as the redundant; then, from equations of static equilibrium, we obtain the reactions at A in terms of R_b as follows:

$$R_a = qL - R_b \qquad M_a = \frac{qL^2}{2} - R_b L \tag{a}$$

Now we can obtain an expression for the bending moment M in terms of R_b:

$$M = R_a x - M_a - \frac{qx^2}{2} = qLx - R_b x - \frac{qL^2}{2} + R_b L - \frac{qx^2}{2}$$

The second-order differential equation of the deflection curve is

$$EIv'' = -M = -qLx + R_b x + \frac{qL^2}{2} - R_b L + \frac{qx^2}{2}$$

* The earliest use of the differential equation for analyzing statically indeterminate beams appears in Navier's book on strength of materials (Ref. 8-1).

and two successive integrations give

$$EIv' = -\frac{qLx^2}{2} + \frac{R_b x^2}{2} + \frac{qL^2 x}{2} - R_b L x + \frac{qx^3}{6} + C_1$$

$$EIv = -\frac{qLx^3}{6} + \frac{R_b x^3}{6} + \frac{qL^2 x^2}{4} - \frac{R_b L x^2}{2} + \frac{qx^4}{24} + C_1 x + C_2$$

There are three unknown quantities in these equations (C_1, C_2, and R_b) and three boundary conditions:

$$v(0) = 0 \qquad v'(0) = 0 \qquad v(L) = 0$$

Application of these conditions to the preceding equations gives $C_1 = 0$, $C_2 = 0$, and

$$R_b = \frac{3qL}{8} \tag{8-1}$$

With the value of the redundant R_b established, we can find the remaining reactions from Eqs. (a):

$$R_a = \frac{5qL}{8} \qquad M_a = \frac{qL^2}{8} \tag{8-2a,b}$$

Also, we can substitute these quantities into the equations for the deflection v, the slope v', and the bending moment M and obtain a complete analysis of the beam.

An alternative way to analyze this beam is to take the reactive moment M_a as the redundant. Then we can express the bending moment M in terms of M_a, substitute the resulting expression into the second-order differential equation, and solve as before. Still another approach is to begin with the fourth-order equation, as illustrated in the next example. Also, the differential equations can be solved with the aid of discontinuity functions if desired.

Example 2

Analyze the fixed-end beam shown in Fig. 8-3 by solving the fourth-order differential equation of the deflection curve (the load equation).

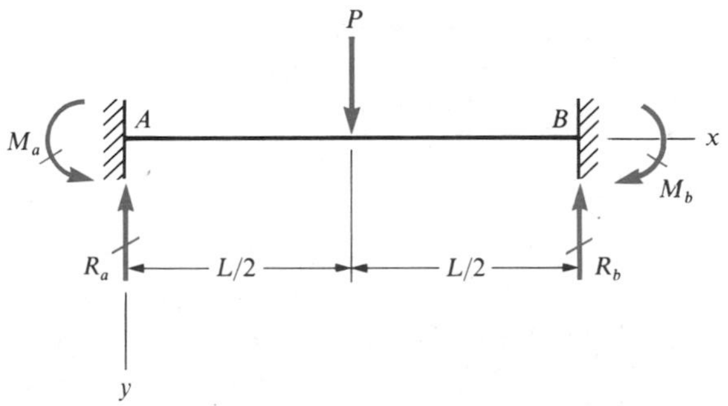

Fig. 8-3 Example 2. Fixed-end beam

Because the concentrated load P acts at the midpoint of the beam, we conclude from symmetry that $M_b = M_a$ and $R_a = R_b = P/2$. Thus, only one redundant quantity (M_a) remains to be determined. No load acts on the beam in the region between $x = 0$ and $x = L/2$, so the differential equation becomes

$$EIv'''' = 0$$

Successive integrations yield

$$EIv''' = C_1 \tag{b}$$

$$EIv'' = C_1x + C_2 \tag{c}$$

$$EIv' = \frac{C_1x^2}{2} + C_2x + C_3 \tag{d}$$

$$EIv = \frac{C_1x^3}{6} + \frac{C_2x^2}{2} + C_3x + C_4 \tag{e}$$

The boundary conditions applicable to the left-hand half of the beam are as follows: First, the shear force throughout this part of the beam is equal to R_a; hence from Eq. (b) we get $C_1 = -P/2$. Next, we observe that the bending moment at $x = 0$ is equal to $-M_a$; therefore, from Eq. (c) we get $C_2 = M_a$. The two conditions on the slope, namely, $v' = 0$ when $x = 0$ and $x = L/2$, yield $C_3 = 0$ and

$$M_a = \frac{PL}{8} \tag{8-3}$$

Thus, the redundant moment M_a has been found. Finally, we have the condition that $v = 0$ when $x = 0$, which gives $C_4 = 0$. Combining these results enables us to write the equation of the deflection curve:

$$v = \frac{Px^2}{48EI}(3L - 4x) \qquad \left(0 \le x \le \frac{L}{2}\right) \tag{8-4}$$

From this equation we may find by differentiation the equations for the slope, the bending moment, and the shear force.

As we observed in this example, the number of boundary conditions is sufficient to evaluate not only the constants of integration but also the redundant reactions. Sometimes it is necessary to set up the differential equations for more than one region of the beam and use conditions of continuity between regions, as was done previously for statically determinate beams. Such analyses are very long and tedious.

8.3 MOMENT-AREA METHOD

Another method for analyzing a statically indeterminate beam is the moment-area method, described in Section 7.5 for finding beam deflections. In the case of an indeterminate beam, the procedure consists of using the two moment-area theorems to obtain the additional equations that are needed to calculate the redundant reactions. These additional equations represent conditions on the slopes and deflections of the beam, and the number of such conditions will always equal the number of redundants.

The analysis of a statically indeterminate beam by the moment-area method begins with the selection of the redundant reactions. Then these reactions are removed from the beam, leaving a statically determinate released structure. Next we place the loads on the released structure and draw the corresponding diagram of bending moment divided by flexural rigidity (that is, the M/EI diagram). In a similar manner, the redundants are applied as loads acting on the released structure, and again the M/EI diagram is drawn. Finally, we invoke the moment-area theorems to provide the necessary equations for calculating the redundant reactions. These equations pertain to the areas and first moments of the M/EI diagrams; the particular relations to be used depend upon the type of beam and the choice of redundants. The following examples illustrate the techniques to be used.

Example 1

Using the moment-area method, determine the reactions for the propped cantilever beam AB shown in Fig. 8-4a.

This beam is statically indeterminate to the first degree; hence we select one reaction as the redundant. If R_b is selected as the redundant and removed, the released structure is a cantilever beam supported at A (Fig. 8-4b). On this beam we place P and R_b as loads, thus producing the bending-moment diagram shown in Fig. 8-4c. Since the beam has constant flexural rigidity EI, the M/EI diagram has the same shape as the moment diagram.

Now we are ready to use the moment-area theorems. Because the slope of the deflection curve at support A is zero (Fig. 8-4a), we note that the tangent to the curve at point A passes through point B. In other words, the offset Δ_{ba} of point B from the tangent at A is zero. Hence it follows from the second moment-area theorem that the first moment of the area of the M/EI diagram between A and B, taken with respect to point B, must equal zero. This relation gives the equation

$$\frac{L}{2}\left(\frac{R_b L}{EI}\right)\left(\frac{2L}{3}\right) - \frac{a}{2}\left(\frac{Pa}{EI}\right)\left(L - \frac{a}{3}\right) = 0$$

from which

$$R_b = \frac{Pa^2}{2L^3}(3L - a) \tag{8-5}$$

Knowing this redundant reaction, we can find from static equilibrium the other two reactions; the results are

$$R_a = \frac{Pb}{2L^3}(3L^2 - b^2) \qquad M_a = \frac{Pab}{2L^2}(L + b) \tag{8-6a,b}$$

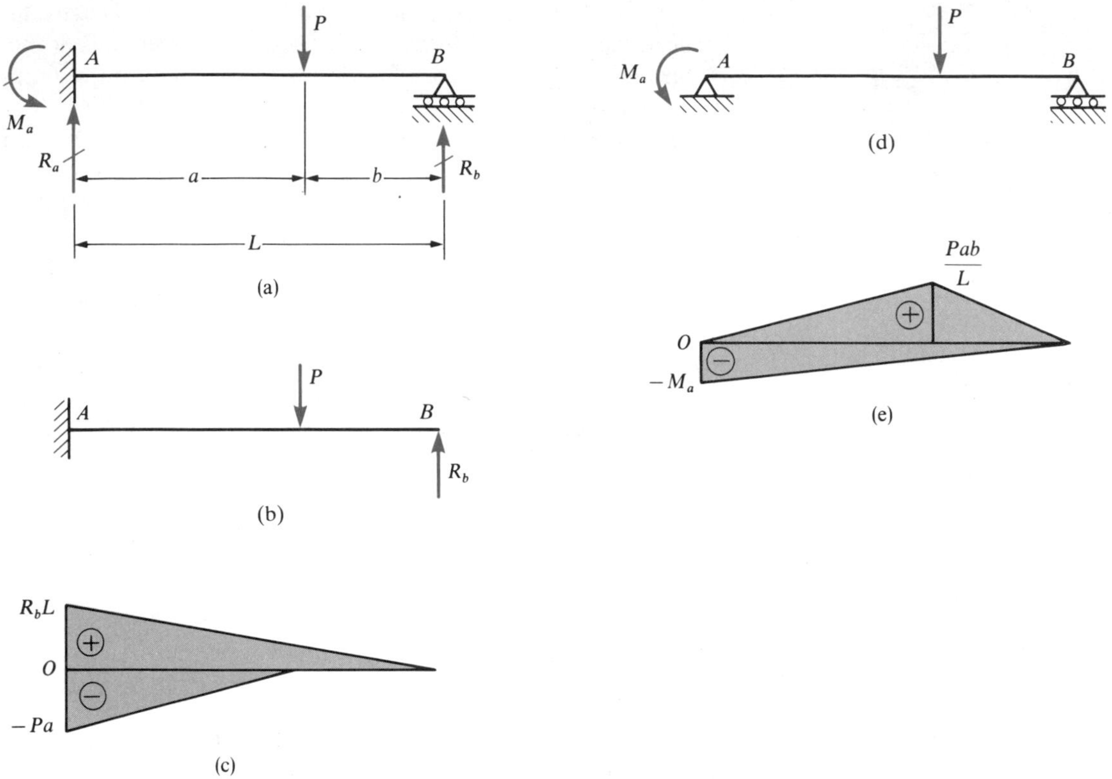

Fig. 8-4 Example 1. Moment-area method

With all reactions known, we can proceed to calculate stresses and deflections as needed.

As an alternative solution, we can solve this same problem by considering the reactive moment M_a as the redundant. In that event, the released structure is a simple beam (Fig. 8-4d), and the corresponding bending-moment diagram due to P and M_a is shown in Fig. 8-4e. Again using the second moment-area theorem and taking the first moment of the area of the M/EI diagram about point B, we get

$$\frac{L}{2}\left(\frac{Pab}{LEI}\right)\left(\frac{L+b}{3}\right)-\frac{L}{2}\left(\frac{M_a}{EI}\right)\left(\frac{2L}{3}\right)=0$$

Solving this equation, we obtain the same result for M_a as before (see Eq. 8-6b).

From the preceding results, we obtain the reactions for a beam with a concentrated load at the midpoint by substituting $a = b = L/2$; thus,

$$R_a=\frac{11P}{16} \qquad R_b=\frac{5P}{16} \qquad M_a=\frac{3PL}{16} \qquad \text{(8-7a,b,c)}$$

Note that the vertical reaction at the fixed support is over twice as large as the reaction at the simple support, although the load is at the midpoint.

Example 2

Determine the reactions for a fixed-end beam loaded by a couple M_0 (Fig. 8-5a). Also, find the deflection of the beam at point C where the couple acts.

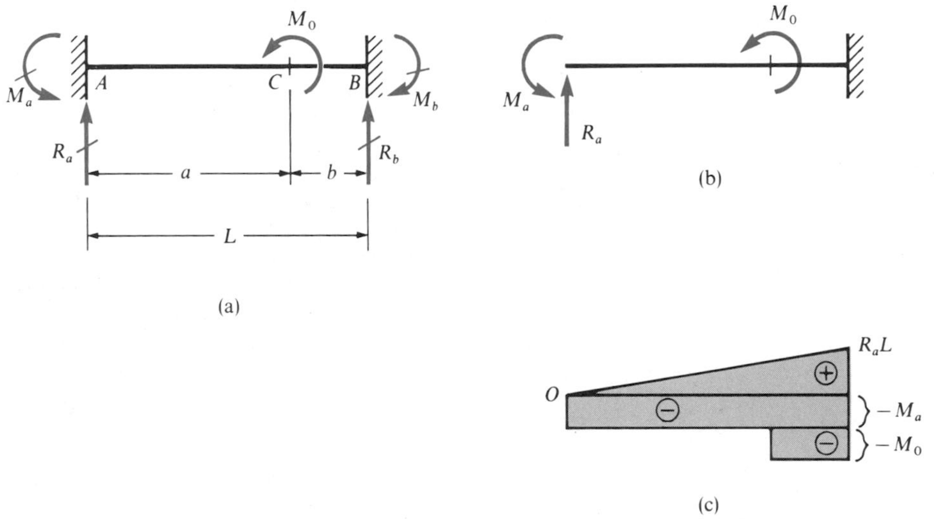

Fig. 8-5 Example 2. Moment-area method

This beam is statically indeterminate to the second degree; hence two redundants must be selected. The three useful possibilities are as follows: R_a and M_a, which gives a released structure in the form of a cantilever beam fixed at B; R_b and M_b, which gives a cantilever beam fixed at A; and M_a and M_b, which gives a simple beam. Let us select the first alternative, taking R_a and M_a as the redundants. Then the released structure is the cantilever beam of Fig. 8-5b. We can easily draw the bending-moment diagram (Fig. 8-5c) for this beam, which is subjected to R_a, M_a, and M_0 as loads. Again, the M/EI diagram has the same shape, because EI is constant.

Two conditions concerning the deflections of the beam are required for finding the two redundants. As a first condition, we note that both ends of the beam have zero slopes; hence the change in slope between A and B is zero. It follows from the first moment-area theorem that the area of the M/EI diagram between A and B must be zero; thus,

$$\frac{L}{2}\left(\frac{R_aL}{EI}\right) - \frac{M_a}{EI}(L) - \frac{M_0}{EI}(b) = 0$$

or

$$R_aL^2 - 2M_aL = 2M_0b \tag{a}$$

The second condition is obtained from the fact that the tangent to the deflection curve at A passes through point B, which means that the first moment of the area of the M/EI diagram between A and B, taken about B, is zero. The resulting equation is

$$\frac{L}{2}\left(\frac{R_aL}{EI}\right)\left(\frac{L}{3}\right) - L\left(\frac{M_a}{EI}\right)\left(\frac{L}{2}\right) - b\left(\frac{M_0}{EI}\right)\left(\frac{b}{2}\right) = 0$$

or

$$R_aL^3 - 3M_aL^2 = 3M_0b^2 \tag{b}$$

Now we can solve simultaneously Eqs. (a) and (b) and obtain the redundants:

$$R_a = \frac{6M_0ab}{L^3} \qquad M_a = \frac{M_0b}{L^2}(2a - b) \tag{8-8a,b}$$

The other two reactions are

$$R_b = -R_a \qquad M_b = \frac{M_0a}{L^2}(a - 2b) \tag{8-9a,b}$$

as found from static equilibrium equations.

The deflection δ_c at the point where the load is applied can be found from the second moment-area theorem. We observe that δ_c is equal to the offset of point C from the tangent at A, because the tangent at A is horizontal. Therefore, δ_c is equal to the first moment of the area of the M/EI diagram between A and C, taken about point C. Referring to Fig. 8-5c, we see that this deflection is calculated as follows:

$$\delta_c = \frac{a}{2}\left(\frac{a}{L}\right)\left(\frac{R_aL}{EI}\right)\left(\frac{a}{3}\right) - a\left(\frac{M_a}{EI}\right)\left(\frac{a}{2}\right) = \frac{R_aa^3}{6EI} - \frac{M_aa^2}{2EI}$$

Substituting the expressions for R_a and M_a (see Eqs. 8-8), we get

$$\delta_c = \frac{M_0a^2b^2(b - a)}{2L^3EI} \tag{8-10}$$

for the deflection under the load.

If the couple M_0 acts at the midpoint of the span ($a = b = L/2$), the reactions are

$$M_a = -M_b = \frac{M_0}{4} \qquad R_a = -R_b = \frac{3M_0}{2L} \tag{8-11a,b}$$

and the deflection at the midpoint is zero.

Example 3

A fixed-end beam AB has its right-hand support lowered vertically without rotation by an amount Δ (Fig. 8-6a). Determine the reactions for this beam.

Let us select the reactions R_b and M_b at support B as the redundants. The corresponding released structure, subjected to loads R_b and M_b, is shown in Fig. 8-6b. The bending-moment diagram is shown in the last part of the figure.

Two equations are required to solve for both R_b and M_b. The first equation expresses the condition that the slope of the beam is zero at each support. Hence the area of the M/EI diagram between A and B equals zero (from the first

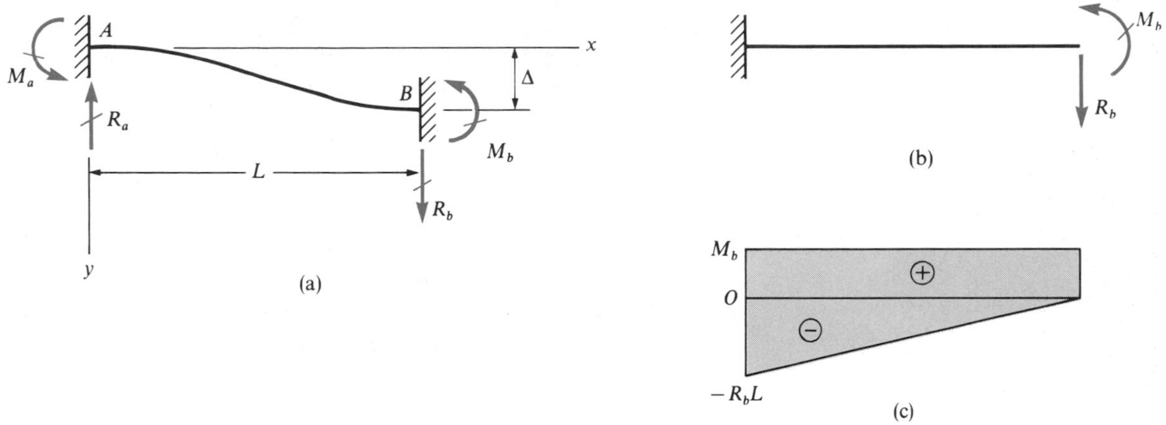

Fig. 8-6 Example 3. Fixed-end beam with a support displacement

moment-area theorem):

$$L\left(\frac{M_b}{EI}\right) - \frac{L}{2}\left(\frac{R_bL}{EI}\right) = 0$$

or

$$2M_b = R_bL \tag{c}$$

The second condition relates to the offset of point B from the tangent at A. Since this offset is equal to Δ, it follows from the second moment-area theorem that Δ is equal to the negative of the first moment of the area of the M/EI diagram between A and B, taken with respect to B:

$$\Delta = -L\left(\frac{M_b}{EI}\right)\left(\frac{L}{2}\right) + \frac{L}{2}\left(\frac{R_bL}{EI}\right)\left(\frac{2L}{3}\right)$$

or

$$2R_bL - 3M_b = \frac{6EI\,\Delta}{L^2} \tag{d}$$

Solving Eqs. (c) and (d), we get

$$R_b = \frac{12EI\Delta}{L^3} \qquad M_b = \frac{6EI\Delta}{L^2} \tag{8-12a,b}$$

From the symmetry of the beam, we note that $R_a = R_b$ and $M_a = M_b$.

8.4 METHOD OF SUPERPOSITION (FLEXIBILITY METHOD)

The method of superposition is of great importance in the analysis of statically indeterminate structures. It is applicable to many different types of structures, such as trusses and frames, in addition to beams, which are our primary concern in this chapter. We have previously used the method in analyzing statically indeterminate systems composed of members in tension and compression (see Section 2.3).

The essence of the superposition method can be described in simple terms. We begin by identifying the statical redundants, as explained in the preceding sections. Then the redundants are removed from the indeterminate beam, leaving a statically determinate released structure. The deflections of the released structure can be found by the techniques described in Chapter 7. In particular, the deflections (either translations or rotations) corresponding to the redundants and caused by the loads can be ascertained. Next, the redundants themselves are visualized as loads acting on the released structure, and the corresponding deflections are calculated. From the principle of superposition, we know that the final deflections due to both the actual loads and the redundants must be equal to the sum of those deflections as calculated separately. In the case of redundant constraints, the corresponding deflections either are zero or have known magnitudes; hence we can write equations of superposition expressing these relationships. Finally, these equations can be solved for the values of the redundant reactions, after which all other reactions can be determined from equations of static equilibrium.

The steps described in the preceding paragraph can be made clearer through an illustration. Let us again analyze a propped cantilever beam supporting a uniform load (see Fig. 8-7a). If the reaction R_b is selected as the statical redundant and the corresponding constraint is removed, we obtain a cantilever beam for the released structure (Fig. 8-7b). The deflection of this beam at end B due to the uniform load is denoted δ'_b, and the deflection caused by the redundant is δ''_b (see Fig. 8-7c). The total deflection δ_b in the original structure, obtained by superimposing the deflections δ'_b and δ''_b, must be zero. This conclusion leads to the following equation:

$$\delta_b = \delta'_b - \delta''_b = 0 \tag{a}$$

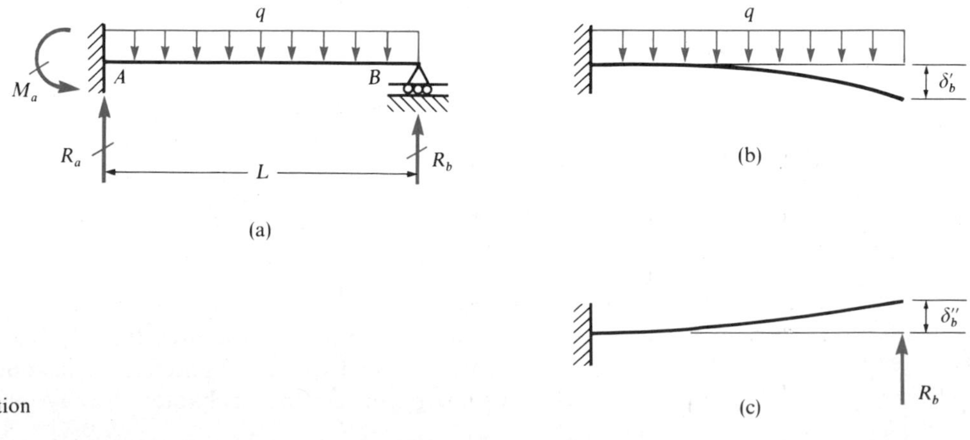

Fig. 8-7 Method of superposition

The minus sign appears in this equation because δ_b' is downward whereas δ_b'' is upward. The deflections δ_b' and δ_b'' due to the uniform load q and the redundant R_b are easily found with the aid of Table G-1 in Appendix G (see Cases 1 and 4). Using the formulas given there, we get from Eq. (a):

$$\delta_b = \frac{qL^4}{8EI} - \frac{R_bL^3}{3EI} = 0$$

from which

$$R_b = \frac{3qL}{8} \tag{8-13}$$

The reaction R_a and the moment M_a can now be found from the equilibrium of the beam; the results are

$$R_a = \frac{5qL}{8} \qquad M_a = \frac{qL^2}{8} \tag{8-14a,b}$$

This same beam (Fig. 8-8a) could have been analyzed in a different way by taking the moment M_a as the redundant; then the released structure is a simple beam (Figs. 8-8b and c). The angle of rotation produced by the uniform load acting on the released structure (Table G-2, Case 1) is

$$\theta_a' = \frac{qL^3}{24EI}$$

and the corresponding angle caused by the redundant M_a (Table G-2, Case 7) is

$$\theta_a'' = \frac{M_aL}{3EI}$$

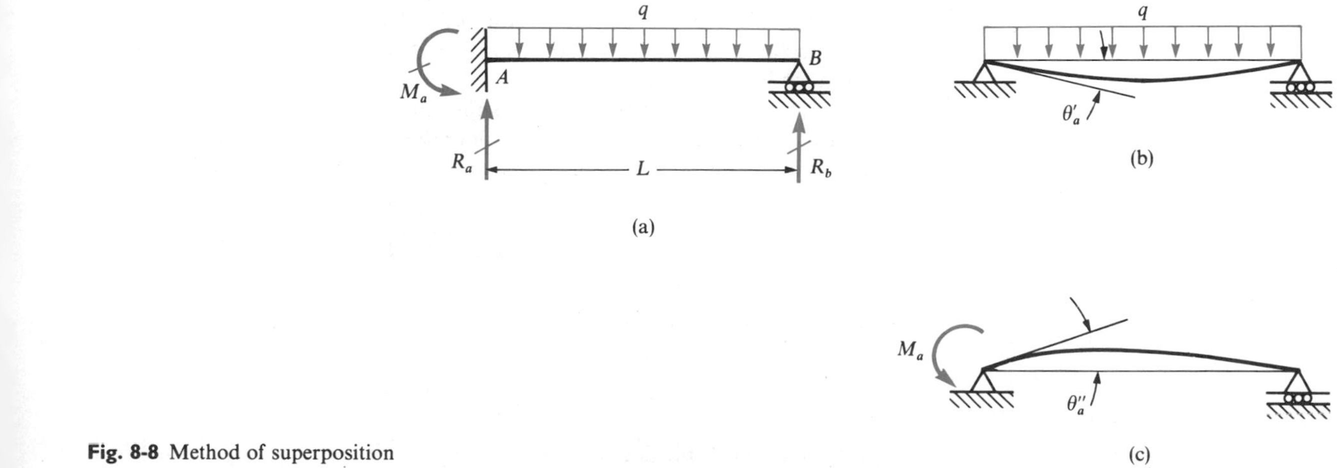

Fig. 8-8 Method of superposition

The equation of superposition, stating that the angle of rotation at support A of the original beam is zero, becomes

$$\theta_a = \theta'_a - \theta''_a = \frac{qL^3}{24EI} - \frac{M_a L}{3EI} = 0 \tag{b}$$

Solving this equation, we get $M_a = qL^2/8$, which agrees with the previous result.

After finding the reactions of a statically indeterminate beam, the calculation of all stress resultants (axial forces, shear forces, and bending moments) presents no further difficulty because equations of static equilibrium are sufficient for this purpose. Furthermore, deflections and slopes can also be obtained at any point, either by using the differential equation of the deflection curve or by using the principle of superposition in conjunction with the deflection formulas listed in Appendix G. In the illustrative examples that follow, and also in the problems, attention will be directed primarily to finding the reactions, which is the key step in the solution.

The method of analysis described in this section is called the **flexibility method** or the **force method** of analysis. The latter name arises from the use of force quantities (forces and couples) as the redundants; the former name is used because the coefficients of the unknowns (terms such as $L/3EI$ in Eq. b) are **flexibilities** (that is, deflections produced by a unit load). The equations of superposition that express conditions imposed on the deflections (see Eqs. a and b) are usually called **equations of compatibility**. The method of superposition, like the moment-area method and the differential equation of the deflection curve, applies only to linearly elastic structures.

Example 1

The reactions for the two-span continuous beam ABC shown in Fig. 8-9a are to be determined by the method of superposition. Note that the beam supports a uniform load of intensity q.

Selecting the middle reaction R_b as the redundant, we see that the released structure is a simple beam (Fig. 8-9b). Under the action of the uniform load, the deflection at point B in the released structure (see Table G-2, Case 1) is

$$\delta'_b = \frac{5q(2L)^4}{384EI} = \frac{5qL^4}{24EI}$$

where L is the length of each span. The deflection in the upward direction produced by the redundant (see Fig. 8-9c) is

$$\delta''_b = \frac{R_b(2L)^3}{48EI} = \frac{R_b L^3}{6EI}$$

as obtained from Table G-2, Case 4. The equation of compatibility pertaining to

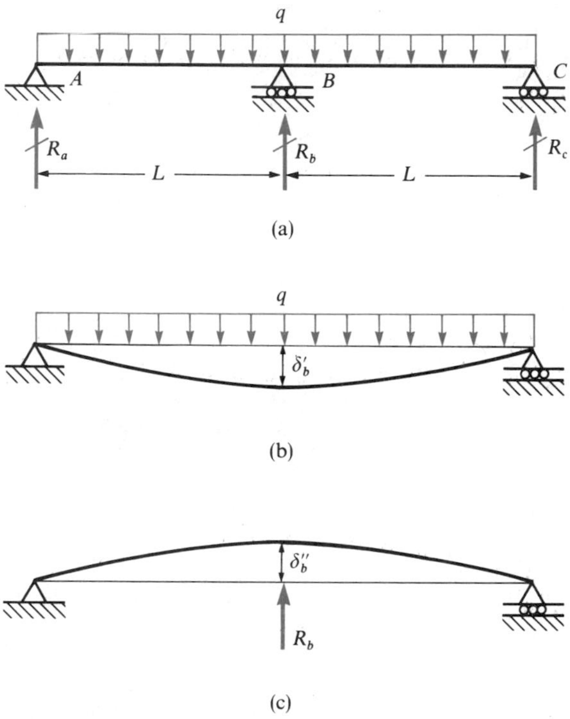

Fig. 8-9 Example 1. Two-span continuous beam

the vertical deflection at point B is

$$\delta_b = \delta_b' - \delta_b'' = \frac{5qL^4}{24EI} - \frac{R_bL^3}{6EI} = 0$$

from which

$$R_b = \frac{5qL}{4} \tag{8-15}$$

The other reactions have the values $R_a = R_c = 3qL/8$, as found from static equilibrium. With the reactions known, we will have no difficulty in finding the bending moments, shear forces, stresses, and deflections.

Example 2

A beam with fixed ends is loaded by a force P acting at the position shown in Fig. 8-10a. Find the reactive forces and moments at the ends of the beam.

Let us select the reactive moments M_a and M_b for the redundants, thereby giving a released structure in the form of a simple beam (Fig. 8-10b). The angles at the ends produced by the load P are obtained from Case 5 of Table G-2:

$$\theta_a' = \frac{Pab(L+b)}{6LEI} \qquad \theta_b' = \frac{Pab(L+a)}{6LEI}$$

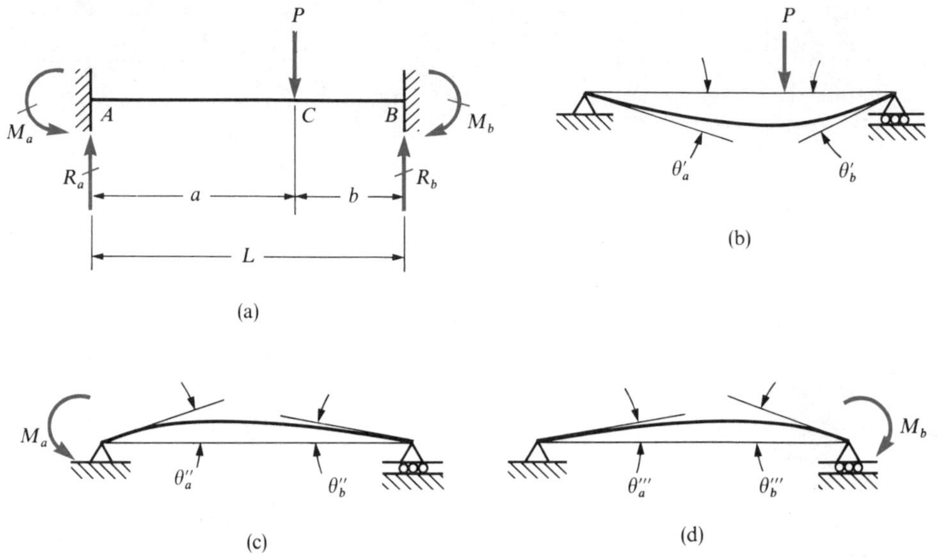

Fig. 8-10 Example 2. Fixed-end beam with a concentrated load

Now we apply the redundant moments M_a and M_b as loads on the released structure (Figs. 8-10c and d). The angles at the ends due to M_a are

$$\theta''_a = \frac{M_a L}{3EI} \qquad \theta''_b = \frac{M_a L}{6EI}$$

and due to M_b are

$$\theta'''_a = \frac{M_b L}{6EI} \qquad \theta'''_b = \frac{M_b L}{3EI}$$

Because the angles of rotation at both ends in the original beam are zero, we have two equations of compatibility:

$$\theta_a = \theta'_a - \theta''_a - \theta'''_a = 0$$

$$\theta_b = \theta'_b - \theta''_b - \theta'''_b = 0$$

When the various expressions for the angles are substituted into these equations, we arrive at two simultaneous equations containing M_a and M_b as unknowns:

$$\frac{M_a L}{3EI} + \frac{M_b L}{6EI} = \frac{Pab(L+b)}{6LEI}$$

$$\frac{M_a L}{6EI} + \frac{M_b L}{3EI} = \frac{Pab(L+a)}{6LEI}$$

The solutions are

$$M_a = \frac{Pab^2}{L^2} \qquad M_b = \frac{Pa^2 b}{L^2} \qquad \text{(8-16a,b)}$$

Using these results and also equations of equilibrium, we obtain the following formulas for the vertical reactions:

$$R_a = \frac{Pb^2}{L^3}(L + 2a) \qquad R_b = \frac{Pa^2}{L^3}(L + 2b) \tag{8-17a,b}$$

To illustrate how the principle of superposition may be used to obtain deflections, let us now calculate the deflection of the fixed-end beam at point C where the load is applied (Fig. 8-10a). The deflection at this point in the released structure under the action of the load P (Fig. 8-10b) is

$$\delta_c' = \frac{Pa^2b^2}{3LEI}$$

from Case 5, Table G-2. The upward deflections at this same point of the released structure due to the couples M_a and M_b (see Figs. 8-10c and d) are

$$\delta_c'' = \frac{M_a ab}{6LEI}(L + b) \qquad \delta_c''' = \frac{M_b ab}{6LEI}(L + a)$$

as found from Case 7, Table G-2. Substituting the values of M_a and M_b from Eqs. (8-16), we get

$$\delta_c'' = \frac{Pa^2b^3}{6L^3EI}(L + b) \qquad \delta_c''' = \frac{Pa^3b^2}{6L^3EI}(L + a)$$

Therefore, the total deflection at point C is

$$\delta_c = \delta_c' - \delta_c'' - \delta_c''' = \frac{Pa^3b^3}{3L^3EI} \tag{8-18}$$

In the special case when the load P is at the middle of the beam, the deflection at the midpoint is

$$\delta_c = \frac{PL^3}{192EI} \tag{8-19}$$

and the reactions are

$$M_a = M_b = \frac{PL}{8} \qquad R_a = R_b = \frac{P}{2} \tag{8-20a,b}$$

Example 3

Find the reactions for a fixed-end beam with a uniform load over part of the span (Fig. 8-11).

Let us isolate an element $q\,dx$ of the load at distance x from the left-hand end of the beam. By treating this element as a concentrated load, we can utilize the formulas derived in the preceding example. Beginning with the formulas for the moments M_a and M_b (Eqs. 8-16), we can replace P with $q\,dx$, a with x, and b with $L - x$; hence the fixed-end moments due to the element of load are

$$dM_a = \frac{qx(L - x)^2\,dx}{L^2} \qquad dM_b = \frac{qx^2(L - x)\,dx}{L^2}$$

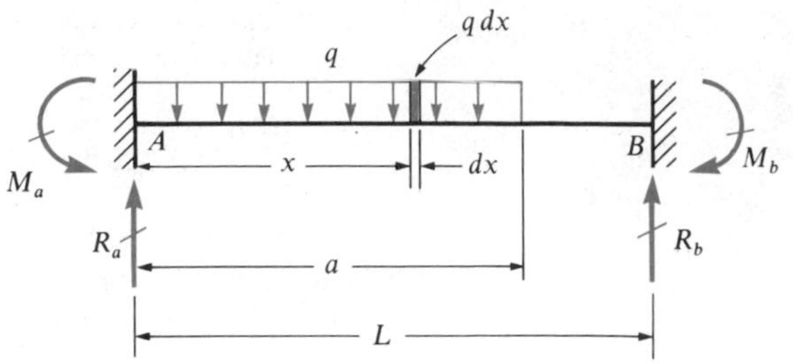

Fig. 8-11 Example 3. Fixed-end beam with a uniform load over part of the span

Integration over the loaded part of the beam yields

$$M_a = \int dM_a = \frac{q}{L^2}\int_0^a x(L-x)^2\,dx = \frac{qa^2}{12L^2}(6L^2 - 8aL + 3a^2) \qquad (8\text{-}21)$$

$$M_b = \int dM_b = \frac{q}{L^2}\int_0^a x^2(L-x)\,dx = \frac{qa^3}{12L^2}(4L - 3a) \qquad (8\text{-}22)$$

Similarly, the vertical reactions at the ends can be found with the aid of Eqs. (8-17):

$$R_a = \frac{q}{L^3}\int_0^a (L-x)^2(L+2x)\,dx = \frac{qa}{2L^3}(2L^3 - 2a^2L + a^3) \qquad (8\text{-}23)$$

$$R_b = \frac{q}{L^3}\int_0^a x^2(3L-2x)\,dx = \frac{qa^3}{2L^3}(2L - a) \qquad (8\text{-}24)$$

Thus, the desired results have been found.

If a fixed-end beam has a uniform load over the entire span (Fig. 8-12), we can obtain the reactions by substituting $a = L$ into the preceding equations, yielding

$$M_a = M_b = \frac{qL^2}{12} \qquad R_a = R_b = \frac{qL}{2} \qquad (8\text{-}25a,b)$$

The reactions at the supports of a beam with fixed ends are called **fixed-end moments** and **fixed-end forces**. They have an important role in several methods of structural analysis.

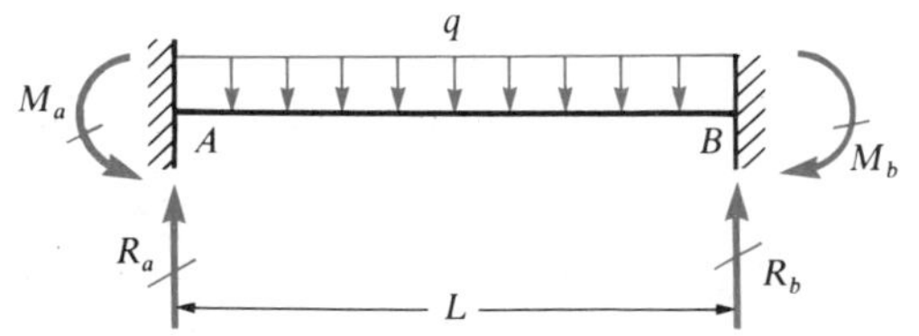

Fig. 8-12 Example 3. Fixed-end beam with a uniform load

Example 4

A beam *ABC* (Fig. 8-13a) is simply supported at *A* and *B* and is hung from a cable *CD* at point *C*. Prior to the application of the uniform load *q*, there is no force in the cable nor is there any slack in the cable. When the load *q* is applied, the beam

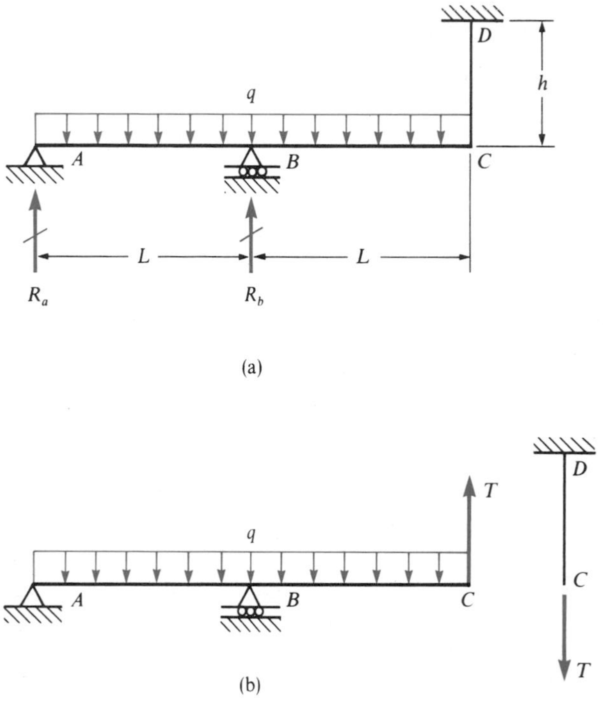

Fig. 8-13 Example 4. Beam *ABC* supported by a cable *CD*

deflects downward at C and a tensile force T develops in the cable. Find the magnitude of this force.

In this analysis it is convenient to select the unknown force T in the cable as the redundant and to cut the structure into two parts (Fig. 8-13b). The released structure then consists of the beam ABC and the cable CD as independent structures, with the force T acting upward on the beam and downward on the cable. The deflection at point C on the beam will consist of two parts, a downward deflection δ_c' due to the uniform load and an upward deflection δ_c'' due to the force T. At the same time, the end of the cable (point C) will displace downward by an amount δ_c''', equal to the elongation of the cable. Therefore, the equation of compatibility, expressing the fact that the downward deflection of the end of the beam is equal to the elongation of the cable, is

$$\delta_c' - \delta_c'' = \delta_c'''$$

Having formulated this equation, we now turn to the task of evaluating the three displacements.

The deflection at the end of the overhang produced by the uniform load can be found from the results given in Example 3 of Section 7.6. Using Eq. (7-54) of that example, and substituting $a = L$, we get

$$\delta_c' = \frac{qL^4}{4EI}$$

where EI is the flexural rigidity of the beam. The deflection of the beam at C due to the force T can be taken from the answer to Problem 7.5-14 by again substituting

$a = L$; hence

$$\delta_c'' = \frac{2TL^3}{3EI}$$

Finally, the stretch of the cable is

$$\delta_c''' = \frac{Th}{EA}$$

where h is the length of the cable and EA is its axial rigidity.

By substituting the three preceding expressions into the equation of compatibility and solving for the force T, we get

$$T = \frac{3qAL^4}{8AL^3 + 12hI} \tag{8-26}$$

Note that in this example the redundant was selected as an internal force quantity instead of an external reaction.

8.5 CONTINUOUS BEAMS

Beams that are continuous over many supports (see Fig. 8-14) are known as **continuous beams** and are commonly encountered in aircraft, buildings, pipelines, bridges, and various kinds of specialized structures. If the loads on a continuous beam are vertical and if there are no axial deformations, then all reactions will be vertical. To represent this behavior schematically, we can consider one of the supports to be a pin support and all the others to be roller supports, as shown in the figure. Then the total number of reactions will be the same as the number of supports, and the degree of statical indeterminacy will be two less than this number. Thus, for the beam shown in Fig. 8-14, there are five vertical reactions, of which three are redundants.

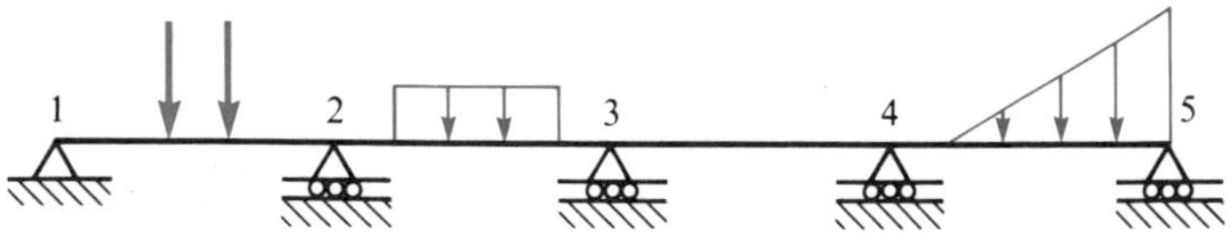

Fig. 8-14 Continuous beam

Although we may analyze a continuous beam by any of the methods described in the preceding sections, only the method of superposition is practical. When using this method, one possibility is to select the reactions at the intermediate supports as the redundants, which means that the released structure is a simple beam. This technique was used in Example 1 of Section 8.4 (see Fig. 8-9) and is satisfactory for beams of only two or three spans. When there are more than two redundants, it is advantageous to select the bending moments in the beam at the intermediate supports, instead of the reactions, as the redundants. This choice greatly simplifies the calculations because it leads to a set of simultaneous equations in which no more than three unknowns appear in each equation regardless of the total number of redundants.

Let us now develop in more detail this procedure for analyzing continuous beams. When the bending moments at the supports are released from the structure, the continuity of the beam at the supports is broken, and the released structure consists of a series of simple beams. Each such beam is subjected to the external loads that normally act on it, together with the two redundant moments at its ends. Under the action of these loads, we can determine the angles of rotation at the ends of each simple beam. The equations of compatibility, which express the fact that at each support the adjacent beams must have the same angle of rotation, provide the necessary equations to solve for the unknown bending moments.

Consider, for example, the portion of a continuous beam shown in Fig. 8-15a. The three consecutive supports are identified as A, B, and C, and the lengths and moments of inertia of the two adjacent spans are denoted by L_a, I_a and L_b, I_b, respectively. Let M_a, M_b, and M_c denote the bending moments at the three supports. The true directions of these moments will depend upon the loads on the beam, but for purposes of this derivation we will assume they are positive (that is, causing compression on the top of the beam). The released structure, consisting of simple beams, is shown in Fig. 8-15b for the two adjacent spans. Each span is loaded by the applied external loads plus the redundant bending moments. These loads produce deflections and rotations of the two simple beams. The angle of rotation of the left-hand beam at support B is indicated in the figure as θ_b', and the rotation of the right-hand beam at this same support is θ_b''. These angles are taken to be positive as shown in the figure (that is, in the same directions as the positive bending moments M_b). Because the axis of the beam is actually continuous across support B, the equation of compatibility is

$$\theta_b' = -\theta_b'' \tag{8-27}$$

which expresses the fact that the angles of rotation of the two simple beams must match each other at B. The next step is to develop appropriate expressions for the angles θ_b' and θ_b'' and substitute into Eq. (8-27).

The bending-moment diagrams associated with the external loads acting on the released structure are shown in Fig. 8-15c. The particular shapes of these diagrams will depend upon the nature of the loads. However, in all instances, we can characterize the bending-moment diagrams by their areas and centroidal distances. Let us denote the areas of the two diagrams as A_a and A_b, respectively. Also, the distance from point A to the centroid of A_a is denoted by $\bar{x}_a$, and the distance from point C to the centroid of A_b is denoted by $\bar{x}_b$. We can use these properties of the

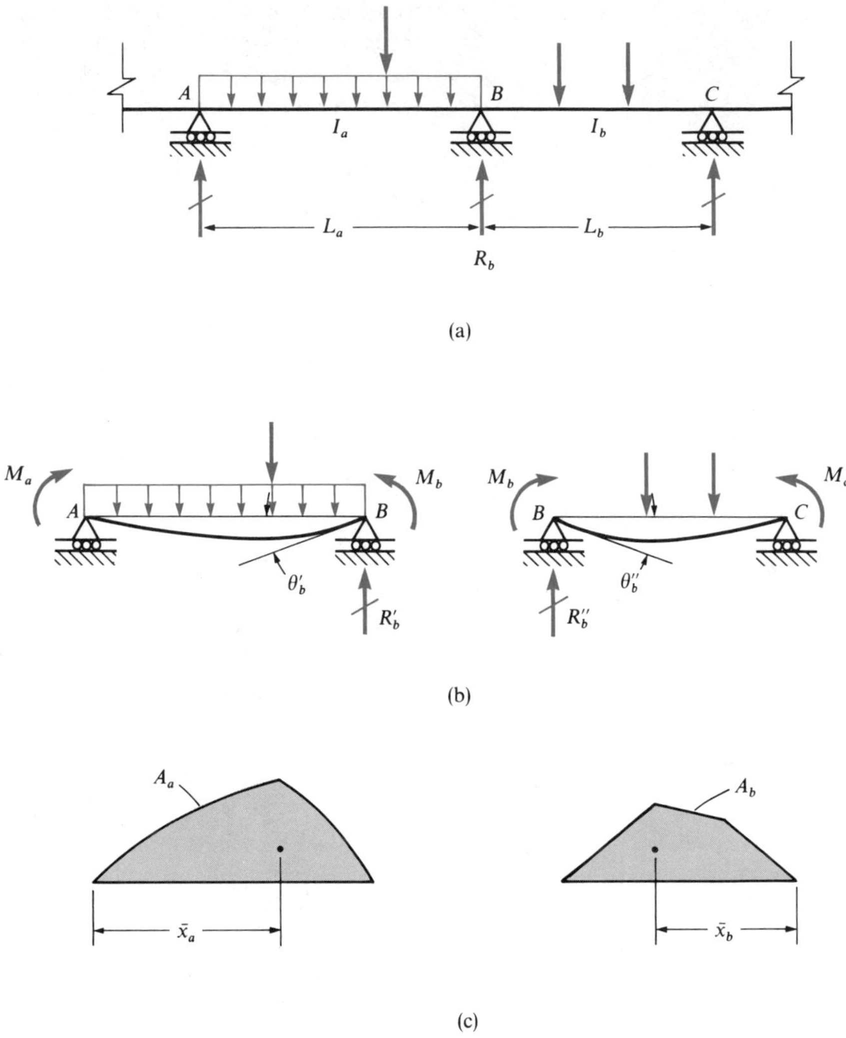

Fig. 8-15 Three-moment equation

bending-moment diagrams to evaluate the angles θ'_b and θ''_b. We note from the second moment-area theorem that the contribution to the angle θ'_b from the external loads on beam AB is

$$\frac{A_a \bar{x}_a}{EI_a L_a}$$

In addition, the moments M_a and M_b contribute to θ'_b in the following amounts:

$$\frac{M_a L_a}{6EI_a} \quad \text{and} \quad \frac{M_b L_a}{3EI_a}$$

Thus, the angle θ_b' is

$$\theta_b' = \frac{M_a L_a}{6EI_a} + \frac{M_b L_a}{3EI_a} + \frac{A_a \bar{x}_a}{EI_a L_a} \tag{a}$$

Considering the right-hand span BC, we obtain in an analogous manner the following equation for θ_b'':

$$\theta_b'' = \frac{M_b L_b}{3EI_b} + \frac{M_c L_b}{6EI_b} + \frac{A_b \bar{x}_b}{EI_b L_b} \tag{b}$$

Thus, the angles of rotation at support B are expressed in terms of the bending moments at the supports and the properties of the M/EI diagrams for the two adjacent spans.

Next, we substitute the expressions for θ_b' and θ_b'' (Eqs. a and b) into the equation of compatibility (Eq. 8-27) and rearrange the terms; in this manner, we obtain the following equation:

$$M_a\left(\frac{L_a}{I_a}\right) + 2M_b\left(\frac{L_a}{I_a} + \frac{L_b}{I_b}\right) + M_c\left(\frac{L_b}{I_b}\right) = -\frac{6A_a\bar{x}_a}{I_a L_a} - \frac{6A_b\bar{x}_b}{I_b L_b} \tag{8-28}$$

This equation is called the **three-moment equation** because it relates three consecutive bending moments in the beam. If both spans have the same moment of inertia I, the three-moment equation simplifies to the following:

$$M_a L_a + 2M_b(L_a + L_b) + M_c L_b = -\frac{6A_a\bar{x}_a}{L_a} - \frac{6A_b\bar{x}_b}{L_b} \tag{8-29}$$

If both spans have the same length L, the equation becomes even simpler:

$$M_a + 4M_b + M_c = -\frac{6}{L^2}(A_a\bar{x}_a + A_b\bar{x}_b) \tag{8-30}$$

The procedure when using the three-moment equation is to write the equation once for each intermediate support of the continuous beam, thus providing as many equations as there are redundant bending moments. Then these equations can be solved simultaneously for the moments.

The terms on the right-hand sides of the three-moment equations can be evaluated from the loads on the beam. For instance, if a uniform load of intensity q acts on span AB, we have

$$A_a = \frac{2}{3}\left(\frac{qL_a^2}{8}\right)(L_a) = \frac{qL_a^3}{12} \qquad \bar{x}_a = \frac{L_a}{2}$$

and
$$A_a\bar{x}_a = \frac{qL_a^4}{24} \tag{8-31}$$

For a concentrated load P at the center of the span, the terms are

$$A_a = \frac{1}{2}\left(\frac{PL_a}{4}\right)(L_a) = \frac{PL_a^2}{8} \qquad \bar{x}_a = \frac{L_a}{2}$$

and
$$A_a\bar{x}_a = \frac{PL_a^3}{16} \tag{8-32}$$

These two examples are sufficient to show that there is no major difficulty in evaluating the terms involving the bending-moment diagrams. Once this step is accomplished, the equations can be formulated by merely writing them down, after which they can be solved for the unknown bending moments.*

Throughout the foregoing discussion, it has been assumed that the two extreme ends of the continuous beam were simply supported. If either or both of these ends should be a **fixed support**, the number of redundant moments will be increased (see Fig. 8-16a). An easy way to handle this situation is to replace the fixed support by an additional span having an infinite moment of inertia (Fig. 8-16b). The effect of this extra span having infinitely large stiffness is to prevent rotation at support 1, which is the same condition imposed by the fixed support. The bending moments found at points 1, 2, and 3 in the continuous beam shown in Fig. 8-16b will

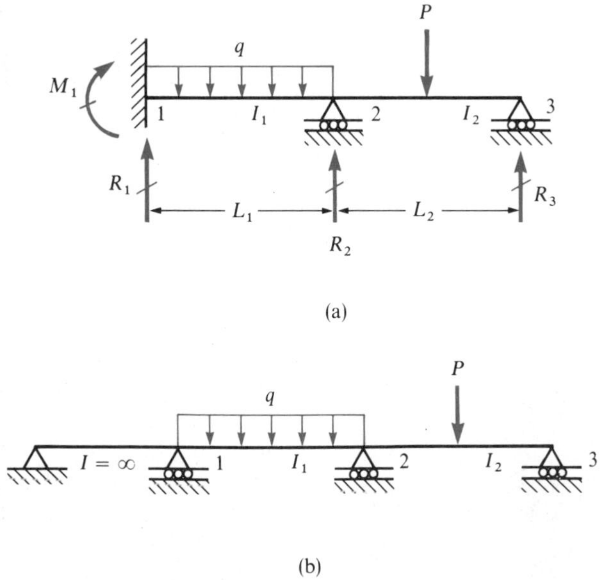

Fig. 8-16 Replacement of a fixed support by a span having infinite moment of inertia

* The three-moment equation was developed by the French engineers B. P. E. Clapeyron and H. Bertot in the middle of the 19th century (see Ref. 8-2).

be identical to those in the original beam. The length assigned to the extra span is not relevant (except that it must be greater than zero) because it always disappears from the three-moment equation.

After the bending moments at the supports of a continuous beam have been found, there is no difficulty in using equations of static equilibrium to find the **reactions**. Again considering two adjacent spans (Fig. 8-15b), let R_b' and R_b'' be the reactions at B for the two simple beams AB and BC. The sum of these reactions gives the total reaction R_b at support B (Fig. 8-15a). The reaction R_b' is made up of three parts: the simple-beam reaction due to the external loads, the reaction due to M_a (equal to M_a/L_a), and the reaction due to M_b (equal to $-M_b/L_a$). Similarly, the reaction R_b'' is equal to the simple-beam reaction due to the external loads plus M_c/L_b and $-M_b/L_b$. Combining all these terms gives the total reaction R_b. This same procedure is used at each support until all reactions are calculated. Of course, if a concentrated load acts on the beam over a support, it will be transmitted directly into the reaction.

Example 1

As an illustration of the use of the three-moment equation, we will analyze the continuous beam shown in Fig. 8-17. The beam has three spans of equal length and constant moment of inertia, and it is loaded in the first and third spans. The concentrated load P is equal to qL.

Equation (8-30) can be used in this example because the moments of inertia and the lengths are the same for all spans. As a preliminary matter, we will determine the terms of the form $A\bar{x}$ that appear on the right-hand side of the equation, for each of the three spans. For span 1-2 this term is $qL^4/24$, as given by Eq. (8-31). For span 2-3 this term becomes zero because there is no load. Finally, for span 3-4 we observe that the bending-moment diagram is a triangle with maximum ordinate equal to $3PL/16$. The area of the triangle is $3PL^2/32$, and its centroidal distance from point 4 is $5L/12$ (see Appendix D, Case 3). Therefore, the term $A\bar{x}$ has the value $5PL^3/128$, which becomes $5qL^4/128$ because P equals qL.

Now we are ready to write the three-moment equations for the interior supports. Considering support 2, we recognize that M_a in the general equation (Eq. 8-30) becomes M_1 (which is zero), M_b becomes M_2, and M_c becomes M_3; thus

$$4M_2 + M_3 = -\frac{qL^2}{4} \tag{c}$$

Similarly, the three-moment equation for support 3 is

$$M_2 + 4M_3 = -\frac{15qL^2}{64} \tag{d}$$

Solving Eqs. (c) and (d), we get the bending moments:

$$M_2 = -\frac{49qL^2}{960} \qquad M_3 = -\frac{11qL^2}{240} \tag{e}$$

By drawing free-body diagrams of each of the three parts of the beam and then

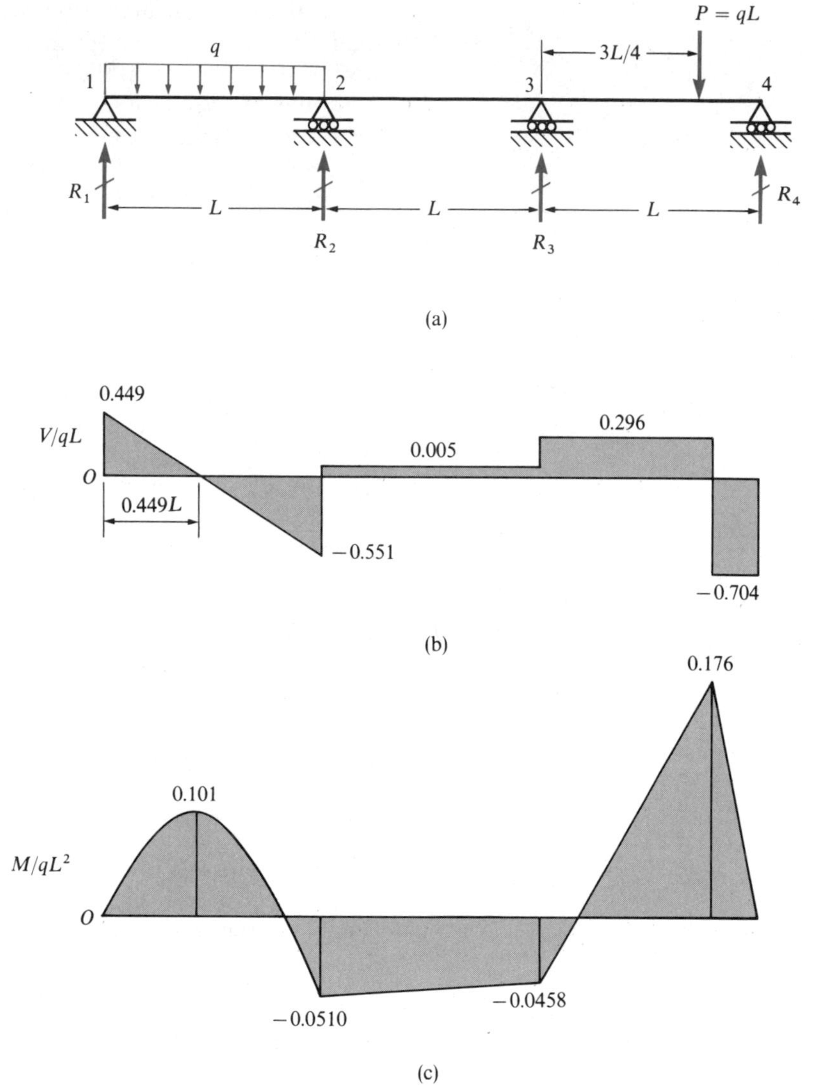

Fig. 8-17 Example 1. Continuous beam

writing equations of static equilibrium, we obtain the reactions: .

$$R_1 = \frac{431qL}{960} \qquad R_2 = \frac{89qL}{160} \qquad R_3 = \frac{93qL}{320} \qquad R_4 = \frac{169qL}{240} \tag{f}$$

With this information we can construct the shear-force and bending-moment diagrams for the entire beam, as shown in Figs. 8-17b and c.

Example 2

Suppose that the supports of a horizontal continuous beam are not on the same level, because the supports moved vertically after the beam was constructed. How can this effect be incorporated into the three-moment equation?

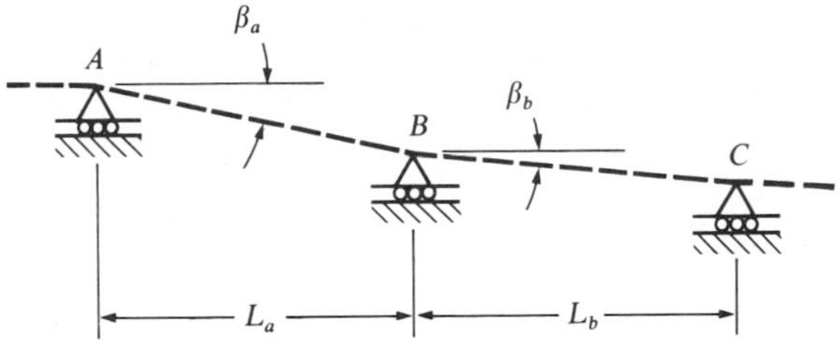

Fig. 8-18 Example 2. Supports at different levels

Let us assume that three consecutive supports A, B, and C are oriented as shown in Fig. 8-18. The dashed lines connecting points A, B, and C do not represent the axis of the beam but are merely straight lines between the points of support (compare with Fig. 8-15a). Let β_a and β_b represent the angles of inclination of these lines, assumed to be positive when the right-hand support is lower than the left-hand support. Returning to Eq. (a) for θ_b', we note that the existing terms will remain in this equation, but in addition we must account for the effect of the angle β_a. The presence of the angle β_a reduces the value of θ_b' so that, instead of Eq. (a), we get

$$\theta_b' = \frac{M_a L_a}{6EI_a} + \frac{M_b L_a}{3EI_a} + \frac{A_a \bar{x}_a}{EI_a L_a} - \beta_a \tag{g}$$

The equation for θ_b'' (see Eq. b) is also altered, as follows:

$$\theta_b'' = \frac{M_b L_b}{3EI_b} + \frac{M_c L_b}{6EI_b} + \frac{A_b \bar{x}_b}{EI_b L_b} + \beta_b \tag{h}$$

Substituting into the equation of compatibility (Eq. 8-27) and rearranging, we obtain a more general form of the three-moment equation:

$$M_a\left(\frac{L_a}{I_a}\right) + 2M_b\left(\frac{L_a}{I_a} + \frac{L_b}{I_b}\right) + M_c\left(\frac{L_b}{I_b}\right) = -\frac{6A_a\bar{x}_a}{I_a L_a} - \frac{6A_b\bar{x}_b}{I_b L_b} + 6E(\beta_a - \beta_b) \tag{8-33}$$

This equation can be used instead of Eq. (8-28) whenever the supports are not on the same level.

*8.6 TEMPERATURE EFFECTS

Temperature changes in a statically indeterminate beam will produce stresses and deflections in the beam. These quantities can be determined by methods similar to those already described for loads on a beam. The most useful method is probably the method of superposition, which we will illustrate by considering the fixed-end beam shown in Fig. 8-19. This beam is assumed to have a temperature T_1 on the upper surface and T_2 on the lower surface, with a linear variation of temperature over the height.

When using the superposition method, we begin by releasing the redundants in order to obtain a statically determinate beam. If we take R_b and M_b as the redundants, the released structure becomes a cantilever

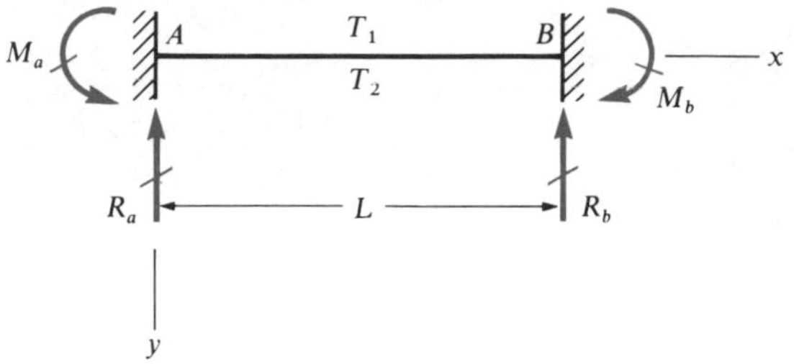

Fig. 8-19 Fixed-end beam with a temperature differential

beam. The deflection and angle of rotation at end B of this cantilever due to the temperature differential are

$$\delta'_b = \frac{\alpha(T_2 - T_1)L^2}{2h} \qquad \theta'_b = \frac{\alpha(T_2 - T_1)L}{h}$$

as obtained in Problem 7.11-2. In these equations, α is the coefficient of thermal expansion and h is the height of the beam. When T_2 is greater than T_1, the deflection δ'_b is upward and the angle of rotation θ'_b is counterclockwise.

The deflection and rotation in the released structure due to R_b are

$$\delta''_b = \frac{R_b L^3}{3EI} \qquad \theta''_b = \frac{R_b L^2}{2EI}$$

and due to M_b are

$$\delta'''_b = -\frac{M_b L^2}{2EI} \qquad \theta'''_b = -\frac{M_b L}{EI}$$

where upward deflection and counterclockwise rotation are positive.

We can now write the compatibility equations as follows:

$$\delta'_b + \delta''_b + \delta'''_b = 0 \qquad \theta'_b + \theta''_b + \theta'''_b = 0$$

Substituting the various deflections and rotations into these equations and solving, we find

$$R_b = 0 \qquad M_b = \frac{\alpha EI(T_2 - T_1)}{h}$$

The fact that R_b is zero could have been anticipated from the symmetry of the beam; if this fact had been utilized at the outset, the preceding solution would have been simplified because only one equation of compatibility would have been needed. Another observation based upon symmetry is that the moment M_a is equal to M_b. Therefore, the final reactions for the fixed-end beam shown in Fig. 8-19 are as follows:

$$R_a = R_b = 0 \qquad M_a = M_b = \frac{\alpha EI(T_2 - T_1)}{h} \tag{8-34}$$

Thus, the beam is subjected to pure bending because of the temperature changes.

Another approach that can be used to analyze the fixed-end beam shown in Fig. 8-19 is the method of solving the differential equation of the deflection curve. For this example, the differential equation becomes

$$EIv'' = -M - \frac{\alpha EI(T_2 - T_1)}{h} \tag{8-35}$$

Note that this equation incorporates the effects of both bending moments (see Eq. 7-10a) and temperature effects (see Eq. 7-72). The expression for the bending moment in the beam is

$$M = R_a x - M_a$$

However, as noted previously, we can see from the symmetry of the beam that no vertical reactions exist; hence we can set R_a equal to zero and obtain

$$EIv'' = M_a - \frac{\alpha EI(T_2 - T_1)}{h}$$

The first integration of this equation yields

$$EIv' = M_a x - \frac{\alpha EI(T_2 - T_1)}{h} x + C_1$$

The two boundary conditions on the slope ($v' = 0$ when $x = 0$ and when $x = L$) give $C_1 = 0$ and $M_a = \alpha EI(T_2 - T_1)/h$, which agrees with Eq. (8-34). Having found the fixed-end moment in this manner, we can consider the statically indeterminate problem as solved because all reactions are known. Also, you can quickly determine the slope v' and the deflection v. The results may surprise you, but they are easily explained.

*8.7 LONGITUDINAL DISPLACEMENTS AT THE ENDS OF A BEAM

When a beam undergoes lateral deflections due to bending, points on the axis of the beam are displaced longitudinally and the ends of the beam move closer together. These longitudinal displacements are illustrated in Fig. 8-20a, which shows a simple beam AB that is pin-supported at one end and free to displace longitudinally at the other. When this beam is bent by loads, end B of the deflection curve will move horizontally through a small distance λ from B to B'. The displacement λ, called the **curvature shortening** of the beam, is equal to the difference between the initial length L of the beam and the length of the chord AB' of the bent beam.

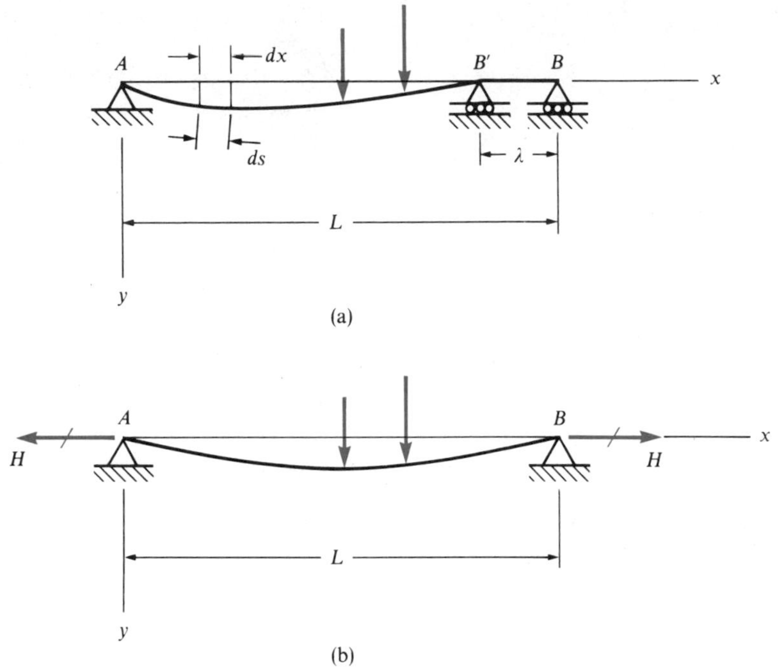

Fig. 8-20 (a) Horizontal displacement λ of the end of a beam, and (b) horizontal reactions H for a beam with immovable supports

To find the curvature shortening, we consider an element of length ds measured along the curved axis of the beam. The projection of this element on the x axis has length dx. The difference between the length ds and the length of its horizontal projection is

$$ds - dx = \sqrt{dx^2 + dv^2} - dx = dx\sqrt{1 + \left(\frac{dv}{dx}\right)^2} - dx \qquad \text{(a)}$$

where v represents the deflection of the beam. Now let us introduce the following binomial series:

$$(1 + t)^{1/2} = 1 + \frac{t}{2} - \frac{t^2}{8} + \frac{t^3}{16} - \cdots \qquad \text{(8-36)}$$

which converges when t is numerically less than one. If t is very small compared to unity, we can disregard the terms involving t^2, t^3, and so on, in comparison with the first two terms, and obtain

$$(1 + t)^{1/2} \approx 1 + \frac{t}{2} \qquad \text{(8-37)}$$

The term $(dv/dx)^2$ in Eq. (a) is ordinarily very small; therefore, we can use Eq. (8-37), and rewrite Eq. (a) as

$$ds - dx = dx\left[1 + \frac{1}{2}\left(\frac{dv}{dx}\right)^2\right] - dx = \frac{1}{2}\left(\frac{dv}{dx}\right)^2 dx$$

If this expression is integrated over the length of the beam, we obtain an expression for the curvature shortening:

$$\lambda = \frac{1}{2}\int_0^L \left(\frac{dv}{dx}\right)^2 dx \tag{8-38}$$

When the equation of the deflection curve of the beam is known, we can use Eq. (8-38) to obtain the displacement λ.

If the ends of the beam are prevented from translating longitudinally, as shown in Fig. 8-20b, then a horizontal reaction H will develop at each end. This force will cause the axis of the beam to elongate as bending occurs. In addition, the force H itself will have an effect upon the bending moments in the beam and hence also upon the deflection curve of the beam. Rather than attempt an exact analysis of this complicated problem, let us obtain an approximate expression for the force H in order to ascertain its importance. A reasonable approximation to the deflection curve of a simple beam is a parabola having the equation

$$v = \frac{4\delta x(L - x)}{L^2} \tag{b}$$

where δ is the deflection at the middle. The curvature shortening λ corresponding to this assumed deflected shape can be found by substituting Eq. (b) into Eq. (8-38) and integrating; the result is

$$\lambda = \frac{8\delta^2}{3L}$$

The force H required to elongate the beam by this amount is

$$H = \frac{EA\lambda}{L} = \frac{8EA\delta^2}{3L^2}$$

and the corresponding axial tensile stress in the beam is

$$\sigma = \frac{H}{A} = \frac{8E\delta^2}{3L^2} \tag{c}$$

The deflection δ at the midpoint of the beam is usually very small compared to the length; for example, the ratio δ/L might be 1/500. Using this value, and also assuming that the material is steel with $E = 30 \times 10^6$ psi, we see from Eq. (c) that the stress σ is only 320 psi. This low value confirms that the axial stress due to H is very small compared to the allowable stress in the material. Furthermore, in reality the ends of the beam cannot be held rigidly, and some small longitudinal displacement always will occur, thereby reducing the axial stress given by Eq. (c). (For a more complete analysis of beams with immovable supports, see Ref. 8-5.)

Another effect of the axial force H is to reduce the deflection of the beam as compared to a simple beam. However, this effect also may be disregarded for ordinary beams. It becomes significant only for extremely long, slender bars carrying large loads.

From the preceding discussions, we conclude that the customary practice of disregarding the effects of any longitudinal restraints and assuming that one end of the beam has a roller support is justified.

PROBLEMS / CHAPTER 8

The problems for Section 8.2 are to be solved by integrating the differential equation of the deflection curve.

8.2-1 Determine the reactions R_a, R_b, and M_a for the propped cantilever beam AB shown in Fig. 8-2 (Example 1 of Section 8.2) by taking the reactive moment M_a as the redundant and using the second-order differential equation. Then obtain the equation of the deflection curve and the location and magnitude of the maximum deflection δ_{max}.

8.2-2 Beginning with the fourth-order differential equation in terms of the intensity q of the distributed load, obtain the reactions for the propped cantilever beam AB of Fig. 8-2 (Example 1 of Section 8.2). Then draw the shear-force and bending-moment diagrams for the beam, labeling all critical ordinates including maximum and minimum values.

8.2-3 Determine the reactions R_a, R_b, M_a, and M_b for the fixed-end beam with a concentrated load P shown in Fig. 8-3 (Example 2 of Section 8.2) by starting with the second-order differential equation in terms of the bending moment. Also, determine the maximum deflection δ_{max} of the beam and draw the shear-force and bending-moment diagrams, labeling all critical ordinates including maximum and minimum values.

8.2-4 A propped cantilever beam AB is subjected to a couple M_0 acting at support B (see figure). Derive the equation of the deflection curve and determine the reactions R_a, R_b, and M_a at the supports. Also, construct the shear-force and bending-moment diagrams, labeling all critical ordinates.

8.2-5 A cantilever beam AB has a fixed support at A (see figure). At end B, the beam is pulled downward by a distance δ_b and held in position by a wire. Determine the equation of the deflection curve in terms of the deflection δ_b. Also, determine the reactions R_a, R_b, and M_a in terms of δ_b.

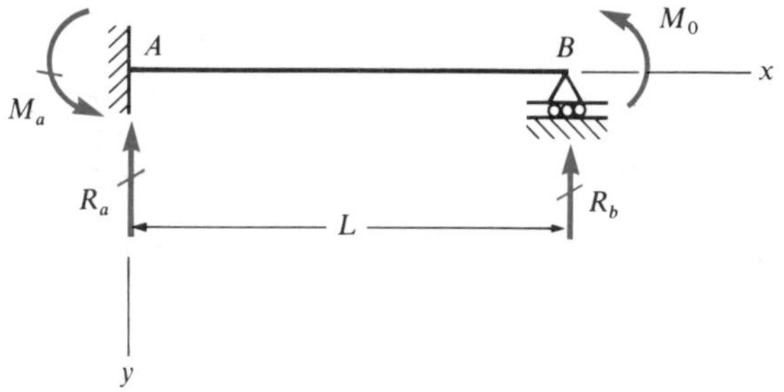

Prob. 8.2-4

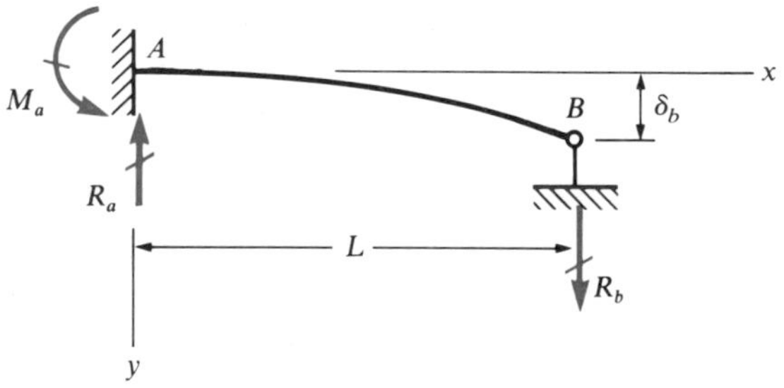

Prob. 8.2-5

8.2-6 Determine the equation of the deflection curve and the reactions R_a, R_b, M_a, and M_b for a fixed-end beam AB subjected to a uniform load of intensity q (see figure).

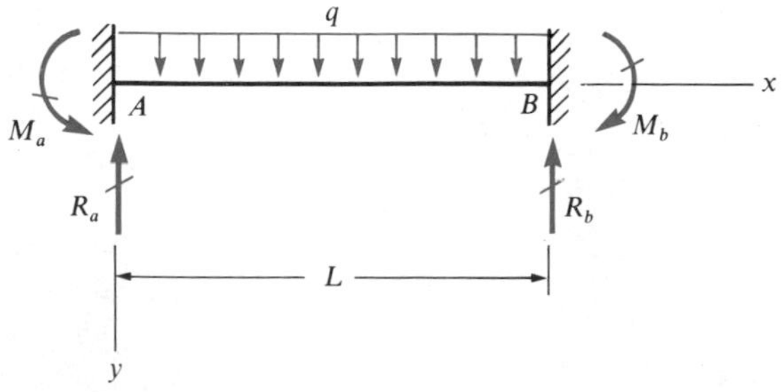

Prob. 8.2-6

8.2-7 Obtain the equation of the deflection curve and the reactions R_a, R_b, and M_a for a propped cantilever beam AB supporting a triangular load of maximum intensity q_0 (see figure).

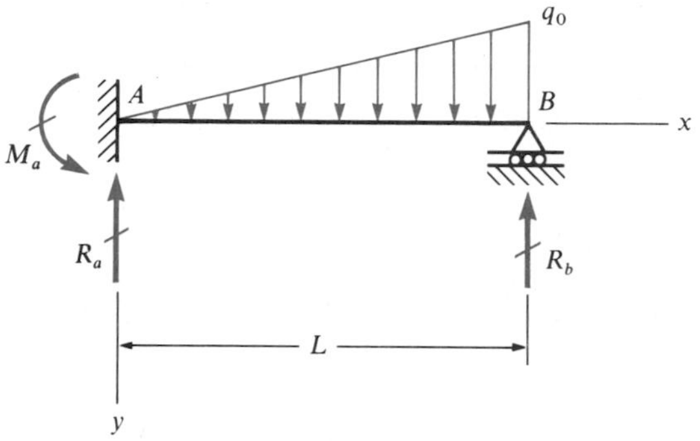

Prob. 8.2-7

8.2-8 A fixed-end beam AB is subjected to a symmetric triangular load of maximum intensity q_0 as shown in the figure. Determine all reactions, the equation of the deflection curve, and the maximum deflection δ_{max} of the beam.

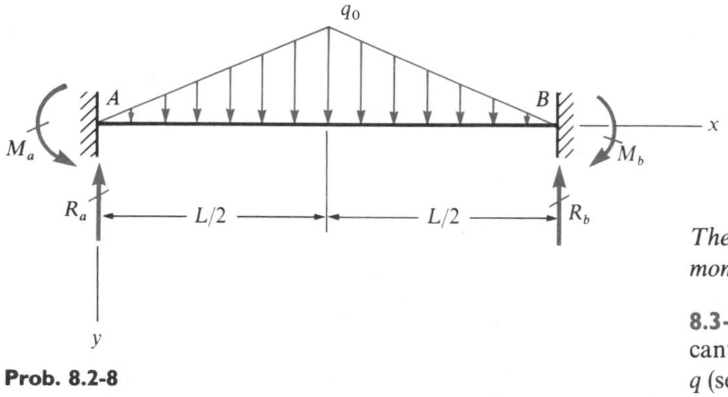

Prob. 8.2-8

8.2-9 Determine all reactions and the equation of the deflection curve for a fixed-end beam AB supporting a triangular load of maximum intensity q_0 (see figure).

8.2-10 A propped cantilever beam AB (see figure) supports a parabolically varying load of intensity $q = q_0(1 - x^2/L^2)$. Derive the equation of the deflection curve and determine the reactions at the supports.

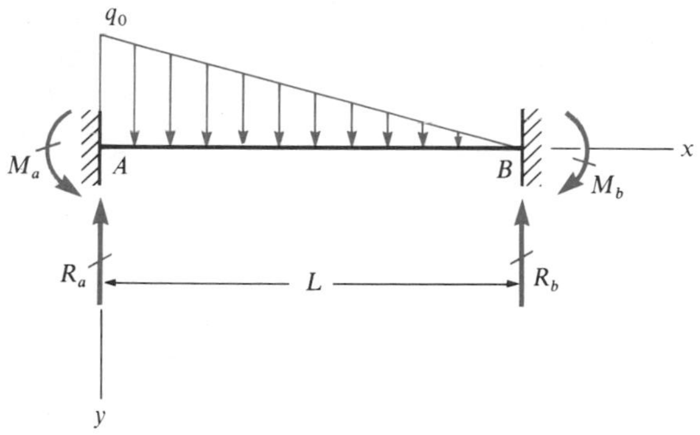

Prob. 8.2-9

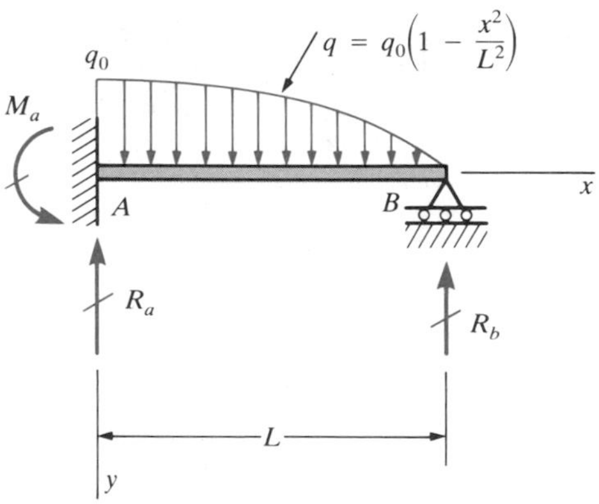

Prob. 8.2-10

The problems for Section 8.3 are to be solved by the moment-area method.

8.3-1 Determine the reactions R_a, R_b, and M_a for a propped cantilever beam AB subjected to a uniform load of intensity q (see figure).

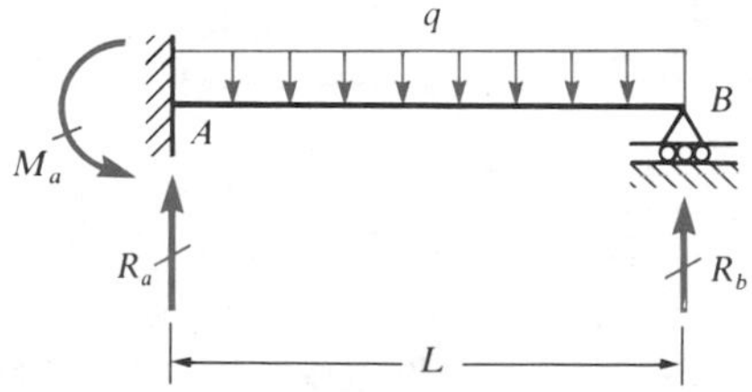

Prob. 8.3-1

8.3-2 Two concentrated loads, each of magnitude P, act at the third points of a propped cantilever beam AB (see figure). Find the reactions R_a, R_b, and M_a of the beam.

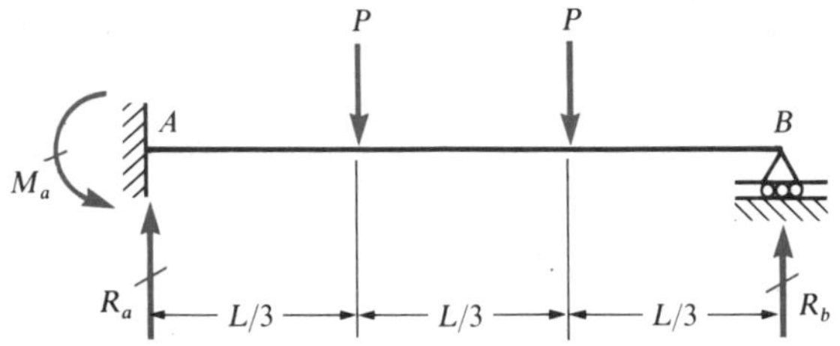

Prob. 8.3-2

8.3-3 Determine the reactions R_a, R_b, and M_a for the propped cantilever beam AB loaded as shown in the figure. Also, draw the shear-force and bending-moment diagrams for the beam, labeling all critical ordinates.

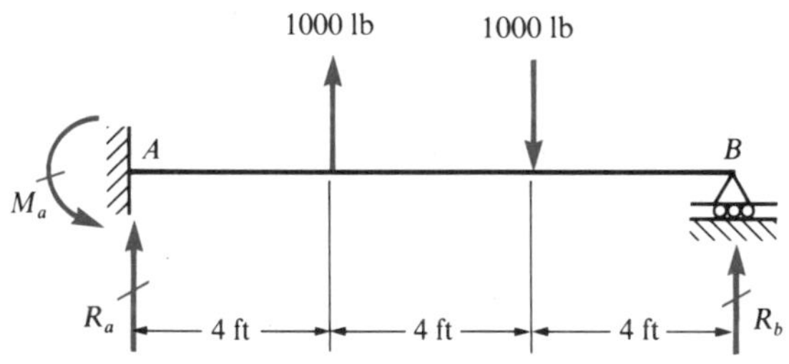

Prob. 8.3-3

8.3-4 The propped cantilever beam AB shown in the figure is subjected to a couple M_0 acting at distance a from the left-hand support. (a) Determine the reactions R_a, R_b, and M_a for this beam. (b) At what distance a_1 from the fixed support should the couple act in order that the deflection curve at support B will be horizontal? (c) At what distance a_2 from the fixed support should the couple act in order that no moment reaction develops at support A?

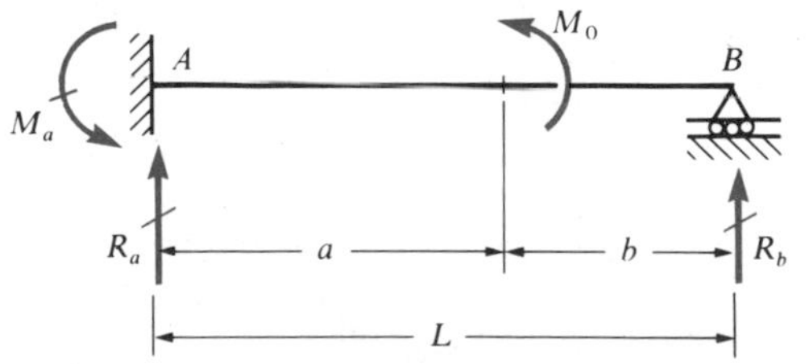

Prob. 8.3-4

8.3-5 Calculate the reactions R_a, R_b, and M_a for the propped cantilever beam with an overhang shown in the figure. (Take the reaction R_b at support B as the redundant.)

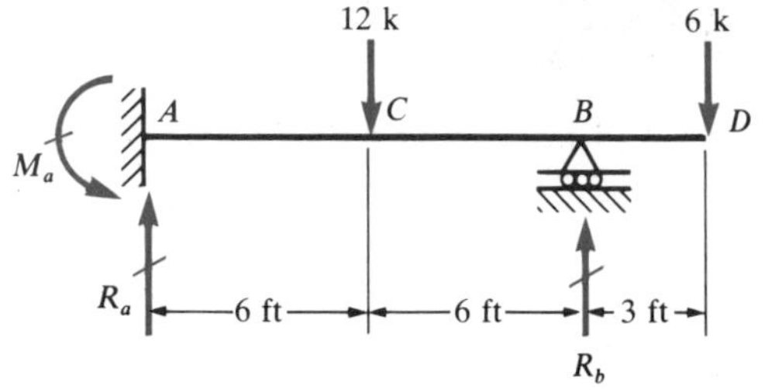

Prob. 8.3-5

8.3-6 Determine all reactions and the deflection δ_{max} at the midpoint for a fixed-end beam carrying a uniform load of intensity q (see figure).

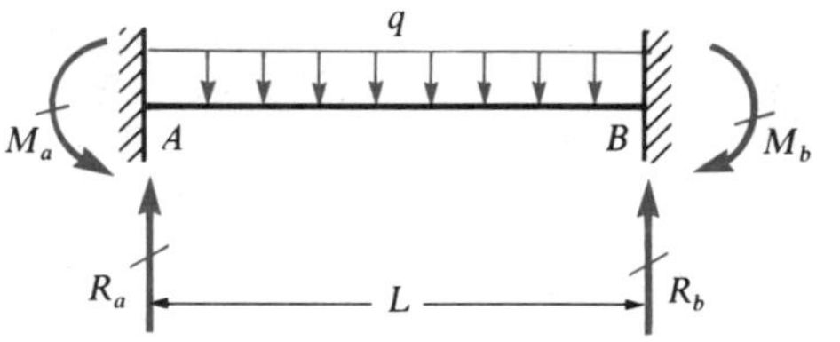

Prob. 8.3-6

8.3-7 Find all reactions for a fixed-end beam AB carrying two concentrated loads P acting at the positions shown in the figure. Also, determine the maximum deflection δ_{max} of the beam.

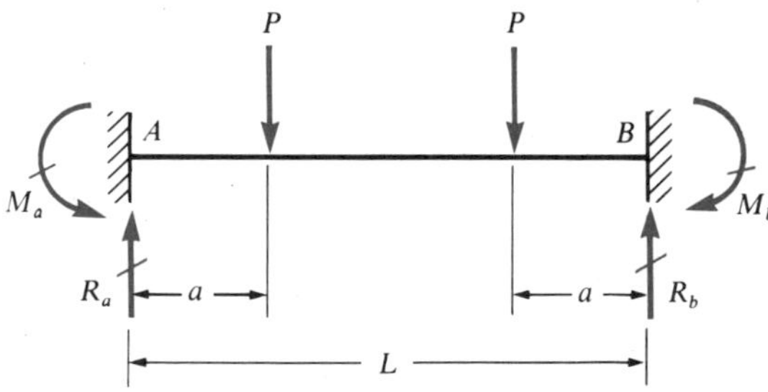

Prob. 8.3-7

8.3-8 Determine all reactions for a fixed-end beam AB due to a symmetric triangular load of maximum intensity q_0 (see figure), taking the reactive moments at the supports as the redundants. Also, determine the maximum deflection δ_{max} of the beam.

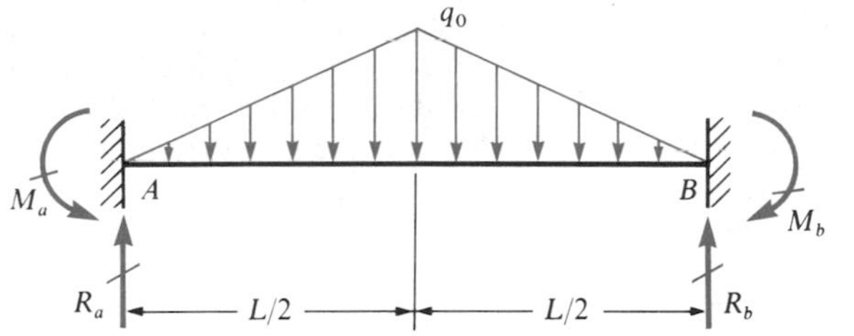

Prob. 8.3-8

8.3-9 Determine all reactions for a fixed-end beam AB with a concentrated load P (see figure). Also, obtain the deflection δ_1 under the load P.

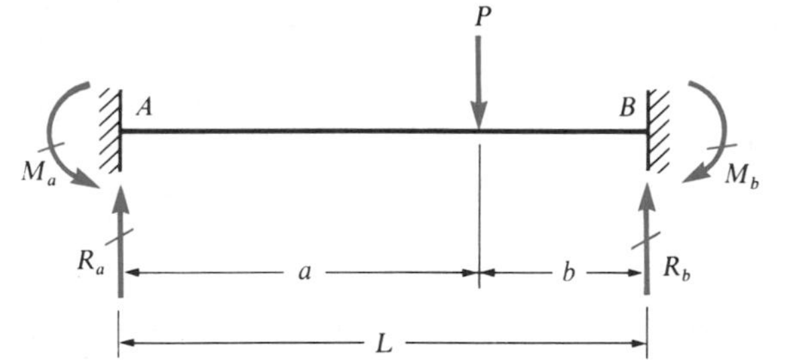

Prob. 8.3-9

8.3-10 A nonprismatic beam AB with fixed ends is subjected to a concentrated load P at the midpoint (see figure). Determine the reactive moments M_a and M_b at the supports and the deflection $\delta_{\max}$ at the midpoint. (Take the reactive moments M_a and M_b as the redundants.)

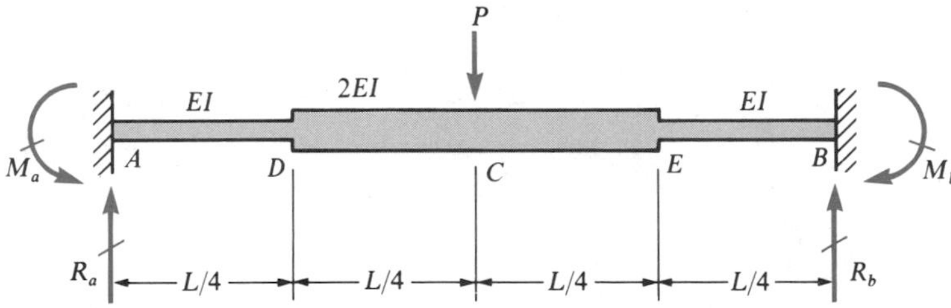

Prob. 8.3-10

8.3-11 Solve the preceding problem if the beam is subjected to two equal loads P acting at points D and E and there is no load at the midpoint of the beam.

The problems for Section 8.4 are to be solved by the method of superposition.

8.4-1 Determine the reactions R_a, R_b, and R_c for the two-span beam ABC with a uniform load of intensity q shown in Fig. 8-9a (Example 1 of Section 8.4) by taking the reaction R_c as the redundant. Also, draw the shear-force and bending-moment diagrams for the beam, labeling all critical ordinates including the maximum and minimum values.

8.4-2 Solve Problem 8.3-2 (a propped cantilever beam with two concentrated loads).

8.4-3 A propped cantilever beam AB supporting a triangular load of maximum intensity q_0 is shown in the figure. Determine all reactions for this beam.

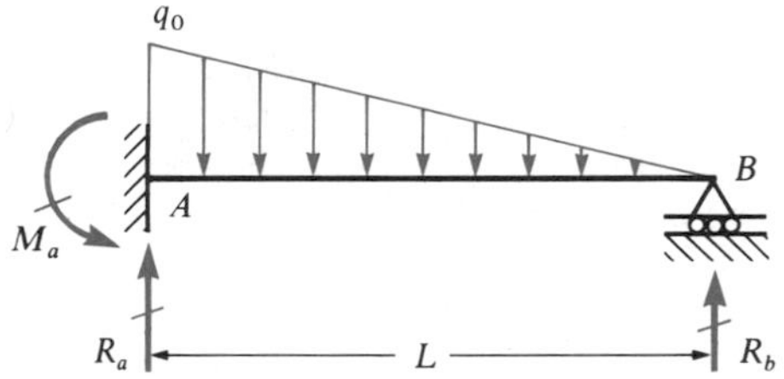

Prob. 8.4-3

8.4-4 Obtain the reactions R_a, R_b, and M_a for the propped cantilever beam with an overhang (see figure). Also, draw the shear-force and bending-moment diagrams for the beam, labeling all critical ordinates.

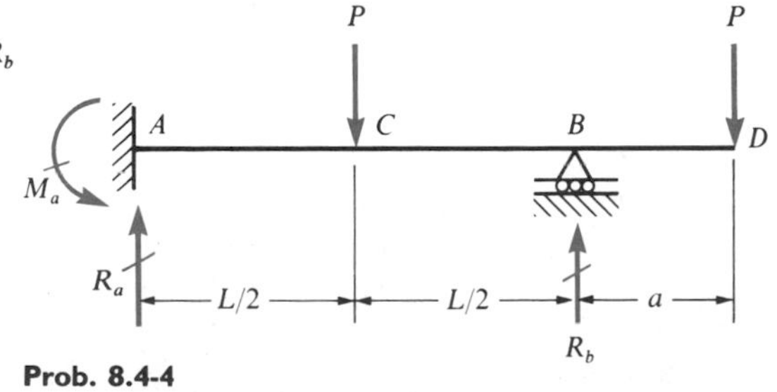

Prob. 8.4-4

8.4-5 Solve Problem 8.3-5 (a propped cantilever beam with an overhang).

8.4-6 A continuous beam ABC with two unequal spans supports a uniform load of intensity q as shown in the figure. Find the reactions R_a, R_b, and R_c for the beam.

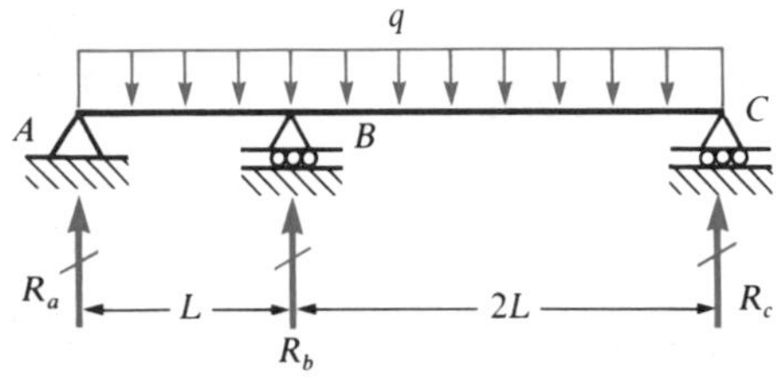

Prob. 8.4-6

8.4-7 Two flat bars AB and CD, lying in a horizontal plane, cross at right angles and jointly support a vertical load P at their midpoints (see figure). The bars are made of the same material and have the same widths, but their lengths may be different. How should the thicknesses of the bars be determined if it is desired to have all four reactions equal?

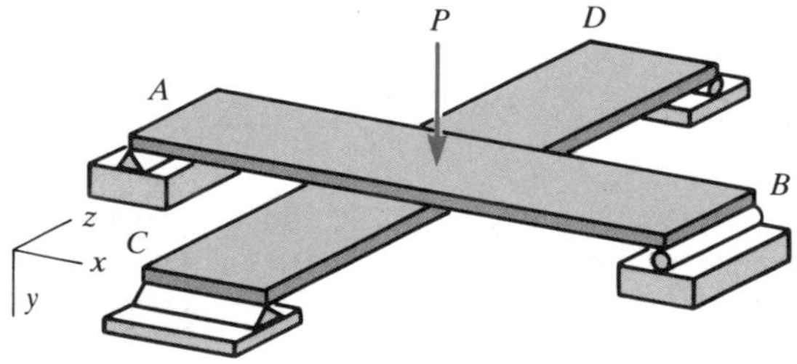

Prob. 8.4-7

8.4-8 The cantilever beam AB shown in the figure is supported by a cable BC at its free end B. A uniform load of intensity q acts on the beam. Before the load q is applied, the cable is taut but has no force in it. Find the force T in the cable due to the load q. (Assume EI is the flexural rigidity of the beam and EA is the axial rigidity of the cable.)

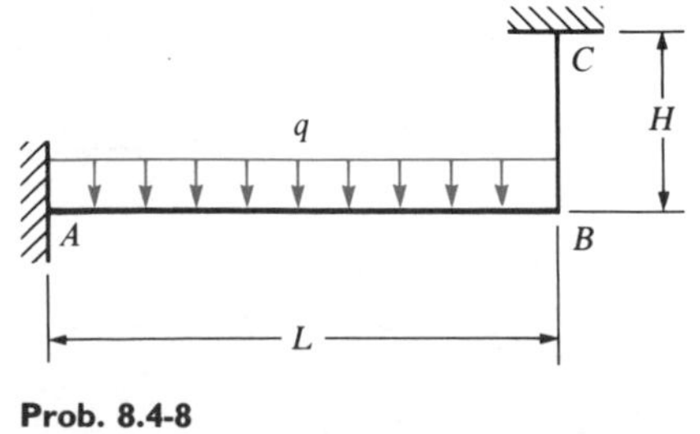

Prob. 8.4-8

8.4-9 Two cantilever beams AB and CD are supported as shown in the figure. A roller fits snugly between the two beams at point D. The upper beam has flexural rigidity EI_1 and the lower beam has flexural rigidity EI_2. Find the force F transmitted between the beams at D.

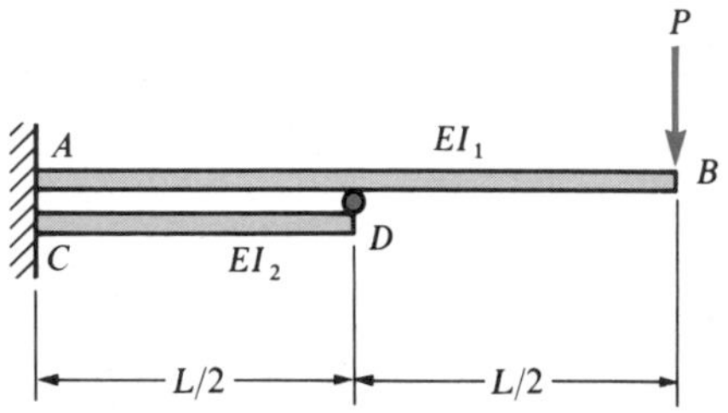

Prob. 8.4-9

8.4-10 A three-span continuous beam $ABCD$ with three equal spans supports a uniform load of intensity q (see figure). Determine all reactions of this beam.

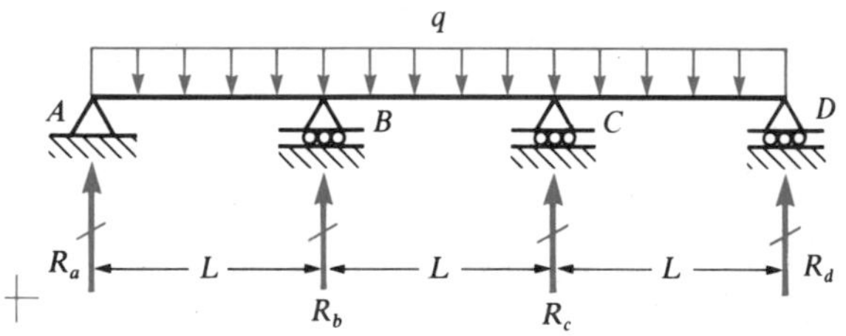

Prob. 8.4-10

8.4-11 The figure shows a nonprismatic, propped cantilever beam AB with flexural rigidity $2EI$ from A to C and EI from C to B. Determine all reactions of the beam due to the uniform load of intensity q.

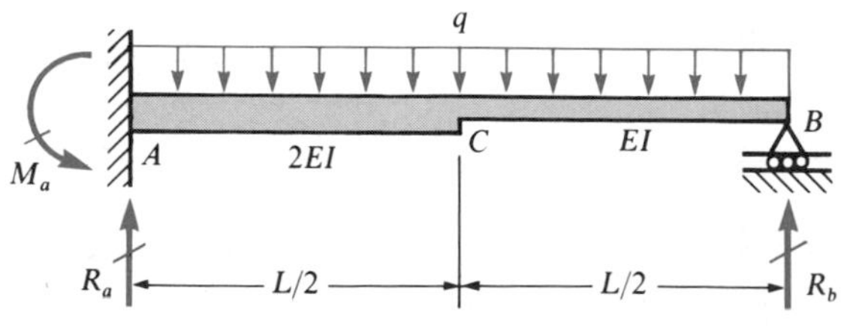

Prob. 8.4-11

8.4-12 The beam ABC shown in the figure has flexural rigidity $EI = 4.0\ \text{MN·m}^2$. When the loads are applied to the beam, the support at B settles vertically downward by a distance of 8.0 mm. Calculate the reaction R_b at support B.

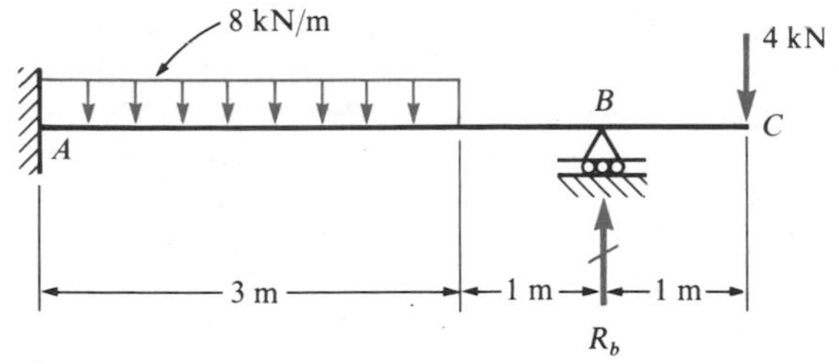

Prob. 8.4-12

8.4-13 A two-span beam ABC rests on supports at A and C when there is no load on the beam (see figure). There is a small gap Δ between the beam and the support at B. When the uniform load of intensity q is applied to the beam, the gap closes and reactions develop at all three supports. What should be the magnitude of the gap Δ in order that all three reactions will be equal?

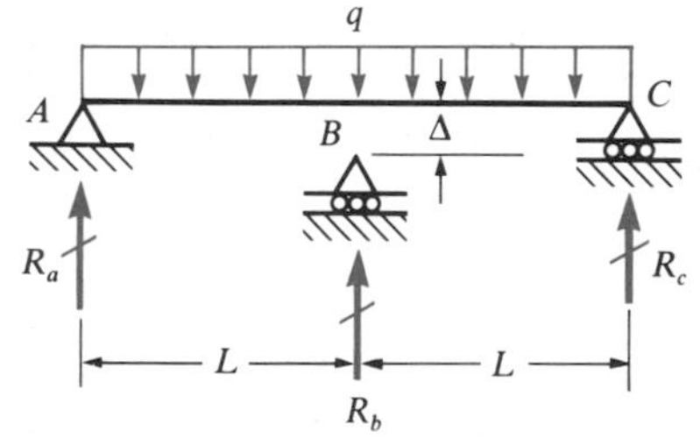

Prob. 8.4-13

8.4-14 Obtain the reactions for the fixed-end beam AB with a concentrated load P shown in Fig. 8-10a (see Example 2 of Section 8.4) by taking the reactions R_b and M_b as the redundants. Also, draw the shear-force and bending-moment diagrams for the beam, labeling all critical ordinates.

8.4-15 Determine all reactions for a fixed-end beam AB with a uniform load acting over part of the span, as shown in Fig. 8-11 (see Example 3 of Section 8.4), by taking the moments M_a and M_b as the redundants.

8.4-16 A fixed-end beam AB has its left-hand support A rotated through a small angle θ (see figure). Determine all reactions for the beam.

8.4-17 One end of a fixed-end beam AB is displaced laterally through a small distance Δ with respect to the other end (see figure). Determine all reactions for the beam.

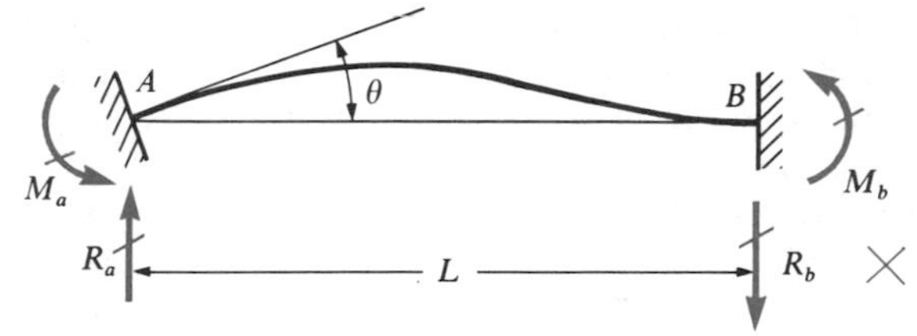

Prob. 8.4-16

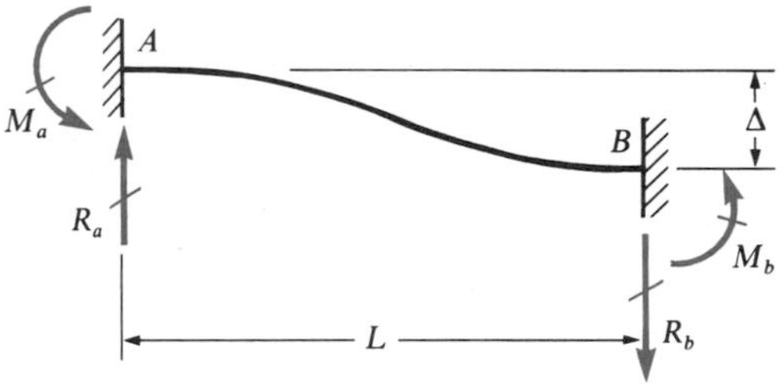

Prob. 8.4-17

8.4-18 A three-span continuous beam $ABCD$ has a uniform load of intensity q acting on the end span AB (see figure). Find all reactions for this beam.

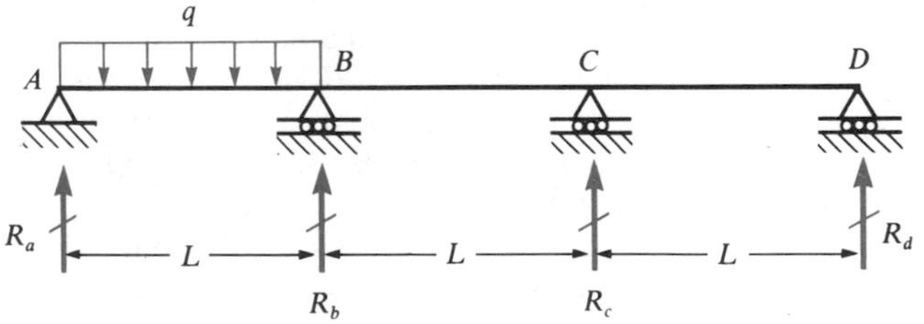

Prob. 8.4-18

8.4-19 Determine all reactions for the two-span continuous beam ABC with a fixed support at A as shown in the figure. Also, construct the shear-force and bending-moment diagrams for this beam, labeling all critical ordinates including the maximum and minimum values.

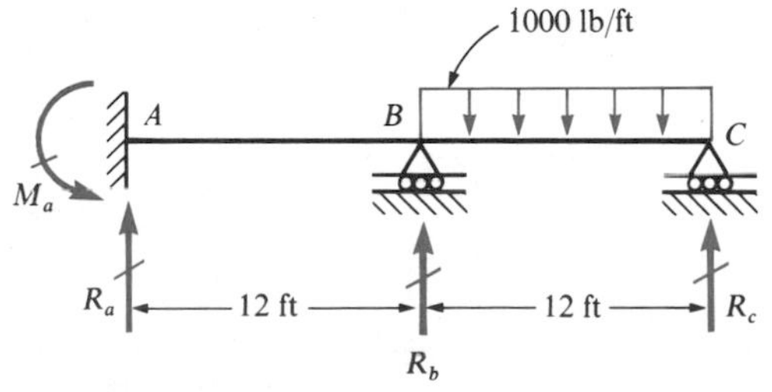

Prob. 8.4-19

*8.4-20 A cantilever beam AB is supported on top of a second cantilever beam CD (see figure). The beams are identical except for their lengths. When there is no load acting, the two beams are in contact but without any pressure between them. When the concentrated load P is applied, what force F is developed between the beams at point D? (Hint: Assuming that the beams are in contact only at points C and D, find the force F. Then, knowing F, obtain the deflection curves for both beams and verify that they touch only at C and D.)

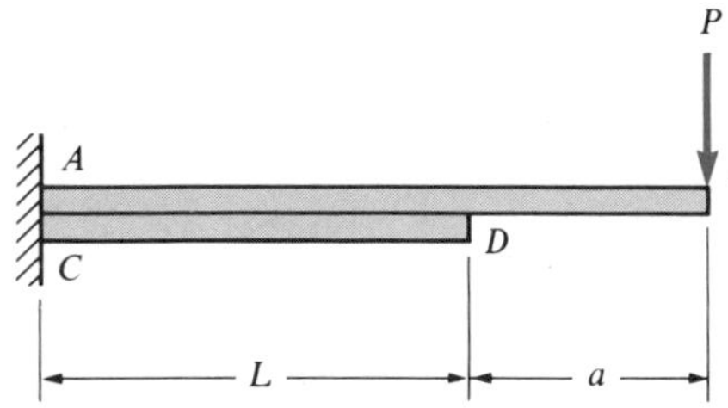

Prob. 8.4-20

*8.4-21 The cantilever beam AB shown in the figure is an S 6 × 12.5 steel I-beam with $E = 30 \times 10^6$ psi. The simple beam DE is a wood beam 4 in. × 12 in. (nominal dimensions) in cross section with $E = 1.5 \times 10^6$ psi. A steel rod AC of diameter 0.25 in., length 10 ft, and $E = 30 \times 10^6$ psi serves as a hanger joining the two beams. The hanger fits snugly between the beams before the uniform load is applied to beam DE. Determine the force F in the hanger and the maximum bending moments M_{ab} and M_{de} in the two beams due to the uniform load, which has intensity $q =$ 400 lb/ft.

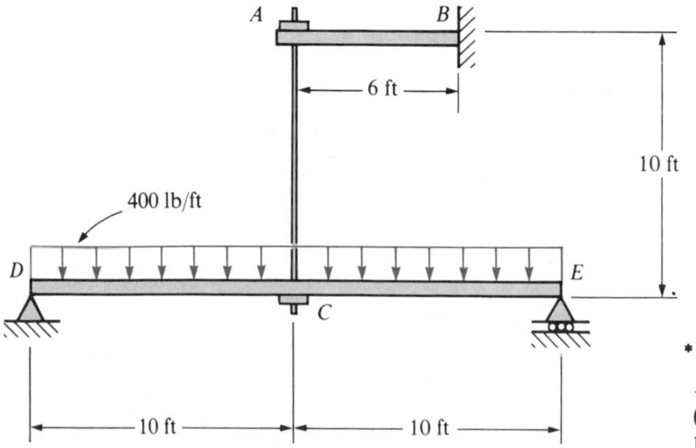

Prob. 8.4-21

**8.4-22 A steel beam AB used in conjunction with an electromagnet in a high-energy physics experiment is securely bolted to rigid supports (see figures). A magnetic field produced by coils C results in a force acting on the beam. The force is trapezoidally distributed with maximum intensity $q_0 = 18$ kN/m. The beam has a rectangular cross section with width $b = 60$ mm and height $h = 20$ mm. Its length between supports is $L = 200$ mm and the dimension c of the load is 50 mm. Determine the maximum bending stress σ_{max} and the maximum deflection δ_{max} for the beam. (Disregard the effects of axial deformations and consider only the effects of bending. Use $E = 200$ GPa.)

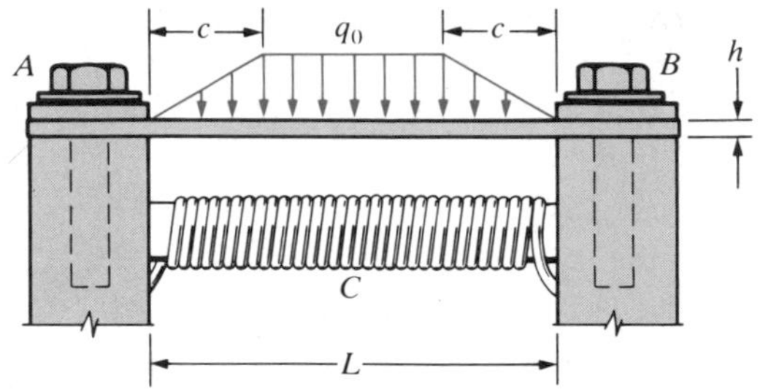

Prob. 8.4-22

*8.4-23 The beam ACB is simply supported at A and B and supported on a spring of stiffness k (force per unit of shortening) at the midpoint C. What should be the stiffness k of the spring in order that the maximum bending moment in the beam (due to the uniform load) will have the smallest possible value?

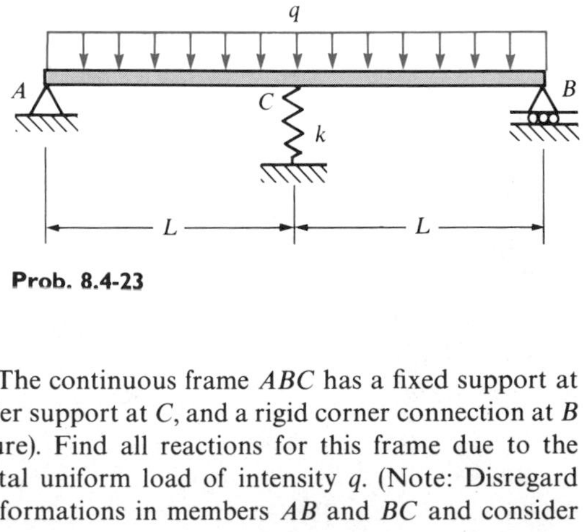

Prob. 8.4-23

*8.4-24 The continuous frame ABC has a fixed support at A, a roller support at C, and a rigid corner connection at B (see figure). Find all reactions for this frame due to the horizontal uniform load of intensity q. (Note: Disregard axial deformations in members AB and BC and consider only the effects of bending. Also, assume that each member has length L and flexural rigidity EI.)

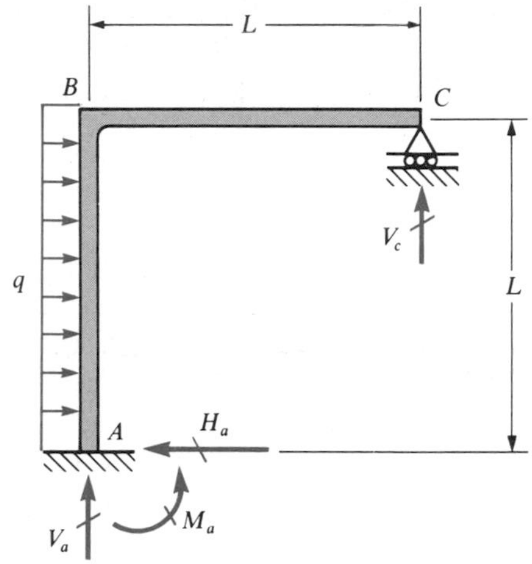

Prob. 8.4-24

***8.4-25** The beam AB with an attached bracket BCD is supported and loaded as shown in the figure. The flexural rigidity EI is the same for all parts of the structure. Determine the horizontal deflection δ_h and vertical deflection δ_v at point D.

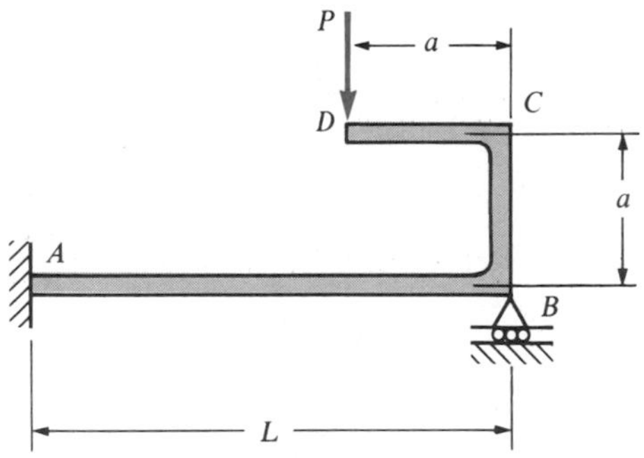

Prob. 8.4-25

The problems for Section 8.5 are to be solved by the three-moment equation.

8.5-1 Solve Problem 8.4-6 (a two-span continuous beam with a uniform load).

8.5-2 Solve Problem 8.4-10 (a three-span continuous beam with a uniform load).

8.5-3 A two-span continuous beam with simple supports is subjected to a uniform load of intensity q throughout its length. Both spans have the same length L and the same flexural rigidity EI. Determine the bending moment M_0 at the middle support assuming that this support settles vertically by a distance Δ when the load q is applied.

8.5-4 A continuous beam $ABCD$ of three spans, each span having length L and flexural rigidity EI, supports a concentrated load P at the midpoint of each span (see figure). Determine the reaction R at the end of the beam. Also, draw the shear-force and bending-moment diagrams for the beam, labeling all critical ordinates.

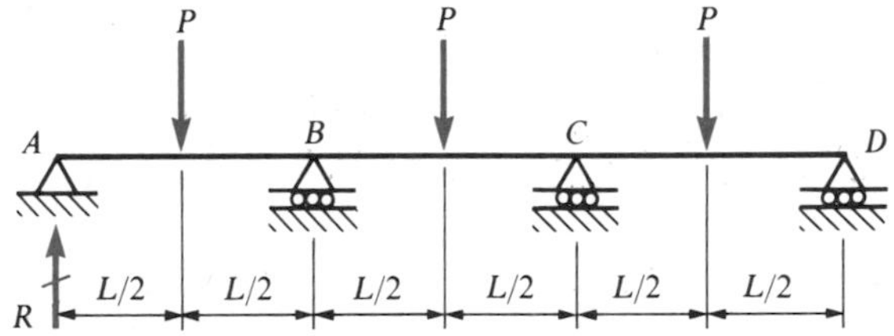

Prob. 8.5-4

8.5-5 Solve Problem 8.4-18 (a three-span continuous beam with a uniform load in one span).

8.5-6 The continuous beam shown in the figure has an overhang at one end. Calculate the reactions R_1, R_2, and R_3 due to the loads shown. Also, draw the shear-force and bending-moment diagrams for the beam, labeling all critical ordinates.

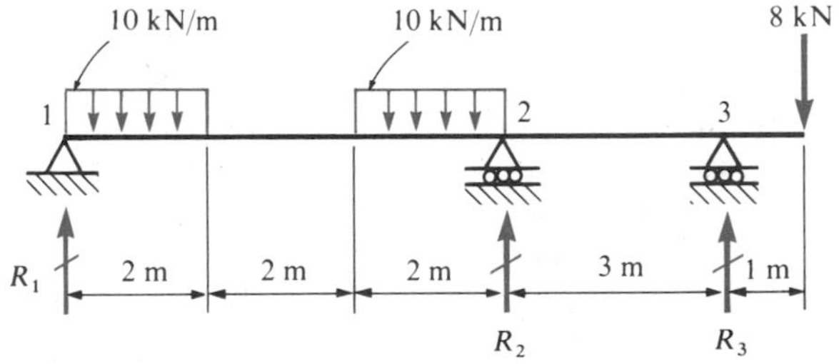

Prob. 8.5-6

8.5-7 A four-span continuous beam has a uniform load of intensity q throughout its entire length (see figure). Assuming that each span is of length L, determine the reaction R at the end of the beam. Also, draw the shear-force and bending-moment diagrams for the beam, labeling all critical ordinates.

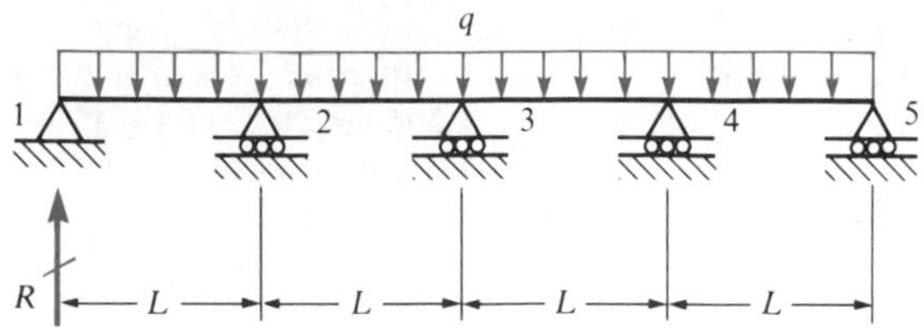

Prob. 8.5-7

8.5-12 A continuous beam on four supports has an overhang at the left-hand end, as shown in the figure. Find the bending moments M_1, M_2, M_3, and M_4 at the supports assuming that $P_1 = 5$ kN, $q = 6$ kN/m, $P_2 = 10$ kN, $b = 3$ m, $I_1 = 46.6 \times 10^6$ mm^4, $I_2 = 43.7 \times 10^6$ mm^4, $L_0 = 3$ m, $L_1 = 6$ m, and $L_2 = 8$ m.

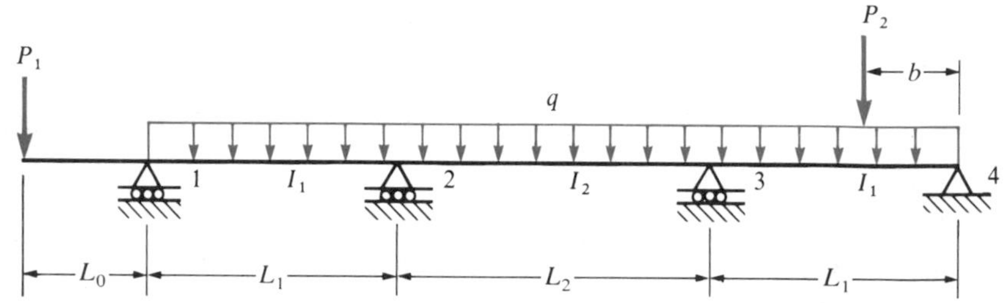

Prob. 8.5-12

8.5-8 Refer to the two-span continuous beam with a fixed support shown in Fig. 8-16a and assume that $L_1 = L_2 = L$, $I_1 = I_2$, and $P = qL$. Determine the bending moments M_1 and M_2 assuming that the load P acts at the midpoint of span 2-3.

8.5-9 Solve Problem 8.4-19 (a two-span continuous beam with a fixed support).

8.5-10 Refer to the two-span continuous beam with a fixed support shown in Fig. 8-16a and assume that $L_1 = 6$ m and $L_2 = 4$ m. Also, the flexural rigidity EI is the same for both spans. The loads have magnitudes $q = 6$ kN/m and $P = 18$ kN. Determine the reactions M_1, R_1, R_2, and R_3 assuming that the load P acts at the midpoint of span 2-3.

8.5-11 Determine the bending moments M_1, M_2, and M_3 at supports 1, 2, and 3, respectively, for the continuous beam with an overhang shown in the figure. Also, draw the shear-force and bending-moment diagrams for the beam, labeling all critical ordinates.

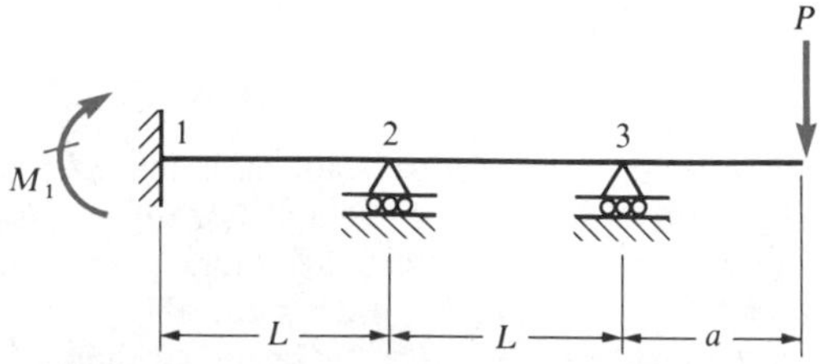

Prob. 8.5-11

8.5-13 The continuous beam shown in the figure supports a uniform load of intensity q over its entire length. Determine the bending moments M_1, M_2, and M_3 at supports 1, 2, and 3, respectively. Also, draw the shear-force and bending-moment diagrams for the beam, labeling all critical ordinates.

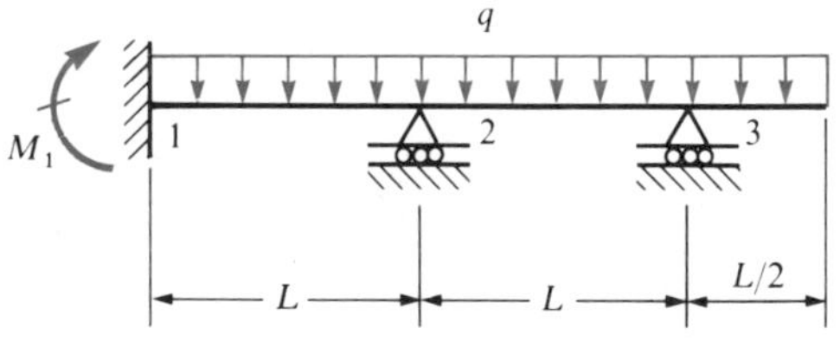

Prob. 8.5-13

8.5-14 Determine the bending moments $M_1, M_2, \ldots, M_8$ at the supports of a continuous beam with seven spans of equal length L when the two end spans are loaded by uniform loads of intensity q.

***8.5-15** The continuous beam shown in the figure has a fixed support at 1 and simple supports at 2, 3, and 4. The moments of inertia in the three spans are $I_1 = 2400$ in.^{4}, $I_2 = 1200$ in.^{4}, and $I_3 = 7200$ in.^{4} The beam supports a uniform load of intensity $q = 6$ k/ft on span 1-2 and a concentrated load $P = 16$ k at the midpoint of span 3-4. Under the action of these loads, the support at 3 settles vertically by an amount $\Delta = 0.08$ in. Find the bending moments M_1, M_2, and M_3 at supports 1, 2, and 3, respectively, assuming $E = 30 \times 10^6$ psi.

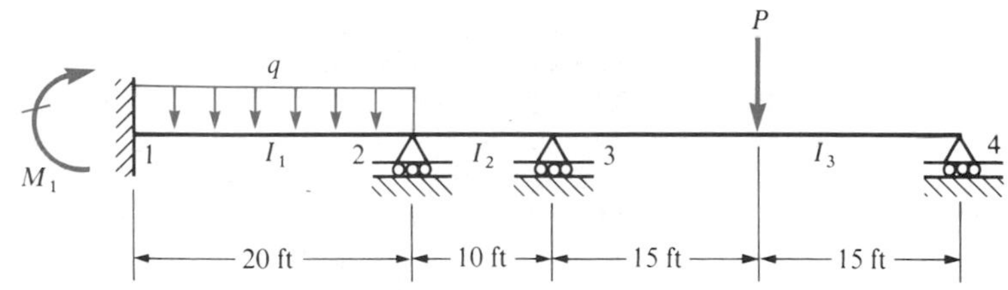

Prob. 8.5-15

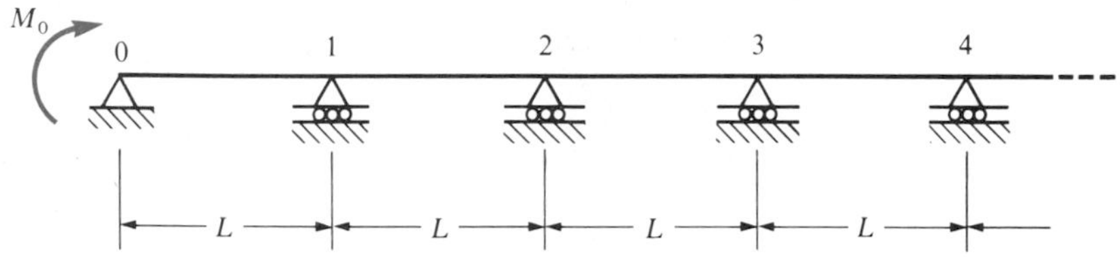

Prob. 8.5-16

***8.5-16** A continuous beam with equal spans and constant flexural rigidity EI extends for an infinite distance to the right, as shown in the figure. The supports are numbered $n = 0, 1, 2, 3, \ldots$. At the left-hand support ($n = 0$), a couple M_0 is applied. Derive a formula for the bending moment M_n at the nth support in terms of M_0. (Hint: Note that the ratio M_{i+1}/M_i of the bending moment at any support to the moment at the preceding support is a constant. Then apply the three-moment equation to two consecutive spans.)

8.6-1 A propped cantilever beam, fixed at the left-hand end A and simply supported at the right-hand end B, is subjected to a temperature differential with temperature T_1 on its upper surface and T_2 on its lower surface (see figure). Find all reactions for the beam.

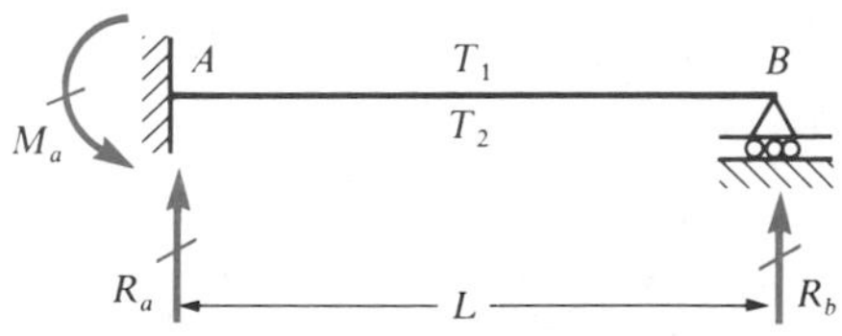

Prob. 8.6-1

8.6-2 A two-span beam ABC is subjected to a temperature differential with temperature T_1 on the upper surface and T_2 on the lower surface (see figure). Determine all reactions for the beam.

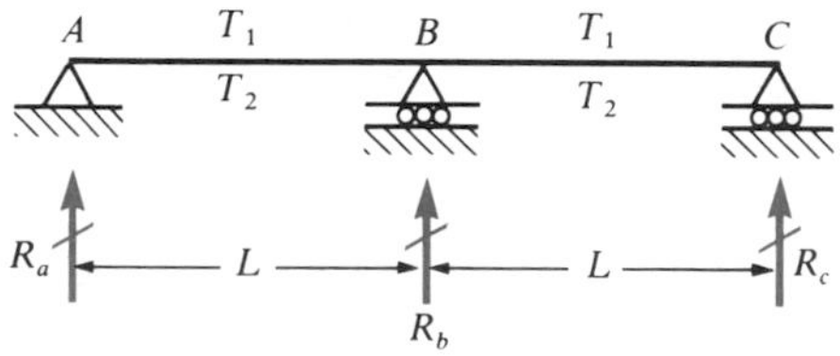

Prob. 8.6-2

8.6-3 A simple beam AB is attached to a cable CD that is taut but without initial tension (see figure). The length of the cable is H and its effective cross-sectional area is A. Find the tensile force S in the cable when the temperature drops T degrees, assuming that the beam and the cable are made of the same material.

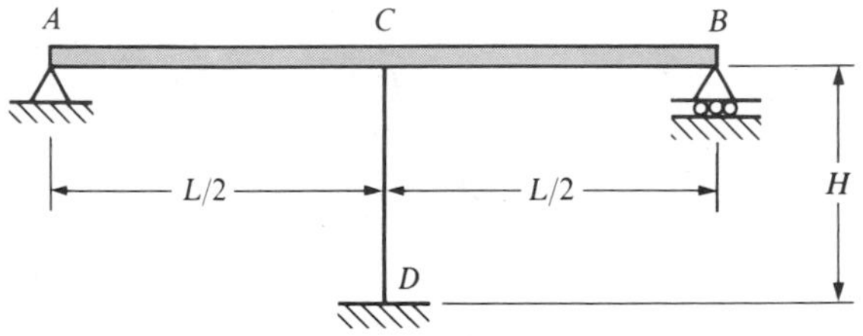

Prob. 8.6-3

8.7-1 Assuming that the deflected shape of a beam with immovable pinned supports (see Fig. 8-20b) is given by the equation $v = \delta \sin \pi x/L$, where δ is the deflection at the midpoint of the beam and L is the length, determine the longitudinal force H at the ends of the beam. For a beam of aluminum alloy with $E = 10 \times 10^6$ psi, calculate the axial tensile stress σ when the ratio of the deflection δ to the length L equals 1/200, 1/400, and 1/600.

8.7-2 Obtain a formula for the curvature shortening λ of a simple beam of length L carrying a uniform load of intensity q. If the beam is a steel beam (modulus of elasticity $E = 200$ GPa, moment of inertia $I = 36 \times 10^6\ \text{mm}^4$, and depth $d = 300$ mm) with $L = 3$ m and $q = 25$ kN/m, what uniform axial stress σ_1 is required to elongate the beam by the amount λ? Compare the stress σ_1 with the maximum bending stress σ_2 in the uniformly loaded simple beam.

8.7-3 A solid circular bar of diameter d and length L is supported as a simple beam. The bar carries a concentrated load P at its midpoint. (a) Derive a formula for the curvature shortening λ of the bar. (b) Determine the uniform axial tensile stress σ_1 required to elongate the bar by the amount λ. (c) Determine the maximum bending stress σ_2 in the bar when it is loaded as a simple beam. (d) For a steel bar ($E = 30 \times 10^6$ psi, $\sigma_{yp} = 42{,}000$ psi) with $d = 1.0$ in. and $L = 50$ in., plot a graph showing σ_1 and σ_2 versus the load P. (From the graph, observe that the effect of curvature shortening becomes relatively more important as the lateral load increases.)

CHAPTER 9

Columns

9.1 BUCKLING AND STABILITY

Structures and machines may fail in a variety of ways, depending upon the materials, kinds of loads, and conditions of support. For instance, ductile members may stretch or bend excessively if overloaded, allowing the structure to come apart or collapse. Fractures can occur from repeated cycles of loading (fatigue failures) or from the overstressing of brittle members. Most of these types of failures are avoided by designing the members so that their maximum stresses and maximum deflections remain within tolerable limits. Thus, the **strength** and **stiffness** of a member are the important criteria in design. These subjects were considered in preceding chapters.

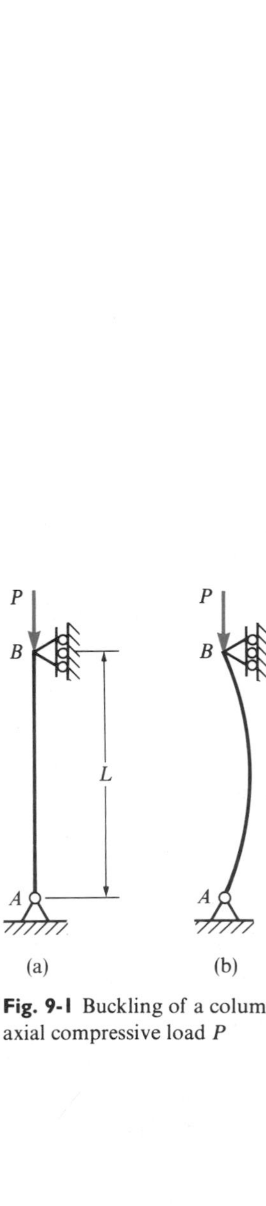

Fig. 9-1 Buckling of a column due to an axial compressive load P

Another type of failure is **buckling**, which is the subject matter of this chapter. We will consider specifically the buckling of **columns** (that is, long, slender structural members loaded axially in compression) (Fig. 9-1a). If such a member is slender, then, instead of failing by direct compression, it may bend and deflect laterally (Fig. 9-1b), and we say that the column has buckled. Under an increasing axial load, the lateral deflections increase too, and eventually the column will collapse completely. Of course, buckling can occur in many different kinds of structures, and it can take a variety of forms. When you step on an empty aluminum can, the thin cylindrical walls buckle under your weight; and, when a large bridge collapsed a few years ago, it was due to buckling of a flat steel plate that wrinkled under compressive stresses.

To illustrate the phenomenon of buckling in an elementary manner, we will consider the **idealized structure** shown in Fig. 9-2a. Member AB is a rigid bar that is pinned at the base and supported by an elastic spring of stiffness β at the top. (The stiffness β is also known as the *spring constant*.)

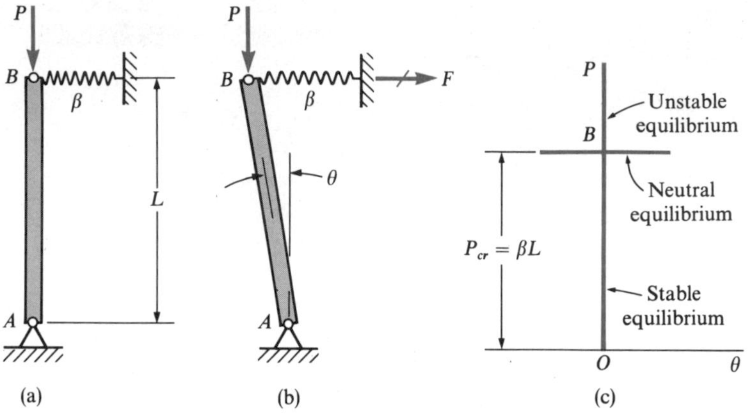

Fig. 9-2 Buckling of a rigid bar supported by a spring

The bar supports a centrally applied load P that is perfectly aligned with the axis of the bar, so that initially the spring has no force in it. Now suppose that the bar is disturbed by some external force so that it rotates slightly through a small angle θ about support A (Fig. 9-2b). If the force P is small, the bar-spring system is **stable** and will return to its initial position when the disturbing force is removed. But if the force P is very large, the bar will continue to rotate and the system will collapse; thus, for a large force, the system is **unstable** and buckles by undergoing large rotations of the bar.

We can analyze the bar-spring system in more detail by considering its static equilibrium. When the bar is rotated slightly (Fig. 9-2b), the spring is elongated by an amount equal to θL, where L is the length of the bar. The corresponding force F in the spring is

$$F = \beta\theta L$$

This force creates a clockwise moment about point A equal to FL, or $\beta\theta L^2$. The tendency of this moment is to return the bar to its original position; hence we shall refer to $\beta\theta L^2$ as the **restoring moment**. The force P creates a counterclockwise moment about A that tends to overturn the bar; thus, the **overturning moment** is $P\theta L$. If the restoring moment exceeds the overturning moment, the system is stable and the bar returns to the initial vertical position; if the overturning moment exceeds the restoring moment, the system is unstable and the bar collapses by rotating through large angles. Therefore, we have the following conditions:

If $P\theta L < \beta\theta L^2$, or $P < \beta L$, the system is stable.
If $P\theta L > \beta\theta L^2$, or $P > \beta L$, the system is unstable.

The transition from a stable to an unstable system occurs when $P\theta L = \beta\theta L^2$, or $P = \beta L$; this value of the load is called the **critical load**:

$$P_{cr} = \beta L \tag{9-1}$$

We see that the system is stable when $P < P_{cr}$ and unstable when $P > P_{cr}$.

As long as P remains less than P_{cr}, the system returns to its initial position and $\theta = 0$. In other words, the bar is in equilibrium only when $\theta = 0$. When P is greater than P_{cr}, the bar is still in equilibrium when $\theta = 0$ (because the bar is in direct compression and there is no force in the spring), but the equilibrium is unstable and cannot be maintained. The slightest disturbance will cause the bar to collapse. At the critical load, the restoring and overturning moments are equal for any small value of θ (note that θ cancels out of the equilibrium equation). Thus, the bar is in equilibrium for any small angle θ; this condition is referred to as **neutral equilibrium**.

These equilibrium relationships are shown in the graph of P versus θ (Fig. 9-2c), in which the two heavy lines represent the equilibrium conditions. Point B, where the equilibrium diagram branches, is called a **bifurcation point**. The horizontal line for neutral equilibrium extends to the left and right of the vertical axis because the angle θ may be clockwise or counterclockwise. The line extends only a short distance, however, because our analysis is based upon the assumption that θ is a small angle. Of course, this assumption is quite valid, because θ is indeed small when buckling begins and the bar departs from the vertical position. (If θ becomes large, the equilibrium line curves away from the horizontal.)

The equilibrium of the bar described in Fig. 9-2 is analogous to that of a ball placed upon a surface (Fig. 9-3). If the surface is concave upward, like the inside of a dish, the equilibrium is stable and the ball always returns to the low point; if it is convex upward, like a dome, then the ball theoretically can be in equilibrium on top of the surface, but the equilibrium is unstable and in reality the ball rolls away. When placed on a flat surface, the ball is in neutral equilibrium and remains wherever it is placed.

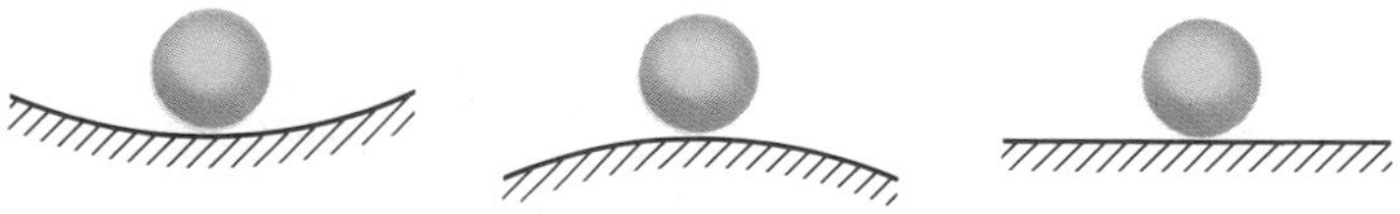

Fig. 9-3 Ball in stable, unstable, and neutral equilibrium

As we will see in the next section, the behavior of an ideal elastic column is analogous to that of the bar-spring system shown in Fig. 9-2. Furthermore, many kinds of buckling and stability problems in both structural and mechanical systems fit this model.

9.2 COLUMNS WITH PINNED ENDS

To investigate the stability behavior of columns, we will begin by considering a slender column with pinned ends (Fig. 9-4a). The column is loaded by a vertical force P that is applied through the centroid of the cross section and aligned with the longitudinal axis of the column. The

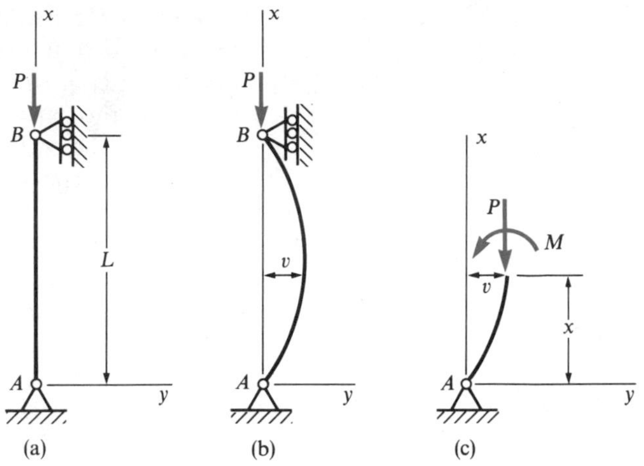

Fig. 9-4 Column with pinned ends: (a) ideal column, (b) buckled shape, and (c) free-body diagram of part of the column

column itself is perfectly straight and is made of a linearly elastic material that follows Hooke's law. Thus, we are going to analyze the behavior of an **ideal column**. The xy plane is a plane of symmetry, and we assume that any bending of the column takes place in that plane (Fig. 9-4b).

When the axial load P has a small value, the column remains straight and undergoes only axial compression. The uniform compressive stresses are obtained from the equation $\sigma = P/A$. This straight form of equilibrium is **stable**, which means that the column returns to the straight position if it is disturbed. For instance, if we apply a small lateral load and cause the column to bend, the deflection will disappear and the column will return to its original position when the lateral load is removed. As the axial load P is gradually increased, we reach a condition of **neutral equilibrium** in which the column may have a bent shape. The corresponding value of the load is the **critical load** P_{cr}. At this load the ideal column may undergo small lateral deflections with no change in the axial force, and a small lateral load will produce a bent shape that does not disappear when the lateral load is removed. Thus, the critical load can maintain the column in static equilibrium either in the straight position or in a slightly bent position. At higher values of the load, the column is **unstable** and will collapse by bending. For the ideal case that we are discussing, the column is in equilibrium in the straight position even when P is greater than P_{cr}. However, the equilibrium is unstable, and the smallest imaginable disturbance will cause the column to deflect sideways; the deflections will increase immediately and the column will collapse. The situation is analogous to balancing a pencil on its point. In theory, the pencil is in equilibrium when so balanced, but the position cannot be maintained.

The behavior of an ideal column compressed by an axial load P (Fig. 9-4) may be summarized as follows:

If $P < P_{cr}$, the column is in stable equilibrium in the straight position.

If $P = P_{cr}$, the column is in neutral equilibrium in either the straight or a slightly bent position.

If $P > P_{cr}$, the column is in unstable equilibrium in the straight position, and hence it buckles.

Of course, actual columns do not behave in this idealized manner because imperfections always exist. Nevertheless, we begin by studying ideal columns because they provide insight into the behavior of real columns.

To determine the critical load and the deflected shape of the buckled column (Fig. 9-4b), we use one of the differential equations of the deflection curve of a beam (see Eqs. 7-10). These equations are applicable to a column because, when buckling occurs, bending moments are developed in the column, which bends as though it were a beam. Although the fourth-order differential equation in terms of load intensity q and the third-order equation in terms of shear force V are suitable for analyzing columns, we will use the second-order equation in terms of bending moment M because its general solution is the simplest. This equation (Eq. 7-10a) is

$$EIv'' = -M \tag{9-2}$$

in which v is the lateral deflection in the y direction. For a column, we orient the x and y axes as shown in Fig. 9-4b, which corresponds to a beam AB that has been rotated 90° from the horizontal. The bending moment M at distance x from end A of the buckled column can be obtained from the free-body diagram shown in Fig. 9-4c. We cut the column at distance x from support A and observe from static equilibrium that a vertical force P and a bending moment M (equal to Pv) must act on the cut cross section. Therefore, the differential equation becomes

$$EIv'' = -M = -Pv$$

or

$$EIv'' + Pv = 0 \tag{9-3}$$

The quantity EI is the flexural rigidity for bending in the xy plane, which we have assumed to be the plane of buckling.

The solution of Eq. (9-3), which is a homogeneous linear differential equation of second order with constant coefficients, gives the deflection v as a function of x. For convenience in writing the general solution of the equation, we introduce the notation

$$k^2 = \frac{P}{EI} \tag{9-4}$$

Then we can rewrite Eq. (9-3) in the form

$$v'' + k^2 v = 0 \tag{9-5}$$

The general solution of this equation is

$$v = C_1 \sin kx + C_2 \cos kx \tag{9-6}$$

where C_1 and C_2 are constants to be evaluated from the boundary conditions (or end conditions) of the column. Note that the number of arbitrary constants (two in this case) must agree with the order of the differential equation. Also, the fact that Eq. (9-6) represents the general solution of Eq. (9-5) may be verified easily by substituting the expression for v into the differential equation and observing that the equation is satisfied.

To evaluate the constants of integration, we use the boundary conditions at the ends:

$$v(0) = 0 \quad \text{and} \quad v(L) = 0$$

The first condition gives $C_2 = 0$, and the second gives

$$C_1 \sin kL = 0 \tag{a}$$

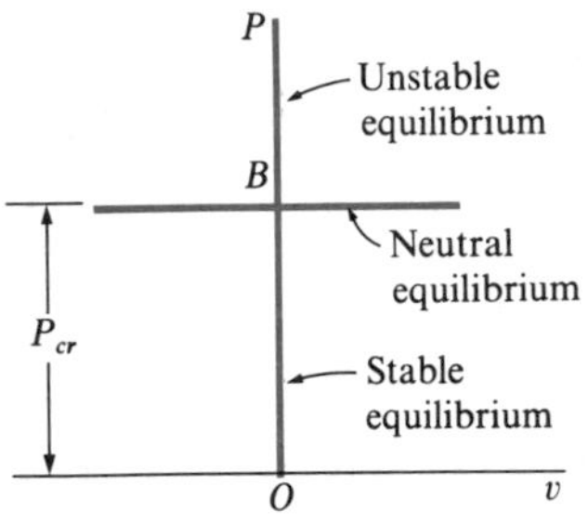

Fig. 9-5 Load-deflection diagram for an ideal, linearly elastic column

From this equation we conclude that either $C_1 = 0$ or $\sin kL = 0$. If $C_1 = 0$, the deflection v is zero and the column remains straight. In that case, Eq. (a) is satisfied for any value of the quantity kL. Therefore, the axial load P also may have any value (see Eq. 9-4). This solution of the differential equation (often called the *trivial solution*) is represented by the vertical axis of the load-deflection diagram shown in Fig. 9-5. This solution corresponds to an ideal column that is in equilibrium (either stable or unstable) under the action of the compressive load P.

The other possibility for satisfying Eq. (a) is to meet the following condition:

$$\sin kL = 0 \tag{b}$$

This equation is satisfied when $kL = 0, \pi, 2\pi, \ldots$. Since $kL = 0$ means that $P = 0$, this solution is not of interest. Therefore, the solutions we will consider are

$$kL = n\pi \qquad n = 1, 2, 3, \ldots \tag{c}$$

or

$$P = \frac{n^2\pi^2 EI}{L^2} \qquad n = 1, 2, 3, \ldots \tag{d}$$

(see Eq. 9-4). This equation gives the values of P that satisfy Eq. (a) and provide solutions (other than the trivial solution) to the differential equation. Thus, the equation of the deflection curve is

$$v = C_1 \sin kx = C_1 \sin \frac{n\pi x}{L} \qquad n = 1, 2, 3, \ldots \tag{e}$$

Only when P has one of the values given by Eq. (d) is it theoretically possible for the column to have a bent shape; for all other values of P, the column is in equilibrium only if it remains straight. Therefore, the values of P given by Eq. (d) are the **critical loads** for the column.

The smallest critical load for the column is obtained when $n = 1$:

$$P_{cr} = \frac{\pi^2 EI}{L^2} \tag{9-7}$$

The corresponding buckled shape (sometimes called a *mode shape*) is

$$v = C_1 \sin\frac{\pi x}{L} \tag{9-8}$$

as shown in Fig. 9-6b. The constant C_1 represents the deflection at the midpoint of the column and may be positive or negative. Therefore, the part of the load-deflection diagram corresponding to P_{cr} is a horizontal straight line (Fig. 9-5). The deflection at this load is undefined, although it must remain small for our equations to be valid. The bifurcation point B is at the critical load; above point B the equilibrium is unstable, and below it is stable. Buckling of a pinned-end column in the first mode ($n = 1$) is called the **fundamental case** of column buckling.

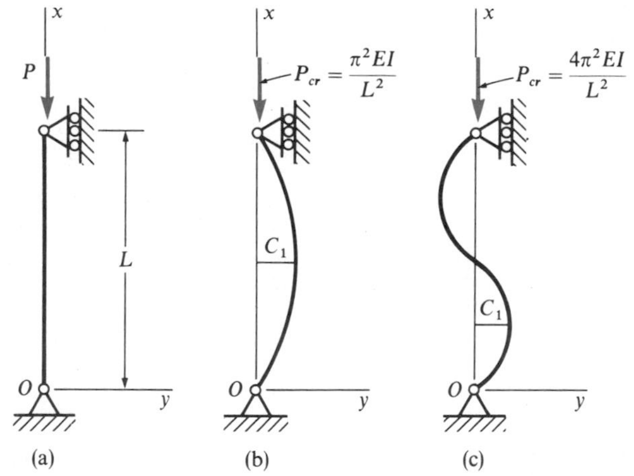

Fig. 9-6 Buckled shapes for an ideal column with pinned ends: (a) initially straight column, (b) buckled shape for $n = 1$, and (c) buckled shape for $n = 2$

The critical load for an ideal elastic column is also known as the **Euler load**. The famous mathematician Leonhard Euler (1707–1783), perhaps the greatest mathematician of all time, was the first person to investigate the bending of a slender column and determine its critical load (in 1744); see Refs. 1-1 through 1-3 and 9-1 through 9-7. For information about Euler's life and works, see the note to Ref. 9-3.

By taking higher values of the index n in Eqs. (d) and (e), we obtain an infinite number of critical loads and corresponding mode shapes. The mode shape for $n = 2$ is pictured in Fig. 9-6c; the corresponding critical load is four times larger than that for the fundamental case. We see that the magnitudes of the critical loads are proportional to the square of n, and the number of half-waves in the buckled shape is equal to n. Such buckled shapes are usually of no practical interest because the column will always buckle when the axial load P reaches its lowest critical value (Eq. 9-7). The only way to obtain higher modes of buckling is to provide lateral support of the column at the inflection points (or nodal points).

The critical load of a column is proportional to the flexural rigidity EI and inversely proportional to the square of the length. However, the strength of the material itself (for instance, the proportional limit) does not appear in the equation for the critical load. Thus, the critical load is not increased by using a stronger material. However, the load can be increased by using a stiffer material (that is, a material with larger modulus of elasticity E). Also, the load can be increased by distributing the material in such a way as to increase the moment of inertia I of the cross section, just as a beam can be made stiffer by increasing the value of I. The moment of inertia is increased by distributing the material away from the centroid of the cross section. Hence hollow tubular members are more economical for columns than are solid members having the same cross-sectional areas. Reducing the wall thickness of a tubular member and increasing its lateral dimensions, while keeping the cross-sectional area constant, increases the critical load because I is increased. This process has a practical limit, however, because eventually the wall itself becomes unstable. Then localized buckling occurs in the form of small corrugations or wrinkles. Thus, we must distinguish between overall buckling of a column, as shown in Fig. 9-6, and local buckling of its parts. The latter requires more detailed investigations, such as those found in books on buckling and stability (Refs. 9-1, 9-2, 9-8, 9-9, and 9-10). In this chapter we consider only overall buckling of columns.

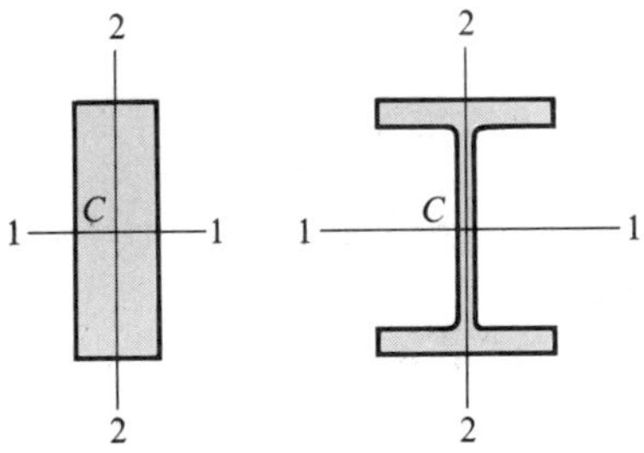

Fig. 9-7 Cross sections of columns showing principal centroidal axes with $I_1 > I_2$

In the preceding analysis we assumed that the xy plane was a plane of symmetry of the column and that buckling took place in that plane (Fig. 9-6). The latter assumption is met if the column has lateral supports perpendicular to the plane of the figure, so that the column is constrained to buckle in the xy plane. If the column is supported only at its ends, so that it is free to buckle in any direction, then bending will occur about the principal centroidal axis having the smaller moment of inertia. For instance, consider the rectangular and wide-flange cross sections shown in Fig. 9-7. In each case, the moment of inertia I_1 is greater than I_2; hence the column will buckle in the 1-1 plane, and the smaller moment of inertia I_2 should be used in the formula for the critical load. If the cross section is square or circular, all centroidal axes have the same moment of inertia and buckling may occur in any longitudinal plane.

EFFECTS OF LARGE DEFLECTIONS, IMPERFECTIONS, AND INELASTIC BEHAVIOR

The equation for the critical load was derived for an ideal column in which the deflections are small, the construction is perfect, and the material

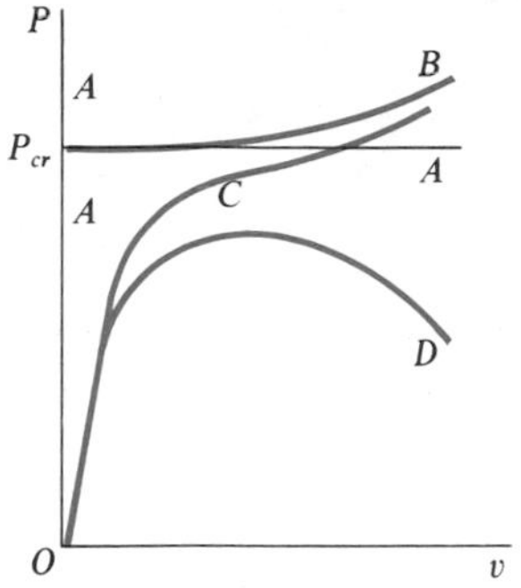

Fig. 9-8 Load-deflection diagram for columns: Lines *A*, ideal elastic column with small deflections; Curve *B*, ideal elastic column with large deflections; Curve *C*, elastic column with imperfections; and Curve *D*, inelastic column with imperfections

follows Hooke's law. As a consequence, we found that the magnitudes of the deflections at buckling were undefined.* Thus, at $P = P_{cr}$, the column may have any small deflection, a condition represented by the horizontal line *A* in the load-deflection diagram of Fig. 9-8. (In this figure, we show only the right-hand half of the diagram, but the two halves are symmetric about the vertical axis.) The theory is limited to small deflections because we used v'' for the curvature. A more exact analysis based upon the exact expression for curvature (see Eq. 7-11) shows that there is no indefiniteness in the magnitudes of the deflections at buckling. Instead, for an ideal elastic column, the load-deflection diagram goes upward in accord with curve *B* of Fig. 9-8. Thus, after an elastic column begins to buckle, a larger and larger load is required to cause an increase in the deflections.

Now suppose that the column is not constructed perfectly; for instance, it could have an imperfection in the form of a small initial curvature, so that the unloaded column is not perfectly straight. Such imperfections produce deflections from the onset of loading, as shown by curve *C* in Fig. 9-8. For small deflections, curve *C* approaches line *A*; as the deflections become large, it approaches curve *B*. The larger the imperfections, the further curve *C* moves to the right. If the column is constructed with great accuracy, curve *C* approaches more closely to the straight lines *A* (one vertical and one horizontal). From lines *A*, *B*, and *C*, we see that the critical load represents the maximum load-carrying capacity of an elastic column for practical purposes, because large deflections usually are not acceptable.

Finally, we consider what happens when the stresses exceed the proportional limit and the column material no longer follows Hooke's law. Of course, the load-deflection diagram is unchanged up to the level of load at which the proportional limit is reached. Then the curve for inelastic behavior (curve *D*) departs from the elastic curve, continues upward, reaches a maximum, and turns downward. Naturally, the detailed shapes of these curves depend upon the material properties and column dimensions, but the general nature of the behavior is typified by the curves shown.

Only extremely slender columns remain elastic up to the critical load P_{cr}. Stockier columns behave inelastically and follow a curve such as *D*. It is important to recognize that the maximum load *P* that can be supported by an inelastic column may be considerably less than the critical load P_{cr}. Furthermore, the descending part of curve *D* represents catastrophic collapse, because it takes smaller and smaller loads to maintain larger and larger deflections. By contrast, the curves for elastic columns are quite stable, because they continue upward as the deflections increase; that is, it takes larger and larger loads to cause an increase in deflection.

* Using mathematical terminology, we solved a *linear eigenvalue problem*. The critical load is an *eigenvalue* and the corresponding buckled mode shape is an *eigenfunction*.

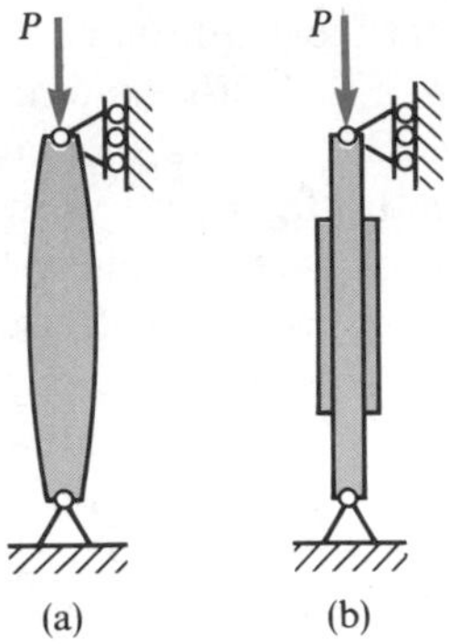

Fig. 9-9 Columns with varying cross sections

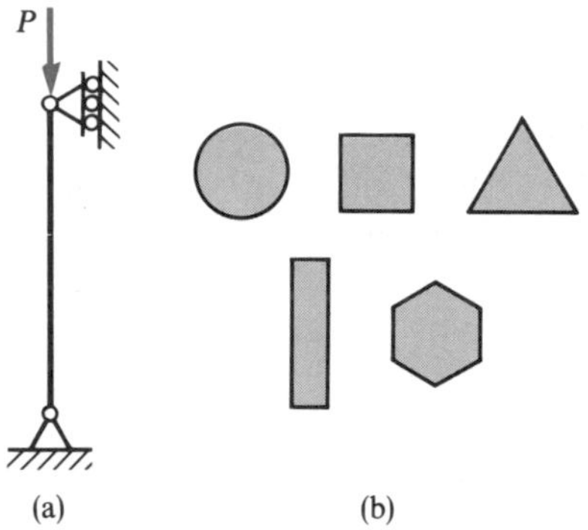

Fig. 9-10 Which is the optimum shape for a prismatic column?

OPTIMUM COLUMNS

Most columns are prismatic members; that is, they have the same cross sections throughout their lengths. Only such columns are analyzed in this chapter. However, the critical load of a column consisting of a given amount of material may be increased by tapering the column so that it has larger cross sections in regions where the bending moments are larger. Consider, for instance, a column of solid circular cross section with pinned ends. A "submarine-shaped" column with appropriately varying cross section (Fig. 9-9a) will have a larger critical load than if the same volume of material is made into a prismatic column. Of course, it is not practical to construct such a column, but prismatic columns sometimes are reinforced over part of their lengths (Fig. 9-9b) in order to approximate the optimum conditions.

Now consider a prismatic column with pinned ends that is free to buckle in any direction (Fig. 9-10a). Also, consider only a solid convex cross section, such as a circle, square, triangle, rectangle, or hexagon (Fig. 9-10b). For a given cross-sectional area, which of these shapes makes the most efficient column? Or, in more precise terms, we wish to know which cross section gives the largest critical load, considering that the critical load is calculated from the formula $P_{cr} = \pi^2 EI/L^2$ using the smallest moment of inertia for the cross section. For most people, intuition suggests that the circular shape is best. However, you can easily prove that a cross section in the shape of an equilateral triangle gives a 21% higher critical load than does a circular cross section of the same area (see Problem 9.2-11). The equilateral triangle also gives higher loads than any of the other shapes; hence it is the optimum cross section. (For a mathematical analysis of optimum column shapes, see Ref. 9-11.)

In practice, columns often are constrained to buckle in only one plane; hence a shape should be chosen that gives a large moment of inertia for bending in that plane. In steel construction, wide-flange sections are commonly used for columns; in both reinforced concrete and wood construction, rectangular and circular sections are used. The shapes used in aircraft and space structures, as well as in machine design, are quite varied, depending upon the particular application. In general, hollow sections are more efficient than solid sections because they provide a larger moment of inertia for the same cross-sectional area.

9.3 COLUMNS WITH OTHER SUPPORT CONDITIONS

Buckling of a column with pinned ends is often called the fundamental case of buckling. However, many other conditions, such as fixed ends, elastic supports, and free ends, are encountered in practice. The critical loads for columns with various kinds of supports can be determined from the differential equation of the deflection curve, in a manner similar to that for a pinned-end column. We begin by drawing free-body diagrams of

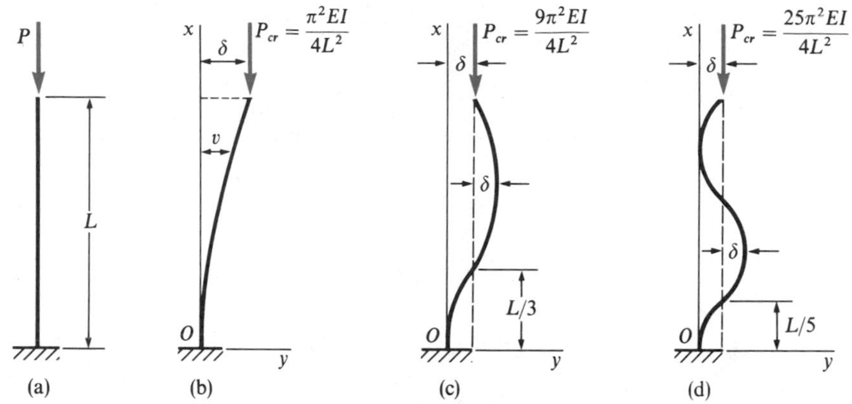

Fig. 9-11 Ideal column fixed at the base and free at the upper end: (a) initially straight column, (b) buckled shape for $n = 1$, (c) buckled shape for $n = 3$, and (d) buckled shape for $n = 5$

the column in order to obtain expressions for the bending moment M. Then we solve the differential equation in terms of the bending moment. Boundary conditions on the deflection v and slope v' are used to evaluate the arbitrary constants and any other unknowns that appear in the solution. The final solution consists of the critical load P_{cr} and the deflected shape of the buckled column.

To illustrate this procedure, let us analyze an ideal elastic column that is fixed at the base, free at the top, and subjected to a vertical axial load P (Fig. 9-11a). This particular column is of historical interest because it is the one first analyzed by Euler in 1744. The deflected column is shown in Fig. 9-11b, and from this figure we can see that the bending moment at distance x from the base is

$$M = -P(\delta - v)$$

where δ is the deflection at the free end. The differential equation of the deflection curve (Eq. 9-2) then becomes

$$EIv'' = -M = P(\delta - v) \tag{a}$$

in which I is the moment of inertia for buckling in the xy plane.

Using the notation $k^2 = P/EI$ (see Eq. 9-4), we can write Eq. (a) in the form

$$v'' + k^2 v = k^2 \delta \tag{b}$$

This equation is another linear differential equation of second order with constant coefficients. However, it is more complicated than the equation for a column with pinned ends (see Eq. 9-5) because it has a nonzero term on the right-hand side. Its general solution consists of two parts: (1) the *homogeneous solution*, which is the solution of the corresponding homogeneous equation obtained by replacing the right-hand side with zero, and (2) the *particular solution*, which is a solution of the equation that produces the actual right-hand side. The homogeneous solution v_H (also called the *complementary solution*) is the same as the solution of

Eq. (9-5); hence

$$v_H = C_1 \sin kx + C_2 \cos kx \tag{c}$$

where C_1 and C_2 are arbitrary constants. When v_H is substituted into the left-hand side of the differential equation (Eq. b), it produces zero. The particular solution is

$$v_P = \delta \tag{d}$$

When v_P is substituted into the left-hand side of the differential equation, it produces the right-hand side. Therefore, the general solution of the equation is the sum of v_H and v_P:

$$v = C_1 \sin kx + C_2 \cos kx + \delta \tag{e}$$

This equation contains three unknowns, C_1, C_2, and δ, and therefore three boundary conditions are needed to obtain the solution.

At the fixed base of the column, we have two conditions:

$$v(0) = 0 \qquad v'(0) = 0$$

The first condition gives

$$C_2 = -\delta \tag{f}$$

To use the second condition, we first differentiate Eq. (e) to obtain the slope:

$$v' = C_1 k \cos kx - C_2 k \sin kx$$

Then the second condition yields $C_1 = 0$. Substituting the values of C_1 and C_2 into the general solution, we get the equation of the deflection curve:

$$v = \delta(1 - \cos kx) \tag{g}$$

This equation gives the shape of the deflection curve, but the amplitude of the deflection is undefined.

The third boundary condition is at the upper end of the column where the deflection v is equal to δ:

$$v(L) = \delta$$

Using this condition with Eq. (g), we get

$$\delta \cos kL = 0 \tag{h}$$

from which we conclude that either $\delta = 0$ or $\cos kL = 0$. If $\delta = 0$, there is no deflection of the bar, and buckling does not occur (Fig. 9-11a). In that case, Eq. (h) will be satisfied for any value of the quantity kL. Therefore, the load P can have any value also. This result is represented by the vertical axis of the load-deflection diagram shown in Fig. 9-5.

The other possibility is that $\cos kL = 0$. In this case, Eq. (h) is satisfied regardless of the value of the deflection δ; thus, δ is undefined and may have any (small) value. The condition $\cos kL = 0$ requires that

$$kL = \frac{n\pi}{2} \qquad n = 1, 3, 5, \ldots \tag{i}$$

The corresponding formula for the critical loads is

$$P_{cr} = \frac{n^2\pi^2 EI}{4L^2} \qquad n = 1, 3, 5, \ldots \tag{9-9}$$

Also, the buckled mode shapes (see Eq. g) are given by the following equation:

$$v = \delta\left(1 - \cos\frac{n\pi x}{2L}\right) \qquad n = 1, 3, 5, \ldots \tag{9-10}$$

The smallest critical load ($n = 1$) is the only load of practical interest:

$$P_{cr} = \frac{\pi^2 EI}{4L^2} \tag{9-11}$$

The corresponding buckled shape is

$$v = \delta\left(1 - \cos\frac{\pi x}{2L}\right) \tag{9-12}$$

which is shown in Fig. 9-11b. As previously mentioned, the deflection δ is undefined, and we obtain a horizontal line on the load-deflection diagram (Fig. 9-5).

By taking higher values of the index n, we obtain an infinite number of critical loads from Eq. (9-9). The corresponding buckled mode shapes have more and more waves in them. When $n = 3$, P_{cr} is nine times larger than for $n = 1$, and the buckled shape is shown in Fig. 9-11c. Similarly, the shape for $n = 5$ is pictured in part (d) of the figure.

EFFECTIVE LENGTHS OF COLUMNS

The critical loads for columns with various support conditions can be related to the critical load of a pinned-end column through the concept of an **effective length**. To explain the idea, let us observe the deflected shape of a column fixed at the base and free at the top (Fig. 9-12a). This column buckles in a curve that is one-fourth of a full sine wave. If we extend the deflection curve (Fig. 9-12b), we see that it becomes the deflection curve for a pinned-end column, or one-half of a sine wave. The effective length L_e is the length of the equivalent pinned-end column, or the distance between points of inflection in the deflection curve. Thus, for the fixed-free column, the effective length is

$$L_e = 2L \tag{j}$$

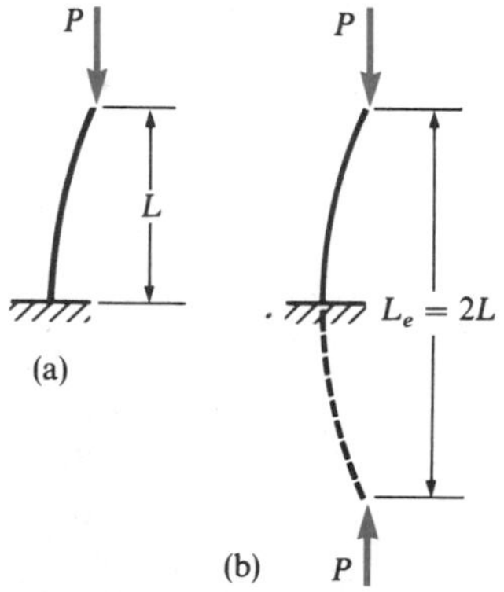

Fig. 9-12 Effective length L_e for a column fixed at the base and free at the upper end

Since the effective length is the length of an equivalent pinned-end column, we can write a general formula for the critical load as follows:

$$P_{cr} = \frac{\pi^2 EI}{L_e^2} \tag{9-13}$$

Substituting $L_e = 2L$, we get the critical load for a fixed-free column (Eq. 9-11).

The effective length is often expressed in terms of an **effective length factor** K:

$$L_e = KL \tag{9-14}$$

Hence the critical load is

$$P_{cr} = \frac{\pi^2 EI}{(KL)^2} \tag{9-15}$$

The factor K equals 2 for a column fixed at the base and free at the top and equals 1 for a pinned-end column.

COLUMN WITH FIXED ENDS

Now consider a column with both ends fixed against rotation (Fig. 9-13a). We assume that the ends of the column are free to move toward each other. Then, when the axial load P is applied at the top, an equal reactive force develops at the base. When buckling occurs, reactive moments M_0 also develop at the supports (Fig. 9-13b). The deflection curve for the first mode of buckling is a trigonometric curve having inflection points at distance $L/4$ from the ends. Thus, the effective length, equal to the distance between inflection points, is

$$L_e = \frac{L}{2} \tag{k}$$

Substituting into Eq. (9-13) gives the critical load:

$$P_{cr} = \frac{4\pi^2 EI}{L^2} \tag{9-16}$$

We see that the critical load for a column with fixed ends is four times that for a column with pinned ends. This result can also be obtained by solving the differential equation of the deflection curve (Problem 9.3-7).

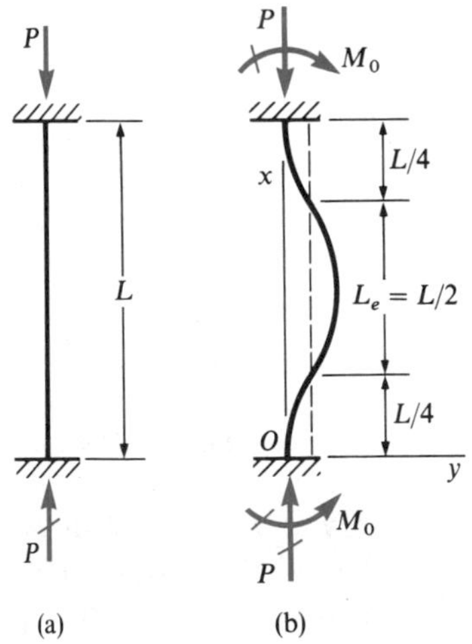

Fig. 9-13 Effective length for a column with both ends fixed against rotation

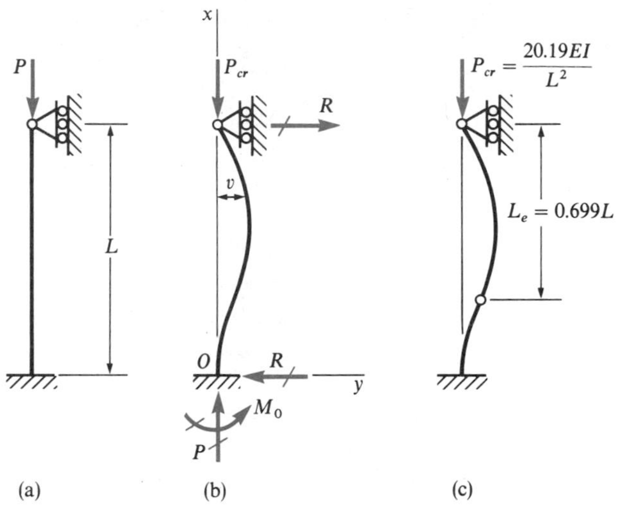

Fig. 9-14 Column fixed at the base and pinned at the top

COLUMN FIXED AT THE BASE AND PINNED AT THE TOP

The critical load and buckled mode shape for a column that is fixed at the base and pinned at the top (Fig. 9-14a) cannot be determined by inspection of the buckled mode shape (Fig. 9-14b) because the location of the point of inflection is not apparent. Hence we must solve the differential equation in order to find P_{cr}.

When the column buckles, horizontal reactive forces R develop at the supports and a reactive couple M_0 develops at the base (Fig. 9-14b). From static equilibrium we know that the horizontal forces are equal in magnitude and opposite in direction and that

$$M_0 = RL$$

The bending moment in the buckled column, at distance x from the base, is

$$M = Pv - R(L - x)$$

By proceeding as in previous analyses, we arrive at the following differential equation:

$$v'' + k^2 v = \frac{R}{EI}(L - x) \tag{l}$$

in which $k^2 = P/EI$.

The general solution of the differential equation is

$$v = C_1 \sin kx + C_2 \cos kx + \frac{R}{P}(L - x) \tag{m}$$

which has three unknown constants (C_1, C_2, and R). The three boundary conditions required are

$$v(0) = 0 \qquad v'(0) = 0 \qquad v(L) = 0$$

Applying these conditions to Eq. (m) yields

$$C_2 + \frac{RL}{P} = 0 \qquad C_1 k - \frac{R}{P} = 0 \qquad C_1 \tan kL + C_2 = 0 \tag{n}$$

All three equations are satisfied if $C_1 = C_2 = R = 0$, in which case we have the trivial solution and the deflection is zero. To obtain the solution for buckling, we must solve the equations in a more general manner. One method of solution is to eliminate R from the first two equations, which yields

$$C_1 kL + C_2 = 0$$

or $C_2 = -C_1 kL$. Now we substitute this expression for C_2 into the third of Eqs. (n) and obtain the **buckling equation**:

$$kL = \tan kL \tag{o}$$

The solution of the buckling equation gives the critical load.

Since the buckling equation is a transcendental equation,* it cannot be solved explicitly. However, the value of kL can be determined by trial and error or by using a computer program for finding roots of equations. The smallest nonzero value of kL that satisfies Eq. (o) is

$$kL = 4.4934 \tag{p}$$

The corresponding critical load is

$$P_{cr} = \frac{20.19\,EI}{L^2} = \frac{2.046\pi^2 EI}{L^2} \tag{9-17}$$

which is between the values of the critical loads for columns with pinned ends and fixed ends (see Eqs. 9-7 and 9-16). The effective length for the column is obtained by comparing Eqs. (9-17) and (9-13); thus, we find

$$L_e = 0.699L \approx 0.7L \tag{q}$$

This length represents the distance from the pinned end of the column to the point of inflection in the buckled shape (Fig. 9-14c).

The equation of the buckled mode shape is obtained by substituting $C_2 = -C_1 kL$ and $R/P = kC_1$ into the general solution (Eq. m):

$$v = C_1[\sin kx - kL \cos kx + k(L - x)] \tag{9-18}$$

* A transcendental function cannot be expressed by a finite number of algebraic operations; hence trigonometric, logarithmic, exponential, and other such functions are transcendental.

in which $k = 4.4934/L$. The term in brackets gives the mode shape for the deflection of the buckled column, but the amplitude of the deflection is undefined because C_1 may have any value (with the limitation that v must remain small).

9.4 COLUMNS WITH ECCENTRIC AXIAL LOADS

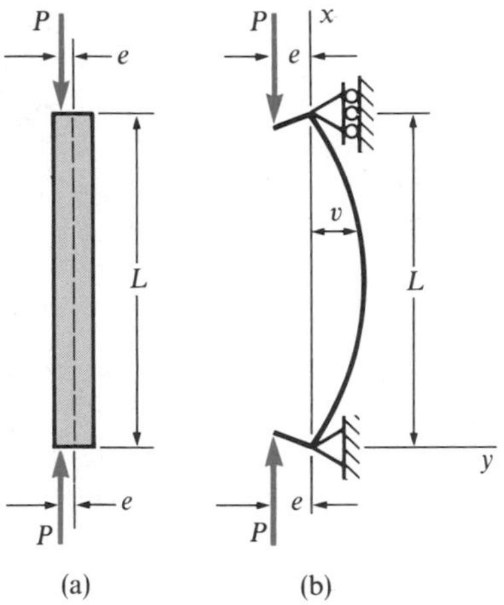

Fig. 9-15 Column with eccentric axial loads

In the preceding sections we analyzed ideal columns for which the axial load P acted at the centroid of the cross section. In such cases the column remains straight until the critical load is reached. Now we will assume that the load is applied with a small eccentricity e from the axis of the column (Fig. 9-15a). As a consequence of the eccentricity, the load P produces bending of the column even when the load is small. Therefore, the column deflects from the onset of loading, and the deflection becomes steadily larger as P increases. In this situation, the allowable load for the column may be determined by the magnitude of the deflection or the bending stress, rather than by the critical load.

To analyze the column, we represent it by the idealized pinned-end column shown in Fig. 9-15b; that is, we assume that the xy plane is a plane of symmetry, the column is initially straight, and the material is linearly elastic. The bending moment in the column at distance x from the lower end is

$$M = P(e + v)$$

where v is the deflection measured from the axis of the column. The differential equation of the deflection curve is

$$EIv'' = -M = -P(e + v)$$

or

$$v'' + k^2 v = -k^2 e \tag{a}$$

in which $k^2 = P/EI$, as before. The general solution of this equation is the sum of the homogeneous solution and the particular solution, as follows:

$$v = C_1 \sin kx + C_2 \cos kx - e \tag{b}$$

in which C_1 and C_2 are constants in the homogeneous solution and $-e$ is the particular solution.

The boundary conditions from which we obtain C_1 and C_2 are obtained from the deflection v at the ends of the column:

$$v(0) = 0 \qquad v(L) = 0$$

These conditions yield

$$C_2 = e \qquad C_1 = \frac{e(1 - \cos kL)}{\sin kL} = e \tan \frac{kL}{2}$$

Therefore, the equation of the deflection curve is

$$v = e\left(\tan\frac{kL}{2}\sin kx + \cos kx - 1\right) \tag{9-19}$$

For a column with a known load P and known eccentricity e, we can use this equation to calculate the deflection at any point. Thus, the condition of the column is quite different from what we encountered when discussing critical loads; in those cases, the magnitude of the deflection either is zero or is undefined, because at the critical load the column is in neutral equilibrium. Now, however, each value of the load P produces a definite value of the deflection. Of course, our results are still limited to small deflections.*

The maximum deflection δ occurs at the midpoint of the column and is obtained by setting x equal to $L/2$ in Eq. (9-19):

$$\delta = v_{max} = v\left(\frac{L}{2}\right) = e\left(\sec\frac{kL}{2} - 1\right) \tag{9-20}$$

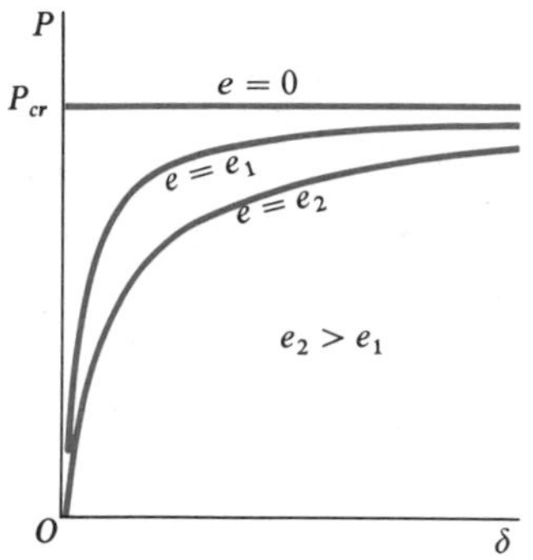

Fig. 9-16 Load-deflection diagram for a column with eccentric axial loads (Fig. 9-15)

As special cases, note that we get $\delta = 0$ if $e = 0$ or if $P = 0$.

A load-deflection diagram for the column can be plotted from Eq. (9-20). We select a particular value e_1 of the eccentricity and then calculate corresponding values of P and δ. The resulting diagram is shown in Fig. 9-16 by the curve labeled $e = e_1$. We note immediately that the deflection δ increases as P increases, but the relationship is nonlinear. Hence we cannot use the principle of superposition for calculating deflections due to more than one load. However, the deflection δ is linear with e, so the curve for $e = e_2$ has the same shape as the curve for e_1 but the abscissas are increased in the ratio e_2/e_1.

As the load P approaches the critical load ($P_{cr} = \pi^2 EI/L^2$), the value of kL approaches π and the secant term in Eq. (9-20) approaches infinity. Therefore, the deflection δ increases without limit as the load approaches P_{cr}, and the horizontal line corresponding to $P = P_{cr}$ in Fig. 9-16 is an asymptote for the curves. In the limit, as e becomes smaller and approaches zero, the curve on the diagram approaches two straight lines, one vertical and one horizontal. Thus, an ideal column with a centrally applied load is the limiting case of a column with an eccentric load. The curves plotted in Fig. 9-16 are mathematically correct, but again we must recall that the differential equation is valid only for small deflections. For large values of δ, the curves must be modified to take into account either the presence of large deflections or inelastic bending effects (see Fig. 9-8).

The reason for the nonlinear relationship between load and deflection, even when the deflections are small and Hooke's law holds, can be understood if we observe that the axial loads P (Fig. 9-15b) are equivalent

* A structural member subjected to both an axial compressive load and lateral bending loads is called a *beam-column*.

to centrally applied loads P plus couples Pe acting at the ends. The couples Pe, if acting alone, would produce bending deflections of the column in the same manner as for a beam. In a beam, the presence of the deflections does not change the action of the loads, and the bending moments are the same whether the deflections exist or not. However, when an axial load acts on the member, the presence of the deflections increases the bending moments produced by the axial forces (because of the additional moments Pv). When the moments are increased, the deflections are further increased; hence the moments increase even more, and so on. Thus, the bending moments depend upon the deflections, and their determination is part of the deflection analysis. This behavior results in a nonlinear relationship between the axial loads and the deflections.

When the axial load P is very small compared to the critical load, the load-deflection relationship can be approximated by a straight line near the origin (Fig. 9-16). The curves have a steep slope near the origin, but the slope is finite. To determine the equation of the initial part of the curves, we use the series expansion for the secant function:

$$\sec t = 1 + \frac{t^2}{2!} + \frac{5t^4}{4!} + \cdots$$

In the present case, the quantity $kL/2$ corresponds to t (see Eq. 9-20). When the axial load is very small compared to the critical load (and hence kL is very small), we can disregard the higher power terms in the series and use only the first two terms:

$$\sec\frac{kL}{2} = 1 + \frac{k^2L^2}{8}$$

Substitution of this expression into Eq. (9-20) yields

$$\delta = \frac{k^2L^2e}{8} = \frac{PeL^2}{8EI} \tag{c}$$

This equation gives the deflection at the middle of a simple beam loaded by couples Pe at the ends (see Case 10, Table G-2, Appendix G). It also gives the slope of the load-deflection curves at the origin:

$$\text{Slope at origin} = \frac{P}{\delta} = \frac{8EI}{eL^2} \tag{d}$$

If $e = 0$, the slope becomes infinite, as expected.

The maximum bending moment in the eccentrically loaded column (Fig. 9-15b) occurs at the midpoint where the deflection is a maximum; it is given by the equation

$$M_{max} = P(e + \delta) = Pe\sec\frac{kL}{2} \tag{9-21}$$

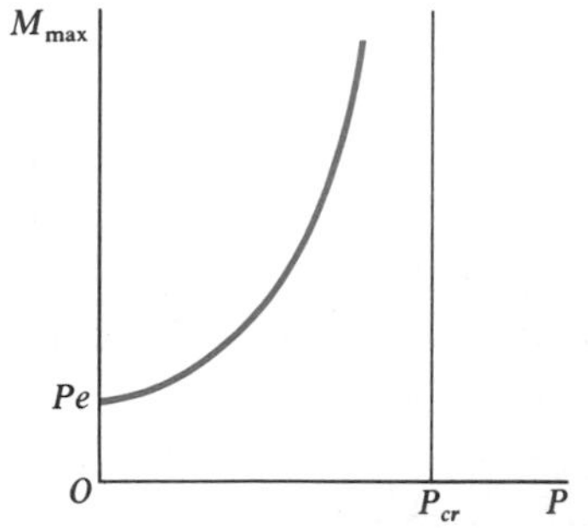

Fig. 9-17 Graph of maximum bending moment for a column with eccentric axial loads (Fig. 9-15)

The manner in which M_{max} varies as a function of the axial load P is shown in Fig. 9-17. When P is small, the maximum moment is equal to Pe, which means that the effect of the deflections is negligible. As P increases, the bending moment grows nonlinearly and theoretically becomes very large as P approaches the critical load. However, as explained before, our equations are not valid when the deflections become large and other effects become important.

9.5 SECANT FORMULA

In the preceding section we analyzed a column subjected to eccentric axial loads P (Fig. 9-15), and we determined the maximum bending moment M_{max} in the column (Eq. 9-21). The stresses in the column are of two kinds: first, the uniformly distributed normal stresses produced by the axial load, and second, the normal stresses produced by the bending moment. Because the material of the column is assumed to follow Hooke's law, the stresses due to the bending moment vary linearly across the section and can be obtained from the flexure formula. Thus, the maximum compressive stress in the column (on the concave side) is

$$\sigma_{max} = \frac{P}{A} + \frac{M_{max}}{S} \tag{a}$$

where A is the cross-sectional area and S is the section modulus. Note that for a column we assume compressive stresses to be positive. Substitution of the expression for M_{max} (see Eq. 9-21) yields

$$\sigma_{max} = \frac{P}{A} + \frac{Pe}{S} \sec\frac{kL}{2} \tag{b}$$

We can put this equation in a more useful form by making three substitutions. First, the section modulus S is replaced by I/c, where c is the distance from the centroidal axis to the extreme point on the concave side of the column. Second, we introduce the notation

$$r = \sqrt{\frac{I}{A}} \tag{9-22}$$

for the **radius of gyration** of the cross section in the plane of bending. Third, we replace k by $\sqrt{P/EI}$. With these substitutions, Eq. (b) becomes

$$\sigma_{max} = \frac{P}{A}\left[1 + \frac{ec}{r^2}\sec\left(\frac{L}{2r}\sqrt{\frac{P}{EA}}\right)\right] \tag{9-23}$$

This equation is known as the **secant formula** for an eccentrically loaded column. It gives the maximum compressive stress in the column as a function of the average compressive stress P/A and two nondimensional ratios, called the **eccentricity ratio** and the **slenderness ratio**:

$$\text{Eccentricity ratio} = \frac{ec}{r^2} \tag{9-24}$$

$$\text{Slenderness ratio} = \frac{L}{r} \tag{9-25}$$

The first of these ratios is a measure of the eccentricity of the load as compared to properties of the cross section, and the second represents the extent to which the column is long and slender. A slenderness ratio of 200 is extremely large for a column.

The secant formula relates the maximum stress in the column to the average stress P/A. If we set a limit on the maximum stress (for example, we might set σ_{max} equal to the yield stress σ_y), then we can calculate the corresponding value of P from the secant formula. Because the equation is transcendental, we must solve it by trial and error. To assist in using the formula, we can plot graphs such as the one shown in Fig. 9-18. This graph is plotted for a maximum stress $\sigma_{max} = 36$ ksi and for steel with modulus of elasticity $E = 30 \times 10^3$ ksi. The abscissa is the slenderness ratio L/r, and the ordinate is the average compressive stress P/A. Curves are plotted for several values of the eccentricity ratio ec/r^2. Of course, the secant formula is valid only when the maximum stress is less than (or, at most,

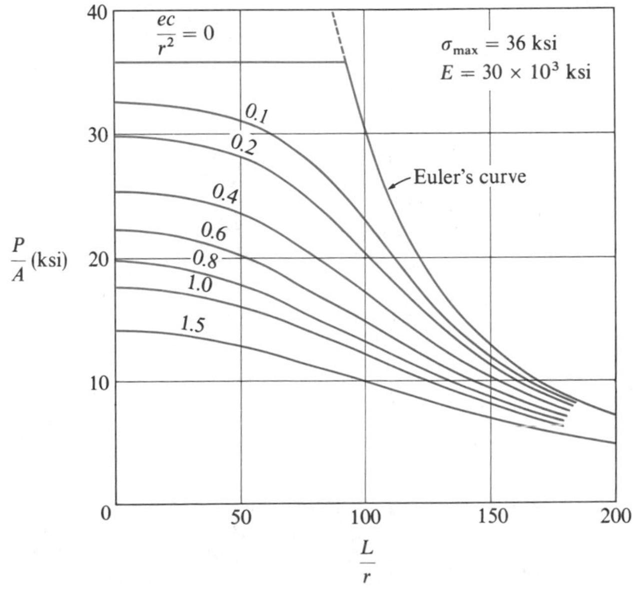

Fig. 9-18 Graph of the secant formula (Eq. 9-23) for $\sigma_{max} = 36$ ksi and $E = 30 \times 10^3$ ksi

equal to) the proportional limit of the material, because the formula was derived using Hooke's law.

When the eccentricity e is zero, the secant formula no longer applies. Instead, we have an ideal column with a centrally applied load. Then the maximum load is the critical load ($P_{cr} = \pi^2 EI/L^2$), and the corresponding critical stress is

$$\sigma_{cr} = \frac{P_{cr}}{A} = \frac{\pi^2 EI}{AL^2} = \frac{\pi^2 E}{(L/r)^2} \tag{9-26}$$

This equation shows that the critical stress is inversely proportional to the square of the slenderness ratio. Equation (9-26) is valid as long as σ_{cr} does not exceed the proportional limit of the material.

Since Eq. (9-26) relates the average stress P/A to the slenderness ratio L/r, we can also plot it on the graph shown in Fig. 9-18. The corresponding curve is labeled **Euler's curve** to distinguish it from the plots of the secant formula.* However, the curves from the secant formula approach Euler's curve as e approaches zero. As already mentioned, Euler's curve is valid only if the stress P/A is less than the proportional limit. In the present case, we are assuming that the maximum stress is 36 ksi and that this value does not exceed the proportional limit. Thus, we construct a horizontal line on the graph at this value of stress; the horizontal line and Euler's curve represent the limits of the secant-formula curves as the eccentricity e approaches zero.

Both the secant formula (Eq. 9-23) and the equation for the critical stress (Eq. 9-26) may be used with columns having other end conditions by replacing the length L with the effective length L_e (described in Section 9.3).

We see from Fig. 9-18 that the load-carrying capacity of a column decreases significantly as the slenderness ratio L/r increases, especially in the middle region of L/r values. Thus, long, slender columns are much less stable than short, stocky columns. The graph also shows that the load-carrying capacity decreases with increasing eccentricity, and this effect is relatively greater for short columns than for long ones.

The load P in the secant formula is the load that produces the maximum stress σ_{max} under ideal, linearly elastic conditions. In design, we usually assign to σ_{max} a limiting stress value, such as the yield stress σ_y. Then the corresponding value of P is the axial load that will produce that maximum stress. The allowable load P_{allow} is obtained by dividing P by a factor of safety n. Although the secant formula gives an excellent theoretical description of column behavior, difficulties arise when using it in practical design because the eccentricity e of the load may not be known accurately.

* Euler's curve is not a common geometric shape. It is sometimes mistakenly called a hyperbola, but hyperbolas are plots of polynomial equations of the second degree in two variables, whereas Euler's curve is a plot of an equation of the third degree in two variables.

Now let us consider an actual column, which inevitably differs from an ideal column because of imperfections such as initial curvature of the axis, imperfect support conditions, and nonhomogeneity of the material. Furthermore, even when the load is supposed to be a centrally applied load, unavoidable eccentricities in its point of application and direction will exist. The extent of these imperfections and eccentricities varies from one column to another and produces scatter in the results of column tests. Of course, all of these imperfections have the effect of subjecting the column to bending in addition to direct compression. Hence actual columns with centrally applied loads depart from the idealized behavior described in Sections 9.2 and 9.3. It is reasonable to assume that the behavior of an imperfect, centrally loaded column is similar to that of an ideal, eccentrically loaded column. Therefore, the secant formula is often used for the design of supposedly straight, centrally loaded columns by choosing an appropriate value of the eccentricity ratio ec/r^2 to account for the effects of all imperfections. However, the value of ec/r^2 must be based upon test results because there is no theoretical way to know what the imperfections are. For instance, a commonly used value of the eccentricity ratio for pinned-end columns in structural-steel design is $ec/r^2 = 0.25$. The use of the secant formula in this manner for columns with centrally applied loads provides a rational means of accounting for the effects of imperfections, rather than allowing for them simply by increasing the factor of safety.

The method for analyzing a column with a centrally applied load by means of the secant formula can be summarized as follows. Assume a value of the eccentricity ratio ec/r^2 based upon test results and other experience. Substitute this value into the secant formula, along with the values of L/r, A, and E. Assign a value to σ_{max} representing the yield stress σ_y of the material (or use the proportional limit). Then solve the secant formula for the load P_y that produces the stress σ_y in the column. This load is always less than the critical load P_{cr} for the column. The allowable load on the column equals the load P_y divided by a factor of safety n.

Example

A steel wide-flange column of W 14 × 82 section (Fig. 9-19) with pinned ends is 25 ft long. It supports a centrally applied load $P_1 = 320$ k and an eccentrically applied load $P_2 = 40$ k; the eccentric load acts on axis 2-2 at a distance of 13.5 in. from the centroid (Fig. 9-19b). Buckling occurs in plane 2-2. (a) Using the secant formula, calculate the maximum compressive stress in the column. (b) If the yield stress for the steel is $\sigma_y = 42{,}000$ psi, what is the factor of safety with respect to initial yielding of the steel?

(a) The two loads P_1 and P_2 acting as shown in Fig. 9-19b are statically equivalent to a single load $P = 360$ k acting with an eccentricity $e = 1.5$ in. (Fig. 9-19c). Using the properties of a W 14 × 82 section from Appendix E,

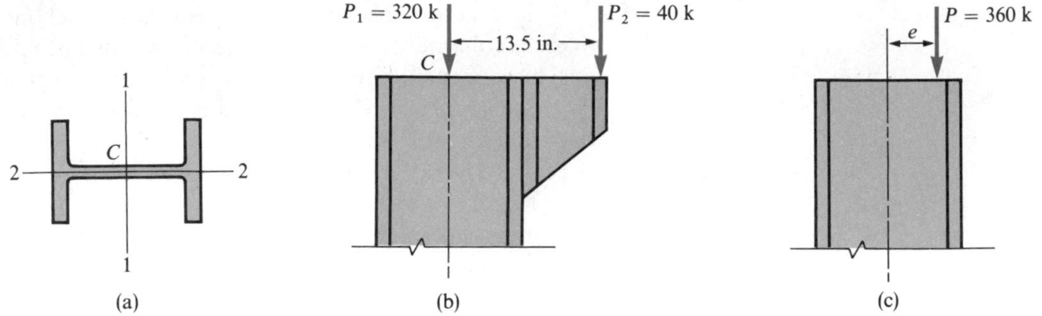

Fig. 9-19 Example. Use of the secant formula for a column with eccentric axial loads

we find

$$\frac{P}{A} = \frac{360 \text{ k}}{24.1 \text{ in.}^2} = 14.94 \text{ ksi} \qquad \frac{L}{r} = \frac{25 \text{ ft}}{6.05 \text{ in.}} = 49.59$$

$$\frac{ec}{r^2} = \frac{eA}{S} = \frac{(1.5 \text{ in.})(24.1 \text{ in.}^2)}{123 \text{ in.}^3} = 0.2939$$

Substituting these values into the secant formula, and also using $E = 30 \times 10^3$ ksi, we get

$$\sigma_{\max} = \frac{P}{A}\left[1 + \frac{ec}{r^2}\sec\left(\frac{L}{2r}\sqrt{\frac{P}{EA}}\right)\right] = 20.10 \text{ ksi}$$

Thus, this stress is the largest compressive stress in the column.

(b) We now wish to determine the load P that will bring the maximum stress up to the yield stress $\sigma_y = 42$ ksi. Since this value of the load is just sufficient to produce initial yielding of the material, we will denote it as P_y. Note that we cannot determine P_y by multiplying P by the ratio $\sigma_y/\sigma_{\max}$; the reason, of course, is that we are dealing with a nonlinear relationship between load and stress. Instead, we substitute $\sigma_{\max} = \sigma_y = 42$ ksi into the secant formula and then solve by trial and error for the corresponding load P_y. Thus, we have to find the value of P_y that satisfies the following equation:

$$42 \text{ ksi} = \frac{P_y}{24.1 \text{ in.}^2}\left[1 + 0.2939 \sec\left(24.79\sqrt{\frac{P_y}{(30{,}000 \text{ ksi})(24.1 \text{ in.}^2)}}\right)\right]$$

or

$$1012 = P_y[1 + 0.2939 \sec(0.02915\sqrt{P_y})]$$

in which P_y has units of kips. The result is

$$P_y = 716 \text{ k}$$

Since the actual load P is 360 k, we obtain

$$n = \frac{P_y}{P} = \frac{716 \text{ k}}{360 \text{ k}} = 1.99$$

as the factor of safety with respect to yielding of the column.

9.6 IMPERFECTIONS IN COLUMNS

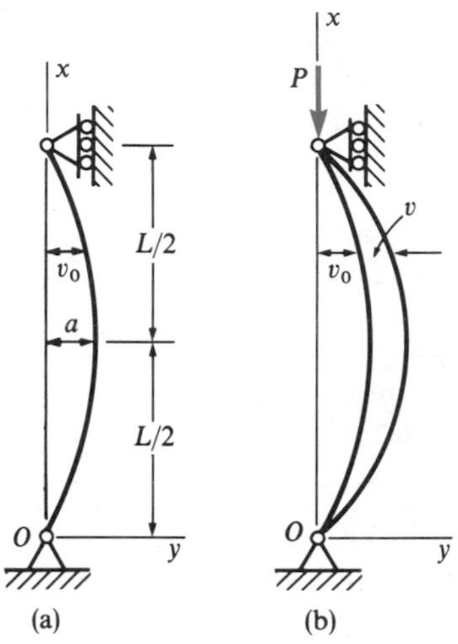

Fig. 9-20 Column with initial deflection v_0

In the discussion at the end of the preceding section, we described how to account for the effects of imperfections in the construction of the column and eccentricities in the line of application of the load by using the secant formula with an assumed value of the eccentricity ratio. Another approach to the problem is to assume that the inaccuracies are equivalent to an initial deflection, or crookedness, of the column. For a column with pinned ends, we may assume that the initial deflection v_0 of the column is a half-wave of a sine curve with maximum deflection equal to a (Fig. 9-20a):

$$v_0 = a \sin \frac{\pi x}{L} \tag{a}$$

This equation gives an initial shape that is a good approximation of any actual shape that a bent member may have. The bending moment in the column when the axial load P is applied centrally (Fig. 9-20b) then becomes

$$M = P(v_0 + v)$$

where v is the additional deflection of the column due to bending. This expression for M can be substituted into the differential equation of the deflection curve, which then can be solved in the manner described previously. The resulting expression for the deflection at the middle of the column is

$$\delta_{\max} = a + v_{\max} = \frac{a}{1 - \alpha} \tag{9-27}$$

where α is the ratio of the axial load P to the critical load for the column:

$$\alpha = \frac{P}{P_{cr}} = \frac{PL^2}{\pi^2 EI} \tag{9-28}$$

Equation (9-27) shows that the axial load causes the initial deflection of the column to increase by the factor $1/(1 - \alpha)$. Because $\alpha < 1$, this factor is always greater than unity. Note that when $\alpha = 0$ the maximum deflection is a, and when $\alpha = 1$ the maximum deflection becomes infinitely large, as expected.

The maximum bending moment in the column is

$$M_{\max} = P\delta_{\max} = \frac{Pa}{1 - \alpha} \tag{b}$$

and the maximum compressive stress is

$$\sigma_{\max} = \frac{P}{A} + \frac{M_{\max} c}{I} = \frac{P}{A}\left[1 + \frac{ac}{r^2(1 - \alpha)}\right]$$

in which $r^2 = I/A$. Substituting for α from Eq. (9-28), we get

$$\sigma_{max} = \frac{P}{A}\left[1 + \frac{\dfrac{ac}{r^2}}{1 - \dfrac{P}{\pi^2 EA}\left(\dfrac{L}{r}\right)^2}\right] \tag{9-29}$$

in which ac/r^2 is the **imperfection ratio**. If we now proceed as with the secant formula (Section 9.5) and set a limit on the maximum stress σ_{max}, we can calculate from Eq. (9-29) the corresponding value of P/A for any given ratio ac/r^2.

Equation (9-29) can be represented graphically in a manner similar to that for the secant formula. The graph of Eq. (9-29) is almost identical to that for the secant formula (Fig. 9-18) except that the term ac/r^2 replaces the eccentricity ratio ec/r^2. When $L/r = 0$, both formulas give the same value of P/A. When $L/r > 0$, the curves obtained from Eq. (9-29) are always slightly above those for the secant formula. In addition, they always are below Euler's curve (except in the limiting case when $a = 0$).

For purposes of calculation, Eq. (9-29) can be rewritten as a quadratic equation with P/A as the unknown:

$$b_1\left(\frac{P}{A}\right)^2 - b_2\left(\frac{P}{A}\right) + \sigma_{max} = 0 \tag{9-30}$$

in which

$$b_1 = \frac{1}{\pi^2 E}\left(\frac{L}{r}\right)^2 \qquad b_2 = 1 + \frac{ac}{r^2} + \frac{\sigma_{max}}{\pi^2 E}\left(\frac{L}{r}\right)^2$$

Thus, if σ_{max} is given, we can solve Eq. (9-30) for P/A. (Use the smaller of the two values obtained from the quadratic formula.) If P/A is given, we can solve Eq. (9-29) for σ_{max}. Of course, P must be less than the critical load P_{cr}. Also, Eqs. (9-29) and (9-30) are valid only for linearly elastic behavior of the material.

The initial deflection a for columns is usually in the range $L/1000$ to $L/400$. If a can be estimated, then ac/r^2 can be calculated and we can use Eq. (9-29) to obtain the maximum stress σ_{max} for any given load P. Conversely, if the maximum stress is given, we can solve Eq. (9-30) for the corresponding load P. (For further discussions of the effects of imperfections, see Refs. 9-1, 9-2, 9-8, and 9-12.)

9.7 ELASTIC AND INELASTIC COLUMN BEHAVIOR

In the preceding sections we discussed the behavior of columns when the material follows Hooke's law. We began by considering an ideal column subjected to a centrally applied load, and we arrived at the concept of a critical load P_{cr}. The behavior of an ideal column is represented by **Euler's**

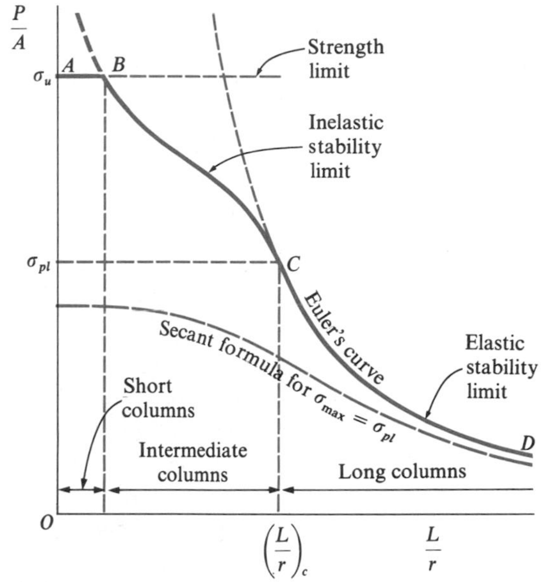

Fig. 9-21 Diagram of average compressive stress P/A versus slenderness ratio L/r

curve on a diagram of average compressive stress P/A versus slenderness ratio L/r (Fig. 9-21). This curve is valid only in the region CD below the proportional limit σ_{pl} of the material. The value of slenderness ratio above which the Euler curve applies is obtained by setting σ_{cr} equal to σ_{pl} in Eq. (9-26) and solving for L/r; thus, letting $(L/r)_c$ represent the critical value of slenderness ratio, we get

$$\left(\frac{L}{r}\right)_c = \sqrt{\frac{\pi^2 E}{\sigma_{\text{pl}}}} \qquad \text{(9-31)}$$

As an example, consider steel with $\sigma_{\text{pl}} = 36$ ksi and $E = 30{,}000$ ksi; then $(L/r)_c = 90.7$.

If we take into account the effects of eccentricities in loading or imperfections in the column, and still assume that the material follows Hooke's law, then we obtain a curve such as the one labeled **secant formula** in Fig. 9-21. This curve is plotted for a maximum stress equal to the proportional limit σ_{pl}. Thus, when comparing the curve for the secant formula with Euler's curve, we must keep in mind the distinctions between them. In the case of Euler's curve, the stress P/A not only is proportional to the applied load P but also is the maximum stress in the column when buckling occurs. Hence as we move from C to D, both the load P and the maximum stress are decreasing in the same proportions. However, for the secant formula, the load P decreases as we move from left to right along the curve, but the maximum stress does not change.

From Euler's curve, we see that columns with large slenderness ratios buckle at low values of the average compressive stress P/A. This condition cannot be improved by using a higher-strength material, because collapse is by instability of the column as a whole and not by failure of the material itself. The critical stress can be raised by reducing L/r or by using a material with higher modulus of elasticity E.

When a compression member is very short, it fails by yielding or crushing of the material; hence there are no buckling or stability considerations. In such a case, we can define a maximum compressive stress σ_u as the failure stress for the material. This stress establishes a **strength limit** for the column, represented by the horizontal line AB in the figure. This stress is much higher than the proportional limit, since it represents the ultimate stress in compression.

Between the regions of short and long columns, there is a range of intermediate slenderness ratios too small for elastic stability to govern and too large for strength considerations alone to govern. Such an intermediate-length column fails by **inelastic buckling**. Instability of the column occurs, but the maximum stresses exceed the proportional limit. The slope of the stress-strain curve diminishes after the proportional limit is passed; hence the critical load for inelastic buckling is less than the Euler load, as explained in the next section. The dividing lines between short, intermediate, and long columns are not precise; nevertheless, it is useful to distinguish between these three types of column behavior. The curve $ABCD$ in Fig. 9-21 represents the maximum load-carrying capacity of a typical column. The diagram applies to columns with any end conditions if the length L is replaced by the effective length L_e.

Tests of columns show reasonably good agreement with curve $ABCD$. When the test results are plotted on the diagram, they generally form a band that lies just below curve $ABCD$. Considerable scatter of test results is to be expected with columns, because they are sensitive to the accuracy of the construction, the alignment of loads, and the support conditions. To account for this variability, we obtain the allowable stress for a column by dividing the maximum stress (curve $ABCD$) by a suitable **factor of safety**, often having a value of about 2. Because imperfections are apt to increase as a column becomes longer, a variable factor of safety (increasing as L/r increases) is sometimes used. In Section 9.9 we will give some typical formulas for allowable stresses.

9.8 INELASTIC BUCKLING

The critical load (or Euler load) for elastic buckling is valid only for long columns, as discussed in the preceding section. For ideal columns of intermediate lengths, the stress in the column exceeds the proportional limit before buckling begins (Fig. 9-21). Hence we need a theory of **inelastic buckling** for calculating critical loads in this range of slenderness ratios.

Let us again consider an ideal pinned-end column subjected to an

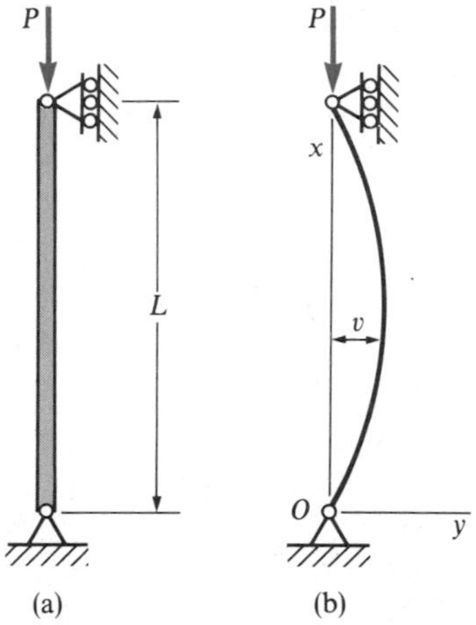

Fig. 9-22 Ideal column that buckles inelastically

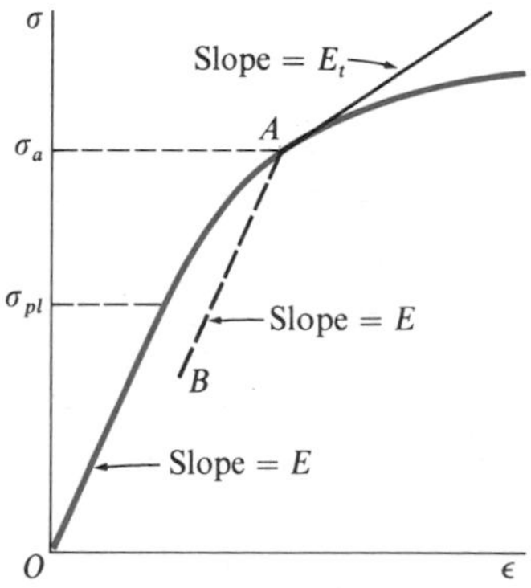

Fig. 9-23 Stress-strain diagram

axial force P (Fig. 9-22a). The column has a slenderness ratio L/r that is less than $(L/r)_c$; hence the axial stress P/A reaches the proportional limit before the elastic critical load is reached (see Fig. 9-21). The compression stress-strain diagram for the material of the column is shown in Fig. 9-23. Let us assume that the stress σ_a in the column is above the proportional limit, so that we are at a point such as A on the stress-strain diagram. Then, if a small increase in stress is produced, the relationship between the increment of stress and the corresponding increment of strain is given by the slope of the stress-strain diagram at point A. This slope, equal to the slope of the tangent line at A, is called the **tangent modulus** and is denoted by E_t; thus,

$$E_t = \frac{d\sigma}{d\epsilon} \tag{9-32}$$

The tangent modulus is a variable property of the material that decreases as the stress increases beyond the proportional limit. When the stress is below the proportional limit, E_t is the same as the elastic modulus E.

According to the **tangent-modulus theory** of inelastic buckling, the column shown in Fig. 9-22a remains straight until the critical load is reached. At that value of load, the column may undergo a small lateral deflection (Fig. 9-22b). The resulting bending stresses are superimposed upon the axial compressive stresses σ_a. Since the column starts bending from the straight position, the bending stresses initially represent only a small increment of stress. Therefore, the relationship between the bending stresses and strains is given by the tangent modulus E_t. Since the strains vary linearly across the section, the initial bending stresses also vary linearly and the expressions for curvature are the same as for elastic bending except that E_t replaces E:

$$\kappa = \frac{1}{\rho} = \frac{d^2v}{dx^2} = -\frac{M}{E_tI} \tag{9-33}$$

Because the bending moment $M = Pv$, the differential equation of the deflection curve is

$$E_tIv'' + Pv = 0 \tag{9-34}$$

This equation has the same form as the equation for elastic buckling (see Eq. 9-3) except that E_t appears in place of E. Therefore, we can solve the equation in the same manner as before and obtain the following equation for the **tangent-modulus load**:

$$P_t = \frac{\pi^2E_tI}{L^2} \tag{9-35}$$

This load represents the critical load for the column according to the tangent-modulus theory. The corresponding critical stress is

$$\sigma_t = \frac{\pi^2 E_t}{(L/r)^2} \tag{9-36}$$

which is similar in form to Eq. (9-26).

Inasmuch as the tangent modulus E_t varies with the compressive stress $\sigma = P/A$ (Fig. 9-23), we usually must find P_t by an iterative procedure. We begin by estimating the value of P_t; this trial value P_1 should be slightly larger than $\sigma_{pl}A$, which is the load when the stress reaches the proportional limit. Knowing P_1, we can calculate the corresponding axial stress $\sigma_1 = P_1/A$ and determine the tangent modulus E_t from the stress-strain diagram. Next, we use Eq. (9-35) to obtain a second estimate of P_t. Let us call this value P_2. If P_2 is close enough to P_1, then we may take the load P_2 as the tangent-modulus load P_t. However, it is more likely that additional cycles of iteration will be required until we reach a calculated load that is in close agreement with the trial load. In this manner, we arrive at a value for the tangent-modulus load.

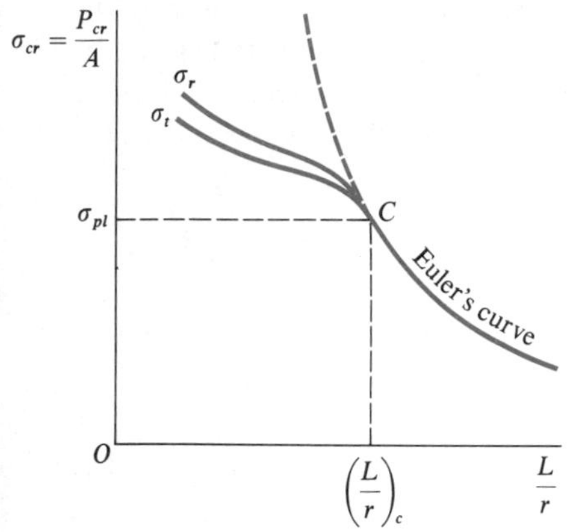

Fig. 9-24 Diagram of critical stress versus slenderness ratio

A diagram showing how the critical stress σ_t, calculated from the tangent-modulus load, varies with the slenderness ratio is given in Fig. 9-24 for a typical metal column. The tangent-modulus formulas may be used for columns with other support conditions by using the effective length L_e for L.

The tangent-modulus theory is distinguished by its simplicity and ease of use. However, it is conceptually deficient in that it does not account for the complete behavior of the column, as follows. When the column shown in Fig. 9-22 first departs from the straight position, the material on the concave side is subjected to compressive bending stresses and the material on the convex side is subjected to tensile bending stresses. These stresses are superimposed on the pre-existing large compressive stresses $\sigma_a = P/A$ at point A on the stress-strain diagram (Fig. 9-23). Hence on the concave side the compressive stresses increase and on the convex side they decrease. When the stress at point A is increased, the stress-strain relation is given by the tangent modulus E_t. However, when the stress is decreased, the material follows the unloading line AB on the stress-strain diagram. This line is parallel to the initial linear part of the diagram, and its slope is equal to the elastic modulus E. Thus, during the initial bending, the column behaves as if it were made of two different materials, a material of modulus E_t on the concave side and a material of modulus E on the convex side. A bending analysis of such a column can be made using the theory for a beam of two materials (see Section 5.11). The results of such analyses show that the column bends as though it had a modulus of elasticity intermediate between the values of E and E_t. This effective modulus is known as the **reduced modulus** E_r, and its value depends not only upon the magnitude of the stress (because E_t depends upon the magnitude of the stress) but also upon the shape of the column cross section. Thus, E_r is more difficult to determine than E_t. Hence we shall not derive formulas for E_r here but, instead, will give one result as an example. If a column has a **rectangular cross section**, the equation for the reduced modulus is

$$E_r = \frac{4EE_t}{(\sqrt{E} + \sqrt{E_t})^2} \tag{9-37}$$

Further discussion of the reduced modulus, which is also known as the *double modulus*, is given in Refs. 9-1, 9-2, and 9-10.

Since the reduced modulus represents an effective modulus that governs the bending of the column when it first departs from the straight position, we can formulate a **reduced-modulus theory** of inelastic buckling. Proceeding in the same manner as for the tangent-modulus theory, we can write an equation for the curvature and then we can write the differential equation of the deflection curve. These equations are the same as Eqs. (9-33) and (9-34) except that E_r appears instead of E_t. Thus, we arrive at the following equation for the reduced-modulus load:

$$P_r = \frac{\pi^2 E_r I}{L^2} \tag{9-38}$$

The corresponding equation for the critical stress is

$$\sigma_r = \frac{\pi^2 E_r}{(L/r)^2} \tag{9-39}$$

To find the reduced-modulus load P_r, we must use a trial-and-error procedure, because E_r depends upon E_t. The critical stress according to the reduced-modulus theory is also shown in Fig. 9-24. Note that the curve for σ_r is above that for σ_t, because E_r is always greater than E_t.

The reduced-modulus theory is difficult to use in practice because E_r depends upon the shape of the cross section as well as the stress-strain curve and must be evaluated for each particular column. (The only exception is a rectangular cross section, for which Eq. 9-37 is available.) Moreover, this theory also has a conceptual defect. In order for the reduced modulus E_r to apply, the material on the convex side of the column must be undergoing a reduction in stress. But such a reduction in stress cannot occur until bending takes place. Therefore, the load P, applied to an ideal straight column, can never reach the reduced-modulus load P_r; to reach that load would require that bending already exists, which is a contradiction.

Thus, neither the tangent-modulus theory nor the reduced-modulus theory is adequate by itself to explain the phenomenon of inelastic buckling, although an understanding of both theories is necessary in order to develop a more complete and logically consistent theory. Such a theory was developed by F. R. Shanley (see the historical note that follows), and it is called the **Shanley theory of inelastic buckling**. This

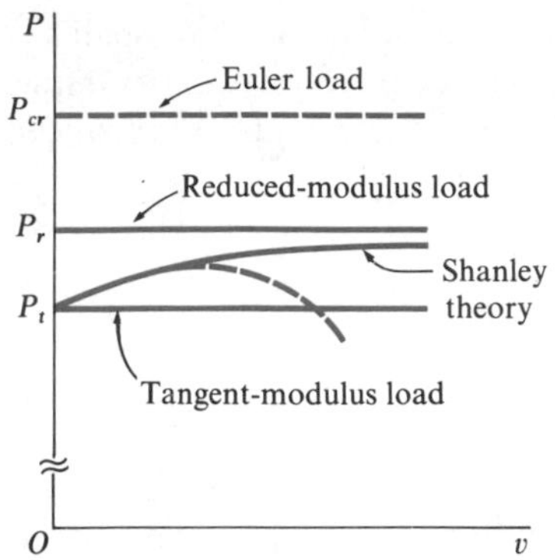

Fig. 9-25 Load-deflection diagram for elastic and inelastic buckling

theory overcomes the difficulties with both the tangent-modulus and reduced-modulus theories by recognizing that it is not possible to have buckling of the Euler type when the column becomes inelastic. In Euler buckling, a critical load is reached at which the column is in neutral equilibrium, represented by a horizontal line on the load-deflection diagram (Fig. 9-25). As already explained, neither the tangent-modulus load P_t nor the reduced-modulus load P_r can represent this type of behavior; in both cases, we are led to contradictions if we try to associate these loads with a condition of neutral equilibrium. Instead, we must think of a column that has an ever-increasing axial load. When the load reaches P_t, bending can begin *provided the load continues to increase.* Under these conditions, bending occurs simultaneously with an increase in load, resulting in a decrease in strain on the convex side of the column. Thus, the effective modulus of the material throughout the cross section becomes greater than E_t, and therefore an increase in load is possible. However, the effective modulus is not as great as E_r, because E_r is based upon full strain reversal on the convex side of the column. In other words, E_r is based upon the amount of strain reversal that exists if the column bends without a change in the axial force, whereas the presence of an increasing axial force means that the reduction in strain is not as great. Thus, instead of neutral equilibrium, where the relationship between load and deflection is undefined, we now have a definite relationship between each value of the load and the corresponding deflection. This behavior is shown by the curve labeled *Shanley theory* in Fig. 9-25. Note that buckling begins at the tangent-modulus load; then the load increases but does not reach the reduced-modulus load until the deflection becomes infinite. However, other effects become important as the deflection increases, and in reality the curve eventually goes downward as shown by the dashed line.

The Shanley concept of inelastic buckling has been verified by numerous investigators and by many tests. However, the maximum load attained in real columns (see the dashed curve in Fig. 9-25) is only slightly above the tangent-modulus load P_t. Fortunately, the tangent-modulus load is very simple to calculate, and for all practical purposes it is reasonable to adopt the tangent-modulus load as the critical load for inelastic buckling of a column.

The preceding discussions of elastic and inelastic buckling are based upon idealized conditions. The theoretical concepts are important in understanding column behavior, but the design of actual columns must take into account additional factors not considered in the theory. For instance, steel columns always contain residual stresses from the rolling process. Because of these stresses, which vary greatly in different parts of the cross section, the stress level required to produce yielding also varies over the cross section. For such reasons a variety of empirical design formulas are used for designing columns. Some of the most commonly used formulas are given in the next section.

HISTORICAL NOTE

The critical-load formulas for elastic buckling originated with Euler in 1744, as explained in the note to Ref. 9-3. Very little additional progress was made for a hundred years. Then, in 1845, A. H. E. Lamarle pointed out that Euler's formula should be used only for slenderness ratios beyond a certain limit and that experimental data should be relied upon for smaller ratios (Ref. 9-13). In 1889, the French engineer A. Considère performed a series of 32 tests on columns (Ref. 9-14). He observed that the stresses on the concave side of the column increased with E_t and the stresses on the convex side decreased with E. Thus, he showed why the Euler formula was not applicable to inelastic buckling, and he stated that the effective modulus was between E and E_t. Although he made no attempt to evaluate the effective modulus, he is responsible for beginning the reduced-modulus theory. In the same year, and quite independently, the German engineer F. Engesser suggested the tangent-modulus theory (Ref. 9-15). He denoted the tangent modulus by the symbol T (equal to $d\sigma/d\epsilon$) and proposed that T be substituted for E in Euler's formula for the critical load. Later, in March 1895, Engesser again presented the tangent-modulus theory (Ref. 9-16), obviously without knowledge of Considère's work. Three months later, Polish-born F. Jasinsky, then a professor in St. Petersburg, pointed out that Engesser's tangent-modulus theory was incorrect, called attention to Considère's work, and presented the reduced-modulus theory (Ref. 9-17). He also stated that the reduced modulus could not be calculated theoretically. In response, and only one month later, Engesser acknowledged the error in the tangent-modulus approach and showed how to obtain the reduced modulus for any cross section (Ref. 9-18). Thus, the tangent-modulus theory is sometimes called the *Engesser theory*, and the reduced-modulus theory is called the *Considère-Engesser theory*.

The reduced-modulus theory was also presented by the famous scientist Theodore von Kármán in 1908 and 1910 (Refs. 9-19 and 9-20), apparently independently of the earlier investigations. In the latter paper, he derived formulas for E_r for rectangular and idealized wide-flange sections (that is, sections without a web). He extended the theory to include the effects of eccentricities on the buckling load, and he showed that the maximum load decreases rapidly as the eccentricity increases. The English elastician R. V. Southwell also presented the reduced-modulus theory. In a paper in 1912 (Ref. 9-22), he derived the theory but used a modified length for the column instead of a reduced modulus for the material. His work appears to be independent of the others, although the basic concepts are the same. The reduced-modulus theory was the accepted theory of inelastic buckling until 1946, when the American aeronautical-engineering professor F. R. Shanley pointed out the logical paradoxes in both theories. In a remarkable one-page paper (Ref. 9-23), Shanley not only explained what was wrong with the generally accepted theories but also proposed his own theory that resolved the paradoxes. In a second paper, five months later, he gave further analyses to support his earlier theory and gave results from tests on columns (Ref. 9-24). Since that time, many other investigators have confirmed and expanded Shanley's concept. The background of the inelastic-buckling theories is described in Refs. 9-1, 9-2, and 9-8. For excellent discussions of the column-buckling problem, see the comprehensive papers by Hoff (Refs. 9-6 and 9-25), and for a historical account, see the paper by Johnston (Ref. 9-7).

9.9 COLUMN DESIGN FORMULAS

In the preceding sections we discussed the load-carrying capacity of columns based only upon theoretical considerations. The next step is to determine allowable loads on columns considering not only the theoretical results but also the behavior of real columns as observed in laboratory tests. A common procedure is to employ empirical design formulas that fit the test data in the inelastic range of column behavior (low values of slenderness ratio) and to use the Euler formula for the critical load in the elastic range (high values of slenderness ratio). A factor of safety must be applied to obtain allowable loads from the maximum loads (or allowable stresses from the maximum stresses). The use of empirical design formulas is quite satisfactory within the limits for which they were established, provided there is adequate experimental data. Thus, the following restrictions should be noted when using a column design formula: (1) The formula is valid only for a particular material. (2) The formula is valid only for a specified range of slenderness ratios. (3) The formula may specify the allowable stress or it may specify the maximum stress; if the latter, a factor of safety must be applied to obtain the allowable stress.

The following examples of column design formulas are applicable to centrally loaded columns. However, many factors besides those discussed here enter into the design of columns, so other references should be consulted before designing a column for a specific application.

Column design using these formulas requires trial-and-error calculations. Knowing the axial compressive load, we begin by estimating the allowable stress σ_{allow} and calculating approximately the required cross-sectional area. Then a section is selected from tables of available sizes. This section is checked with the aid of the appropriate design formulas to see if it is adequate to support the load. If it is not, then a heavier section is selected and the process repeated; if it appears to be overdesigned, a lighter section is selected and checked.

STRUCTURAL STEEL

For the design of centrally loaded, structural-steel columns, the Structural Stability Research Council (SSRC) proposed design formulas that are in common use today (see Ref. 9-8). The SSRC formulas give the maximum stress or critical stress in the column (that is, the stress obtained by dividing the assumed maximum load that the column can carry by the cross-sectional area).

When the slenderness ratio L/r is large, the maximum stress is based upon the Euler load:

$$\sigma_{max} = \frac{\pi^2 E}{(KL/r)^2} \tag{a}$$

where the effective length KL (see Eqs. 9-14 and 9-15) is used so that the formula may be applied to a variety of support conditions. Equation (a) is valid only when the stresses in the column do not exceed the proportional limit σ_{pl}, as discussed previously (see Fig. 9-21). For structural steel, we normally assume that the proportional limit is equal to the yield stress σ_y. However, rolled sections (such as wide-flange sections) have large residual stresses in them; the compressive residual stresses may be as large as one-half the yield stress. Hence for such a column, the proportional limit is

$$\sigma_{pl} = \sigma_y - \sigma_{rc} = 0.5\sigma_y \tag{b}$$

where σ_{rc} represents the compressive residual stress in the column and is assumed to be equal to $0.5\sigma_y$.

To determine the smallest slenderness ratio for which Eq. (a) is applicable, we set σ_{max} equal to σ_{pl} (Eq. b) and solve for the corresponding value of KL/r, denoted by $(KL/r)_c$:

$$\left(\frac{KL}{r}\right)_c = \sqrt{\frac{2\pi^2 E}{\sigma_y}} \tag{9-40}$$

If the effective slenderness ratio is equal to or larger than $(KL/r)_c$, the Euler formula for the critical stress (Eq. a) may be used. This equation may be expressed in a convenient nondimensional form by dividing by the yield stress σ_y and also substituting from Eq. (9-40):

$$\frac{\sigma_{max}}{\sigma_y} = \frac{\pi^2 E}{\sigma_y\left(\frac{KL}{r}\right)^2} = \frac{\left(\frac{KL}{r}\right)_c^2}{2\left(\frac{KL}{r}\right)^2} \qquad \frac{KL}{r} \geq \left(\frac{KL}{r}\right)_c \tag{9-41}$$

This equation is plotted in Fig. 9-26 and labeled *Euler's curve.*

For the region of inelastic buckling, that is, the region where $KL/r \leq (KL/r)_c$, the maximum stress is given by a parabolic curve:

$$\frac{\sigma_{max}}{\sigma_y} = 1 - \frac{\left(\frac{KL}{r}\right)^2}{2\left(\frac{KL}{r}\right)_c^2} \qquad \frac{KL}{r} \leq \left(\frac{KL}{r}\right)_c \tag{9-42}$$

This empirical formula is also plotted in Fig. 9-26. The curve has a horizontal tangent at $KL/r = 0$, where the maximum stress is equal to σ_y.

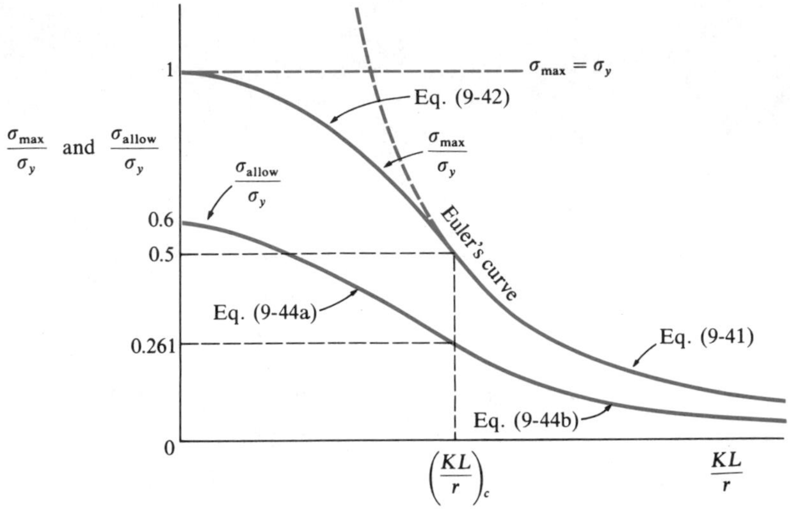

Fig. 9-26 Design formulas for structural-steel columns

At $KL/r = (KL/r)_c$, the curve merges smoothly with Euler's curve (both curves have the same slope at the point where they meet).

The American Institute of Steel Construction (AISC), an organization that prepares specifications for structural steel design, has adopted the formulas for σ_{max} proposed by the SSRC.* To obtain allowable stresses, the AISC specifies that the maximum stresses be divided by the following factors of safety:

$$n_1 = \frac{5}{3} + \frac{3\left(\frac{KL}{r}\right)}{8\left(\frac{KL}{r}\right)_c} - \frac{\left(\frac{KL}{r}\right)^3}{8\left(\frac{KL}{r}\right)_c^3} \qquad \frac{KL}{r} \le \left(\frac{KL}{r}\right)_c \qquad \text{(9-43a)}$$

$$n_2 = \frac{23}{12} \approx 1.92 \qquad \frac{KL}{r} \ge \left(\frac{KL}{r}\right)_c \qquad \text{(9-43b)}$$

Thus, the factor of safety is 5/3 when $KL/r = 0$ and gradually increases to 23/12 when $KL/r = (KL/r)_c$. For higher slenderness ratios, the factor of safety remains constant at that value. The AISC formulas for the allowable stresses are obtained by dividing the maximum stresses σ_{max} by the appropriate factors of safety; thus,

$$\frac{\sigma_{allow}}{\sigma_y} = \frac{1}{n_1}\left[1 - \frac{\left(\frac{KL}{r}\right)^2}{2\left(\frac{KL}{r}\right)_c^2}\right] \qquad \frac{KL}{r} \le \left(\frac{KL}{r}\right)_c \qquad \text{(9-44a)}$$

* The specifications of the AISC are given in Ref. 5-4.

$$\frac{\sigma_{\text{allow}}}{\sigma_y} = \frac{\left(\frac{KL}{r}\right)_c^2}{2n_2\left(\frac{KL}{r}\right)^2} \qquad \frac{KL}{r} \geq \left(\frac{KL}{r}\right)_c \tag{9-44b}$$

These equations for the allowable stresses are also plotted in Fig. 9-26. They may be used with either USCS or SI units. The maximum value of KL/r permitted by the AISC specifications is 200; also, the modulus of elasticity E is specified as 29×10^6 psi. Finally, we note that the AISC uses the symbols C_c for $(KL/r)_c$, F_a for σ_{allow}, and F_y for σ_y.

ALUMINUM

The design of aluminum columns is in accord with the specifications of the Aluminum Association (Ref. 5-5) and follows a pattern similar to that for structural steel columns. Euler's curve is used for large values of slenderness ratio. For smaller slenderness ratios, two straight lines, one horizontal and one inclined, are used (Fig. 9-27). The slenderness ratios that separate short, intermediate, and long columns are denoted by S_1 and S_2. The formulas for the allowable stresses are as follows:

$$\sigma_{\text{allow}} = \frac{\sigma_y}{k_c n_y} \qquad 0 \leq \frac{L}{r} \leq S_1 \tag{9-45a}$$

$$\sigma_{\text{allow}} = \frac{1}{n_u}\left(B_c - D_c\frac{L}{r}\right) \qquad S_1 \leq \frac{L}{r} \leq S_2 \tag{9-45b}$$

$$\sigma_{\text{allow}} = \frac{\pi^2 E}{n_u\left(\frac{L}{r}\right)^2} \qquad S_2 \leq \frac{L}{r} \tag{9-45c}$$

In these equations, the length L is defined as the distance between points of lateral support, or twice the length of a cantilever column. Thus, L may be interpreted as the effective length. The stress σ_y is the compressive yield

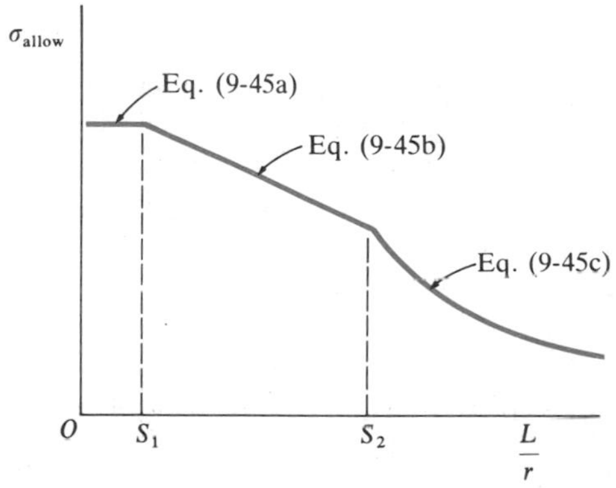

Fig. 9-27 Design formulas for aluminum columns

stress (0.2% offset), k_c is a constant, n_y is the factor of safety with respect to the yield stress, n_u is the factor of safety with respect to the ultimate stress, and B_c, D_c, S_1, and S_2 are constants. The values of these quantities vary according to the alloy, temper, and usage. Numerous aluminum alloys and tempers are available, so the Aluminum Association specifications give tables of the various constants according to the material. As examples, the following are some typical formulas for alloys used in buildings and aircraft structures.

1. Alloy 2014-T6

$$\sigma_{\text{allow}} = 28 \text{ ksi} \qquad 0 \le \frac{L}{r} \le 12 \tag{9-46a}$$

$$\sigma_{\text{allow}} = 30.7 - 0.23\left(\frac{L}{r}\right) \text{ksi} \qquad 12 \le \frac{L}{r} \le 55 \tag{9-46b}$$

$$\sigma_{\text{allow}} = \frac{54{,}000 \text{ ksi}}{\left(\dfrac{L}{r}\right)^2} \qquad 55 \le \frac{L}{r} \tag{9-46c}$$

2. Alloy 6061-T6

$$\sigma_{\text{allow}} = 19 \text{ ksi} \qquad 0 \le \frac{L}{r} \le 9.5 \tag{9-47a}$$

$$\sigma_{\text{allow}} = 20.2 - 0.126\left(\frac{L}{r}\right) \text{ksi} \qquad 9.5 \le \frac{L}{r} \le 66 \tag{9-47b}$$

$$\sigma_{\text{allow}} = \frac{51{,}000 \text{ ksi}}{\left(\dfrac{L}{r}\right)^2} \qquad 66 \le \frac{L}{r} \tag{9-47c}$$

These formulas give the allowable stresses; hence they incorporate the safety factors, which vary from 1.65 to 2.20.

WOOD

The design of wood structural members is governed by the *National Design Specification for Wood Construction*, published by the National Forest Products Association (Ref. 5-7). The formulas for allowable stresses in rectangular wood columns for short, intermediate, and long columns are as follows:

$$\sigma_{\text{allow}} = F_c \qquad 0 \le \frac{L}{d} \le 11 \tag{9-48a}$$

$$\sigma_{\text{allow}} = F_c\left[1 - \frac{1}{3}\left(\frac{L/d}{K}\right)^4\right] \qquad 11 < \frac{L}{d} \le K \tag{9-48b}$$

$$\sigma_{\text{allow}} = \frac{0.3E}{(L/d)^2} = \frac{2F_c}{3}\left(\frac{K}{L/d}\right)^2 \qquad K \le \frac{L}{d} \le 50 \qquad \text{(9-48c)}$$

The stress F_c is the design value in compression parallel to the grain as modified for duration of loading and other service conditions; typical values of F_c for structural lumber are in the range of 700 to 1800 psi. The slenderness ratio L/d is the effective length L of the column for buckling in a principal plane divided by the cross-sectional dimension d in that same plane. The factor K is the slenderness ratio that separates intermediate from long columns:

$$K = \sqrt{\frac{0.45E}{F_c}} \qquad \text{(9-49)}$$

Values of K are usually in the range from 18 to 30. The modulus of elasticity E, like the compressive stress F_c, varies with species and grade, but typical values are in the range of 1×10^6 to 1.8×10^6 psi. A graph of Eqs. (9-48a, b, and c) is shown in Fig. 9-28 for $E/F_c = 1500$. Equations (9-48) may be used with either USCS or SI units.

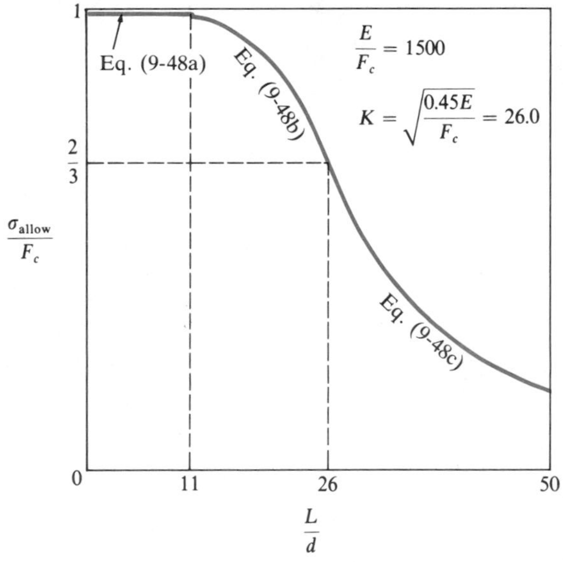

Fig. 9-28 Design formulas for rectangular wood columns with $E/F_c = 1500$

The column formulas for steel, aluminum, and wood given in this section are for use in solving problems, but they should not be used by inexperienced persons for the design of actual columns, because they represent only a small part of the complete design process for columns. All design formulas taken from specifications and codes require informed judgment in their use. There are many cases of structures that "met the building code" but nevertheless collapsed or failed to perform adequately. In addition, codes specify numerous other limitations that are not described here.

PROBLEMS / CHAPTER 9

The problems for Section 9.1 are to be solved using the assumptions that the bars are rigid, the springs are linearly elastic, and the displacements and angles of rotation are small.

9.1-1 Determine the critical load P_{cr} for the bar-spring system shown in the figure. The bar is free at B and supported at A by a rotational spring of stiffness α, that is, $M = \alpha\theta$, where M is the moment acting on the spring and θ is the angle of rotation of the bar.

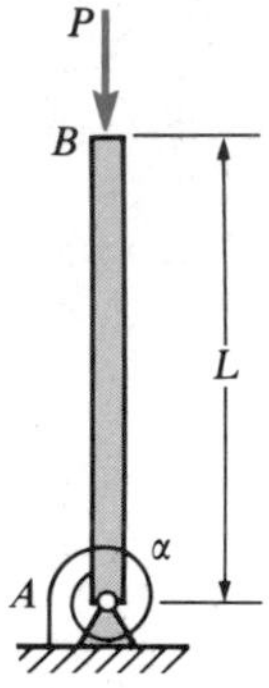

Prob. 9.1-1

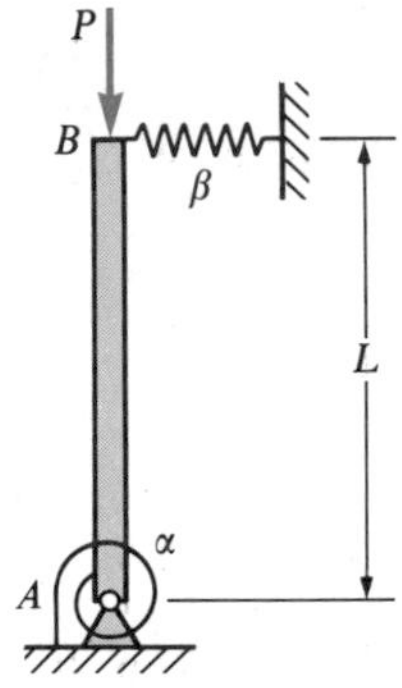

Prob. 9.1-2

9.1-2, 9.1-3, 9.1-4, and 9.1-5 Determine the critical load P_{cr} for the bar-spring system shown in the figure.

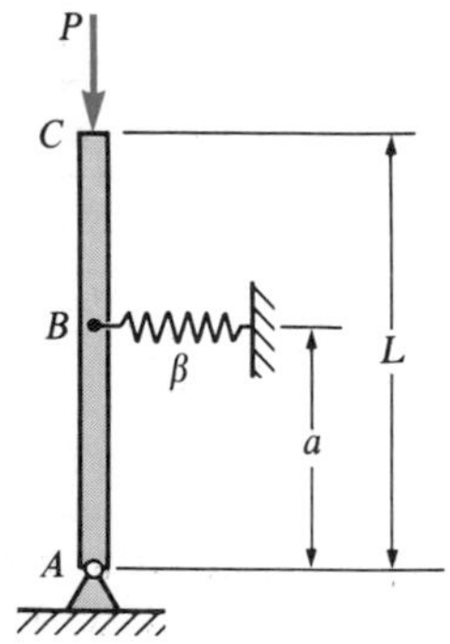

Prob. 9.1-3

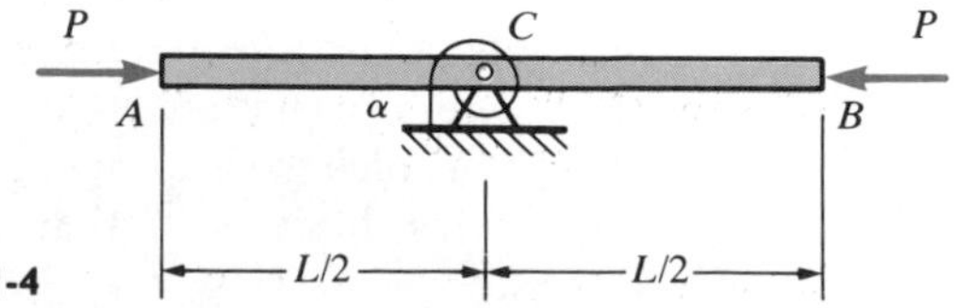

Prob. 9.1-4

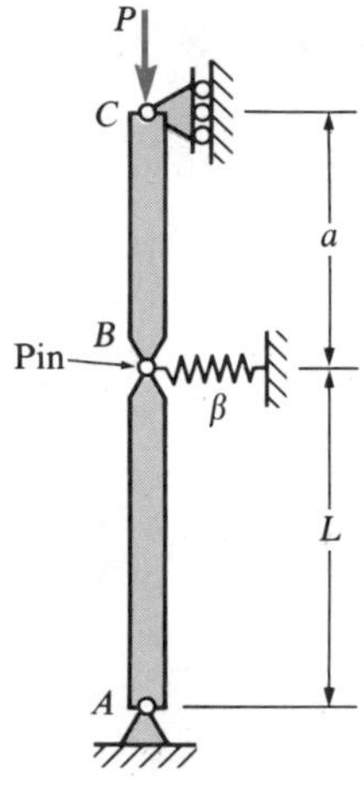

Prob. 9.1-5

The problems for Section 9.2 are to be solved using the assumptions of ideal, slender, prismatic, elastic columns. Buckling occurs in the plane of the figure unless otherwise stated.

9.2-1 Calculate the critical load P_{cr} for a W 10 × 60 steel column having length $L = 18$ ft and $E = 30 \times 10^6$ psi. Assume that the column has pinned ends and may buckle in any direction.

9.2-2 Solve the preceding problem for a W 12 × 87 steel column having length $L = 22$ ft.

9.2-3 A pinned-end column must support an axial load $P = 185$ k with a factor of safety $n = 2.5$ with respect to buckling. The column has length $L = 15$ ft and may buckle in any direction. Select from Table E-1 in Appendix E the lightest steel wide-flange section ($E = 30 \times 10^6$ psi) that will support the load.

9.2-4 Solve the preceding problem for a column with $P = 280$ k and $L = 20$ ft.

9.2-5 Three identical circular rods, each of radius r and length L, are placed together to form a compression member (see figure for cross section). Assuming pinned-end

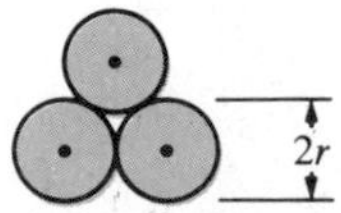

Prob. 9.2-5

conditions, determine the critical load P_{cr} as follows: (a) the rods are securely joined by epoxy so that they function as a single member, and (b) the rods act independently.

9.2-6 A pinned-end strut of aluminum ($E = 73$ GPa) with length $L = 2$ m is constructed of circular tubing with outside diameter $d = 50$ mm. The strut must resist an axial load $P = 15$ kN with a factor of safety $n = 2$ with respect to buckling. Determine the required thickness t of the tube.

9.2-7 Solve the preceding problem for a steel pipe section ($E = 30 \times 10^6$ psi) with length $L = 22$ ft, outside diameter $d = 5.5$ in., and $P = 30$ k.

9.2-8 The cross section of a column built up of two steel I-beams (S 8 × 18.4 sections) is shown in the figure. The sections are connected by spacer bars (represented by the dashed lines) to ensure that they act together as a single column. The column has pinned ends and may buckle in any direction. Assuming $E = 30 \times 10^6$ psi and $L = 20$ ft, calculate the critical load P_{cr} for the column.

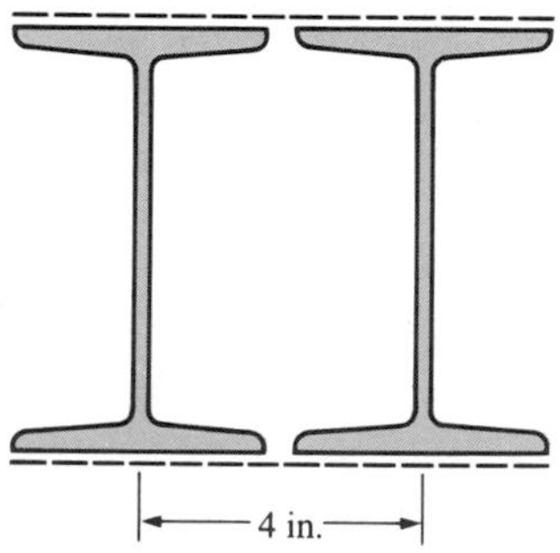

Prob. 9.2-8

9.2-9 An equal-leg angle section of length $L = 9$ ft is to be used as a pinned-end strut. The angle is supported only at the ends; hence it is free to buckle in any direction. The axial load on the angle is $P = 22$ k, and a factor of safety $n = 2.5$ against buckling is required. Select from Table E-4 in Appendix E the lightest steel angle section ($E = 30 \times 10^6$ psi) that is satisfactory.

9.2-10 Solve the preceding problem for an angle with $L = 8$ ft and $P = 60$ k.

9.2-11 Three pinned-end columns of the same material have the same length and the same cross-sectional area. The columns are free to buckle in any direction. The columns have cross sections as follows: (1) equilateral triangle, (2) square, and (3) circle. Determine the ratios $P_1 : P_2 : P_3$ of the critical loads for these columns.

9.2-12 A rectangular column with cross-sectional dimensions b and h is pin-supported at ends A and C (see figure). The column is restrained in the plane of the figure at midheight but is free to deflect perpendicular to the plane of the figure (except at ends A and C). Determine the ratio h/b such that the critical load is the same for buckling in the two principal planes of the column.

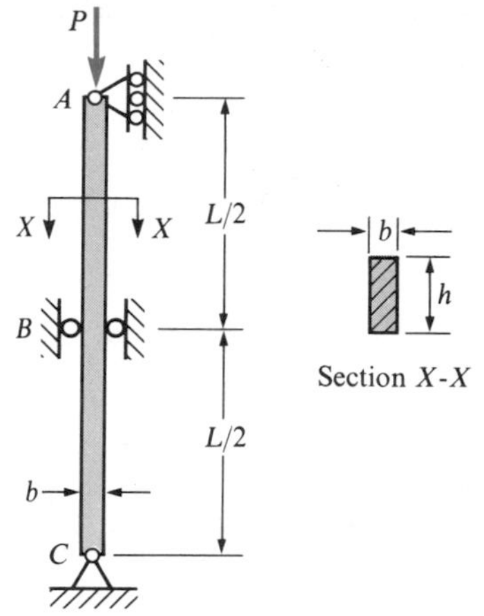

Prob. 9.2-12

9.2-13 A horizontal bar AB is supported by a pinned-end column CD as shown in the figure. The column is a steel bar ($E = 30 \times 10^6$ psi) of square cross section having length $L = 10$ ft and side dimensions $b = 2$ in. Calculate the allowable load Q if the factor of safety with respect to buckling of the column is $n = 3$.

9.2-14 A horizontal bar AB is supported by a pinned-end column CD as shown in the figure. The column is a steel

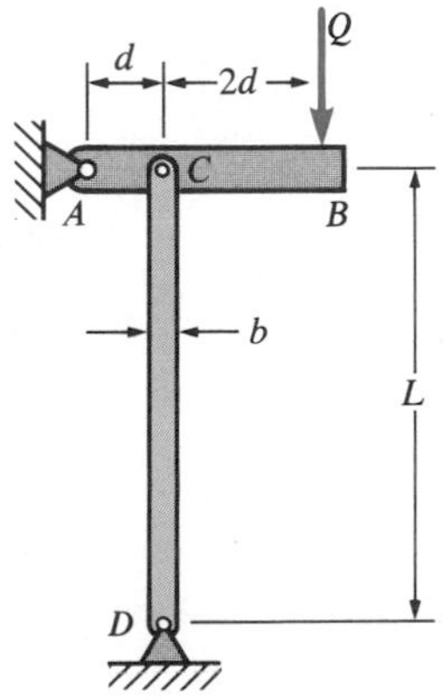

Probs. 9.2-13 and 9.2-14

bar ($E = 200$ GPa) of square cross section having length $L = 3$ m and side dimensions $b = 50$ mm. Calculate the allowable load Q if the factor of safety with respect to buckling of the column is $n = 2.5$.

9.2-15 A horizontal bar AB is pin-supported at end A and carries a load Q at end B, as shown in the figure. It is supported at C and D by two identical pinned-end columns of length L. Each column has flexural rigidity EI. At what load Q_{cr} does the system collapse by buckling of the columns?

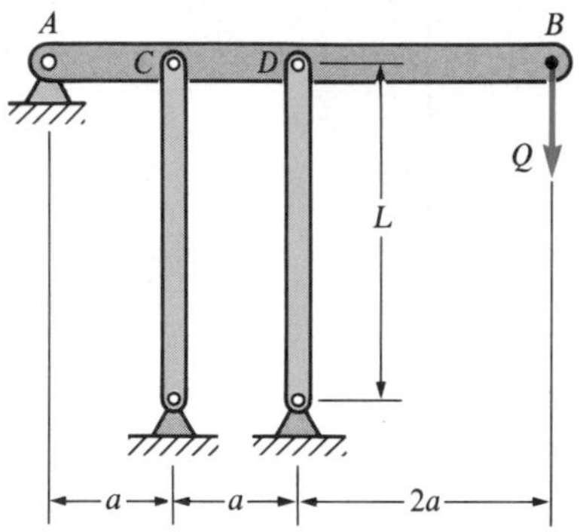

Prob. 9.2-15

9.2-16 A slender bar AB with pinned ends is held between immovable supports (see figure). What increase ΔT in the temperature of the bar will produce buckling?

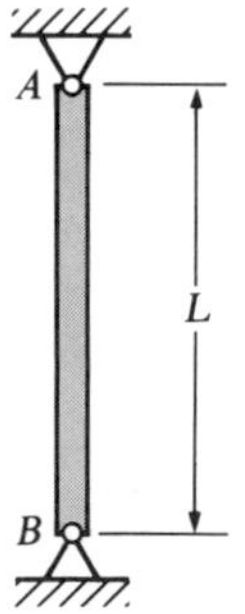

Prob. 9.2-16

9.2-17 A long slender column ABC is pinned at ends A and C and compressed by an axial force P (see figure). At the midpoint B, lateral support is provided to prevent deflection in the plane of the figure; however, deflection is free to occur perpendicular to the plane of the figure. The column is a steel wide-flange section (W 10 × 45) with $E = 30 \times 10^6$ psi. The distance between supports is $L = 18$ ft. Calculate the allowable load P using a factor of safety $n = 2.5$, taking into account the possibility of buckling about either principal centroidal axis 1-1 or 2-2.

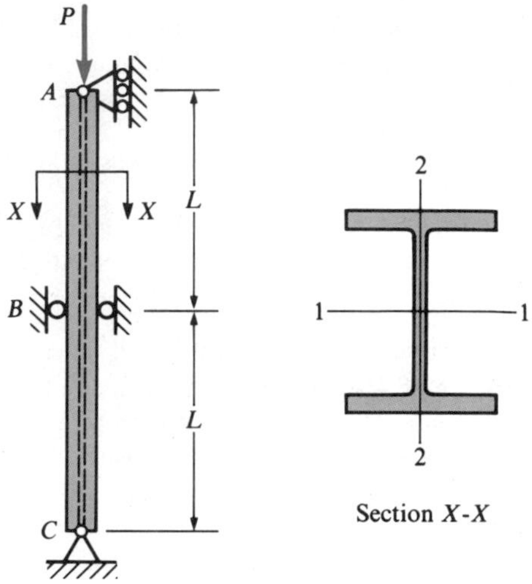

Prob. 9.2-17

9.2-18 The truss ABC shown in the figure supports a vertical load P. Each member is a slender circular steel pipe ($E = 200$ GPa) with outside diameter 100 mm and wall thickness 6.0 mm. Determine the allowable load P_{allow} if the factor of safety with respect to buckling of the members is $n = 2$ and the distance between supports is $L = 7$ m.

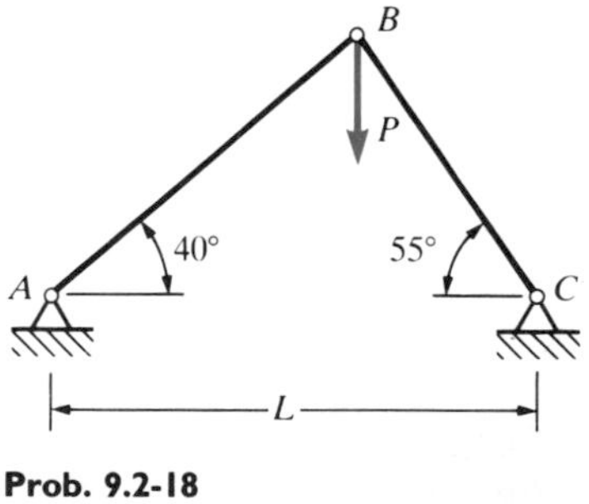

Prob. 9.2-18

***9.2-19** A pin-connected truss ABC (see figure) is composed of two bars of the same material with identical cross sections. A load P is applied at joint B at an angle θ from the prolongation of line AB. The angle θ may be varied from 0 to 90°. Assuming that collapse occurs by buckling of the bars, obtain a formula for the angle θ such that the load P will be a maximum.

***9.2-20** A truss ABC supports a load W at joint B, as shown in the figure. The length L_1 of member AB is fixed, but the

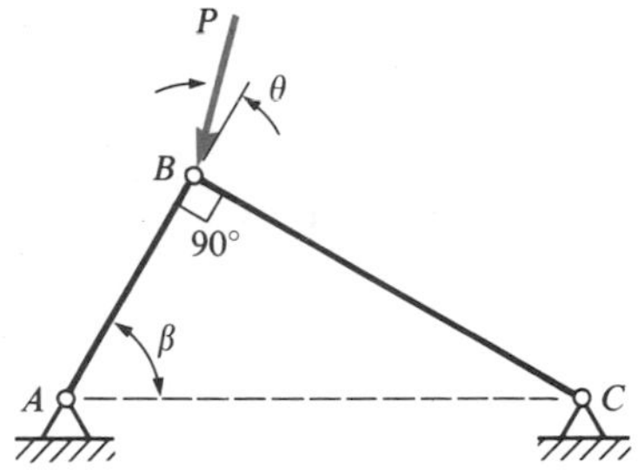

Prob. 9.2-19

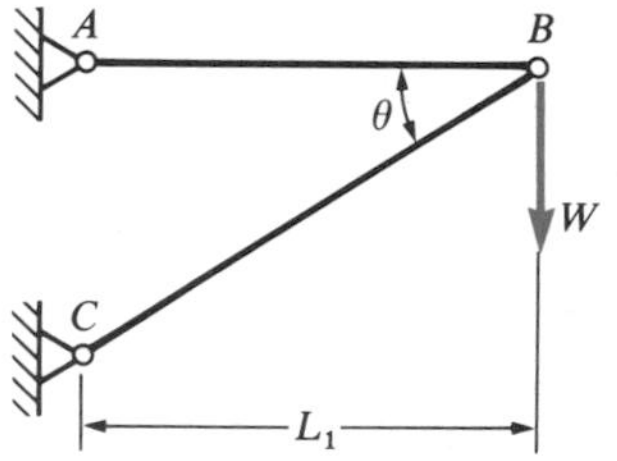

Prob. 9.2-20

length of strut BC varies as the angle θ is changed. Strut BC has a solid circular cross section. Assuming that collapse occurs by buckling of the strut, determine the angle θ for minimum weight of the strut.

The problems for Section 9.3 are to be solved using the assumptions of ideal, slender, prismatic, elastic columns. Buckling occurs in the plane of the figure unless otherwise stated.

9.3-1 A steel wide-flange column ($E = 30 \times 10^6$ psi) of W 12 × 50 shape has a length $L = 20$ ft. It is supported only at the ends and may buckle in any direction. Calculate the critical load P_{cr} for the following support conditions: (1) pinned-pinned, (2) fixed-free, (3) fixed-pinned, and (4) fixed-fixed.

9.3-2 Solve the preceding problem for a W 8 × 35 section with length $L = 18$ ft.

9.3-3 An aluminum pipe column ($E = 10{,}000$ ksi) has length $L = 10$ ft. The inside and outside diameters are 5.2 in. and 6.0 in., respectively. The column is supported only at the ends and may buckle in any direction. Calculate the critical load P_{cr} for the following support conditions: (1) pinned-pinned, (2) fixed-free, (3) fixed-pinned, and (4) fixed-fixed.

9.3-4 A pipe column AB is fixed at the base and pinned at the top to a horizontal bar supporting a load $Q = 200$ kN (see figure). Determine the required thickness t of the pipe if its outside diameter d is 110 mm and the desired factor of safety with respect to buckling is $n = 3$. The pipe is made of aluminum with $E = 72$ GPa.

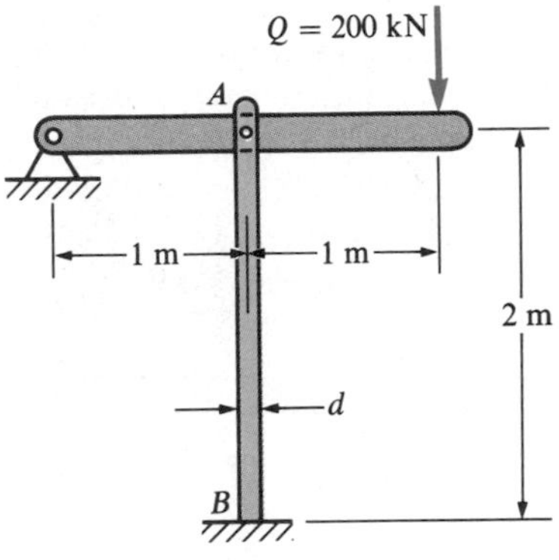

Prob. 9.3-4

9.3-5 The steel roof beams of a warehouse are supported by standard steel pipe columns (see figure) having outside diameter $d = 4.00$ in. and inside diameter 3.548 in. If the columns have length $L = 16$ ft and $E = 30 \times 10^6$ psi, calculate the critical load P_{cr} using the following assumptions: (a) the column has pinned ends; (b) the column has fixed ends; and (c) the base of the column is fixed, the top of the column is rigidly attached to the beam and does not rotate, and the beam is free to translate horizontally.

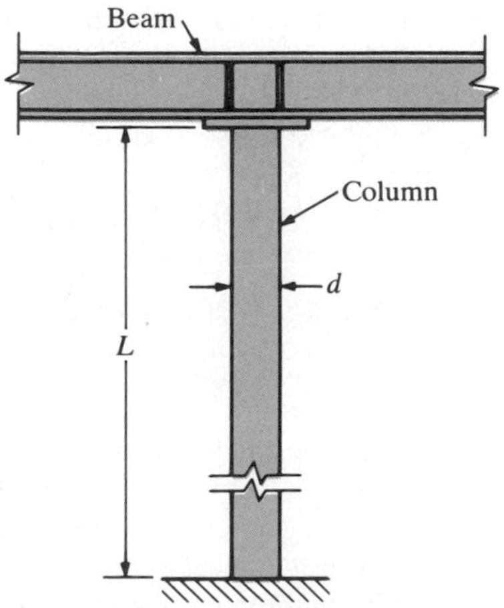

Prob. 9.3-5

9.3-6 The horizontal bar shown in the figure is supported by columns AB and CD. The bar is prevented from moving horizontally by the roller support at its left-hand end. Each column is pinned at its upper end to the horizontal bar, but at the lower ends, support A is fixed and support D is pinned. Both columns are solid steel bars ($E = 200$ GPa) of square cross section with width equal to 16 mm. A load Q acts at distance a from column AB. (a) If the distance $a = 0.4$ m, what is the critical value Q_{cr} of the load? (b) If the distance a can be varied between 0 and 1 m, what is the maximum possible value of Q_{cr}? (c) What is the corresponding value of the distance a?

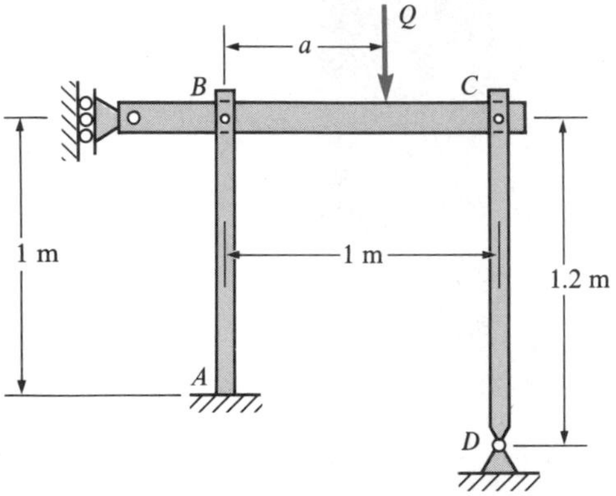

Prob. 9.3-6

9.3-7 Determine the critical load P_{cr} and the equation of the buckled mode shape for a column with fixed ends (see Fig. 9-13) by solving the differential equation of the deflection curve.

***9.3-8** By solving the differential equation of the deflection curve, obtain the following buckling equation for a column that is fixed at the base and supported at the top by a linearly elastic spring of stiffness β (see figure):

$$\tan kL - kL + k^3L^3\left(\frac{EI}{\beta L^3}\right) = 0$$

For the particular case of $\beta = 3EI/L^3$, calculate the critical load P_{cr}.

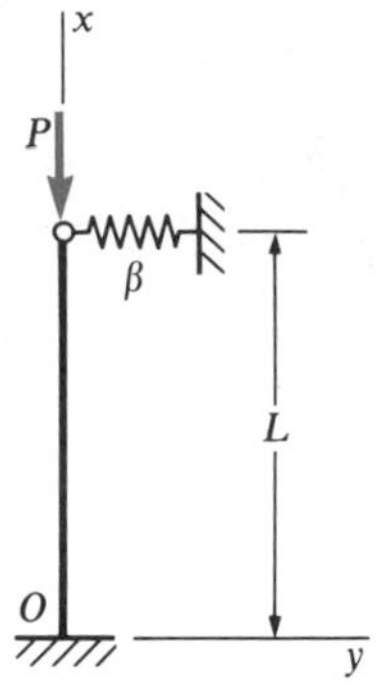

Prob. 9.3-8

****9.3-9** A uniform beam AB of length $L = 25$ ft and weight $W = 2000$ lb is to be supported at an angle $\alpha = 50°$ to the horizontal by a strut CD of length $S = 10$ ft (see figure). The strut is a solid wood member of square cross section ($E = 1{,}500{,}000$ psi) placed in such a position that it carries the smallest possible compressive force. What is the minimum permissible width b of the strut if it is required to have a factor of safety of 3.0 with respect to buckling? (For purposes of analysis, disregard the weight of the strut and assume pinned conditions at A, C, and D.)

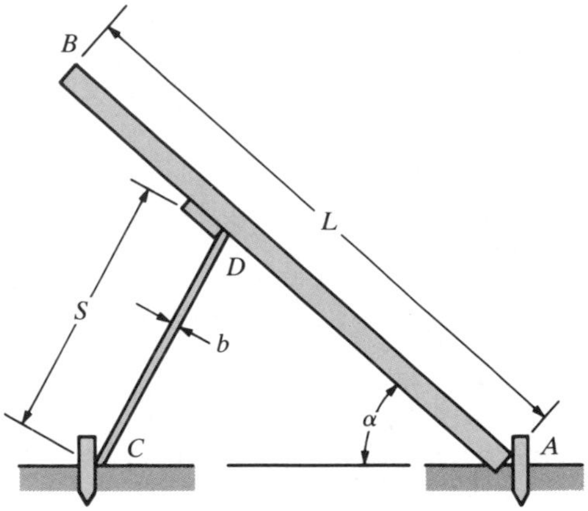

Prob. 9.3-9

****9.3-10** A column AB of length L and flexural rigidity EI is fixed at the base and supports a container C that is attached rigidly to the top of the column, as shown in the figure. The

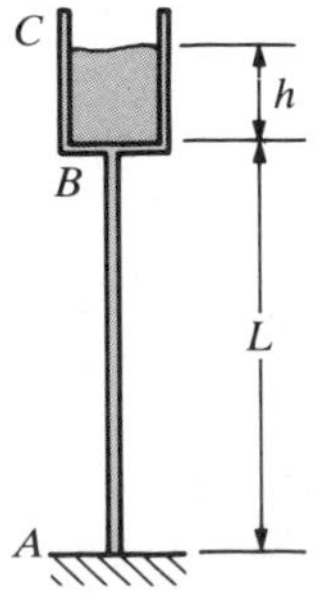

Prob. 9.3-10

container, which is a circular cylinder of base area A_c, is slowly filled with sand of specific weight γ (weight per unit volume). As filling occurs, the surface of the sand is kept level. (a) Derive the following equation for determining the height h of the sand at which the column buckles:

$$\left(\frac{\alpha h}{L}\right)^{3/2} \tan\left(\frac{\alpha h}{L}\right)^{1/2} - 2\alpha = 0 \qquad \text{(a)}$$

in which α is a nondimensional parameter defined as

$$\alpha = \frac{\gamma A_c L^3}{EI}$$

Note that Eq. (a) can be solved numerically for the ratio h/L if the value of α is given. (b) For the particular case of $\alpha = 4$, calculate the ratio h/L.

The problems for Section 9.4 are to be solved using the assumptions of ideal, slender, prismatic, elastic columns. Bending occurs in the plane of the figure unless otherwise stated.

9.4-1 Obtain an expression for the bending moment M in the column with eccentric axial loads shown in Fig. 9-15b. Then plot a bending-moment diagram for the column for an axial load $P = 0.2P_{cr}$. (Suggestion: Plot the bending moment M, expressed in nondimensional form as M/Pe, as the ordinate to the diagram, and plot x/L as the abscissa.)

9.4-2 The column shown in the figure is fixed at the base and free at the upper end. A compressive load P acts at the top of the column with an eccentricity e from the axis of the column. Beginning with the differential equation, derive formulas for: (a) the maximum deflection δ of the column, and (b) the maximum bending moment M_{max} in the column.

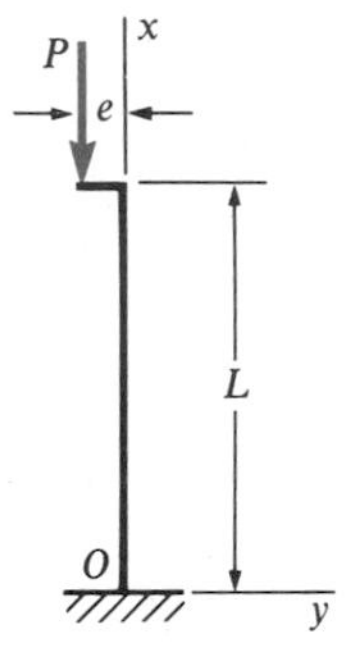

Prob. 9.4-2

9.4-3 A steel bar of square cross section (2 in. × 2 in.) and length $L = 6$ ft is eccentrically compressed by axial loads $P = 12$ k (see Fig. 9-15). The forces are applied on the edge of the cross section at the middle of one side. Calculate the maximum deflection δ of the bar and the maximum bending moment M_{max}, assuming $E = 30 \times 10^6$ psi.

9.4-4 An aluminum bar of rectangular cross section (30 mm × 60 mm) and length $L = 1.0$ m is eccentrically compressed by axial loads $P = 14$ kN (see Fig. 9-15). The forces are applied at the middle of the long edge of the cross section. Calculate the maximum deflection δ of the bar and the maximum bending moment M_{max}, assuming $E = 70$ GPa.

9.4-5 An aluminum hollow box column of square cross section is fixed at the base and free at the top (see figure). The width of each side is $b = 6$ in. and the thickness $t = 0.5$ in. A compressive load $P = 45$ k acts on the outer edge of the column at the midpoint of one side. What is the longest permissible length L of the column if the deflection at the top is not to exceed 4.0 in.? (Assume $E = 10 \times 10^3$ ksi.)

9.4-6 An aluminum hollow box column of square cross section is fixed at the base and free at the top (see figure). The width of each side is $b = 150$ mm and the thickness $t = 10$ mm. A compressive load $P = 160$ kN acts on the outer edge of the column at the midpoint of one side. What is the longest permissible length L of the column if the deflection at the top is not to exceed 100 mm? (Assume $E = 70$ GPa.)

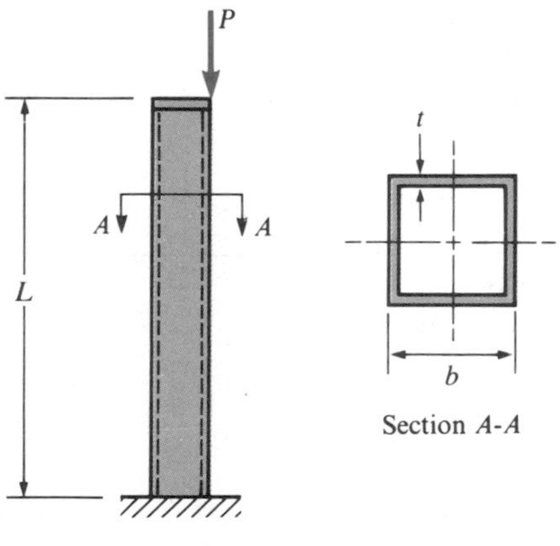

Probs. 9.4-5 and 9.4-6

9.4-7 A steel I-beam (S 8 × 23 section) used as a column supports a load P applied on axis 1-1 at the position shown in the figure (that is, the eccentricity equals half the width of the flange). The column is fixed at the base and free at the top and is 15 ft long. Assuming $E = 30 \times 10^3$ ksi, determine the load P that will produce a maximum bending moment of 40 in.-k in the column.

9.4-8 Solve the preceding problem for a 16-ft long column of S 10 × 35 shape if the maximum bending moment is 60 in.-k.

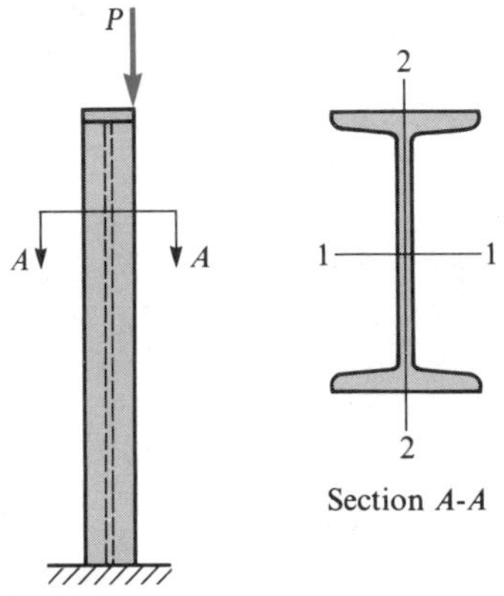

Probs. 9.4-7 and 9.4-8

9.4-9 A frame $ABCD$ is built of steel wide-flange sections (W 8 × 21; $E = 30 \times 10^6$ psi) and subjected to triangularly distributed loads of maximum intensity q_0 acting along members AB and CD (see figure). The distance between supports is $L = 20$ ft and the height of the frame is $h = 4$ ft. (a) Calculate the intensity of load q_0 required to produce a bending moment of 80 in.-k in member BC. (b) If the load q_0 is reduced to one-half its value in part (a), what is the maximum bending moment M_{max} in member BC?

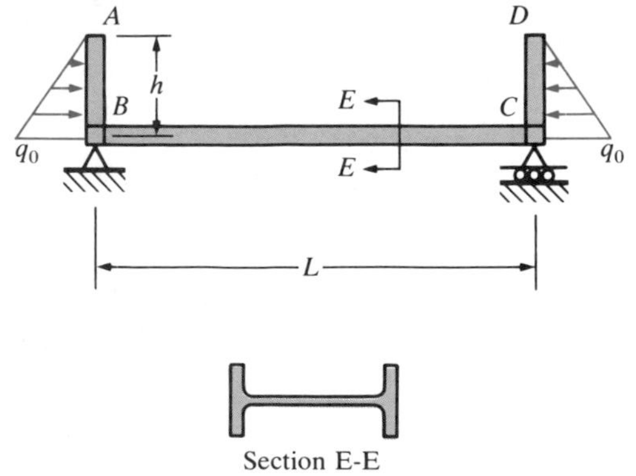

Prob. 9.4-9

The problems for Section 9.5 are to be solved using the assumptions of ideal, slender, prismatic, elastic columns. Bending occurs in the plane of the figure unless otherwise stated.

9.5-1 A steel bar ($E = 30 \times 10^3$ ksi) has a square cross section of width $b = 2$ in. on each side (see figure). The bar has length $L = 6$ ft and is supported as a pinned-end column. (a) Assume that the bar is compressed by a centrally applied force P. What is the critical stress σ_{cr} in the bar when the load P reaches the critical load? (b) If a load $P = 8$ k acts at the middle of one side ($e = 1$ in.), what is the maximum stress σ_{max} in the bar? (c) What load P_1, applied at the middle of one side with an eccentricity of 1 in., will produce a maximum stress $\sigma_1 = 12$ ksi in the bar?

9.5-2 A steel bar ($E = 200$ GPa) has a square cross section of width $b = 40$ mm on each side (see figure). The bar has length $L = 1.5$ m and is supported as a pinned-end column. (a) Assume that the bar is compressed by a centrally applied force P. What is the critical stress σ_{cr} in the bar when the load P reaches the critical load? (b) If a load $P = 10$ kN acts at the middle of one side ($e = 20$ mm), what is the maximum stress σ_{max} in the bar? (c) What load P_1, applied at the middle of one side with an eccentricity of 20 mm, will produce a maximum stress of 60 MPa in the bar?

9.5-3 A pinned-end column of length $L = 5$ ft is constructed of steel pipe ($E = 30 \times 10^3$ ksi) having inside diameter $d_1 = 2.0$ in. and outside diameter $d_2 = 2.2$ in. (see figure). A compressive load P is applied at the middle of the wall of the pipe ($e = 1.05$ in.). (a) If the load $P = 2.0$ k, what is the maximum stress σ_{max} in the pipe? (b) What is the allowable load P_{allow} if a factor of safety $n = 2$ with respect to yielding of the material is required? (Assume $\sigma_y = 42$ ksi.)

9.5-4 A pinned-end column of length $L = 2.5$ m is constructed of steel pipe ($E = 210$ GPa) having inside diameter $d_1 = 60$ mm and outside diameter $d_2 = 70$ mm (see figure). A compressive load P is applied at the middle of the wall of the pipe ($e = 32.5$ mm). (a) If the load $P = 10$ kN, what is the maximum stress σ_{max} in the pipe? (b) What is the allowable load P_{allow} if a factor of safety $n = 2$ with respect to yielding of the material is required? (Assume $\sigma_y = 250$ MPa.)

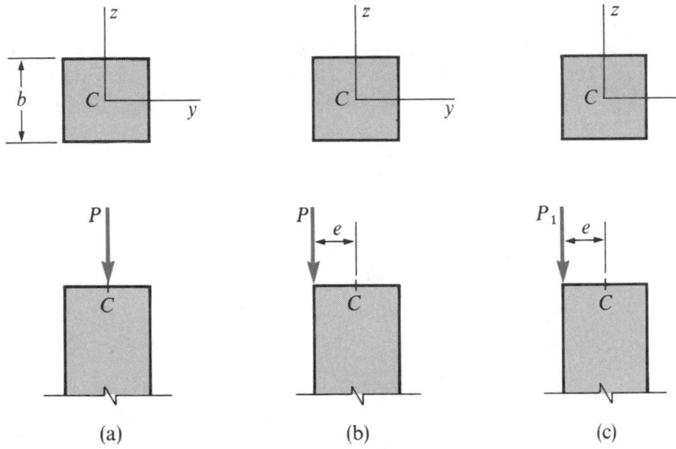

Probs. 9.5-1 and 9.5-2

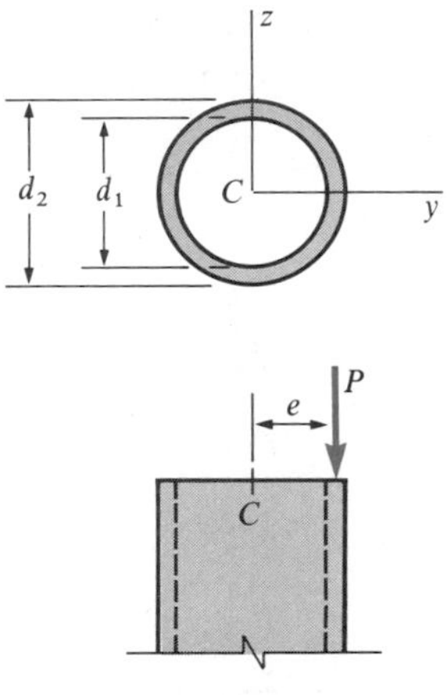

Probs. 9.5-3 and 9.5-4

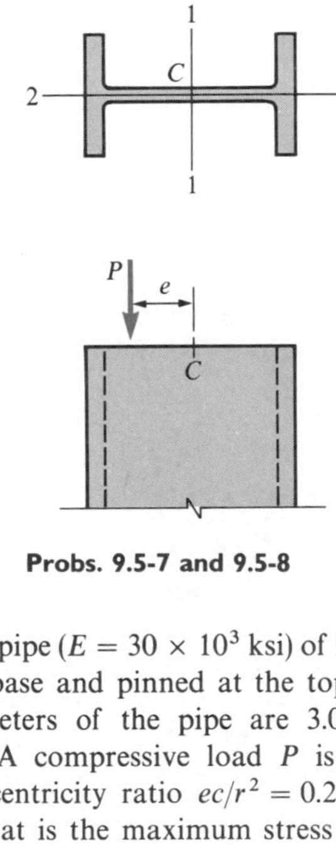

Probs. 9.5-7 and 9.5-8

9.5-5 A steel pipe ($E = 30 \times 10^3$ ksi) of length $L = 12$ ft is fixed at the base and pinned at the top. The inside and outside diameters of the pipe are 3.0 in. and 3.5 in., respectively. A compressive load P is applied with an estimated eccentricity ratio $ec/r^2 = 0.25$. (a) If the load $P = 10$ k, what is the maximum stress σ_{max} in the pipe? (b) Determine the allowable load P_{allow} if the factor of safety with respect to yielding of the material is $n = 2.5$. (Assume $\sigma_y = 36$ ksi.)

9.5-6 A steel pipe ($E = 200$ GPa) of length $L = 3.5$ m is fixed at the base and pinned at the top. The inside and outside diameters of the pipe are 75 mm and 90 mm, respectively. A compressive load P is applied with an estimated eccentricity ratio $ec/r^2 = 0.4$. (a) If the load $P = 50$ kN, what is the maximum stress σ_{max} in the pipe? (b) Determine the allowable load P_{allow} if the factor of safety with respect to yielding of the material is $n = 2.5$. (Assume $\sigma_y = 250$ MPa.)

9.5-7 A steel column ($E = 30 \times 10^3$ ksi) with pinned ends and length $L = 16$ ft is constructed of a W 8 × 35 wide-flange section (see figure). A compressive axial load P acts on axis 2-2 with an eccentricity $e = 2.5$ in. (a) If $P = 100$ k, determine the maximum compressive stress σ_{max} in the column. (b) What load P_y will produce yielding of the material if $\sigma_y = 36$ ksi?

9.5-8 A steel column ($E = 30 \times 10^3$ ksi) that is fixed at the base and free at the top is constructed of a W 10 × 60 wide-flange section (see figure). The column is 12 ft long. A compressive axial load P acts on axis 2-2 with an eccentricity $e = 2.0$ in. (a) If $P = 120$ k, determine the maximum compressive stress σ_{max} in the column. (b) Determine the allowable load P_{allow} if $\sigma_y = 42$ ksi and the factor of safety with respect to yielding of the material is $n = 2.0$.

9.5-9 A pinned-end column with length $L = 30$ ft is constructed of a steel ($E = 30 \times 10^3$ ksi) W 14 × 53 wide-flange section (see figure). The column is subjected to a centrally applied load $P_1 = 150$ k and an eccentric load $P_2 = 60$ k acting on axis 2-2 at the edge of the column. (a) Calculate the maximum compressive stress σ_{max} in the column. (b) What is the factor of safety against yielding of the material if $\sigma_y = 36$ ksi?

9.5-10 A pinned-end column with length $L = 32$ ft is constructed of a steel ($E = 30 \times 10^3$ ksi) W 12 × 50 wide-flange section (see figure). The column is subjected to a centrally applied load $P_1 = 140$ k and an eccentric load

$P_2 = 50$ k acting on axis 2-2 at the edge of the column. (a) Calculate the maximum compressive stress σ_{max} in the column. (b) What is the factor of safety against yielding of the material if $\sigma_y = 42$ ksi?

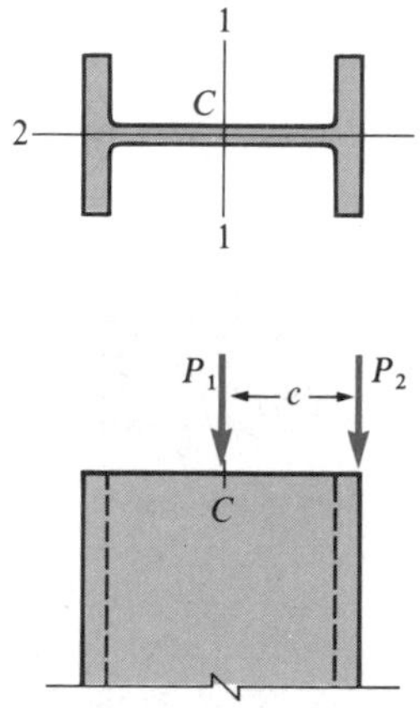

Probs. 9.5-9 and 9.5-10

9.5-11 A W 16 × 57 steel section is used as a column with pinned supports. It carries an eccentric axial load $P = 180$ k acting on axis 1-1, so that buckling occurs in the 1-1 plane (see figure). The steel has modulus of elasticity $E = 30 \times 10^3$ ksi and yield stress $\sigma_y = 36$ ksi. Also, the eccentricity ratio is assumed to be $ec/r^2 = 0.2$. What is the longest length L that the column can have if a factor of safety of 2.0 is required with respect to yielding?

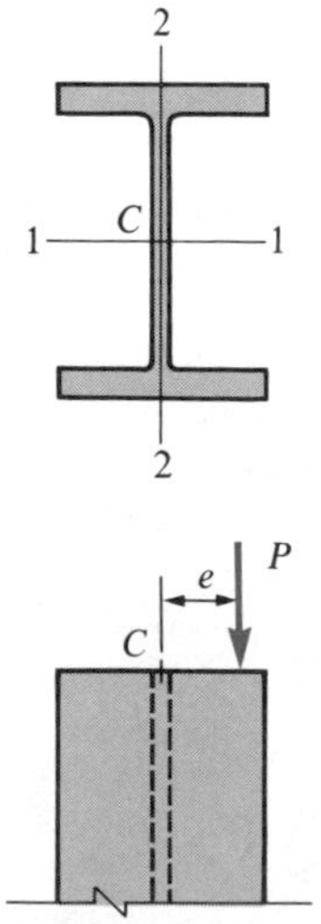

Prob. 9.5-11

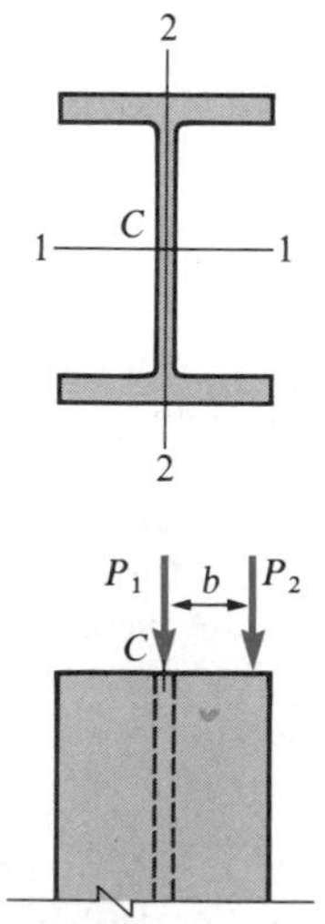

Prob. 9.5-12

9.5-12 A steel column (W 12 × 87 wide-flange section) has pinned ends and supports two loads; load P_1 acts at the centroid C and load P_2 acts on axis 1-1 at distance $b =$ 7.5 in. from the centroid (see figure). The column has length $L = 16$ ft, $E = 30 \times 10^3$ ksi, and $\sigma_y = 42$ ksi. If $P_1 = 150$ k, what is the largest permissible value of P_2 in order to maintain a factor of safety of 2.5 with respect to yielding?

The problems for Section 9.6 are to be solved using the assumptions of ideal, slender, prismatic, elastic columns with an initial deflection given by a half-wave of a sine curve. Bending occurs in the plane of the figure unless otherwise stated.

9.6-1 Consider a pinned-end column (Fig. 9-20a) with an initial deflection v_0 in the form of a half-wave of a sine curve:

$$v_0 = a \sin \frac{\pi x}{L}$$

Derive Eq. (9-27) for the maximum deflection δ_{max} of this column by solving the differential equation of the deflection curve.

9.6-2 A solid circular steel bar ($E = 200$ GPa) has a diameter $d = 40$ mm and length $L = 0.8$ m. The bar is pin-supported at the ends and is assumed to have an initial deflection at the midpoint equal to 1/500 of the length. The axial load on the bar is $P = 50$ kN. (a) Calculate the maximum stress σ_{max} in the bar. (b) Calculate the maximum allowable load P_{allow} if the factor of safety with respect to yielding is $n = 2$ and $\sigma_y = 250$ MPa.

9.6-3 A solid circular aluminum bar of diameter $d = 2$ in. and length $L = 40$ in. is pin-supported at the ends. The bar is assumed to have an initial deflection a at the midpoint equal to 1/800 of the length. The axial load on the bar is $P = 20$ k and $E = 10 \times 10^3$ ksi. (a) Determine the maximum stress σ_{max} in the bar. (b) Determine the factor of safety n with respect to yielding if $\sigma_y = 40$ ksi.

9.6-4 A W 10 × 45 steel column ($E = 30 \times 10^3$ ksi) with pinned ends and length $L = 15$ ft supports an axial load $P = 150$ k. The column is supported only at the ends; hence it buckles by bending about axis 2-2 (see figure). Assume that the column has an initial deflection at the midpoint equal to 1/900 of its length ($L/a = 900$). (a) Determine the maximum stress σ_{max} in the column. (b) Determine the factor of safety n with respect to yielding if $\sigma_y = 36$ ksi.

9.6-5 A W 8 × 35 steel column ($E = 30 \times 10^3$ ksi) with pinned ends and length $L = 12$ ft supports an axial load

$P = 80$ k. The column is supported only at the ends; hence it buckles by bending about axis 2-2 (see figure). Assume that the column has an initial deflection at the midpoint of 0.3 in. (a) What is the maximum stress σ_{max} in the column? (b) What is the maximum allowable load P_{allow} if the factor of safety with respect to yielding is $n = 2.2$ and $\sigma_y = 36$ ksi?

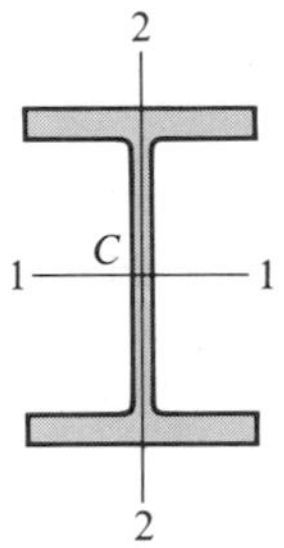

Probs. 9.6-4 and 9.6-5

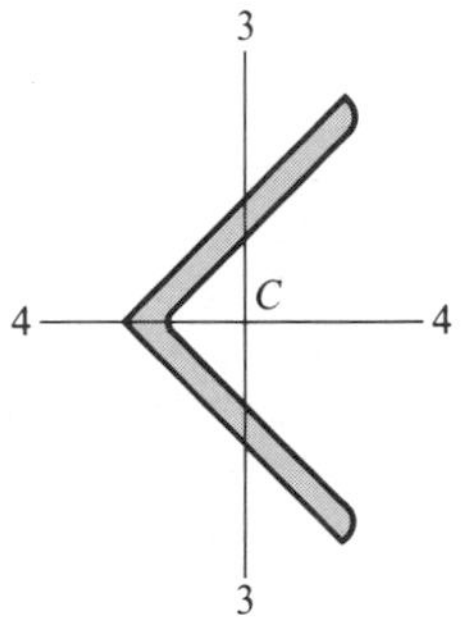

Prob. 9.6-6

9.6-6 An L 3 × 3 × 1/4 angle section is used as a compression member with pinned ends (see figure). It is supported only at the ends; hence buckling occurs by bending about axis 3-3. Data for the angle section are as follows: $L = 4$ ft, $E = 30 \times 10^3$ ksi, and $\sigma_y = 36$ ksi. Assume that the initial deflection at the midpoint of the angle is 1/800 of the length. Calculate the allowable axial load P_{allow} if the factor of safety with respect to yielding is $n = 2.5$.

The problems for Section 9.9 are to be solved assuming that the axial loads are centrally applied at the ends of the columns. The columns are pin-supported at the ends and may buckle in any direction.

9.9-1 Determine the allowable axial load P for a W 10 × 45 steel wide-flange column with pinned ends for each of the following lengths: $L = 8$ ft, 16 ft, 24 ft, and 32 ft. (Assume $E = 29 \times 10^3$ ksi and $\sigma_y = 36$ ksi.)

9.9-2 Determine the allowable axial load P for a W 10 × 60 steel wide-flange column with pinned ends for each of the following lengths: $L = 10$ ft, 20 ft, 30 ft, and 40 ft. (Assume $E = 29 \times 10^3$ ksi and $\sigma_y = 36$ ksi.)

9.9-3 Calculate the allowable axial load P for a W 12 × 50 steel wide-flange column with pinned ends for each of the following lengths: $L = 8$ ft, 16 ft, 24 ft, and 32 ft. (Assume $E = 29 \times 10^3$ ksi and $\sigma_y = 50$ ksi.)

9.9-4 Calculate the allowable axial load P for a W 12 × 87 steel wide-flange column with pinned ends for each of the following lengths: $L = 10$ ft, 20 ft, 30 ft, and 40 ft. (Assume $E = 29 \times 10^3$ ksi and $\sigma_y = 50$ ksi.)

9.9-5 Determine the allowable axial load P for a steel pipe column with pinned ends having outside diameter 4.5 in. and wall thickness 0.237 in. for each of the following lengths: $L = 6$ ft, 12 ft, 18 ft, and 24 ft. (Assume $E = 29 \times 10^3$ ksi and $\sigma_y = 36$ ksi.)

9.9-6 Calculate the allowable axial load P for a steel pipe column with pinned ends having outside diameter 140 mm and wall thickness 7 mm for each of the following lengths: $L = 2$ m, 4 m, 6 m, and 8 m. (Assume $E = 200$ GPa and $\sigma_y = 250$ MPa.)

9.9-7 Determine the allowable axial load P for a steel pipe column with pinned ends having outside diameter 6.625 in. and wall thickness 0.432 in. for each of the following lengths: $L = 8$ ft, 16 ft, 24 ft, and 32 ft. (Assume $E = 29 \times 10^3$ ksi and $\sigma_y = 36$ ksi.)

9.9-8 Calculate the allowable axial load P for a steel pipe column with pinned ends having outside diameter 220 mm and wall thickness 12 mm for each of the following lengths: $L = 2.5$ m, 5 m, 7.5 m, and 10 m. (Assume $E = 200$ GPa and $\sigma_y = 250$ MPa.)

9.9-9 A steel pipe column with pinned ends supports an axial load $P = 21$ k. The pipe has outside and inside diameters of 3.5 in. and 2.9 in., respectively. What is the longest permissible length L of the column if $E = 29 \times 10^3$ ksi and $\sigma_y = 36$ ksi?

9.9-10 A W 8 × 28 steel wide-flange column with pinned ends carries an axial load P. What is the longest permissible length L of the column if $P = 50$ k and if $P = 100$ k? (Assume $E = 29 \times 10^3$ ksi and $\sigma_y = 36$ ksi.)

9.9-11 A W 8 × 35 steel wide-flange column with pinned ends carries an axial load P. What is the longest permissible length L of the column if $P = 75$ k and if $P = 150$ k? (Assume $E = 29 \times 10^3$ ksi and $\sigma_y = 36$ ksi.)

9.9-12 What is the minimum permissible diameter d of a solid steel column of circular cross section with pinned ends and length 2 m if it must support an axial load $P = 800$ kN? (Assume $E = 200$ GPa and $\sigma_y = 250$ MPa.)

9.9-13 What is the minimum permissible width b of a solid steel column of square cross section with pinned ends and length $L = 5$ ft if it must support an axial load $P = 56$ k? (Assume $E = 29 \times 10^3$ ksi and $\sigma_y = 36$ ksi.)

9.9-14 Determine the allowable axial load P for an aluminum pipe column (alloy 2014-T6) with pinned ends for each of the following lengths: $L = 4$ ft, 8 ft, 12 ft, and 16 ft. The column has outside diameter 5.563 in. and inside diameter 4.813 in.

9.9-15 Solve the preceding problem for an aluminum pipe with outside diameter 6.625 in. and inside diameter 6.065 in.

9.9-16 A square aluminum bar of width 2 in. is loaded in compression by a force P. The bar has pinned ends and is constructed of alloy 2014-T6. What is the longest permissible length L of the bar if $P = 90$ k and if $P = 30$ k?

9.9-17 What is the minimum permissible diameter d of a solid aluminum strut (alloy 2014-T6) of circular cross section with pinned ends and length 24 in. if it must support an axial load $P = 12$ k?

9.9-18 A square wood column of width 120 mm (actual dimensions) is supported with pinned ends. Determine the longest permissible length L of the column for each of the following axial loads: $P = 90$ kN, 60 kN, and 30 kN. (Assume $E = 12$ GPa and $F_c = 8$ MPa.)

9.9-19 Determine the allowable axial load P for a 6 in. × 6 in. wood column (actual dimensions 5.5 in. × 5.5 in., see Appendix F) with pinned ends for each of the following lengths: $L = 5$ ft, 10 ft, 15 ft, and 20 ft. (Assume $E = 1.8 \times 10^6$ psi and $F_c = 1200$ psi.)

9.9-20 What is the minimum permissible width b of a square wood column with pinned ends and length 3 m if it must support an axial load $P = 150$ kN? (Assume $E = 12$ GPa and $F_c = 9$ MPa.)

9.9-21 Determine the allowable axial load P for a 4 in. × 8 in. wood column (actual dimensions 3.5 in. × 7.25 in.; see Appendix F) with pinned ends for each of the following lengths: $L = 6$ ft, 8 ft, 10 ft, and 12 ft. (Assume $E = 1.6 \times 10^6$ psi and $F_c = 1000$ psi.)

CHAPTER 10

Energy Methods

10.1 INTRODUCTION

The concepts of work and strain energy were introduced in earlier chapters, and expressions for the strain energy stored in linearly elastic structures subjected to axial loads, torsion, and bending were derived. Now we will consider some additional concepts, including virtual work and complementary energy. The principle of virtual work leads to the unit-load method, which is a powerful method for finding displacements of structures of all kinds. Strain energy and complementary energy provide methods for analyzing both linear and nonlinear structures, whether statically determinate or indeterminate. We will consider only simple trusses, beams, and frames in this chapter, but the concepts are applicable to more complex structures. Thus, this chapter provides an introduction to principles of applied mechanics that are fundamental to advanced structural analysis.

10.2 PRINCIPLE OF VIRTUAL WORK

The principle of virtual displacements and the principle of virtual work are usually introduced in the study of statics, in which they are used to solve problems of static equilibrium. The word *virtual* means that the quantities are hypothetical and do not exist in a real or physical sense. Thus, a **virtual displacement** is an imaginary displacement that is arbitrarily imposed upon a structural or mechanical system; it is not an actual displacement, such as the deflection caused by a load acting on a structure. The work done by the real forces during a virtual displacement is called **virtual work**.

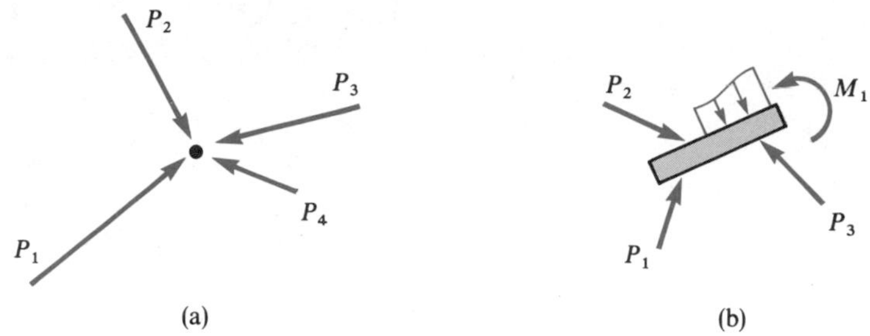

Fig. 10-1 Principle of virtual displacements for: (a) a particle, and (b) a rigid body.

When a single particle is acted upon by a set of forces in static equilibrium (Fig. 10-1a), we may impart to the particle a virtual displacement consisting of a translation of the particle in any direction. During this virtual displacement the virtual work done by the forces must be zero because the forces are in equilibrium. This seemingly simple statement is the **principle of virtual displacements**. As shown in the study of statics, this principle may be used instead of the more familiar equations of equilibrium for the purpose of solving statics problems.

The principle of virtual displacements is also applicable to a rigid body that is held in equilibrium by the action of a set of loads that may include forces, couples, and distributed loads (Fig. 10-1b). The rigid body may be given a virtual displacement consisting of a translation in any direction, a rotation about any axis, or a combination of translation and rotation. In all instances the virtual work done by the forces will be zero if the body is in equilibrium. Usually we must restrict the virtual displacement to a very small displacement in order that the lines of action of the forces will not be altered during the virtual displacement.*

For use in finding displacements of deformable structures, we must extend the principle of virtual displacements to take into account not only the virtual work of the external forces, or loads, but also the virtual work associated with the internal forces, or stress resultants. To show how this is accomplished, let us consider the beam shown in Fig. 10-2a, although any other structure could be used. This structure is assumed to be loaded in a

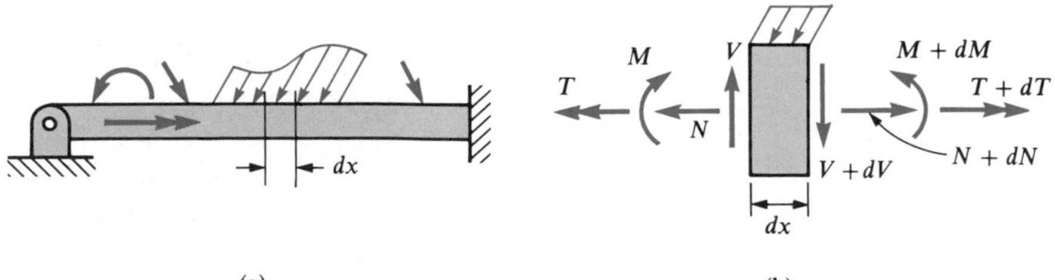

Fig. 10-2 Derivation of principle of virtual work

* The principle of virtual displacements was first formulated by John Bernoulli, 1667–1748 (see Ref. 7-1).

completely general fashion by forces, bending couples, torques, and distributed loads. The structure is at rest and in equilibrium under the action of these loads. At any cross section of the structure, stress resultants may exist in the form of an axial force N, bending moment M, shear force V, and twisting couple T, as shown on the left-hand face of an element of length dx cut out from the member (Fig. 10-2b). On the opposite face of the element, the stress resultants may have changed by small amounts; hence they are shown as $N + dN$, $M + dM$, $V + dV$, and $T + dT$.

Assume now that the beam (Fig. 10-2a) is given a virtual deformation consisting of a small change in its deflected shape. This virtual deformation is imposed upon the beam in some unspecified manner, and it is completely independent of the fact that the structure has already been subjected to real deflections caused by the loads acting on it. Such real deflections have definite magnitudes dictated by the nature of both the loads and the structure. The virtual deformation, however, represents an additional deformation that is imparted arbitrarily to the structure. The only restriction on the virtual deformation is that it must represent a deformed shape that could actually occur physically; that is, the virtual change in shape must be compatible with the conditions of support for the structure and must maintain continuity between elements of the structure. As a consequence of the virtual deformation of the beam, points on the axis of the beam will undergo virtual displacements, such as deflections in the vertical direction.

During the virtual deformation each element of the beam will be displaced to a new location and also will be deformed in shape. Therefore, the forces acting on an element (both stress resultants and external loads, as shown in Fig. 10-2b) will perform virtual work. Let us denote the total amount of this virtual work by dW_e. This work may be considered to be composed of two parts: (1) the work dW_r due to the displacement (both translation and rotation) of the element as a rigid body, and (2) the work dW_d associated with the deformation of the element. Thus,

$$dW_e = dW_r + dW_d$$

Because the element is in equilibrium, the virtual work dW_r done by the forces (both external and internal) during the displacement of the element as a rigid body must be zero, and so the preceding equation reduces to

$$dW_e = dW_d \tag{a}$$

This equation states that the total virtual work done by all forces acting on the element during its virtual displacement is equal to the virtual work done by those same forces during only the virtual deformation of the element. Summing up the virtual-work terms in Eq. (a) for all elements of the structure, we get

$$\int dW_e = \int dW_d \tag{b}$$

in which it is understood that the integrations are performed over the entire structure.

The integrals in Eq. (b) can be interpreted as follows. The integral on the left-hand side of the equation is the total virtual work done during the virtual deformation of the structure by all the forces, both loads and stress resultants, acting on all faces of all elements, of which the element shown in Fig. 10-2b is typical. We note, however, that the sides of each element are in direct contact with the sides of adjacent elements. Therefore, the virtual work of the stress resultants acting on one element exactly cancels the virtual work of the equal and opposite stress resultants acting on adjacent elements. The only remaining virtual work is the work of the external forces acting on the external boundaries of the elements (such as the top and bottom of the element in Fig. 10-2b). Thus, we conclude that the integral on the left-hand side of Eq. (b) is equal to the virtual work of the external forces acting on the structure. We will refer to this quantity as the **external virtual work** and designate it by W_{ext}.

The term on the right-hand side of Eq. (b) was obtained by integrating the virtual work associated with the deformation of an element. This work includes the effects of all forces acting on the element, both stress resultants and external forces. However, when an element deforms (for example, when it elongates) only the stress resultants perform work. Thus, the second term in Eq. (b) actually represents the virtual work of the stress resultants alone. This virtual work is equal to the work done by the stress resultants when the elements on which they act are deformed virtually. The total amount of this virtual work, obtained by summing over all elements, is called the **internal virtual work** and is denoted by W_{int}. Thus, Eq. (b) becomes

$$W_{ext} = W_{int} \tag{10-1}$$

This equation represents the **principle of virtual work**, and it may be stated as follows:

> *If a deformable structure in equilibrium under the action of a system of loads is given a small virtual deformation, then the virtual work done by the external forces (or loads) is equal to the virtual work done by the internal forces (or stress resultants).*

The principle of virtual work is extremely general and has many applications in structural analysis and applied mechanics. We will use it in the next section to derive the unit-load method for finding deflections. However, before proceeding to use the principle, it is important to emphasize some of its features. First, the virtual deformation or virtual displacement must be compatible with the supports of the structure and must maintain the continuity of the structure. Second, the deformations and displacements must be sufficiently small so that the geometry of the structure is not altered, allowing all calculations to be based upon the initial configuration of the structure. Except for these restrictions, the virtual change in shape may be arbitrarily imposed on the structure.

However, the virtual deformations should not be confused with the deformed shape of the structure caused by the real loads. Finally, a review of the development of the principle will show that the properties of the material of the structure never entered the derivation. Therefore, the principle of virtual work applies to all structures irrespective of whether the material behaves linearly or nonlinearly, elastically or inelastically.

Now consider the evaluation of the terms for external and internal virtual work appearing in Eq. (10-1). The external work W_{ext} is the work performed by the loads acting on the structure during the virtual displacement. Since these loads are acting at their full values when the virtual displacement is imposed, the virtual work they perform is simply the product of load and displacement. Specifically, the virtual work of a concentrated load is the product of the force and the virtual displacement of its point of application. In such a case the positive direction of the displacement must be taken in the same direction as the positive direction of the force. If the load is a couple, then the virtual work is the product of the moment of the couple and the virtual angle through which it rotates.

The evaluation of the internal work W_{int} of the stress resultants is more involved. The virtual work done by the stress resultants acting on an element (Fig. 10-2b) depends upon the deformation of the element during the virtual displacement. The various kinds of small virtual deformations are shown in Fig. 10-3. Part (a) of the figure shows a virtual deformation consisting of a uniform axial strain in the element; thus, the length of the element has increased by an amount denoted by $d\delta$. This virtual deformation results in virtual work $(N + dN)\,d\delta$ being done by the axial force (Fig. 10-2b), but no virtual work is done by the bending moment, shear force, or twisting couple. Again we should note that nothing is said at this time about the cause of the virtual deformation $d\delta$; it is clearly not caused by the axial force N itself.

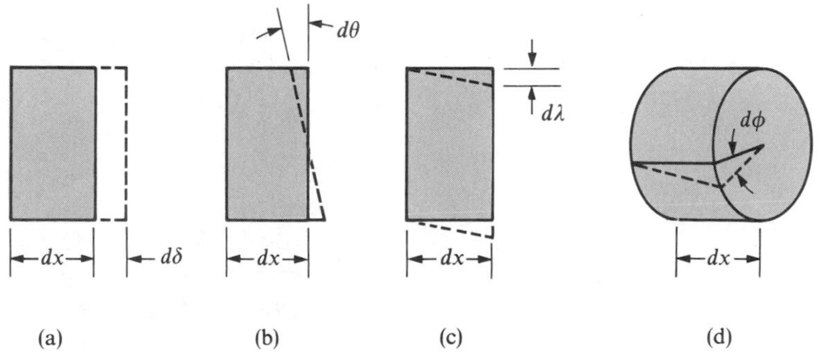

Fig. 10-3 Virtual deformations of an element of a structure: (a) axial, (b) flexural, (c) shearing, and (d) torsional

The next type of virtual displacement is a flexural deformation consisting of a relative rotation $d\theta$ between two sides of the element (Fig. 10-3b). During such a virtual deformation, the only stress resultant that does work is the bending moment, and its virtual work is $(M + dM)\,d\theta$.

A shearing virtual deformation and a torsional virtual deformation are shown in Figs. 10-3c and d, respectively. The former consists of a lateral translation $d\lambda$ of one side of the element with respect to the other, and the latter consists of a relative rotation $d\phi$ about the longitudinal axis. The virtual-work terms associated with these virtual deformations are $(V + dV)\,d\lambda$ and $(T + dT)\,d\phi$.

Each of the four preceding expressions for internal virtual work can be simplified by disregarding the product of two differentials (such as $dN\,d\delta$) in comparison with the product of the finite term and the differential (such as $N\,d\delta$). Thus, the expression for the internal virtual work done by the stress resultants acting on one element of the structure becomes

$$N\,d\delta + M\,d\theta + V\,d\lambda + T\,d\phi$$

Integrating over all elements of the structure gives the complete expression for the internal virtual work:

$$W_{\text{int}} = \int N\,d\delta + \int M\,d\theta + \int V\,d\lambda + \int T d\phi \qquad (10\text{-}2)$$

In this expression the stress resultants N, M, V, and T are the actual stress resultants in the structure due to the real loads, whereas the deformations $d\delta$, $d\theta$, $d\lambda$, and $d\phi$ are the fictitious deformations associated with the virtual displacement of the structure. In the next section we will develop a method for finding displacements of structures based upon Eqs. (10-1) and (10-2).

10.3 UNIT-LOAD METHOD FOR CALCULATING DISPLACEMENTS

The determination of displacements in structures is an essential part of structural and mechanical design. We already have discussed methods for finding displacements of beams, shafts, and other simple structures in previous chapters. However, those methods are limited to individual structural elements or to structures containing only a few members. In this section, we will use the principle of virtual work to derive the **unit-load method**, which is widely used for finding displacements of all types of structures. Furthermore, the method is suitable for finding not only displacements of a point in a structure but also other displacement quantities, such as the rotation of a member or the relative displacement between two points. Theoretically, the unit-load method may be used for either statically determinate or indeterminate structures, although for practical purposes the method is limited to determinate structures because its use requires that the stress resultants be known throughout the structure.

Because the basic equation of the unit-load method may be derived from the principle of virtual work, the method itself is sometimes called the **method of virtual work**. It is also known as the **dummy-load method** and the **Maxwell-Mohr method**. The former name arose because the method requires the use of a fictitious, or dummy, load (that is, the unit load), and the latter name is used because J. C. Maxwell in 1864 and O. Mohr in 1874 independently developed the method (Refs. 10-1 and 10-4).

Two loading systems must be considered when using the unit-load method to find the displacements of a structure. The first system consists of the structure subjected to the actual loads, temperature changes, or other causes that are responsible for producing the displacement to be calculated. The second system consists of a unit load acting alone on the structure. The unit load is a fictitious, or dummy, load that is introduced solely for the purpose of calculating a displacement Δ of the structure due to the actual loads. The unit load must **correspond** to the desired displacement Δ. A load that corresponds to a displacement is a load that acts on the structure at the point where the displacement is to be determined and has its positive direction in the same direction as a positive displacement.

The term **displacement** is used here in a generalized sense; thus, the displacement Δ may be a translation, a rotation, a relative displacement, or a relative rotation. If the displacement to be calculated is a translation, then the corresponding unit load is a force acting at the point where the translation occurs and having its positive direction in the positive direction of the translation. If the displacement to be calculated is a rotation, then the unit load is a couple acting at the point on the structure where the rotation occurs; the positive sense of the unit couple must be the same as the positive sense of the rotation. If the displacement is the relative translation of two points along the line joining them, the unit load will consist of two collinear and oppositely directed forces acting at the two points. All of these possibilities are illustrated later by examples.

The unit load acting on the structure produces reactions at the supports and stress resultants within the members. Let us designate these stress resultants by the symbols N_U, M_U, V_U, and T_U. These quantities, in combination with the unit load and the reactions, constitute a force system that is in equilibrium. According to the principle of virtual work, if the structure is given a small virtual deformation (or change in shape), then the virtual work of the external forces is equal to the virtual work of the internal forces (see Eq. 10-1). Now we come to the key step in the development of the unit-load method; namely, we must make the proper choice of the virtual deformation. Let us take the actual deformations of the structure caused by the first system of loading as the virtual deformations to be imposed upon the second system (the structure with the unit load). During this virtual deformation the only external virtual work is the work performed by the unit load itself, because it is the only external load on the structure. This virtual work is the product of the unit load and the displacement Δ through which it moves; thus,

$$W_{\text{ext}} = 1 \cdot \Delta \tag{a}$$

where Δ represents the displacement of the structure due to the actual loads (recall that the unit load was intentionally selected to correspond to Δ).

The internal virtual work is the work performed by the stress resultants (N_U, M_U, V_U, and T_U) when the elements of the structure are deformed virtually. However, the virtual deformations are chosen to be the same as the actual deformations that occur in the structure supporting the real loads. Denoting those deformations by $d\delta$, $d\theta$, $d\lambda$, and $d\phi$, as shown in Fig. 10-3, we get for the internal work (see Eq. 10-2)

$$W_{\text{int}} = \int N_U\, d\delta + \int M_U\, d\theta + \int V_U\, d\lambda + \int T_U\, d\phi \tag{b}$$

Finally, by equating external and internal work (see Eqs. a and b), we obtain the **equation of the unit-load method**:

$$\Delta = \int N_U\, d\delta + \int M_U\, d\theta + \int V_U\, d\lambda + \int T_U\, d\phi \tag{10-3}$$

In this equation, Δ represents the displacement to be calculated, which may be a translation, a rotation, or a relative displacement. The stress resultants N_U, M_U, V_U, and T_U represent the axial force, bending moment, shear force, and twisting couple caused by the unit load corresponding to Δ, and $d\delta$, $d\theta$, $d\lambda$, and $d\phi$ represent deformations caused by the actual loads. Because the unit load has been divided from the left-hand side of Eq. (10-3), leaving only the term Δ, it is necessary to consider the quantities N_U, M_U, V_U, and T_U as having the dimensions of force or moment per unit of the applied unit load.

The equation of the unit-load method (Eq. 10-3) is quite general and is not subject to any restrictions concerning linear behavior of the material or the structure. In other words, the principle of superposition need not be valid in order to use Eq. (10-3). However, the deformations $d\delta$, $d\theta$, and so on, must be small in order not to change the geometry of the structure, as explained in the preceding section.

The most common situation occurs when the material of the structure follows Hooke's law and the structure behaves linearly. In this case we can readily obtain expressions for the deformations $d\delta$, $d\theta$, $d\lambda$, and $d\phi$ caused by the real loads acting on the structure. If we denote the stress resultants in the structure due to the real loads as N_L, M_L, V_L, and T_L, then the deformations of an element are

$$d\delta = \frac{N_L\, dx}{EA} \qquad d\theta = \frac{M_L\, dx}{EI} \qquad d\lambda = \frac{f_s V_L\, dx}{GA} \qquad d\phi = \frac{T_L\, dx}{GI_p} \tag{c}$$

The first of these equations gives the elongation of an element (Fig. 10-3a) due to the axial force N_L. Similarly, the remaining three terms give the deformations associated with flexure, shear, and torsion (Figs. 10-3b, c, and d). The expressions for $d\delta$, $d\theta$, and $d\phi$ are based upon equations derived in earlier chapters (see Eqs. 2-1, 7-6, and 3-8). The expression for $d\lambda$ is derived later in Section 10.9 on shear deflections in beams. Values of the term f_s, called the *form factor for shear*, are listed in Table 10-4 in Section 10.9

Substituting the four expressions (c) into Eq. (10-3) gives the equation of the **unit-load method for linearly elastic structures** in the following form:

$$\Delta = \int \frac{N_U N_L \, dx}{EA} + \int \frac{M_U M_L \, dx}{EI} + \int \frac{f_s V_U V_L \, dx}{GA} + \int \frac{T_U T_L \, dx}{GI_p} \qquad (10\text{-}4)$$

This equation can be used to find the displacement Δ at any point of a structure when the material is linearly elastic and the principle of superposition is valid. Each integral in the equation represents the contribution of one type of deformation to the total displacement. Thus, the first integral gives the contribution to the displacement Δ from the effects of axial deformations, the second term gives the contribution from flexural deformations, and so forth, for the remaining terms.

The **sign conventions** used for the stress resultants appearing in Eq. (10-4) must be consistent; thus, the axial forces N_U and N_L must be obtained according to the same sign convention, as must M_U and M_L, V_U and V_L, and T_U and T_L. Only under these conditions will the displacement Δ have the same positive sense as the unit load.

The procedure for calculating a displacement by means of the unit-load method using Eq. (10-4) may be summarized as follows: (1) determine the stress resultants N_L, M_L, V_L, and T_L in the structure due to the actual loads; (2) place a unit load on the structure corresponding to the displacement Δ that is to be found; (3) determine the stress resultants N_U, M_U, V_U, and T_U due to the unit load; (4) form the terms shown in Eq. (10-4) and integrate each term for the entire structure; and (5) sum the results to obtain the displacement Δ. These steps are illustrated in the examples given later.

Depending upon the type of structure, we can anticipate that some of the terms in Eq. (10-4) will not be needed. For example, if a truss with pinned joints has loads acting only at the joints, there will be no flexural, shearing, or torsional deformations in the members and only the first term in Eq. (10-4) is required. Furthermore, the axial forces in the members will be constant throughout the lengths of the members; hence if the members are prismatic, the integration for one member yields $N_U N_L L/EA$, where L is the length of the member. Then a summation for all members yields the following unit-load equation for a **truss**:

$$\Delta = \sum \frac{N_U N_L L}{EA} \tag{10-5}$$

This equation shows that the displacement Δ at any joint of the truss can be found by the following procedure: (1) determine the axial forces N_U and N_L in all members due to the unit load and the actual loads, respectively; (2) form the expression $N_U N_L L/EA$ for each member; and (3) add these expressions for all members to obtain the displacement.

Only flexural deformations are apt to be important in a beam or plane frame. Therefore, the unit-load equation for a **beam** is

$$\Delta = \int \frac{M_U M_L \, dx}{EI} \tag{10-6}$$

This expression can be integrated for each member of the structure, and then the resulting terms can be summed for all members of the structure.

In general, it is possible to calculate displacements of structures by using any appropriate combination of terms from Eq. (10-4), depending upon the type of structure and loading.

TEMPERATURE EFFECTS

If the displacements are caused by effects other than loads, such as temperature changes, then we must use the appropriate expressions for $d\delta$, $d\theta$, $d\lambda$, and $d\phi$ in place of the expressions given in Eqs. (c), which are for the effects of loads only. For instance, an increase in temperature produces an increase in length (see Fig. 10-3a) given by the equation

$$d\delta = \alpha(\Delta T)\, dx$$

in which α is the coefficient of thermal expansion and ΔT is the increase in temperature (see Eq. 2-19). Then the unit-load equation (Eq. 10-3) takes the form

$$\Delta = \int N_U \alpha(\Delta T)\, dx \tag{10-7}$$

This equation may be used even when the temperature change ΔT varies along the axes of the members; it is necessary only to express ΔT as a function of x and then integrate. However, for the common case in which the temperature change is constant along the length of each member, we can replace the expression in Eq. (10-7) by a summation that includes all members:

$$\Delta = \Sigma\, N_U \alpha L(\Delta T) \tag{10-8}$$

where L is the length of a member. Thus, the term $N_U \alpha L(\Delta T)$ must be evaluated for each member of the structure, and then these terms must be summed to obtain the displacement.

When the temperature varies linearly from top to bottom of the member but is constant along the length, the deformation is of the type pictured in Fig. 10-3b in which the deformation $d\theta$ is

$$d\theta = \frac{\alpha(T_2 - T_1)\,dx}{h}$$

as given by Eq. (7-71) with h being the height of the beam, T_2 the temperature on the lower surface, and T_1 the temperature on the upper surface. Thus, under these conditions the unit-load equation is

$$\Delta = \int \frac{M_U \alpha(T_2 - T_1)\,dx}{h} \tag{10-9}$$

In these equations $d\theta$ is assumed to be positive when the top part of the beam shortens and the lower part lengthens, as pictured in Fig. 10-3b. Therefore, the bending moment M_U must be taken as positive for the same conditions; this means that M_U is positive when it produces compression on the top of the beam.

EVALUATION OF PRODUCT INTEGRALS

When evaluating the integrals in Eq. (10-4), we usually are dealing with members for which the material properties and cross-sectional dimensions are constant from one end of the member to the other. Therefore, the rigidities EA, EI, GA/f_s, and GI_p may be placed outside the integral signs. The remaining integrals are in the form of a product, such as

$$\int M_U M_L\,dx \tag{10-10}$$

These **product integrals** must be evaluated over the length of each member and then added for all members. For any particular member, each quantity (such as M_U or M_L) is a function of the distance x measured along the axis of the member; specifically, the quantity may be constant along the length, may vary linearly along the length, or may be a function of higher order, such as quadratic or cubic. To save time when performing calculations, these product integrals can be evaluated in advance and the results tabulated for ready use. A compilation of product integrals, covering the most commonly encountered functions, is given in Table 10-1. The table is presented in terms of the functions M_U and M_L, but it is apparent that these functions can be replaced by others, such as V_U and V_L or T_U and T_L. Illustrations of the use of the table are given in some of the examples.

Table 10-1 Values of product integrals $\int_0^L M_U M_L \, dx$

M_U \ M_L	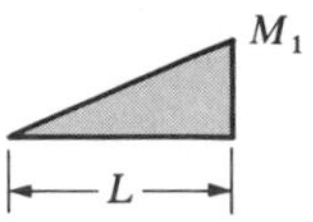	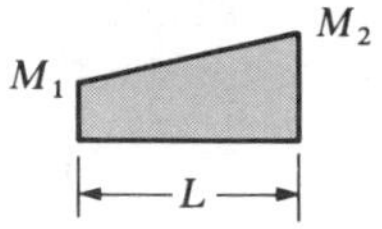	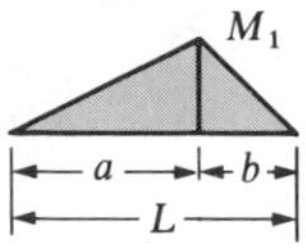	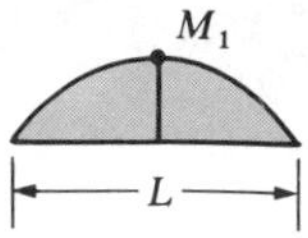
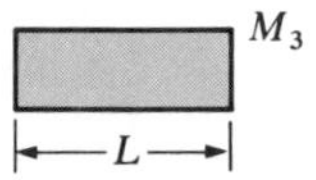	$\frac{L}{2} M_1 M_3$	$\frac{L}{2}(M_1 + M_2)M_3$	$\frac{L}{2} M_1 M_3$	$\frac{2L}{3} M_1 M_3$
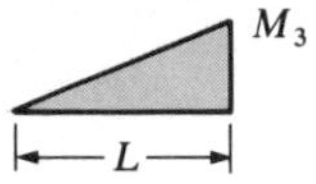	$\frac{L}{3} M_1 M_3$	$\frac{L}{6}(M_1 + 2M_2)M_3$	$\frac{L}{6}\left(1 + \frac{a}{L}\right) M_1 M_3$	$\frac{L}{3} M_1 M_3$
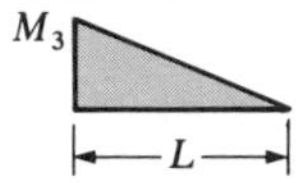	$\frac{L}{6} M_1 M_3$	$\frac{L}{6}(2M_1 + M_2)M_3$	$\frac{L}{6}\left(1 + \frac{b}{L}\right) M_1 M_3$	$\frac{L}{3} M_1 M_3$
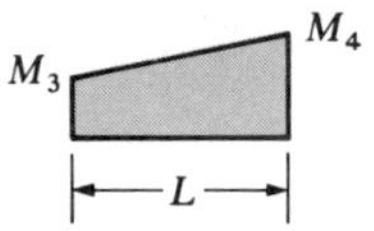	$\frac{L}{6} M_1(M_3 + 2M_4)$	$\frac{L}{6} M_1(2M_3 + M_4)$ $+ \frac{L}{6} M_2(M_3 + 2M_4)$	$\frac{L}{6}\left(1 + \frac{b}{L}\right) M_1 M_3$ $+ \frac{L}{6}\left(1 + \frac{a}{L}\right) M_1 M_4$	$\frac{L}{3} M_1(M_3 + M_4)$
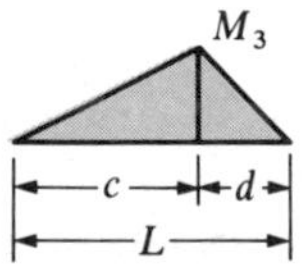	$\frac{L}{6}\left(1 + \frac{c}{L}\right) M_1 M_3$	$\frac{L}{6}\left(1 + \frac{d}{L}\right) M_1 M_3$ $+ \frac{L}{6}\left(1 + \frac{c}{L}\right) M_2 M_3$	For $c \le a$: $\frac{L}{3} M_1 M_3$ $- \frac{L(a - c)^2}{6ad} M_1 M_3$	$\frac{L}{3}\left(1 + \frac{cd}{L^2}\right) M_1 M_3$
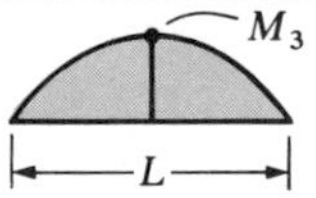	$\frac{L}{3} M_1 M_3$	$\frac{L}{3}(M_1 + M_2)M_3$	$\frac{L}{3}\left(1 + \frac{ab}{L^2}\right) M_1 M_3$	$\frac{8L}{15} M_1 M_3$
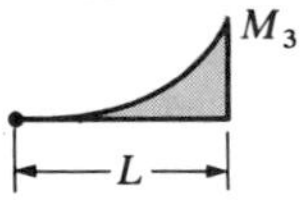	$\frac{L}{4} M_1 M_3$	$\frac{L}{12}(M_1 + 3M_2)M_3$	$\frac{L}{12}\left(1 + \frac{a}{L} + \frac{a^2}{L^2}\right) M_1 M_3$	$\frac{L}{5} M_1 M_3$

Note: All curves are second-degree parabolas with vertices shown by heavy dots.

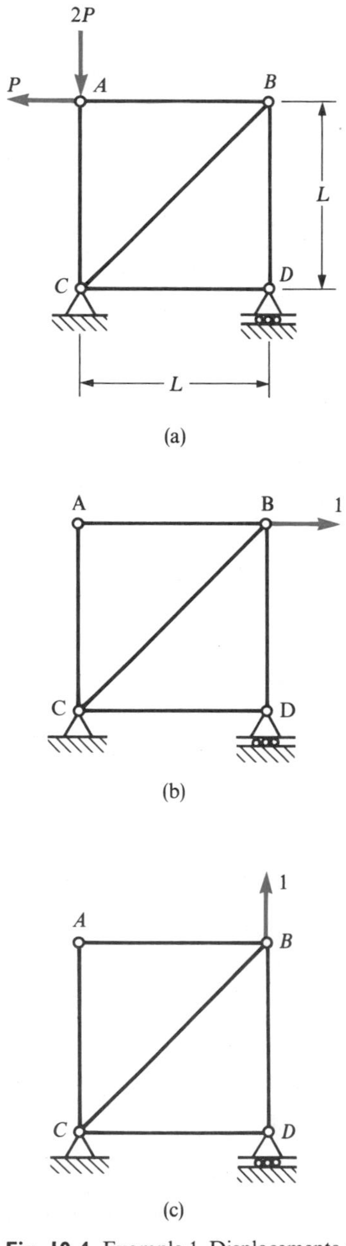

Fig. 10-4 Example 1. Displacements of a truss by the unit-load method

Example 1

The truss pictured in Fig. 10-4a is subjected to loads P and $2P$ at joint A. All members of the truss are assumed to be prismatic and to have the same axial rigidity EA. Calculate the horizontal and vertical displacements of joint B of the truss using the unit-load method.

Because the loads act only at the joints, the axial force in each member is constant throughout the length of the member. Therefore, we can use Eq. (10-5) to determine the desired deflections. It is helpful to record the calculations in a systematic manner, as shown in Table 10-2. The first two columns in the table identify the members of the truss and their lengths. The axial forces N_L, obtained by static-equilibrium analysis of the truss shown in Fig. 10-4a, are listed in column 3 of the table (tensile forces are positive).

Table 10-2 Calculations for Example 1

(1) Member	(2) Length	(3) N_L	(4) N_U	(5) $N_U N_L L$	(6) N_U	(7) $N_U N_L L$
AB	L	P	0	0	0	0
AC	L	$-2P$	0	0	0	0
BD	L	P	-1	$-PL$	1	PL
CD	L	0	0	0	0	0
CB	$\sqrt{2}L$	$-\sqrt{2}P$	$\sqrt{2}$	$-2.828PL$	0	0
				$-3.828PL$		PL

In order to find the horizontal displacement δ_h of joint B, we introduce a horizontal unit load on the structure at B (see Fig. 10-4b). The axial forces N_U produced by this unit load are given in column 4 of the table; again, tensile forces are positive. Next, the products $N_U N_L L$ are calculated for each member and summed (column 5). Dividing this result by EA gives the desired displacement (see Eq. 10-5):

$$\delta_h = -3.828\frac{PL}{EA}$$

The negative sign in this expression means that the displacement is in the direction opposite to the direction of the unit load; that is, toward the left.

The same general procedure is used to find the vertical displacement δ_v of joint B. The corresponding unit load (taken as positive when upward) is portrayed in Fig. 10-4c, and the axial forces N_U for this loading condition are listed in column 6 of the table. In the last column, the products $N_U N_L L$ are calculated and summed. Finally, dividing the sum by EA yields

$$\delta_v = \frac{PL}{EA}$$

Because this result is positive, we know that the vertical displacement of joint B produced by the loads P and $2P$ is upward.

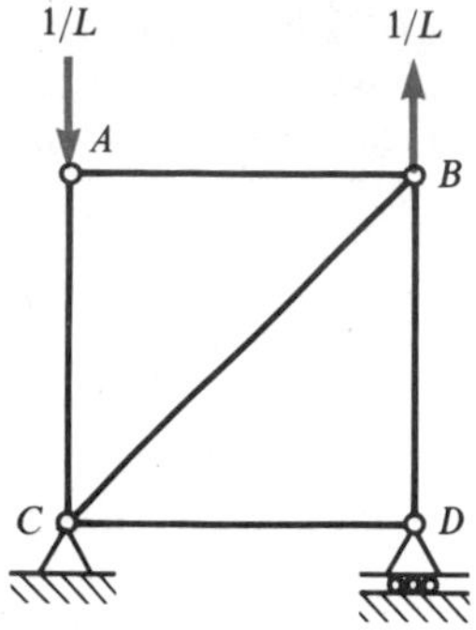

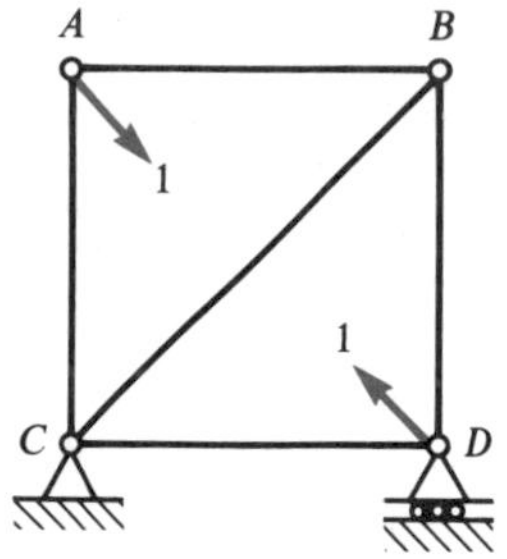

Fig. 10-5 Example 2

Example 2

In this example we again consider the truss shown in Fig. 10-4a. However, rather than find the displacement of a joint, let us determine the angle of rotation of member AB and the change in distance between joints A and D.

The unit load corresponding to a rotation is a unit couple. In this example we create a unit couple consisting of equal and opposite forces acting at the ends of member AB (see Fig. 10-5a). Each force has a magnitude equal to the unit couple divided by the length of bar AB. We can easily show that the displacement corresponding to this couple is the counterclockwise rotation of bar AB; we merely need to note that the external work done by these two forces during the virtual deformation of the truss is

$$W_{\text{ext}} = \frac{1}{L}(\delta_a) + \frac{1}{L}(\delta_b) = \frac{1}{L}(\delta_a + \delta_b) \tag{d}$$

where δ_a is the downward displacement of joint A and δ_b is the upward displacement of joint B. The units of the terms appearing in Eq. (d) are consistent because each "1" represents the unit couple and has units of length times force. The sum of the two displacements δ_a and δ_b, when divided by the length L of the member, is the angle of rotation of member AB:

$$\theta_{ab} = \frac{\delta_a + \delta_b}{L} \tag{e}$$

From Eqs. (d) and (e) we get

$$W_{\text{ext}} = 1 \cdot \theta_{ab} \tag{f}$$

as expected. Thus, the application of a unit couple in the form of the two forces shown in Fig. 10-5a will enable us to find the angle of rotation θ_{ab}.

The calculations for this example are given in Table 10-3; the first three columns are the same as in Table 10-2, but the last four columns differ because we are calculating different displacements. Column 4 contains the axial forces N_U for the loads shown in Fig. 10-5a, and column 5 gives the products $N_U N_L L$. From the sum of those terms we get

$$\theta_{ab} = \frac{3P}{EA}$$

in which a positive result means that the bar rotates counterclockwise. Thus, the

Table 10-3 Calculations for Example 2

(1) Member	(2) Length	(3) N_L	(4) N_U	(5) $N_U N_L L$	(6) N_U	(7) $N_U N_L L$
AB	L	P	0	0	$-1/\sqrt{2}$	$-0.707PL$
AC	L	$-2P$	$-1/L$	$2P$	$-1/\sqrt{2}$	$1.414PL$
BD	L	P	$1/L$	P	$-1/\sqrt{2}$	$-0.707PL$
CD	L	0	0	0	$-1/\sqrt{2}$	0
CB	$\sqrt{2}L$	$-\sqrt{2}P$	0	0	1	$-2PL$
				$3P$		$-2PL$

desired angle of rotation produced by the loads P and $2P$ has been found. A similar procedure can be followed to determine the angle of rotation for any other member.

Next, let us consider the change in distance between points A and D of the truss shown in Fig. 10-4a. In other words, we wish to determine the relative translation δ_{ad} of joints A and D along the line that joins them. The corresponding unit load consists of two equal and opposite unit forces acting along the line from A to D (see Fig. 10-5b). The resulting axial forces N_U in the truss are listed in column 6 of Table 10-3, and the products $N_U N_L L$ are listed in column 7. Thus, the relative translation of joints A and D is

$$\delta_{ad} = -\frac{2PL}{EA}$$

in which the minus sign indicates that the distance between points A and D has increased (that is, the relative translation is opposite to the sense of the unit loads).

Example 3

Let us consider the effects of a uniform temperature change in one of the bars of the truss shown in Fig. 10-4a. Assume that member BD has its temperature increased uniformly by an amount ΔT, so that the length of the member is increased by $\alpha(\Delta T)L$. We wish to calculate the horizontal displacement of joint B of the truss.

To obtain the displacement of a joint, we use Eq. (10-8). However, because only one bar of the truss has a change in length due to a temperature change, there is only one term in the summation. When finding the horizontal displacement of joint B due to the temperature change, we use the unit load shown in Fig. 10-4b, and therefore we take N_U from column 4 of Table 10-2. Thus, the force in bar BD is $N_U = -1$, and the horizontal displacement (from Eq. 10-8) is

$$\delta_h = -\alpha L(\Delta T)$$

in which the minus sign means that the displacement is toward the left (that is, opposite to the direction of the unit load). By a similar procedure we can easily calculate any other joint displacements caused by the temperature change.

Example 4

A prismatic cantilever beam AB supporting a uniform load of intensity q acting over part of the span is shown in Fig. 10-6a. We wish to determine the deflection δ and angle of rotation θ at the free end B.

When finding beam deflections, we use the form of the unit-load equation that considers only the effects of flexural deformations (Eq. 10-6). If we take the origin of coordinates at the left end A and measure x to the right (Fig. 10-6a), the equations for the bending moment M_L due to the load are

$$M_L = -\frac{q}{2}(a - x)^2 \qquad (0 \le x \le a)$$

$$M_L = 0 \qquad (a \le x \le L)$$

in which a positive bending moment causes compression on the top of the beam.

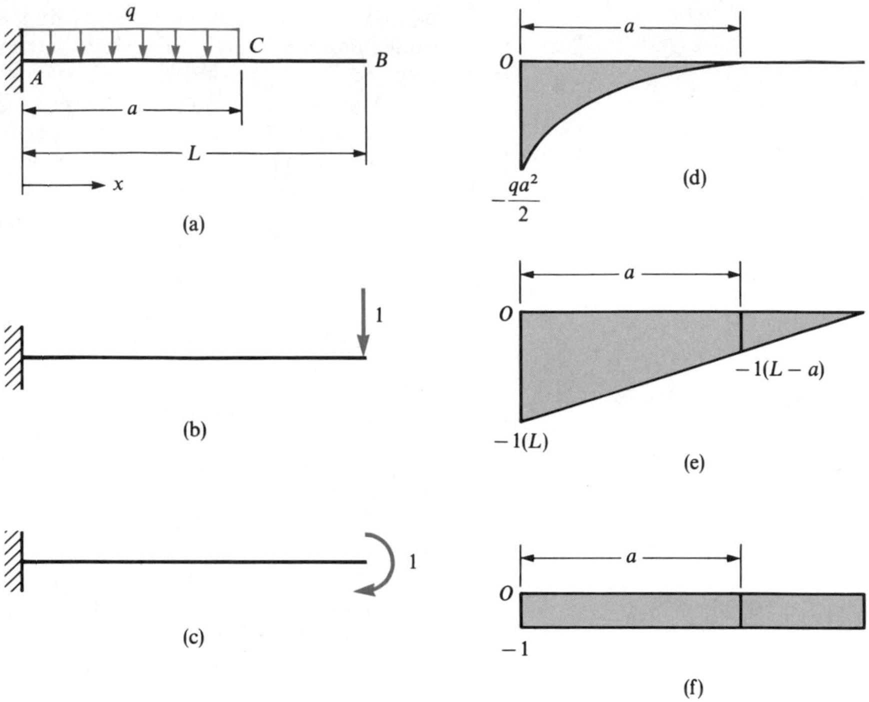

Fig. 10-6 Example 4. Deflection and angle of rotation of a beam by the unit-load method

The unit load corresponding to a downward deflection δ is shown in Fig. 10-6b; this load produces a bending moment as follows:

$$M_U = -1(L - x) \qquad (0 \le x \le L)$$

Substituting M_U and M_L into Eq. (10-6) and integrating, we obtain the deflection of point B:

$$\delta = \frac{1}{EI}\int_0^a (-1)(L - x)\left(-\frac{q}{2}\right)(a - x)^2\,dx = \frac{qa^3}{24EI}(4L - a)$$

This result is positive; hence the deflection is downward.

The procedure for finding the angle of rotation θ is similar except that the unit load is a unit couple, as shown in Fig. 10-6c. The bending moment due to this load is

$$M_U = -1 \qquad (0 \le x \le L)$$

and the unit-load equation becomes

$$\theta = \frac{1}{EI}\int_0^a (-1)\left(-\frac{q}{2}\right)(a - x)^2\,dx = \frac{qa^3}{6EI}$$

This result is positive, which indicates that the angle θ has the same sense (clockwise) as the unit couple.

An alternative procedure for finding the displacements is to use Table 10-1 for the product integrals. The bending-moment diagram for the load q is a parabola over part AC of the beam, as shown in Fig. 10-6d. The moment diagrams in this region of the beam for the unit loads are a trapezoid and a rectangle, respectively (see Figs. 10-6e and f). Taking the case of a parabola and trapezoid from Table 10-1, we get for the value of the product integral

$$\frac{a}{12}[-1(L-a)-3(L)]\left(-\frac{qa^2}{2}\right)=\frac{qa^3}{24}(4L-a)$$

Now dividing by EI, we obtain the same expression for the deflection δ as before. Also, the case of a parabola and a rectangle* yields

$$\frac{a}{12}(-4)\left(-\frac{qa^2}{2}\right)=\frac{qa^3}{6}$$

which leads to the previous result for the angle of rotation θ. The use of the table of product integrals in conjunction with bending-moment diagrams is often faster than writing the equations for the bending moments and then integrating.

Example 5

A beam ABC with a span of length L and an overhang of length b (Fig. 10-7a) undergoes a change in temperature such that the top surface of the beam is at temperature T_1 while the lower surface is at temperature T_2. Calculate the vertical deflection δ_c of the end of the overhanging portion of the beam.

The unit load corresponding to δ_c is pictured in Fig. 10-7b. The bending moment caused by this unit load, for region AB, is

$$M_U=-1\left(\frac{b}{L}\right)x \qquad (0 \le x \le L)$$

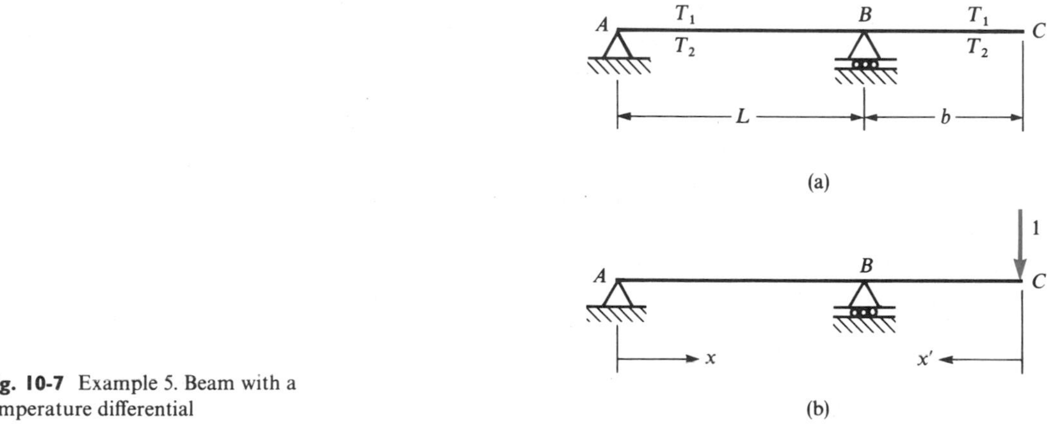

Fig. 10-7 Example 5. Beam with a temperature differential

* Note that the formulas for a rectangle can be obtained from Table 10-1 by taking a trapezoid with $M_1 = M_2$.

where x is the distance measured from A toward the right. Also, the bending moment in region BC is

$$M_U = -1(x') \qquad (0 \le x' \le b)$$

where x' is measured from C toward the left. In both parts of the beam, the bending moment M_U is negative because it causes tension on the top of the beam.

Knowing the bending moments, we can now substitute directly into Eq. (10-9) and obtain the deflection:

$$\delta_c = \int_0^L -1\left(\frac{b}{L}\right)(x)\frac{\alpha(T_2 - T_1)}{h}dx + \int_0^b -1(x')\frac{\alpha(T_2 - T_1)}{h}dx'$$

from which

$$\delta_c = \frac{\alpha b(T_1 - T_2)(L + b)}{2h} \tag{g}$$

where h is the height of the beam and α is the coefficient of thermal expansion. If the calculated value of δ_c is negative, then the deflection is upward; this condition exists whenever T_2 is greater than T_1.

Example 6

The plane frame ABC shown in Fig. 10-8a has a fixed support at point A and carries a vertical load P at the free end C. The members AB and BC are joined rigidly at B, and both members of the frame have constant flexural rigidity EI. Determine the horizontal deflection δ_h, the vertical deflection δ_v, and the angle of rotation θ at point C.

The bending moments M_L caused by the load P are portrayed in Fig. 10-8b, in which the bending-moment diagrams are plotted on the sides of the members that are in tension. The unit loads corresponding to the horizontal deflection, vertical deflection, and angle of rotation at point C are shown in the last three parts of the figure. In each case the complete bending-moment diagram for M_U is also shown, again plotted on the tension sides of the members.

Knowing the bending moments, we can now obtain the deflections either by integrating (see Eq. 10-6) or by using the formulas for the product integrals listed in Table 10-1. The latter approach seems simpler in this example. For instance, to find δ_h we take M_U from Fig. 10-8c and M_L from Fig. 10-8b. Then using Table 10-1 we get

$$\frac{H}{2}(1)(H)(PL)$$

Dividing this expression by EI, we get

$$\delta_h = \frac{PLH^2}{2EI} \tag{h}$$

for the horizontal deflection of point C.

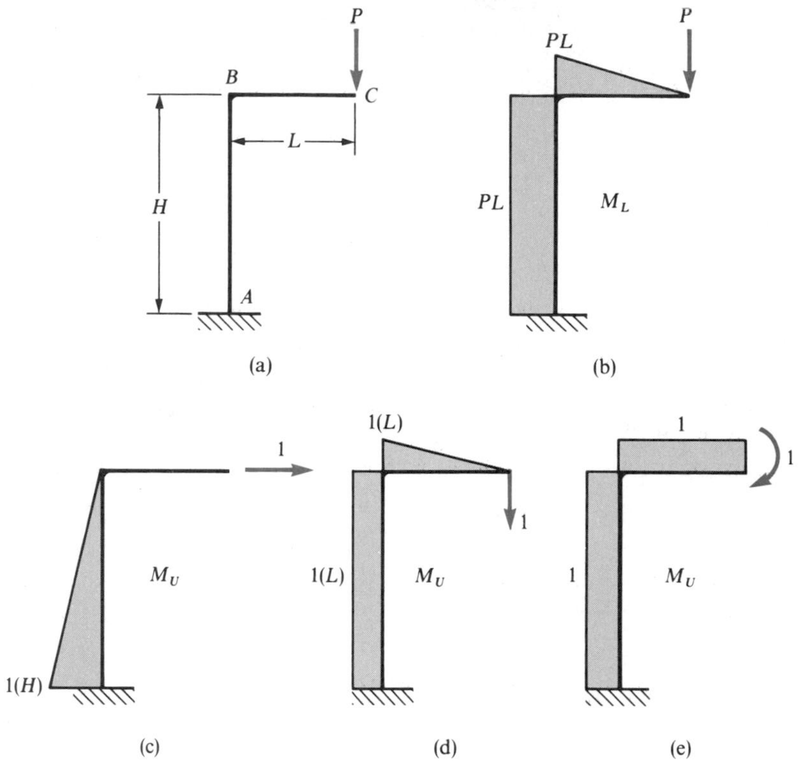

Fig. 10-8 Examples 6 and 7. Displacements of a plane frame

The vertical deflection is calculated in a similar manner from the diagrams of Figs. 10-8b and d. For member BC we have the case of two triangles, and for member AB the case of two rectangles; thus, from Table 10-1 we find

$$\frac{L}{3}(1)(L)(PL) + \frac{H}{2}(2L)(PL) = \frac{PL^3}{3} + PL^2H$$

and the vertical deflection is

$$\delta_v = \frac{PL^2(L + 3H)}{3EI} \tag{i}$$

Finally, using Figs. 10-8b and e and Table 10-1, we obtain

$$\frac{L}{2}(PL)(1) + \frac{H}{2}(2)(PL) = \frac{PL^2}{2} + PLH$$

from which

$$\theta = \frac{PL(L + 2H)}{2EI} \tag{j}$$

Thus, the desired displacements of the frame have been determined.

Example 7

Referring to the plane frame described in Example 6, let us determine the additional contributions to the deflections δ_h, δ_v, and θ caused by axial deformations in the members. Assume that both members have constant axial rigidity EA.

The additional deflections will be found with the aid of Eq. (10-5). The axial forces N_L in that equation are the axial forces caused by the load P shown in Fig. 10-8a; the only such force is $N_L = -P$ in member AB. To find the horizontal deflection δ_h, we take the axial forces N_U from Fig. 10-8c. Since the only such force is $N_U = 1$ in member BC, we conclude that the horizontal deflection δ_h is not affected by the presence of axial deformations.

To obtain the vertical deflection δ_v, we observe from Fig. 10-8d that the only axial force caused by the unit load is $N_U = -1$ for member AB. It follows that the vertical deflection due to axial deformations (from Eq. 10-5) is

$$\sum \frac{N_U N_L L}{EA} = \frac{(-1)(-P)(H)}{EA} = \frac{PH}{EA}$$

This quantity can be added to the earlier result obtained in Example 6 to give the total vertical deflection of joint C:

$$\delta_v = \frac{PL^2(L + 3H)}{3EI} + \frac{PH}{EA} \tag{k}$$

When numerical values are substituted into this expression, we find that the last term, representing the effects of axial deformations, is extremely small compared to the first term. For this reason it is common practice to consider only the effects of flexural deformations when analyzing plane frames and to disregard the contribution of the axial deformations.

To complete this example, we consider the angle of rotation θ and observe that the corresponding unit load (Fig. 10-8e) produces no axial forces in the members. Hence, the angle θ is unaffected by the presence of axial deformations.

Example 8

A curved bar AB has a centerline in the form of a quarter circle of radius R, as shown in Fig. 10-9a. The bar has a fixed support at A and carries a vertical load P at the free end B. Obtain an expression for the horizontal deflection δ_h of point B.

The unit-load equation for finding deflections of the curved bar can be expressed in the general form

$$\Delta = \int \frac{M_U M_L \, ds}{EI} \tag{10-11}$$

where ds, equal to $R\,d\theta$, is the length of an element mn of the bar. It is assumed that the bar is thin compared to the radius R so that the preceding formula, originally derived for the bending of a straight bar, can be used. Also, only the effects of flexural deformations are considered in Eq. (10-11).

The bending moment M_L caused by the load P is $M_L = -PR\cos\theta$, where a positive moment is assumed to cause compression on the outside of the curved

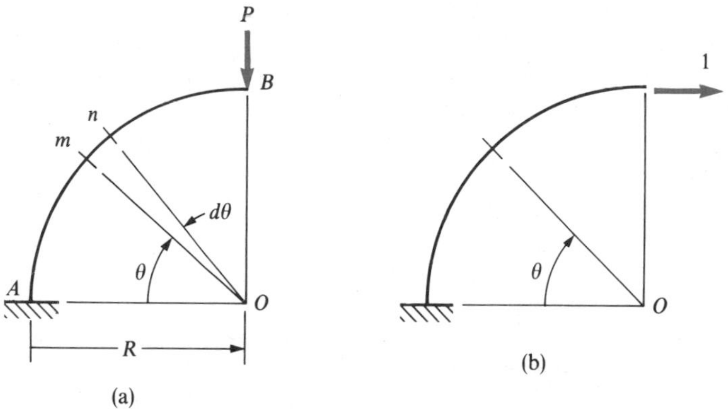

Fig. 10-9 Example 8. Deflection of a curved bar

bar. Also, the bending moment M_U due to a horizontal unit load (Fig. 10-9b) is $M_U = -R(1 - \sin\theta)$. Substituting for M_U and M_L in Eq. (10-11) and integrating, we obtain

$$\delta_h = \frac{1}{EI}\int_0^{\pi/2}(-R)(1 - \sin\theta)(-PR\cos\theta)R\,d\theta = \frac{PR^3}{2EI} \tag{1}$$

Thus, the horizontal deflection of point B has been found by the unit-load method.

10.4 RECIPROCAL THEOREMS

The reciprocal theorems are important concepts in applied mechanics and structural analysis. They apply only to linearly elastic structures (that is, structures for which the principle of superposition is valid). Thus, two basic conditions must be satisfied: (1) the material must follow Hooke's law, and (2) the displacements must be small enough that all calculations can be based upon the undeformed geometry of the structure. We will use the concepts of work and strain energy to derive the theorems.

RECIPROCAL-DISPLACEMENT THEOREM

To illustrate the reciprocal-displacement theorem, let us take as an example a cantilever beam AB subjected to a concentrated load P at the free end (Fig. 10-10a). We can obtain the deflection at the midpoint C of this beam from the formulas in Appendix G; thus

$$\delta_{cb} = \frac{5PL^3}{48EI}$$

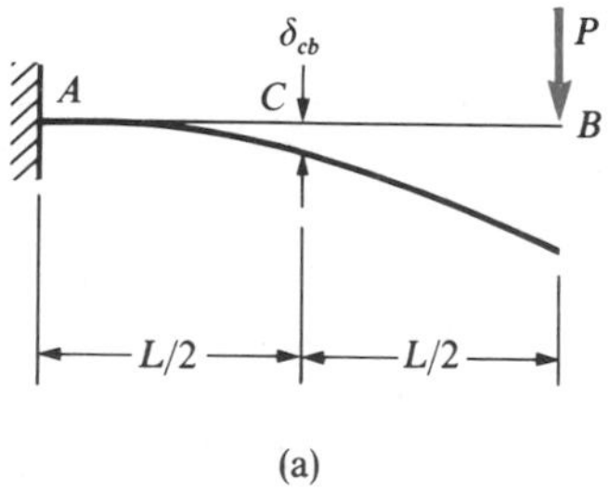

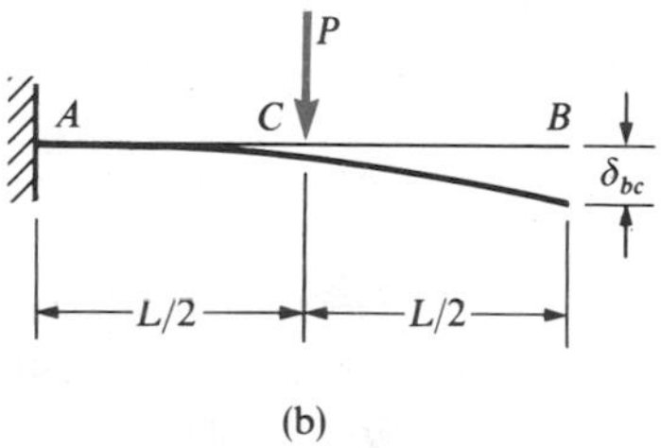

Fig. 10-10 Reciprocal-displacement theorem

The subscript notation used with the symbol δ is based upon the following scheme: the first subscript denotes the point at which the deflection occurs, and the second subscript denotes the point at which the load is applied. Thus, the symbol δ_{cb} identifies the deflection at point C caused by a load acting at point B.

Now let us consider the same cantilever beam subjected to a load P acting at the midpoint C (Fig. 10-10b). In this case we wish to find the deflection at the free end B, denoted by the symbol δ_{bc}. Again referring to the formulas in Appendix G, we find

$$\delta_{bc} = \frac{5PL^3}{48EI}$$

Thus, we observe that the deflection at C due to the load acting at B is equal to the deflection at B due to the load acting at C. This statement is an example of the **reciprocal-displacement theorem**.

To prove the theorem in more general terms, let us consider a structure of any type (that is, a truss, beam, three-dimensional body of arbitrary shape, and so on). For convenience a simple beam will be discussed (Fig. 10-11), but any other structure would be satisfactory. Also, let us consider two states of loading acting on the structure. In the first state of loading, a force P acts at any point A (Fig. 10-11a); in the second state, the same load P acts at any other point B (Fig. 10-11b). The deflections at points A and B for the first state of loading are denoted by δ_{aa} and δ_{ba}, respectively, in accordance with the subscript notation already described. In an analogous manner, the deflections for the second state of loading are identified as δ_{ab} and δ_{bb}.

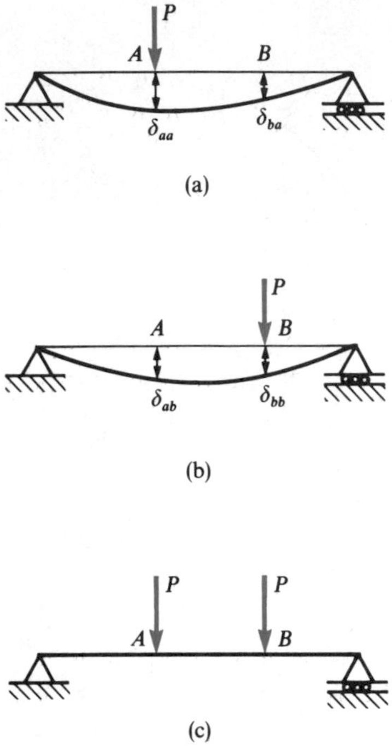

Fig. 10-11 Reciprocal-displacement theorem

The deflections of the beam may be described by using the notion of **correspondence** between loads and displacements (see Section 10.3). For instance, the deflections δ_{aa} and δ_{ab} both correspond to the load P acting at A (Fig. 10-11a). To clarify this idea, recall that the displacement corresponding to a force is the displacement at the point where the force acts; furthermore, this displacement is measured along the line of action of the force and is positive in the direction of the force. However, the displacement need not be *caused* by the force to which it corresponds. In our example (Fig. 10-11), deflection δ_{aa} is caused by the force P from the first state of loading and the deflection δ_{ab} is caused by the force P from the second state of loading. Nevertheless, both deflections correspond to the load P of Fig. 10-11a. Similarly, the deflections δ_{ba} and δ_{bb} both correspond to the load P from the second state of loading, although δ_{ba} is caused by the first load and δ_{bb} is caused by the second load. This concept of correspondence as a means of identifying a displacement will be very useful in subsequent discussions.

Now returning to the derivation of the reciprocal-displacement theorem, let us suppose that both forces P act simultaneously on the beam (Fig. 10-11c). If the material of the beam is linearly elastic and if the deflections are small, we can use the principle of superposition to obtain the deflections for this beam. The deflection corresponding to the load P acting at A is $\delta_{aa} + \delta_{ab}$, and the deflection corresponding to the load P acting at B is $\delta_{ba} + \delta_{bb}$. Knowing these deflections, we can easily calculate the work done by the two loads P as they are slowly and simultaneously applied to the beam. This work, equal to the total strain energy U of the beam, is

$$U = \frac{1}{2}P(\delta_{aa} + \delta_{ab}) + \frac{1}{2}P(\delta_{ba} + \delta_{bb}) \tag{a}$$

as obtained from Clapeyron's theorem (Eq. 2-32).

The total strain energy of the beam subjected to two loads (Fig. 10-11c) does not depend upon the order in which the two loads are applied. Because the beam behaves linearly, the strain energy must be the same when the two loads are applied simultaneously and when one load is applied before the other load. The final state of the beam is the same in both cases. Let us assume that the load at A is applied first, followed by the load at B. Then the strain energy of the beam during the application of the first load is

$$\frac{1}{2}P\delta_{aa} \tag{b}$$

because this load causes the deflection δ_{aa}, shown in Fig. 10-11a. When the second load is applied, an additional deflection results at B equal to δ_{bb}; hence the second load does work equal to

$$\frac{1}{2}P\delta_{bb} \tag{c}$$

and an equal amount of strain energy is developed in the beam. However, we must not overlook the fact that, while the load at B is being applied, the load P acting at A undergoes an additional deflection δ_{ab}. The corresponding amount of work done by that load is

$$P\delta_{ab} \tag{d}$$

Thus, this amount of additional strain energy is produced. Equation (d) does not contain the factor 1/2 because the force P remains constant during the time that the additional deflection δ_{ab} occurs. Summing Eqs. (b), (c), and (d), we get the total strain energy for the case when one load is applied before the other:

$$U = \frac{1}{2}P\delta_{aa} + \frac{1}{2}P\delta_{bb} + P\delta_{ab} \tag{e}$$

This amount of strain energy must be equal to the strain energy produced when the two loads are applied simultaneously (Eq. a). Equating the two expressions for strain energy leads to the following result:

$$\delta_{ab} = \delta_{ba} \tag{10-12}$$

This equation represents the **reciprocal-displacement theorem**, which may be stated as follows:

> *The displacement at A due to a load acting at B is equal to the displacement at B due to the same load acting at A.*

The positive directions of the displacements are understood to be the same as the positive directions of the corresponding loads.

The reciprocal-displacement theorem is also applicable if one load is a force and the other is a couple, or if both loads are couples. To illustrate the former possibility, consider again a simple beam subjected to two states of loading (Fig. 10-12). The first loading consists of a couple M acting at point A (Fig. 10-12a). The displacement corresponding to M in the first beam is the angle of rotation θ_{aa}, and in the second beam it is the angle θ_{ab}. The second loading is the force P acting at B. Repeating the same steps as in the preceding derivation, we get the following expression for the strain energy of the beam when the two loads M and P are applied simultaneously:

$$U = \frac{1}{2}M(\theta_{aa} + \theta_{ab}) + \frac{1}{2}P(\delta_{ba} + \delta_{bb})$$

When the couple M is applied first, followed by the force P, the strain energy is

$$U = \frac{1}{2}M\theta_{aa} + \frac{1}{2}P\delta_{bb} + M\theta_{ab}$$

Equating these two expressions for strain energy yields

$$M\theta_{ab} = P\delta_{ba} \qquad (10\text{-}13)$$

If the loads M and P are numerically equal, then θ_{ab} and δ_{ba} also will be numerically equal. Therefore, for this case we may state the reciprocal-displacement theorem as follows:

> *The angle of rotation at A due to a force acting at B is equal numerically to the displacement at B due to a couple acting at A if the force and couple are numerically equal.*

Of course, consistent units must be used for all quantities.

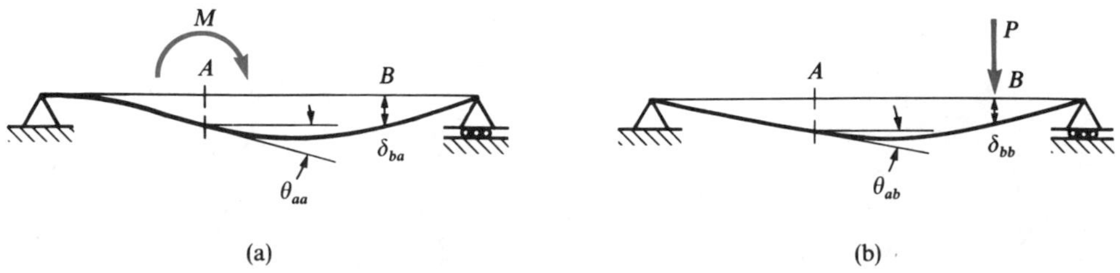

Fig. 10-12 Reciprocal-displacement theorem

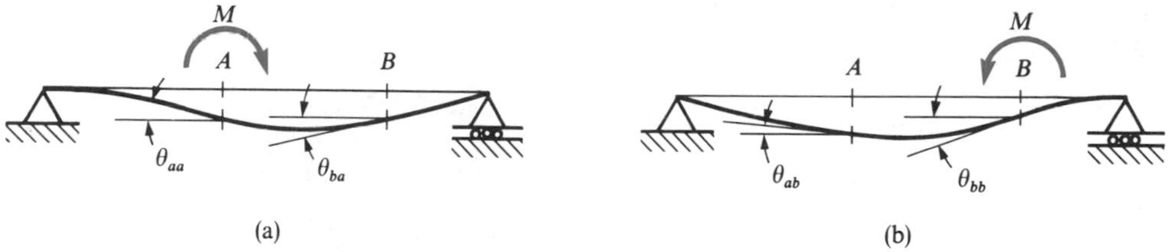

Fig. 10-13 Reciprocal-displacement theorem

If both loads acting on the structure consist of couples M (Fig. 10-13), then we find

$$\theta_{ab} = \theta_{ba} \tag{10-14}$$

In this situation the theorem is as follows:

> *The angle of rotation at A due to a couple acting at B is equal to the angle of rotation at B due to the same couple acting at A.*

The positive senses of the angles are understood to be the same as the positive senses of the corresponding couples.

The preceding derivations of the reciprocal-displacement theorem used a simple beam for illustrative purposes. Any other type of structure, such as a truss, frame, or massive body, could have been used, because the derivation was based only on strain-energy considerations and the principle of superposition. Thus, the theorem is quite general and applies to structures having axial, flexural, shearing, or torsional deformations. The only restriction on the use of the theorem is that the principle of superposition must be valid, which requires that the structure be linearly elastic. The theorem was first derived by J. C. Maxwell and published in 1864 (see Ref. 10-1); it is often called **Maxwell's reciprocal theorem**.

RECIPROCAL-WORK THEOREM

The reciprocal-work theorem is much more general than the reciprocal-displacement theorem and contains the latter as a special case. To derive the theorem, let us consider any linearly elastic body for which the principle of superposition holds (Fig. 10-14). This body could represent a beam, truss, frame, or other kind of structure. Two states of loading on the structure must be considered. In the first state (Fig. 10-14a) there are m loads $P_1, P_2, \ldots, P_m$; in the second state (Fig. 10-14b) there are n loads Q_1, $Q_2, \ldots, Q_n$. The displacements in the first state are designated by the symbol δ, with the subscript identifying the particular load to which each displacement corresponds. For instance, δ_{Q_2} represents the displacement corresponding to the force Q_2. It follows that this displacement must be measured in the direction of the force Q_2, although there is no implication that the point where Q_2 is applied moves *only* in the direction of Q_2. This

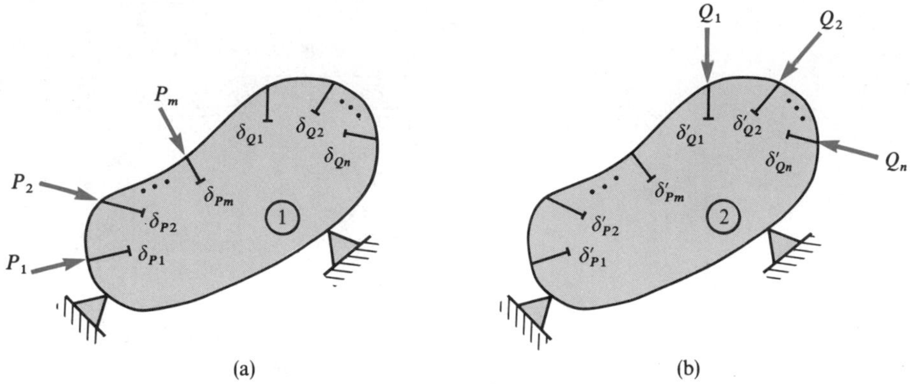

Fig. 10-14 Reciprocal-work theorem

point may also have a component of displacement normal to the direction of δ_{Q_2}, but this component does not enter into our discussion because it does not correspond to Q_2.

In the second state of loading (Fig. 10-14b), the situation is similar; the loads Q cause displacements, some of which correspond to the forces P and some of which correspond to the forces Q. These displacements are denoted by the symbol δ', and again the subscripts denote the forces to which the displacements correspond.

To derive the reciprocal-work theorem, we use the same ideas about strain energy that we used in deriving the reciprocal-displacement theorem. If both load systems P and Q are applied simultaneously to the body, the strain energy (equal to the work done by the forces) is

$$U = \frac{1}{2}P_1(\delta_{P_1} + \delta'_{P_1}) + \frac{1}{2}P_2(\delta_{P_2} + \delta'_{P_2}) + \cdots + \frac{1}{2}P_m(\delta_{P_m} + \delta'_{P_m})$$
$$+ \frac{1}{2}Q_1(\delta_{Q_1} + \delta'_{Q_1}) + \frac{1}{2}Q_2(\delta_{Q_2} + \delta'_{Q_2}) + \cdots + \frac{1}{2}Q_n(\delta_{Q_n} + \delta'_{Q_n}) \quad \text{(f)}$$

This strain energy must be the same as the strain energy obtained when we apply first the entire P-load system and then the entire Q-load system. When the loads P are applied alone, the strain energy is

$$\frac{1}{2}P_1\delta_{P_1} + \frac{1}{2}P_2\delta_{P_2} + \cdots + \frac{1}{2}P_m\delta_{P_m} \quad \text{(g)}$$

When the second set of loads is applied, we get the following amount of strain energy due to the work done by the loads Q:

$$\frac{1}{2}Q_1\delta'_{Q_1} + \frac{1}{2}Q_2\delta'_{Q_2} + \cdots + \frac{1}{2}Q_n\delta'_{Q_n} \quad \text{(h)}$$

At the same time, we get the following additional amount of strain energy due to the work done by the loads P:

$$P_1\delta'_{P_1} + P_2\delta'_{P_2} + \cdots + P_m\delta'_{P_m} \quad \text{(i)}$$

Therefore, the total strain energy (for the case when the loads P are applied first, followed by the loads Q) is the sum of expressions (g), (h), and (i). Equating this sum with the strain energy associated with simultaneous application of the loads (see Eq. f) gives

$$P_1\delta'_{P_1} + P_2\delta'_{P_2} + \cdots + P_m\delta'_{P_m} = Q_1\delta_{Q_1} + Q_2\delta_{Q_2} + \cdots + Q_n\delta_{Q_n}$$

or

$$\sum_{i=1}^{m} P_i\delta'_{P_i} = \sum_{j=1}^{n} Q_j\delta_{Q_j} \tag{10-15}$$

The expression on the left-hand side of this equation is the sum of the products of the P forces and their corresponding displacements caused by the Q forces. On the right-hand side appears the sum of the products of the Q forces and their corresponding displacements caused by the P forces. This equation represents the **reciprocal-work theorem**, and it may be stated as follows:

> *The work done by the forces in the first state of loading when they move through their corresponding displacements in the second state of loading is equal to the work done by the forces in the second state of loading when they move through their corresponding displacements in the first state of loading.*

The reciprocal-work theorem applies to both forces and couples. For instance, P_i may represent either a force or a couple, and then the corresponding displacement δ_{P_i} is either a translation or an angle of rotation, respectively.

Although in the derivation of the reciprocal-work theorem we portrayed the two sets of forces as acting at different points of the structure (see Fig. 10-14), it is not necessary that they do so. The force Q_1, for example, could act at the same point on the body as one of the P forces; it could even act in the same direction and have the same magnitude. In other words, no restriction is imposed on either the P forces or the Q forces as to the number of forces, the locations of their points of application, or their directions. This generality makes the reciprocal-work theorem a very useful principle in structural mechanics. As in the case of the reciprocal-displacement theorem, the reciprocal-work theorem is valid only for structures for which the principle of superposition may be used. The theorem was derived by E. Betti (Ref. 10-5) and Lord Rayleigh (Refs. 10-6 through 10-8), and it is often called the **Betti-Rayleigh reciprocal theorem**.

It is evident that the reciprocal-displacement theorem is a special case of the reciprocal-work theorem. For instance, we can apply the reciprocal-work theorem to the two cases of loading shown in Figs. 10-11a and b; then we obtain $P\delta_{ab} = P\delta_{ba}$, which immediately gives Eq. (10-12) for the reciprocal-displacement theorem. Similarly, application of the work theorem to the two loading conditions pictured in Fig. 10-12 gives $M\theta_{ab} = P\delta_{ba}$, which is the same as Eq. (10-13). Finally, we can get Eq. (10-14) by applying the reciprocal-work theorem to the two states of loading shown in Fig. 10-13.

10.5 STRAIN ENERGY AND COMPLEMENTARY ENERGY

The dual concepts of strain energy and complementary energy provide the bases for some extremely powerful methods of analysis. In their most general form, these methods are applicable to both linear and nonlinear structures. Our approach will be to describe the concepts for nonlinear structures and then to consider linear structures as a special case. We used this procedure when discussing the unit-load method in Section 10.3, where we first derived Eq. (10-3) for nonlinear structures and then specialized to Eq. (10-4) for linear structures.*

Nonlinearities in the behavior of a structure have two primary causes. The most obvious cause is a material having a nonlinear stress-strain curve; in this case we refer to the structure as having **material nonlinearities**. The second cause of nonlinearities is a change in the geometry of the structure. This situation occurs whenever the displacements of the structure alter the action of the applied loads or the reactions. An example is a column with an eccentric axial load (Section 9.4), for which we observed that lateral deflections of the column (even though small in magnitude) have a significant effect upon the deflections and bending moments in the column. Another example is a structure with large deflections, such as the beam described in Section 7.12. In both of these examples, the material of the beam is assumed to follow Hooke's law, but, because of the changing geometry of the deformed structure, the deflections and stress resultants are related nonlinearly to the applied loads. These examples are illustrations of **geometric nonlinearities**. When analyzing a nonlinear structure, it is important to remember that the principle of superposition is not applicable except in very special ways.

Independently of whether material or geometric nonlinearities are present, we assume that the material of the structure is **elastic**. This requirement is necessary in order that the various energy principles, including conservation of energy, will be valid.

To illustrate the energy concepts, let us consider a prismatic bar subjected to an axial force P that produces a uniformly distributed stress $\sigma = P/A$ (Fig. 10-15a). The strain in the bar is $\epsilon = \delta/L$, where δ is the elongation of the bar and L is the length. The material of the bar is assumed to have the nonlinear stress-strain curve shown in Fig. 10-15b. The load-displacement diagram (Fig. 10-15c) has the same shape as the stress-strain curve.

The **work** done by the load P during loading of the bar is

$$W = \int_0^{\delta} P_1 \, d\delta_1$$

* The expression **linear structure** is often used for a structure that behaves in a linearly elastic manner and for which the principle of superposition is applicable.

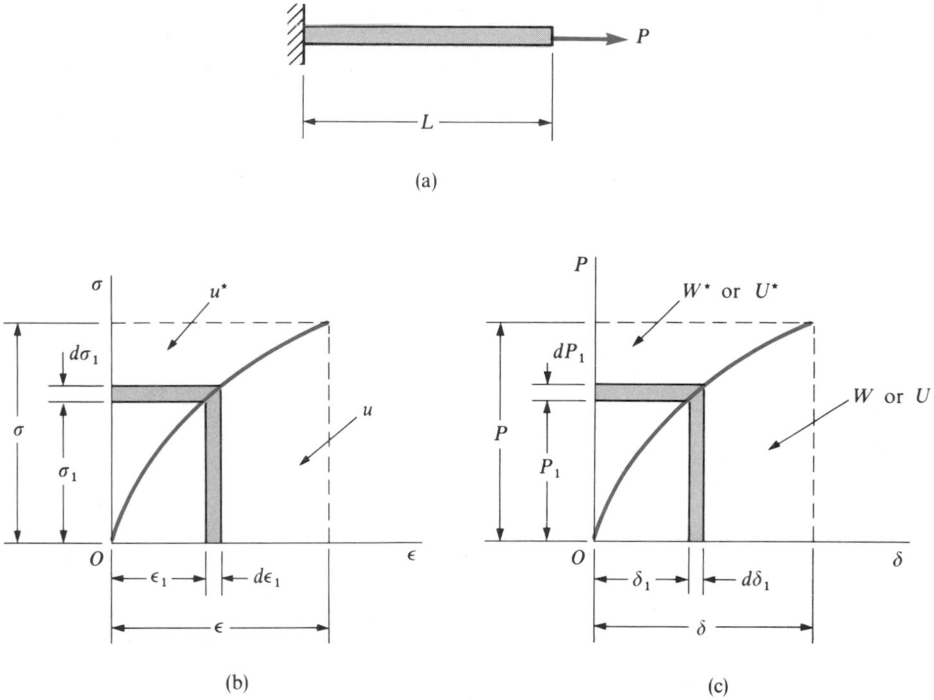

Fig. 10-15 Strain energy and complementary energy

In this equation, P_1 represents any value of the axial load between zero and the maximum value P; δ_1 is the corresponding elongation; and δ is the maximum elongation. Because the bar behaves elastically and because we are disregarding any losses of energy during loading and unloading (in other words, because we are dealing with a conservative system), all of the work done by the load will be stored in the bar in the form of elastic strain energy that can be recovered during unloading. Therefore, the **strain energy** is equal to the work W, as follows:

$$U = W = \int_0^\delta P_1 \, d\delta_1 \tag{10-16}$$

The integral in this equation may be interpreted geometrically as the area below the load-displacement curve shown in Fig. 10-15c.

The **strain energy density** u (that is, the strain energy per unit volume of material) can be obtained by considering a differential volume element of unit dimensions subjected to a stress σ_1 and a strain ϵ_1; thus,

$$u = \int_0^\epsilon \sigma_1 \, d\epsilon_1 \tag{10-17}$$

where $\epsilon = \delta/L$ is the maximum strain. The integral in Eq. (10-17) represents the area below the stress-strain curve in Fig. 10-15b. This same expression for u can be obtained from the expression for the total strain energy (Eq. 10-16) by dividing by the volume V of the bar (equal to AL) and noting that $\sigma_1 = P_1/A$ and $d\epsilon_1 = d\delta_1/L$. Conversely, the total strain energy U can be obtained from the strain energy density by integrating as follows:

$$U = \int u\,dV \tag{10-18}$$

where dV is an element of volume and the integration is performed throughout the volume of the bar.

In the special case when the stress-strain curve follows Hooke's law, so that $\sigma_1 = E\epsilon_1$ (and hence $P_1 = EA\delta_1/L$), the expressions for U and u (Eqs. 10-16 and 10-17) yield

$$U = \frac{EA\delta^2}{2L} \tag{10-19}$$

$$u = \frac{E\epsilon^2}{2} \tag{10-20}$$

These expressions are the same as those derived in Section 2.6 (see Eqs. 2-33b and 2-37b).

Let us now define another type of work for the prismatic bar shown in Fig. 10-15. This new kind of work is called **complementary work** $W^\star$ and is defined as follows:

$$W^\star = \int_0^P \delta_1\,dP_1$$

The complementary work is represented by the area between the load-displacement curve and the load axis (Fig. 10-15c). It does not have a physical meaning as does the work W, but note that

$$W + W^\star = P\delta \tag{10-21}$$

Thus, in a geometric sense, the work $W^\star$ is the complement of the work W because it completes the rectangle shown in Fig. 10-15c.

The **complementary energy** $U^\star$ of the bar is equal to the complementary work of the applied load, so that

$$U^\star = W^\star = \int_0^P \delta_1\,dP_1 \tag{10-22}$$

The **complementary energy density** $u^\star$ (or the complementary energy per unit volume) is obtained by considering a volume element subjected to the stress σ_1 and strain ϵ_1, in a manner analogous to that used in defining the strain energy density; thus,

$$u^\star = \int_0^\sigma \epsilon_1 \, d\sigma_1 \tag{10-23}$$

The complementary energy density is equal to the area between the stress-strain curve and the stress axis (Fig. 10-15b). Also, the total complementary energy of the bar may be obtained from $u^\star$ by integration:

$$U^\star = \int u^\star \, dV \tag{10-24}$$

Sometimes the complementary energy is called **stress energy** in order to maintain the analogy with strain energy.

Again considering the special case of a material that follows Hooke's law ($\epsilon_1 = \sigma_1/E$ and $\delta_1 = P_1L/EA$), we substitute into Eqs. (10-22) and (10-23) to obtain the following expressions for $U^\star$ and $u^\star$:

$$U^\star = \frac{P^2L}{2EA} \tag{10-25}$$

$$u^\star = \frac{\sigma^2}{2E} \tag{10-26}$$

Note that the complementary energy is expressed in terms of the load and that the strain energy (Eq. 10-19) is expressed in terms of the displacement. This distinction is inherent in the definitions of U and $U^\star$ (Eqs. 10-16 and 10-22, respectively). However, when Hooke's law holds, the strain and complementary energies are equal, and we obtain

$$U = U^\star = \frac{EA\delta^2}{2L} = \frac{P^2L}{2EA} \tag{10-27}$$

$$u = u^\star = \frac{E\epsilon^2}{2} = \frac{\sigma^2}{2E} \tag{10-28}$$

Despite these equalities for the special case of a linearly elastic member, the conceptual distinctions between U and $U^\star$ must be maintained.

Although the preceding equations for strain energy and complementary energy were derived for a bar in tension, they can be extended readily to include other cases of loading of a bar, such as compression, torsion, and bending. Thus, we may consider the load-displacement curve shown in Fig. 10-15c to represent the relationship between a load and its corresponding displacement for any type of member. If the load is a couple M with a corresponding displacement θ, we merely replace P and δ with M and θ, respectively.

Next, consider an elastic structure composed of more than one member but supporting a single load P, such as the simple truss ABC shown in Fig. 10-16a. We assume that the material has the stress-strain

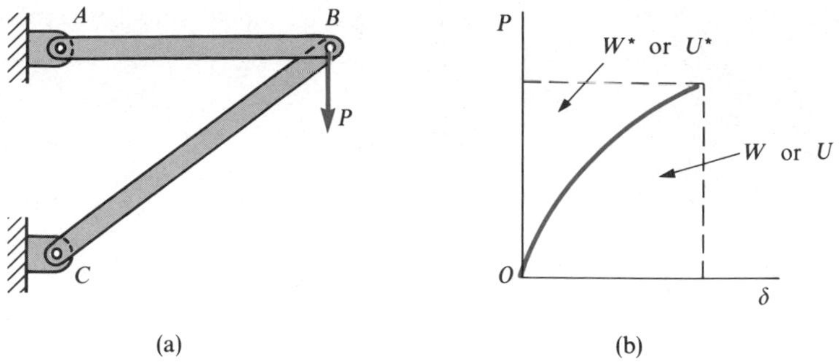

Fig. 10-16 Truss with material nonlinearity

curve shown in Fig. 10-15b. Consequently, the diagram of the load P versus the corresponding displacement δ (that is, the vertical displacement of joint B) will also be nonlinear (Fig. 10-16b). Shown on this load-displacement diagram are the work W and complementary work $W^\star$ of the load P. These quantities are equal, respectively, to the strain energy U and complementary energy $U^\star$ of the structure.

Now consider the individual members of the structure. Each member has both strain energy and complementary energy; these quantities can be found easily because we know the axial force in each member (from static equilibrium) and we know the stress-strain relationship. The sum U_m of the strain energies for the members equals the work W of the load P, by the principle of conservation of energy for an elastic structure (thus, $U_m = U$). This conclusion holds even when the structure is geometrically nonlinear; for instance, it holds when the displacements are large. However, the sum $U_m^\star$ of the complementary energies of the members equals the complementary work $W^\star$ of the load P only if the structure is geometrically linear (Ref. 10-14). Thus, complementary energy is conserved and $U_m^\star = U^\star$ when material nonlinearities are present but not when geometric nonlinearities exist. A necessary condition for geometric linearity is that the displacements be small, but this condition is not sufficient, as shown in Example 2 at the end of the section. If geometric nonlinearities are present, we define the complementary energy $U^\star$ of the structure as being equal to the complementary work $W^\star$, but this energy is larger than the total complementary energy $U_m^\star$ of the individual members. Under these conditions, complementary energy is not conserved and some of it is "lost" during the loading process.

When more than one load acts on the structure, the total work done by the loads can be obtained by summation. The loads must be applied simultaneously and must be maintained in the same proportions to one another during the loading process. The work and the complementary work of the individual loads can be added to give the total work W and total complementary work $W^\star$, as follows:

$$W = \sum_{i=1}^{n} \int_0^{\delta_i} P\, d\delta \qquad W^\star = \sum_{i=1}^{n} \int_0^{P_i} \delta\, dP \qquad \text{(10-29a,b)}$$

In these equations, P_i and δ_i are the maximum values of the ith load and its corresponding displacement, respectively; P and δ represent intermediate values of those same quantities (between zero and the maximum values); and n is the total number of loads. The work W given by Eq. (10-29a) equals the strain energy U of the structure; it also equals the sum U_m of the strain energies of the individual members, regardless of whether the structure is materially or geometrically nonlinear. The complementary work $W^\star$ (Eq. 10-29b), which equals the complementary energy $U^\star$ of the structure, is equal to the sum $U_m^\star$ of the complementary energies of the members only if no geometric nonlinearities exist, as stated previously. Assuming that we can obtain the load-displacement relations, we can substitute into Eqs. (10-29) and evaluate both W and $W^\star$. The former quantity will be expressed in terms of the displacements, and the latter will be in terms of the loads. However, since the relationships between the loads and their corresponding displacements usually are not known until the analysis of the structure is completed, the usual procedure is to evaluate the strain energy and complementary energy on a member-by-member basis (using Eqs. 10-16 and 10-22) and then to sum the energies to obtain U and $U^\star$ for the entire structure. This method for obtaining U is valid for structures with either kind of nonlinearity, but for $U^\star$ only material nonlinearities are permitted.

If the structure behaves linearly, the work W and the complementary work $W^\star$ of the loads are equal. Furthermore, each of these work terms is equal to the strain energy and the complementary energy of the structure ($W = W^\star = U = U^\star$). Since the work of the ith load is $P_i\delta_i/2$, we obtain the following equation for U and $U^\star$:

$$U = U^\star = \sum_{i=1}^{n} \frac{P_i\delta_i}{2} = \frac{P_1\delta_1}{2} + \frac{P_2\delta_2}{2} + \cdots + \frac{P_n\delta_n}{2} \qquad \text{(10-30)}$$

This equation expresses U and $U^\star$ in terms of the applied loads and their corresponding displacements.

We can also express U and $U^\star$ in terms of the displacements only or in terms of the loads only, simply by substituting for one quantity in terms of the other. Let us demonstrate this process symbolically in order to show the nature of the resulting expressions. Because the structure behaves linearly, the loads may be expressed as linear combinations of the displacements, in the following general manner:

$$\begin{aligned} P_1 &= a_{11}\delta_1 + a_{12}\delta_2 + \cdots + a_{1n}\delta_n \\ P_2 &= a_{21}\delta_1 + a_{22}\delta_2 + \cdots + a_{2n}\delta_n \\ &\cdots \\ P_n &= a_{n1}\delta_1 + a_{n2}\delta_2 + \cdots + a_{nn}\delta_n \end{aligned} \qquad \text{(10-31)}$$

where the coefficients $a_{11}, a_{12}, \ldots, a_{nn}$ are constants that depend only upon the properties of the structure. If these relations are substituted into Eq. (10-30), the strain and complementary energies are expressed as a function of the displacements only. The general form of the resulting equation is as follows:

$$\begin{aligned} U = U^\star = {} & b_{11}\delta_1^2 + b_{12}\delta_1\delta_2 + \cdots + b_{1n}\delta_1\delta_n \\ & + b_{21}\delta_2\delta_1 + b_{22}\delta_2^2 + \cdots + b_{2n}\delta_2\delta_n \\ & \ldots \\ & + b_{n1}\delta_n\delta_1 + b_{n2}\delta_n\delta_2 + \cdots + b_{nn}\delta_n^2 \end{aligned} \tag{10-32}$$

where the b's are new constants obtained from the a's. The expression on the right-hand side is called a **quadratic form**, which is a homogeneous polynomial of the second degree.

In addition to expressing the loads in terms of the displacements, we can also express the displacements as linear combinations of the loads:

$$\begin{aligned} \delta_1 &= c_{11}P_1 + c_{12}P_2 + \cdots + c_{1n}P_n \\ \delta_2 &= c_{21}P_1 + c_{22}P_2 + \cdots + c_{2n}P_n \\ & \ldots \\ \delta_n &= c_{n1}P_1 + c_{n2}P_2 + \cdots + c_{nn}P_n \end{aligned} \tag{10-33}$$

These equations can be obtained from Eqs. (10-31) by solving them simultaneously, hence the c's also are constants that depend only upon the properties of the structure. When Eqs. (10-33) are substituted into Eq. (10-30), we get another quadratic form for U and $U^\star$; thus,

$$\begin{aligned} U = U^\star = {} & d_{11}P_1^2 + d_{12}P_1P_2 + \cdots + d_{1n}P_1P_n \\ & + d_{21}P_2P_1 + d_{22}P_2^2 + \cdots + d_{2n}P_2P_n \\ & \ldots \\ & + d_{n1}P_nP_1 + d_{n2}P_nP_2 + \cdots + d_{nn}P_n^2 \end{aligned} \tag{10-34}$$

where the d's are constants obtained from the c's. These results show that in the case of a linearly elastic structure, the strain energy and the complementary energy may be expressed either as a quadratic function of the displacements or as a quadratic function of the loads. (Additional discussions of strain energy and complementary energy, with many examples, are given in Refs. 10-14 and 10-15.)

Example 1

Assume that a certain structure with material nonlinearities is acted on by a force P that produces a corresponding displacement δ given by the equation $\delta = CP^2$, where C is a constant (see Fig. 10-17). Determine the strain energy and complementary energy for this structure.

The strain energy is found from Eq. (10-16) by substituting $P_1 = \sqrt{\delta_1/C}$; thus,

$$U = \int_0^\delta \sqrt{\frac{\delta_1}{C}}\, d\delta_1 = \frac{2}{3}\sqrt{\frac{\delta^3}{C}} \tag{10-35}$$

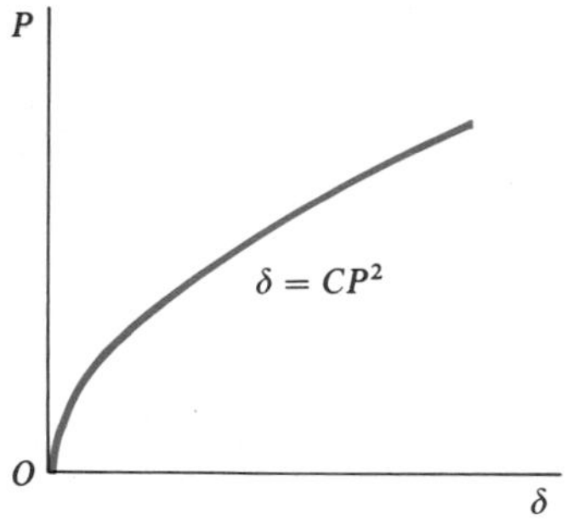

Fig. 10-17 Example 1. Nonlinear load-displacement diagram

The complementary energy, from Eq. (10-22), is

$$U^\star = \int_0^P CP_1^2\,dP_1 = \frac{CP^3}{3} \tag{10-36}$$

Note that the strain energy is expressed in terms of the displacement and the complementary energy is expressed in terms of the load. These forms for the energies are consistent with the nature of the definitions of U and $U^\star$; furthermore, we will see in subsequent sections that these forms are the most useful when finding displacements and analyzing structures. Of course, in some cases it is quite possible to express strain energy in terms of loads and complementary energy in terms of displacements. Such a result could be accomplished for this example by substituting from the original load-displacement relationship $\delta = CP^2$ into Eqs. (10-35) and (10-36). Also, note that U and $U^\star$ are not equal for this nonlinear structure (in this particular example, $U = 2U^\star$).

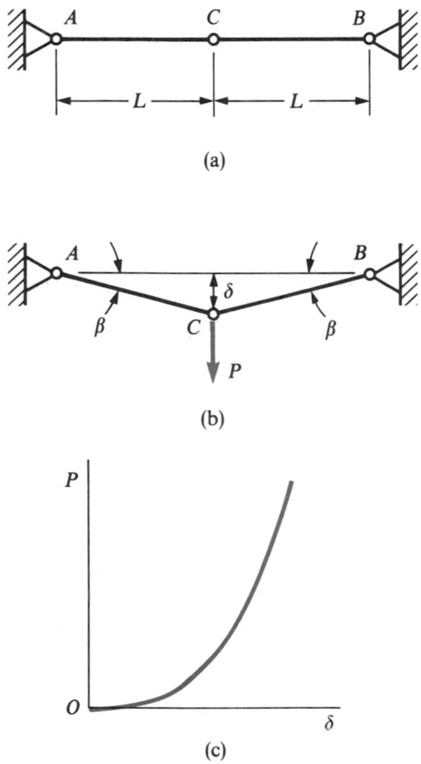

Fig. 10-18 Example 2. Structure with geometric nonlinearity

Example 2

A structure consisting of two horizontal bars AC and CB, each of length L, is shown in Fig. 10-18a. The bars have pinned supports and are linked together at C. The material of the bars is linearly elastic, and each bar has constant axial rigidity EA. If a vertical load P is applied at C, the bars must rotate because they are incapable of supporting the load when they are horizontal. Thus, joint C moves downward and tensile forces develop in the bars. When an equilibrium position is reached, the structure has a displacement δ at joint C (Fig. 10-18b). Determine the strain energy and complementary energy for this structure, assuming that the displacement δ remains small.

To determine the strain energy U, we will obtain the strain energy in one member and then multiply by 2. The strain energy of one member (since Hooke's law holds for the material) is obtained from the formula $EA\,\Delta^2/2L'$ (see Eq. 10-19), in which Δ is the elongation and L' is the length of the member. To find the length L', we note from Fig. 10-18b that

$$L' = \frac{L}{\cos\beta} \tag{a}$$

where β is the small angle shown in the figure. The series expression for $\cos\beta$ is

$$\cos\beta = 1 - \frac{\beta^2}{2!} + \frac{\beta^4}{4!} - \cdots \tag{b}$$

Since β is a small quantity, we can disregard (in comparison to unity) all terms containing powers of β. Then we can take $\cos\beta$ equal to 1 and L' equal to L. To obtain the elongation Δ in terms of the displacement δ, we follow a similar procedure. First, we note that

$$L' = L + \Delta = \frac{L}{\cos\beta} = L\sec\beta \tag{c}$$

The series expression for $\sec\beta$ is

$$\sec\beta = 1 + \frac{\beta^2}{2!} + \frac{5\beta^4}{4!} + \cdots \tag{d}$$

Substituting from this equation into Eq. (c), we get

$$L + \Delta = L\left(1 + \frac{\beta^2}{2!} + \frac{5\beta^4}{4!} + \cdots\right)$$

or

$$\Delta = \frac{L\beta^2}{2}\left(1 + \frac{5\beta^2}{12} + \cdots\right)$$

We can disregard all terms within the parentheses except the first, inasmuch as the angle β has a small value; therefore, we can assume that

$$\Delta = \frac{L\beta^2}{2} \tag{e}$$

Next, we observe from Fig. 10-18b that $\tan\beta = \delta/L$. However, for small angles we can replace $\tan\beta$ by β and obtain

$$\beta = \frac{\delta}{L} \tag{f}$$

Combining Eqs. (e) and (f), we get

$$\Delta = \frac{\delta^2}{2L} \tag{g}$$

Thus, the total strain energy of the members is

$$U_m = (2)\frac{EA\,\Delta^2}{2L'} = \frac{EA\delta^4}{4L^3} \tag{10-37}$$

This analysis is valid as long as δ remains small in comparison to the length L.

The complementary energy of the members is equal to the strain energy of the members because the material follows Hooke's law:

$$U_m^\star = \frac{EA\delta^4}{4L^3} \tag{h}$$

However, this structure is geometrically nonlinear because the displacement of the structure alters the action of the forces; hence $U_m^\star$ does not represent the total complementary energy $U^\star$. To show this fact, let us obtain the load-displacement equation for the structure and then evaluate both U and $U^\star$ from the work of the load P (see Eqs. 10-16 and 10-22). From static equilibrium at joint C (Fig. 10-18b), we obtain the tensile forces T in the members:

$$T = \frac{P}{2\sin\beta}$$

Replacing $\sin\beta$ with β (because β is a small angle), and also using Eq. (f), we obtain

$$T = \frac{PL}{2\delta}$$

The elongation Δ of one bar is related to the axial force T as follows:

$$\Delta = \frac{TL}{EA} = \frac{PL}{2\delta}\left(\frac{L}{EA}\right) = \frac{PL^2}{2EA\delta} \tag{i}$$

Now we can eliminate Δ between Eqs. (g) and (i) and obtain the load-displacement relationship in either of the following forms:

$$P = \frac{EA\delta^3}{L^3} \qquad \delta = \sqrt[3]{\frac{PL^3}{EA}} \tag{10-38a,b}$$

A graph of these equations is sketched in Fig. 10-18c. It is important to note that the structure is geometrically nonlinear, even though the material itself follows Hooke's law and the displacements are small.

Now we can calculate the strain energy of the structure from Eq. (10-16):

$$U = \int_0^\delta P_1\, d\delta_1 = \int_0^\delta \frac{EA\delta_1^3}{L^3}\, d\delta_1 = \frac{EA\delta^4}{4L^3} \tag{10-39}$$

which agrees with Eq. (10-37). The complementary energy is found from Eq. (10-22):

$$U^\star = \int_0^P \delta_1\, dP_1 = \int_0^P \sqrt[3]{\frac{P_1 L^3}{EA}}\, dP_1 = \frac{3P^{4/3}L}{4\sqrt[3]{EA}} \tag{10-40}$$

Note again that the strain energy is expressed in terms of the displacement and the complementary energy is expressed in terms of the load.

In order to compare the complementary energy $U^\star$ with the complementary energy $U_m^\star$ of the members (Eq. h), we substitute from Eq. (10-38a) into Eq. (10-40), thus obtaining an alternative expression for the complementary energy:

$$U^\star = \frac{3EA\delta^4}{4L^3} \tag{10-41}$$

This result is larger than $U_m^\star$, which demonstrates that, when a structure has geometric nonlinearities, complementary energy is not conserved.

Example 3

A cantilever beam of length L and rectangular cross section (width b, height h) carries a concentrated load P at the free end (see Fig. 10-19). The stress-strain curve of the material in tension is represented by the equation $\sigma = B\sqrt{\epsilon}$, where B is a constant; the curve has the same shape in compression. Determine the strain energy and complementary energy for this beam.

In this example the stresses and strains vary throughout the volume of the beam; hence it is necessary to determine the strain energy density u and complementary energy density $u^\star$ and then integrate to obtain U and $U^\star$. These quantities can be found if we know the stress σ and strain ϵ at every point in the beam.

We begin the analysis by denoting the stress and strain at the top surface of the beam as σ_0 and ϵ_0, respectively. Because of the symmetry of the cross section (Fig. 10-19b), the corresponding quantities at the bottom surface have the same

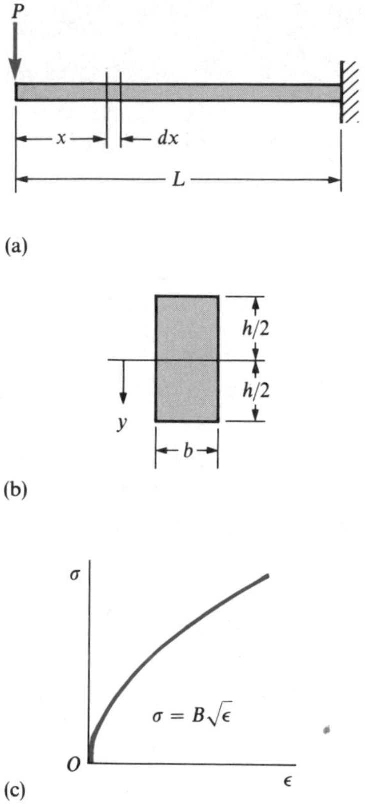

Fig. 10-19 Example 3. Cantilever beam with material nonlinearity

numerical values but opposite signs. The strain at any point in the cross section, distance y from the neutral axis, is

$$\epsilon = -\frac{2y\epsilon_0}{h} \tag{j}$$

The stress at the same point is

$$\sigma = B\sqrt{\epsilon} = -B\sqrt{\frac{2y\epsilon_0}{h}} \tag{k}$$

The bending moment M is related to the stresses by the familiar equation of statics:

$$M = \int \sigma y\, dA = 2\int_0^{h/2} -B\sqrt{\frac{2y\epsilon_0}{h}}\, yb\, dy$$

$$= -\frac{1}{5}B\sqrt{\epsilon_0}\, bh^2$$

Now we equate the preceding expression for M to $-Px$ and solve for ϵ_0:

$$\epsilon_0 = \frac{25P^2x^2}{B^2b^2h^4} \tag{l}$$

Finally, we can substitute ϵ_0 into Eqs. (j) and (k) and obtain the strains and stresses in the beam:

$$\epsilon = -\frac{50P^2x^2y}{B^2b^2h^5} \tag{m}$$

$$\sigma = -\frac{5\sqrt{2}\,Pxy^{1/2}}{bh^{5/2}} \tag{n}$$

By substituting the appropriate values of x and y into these equations, we can determine the strain and stress at any point in the beam.

The strain energy density (Eq. 10-17) is

$$u = \int_0^{\epsilon} \sigma_1\, d\epsilon_1 = \int_0^{\epsilon} B\sqrt{\epsilon_1}\, d\epsilon_1 = \frac{2B\epsilon^{3/2}}{3}$$

Substituting for the strain ϵ (see Eq. m), and using only absolute values, we get

$$u = \frac{500\sqrt{2}\,P^3x^3y^{3/2}}{3B^2b^3h^{15/2}} \tag{o}$$

This expression gives the strain energy density as a function of the coordinates x and y. The strain energy U of the entire beam is found by integrating u throughout the volume of the beam. In setting up the integral, we will let x vary from 0 to L and y from 0 to $h/2$; then we will multiply by 2 to account for the two halves of the beam. Thus, the equation for the strain energy U is

$$U = \int u\, dV = 2\int_0^L \left[\int_0^{h/2} ub\, dy\right] dx$$

Substituting from Eq. (o) for u, we get

$$U = \frac{1000\sqrt{2}\,P^3}{3B^2b^2h^{15/2}}\int_0^L\left[\int_0^{h/2} y^{3/2}\,dy\right]x^3\,dx = \frac{25P^3L^4}{3B^2b^2h^5} \tag{10-42}$$

The work W done by the load P is given by this same expression.

The complementary energy density (Eq. 10-23) is

$$u^\star = \int_0^\sigma \epsilon_1\,d\sigma_1 = \int_0^\sigma \frac{\sigma_1^2}{B^2}\,d\sigma_1 = \frac{\sigma^3}{3B^2}$$

Substituting for the stress σ (see Eq. n), and again using only absolute values, we get

$$u^\star = \frac{250\sqrt{2}\,P^3x^3y^{3/2}}{3B^2b^3h^{15/2}} \tag{p}$$

which expresses the complementary energy density as a function of x and y. Now we can integrate throughout the volume of the beam, in the same manner as for strain energy:

$$U^\star = \int u^\star\,dV = 2\int_0^L\left[\int_0^{h/2} u^\star b\,dy\right]dx$$

Substitution of Eq. (p) for $u^\star$ gives

$$U^\star = \frac{500\sqrt{2}\,P^3}{3B^2b^2h^{15/2}}\int_0^L\left[\int_0^{h/2} y^{3/2}\,dy\right]x^3\,dx = \frac{25P^3L^4}{6B^2b^2h^5} \tag{10-43}$$

Thus, we have expressions for the strain energy and complementary energy of this beam. Note that the strain energy is twice the complementary energy, because the area below the stress-strain curve (Fig. 10-19c) is twice the area between the curve and the vertical axis.

10.6 STRAIN-ENERGY METHODS

We will now develop a strain-energy theorem that is important in structural mechanics. Consider a structure subjected to n loads $P_1, P_2, \ldots, P_n$ that produce corresponding displacements $\delta_1, \delta_2, \ldots, \delta_n$. As in previous discussions, it is understood that P and δ represent forces and corresponding displacements in a generalized sense, thereby including the possibility of a force and a translation, a couple and a rotation, a pair of forces and a relative displacement, or a pair of couples and a relative rotation. It is also understood that the structure may behave nonlinearly, which means that the relationship between any force P_i and the corresponding displacement δ_i is typified by the load-displacement diagram of Fig. 10-15c.

The strain energy U of the structure is equal to the work W done by the loads during their application, as discussed in the preceding section.

Each force P_i theoretically can be expressed as a function of its corresponding displacement δ_i through the use of the appropriate load-displacement relationship. These expressions for the loads can be substituted into the expression for work (see Eq. 10-29a), thereby obtaining the strain energy U as a function of the displacements δ_i. With U expressed in terms of these displacements, we can evaluate the change in strain energy when one displacement δ_i is increased by a small amount $d\delta_i$ while all other displacements are held constant. This increase in strain energy, denoted by dU, is given by the expression

$$dU = \frac{\partial U}{\partial \delta_i} d\delta_i$$

where the partial derivative $\partial U / \partial \delta_i$ is the rate of change of the strain energy with respect to δ_i. We observe that when the displacement δ_i is increased by the small amount $d\delta_i$, work is done by the corresponding force P_i but not by any of the other forces because the other displacements are not changed. This work, which is equal to $P_i\, d\delta_i$, is equal to the increase in strain energy stored in the structure:

$$dU = P_i\, d\delta_i$$

Equating the preceding expressions for dU gives

$$P_i = \frac{\partial U}{\partial \delta_i} \qquad (10\text{-}44)$$

This equation states that the partial derivative of the strain energy with respect to any displacement δ_i is equal to the corresponding force P_i, provided that the strain energy is expressed as a function of the displacements. The equation is called **Castigliano's first theorem**, after the Italian engineer who derived and applied the theorem in his famous book, published in 1879 (see Refs. 10-16 through 10-20). The theorem can also be derived directly from the definition of strain energy (Eq. 10-16) by taking the derivatives of both sides of the equation.

As an illustration of Castigliano's first theorem, let us consider the nonlinear structure described in Example 1 of the preceding section. The strain energy of the structure is

$$U = \frac{2}{3}\sqrt{\frac{\delta^3}{C}}$$

as given by Eq. (10-35). Note that this equation gives U as a function of the displacement δ corresponding to the load P. Applying Castigliano's first theorem, we get

$$P = \frac{dU}{d\delta} = \sqrt{\frac{\delta}{C}}$$

from which $\delta = CP^2$, which is the correct relation between load and displacement.

In a similar manner, we can apply Castigliano's first theorem to the geometrically nonlinear structure shown in Fig. 10-18 and described in Example 2 of the preceding section. The strain energy of the structure (in terms of the displacement δ) is

$$U = \frac{EA\delta^4}{4L^3}$$

(see Eqs. 10-37 and 10-39). Thus, Castigliano's first theorem yields

$$P = \frac{dU}{d\delta} = \frac{EA\delta^3}{L^3}$$

This result agrees with the previous analysis (see Eq. 10-38a).

Castigliano's first theorem is the basis for a method of structural analysis based upon strain energy. To explain the method, let us assume that we have a nonlinear structure with n unknown joint displacements $D_1, D_2, \ldots, D_n$. We use the symbol D for these displacements to distinguish them from the displacements δ_i, which are more general quantities. Whereas δ_i and P_i may represent any displacement and corresponding force, the displacements D are joint displacements and represent the unknown quantities in the analysis. Furthermore, for practical purposes it is necessary that every joint displacement be included as one of the D's. An additional restriction is that all loads on the structure must act at the joints. Thus, each load corresponds to a displacement. The loads will be denoted by $P_1, P_2, \ldots, P_n$ corresponding to $D_1, D_2, \ldots, D_n$, respectively.

As already explained, it is theoretically possible to express the strain energy U of the structure in terms of the unknown joint displacements D_1, $D_2, \ldots, D_n$. If U is expressed in this manner, we can apply Castigliano's first theorem with respect to each displacement and thereby obtain a set of n simultaneous equations:

$$P_1 = \frac{\partial U}{\partial D_1} \qquad P_2 = \frac{\partial U}{\partial D_2} \qquad \cdots \qquad P_n = \frac{\partial U}{\partial D_n} \tag{10-45}$$

If we examine a typical equation, say the ith equation, we find that it consists of terms on the right-hand side that contain the n joint displacements $D_1, D_2, \ldots, D_n$ as unknown quantities. Because these terms sum up to the load P_i itself, we conclude that the equation represents an equilibrium condition for forces corresponding to P_i. Thus, Eqs. (10-45) are a set of equilibrium equations that can be solved simultaneously for the joint displacements $D_1, D_2, \ldots, D_n$ in terms of the loads $P_1, P_2, \ldots, P_n$. Knowing the joint displacements, we can calculate other quantities, such as reactions and stress resultants, from the displacements. Thus, the analysis of the structure is completed.

The strain-energy method described in the preceding paragraph utilizes joint displacements as unknowns and requires the solution of equilibrium equations. It is called the **displacement method** of analysis and may be used to analyze both linear and nonlinear structures. In the case of a linear structure, the method is called the **stiffness method**.

Let us consider in more detail the special case of a linear structure (that is, a linearly elastic structure for which the principle of superposition is valid). In this case the strain energy U is a quadratic function of the displacements, as explained in the preceding section. Therefore, when there are n unknown joint displacements $D_1, D_2, \ldots, D_n$ and n corresponding loads $P_1, P_2, \ldots, P_n$, the general form for the strain energy U is

$$\begin{aligned} U = {} & a_{11}D_1^2 + a_{12}D_1D_2 + \cdots + a_{1n}D_1D_n \\ & + a_{21}D_2D_1 + a_{22}D_2^2 + \cdots + a_{2n}D_2D_n \\ & \ldots \\ & + a_{n1}D_nD_1 + a_{n2}D_nD_2 + \cdots + a_{nn}D_n^2 \end{aligned}$$

where the coefficients a_{11}, a_{12}, and so on, are new constants that depend only upon the properties of the structure. Applying Castigliano's first theorem (Eq. 10-44), we obtain the equations of equilibrium (Eqs. 10-45) in the following form:

$$\begin{aligned} P_1 &= \frac{\partial U}{\partial D_1} = S_{11}D_1 + S_{12}D_2 + \cdots + S_{1n}D_n \\ P_2 &= \frac{\partial U}{\partial D_2} = S_{21}D_1 + S_{22}D_2 + \cdots + S_{2n}D_n \\ & \ldots \\ P_n &= \frac{\partial U}{\partial D_n} = S_{n1}D_1 + S_{n2}D_2 + \cdots + S_{nn}D_n \end{aligned} \tag{10-46}$$

in which the coefficients S_{11}, S_{12}, and so on, are constants that depend upon the a's (that is, upon the properties of the structure). These coefficients are known as **stiffness coefficients**, or **stiffnesses**, and Eqs. (10-46) are the **equilibrium equations** of the stiffness method of analysis.

An important relationship between the strain energy of the structure and the stiffnesses can be obtained by taking partial derivatives of Eqs. (10-46), as follows:

$$\frac{\partial P_1}{\partial D_1} = S_{11} \qquad \frac{\partial P_1}{\partial D_2} = S_{12} \qquad \cdots \qquad \frac{\partial P_1}{\partial D_n} = S_{1n} \tag{a}$$

$$\frac{\partial P_2}{\partial D_1} = S_{21} \qquad \frac{\partial P_2}{\partial D_2} = S_{22} \qquad \cdots \qquad \frac{\partial P_2}{\partial D_n} = S_{2n} \tag{b}$$

and so forth for the remaining equations. Inasmuch as P_1 is equal to $\partial U/\partial D_1$, Eqs. (a) can be written as

$$S_{11} = \frac{\partial^2 U}{\partial D_1^2} \qquad S_{12} = \frac{\partial^2 U}{\partial D_2\,\partial D_1} \qquad \cdots \qquad S_{1n} = \frac{\partial^2 U}{\partial D_n\,\partial D_1}$$

In a similar manner, Eqs. (b) become

$$S_{21} = \frac{\partial^2 U}{\partial D_1\, \partial D_2} \qquad S_{22} = \frac{\partial^2 U}{\partial D_2^2} \qquad \cdots \qquad S_{2n} = \frac{\partial^2 U}{\partial D_n\, \partial D_2}$$

From the pattern of derivatives that emerges from these expressions, we conclude that the stiffness coefficients are related to the strain energy by the following general equation:

$$S_{ij} = \frac{\partial^2 U}{\partial D_j\, \partial D_i} \tag{10-47}$$

Thus, whenever the strain energy U is expressed as a quadratic function of the unknown displacements $D_1, D_2, \ldots, D_n$, we can immediately get the stiffnesses for the structure by the process of differentiation. Furthermore, since the order of differentiation of U is immaterial, we obtain the reciprocal theorem for stiffnesses:

$$S_{ij} = \frac{\partial^2 U}{\partial D_j\, \partial D_i} = \frac{\partial^2 U}{\partial D_i\, \partial D_j} = S_{ji} \tag{10-48}$$

Of course, the preceding discussion of stiffnesses is applicable only to linearly elastic structures. For a thorough discussion of the stiffness method, textbooks on structural analysis should be consulted (for example, see Ref. 10-22).

Example 1

The truss ABC shown in Fig. 10-20a supports a vertical load P at joint B. Bars AB and BC have constant cross-sectional area A. The stress-strain relationship for the material in tension is $\sigma = B\sqrt{\epsilon}$, where B is a constant (see Fig. 10-20b). The shape of the stress-strain curve is the same in both tension and compression. Analyze this materially nonlinear truss by the displacement method, using Castigliano's first theorem.

Two unknown displacements may be identified at joint B of this structure: a horizontal displacement D_1 and a vertical displacement D_2 (Fig. 10-20). The strain energy of the structure must be expressed in terms of these displacements in order

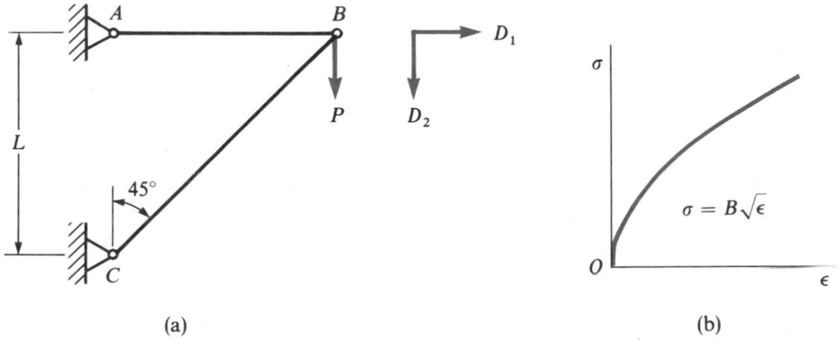

Fig. 10-20 Example 1. Truss of material with a nonlinear stress-strain curve

to obtain the equations of equilibrium by the use of Castigliano's first theorem. Thus, the strain energy is determined as follows. Considering only the displacement D_1, we note that bar AB elongates by an amount equal to D_1 and bar BC elongates by an amount equal to $D_1/\sqrt{2}$. Also, due to the displacement D_2 alone, bar BC shortens by an amount equal to $D_2/\sqrt{2}$, and bar AB does not change in length. Thus, the total elongation of bar AB is D_1 and the total shortening of bar BC is $(D_2 - D_1)/\sqrt{2}$. Hence the strains in the bars are

$$\epsilon_{ab} = \frac{D_1}{L} \qquad \text{(elongation)} \tag{c}$$

$$\epsilon_{bc} = \frac{D_2 - D_1}{2L} \qquad \text{(shortening)} \tag{d}$$

The strain energy density in each bar can be found from Eq. (10-17):

$$u_{ab} = \int_0^{\epsilon_{ab}} \sigma_1\, d\epsilon_1 = \int_0^{\epsilon_{ab}} B\sqrt{\epsilon_1}\, d\epsilon_1 = \frac{2B}{3}\left(\frac{D_1}{L}\right)^{3/2}$$

$$u_{bc} = \int_0^{\epsilon_{bc}} \sigma_1\, d\epsilon_1 = \int_0^{\epsilon_{bc}} B\sqrt{\epsilon_1}\, d\epsilon_1 = \frac{2B}{3}\left(\frac{D_2 - D_1}{2L}\right)^{3/2}$$

Because the stress and strain are constant throughout the volume of each bar, we can get the total strain energy of each bar by multiplying the strain energy density by the volume of the bar. Then these strain energies can be added to give the total strain energy U of the structure:

$$U = U_{ab} + U_{bc} = u_{ab}AL + u_{bc}AL\sqrt{2}$$

from which

$$U = \frac{AB}{3\sqrt{L}}[2D_1^{3/2} + (D_2 - D_1)^{3/2}] \tag{10-49}$$

where A is the cross-sectional area of each bar. This equation gives the strain energy in terms of the unknown joint displacements D_1 and D_2.

The equations of equilibrium can now be obtained by using Castigliano's first theorem (see Eqs. 10-44 and 10-45). The quantity P_1 in Eqs. (10-45) represents the load corresponding to D_1 and is equal to zero, while P_2 represents the load corresponding to D_2 and is equal to P. Thus, we get

$$P_1 = \frac{\partial U}{\partial D_1} = \frac{AB}{2\sqrt{L}}[2D_1^{1/2} - (D_2 - D_1)^{1/2}] = 0$$

$$P_2 = \frac{\partial U}{\partial D_2} = \frac{AB}{2\sqrt{L}}(D_2 - D_1)^{1/2} = P$$

Simplifying these two equations, we get

$$2D_1^{1/2} - (D_2 - D_1)^{1/2} = 0 \tag{e}$$

$$(D_2 - D_1)^{1/2} = \frac{2P\sqrt{L}}{AB} \tag{f}$$

Solving simultaneously the preceding equations, we obtain the joint displacements:

$$D_1 = \frac{P^2L}{A^2B^2} \qquad D_2 = \frac{5P^2L}{A^2B^2} \tag{10-50a,b}$$

This step completes the essential part of the analysis by the displacement method, because, now that we know D_1 and D_2, we can calculate all bar forces and reactions.

To illustrate these final calculations, let us determine the forces in the bars. The strains in the bars are obtained by substituting for D_1 and D_2 (Eqs. 10-50a and b) into Eqs. (c) and (d):

$$\epsilon_{ab} = \frac{D_1}{L} = \frac{P^2}{A^2B^2} \qquad \text{(elongation)}$$

$$\epsilon_{bc} = \frac{D_2 - D_1}{2L} = \frac{2P^2}{A^2B^2} \qquad \text{(shortening)}$$

Next, the stresses are determined from the stress-strain law:

$$\sigma_{ab} = B\sqrt{\epsilon_{ab}} = \frac{P}{A} \qquad \text{(tension)}$$

$$\sigma_{bc} = B\sqrt{\epsilon_{bc}} = \frac{\sqrt{2}\,P}{A} \qquad \text{(compression)}$$

Finally, the axial forces N in the bars are

$$N_{ab} = \sigma_{ab}A = P \qquad \text{(tension)}$$

$$N_{bc} = \sigma_{bc}A = \sqrt{2}\,P \qquad \text{(compression)}$$

These results are readily verified by static equilibrium.

In this example we have intentionally described all steps in the solution in order to illustrate the general concepts of the displacement method and the use of Castigliano's first theorem, even though the structure is very simple and could have been analyzed much more easily as a statically determinate structure. The use of the displacement method required the solution of two simultaneous equations because there are two unknown joint displacements. However, because the structure is statically determinate, it can also be analyzed as follows: (1) find the forces in the bars from static equilibrium, (2) find the stresses in the bars by dividing the forces by the cross-sectional areas, (3) obtain the strains in the bars from the stress-strain diagram, (4) determine the elongations of the bars from the strains, and (5) obtain the displacements D_1 and D_2 of joint B from the elongations, using either geometry or the unit-load method (see Example 3 of Section 10.7).

Example 2

A truss consisting of four bars that meet at a common joint E is shown in Fig. 10-21a. All bars are constructed of the same linearly elastic material with modulus of elasticity E. Each bar has length L and cross-sectional area A, and

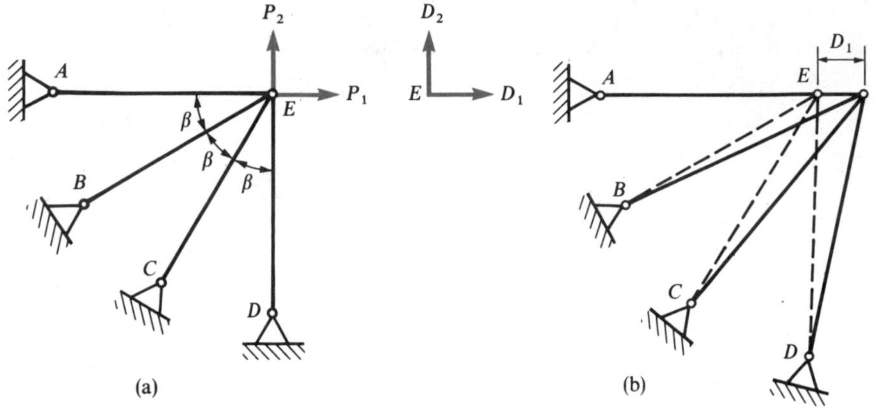

Fig. 10-21 Example 2. Truss of material that follows Hooke's law

the angle β equals 30°. Loads P_1 and P_2 act on the truss at joint E. Analyze this statically indeterminate truss by using Castigliano's first theorem and the stiffness method.

This truss has two unknown joint displacements, namely, the horizontal and vertical translations D_1 and D_2 at joint E (Fig. 10-21a). To express the strain energy U as a function of D_1 and D_2, we begin by assuming that D_1 occurs alone (Fig. 10-21b). Under these conditions the elongations of the bars are as follows:

$$\Delta_{ae} = D_1 \qquad \Delta_{be} = \frac{\sqrt{3}\,D_1}{2} \qquad \Delta_{ce} = \frac{D_1}{2} \qquad \Delta_{de} = 0$$

as can be readily verified from the geometry of the figure. When the displacement D_2 occurs alone, the elongations of the members are

$$\Delta_{ae} = 0 \qquad \Delta_{be} = \frac{D_2}{2} \qquad \Delta_{ce} = \frac{\sqrt{3}\,D_2}{2} \qquad \Delta_{de} = D_2$$

Therefore, the elongations of the bars when both D_1 and D_2 occur simultaneously are as follows:

$$\Delta_{ae} = D_1 \qquad \Delta_{be} = \frac{\sqrt{3}\,D_1 + D_2}{2} \qquad \Delta_{ce} = \frac{D_1 + \sqrt{3}\,D_2}{2} \qquad \Delta_{de} = D_2 \tag{g}$$

The strain energy of each bar can be evaluated from its elongation (see Eq. 10-19), and then the total strain energy U of the truss can be obtained by summing the energies for all four bars:

$$U = \frac{EA}{2L}D_1^2 + \frac{EA}{2L}\left(\frac{\sqrt{3}\,D_1 + D_2}{2}\right)^2 + \frac{EA}{2L}\left(\frac{D_1 + \sqrt{3}\,D_2}{2}\right)^2 + \frac{EA}{2L}D_2^2$$

or

$$U = \frac{EA}{2L}(2D_1^2 + \sqrt{3}\,D_1 D_2 + 2D_2^2) \tag{10-51}$$

Note that this expression gives the strain energy as a quadratic function of the displacements.

Applying Castigliano's first theorem, we get the following equations of equilibrium (see Eqs. 10-46):

$$P_1 = \frac{2EA}{L} D_1 + \frac{\sqrt{3}\,EA}{2L} D_2 \tag{h}$$

$$P_2 = \frac{\sqrt{3}\,EA}{2L} D_1 + \frac{2EA}{L} D_2 \tag{i}$$

These equations can be solved for the joint displacements; the results are

$$D_1 = \frac{2L}{13EA}(4P_1 - \sqrt{3}\,P_2) \qquad D_2 = \frac{2L}{13EA}(-\sqrt{3}\,P_1 + 4P_2)$$

Finally, the axial forces N in the bars of the truss can be calculated. The procedure is to find the elongations Δ of the bars by substituting the joint displacements D_1 and D_2 into Eqs. (g), and then to find the bar forces from the elongations by using the equation $N = EA\,\Delta/L$. The results are

$$N_{ae} = \frac{8P_1}{13} - \frac{2\sqrt{3}\,P_2}{13} \qquad N_{be} = \frac{3\sqrt{3}\,P_1}{13} + \frac{P_2}{13}$$

$$N_{ce} = \frac{P_1}{13} + \frac{3\sqrt{3}\,P_2}{13} \qquad N_{de} = -\frac{2\sqrt{3}\,P_1}{13} + \frac{8P_2}{13}$$

Thus, the complete analysis of the statically indeterminate truss has been performed by the stiffness method.

Example 3

The statically indeterminate frame ABC shown in Fig. 10-22a has a fixed support at A and a pin support at C. The load on the structure is a couple M_0 acting at joint B. The material of the structure follows Hooke's law. Members AB and BC have length L and flexural rigidity EI. Determine the angles of rotation D_1 and D_2 at joints B and C, respectively (Fig. 10-22b).

The only unknown joint displacements for this frame are the angles of rotation D_1 and D_2, inasmuch as we are disregarding the effects of axial deformations (and therefore the lengths of the members do not change). The strain energy of the frame must be expressed in terms of the unknown displacements D_1 and D_2. To accomplish this step for a frame, it is helpful to imagine that the unknown joint displacements are imposed on the structure by the addition of restraints corresponding to those displacements (see Fig. 10-22c). Then each member of the frame is in the same condition as a fixed-end beam with rotations imposed at the ends. If we can obtain a formula for the strain energy stored in such a beam, we can use this formula to obtain the strain energy of the frame.

Consider the beam AB shown in Fig. 10-23. The ends A and B are assumed to be rotated through angles θ_1 and θ_2, respectively. Our objective is to obtain a formula for the strain energy U of this beam in terms of θ_1 and θ_2. One method is to derive the equation of the deflection curve by solving the fourth-order differential equation (Eq. 7-10c):

$$EIv'''' = q = 0$$

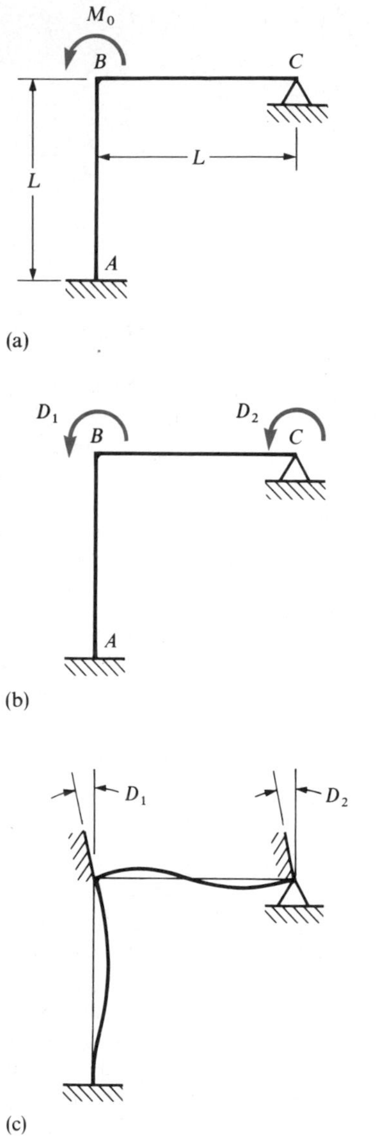

Fig. 10-22 Example 3. Frame of material that follows Hooke's law

Four successive integrations of this equation yield

$$EIv = \frac{C_1x^3}{6} + \frac{C_2x^2}{2} + C_3x + C_4$$

The boundary conditions are as follows:

$$v'(0) = -\theta_1 \qquad v'(L) = -\theta_2 \qquad v(0) = 0 \qquad v(L) = 0$$

Application of these conditions yields the following equation of the deflection curve:

$$v = -\frac{x^3}{L^2}(\theta_1 + \theta_2) + \frac{x^2}{L}(2\theta_1 + \theta_2) - \theta_1 x$$

The strain energy of the beam can now be calculated (see Eq. 7-59b):

$$U = \frac{EI}{2}\int_0^L \left(\frac{d^2v}{dx^2}\right)^2 dx = \frac{2EI}{L}(\theta_1^2 + \theta_1\theta_2 + \theta_2^2) \tag{10-52}$$

Another way to find the strain energy of the beam is to determine the work done by the moments M_1 and M_2 acting at the ends of the beam (see Fig. 10-23). When these moments act on a simple beam, the angles at the ends are

$$\theta_1 = \frac{M_1L}{3EI} - \frac{M_2L}{6EI} \qquad \theta_2 = -\frac{M_1L}{6EI} + \frac{M_2L}{3EI}$$

(see Case 7 of Table G-2 in Appendix G). These equations can be solved simultaneously for M_1 and M_2:

$$M_1 = \frac{2EI}{L}(2\theta_1 + \theta_2) \qquad M_2 = \frac{2EI}{L}(\theta_1 + 2\theta_2) \tag{10-53a,b}$$

The strain energy of the beam, from Eq. (10-30), is

$$U = \frac{M_1\theta_1}{2} + \frac{M_2\theta_2}{2}$$

Substituting from Eqs. (10-53), we obtain the same expression for U as before (see Eq. 10-52).

Now we are ready to determine the strain energy of the frame shown in Fig. 10-22 in terms of the displacements D_1 and D_2. The procedure is to apply Eq. (10-52) to each member and sum the results. For member AB we have $\theta_1 = 0$ and $\theta_2 = D_1$, and for member BC we have $\theta_1 = D_1$ and $\theta_2 = D_2$. Thus, the strain

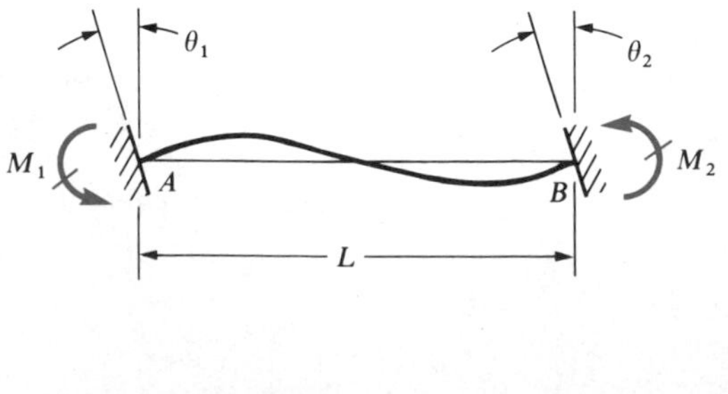

Fig. 10-23 Strain energy in a beam with rotations θ_1 and θ_2 at the ends

energy is

$$U = \frac{2EI}{L}(D_1^2) + \frac{2EI}{L}(D_1^2 + D_1D_2 + D_2^2) = \frac{2EI}{L}(2D_1^2 + D_1D_2 + D_2^2)$$

From Castigliano's first theorem we now get the following equations of equilibrium:

$$M_0 = \frac{\partial U}{\partial D_1} = \frac{2EI}{L}(4D_1 + D_2)$$

$$0 = \frac{\partial U}{\partial D_2} = \frac{2EI}{L}(D_1 + 2D_2)$$

These equations can be solved for the joint displacements D_1 and D_2:

$$D_1 = \frac{M_0L}{7EI} \qquad D_2 = -\frac{M_0L}{14EI}$$

Thus, the joint rotations have been found for the frame. As a final step in the solution, the moments at the ends of each member could be determined by using Eqs. (10-53).

This example is well suited to the use of strain energy and Castigliano's first theorem because the load M_0 corresponds to one of the unknown joint displacements. The only other possible load on the structure is a couple corresponding to D_2, inasmuch as one of the requirements of this method of analysis is that each load must correspond to an unknown joint displacement. This fact raises the question of how to analyze a structure when there are loads at other locations, such as a concentrated load acting at the midpoint of a member. One possibility is to consider every point of loading to be a joint of the structure, thus introducing additional unknown joint displacements that correspond to the loads. The disadvantage of such an approach is that it greatly increases the number of equations of equilibrium to be solved. A method preferred by most structural analysts is to replace any loads that act between the joints by a set of statically equivalent loads acting at the joints. Of course, the use of equivalent joint loads is possible only when the principle of superposition is valid. The technique for determining the equivalent loads is not difficult to use; however, we will not explain it here. Instead, readers should consult books on structural analysis for further details (for example, see Ref. 10-22).

10.7 COMPLEMENTARY-ENERGY METHODS

In the preceding section we dealt with several important principles pertaining to strain energy. Now we turn to a consideration of some analogous principles relating to complementary energy. It was pointed out in earlier discussions that strain energy is usually expressed as a function of displacements, whereas complementary energy is usually expressed in terms of forces. Thus, it is quite natural that strain energy is related to the displacement and stiffness methods of analysis, whereas complementary energy is related to the **force** and **flexibility methods**, as explained in the following discussions.

To derive a basic theorem relating to complementary energy, let us consider again a nonlinear structure subjected to n loads $P_1, P_2, \ldots, P_n$ that produce corresponding displacements $\delta_1, \delta_2, \ldots, \delta_n$. As in our earlier discussions, it is understood that P and δ represent forces and corresponding displacements in a generalized sense. The complementary energy $U^\star$ of the structure is defined as being equal to the complementary work $W^\star$ of the loads (see the discussion in the middle of Section 10.5). When evaluating this work, we express the displacements in terms of the loads and then integrate, as shown in Eq. (10-29b). The resulting expression for $U^\star$ is a function of the loads $P_1, P_2, \ldots, P_n$. Then, if we imagine that one load, say P_i, is given a small increase dP_i while the other loads are unchanged, the complementary energy will be increased by a small amount $dU^\star$, given by the following equation:

$$dU^\star = \frac{\partial U^\star}{\partial P_i} dP_i$$

This equation states that the increase in $U^\star$ is equal to the rate of change of $U^\star$ with respect to P_i multiplied by the increment in P_i.

Another way to obtain an expression for $dU^\star$ is to consider the complementary work of the loads when the force P_i is increased by the amount dP_i. This complementary work is the same as the increase $dU^\star$ in the complementary energy of the structure. The only load that does any complementary work is P_i itself, because the other forces are not changed. Therefore, the increase in complementary energy is the product of the displacement δ_i and the increment dP_i of the load:

$$dU^\star = \delta_i \, dP_i$$

Equating the two preceding expressions for $dU^\star$, we obtain

$$\delta_i = \frac{\partial U^\star}{\partial P_i} \tag{10-54}$$

This equation states that the partial derivative of the complementary energy with respect to any load P_i is equal to the corresponding displacement δ_i, provided that the complementary energy is expressed as a function of the loads. The equation is called the **Crotti-Engesser theorem**, after the Italian engineer Francesco Crotti who derived it in 1878 (Refs. 10-23 and 10-24) and the German engineer Friedrich Engesser who independently obtained it in 1889 (Ref. 10-25). The theorem can also be obtained directly from the definition of complementary energy (see Eqs. 10-22 and 10-29b) by taking the derivatives of both sides of the equations.

The Crotti-Engesser theorem provides a remarkable analogy with Castigliano's first theorem (Eq. 10-44). In the case of the Crotti-Engesser theorem, we express the complementary energy as a function of the loads and then obtain the corresponding displacements by taking derivatives with respect to the loads, whereas, in the case of Castigliano's first theorem, we express the strain energy as a function of the displacements and then obtain the corresponding loads by taking derivatives with respect to the displacements. Both theorems are very general and apply to structures that behave nonlinearly. However, as explained in Section 10.5, the complementary energy $U^\star$ of the structure must be obtained from the complementary work of the loads. If no geometric nonlinearities exist, the complementary energy $U^\star$ is equal to the sum of the complementary energies of the members; hence the member energies can be used to obtain $U^\star$. If there are geometric nonlinearities, then the complementary energy of the members is less than the complementary energy $U^\star$ of the structure.*

To illustrate the Crotti-Engesser theorem, let us again consider the materially nonlinear structure described in Example 1 of Section 10.5. The complementary energy of the structure (from Eq. 10-36) is

$$U^\star = \frac{CP^3}{3}$$

Note that this equation gives $U^\star$ as a function of the load P corresponding to the displacement δ. From the Crotti-Engesser theorem we get

$$\delta = \frac{dU^\star}{dP} = CP^2$$

which is the correct relation between load and displacement.

As another example, consider the geometrically nonlinear structure shown in Fig. 10-18 and described in Example 2 of Section 10.5. The complementary energy of the structure (in terms of the load P) is

$$U^\star = \frac{3P^{4/3}L}{4\sqrt[3]{EA}}$$

(see Eq. 10-40). Thus, the Crotti-Engesser theorem yields

$$\delta = \frac{dU^\star}{dP} = \sqrt[3]{\frac{PL^3}{EA}}$$

which agrees with the previous analysis (see Eq. 10-38b).

A third example is the materially nonlinear cantilever beam of Example 3 of Section 10.5, for which the complementary energy is

$$U^\star = \frac{25P^3L^4}{6B^2b^2h^5}$$

* In the special case of a structure that behaves linearly, the complementary energy is equal to the strain energy and the Crotti-Engesser theorem reduces to Castigliano's second theorem (see Section 10.8).

(see Eq. 10-43). We can now determine the deflection δ at the end of the beam:

$$\delta = \frac{dU^\star}{dP} = \frac{25P^2L^4}{2B^2b^2h^5} \tag{10-55}$$

Note that this deflection varies nonlinearly with the load P.

Example 1

The truss ABC shown in Fig. 10-20a supports a vertical load P at joint B. The material of the truss has a stress-strain relationship in both tension and compression that is given by the equation $\sigma = B\sqrt{\epsilon}$, where B is a constant. Determine the vertical displacement δ_v of joint B.

The first step in finding the displacement by means of the Crotti-Engesser theorem is to obtain the complementary energy of the structure in terms of the force P. Because the truss has no geometric nonlinearities, we can obtain the total complementary energy $U^\star$ by summing the energies of the two bars:

$$U^\star = U^\star_{ab} + U^\star_{bc} \tag{a}$$

The complementary energy of an individual bar can be found by multiplying the complementary energy density by the volume of the bar, inasmuch as the stress and strain remain constant throughout the bar. Therefore, we have

$$U^\star_{ab} = u^\star_{ab}AL \qquad U^\star_{bc} = u^\star_{bc}AL\sqrt{2} \tag{b}$$

where A is the cross-sectional area of each bar. The complementary energy density of each bar can be obtained from Eq. (10-23), as follows:

$$u^\star_{ab} = \int_0^{\sigma_{ab}} \epsilon_1\, d\sigma_1 = \int_0^{\sigma_{ab}} \frac{\sigma_1^2}{B^2}\, d\sigma_1 = \frac{\sigma_{ab}^3}{3B^2} \tag{c}$$

$$u^\star_{bc} = \int_0^{\sigma_{bc}} \epsilon_1\, d\sigma_1 = \int_0^{\sigma_{bc}} \frac{\sigma_1^2}{B^2}\, d\sigma_1 = \frac{\sigma_{bc}^3}{3B^2} \tag{d}$$

where σ_{ab} and σ_{bc} are the stresses in the bars. Now we can obtain the total complementary energy of the truss from Eqs. (a) through (d); the result is

$$U^\star = \frac{AL}{3B^2}(\sigma_{ab}^3 + \sqrt{2}\,\sigma_{bc}^3) \tag{e}$$

The load P acting on the truss (Fig. 10-20a) produces stresses in the bars that can be found from equilibrium:

$$\sigma_{ab} = \frac{P}{A} \qquad \sigma_{bc} = \frac{\sqrt{2}\,P}{A} \tag{f}$$

in which we consider only the absolute values of the stresses. Therefore, the complementary energy of the truss (Eq. e) is

$$U^\star = \frac{5P^3L}{3A^2B^2} \tag{10-56}$$

This equation expresses the complementary energy as a function of the load P.

The next step is to apply the Crotti-Engesser theorem to obtain the displacement corresponding to P, as follows:

$$\delta_v = \frac{dU^\star}{dP} = \frac{5P^2L}{A^2B^2} \tag{10-57}$$

This equation gives the vertical displacement δ_v of joint B of the truss.

To find the horizontal displacement δ_h of joint B, it is necessary to add a horizontal load Q corresponding to that displacement. We can then repeat the procedure for finding the complementary energy $U^\star$, the difference being that $U^\star$ is now a function of both P and Q. Then we obtain the horizontal displacement δ_h by applying the Crotti-Engesser theorem and taking the derivative with respect to Q. The resulting expression for δ_h contains both P and Q. Setting Q equal to zero gives the equation for the horizontal displacement due to P acting alone (see Problem 10.7-2).

UNIT-LOAD METHOD

The direct use of the Crotti-Engesser theorem in finding displacements of structures, as illustrated in the preceding example, requires that the complementary energy of the structure be determined as a function of the loads. Then the derivatives of the complementary energy must be evaluated in order to obtain the desired displacements. This procedure is quite lengthy, but fortunately a more practical method is available, as will now be explained.

To derive the method in the simplest manner, we will discuss a truss structure in which the only stress resultants are the axial forces. Let us consider an individual member of the truss and assume that it has length L and cross-sectional area A and that it is subjected to an axial force N. Then the complementary energy $U^\star$ of this member is

$$U^\star = \int_0^N \epsilon_1 L\, dN_1 \tag{g}$$

as obtained from Eq. (10-22) with $\delta_1 = \epsilon_1 L$, where ϵ_1 is the uniform strain in the bar due to the axial force N_1. We may note that, as the axial force N_1 increases from its initial value of zero to its maximum value N, the strain ϵ_1 increases from zero to a maximum value ϵ. Similarly, the stress σ_1 in the bar increases from zero to a maximum value σ. The relationship between the stress σ_1 and the strain ϵ_1 is assumed to be represented by a nonlinear stress-strain diagram such as that shown in Fig. 10-15b.

It is convenient to define a function F representing the complementary energy per unit length of the bar; thus, dividing Eq. (g) by the length L, we obtain

$$F = \int_0^N \epsilon_1\, dN_1 \tag{h}$$

The quantity F, which is a function of the axial force N, can be evaluated with the aid of the stress-strain curve. (Note that the stress-strain curve gives the relationship between ϵ_1 and σ_1, but, since $N_1 = \sigma_1 A$, we also know the relationship between ϵ_1 and N_1. Thus, the integral in Eq. h can be evaluated.)

Now consider a truss composed of many members and subjected to loads $P_1, P_2, \ldots, P_n$ that produce corresponding displacements $\delta_1, \delta_2, \ldots, \delta_n$. Also, let us consider a typical element of the truss between two cross sections a distance dx apart (Fig. 10-3a). The axial force acting on this element is N, and the corresponding elongation is $d\delta$ (equal to $\epsilon\, dx$). Therefore, the complementary energy of the element is $dU^\star = F\, dx$, where F is given by Eq. (h). The complementary energy of the entire structure is

$$U^\star = \int F\, dx \tag{i}$$

where it is understood that the integration extends along the axes of all members of the structure. Since we are obtaining $U^\star$ by summing the complementary energies of the members, we are restricted to structures with only material nonlinearities.

To obtain the displacement δ_i corresponding to the load P_i, we apply the Crotti-Engesser theorem to Eq. (i), thereby obtaining

$$\delta_i = \frac{\partial U^\star}{\partial P_i} = \frac{\partial}{\partial P_i}\left[\int F\, dx\right]$$

Differentiating under the integral sign, we get

$$\delta_i = \int \frac{\partial F}{\partial P_i}\, dx$$

Now consider the derivative $\partial F/\partial P_i$. The quantity F is a function of the axial force N, as already explained in connection with Eq. (h). The axial force N is a function of the applied loads $P_1, P_2, \ldots, P_n$. Therefore, F is a function of P_i through the intermediate variable N. Hence we can rewrite the preceding equation in the form

$$\delta_i = \int \frac{\partial F}{\partial N}\,\frac{\partial N}{\partial P_i}\, dx \tag{j}$$

Each of the derivatives in this expression has a simple physical interpretation. The derivative $\partial F/\partial N$ is equal to ϵ (see Eq. h).* The derivative $\partial N/\partial P_i$ represents the value of the axial force N caused by a unit value

* From calculus we know that

$$\frac{\partial}{\partial a}\int_0^a f(x)\, dx = f(a)$$

Applying this relationship to Eq. (h), we have $a = N$, $f(x) = \epsilon_1$, $dx = dN_1$, and $f(a) = \epsilon$.

of the load P_i. Thus, in the notation of the unit-load method (Section 10.3), this derivative is equal to N_U, which is defined as the axial force due to a unit load corresponding to the displacement δ_i. Replacing the derivatives in Eq. (j) with ϵ and N_U, respectively, we obtain

$$\delta_i = \int \epsilon N_U \, dx = \int N_U \, d\delta \tag{k}$$

where $d\delta$, equal to $\epsilon \, dx$, is the elongation of an element (see Fig. 10-3a). The preceding equation is the fundamental equation of the unit-load method (Eq. 10-3) for the case in which only axial deformations are considered.

Similar derivations can be made for structures in which the effects of flexural, shearing, and torsional deformations are considered, and therefore we conclude that the use of complementary energy and the Crotti-Engesser theorem leads directly to the unit-load method as expressed by Eq. (10-3). The method is a very efficient means of determining displacements, and it is valid for structures that have material nonlinearities (but not geometric nonlinearities).*

Example 2

A cantilever beam of length L and rectangular cross section carries a concentrated load P at the free end (see Fig. 10-19 and Example 3 of Section 10.5). The stress-strain relationship of the material is represented by the equation $\sigma = B\sqrt{\epsilon}$, where B is a constant. Determine the deflection δ at the free end of the beam by using the unit-load method.

Because only flexural deformations are important in this beam, we will use only the second term in the unit-load equation (Eq. 10-3). The bending moment M_U is the moment caused by a unit load corresponding to the deflection δ; therefore, the moment M_U is numerically equal to $1 \cdot x$, where x is the distance measured from the free end of the beam (Fig. 10-19a). The deformation $d\theta$ is numerically equal to

$$\frac{\epsilon_0 \, dx}{h/2}$$

where ϵ_0 is the strain at the outer surface of the beam. For the beam under discussion, the strain ϵ_0 is given by Eq. (l) of Section 10.5, and therefore the deformation $d\theta$ is

$$d\theta = \frac{50P^2x^2}{B^2b^2h^5} dx \tag{l}$$

Substituting into the unit-load equation, we obtain the following expression for the deflection:

$$\delta = \int M_U \, d\theta = \int_0^L 1(x)\left(\frac{50P^2x^2}{B^2b^2h^5}\right) dx = \frac{25P^2L^4}{2B^2b^2h^5} \tag{m}$$

* The case of linear structures is discussed later in Section 10.8.

This result agrees with the deflection found previously (see Eq. 10-55). However, the unit-load method provides an easier solution than does the earlier example, because it is not necessary to go through the lengthy process of evaluating the complementary energy (see the derivation of Eq. 10-43).

Example 3

The truss ABC shown in Fig. 10-20a supports a vertical load P at joint B. The material of the truss has a stress-strain relationship given by the equation $\sigma = B\sqrt{\epsilon}$, where B is a constant. Determine the vertical displacement δ_v of joint B by using the unit-load method.

Since axial forces are the only stress resultants in the truss, we will use only the first term of Eq. (10-3). The unit load corresponding to the displacement δ_v is a vertical load at joint B. The corresponding axial forces N_U in the two members of the truss are

$$(N_U)_{ab} = 1 \qquad (N_U)_{bc} = -\sqrt{2}$$

The strains in the bars are obtained from the stress-strain relationship:

$$\epsilon_{ab} = \frac{\sigma_{ab}^2}{B^2} \qquad \epsilon_{bc} = -\frac{\sigma_{bc}^2}{B^2}$$

The stresses are $\sigma_{ab} = P/A$ and $\sigma_{bc} = -\sqrt{2}P/A$, as obtained from equilibrium. Substituting these stresses into the expressions for the strains, we get

$$\epsilon_{ab} = \frac{P^2}{A^2B^2} \qquad \epsilon_{bc} = -\frac{2P^2}{A^2B^2}$$

Finally, we can substitute the values for N_U and $d\delta$ (equal to $\epsilon\,dx$) into the unit-load equation (Eq. 10-3) and obtain the displacement.

$$\delta_v = \int N_U\,d\delta = \int_0^L (1)\left(\frac{P^2}{A^2B^2}\right)dx + \int_0^{\sqrt{2}L} (-\sqrt{2})\left(-\frac{2P^2}{A^2B^2}\right)dx$$

$$= \frac{5P^2L}{A^2B^2} \tag{n}$$

This displacement is the same as the displacement obtained earlier in Example 1 of this section (see Eq. 10-57). However, we again observe the simplicity of the solution by the unit-load method. The advantage of the unit-load method is even more pronounced if the displacement to be calculated does not correspond to one of the actual loads on the structure (see Problem 10.7-2).

FORCE METHOD

The use of complementary energy and the Crotti-Engesser theorem leads to another important method for analyzing structures. This method, called the *force method*, is based upon the concept of statical indeterminacy and utilizes statical redundants as the unknown quantities in the analysis. The fact that force quantities (stress resultants and reactions) are the unknowns is consistent with the requirement that the complementary energy must be expressed as a function of the forces in order to apply the Crotti-Engesser theorem.

Let us consider a materially nonlinear structure having n degrees of statical indeterminacy. After selecting the redundant quantities X_1, $X_2, \ldots, X_n$, we then release these redundants by making suitable releases in the structure. For instance, a support may be removed if the redundant is a reaction, or a pin may be inserted if the redundant is a bending moment. The resulting released structure must be statically determinate and immovable. Next, we visualize the released structure as being subject not only to the actual loads but also to the redundants themselves. In other words, the redundant actions must be treated as loads acting on the released structure. Then the complementary energy $U^\star$ of the released structure can be evaluated in the usual manner. The only new feature lies in the fact that the complementary energy of the released structure is a function of both the loads and the redundants. According to the Crotti-Engesser theorem, we can obtain the displacements of the released structure corresponding to the redundants by taking partial derivatives of the complementary energy with respect to those redundants. Let us denote by the symbols D_1, $D_2, \ldots, D_n$ the displacements of the original structure corresponding to the redundants X_1, $X_2, \ldots, X_n$, respectively. Then the Crotti-Engesser theorem (Eq. 10-54) yields n simultaneous equations, as follows:

$$D_1 = \frac{\partial U^\star}{\partial X_1} \qquad D_2 = \frac{\partial U^\star}{\partial X_2} \qquad \cdots \qquad D_n = \frac{\partial U^\star}{\partial X_n} \tag{10-58}$$

Each of these equations has the same general form. If we examine the ith equation, we find that the right-hand side consists of terms containing both the statical redundants and the loads, with the redundants being the unknown quantities. These terms add up to the true displacement D_i of the original structure, which will be zero whenever the corresponding redundant X_i is a stress resultant or a support reaction that does not undergo any displacement. Thus, we see that Eqs. (10-58) actually represent compatibility conditions for displacements corresponding to the redundants. These equations can be solved simultaneously for the redundants in terms of the loads, and then the remaining reactions and stress resultants can be determined by static equilibrium.

The complementary-energy method of analysis described in the preceding paragraph utilizes statical redundants as the unknown quantities and requires the solution of compatibility equations. It is called the **force method** of analysis, and it applies to materially nonlinear structures. In the case of a linear structure, the method is known as the **flexibility method**.*

* The flexibility method is discussed further in Section 10.8.

The force method, as represented by Eqs. (10-58), is analogous to the displacement method, represented by Eqs. (10-45). In the force method we express the complementary energy as a function of statical redundants, and then we use the Crotti-Engesser theorem to obtain equations of compatibility that can be solved for the redundant forces. In the displacement method we express the strain energy as a function of unknown joint displacements, and then we use Castigliano's first theorem to obtain equations of equilibrium that can be solved for the joint displacements.

Now consider the particular case of the force method in which there are no displacements in the original structure corresponding to the statical redundants. As already mentioned, this situation will exist if the redundants are stress resultants or reactions at supports that do not move. Under these conditions the displacements $D_1, D_2, \ldots, D_n$ in the compatibility equations (Eqs. 10-58) are zero. Then the equations simplify to the following:

$$\frac{\partial U^*}{\partial X_1} = 0 \qquad \frac{\partial U^*}{\partial X_2} = 0 \qquad \cdots \qquad \frac{\partial U^*}{\partial X_n} = 0 \tag{10-59}$$

These equations represent the conditions for a stationary value of the complementary energy. For a structure in stable equilibrium, the stationary value is actually a minimum value; hence Eqs. (10-59) represent the **principle of minimum complementary energy**. This principle states that the statical redundants $X_1, X_2, \ldots, X_n$ for a stable structure have values such that the complementary energy is a minimum, provided that there are no displacements corresponding to the redundants in the original structure.*

Example 4

The truss shown in Fig. 10-24a is constructed of a material having a stress-strain relationship given by the equation $\sigma = B\sqrt{\epsilon}$, where B is a constant. Find the axial forces in all three bars by using complementary energy and the force method.

Assuming that the reactive force at joint B is the statical redundant X, we obtain the released structure shown in Fig. 10-24b. The axial forces in the bars of the released structure, from equilibrium, are as follows:

$$N_{ad} = N_{cd} = \frac{P - X}{\sqrt{2}} \qquad N_{bd} = X \tag{o}$$

* The principle of minimum complementary energy was formulated first by Crotti (Refs. 10-23 and 10-24) and later by Engesser (Ref. 10-25). If the structure behaves linearly, the complementary energy and the strain energy are equal; then the principle of minimum complementary energy reduces to the principle of minimum strain energy (see Section 10.8).

The complementary-energy methods have provided the basis for significant progress in structural mechanics. The reader who is interested in additional study of the methods should consult other sources, such as Refs. 10-14, 10-15, 10-26, and 10-27. The history of complementary-energy methods was traced by Oravas and McLean (Ref. 2-4); other historical comments are given in Refs. 10-15, 10-28, and 10-29.

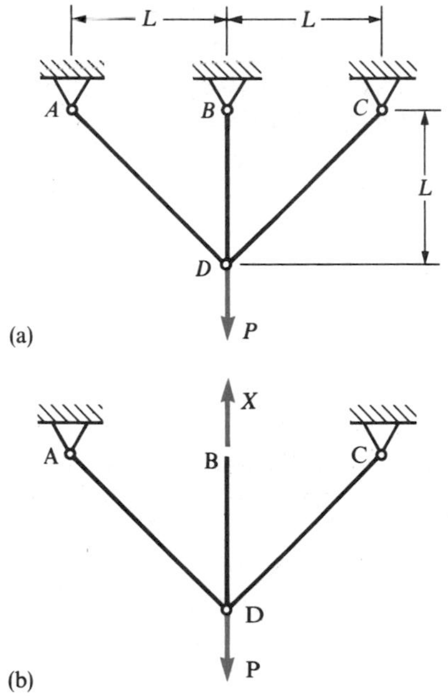

Fig. 10-24 Example 4. Force method

The corresponding stresses are

$$\sigma_{ad} = \sigma_{cd} = \frac{P - X}{\sqrt{2}\,A} \qquad \sigma_{bd} = \frac{X}{A}$$

in which A is the cross-sectional area of each bar.

The complementary energy density of bar BD is obtained as follows (see Eq. 10-23):

$$u_{bd}^{\star} = \int_0^{\sigma_{bd}} \epsilon\, d\sigma = \int_0^{\sigma_{bd}} \frac{\sigma^2}{B^2}\, d\sigma = \frac{\sigma_{bd}^3}{3B^2} = \frac{1}{3B^2}\left(\frac{X}{A}\right)^3$$

The complementary energy density of bars AD and CD can be obtained in a similar manner; the results are

$$u_{ad}^{\star} = u_{cd}^{\star} = \frac{1}{3B^2}\left(\frac{P - X}{\sqrt{2}\,A}\right)^3$$

Finally, we can multiply the complementary energy density of each bar by the volume of the bar, and then we can sum the results to obtain the total complementary energy of the released structure:

$$U^{\star} = \frac{L}{3A^2B^2}[(P - X)^3 + X^3]$$

This expression for $U^{\star}$ is a nonlinear function of both the load P and the redundant X.

The displacement in the original structure corresponding to the redundant X is zero because there is no displacement at support B. Therefore, the Crotti-Engesser theorem, when applied to the redundant X, gives the following equation (see Eqs. 10-58 and 10-59):

$$\frac{dU^{\star}}{dX} = \frac{L}{3A^2B^2}[3(P - X)^2(-1) + 3X^2] = 0$$

from which we get $X = P/2$. Substituting this value for X into Eqs. (o) yields

$$N_{ad} = N_{cd} = \frac{P}{2\sqrt{2}} \qquad N_{bd} = \frac{P}{2} \tag{p}$$

Thus, the axial forces in the bars of this statically indeterminate, nonlinear truss have been found by the force method.

10.8 CASTIGLIANO'S SECOND THEOREM

In the preceding section we discussed the methods by which complementary energy may be used to find deflections and to analyze structures. We emphasized that the complementary-energy methods apply to structures with nonlinear behavior. Let us now limit our discussion to structures that behave linearly and for which the principle of superposition applies.

Under these conditions the complementary energy $U^\star$ and the strain energy U of the structure are equal.

Now assume that a linear structure is subjected to loads $P_1, P_2, \ldots, P_n$ and that these loads produce corresponding displacements $\delta_1, \delta_2, \ldots, \delta_n$. Then both U and $U^\star$ may be expressed as quadratic functions of the loads (see Eq. 10-34). Also, we can replace $U^\star$ by U in the Crotti-Engesser theorem (Eq. 10-54) and obtain

$$\delta_i = \frac{\partial U}{\partial P_i} \qquad (10\text{-}60)$$

This equation is known as **Castigliano's second theorem**, and it may be stated as follows: For a linear structure the partial derivative of the strain energy with respect to any load P_i is equal to the corresponding displacement δ_i provided that the strain energy is expressed as a function of the loads. (For the history of this theorem, see Refs. 10-16 through 10-20.)

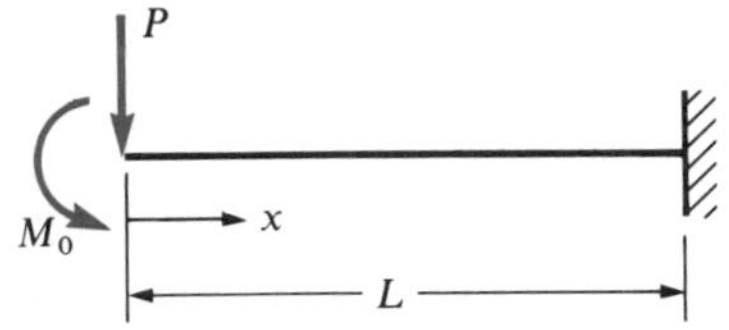

Fig. 10-25 Displacements of a beam by Castigliano's second theorem

As an application of Castigliano's second theorem, let us consider a cantilever beam subjected to a load P and a couple M_0 acting at the free end (Fig. 10-25). The beam behaves linearly and has constant flexural rigidity EI. The strain energy of the beam can be found from Eq. (7-59a), which is repeated here:

$$U = \int \frac{M^2\,dx}{2EI} \qquad (10\text{-}61)$$

In this equation, M represents the bending moment at any cross section. For the beam shown in Fig. 10-25, the bending moment at distance x from the free end is $M = -Px - M_0$. Substituting this expression for M into Eq. (10-61), we get

$$U = \frac{1}{2EI}\int_0^L (-Px - M_0)^2\,dx = \frac{P^2L^3}{6EI} + \frac{PM_0L^2}{2EI} + \frac{M_0^2 L}{2EI} \qquad \text{(a)}$$

This equation expresses the strain energy U as a quadratic function of the loads P and M_0.

In order to obtain the vertical deflection δ at the free end of the beam, we use Castigliano's second theorem and take the partial derivative of U with respect to P; thus,

$$\delta = \frac{\partial U}{\partial P} = \frac{PL^3}{3EI} + \frac{M_0L^2}{2EI}$$

In a similar manner, we can find the angle of rotation θ at the free end of the beam by taking the partial derivative of U with respect to M_0:

$$\theta = \frac{\partial U}{\partial M_0} = \frac{PL^2}{2EI} + \frac{M_0L}{EI}$$

The sign conventions for δ and θ are as follows. A positive sign for δ means that the deflection is in the same direction as the load P (that is, downward) and a positive sign for θ means that the angle of rotation has the same sense as the couple M_0 (that is, counterclockwise). As expected, both δ and θ are linear functions of the loads.

It is apparent that Castigliano's second theorem can be used only when finding displacements that correspond to loads acting on the structure, as is also the case for the Crotti-Engesser theorem. If the desired displacement does not correspond to a load, we must place a fictitious load on the structure corresponding to that displacement. Then we may calculate that displacement by using Castigliano's second theorem; the resulting expression for the displacement is in terms of both the actual loads and the fictitious load. By setting the fictitious load equal to zero, the displacement due to the actual loads is obtained.

UNIT-LOAD METHOD

The process of finding displacements by a direct application of Castigliano's second theorem is cumbersome if more than two loads act on the structure. The reason is that the evaluation of the strain energy and its square is very lengthy. For instance, suppose that four loads act on the cantilever beam shown in Fig. 10-25, rather than two loads. Then the evaluation of the equation for the strain energy, analogous to Eq. (a), requires squaring a four-term expression, and the final result for U contains ten terms.

A considerable simplification in finding displacements is obtained if we apply Castigliano's second theorem *before* integrating the expression for the bending moment. To illustrate this point, consider a beam or frame in which only flexural deformations are important; then the strain energy U is given by Eq. (10-61). To obtain the deflection δ_i corresponding to the load P_i, we must take the partial derivative of U with respect to P_i. Thus, differentiating under the integral sign, we obtain

$$\delta_i = \frac{\partial U}{\partial P_i} = \frac{\partial}{\partial P_i}\int \frac{M^2\,dx}{2EI} = \int \frac{M}{EI}\frac{\partial M}{\partial P_i}\,dx \tag{b}$$

The partial derivative $\partial M/\partial P_i$ represents the value of the bending moment M caused by a unit value of the load P_i. Thus, this derivative is equal to M_U, which is the bending moment in the structure due to a unit load corresponding to the desired displacement. The moment M appearing under the integral sign in Eq. (b) is the bending moment due to the actual loads on the structure; hence this moment is equal to M_L. Using this notation, we see that Eq. (b) becomes

$$\delta_i = \int \frac{M_U M_L\,dx}{EI}$$

This equation is the unit-load equation (Eq. 10-4) for the case when only flexural deformations are considered.

Similar derivations can be made for the effects of axial, shearing, and torsional deformations. We conclude, therefore, that the unit-load method as applied to linear structures (Eq. 10-4) can be derived directly from Castigliano's second theorem. This result is not surprising since we have already shown that the more general equation of the unit-load method (Eq. 10-3), which is applicable to nonlinear structures, can be derived from the Crotti-Engesser theorem. As already pointed out, the unit-load method is an efficient method for determining displacements of structures, and ordinarily it should be used instead of Castigliano's second theorem.

FLEXIBILITY METHOD

In the preceding section we used complementary energy and the Crotti-Engesser theorem to derive the force method of structural analysis. A special case of the force method, called the **flexibility method**, occurs when the structure behaves linearly. Under such conditions the strain energy of the released structure, which is equal to the complementary energy, may be expressed as a quadratic function of both the loads and the statical redundants $X_1, X_2, \ldots, X_n$. Then the application of Castigliano's second theorem leads to the following simultaneous equations:

$$D_1 = \frac{\partial U}{\partial X_1} \qquad D_2 = \frac{\partial U}{\partial X_2} \qquad \cdots \qquad D_n = \frac{\partial U}{\partial X_n} \qquad (10\text{-}62)$$

in which $D_1, D_2, \ldots, D_n$ represent the displacements in the original structure corresponding to the redundants. The general form of each of these equations is the same. The right-hand sides consist of terms that represent displacements corresponding to the redundants; some of the displacements are due to the redundants themselves and others are due to the actual loads on the structure. These terms add up to the true displacements in the original structure, which are zero whenever the corresponding redundants are either stress resultants or reactions that do not undergo any displacements. Therefore, Eqs. (10-62) are equations of compatibility that can be solved simultaneously for the redundants in terms of the loads. Furthermore, Eqs. (10-62) are a special case of the compatibility equations for the force method (Eqs. 10-58), because we can use the strain energy U in place of the complementary energy U^* when a structure behaves linearly. Equations (10-62) are the equations of the flexibility method, which is the counterpart of the stiffness method.*

A special case arises whenever the redundants are either stress resultants or reactions that do not undergo any displacements. In such cases the displacements in the original structure corresponding to the redun-

* The flexibility method was originated by J. C. Maxwell in 1864 and O. Mohr in 1874 (Refs. 10-1, 10-13, and 10-30), and it is often called the *Maxwell-Mohr method*. This commonly used method of structural analysis is discussed in detail in textbooks on structural analysis (for example, see Ref. 10-22).

dants are zero, and the equations of the flexibility method (Eqs. 10-62) become

$$\frac{\partial U}{\partial X_1} = 0 \qquad \frac{\partial U}{\partial X_2} = 0 \qquad \cdots \qquad \frac{\partial U}{\partial X_n} = 0 \tag{10-63}$$

These equations represent the conditions for a stationary value of the strain energy. If the structure is in stable equilibrium, the stationary value is actually a minimum value. Therefore, Eqs. (10-63) represent the **principle of minimum strain energy**, which states that for a linear structure the redundant quantities $X_1, X_2, \ldots, X_n$ have values such that the strain energy is a minimum, provided that there are no displacements corresponding to the redundants in the original structure. The principle of minimum strain energy is a special case (for linear structures) of the more general principle of minimum complementary energy (Eqs. 10-59).*

In order to show in more detail the general form of the equations of the flexibility method (Eqs. 10-62), let us take a specific illustration. Assume that a certain structure is statically indeterminate to the second degree and that it supports two loads P_1 and P_2. Then the strain energy of the released structure, which is a quadratic function of the two redundants X_1 and X_2 and the loads P_1 and P_2, will have the following general form (see Eq. 10-34):

$$\begin{aligned} U = {} & a_{11}X_1^2 + a_{12}X_1X_2 + a_{13}X_1P_1 + a_{14}X_1P_2 \\ & + a_{21}X_2X_1 + a_{22}X_2^2 + a_{23}X_2P_1 + a_{24}X_2P_2 \\ & + a_{31}P_1X_1 + a_{32}P_1X_2 + a_{33}P_1^2 + a_{34}P_1P_2 \\ & + a_{41}P_2X_1 + a_{42}P_2X_2 + a_{43}P_2P_1 + a_{44}P_2^2 \end{aligned}$$

where the coefficients $a_{11}, a_{12}, \ldots, a_{44}$ are constants that depend only upon the properties of the structure. The application of Castigliano's second theorem to the preceding expression for U gives the equations of compatibility (Eqs. 10-62):

$$D_1 = \frac{\partial U}{\partial X_1} = F_{11}X_1 + F_{12}X_2 + b_{11}P_1 + b_{12}P_2$$

$$D_2 = \frac{\partial U}{\partial X_2} = F_{21}X_1 + F_{22}X_2 + b_{21}P_1 + b_{22}P_2$$

in which the coefficients F and b are new constants that are obtained from the a's and that depend only upon the properties of the structure. The constants F_{11}, F_{12}, F_{21}, and F_{22} are coefficients of the unknown redundants and are known as the **flexibility coefficients**, or **flexibilities**,

* The principle of minimum strain energy is also known as the *principle of least work*. This name was used by L. F. Ménabréa, who first stated the principle in 1858 although without a satisfactory proof, and by Castigliano, who proved it in 1873 (see Refs. 10-16 through 10-20 and Ref. 10-31).

of the released structure. The remaining terms depend only upon the loads P_1 and P_2 and the properties of the structure. These terms are the displacements of the released structure corresponding to the redundants and caused by the loads.

Suppose we now take additional partial derivatives of the preceding equations, as follows:

$$\frac{\partial D_1}{\partial X_1} = \frac{\partial^2 U}{\partial X_1^2} = F_{11} \qquad \frac{\partial D_1}{\partial X_2} = \frac{\partial^2 U}{\partial X_2\,\partial X_1} = F_{12}$$

$$\frac{\partial D_2}{\partial X_1} = \frac{\partial^2 U}{\partial X_1\,\partial X_2} = F_{21} \qquad \frac{\partial D_2}{\partial X_2} = \frac{\partial^2 U}{\partial X_2^2} = F_{22}$$

From these relationships we can readily discern the following general equation for the flexibilities:

$$F_{ij} = \frac{\partial^2 U}{\partial X_j\,\partial X_i} \tag{10-64}$$

This equation shows that the flexibilities for the released structure can be obtained from the strain energy U by the process of differentiation whenever the strain energy is expressed as a quadratic function of the statical redundants $X_1, X_2, \ldots, X_n$. Also, because the order of differentiation of U is immaterial, we can write

$$F_{ij} = \frac{\partial^2 U}{\partial X_j\,\partial X_i} = \frac{\partial^2 U}{\partial X_i\,\partial X_j} = F_{ji} \tag{10-65}$$

Thus, the reciprocal theorem for flexibilities is established from strain-energy considerations.

OTHER METHODS OF STRUCTURAL ANALYSIS

In addition to the flexibility method and the stiffness method, numerous other methods of structural analysis are of importance to engineers. For example, *matrix methods of structural analysis* are in common usage. The matrix methods consist of the flexibility and stiffness methods with the refinement that all equations are presented in matrix form and all derivations and calculations are performed with the aid of matrix algebra. The use of matrices makes the methods more systematic, provides compactness of presentation, and is ideally suited for computer programming.* The most general method for the numerical analysis of structures (including plates, shells, and solid continua) is the *finite-element method* (Ref. 10-32). These advanced methods of analysis require a thorough understanding of mechanics of materials, including the concepts of strain energy and complementary energy.

* Matrix methods of structural analysis are described in textbooks on structural theory (for example, see Ref. 10-22). A working knowledge of matrix algebra can be obtained from engineering mathematics books (for example, see Ref. 6-1).

*10.9 SHEAR DEFLECTIONS OF BEAMS

Because the effects of shear deformations on the deflections of beams are usually relatively small compared to the effects of flexural deformations, it is common practice to disregard them. However, if greater accuracy is required, the shear deflections can be determined and added to the bending deflections. In this section we will use the unit-load method for this purpose.

The unit-load method for obtaining beam deflections is based upon the principle of virtual work (see Sections 10.2 and 10.3). The method requires that expressions for both external and internal work be obtained. Because the unit load is the only load on the structure, the expression for the external work is $W_{\text{ext}} = 1 \cdot \Delta$, as given by Eq. (a) of Section 10.3. In that equation, Δ represents the desired deflection due to the actual loads on the beam and the 1 represents the unit load corresponding to that deflection.

The internal virtual work is the work performed by the stresses in the beam (the stresses are caused by the unit load) as those stresses move through the deformations caused by the actual loads. Previously, we considered four possible stress resultants: axial force, bending moment, shear force, and torque. Because we are now limiting our attention to shear deflections of beams, we will omit the effects of axial forces and torques. Also, in the previous derivation we obtained the internal virtual work by multiplying each stress resultant by the appropriate deformation quantity for a beam element; in this way we obtained Eq. (b) of Section 10.3. Now, however, we will deal directly with the stresses in the beam and integrate over the volume of the beam to obtain the internal virtual work.

Let us consider a differential element of dimensions dx, dy, dz (Fig. 10-26a) from the interior of a beam subjected to a unit load. Acting on the sides of this element are normal stresses σ and shear stresses τ (Fig. 10-26b) caused by the bending moment M_U and the shear force V_U produced by the unit load. These stresses can be calculated from the flexure and shear formulas:

$$\sigma = \frac{M_U y}{I} \qquad \tau = \frac{V_U Q}{Ib}$$

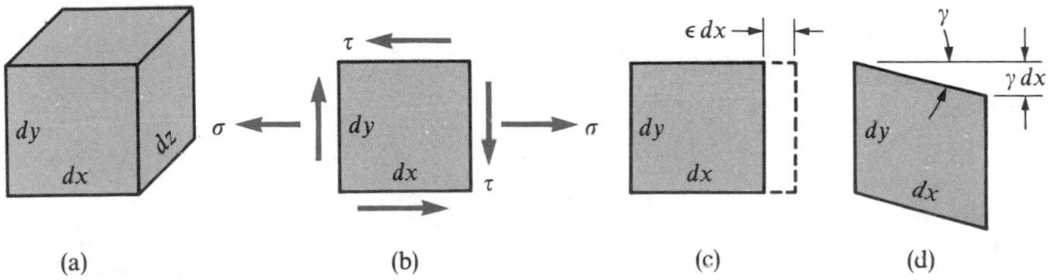

Fig. 10-26 Element from a beam

When using the unit-load method, the virtual deformations imparted to the element are chosen to be the same as the deformations caused by the actual loads. These loads produce bending moments M_L and shear forces V_L. The corresponding deformations consist of an extension (Fig. 10-26c) caused by the moment M_L and a shear distortion (Fig. 10-26d) caused by the shear force V_L. The extensional strain ϵ and the shear strain γ associated with these deformations are

$$\epsilon = \frac{M_L y}{EI} \qquad \gamma = \frac{V_L Q}{GIb}$$

Therefore, the internal virtual work of the stresses σ and τ acting on the differential element is

$$\begin{aligned} dW_{\text{int}} &= (\sigma\, dy\, dz)(\epsilon\, dx) + (\tau\, dy\, dz)(\gamma\, dx) \\ &= \frac{M_U M_L y^2}{EI^2} dx\, dy\, dz + \frac{V_U V_L Q^2}{GI^2 b^2} dx\, dy\, dz \end{aligned}$$

The total internal work is obtained by integrating the preceding expression throughout the volume of the beam; thus,

$$W_{\text{int}} = \int \frac{M_U M_L y^2}{EI^2} dx\, dy\, dz + \int \frac{V_U V_L Q^2}{GI^2 b^2} dx\, dy\, dz$$

This expression may be simplified by noting that, at a given cross section of the beam, the following quantities are constants: M_U, M_L, V_U, V_L, E, G, and I. Therefore, we can separate each of the preceding integrals into an integration over the cross-sectional area and an integration along the axis of the beam, as follows:

$$W_{\text{int}} = \int_L \frac{M_U M_L}{EI^2}\left[\int_A y^2\, dy\, dz\right]dx + \int_L \frac{V_U V_L}{GI^2}\left[\int_A \frac{Q^2}{b^2} dy\, dz\right]dx \qquad \text{(a)}$$

where the symbols L and A denote that the integration is to be carried out over the length L of the beam and over the cross-sectional area A, respectively.

The first term in brackets in Eq. (a) is the moment of inertia I, a property of the cross section. The second term in brackets is also dependent only upon the cross-sectional dimensions of the beam. Therefore, we define a new cross-sectional property f_s, called the **form factor for shear**, as follows:

$$f_s = \frac{A}{I^2}\int_A \frac{Q^2}{b^2} dA \qquad \text{(10-66)}$$

where $dA = dy\, dz$ represents an element of area in the cross section of the beam. The form factor is a dimensionless quantity that can be evaluated for each particular shape of beam, as illustrated later.

Replacing the two terms in brackets in Eq. (a) with I and $f_s I^2/A$, respectively, we get the expression for internal virtual work:

$$W_{\text{int}} = \int \frac{M_U M_L\,dx}{EI} + \int \frac{f_s V_U V_L\,dx}{GA} \tag{b}$$

Finally, the **unit-load equation** for the deflection Δ is obtained by equating external work (equal to $1 \cdot \Delta$) with internal work (Eq. b):

$$\Delta = \int \frac{M_U M_L\,dx}{EI} + \int \frac{f_s V_U V_L\,dx}{GA} \tag{10-67}$$

Equation (10-67) may be used to find beam deflections when considering the effects of both bending moments and shear forces (compare with Eq. 10-4). The term GA/f_s is called the **shearing rigidity** of the beam.

The form factor for shear must be evaluated for each particular shape of cross section. For example, if the cross section is rectangular with width b and height h, the expression for the first moment Q (see Fig. 5-23b and Eq. c of Section 5.5) is

$$Q = \frac{b}{2}\left(\frac{h^2}{4} - y_1^2\right) \tag{c}$$

Also, for a rectangular beam, the quantity A/I^2 equals $144/bh^5$. Therefore, the form factor is

$$f_s = \frac{144}{bh^5}\int_{-h/2}^{h/2} \frac{1}{4}\left(\frac{h^2}{4} - y_1^2\right)^2 b\,dy_1 = \frac{6}{5} \tag{d}$$

In an analogous manner the form factors for other cross sections can be calculated. For instance, the form factor for a solid circular section equals 10/9 and for a thin tubular section equals 2. For an I-beam or box beam, we can assume that the shear stress is uniformly distributed over the height of the web and is approximately equal to the shear force divided by the area of the web (see Section 5.6 and Fig. 5-29). These assumptions lead to the conclusion that the form factor for such a beam is A/A_{web}, a ratio that typically is in the range of 2 to 5. A listing of values of the form factor f_s is given in Table 10-4.

To illustrate the determination of shear deflections in a beam by using the unit-load method, let us obtain the deflection at the midpoint of a uniformly loaded beam with simple supports. With the coordinate distance x measured from the left-hand support of the beam, the expressions for the bending moment and shear force due to the actual load are

$$M_L = \frac{qLx}{2} - \frac{qx^2}{2} \qquad V_L = \frac{qL}{2} - qx$$

where L is the length of the beam and q is the intensity of the uniform load.

Table 10-4 Form factor f_s for shear

	Section	f_s
	Rectangle	$\frac{6}{5}$
	Circle	$\frac{10}{9}$
	Thin tube	2
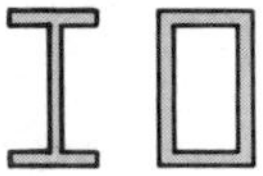	I-section or box section	$\frac{A}{A_{\text{web}}}$

The unit load acting at the middle of the beam produces the following bending moments and shear forces:

$$M_U = \frac{1(x)}{2} \qquad V_U = \frac{1}{2} \qquad \left(0 \le x \le \frac{L}{2}\right)$$

Substitution of the preceding quantities into Eq. (10-67) yields the following expression for the deflection δ at the midpoint of the beam:

$$\delta = \frac{2}{EI}\int_0^{L/2} \frac{x}{2}\left(\frac{qLx}{2} - \frac{qx^2}{2}\right)dx + \frac{2f_s}{GA}\int_0^{L/2} \frac{1}{2}\left(\frac{qL}{2} - qx\right)dx$$

$$= \frac{5qL^4}{384EI}\left(1 + \frac{48f_sEI}{5GAL^2}\right) \qquad (10\text{-}68)$$

If we use the unit-load method for the case of a concentrated load P acting at the midpoint of a simple beam, we get for the deflection at the middle

$$\delta = \frac{PL^3}{48EI}\left(1 + \frac{12f_sEI}{GAL^2}\right) \qquad (10\text{-}69)$$

Finally, for a cantilever beam with a concentrated load P at the free end, the unit-load method yields

$$\delta = \frac{PL^3}{3EI}\left(1 + \frac{3f_sEI}{GAL^2}\right) \qquad (10\text{-}70)$$

as the deflection at the free end.

The relative importance of the shear deformations can be determined in each particular beam by evaluating numerically the last term in Eqs. (10-68), (10-69), and (10-70). For solid beams, such as beams of rectangular cross section, the shear term usually represents only a few percent increase in the deflection.

STRAIN ENERGY OF SHEAR

The strain energy of shear in an element of a beam subjected to shear stresses τ (see Figs. 10-26b and d) is equal to $u\,dx\,dy\,dz$, where u is the strain energy of shear per unit volume. The strain energy per unit volume is $u = \tau^2/2G$, as given by Eq. (3-37a). Thus, we can express the strain energy of shear dU_s in terms of the shear stress τ as follows:

$$dU_s = \frac{\tau^2}{2G}\,dx\,dy\,dz$$

Since the shear stress τ is equal to VQ/Ib, the strain energy of shear becomes

$$dU_s = \frac{V^2Q^2}{2GI^2b^2}\,dx\,dy\,dz$$

Integrating this expression throughout the volume of the beam gives the total strain energy of shear:

$$U_s = \int_L \frac{V^2}{2GI^2}\left[\int_A \frac{Q^2}{b^2}\,dy\,dz\right]dx$$

where, as before, the symbols L and A indicate that the integrals are evaluated along the length of the beam and over the cross-sectional area, respectively. The term in brackets is equal to $f_s I^2/A$, as shown by comparison with Eq. (10-66). Therefore, the expression for U_s simplifies to

$$U_s = \int \frac{f_s V^2\,dx}{2GA} \tag{10-71}$$

This equation gives the shear strain energy of a beam in terms of the shear force V. In most situations, the shear strain energy is only a few percent of the bending strain energy (Eq. 10-61) and can be disregarded.

PROBLEMS / CHAPTER 10

The problems for Section 10.3 are to be solved by the unit-load method.

10.3-1 A truss ABC consisting of two members (see figure) supports a vertical load P at joint B. Each member has axial rigidity EA. Assuming that $\beta = 60°$, determine the following displacements: (a) the vertical displacement δ_v of joint B, (b) the horizontal displacement δ_h of joint B, and (c) the angle of rotation θ_{ab} of member AB.

10.3-2 Solve the preceding problem assuming that member BC is subjected to a uniform temperature increase ΔT and $P = 0$.

10.3-3 Solve Problem 10.3-1 for $\beta = 30°$.

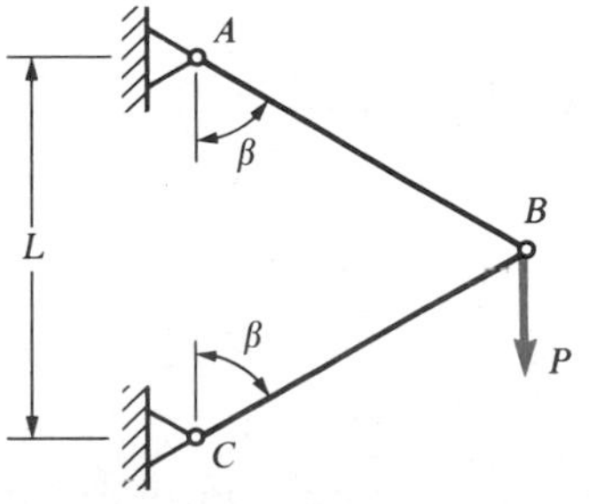

Probs. 10.3-1, 10.3-2, and 10.3-3

10.3-4 Determine the vertical displacement δ_v and horizontal displacement δ_h of joint B of the truss shown in the figure if both members have axial rigidity EA and $\beta = 30°$.

10.3-5 Solve the preceding problem for $\beta = 45°$.

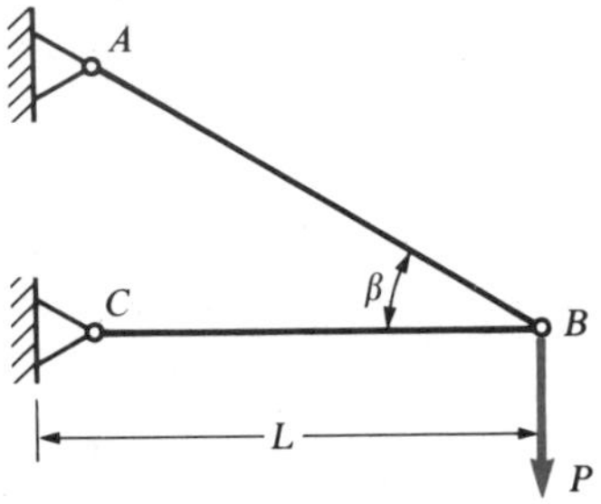

Probs. 10.3-4 and 10.3-5

10.3-6 The four outside bars of the square truss shown in the figure have their temperatures lowered by an amount ΔT. What is the change in distance between joints B and D?

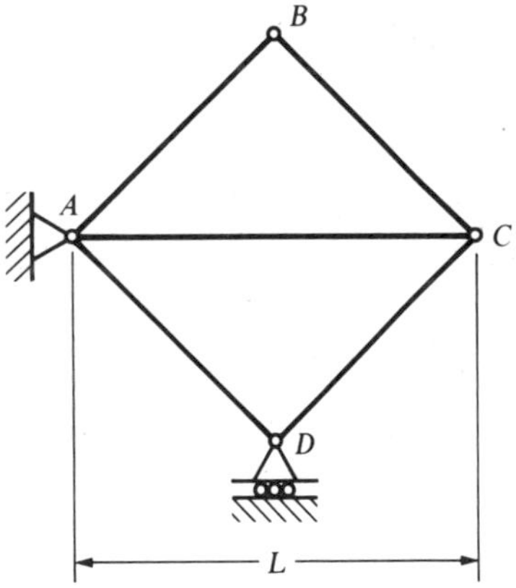

Prob. 10.3-6

10.3-7 A symmetric truss $ABCD$ has height $H = 6$ ft and span length $L = 16$ ft (see figure). A load $P = 18$ k acts vertically at joint D. The cross-sectional area of each tension member is $A_t = 2.5$ in.^{2} and of each compression member is $A_c = 5.0$ in.^{2} The truss is made of steel with $E = 30 \times 10^3$ ksi. Calculate the horizontal displacement δ_h of joint C and the vertical displacement δ_v of joint D.

10.3-8 Solve the preceding problem for a truss with the following data: $H = 2$ m, $L = 4$ m, $P = 120$ kN, $A_t = 1200$ mm^2, $A_c = 3000$ mm^2, and $E = 210$ GPa.

10.3-9 Refer to the truss $ABCD$ described in Problem 10.3-7 and assume $P = 0$. To eliminate the appearance of sagging when the truss is eventually built and loaded, the truss is to be constructed with an upward displacement, called *camber*, of joint D. By what amount ΔL should members AB and BC be increased in length (as compared to their theoretical lengths of 10 ft) in order that the camber of joint D will be 1.0 in.?

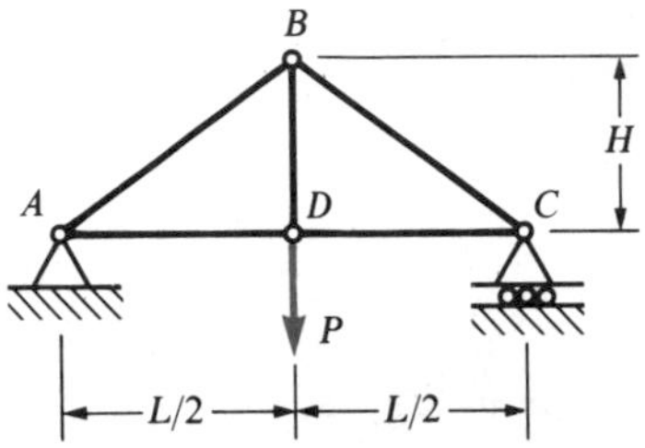

Probs. 10.3-7, 10.3-8, and 10.3-9

10.3-10 The truss shown in the figure is constructed of nine bars, each having axial rigidity EA. Loads P and $2P$ act at joints D and B, respectively. Determine the following quantities: (a) the vertical displacement δ_v of joint E, and (b) the increase δ_{ae} in the distance between joints A and E.

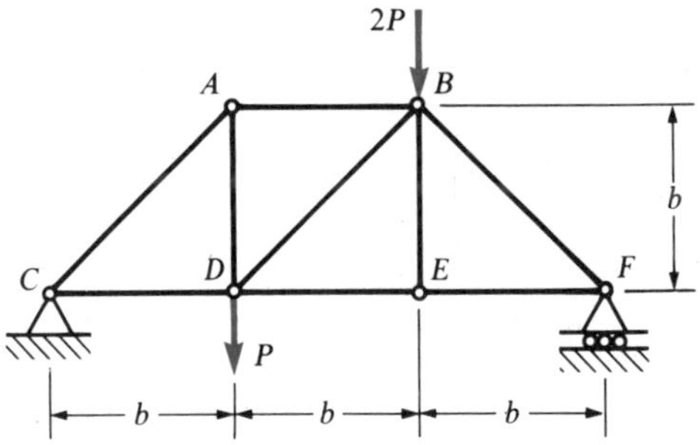

Prob. 10.3-10

10.3-11 A steel truss supports three loads P_1, P_2, and P_3 as shown in the figure. The cross-sectional areas of the members are as follows:

Members AB and CD:	Area $= A_1$
Members AE, EF, FG, and GD:	Area $= A_2$
Member BC:	Area $= A_3$
Members BE, BF, CF, and CG:	Area $= A_4$

Calculate the vertical and horizontal displacements δ_v and δ_h of joint F. Use the following data: $P_1 = P_2 = 8$ k, $P_3 = 4$ k, $A_1 = 6$ in.^{2}, $A_2 = 3$ in.^{2}, $A_3 = 4$ in.^{2}, $A_4 = 2$ in.^{2}, $L = 48$ ft, $H = 9$ ft, and $E = 30 \times 10^3$ ksi.

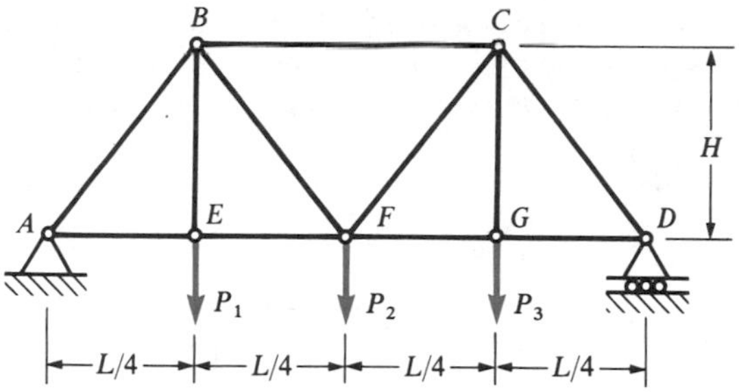

Probs. 10.3-11, 10.3-12, and 10.3-13

10.3-12 Solve the preceding problem using the following data: $P_1 = 32$ kN, $P_2 = 40$ kN, $P_3 = 16$ kN, $A_1 = 4000$ mm^2, $A_2 = 2000$ mm^2, $A_3 = 3000$ mm^2, $A_4 = 1200$ mm^2, $L = 12$ m, $H = 4$ m, and $E = 200$ GPa.

10.3-13 Refer to the truss described in Problem 10.3-11 and calculate the angle of rotation θ_{bc} of member BC and the decrease δ_{bg} in the distance between joints B and G.

10.3-14 A simple beam AB of length L and flexural rigidity EI supports a concentrated load P at the midpoint C (see figure). Determine the deflection δ_c at the load and the angle of rotation θ_a at the left-hand support.

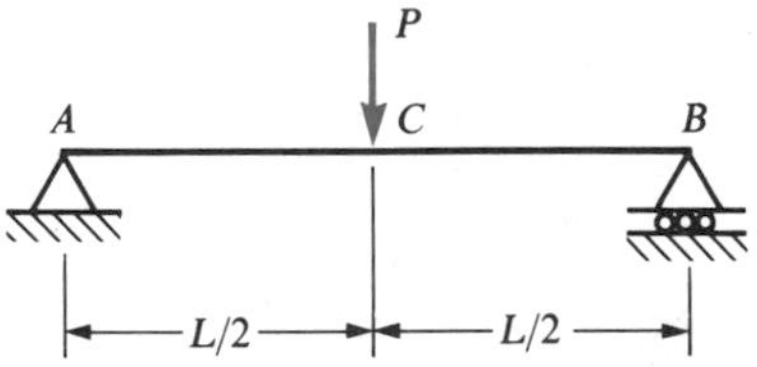

Prob. 10.3-14

10.3-15 A cantilever beam AB of length L and flexural rigidity EI supports a uniform load of intensity q throughout its length (see figure). Determine the deflection δ_b and angle of rotation θ_b at the free end B of the beam.

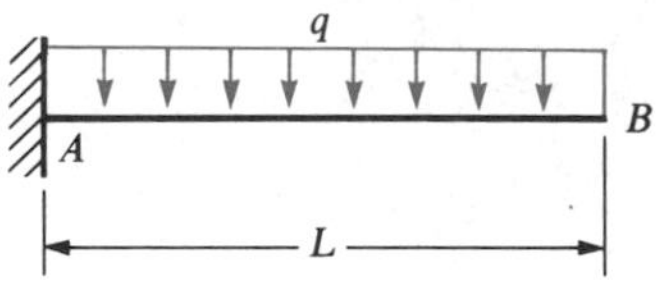

Prob. 10.3-15

10.3-16 A simple beam AB of length L and flexural rigidity EI supports three concentrated loads P as shown in the figure. Determine the deflection δ_c at the midpoint of the beam.

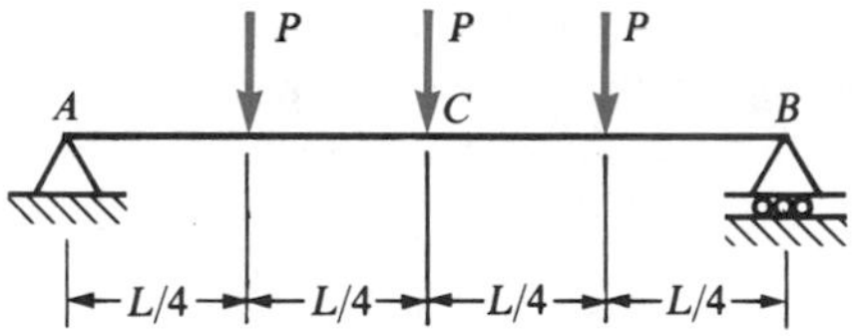

Prob. 10.3-16

10.3-17 A cantilever beam AB of length L and flexural rigidity EI supports two concentrated loads P as shown in the figure. Determine the deflections δ_c, δ_d, and δ_b at points C, D, and B, respectively.

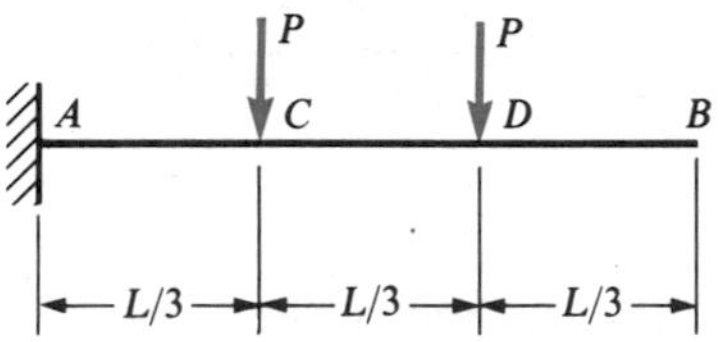

Prob. 10.3-17

10.3-18 A simple beam AB with an overhang BC (see figure) is subjected to a load P at the free end. The flexural rigidity EI is the same for both parts of the beam. Determine the deflection δ_c and angle of rotation θ_c at the point where the load is applied.

10.3-19 Refer to the beam described in the preceding problem (see figure). Determine the deflection δ at the midpoint of span AB and the angle of rotation θ at support A.

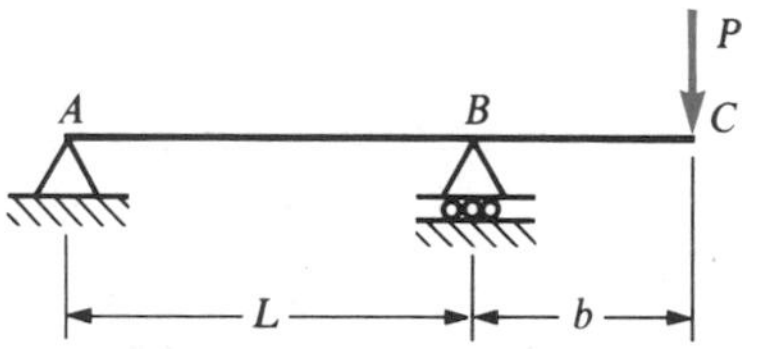

Probs. 10.3-18 and 10.3-19

10.3-20 A prismatic beam $ABCD$ with overhangs at each end is shown in the figure. The beam undergoes a change in temperature such that the top surface of the beam is at

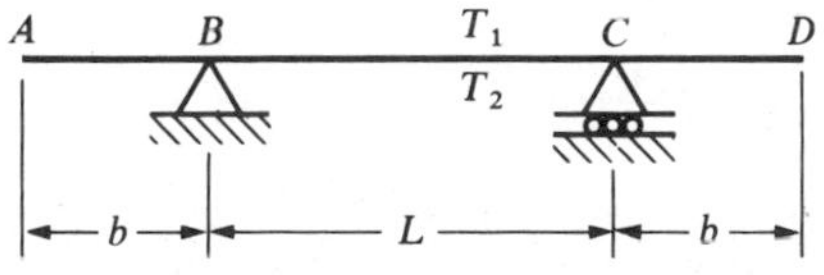

Prob. 10.3-20

temperature T_1 and the bottom surface is at temperature T_2. The height of the beam is h. Determine the deflection δ at the midpoint of the span, the angle of rotation θ_a at the free end A, and the angle of rotation θ_b at support B.

10.3-21 A cantilever beam AB of length L and flexural rigidity EI supports a uniform load of intensity q over one-half of the length (see figure). Determine the deflections δ_b and δ_c at points B and C, respectively.

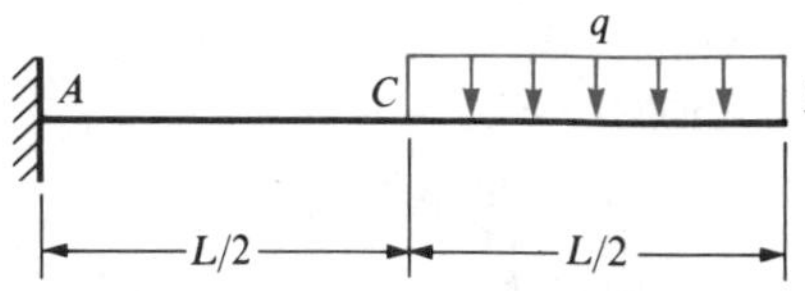

Prob. 10.3-21

10.3-22 A simple beam AB of length L and flexural rigidity EI supports a triangularly distributed load of maximum intensity q_0 (see figure). Determine the angles of rotation θ_a and θ_b at the supports.

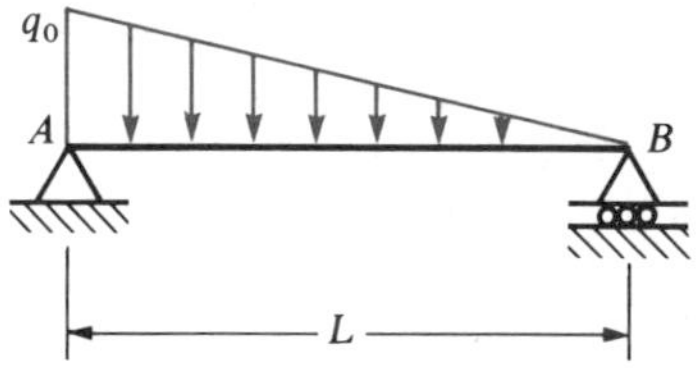

Prob. 10.3-22

10.3-23 A frame ABC has a pin support at A and a roller support at C (see figure). Members AB and BC each have length L and flexural rigidity EI. A horizontal load P acts at joint B. Determine the horizontal deflection δ at joint B.

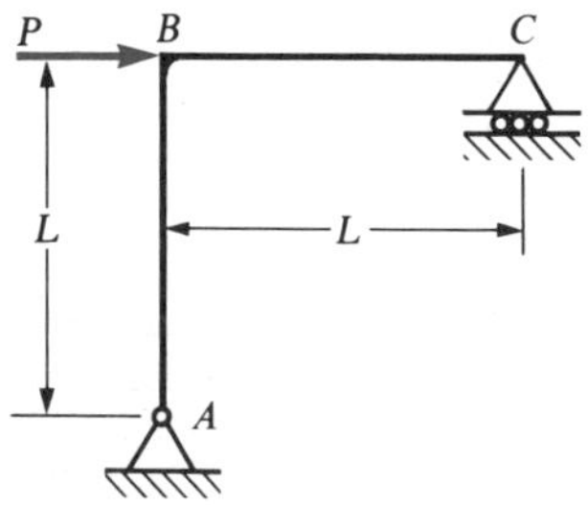

Prob. 10.3-23

10.3-24 A rectangular frame $ABCD$ has a roller support at A and a pinned support at D (see figure). The flexural rigidity of the vertical members is EI_1 and of the horizontal member is EI_2. The load on the frame consists of a force P acting at the midpoint of member BC. Determine the horizontal deflection δ_h and the angle of rotation θ at point A.

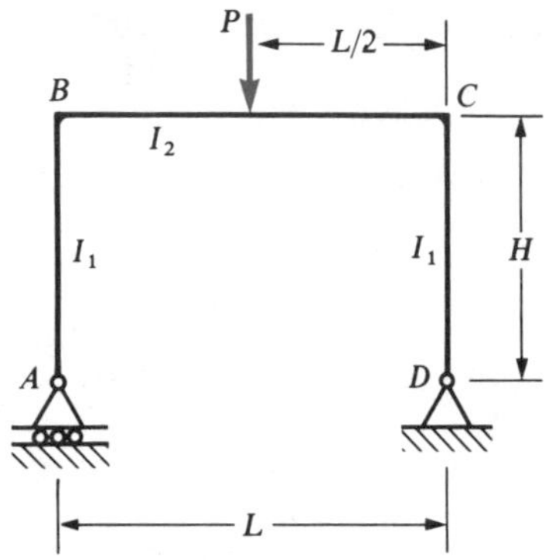

Prob. 10.3-24

10.3-25 The frame ABC shown in the figure is fixed at support A and free at end C. Members AB and BC, which are perpendicular to each other, have length L and flexural rigidity EI. A horizontal load P acts at point C. Determine the horizontal and vertical deflections δ_h and δ_v at C.

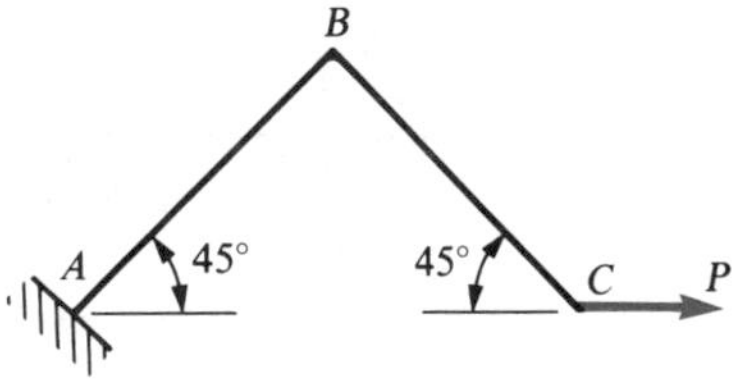

Prob. 10.3-25

10.3-26 A rectangular frame $ABCD$ is fixed at support A and free at end D, as shown in the figure. All three members

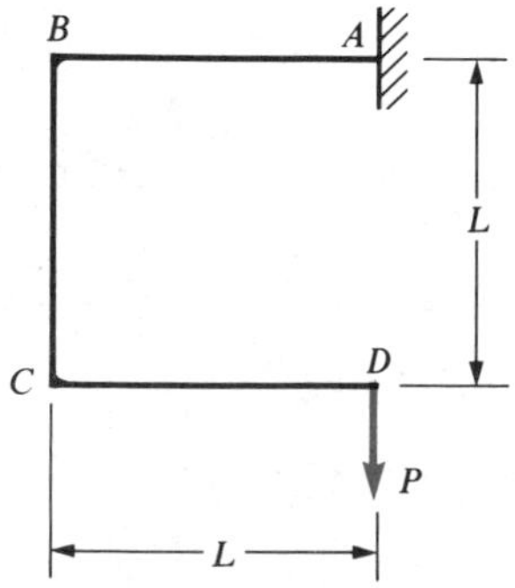

Prob. 10.3-26

have length L and flexural rigidity EI. A vertical load P acts at D. Determine the horizontal deflection δ_h, the vertical deflection δ_v, and the angle of rotation θ at the free end.

10.3-27 The Z-shaped frame $ABCD$ shown in the figure is fixed at support D and free at end A. The flexural rigidity EI is the same for all members. Determine the vertical deflection δ_v and the angle of rotation θ at point A due to the vertical load P.

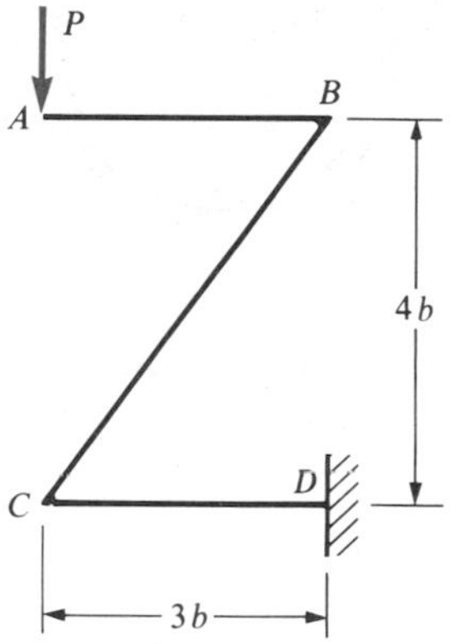

Prob. 10.3-27

10.3-28 The frame ABC shown in the figure is loaded by forces P at points A and C. Members AB and BC are identical and have length L, flexural rigidity EI, and axial rigidity EA. Determine the increase Δ in the distance between points A and C (due to the forces P), considering the effects of both flexural and axial deformations. Check the result for the special cases of $\beta = 0$ and $\beta = 90°$.

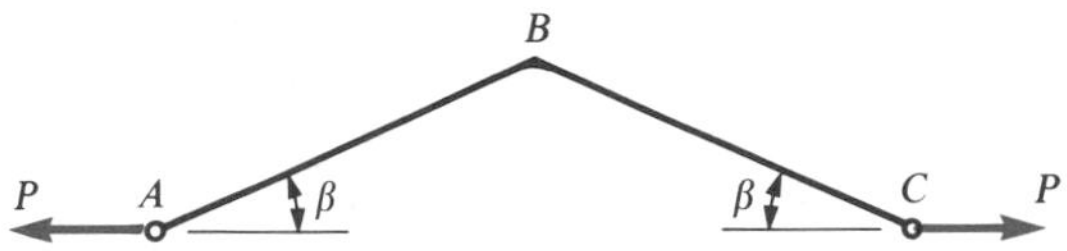

Prob. 10.3-28

10.3-29 A frame ABC is fixed at A and free at C (see figure). The frame undergoes a temperature change such that the temperature on the left-hand and top surfaces becomes T_1 and on the right-hand and bottom surfaces becomes T_2. Each member has the same cross-sectional shape (h = height of cross section) and is made of the same material (α = coefficient of thermal expansion). Determine the horizontal deflection δ_h, the vertical deflection δ_v, and the angle of rotation θ at point C.

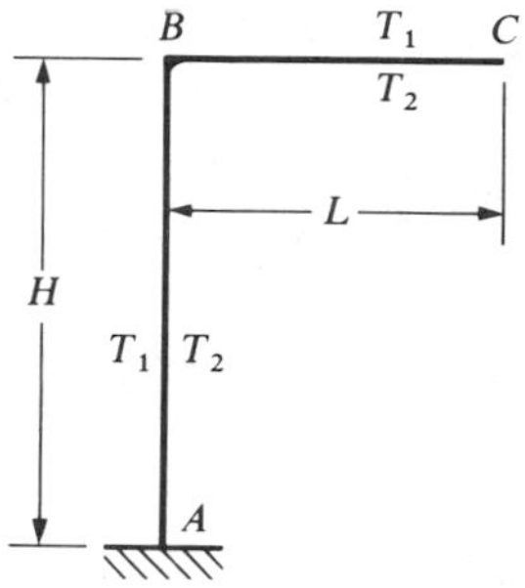

Prob. 10.3-29

10.3-30 A slender curved bar AB has a centerline in the form of a quarter circle of radius R, as shown in the figure. The bar is fixed at support A and free at B. A horizontal load P acts at the free end. Determine the horizontal deflection δ_h, the vertical deflection δ_v, and the angle of rotation θ at the free end.

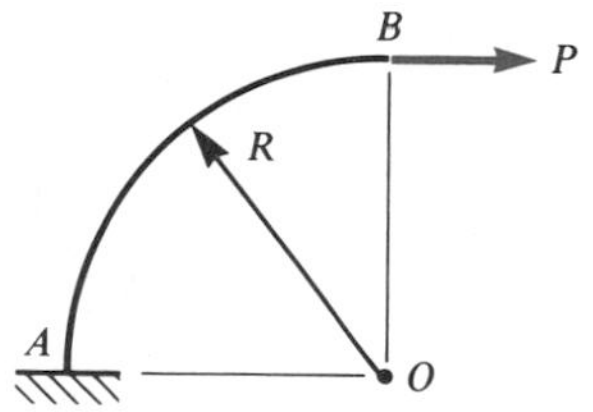

Prob. 10.3-30

10.3-31 A slender curved bar AB of semicircular shape is fixed at support A and free at B (see figure). A vertical load P acts at the free end. Determine the horizontal deflection δ_h, the vertical deflection δ_v, and the angle of rotation θ at the free end.

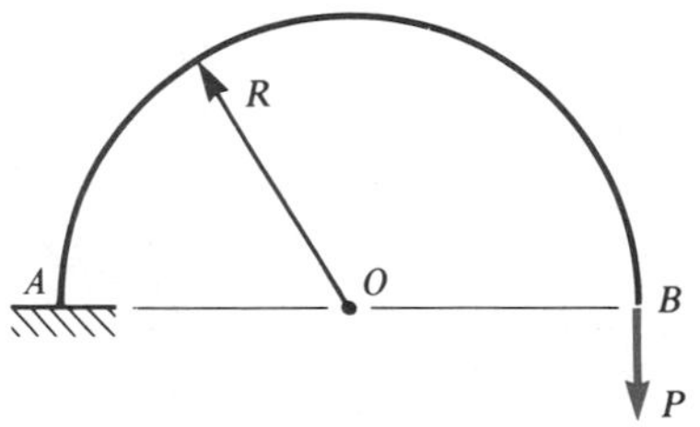

Prob. 10.3-31

10.3-32 A slender curved bar ACB of semicircular shape is pin-supported at A and roller-supported at B (see figure on p. 700). A vertical load P acts at the midpoint C. Determine the vertical deflection δ_c at C and the horizontal deflection δ_b at B.

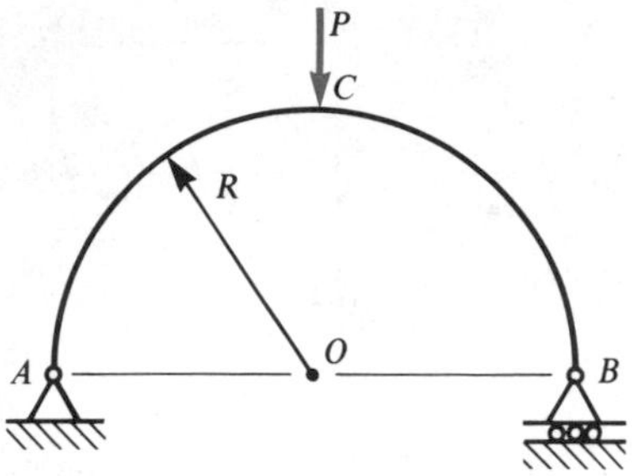

Prob. 10.3-32

10.3-33 A slender frame $ABCD$ consists of a semicircular part BC of radius R and two straight parts AB and CD of length L (see figure). All parts have flexural rigidity EI. Determine the increase Δ in the distance between points A and D due to the loads P.

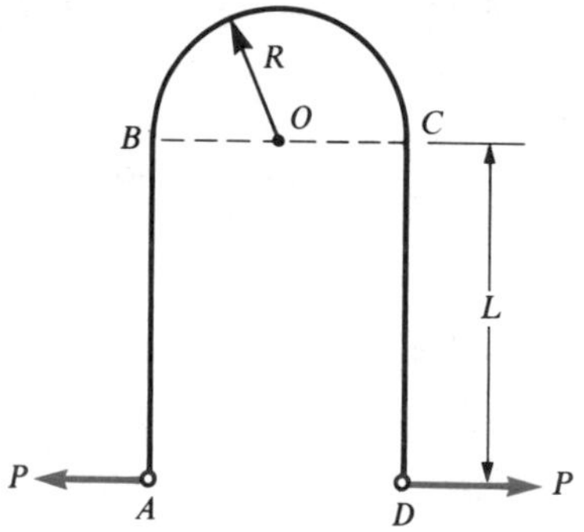

Prob. 10.3-33

10.3-34 A slender bar ABC consists of a straight part AB and a quarter-circle part BC (see figure). Both parts have flexural rigidity EI. A vertical load P acts at C. Determine the horizontal deflection δ_h, the vertical deflection δ_v, and the angle of rotation θ at point C.

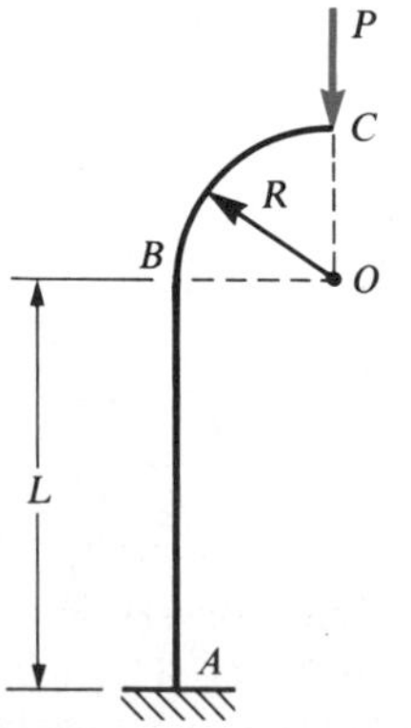

Prob. 10.3-34

10.3-35 A slender circular ring of average radius R and flexural rigidity EI is cut through at a certain point and spread open by placing a small block in the gap (see figure). Assuming that the width of the block is e, find the maximum bending moment M_{max} in the ring.

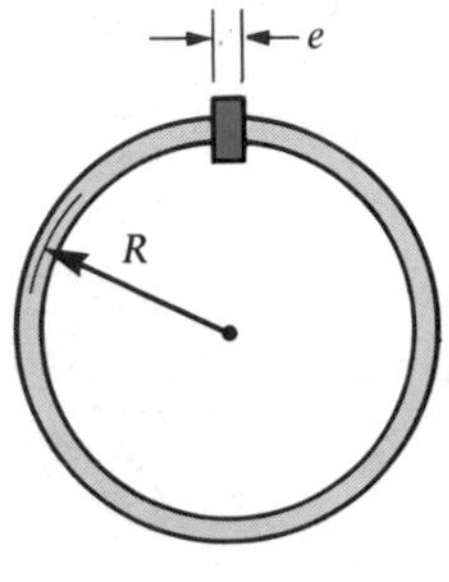

Prob. 10.3-35

10.3-36 A slender curved bar AB lying in a horizontal plane with its centerline forming a quarter circle of radius R is subjected to a vertical load P at the free end B (see figure). The bar has flexural rigidity EI and torsional rigidity GI_p. Determine the vertical deflection δ_v and the angle of twist ϕ at end B.

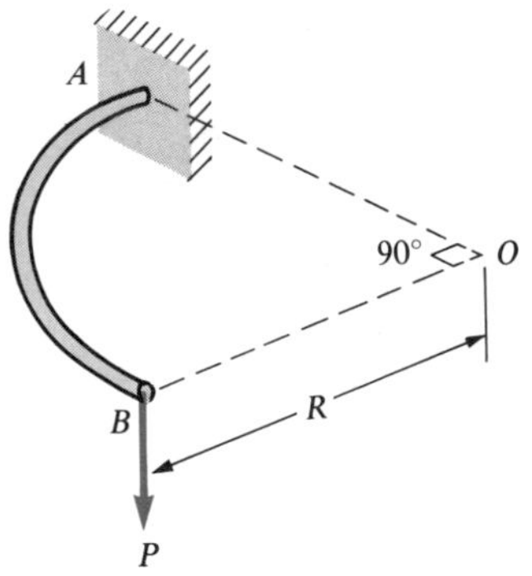

Prob. 10.3-36

10.3-37 A horizontal bracket $ABCD$ is fixed at support A and free at end D (see figure). The bracket is constructed from a circular pipe of constant cross section (EI = flexural rigidity; GI_p = torsional rigidity). Let L denote the lengths of members AB and CD, and let b denote the length of member BC. A vertical load P acts at the free end D. (a) Determine the vertical deflection δ_v and the angle of twist ϕ at D. (b) Calculate numerical values for δ_v and ϕ for an aluminum pipe with the following data: $P = 200$ lb, $L = 4$ ft, $b = 3$ ft, $I = 8.0$ in.^{4}, $E = 10 \times 10^6$ psi, and $G = 3.8 \times 10^6$ psi.

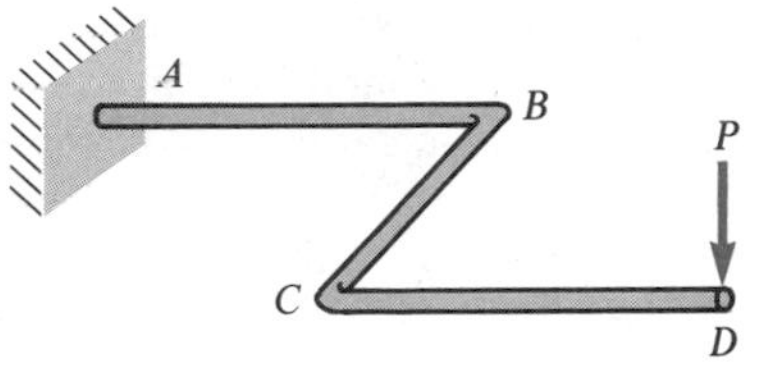

Prob. 10.3-37

*10.3-38 A square frame $ABCD$ has a cut in the middle of side AD (see figure). Each side of the square has length L, and each member has flexural rigidity EI and torsional rigidity GI_p. Equal and opposite forces P, with their lines of action perpendicular to the plane of the frame, are applied at the cut ends of the frame. Determine the gap Δ between the cut ends (in the direction perpendicular to the plane of the frame).

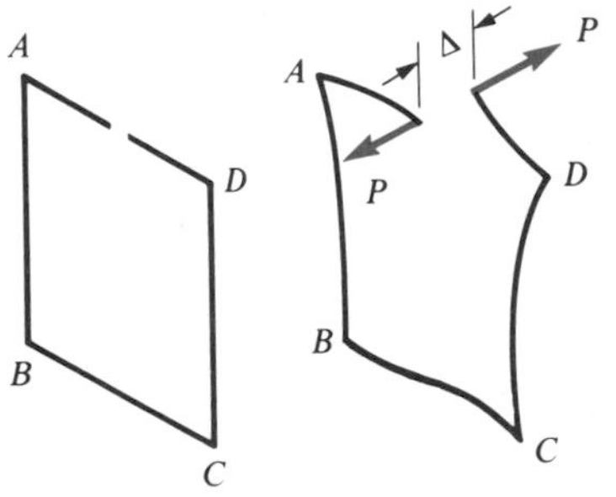

Prob. 10.3-38

*10.3-39 The "staircase" frame shown in the figure has a fixed support at the upper end and is free at the lower end. Every member of the frame has length L and flexural rigidity EI. A vertical load P acts at the free end. Determine the vertical deflection δ at the free end, assuming that there are n "steps."

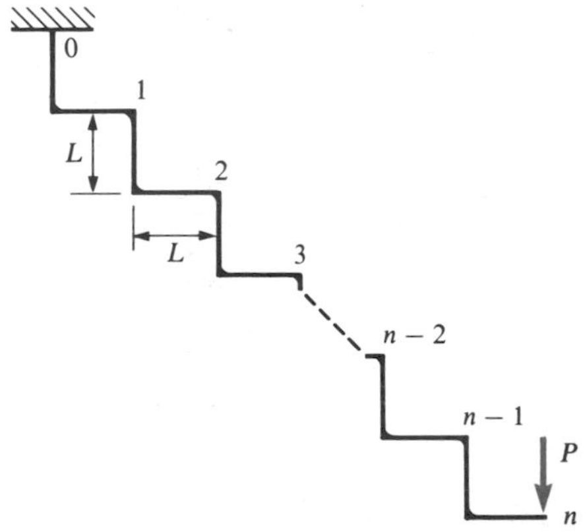

Prob. 10.3-39

10.4-1 Derive the reciprocal-displacement theorem $\delta_{ab} = \delta_{ba}$ (see Eq. 10-12) by applying the unit-load method (Eq. 10-6) to the simple beams shown in Figs. 10-11a and b. (Note that points A and B may be selected anywhere along the beam.)

10.4-2 A simple beam AB with an overhang BD is subjected to two loading conditions (see figure). The first load is a vertical force P acting at C and the second is a vertical force P acting at D. Show that $\delta_{dc} = \delta_{cd}$ by evaluating both deflections by the unit-load method.

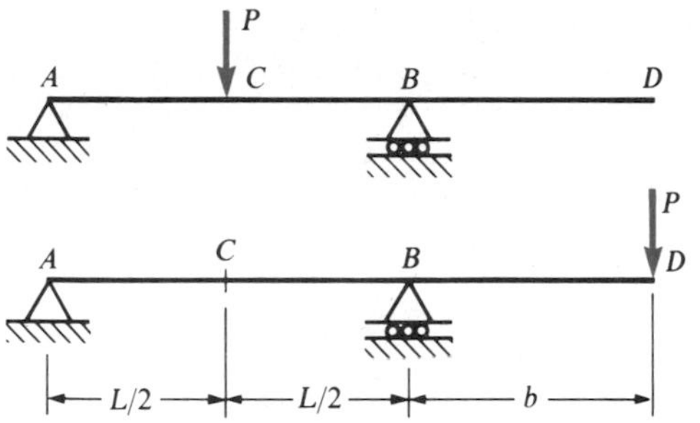

Prob. 10.4-2

10.4-3 A frame ABC of constant cross section is subjected to two loading conditions (see figure). The first load is a vertical force P acting at C and the second is a couple M acting at B. Show that $M\theta_{bc} = P\delta_{cb}$ (see Eq. 10-13), where θ_{bc} is the angle of rotation at B due to the load P and δ_{cb} is the deflection at C due to the couple M. (Use the unit-load method for evaluating the displacements.)

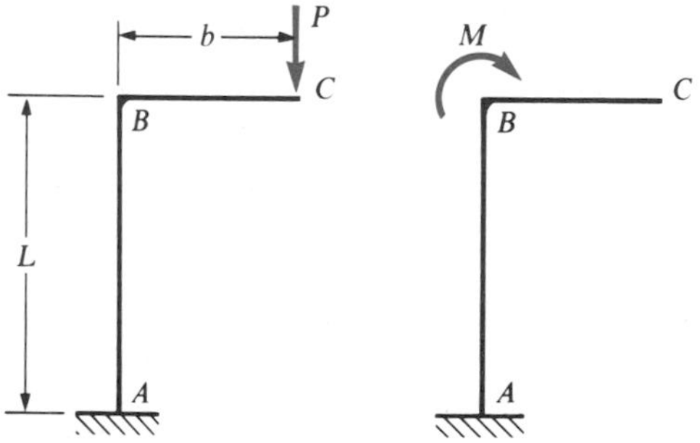

Prob. 10.4-3

10.4-4 A simple beam AB is subjected to two loading conditions; the first is a couple M acting at point C and the second is a couple M acting at D (see figure on p. 702). Show that $\theta_{dc} = \theta_{cd}$ by evaluating both angles of rotation by the unit-load method.

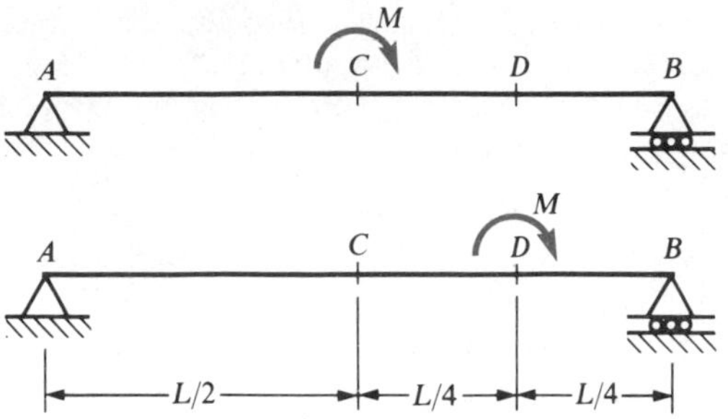

Prob. 10.4-4

10.4-5 A statically indeterminate truss consisting of 12 bars made of the same material and having the same cross section is shown in the figure. A vertical force P acts at joint A. By using the reciprocal-displacement theorem, determine the horizontal displacement δ_h of joint A.

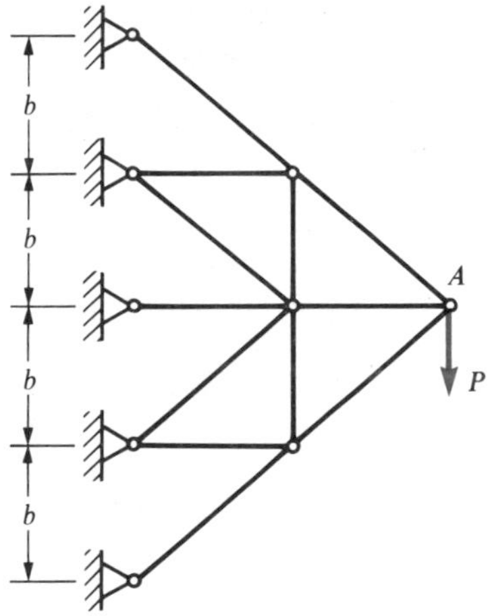

Prob. 10.4-5

10.4-6 A statically indeterminate frame $ABCDE$ is subjected to a horizontal load P (see figure). Members AB and DE are identical. By using the reciprocal-displacement theorem, determine the vertical deflection δ_c at point C. (Consider only the effects of flexural deformations.)

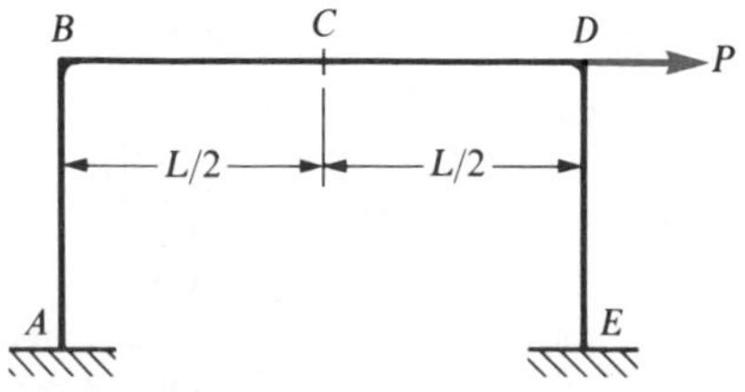

Prob. 10.4-6

10.4-7 A two-bar truss ABC (see figure) is subjected to two loading conditions, a horizontal force P acting at joint B and a vertical force P acting at joint B. Each member has axial rigidity EA. Show that $\delta_{12} = \delta_{21}$, where δ_{12} is the horizontal displacement due to the vertical force and δ_{21} is the vertical displacement due to the horizontal force. (Use the unit-load method for evaluating the displacements.)

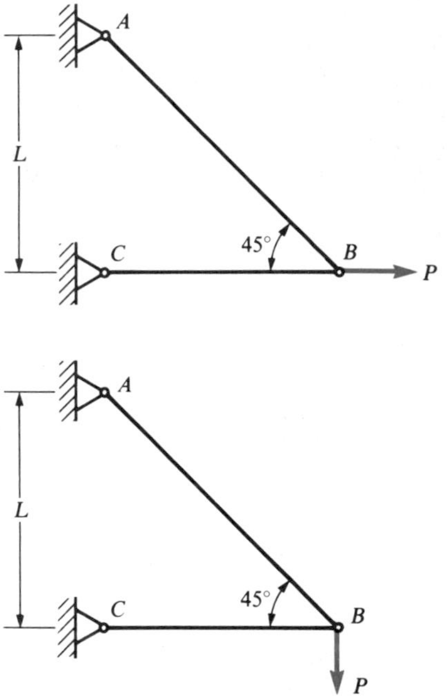

Prob. 10.4-7

10.4-8 A cantilever beam of length L, fixed at the left-hand end and free at the other end, is subjected to two states of loading (see figure). In the first state of loading, a force P_1 acts downward at the midpoint of the beam and a clockwise couple M_1 acts at the free end. In the second state, a

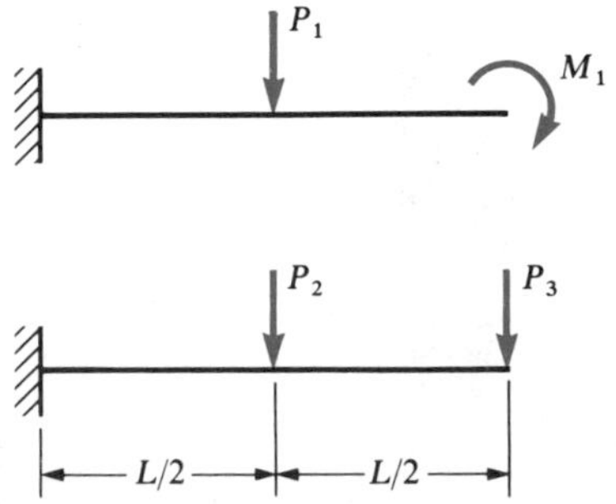

Prob. 10.4-8

force P_2 acts downward at the midpoint and a force P_3 acts downward at the free end. Verify the reciprocal-work theorem for these two loading cases by evaluating the displacements in both cases and substituting into Eq. (10-15).

10.4-9 A simple beam of length L is subjected to two states of loading (see figure). In the first state of loading, a force P_1 acts downward at distance $L/3$ from the left-hand support. In the second state, two equal counterclockwise couples M_0 act on the beam, one at the left-hand end and the other at the midpoint. Verify the reciprocal-work theorem for these two loading cases by evaluating the displacements in both cases and substituting into Eq. (10-15).

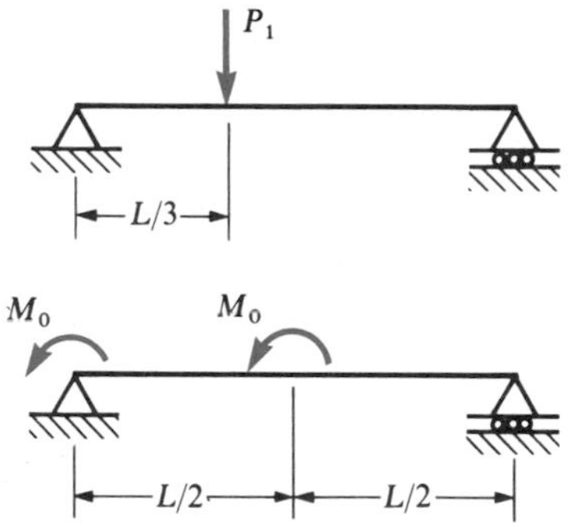

Prob. 10.4-9

***10.4-10** A simple beam of length L is subjected to two states of loading (see figure). The first state of loading consists of a uniform load of intensity q acting over the entire span of the beam. The second state consists of a concentrated load P acting at the midpoint of the beam. Verify the reciprocal-work theorem for these two loading cases by evaluating the displacements in both cases and substituting into Eq. (10-15). (Hint: In the case of a uniform load, the corresponding displacement is the area of the deflection diagram.)

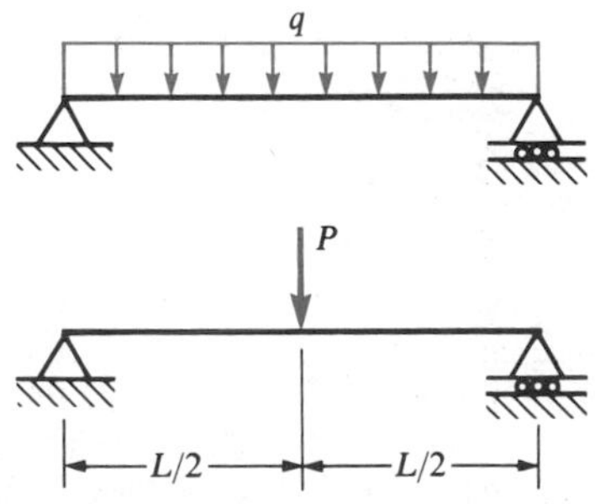

Prob. 10.4-10

The problems for Section 10.6 are to be solved using strain-energy methods.

10.6-1 The symmetric truss shown in the figure consists of three bars meeting at joint D. All bars have the same axial rigidity EA, and the middle bar has length L. The only load on the truss is the vertical force P acting at joint D. (a) Express the strain energy U of the structure as a function of the vertical displacement δ at D. (b) Using Castigliano's first theorem, determine the displacement δ. (c) Determine the axial forces N_{ad}, N_{bd}, and N_{cd} in the bars of the truss.

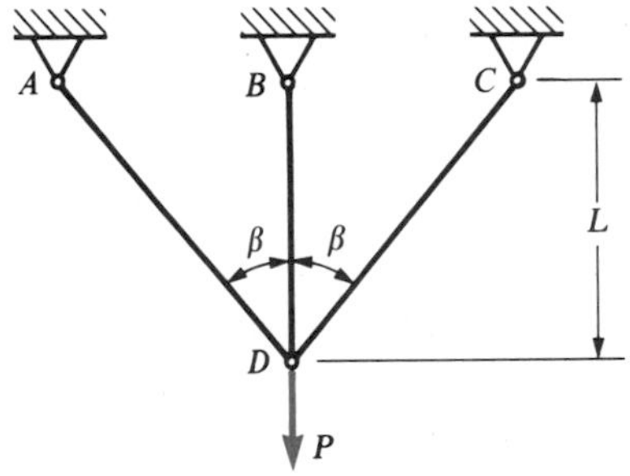

Prob. 10.6-1

10.6-2 A simple truss ABC is shown in the figure. Assume that both bars have the same axial rigidity EA and that the length of member AB is L. Denote the horizontal displacement of joint B by D_1 (positive to the right) and denote the vertical displacement by D_2 (positive downward). (a) Express the strain energy U of the structure as a function of the displacements D_1 and D_2. (b) Determine D_1 and D_2 by using Castigliano's first theorem.

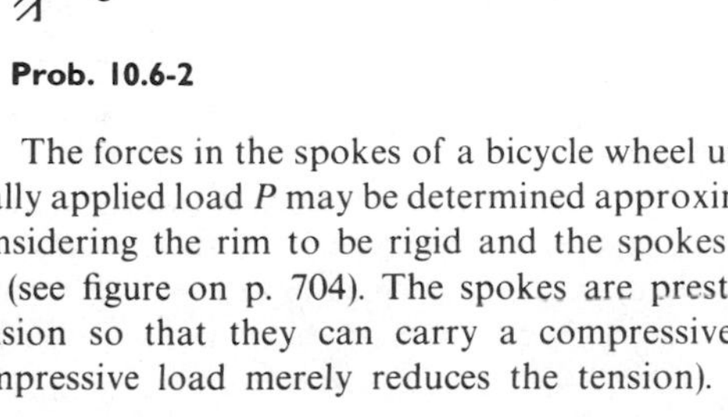

Prob. 10.6-2

10.6-3 The forces in the spokes of a bicycle wheel under a statically applied load P may be determined approximately by considering the rim to be rigid and the spokes to be radial (see figure on p. 704). The spokes are prestressed in tension so that they can carry a compressive load (a compressive load merely reduces the tension). Using

Castigliano's first theorem, determine the downward displacement δ of the hub due to the load P if there are 32 equally spaced spokes having cross-sectional area A, length L, and modulus of elasticity E.

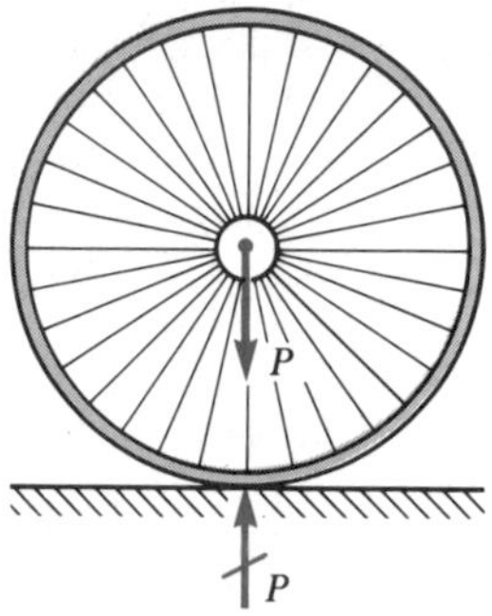

Prob. 10.6-3

10.6-4 A truss consisting of two identical members AB and BC supports a vertical load P (see figure). Each bar has cross-sectional area A and length L. The stress-strain relationship for the material is $\sigma^n = B\epsilon$, where n and B are constants. (a) Express the strain energy U of the truss as a function of the vertical displacement δ at joint B. (b) Using Castigliano's first theorem, determine the displacement δ.

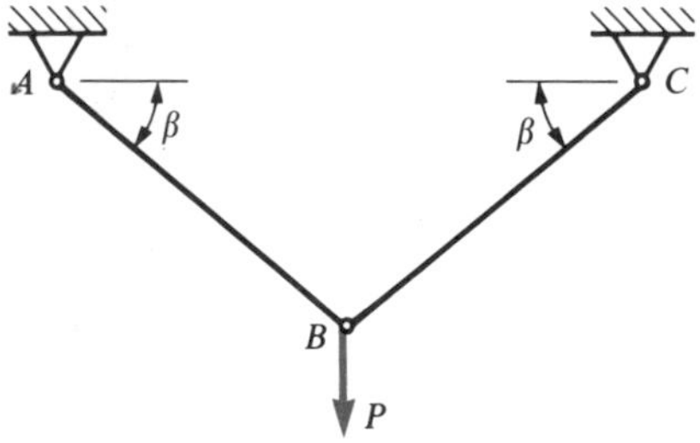

Probs. 10.6-4 and 10.7-1

The problems for Section 10.7 are to be solved using complementary-energy methods.

10.7-1 A truss consisting of two identical members AB and BC supports a vertical load P (see figure). Each bar has cross-sectional area A and length L. The stress-strain relationship for the material is $\sigma^n = B\epsilon$, where n and B are constants. (a) Determine the vertical displacement δ of joint B by evaluating the complementary energy of the structure and using the Crotti-Engesser theorem. (b) Determine the displacement δ by the unit-load method.

10.7-2 The truss ABC shown in Fig. 10-20 supports a vertical load P at joint B. Each bar has cross-sectional area A. The stress-strain relationship for the material is $\sigma = B\sqrt{\epsilon}$, where B is a constant. (a) Determine the horizontal displacement δ_h of joint B by evaluating the complementary energy of the structure and using the Crotti-Engesser theorem. (b) Determine the displacement δ_h by the unit-load method.

10.7-3 A vertical prismatic bar of length L and weight γ per unit volume hangs from one end. The stress-strain relationship for the material is $\sigma^n = B\epsilon$, where n and B are constants. (a) Determine the elongation δ of the bar by evaluating the complementary energy and using the Crotti-Engesser theorem. (b) Determine the elongation δ by the unit-load method.

The problems for Section 10.8 are to be solved using Castigliano's second theorem.

10.8-1 Obtain the deflection δ_c at the free end of the overhanging beam ABC shown in the figure. The beam has constant flexural rigidity EI and supports a concentrated load P at end C.

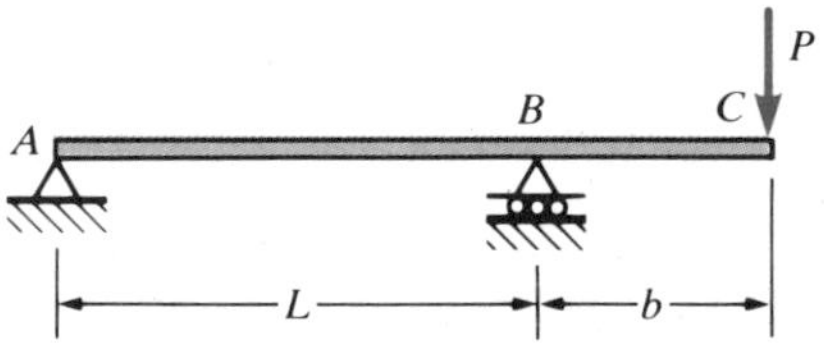

Prob. 10.8-1

10.8-2 The simple beam AB shown in the figure has constant flexural rigidity EI and is subjected to a force P acting at distance $L/4$ from one end. Determine the deflection δ_c at the point where the load is applied.

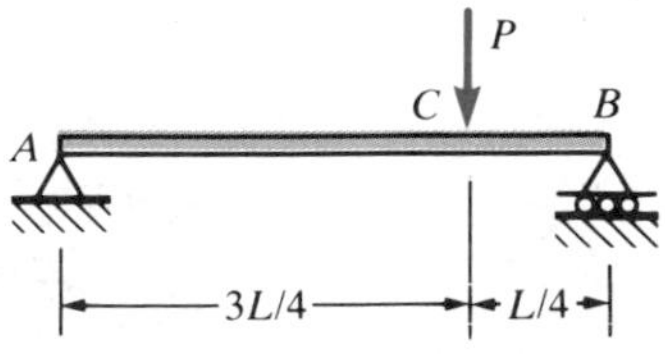

Prob. 10.8-2

10.8-3 The cantilever bracket ABC (see figure) is loaded by a horizontal force P. The arms AB and BC of the bracket have lengths L and b, respectively, and constant flexural rigidity EI. Obtain the formula for the horizontal deflection δ_c at end C.

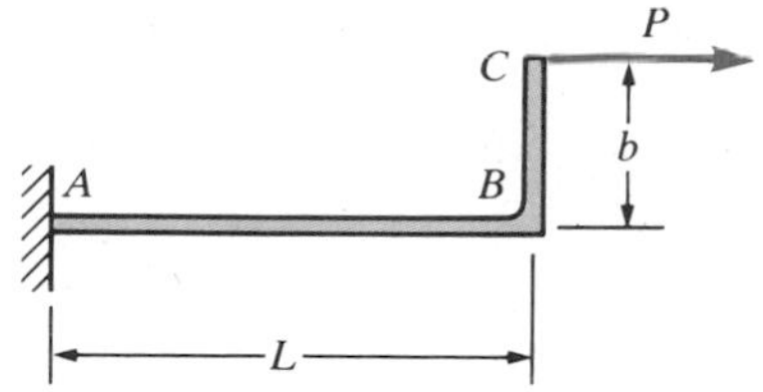

Prob. 10.8-3

10.8-4 The truss ABC shown in the figure supports a vertical load P at joint B. Both bars have axial rigidity EA. Obtain the vertical displacement δ_b at joint B.

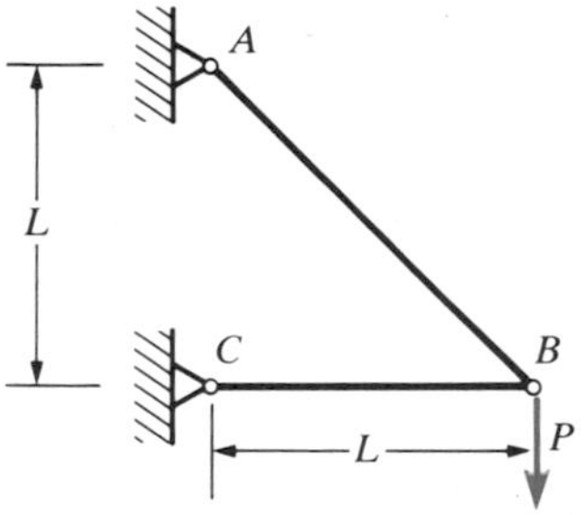

Prob. 10.8-4

10.8-5 A horizontal load P acts on the two-bar truss shown in the figure. Both bars have axial rigidity EA. Obtain the horizontal displacement δ_b of joint B.

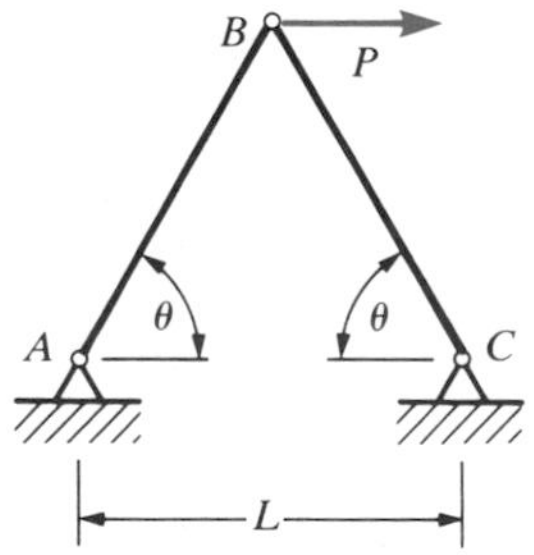

Prob. 10.8-5

10.8-6 A thin semicircular bar AB of radius R and constant flexural rigidity EI is supported and loaded as shown in the figure. Determine the horizontal deflection δ_b of point B.

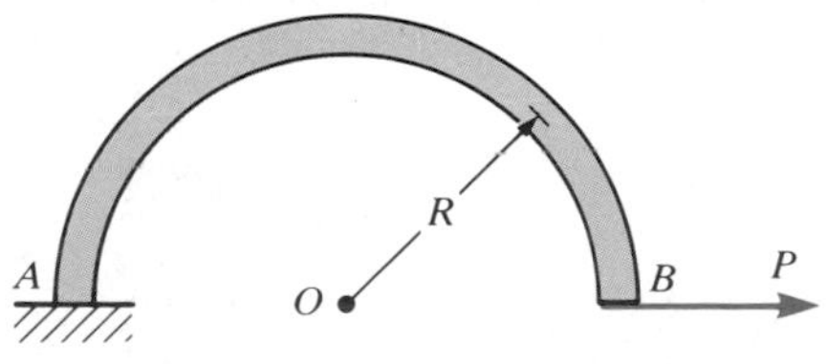

Prob. 10.8-6

10.8-7 The cantilever beam ABC shown in the figure has constant flexural rigidity EI and carries two concentrated loads, P_1 and P_2. Determine the deflection δ_b at point B. (Hint: Before integrating, apply Castigliano's theorem by differentiating under the integral sign.)

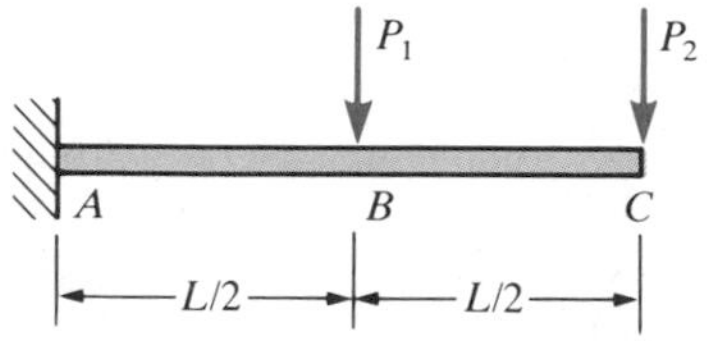

Prob. 10.8-7

10.8-8 A simple beam AB supports a uniform load of intensity q over the left-hand half of the span and a couple M_0 at end B (see figure). The flexural rigidity of the beam is EI. Determine the angle of rotation θ_b at end B. (Hint: Before integrating, apply Castigliano's theorem by differentiating under the integral sign.)

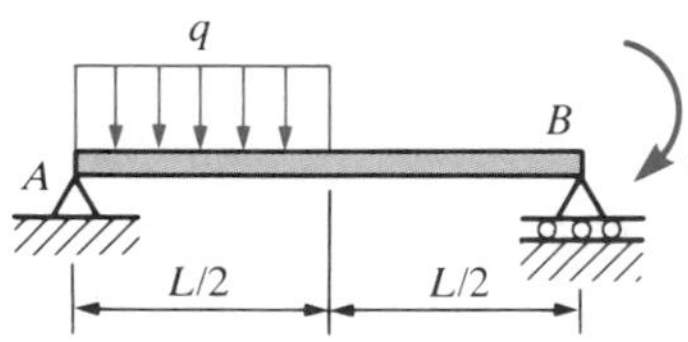

Prob. 10.8-8

10.8-9 A simple beam AB of constant flexural rigidity EI supports a uniform load of intensity q (see figure). Obtain the formula for the maximum deflection δ_{max} of the beam. (Hints: Add a concentrated load at the midpoint of the beam. Also, before integrating, apply Castigliano's theorem by differentiating under the integral sign.)

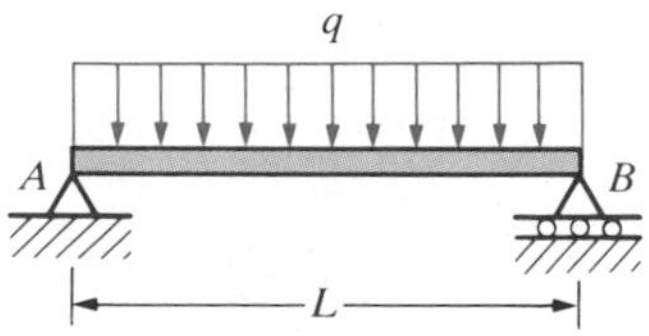

Prob. 10.8-9

10.8-10 A cantilever beam AB of constant flexural rigidity EI supports a uniform load of intensity q acting over one-half of the length (see figure on p. 706). Determine the angle of rotation θ_a at the free end A. (Hints: Add a load in the form of a couple M_0 acting at end A. Also, before integrating, apply Castigliano's theorem by differentiating under the integral sign.)

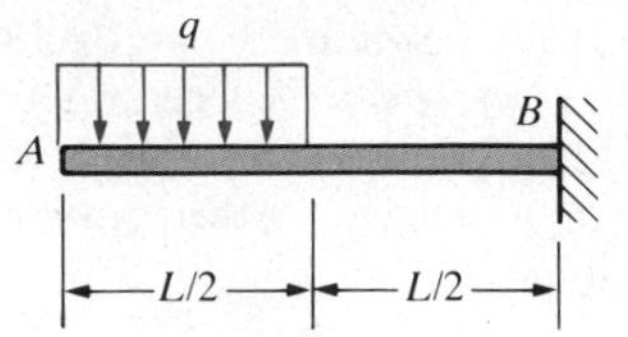

Prob. 10.8-10

10.8-11 The truss $ABCD$ shown in the figure is loaded by a vertical force P at joint B and by a horizontal force P at joint C. The axial rigidity of every member is EA. Determine the vertical displacement δ_c of joint C. (Hint: Add a vertical load Q at joint C.)

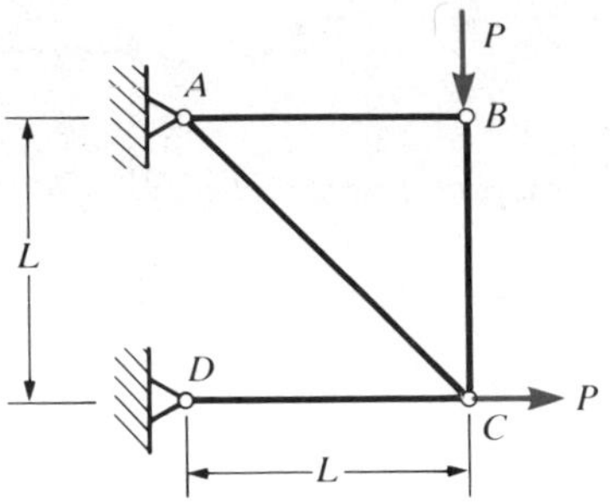

Prob. 10.8-11

References and Historical Notes

1-1 Timoshenko, S. P., *History of Strength of Materials*, Dover Publications, Inc., New York, 1983, 452 pages (originally published by McGraw-Hill Book Co., Inc., New York, 1953). (Note: Stephen P. Timoshenko, 1878–1972, was a famous scientist, engineer, and teacher. Born in Russia, he came to the United States in 1922. He was a researcher with the Westinghouse Research Laboratory, a professor at the University of Michigan, and later a professor at Stanford University, where he retired in 1944. Timoshenko made many original contributions, both theoretical and experimental, to the field of applied mechanics, and he wrote twelve pioneering textbooks that revolutionized the teaching of mechanics in the United States. These books, which were published in as many as 5 editions and translated into as many as 35 languages, covered the subjects of statics, dynamics, mechanics of materials, vibrations, structural theory, stability, elasticity, plates, and shells.)

1-2 Todhunter, I., and Pearson, K., *A History of the Theory of Elasticity and of the Strength of Materials*, Vols. I and II, Dover Publications, Inc., New York, 1960 (originally published by the Cambridge University Press in 1886 and 1893). (Note: Isaac Todhunter, 1820–1884, and Karl Pearson, 1857–1936, were English mathematicians and educators. Pearson was especially famous for his many original contributions to statistics.)

1-3 Love, A. E. H., *A Treatise on the Mathematical Theory of Elasticity*, 4th Ed., Dover Publications, Inc., New York, 1944, 643 pages (originally published by the Cambridge University Press in 1927); see "Historical Introduction," pp. 1–31. (Note: Augustus Edward Hough Love, 1863–1940, was an outstanding English elastician who taught at Oxford University. He analyzed seismic surface waves, which are now called *Love waves* by geophysicists.)

1-4 See Ref. 1-1, p. 88 and Ref. 1-2, Vol. I, pp. 10, 533, and 873. (Note: Jacob Bernoulli, 1654–1705, is also known by the names James, Jacques, and Jakob; Ref. 1-2 uses James. He was a member of the famous family of mathematicians and scientists of Basel, Switzerland, and he did important work in connection with elastic curves of beams. He also developed polar coordinates and became famous for his work in theory of probability, analytic geometry, and other fields. Jean Victor Poncelet, 1788–1867, was a Frenchman who fought in Napoleon's campaign against Russia and was given up for dead on the battlefield. He survived, was taken prisoner, and later returned to France to continue his work in mathematics. His major contributions to mathematics are in geometry, and in mechanics he is best known for his work on properties of materials and dynamics.)

1-5 See Ref. 1-1, pp. 17–20 and Ref. 1-2, Vol. I, p. 5. (Note: Robert Hooke, 1635–1703, was an English scientist who performed many experiments with elastic bodies and developed methods for improvement in timepieces; he also formulated the laws of gravitation independently of Newton, with whom he was a contemporary. Upon the founding of the Royal Society of London in 1662, he was appointed its first curator.)

1-6 Hooke, Robert, *De Potentiâ Restitutiva*, John Martyn, London, 1678.

1-7 See Ref. 1-1, pp. 90–98 and Ref. 1-2, Vol. I, pp. 80–86. (Note: Thomas Young, 1773–1829, was an outstanding English scientist who did pioneering work in optics, sound, impact, and other subjects.)

1-8 Young, Thomas, *A Course of Lectures on Natural Philosophy and the Mechanical Arts*, Vols. 1 and 2, London, 1807.

1-9 See Ref. 1-1, pp. 111–114; Ref. 1-2, Vol. I, pp. 208–318; and Ref. 1-3, p. 13. (Note: Siméon Denis Poisson, 1781–1840, was a great French mathematician. He made many contributions to both mathematics and mechanics, and his name is preserved in numerous ways besides Poisson's ratio. For instance, we have Poisson's equation in partial differential equations and the Poisson distribution in theory of probability. Based upon his own theory for the behavior of materials, he calculated the lateral strain of a bar in tension and found it to be one-fourth of the longitudinal strain.)

2-1 Timoshenko, S. P., and Goodier, J. N., *Theory of Elasticity*, 3rd Ed., McGraw-Hill Book Co., Inc., New York, 1970, p. 110. (Note: James Norman Goodier, 1905–1969, was well known for his research contributions to theory of elasticity, stability, wave propagation in solids, and other branches of applied mechanics. Born in England, he studied at Cambridge University and the University of Michigan. He was a professor at Cornell University and later at Stanford University, where he headed the program in applied mechanics.)

2-2 Weaver, W., Jr., and Gere, J. M., *Matrix Analysis of Framed Structures*, 3rd Ed., Van Nostrand Reinhold Co., New York, 1990. (Note: The flexibility and stiffness methods are explained in this book.)

2-3 See Ref. 1-1, p. 36 and Ref. 2-4, p. 650. (Note: Leonhard Euler, 1707–1783, was a famous Swiss mathematician, perhaps the greatest mathematician of all time. See the note to Ref. 9-3 for information about his life.)

2-4 Oravas, G. A., and McLean, L., "Historical development of energetical principles in elastomechanics," *Applied Mechanics Reviews*, Part I, vol. 19, no. 8, August 1966, pp. 647–658 and Part II, vol. 19, no. 11, November 1966, pp. 919–933.

2-5 See Ref. 1-1, p. 75; Ref. 1-2, Vol. I, p. 146; and Ref. 2-4, p. 652. (Note: Louis Marie Henri Navier, 1785–1836, a famous French mathematician and engineer, was one of the founders of the mathematical theory of elasticity. He contributed to beam, plate, and shell theory, to theory of vibrations, and to the theory of viscous fluids.)

2-6 Piobert, G., Morin, A.-J., and Didion, I., "Commission des Principes du Tir," *Mémorial de l'Artillerie*, vol. 5, 1842, pp. 501–552. (Note: This paper describes experiments made by firing artillery projectiles against iron plating. On page 505 appears the description of the markings that are the slip bands. The description is quite brief, and there is no indication that the authors attributed the markings to inherent material characteristics. Guillaume Piobert, 1793–1871, was a French general and mathematician who made many studies of ballistics; when this paper was written, he was a captain in the artillery.)

2-7 Lüders, W., "Ueber die Äusserung der elasticität an stahlartigen Eisenstäben und Stahlstäben, und über eine beim Biegen solcher Stäbe beobachtete Molecularbewegung," *Dingler's Polytechnisches Journal*, vol. 155, 1860, pp. 18–22. (Note: This paper clearly describes and illustrates the bands that appear on the surface of a polished steel specimen during yielding. Of course, these bands are only the surface manifestation of three-dimensional zones of deformation; hence the zones should probably be characterized as "wedges" rather than bands. Nevertheless, the name *Lüders' bands* is usually given to the markings, although sometimes they are called *Piobert's bands*. Further discussions of the bands, with photos and bibliographies, can be found in Refs. 2-8 and 2-9. We are still seeking information about Lüders himself.)

2-8 Fell, E. W., "The Piobert effect in iron and soft steel," *The Journal of the Iron and Steel Institute*, vol. 132, no. 2, 1935, pp. 75–91.

2-9 Turner, T. H., and Jevons, J. D., "The detection of strain in mild steels," *The Journal of the Iron and Steel Institute*, vol. 111, no. 1, 1925, pp. 169–189.

2-10 See Ref. 1-1, pp. 118 and 288; Ref. 1-2, Vol. I, p. 578; and Ref. 1-2, Vol. II, p. 418. (Note: Benoit Paul Emile Clapeyron, 1799–1864, was a famous French structural engineer and bridge designer; he taught engineering at the École des Ponts et Chaussées in Paris. It appears that Clapeyron's theorem, which states that the work of the external loads acting on a linearly elastic body is equal to the strain energy, was first published in 1833.)

2-11 See Ref. 1-1, p. 88. (Note: Poncelet investigated longitudinal vibrations of a bar due to impact loads. See Ref. 1-4 for additional information about his life and works.)

2-12 Ramberg, W. A., and Osgood, W. R., "Description of stress-strain curves by three parameters," *National Advisory Committee for Aeronautics*, Technical Note No. 902, July 1943.

2-13 See Ref. 1-1, pp. 229–242; Ref. 1-2, Vol. I, pp. 833–872, Vol. II, Part I, pp. 1–286, Vol. II, Part II, pp. 1–51; and Ref. 2-1, pp. 39–40. (Note: Barré de Saint-Venant,

1797–1886, is generally recognized as the most outstanding elastician of all time. Born near Paris, he studied briefly at the École Polytechnique and later graduated from the École des Ponts et Chaussées. His later professional career suffered greatly from his refusal, as a matter of conscience and politics, to join his schoolmates in preparing for the defense of Paris in March 1814, just prior to Napoleon's abdication. As a consequence, his achievements received greater recognition in other countries than they did in France. Some of his most famous contributions are the formulation of the fundamental equations of elasticity and the development of the exact theories of bending and torsion. He also developed theories for plastic deformations and vibrations. His full name was Adéhmar Jean Claude Barré, Count de Saint-Venant.)

2-14 Roark, R. J., and Young, W. C., *Formulas for Stress and Strain*, 6th Ed., McGraw-Hill Book Co., Inc., 1989.

3-1 See Ref. 1-1, pp. 51–53, 82, and 92, and Ref. 1-2, Vol. I, p. 69. (Note: The relationship between torque and angle of twist in the case of a circular bar was correctly established in 1784 by Charles Augustin de Coulomb, 1736–1806, a famous French scientist. Coulomb made contributions in electricity and magnetism, viscosity of fluids, friction, beam bending, retaining walls and arches, torsion and torsional vibrations, and other subjects; see Ref. 1-1, pp. 47–54. Thomas Young, in his book of 1807, Ref. 1-8, observed that the applied torque is balanced by the shear stresses on the cross section and that the shear stresses are proportional to the distance from the axis. The French engineer Alphonse J. C. B. Duleau, 1789–1832, performed tests on bars in torsion and also developed a theory for circular bars; see Ref. 1-1, p. 82.)

3-2 See Ref. 1-1, pp. 229–237 and Ref. 1-2, Vol. II, Part II, pp. 1–51. (Note: Saint-Venant's conclusive memoir on torsion, published in 1855, is described in these references. See Ref. 2-13 for additional information about his life and works.)

3-3 See Ref. 1-1, p. 216 and note to Ref. 1-9.

3-4 Bredt, R., "Kritische Bemerkungen zur Drehungselastizität," *Zeitschrift des Vereines Deutscher Ingenieure*, vol. 40, 1896, pp. 785–790 and 813–817. (Note: Rudolph Bredt, 1842–1900, was a German engineer. He studied in Karlsruhe and Zürich and then worked for a while in Crewe, England, at a train factory, where he learned about the design and construction of cranes. This experience formed the basis for his later work as a crane manufacturer in Germany. His theory of torsion was developed in connection with the design of box-girder cranes.)

3-5 Timoshenko, S. P., and Gere, J. M., *Theory of Elastic Stability*, 2nd Ed., McGraw-Hill Book Co., Inc., 1961, pp. 500–509.

5-1 See Ref. 2-1, pp. 42 and 48.

5-2 Galilei, Galileo, *Dialogues Concerning Two New Sciences*, translated from the Italian and Latin into English by Henry Crew and Alfonso De Salvio, The Macmillan Company, New York, 1933 (translation first published in 1914.) (Note: *Two New Sciences* was published in 1638 by Louis Elzevir in Leida, now Leiden, Netherlands. This book represented the culmination of Galileo's work on dynamics and mechanics of materials. It can truly be said that these two subjects, as we know them today, began with Galileo and the publication of this book.

Galileo Galilei was born in Pisa in 1564. He made many famous experiments and discoveries, including those on falling bodies and pendulums that initiated the science of dynamics. Galileo was an eloquent lecturer and attracted students from many countries. He pioneered in astronomy and developed a telescope with which he made many astronomical discoveries, including the mountainous character of the moon, Jupiter's satellites, the phases of Venus, and sunspots. Because his scientific views of the solar system were contrary to theology, he was condemned and spent the last years of his life in seclusion in Florence; during this period he wrote *Two New Sciences*. Galileo died in 1642 and was buried in Florence.)

5-3 See Ref. 1-1, pp. 11–47 and 135–141, and Ref. 1-2. (Note: The history of beam theory is described in these references. Edme Mariotte, 1620–1684, was a French physicist who made developments in dynamics, hydrostatics, optics, and mechanics. He made tests on beams and developed a theory for calculating load-carrying capacity; his theory was an improvement on Galileo's work, but still not correct. Jacob Bernoulli, 1654–1705, who is described in the note to Ref. 1-4, first determined that the curvature is proportional to the bending moment. However, his constant of proportionality was incorrect. Leonhard Euler, 1707–1783, obtained the differential equation of the deflection curve of a beam and used it to solve many problems of both large and small deflections. His life and work are described in the note to Ref. 9-3. The first person to obtain the distribution of stresses in a beam and correctly relate the stresses to the bending moment probably was Antoine Parent, 1666–1716, a French physicist and mathematician. Later, a rigorous investigation of strains and stresses in beams was made by Saint-Venant, 1797–1886; see note to Ref. 2-13).

5-4 *Manual of Steel Construction*, latest edition, American Institute of Steel Construction, Inc., 400 North Michigan Avenue, Chicago, Illinois 60611.

5-5 *Aluminum Construction Manual*, latest edition, Section 1, "Specifications for Aluminum Structures," The Aluminum Association, Inc., 900 19th Street, N.W., Washington, D.C. 20006.

5-6 *Aluminum Construction Manual*, latest edition, Section 3, "Engineering Data for Aluminum Structures," *ibid.*

5-7 *National Design Specification for Wood Construction*, latest edition, National Forest Products Association, 1619 Massachusetts Avenue, N.W., Washington, D.C., 20036.

5-8 *Wood Structural Design Data*, latest edition, *ibid.*

5-9 See Ref. 3-5, pp. 251–277, for a discussion of lateral buckling of beams.

5-10 See Ref. 1-1, pp. 141–144 and Ref. 1-2, Vol. II, Part I, pp. 641–642. (Note: D. J. Jourawski, 1821–1891, was a Russian bridge and railway engineer who developed the now widely used approximate theory for shear stresses in beams. In 1844, only two years after he graduated from the Institute of Engineers of Ways of Communication in St. Petersburg, he was assigned the task of designing and constructing a major bridge on the first railway line from Moscow to St. Petersburg. He noticed that some of the large timber beams split longitudinally in the centers of the cross sections, where he knew the bending stresses were zero. Jourawski drew free-body diagrams and quickly discovered the existence of horizontal shear stresses in the beams. He derived the shear formula and applied his theory to various shapes of beams. While the exact theory for shear stresses in beams was given by Saint-Venant, it is useful in only a very few practical cases, so it appears that the critical remarks about Jourawski's theory made by Todhunter and Pearson on p. 642 of Ref. 1-2 are unjustified. Jourawski's paper on shear in beams is cited in Ref. 5-11. His name is also transliterated as Dimitrii Ivanovich Zhuravskii.)

5-11 Jourawski, D. J., "Sur la résistance d'un corps prismatique . . . ," *Annales des Ponts et Chaussées*, Mémoires et Documents, 3rd Series, vol. 12, Part 2, 1856, pp. 328–351.

5-12 Zaslavsky, A., "On the limitations of the shearing stress formula," *International Journal of Mechanical Engineering Education*, vol. 8, no. 1, 1980, pp. 13–19.

5-13 See Ref. 2-1, pp. 358–359.

5-14 Maki, A. C., and Kuenzi, E. W., "Deflection and stresses of tapered wood beams," Research Paper FPL 34, U. S. Forest Service, Forest Products Laboratory, Madison, Wisconsin, September 1965, 54 pages.

5-15 See Ref. 1-1, p. 147. (Note: The French engineer Jacques Antoine Charles Bresse, 1822–1883, is best known for his work with curved bars and arches.)

5-16 Timoshenko, S. P., "Use of stress functions to study flexure and torsion of prismatic bars," (in Russian), St. Petersburg, 1913 (reprinted in Vol. 82 of the *Memoirs of the Institute of Ways of Communication*, pp. 1–21). (Note: In this paper, the point in the cross section of a beam through which a concentrated force should act in order to eliminate rotation was found. Thus, it appears that this work contains the first determination of a shear center. The particular beam under investigation had a solid semicircular cross section; see Ref. 5-17. In 1909, C. Bach tested channel beams and observed the twisting that occurs when a load parallel to the plane of the web is applied to the beam (Refs. 5-18 and 5-19). He also observed that the rotation changed as the load was translated laterally, but apparently he did not locate the shear center. In 1917, A. A. Griffith and G. I. Taylor used the soap-film method to investigate bending; they located the shear center, which they called the *flexural center*, for several shapes of structural sections (Ref. 5-20). The general approximate solution for the shear center of a thin-walled bar of open cross section was developed by R. Maillart, who explained the practical significance of the shear center for structural shapes; see Ref. 5-21. It also appears that Maillart introduced the name *shear center*. Further contributions to the development of the shear-center concept are described in Refs. 5-22 through 5-26. A comprehensive discussion of the shear center, as well as a general presentation of the problem of bending and torsion of beams, is given in Ref. 5-27, and some historical comments concerning the shear center can be found in Refs. 5-28 and 5-29.)

5-17 See Ref. 2-1, pp. 371–373.

5-18 Bach, C., "Versuche über die tatsächliche Widerstandsfähigkeit von Balkan mit [-förmigem Querschnitt," *Zeitschrift des Vereines Deutscher Ingenieure*, vol. 53, no. 44, 1909, pp. 1790–1795.

5-19 Bach, C., "Versuche über die tatsächliche Widerstandsfähigkeit von Trägern mit [-förmigem Querschnitt," *Zeitschrift des Vereines Deutscher Ingenieure*, vol. 54, no. 10, 1910, pp. 382–387.

5-20 Griffith, A. A., and Taylor, G. I., "The problem of flexure and its solution by the soap-film method," *Technical Report of the Advisory Committee for Aeronautics*, 1917–1918, vol. 3, published in 1921 by His Majesty's Stationery Office, London (Reports and Memoranda, No. 399, November 1917), pp. 950–969.

5-21 Maillart, R., "Zur Frage der Biegung," *Schweizerische Bauzeitung*, vol. 77, 1921, pp. 195–197; "Bemerkungen zur Frage der Biegung," *ibid.*, vol. 78, 1921, pp.18–19; "Ueber Drehung und Biegung," *ibid.*, vol. 79, 1922, pp. 254–

257; "Der Schubmittelpunkt," *ibid.*, vol. 83, 1924, pp. 109–111; "Zur Frage des Schubmittelpunktes," *ibid.*, vol. 83, 1924, pp. 176–177.

5-22 Osgood, W. R., "The center of shear again," *Journal of Applied Mechanics*, vol. 10, no. 2, June 1943, pp. A-62 to A-64 (*Transactions of the American Society of Mechanical Engineers*, vol. 65, 1943).

5-23 Mandel, J., "Détermination du centre de torsion à l'aide du théorème de réciprocité," *Annales des Ponts et Chaussées*, vol. 118, 1948, pp. 271–290.

5-24 Jacobs, J. A., "The centre of shear of aerofoil sections," *Journal of the Royal Aeronautical Society*, vol. 57, 1953, pp. 235–237.

5-25 Duncan, W. J., "The flexural centre or centre of shear," *Journal of the Royal Aeronautical Society*, vol. 57, 1953, pp. 594–597.

5-26 Koiter, W. T., "The flexural centre or centre of shear," *Journal of the Royal Aeronautical Society*, vol. 58, 1954, pp. 64–65.

5-27 Timoshenko, S. P., "Theory of bending, torsion, and buckling of thin-walled members of open cross section," *Journal of the Franklin Institute*, vol. 239, no. 3, March 1945, pp. 201–219; no. 4, April 1945, pp. 249–268; and no. 5, May 1945, pp. 343–361.

5-28 See Ref. 1-1, p. 401.

5-29 Nowinski, J. L., "Theory of thin-walled bars," *Applied Mechanics Surveys*, edited by Abramson, H. N., et al., Spartan Books, Washington, D.C., 1966, pp. 325–338.

5-30 Horne, M. R., *Plastic Theory of Structures*, 2nd Ed., Pergamon Press, Oxford, 1979, 179 pages.

5-31 Massonnet, C. E., and Save, M. A., *Plastic Analysis and Design*, Vol. I: Beams and Frames, Blaisdell Publishing Co., New York, 1965, 379 pages.

5-32 Neal, B. G., *The Plastic Methods of Structural Analysis*, 3rd Ed., Chapman and Hall, London, 1977, 205 pages.

6-1 Gere, J. M., and Weaver, W., *Matrix Algebra for Engineers*, 2nd Ed., PWS-KENT Publishing Company, Boston, Massachusetts, 1983, 232 pages.

6-2 See Ref. 1-1, pp. 107–111. (Note: Augustin Louis Cauchy, 1789–1857, was one of the greatest mathematicians. Born in Paris, he entered the École Polytechnique at the age of 16, where he studied under Lagrange, Laplace, Fourier, and Poisson. He was quickly recognized for his mathematical prowess, and at age 27 he became a professor at the École and a member of the Academy of Sciences. His major works in pure mathematics were in group theory, number theory, series, integration, differential equations, and analytical functions. In applied mathematics, he introduced the concept of stress as we know it today, developed the equations of theory of elasticity, and introduced the notion of principal stresses and principal strains. An entire chapter is devoted to his work on theory of elasticity in Ref. 1-2; see Vol. I, pp. 319–376.)

6-3 See Ref. 1-1, pp. 229–242. (Note: Saint-Venant was a pioneer in many aspects of theory of elasticity, and Todhunter and Pearson dedicate their book, *A History of the Theory of Elasticity*, Ref. 1-2, to him. For further information about Saint-Venant, see Ref. 2-13.)

6-4 See Ref. 1-1, pp. 197–202, and Ref. 1-2, Vol. II, Part I, pp. 86 and 287–322. (Note: William John Macquorn Rankine, 1820–1872, was born in Edinburgh, Scotland, and taught engineering at Glasgow University. He derived the stress transformation equations in 1852 and made many other contributions to theory of elasticity and applied mechanics. His engineering subjects included arches, retaining walls, and structural theory. He also achieved scientific fame for his work with fluids, light, sound, and behavior of crystals.)

6-5 See Ref. 1-1, pp. 283–288, Ref. 6-6, and Ref. 6-7. (Note: The famous German civil engineer Otto Christian Mohr, 1835–1918, was both a theoretician and a practical designer. He was a professor at the Stuttgart Polytechnikum and later at the Dresden Polytechnikum. He developed the circle of stress in 1882, Ref. 6-6; it is also described in his book, Ref. 6-7, pp. 187–219. Mohr made numerous contributions to the theory of structures, including the Williot-Mohr diagram for truss displacements, the moment-area method for beam deflections, and the Maxwell-Mohr method for analyzing statically indeterminate structures.)

6-6 Mohr, O., "Über die Darstellung des Spannungszustandes und des Deformationszustandes eines Körperelementes," *Zivilingenieur*, 1882, p. 113.

6-7 Mohr, O., *Abhandlungen aus dem Gebiete der technischen Mechanik*, Wilhelm Ernst and Sohn, Berlin, 1906, 459 pages.

6-8 See Ref. 1-1, pp. 190–197. (Note: Karl Culmann, 1821–1881, was a famous German bridge and railway engineer. In 1849–1850 he spent two years traveling in England and the United States to study bridges, which he later wrote about in Germany. He designed numerous bridge structures in Europe, and in 1855 he became professor of structures at the newly organized Zürich Polytechnicum. Culmann made many developments in graphical methods and wrote the first book on graphic statics, published in Zürich in 1866. Stress trajectories are one of the original topics presented in this book. For additional information about Culmann and the history of bridges, see Ref. 6-9.)

6-9 Kuzmanovic, B. O., "History of the theory of bridge structures," *Proceedings of the American Society of Civil Engineers, Journal of the Structural Division*, vol. 103, no. ST5, May 1977, pp. 1095–1111.

6-10 Hetényi, M., Editor, *Handbook of Experimental Stress Analysis*, John Wiley and Sons, Inc., New York, 1960.

6-11 Dally, J. W., and Riley, W. F., *Experimental Stress Analysis*, 2nd Ed., McGraw-Hill Book Co., Inc., New York, 1978, 571 pages.

7-1 See Ref. 1-1, pp. 27 and 30–36. (Note: The work of Jacob Bernoulli, Euler, and many others with respect to elastic curves is also described in Ref. 1-2. Another member of the Bernoulli family, Daniel Bernoulli, 1700–1782, proposed to Euler that he obtain the differential equation of the deflection curve by minimizing the strain energy, which Euler did. Daniel Bernoulli, a nephew of Jacob Bernoulli, is renowned for his work in hydrodynamics, kinetic theory of gases, beam vibrations, and other subjects. His father, John Bernoulli, 1667–1748, a younger brother of Jacob, was an equally famous mathematician and scientist who first formulated the principle of virtual displacements, solved the problem of the brachystochrone, and established the rule for obtaining the limiting value of a fraction when both the numerator and denominator tend to zero. He communicated this last rule to G. F. A. de l'Hôpital, 1661–1704, who wrote the first book on calculus and included this theorem, which today we know as *L'Hôpital's rule*; see Ref. 7-2. Daniel's nephew, Jacob Bernoulli, 1759–1789, who is also known as James or Jacques, was a pioneer in the theory of plate bending and plate vibrations. Much interesting information about the many prominent members of the Bernoulli family, as well as other pioneers in mechanics and mathematics, can be found in books on the history of mathematics; for instance, see Refs. 7-3, 7-4, and 7-5.)

7-2 Struik, D. J., "The origin of L'Hôpital's rule," *The Mathematics Teacher*, vol. 56, no. 4, April 1963, pp. 257–260.

7-3 Newman, J. R., *The World of Mathematics*, Vols. 1-4, Simon and Schuster, New York, 1956, 2469 pages.

7-4 Struik, D. J., *A Concise History of Mathematics*, 4th Ed., Dover Publications, Inc., New York, 1987.

7-5 Cajori, F., *A History of Mathematics*, 4th Ed., Chelsea Publishing Co., New York, 1985.

7-6 See Ref. 1-1, p. 137.

7-7 Saint-Venant, Barré de, notes and appendices to the third edition of the book by Navier, *Résumé des Leçons données à l'école des ponts et chaussées sur l'application de la mécanique à l'établissement des constructions et des machines*, 1st Part, "De la Résistance des corps solides," Paris, 1864, p. 72.

7-8 See Ref. 1-1, p. 284.

7-9 Mohr, O., "Beitrag zur Theorie der Holz- und Eisen-Constructionen," *Zeitschrift des Architekten- und Ingenieur-Vereins zu Hannover*, vol. 14, 1868, pp. 19–51.

7-10 Greene, C. E., *Graphical Method for the Analysis of Bridge Trusses*, D. Van Nostrand Co, Inc., New York, 1875. (Note: The moment-area method was discovered independently by Greene, a professor at the University of Michigan, in 1873. He began teaching it to students in that year. The method is described under the name *area-moment method* on pp. 35–40 of this book.)

7-11 Macaulay, W. H., "Note on the deflection of beams," *The Messenger of Mathematics*, vol. XLVIII, May 1918–April 1919, Cambridge, 1919, pp. 129–130. (Note: William Herrick Macaulay, 1853–1936, was an English mathematician and Fellow of King's College, Cambridge. In this paper he defined "by $\{f(x)\}_a$ a function of x which is zero when x is less than a and equal to $f(x)$ when x is equal to or greater than a." Then he showed how to use this function when finding beam deflections. Unfortunately, he did not give any references to the earlier work of Föppl and Clebsch; see Refs. 7-12 through 7-15.)

7-12 Clebsch, A., *Theorie der Elasticität fester Körper*, B. G. Teubner, Leipzig, 1862, 424 pages. (Translated into French and annotated by Saint-Venant, *Théorie de l'Élasticité des Corps Solides*, Paris, 1883. Saint-Venant's notes increased Clebsch's book threefold in size.) (Note: The method of finding beam deflections by integrating across points of discontinuity was presented first in this book; see Ref. 1-1, pp. 258–259 and Ref. 7-15. Rudolf Friedrich Alfred Clebsch, 1833–1872, was a German mathematician and scientist. He was a professor of engineering at the Karlsruhe Polytechnicum and later a professor of mathematics at Göttingen University.)

7-13 Föppl, A., *Vorlesungen über technische Mechanik*, Vol. III: Festigkeitslehre, B. G. Teubner, Leipzig, 1897. (Note: In this book, Föppl extended Clebsch's method for finding beam deflections. August Föppl, 1854–1924, was a German mathematician and engineer. He was a professor at the University of Leipzig and later at the Polytechnic Institute of Munich. For a fascinating account of his life, see Ref. 7-14.)

7-14 Oravas, G. A., Introduction to *Drang und Zwang*, by A. Föppl and L. Föppl, Vol. 1, 3rd Ed., Johnson Reprint Corporation, New York, 1969 (a reprint of the 3rd Ed., 1941, with a new biographical introduction by Oravas; 1st Ed. published by R. Oldenbourg Verlag, Munich, in 1920). (Note: Carl Ludwig Föppl, 1887–1976, was the younger son of August Föppl. His older brother, Otto Föppl, born 1885, also coauthored a book with his father. Both sons were well known for their work in applied mechanics.)

7-15 Pilkey, W. D., "Clebsch's method for beam deflections," *Journal of Engineering Education*, vol. 54, no. 5, January 1964, pp. 170–174. (Note: This paper describes Clebsch's method and gives a very complete historical account, with many references. For extensions of the method, see Refs. 7-16 and 7-17.)

7-16 Weissenburger, J. T., "Integration of discontinuous expressions arising in beam theory," *American Institute of Aeronautics and Astronautics Journal*, vol. 2, no. 1, January 1964, pp. 106–108.

7-17 Wittrick, W. H., "A generalization of Macaulay's method with applications in structural mechanics," *American Institute of Aeronautics and Astronautics Journal*, vol. 3, no. 2, February 1965, pp. 326–330.

7-18 See Ref. 1-1, pp. 25, 30–36, 39–40. (Note: The basic equation relating curvature and bending moment originated with Jacob Bernoulli, 1654–1705, although he did not evaluate correctly the constant of proportionality. Nevertheless, his work must be considered as the first contribution to solving problems of large deflections of beams. Later, acting upon a suggestion by Daniel Bernoulli, Euler rederived the differential equation of the deflection curve and solved various problems of the elastica; see Ref. 1-1, p. 27; Ref. 1-2, Vol. I, pp. 30 and 34; and Ref. 1-3, p. 3. Euler's famous paper on elastic curves is cited in Ref. 7-19. Joseph Louis Lagrange, 1736–1813, was a famous French mathematician; he was born in Turin of French and Italian ancestry. Lagrange, who first formulated the principle of virtual work and who made major contributions to dynamics, apparently was the next person to investigate the elastica. He considered a cantilever beam with a load at the end; see Ref. 1-1, pp. 39–40 and Ref. 1-2, Vol. I, pp. 58–61. Lagrange's paper is cited in Ref. 7-20 and brief biographies of him can be found in Ref. 7-4, p. 132 and Ref. 7-5, p. 250. Another early investigator in elasticity theory was Giovanni Antonio Amaedo Plana, 1781–1864, a nephew of Lagrange, who corrected a mistake in Lagrange's investigation of the elastica; see Ref. 1-2, Vol. I, pp. 89–90. Plana's paper is cited in Ref. 7-21, and biographical information is given in Ref. 7-5. An investigation of the elastica by variational methods was made by Max Born in his dissertation; see Ref. 2-4, pp. 927–928 and 932, Ref. 7-22, and Ref. 7-23, p. xxviii. Max Born, 1882–1970, was the great physicist who founded modern wave mechanics and did important research in quantum mechanics and relativity. In his dissertation he established new variational principles in elasticity.)

7-19 Euler, L., "Methodus inveniendi lineas curvas maximi minimive proprietate gaudentes . . . ," Appendix I, "De curvis elasticis," Bousquet, Lausanne and Geneva, 1744. (English translation: Oldfather, W. A., Ellis, C. A., and Brown, D. M., *Isis*, vol. 20, 1933, pp. 72–160. Also, republished in *Leonhardi Euleri Opera Omnia*, series 1, vol. 24, 1952.)

7-20 Lagrange, J. L., "Sur la force des ressorts pliés," *Mémoires de l'Académie Royale des Sciences et Belles-Lettres de Berlin*, vol. 25, 1771. (Reprinted in "Oeuvres de Lagrange," Gauthier-Villars, Paris, vol. 3, 1869, pp. 77–110.)

7-21 Plana, G. A. A., "Équation de la courbe formée par une lame élastique," *Memoirs of the Royal Society of Turin*, vol. 18, 1809–1810, pp. 123–155.

7-22 Born, M., "Untersuchungen über der Stabilität der elastichen Linie in Ebene und Raum unter verschiedenen Grenzbedingungen," Dissertation, Göttingen, 1906.

7-23 Oravas, G. A., "Historical Review of Extremum Principles in Elastomechanics," an introductory section (pp. xx–xlvi) of the book, *The Theory of Equilibrium of Elastic Systems and Its Applications*, by C. A. P. Castigliano, translated by E. S. Andrews, Dover Publications, Inc., New York, 1966. (Note: Castigliano's book was originally published in Turin in 1879; subsequently, it was translated into German in 1886 and into English in 1919. The English translation by E. S. Andrews was published under the title *Elastic Stresses in Structures* by Scott, Greenwood, and Son, London. The Dover edition is a republication of the English translation but with the addition of historical material by Oravas; see Refs. 10-16 through 10-20.)

7-24 See Ref. 1-3, pp. 401–412.

7-25 See Ref. 3-5, pp. 76–82.

7-26 Southwell, R. V., *An Introduction to the Theory of Elasticity*, 2nd Ed., Oxford University Press, London, 1941, pp. 429–436. (Note: Richard Vynne Southwell, 1888–1970, was an English mathematician, engineer, and educator. He developed relaxation methods for solving problems in various branches of engineering, including applied mechanics.)

7-27 Frisch-Fay, R., *Flexible Bars*, Butterworth and Co., Ltd., 1962, 220 pages.

7-28 Eisley, J. G., "Nonlinear deformation of elastic beams, rings, and strings," *Applied Mechanics Reviews*, vol. 16, no. 9, September 1963, pp. 677–680.

7-29 Rojahn, C., "Large deflections of elastic beams," thesis for the Degree of Engineer, Stanford University, June 1968.

7-30 Bisshopp, K. E., and Drucker, D. C., "Large deflection of cantilever beams," *Quarterly of Applied Mathematics*, vol. 3, 1945, pp. 272–275.

7-31 Barten, H. J., "On the deflection of a cantilever beam," *Quarterly of Applied Mathematics*, vol. 2, 1944, pp. 168–171, and *ibid.*, vol. 3, 1945, pp. 275–276. (Note: The second article corrects a mistake in the first one.)

7-32 Rohde, F. V., "Large deflections of a cantilever beam with uniformly distributed load," *Quarterly of Applied Mathematics*, vol. 11, 1953, pp. 337–338.

8-1 Navier, L. M. H., *Résumé des leçons données à l'école des ponts et chaussées sur l'application de la mécanique à l'établissement des constructions et des machines*, 1st Ed., 1826; 2nd Ed., 1833; 3rd Ed. (with notes and appendices by Saint-Venant), Paris, 1864. (Note: Navier prepared the first two editions; the third edition was prepared by Saint-Venant after Navier's death. Saint-Venant's additions are ten times the volume of the original book. For discussions of this famous book, see Ref. 1-1, pp. 73–77 and 232–233, and Ref. 1-2, Vol. I, pp. 144–146, Vol. II, Part I, pp. 105–135.

8-2 See Ref. 1-1, pp. 144–146. (Note: B. P. E. Clapeyron, 1799–1864, was a French engineer who developed the three-moment equation in connection with his work on bridges. He is also known for his theorems on work and strain energy; see Ref. 2-10. Clapeyron's paper giving the three-moment equation is cited in Ref. 8-3. Bertot, another French engineer, was the first to publish the three-moment equation in its present-day form; see his paper cited in Ref. 8-4. While Bertot's paper was published prior to Clapeyron's, it seems that Clapeyron used the method several years earlier and is the true originator of the equation, as pointed out in Ref. 1-1.)

8-3 Clapeyron, B. P. E., "Calcul d'une poutre élastique reposant librement sur des appuis inégalement espacés," *Comptes Rendus*, vol. 45, 1857, pp. 1076–1080.

8-4 Bertot, H., *Mémoires et compte-rendu des travaux de la Société des Ingénieurs Civils*, vol. 8, 1855, p. 278.

8-5 Zaslavsky, A., "Beams on immovable supports," *Publications of the International Association for Bridge and Structural Engineering*, vol. 25, 1965, pp. 353–362.

9-1 Timoshenko, S. P., and Gere, J. M., *Theory of Elastic Stability*, 2nd Ed., McGraw-Hill Book Company, Inc., New York, 1961, 541 pages.

9-2 Bleich, F., *Buckling Strength of Metal Structures*, McGraw-Hill Book Company, Inc., New York, 1952, 508 pages.

9-3 Euler, L., "Methodus inveniendi lineas curvas maximi minimive proprietate gaudentes...," Appendix I, "De curvis elasticis," Bousquet, Lausanne and Geneva, 1744. (English translation: Oldfather, W. A., Ellis, C. A., and Brown, D. M., *Isis*, vol. 20, 1933, pp. 72–160. Also, republished in *Leonhardi Euleri Opera Omnia*, series 1, vol. 24, 1952.)

Note: Leonhard Euler, 1707–1783, made many remarkable contributions to mathematics and mechanics. He was the most productive mathematician of all time (Ref. 7-4, p. 120) and Newman refers to him as "a hero for mathematicians" (Ref. 7-3, p. 150). His name appears repeatedly in present-day textbooks; for instance, in mechanics we have Euler's equations of motion of a rigid body, Euler's angles, Euler's equations of fluid flow, the Euler load in column buckling, and much more; and in mathematics we encounter the famous Euler constant, as well as Euler's numbers, the Euler identities ($e^{i\theta} = \cos\theta + i\sin\theta$, etc.) and Euler's formula ($e^{i\pi} + 1 = 0$), Euler's differential equation, Euler's equation of a variational problem, Euler's quadrature formula, the Euler summation formula, Euler's theorem on homogeneous functions, Euler's integrals, and even Euler squares (square arrays of numbers possessing special properties).

In applied mechanics Euler was the first to derive the formula for the critical buckling load of an ideal, slender column and the first to solve the problem of the elastica. This work was published in 1744, as cited. He dealt with a column that is fixed at the base and free at the top. Later, he extended his work on columns (Ref. 9-4). Euler's numerous books include treatises on celestial mechanics, dynamics, and hydromechanics, and his papers also include subjects such as vibrations of beams and plates.

In the field of mathematics, Euler made outstanding contributions to trigonometry, algebra, number theory, differential and integral calculus, infinite series, analytic geometry, differential equations, calculus of variations, and many other subjects. He was the first to conceive of trigonometric values as the ratios of numbers and the first to present the equation $e^{i\theta} = \cos\theta + i\sin\theta$. Within his books on mathematics, all of which were classical references for many generations later, we find the first development of the calculus of variations as well as such intriguing items as the proof of Fermat's "last theorem" for

$n = 3$ and $n = 4$. He solved the famous problem of the seven bridges of Königsberg, a problem of topology, another field in which he pioneered.

Euler was born near Basel, Switzerland, and attended the University of Basel, where he studied under John Bernoulli (1667–1748). From 1727 to 1741 he lived and worked in St. Petersburg, where he established a great reputation as a mathematician. In 1741 he moved to Berlin upon the invitation of Frederick the Great, King of Prussia. He continued his mathematical research in Berlin until the year 1766, when he returned to St. Petersburg at the request of Catherine II, Empress of Russia. Euler continued to be prolific until his death in St. Petersburg at the age of 76; during this final period of his life he wrote more than 400 papers. In his entire lifetime, the number of books and papers written by Euler totaled 886; he left many manuscripts at his death and they continued to be published by the Russian Academy of Sciences in St. Petersburg for 47 years afterward. All this in spite of the fact that one of his eyes went blind in 1735 and the other in 1766! The story of Euler's life is told in Ref. 1-1, pp. 28–30 and Ref. 7-3, pp. 148–151; some of his contributions to mechanics are described in Ref. 1-1, pp. 30–36.

9-4 Euler, L., "Sur la force des colonnes," *Histoire de L'Académie Royale des Sciences et Belles Lettres*, 1757, published in *Memoires* of the Academie, vol. 13, Berlin, 1759, pp. 252–282. (Note: See Ref. 9-5 for a translation and discussion of this paper.)

9-5 Van den Broek, J. A., "Euler's classic paper 'On the strength of columns,'" *American Journal of Physics*, vol. 15, no. 4, July–August 1947, pp. 309–318.

9-6 Hoff, N. J., "Buckling and Stability," The Forty-First Wilbur Wright Memorial Lecture, *Journal of the Royal Aeronautical Society*, vol. 58, January 1954, pp. 3–52.

9-7 Johnston, B. G., "Column buckling theory: Historical highlights," *Journal of Structural Engineering*, Structural Division, American Society of Civil Engineers, vol. 109, no. 9, September 1983, pp. 2086–2096.

9-8 Johnston, B. G., editor, *Guide to Stability Design Criteria for Metal Structures*, 3rd Ed., John Wiley and Sons, New York, 1976, 616 pages.

9-9 Brush, D. O., and Almroth, B. O., *Buckling of Bars, Plates, and Shells,* McGraw-Hill Book Co., New York, 1975, 379 pages.

9-10 Chajes, A., *Principles of Structural Stability Theory*, Prentice-Hall, Inc., Englewood Cliffs, New Jersey, 1974, 336 pages.

9-11 Keller, J. B., "The shape of the strongest column," *Archive for Rational Mechanics and Analysis*, vol. 5, no. 4, 1960, pp. 275–285.

9-12 Young, D. H., "Rational design of steel columns," *Transactions of the American Society of Civil Engineers*, vol. 101, 1936, pp. 422–451. (Note: Donovan Harold Young, 1904–1980, was a well-known engineering educator. He was a professor at the University of Michigan and later at Stanford University. His five textbooks in the field of applied mechanics, written with S. P. Timoshenko, were translated into many languages and used throughout the world.)

9-13 Lamarle, A. H. E., "Mémoire sur la flexion du bois," *Annales des Travaux Publiques de Belgique*, Part 1, vol. 3, 1845, pp. 1–64, and Part 2, vol. 4, 1846, pp. 1–36. (Note: See Ref. 1-1, p. 208. Anatole Henri Ernest Lamarle, 1806–1875, was an engineer and professor. He was born in Calais, studied in Paris, and became a professor at the University of Gand, or Ghent, Belgium.)

9-14 Considère, A., "Résistance des pièces comprimées," *Congrès International des Procédés de Construction*, Paris, September 9–14, 1889, proceedings published by Librairie Polytechnique, Paris, vol. 3, 1891, p. 371. (Note: Armand Gabriel Considère, 1841–1914, was a French engineer.)

9-15 Engesser, F., "Ueber die Knickfestigkeit gerader Stäbe," *Zeitschrift für Architektur und Ingenieurwesen*, vol. 35, no. 4, 1889, pp. 455–462. (Note: Friedrich Engesser, 1848–1931, was a German railway and bridge engineer. Later, he became a professor at the Karlsruhe Polytechnical Institute, where he made important advances in the theory of structures, especially in buckling and energy methods. See Ref. 1-1, pp. 292 and 297–299.)

9-16 Engesser, F., "Knickfragen," *Schweizerische Bauzeitung*, vol. 25, no. 13, March 30, 1895, pp. 88–90.

9-17 Jasinski, F., "Noch ein Wort zu den 'Knickfragen,'" *Schweizerische Bauzeitung*, vol. 25, no. 25, June 22, 1895, pp. 172–175. (Note: Félix S. Jasinski, 1856–1899, was born in Warsaw and studied in Russia. He became a professor at the Institute of Engineers of Ways of Communication in St. Petersburg, now Leningrad.)

9-18 Engesser, F., "Ueber Knickfragen," *Schweizerische Bauzeitung*, vol. 26, no. 4, July 27, 1895, pp. 24–26.

9-19 von Kármán, T., "Die Knickfestigkeit gerader Stäbe," *Physikalische Zeitschrift*, vol. 9, no. 4, 1908, pp. 136–140. (Note: This paper also appears in Vol. I of Ref. 9-21. Theodore von Kármán, 1881–1963, was born in Hungary and later worked at the University of Göttingen in the field of aerodynamics. After coming to the United States in 1929, he founded the Jet Propulsion Laboratory and pioneered in aircraft and rocket problems. His research also included inelastic buckling of columns and stability of shells.)

9-20 von Kármán, T., "Untersuchungen über Knickfestigkeit," *Mitteilungen über Forschungsarbeiten auf dem Gebiete des Ingenieurwesens, Verein Deutscher Ingenieure*, Berlin, Heft 81, 1910. (Note: This paper also appears in Ref. 9-21.)

9-21 *Collected Works of Theodore von Kármán*, Vols. I-IV, Butterworths Scientific Publications, London, 1956.

9-22 Southwell, R. V., "The strength of struts," *Engineering*, vol. 94, August 23, 1912, pp. 249–250.

9-23 Shanley, F. R., "The column paradox," *Journal of the Aeronautical Sciences*, vol. 13, no. 12, December 1946, p. 678. (Note: Francis Reynolds Shanley, 1904–1968, was a professor of aeronautical engineering at the University of California, Los Angeles.)

9-24 Shanley, F. R., "Inelastic column theory," *Journal of the Aeronautical Sciences*, vol. 14, no. 5, May 1947, pp. 261–267.

9-25 Hoff, N. J., "The idealized column," *Ingenieur-Archiv*, vol. 28, 1959 (Festschrift Richard Grammel), pp. 89–98.

10-1 Maxwell, J. C., "On the calculation of the equilibrium and stiffness of frames," *Philosophical Magazine*, series 4, vol. 27, 1864, pp. 294–299. (Republished in *The Scientific Papers of James Clerk Maxwell*, Cambridge University Press, Vol. 1, 1890, pp. 598–604; see Ref. 10-3.) (Note: James Clerk Maxwell, 1831–1879, derived the unit-load method for finding displacements of elastic trusses in this paper. The paper also contains the reciprocal-displacement theorem and the flexibility method as applied to statically indeterminate trusses. Ten years later Otto Mohr rediscovered Maxwell's method for finding displacements, Ref. 10-4, and also his method for analyzing statically indeterminate trusses, Ref. 10-13. Maxwell performed other important work in structural mechanics, including the development of the photoelastic method of stress analysis and numerous investigations in the theory of elasticity. For an account of this work and a brief biography, see Ref. 1-1, pp. 202–208 and 268–275. Maxwell is best known for his scientific work in optics, the kinetic theory of gases, electricity, and magnetism. For the life story of this famous mathematician and physicist, see Ref. 10-2.)

10-2 Campbell, L., and Garnett, W., *The Life of James Clerk Maxwell*, Macmillan and Co., London, 1882.

10-3 Maxwell, J. C., *The Scientific Papers of James Clerk Maxwell*, Vols. 1 and 2, edited by W. D. Niven, Cambridge University Press, 1890.

10-4 Mohr, O., "Beitrag zur Theorie der Bogenfachwerksträger," *Zeitschrift des Architekten- und Ingenieur-Vereins zu Hannover*, vol. 20, no. 2, 1874, pp. 223–238. (Note: This paper presents the unit-load method for finding displacements of trusses.)

10-5 Betti, E., "Teoria della Elasticita," *Il Nuovo Cimento*, series 2, vols. 7 and 8, 1872. (Note: Enrico Betti, 1823–1892, was an Italian mathematician and engineer.)

10-6 Rayleigh, Lord, "Some general theorems relating to vibrations," *Proceedings of the London Mathematical Society*, vol. 4, 1873, pp. 357-368. (Republished in *Scientific Papers*, by John William Strutt, Vol. 1, Cambridge University Press, 1899, pp. 170–181.)

Note: In this paper, Lord Rayleigh presented the reciprocal theorem for a vibrating mechanical system acted upon by harmonically varying forces at two different points. He obtained the static reciprocal-work theorem, for the case of two forces, by letting the period of the forces become infinitely large. In a paper published in 1874 and 1875 (Ref. 10-7), Rayleigh obtained the reciprocal theorem for both flexibilities and stiffnesses, although he did not use that terminology. Later, in his book, *The Theory of Sound*, first published in 1877 (Ref. 10-8), he gave an explicit statement and derivation of the reciprocal-work theorem for two sets of corresponding forces and displacements, each set containing any number of forces.

Lord Rayleigh (1842–1919), who was born John William Strutt, was a famous British physicist. His best-known research pertained to sound, light, and electricity, and his book *The Theory of Sound* is a classic that is still used widely. His scientific papers fill six volumes and have been republished twice (Refs. 10-9 and 10-10). Lord Rayleigh studied at Trinity College, Cambridge University, under the mathematicians E. J. Routh and G. G. Stokes. He performed much experimental work in addition to his theoretical studies. He became professor at the Cavendish Laboratory in Cambridge in 1879, and later he served as Professor of Natural Philosophy at the Royal Institution of Great Britain. He helped organize the National Physical Laboratory and was President of the Advisory Committee on Aeronautics from its beginning in 1909 until his death. In addition, Lord Rayleigh was President of the Royal Society from 1905 to 1908 and Chancellor of Cambridge University from 1908 until his death. He received many honors, including the Order of Merit in 1902 and the Nobel Prize in physics in 1904, the latter for his joint work with Sir William Ramsay in discovering the first of the rare gases, argon, in

1895. Among his many important discoveries, we should perhaps note that Lord Rayleigh was the first to explain why the sky is blue. Rayleigh visited the United States twice, the second time in 1884, when he attended Lord Kelvin's famous "Baltimore lectures." Biographies of Lord Rayleigh can be found in Refs. 1-1, 10-8, 10-11, and 10-12.

10-7 Rayleigh, Lord, "A statical theorem," *Philosophical Magazine*, 4th series, vol. 48, 1874, pp. 452–456; 4th series, vol. 49, 1875, pp. 183–185. (Republished in *Scientific Papers*, by John William Strutt, Vol. 1, Cambridge University Press, 1899, pp. 223–229.)

10-8 Strutt, John William (Baron Rayleigh), *The Theory of Sound*, Vols. 1 and 2, 2nd Ed., Dover Publications, Inc., 1945. (Note: This republication contains a historical introduction and biographical sketch by Robert Bruce Lindsay. The first edition originally was published as follows: Vol. 1, 1877, Vol. 2, 1878. The second edition originally was published as follows: Vol. 1, 1894, Vol. 2, 1896; both editions by Macmillan and Co., Ltd., London.)

10-9 Strutt, John William (Baron Rayleigh), *Scientific Papers*, Cambridge University Press, Vol. 1, 1899; Vol. 2, 1900; Vol. 3, 1902; Vol. 4, 1903; Vol. 5, 1912; Vol. 6, 1920. (Republished in 1964 by Dover Publications, Inc.; Ref. 10-10.)

10-10 Rayleigh, Lord, *Scientific Papers*, Dover Publications, Inc., New York, 1964. (A republication of the six volumes originally published by the Cambridge University Press from 1899 to 1920; however, it is bound in three volumes.)

10-11 Strutt, Robert John (Fourth Baron Rayleigh), *John William Strutt, Third Baron Rayleigh*, Edward Arnold and Co., London, 1924, 403 pages. (A biography of Lord Rayleigh by his eldest son.)

10-12 Strutt, Robert John (Fourth Baron Rayleigh), *Life of John William Strutt, Third Baron Rayleigh*, University of Wisconsin Press, 1968, 439 pages. (A republication of the biography originally published in 1924 by Edward Arnold and Co., Ref. 10-11, but with annotations by the author and a foreword by John N. Howard.)

10-13 Mohr, O., "Beitrag zur Theorie des Fachwerks," *Zeitschrift des Architekten- und Ingenieur-Vereins zu Hannover*, vol. 20, no. 4, 1874, pp. 509–526, and vol. 21, no. 1, 1875, pp. 17–38. (Note: This paper presents the flexibility method for statically indeterminate trusses.)

10-14 Charlton, T. M., *Energy Principles in Theory of Structures*, Oxford University Press, London, 1973, 118 pages.

10-15 Gregory, M. S., *An Introduction to Extremum Principles*, Butterworth and Co., London, 1969, 196 pages.

10-16 Castigliano, A., *Théorie de l'équilibre des systèmes élastiques et ses applications*, A. F. Negro, Turin, 1879, 480 pages. (Note: In this book Castigliano presented in very complete form many concepts and principles of structural analysis. Although Castigliano was Italian, he wrote this book in French in order to gain a wider audience for his work. It was translated into both German and English, Refs. 10-17 and 10-18. The English translation was republished in 1966 by Dover Publications and is especially valuable because of the introductory material by Gunhard A. Oravas; see Refs. 10-19 and 7-23.

Castigliano's first and second theorems appear on pp. 15–16 of the 1966 edition of his book. He identified them as Part 1 and Part 2 of the "Theorem of the Differential Coefficients of the Internal Work." His statements of the theorem are as follows:

"Part 1. If the internal work of a framed structure is expressed as a function of the relative displacements of the external forces applied at its nodes, the resulting expression is such that its differential coefficients with regard to these displacements give the values of the corresponding forces.

"Part 2. If, on the contrary, the internal work of a framed structure is expressed as a function of the external forces, the resulting expression is such that its differential coefficients give the relative displacements of their points of application."

Today these propositions are called Castigliano's first and second theorems, respectively, rather than Parts 1 and 2 of a single theorem. After presenting these theorems, Castigliano proceeded in this book to derive them and apply them to many situations. In mathematical form, they appear in his book as

$$F_p = \frac{dW_i}{dr_p} \quad \text{and} \quad r_p = \frac{dW_i}{dF_p}$$

where W_i is the internal work (or strain energy), F_p represents any one of the external forces, and r_p is the displacement of the point of application of F_p.

Castigliano did not claim complete originality for the first theorem, although he stated in the Preface to his book that his presentation and proof were more general than anything published previously. The second theorem was original with him and was part of his thesis for the civil engineering degree at the Polytechnic Institute of Turin in 1873; see Ref. 10-20.

The "principle of least work" was also proved by Castigliano in his thesis published in 1873, and it appears in this book. The story of the controversy between Castigliano and Ménabréa about the authorship of this principle is described in Oravas' introduction to the 1966 edition, Ref. 10-19.

Carlo Alberto Pio Castigliano was born of a poor family in Asti in 1847 and died of pneumonia in 1884, while at the height of his productivity. The story of his life is told by Oravas in the introduction to the 1966 edition, and a bibliography of Castigliano's works and a list of his honors and awards are also given there. His contributions are also documented in Refs. 2-4, 1-1, and 10-21. He used the name Alberto Castigliano when signing his writings.)

10-17 Hauff, E., *Theorie des Gleichgewichtes elastischer Systeme und deren Anwendung*, Carl Gerold's Sohn, Vienna, 1886. (A translation of Castigliano's book, Ref. 10-16.)

10-18 Andrews, E. S., *Elastic Stresses in Structures*, Scott, Greenwood and Son, London, 1919. (A translation of Castigliano's book, Ref. 10-16.)

10-19 Castigliano, C. A. P., *The Theory of Equilibrium of Elastic Systems and Its Applications*, translated by E. S. Andrews with a new introduction and biographical portrait section by G. A. Oravas, Dover Publications, Inc., New York, 1966. (A republication of Ref. 10-18.)

10-20 Castigliano, A., "Intorno ai sistemi elastici," thesis presented to the Reale Scuola d'Applicazione degli Ingegneri in Torino for the civil engineering degree in 1873, published by Vincenzo Bona, Turin, 1873, 52 pages.

10-21 Grüning, M., "Theorie der Baukonstruktionen," *Encyklopädie der Mathematischen Wissenschaft*, Leipzig, Vol. 4, Part 4, 1907–1914, pp. 419–534.

10-22 Weaver, W., Jr., and Gere, J. M., *Matrix Analysis of Framed Structures*, 3rd Ed., Van Nostrand Reinhold Co., New York, 1990.

10-23 Crotti, F., "Esposizione del Teorema Castigliano e suo raccordo colla teoria dell' elasticità," *Atti del Collegio degli Ingegneri ed Architetti in Milano*, vol. 11, sect. 4, part 2, 1878, p. 225. (Also published in *Il Politecnico*, vol. 27, 1879, p. 45). (Note: Francesco Crotti, 1839–1896, was an Italian railroad engineer and a friend of Castigliano. In this paper he first derived the Crotti-Engesser theorem, although his later researches pertaining to complementary energy were much more extensive. Crotti formulated all of the basic theorems and principles pertaining to complementary energy in a series of papers beginning in 1877 and culminating in an important treatise in 1888, Ref. 10-24. Crotti's work is described on pp. 922–925 of Ref. 2-4 and on pp. xxv–xxvii of Ref. 7-23. Brief information about Crotti's life is given on p. xliii of Ref. 7-23.)

10-24 Crotti, F., *La Teoria dell' elasticità né suoi Principi Fondamentali e nelle sue Applicazioni Practiche alle Costruzioni*, Ulrico Hoepli, Milan, 1888.

10-25 Engesser, F., "Ueber statisch unbestimmte Träger bei beliebigem Formänderungs—Gesetze und über den Satz von der kleinsten Ergänzungsarbeit," *Zeitschrift des Architekten- und Ingenieur-Vereins zu Hannover*, vol. 35, 1889, pp. 733–744. (Note: In this paper Friedrich Engesser, 1848–1931, introduced the term *complementary work* and derived the Crotti-Engesser theorem. He was apparently unaware of Crotti's more general approach; see pp. 925–926 of Ref. 2-4 and p. xxvii of Ref. 7-23.)

10-26 Oden, J. T., and Ripperger, E. A., *Mechanics of Elastic Structures*, 2nd Ed., McGraw-Hill Book Co., Inc., New York, 1981.

10-27 Hoff, N. J., *The Analysis of Structures*, John Wiley and Sons, Inc., New York, 1956, 493 pages.

10-28 Westergaard, H. M., "On the method of complementary energy," *Transactions of the American Society of Civil Engineers*, vol. 107, 1942, pp. 765–793.

10-29 Westergaard, H. M., "One hundred fifty years advance in structural analysis," *Transactions of the American Society of Civil Engineers*, vol. 94, 1930, pp. 226–240.

10-30 Charlton, T. M., "Maxwell, Jenkin and Cotterill and the theory of statically-indeterminate structures," *Notes and Records of the Royal Society of London*, vol. 26, no. 2, December 1971, pp. 233–246.

10-31 Ménabréa, L. F., "Nouveau principe sur la distribution des tensions dans les systèmes élastiques," *Comptes Rendus*, vol. 46, 1858, pp. 1056–1060. (Note: In this paper Ménabréa set forth the concept that the redundant forces in the bars of a truss have such values as to make the strain energy a minimum, but he did not prove the idea correctly; see Ref. 7-23, p. xxii, Ref. 2-4, p. 655, and Ref. 1-1, p. 289. Luigi Frederico Ménabréa, 1809–1896, was an Italian nobleman, general, and engineer. His biography can be found on p. xl of Ref. 7-23.)

10-32 Weaver, W., Jr., and Johnston, P. R., *Finite Elements for Structural Analysis*, Prentice-Hall, Inc., Englewood Cliffs, New Jersey, 1984, 403 pages.

APPENDIX A

Systems of Units

A.1 INTRODUCTION

Numerous systems of measurement have been devised over the centuries for use in engineering and scientific work. The most recent and universal system is the International System of Units (Système International d'Unités), referred to as SI. Other important systems are the U.S. Customary System (USCS), the British Imperial System of Units, and the metric system.

The International System (SI), like the metric system, is based upon mass, length, and time as fundamental quantities. It is called an **absolute system** of units because measurements of the three fundamental quantities are independent of the locations at which the measurements are made. In contrast, the U.S. Customary System and the British Imperial System are based upon force, length, and time. The unit of force (pound) in these systems is defined as the weight of a certain standard mass; since weight depends upon gravitational attraction and varies with location, the USCS and the Imperial System are called **gravitational systems**. In an absolute system, force is a derived quantity (obtained from mass through the use of Newton's second law, $F = Ma$); in a gravitational system, the situation is reversed, with mass being the derived quantity.

The metric system is the predecessor of SI, but the two systems are not identical. Many of the units of the metric system are the same in SI, but SI has important features not previously part of any metric system. Thus, SI is an improved and modernized metric system.

The fundamental unit of mass in SI is the **kilogram**, and the corresponding unit of force is the **newton.*** A newton is defined as the force

* The weight of a small apple is approximately one newton.

required to impart an acceleration of one meter per second squared to a mass of one kilogram. In the USCS, the fundamental unit of force is the **pound**, and the corresponding unit of mass is the **slug**. The latter is defined as the mass that will be accelerated one foot per second squared when acted upon by a force of one pound.

A.2 SI UNITS

The International System of Units has several **base units** from which all other units are derived. The base units of importance in mechanics are the kilogram (kg) for mass, meter (m) for length, second (s) for time, and kelvin (K) for temperature. Other units needed in mechanics are derived from the base units. For instance, from the equation $F = Ma$ we can derive the unit of force:

$$1 \text{ newton} = (1 \text{ kilogram})(1 \text{ meter/second}^2)$$

Thus, the newton (N) is given in terms of base units by the formula

$$1 \text{ N} = 1 \text{ kg}\cdot\text{m/s}^2$$

The symbols and formulas for SI units of importance in mechanics are listed in Table A-1. Note that some derived units have special names; for instance, force is measured in newtons (N) and pressure in pascals (Pa). The units of other quantities, such as area and moment, have no special names and must be expressed in terms of base units and other derived units.

The **weight** of an object is the force of gravity on that object; thus, weight is measured in newtons. Unfortunately, the term *weight* has other meanings in popular usage, and hence its meaning is not always clear. For instance, we read that an astronaut becomes "weightless" in orbit and that a body "weighs less" when submerged in water than it does in air. However, the force of gravity is still acting on the astronaut, and the force of gravity on a body is the same whether the body is in air or under water. We even find that "weight" is sometimes used by the general public as synonymous with "mass." Therefore, to avoid confusion in engineering work, we always use the term *weight* to mean **force of gravity**. Since this force depends upon the location of the object with respect to the earth, weight is not an invariant property. In contrast, the mass of an object does not change with position.

The **acceleration of gravity**, denoted by the letter g, varies with elevation and latitude on the earth, but for ordinary purposes we can assume the following value:

$$g = 9.81 \text{ m/s}^2 \qquad \text{(A-1)}$$

anywhere on the surface of the earth. The weight and the mass of a body are related by the equation

$$W = Mg \tag{A-2}$$

where W is the weight in newtons, M is the mass in kilograms, and g is given by Eq. (A-1). Therefore, a body having a mass of 1 kilogram has a weight of 9.81 newtons.

Prefixes can be attached to SI units to form multiples and submultiples (see Table A-2). The use of prefixes avoids unusually large or small numbers. For instance, we may find it convenient to express the diameter of a bar in millimeters (mm) and the magnitude of a load on a beam in kilonewtons (kN).

Temperature is expressed in SI by a unit called the kelvin (K), but for common purposes it is customary to use the degree Celsius (°C). The temperature interval is the same for kelvins and degrees Celsius (that is, one kelvin equals one degree Celsius), but the scales have different origins. The origin of the Kelvin scale is at absolute zero temperature, whereas the origin of the Celsius scale (formerly called the centigrade scale in the United States) is at the freezing point of water. The Celsius scale registers 100 at the boiling point of water under standard conditions. The relationship between the scales is defined as follows:

Temperature in degrees Celsius = temperature in kelvins − 273.15

or

$$T(°\text{C}) = T(\text{K}) - 273.15 \tag{A-3}$$

where T denotes the temperature. When working only with temperature *intervals*, either scale can be used.*

A.3 U.S. CUSTOMARY UNITS

The system of units commonly used in the United States came from the British Imperial System, although the two systems are not identical. These units have never been officially adopted by the U.S. government in a legal sense; hence for lack of a better name, they are called the "customary" units. In this system, the base units of relevance to mechanics are the pound (lb) for force, foot (ft) for length, second (s) for time, and degree Fahrenheit (°F) for temperature.

The unit of mass can be derived from the equation $M = F/a$:

$$1 \text{ slug} = \frac{1 \text{ pound}}{1 \text{ ft/s}^2}$$

We see that the slug is given in terms of base units by the formula

$$1 \text{ slug} = 1 \text{ lb}\cdot\text{s}^2/\text{ft}$$

* SI is governed by very detailed rules and recommendations; for complete information, see the following document: *ASTM Standard for Metric Practice*, Publication E 380-86, American Society for Testing and Materials, 1916 Race Street, Philadelphia, Pennsylvania 19103.

Table A-1 Principal units used in mechanics

Quantity	International System (SI)			U.S. Customary System (USCS)		
	Unit	Symbol	Formula	Unit	Symbol	Formula
Acceleration, angular	radian per second squared		rad/s^2	radian per second squared		rad/s^2
Acceleration, linear	meter per second squared		m/s^2	foot per second squared		ft/s^2
Area	square meter		m^2	square foot		ft^2
Density (mass)	kilogram per cubic meter		kg/m^3	slug per cubic foot		$slug/ft^3$
Energy	joule	J	N·m	foot-pound		ft-lb
Force	newton	N	$kg \cdot m/s^2$	pound	lb	(base unit)
Frequency	hertz	Hz	s^{-1}	hertz	Hz	s^{-1}
Impulse, angular	newton meter second		N·m·s			ft-lb-s
Impulse, linear	newton second		N·s	pound-second		lb-s
Intensity of force	newton per meter		N/m	pound per foot		lb/ft
Length	meter	m	(base unit)	foot	ft	(base unit)
Mass	kilogram	kg	(base unit)	slug		$lb\text{-}s^2/ft$
Moment of a force; torque	newton meter		N·m	foot-pound		ft-lb
Moment of inertia (mass)	kilogram meter squared		$kg \cdot m^2$	slug foot squared		$slug\text{-}ft^2$
Moment of inertia (second moment of area)	meter to fourth power		m^4	inch to fourth power		$in.^4$
Power	watt	W	J/s	foot-pound per second		ft-lb/s
Pressure	pascal	Pa	N/m^2	pound per square foot	psf	lb/ft^2
Section modulus	meter to third power		m^3	inch to third power		$in.^3$
Specific weight (weight density)	newton per cubic meter		N/m^3	pound per cubic foot	pcf	lb/ft^3
Stress	pascal	Pa	N/m^2	pound per square inch	psi	$lb/in.^2$

(Continued)

Table A-I (Continued)

Quantity	International System (SI)			U.S. Customary System (USCS)		
	Unit	Symbol	Formula	Unit	Symbol	Formula
Time	second	s	(base unit)	second	s	(base unit)
Velocity, angular	radian per second		rad/s	radian per second		rad/s
Velocity, linear	meter per second		m/s	foot per second	fps	ft/s
Volume (liquids)	liter	L	10^{-3} m^3	gallon	gal.	231 in.^{3}
Volume (solids)	cubic meter		m^3	cubic foot	cf	ft^3
Work	joule	J	N·m	foot-pound		ft-lb

Note 1. Relationships pertaining to various SI units:
1 joule (J) = 1 newton meter (N·m) = 1 watt second (W·s)
1 hertz (Hz) = 1 cycle per second (cps)
1 watt (W) = 1 joule per second (J/s)
1 pascal (Pa) = 1 newton per meter squarcd (N/m^2)
1 liter (L) = 0.001 cubic meter (10^{-3} m^3) = 1000 cubic centimeters (1000 cm^3)

Note 2. Relationships between some commonly used metric units and SI units:
1 gal = 1 centimeter per second squared (cm/s^2); for example, g = 981 gals
1 are (a) = 100 square meters (m^2)
1 hectare (ha) = 10,000 square meters (m^2)
1 erg = 10^{-7} joules (J)
1 kilowatt-hour (kWh) = 3.6 megajoules (MJ)
1 dyne = 10^{-5} newtons (N)
1 kilopond (kp) = 9.81 newtons (N)
1 centimeter (cm) = 10^{-2} meters (m)
1 gram (g) = 10^{-3} kilograms (kg)
1 metric ton (t) = 1 megagram (Mg) = 1000 kilograms (kg)
1 watt (W) = 10^7 ergs per second (erg/s)
1 stere = 1 cubic meter (m^3)

Note 3. Some additional USCS and Imperial units:
1 kilowatt-hour (kWh) = 2,655, 220 foot-pounds (ft-lb)
1 British thermal unit (Btu) = 778.171 foot-pounds (ft-lb)
1 kip (k) = 1000 pounds (lb)
1 ounce (oz) = 1/16 pound (lb)
1 ton = 2000 pounds (lb)
1 Imperial ton (or long ton) = 2240 pounds (lb)
1 inch (in.) = 1/12 foot (ft)
1 mil = 0.001 inch (in.)
1 yard (yd) = 3 feet (ft)
1 mile = 5280 feet (ft)
1 mechanical horsepower (hp) = 550 foot-pounds per second (ft-lb/s)
1 kilowatt (kW) = 737.562 foot-pounds per second (ft-lb/s)
1 kilowatt (kW) = 1.34102 horsepower (hp)
1 pound per square inch (psi) = 144 pounds per square foot (psf)
1 revolution per minute (rpm) = $2\pi/60$ radians per second (rad/s)
1 mile per hour (mph) = 22/15 feet per second (fps)
1 quart (qt) = 1/4 gallon (gal.)
1 cubic foot (cf) = 576/77 gallons = 7.48052 gallons (gal.)
1 Imperial gallon = 277.420 cubic inches (in.^{3})

Note 4. The SI unit of work and energy (joule) is pronounced "jool."

Note 5. The international symbol for liter is lowercase l, but to avoid confusion with the numeral 1, uppercase L is recommended by the National Bureau of Standards for use in the United States.

Table A-2 SI prefixes

Prefix	Symbol	Multiplication factor
tera	T	$10^{12} = 1\ 000\ 000\ 000\ 000$
giga	G	$10^{9} = 1\ 000\ 000\ 000$
mega	M	$10^{6} = 1\ 000\ 000$
kilo	k	$10^{3} = 1\ 000$
hecto	h	$10^{2} = 100$
deka	da	$10^{1} = 10$
deci	d	$10^{-1} = 0.1$
centi	c	$10^{-2} = 0.01$
milli	m	$10^{-3} = 0.001$
micro	μ	$10^{-6} = 0.000\ 001$
nano	n	$10^{-9} = 0.000\ 000\ 001$
pico	p	$10^{-12} = 0.000\ 000\ 000\ 001$

Notes: The use of the prefixes hecto, deka, deci, and centi is not recommended in SI, because the powers of 10 are not multiples of 3. The prefixes giga and pico are properly pronounced "jig-uh" and "pea-ko," respectively, although "gig-uh" is often heard.

In order to obtain the mass of an object of known weight, we first note that the **acceleration of gravity** at the surface of the earth in the USCS is

$$g = 32.2 \text{ ft/s}^2 \tag{A-4}$$

Next, we rewrite Eq. (A-2) in the form

$$M = \frac{W}{g} \tag{A-5}$$

where M is the mass in slugs, W is the weight in pounds, and g is given by Eq. (A-4). From these two equations, we conclude that an object having a mass of 1 slug will weigh 32.2 pounds at the earth's surface. Another unit of mass is the pound mass (lb mass), which is the mass of an object weighing 1 pound. Thus, 1 pound mass equals 1/32.2 slug.

The symbols and formulas for various USCS units are listed in Table A-1. One of the units mentioned in Note 3 to the table has a metric origin, namely the kilopound, or kip, equal to 1000 pounds.

The customary unit for **temperature** is the degree Fahrenheit (°F). The Fahrenheit scale registers 32 at the freezing point of water and 212 at the boiling point. Thus, each Fahrenheit degree is 5/9 of a kelvin or a degree Celsius. The conversion formulas between the temperature scales are as follows:

$$T(^\circ\mathrm{F}) = \frac{9}{5}T(^\circ\mathrm{C}) + 32 \qquad \text{(A-6)}$$

$$T(^\circ\mathrm{F}) = \frac{9}{5}T(\mathrm{K}) - 459.67 \qquad \text{(A-7)}$$

$$T(^\circ\mathrm{C}) = \frac{5}{9}[T(^\circ\mathrm{F}) - 32] \qquad \text{(A-8)}$$

$$T(\mathrm{K}) = \frac{5}{9}[T(^\circ\mathrm{F}) - 32] + 273.15 \qquad \text{(A-9)}$$

As before, T denotes the temperature on the appropriate scale.

A.4 CONVERSIONS

Quantities given in either of the unit systems can be converted quickly to the other system by means of the **conversion factors** listed in Table A-3. If the given number has USCS units, it can be converted to SI units by multiplying by the conversion factor. For instance, from the first line of the table we see that an acceleration of 1 foot per second squared converts to 0.3048 meters per second squared. If the actual acceleration is 12.9 feet per second squared, then the conversion is performed as follows:

$$(12.9\ \mathrm{ft/s^2})(0.3048) = 3.93\ \mathrm{m/s^2}$$

To reverse the process (that is, to convert from SI to customary units), the number in SI units is divided by the factor. As an example, suppose that the acceleration is 9.81 meters per second squared. To convert to customary units, proceed as follows:

$$\frac{(9.81\ \mathrm{m/s^2})}{0.3048} = 32.2\ \mathrm{ft/s^2}$$

The factor 0.3048 happens to be exact, but in most cases the conversion factor listed in the table is given to six significant digits in the first column and to three significant digits in the second column.

Table A-4 gives a few important **physical properties** in both SI and USCS units.

Table A-3 Conversion of USCS units to SI units

USCS unit		Times conversion factor: Accurate	Times conversion factor: Practical	Equals SI unit	
Acceleration					
foot per second squared	ft/s^2	0.3048*	0.305	meter per second squared	m/s^2
inch per second squared	$in./s^2$	0.0254*	0.0254	meter per second squared	m/s^2
Area					
square foot	ft^2	0.09290304*	0.0929	square meter	m^2
square inch	$in.^2$	645.16*	645	square millimeter	mm^2
Density (mass)					
slug per cubic foot	$slug/ft^3$	515.379	515	kilogram per cubic meter	kg/m^3
Energy; work					
foot-pound	ft-lb	1.35582	1.36	joule	J
kilowatt-hour	kWh	3.6*	3.6	megajoule	MJ
British thermal unit	Btu	1055.06	1055	joule	J
Force					
pound	lb	4.44822	4.45	newton	N
kip (1000 pounds)	k	4.44822	4.45	kilonewton	kN
Intensity of force					
pound per foot	lb/ft	14.5939	14.6	newton per meter	N/m
kip per foot	k/ft	14.5939	14.6	kilonewton per meter	kN/m
Length					
foot	ft	0.3048*	0.305	meter	m
inch	in.	25.4*	25.4	millimeter	mm
mile		1.609344*	1.61	kilometer	km
Mass					
slug		14.5939	14.6	kilogram	kg
Moment of a force; torque					
foot-pound	ft-lb	1.35582	1.36	newton meter	N·m
inch-pound	in.-lb	0.112985	0.113	newton meter	N·m
foot-kip	ft-k	1.35582	1.36	kilonewton meter	kN·m
inch-kip	in.-k	0.112985	0.113	kilonewton meter	kN·m
Moment of inertia (mass)					
slug foot squared		1.35582	1.36	kilogram meter squared	$kg{\cdot}m^2$

(Continued)

Table A-3 (Continued)

USCS unit		Times conversion factor		Equals SI unit	
		Accurate	Practical		
Moment of inertia (second moment of area)					
inch to fourth power	in.^{4}	416,231	416,000	millimeter to fourth power	mm^4
inch to fourth power	in.^{4}	0.416231×10^{-6}	0.416×10^{-6}	meter to fourth power	m^4
Power					
foot-pound per second	ft-lb/s	1.35582	1.36	watt	W
foot-pound per minute	ft-lb/min	0.0225970	0.0226	watt	W
horsepower (550 foot-pounds per second)	hp	745.701	746	watt	W
Pressure; stress					
pound per square foot	psf	47.8803	47.9	pascal (N/m^2)	Pa
pound per square inch	psi	6894.76	6890	pascal	Pa
kip per square foot	ksf	47.8803	47.9	kilopascal	kPa
kip per square inch	ksi	6894.76	6890	kilopascal	kPa
Section modulus					
inch to third power	in.^{3}	16,387.1	16,400	millimeter to third power	mm^3
inch to third power	in.^{3}	16.3871×10^{-6}	16.4×10^{-6}	meter to third power	m^3
Specific weight (weight density)					
pound per cubic foot	lb/ft^3	157.087	157	newton per cubic meter	N/m^3
pound per cubic inch	lb/in.^{3}	271.447	271	kilonewton per cubic meter	kN/m^3
Velocity					
foot per second	ft/s	0.3048*	0.305	meter per second	m/s
inch per second	in./s	0.0254*	0.0254	meter per second	m/s
mile per hour	mph	0.44704*	0.447	meter per second	m/s
mile per hour	mph	1.609344*	1.61	kilometer per hour	km/h
Volume					
cubic foot	ft^3	0.0283168	0.0283	cubic meter	m^3
cubic inch	in.^{3}	16.3871×10^{-6}	16.4×10^{-6}	cubic meter	m^3
cubic inch	in.^{3}	16.3871	16.4	cubic centimeter	cm^3
gallon	gal.	3.78541	3.79	liter	L
gallon	gal.	0.00378541	0.00379	cubic meter	m^3

* Exact conversion factor

Note: To convert from SI units to USCS units, *divide* by the conversion factor.

Table A-4 Physical properties in SI and USCS units

Property	SI	USCS
Water (fresh)		
specific weight	9.81 kN/m^3	62.4 lb/ft^3
mass density	1000 kg/m^3	1.94 $slugs/ft^3$
Sea water		
specific weight	10.0 kN/m^3	63.8 lb/ft^3
mass density	1020 kg/m^3	1.98 $slugs/ft^3$
Aluminum		
specific weight	26.6 kN/m^3	169 lb/ft^3
mass density	2710 kg/m^3	5.26 $slugs/ft^3$
Steel		
specific weight	77.0 kN/m^3	490 lb/ft^3
mass density	7850 kg/m^3	15.2 $slugs/ft^3$
Reinforced concrete		
specific weight	23.6 kN/m^3	150 lb/ft^3
mass density	2400 kg/m^3	4.66 $slugs/ft^3$
Acceleration of gravity (on the earth's surface)		
Recommended value	9.81 m/s^2	32.2 ft/s^2
Standard international value	9.80665 m/s^2	32.1740 ft/s^2
Atmospheric pressure (at sea level)		
Recommended value	101 kPa	14.7 psi
Standard international value	101.325 kPa	14.6959 psi

APPENDIX B

Significant Digits

An important aspect of engineering calculations is the relative accuracy of all numerical values. Because calculations are usually performed by electronic calculators and computers, the possibility exists that numerical results will be reported to an accuracy that is not justified. As an example, suppose that a computation produces the result $R = 6287.46$ lb for the reaction of a beam. To present the result in this form is misleading, because it implies that the reaction is known to the nearest 1/100 of a pound even though the magnitude is over 6000 pounds. Thus, it implies an accuracy of approximately 1/600,000 and a precision of 0.01 lb, neither of which is justified. The accuracy of the calculated reaction depends upon how accurately the loads, dimensions, and other data used in the calculations are known. In reality, the accuracy is likely to be 1/1000 or 1/100 in this example. Therefore, the reaction would be known to the nearest 10 pounds, or possibly only to the nearest 100 pounds, and should be stated either as 6290 or 6300 lb. Under these circumstances, the last few digits in the original number have no justification. In order to decide how accurately to state a given numerical value, it is common practice to make use of significant digits, as described in the next section.

B.1 SIGNIFICANT DIGITS

In everyday practice, the accuracy of a number is indicated by the significant digits used in recording the value. A significant digit is a digit from 1 to 9 or any zero not used to show the position of the decimal point; for instance, the numbers 316, 7.23, 3.70, and 0.00347 each have three significant digits. However, the accuracy of a number such as 52,000 is not apparent. It may have two significant digits, with the three zeros serving only to locate the decimal point, or it may have three, four, or five significant digits if one or more of the zeros is accurate. By using powers of ten, the accuracy of a number such as 52,000 can be made clear. When written as 52×10^3, the number is understood to have two significant digits; when written as 520×10^2 or 52.0×10^3, it has three significant digits. For convenience in converting between SI units, it is desirable that the exponents of 10 be in multiples of three.

As a general rule, the precision of a number is plus or minus one-half of the unit corresponding to the last significant digit. Thus, the number 316 is understood to have a precision of ± 0.5 and 7.23 has a precision of ± 0.005.

When a number is obtained by calculation, its accuracy depends upon the accuracy of the numbers used in performing the calculations. A rule of thumb that serves for calculations involving multiplication and division is the following: The number of significant digits in the calculated result is the same as the least number of significant digits in any of the numbers used in the calculation. As an illustration, consider the product (2743.1)(31.6). When multiplied on a calculator, the result is 86,681.960 when recorded to eight digits. However, to write this number with eight digits is misleading, because it implies much more accuracy than is warranted by the original numbers. Inasmuch as the number 31.6 has three significant digits, the proper way to write the result is either 86,700 or 86.7×10^3.

For calculations involving addition or subtraction of a column of numbers, the last significant digit in the result is found in the last column of digits that has significant digits in all of the numbers being added or subtracted; that is, the last significant digit in the result corresponds to the last digit in the least precise number. To make this notion clear, consider the following examples:

	142.734	127.58	945,000
	+ 9.8	− 6	+ 11,230
From calculator:	152.534	121.58	956,230
Write as:	152.5	122	956,000

In the first example, the number 142.734 has six significant digits and the number 9.8 has two. When added, the result has four significant digits because all digits in the result to the right of the column containing the 8 are meaningless. In the second example, the number 6 is assumed to be accurate to only one significant digit (that is, it is not an exact number). Therefore, the final result is accurate to three significant digits and is recorded as 122. In the third example, the number 945,000 is assumed to be accurate to three significant digits; hence any digits in the final result to the right of the comma are meaningless.

In engineering problems, all calculations can be performed to a high degree of numerical accuracy with a computer or calculator (usually carrying eight or more digits), but the user must recognize that most of the digits are superfluous and have no physical meaning. When reporting final results, the number of digits should be limited to those that are significant. In mechanics of materials, the data for a problem (such as the dimensions and loads) are usually accurate to only two or three significant digits, and the final results should be reported to a comparable accuracy.

Although the use of significant digits is convenient in engineering work, it should be recognized that they do not give a precise indication of the accuracy of a number. To illustrate this point, consider the numbers 999 and 101, each of which has three significant digits. In the first case, the number has an accuracy of 1/999, or 0.1%; in the second case, the accuracy is 1/101, or 1.0%. This disparity in accuracy can be reduced by using one additional significant digit for numbers beginning with the digit 1.

For consistency throughout this book, we generally follow the common engineering rule that numbers beginning with the digits 2 through 9 are recorded to three significant digits and numbers beginning with the digit 1 are recorded to four significant digits. Intermediate calculations are sometimes recorded to additional digits in order to preserve numerical accuracy.

Some of the numbers entering into a typical engineering calculation are exact (for example, the number π, or the number 48 in the formula $PL^3/48EI$ for a beam deflection). Exact numbers can be considered to have an infinite number of significant digits; hence they do not affect the accuracy of the calculated result.

B.2 ROUNDING OFF NUMBERS

The process of discarding the insignificant digits and keeping only the significant ones is called *rounding off*. To illustrate the process, let us assume that a number is to be rounded off to three significant digits. Then the following rules apply: (a) If the fourth digit is less than 5, the first three digits are left unchanged and all following digits are dropped or set equal to zero. For example, 22.34 rounds off to 22.3 and 576,234 rounds off to 576,000. (b) If the fourth digit is greater than 5, or if the fourth digit is 5 and it is followed by at least one digit other than zero, then the third digit is increased by 1 and all following digits are dropped or set equal to zero. For example, 31.56 rounds off to 31.6; 28,172 rounds off to 28,200; and 1.755001 rounds off to 1.76. (c) Finally, if the fourth digit is exactly 5, then we round off to the even digit. For instance, 22.45 rounds off to 22.4 and 67.75 rounds off to 67.8. Because the occurrence of even and odd digits is more or less random, the use of the preceding rule means that we round off upward about as often as we round off downward, thereby reducing the chances of accumulating round-off errors. (Instead of following the "even" rule, most calculators round off upward when the last digit is exactly 5.)

The rules illustrated in the preceding paragraph for rounding off to three significant digits apply in the same fashion when rounding off to any other number of significant digits.

APPENDIX C

Centroids and Moments of Inertia of Plane Areas

C.1 CENTROIDS OF AREAS

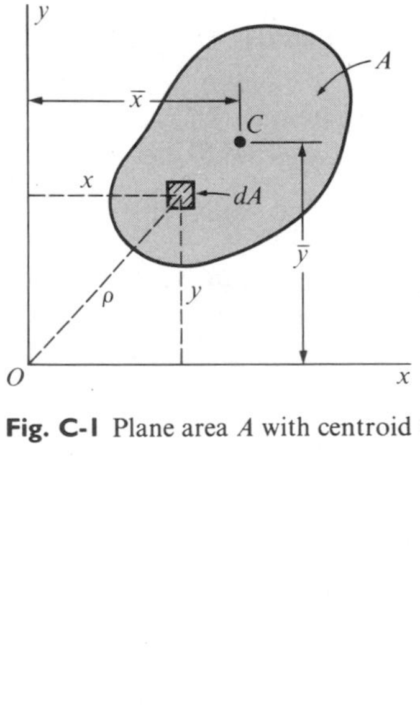

Fig. C-1 Plane area A with centroid C

The location of the **centroid** of a plane area is an important geometric property of the area.* To determine the coordinates of the centroid, let us refer to the area A and the xy coordinate system shown in Fig. C-1. A differential element of area dA, with coordinates x and y, is shown in the figure. The total area A is defined as the following integral:

$$A = \int dA \tag{C-1}$$

In addition, the **first moments** of the area about the x and y axes, respectively, are defined as

$$Q_x = \int y\,dA \qquad Q_y = \int x\,dA \tag{C-2}$$

The coordinates $\bar{x}$ and $\bar{y}$ of the centroid C (Fig. C-1) are equal to the first

* In accord with common terminology, we use the term "area" to mean *plane surface*. Strictly speaking, area is a measure of the size of a surface and is not the same thing as the surface itself.

moments divided by the area itself:

$$\bar{x} = \frac{Q_y}{A} = \frac{\int x\,dA}{\int dA} \qquad \bar{y} = \frac{Q_x}{A} = \frac{\int y\,dA}{\int dA} \tag{C-3}$$

If the boundaries of the area are defined by simple mathematical expressions, we can evaluate the integrals appearing in Eqs. (C-3) in closed form and thereby obtain formulas for $\bar{x}$ and $\bar{y}$. A table of formulas obtained in this manner is given in Appendix D.

If an area is **symmetric about an axis**, the centroid must lie on that axis because the first moment about an axis of symmetry equals zero. For example, the centroid of the singly symmetric area shown in Fig. C-2 must lie on the x axis, which is the axis of symmetry; hence only one distance must be calculated in order to locate C. If an area has two axes of symmetry, as illustrated in Fig. C-3, the position of the centroid can be determined by inspection because it lies at the intersection of the axes of symmetry.

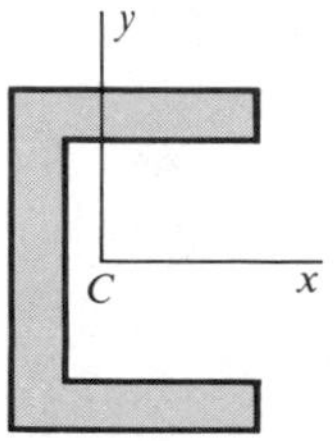

Fig. C-2 Area with one axis of symmetry

An area of the type shown in Fig. C-4 is **symmetric about a point**. It has no axes of symmetry, but there is a point (called the **center of symmetry**) such that every line in the area drawn through that point is symmetric about that point. The centroid coincides with the center of symmetry and can be located by inspection.

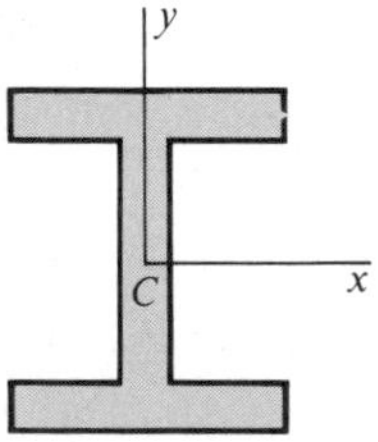

Fig. C-3 Area with two axes of symmetry

If the boundaries of the area are irregular curves not defined by mathematical expressions, then we can evaluate approximately the integrals in Eqs. (C-3) by numerical methods. The simplest procedure is to divide the area into small elements of area ΔA_i and replace the integrations with summations:

$$A = \sum_{i=1}^{n} \Delta A_i \qquad Q_x = \sum_{i=1}^{n} y_i \Delta A_i \qquad Q_y = \sum_{i=1}^{n} x_i \Delta A_i \tag{C-4}$$

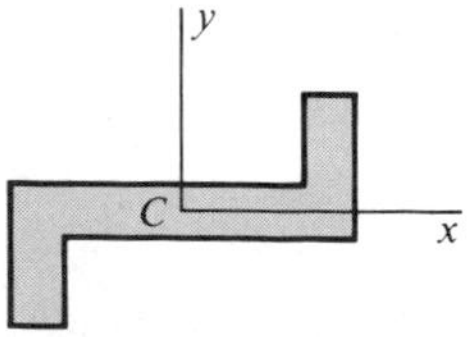

Fig. C-4 Area that is symmetric about a point

in which n is the total number of elements of area, y_i is the y coordinate of the centroid of area ΔA_i, and x_i is the x coordinate of the centroid of area ΔA_i. The accuracy of the calculations for $\bar{x}$ and $\bar{y}$ depends upon how closely the selected elements fit the actual area.

Example

A parabolic semisegment OAB is bounded by the x axis, the y axis, and a parabolic curve having its vertex at A (Fig. C-5). The equation of the curve is

$$y = f(x) = h\left(1 - \frac{x^2}{b^2}\right) \tag{a}$$

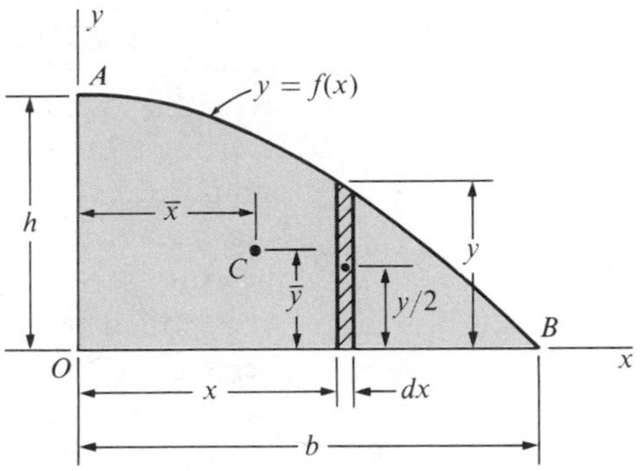

Fig. C-5 Example. Centroid of a parabolic semisegment

in which b is the base and h is the height of the semisegment. Locate the centroid C of the semisegment.

To carry out this analysis, we will select an element of area dA in the form of a thin strip of width dx and height y, as shown in the figure. The area of this element is

$$dA = y\,dx = h\left(1 - \frac{x^2}{b^2}\right)dx$$

and, therefore, the area of the semisegment is

$$A = \int dA = \int_0^b h\left(1 - \frac{x^2}{b^2}\right)dx = \frac{2bh}{3} \qquad \text{(b)}$$

The first moment of the element of area about any axis can be obtained by multiplying the area of the element by the distance from its centroid to the axis. Since the x and y coordinates of the centroid of the element are x and $y/2$, respectively, the first moments are

$$Q_x = \int \frac{y}{2}\,dA = \int_0^b \frac{h^2}{2}\left(1 - \frac{x^2}{b^2}\right)^2 dx = \frac{4bh^2}{15}$$

$$Q_y = \int x\,dA = \int_0^b hx\left(1 - \frac{x^2}{b^2}\right)dx = \frac{b^2h}{4}$$

Now we can determine the coordinates of the centroid C as follows:

$$\bar{x} = \frac{Q_y}{A} = \frac{3b}{8} \qquad \bar{y} = \frac{Q_x}{A} = \frac{2h}{5} \qquad \text{(c)}$$

The centroid may also be located by taking the element of area dA as a horizontal strip of height dy and width

$$x = b\sqrt{1 - \frac{y}{h}} \qquad \text{(d)}$$

which is obtained by solving Eq. (a) for x in terms of y. Yet another possibility is to take the element as a rectangle of width dx and height dy; in this case, the expressions for A, Q_x, and Q_y are in the form of double integrals.

C.2 CENTROIDS OF COMPOSITE AREAS

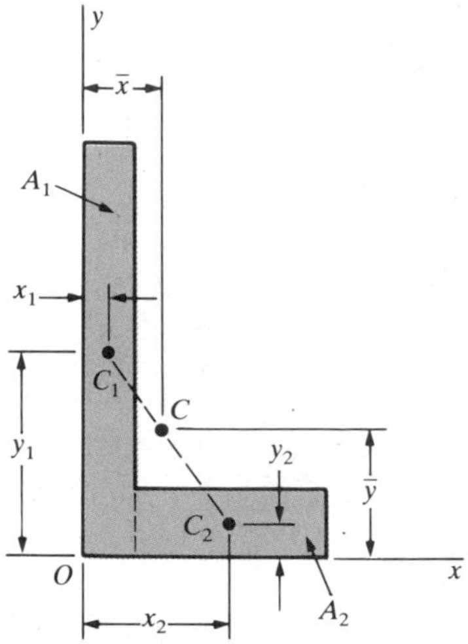

Fig. C-6 Centroid of a composite area consisting of two parts

In engineering work we frequently need to locate the centroid of an area composed of several parts, each part having a familiar geometric shape (such as a rectangle or a triangle). Examples of such **composite areas** are the cross sections of beams, which often are composed of rectangular areas (for instance, see Figs. C-2, C-3, and C-4). The area and first moments of a composite area may be calculated by summing the corresponding properties of its parts:

$$A = \sum_{i=1}^{n} A_i \qquad Q_x = \sum_{i=1}^{n} y_i A_i \qquad Q_y = \sum_{i=1}^{n} x_i A_i \qquad \text{(C-5)}$$

in which A_i is the area of the ith part, x_i and y_i are the coordinates of the centroid of the ith part, and n is the number of parts. When using these formulas, it is possible to treat the absence of an area as a "negative area"; for instance, this concept is useful when a hole exists in the figure. After finding A, Q_x, and Q_y from Eqs. (C-5), we can determine the coordinates of the centroid from Eqs. (C-3).

To illustrate the procedure, let us consider the special case of a composite area that can be divided conveniently into only two parts. The L-shaped area shown in Fig. C-6 is of this kind, because it can be divided into two rectangles of areas A_1 and A_2. These rectangles have centroids C_1 and C_2 with known coordinates (x_1, y_1) and (x_2, y_2), respectively. Thus, we obtain the following expressions from Eqs. (C-5):

$$A = A_1 + A_2 \qquad Q_x = y_1 A_1 + y_2 A_2 \qquad Q_y = x_1 A_1 + x_2 A_2$$

Therefore, the coordinates of the centroid C are

$$\bar{x} = \frac{Q_y}{A} = \frac{x_1 A_1 + x_2 A_2}{A_1 + A_2} \qquad \bar{y} = \frac{Q_x}{A} = \frac{y_1 A_1 + y_2 A_2}{A_1 + A_2} \qquad \text{(C-6)}$$

(see Eqs. C-3). When an area is divided into only two parts, the centroid C of the entire area lies on the line joining the centroids C_1 and C_2 of the parts, as shown in the figure.

Example

The cross-section of a beam constructed of a W 18 × 71 section with a 6 × $\frac{1}{2}$ in. cover plate welded to the top flange and a C 10 × 30 channel section welded to the bottom flange is shown in Fig. C-7. Locate the centroid C of the cross-sectional area.

Let us denote the centroids of the cross-sectional areas of the cover plate, the wide-flange section, and the channel section as C_1, C_2, and C_3, respectively. Also, the corresponding areas are

$$A_1 = (6 \text{ in.})(0.5 \text{ in.}) = 3.0 \text{ in.}^2$$

$$A_2 = 20.8 \text{ in.}^2 \qquad A_3 = 8.82 \text{ in.}^2$$

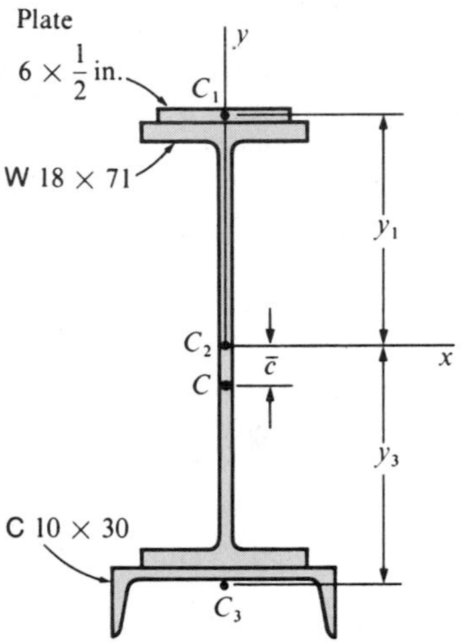

Fig. C-7 Example

in which the areas A_2 and A_3 are obtained from Tables E-1 and E-3, Appendix E. If the x and y axes are taken with origin at C_2, the centroidal distances for the three areas are

$$y_1 = \frac{18.47 \text{ in.}}{2} + \frac{0.5 \text{ in.}}{2} = 9.485 \text{ in.}$$

$$y_2 = 0 \qquad y_3 = \frac{18.47 \text{ in.}}{2} + 0.649 \text{ in.} = 9.884 \text{ in.}$$

in which the pertinent dimensions of the sections also are obtained from Tables E-1 and E-3.

The area A and first moment Q_x of the entire cross section are

$$A = A_1 + A_2 + A_3 = 32.62 \text{ in.}^2$$

$$Q_x = y_1 A_1 + y_2 A_2 - y_3 A_3$$

$$= (9.485 \text{ in.})(3.0 \text{ in.}^2) + 0 - (9.884 \text{ in.})(8.82 \text{ in.}^2)$$

$$= -58.72 \text{ in.}^3$$

The coordinate $\bar{y}$ to the centroid C is obtained from the following equation:

$$\bar{y} = \frac{Q_x}{A} = -\frac{58.72 \text{ in.}^3}{32.62 \text{ in.}^2} = -1.80 \text{ in.}$$

Since $\bar{y}$ is positive in the direction of the positive y axis, the minus sign means that the centroid C is located below the x axis (see figure). Thus, the distance $\bar{c}$ between the x axis and the centroid C is

$$\bar{c} = -\bar{y} = 1.80 \text{ in.}$$

Note that the position of the reference axis (the x axis) is arbitrary; however, in this example we placed it through the centroid of the wide-flange section for convenience in the calculations.

C.3 MOMENTS OF INERTIA OF AREAS

The **moments of inertia** of a plane area (see Fig. C-1) with respect to the x and y axes, respectively, are defined by the integrals

$$I_x = \int y^2 \, dA \qquad I_y = \int x^2 \, dA \tag{C-7}$$

in which x and y are the coordinates of the differential element of area dA. Because dA is multiplied by the square of the distance, moments of inertia are also called **second moments** of the area.

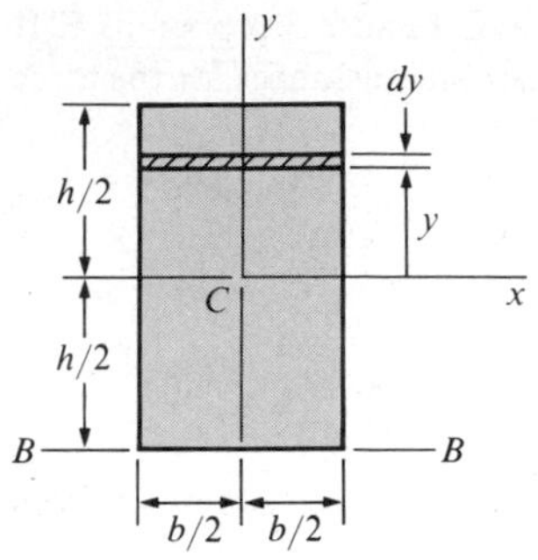

Fig. C-8 Moments of inertia of a rectangle

To illustrate how moments of inertia are obtained by integration, let us consider the rectangle shown in Fig. C-8. The x and y axes have their origin at the centroid C. For convenience, we use an element of area in the form of a thin strip of width b and height dy, so that $dA = b\,dy$. Then the moment of inertia with respect to the x axis is

$$I_x = \int_{-h/2}^{h/2} y^2 b\,dy = \frac{bh^3}{12} \tag{a}$$

In a similar manner, we can use an element of area dA in the form of a vertical strip and obtain the moment of inertia with respect to the y axis:

$$I_y = \int_{-b/2}^{b/2} x^2 h\,dx = \frac{hb^3}{12} \tag{b}$$

If different axes are selected, then the moments of inertia will have different values. For instance, let us consider axis BB at the base of the rectangle. In that case, we define y as the distance from axis BB to the element of area dA. Then the calculations for the moment of inertia proceed as follows:

$$I_{BB} = \int y^2\,dA = \int_0^h y^2 b\,dy = \frac{bh^3}{3} \tag{c}$$

Note that the moment of inertia is larger with respect to axis BB than with respect to the centroidal x axis. In general, the moment of inertia increases as the reference axis is moved parallel to itself farther from the centroid. Regardless of the axes that are selected, moments of inertia always are positive quantities because the coordinates x and y are squared (see Eqs. C-7).

The moment of inertia of a composite area with respect to a particular axis is the sum of the moments of inertia of its parts with respect to that same axis. An example is the hollow box section shown in Fig. C-9a. The x axis is an axis of symmetry through the centroid C. The moment of inertia with respect to the x axis is equal to the difference

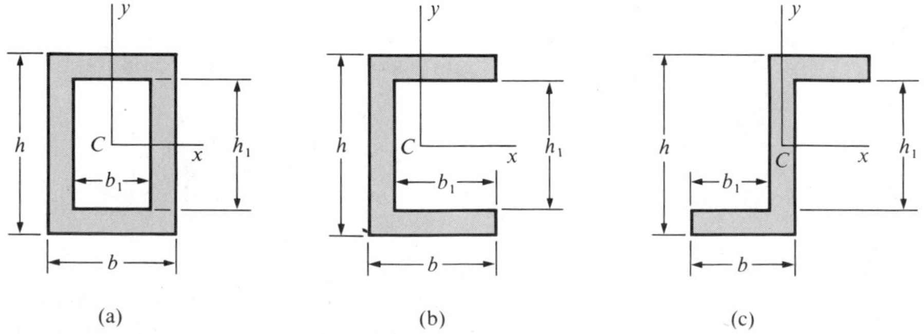

Fig. C-9 Composite areas

between the moments of inertia of the two rectangles:

$$I_x = \frac{bh^3}{12} - \frac{b_1 h_1^3}{12} \tag{d}$$

This same formula applies to the channel section and the Z-section shown in parts (b) and (c), respectively, of the figure. A similar technique can be used to obtain the moment of inertia I_y for all three sections. However, in the case of the channel section and Z-section, I_y is more easily obtained by using the parallel-axis theorem, which is described in the next section.

Moments of inertia are tabulated in Appendix D for many common areas. The use of the formulas in the table in conjunction with the parallel-axis theorem makes it possible to obtain moments of inertia for most other shapes that are encountered. However, if an area is of irregular shape, we can always obtain its moments of inertia by numerical methods. The procedure is to divide the area into small elements of area ΔA, multiply each such area by the square of its distance from the axis, and then sum the products.

RADIUS OF GYRATION

A distance known as the **radius of gyration** of an area is used occasionally in mechanics. It is defined as the square root of the moment of inertia divided by the area itself; thus,

$$r_x = \sqrt{\frac{I_x}{A}} \qquad r_y = \sqrt{\frac{I_y}{A}} \tag{C-8}$$

where r_x and r_y denote the radii of gyration with respect to the x and y axes, respectively. Since I has units of length to the fourth power and A has units of length to the second power, radius of gyration has units of length. We may consider the radius of gyration of an area to be the distance from the axis at which the entire area could be concentrated and still have the same moment of inertia as the original area.

Example

Determine the moments of inertia I_x and I_y for the parabolic semisegment OAB shown in Fig. C-5. The equation of the parabolic boundary is

$$y = f(x) = h\left(1 - \frac{x^2}{b^2}\right)$$

as given in the example of Section C.1.

Let us use the element of area dA shown as a vertical strip in Fig. C-5:

$$dA = y\,dx = h\left(1 - \frac{x^2}{b^2}\right)dx$$

Since every point in this element of area is at the same distance x from the y axis, the moment of inertia of the element with respect to the y axis is $x^2\,dA$. Therefore, the moment of inertia of the entire area with respect to the y axis is determined as follows:

$$I_y = \int x^2\,dA = \int_0^b x^2 h\left(1 - \frac{x^2}{b^2}\right)dx = \frac{2hb^3}{15} \qquad \text{(e)}$$

To obtain the moment of inertia with respect to the x axis, we note that the element of area dA has a moment of inertia equal to

$$\frac{1}{3}(dx)y^3$$

with respect to the x axis (see Eq. c). Hence, the moment of inertia of the entire area is

$$I_x = \int_0^b \frac{y^3}{3}dx = \int_0^b \frac{h^3}{3}\left(1 - \frac{x^2}{b^2}\right)^3 dx = \frac{16bh^3}{105} \qquad \text{(f)}$$

These same results can be obtained by using an element in the form of a horizontal strip or by using a rectangular element of area $dA = dx\,dy$ and performing a double integration.

C.4 PARALLEL-AXIS THEOREM FOR MOMENTS OF INERTIA

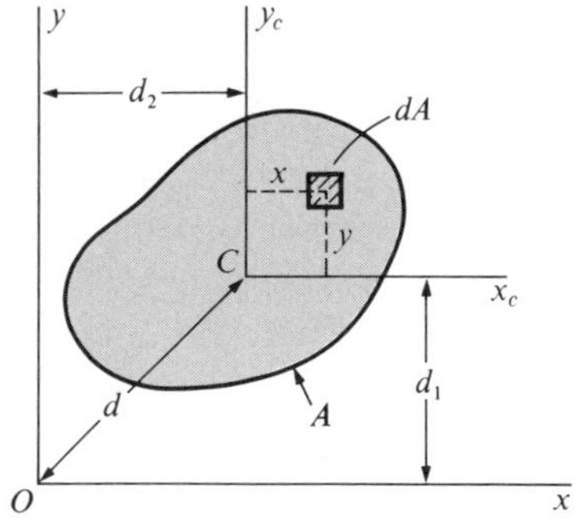

Fig. C-10 Derivation of parallel-axis theorems

The moment of inertia of an area with respect to any axis in the plane of the area is related to the moment of inertia with respect to a parallel centroidal axis by the **parallel-axis theorem**. To derive this extremely useful theorem, we will consider the area shown in Fig. C-10. Assume that the $x_c y_c$ axes have their origin at the centroid C of the area. The xy axes are parallel to the $x_c y_c$ axes and have their origin at any point O. The distances between corresponding axes are d_1 and d_2. From the definition of moment of inertia, we obtain the following equation for the moment of inertia I_x with respect to the x axis:

$$I_x = \int (y + d_1)^2\,dA$$
$$= \int y^2\,dA + 2d_1 \int y\,dA + d_1^2 \int dA$$

The first integral on the right-hand side is the moment of inertia I_{x_c} with respect to the x_c axis; the second integral vanishes because the x_c axis passes through the centroid; and the third integral is the area A of the figure. Therefore, the preceding equation reduces to

$$I_x = I_{x_c} + Ad_1^2 \qquad \text{(C-9a)}$$

Proceeding in the same manner for the y axis, we obtain

$$I_y = I_{y_c} + Ad_2^2 \tag{C-9b}$$

Equations (C-9) represent the parallel-axis theorem for moments of inertia: *The moment of inertia of an area with respect to any axis in its plane is equal to the moment of inertia with respect to a parallel centroidal axis plus the product of the area and the square of the distance between the two axes.*

From the parallel-axis theorem, we see that the moment of inertia increases as the axis is moved parallel to itself farther from the centroid. Hence the moment of inertia about a centroidal axis is the least moment of inertia of the area (for a given direction of the axes).

The parallel-axis theorem is extremely useful for finding moments of inertia, especially for composite areas. When using the theorem, we must always keep in mind that one of the two parallel axes must be a centroidal axis. To illustrate this point, consider again the rectangle shown in Fig. C-8. Knowing that the moment of inertia about the x axis, which is through the centroid, is equal to $bh^3/12$ (see Eq. a of Section C.3), we can quickly determine the moment of inertia I_{BB} about the base of the rectangle:

$$I_{BB} = I_x + Ad^2 = \frac{bh^3}{12} + bh\left(\frac{h}{2}\right)^2$$
$$= \frac{bh^3}{3}$$

This result agrees with the result obtained previously by integration (Eq. c of Section C.3).

If it is necessary to find the moment of inertia I_1 about a noncentroidal axis when the moment of inertia I_2 about another noncentroidal (and parallel) axis is known, we must apply the parallel-axis theorem twice. First, we use the theorem to find the centroidal moment of inertia from the known moment of inertia I_2. Then we use the theorem a second time to find I_1 from the centroidal moment of inertia.

In the case of a composite area, we can find the moment of inertia (with respect to a particular axis) of each part of the area and then add these moments of inertia to obtain I for the entire area. For instance, consider the Z-section shown in Fig. C-9c and assume that I_y is to be calculated. Then we can divide the area into three rectangles, each of which has a centroid that can be located by inspection. The moments of inertia of the rectangles, with respect to axes through their centroids parallel to the y axis, can be obtained from the general formula $I = bh^3/12$. Then we can use the parallel-axis theorem to calculate the moments of inertia with respect to the y axis. Adding these moments of inertia, we obtain I_y for the entire area.

Example 1

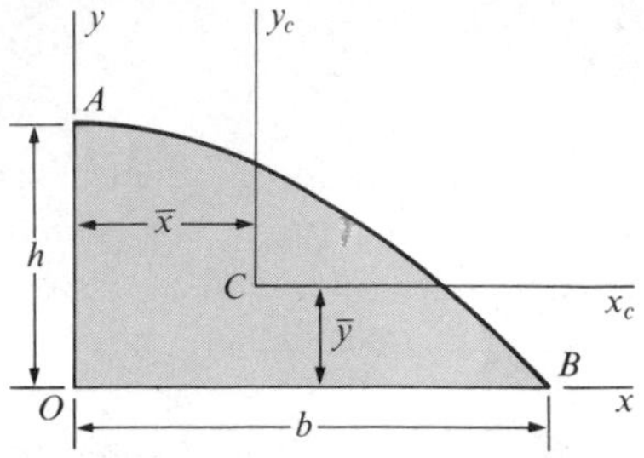

Fig. C-11 Example 1. Parallel-axis theorem

Determine the centroidal moments of inertia I_{x_c} and I_{y_c} for the parabolic semisegment OAB shown in Fig. C-11. The area of the semisegment is $2bh/3$, and the centroid C has coordinates $\bar{x} = 3b/8$ and $\bar{y} = 2h/5$; also, the moments of inertia with respect to the x and y axes are

$$I_x = \frac{16bh^3}{105} \qquad I_y = \frac{2hb^3}{15}$$

(see Case 18, Appendix D).

To obtain the moment of inertia with respect to the x_c axis, we write the parallel-axis theorem as follows:

$$I_x = I_{x_c} + A\bar{y}^2$$

Therefore, the moment of inertia I_{x_c} is

$$I_{x_c} = I_x - A\bar{y}^2 = \frac{16bh^3}{105} - \frac{2bh}{3}\left(\frac{2h}{5}\right)^2 = \frac{8bh^3}{175} \tag{a}$$

In a similar manner, we obtain the moment of inertia with respect to the y_c axis:

$$I_{y_c} = I_y - A\bar{x}^2 = \frac{2hb^3}{15} - \frac{2bh}{3}\left(\frac{3b}{8}\right)^2 = \frac{19hb^3}{480} \tag{b}$$

Thus, the required centroidal moments of inertia have been found.

Example 2

Determine the moment of inertia I_c with respect to a horizontal axis through the centroid C of the beam cross section described in the example of Section C.2 and shown in Fig. C-7. (This moment of inertia is required when calculating the stresses and deflections of the beam.)

To find the centroidal moment of inertia I_c of the composite area, we consider the area in three parts: (1) the cover plate, (2) the wide-flange section, and (3) the channel section. Also, we take the following properties and dimensions from the example of Section C.2:

$$A_1 = 3.0 \text{ in.}^2 \qquad A_2 = 20.8 \text{ in.}^2 \qquad A_3 = 8.82 \text{ in.}^2$$

$$y_1 = 9.485 \text{ in.} \qquad y_3 = 9.884 \text{ in.} \qquad \bar{c} = 1.80 \text{ in.}$$

The moments of inertia of the three parts with respect to their own centroids are as follows:

$$I_1 = \frac{bh^3}{12} = \frac{1}{12}(6.0 \text{ in.})(0.5 \text{ in.})^3 = 0.063 \text{ in.}^4$$

$$I_2 = 1170 \text{ in.}^4$$

$$I_3 = 3.94 \text{ in.}^4$$

in which I_2 and I_3 are obtained from Tables E-1 and E-3, respectively.

Now we can use the parallel-axis theorem to calculate the moments of inertia about an axis through C for the three parts of the cross section:

$$I_{c1} = I_1 + A_1(y_1 + \bar{c})^2 = 0.063 + 3.0(11.28)^2 = 382 \text{ in.}^4$$

$$I_{c2} = I_2 + A_2\bar{c}^2 = 1170 + 20.8(1.80)^2 = 1237 \text{ in.}^4$$

$$I_{c3} = I_3 + A_3(y_3 - \bar{c})^2 = 3.94 + 8.82(8.084)^2 = 580 \text{ in.}^4$$

The sum of these moments of inertia is

$$I_c = I_{c1} + I_{c2} + I_{c3} = 2200 \text{ in.}^4$$

which is the centroidal moment of inertia of the entire cross section.

C.5 POLAR MOMENTS OF INERTIA

The moments of inertia discussed in the preceding sections are defined with respect to axes lying in the plane of the area itself, such as the x and y axes in Fig. C-1. Now let us consider an axis perpendicular to the plane of the area and intersecting the plane at the origin O. The moment of inertia with respect to this axis is called the **polar moment of inertia I_p**; it is defined as the integral

$$I_p = \int \rho^2 \, dA \qquad \text{(C-10)}$$

in which ρ is the distance from point O to the element of area dA (see Fig. C-1). Inasmuch as $\rho^2 = x^2 + y^2$, where x and y are the rectangular coordinates of the element dA, we get the following expression for I_p:

$$I_p = \int \rho^2 \, dA = \int (x^2 + y^2) \, dA$$

Therefore, we obtain the relationship

$$I_p = I_x + I_y \qquad \text{(C-11)}$$

This equation shows that the polar moment of inertia with respect to an axis perpendicular to the plane of the figure at point O is equal to the sum of the moments of inertia with respect to any two perpendicular axes x and y passing through that same point and lying in the plane of the figure. For simplicity, we usually will refer to I_p as the polar moment of inertia with respect to point O.

The calculation of polar moments of inertia is greatly facilitated by using the **parallel-axis theorem for polar moments of inertia**. We can derive this theorem by referring again to Fig. C-10. Let us denote the polar moments of inertia about the origin O and the centroid C by I_{p_o} and I_{p_c}, respectively; then, we can write the following equations:

$$I_{p_o} = I_x + I_y \qquad I_{p_c} = I_{x_c} + I_{y_c} \tag{a}$$

(see Eq. C-11). Next, let us introduce the parallel-axis theorems derived in Section C.4 (see Eqs. C-9):

$$I_x = I_{x_c} + Ad_1^2 \qquad I_y = I_{y_c} + Ad_2^2 \tag{b}$$

Adding the last two equations, we get

$$I_x + I_y = I_{x_c} + I_{y_c} + A(d_1^2 + d_2^2)$$

Now substituting from Eqs. (a), and also noting that $d^2 = d_1^2 + d_2^2$ (see Fig. C-10), we obtain

$$I_{p_o} = I_{p_c} + Ad^2 \tag{C-12}$$

This equation represents the parallel-axis theorem for polar moments of inertia: *The polar moment of inertia of an area with respect to any point O in its plane is equal to the polar moment of inertia with respect to the centroid C plus the product of the area and the square of the distance between points O and C.*

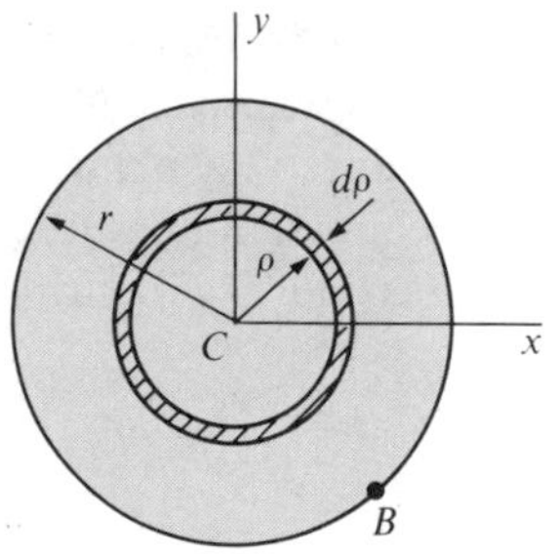

Fig. C-12 Polar moment of inertia of a circle

To illustrate the determination of polar moments of inertia, consider a circle of radius r (Fig. C-12). Let us take an element of area dA in the form of a thin ring of radius ρ and thickness $d\rho$; thus, $dA = 2\pi\rho\, d\rho$. Since every point in this element is at the same distance ρ from the center C of the circle, the polar moment of inertia of the element with respect to C is $\rho^2\, dA$, or $2\pi\rho^3\, d\rho$. To obtain the polar moment of inertia for the entire circle, we integrate as follows:

$$I_p = \int \rho^2\, dA = \int_0^r 2\pi\rho^3\, d\rho = \frac{\pi r^4}{2} \tag{c}$$

The polar moment of inertia with respect to any point B on the circumference can be obtained from the parallel-axis theorem:

$$I_{p_b} = I_{p_c} + Ad^2 = \frac{\pi r^4}{2} + \pi r^2(r^2) = \frac{3\pi r^4}{2} \tag{d}$$

As an incidental matter, we can easily obtain the moment of inertia of a circle with respect to a diameter by using Eq. (C-11). For the circle in Fig. C-12, we get

$$I_x = I_y = \frac{I_p}{2} = \frac{\pi r^4}{4} \tag{e}$$

A circle is obviously a special case in which the polar moment of inertia can easily be determined by integration. However, most of the shapes encountered in engineering work do not lend themselves to this technique. Instead, the polar moment of inertia is usually obtained by summing the moments of inertia about two perpendicular axes (Eq. C-11). The latter moments of inertia are found by the methods described in Sections C.3 and C.4.

C.6 PRODUCTS OF INERTIA

The product of inertia of a plane area is a property that is defined with respect to a set of perpendicular axes lying in the plane of the area. Thus, we refer again to the area A shown in Fig. C-1, and we define the **product of inertia** with respect to the x and y axes as follows:

$$I_{xy} = \int xy\,dA \qquad \text{(C-13)}$$

From this definition we see that each element of area dA is multiplied by the product of its coordinates. As a consequence, products of inertia may be positive, negative, or zero, depending upon the position of the xy axes with respect to the area. If the area lies entirely in the first quadrant of the axes (as in Fig. C-1), then the product of inertia is positive because every element dA has positive coordinates x and y. If the area lies entirely in the second quadrant, the product of inertia is negative because every element has a positive y coordinate and a negative x coordinate. Similarly, areas entirely within the third and fourth quadrants have positive and negative products of inertia, respectively. When the area is located in more than one quadrant, the sign of the product of inertia depends upon the distribution of the area within the quadrants.

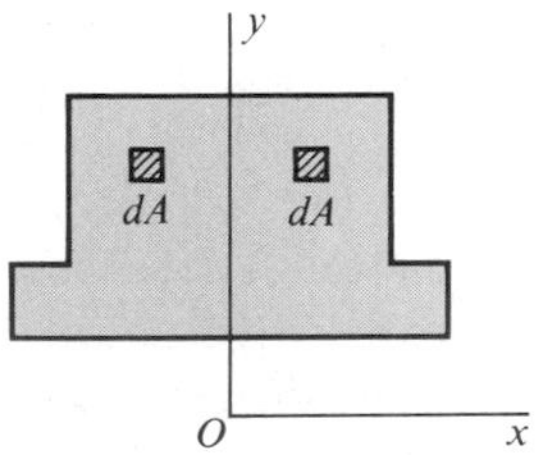

Fig. C-13 The product of inertia equals zero when one axis is an axis of symmetry.

An important special case arises when one of the axes is an axis of symmetry of the area. For illustration, consider the area shown in Fig. C-13, which is symmetric about the y axis. For every element dA having coordinates x and y, there exists an equal and symmetrically located element dA having the same y coordinate but an x coordinate of opposite sign. Therefore, the products $xy\,dA$ cancel each other and the integral in Eq. (C-13) vanishes. Thus, *the product of inertia of an area is zero with respect to any pair of axes in which at least one axis is an axis of symmetry.* For instance, the product of inertia I_{xy} equals zero for the areas shown in Figs. C-7, C-8, C-9a, C-9b, and C-12. In contrast, the product of inertia I_{xy} has a positive nonzero value for the areas shown in Figs. C-6, C-9c, and C-11. These conclusions are valid for products of inertia with respect to the particular xy axes shown in the figures; of course, the product of inertia of an area is changed when the axes are shifted to another position.

The products of inertia of an area with respect to parallel sets of axes are related by a parallel-axis theorem that is analogous to the corresponding theorems for moments of inertia and polar moments of inertia. To obtain this theorem, let us consider again the area shown in Fig. C-10, and let us denote by $I_{x_c y_c}$ the product of inertia with respect to the centroidal $x_c y_c$ axes. The product of inertia I_{xy} with respect to any set of axes, parallel to the $x_c y_c$ axes, is

$$
\begin{aligned}
I_{xy} &= \int (x + d_2)(y + d_1)\,dA \\
&= \int xy\,dA + d_1 \int x\,dA + d_2 \int y\,dA + d_1 d_2 \int dA
\end{aligned}
$$

in which d_1 and d_2 are the coordinates of the centroid C with respect to the xy axes (thus, d_1 and d_2 may have positive or negative values). The first integral in the last expression is the product of inertia $I_{x_c y_c}$ with respect to the centroidal axes; the second and third integrals vanish because they are the first moments of the area with respect to the centroidal axes; and the last integral is the area A. Hence the preceding equation reduces to

$$I_{xy} = I_{x_c y_c} + A d_1 d_2 \tag{C-14}$$

This equation represents the **parallel-axis theorem** for products of inertia: *The product of inertia of an area with respect to any pair of axes in its plane is equal to the product of inertia with respect to parallel centroidal axes plus the product of the area and the coordinates of the centroid with respect to the pair of axes.*

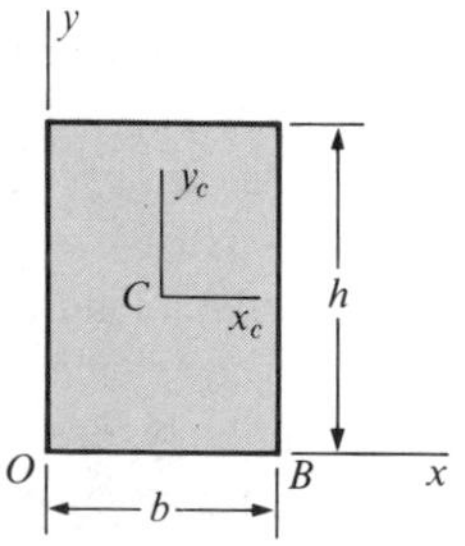

Fig. C-14 Parallel-axis theorem for products of inertia

To demonstrate the use of the parallel-axis theorem, let us determine the product of inertia of a rectangle with respect to xy axes with origin at a corner (Fig. C-14). The product of inertia with respect to the centroidal axes x_c and y_c is zero because they are axes of symmetry. Also, the coordinates of the centroid with respect to the xy axes are

$$d_1 = \frac{h}{2} \qquad d_2 = \frac{b}{2}$$

Substituting into Eq. (C-14), we obtain

$$I_{xy} = I_{x_c y_c} + A d_1 d_2 = 0 + bh\left(\frac{h}{2}\right)\left(\frac{b}{2}\right) = \frac{b^2 h^2}{4} \tag{a}$$

The product of inertia is positive because the entire area lies in the first quadrant. If the xy axes are translated horizontally so that the origin moves to point B at the lower right-hand corner of the rectangle (Fig. C-14), the entire area lies in the second quadrant and the product of inertia becomes $-b^2h^2/4$.

Example 1

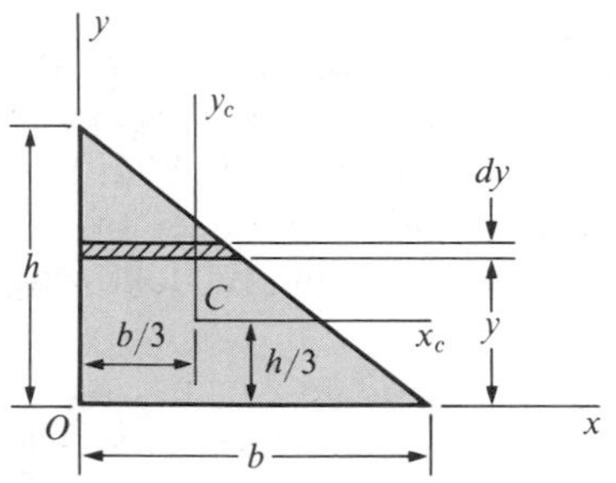

Fig. C-15 Example 1

Determine the product of inertia I_{xy} of a right triangle (Fig. C-15) with respect to xy axes with origin at the 90° vertex, and then determine the product of inertia $I_{x_c y_c}$ with respect to parallel centroidal axes.

Let us consider an element of area dA in the form of a thin strip of height equal to dy and width equal to

$$\frac{(h-y)b}{h}$$

(see Fig. C-15). The product of inertia of this strip (which may be treated as a rectangle) with respect to axes through its own centroid and parallel to the xy axes is zero (from symmetry). Therefore, its product of inertia dI_{xy} with respect to the xy axes (from the parallel-axis theorem) is

$$dI_{xy} = dA d_1 d_2 = \left[\frac{(h-y)b\,dy}{h}\right][y]\left[\frac{(h-y)b}{2h}\right] = \frac{b^2}{2h^2}(h-y)^2 y\,dy$$

The product of inertia I_{xy} of the triangle is obtained by integration:

$$I_{xy} = \int dI_{xy} = \frac{b^2}{2h^2}\int_0^h (h-y)^2 y\,dy = \frac{b^2h^2}{24} \qquad \text{(b)}$$

The product of inertia with respect to the parallel centroidal axes may now be determined from the parallel-axis theorem:

$$I_{x_c y_c} = I_{xy} - A d_1 d_2 = \frac{b^2h^2}{24} - \frac{bh}{2}\left(\frac{h}{3}\right)\left(\frac{b}{3}\right) = -\frac{b^2h^2}{72} \qquad \text{(c)}$$

These results are given in Cases 7 and 6, respectively, of Appendix D.

Example 2

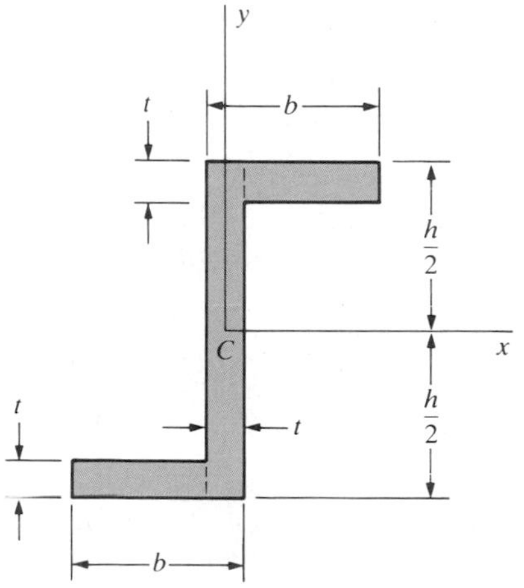

Fig. C-16 Example 2

Determine the product of inertia I_{xy} of the Z-section shown in Fig. C-16.

We divide the section into three rectangles, a web rectangle of thickness t and height h and two flange rectangles of width $b-t$ and thickness t. The product of inertia of the web rectangle is zero (from symmetry). The product of inertia $(I_{xy})_1$ of the upper flange rectangle is determined by using the parallel-axis theorem:

$$(I_{xy})_1 = I_{x_c y_c} + A d_1 d_2 = 0 + (b-t)(t)\left(\frac{h}{2} - \frac{t}{2}\right)\left(\frac{b}{2}\right) = \frac{bt}{4}(h-t)(b-t)$$

The product of inertia of the lower flange rectangle is the same; hence the product of inertia of the entire Z-section is

$$I_{xy} = \frac{bt}{2}(h-t)(b-t) \qquad \text{(d)}$$

Note that this product of inertia is positive because the flanges lie in the first and third quadrants.

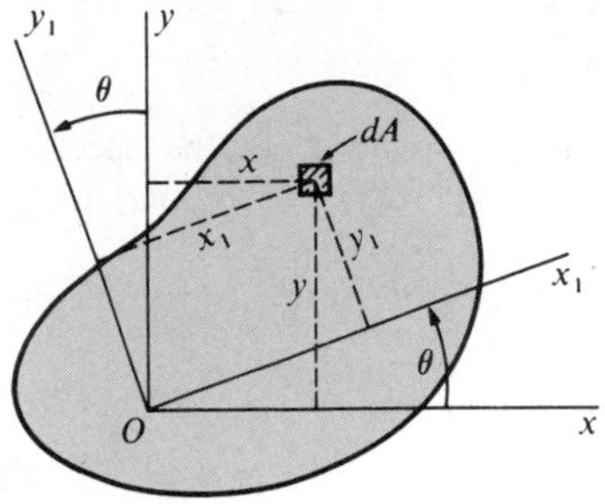

Fig. C-17 Rotation of axes

C.7 ROTATION OF AXES

The moments of inertia of a plane area depend upon the location of the origin and the orientation of the reference axes. For a given origin, the moments and product of inertia vary as the axes are rotated about that origin. The manner in which they vary, and the magnitudes of the maximum and minimum values, are discussed in this and the following section.

Let us consider the plane area shown in Fig. C-17, and let us assume that the xy axes are a pair of arbitrarily located reference axes. The moments and products of inertia with respect to these axes are

$$I_x = \int y^2 \, dA \qquad I_y = \int x^2 \, dA \qquad I_{xy} = \int xy \, dA \tag{a}$$

in which x and y are the coordinates of an element of area dA. The x_1y_1 axes have the same origin but are rotated through a counterclockwise angle θ with respect to the xy axes. The moments and product of inertia with respect to the x_1y_1 axes are denoted by I_{x_1}, I_{y_1}, and $I_{x_1y_1}$, respectively. To obtain these quantities, we need the coordinates x_1 and y_1 of the element dA; these coordinates may be expressed in terms of the xy coordinates and the angle θ, as follows:

$$x_1 = x\cos\theta + y\sin\theta \qquad y_1 = y\cos\theta - x\sin\theta \tag{C-15}$$

Then the moment of inertia with respect to the x_1 axis is

$$I_{x_1} = \int y_1^2 \, dA = \int (y\cos\theta - x\sin\theta)^2 \, dA$$

$$= \cos^2\theta \int y^2 \, dA + \sin^2\theta \int x^2 \, dA - 2\sin\theta\cos\theta \int xy \, dA$$

or, by using Eqs. (a),

$$I_{x_1} = I_x\cos^2\theta + I_y\sin^2\theta - 2I_{xy}\sin\theta\cos\theta \tag{b}$$

Now we introduce the following trigonometric identities:

$$\cos^2\theta = \frac{1}{2}(1 + \cos 2\theta) \qquad \sin^2\theta = \frac{1}{2}(1 - \cos 2\theta)$$

$$2\sin\theta\cos\theta = \sin 2\theta$$

Then Eq. (b) becomes

$$I_{x_1} = \frac{I_x + I_y}{2} + \frac{I_x - I_y}{2}\cos 2\theta - I_{xy}\sin 2\theta \tag{C-16}$$

In a similar manner, we can obtain the product of inertia with respect to the x_1y_1 axes:

$$I_{x_1y_1} = \int x_1 y_1 \, dA = \int (x\cos\theta + y\sin\theta)(y\cos\theta - x\sin\theta)\, dA$$
$$= (I_x - I_y)\sin\theta\cos\theta + I_{xy}(\cos^2\theta - \sin^2\theta)$$

Again using the trigonometric identities, we obtain

$$I_{x_1y_1} = \frac{I_x - I_y}{2}\sin 2\theta + I_{xy}\cos 2\theta \tag{C-17}$$

Thus, Eqs. (C-16) and (C-17) give the moment of inertia I_{x_1} and the product of inertia $I_{x_1y_1}$ with respect to the rotated axes in terms of the moments and product of inertia for the original axes. These equations are called the **transformation equations for moments and products of inertia.** Note that they have the same form as the transformation equations for plane stress (Eqs. 6-4, Section 6.2), with I_{x_1} corresponding to σ_{x_1}, $-I_{x_1y_1}$ corresponding to $\tau_{x_1y_1}$, I_x corresponding to σ_x, I_y corresponding to σ_y, and $-I_{xy}$ corresponding to τ_{xy}. Therefore, we can also analyze moments and products of inertia by the use of Mohr's circle (see Section 6.4).

The moment of inertia I_{y_1} may be obtained by the same procedure that we used for I_{x_1} and $I_{x_1y_1}$ or by replacing θ by $\theta + 90°$ in Eq. (C-16); the result is

$$I_{y_1} = \frac{I_x + I_y}{2} - \frac{I_x - I_y}{2}\cos 2\theta + I_{xy}\sin 2\theta \tag{C-18}$$

Taking the sum of I_{x_1} and I_{y_1}, we find

$$I_{x_1} + I_{y_1} = I_x + I_y \tag{C-19}$$

This equation shows that the sum of the moments of inertia with respect to a pair of axes remains constant as the axes are rotated about the origin. This sum is the polar moment of inertia of the area with respect to the origin.

C.8 PRINCIPAL AXES

The transformation equations for moments and products of inertia of an area were derived in the preceding section (see Eqs. C-16 and C-17). These equations show how the inertia quantities vary as the angle of rotation θ varies. Of special interest are the maximum and minimum values of the moment of inertia; such values are known as **principal moments of inertia**, and the corresponding axes are known as **principal axes.**

To find the values of the angle θ that make the moment of inertia I_{x_1} a maximum or a minimum, we take the derivative with respect to θ of the expression on the right-hand side of Eq. (C-16) and set it equal to zero:

$$(I_x - I_y)\sin 2\theta + 2I_{xy}\cos 2\theta = 0 \qquad \text{(a)}$$

Solving for θ from this equation, we get

$$\tan 2\theta_p = -\frac{2I_{xy}}{I_x - I_y} \qquad \text{(C-20)}$$

in which θ_p denotes the angle defining a principal axis. This same result is obtained if we take the derivative of I_{y_1} (Eq. C-18). Two values of the angle $2\theta_p$ in the range from 0 to 360° are obtained from Eq. (C-20); these values differ by 180°. The corresponding two values of θ_p differ by 90° and define the two perpendicular principal axes, of which one corresponds to the maximum moment of inertia and the other corresponds to the minimum moment of inertia.

Now let us examine the variation in the product of inertia $I_{x_1y_1}$ as θ varies (see Eq. C-17). If $\theta = 0$, we get $I_{x_1y_1} = I_{xy}$, as expected. If $\theta = 90°$, we obtain $I_{x_1y_1} = -I_{xy}$. Thus, during a 90° rotation the product of inertia changes sign, which means that for some orientation of the axes, the product of inertia must vanish. To determine this orientation, we set $I_{x_1y_1}$ (Eq. C-17) equal to zero:

$$(I_x - I_y)\sin 2\theta + 2I_{xy}\cos 2\theta = 0$$

This equation is the same as Eq. (a), which defines the angle θ_p to the principal axes. Therefore, we conclude that *the product of inertia is zero for the principal axes.*

In Section C.6 we showed that the product of inertia of an area with respect to a pair of axes equals zero if at least one of the axes is an axis of symmetry. It follows that if an area has an axis of symmetry, that axis and any axis perpendicular to it constitute a set of principal axes.

The preceding observations may be summarized as follows: (1) principal axes through an origin O are a pair of orthogonal axes for which the moments of inertia are a maximum and a minimum; (2) the orientation of the principal axes is given by the angle θ_p obtained from Eq. (C-20); (3) the product of inertia is zero for principal axes; and (4) an axis of symmetry is always a principal axis.

PRINCIPAL POINTS

Now consider a pair of principal axes with origin at a given point O. If there exists a different pair of principal axes through that same point, then it follows that every pair of axes through that point is a set of principal axes. Furthermore, the moment of inertia must be constant as the angle θ is varied. These conclusions follow from the nature of the transformation

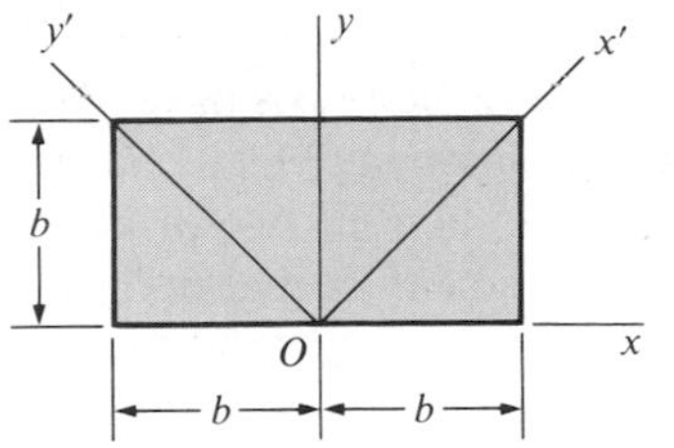

Fig. C-18 Illustration in which every axis through point O is a principal axis

equation for I_{x_1} (Eq. C-16). Because this equation contains trigonometric functions of the angle 2θ, there is one maximum value and one minimum value of I_{x_1} as 2θ varies through a range of 360°, or as θ varies through a range of 180°. If a second maximum exists, then the only possibility is that I_{x_1} remains constant, which means that every pair of axes is a set of principal axes and all moments of inertia are the same. A point so located that every axis through the point is a principal axis, and hence the moments of inertia are the same for all axes through the point, is called a **principal point**.

An illustration of this situation is the rectangle of width $2b$ and height b shown in Fig. C-18. The xy axes, with origin at point O, are principal axes of the rectangle because the y axis is an axis of symmetry. The $x'y'$ axes, with the same origin, are also principal axes because the product of inertia $I_{x'y'}$ equals zero (because the triangles are symmetrically located with respect to the x' and y' axes). It follows that every pair of axes through O is a set of principal axes and every moment of inertia is the same (and equal to $2b^4/3$). Therefore, point O is a principal point for the rectangle. (A second principal point is located where the y axis intersects the upper side of the rectangle.)

A useful corollary of the proposition described in the preceding two paragraphs applies to axes through the centroid C of an area. Consider an area having two *different* pairs of centroidal axes such that at least one axis in each pair is an axis of symmetry. In other words, there exist two different axes of symmetry that are not perpendicular to each other. Then it follows that the centroid is a principal point. Two examples, a square and an equilateral triangle, are shown in Fig. C-19. In each case the xy axes are principal axes because at least one of the two axes is an axis of symmetry. In addition, a second pair of axes (the $x'y'$ axes) has at least one axis of symmetry. It follows that both the xy and $x'y'$ axes are principal axes, and therefore every axis through C is a principal axis and has the same moment of inertia.

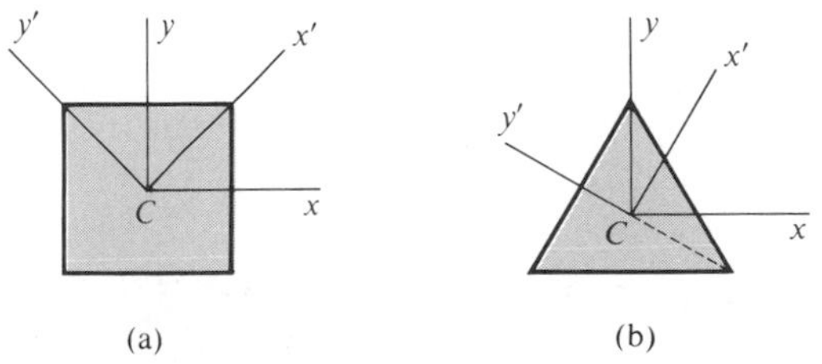

Fig. C-19 Areas for which every centroidal axis is a principal axis and the centroid C is a principal point: (a) a square, and (b) an equilateral triangle

If an area has three different axes of symmetry, the conditions described in the preceding paragraph are automatically fulfilled. Therefore, we conclude that if an area has three or more axes of symmetry, the centroid is a principal point. These conditions are fulfilled for a circle, for regular polygons (equilateral triangle, square, regular pentagon, regular hexagon, and so on), and for many other symmetric shapes.

In general, every plane area has two principal points. These points lie equidistant from the centroid on the principal centroidal axis having the larger principal moment of inertia. An exception occurs when the two principal centroidal moments of inertia are equal; then the two principal points merge at the centroid, which becomes the sole principal point.

PRINCIPAL MOMENTS OF INERTIA

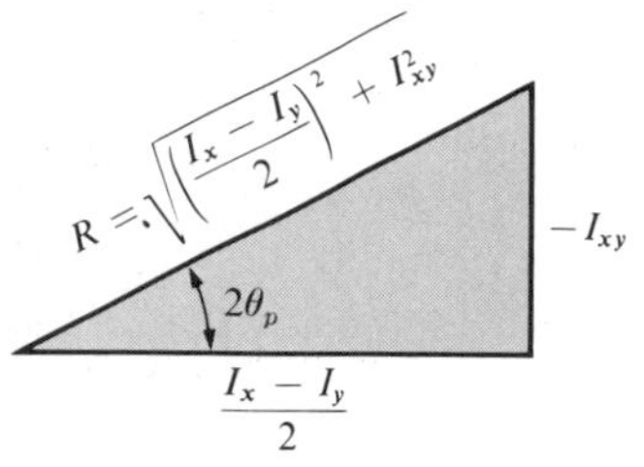

Fig. C-20

Let us now consider the determination of the principal moments of inertia, assuming that I_x, I_y, and I_{xy} are known. One method is to determine the two values of θ_p (differing by 90°) from Eq. (C-20) and then substitute these values into Eq. (C-16) for I_{x_1}. The resulting two values are the principal moments of inertia, denoted by I_1 and I_2. The advantage of this method is that we know which of the two principal angles θ_p corresponds to each principal moment of inertia.

It is also possible to obtain general formulas for the principal moments of inertia. We note from Eq. (C-20) and Fig. C-20 that

$$\cos 2\theta_p = \frac{I_x - I_y}{2R} \qquad \sin 2\theta_p = \frac{-I_{xy}}{R} \qquad \text{(C-21a,b)}$$

in which

$$R = \sqrt{\left(\frac{I_x - I_y}{2}\right)^2 + I_{xy}^2} \qquad \text{(C-22)}$$

When evaluating R, we always take the positive square root. Now we substitute the expressions for $\cos 2\theta_p$ and $\sin 2\theta_p$ into Eq. (C-16) and obtain the algebraically larger of the two principal moments of inertia, denoted by I_1:

$$I_1 = \frac{I_x + I_y}{2} + \sqrt{\left(\frac{I_x - I_y}{2}\right)^2 + I_{xy}^2} \qquad \text{(C-23a)}$$

The smaller principal moment of inertia, denoted by I_2, may be obtained from the equation

$$I_1 + I_2 = I_x + I_y$$

(see Eq. C-19). Substituting the expression for I_1 into this equation and solving for I_2, we get

$$I_2 = \frac{I_x + I_y}{2} - \sqrt{\left(\frac{I_x - I_y}{2}\right)^2 + I_{xy}^2} \qquad \text{(C-23b)}$$

Equations (C-23) provide a convenient way to calculate the principal moments of inertia.

Example

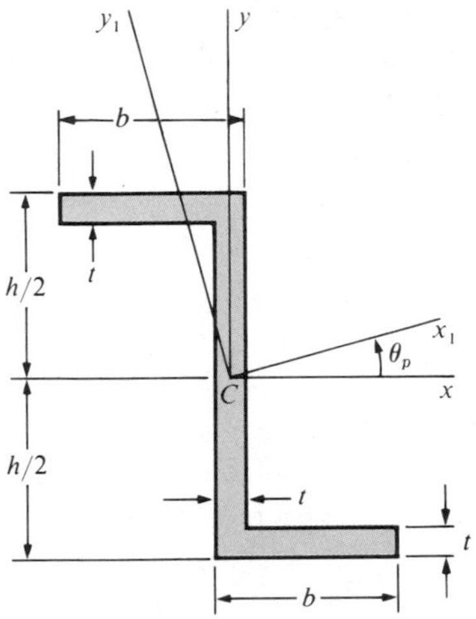

Fig. C-21 Example. Principal axes of a Z-section

Determine the orientations of the principal centroidal axes and the magnitudes of the principal centroidal moments of inertia for the cross-sectional area of the Z-section shown in Fig. C-21. Use the following numerical data: $h = 8$ in., $b = 3.5$ in., and $t = 0.55$ in.

Let us use the xy axes (Fig. C-21) as the reference axes through the centroid C. The moments and product of inertia with respect to these axes can be obtained by dividing the area into three rectangles and using the parallel-axis theorems. The results of such calculations are as follows:

$$I_x = 68.57 \text{ in.}^4 \qquad I_y = 12.40 \text{ in.}^4 \qquad I_{xy} = -21.15 \text{ in.}^4$$

Substituting these values into the equation for the angle θ_p (Eq. C-20), we get

$$\tan 2\theta_p = -\frac{2I_{xy}}{I_x - I_y} = 0.7532 \qquad 2\theta_p = 37.0^\circ \text{ and } 217.0^\circ$$

Thus, the two values of θ_p are

$$\theta_p = 18.5^\circ \text{ and } 108.5^\circ$$

Using these values of θ_p in the transformation equation for I_{x_1} (Eq. C-16), we find $I_{x_1} = 75.65$ in.^{4} and 5.33 in.^{4}, respectively. These same values are obtained if we substitute into Eqs. (C-23). Thus, the principal moments of inertia and the angles to the corresponding axes are:

$$I_1 = 75.6 \text{ in.}^4 \qquad \theta_{p_1} = 18.5^\circ$$

$$I_2 = 5.3 \text{ in.}^4 \qquad \theta_{p_2} = 108.5^\circ$$

The principal axes are shown in Fig. C-21 as the x_1y_1 axes.

PROBLEMS / APPENDIX C

The problems for Section C.1 are to be solved by integration.

C.1-1 Determine the distances $\bar{x}$ and $\bar{y}$ to the centroid C of a right triangle having base b and altitude h (see Case 6, Appendix D).

C.1-2 Determine the distance $\bar{y}$ to the centroid C of a trapezoid having bases a and b and altitude h (see Case 8, Appendix D).

C.1-3 Determine the distance $\bar{y}$ to the centroid C of a semicircle of radius r (see Case 11, Appendix D).

C.1-4 Determine the distances $\bar{x}$ and $\bar{y}$ to the centroid C of a parabolic spandrel of base b and height h (see Case 19, Appendix D).

C.1-5 Determine the distances $\bar{x}$ and $\bar{y}$ to the centroid C of a semisegment of nth degree having base b and height h (see Case 20, Appendix D).

The problems for Section C.2 are to be solved using the formulas for composite areas.

C.2-1 Determine the distance $\bar{y}$ to the centroid C of a trapezoid having bases a and b and altitude h (see Case 8, Appendix D) by dividing the trapezoid into two triangles.

C.2-2 One quarter of a square of side a is removed (see figure on next page). What are the coordinates $\bar{x}$ and $\bar{y}$ of the centroid C of the remaining area?

C.2-3 Calculate the distance $\bar{y}$ to the centroid C of the channel section shown in the figure on the next page if $a = 6$ in., $b = 1$ in., and $c = 2$ in.

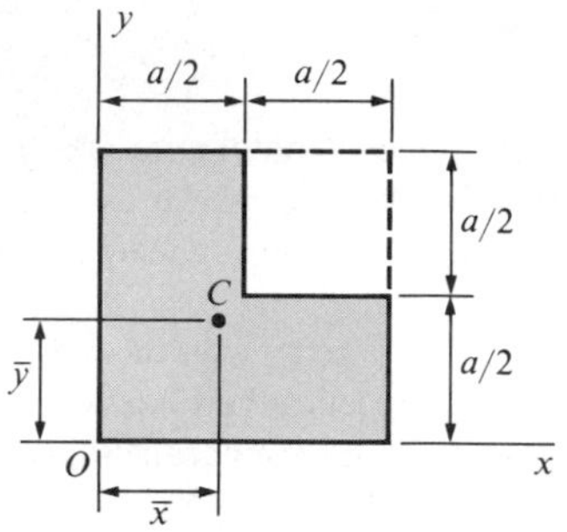

Probs. C.2-2 and C.4-2

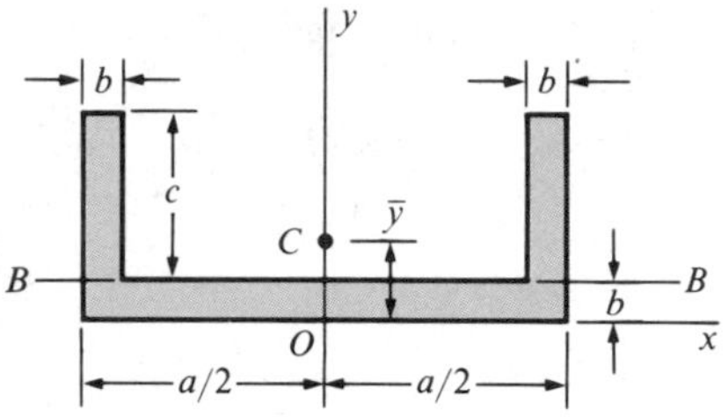

Probs. C.2-3, C.2-4, and C.4-3

C.2-4 What must be the relationship between the dimensions a, b, and c of the channel section shown in the figure for the preceding problem in order that the centroid C will lie on line BB?

C.2-5 The cross section of a beam constructed of a W 24 × 162 section with an 8 × $\frac{3}{4}$ in. cover plate welded to the top flange is shown in the figure. Determine the distance $\bar{y}$ from the base of the beam to the centroid C of the cross-sectional area.

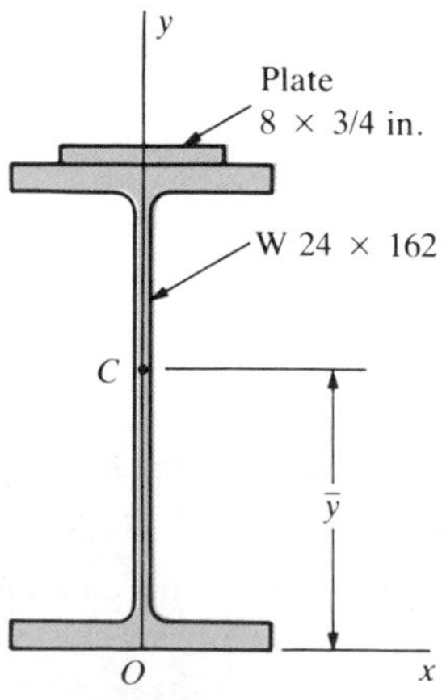

Probs. C.2-5 and C.4-5

C.2-6 Determine the distance $\bar{y}$ to the centroid C of the composite area shown in the figure.

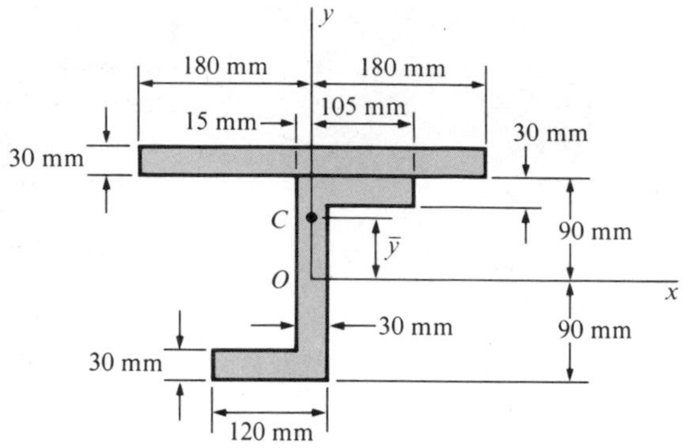

Probs. C.2-6, C.4-6, and C.6-6

C.2-7 Determine the coordinates $\bar{x}$ and $\bar{y}$ of the centroid C of the L-shaped area shown in the figure.

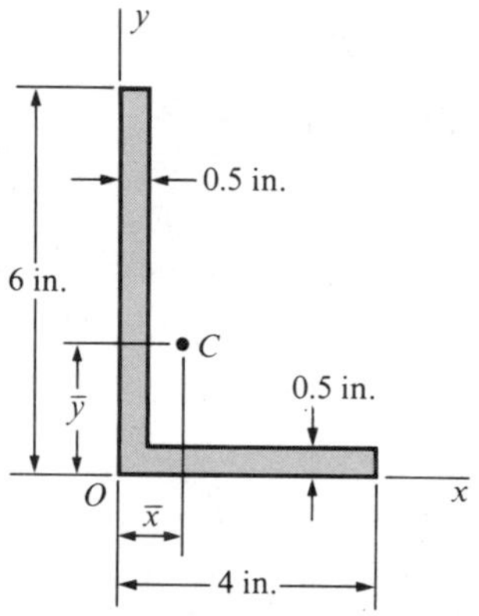

Probs. C.2-7, C.3-7, C.4-7, and C.6-7

C.2-8 Determine the coordinates $\bar{x}$ and $\bar{y}$ of the centroid C of the area shown in the figure.

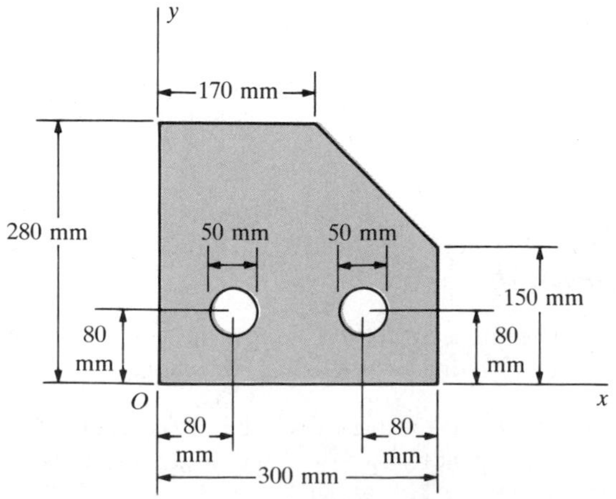

Prob. C.2-8

Problems C.3-1 through C.3-4 are to be solved by integration.

C.3-1 Determine the moment of inertia I_x of a triangle of base b and altitude h with respect to its base (see Case 4, Appendix D).

C.3-2 Determine the moment of inertia I_{BB} of a trapezoid having bases a and b and altitude h with respect to its base (see Case 8, Appendix D).

C.3-3 Determine the moment of inertia I_x of a parabolic spandrel of base b and height h with respect to its base (see Case 19, Appendix D).

C.3-4 Determine the moment of inertia I_x of a circle of radius r with respect to a diameter (see Case 9, Appendix D).

Problems C.3-5 through C.3-9 are to be solved by considering the area to be a composite area.

C.3-5 Determine the moment of inertia I_{BB} of a rectangle having sides of lengths b and h with respect to a diagonal of the rectangle (see Case 2, Appendix D).

C.3-6 Calculate the moment of inertia I_x for the composite circular area shown in the figure. The origin of the axes is at the center of the concentric circles, and the three diameters are 50, 100, and 150 mm.

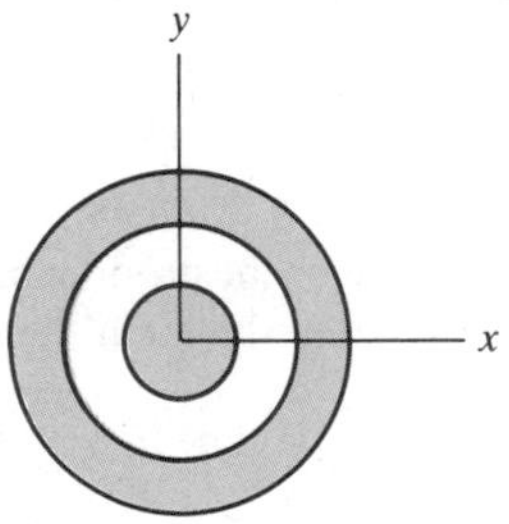

Prob. C.3-6

C.3-7 Calculate the moments of inertia I_x and I_y with respect to the x and y axes, respectively, for the L-shaped area shown in the figure for Prob. C.2-7.

C.3-8 A semicircular area of radius 150 mm has a rectangular cutout of dimensions 50 mm × 100 mm (see figure). Calculate the moments of inertia I_x and I_y with respect to the x and y axes, respectively. Then calculate the corresponding radii of gyration r_x and r_y.

C.3-9 Calculate the moments of inertia I_1 and I_2 of a W 16 × 100 section using the cross-sectional dimensions given in Table E-1, Appendix E. (Disregard the cross-sectional areas of the fillets.) Also, calculate the corresponding radii of gyration r_1 and r_2, respectively.

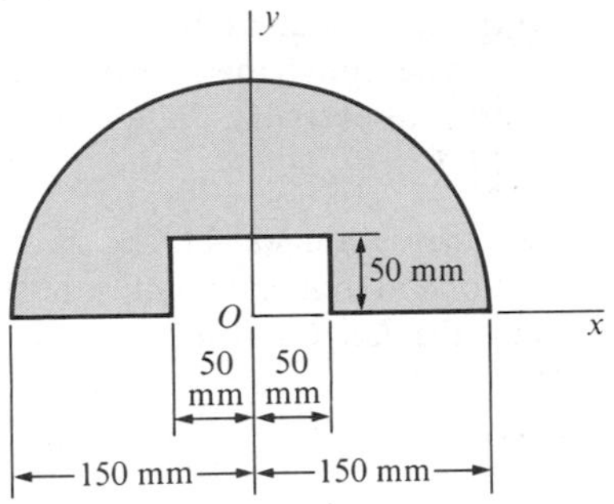

Prob. C.3-8

The problems for Section C.4 are to be solved by using the parallel-axis theorem.

C.4-1 Calculate the moment of inertia I_b of a W 12 × 50 section with respect to its base. (Use data from Table E-1, Appendix E.)

C.4-2 Determine the moment of inertia I_c with respect to an axis through the centroid C parallel to one of the sides for the figure described in Prob. C.2-2.

C.4-3 For the channel section described in Prob. C.2-3, calculate the centroidal moments of inertia I_{x_c} and I_{y_c} with respect to axes through the centroid C such that the x_c axis is parallel to the x axis and the y_c axis coincides with the y axis.

C.4-4 The moment of inertia with respect to axis 1-1 of the scalene triangle shown in the figure is 90×10^3 mm^4. Calculate its moment of inertia I_2 with respect to axis 2-2.

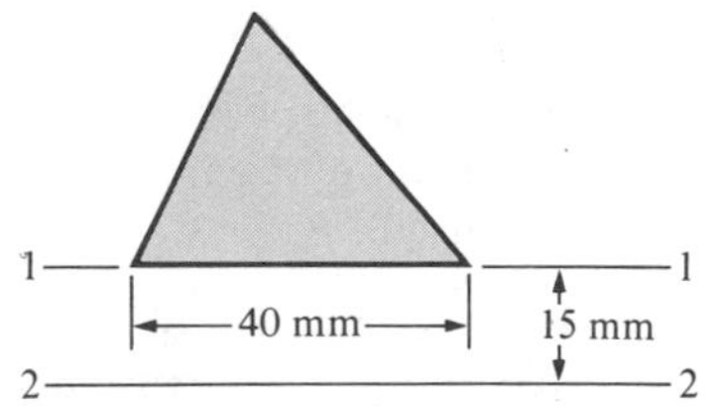

Prob. C.4-4

C.4-5 For the beam cross section described in Prob. C.2-5, calculate the centroidal moments of inertia I_{x_c} and I_{y_c} with respect to axes through the centroid C such that the x_c axis is parallel to the x axis and the y_c axis coincides with the y axis.

C.4-6 Calculate the moment of inertia I_c with respect to an axis through the centroid C parallel to the x axis for the composite area shown in Prob. C.2-6.

C.4-7 Calculate the centroidal moments of inertia I_{x_c} and I_{y_c} with respect to axes through the centroid C and parallel to the x and y axes, respectively, for the L-shaped area shown in Prob. C.2-7.

C.4-8 The wide-flange beam section shown in the figure has a total height of 250 mm and a constant thickness of 15 mm. Determine the flange width b if it is required that the centroidal moments of inertia I_x and I_y be in the ratio 3 to 1, respectively.

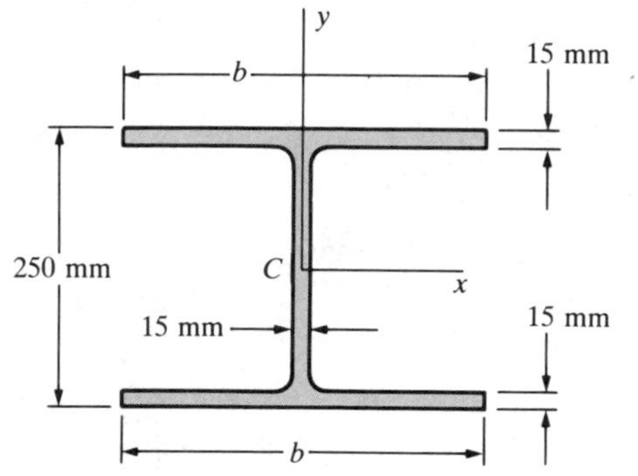

Prob. C.4-8

C.5-1 Determine the polar moment of inertia I_p of an isosceles triangle of base b and altitude h with respect to its apex (see Case 5, Appendix D).

C.5-2 Determine the polar moment of inertia I_{p_c} with respect to the centroid C for a circular sector (see Case 14, Appendix D).

C.5-3 Determine the polar moment of inertia I_p of a W 8 × 21 section with respect to one of its outermost corners.

C.5-4 Obtain a formula for the polar moment of inertia I_p with respect to the midpoint of the hypotenuse for a right triangle of base b and height h (see Case 6, Appendix D).

C.5-5 Determine the polar moment of inertia I_{p_c} with respect to the centroid C for a quarter-circular spandrel (see Case 13, Appendix D).

***C.5-6** Calculate the polar moment of inertia I_{p_c} with respect to the centroid C for the hollow section of constant thickness shown in the figure.

C.6-1 Using integration, determine the product of inertia I_{xy} for the parabolic semisegment shown in Fig. C-5 (see also Case 18, Appendix D).

C.6-2 Using integration, determine the product of inertia I_{xy} for the quarter-circular spandrel shown in Case 13, Appendix D.

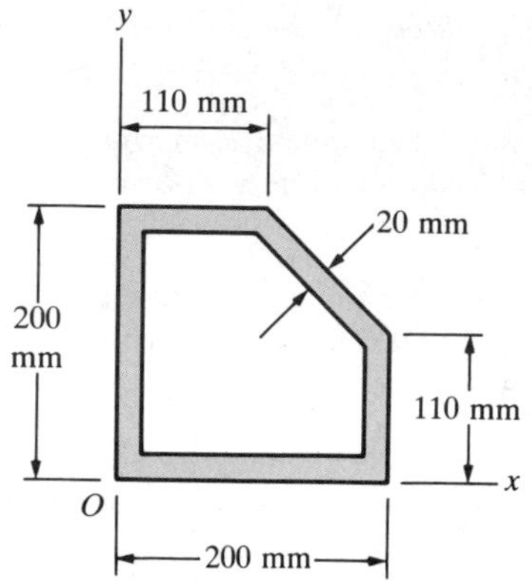

Probs. C.5-6 and C.6-8

C.6-3 Find the relationship between the radius r and the distance b for the composite area shown in the figure in order that the product of inertia I_{xy} will be zero.

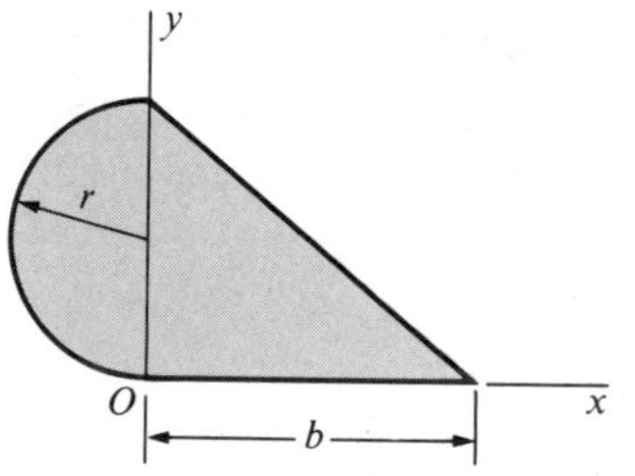

Prob. C.6-3

C.6-4 Obtain a formula for the product of inertia I_{xy} of the symmetrical L-shaped area shown in the figure.

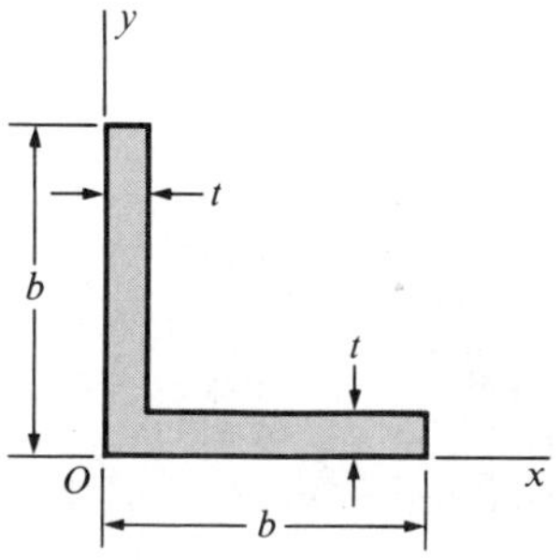

Prob. C.6-4

C.6-5 Calculate the product of inertia I_{12} with respect to the centroidal axes 1-1 and 2-2 for an L 6 × 6 × 1 in. angle section (see Table E-4, Appendix E). (Disregard the cross-sectional areas of the fillet and rounded corners.)

C.6-6 Calculate the product of inertia I_{xy} for the composite area shown in Prob. C.2-6.

C.6-7 Determine the product of inertia $I_{x_c y_c}$ with respect to centroidal axes x_c and y_c parallel to the x and y axes, respectively, for the L-shaped area shown in Prob. C.2-7.

***C.6-8** Calculate the product of inertia I_{xy} for the hollow section of constant thickness shown in the figure for Prob. C.5-6.

The problems for Section C.7 are to be solved by using the transformation equations for moments and products of inertia.

C.7-1 Determine the moments of inertia I_{x_1} and I_{y_1} and the product of inertia $I_{x_1 y_1}$ for the square shown in the figure. Note that the $x_1 y_1$ axes are centroidal axes rotated through an angle θ with respect to the xy axes.

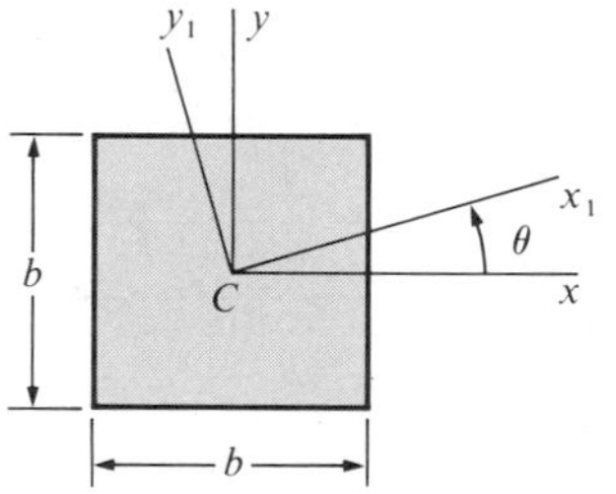

Prob. C.7-1

C.7-2 Determine the moments and product of inertia with respect to the $x_1 y_1$ axes for the rectangle shown in the figure. Note that the x_1 axis is a diagonal of the rectangle.

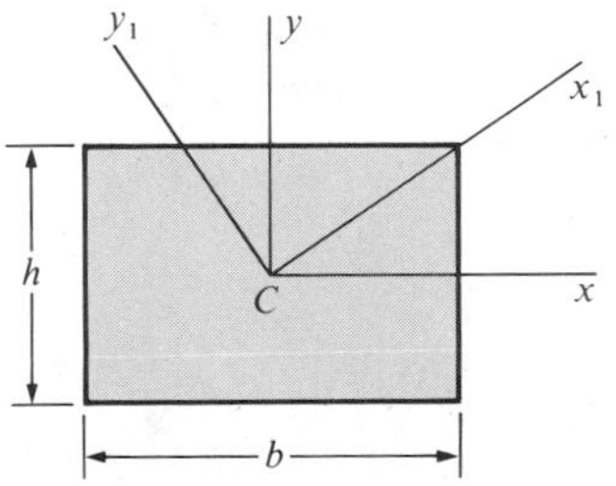

Prob. C.7-2

C.7-3 Calculate the moment of inertia I_d for a W 12 × 50 wide-flange section with respect to a diagonal axis passing through the centroid and two outside corners of the flanges. Use the dimensions and properties given in Table E-1.

C.7-4 Calculate the moments of inertia I_{x_1} and I_{y_1} and the product of inertia $I_{x_1 y_1}$ for the L-shaped area shown in the figure if $a = 150$ mm, $b = 100$ mm, $t = 15$ mm, and $\theta = 30°$.

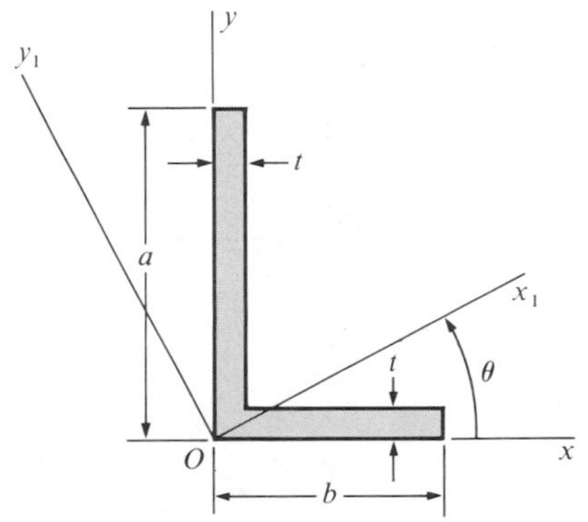

Probs. C.7-4 and C.8-4

C.7-5 Calculate the moments of inertia I_{x_1} and I_{y_1} and the product of inertia $I_{x_1 y_1}$ with respect to the $x_1 y_1$ axes for the Z-section shown in the figure if $b = 3$ in., $h = 4$ in., $t = 0.5$ in., and $\theta = 60°$.

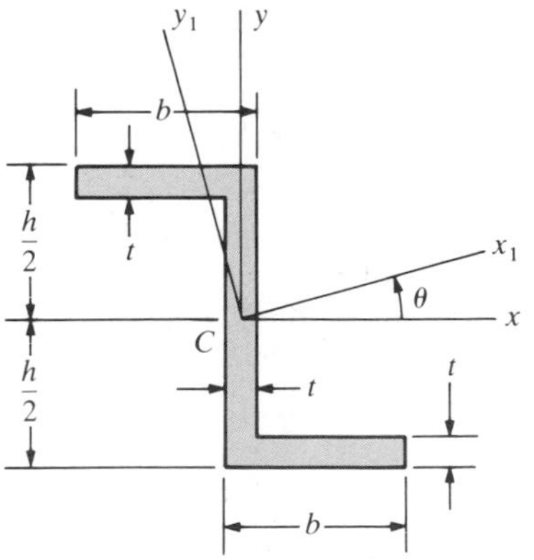

Probs. C.7-5, C.7-6, C.8-5, and C.8-6

C.7-6 Solve the preceding problem if $b = 80$ mm, $h = 120$ mm, $t = 12$ mm, and $\theta = 30°$.

C.8-1 An ellipse with major axis of length $2a$ and minor axis of length $2b$ is shown in the figure on the next page. (a) Determine the distance c from the centroid C of the ellipse to the principal points P on the minor axis (y axis). (b) For what ratio a/b do the principal points lie on the circumference of the ellipse? (c) For what ratios do they lie inside the ellipse?

C.8-2 Demonstrate that the two points P_1 and P_2, located as shown in the figure on the next page, are the principal points of the isosceles right triangle.

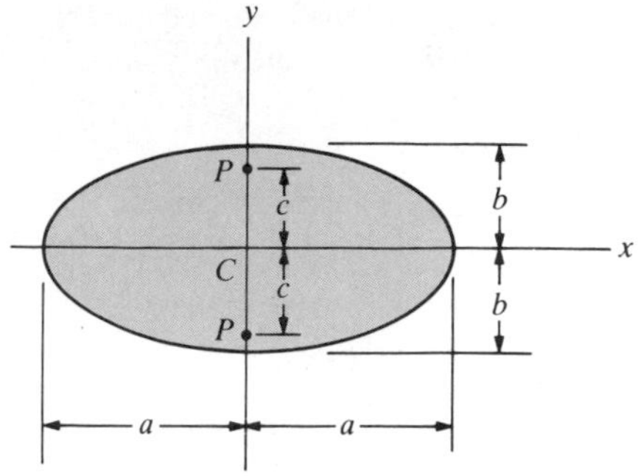

Prob. C.8-1

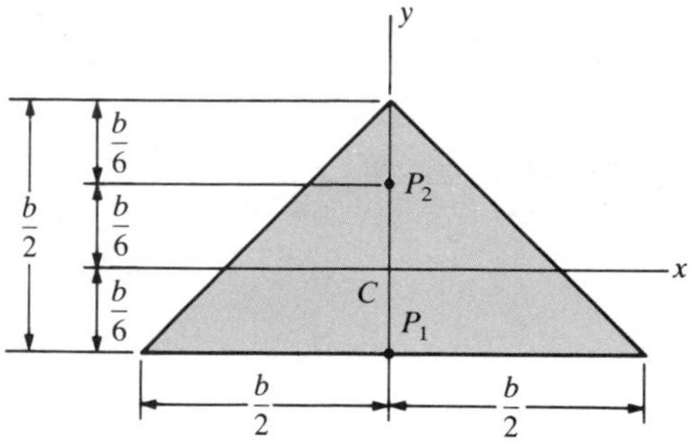

Prob. C.8-2

C.8-3 Determine the angles θ_{p_1} and θ_{p_2} defining the orientations of the principal axes through the origin O for the right triangle shown in Fig. C-15 if $b = 6$ in. and $h = 8$ in. Also, calculate the corresponding principal moments of inertia I_1 and I_2.

C.8-4 Determine the angles θ_{p_1} and θ_{p_2} defining the orientations of the principal axes through the origin O and the corresponding principal moments of inertia I_1 and I_2 for the L-shaped area described in Prob. C.7-4 ($a = 150$ mm, $b = 100$ mm, and $t = 15$ mm).

C.8-5 Determine the angles θ_{p_1} and θ_{p_2} defining the orientations of the principal axes through the centroid C and the corresponding principal centroidal moments of inertia I_1 and I_2 for the Z-section described in Prob. C.7-5 ($b = 3$ in., $h = 4$ in., and $t = 0.5$ in.).

C.8-6 Solve the preceding problem for the Z-section described in Prob. C.7-6 ($b = 80$ mm, $h = 120$ mm, and $t = 12$ mm).

C.8-7 Determine the angles θ_{p_1} and θ_{p_2} defining the orientations of the principal axes through the centroid C for the right triangle shown in Fig. C-15 if $h = 2b$. Also, determine the corresponding principal centroidal moments of inertia I_1 and I_2.

C.8-8 Determine the angles θ_{p_1} and θ_{p_2} defining the orientations of the principal axes through the centroid C and the corresponding principal centroidal moments of inertia I_1 and I_2 for the Z-section shown in Fig. C-16 if $b = 80$ mm, $h = 150$ mm, and $t = 10$ mm.

C.8-9 Solve the preceding problem if $b = 3$ in., $h = 6$ in., and $t = \frac{3}{8}$ in.

C.8-10 Determine the angles θ_{p_1} and θ_{p_2} defining the orientations of the principal centroidal axes and the corresponding principal moments of inertia I_1 and I_2 for the L-shaped area shown in the figure if $a = 80$ mm, $b = 150$ mm, and $t = 16$ mm.

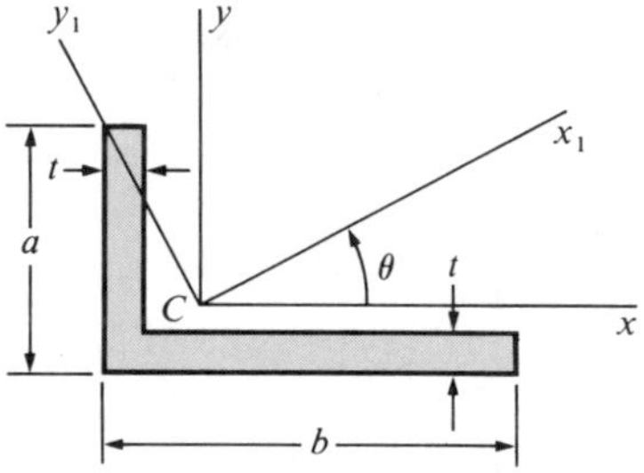

Probs. C.8-10 and C.8-11

C.8-11 Solve the preceding problem if $a = 3$ in., $b = 6$ in., and $t = \frac{5}{8}$ in.

***C.8-12** Determine the coordinates x_1, y_1 and x_2, y_2, respectively, of the two principal points P_1 and P_2 of the right triangle shown in the figure. (Note that the height of the triangle is twice the length of the base.) What is the moment of inertia I with respect to any axis through either of these points?

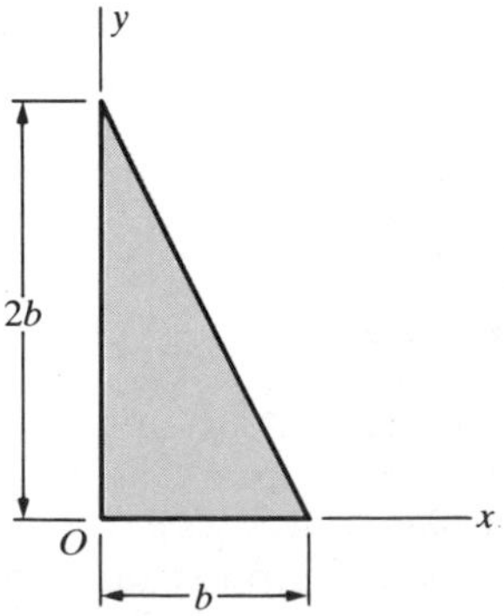

Prob. C.8-12

APPENDIX D

Properties of Plane Areas

Notation: A = area
$\bar{x}, \bar{y}$ = distances to centroid C
I_x, I_y = moments of inertia with respect to the x and y axes, respectively
I_{xy} = product of inertia with respect to the x and y axes
$I_p = I_x + I_y$ = polar moment of inertia
I_{BB} = moment of inertia with respect to axis B-B

1

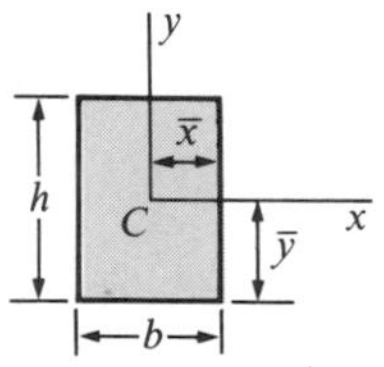

Rectangle (Origin of axes at centroid)

$$A = bh \qquad \bar{x} = \frac{b}{2} \qquad \bar{y} = \frac{h}{2}$$

$$I_x = \frac{bh^3}{12} \qquad I_y = \frac{hb^3}{12}$$

$$I_{xy} = 0 \qquad I_p = \frac{bh}{12}(h^2 + b^2)$$

2

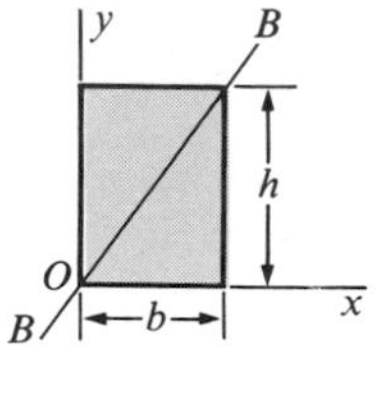

Rectangle (Origin of axes at corner)

$$I_x = \frac{bh^3}{3} \qquad I_y = \frac{hb^3}{3}$$

$$I_{xy} = \frac{b^2h^2}{4} \qquad I_p = \frac{bh}{3}(h^2 + b^2)$$

$$I_{BB} = \frac{b^3h^3}{6(b^2 + h^2)}$$

3

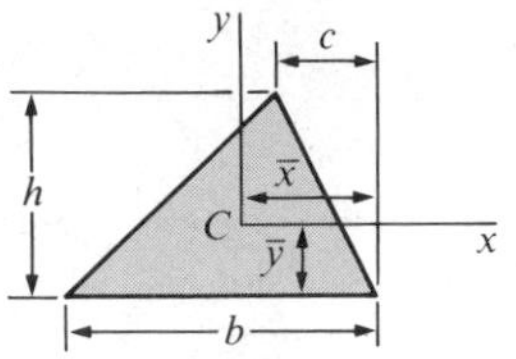

Triangle (Origin of axes at centroid)

$$A = \frac{bh}{2} \qquad \bar{x} = \frac{b + c}{3} \qquad \bar{y} = \frac{h}{3}$$

$$I_x = \frac{bh^3}{36} \qquad I_y = \frac{bh}{36}(b^2 - bc + c^2)$$

$$I_{xy} = \frac{bh^2}{72}(b - 2c) \qquad I_p = \frac{bh}{36}(h^2 + b^2 - bc + c^2)$$

4

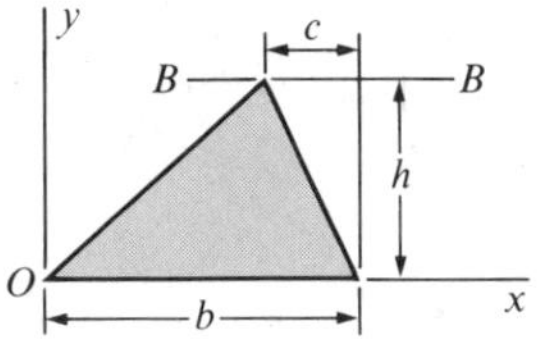

Triangle (Origin of axes at vertex)

$$I_x = \frac{bh^3}{12} \qquad I_y = \frac{bh}{12}(3b^2 - 3bc + c^2)$$

$$I_{xy} = \frac{bh^2}{24}(3b - 2c) \qquad I_{BB} = \frac{bh^3}{4}$$

5

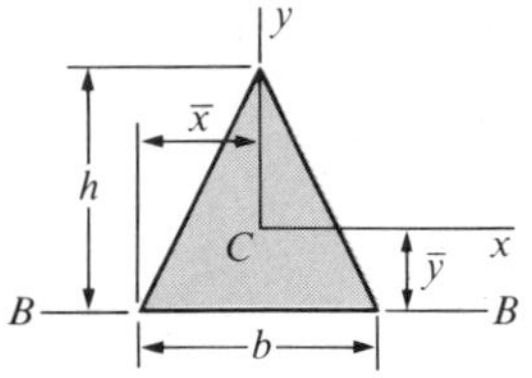

Isosceles triangle (Origin of axes at centroid)

$$A = \frac{bh}{2} \qquad \bar{x} = \frac{b}{2} \qquad \bar{y} = \frac{h}{3}$$

$$I_x = \frac{bh^3}{36} \qquad I_y = \frac{hb^3}{48} \qquad I_{xy} = 0$$

$$I_p = \frac{bh}{144}(4h^2 + 3b^2) \qquad I_{BB} = \frac{bh^3}{12}$$

(Note: For an equilateral triangle, $h = \sqrt{3}\,b/2$.)

6

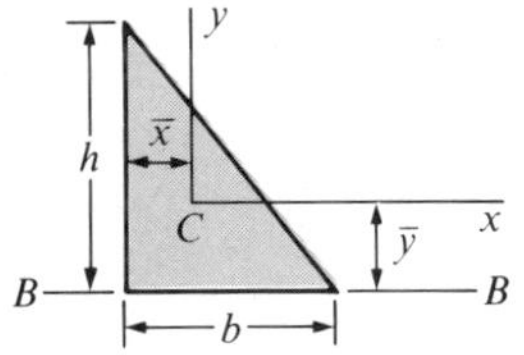

Right triangle (Origin of axes at centroid)

$$A = \frac{bh}{2} \qquad \bar{x} = \frac{b}{3} \qquad \bar{y} = \frac{h}{3}$$

$$I_x = \frac{bh^3}{36} \qquad I_y = \frac{hb^3}{36} \qquad I_{xy} = -\frac{b^2h^2}{72}$$

$$I_p = \frac{bh}{36}(h^2 + b^2) \qquad I_{BB} = \frac{bh^3}{12}$$

7

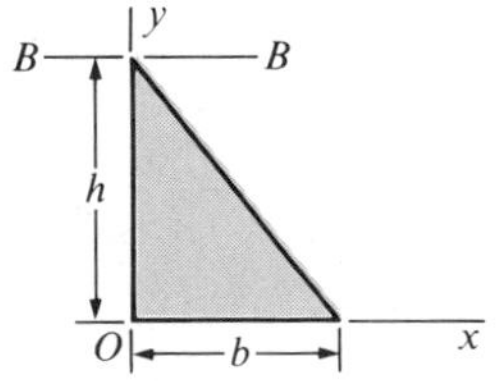

Right triangle (Origin of axes at vertex)

$$I_x = \frac{bh^3}{12} \qquad I_y = \frac{hb^3}{12} \qquad I_{xy} = \frac{b^2h^2}{24}$$

$$I_p = \frac{bh}{12}(h^2 + b^2) \qquad I_{BB} = \frac{bh^3}{4}$$

8

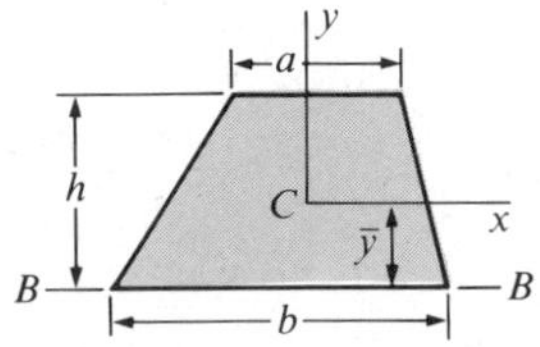

Trapezoid (Origin of axes at centroid)

$$A = \frac{h(a + b)}{2} \qquad \bar{y} = \frac{h(2a + b)}{3(a + b)}$$

$$I_x = \frac{h^3(a^2 + 4ab + b^2)}{36(a + b)} \qquad I_{BB} = \frac{h^3(3a + b)}{12}$$

9 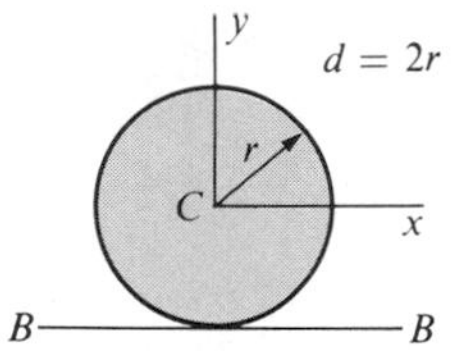

Circle (Origin of axes at center)

$$A = \pi r^2 = \frac{\pi d^2}{4} \qquad I_x = I_y = \frac{\pi r^4}{4} = \frac{\pi d^4}{64}$$

$$I_{xy} = 0 \qquad I_p = \frac{\pi r^4}{2} = \frac{\pi d^4}{32} \qquad I_{BB} = \frac{5\pi r^4}{4} = \frac{5\pi d^4}{64}$$

10 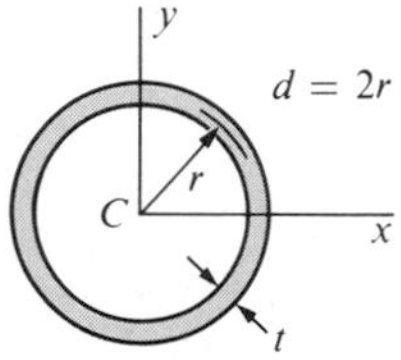

Circular ring (Origin of axes at center)
Approximate formulas for case when t is small

$$A = 2\pi r t = \pi d t \qquad I_x = I_y = \pi r^3 t = \frac{\pi d^3 t}{8}$$

$$I_{xy} = 0 \qquad I_p = 2\pi r^3 t = \frac{\pi d^3 t}{4}$$

11 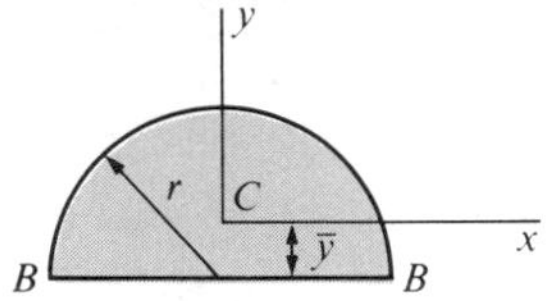

Semicircle (Origin of axes at centroid)

$$A = \frac{\pi r^2}{2} \qquad \bar{y} = \frac{4r}{3\pi}$$

$$I_x = \frac{(9\pi^2 - 64)r^4}{72\pi} \approx 0.1098 r^4 \qquad I_y = \frac{\pi r^4}{8}$$

$$I_{xy} = 0 \qquad I_{BB} = \frac{\pi r^4}{8}$$

12 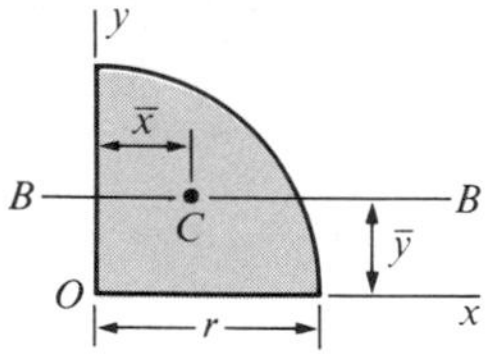

Quarter circle (Origin of axes at center of circle)

$$A = \frac{\pi r^2}{4} \qquad \bar{x} = \bar{y} = \frac{4r}{3\pi}$$

$$I_x = I_y = \frac{\pi r^4}{16} \qquad I_{xy} = \frac{r^4}{8}$$

$$I_{BB} = \frac{(9\pi^2 - 64)r^4}{144\pi} \approx 0.05488 r^4$$

13 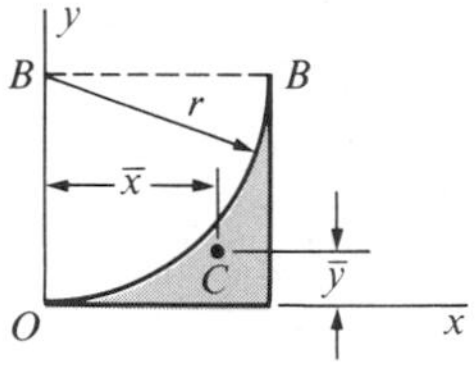

Quarter-circular spandrel (Origin of axes at point of tangency)

$$A = \left(1 - \frac{\pi}{4}\right) r^2 \qquad \bar{x} = \frac{2r}{3(4 - \pi)} \approx 0.7766 r \qquad \bar{y} = \frac{(10 - 3\pi)r}{3(4 - \pi)} \approx 0.2234 r$$

$$I_x = \left(1 - \frac{5\pi}{16}\right) r^4 \approx 0.01825 r^4 \qquad I_y = I_{BB} = \left(\frac{1}{3} - \frac{\pi}{16}\right) r^4 \approx 0.1370 r^4$$

14 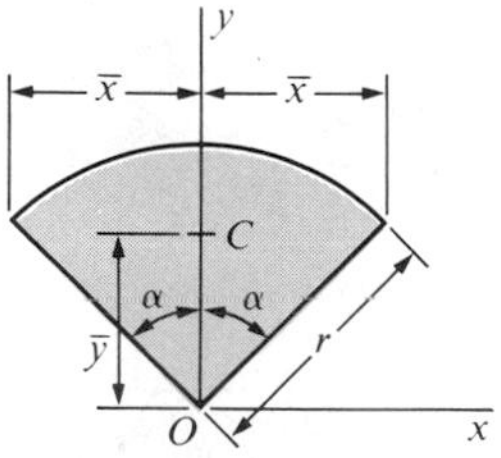

Circular sector (Origin of axes at center of circle)
α = angle in radians ($\alpha \leq \pi/2$)

$$A = \alpha r^2 \qquad \bar{x} = r \sin\alpha \qquad \bar{y} = \frac{2r \sin\alpha}{3\alpha}$$

$$I_x = \frac{r^4}{4}(\alpha + \sin\alpha\cos\alpha) \qquad I_y = \frac{r^4}{4}(\alpha - \sin\alpha\cos\alpha)$$

$$I_{xy} = 0 \qquad I_p = \frac{\alpha r^4}{2}$$

15 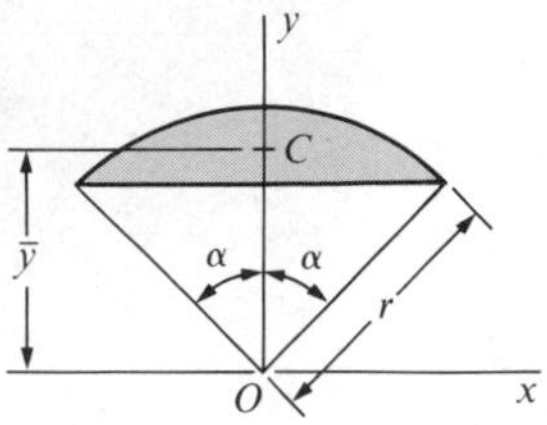

Circular segment (Origin of axes at center of circle)

α = angle in radians $\quad (\alpha \leq \pi/2)$

$$A = r^2(\alpha - \sin\alpha\cos\alpha) \qquad \bar{y} = \frac{2r}{3}\left(\frac{\sin^3\alpha}{\alpha - \sin\alpha\cos\alpha}\right)$$

$$I_x = \frac{r^4}{4}(\alpha - \sin\alpha\cos\alpha + 2\sin^3\alpha\cos\alpha) \qquad I_{xy} = 0$$

$$I_y = \frac{r^4}{12}(3\alpha - 3\sin\alpha\cos\alpha - 2\sin^3\alpha\cos\alpha)$$

16 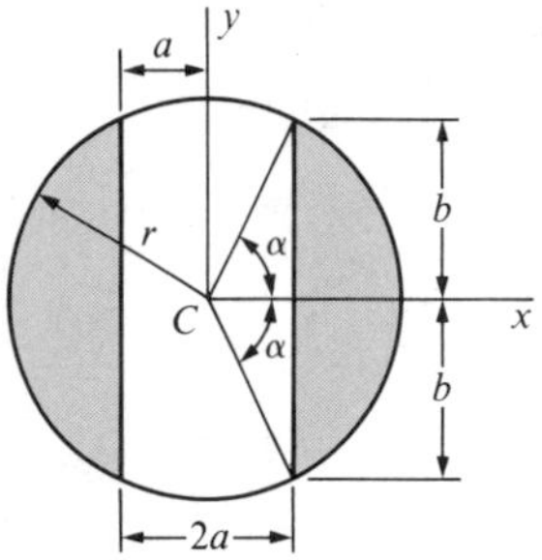

Circle with core removed (Origin of axes at center of circle)

α = angle in radians $\quad (\alpha \leq \pi/2)$

$$\alpha = \arccos\frac{a}{r} \qquad b = \sqrt{r^2 - a^2} \qquad A = 2r^2\left(\alpha - \frac{ab}{r^2}\right)$$

$$I_x = \frac{r^4}{6}\left(3\alpha - \frac{3ab}{r^2} - \frac{2ab^3}{r^4}\right) \qquad I_y = \frac{r^4}{2}\left(\alpha - \frac{ab}{r^2} + \frac{2ab^3}{r^4}\right) \qquad I_{xy} = 0$$

17 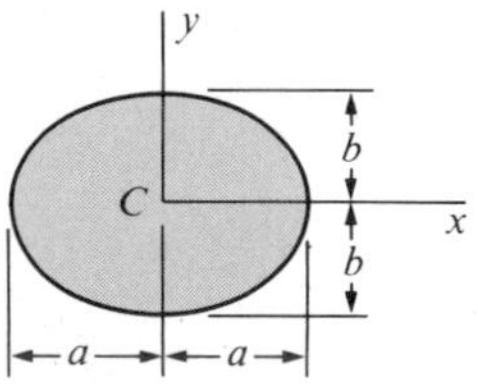

Ellipse (Origin of axes at centroid)

$$A = \pi ab \qquad I_x = \frac{\pi ab^3}{4} \qquad I_y = \frac{\pi ba^3}{4}$$

$$I_{xy} = 0 \qquad I_P = \frac{\pi ab}{4}(b^2 + a^2)$$

$$\text{Circumference} \approx \pi[1.5(a + b) - \sqrt{ab}] \qquad a/3 \leq b \leq a$$

$$\approx 4.17b^2/a + 4a \qquad 0 \leq b \leq a/3$$

18 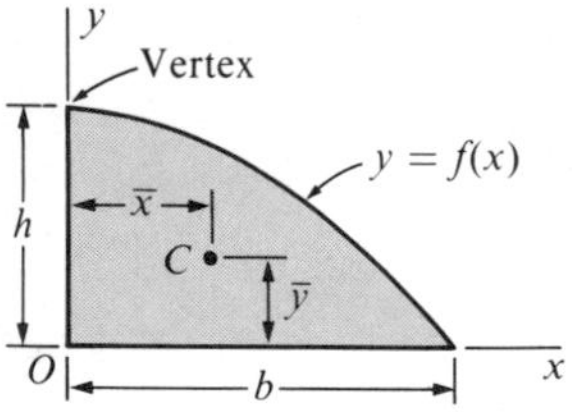

Parabolic semisegment (Origin of axes at corner)

$$y = f(x) = h\left(1 - \frac{x^2}{b^2}\right)$$

$$A = \frac{2bh}{3} \qquad \bar{x} = \frac{3b}{8} \qquad \bar{y} = \frac{2h}{5}$$

$$I_x = \frac{16bh^3}{105} \qquad I_y = \frac{2hb^3}{15} \qquad I_{xy} = \frac{b^2h^2}{12}$$

19 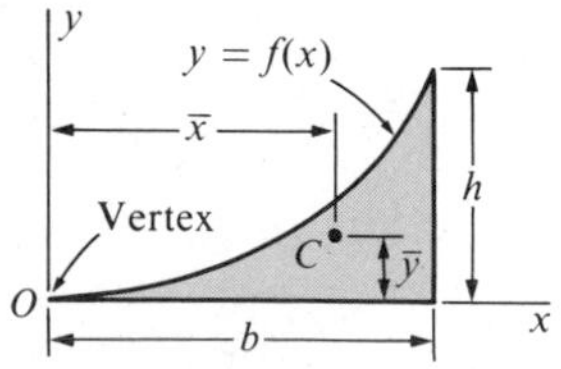

Parabolic spandrel (Origin of axes at vertex)

$$y = f(x) = \frac{hx^2}{b^2}$$

$$A = \frac{bh}{3} \qquad \bar{x} = \frac{3b}{4} \qquad \bar{y} = \frac{3h}{10}$$

$$I_x = \frac{bh^3}{21} \qquad I_y = \frac{hb^3}{5} \qquad I_{xy} = \frac{b^2h^2}{12}$$

20

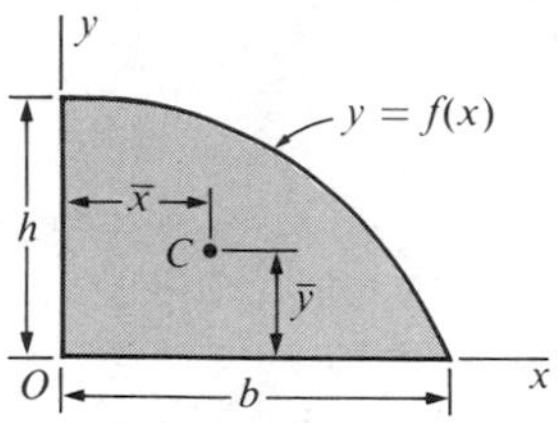

Semisegment of *n*th degree (Origin of axes at corner)

$$y = f(x) = h\left(1 - \frac{x^n}{b^n}\right) \qquad n > 0$$

$$A = bh\left(\frac{n}{n+1}\right) \qquad \bar{x} = \frac{b(n+1)}{2(n+2)} \qquad \bar{y} = \frac{hn}{2n+1}$$

$$I_x = \frac{2bh^3n^3}{(n+1)(2n+1)(3n+1)}$$

$$I_y = \frac{hb^3n}{3(n+3)} \qquad I_{xy} = \frac{b^2h^2n^2}{4(n+1)(n+2)}$$

21

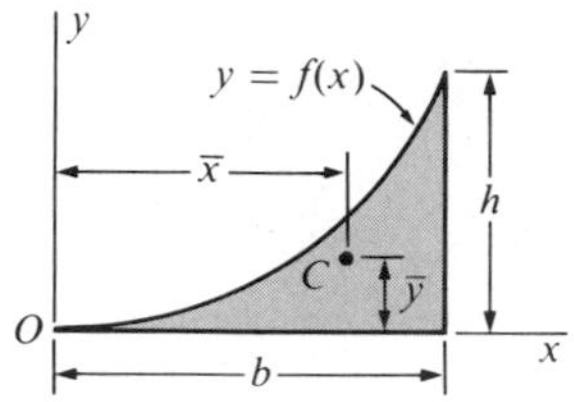

Spandrel of *n*th degree (Origin of axes at point of tangency)

$$y = f(x) = \frac{hx^n}{b^n} \qquad n > 0$$

$$A = \frac{bh}{n+1} \qquad \bar{x} = \frac{b(n+1)}{n+2} \qquad \bar{y} = \frac{h(n+1)}{2(2n+1)}$$

$$I_x = \frac{bh^3}{3(3n+1)} \qquad I_y = \frac{hb^3}{n+3} \qquad I_{xy} = \frac{b^2h^2}{4(n+1)}$$

22

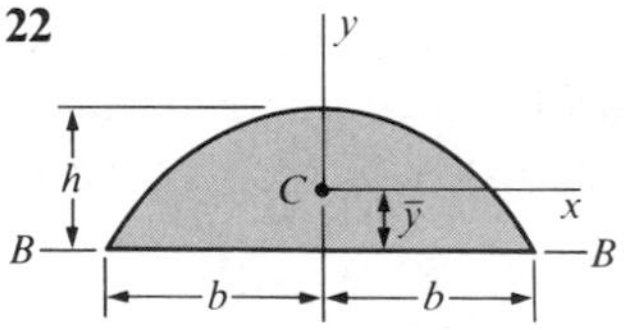

Sine wave (Origin of axes at centroid)

$$A = \frac{4bh}{\pi} \qquad \bar{y} = \frac{\pi h}{8}$$

$$I_x = \left(\frac{8}{9\pi} - \frac{\pi}{16}\right)bh^3 \approx 0.08659bh^3 \qquad I_y = \left(\frac{4}{\pi} - \frac{32}{\pi^3}\right)hb^3 \approx 0.2412hb^3$$

$$I_{xy} = 0 \qquad I_{BB} = \frac{8bh^3}{9\pi}$$

APPENDIX E

Properties of Selected Structural-Steel Shapes

In the following tables, the properties of a few structural-steel shapes are presented as an aid to the reader in solving problems in the text. These tables were compiled from the extensive tables found in the *Manual of Steel Construction*, published by the American Institute of Steel Construction, Inc. (Ref. 5-4).

Notation:

I = moment of inertia
S = section modulus
$r = \sqrt{I/A}$ = radius of gyration

Table E-1 **Properties of W shapes (wide-flange sections) (abridged list)**

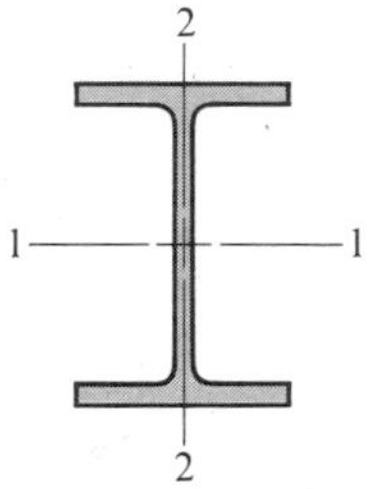

Designation	Weight per foot	Area	Depth	Web thickness	Flange		Axis 1-1			Axis 2-2		
					Width	Thickness	I	S	r	I	S	r
	lb	in.^{2}	in.	in.	in.	in.	in.^{4}	in.^{3}	in.	in.^{4}	in.^{3}	in.
W 30 × 211	211	62.0	30.94	0.775	15.105	1.315	10300	663	12.9	757	100	3.49
W 30 × 132	132	38.9	30.31	0.615	10.545	1.000	5770	380	12.2	196	37.2	2.25
W 24 × 162	162	47.7	25.00	0.705	12.955	1.220	5170	414	10.4	443	68.4	3.05
W 24 × 94	94	27.7	24.31	0.515	9.065	0.875	2700	222	9.87	109	24.0	1.98
W 18 × 119	119	35.1	18.97	0.655	11.265	1.060	2190	231	7.90	253	44.9	2.69
W 18 × 71	71	20.8	18.47	0.495	7.635	0.810	1170	127	7.50	60.3	15.8	1.70
W 16 × 100	100	29.4	16.97	0.585	10.425	0.985	1490	175	7.10	186	35.7	2.51
W 16 × 77	77	22.6	16.52	0.455	10.295	0.760	1110	134	7.00	138	26.9	2.47
W 16 × 57	57	16.8	16.43	0.430	7.120	0.715	758	92.2	6.72	43.1	12.1	1.60
W 16 × 31	31	9.12	15.88	0.275	5.525	0.440	375	47.2	6.41	12.4	4.49	1.17
W 14 × 120	120	35.3	14.48	0.590	14.670	0.940	1380	190	6.24	495	67.5	3.74
W 14 × 82	82	24.1	14.31	0.510	10.130	0.855	882	123	6.05	148	29.3	2.48
W 14 × 53	53	15.6	13.92	0.370	8.060	0.660	541	77.8	5.89	57.7	14.3	1.92
W 14 × 26	26	7.69	13.91	0.255	5.025	0.420	245	35.3	5.65	8.91	3.54	1.08
W 12 × 87	87	25.6	12.53	0.515	12.125	0.810	740	118	5.38	241	39.7	3.07
W 12 × 50	50	14.7	12.19	0.370	8.080	0.640	394	64.7	5.18	56.3	13.9	1.96
W 12 × 35	35	10.3	12.50	0.300	6.560	0.520	285	45.6	5.25	24.5	7.47	1.54
W 12 × 14	14	4.16	11.91	0.200	3.970	0.225	88.6	14.9	4.62	2.36	1.19	0.753
W 10 × 60	60	17.6	10.22	0.420	10.080	0.680	341	66.7	4.39	116	23.0	2.57
W 10 × 45	45	13.3	10.10	0.350	8.020	0.620	248	49.1	4.32	53.4	13.3	2.01
W 10 × 30	30	8.84	10.47	0.300	5.810	0.510	170	32.4	4.38	16.7	5.75	1.37
W 10 × 12	12	3.54	9.87	0.190	3.960	0.210	53.8	10.9	3.90	2.18	1.10	0.785
W 8 × 35	35	10.3	8.12	0.310	8.020	0.495	127	31.2	3.51	42.6	10.6	2.03
W 8 × 28	28	8.25	8.06	0.285	6.535	0.465	98.0	24.3	3.45	21.7	6.63	1.62
W 8 × 21	21	6.16	8.28	0.250	5.270	0.400	75.3	18.2	3.49	9.77	3.71	1.26
W 8 × 15	15	4.44	8.11	0.245	4.015	0.315	48.0	11.8	3.29	3.41	1.70	0.876

Note: Axes 1-1 and 2-2 are principal centroidal axes.

Table E-2 **Properties of S shapes (I-beam sections) (abridged list)**

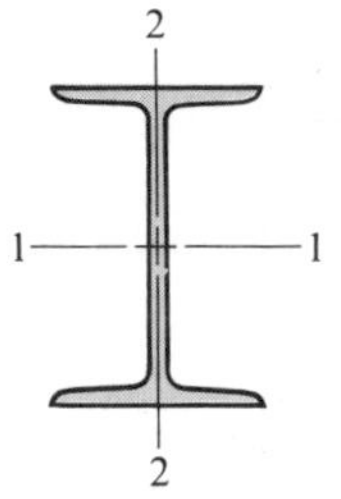

Designation	Weight per foot	Area	Depth	Web thickness	Flange		Axis 1-1			Axis 2-2		
					Width	Average thickness	*I*	*S*	*r*	*I*	*S*	*r*
	lb	in.^{2}	in.	in.	in.	in.	in.^{4}	in.^{3}	in.	in.^{4}	in.^{3}	in.
S 24 × 100	100	29.3	24.00	0.745	7.245	0.870	2390	199	9.02	47.7	13.2	1.27
S 24 × 80	80	23.5	24.00	0.500	7.000	0.870	2100	175	9.47	42.2	12.1	1.34
S 20 × 96	96	28.2	20.30	0.800	7.200	0.920	1670	165	7.71	50.2	13.9	1.33
S 20 × 75	75	22.0	20.00	0.635	6.385	0.795	1280	128	7.62	29.8	9.32	1.16
S 18 × 70	70	20.6	18.00	0.711	6.251	0.691	926	103	6.71	24.1	7.72	1.08
S 18 × 54.7	54.7	16.1	18.00	0.461	6.001	0.691	804	89.4	7.07	20.8	6.94	1.14
S 15 × 50	50	14.7	15.00	0.550	5.640	0.622	486	64.8	5.75	15.7	5.57	1.03
S 15 × 42.9	42.9	12.6	15.00	0.411	5.501	0.622	447	59.6	5.95	14.4	5.23	1.07
S 12 × 50	50	14.7	12.00	0.687	5.477	0.659	305	50.8	4.55	15.7	5.74	1.03
S 12 × 35	35	10.3	12.00	0.428	5.078	0.544	229	38.2	4.72	9.87	3.89	0.980
S 10 × 35	35	10.3	10.00	0.594	4.944	0.491	147	29.4	3.78	8.36	3.38	0.901
S 10 × 25.4	25.4	7.46	10.00	0.311	4.661	0.491	124	24.7	4.07	6.79	2.91	0.954
S 8 × 23	23	6.77	8.00	0.441	4.171	0.426	64.9	16.2	3.10	4.31	2.07	0.798
S 8 × 18.4	18.4	5.41	8.00	0.271	4.001	0.426	57.6	14.4	3.26	3.73	1.86	0.831
S 6 × 17.25	17.25	5.07	6.00	0.465	3.565	0.359	26.3	8.77	2.28	2.31	1.30	0.675
S 6 × 12.5	12.5	3.67	6.00	0.232	3.332	0.359	22.1	7.37	2.45	1.82	1.09	0.705
S 4 × 9.5	9.5	2.79	4.00	0.326	2.796	0.293	6.79	3.39	1.56	0.903	0.646	0.569
S 4 × 7.7	7.7	2.26	4.00	0.193	2.663	0.293	6.08	3.04	1.64	0.764	0.574	0.581

Note: Axes 1-1 and 2-2 are principal centroidal axes.

Table E-3 **Properties of channel sections (C shapes) (abridged list)**

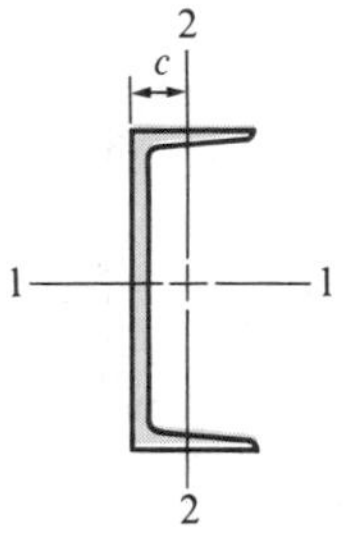

Designation	Weight per foot	Area	Depth	Web thickness	Flange		Axis 1-1			Axis 2-2			
					Width	Average thickness	I	S	r	I	S	r	c
	lb	in.^{2}	in.	in.	in.	in.	in.^{4}	in.^{3}	in.	in.^{4}	in.^{3}	in.	in.
C 15 × 50	50.0	14.7	15.00	0.716	3.716	0.650	404	53.8	5.24	11.0	3.78	0.867	0.798
C 15 × 40	40.0	11.8	15.00	0.520	3.520	0.650	349	46.5	5.44	9.23	3.37	0.886	0.777
C 15 × 33.9	33.9	9.96	15.00	0.400	3.400	0.650	315	42.0	5.62	8.13	3.11	0.904	0.787
C 12 × 30	30.0	8.82	12.00	0.510	3.170	0.501	162	27.0	4.29	5.14	2.06	0.763	0.674
C 12 × 25	25	7.35	12.00	0.387	3.047	0.501	144	24.1	4.43	4.47	1.88	0.780	0.674
C 12 × 20.7	20.7	6.09	12.00	0.282	2.942	0.501	129	21.5	4.61	3.88	1.73	0.799	0.698
C 10 × 30	30.0	8.82	10.00	0.673	3.033	0.436	103	20.7	3.42	3.94	1.65	0.669	0.649
C 10 × 25	25	7.35	10.00	0.526	2.886	0.436	91.2	18.2	3.52	3.36	1.48	0.676	0.617
C 10 × 20	20.0	5.88	10.00	0.379	2.739	0.436	78.9	15.8	3.66	2.81	1.32	0.692	0.606
C 10 × 15.3	15.3	4.49	10.00	0.240	2.600	0.436	67.4	13.5	3.87	2.28	1.16	0.713	0.634
C 8 × 18.75	18.75	5.51	8.00	0.487	2.527	0.390	44.0	11.0	2.82	1.98	1.01	0.599	0.565
C 8 × 13.75	13.75	4.04	8.00	0.303	2.343	0.390	36.1	9.03	2.99	1.53	0.854	0.615	0.553
C 8 × 11.5	11.5	3.38	8.00	0.220	2.260	0.390	32.6	8.14	3.11	1.32	0.781	0.625	0.571
C 6 × 13	13.0	3.83	6.00	0.437	2.157	0.343	17.4	5.80	2.13	1.05	0.642	0.525	0.514
C 6 × 10.5	10.5	3.09	6.00	0.314	2.034	0.343	15.2	5.06	2.22	0.866	0.564	0.529	0.499
C 6 × 8.2	8.2	2.40	6.00	0.200	1.920	0.343	13.1	4.38	2.34	0.693	0.492	0.537	0.511
C 4 × 7.25	7.25	2.13	4.00	0.321	1.721	0.296	4.59	2.29	1.47	0.433	0.343	0.450	0.459
C 4 × 5.4	5.4	1.59	4.00	0.184	1.584	0.296	3.85	1.93	1.56	0.319	0.283	0.449	0.457

Notes: 1. Axes 1-1 and 2-2 are principal centroidal axes.
2. The distance c is measured from the centroid to the back of the web.
3. For axis 2-2, the tabulated value of S is the smaller of the two section moduli for this axis.

Table E-4 **Properties of angle sections with equal legs (L shapes) (abridged list)**

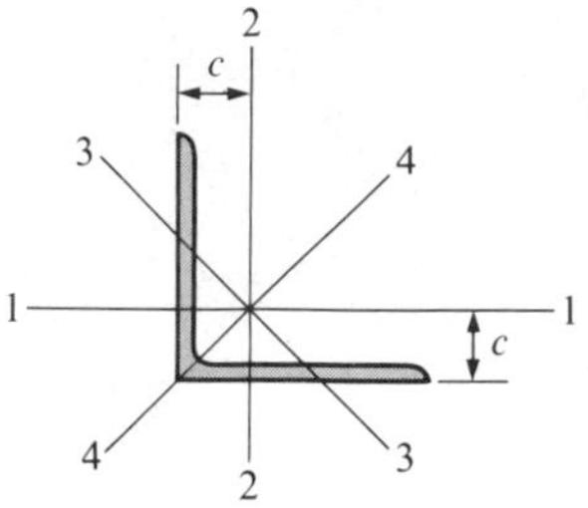

Designation	Weight per foot	Area	Axis 1-1 and Axis 2-2				Axis 3-3
			I	S	r	c	r_{min}
in.	lb	in.^{2}	in.^{4}	in.^{3}	in.	in.	in.
L $8 \times 8 \times 1$	51.0	15.0	89.0	15.8	2.44	2.37	1.56
L $8 \times 8 \times \frac{3}{4}$	38.9	11.4	69.7	12.2	2.47	2.28	1.58
L $8 \times 8 \times \frac{1}{2}$	26.4	7.75	48.6	8.36	2.50	2.19	1.59
L $6 \times 6 \times 1$	37.4	11.0	35.5	8.57	1.80	1.86	1.17
L $6 \times 6 \times \frac{3}{4}$	28.7	8.44	28.2	6.66	1.83	1.78	1.17
L $6 \times 6 \times \frac{1}{2}$	19.6	5.75	19.9	4.61	1.86	1.68	1.18
L $5 \times 5 \times \frac{7}{8}$	27.2	7.98	17.8	5.17	1.49	1.57	0.973
L $5 \times 5 \times \frac{1}{2}$	16.2	4.75	11.3	3.16	1.54	1.43	0.983
L $5 \times 5 \times \frac{3}{8}$	12.3	3.61	8.74	2.42	1.56	1.39	0.990
L $4 \times 4 \times \frac{3}{4}$	18.5	5.44	7.67	2.81	1.19	1.27	0.778
L $4 \times 4 \times \frac{1}{2}$	12.8	3.75	5.56	1.97	1.22	1.18	0.782
L $4 \times 4 \times \frac{3}{8}$	9.8	2.86	4.36	1.52	1.23	1.14	0.788
L $3\frac{1}{2} \times 3\frac{1}{2} \times \frac{3}{8}$	8.5	2.48	2.87	1.15	1.07	1.01	0.687
L $3\frac{1}{2} \times 3\frac{1}{2} \times \frac{1}{4}$	5.8	1.69	2.01	0.794	1.09	0.968	0.694
L $3 \times 3 \times \frac{1}{2}$	9.4	2.75	2.22	1.07	0.898	0.932	0.584
L $3 \times 3 \times \frac{1}{4}$	4.9	1.44	1.24	0.577	0.930	0.842	0.592

Notes: 1. Axes 1-1 and 2-2 are centroidal axes parallel to the legs.
2. The distance c is measured from the centroid to the back of the legs.
3. For axes 1-1 and 2-2, the tabulated value of S is the smaller of the two section moduli for those axes.
4. Axes 3-3 and 4-4 are principal centroidal axes.
5. The moment of inertia for axis 3-3, which is the smaller of the two principal moments of inertia, can be found from the equation $I_{33} = Ar_{min}^2$.
6. The moment of inertia for axis 4-4, which is the larger of the two principal moments of inertia, can be found from the equation $I_{44} + I_{33} = I_{11} + I_{22}$.

Table E-5 Properties of angle sections with unequal legs (L shapes) (abridged list)

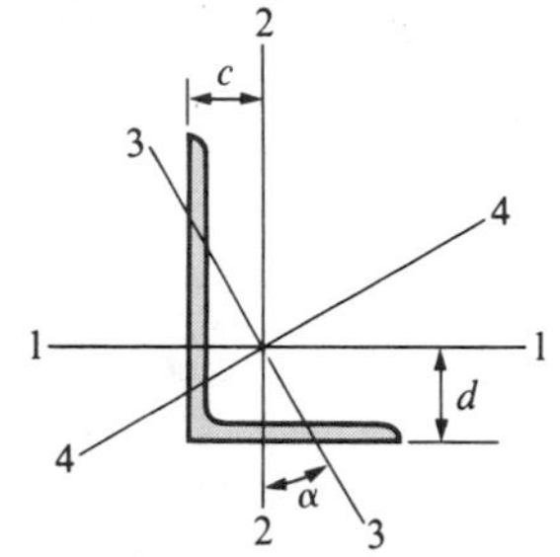

Designation	Weight per foot	Area	Axis 1-1				Axis 2-2				Axis 3-3	
			I	S	r	d	I	S	r	c	r_{min}	tan α
in.	lb	in.^{2}	in.^{4}	in.^{3}	in.	in.	in.^{4}	in.^{3}	in.	in.	in.	
L 8 × 6 × 1	44.2	13.00	80.8	15.1	2.49	2.65	38.8	8.92	1.73	1.65	1.28	0.543
L 8 × 6 × $\frac{1}{2}$	23.0	6.75	44.3	8.02	2.56	2.47	21.7	4.79	1.79	1.47	1.30	0.558
L 7 × 4 × $\frac{3}{4}$	26.2	7.69	37.8	8.42	2.22	2.51	9.05	3.03	1.09	1.01	0.860	0.324
L 7 × 4 × $\frac{1}{2}$	17.9	5.25	26.7	5.81	2.25	2.42	6.53	2.12	1.11	0.917	0.872	0.335
L 6 × 4 × $\frac{3}{4}$	23.6	6.94	24.5	6.25	1.88	2.08	8.68	2.97	1.12	1.08	0.860	0.428
L 6 × 4 × $\frac{1}{2}$	16.2	4.75	17.4	4.33	1.91	1.99	6.27	2.08	1.15	0.987	0.870	0.440
L 5 × $3\frac{1}{2}$ × $\frac{3}{4}$	19.8	5.81	13.9	4.28	1.55	1.75	5.55	2.22	0.977	0.996	0.748	0.464
L 5 × $3\frac{1}{2}$ × $\frac{1}{2}$	13.6	4.00	9.99	2.99	1.58	1.66	4.05	1.56	1.01	0.906	0.755	0.479
L 5 × 3 × $\frac{1}{2}$	12.8	3.75	9.45	2.91	1.59	1.75	2.58	1.15	0.829	0.750	0.648	0.357
L 5 × 3 × $\frac{1}{4}$	6.6	1.94	5.11	1.53	1.62	1.66	1.44	0.614	0.861	0.657	0.663	0.371
L 4 × $3\frac{1}{2}$ × $\frac{1}{2}$	11.9	3.50	5.32	1.94	1.23	1.25	3.79	1.52	1.04	1.00	0.722	0.750
L 4 × $3\frac{1}{2}$ × $\frac{1}{4}$	6.2	1.81	2.91	1.03	1.27	1.16	2.09	0.808	1.07	0.909	0.734	0.759
L 4 × 3 × $\frac{1}{2}$	11.1	3.25	5.05	1.89	1.25	1.33	2.42	1.12	0.864	0.827	0.639	0.543
L 4 × 3 × $\frac{3}{8}$	8.5	2.48	3.96	1.46	1.26	1.28	1.92	0.866	0.879	0.782	0.644	0.551
L 4 × 3 × $\frac{1}{4}$	5.8	1.69	2.77	1.00	1.28	1.24	1.36	0.599	0.896	0.736	0.651	0.558

Notes: 1. Axes 1-1 and 2-2 are centroidal axes parallel to the legs.
2. The distances c and d are measured from the centroid to the backs of the legs.
3. For axes 1-1 and 2-2, the tabulated value of S is the smaller of the two section moduli for those axes.
4. Axes 3-3 and 4-4 are principal centroidal axes.
5. The moment of inertia for axis 3-3, which is the smaller of the two principal moments of inertia, can be found from the equation $I_{33} = Ar_{min}^2$.
6. The moment of inertia for axis 4-4, which is the larger of the two principal moments of inertia, can be found from the equation $I_{44} + I_{33} = I_{11} + I_{22}$.

APPENDIX F

Section Properties of Structural Lumber

Properties of surfaced lumber (abridged list)

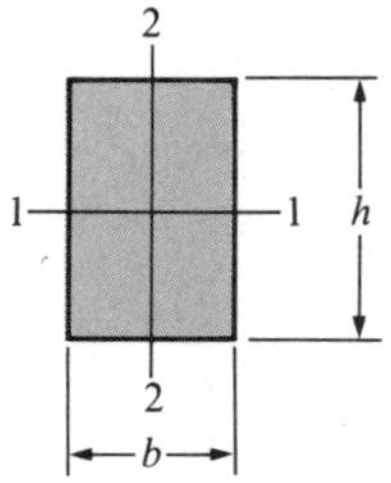

Nominal dimensions $b \times h$	Net dimensions $b \times h$	Area $A = bh$	Axis 1-1		Axis 2-2		Weight per linear foot (specific weight = 35 lb/ft^3)
			Moment of inertia $I_1 = \frac{bh^3}{12}$	Section modulus $S_1 = \frac{bh^2}{6}$	Moment of inertia $I_2 = \frac{hb^3}{12}$	Section modulus $S_2 = \frac{hb^2}{6}$	
in.	in.	in.^{2}	in.^{4}	in.^{3}	in.^{4}	in.^{3}	lb
2 × 4	1.5 × 3.5	5.25	5.36	3.06	0.98	1.31	1.3
2 × 6	1.5 × 5.5	8.25	20.80	7.56	1.55	2.06	2.0
2 × 8	1.5 × 7.25	10.88	47.63	13.14	2.04	2.72	2.6
2 × 10	1.5 × 9.25	13.88	98.93	21.39	2.60	3.47	3.4
2 × 12	1.5 × 11.25	16.88	177.98	31.64	3.16	4.22	4.1
3 × 4	2.5 × 3.5	8.75	8.93	5.10	4.56	3.65	2.1
3 × 6	2.5 × 5.5	13.75	34.66	12.60	7.16	5.73	3.3
3 × 8	2.5 × 7.25	18.13	79.39	21.90	9.44	7.55	4.4
3 × 10	2.5 × 9.25	23.13	164.89	35.65	12.04	9.64	5.6
3 × 12	2.5 × 11.25	28.13	296.63	52.73	14.65	11.72	6.8
4 × 4	3.5 × 3.5	12.25	12.51	7.15	12.51	7.15	3.0
4 × 6	3.5 × 5.5	19.25	48.53	17.65	19.65	11.23	4.7
4 × 8	3.5 × 7.25	25.38	111.15	30.66	25.90	14.80	6.2
4 × 10	3.5 × 9.25	32.38	230.84	49.91	33.05	18.89	7.9
4 × 12	3.5 × 11.25	39.38	415.28	73.83	40.20	22.97	9.6
6 × 6	5.5 × 5.5	30.25	76.3	27.7	76.3	27.7	7.4
6 × 8	5.5 × 7.5	41.25	193.4	51.6	104.0	37.8	10.0
6 × 10	5.5 × 9.5	52.25	393.0	82.7	131.7	47.9	12.7
6 × 12	5.5 × 11.5	63.25	697.1	121.2	159.4	58.0	15.4
8 × 8	7.5 × 7.5	56.25	263.7	70.3	263.7	70.3	13.7
8 × 10	7.5 × 9.5	71.25	535.9	112.8	334.0	89.1	17.3
8 × 12	7.5 × 11.5	86.25	950.5	165.3	404.3	107.8	21.0

Note: Axes 1-1 and 2-2 are principal centroidal axes.

APPENDIX G

Deflections and Slopes of Beams

Table G-1 Deflections and slopes of cantilever beams

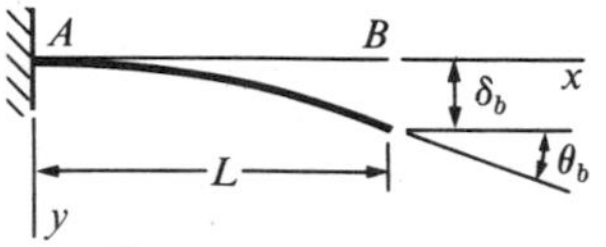

EI = constant

v = deflection in y direction

$v' = dv/dx$ = slope of deflection curve

$\delta_b = v(L)$ = deflection at right end of beam

$\theta_b = v'(L)$ = angle at right end of beam

1

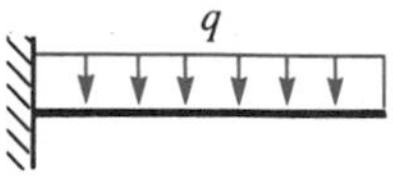

$$v = \frac{qx^2}{24EI}(6L^2 - 4Lx + x^2)$$

$$v' = \frac{qx}{6EI}(3L^2 - 3Lx + x^2)$$

$$\delta_b = \frac{qL^4}{8EI} \qquad \theta_b = \frac{qL^3}{6EI}$$

2

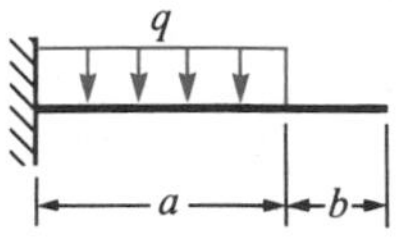

$$v = \frac{qx^2}{24EI}(6a^2 - 4ax + x^2) \qquad 0 \le x \le a$$

$$v' = \frac{qx}{6EI}(3a^2 - 3ax + x^2) \qquad 0 \le x \le a$$

$$v = \frac{qa^3}{24EI}(4x - a) \qquad v' = \frac{qa^3}{6EI} \qquad a \le x \le L$$

At $x = a$: $\quad v = \dfrac{qa^4}{8EI} \qquad v' = \dfrac{qa^3}{6EI}$

$$\delta_b = \frac{qa^3}{24EI}(4L - a) \qquad \theta_b = \frac{qa^3}{6EI}$$

3

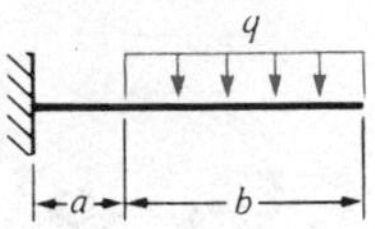

$$v = \frac{qbx^2}{12EI}(3L + 3a - 2x) \qquad 0 \le x \le a$$

$$v' = \frac{qbx}{2EI}(L + a - x) \qquad 0 \le x \le a$$

$$v = \frac{q}{24EI}(x^4 - 4Lx^3 + 6L^2x^2 - 4a^3x + a^4) \qquad a \le x \le L$$

$$v' = \frac{q}{6EI}(x^3 - 3Lx^2 + 3L^2x - a^3) \qquad a \le x \le L$$

$$\text{At } x = a\text{:} \quad v = \frac{qa^2b}{12EI}(3L + a) \qquad v' = \frac{qabL}{2EI}$$

$$\delta_b = \frac{q}{24EI}(3L^4 - 4a^3L + a^4) \qquad \theta_b = \frac{q}{6EI}(L^3 - a^3)$$

4

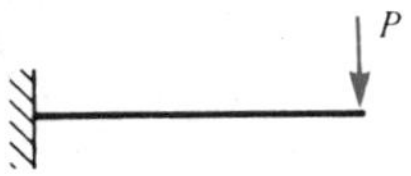

$$v = \frac{Px^2}{6EI}(3L - x) \qquad v' = \frac{Px}{2EI}(2L - x)$$

$$\delta_b = \frac{PL^3}{3EI} \qquad \theta_b = \frac{PL^2}{2EI}$$

5

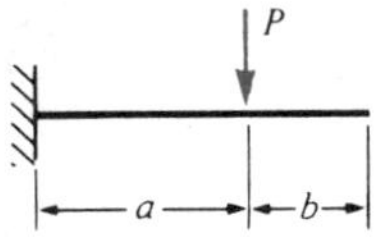

$$v = \frac{Px^2}{6EI}(3a - x) \qquad v' = \frac{Px}{2EI}(2a - x) \qquad 0 \le x \le a$$

$$v = \frac{Pa^2}{6EI}(3x - a) \qquad v' = \frac{Pa^2}{2EI} \qquad a \le x \le L$$

$$\text{At } x = a\text{:} \quad v = \frac{Pa^3}{3EI} \qquad v' = \frac{Pa^2}{2EI}$$

$$\delta_b = \frac{Pa^2}{6EI}(3L - a) \qquad \theta_b = \frac{Pa^2}{2EI}$$

6

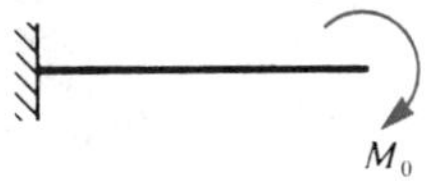

$$v = \frac{M_0x^2}{2EI} \qquad v' = \frac{M_0x}{EI}$$

$$\delta_b = \frac{M_0L^2}{2EI} \qquad \theta_b = \frac{M_0L}{EI}$$

7

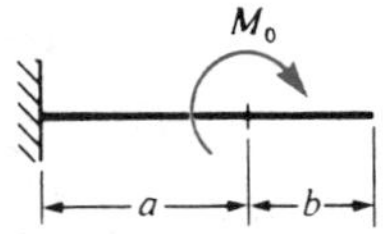

$$v = \frac{M_0x^2}{2EI} \qquad v' = \frac{M_0x}{EI} \qquad 0 \le x \le a$$

$$v = \frac{M_0a}{2EI}(2x - a) \qquad v' = \frac{M_0a}{EI} \qquad a \le x \le L$$

$$\text{At } x = a\text{:} \quad v = \frac{M_0a^2}{2EI} \qquad v' = \frac{M_0a}{EI}$$

$$\delta_b = \frac{M_0a}{2EI}(2L - a) \qquad \theta_b = \frac{M_0a}{EI}$$

8 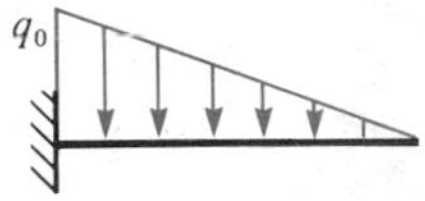

$$v = \frac{q_0 x^2}{120LEI}(10L^3 - 10L^2x + 5Lx^2 - x^3)$$

$$v' = \frac{q_0 x}{24LEI}(4L^3 - 6L^2x + 4Lx^2 - x^3)$$

$$\delta_b = \frac{q_0 L^4}{30EI} \qquad \theta_b = \frac{q_0 L^3}{24EI}$$

9 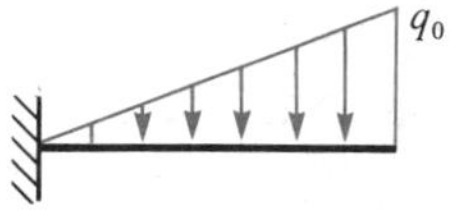

$$v = \frac{q_0 x^2}{120LEI}(20L^3 - 10L^2x + x^3)$$

$$v' = \frac{q_0 x}{24LEI}(8L^3 - 6L^2x + x^3)$$

$$\delta_b = \frac{11q_0 L^4}{120EI} \qquad \theta_b = \frac{q_0 L^3}{8EI}$$

10 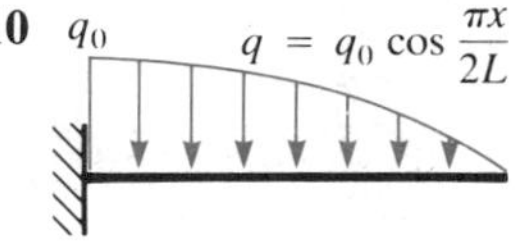

$$v = \frac{q_0 L}{3\pi^4 EI}\left(48L^3 \cos\frac{\pi x}{2L} - 48L^3 + 3\pi^3 Lx^2 - \pi^3 x^3\right)$$

$$v' = \frac{q_0 L}{\pi^3 EI}\left(2\pi^2 Lx - \pi^2 x^2 - 8L^2 \sin\frac{\pi x}{2L}\right)$$

$$\delta_b = \frac{2q_0 L^4}{3\pi^4 EI}(\pi^3 - 24) \qquad \theta_b = \frac{q_0 L^3}{\pi^3 EI}(\pi^2 - 8)$$

Table G-2 Deflections and slopes of simple beams

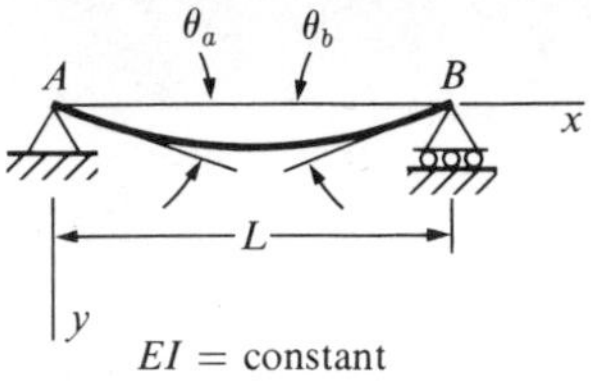

v = deflection in y direction

$v' = dv/dx$ = slope of deflection curve

$\delta_c = v(L/2)$ = deflection at center of beam

x_1 = distance from A to point of maximum deflection

$\delta_{\max} = v_{\max}$ = maximum deflection

$\theta_a = v'(0)$ = angle at left end of beam

$\theta_b = -v'(L)$ = angle at right end of beam

1

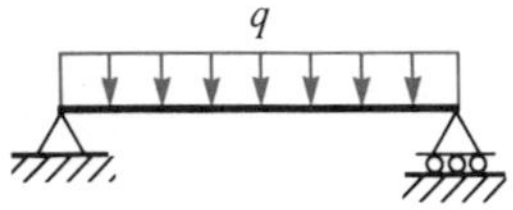

$$v = \frac{qx}{24EI}(L^3 - 2Lx^2 + x^3)$$

$$v' = \frac{q}{24EI}(L^3 - 6Lx^2 + 4x^3)$$

$$\delta_c = \delta_{\max} = \frac{5qL^4}{384EI} \qquad \theta_a = \theta_b = \frac{qL^3}{24EI}$$

2

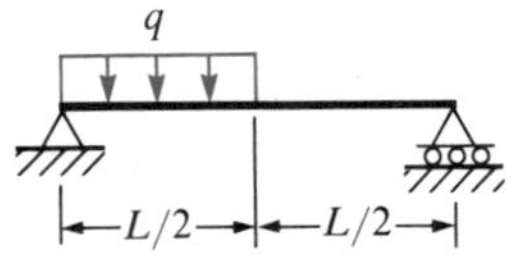

$$v = \frac{qx}{384EI}(9L^3 - 24Lx^2 + 16x^3) \qquad 0 \le x \le \frac{L}{2}$$

$$v' = \frac{q}{384EI}(9L^3 - 72Lx^2 + 64x^3) \qquad 0 \le x \le \frac{L}{2}$$

$$v = \frac{qL}{384EI}(8x^3 - 24Lx^2 + 17L^2x - L^3) \qquad \frac{L}{2} \le x \le L$$

$$v' = \frac{qL}{384EI}(24x^2 - 48Lx + 17L^2) \qquad \frac{L}{2} \le x \le L$$

$$\delta_c = \frac{5qL^4}{768EI} \qquad \theta_a = \frac{3qL^3}{128EI} \qquad \theta_b = \frac{7qL^3}{384EI}$$

3

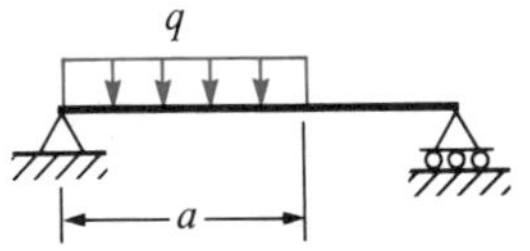

$$v = \frac{qx}{24LEI}(a^4 - 4a^3L + 4a^2L^2 + 2a^2x^2 - 4aLx^2 + Lx^3) \qquad 0 \le x \le a$$

$$v' = \frac{q}{24LEI}(a^4 - 4a^3L + 4a^2L^2 + 6a^2x^2 - 12aLx^2 + 4Lx^3) \qquad 0 \le x \le a$$

$$v = \frac{qa^2}{24LEI}(-a^2L + 4L^2x + a^2x - 6Lx^2 + 2x^3) \qquad a \le x \le L$$

$$v' = \frac{qa^2}{24LEI}(4L^2 + a^2 - 12Lx + 6x^2) \qquad a \le x \le L$$

$$\theta_a = \frac{qa^2}{24LEI}(2L - a)^2 \qquad \theta_b = \frac{qa^2}{24LEI}(2L^2 - a^2)$$

4

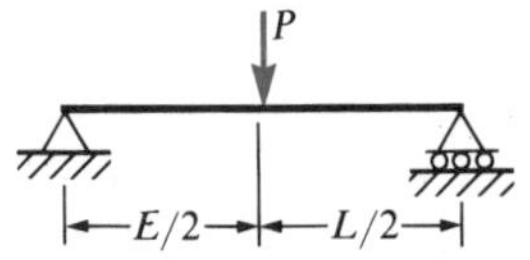

$$v = \frac{Px}{48EI}(3L^2 - 4x^2) \qquad 0 \le x \le \frac{L}{2}$$

$$v' = \frac{P}{16EI}(L^2 - 4x^2) \qquad 0 \le x \le \frac{L}{2}$$

$$\delta_c = \delta_{\max} = \frac{PL^3}{48EI} \qquad \theta_a = \theta_b = \frac{PL^2}{16EI}$$

5

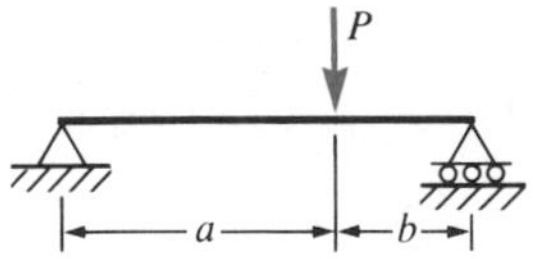

$$v = \frac{Pbx}{6LEI}(L^2 - b^2 - x^2) \qquad 0 \le x \le a$$

$$v' = \frac{Pb}{6LEI}(L^2 - b^2 - 3x^2) \qquad 0 \le x \le a$$

$$\theta_a = \frac{Pab(L + b)}{6LEI}$$

$$\theta_b = \frac{Pab(L + a)}{6LEI}$$

$$\text{If } a \ge b, \quad \delta_c = \frac{Pb(3L^2 - 4b^2)}{48EI}$$

$$\text{If } a \ge b, \quad x_1 = \sqrt{\frac{L^2 - b^2}{3}} \quad \text{and} \quad \delta_{max} = \frac{Pb(L^2 - b^2)^{3/2}}{9\sqrt{3}\,LEI}$$

6

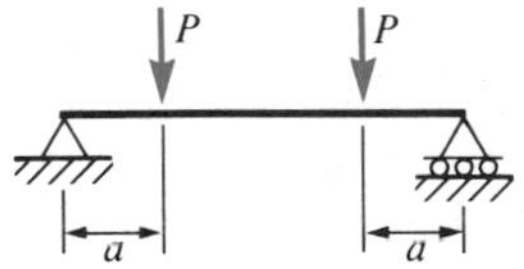

$$v = \frac{Px}{6EI}(3aL - 3a^2 - x^2) \qquad 0 \le x \le a$$

$$v' = \frac{P}{2EI}(aL - a^2 - x^2) \qquad 0 \le x \le a$$

$$v = \frac{Pa}{6EI}(3Lx - 3x^2 - a^2) \qquad a \le x \le L - a$$

$$v' = \frac{Pa}{2EI}(L - 2x) \qquad a \le x \le L - a$$

$$\theta_a = \theta_b = \frac{Pa(L - a)}{2EI}$$

$$\delta_c = \delta_{max} = \frac{Pa}{24EI}(3L^2 - 4a^2)$$

7

$$v = \frac{M_0 x}{6LEI}(2L^2 - 3Lx + x^2)$$

$$v' = \frac{M_0}{6LEI}(2L^2 - 6Lx + 3x^2)$$

$$\delta_c = \frac{M_0 L^2}{16EI} \qquad \theta_a = \frac{M_0 L}{3EI} \qquad \theta_b = \frac{M_0 L}{6EI}$$

$$x_1 = L\left(1 - \frac{\sqrt{3}}{3}\right) \quad \text{and} \quad \delta_{max} = \frac{M_0 L^2}{9\sqrt{3}\,EI}$$

8

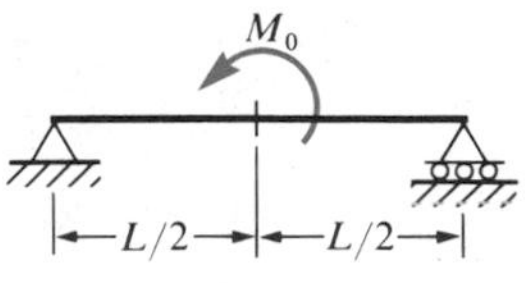

$$v = \frac{M_0 x}{24LEI}(L^2 - 4x^2) \qquad 0 \le x \le \frac{L}{2}$$

$$v' = \frac{M_0}{24LEI}(L^2 - 12x^2) \qquad 0 \le x \le \frac{L}{2}$$

$$\delta_c = 0 \qquad \theta_a = \frac{M_0 L}{24EI} \qquad \theta_b = -\frac{M_0 L}{24EI}$$

9

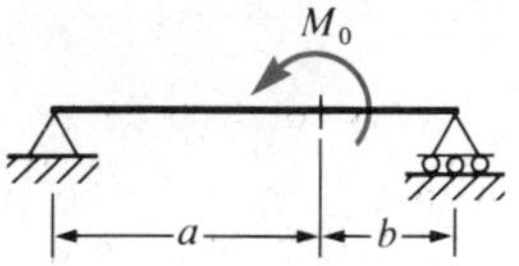

$$v = \frac{M_0 x}{6LEI}(6aL - 3a^2 - 2L^2 - x^2) \qquad 0 \le x \le a$$

$$v' = \frac{M_0}{6LEI}(6aL - 3a^2 - 2L^2 - 3x^2) \qquad 0 \le x \le a$$

$$\text{At } x = a: \quad v = \frac{M_0 ab}{3LEI}(2a - L)$$

$$\text{At } x = a: \quad v' = \frac{M_0}{3LEI}(3aL - 3a^2 - L^2)$$

$$\theta_a = \frac{M_0}{6LEI}(6aL - 3a^2 - 2L^2) \qquad \theta_b = \frac{M_0}{6LEI}(3a^2 - L^2)$$

10

$$v = \frac{M_0 x}{2EI}(L - x)$$

$$v' = \frac{M_0}{2EI}(L - 2x)$$

$$\delta_c = \delta_{max} = \frac{M_0 L^2}{8EI} \qquad \theta_a = \theta_b = \frac{M_0 L}{2EI}$$

11

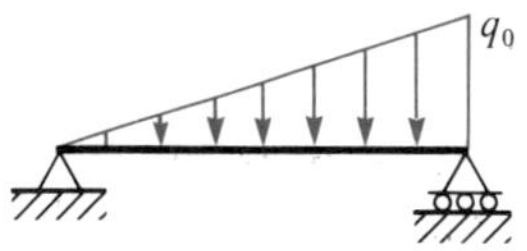

$$v = \frac{q_0 x}{360LEI}(7L^4 - 10L^2x^2 + 3x^4)$$

$$v' = \frac{q_0}{360LEI}(7L^4 - 30L^2x^2 + 15x^4)$$

$$\delta_c = \frac{5q_0 L^4}{768EI} \qquad \theta_a = \frac{7q_0 L^3}{360EI} \qquad \theta_b = \frac{q_0 L^3}{45EI}$$

$$x_1 = 0.5193\,L \qquad \delta_{max} = 0.00652\frac{q_0 L^4}{EI}$$

12

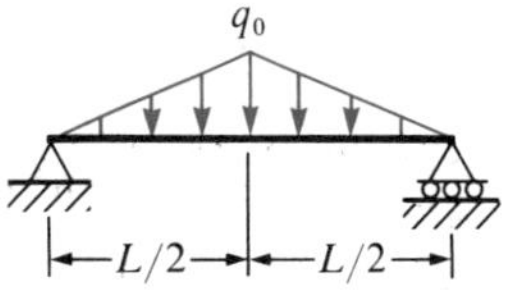

$$v = \frac{q_0 x}{960LEI}(5L^2 - 4x^2)^2 \qquad 0 \le x \le \frac{L}{2}$$

$$v' = \frac{q_0}{192LEI}(5L^2 - 4x^2)(L^2 - 4x^2) \qquad 0 \le x \le \frac{L}{2}$$

$$\delta_c = \delta_{max} = \frac{q_0 L^4}{120EI} \qquad \theta_a = \theta_b = \frac{5q_0 L^3}{192EI}$$

13

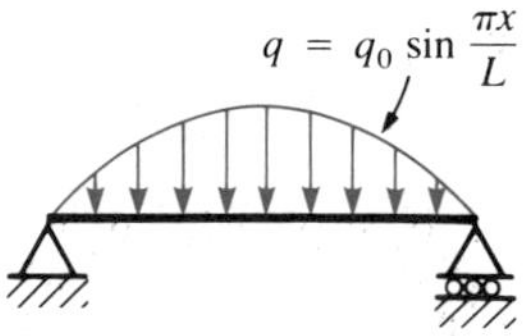

$$v = \frac{q_0 L^4}{\pi^4 EI}\sin\frac{\pi x}{L}$$

$$v' = \frac{q_0 L^3}{\pi^3 EI}\cos\frac{\pi x}{L}$$

$$\delta_c = \delta_{max} = \frac{q_0 L^4}{\pi^4 EI} \qquad \theta_a = \theta_b = \frac{q_0 L^3}{\pi^3 EI}$$

APPENDIX H

Mechanical Properties of Materials

Note: Properties of materials vary greatly depending upon manufacturing processes, chemical composition, internal defects, temperature, dimensions of test specimens, and other factors. Hence the tabulated data are representative of the material but not necessarily suitable for a specific application. In many cases, a range of values is given in the table in order to show some of the possible variations in properties. Unless otherwise indicated, the moduli of elasticity E and other mechanical properties are for materials in tension.

Table H-1 Specific weights and mass densities

Material	Specific Weight γ		Mass density ρ	
	lb/ft³	kN/m³	slugs/ft³	kg/m³
Aluminum (pure)	169	26.6	5.26	2710
Aluminum alloys	160–180	26–28	5.2–5.4	2600–2800
2014-T6	175	28	5.4	2800
6061-T6	170	26	5.2	2700
7075-T6	175	28	5.4	2800
Brass	520–540	82–85	16–17	8400–8600
Red brass (80% Cu, 20% Zn)	540	85	17	8600
Naval brass	525	82	16	8400
Bronze	510–550	80–86	16–17	8200–8800
Manganese bronze	520	82	16	8300
Cast iron	435–460	68–72	13–14	7000–7400
Concrete				
Plain	145	23	4.5	2300
Reinforced	150	24	4.7	2400
Lightweight	70–115	11–18	2.2–3.6	1100–1800
Copper	556	87	17	8900
Glass	150–180	24–28	4.7–5.4	2400–2800
Magnesium (pure)	109	17	3.4	1750
Alloys	110–114	17–18	3.4–3.5	1760–1830
Monel (67% Ni, 30% Cu)	550	87	17	8800
Nickel	550	87	17	8800
Nylon	70	11	2.2	1100
Rubber	60–80	9–13	1.9–2.5	960–1300
Steel	490	77.0	15.2	7850
Stone				
Granite	165	26	5.1	2600
Limestone	125–180	20–28	3.9–5.6	2000–2900
Marble	165–180	26–28	5.1–5.6	2600–2900
Titanium	280	44	8.7	4500
Tungsten	1200	190	37	1900
Wood (air dry)				
Douglas fir	30–35	4.7–5.5	0.9–1.1	480–560
Oak	40–45	6.3–7.1	1.2–1.4	640–720
Southern pine	35–40	5.5–6.3	1.1–1.2	560–640
Wrought iron	460–490	72–77	14–15	7400–7800

Table H-2 Moduli of elasticity and Poisson's ratios

Material	Modulus of elasticity E		Shear modulus of elasticity G		Poisson's ratio ν
	ksi	GPa	ksi	GPa	
Aluminum (pure)	10,000	70	3,800	26	0.33
Aluminum alloys	10,000–11,400	70–79	3,800–4,300	26–30	0.33
2014-T6	10,600	73	4,000	28	0.33
6061-T6	10,000	70	3,800	26	0.33
7075-T6	10,400	72	3,900	27	0.33
Brass	14,000–16,000	96–110	5,200–6,000	36–41	0.34
Red brass (80% Cu, 20% Zn)	15,000	100	5,600	39	0.34
Naval brass	15,000	100	5,600	39	0.34
Bronze	14,000–17,000	96–120	5,200–6,300	36–44	0.34
Manganese bronze	15,000	100	5,600	39	0.34
Cast iron	12,000–25,000	83–170	4,600–10,000	32–69	0.2–0.3
Concrete (compression)	2500–4500	17–31			0.1–0.2
Copper (pure)	16,000–18,000	110–120	5,800–6,800	40–47	0.33–0.36
Beryllium copper (hard)	18,000	120	6,800	47	0.33
Glass	7,000–12,000	48–83	2,800–5,000	19–34	0.20–0.27
Magnesium (pure)	6,000	41	2,200	15	0.35
Alloys	6,500	45	2,400	17	0.35
Monel (67% Ni, 30% Cu)	25,000	170	9,500	66	0.32
Nickel	30,000	210	11,400	80	0.31
Nylon	300–400	2.1–2.8			0.4
Rubber	0.1–0.6	0.0007–0.004	0.03–0.2	0.0002–0.001	0.45–0.50
Steel	28,000–30,000	190–210	10,800–11,800	75–80	0.27–0.30
Stone (compression)					
Granite	6,000–10,000	40–70			0.2–0.3
Limestone	3,000–10,000	20–70			0.2–0.3
Marble	7,000–14,000	50–100			0.2–0.3
Titanium (pure)	15,500	110	5,800	40	0.33
Alloys	15,000–17,000	100–120	5,600–6,400	39–44	0.33
Tungsten	50,000–55,000	340–380	21,000–23,000	140–160	0.2
Wood (bending)					
Douglas fir	1,600–1,900	11–13			
Oak	1,600–1,800	11–12			
Southern pine	1,600–2,000	11–14			
Wrought iron	28,000	190	10,800	75	0.3

Table H-3 Mechanical properties

Material	Yield stress σ_y		Ultimate stress σ_u		Percent elongation (2 in. gage length)
	ksi	MPa	ksi	MPa	
Aluminum (pure)	3	20	10	70	60
Aluminum alloys	5–70	35–500	15–80	100–550	1–45
2014-T6	60	410	70	480	13
6061-T6	40	270	45	310	17
7075-T6	70	480	80	550	11
Brass	10–80	70–550	30–90	200–620	4–60
Red brass (80% Cu, 20% Zn); hard	70	470	85	590	4
Red brass (80% Cu, 20% Zn); soft	13	90	43	300	50
Naval brass; hard	60	410	85	590	15
Naval brass; soft	25	170	59	410	50
Bronze	12–100	82–690	30–120	200–830	5–60
Manganese bronze; hard	65	450	90	620	10
Manganese bronze; soft	25	170	65	450	35
Cast iron (tension)	17–42	120–290	10–70	69–480	0–1
Cast iron (compression)			50–200	340–1,400	
Concrete (compression)			1.5–10	10–70	
Copper					
Hard-drawn	48	330	55	380	10
Soft (annealed)	8	55	33	230	50
Beryllium copper (hard)	110	760	120	830	4
Glass			5–150	30–1,000	
Plate glass			10	70	
Glass fibers			1,000–3,000	7,000–20,000	
Magnesium (pure)	3–10	20–70	15–25	100–170	5–15
Alloys	12–40	80–280	20–50	140–340	2–20
Monel (67% Ni, 30% Cu)	25–160	170–1,100	65–170	450–1,200	2–50
Nickel	20–90	140–620	45–110	310–760	2–50
Nylon			6–10	40–70	50
Rubber	0.2–1.0	1–7	1–3	7–20	100–800
Steel					
High-strength	50–150	340–1,000	80–180	550–1,200	5–25
Machine	50–100	340–700	80–125	550–860	5–25
Spring	60–240	400–1,600	100–270	700–1,900	3–15
Stainless	40–100	280–700	60–150	400–1,000	5–40
Tool	75	520	130	900	8

(Continued)

Table H-3 (Continued)

Material	Yield stress σ_y		Ultimate stress σ_u		Percent elongation (2 in. gage length)
	ksi	MPa	ksi	MPa	
Steel, structural	30–100	200–700	50–120	340–830	10–40
ASTM-A36	36	250	60	400	30
ASTM-A572	50	340	70	500	20
ASTM-A514	100	700	120	830	15
Steel wire	40–150	280–1,000	80–200	550–1,400	5–40
Stone (compression)					
Granite			10–40	70–280	
Limestone			3–30	20–200	
Marble			8–25	50–180	
Titanium (pure)	60	400	70	500	25
Alloys	110–130	760–900	130–140	900–970	10
Tungsten			200–600	1,400–4,000	0–4
Wood (bending)					
Douglas fir	5–8	30–50	8–12	50–80	
Oak	6–9	40–60	8–14	50–100	
Southern pine	6–9	40–60	8–14	50–100	
Wood (compression parallel to grain)					
Douglas fir	4–8	30–50	6–10	40–70	
Oak	4–6	30–40	5–8	30–50	
Southern pine	4–8	30–50	6–10	40–70	
Wrought iron	30	210	50	340	35

Table H-4 Coefficients of thermal expansion

Material	Coefficient of thermal expansion α	
	$10^{-6}/°F$	$10^{-6}/°C$
Aluminum and aluminum alloys	13	23
Brass	10.6–11.8	19.1–21.2
Red brass	10.6	19.1
Naval brass	11.7	21.1
Bronze	9.9–11.6	18–21
Manganese bronze	11	20
Cast iron	5.5–6.6	9.9–12.0
Concrete	4–8	7–14
Copper	9.2–9.8	16.6–17.6
Beryllium copper	9.4	17.0
Glass	3–6	5–11
Magnesium (pure)	14.0	25.2
Alloys	14.5–16.0	26.1–28.8
Monel (67% Ni, 30% Cu)	7.7	14
Nickel	7.2	13
Nylon	40–60	75–100
Rubber	70–110	130–200
Steel	5.5–9.9	10–18
High-strength	8.0	14
Stainless	9.6	17
Structural	6.5	12
Stone	3–5	5–9
Titanium (alloys)	4.5–5.5	8–10
Tungsten	2.4	4.3
Wrought iron	6.5	12

Answers to Selected Problems

CHAPTER 1

1.2-1 $\sigma = 10{,}700$ psi **1.2-2** $P = 19.6$ kN

1.2-3 $\sigma_c = 1590$ psi **1.2-4** $\sigma_{\max} = 34.1$ MPa

1.2-5 $\sigma = 15.3$ ksi, $\epsilon = 560 \times 10^{-6}$

1.2-6 (a) $\delta = 0.45$ mm; (b) $P = 13.0$ kN

1.2-7 $\sigma_{ab} = 3.80$ ksi, $\sigma_{bc} = 9.22$ ksi

1.2-8 (a) $\bar{x} = 0.45$ m, $\bar{y} = 0.6$m; (b) $\sigma_c = 25.0$ MPa

1.2-9 $\sigma_t = 54{,}000$ psi **1.2-10** $\sigma_t = 86.8$ MPa

1.2-11 (a) $\sigma_{ab} = Q/(A_c \tan\alpha \tan\beta)$;
(b) $\sigma_{ab} = 68{,}300$ psi, $P = 8130$ lb

1.2-12 (a) $\sigma_{ab} = 125$ MPa (tension),
$\sigma_{bc} = 60$ MPa (compression); (b) $\epsilon_{ab} = 520 \times 10^{-6}$;
(c) $\epsilon_{bc} = 248 \times 10^{-6}$

1.2-13 $\sigma_x = \gamma\omega^2(L^2 - x^2)/2g$, $\sigma_{\max} = \gamma\omega^2 L^2/2g$

1.3-1 (a) $L = 11{,}800$ ft; (b) $L = 13{,}500$ ft

1.3-2 (a) 8.2%, 8.16%, brittle; (b) 26.0%, 37.9%, ductile;
(c) 41.2%, 74.9%, ductile

1.3-3 65 ksi, 30×10^3 ksi, 69 ksi, 127 ksi, 21%, 46%

1.3-4 42 MPa, 2.4 GPa, 50 MPa, brittle

1.4-1 0.22 in. longer **1.4-2** 4.0 mm longer

1.4-3 0.070 in. **1.4-4** 1.8 mm **1.5-1** $E = 30.6 \times 10^6$ psi

1.5-2 $P = 450$ N **1.5-3** $P = 152$ k

1.5-4 (a) $E = 105$ GPa; (b) $\Delta d = 0.0103$ mm;
(c) $e = 0.000973$

1.5-5 (a) $E = 25.0 \times 10^6$ psi, $v = 0.32$; (b) $e = 0.000235$;
(c) Monel metal

1.5-6 $v = 2/7$ **1.5-7** (a) $E \approx 10{,}700$ ksi; (b) $P \approx 15$ k

1.5-8 (a) $\delta = 0.624$ mm; (b) $\Delta d = 0.0103$ mm;
(c) $\Delta t = 0.000749$ mm

1.5-9 Slope $= \dfrac{b(1 - v\sigma/E)}{L(1 + \sigma/E)}$

1.5-10 $\Delta V = WL(1 - 2v)/2E$ **1.6-1** $\tau_{\text{aver}} = 450$ psi

1.6-2 $\tau_{\text{aver}} = 91$ MPa **1.6-3** $\tau_{\text{aver}} = 3400$ psi

1.6-4 $P = 148$ kN **1.6-5** $\tau_{\text{aver}} = 75$ psi

1.6-6 $P = 70.4$ kN

1.6-7 (a) $\tau = 9780$ psi; (b) $\sigma_b = 17{,}100$ psi

1.6-8 $\tau_{\text{aver}} = 53$ MPa **1.6-9** $\sigma_b = 7650$ psi, $\tau = 8910$ psi

1.6-10 $\tau_{\text{aver}} = 1.59$ MPa **1.6-11** $\tau_{\text{aver}} = 12.5$ ksi

1.6-12 $G = 2.0$ MPa

1.6-13 (a) $\gamma_{\text{aver}} = 0.0040$; (b) $V = 89.6$ k

1.6-14 (a) $\gamma = 0.312$; (b) $\delta = 3.12$ mm

1.6-15 $\tau = \dfrac{4PL}{bc(2d + b)}$ **1.6-16** $\tau = 10.0$ MPa

1.6-17 (a) $\tau = \dfrac{P}{2\pi rh}$; (b) $\delta = \dfrac{P}{2\pi hG}\ln\dfrac{b}{d}$ **1.7-1** $d = 5.70$ in.

1.7-2 $d = 14.1$ mm **1.7-3** $d = 0.616$ in. **1.7-4** 12 bolts

1.7-5 (a) $P = \sigma_c A\sqrt{1-(R/L)^2}$; (b) $P = 2.4$ k

1.7-6 $P = 68.9$ kN **1.7-7** $P = 7360$ lb

1.7-8 $P = 25.4$ kN **1.7-9** $d = 0.620$ in.

1.7-10 $d = 310$ mm

1.7-11 (a) $d_P = 1.60$ in.; (b) $t_G = 0.267$ in.; (c) $d_B = 0.81$ in.

1.7-12 $d = 24.7$ mm

1.7-13 $A = 2M_1L\omega^2/(2\sigma_t - \rho L^2\omega^2)$

1.7-14 $\theta = 45°$ **1.7-15** $\theta = 54.7°$ **1.7-16** $\theta = 38.7°$

1.7-17 $P = 316$ lb

CHAPTER 2

2.2-1 42.554 in. **2.2-2** $t = 11.3$ mm **2.2-3** 1.65 in.

2.2-4 0.400 m **2.2-5** $E = 23.6 \times 10^6$ psi

2.2-6 $\delta = 4W/3k$ **2.2-7** $x = LL_1/(L_1 + L_2)$

2.2-8 $\delta = \dfrac{PL}{\pi E}\left(\dfrac{1}{d_1^2} + \dfrac{1}{d_2^2}\right)$ **2.2-9** $\delta_c = 0.157$ in.

2.2-10 $\delta = 1.02$ mm (elongation)

2.2-11 $\delta = 0.053$ in. (elongation) **2.2-12** $P_1 = 39$ kN

2.2-13 (a) 0.0736 in.; (b) 0.0627 in. **2.2-14** $\delta = 5.54$ mm

2.2-15 $\dfrac{P_2}{P_1} = \dfrac{L_3}{L_4}\left(1 + \dfrac{A_1L_2}{A_2L_1}\right)$

2.2-16 (a) $L_1 = 0.75$ m, $L_2 = 0.25$ m; (b) $\delta = 0.473$ mm

2.2-17 $P = 36{,}700$ lb

2.2-18 (a) $\delta = 0.0212$ mm to the left; (b) $x = 3.38$ m

2.2-19 $\delta = WL/2EA$ **2.2-20** 0.0069 mm

2.2-21 $\delta = fL^2/2EA$ **2.2-22** $d = 23.9$ mm

2.2-23 $P = 1190$ lb **2.2-24** 0.800 mm

2.2-25 $\delta = PL/2Ed^2$

2.2-26 (a) $\delta = \dfrac{PL}{Et(b_2 - b_1)}\ln\dfrac{b_2}{b_1}$; (b) $\delta = 0.085$ mm

2.2-27 $\delta = \gamma L^2/6E$ **2.2-28** $\delta = \dfrac{\rho\omega^2L^2}{6E}(3R + 2L)$

2.2-29 (a) $\delta = \dfrac{qL^3}{8hEA}\left(1 + \dfrac{16h^2}{3L^2}\right)$; $\delta = 11.1$ ft

2.3-1 1/4

2.3-2 $\delta = 2.39$ mm **2.3-3** $P_{max} = 903$ k

2.3-4 $P_{max} = 840$ kN **2.3-5** $P_{max} = 560$ k

2.3-6 $s = PL/6EA$

2.3-7 $R_a = b_2A_1P/(b_1A_2 + b_2A_1)$,
$R_b = b_1A_2P/(b_1A_2 + b_2A_1)$

2.3-8 $\sigma_a = 10.6$ MPa (comp.), $\sigma_s = 60.6$ MPa (tens.)

2.3-9 $\delta_d = 7PL/40EA_1$ (to the right),
$R_a = 7P/10$, $R_b = 3P/10$

2.3-10 $P = 4440$ N **2.3-11** $\sigma = 4000$ psi (comp.)

2.3-12 (a) $e = \dfrac{b(E_2 - E_1)}{2(E_2 + E_1)}$; (b) $P_1 = \dfrac{PE_1}{E_1 + E_2}$, $P_2 = \dfrac{PE_2}{E_1 + E_2}$

2.3-13 $x = L/4$ **2.3-14** $\sigma_2 = 68.0$ MPa

2.3-15 $d_2 = 0.492$ in., $L_2 = 60$ in.

2.3-16 $T_c = 2PaL/(2a^2 + b^2)$, $T_d = PbL/(2a^2 + b^2)$

2.3-17 $\delta_c = 0.00704$ in.

2.3-18 If $x = 0$, $F_a = F_b = F_c = W$; If $x = b$, $F_a = 0$,
$F_b = W$, $F_c = 2W$; If $x = 3b/2$, $F_a = F_b = 0$, $F_c = 3W$

2.3-19 $\delta_b = P(a + b)^2/(k_1a^2 + k_2b^2)$ **2.3-20** $T = 8P/11$

2.3-21 $T_1 = 1.868P$, $T_2 = 1.401P$ **2.3-22** $P_{allow} = 2.5$ kN

2.3-23 $h = 1.538b$

2.3-24 $\sigma_s = 49.1$ MPa, $\sigma_b = 23.4$ MPa, $\sigma_c = 28.0$ MPa

2.4-1 $d = 95.53$ ft **2.4-2** 55.9 °C and 10.4 °C

2.4-3 $\sigma = 14{,}600$ psi

2.4-4 $\delta = L(\Delta T)[\alpha_m + (b/a)(\alpha_m - \alpha_t)]$;
yes, because δ is proportional to ΔT

2.4-5 $\Delta T = P/[E_cA_c(\alpha_c - \alpha_s)]$ **2.4-6** $\sigma_c = E\alpha(\Delta T_1)/3$

2.4-7 $\Delta T = 57.0$ °F

2.4-8 (a) $\sigma = 98.8$ MPa; (b) $T = 33.6$ °C

2.4-9 $\delta = 0.148$ in. **2.4-10** $\sigma = 71.5$ MPa

2.4-11 $\sigma = 19.2$ ksi (tension)

2.4-12 (a) $P = 46.3$ kN; (b) $\sigma = 38.6$ MPa;
(c) $\delta = 0.43$ mm to the left

2.4-13 (a) $\sigma = \dfrac{2E\alpha(\Delta T)A_2}{A_1 + A_2}$; (b) $\delta = \dfrac{\alpha(\Delta T)(L)(A_2 - A_1)}{2(A_1 + A_2)}$

2.4-14 $P = 1.80$ MN

2.4-15 (a) $\Delta T = 190$ °F; (b) $\sigma = 9680$ psi (comp.)

2.4-16 $\Delta T = 46.8$ °C **2.4-17** $\tau_{aver} = 8070$ psi

2.4-18 (a) $T_a = 1600$ N, $T_b = 800$ N;
(b) $T_a = 1920$ N, $T_b = 154$ N; (c) $\Delta T = 161$ °C

2.4-19 $\sigma_s = 24.2$ ksi (tension), $\sigma_c = 12.1$ ksi (comp.)

2.4-20 $\sigma_s = 600$ MPa, $\sigma_c = 20$ MPa

2.4-21 If $0 \leq h/L \leq 0.25$, $P_a = 1600$ lb, $P_b = 0$;
If $h/L = 1.0$, $P_a = 2800$ lb, $P_b = 1200$ lb

2.4-22 $\sigma_s = \dfrac{(\alpha_c - \alpha_s)(\Delta T)E_s E_c}{E_c + 2E_s}$ (tension), $\sigma_c = 2\sigma_s$ (comp.)

2.4-23 Rope B should be 2.5 in. shorter than rope A

2.4-24 $\sigma_p = 24.0$ MPa **2.5-1** $\tau_{max} = 14{,}500$ psi

2.5-2 $P = 353$ kN **2.5-3** $P = 72.0$ k

2.5-4 $d = 40.3$ mm

2.5-5 (a) For $\theta = 45°$: $\sigma_\theta = -7.5$ ksi, $\tau_\theta = 7.5$ ksi;
(b) For $\theta = 35°$: $\sigma_\theta = -10.1$ ksi, $\tau_\theta = 7.0$ ksi

2.5-6 (a) For $\theta = 45°$: $\sigma_\theta = 50$ MPa, $\tau_\theta = -50$ MPa;
(b) For $\theta = 65°$: $\sigma_\theta = 17.9$ MPa, $\tau_\theta = -38.3$ MPa

2.5-7 $\Delta T = 60.2$ °F

2.5-8 (a) $\sigma_\theta = 3.21$ MPa, $\tau_\theta = -6.89$ MPa; (b) $\alpha = 50.2°$

2.5-9 $\tau = 1150$ psi **2.5-10** $\Delta T = 31.3$ °C

2.5-11 $b_{min} = 3.97$ in.

2.5-12 $\sigma_\theta = 67.5$ MPa, $\tau_\theta = -39.0$ MPa

2.5-13 $\theta = 35.26°$, $\tau_\theta = -8485$ psi, $\sigma_{max} = 18{,}000$ psi,
$\tau_{max} = 9000$ psi

2.5-14 For $\theta = 40°$: $\sigma_\theta = -46.9$ MPa, $\tau_\theta = 39.4$ MPa

2.5-15 $\theta = 70°$ **2.5-16** $\theta = 30.96°$, $P_{max} = 18.4$ kN

2.6-1 (a) $U = 5P^2L/16EA$; (b) $U = 3.75$ in.-lb

2.6-2 (a) $U = P^2L/EA$; (b) $U = 3.75$ J

2.6-3 Aluminum: $u_r = 119$ psi, $u_w = 1210$ in.

2.6-4 (a) $U_1 = P^2L/2EA$; (b) $U_2 = Q^2L/4EA$;
(c) $U = L(2P^2 + 2PQ + Q^2)/4EA$

2.6-5 (a) $U = 5EA\delta^2/3b$; (b) $\delta = 3Pb/10EA$

2.6-6 $U = \pi\gamma^2 d^2 L^3/360E$

2.6-7 (a) $U = 2P^2L/\pi E d_1 d_2$; (b) $\delta = 4PL/\pi E d_1 d_2$

2.6-8 (a) $U = \dfrac{P^2L}{2Et(b_2 - b_1)} \ln\dfrac{b_2}{b_1}$; (b) $\delta = \dfrac{PL}{Et(b_2 - b_1)} \ln\dfrac{b_2}{b_1}$

2.6-9 $U = 2\rho^2 A\omega^4 L^5/15E$

2.6-10 (a) $U_1 = 2P_1^2 a/EA$; (b) $U_2 = 21P_2^2 a/2EA$;
(c) $U = a(12P_1^2 + 32P_1P_2 + 63P_2^2)/6EA$

2.6-11 (a) $P = 72.2$ k; (b) $\delta = 0.05026$ in.;
(c) $U = 2330$ in.-lb;
(d) U is not equal to $P\delta/2$ because the load-displacement diagram is not linear.

2.6-12 (a) $U = P^2L/2\sqrt{2}EA$; (b) $\delta = PL/\sqrt{2}EA$

2.6-13 (a) $U = 1.457P^2L/EA$; (b) $\delta = 2.914PL/EA$

2.6-14 (a) $U = P^2L/(4EA \sin\beta \cos^2\beta)$; (b) $\beta = 35.26°$;
(c) $3\sqrt{3}PL/4EA$

2.6-15 $U = (4k^2s^2 - 4Wks + 3W^2)/12k$

2.6-16 $U = 134$ in.-lb **2.7-1** $v = 39.3$ ft/s

2.7-2 (a) Impact factor $= 1 + (1 + 2EA/W)^{1/2}$;
(b) 5.58

2.7-3 (a) $\sigma_{max} = 2340$ psi; (b) $\sigma_{max} = 2100$ psi

2.7-4 $h = 650$ mm **2.7-5** $h = 7.14$ in. **2.7-6** $\sigma_2 = \sigma_1/2$

2.7-7 $\sigma_{max} = 77.0$ ksi **2.7-8** $\sigma_{max} = 640$ MPa

2.7-9 $\delta_{max} = 4.39$ in. **2.7-10** $L = 25.6$ m

2.7-11 $\delta_{max} = 8.01$ in. **2.7-12** $\delta_{max} = 193$ mm

2.7-13 $\sigma_{max} = 4920$ psi

2.7-14 $\delta_{max} = 0.267$ mm, $\sigma_{max} = 164$ MPa

2.7-15 $\delta_{max} = 0.0123$ in., $\sigma_{max} = 27.5$ ksi

2.7-16 $\delta_{max} = 0.277$ mm, $\sigma_{max} = 27.7$ MPa

2.8-2 For $P = 3.2$ kN: $\delta_b = 4.85$ mm;
For $P = 4.8$ kN: $\delta_b = 17.32$ mm

2.8-3 $n = 204$ rpm **2.8-4** $\delta = 528$ mm

2.8-5 (c) $\delta = 5.6$ in.

2.8-6 (a) $P_u = 5\sigma_y A$; (b) P_u is not changed

2.8-7 $P_u = \sigma_y(A_1 + A_2)$ **2.8-8** $P_u = 88.4$ kN

2.8-9 $P_u = 22.2$ k **2.8-10** $P_u = 3.77\sigma_y A$

2.8-11 $P_u = \sigma_y A_b$ **2.8-12** $P_u = \sqrt{2}\,\sigma_y A$

2.8-13 $P_y = \sigma_y A$, $\delta_y = 3\sigma_y L/2E$; $P_u = 4\sigma_y A/3$, $\delta_u = 3\sigma_y L/E$

2.8-14 (a) $P_y = \sigma_y A$, $\delta_y = \sigma_y L/E$;
(b) $P_u = 5\sigma_y A/4$, $\delta_u = 2\sigma_y L/E$

2.8-15 (a) $W_y = 8000$ lb, $\delta_y = 6.0$ in.;
(b) $W_u = 12{,}000$ lb, $\delta_u = 10.0$ in.

2.8-16 $P_y = 624$ kN, $\delta_y = 0.475$ mm;
$P_u = 925$ kN, $\delta_u = 0.775$ mm

2.8-17 $F_1 = 1155$ lb, $F_2 = 690$ lb, $\delta = 0.242$ in.

2.10-1 (a) $\sigma_{max} = 6.2$ ksi and 6.9 ksi;
(b) $\sigma_{max} = 11.0$ ksi and 9.0 ksi

2.10-2 (a) $\sigma_{max} = 26$ MPa and 29 MPa;
(b) $\sigma_{max} = 25$ MPa and 22 MPa

2.10-3 $P_{max} = \sigma_t bt/3$

2.10-4 No, it makes it weaker; $P_1 = 25$ kN; $P_2 = 14$ kN

2.10-5 (a) $P_{max} = 850$ lb; (b) Make d smaller

2.10-6 (a) $P_{max} = 18$ kN; (b) Make d smaller

2.10-7 $d_{max} = 0.51$ in.

CHAPTER 3

3.2-1 $L = 43.0$ in. **3.2-2** $L = 5.03$ m

3.2-3 $\tau_{max} = 3450$ psi, $\gamma_{max} = 0.000909$

3.2-4 $T = 76.3$ N·m **3.2-5** $T = 270$ in.-lb, $\phi = 3.85°$

3.2-6 $\tau_{max} = 54.0$ MPa

3.2-7 $\tau_{max} = 4850$ psi, $G = 4.00 \times 10^6$ psi

3.2-8 $G = 28.0$ GPa **3.2-9** $\tau_1 = 4740$ psi, $\tau_2 = 6830$ psi

3.2-10 $T = 9140$ N·m **3.2-11** $d = 2.75$ in.

3.2-12 (a) $\phi = 4.91°$; (b) $d = 76.8$ mm

3.2-13 (a) $d_H/d_S = 1.279$; (b) $W_H/W_S = 0.454$

3.2-14 (a) Percent area removed $= 100\,\beta^2$;
(b) Percent reduction in torque $= 100\,\beta^4$

3.3-1 (a) $\tau_{max} = 4890$ psi; (b) $\phi = 0.87°$

3.3-2 (a) $\tau_{max} = 66.0$ MPa; (b) $\phi = 2.44°$

3.3-3 $T = 1810$ in.-lb **3.3-4** $d = 28.7$ mm

3.3-5 $d_{ab} = 1.60$ in., $d_{bc} = 1.78$ in., $d_{cd} = 1.53$ in.

3.3-6 $d = 81.3$ mm

3.3-7 $d_{ab} = 1.67$ in., $d_{bc} = 1.83$ in., $d_{cd} = 1.61$ in.

3.3-8 $d_b/d_a = 1.45$ **3.3-9** $\phi = 3TL/2\pi G t d_a^3$

3.3-10 $\phi = qL^2/2GI_p$ **3.3-11** $\phi = q_0L^2/6GI_p$

3.3-12 $\phi = WcL/12GI_p$ **3.4-1** 6760 psi

3.4-2 $\sigma_{max} = 42.0$ MPa, $T = 7.73$ kN·m

3.4-3 $d = 1.43$ in.

3.4-4 $\sigma_{max} = 60.0$ MPa, $T = 23.5$ kN·m

3.4-5 $d = 0.60$ in. **3.5-1** $\gamma = 690 \times 10^{-6}$

3.5-2 $\gamma_{max} = 2260 \times 10^{-6}$, $\epsilon_{max} = 1130 \times 10^{-6}$

3.5-3 $G = 11.5 \times 10^6$ psi **3.5-4** $G = 8T/\pi d^3\epsilon$

3.5-5 $\epsilon = 904 \times 10^{-6}$, $\gamma = 1810 \times 10^{-6}$

3.6-1 $P = 20{,}820$ ft-lb/sec **3.6-2** $P = 18.9$ kW

3.6-3 $H = 6560$ hp **3.6-4** $d = 91.3$ mm

3.6-5 $d = 3.11$ in. **3.6-6** $d = 61.3$ mm

3.6-7 $d = 4.28$ in. **3.6-8** $P = 182$ kW

3.6-9 (a) $d = 4.59$ in.; (b) 4.5 in. **3.6-10** $d_1 = 1.221d$

3.6-11 (a) $d = 2.19$ in.; (b) $d = 2\text{-}1/8$ in.

3.6-12 $d = 120$ mm **3.6-13** $d = 3.90$ in.

3.7-1 $T_a = [T_1(b + c) + T_2c]/L$, $T_b = [T_1a + T_2(a + b)]/L$

3.7-2 $T_a = T_d = T_0/3$, $\phi_b = T_0L/9GI_p$, $\phi_m = 0$

3.7-3 $P = 895$ lb **3.7-4** $\phi = 2b\tau_{allow}/Gd$

3.7-5 $T = 1170$ in.-lb **3.7-6** $T = 814$ N·m

3.7-7 $a = d_aL/(d_a + d_b)$, $b = d_bL/(d_a + d_b)$

3.7-8 $a/L = (d_1/d_2)^4$ **3.7-9** $x = L(3 - I_{pb}/I_{pa})/4$

3.7-10 $T = 2GI_pbck/(GI_p + 2c^2kL)$

3.7-11 (a) $\tau_t = 4750$ psi, $\tau_b = 3160$ psi; (b) $\phi = 0.41°$;
(c) $k = 2.52 \times 10^6$ in.-lb/rad

3.7-12 $T_a = q_0L/6$, $T_b = q_0L/3$

3.7-13 $T_a = q_0L/12$, $T_b = q_0L/4$

3.7-14 $\tau_s = 82.4$ MPa, $\tau_b = 26.8$ MPa

3.7-15 (a) $\tau_s = 13{,}710$ psi, $\tau_b = 5{,}560$ psi; (b) $\phi = 1.71°$

3.7-16 $T = 5130$ N·m **3.7-17** $T = 506$ in.-k

3.7-18 (a) $T_1 = 8.57$ kN·m; (b) $T_2 = 14.66$ kN·m;
(c) $T_3 = 4.71$ kN·m

3.8-1 $U = 56.5$ in.-lb **3.8-2** $U = 388$ J

3.8-3 $U = 10.0$ in.-lb **3.8-4** $U = q_0^2L^3/40GI_p$

3.8-5 (a) $U_1 = T^2L/2GI_p$; (b) $U_2 = q^2L^3/6GI_p$;
(c) $U_3 = L(3T^2 + 3TqL + q^2L^2)/6GI_p$

3.8-6 (a) $U = \dfrac{T^2L(d_a + d_b)}{\pi G t\, d_a^2 d_b^2}$; (b) $\phi = \dfrac{2TL(d_a + d_b)}{\pi G t\, d_a^2 d_b^2}$

3.8-7 $U = \dfrac{19T_0^2L}{32GI_p}$ **3.8-8** $\phi = \dfrac{L_1L_2T}{G(L_1I_{p2} + L_2I_{p1})}$

3.8-9 $U = \dfrac{\beta^2 G\, I_{pa}I_{pb}}{2(L_aI_{pb} + L_bI_{pa})}$

3.9-1 (a) $\tau_{max} = 7640$ psi; (b) $\tau_{max} = 8320$ psi

3.9-2 $\phi_1/\phi_2 = 1 + 1/4\beta^2$ **3.9-3** $U_1/U_2 = 2$

3.9-4 (a) $\tau = 20.0$ MPa; (b) $\phi = 0.771°$

3.9-5 (a) $t = 0.139$ in.; (b) $t = 0.144$ in.

3.9-6 $T = 7790$ N·m, $\theta = 0.827°$ per meter

3.9-7 $\tau = 6370$ psi, $\theta = 0.0128°$ per inch

3.9-8 $\tau = 52.5$ MPa, $\phi = 0.855°$ **3.9-9** $t = 0.140$ in.

3.9-10 $\tau = T\sqrt{3}/9b^2t$, $\theta = 2T/9Gb^3t$

3.9-11 $\tau = \dfrac{2T(1 + \beta)^2}{tL_m^2\beta}$ **3.9-12** $\theta = \dfrac{4T(1 + \beta)^4}{GtL_m^3\beta^2}$

3.9-13 $\phi = \dfrac{2TL}{\pi G t}\left(\dfrac{d_a + d_b}{d_a^2 d_b^2}\right)$ **3.10-1** $T_{max} \approx 16{,}000$ in.-lb

3.10-2 $R_{min} \approx 4$ mm

3.10-3 For $D_1 = 0.8$ in.: $\tau_{max} = 12{,}800$ psi

3.10-4 (a) $D_2 \approx 116$ mm; (b) lower limit

3.10-5 $D_1 = 1.31$ in. **3.11-2** $\tau_{max} = \dfrac{Tr}{I_p}\left(\dfrac{3n+1}{4n}\right)$

3.11-3 $\dfrac{T_u}{T_y} = \dfrac{4(1-\beta^3)}{3(1-\beta^4)}$ with $\beta = \dfrac{r_1}{r_2}$

3.11-4 $T = \dfrac{T_y}{3}\left[4 - \left(\dfrac{\gamma_y}{\gamma_{max}}\right)^3\right]$ for $T_y \leq T \leq T_u$

3.11-5 (a) $T = 1161$ in.-k; (b) $T_u = 1188$ in.-k; (c) $\gamma_u = 0.004$

3.11-6 $\tau_{max} = 50.8$ MPa

CHAPTER 4

4.2-1 $V = 1.5$ k, $M = 35.0$ ft-k

4.2-2 $V = 10$ kN, $M = -17$ kN·m

4.2-3 $V = 2Pa/b$, $M = 0$

4.2-4 $V = 1.25$ kN, $M = 9.75$ kN·m

4.2-5 $V = -2060$ lb, $M = -7960$ ft-lb

4.2-6 $V = -1.25$ kN, $M = -6.5$ kN·m

4.2-7 $M_{max} = 480$ ft-lb

4.2-8 $V = -1.67$ kN, $M = 20.0$ kN·m

4.2-9 $N = P\sin\theta$, $V = P\cos\theta$, $M = Pr\sin\theta$

4.2-10 (a) $M = qx(4b - 3x)/6$, (b) $M_{max} = 2qb^2/9$

4.2-11 $P = 10$ k

4.2-12 $N = 24$ kN (comp.), $V = 8.0$ kN, $M = 56.0$ kN·m

4.2-13 (a) $V_b = 4800$ lb, $M_b = 7200$ ft-lb; (b) $V_m = 0$, $M_m = 16{,}800$ ft-lb

4.2-14 $a/L = 1/4$ **4.2-15** $V = 8.4$ k, $M = -34.0$ ft-k

4.2-16 $V_{max} = 229wL^2\alpha/18g$, $M_{max} = 617wL^3\alpha/45g$

4.4-1 $V_{max} = P$, $M_{max} - Pa$

4.4-2 $V_{max} = qL/2$, $M_{max} = -3qL^2/8$

4.4-3 $V_{max} = M_0/L$, $M_{pos} = M_0 a/L$

4.4-4 $V_{max} = -7P/12$, $M_{max} = 7PL/36$

4.4-5 $V_{max} = P/2$, $M_{max} = 3PL/8$

4.4-6 $V_{max} = -14.4$ kN, $M_{max} = 46.1$ kN·m

4.4-7 $V_{max} = -3M_1/L$, $M_{max} = M_1$

4.4-8 $V_{max} = P$, $M_{max} = -Pa$

4.4-9 $V_{max} = 1.0$ k, $M_{max} = -8.0$ ft-k

4.4-10 $V_{max} = -5.0$ kN, $M_{max} = -8.0$ kN·m

4.4-11 $V_{pos} = 5.5$ k, $M_{max} = 36.1$ ft-k

4.4-12 $V_{max} = 10$ kN, $M_{max} = -22$ kN·m

4.4-13 $V_{pos} = 2Pa/b$, $M_{max} = Pa$

4.4-14 $V_{pos} = 5.25$ kN, $M_{max} = 11.62$ kN·m

4.4-15 $V_{max} = 2940$ lb, $M_{pos} = 8640$ ft-lb

4.4-16 $V_{max} = 9.0$ kN, $M_{max} = -9.0$ kN·m

4.4-17 $V_{max} = -q_0 L/2$, $M_{max} = -q_0 L^2/6$

4.4-18 $a = 0.586L$, $V_{max} = 0.293qL$, $M_{max} = 0.0214qL^2$

4.4-19 $V_{max} = -7q_0 L/18$, $M_{pos} = 2\sqrt{2}\,q_0 L^2/81$

4.4-20 $V_{max} = 13.0$ kN, $M_{max} = -28.0$ kN·m

4.4-21 $V_{max} = -6.2$ k, $M_{max} = -34.5$ ft-k

4.4-22 $V_{max} = -q_0 L/3$, $M_{max} = q_0 L^2/9\sqrt{3}$

4.4-23 $V_{max} = -7000$ lb, $M_{max} = 22{,}360$ ft-lb

4.4-24 $V_{neg} = -q_0 L/12$, $M_{max} = 0.01604 q_0 L^2$

4.4-25 $V_{max} = -165$ lb, $M_{max} = 2722$ in.-lb

4.4-26 $V_{max} = -12.5$ kN, $M_{max} = 20.0$ kN·m

4.4-27 $V_{max} = -1560$ lb, $M_{max} = -2880$ ft-lb

4.4-28 $V_{max} = 32.97$ kN, $M_{max} = 61.15$ kN·m

4.4-29 $V_{max} = 12.5$ k, $M_{pos} = 78.12$ ft-k

4.4-30 $V_{max} = -12.0$ kN, $M_{max} = -24.0$ kN·m

4.4-31 $b = L/3$, $V_{max} = 5q_0 L/36$, $M_{max} = -q_0 L^2/81$

4.4-32 $V_{max} = qL$, $M_{max} = -3qL^2/8$

4.4-33 $V_{max} = -q_0 L/3$, $M_{max} = 3q_0 L^2/64$

4.4-34 $V_{max} = 2.5$ kN, $M_{max} = 5.0$ kN·m

4.4-35 $M_{max} = 20{,}000$ ft-lb

4.4-36 $M_{max} = 30$ kN·m

4.4-37 $V_{max} = -9q_0 L/32$, $M_{max} = 3q_0 L^2/64$

4.4-38 $q_a = 2P/3L$, $V_{max} = 11P/9$, $M_{max} = -3PL/1$

4.4-39 (a) $x = 0$ and $x = L - d$, $V_{max} = P(2 - d/L)$, (b) $x = (2L - d)/4$, $M_{max} = P(L - d/2)^2/2L$

4.4-40 (a) $x = 6.4$ m, $V_{max} = 16.8$ kN; (b) $x = 2.67$ m, $M_{max} = 31.4$ kN·m

4.4-41 $x = 24$ ft, $M_{max} = 535$ ft-k

4.4-42 $M_{max} = 0.560 \text{ kN·m}$

4.4-43 $M_{max} = \frac{PL}{8}\left(\frac{n+2}{n+1}\right)$ for n even;

$M_{max} = \frac{PL}{8}\left(\frac{n+1}{n}\right)$ for n odd

4.4-44 (a) $a/L = 0.4749$; (b) $M_{max} = 0.01785 q_0 L^2$

CHAPTER 5

5.2-1 $\epsilon_{max} = 1560 \times 10^{-6}$ **5.2-2** $L_{min} = 3.93$ m

5.2-3 $\rho = 2500$ in., $\kappa = 0.0004$ in.$^{-1}$, $\delta = 0.461$ in.

5.2-4 (a) $\rho = 534$ mm; (b) $\epsilon = 0.00187$

5.3-1 (a) $\sigma_{max} = 58{,}500$ psi; (b) Decrease

5.3-2 (a) $\sigma_{max} = 188$ MPa; (b) Increase

5.3-3 (a) $\sigma_{max} = 42{,}950$ psi; (b) Increase

5.3-4 $\sigma_{max} = 7.29$ MPa **5.3-5** $\sigma_{max} = 1760$ psi

5.3-6 $\sigma_{max} = 35.4$ MPa **5.3-7** $\sigma_{max} = 11.6$ ksi

5.3-8 (a) $\sigma_{max} = 160$ MPa; (b) $\sigma_c = 80$ MPa

5.3-9 (a) $\sigma_{max} = 18$ ksi; (b) $\Delta b = 0.00036$ in. (increase)

5.3-10 $\sigma_{max} = 2.04$ MPa **5.3-11** $\sigma_{max} = 1010$ psi

5.3-12 (a) $\sigma_{max} = 768\ M/65\pi d^3$; (b) $\sigma_{max} = 32\sqrt{2}\ M/\pi d^3$; (c) $\sigma_{max} = 8M/5b^3$

5.3-13 (a) $\sigma_t = 16\ M/b^3$; (b) $\sigma_t = 768\ M/(9\pi^2 - 64)d^3$; (c) $\sigma_t = 168\ M/37bh^2$

5.3-14 (a) $\sigma_c = 32\ M/b^3$; (b) $\sigma_c = 192(3\pi - 4)M/(9\pi^2 - 64)d^3$; (c) $\sigma_c = 192\ M/37bh^2$

5.3-15 $\sigma_{max} = 21.4$ ksi

5.3-16 (a) $\sigma_{max} = 4.61$ MPa; (b) $\sigma_{max} = 97.2$ kPa; (c) $\sigma_{max} = 4.61$ MPa

5.3-17 $\sigma_t = 4{,}240$ psi, $\sigma_c = 14{,}900$ psi

5.3-18 $\sigma_{max} = 3\rho L^2 a_0/t$

5.3-19 $\sigma_t = 8.15$ ksi, $\sigma_c = 13.59$ ksi

5.3-20 $\sigma_t = 41.0$ MPa, $\sigma_c = 86.3$ MPa

5.3-21 $\sigma_t = 8190$ psi, $\sigma_c = 14{,}600$ psi

5.3-22 $\sigma_t = 141.3$ MPa, $\sigma_c = 55.9$ MPa

5.3-23 $d = 3$ ft, $\sigma_{max} = 171$ psi; $d = 6$ ft, $\sigma_{max} = 830$ psi

5.4-1 W 8 × 21 **5.4-2** W 8 × 21 **5.4-3** S 8 × 18.4

5.4-4 $d_{min} = 15.4$ mm **5.4-5** $d_{min} = 8.19$ in.

5.4-6 $d_{min} = 19.9$ mm

5.4-7 (a) $S = 15.41$ in.^{3}; (b) S 8 × 23 **5.4-8** $b = 136$ mm

5.4-9 $q = 496$ lb/ft

5.4-10 (a) $b = 149$ mm; (b) = 151 mm

5.4-11 $b = d/\sqrt{3}$, $h = d\sqrt{2/3}$ **5.4-12** $t = 10$ mm

5.4-13 $b = 6$ in. **5.4-14** $b = 259$ mm

5.4-15 1:1.260:1.408 **5.4-16** $r = 1.07$

5.4-17 $S = 12.86$ in.^{3}; 2 × 8 in. joist **5.4-18** $s = 15.6$ in.

5.4-19 $s = 72.0$ in. **5.4-20** $q = 13.4$ kN/m

5.4-21 $t_{min} = \sqrt{8\gamma h^3/\sigma_{allow} \sin^2\alpha}$

5.4-22 (a) Stronger when bent about axis 1-1; (b) $s = 1.184$

5.4-23 (a) $\beta = 1/9$; (b) 5.35%

5.5-2 (a) $\tau_{max} = (h/L)\sigma_{max}$; (b) $\tau_{max} = (h/2L)\sigma_{max}$

5.5-3 $\tau_{max} = 101$ psi, $\sigma_{max} = 729$ psi

5.5-4 $\tau_{max} = 189$ kPa, $\sigma_{max} = 15.1$ MPa

5.5-5 $\tau_{max} = 3000$ psi **5.5-6** $\tau_{max} = 750$ kPa

5.5-7 $P = 900$ lb **5.5-8** $b = 172$ mm

5.5-9 $L_0 = h\sigma_{allow}/2\tau_{allow}$ **5.5-10** $L_0 = 2.46$ m

5.5.11 $P = 6870$ lb

5.5-12 (a) $P = 402$ N; (b) $P = 448$ N

5.5-13 (a) $P = 1050$ lb; (b) $P = 1430$ lb

5.5-14 (a) $W = 28.6$ kN; (b) $W = 41.3$ kN

5.5-15 $w_1 = 118$ lb/ft^2, $w_2 = 320$ lb/ft^2, $w_3 = 121$ lb/ft^2, $w_4 = 86.9$ lb/ft^2

5.6-1 (a) $\tau_{max} = 5800$ psi; (b) $\tau_{aver} = 5710$ psi

5.6-2 (a) $\tau_{max} = 28.4$ MPa; (b) $\tau_{aver} = 27.4$ MPa

5.6-3 $\tau_{max} = 2630$ psi; (b) $\tau_{aver} = 2520$ psi

5.6-4 (a) $\tau_{max} = 41.8$ MPa, $\tau_{min} = 30.7$ MPa; (b) $\tau_{aver} = 39.5$ MPa; (c) $V_{web} = 261$ kN

5.6-5 (a) $\tau_{max} = 2920$ psi, $\tau_{min} = 2160$ psi; (b) $\tau_{aver} = 2750$ psi; (c) $V_{web} = 31.0$ k

5.6-6 (a) $q = 144$ kN/m; (b) $q = 174$ kN/m

5.6-7 (a) $q = 1.16$ k/ft; (b) $q = 2.83$ k/ft

5.6-8 $\tau_{max} = 7.63$ MPa, $\tau_{min} = 5.53$ MPa

5.6-9 $\tau_{max} = 3070$ psi, $\tau_{min} = 2060$ psi

5.6-10 $\tau_{max} = 19.5$ MPa **5.6-11** $\tau_{max} = 1530$ psi

5.6-12 $\sigma_{max} = 56.2$ MPa, $\tau_{max} = 1.50$ MPa

5.6-13 (a) $\tau_{max} = 3810$ psi, $\tau_{min} = 3200$ psi; (b) $\tau_{aver} = 4000$ psi; (c) $V_{web} = 14{,}600$ lb

5.6-14 $\tau_{max} = 5.53$ MPa **5.7-1** $F = 1900$ lb/in.

5.7-2 $V = 1.35$ MN **5.7-3** $s = 3.65$ in.

5.7-4 $V = 10.7$ kN **5.7-5** $V = 1630$ lb

5.7-6 $s = 85.1$ mm **5.7-7** (a) $s = 2.77$ in.; (b) $s = 1.85$ in.

5.7-8 $s = 5.42$ in. **5.7-9** $s = 6.03$ in. **5.7-10** $V = 11.8$ k

5.8-1 $x = L/3$, $\sigma_{max} = 256PL/81\pi d_a^3$, $\sigma_{max}/\sigma_b = 125/81$

5.8-2 $1 \le d_b/d_a \le 3/2$

5.8-3 $x = L/4$, $\sigma_{max} = 4PL/9h_a^3$, $\sigma_{max}/\sigma_b = 2$

5.8-4 At $x = \dfrac{L}{4}$: $\tau = \dfrac{24P}{25bh_a}\left(1 - \dfrac{32y_1^2}{25h_a^2}\right)$

5.8-5 $x = 8$ in., $\sigma_{max} = 1250$ psi, $\sigma_{max}/\sigma_b = 25/24$

5.8-6 $x = 0.45$ m, $\sigma_{max} = 268$ MPa

5.8-7 At $x = 0$, $\tau_{max} = 30$ psi and $\tau_{min} = 22.5$ psi; at $x = 10$ in., $\tau_{max} = 31.2$ psi and $\tau_{min} = 14.4$ psi

5.8-8 (a) $d_b/d_a = 1.9059$; (b) $V_{min} = 2.405L(PL/\sigma_{allow})^{2/3}$; (c) $V_{min}/V_{prismatic} = 0.6518$

5.8-9 $b = 6Px/h^2\sigma_{allow}$ **5.8-10** $h = x\sqrt{3q/b\sigma_{allow}}$

5.8-11 $h = \left[\dfrac{q_0L^2}{4b\sigma_{allow}}\left(1 - \dfrac{8x^3}{L^3}\right)\right]^{1/2}$

5.8-12 $h = \left[\dfrac{3PL}{2b\sigma_{allow}}\left(1 - \dfrac{4x^2}{L^2}\right)\right]^{1/2}$

5.9-1 $\sigma_t = 8P/b^2$, $\sigma_c = -4P/b^2$

5.9-2 $\sigma_t = 9.11P/b^2$, $\sigma_c = -6.36P/b^2$

5.9-3 $t_{min} = 0.477$ in. **5.9-4** $T_{allow} = 99.9$ kN

5.9-5 $\sigma_t = 3830$ psi, $\sigma_c = -3950$ psi

5.9-6 $\sigma_t = 8.33$ MPa, $\sigma_c = -8.74$ MPa

5.9-7 $\alpha = (d_2^2 + d_1^2)/4hd_2$ **5.9-8** $d = 66.2$ mm

5.9-9 $b = 19.3$ in.

5.9-10 (a) $\sigma_t = 826$ kPa, $\sigma_c = -918$ kPa; (b) $d_{max} = 0.750$ m

5.9-11 (a) $\sigma_t = 87.6$ psi, $\sigma_c = -99.6$ psi; (b) $d_{max} = 28.9$ in.

5.9-12 $d_{max} = (ht^2\gamma_c/\gamma_w)^{1/3}$ **5.9-13** $W = 7560$ lb

5.9-14 $s = h^2/[12(L - x)]$

5.9-15 (a) $y_0 = 0.928$ in.; (b) $P_{max} = 29{,}600$ lb

5.9-16 (a) $y_0 = 0.847$ in.; (b) $P_{max} = 32{,}300$ lb

5.9-17 (a) $0.765z_0 - 0.264y_0 = -1$; (b) $\sigma_t = 7170$ psi, $\sigma_c = -10{,}580$ psi

5.9-18 (a) $0.644z_0 - 0.217y_0 = -1$; (b) $\sigma_t = 6670$ psi, $\sigma_c = -9790$ psi

5.9-19 (a) $r_c = (r_1^2 + r_2^2)/4r_2$

5.9-20 Rhombus with diagonals of lengths 5.90 in. and 1.61 in.

5.9-21 Rhombus with diagonals of lengths 8.79 in. and 1.90 in.

5.9-22 Square with sides of length $0.3200b$

5.9-23 Equilateral triangle with same centroid and sides of length $b/4$

5.9-24 $c_{max} = 0.172b$

5.10-1 (a) $d = 0.50$ in., $\sigma_{max} = 15{,}500$ psi; (b) $R = 0.10$ in., $\sigma_{max} \approx 49{,}000$ psi

5.10-2 (a) $d = 16$ mm, $\sigma_{max} = 81$ MPa; (b) $R = 4$ mm, $\sigma_{max} = 200$ MPa

5.10-3 $b_{min} = 0.24$ in. **5.10-4** $b_{min} = 33$ mm

5.10-5 (a) $R_{min} = 0.45$ in.; (b) $d_{max} = 4.13$ in.

5.11-1 $\sigma_s = 13{,}700$ psi, $\sigma_w = 660$ psi

5.11-2 $\sigma_s = 62.3$ MPa, $\sigma_w = 2.3$ MPa

5.11-3 $M_{max} = 1030$ in.-k **5.11-4** $M_{max} = 67.9$ kN·m

5.11-5 $M_{max} = 207$ in.-k **5.11-6** $M = \dfrac{\pi d^3\sigma_s}{512}\left(15 + \dfrac{E_a}{E_s}\right)$

5.11-7 $t = 0.525$ in.

5.11-8 $\sigma_s = 41.8$ MPa (tension), $\sigma_w = 5.6$ MPa (compression)

5.11-9 $\sigma_s = 8490$ psi (tension), $\sigma_w = 1510$ psi (compression)

5.11-10 $\sigma_s = 93.5$ MPa

5.11-11 $\sigma_a = 4120$ psi, $\sigma_c = 5230$ psi

5.11-12 $S = 50.6\ \text{mm}^3$; Metal A

5.11-13 $M_{max} = 179$ in.-k **5.11-14** $M_{max} = 55.1$ kN·m

5.12-2 $\sigma_{max} = 7.43$ MPa **5.12-3** $\sigma_{max} = 1790$ psi

5.12-4 $\sigma_{max} = 14{,}780$ psi **5.12-5** $\sigma_{max} = 15{,}060$ psi

5.12-6 $\sigma_{max} = 10{,}760$ psi

5.12-7 $\sigma_a = -\sigma_e = 4930$ psi, $\sigma_b = -\sigma_d = -900$ psi

5.12-8 $\sigma_a = -\sigma_e = 8220$ psi, $\sigma_b = -\sigma_d = -1170$ psi

5.12-9 $\sigma_{max} = 712$ psi **5.12-10** $\sigma_{max} = 5.95$ MPa

5.12-11 $\sigma_a = 48{,}450 \sin\theta + 3{,}870 \cos\theta$ (psi), $\beta = \arctan(37.54 \tan\theta)$

5.13-1 (a) $s = 2.453$ in.; (b) $\sigma_t = 7520$ psi, $\sigma_c = -6330$ psi

5.13-2 (a) $s = 145.3$ mm;
(b) $\sigma_t = 37.8$ MPa, $\sigma_c = -30.2$ MPa

5.13-3 (a) $s = 4.705$ in.; (b) $\sigma_t = 3140$ psi, $\sigma_c = -2670$ psi

5.13-4 (a) $s = 112.9$ mm;
(b) $\sigma_t = 35.9$ MPa, $\sigma_c = -30.0$ MPa

5.14-1 $\tau_{max} = 1830$ psi, $\tau_b = 168$ psi

5.14-2 $\tau_{max} = 17.1$ MPa, $\tau_b = 1.1$ MPa

5.14-3 $\tau_{max} = 1830$ psi, $\tau_b = 172$ psi

5.14-4 $\tau_{max} = \dfrac{3Pb_1^2}{2(t_1 b_1^3 + t_2 b_2^3)}$

5.14-5 (a) $\tau_{max} = \dfrac{3P}{2bt}$; (b) $\tau = \dfrac{3P(b^2 - 4s^2)}{2tb^3}$

5.14-6 (a) $\tau_{max} = \dfrac{3P}{2\sqrt{2}\,tb}$; (b) $\tau = \dfrac{6Ps(b - s)}{\sqrt{2}\,tb^3}$

5.14-7 $\tau_{max} = 1370$ psi **5.14-8** $\tau_{max} = 7.56$ MPa

5.14-9 $\tau_{max} = 1890$ psi **5.14-10** $\tau_{max} = 31.3$ MPa

5.15-1 $e_0 = 0.494$ in. **5.15-2** $e_0 = 0.604$ in.

5.15-12 $e = 23.0$ mm **5.17-1** $f = 16/3\pi \approx 1.70$

5.17-2 $f = 2$

5.17-3 (a) $f = \dfrac{16r_2(r_2^3 - r_1^3)}{3\pi(r_2^4 - r_1^4)}$; (b) $f = \dfrac{4}{\pi} \approx 1.27$

5.17-4 $M_y = 24.0$ kN·m, $M_p = 36.0$ kN·m,
$Z = 144 \times 10^3$ mm^3

5.17-5 $M_y = 192$ in.-k, $M_p = 288$ in.-k, $Z = 8.0$ in.^{3}

5.17-6 $M_p = 298$ kN·m, $Z = 1.028 \times 10^6$ mm^3, $f = 1.14$

5.17-7 $M_p = 1897$ in.-k, $Z = 52.7$ in.^{3}, $f = 1.15$

5.17-8 $Z = 130.4$ in.^{3}, $f = 1.12$

5.17-9 $Z = 54.0$ in.^{3}, $f = 1.12$

5.17-10 $Z = 26.7$ in.^{3}, $f = 1.12$

5.17-11 $M_p = 6130$ in.-k, $Z = 170.3$ in.^{3}, $f = 1.26$

5.17-12 $M_p = 648$ kN·m, $Z = 2.816 \times 10^6$ mm^3,
$f = 1.27$

5.17-13 $Z = 53.25$ in.^{3}, $f = 1.77$

5.17-14 $Z = 440.6 \times 10^3$ mm^3, $f = 1.81$

5.17-15 $M_p = 1310$ in.-k

5.17-16 $M = 2890$ in.-k, $\kappa = 190 \times 10^{-6}$ in.$^{-1}$

5.17-17 $M = 2410$ in.-k, $\kappa = 220 \times 10^{-6}$ in.$^{-1}$

5.17-18 $M = 918$ in.-k, $\kappa = 337 \times 10^{-6}$ in.$^{-1}$

5.17-19 $M/M_y = 1.071$, $\kappa/\kappa_y = 1.120$, $f = 1.203$

5.17-20 $M/M_y = 1.072$, $\kappa/\kappa_y = 1.111$, $f = 1.267$

5.17-21 $M/M_y = 2 - 2(\kappa_y/\kappa)^2 + (\kappa_y/\kappa)^3$ for $\kappa \geq \kappa_y$

5.17-22 $\dfrac{M}{M_y} = \dfrac{2\kappa}{3\pi\kappa_y}\left[3\arcsin\dfrac{\kappa_y}{\kappa} + \dfrac{\kappa_y}{\kappa}\left(5 - \dfrac{2\kappa_y^2}{\kappa^2}\right)\sqrt{1 - \dfrac{\kappa_y^2}{\kappa^2}}\right]$
for $\kappa \geq \kappa_y$

5.18-1 (a) $h_1 = 4.00$ in.; (b) $\sigma_t = 2400$ psi, $\sigma_c = -4800$ psi

5.18-2 (a) $h_1 = 65.40$ mm; (b) $\sigma_t = 11.5$ MPa,
$\sigma_c = -21.7$ MPa

5.18-3 (a) $M = 379.3$ in.-k; (b) $\rho = 882$ in.

5.18-4 (a) $M = 22.27$ kN·m; (b) $\rho = 18.0$ m

5.18-5 $\sigma_{max} = 25{,}590$ psi **5.18-6** $\sigma_{max} = 172$ MPa

5.18-8 $M = \dfrac{bh^2 B_1 \epsilon_1}{6}\left(1 - \dfrac{3B_2\epsilon_1}{4B_1}\right)$

5.18-9 $M = 2540$ in.-k

5.18-11 $M = \dfrac{\sigma_1 bh^2}{6}\left[\dfrac{3m(m + 3)}{2(m + 1)(m + 2)}\right]$

CHAPTER 6

6.2-1 $\sigma_{x_1} = 8690$ psi, $\tau_{x_1y_1} = -2000$ psi,
$\sigma_{y_1} = 3310$ psi

6.2-2 $\sigma_{x_1} = 118.5$ MPa, $\tau_{x_1y_1} = -15.3$ MPa,
$\sigma_{y_1} = 31.5$ MPa

6.2-3 $\sigma_{x_1} = -18{,}410$ psi, $\tau_{x_1y_1} = -11{,}200$ psi,
$\sigma_{y_1} = 5610$ psi

6.2-4 $\sigma_{x_1} = -22.8$ MPa, $\tau_{x_1y_1} = -21.3$ MPa,
$\sigma_{y_1} = 58.8$ MPa

6.2-5 $\sigma_w = 80$ psi, $\tau_w = 960$ psi

6.2-6 $\sigma_w = 8.40$ MPa, $\tau_w = -4.27$ MPa

6.2-7 $\theta = \pm 26.57°$ **6.2-8** $\theta = \pm 60°$

6.2-9 $\sigma_x = -29{,}540$ psi, $\sigma_y = -10{,}060$ psi,
$\tau_{xy} = -7{,}080$ psi

6.2-10 $\sigma_x = 29.5$ MPa, $\sigma_y = -9.5$ MPa,
$\tau_{xy} = -11.8$ MPa

6.2-11 $\sigma_y = 3470$ psi, $\tau_{xy} = 1710$ psi

6.2-12 $\sigma_y = 43.7$ MPa, $\tau_{xy} = 20.5$ MPa

6.2-13 $\theta_1 = 28.26°$, $\sigma_b = 2000$ psi, $\tau_b = -6800$ psi

6.2-14 $\theta_1 = 44.56°$, $\sigma_b = -35$ MPa, $\tau_b = 32.9$ MPa

6.2-15 $\theta_1 = 22.4°$ **6.2-16** $\theta_1 = 60.3°$

6.3-1 $\sigma_1 = 9350$ psi, $\sigma_2 = 2650$ psi, $\theta_{p_1} = 31.7°$,
$\tau_{max} = 3350$ psi

6.3-2 $\sigma_1 = 121$ MPa, $\sigma_2 = 29$ MPa, $\theta_{p_1} = 20.3°$, $\tau_{max} = 46.1$ MPa

6.3-3 $\sigma_1 = 10{,}020$ psi, $\sigma_2 = -22{,}820$ psi, $\theta_{p_1} = -8.5°$, $\tau_{max} = 16{,}420$ psi

6.3-4 $\sigma_1 = 64.0$ MPa, $\sigma_2 = -28.0$ MPa, $\theta_{p_1} = -21.2°$, $\tau_{max} = 46.0$ MPa

6.3-5 $\sigma_1 = 16{,}230$ psi, $\sigma_2 = 1{,}770$ psi, $\theta_{p_1} = 7.21°$, $\tau_{max} = 7{,}230$ psi

6.3-6 $\sigma_1 = 86.0$ MPa, $\sigma_2 = 16.0$ MPa, $\theta_{p_1} = 26.6°$, $\tau_{max} = 35.0$ MPa

6.3-7 $\sigma_1 = 8{,}310$ psi, $\sigma_2 = -28{,}810$ psi, $\theta_{p_1} = -12.8°$, $\tau_{max} = 18{,}560$ psi

6.3-8 $\sigma_1 = 34.4$ MPa, $\sigma_2 = -114.4$ MPa, $\theta_{p_1} = -20.6°$, $\tau_{max} = 74.4$ MPa

6.3-9 $\sigma_1 = 45{,}810$ psi, $\sigma_2 = -25{,}810$ psi, $\theta_{p_1} = 117.0°$, $\tau_{max} = 35{,}810$ psi

6.3-10 $\sigma_1 = 135.0$ MPa, $\sigma_2 = -87.0$ MPa, $\theta_{p_1} = 99.5°$, $\tau_{max} = 111.0$ MPa

6.3-11 $\sigma_1 = -3350$ psi, $\sigma_2 = -22{,}050$ psi, $\theta_{p_1} = 77.7°$, $\tau_{max} = 9350$ psi

6.3-12 $\sigma_1 = -17.3$ MPa, $\sigma_2 = -86.7$ MPa, $\theta_{p_1} = 61.6°$, $\tau_{max} = 34.7$ MPa

6.3-13 $\sigma_1 = 22.8$ ksi, $\sigma_2 = -32.3$ ksi, $\theta_{p_1} = 58.2°$, $\tau_{max} = 27.6$ ksi

6.3-14 $\sigma_1 = 74.5$ MPa, $\sigma_2 = -36.5$ MPa, $\theta_{p_1} = 10.0°$, $\tau_{max} = 55.5$ MPa

6.3-15 $\tau_{max} = \pm 8950$ psi

6.3-16 $\tau_{max} = \pm 120$ MPa

6.3-17 (a) $\sigma_y = -2457$ psi; (b) $\sigma_2 = -3237$ psi, $\theta_{p_1} = -15.0°$

6.3-18 (a) $\sigma_y = 9.86$ MPa; (b) $\sigma_2 = -84.7$ MPa, $\theta_{p_1} = 67.5°$

6.3-19 (a) $\sigma_x = 6729$ psi, $\sigma_y = -2729$ psi; (b) $\sigma_2 = -3600$ psi, $\theta_{p_1} = -16.2°$

6.3-20 (a) $\sigma_x = -17.67$ MPa, $\sigma_y = 45.67$ MPa; (b) $\sigma_2 = -24.0$ MPa, $\theta_{p_1} = 106.8°$

6.4-5 (a) $\sigma_{x_1} = 9000$ psi, $\tau_{x_1y_1} = -5200$ psi; (b) $\tau_{max} = 6000$ psi

6.4-6 (a) $\sigma_{x_1} = -67.5$ MPa, $\tau_{x_1y_1} = 39.0$ MPa; (b) $\tau_{max} = 45.0$ MPa

6.4-7 (a) $\sigma_{x_1} = 2100$ psi, $\tau_{x_1y_1} = -3640$ psi; (b) $\sigma_1 = 4200$ psi, $\sigma_2 = -4200$ psi, $\theta_{p_1} = 45°$

6.4-8 (a) $\sigma_{x_1} = -18.0$ MPa, $\tau_{x_1y_1} = 31.2$ MPa; (b) $\sigma_1 = 36.0$ MPa, $\sigma_2 = -36.0$ MPa, $\theta_{p_1} = 135°$

6.4-9 (a) $\sigma_{x_1} = 3270$ psi, $\tau_{x_1y_1} = -1770$ psi; (b) $\tau_{max} = 2500$ psi

6.4-10 (a) $\sigma_{x_1} = -72.7$ MPa, $\tau_{x_1y_1} = 17.7$ MPa; (b) $\tau_{max} = 25$ MPa

6.5-1 $\sigma_x = 31{,}600$ psi, $\sigma_y = -14{,}510$ psi

6.5-2 $\sigma_x = 114.1$ MPa, $\sigma_y = 60.2$ MPa

6.5-3 (a) $\epsilon_z = -\nu(\epsilon_x + \epsilon_y)/(1 - \nu)$; (b) $\epsilon_z = -55.7 \times 10^{-6}$

6.5-4 $F = \nu P$

6.5-5 (a) $\gamma_{max} = 1900 \times 10^{-6}$; (b) $\Delta t = -0.00014$ in., $\Delta V = 0.0874$ in.^{3}

6.5-6 (a) $\gamma_{max} = 715 \times 10^{-6}$; (b) $\Delta t = -0.0021$ mm, $\Delta V = 896$ mm^3

6.5-7 $\Delta V = -0.0253$ in.^{3}, $U = 13.3$ in.-lb

6.5-8 $\Delta V = -30.7$ mm^3, $U = 2.38$ J

6.5-9 $\Delta V = 0.0481$ in.^{3}, $U = 660$ in.-lb

6.5-10 $\Delta V = 2000$ mm^3, $U = 58.5$ J

6.5-11 $\sigma_x = 30{,}000$ psi or $\sigma_x = -30{,}000$ psi

6.5-12 $\sigma_x = 75$ MPa or $\sigma_x = -75$ MPa

6.5-13 (a) $\Delta ac = 0.00284$ in.; (b) $\Delta bd = -0.0000213$ in.; (c) $\Delta t = -0.0000907$ in.; (d) $e = 170.7 \times 10^{-6}$; (e) $U = 136$ in.-lb

6.6-1 $p = 2790$ psi **6.6-2** $t = 7.5$ mm

6.6-3 $f = 28{,}800$ lb/in. of weld

6.6-4 (b) $\tau_{max} = 21.6$ MPa **6.6-5** (b) $\tau_{max} = 4750$ psi

6.6-6 $h = 16.3$ m **6.6-7** $t_{min} = 0.0867$ in.

6.6-8 (a) $t_{min} = 5.00$ mm; (b) $t_{min} = 3.75$ mm

6.6-9 (a) $\sigma = 4000$ psi; (b) $\sigma_c = 8000$ psi; (c) $\sigma_w = 4000$ psi

6.6-10 (a) $\sigma_1 = 51.0$ MPa, $\sigma_2 = 25.5$ MPa; (b) $\tau_{max} = 12.75$ MPa; (c) $\tau_{max} = 25.5$ MPa

6.6-11 (a) $\sigma_1 = 12{,}000$ psi, $\sigma_2 = 6{,}000$ psi; (b) $\tau_{max} = 3{,}000$ psi; (c) $\tau_{max} = 6{,}000$ psi

6.6-12 $F = 3\pi pr^2$

6.6-13 (a) $\sigma_1 = 6000$ psi, $\sigma_2 = 3000$ psi; (b) $\tau = 1500$ psi; (c) $\tau = 3100$ psi; (d) $\sigma = 3200$ psi and 5800 psi, $\tau = 750$ psi

6.6-14 (a) $\sigma_1 = 80$ MPa, $\sigma_2 = 40$ MPa; (b) $\tau = 20$ MPa; (c) $\tau = 41.2$ MPa; (d) $\sigma = 50$ MPa and 70 MPa, $\tau = 17.3$ MPa

6.6-15 $p = 1000$ psi

6.6-16 $p_{max} = 10.0$ MPa

6.7-1 $\sigma_t = 8870$ psi, $\sigma_c = -2510$ psi, $\tau_{max} = 5690$ psi

6.7-2 $\sigma_t = 8.0$ MPa, $\sigma_c = -72.0$ MPa, $\tau_{max} = 40.0$ MPa

6.7-3 (a) $\sigma_{max} = 10{,}870$ psi, $\tau_{max} = 5470$ psi; (b) $P_{allow} = 395$ lb

6.7-4 $\tau_a = 86.8$ MPa, $\tau_b = 22.6$ MPa, $\tau_c = 26.4$ MPa

6.7-5 $\sigma_t = 4320$ psi, $\sigma_c = -1870$ psi, $\tau_{max} = 3100$ psi

6.7-6 $\sigma_t = 6.1$ MPa, $\sigma_c = -65.1$ MPa, $\tau_{max} = 35.6$ MPa

6.7-7 $\sigma_{max} = 8550$ psi, $\tau_{max} = 5670$ psi

6.7-8 $\phi_{max} = 0.288$ rad

6.7-9 (a) $\sigma_x = 3600$ psi, $\sigma_y = 7200$ psi, $\tau_{xy} = -1990$ psi; (b) $\sigma_{max} = 8080$ psi, $\tau_{max} = 2680$ psi

6.7-10 $P = 34.1$ kN **6.7-11** $d = 1.00$ in.

6.7-12 $\sigma_t = 29.2qR^2/d^3$, $\sigma_c = -8.8qR^2/d^3$, $\tau_{max} = 19.0qR^2/d^3$

6.8-1 $\sigma_1 = 200$ psi, $\sigma_2 = -4520$ psi, $\theta_{p_1} = 78.2°$, $\tau_{max} = 2360$ psi

6.8-2 $\sigma_1 = 69.0$ MPa, $\sigma_2 = -1.5$ MPa, $\theta_{p_1} = 8.3°$, $\tau_{max} = 35.2$ MPa

6.8-3 (a) $\sigma_1 = 0$, $\sigma_2 = -11{,}090$ psi, $\tau_{max} = 5{,}540$ psi; (b) $\sigma_1 = 1{,}620$ psi, $\sigma_2 = -11{,}630$ psi, $\tau_{max} = 6{,}620$ psi

6.8-4 (a) $\sigma_1 = 0$, $\sigma_2 = -82.7$ MPa, $\tau_{max} = 41.3$ MPa; (b) $\sigma_1 = 4.5$ MPa, $\sigma_2 = -76.1$ MPa, $\tau_{max} = 40.3$ MPa

6.8-5 $P = 2600$ lb

6.8-6 (a) $\sigma_1 = 0$, $\sigma_2 = -72.4$ MPa, $\tau_{max} = 36.2$ MPa; (b) $\sigma_1 = 20.2$ MPa, $\sigma_2 = -0.3$ MPa, $\tau_{max} = 10.3$ MPa

6.8-7 Top: $\sigma_1 = 0$, $\sigma_2 = -15{,}000$ psi, $\tau_{max} = 7500$ psi; N.A.: $\sigma_1 = 1125$ psi, $\sigma_2 = -1125$ psi, $\tau_{max} = 1125$ psi

6.8-8 Top: $\sigma_1 = 0$, $\sigma_2 = -225$ MPa, $\tau_{max} = 112$ MPa; N.A.: $\sigma_1 = 45$ MPa, $\sigma_2 = -45$ MPa, $\tau_{max} = 45$ MPa

6.9-1 $\tau_{max} = 8000$ psi, $\Delta a = 0.00632$ in., $\Delta b = -0.00313$ in., $\Delta c = -0.00100$ in., $\Delta V = 0.00883$ in.^{3}, $U = 549$ in.-lb

6.9-2 $\tau_{max} = 7$ MPa, $\Delta a = -0.03400$ mm, $\Delta b = -0.01065$ mm, $\Delta c = -0.00675$ mm, $\Delta V = -1418$ mm^3, $U = 31.7$ J

6.9-3 $\sigma_x = -6610$ psi, $\sigma_y = \sigma_z = -3420$ psi, $\tau_{max} = 1600$ psi, $\Delta V = -0.0130$ in.^{3}, $U = 37.2$ in.-lb

6.9-4 $\sigma_x = -56.6$ MPa, $\sigma_y = \sigma_z = -38.9$ MPa, $\tau_{max} = 8.9$ MPa, $\Delta V = -140$ mm^3, $U = 3.41$ J

6.9-5 (a) $p = 4\nu F/[\pi d^2(1-\nu)]$; (b) $p = 208$ psi (compression)

6.9-6 (a) $p = \nu p_0$; (b) $e = -p_0(1+\nu)(1-2\nu)/E$; (c) $e = -p_0(1+\nu)[1-2\nu+\nu p_0(1-\nu^2)/E]/E$; (d) $p = 1.32$ MPa (compression), $e = -0.286$

6.9-7 $p = 100{,}000$ psi, $K = 25 \times 10^6$ psi, $U = 22{,}600$ in.-lb

6.9-8 $d = 2700$ m; Increase $= 0.05403\%$

6.9-9 $\epsilon_0 = 276 \times 10^{-6}$, $e = 828 \times 10^{-6}$, $u = 4.97$ psi

6.11-1 $\epsilon_{x_1} = 461 \times 10^{-6}$, $\gamma_{x_1y_1} = 225 \times 10^{-6}$, $\epsilon_{y_1} = 239 \times 10^{-6}$

6.11-2 $\epsilon_{x_1} = 351 \times 10^{-6}$, $\gamma_{x_1y_1} = -490 \times 10^{-6}$, $\epsilon_{y_1} = -101 \times 10^{-6}$

6.11-3 $\epsilon_1 = 554 \times 10^{-6}$, $\epsilon_2 = 66 \times 10^{-6}$, $\theta_{p_1} = -22.9°$, $\gamma_{max} = 448 \times 10^{-6}$

6.11-4 $\epsilon_1 = 172 \times 10^{-6}$, $\epsilon_2 = -502 \times 10^{-6}$, $\theta_{p_1} = 163.9°$, $\gamma_{max} = 674 \times 10^{-6}$

6.11-5 (a) $\epsilon_{x_1} = 202 \times 10^{-6}$, $\gamma_{x_1y_1} = -569 \times 10^{-6}$, $\epsilon_{y_1} = 348 \times 10^{-6}$; (b) $\epsilon_1 = 568 \times 10^{-6}$, $\epsilon_2 = -18 \times 10^{-6}$, $\theta_{p_1} = 22.8°$; (c) $\gamma_{max} = 587 \times 10^{-6}$

6.11-6 (a) $\epsilon_{x_1} = -385 \times 10^{-6}$, $\gamma_{x_1y_1} = 690 \times 10^{-6}$, $\epsilon_{y_1} = -1165 \times 10^{-6}$; (b) $\epsilon_1 = -254 \times 10^{-6}$, $\epsilon_2 = -1296 \times 10^{-6}$, $\theta_{p_1} = 65.7°$; (c) $\gamma_{max} = 1041 \times 10^{-6}$

6.11-7 (a) $\epsilon_{x_1} = -756 \times 10^{-6}$, $\gamma_{x_1y_1} = 868 \times 10^{-6}$, $\epsilon_{y_1} = 267 \times 10^{-6}$; (b) $\epsilon_1 = 426 \times 10^{-6}$, $\epsilon_2 = -916 \times 10^{-6}$, $\theta_{p_1} = 99.8°$; (c) $\gamma_{max} = 1340 \times 10^{-6}$

6.11-8 (a) $\epsilon_{x_1} = -1470 \times 10^{-6}$, $\gamma_{x_1y_1} = -717 \times 10^{-6}$, $\epsilon_{y_1} = -907 \times 10^{-6}$; (b) $\epsilon_1 = -732 \times 10^{-6}$, $\epsilon_2 = -1644 \times 10^{-6}$, $\theta_{p_1} = 166°$; (c) $\gamma_{max} = 911 \times 10^{-6}$

6.11-9 $\epsilon_1 = 551 \times 10^{-6}$, $\epsilon_2 = -111 \times 10^{-6}$, $\theta_{p_1} = 12.5°$, $\gamma_{max} = 662 \times 10^{-6}$

6.11-10 $\epsilon_1 = 332 \times 10^{-6}$, $\epsilon_2 = -182 \times 10^{-6}$, $\theta_{p_1} = 12.0°$, $\gamma_{max} = 515 \times 10^{-6}$

6.11-11 $\alpha = 30°$, $P = 30$ k

6.11-12 $\epsilon_x = \epsilon_a$, $\epsilon_y = (2\epsilon_b + 2\epsilon_c - \epsilon_a)/3$, $\gamma_{xy} = 2(\epsilon_b - \epsilon_c)/\sqrt{3}$

CHAPTER 7

7.3-2 $q = q_0 x/L$ **7.3-3** $q = q_0 \sin \pi x/L$

7.3-4 $h_2/h_1 = 3/2$ **7.3-5** $\delta = 0.182$ in., $\theta = 0.199°$

7.3-6 $\delta/L = 1/250$ **7.3-7** $L = 10$ ft

7.3-8 $h = 96$ mm **7.3-9** $\delta = 0.680$ in.

7.3-10 $\theta_a/\theta_b = (2 - a/L)/(1 + a/L)$

7.3-11 $\dfrac{\delta_c}{\delta_{max}} = \dfrac{3\sqrt{3}\,s}{16t^{3/2}}$ in which $s = -1 + \dfrac{8a}{L} - \dfrac{4a^2}{L^2}$, $t = \dfrac{2a}{L} - \dfrac{a^2}{L^2}$

7.3-15 $v = mx^2(3L - x)/6EI$, $\delta_b = mL^3/3EI$, $\theta_b = mL^2/2EI$

7.3-18 $\delta_b = 41qL^4/384EI$, $\delta_c = 7qL^4/192EI$

7.4-4 $\delta_b = 2q_0L^4(\pi^3 - 24)/3\pi^4EI$ **7.4-5** $\delta = q_0L^4/\pi^4EI$

7.4-6 $\delta_b = 19q_0L^4/360EI$, $\theta_b = q_0L^3/15EI$

7.4-10 $\delta_b = 41qL^4/384EI$, $\delta_c = 7qL^4/192EI$

7.4-11 $\delta_c = 11qL^4/384EI$, $\theta_c = qL^3/16EI$

7.5-4 $\theta_b = \dfrac{PL^2}{2EI} - \dfrac{M_0L}{EI}$, $\delta_b = \dfrac{PL^3}{3EI} - \dfrac{M_0L^2}{2EI}$

7.5-5 $\theta_b = 7qL^3/162EI$, $\delta_b = 23qL^4/648EI$

7.5-6 $\delta_b = 11.97$ mm, $\delta_c = 4.05$ mm

7.5-7 $\delta_b = 0.448$ in., $\delta_c = 0.139$ in. **7.5-10** $P = 53.0$ kN

7.5-11 $\theta_a = Pa(L - a)(L - 2a)/6LEI$, $\delta_1 = Pa^2(L - 2a)^2/6LEI$, $\delta_2 = 0$

7.5-12 $\theta_a = M_0L/6EI$, $\theta_b = 0$, $\delta_{max} = M_0L^2/27EI$

7.5-13 $\delta_{max} = 0.859$ in. **7.5-14** $\delta_c = Pa^2(L + a)/3EI$

7.5-15 $P/Q = 4$ **7.5-16** $\delta_c = qL^4/128EI$

7.5-17 $\theta_b = PL^2/12EI$, $\delta_a = PL^3/12EI$, $\delta_e = 0$

7.5-18 $P = 3qL/4$ **7.6-1** $\delta_b = 2PL^3/9EI$

7.6-2 $\delta = 19PL^3/384EI$

7.6-3 (a) $a/L = 2/3$; (b) $a/L = 1/2$ **7.6-4** $y = -Px^3/3EI$

7.6-5 $\theta_b = 7qL^3/162EI$, $\delta_b = 23qL^4/648EI$

7.6-6 $\delta_b = 11.97$ mm, $\delta_c = 4.05$ mm

7.6-7 $\delta_b = 0.448$ in., $\delta_c = 0.139$ in.

7.6-8 $\theta_b = q_0L^3/10EI$, $\delta_b = 13q_0L^4/180EI$

7.6-9 $\delta_2/\delta_1 = n^2$

7.6-10 $\theta_a = q(L^3 - 6La^2 + 4a^3)/24EI$, $\delta = q(5L^4 - 24L^2a^2 + 16a^4)/384EI$

7.6-11 $\theta_a = Pa(L - a)(L - 2a)/6LEI$, $\delta_1 = Pa^2(L - 2a)^2/6LEI$, $\delta_2 = 0$

7.6-12 $\theta_a = M_0L/6EI$, $\theta_b = 0$, $\delta_{max} = M_0L^2/27EI$

7.6-13 $\delta = 3q_0L^4/1280EI$

7.6-14 $\delta_b = L^2(QL - 3Pa)/48EI$, $P/Q = L/3a$

7.6-15 $\delta_a = \dfrac{Pa^2(L + a)}{3EI} - \dfrac{QaL^2}{16EI}$, $\dfrac{P}{Q} = \dfrac{3L^2}{16a(L + a)}$

7.6-16 $\theta_a = qL(L^2 - 2a^2)/24EI$

7.6-17 $\theta_b = PL^2/12EI$, $\delta_a = PL^3/12EI$, $\delta_e = 0$

7.6-18 $P = 3qL/4$ **7.6-19** $\delta = 19WL^3/31{,}104EI$

7.6-20 $\theta_a = 881qL^3/10{,}368EI$ **7.6-21** $\delta = 5Pb^3/2EI$

7.6-22 $\delta_e = 5Pb^3/3EI$ **7.6-23** $\delta_c = 0.300$ in.

7.6-24 $k = 12EI(1 + L/b)/L^3$

7.6-25 $\delta_c = 39PL^3/1024EI$

7.6-26 $\delta_h = Pcb^2/2EI$, $\delta_v = Pc^2(c + 3b)/3EI$

7.6-27 $\delta = PL^2(2L + 3a)/3EI$

7.7-1 (a) $\delta_b = PL^3(7/I_2 + 1/I_1)/24E$; (b) $r = (1 + 7I_1/I_2)/8$

7.7-2 $\delta_b = 3qL^4/64EI$

7.7-3 $\delta_b = qL^4(1 + 15I_1/I_2)/128EI_1$

7.7-4 $\theta_a = \dfrac{PL^2}{64EI_1} + \dfrac{3PL^2}{64EI_2}$, $\delta_c = \dfrac{PL^3}{384EI_1} + \dfrac{7PL^3}{384EI_2}$

7.7-5 $\theta_a = 13qL^3/576EI$, $\delta_c = 53qL^4/9216EI$

7.7-6 $\delta_b = 5.371PL^3/Ebd_a^3$

7.7-7 $\delta_b = 6PL^3t/[Eb(d_b - d_a)^3]$ in which $t = (d_b/d_a - 3)(d_b/d_a - 1) + 2\ln d_b/d_a$

7.7-8 $\delta_b = 1.388PL^3/Etd_a^3$ **7.7-9** $\delta_c = 11PL^3/64Ebh^3$

7.8-1 (a) $U = P^2L^3/96EI$, (b) $\delta = PL^3/48EI$

7.8-2 (a) $U = P^2a^2(L + a)/6EI$; (b) $\delta_c = Pa^2(L + a)/3EI$

7.8-3 (a) $U = q^2L^5/40EI$, (b) $U = 238$ in.-lb

7.8-4 $U_2/U_1 = n^5$ **7.8-5** $U = 4bhL\sigma_{max}^2/45E$

7.8-6 $U = 32EI\delta^2/L^3$ **7.8-7** $U = \pi^4EI\delta^2/4L^3$

7.8-8 $U = P^2L^3/96EI + PM_0L^2/16EI + M_0^2L/6EI$

7.8-9 $\sigma_{max} = \sqrt{18WEh/AL}$

7.8-10 $\sigma_{max}/\sigma_{st} = 1 + (1 + 2h/\delta_{st})^{1/2}$

7.8-11 $\delta_{max} = 0.5937$ in., $\sigma_{max} = 24{,}000$ psi

7.8-12 $d = 280$ mm **7.8-13** W 14 × 53

7.8-14 $h = 0.360$ m **7.8-15** $R = \sqrt{3EIWr^2\omega^2/2gL^3}$

7.9-1 $q(x) = -P\langle x\rangle^{-1} + Pa\langle x\rangle^{-2} + P\langle x - a\rangle^{-1}$

7.9-2 $q(x) = -qb\langle x\rangle^{-1} + \dfrac{qb}{2}(2a + b)\langle x\rangle^{-2} + q\langle x - a\rangle^0 - q\langle x - L\rangle^0$

7.9-3 $q(x) = -16\langle x\rangle^{-1} + 864\langle x\rangle^{-2} + \dfrac{1}{6}\langle x\rangle^0 - \dfrac{1}{6}\langle x - 72\rangle^0 + 4\langle x - 108\rangle^{-1}$, x = in., q = k/in.

7.9-4 $q(x) = -\dfrac{Pb}{L}\langle x\rangle^{-1} + P\langle x - a\rangle^{-1} - \dfrac{Pa}{L}\langle x - L\rangle^{-1}$

7.9-5 $q(x) = -\dfrac{M_0}{L}\langle x\rangle^{-1} + M_0\langle x - a\rangle^{-2} + \dfrac{M_0}{L}\langle x - L\rangle^{-1}$

7.9-6 $q(x) = -P\langle x\rangle^{-1} + P\langle x - a\rangle^{-1} + P\langle x - L + a\rangle^{-1} - P\langle x - L\rangle^{-1}$

7.9-7 $q(x) = -7.692\langle x\rangle^{-1} + 240\langle x\rangle^{-2} + 18\langle x - 192\rangle^{-1} - 10.308\langle x - 312\rangle^{-1}$, x = in., q = k/in.

7.9-8 $q(x) = -\dfrac{qa}{2L}(2L - a)\langle x\rangle^{-1} + q\langle x\rangle^0 - q\langle x - a\rangle^0 - \dfrac{qa^2}{2L}\langle x - L\rangle^{-1}$

7.9-9 $q(x) = -\dfrac{2q_0L}{27}\langle x\rangle^{-1} + \dfrac{3q_0}{L}\left\langle x - \dfrac{L}{3}\right\rangle^1 - \dfrac{3q_0}{L}\left\langle x - \dfrac{2L}{3}\right\rangle^1 - q_0\left\langle x - \dfrac{2L}{3}\right\rangle^0 - \dfrac{5q_0L}{54}\langle x - L\rangle^{-1}$

7.9-10 $q(x) = -180\langle x\rangle^{-1} + 20\langle x\rangle^0 - 20\langle x - 10\rangle^0 + 120\langle x - 15\rangle^{-1} - 140\langle x - 20\rangle^{-1}$, x = m, q = kN/m

7.9-11 $q(x) = 3\langle x\rangle^{-1} + 144\langle x - 72\rangle^{-2} - 11\langle x - 144\rangle^{-1} + 8\langle x - 216\rangle^{-1}$, x = in., q = k/in.

7.9-12 $q(x) = 2.4\langle x\rangle^{-1} + 10\langle x - 1.2\rangle^1 - 10\langle x - 2.4\rangle^1 - 24\langle x - 2.4\rangle^{-1} - 12\langle x - 3.6\rangle^0$, x = m, q = kN/m

7.10-1 $EIv = Px^2(3a - x)/6 + P\langle x - a\rangle^3/6$; $\theta_b = Pa^2/2EI$; $\delta_b = Pa^2(3L - a)/6EI$

7.10-2 $EIv = qbx^2(3L + 3a - 2x)/12 + q\langle x - a\rangle^4/24$; $\theta_b = q(L^3 - a^3)/6EI$; $\delta_b = q(3L^4 - 4a^3L + a^4)/24EI$

7.10-3 $EIv = (x^4 - 384x^3 + 62{,}208x^2 - \langle x - 72\rangle^4)/144$, x = in., v = in., E = ksi, I = in.^{4}; $\theta_b = 0.00749$ rad; $\delta_b = 0.581$ in.

7.10-4 $EIv = Pbx(L^2 - b^2 - x^2)/6L + P\langle x - a\rangle^3/6$; $\theta_a = Pab(L + b)/6LEI$; $\delta_d = Pa^2b^2/3LEI$

7.10-5 $EIv = M_0x(6aL - 3a^2 - 2L^2 - x^2)/6L + M_0\langle x - a\rangle^2/2$; $\theta_a = M_0(6aL - 3a^2 - 2L^2)/6LEI$; $\delta_d = M_0ab(2a - L)/3LEI$

7.10-6 $EIv = Px(3aL - 3a^2 - x^2)/6 + P\langle x - a\rangle^3/6 + P\langle x - L + a\rangle^3/6$; $\theta_a = Pa(L - a)/2EI$; $\delta_d = Pa^2(3L - 4a)/6EI$

7.10-7 $EIv = -1.282x^3 + 120x^2 + 70{,}740x + 3\langle x - 192\rangle^3$, x = in., v = in., E = ksi, I = in.^{4}; $\theta_a = 0.00327$ rad; $\delta_d = 0.414$ in.

7.10-8 $EIv = qx[a^2(2L - a)^2 - 2a(2L - a)x^2 + Lx^3]/24L - q\langle x - a\rangle^4/24$; $\theta_b = qa^2(2L^2 - a^2)/24LEI$; $\delta_d = qa^3(4L^2 - 7aL + 3a^2)/24LEI$

7.10-9 $EIv = 47q_0L^3x/4860 - q_0Lx^3/81 + q_0\langle x - L/3\rangle^5/40L - q_0\langle x - 2L/3\rangle^5/40L - q_0\langle x - 2L/3\rangle^4/24$; $\theta_b = 101q_0L^3/9720EI$; $\delta_d = 121q_0L^4/43{,}740EI$

7.10-10 $EIv = 5625x - 30x^3 + 5x^4/6 - 5\langle x - 10\rangle^4/6 + 20\langle x - 15\rangle^3$, x = m, v = m, E = kPa, I = m^4; $\theta_b = 0.0107$ rad; $\delta_d = 47.7$ mm

7.10-11 $EIv = x^3/2 - 12{,}960x + 72\langle x - 72\rangle^2 - 11\langle x - 144\rangle^3/6$, x = in., v = in., $EI = 8.4 \times 10^6$ k-in.^{2}; $\delta_c = -0.0889$ in. upward; $\delta_d = 0.363$ in. downward

7.10-12 $EIv = 0.4x^3 - 2.390x + \langle x - 1.2\rangle^5/12 - \langle x - 2.4\rangle^5/12 - 4\langle x - 2.4\rangle^3$, x = m, v = m, E = kPa, I = m^4; $\delta_c = -0.726$ mm upward; $\delta_d = 3.19$ mm downward

7.11-1 $v = \alpha(T_2 - T_1)(x)(L - x)/2h$, $\theta = \alpha L(T_2 - T_1)/2h$, $\delta = \alpha L^2(T_2 - T_1)/8h$

7.11-2 $v = -\alpha(T_2 - T_1)(x^2)/2h$, $\theta = -\alpha L(T_2 - T_1)/h$, $\delta = -\alpha L^2(T_2 - T_1)/2h$

7.11-3 $\delta_c = -\alpha a(L + a)(T_2 - T_1)/2h$

7.11-4 $\delta_{max} = \alpha T_0L^3/9\sqrt{3}\,h$

CHAPTER 8

8.2-1 $R_a = 5qL/8$, $R_b = 3qL/8$, $M_a = qL^2/8$, $x_1 = 0.5785L$, $\delta_{max} = 0.005416qL^4/EI$

8.2-2 $R_a = 5qL/8$, $R_b = 3qL/8$, $M_a = qL^2/8$, Max. pos. $M = 9qL^2/128$

8.2-3 $R_a = R_b = P/2$, $M_a = M_b = PL/8$, $\delta_{max} = PL^3/192EI$

8.2-4 $R_a = -R_b = 3M_0/2L$, $M_a = M_0/2$, $v = M_0x^2(L - x)/4LEI$

8.2-5 $v = \delta_bx^2(3L - x)/2L^3$, $R_a = R_b = 3EI\delta_b/L^3$, $M_a = 3EI\delta_b/L^2$

8.2-6 $R_a = R_b = qL/2$, $M_a = M_b = qL^2/12$, $v = qx^2(L - x)^2/24EI$

8.2-7 $R_a = 9q_0L/40$, $R_b = 11q_0L/40$, $M_a = 7q_0L^2/120$, $v = q_0x^2(7L^3 - 9L^2x + 2x^3)/240LEI$

8.2-8 $R_a = R_b = q_0L/4$, $M_a = M_b = 5q_0L^2/96$, $v = q_0x^2(25L^3 - 40L^2x + 16x^3)/960LEI$ for $0 \le x \le L/2$, $\delta_{max} = 7q_0L^4/3840EI$

8.2-9 $R_a = 7q_0L/20$, $R_b = 3q_0L/20$, $M_a = q_0L^2/20$, $M_b = q_0L^2/30$, $v = q_0x^2(3L^3 - 7L^2x + 5Lx^2 - x^3)/120LEI$

8.2-10 $v = q_0x^2(33L^4 - 61L^3x + 30L^2x^2 - 2x^4)/720L^2EI$, $R_a = 61q_0L/120$, $R_b = 19q_0L/120$, $M_a = 11q_0L^2/120$

8.3-1 $R_a = 5qL/8$, $R_b = 3qL/8$, $M_a = qL^2/8$

8.3-2 $R_a = 4P/3$, $R_b = 2P/3$, $M_a = PL/3$

8.3-3 $R_a = -370.4$ lb (downward), $R_b = 370.4$ lb (upward), $M_a = -444.4$ ft-lb (clockwise), $M_{max} = 1481$ ft-lb

8.3-4 (a) $R_a = -R_b = 3M_0a(L + b)/2L^3$, $M_a = -M_0(2L^2 - 6aL + 3a^2)/2L^2$; (b) $a_1 = 2L/3$; (c) $a_2 = L(1 - 1/\sqrt{3})$

8.3-5 $R_a = 6$ k, $R_b = 12$ k, $M_a = 18$ ft-k

8.3-6 $R_a = R_b = qL/2$, $M_a = M_b = qL^2/12$, $\delta_{max} = qL^4/384EI$

8.3-7 $R_a = R_b = P$, $M_a = M_b = Pa(L - a)/L$, $\delta_{max} = Pa^2(3L - 4a)/24EI$

8.3-8 $R_a = R_b = q_0L/4$, $M_a = M_b = 5q_0L^2/96$, $\delta_{max} = 7q_0L^4/3840EI$

8.3-9 $R_a = Pb^2(L + 2a)/L^3$, $R_b = Pa^2(L + 2b)/L^3$, $M_a = Pab^2/L^2$, $M_b = Pa^2b/L^2$, $\delta_1 = Pa^3b^3/3L^3EI$

8.3-10 $M_a = M_b = 5PL/48$, $\delta_{max} = 11PL^3/3072EI$

8.3-11 $M_a = M_b = PL/6$, $\delta_{max} = PL^3/256EI$

8.4-1 $R_a = R_c = 3qL/8$, $R_b = 5qL/4$

8.4-2 $R_a = 4P/3$, $R_b = 2P/3$, $M_a = PL/3$

8.4-3 $R_a = 2q_0L/5$, $R_b = q_0L/10$, $M_a = q_0L^2/15$

8.4-4 $R_a = P(11L - 24a)/16L$, $R_b = 3P(7L + 8a)/16L$, $M_a = P(3L - 8a)/16$

8.4-5 $R_a = 6$ k, $R_b = 12$ k, $M_a = 18$ ft-k

8.4-6 $R_a = qL/8$, $R_b = 33qL/16$, $R_c = 13qL/16$

8.4-7 Thicknesses are proportional to the lengths.

8.4-8 $T = 3qAL^4/(8AL^3 + 24HI)$

8.4-9 $F = 5PI_2/(2I_1 + 2I_2)$

8.4-10 $R_a = R_d = 2qL/5$, $R_b = R_c = 11qL/10$

8.4-11 $R_a = 31qL/48$, $R_b = 17qL/48$, $M_a = 7qL^2/48$

8.4-12 $R_b = 9.48$ kN

8.4-13 $\Delta = 7qL^4/72EI$

8.4-14 $R_a = Pb^2(L + 2a)/L^3$, $R_b = Pa^2(L + 2b)/L^3$, $M_a = Pab^2/L^2$, $M_b = Pa^2b/L^2$

8.4-15 $R_a = qa(2L^3 - 2a^2L + a^3)/2L^3$, $R_b = qa^3(2L - a)/2L^3$, $M_a = qa^2(6L^2 - 8aL + 3a^2)/12L^2$, $M_b = qa^3(4L - 3a)/12L^2$

8.4-16 $R_a = R_b = 6EI\theta/L^2$, $M_a = 2M_b = 4EI\theta/L$

8.4-17 $R_a = R_b = 12EI\Delta/L^3$, $M_a = M_b = 6EI\Delta/L^2$

8.4-18 $R_a = 13qL/30$, $R_b = 13qL/20$, $R_c = -qL/10$, $R_d = qL/60$

8.4-19 $R_a = -1286$ lb, $R_b = 8143$ lb, $R_c = 5143$ lb, $M_a = -5143$ ft-lb, $M_{max} = 13{,}220$ ft-lb

8.4-20 $F = P(2L + 3a)/4L$

8.4-21 $F = 3160$ lb, $M_{ab} = 18{,}960$ ft-lb, $M_{de} = 7320$ ft-lb

8.4-22 $M_a = M_b = 19q_0L^2/256$, $\sigma_{max} = 13.4$ MPa, $\delta_{max} = 19q_0L^4/7680EI = 0.00891$ mm

8.4-23 $k = 89.63EI/L^3$

8.4-24 $H_a = qL$, $V_c = -V_a = qL/8$, $M_a = 3qL^2/8$

8.4-25 $\delta_h = Pa^2(L + 2a)/4EI$, $\delta_v = Pa^2(3L + 16a)/12EI$

8.5-1 $R_a = qL/8$, $R_b = 33qL/16$, $R_c = 13qL/16$

8.5-2 $R_a = R_d = 2qL/5$, $R_b = R_c = 11qL/10$

8.5-3 $M_0 = -qL^2/8 + 3EI\Delta/L^2$

8.5-4 $R = 7P/20$

8.5-5 $R_a = 13qL/30$, $R_b = 13qL/20$, $R_c = -qL/10$, $R_d = qL/60$

8.5-6 $R_1 = 17.63$ kN, $R_2 = 24.44$ kN, $R_3 = 5.93$ kN

8.5-7 $R = 11qL/28$

8.5-8 $M_1 = -3qL^2/56$, $M_2 = -qL^2/7$

8.5-9 $R_a = -1286$ lb, $R_b = 8143$ lb, $R_c = 5143$ lb, $M_a = -5143$ ft-lb, $M_{max} = 13{,}220$ ft-lb

8.5-10 $M_1 = -19.06$ kN · m, $R_1 = 18.53$ kN, $R_2 = 30.44$ kN, $R_3 = 5.03$ kN

8.5-11 $M_1 = -Pa/7$, $M_2 = 2Pa/7$, $M_3 = -Pa$

8.5-12 $M_1 = -15$ kN · m, $M_2 = -25.52$ kN · m, $M_3 = -36.48$ kN · m, $M_4 = 0$

8.5-13 $M_1 = -5qL^2/56$, $M_2 = -qL^2/14$, $M_3 = -qL^2/8$

8.5-14 $M_1 = M_8 = 0$, $M_2 = M_7 = -19qL^2/284$, $M_3 = M_6 = 5qL^2/284$, $M_4 = M_5 = -qL^2/284$

8.5-15 $M_1 = -234.6$ ft-k, $M_2 = -130.9$ ft-k, $M_3 = 58.1$ ft-k

8.5-16 $M_n = M_0(\sqrt{3} - 2)^n$

8.6-1 $M_a = 3\alpha EI(T_2 - T_1)/2h$, $R_a = -R_b = M_a/L$

8.6-2 $R_a = R_c = -3\alpha EI(T_2 - T_1)/2hL$, $R_b = 3\alpha EI(T_2 - T_1)/hL$

8.6-3 $S = 48EIAH\alpha T/(AL^3 + 48IH)$

8.7-1 $H = \pi^2EA\delta^2/4L^2$; $\sigma = 617$ psi, 154 psi, and 69 psi

8.7-2 $\lambda = 17q^2L^7/40{,}320E^2I^2$, $\sigma_1 = 741$ kPa, $\sigma_2 = 117$ MPa

8.7-3 (a) $\lambda = 64P^2L^5/15\pi^2E^2d^8$;
(b) $\sigma_1 = 64P^2L^4/15\pi^2Ed^8$; (c) $\sigma_2 = 8PL/\pi d^3$

CHAPTER 9

9.1-1 $P_{cr} = \alpha/L$ **9.1-2** $P_{cr} = \beta L + \alpha/L$

9.1-3 $P_{cr} = \beta a^2/L$ **9.1-4** $P_{cr} = \alpha/L$

9.1-5 $P_{cr} = \beta aL/(L + a)$ **9.2-1** $P_{cr} = 736$ k

9.2-2 $P_{cr} = 1020$ k **9.2-3** $I = 50.6$ in.^{4}, W 10 × 45

9.2-4 $I = 136$ in.^{4}, W 16 × 77

9.2-5 (a) $P_{cr} = 11\pi^3Er^4/4L^2$; (b) $P_{cr} = 3\pi^3Er^4/4L^2$

9.2-6 $t = 4.45$ mm **9.2-7** $t = 0.248$ in.

9.2-8 $P_{cr} = 261$ k **9.2-9** $I = 2.17$ in.^{4}; L 5 × 5 × 3/8

9.2-10 $I = 4.67$ in.^{4}, L 6 × 6 × 1/2

9.2-11 $P_1:P_2:P_3 = 1.209:1.047:1.000$

9.2-12 $h/b = 2$ **9.2-13** $Q = 3.05$ k

9.2-14 $Q = 15.2$ kN **9.2-15** $Q_{cr} = 3\pi^2EI/4L^2$

9.2-16 $\Delta T = \pi^2I/\alpha AL^2$ **9.2-17** $P = 136$ k

9.2-18 $P_{allow} = 102$ kN **9.2-19** $\theta = \arctan(\cot^2\beta)$

9.2-20 $\theta = 26.57°$ **9.3-1** $P_{cr} = 289$ k, 72 k, 592 k, 1160 k

9.3-2 $P_{cr} = 270$ k, 68 k, 553 k, 1080 k

9.3-3 $P_{cr} = 190$ k, 47.5 k, 389 k, 760 k

9.3-4 $t = 7.84$ mm **9.3-5** $P_{cr} = 38.5$ k, 154 k, 38.5 k

9.3-6 (a) $Q_{cr} = 18.7$ kN; (b) $Q_{cr} = 29.5$ kN; (c) $a = 0.254$ m

9.3-7 $P_{cr} = \dfrac{4\pi^2EI}{L^2}$, $v = \dfrac{\delta}{2}\left(1 - \cos\dfrac{2\pi x}{L}\right)$

9.3-8 $P_{cr} = 4.856\,EI/L^2$ **9.3-9** $b = 2.69$ in.

9.3-10 For $\alpha = 4$, $h/L = 0.424$

9.4-1 $M/Pe = 0.8465\sin(1.4050x/L) + \cos(1.4050x/L)$

9.4-2 (a) $\delta = e(\sec kL - 1)$; (b) $M_{max} = Pe\sec kL$

9.4-3 $\delta = 0.232$ in., $M_{max} = 14.8$ in.-k

9.4-4 $\delta = 3.28$ mm, $M_{max} = 256$ N · m

9.4-5 $L = 10.5$ ft **9.4-6** $L = 3.20$ m **9.4-7** $P = 6.17$ k

9.4-8 $P = 9.38$ k

9.4-9 (a) $q_0 = 2230$ lb/ft; (b) $M_{max} = 37.7$ in.-k

9.5-1 (a) $\sigma_{cr} = 19.0$ ksi; (b) $\sigma_{max} = 8.87$ ksi; (c) $P_1 = 10.5$ k

9.5-2 (a) $\sigma_{cr} = 117$ MPa; (b) $\sigma_{max} = 26.3$ MPa;
(c) $P_1 = 21.4$ kN

9.5-3 (a) $\sigma_{max} = 9.93$ ksi; (b) $P_{allow} = 3.55$ k

9.5-4 (a) $\sigma_{max} = 32.3$ MPa; (b) $P_{allow} = 29.0$ kN

9.5-5 (a) $\sigma_{max} = 5.03$ ksi; (b) $P_{allow} = 22.2$ k

9.5-6 (a) $\sigma_{max} = 37.3$ MPa; (b) $P_{allow} = 105$ kN

9.5-7 (a) $\sigma_{max} = 18.8$ ksi; (b) $P_y = 181$ k

9.5-8 (a) $\sigma_{max} = 10.9$ ksi; (b) $P_{allow} = 200$ k

9.5-9 (a) $\sigma_{max} = 20.2$ ksi; (b) $n = 1.67$

9.5-10 (a) $\sigma_{max} = 19.5$ ksi; (b) $n = 1.87$

9.5-11 $L = 12.7$ ft

9.5-12 $P_2 = 36.2$ k

9.6-2 (a) $\sigma_{max} = 54.4$ MPa; (b) $P_{allow} = 96.1$ kN

9.6-3 (a) $\sigma_{max} = 8.53$ ksi; (b) $n = 2.17$

9.6-4 (a) $\sigma_{max} = 14.5$ ksi; (b) $n = 2.06$

9.6-5 (a) $\sigma_{max} = 10.4$ ksi; (b) $P_{allow} = 113$ k

9.6-6 $P_{allow} = 14.3$ k

9.9-1 $P = 247$ k, 180 k, 96.7 k, 54.4 k

9.9-2 $P = 328$ k, 243 k, 134 k, 75.3 k

9.9-3 $P = 360$ k, 223 k, 102 k, 57.2 k

9.9-4 $P = 664$ k, 496 k, 278 k, 156 k

9.9-5 $P = 58.9$ k, 43.0 k, 23.2 k, 13.0 k

9.9-6 $P = 385$ kN, 300 kN, 186 kN, 104 kN

9.9-7 $P = 159$ k, 122 k, 72.9 k, 41.0 k

9.9-8 $P = 1070$ kN, 906 kN, 692 kN, 438 kN

9.9-9 $L = 13.9$ ft **9.9-10** $L = 21.2$ ft, 14.4 ft

9.9-11 $L = 24.2$ ft, 14.7 ft **9.9-12** $d = 98.3$ mm

9.9-13 $b = 2.08$ in.

9.9-14 $P = 151$ k, 114 k, 53.8 k, 30.3 k

9.9-15 $P = 144$ k, 116 k, 73.3 k, 41.2 k

9.9-16 $L = 20.6$ in., 49.0 in. **9.9-17** $d = 1.27$ in.

9.9-18 $L = 2.81$ m, 3.53 m, 4.99 m

9.9-19 $P = 36.3$ k, 30.3 k, 15.2 k, 8.6 k

9.9-20 $b = 143$ mm

9.9-21 $P = 22.5$ k, 16.2 k, 10.4 k, 7.2 k

CHAPTER 10

10.3-1 (a) $\delta_v = 2PL/EA$ (downward); (b) $\delta_h = 0$;
(c) $\theta_{ab} = \sqrt{3}\,P/EA$ (clockwise)

10.3-2 (a) $\delta_v = -\alpha L(\Delta T)$ (upward); (b) $\delta_h = \alpha L(\Delta T)/\sqrt{3}$ (to the right); (c) $\theta_{ab} = -2\alpha(\Delta T)/\sqrt{3}$ (counterclockwise)

10.3-3 (a) $\delta_v = 2PL/3\sqrt{3}\,EA$ (downward); (b) $\delta_h = 0$; (c) $\theta_{ab} = P/3EA$ (clockwise)

10.3-4 $\delta_v = 7.62PL/EA$ (downward), $\delta_h = 1.73PL/EA$ (to the left)

10.3-5 $\delta_v = 3.83PL/EA$ (downward), $\delta_h = PL/EA$ (to the left)

10.3-6 $\delta_{bd} = 2\alpha L(\Delta T)$

10.3-7 $\delta_h = 0.0307$ in. (to the right), $\delta_v = 0.0578$ in. (downward)

10.3-8 $\delta_h = 0.952$ mm (to the right), $\delta_v = 1.967$ mm (downward)

10.3-9 $\Delta L = 0.6$ in. (increase in length)

10.3-10 (a) $\delta_v = 8.27Pb/EA$ (downward); (b) $\delta_{ae} = 1.85Pb/EA$ (increase)

10.3-11 $\delta_v = 0.178$ in. (downward), $\delta_h = 0.047$ in. (to the right)

10.3-12 $\delta_v = 1.81$ mm (downward), $\delta_h = 0.540$ mm (to the right)

10.3-13 $\theta_{bc} = 58.3 \times 10^{-6}$ rad (counterclockwise), $\delta_{bg} = 0.00632$ in. (decrease in distance)

10.3-14 $\delta_c = PL^3/48EI$ (downward), $\theta_a = PL^2/16EI$ (clockwise)

10.3-15 $\delta_b = qL^4/8EI$ (downward), $\theta_b = qL^3/6EI$ (clockwise)

10.3-16 $\delta_c = 19PL^3/384EI$ (downward)

10.3-17 $\delta_c = 7PL^3/162EI$, $\delta_d = 7PL^3/54EI$, $\delta_b = 2PL^3/9EI$

10.3-18 $\delta_c = Pb^2(L+b)/3EI$, $\theta_c = Pb(2L+3b)/6EI$

10.3-19 $\delta = PbL^2/16EI$ (upward), $\theta = PbL/6EI$ (counterclockwise)

10.3-20 $\delta = \alpha(T_2 - T_1)L^2/8h$ (downward when $T_2 > T_1$), $\theta_a = \alpha(T_2 - T_1)(L+2b)/2h$ (clockwise), $\theta_b = \alpha(T_2 - T_1)L/2h$ (clockwise)

10.3-21 $\delta_b = 41qL^4/384EI$, $\delta_c = 7qL^4/192EI$

10.3-22 $\theta_a = q_0L^3/45EI$, $\theta_b = 7q_0L^3/360EI$

10.3-23 $\delta = 2PL^3/3EI$

10.3-24 $\delta_h = PHL^2/8EI_2$ (to the left), $\theta = PL^2/16EI_2$ (clockwise)

10.3-25 $\delta_h = PL^3/3EI$ (to the right), $\delta_v = PL^3/2EI$ (upward)

10.3-26 $\delta_h - PL^3/EI$ (to the left), $\delta_v = 5PL^3/3EI$ (downward), $\theta = 2PL^2/EI$ (clockwise)

10.3-27 $\delta_v = 33Pb^3/EI$ (downward), $\theta = 33Pb^2/2EI$ (counterclockwise)

10.3-28 $\Delta = \dfrac{2PL^3\sin^2\beta}{3EI} + \dfrac{2PL\cos^2\beta}{EA}$

10.3-29 $\delta_h = \alpha(T_2 - T_1)H^2/2h$ (to the left), $\delta_v = \alpha(T_2 - T_1)(L)(L+2H)/2h$ (upward), $\theta = \alpha(T_2 - T_1)(L+H)/h$ (counterclockwise)

10.3-30 $\delta_h = PR^3(3\pi - 8)/4EI$ (to the right), $\delta_v = PR^3/2EI$ (downward), $\theta = PR^2(\pi - 2)/2EI$ (clockwise)

10.3-31 $\delta_h = 2PR^3/EI$ (to the left), $\delta_v = 3\pi PR^3/2EI$ (downward), $\theta = \pi PR^2/EI$ (clockwise)

10.3-32 $\delta_c = PR^3(3\pi - 8)/8EI$ (downward), $\delta_b = PR^3/2EI$ (to the right)

10.3-33 $\Delta = \dfrac{2PL^3}{3EI} + \dfrac{PR}{2EI}(2\pi L^2 + 8LR + \pi R^2)$

10.3-34 $\delta_h = PR(L+R)^2/2EI$ (to the right), $\delta_v = PR^2(4L + \pi R)/4EI$ (downward), $\theta = PR(L+R)/EI$ (clockwise)

10.3-35 $M_{max} = 2EIe/3\pi R^2$

10.3-36 $\delta_v = \dfrac{\pi PR^3}{4EI} + \dfrac{(3\pi - 8)PR^3}{4GI_p}$, $\phi = \dfrac{\pi PR^2}{4EI} + \dfrac{(\pi - 4)PR^2}{4GI_p}$

10.3-37 (a) $\delta_v = \dfrac{P}{3EI}(8L^3 + b^3) + \dfrac{PLb}{GI_p}(L+b)$, $\phi = \dfrac{Pb^2}{2EI} + \dfrac{PLb}{GI_p}$; (b) $\delta_v = 1.25$ in., $\phi = 0.00730$ rad

10.3-38 $\Delta = 5PL^3/6EI + 3PL^3/2GI_p$

10.3-39 $\delta = PL^3n(4n^2 + 3n + 1)/6EI$

10.4-2 $\delta_{dc} = \delta_{cd} = -PbL^2/16EI$

10.4-3 $M\theta_{bc} = P\delta_{cb} = MPbL/EI$

10.4-4 $\theta_{dc} = \theta_{cd} = -ML/96EI$ **10.4-5** $\delta_h = 0$

10.4-6 $\delta_c = 0$ **10.4-7** $\delta_{12} = \delta_{21} = -PL/EA$

10.6-1 (a) $U = EA\delta^2(1 + 2\cos^3\beta)/2L$; (b) $\delta = PL/[EA(1 + 2\cos^3\beta)]$; (c) $N_{ad} = N_{cd} = (P\cos^2\beta)/(1 + 2\cos^3\beta)$, $N_{bd} = P/(1 + 2\cos^3\beta)$

10.6-2 (a) $U = \dfrac{EA}{2L}[D_1^2 + (\cos\theta)(D_1\cos\theta - D_2\sin\theta)^2]$;

(b) $D_1 = \frac{PL}{EA}\left(\frac{1}{\tan\theta}\right)$, $D_2 = \frac{PL}{EA}\left(\frac{1+\cos^3\theta}{\sin^2\theta\cos\theta}\right)$

10.6-3 $\delta = PL/16EA$

10.6-4 (a) $U = \frac{2nAL}{1+n}\left(\frac{\delta\sin\beta}{L}\right)\left(\frac{\delta B\sin\beta}{L}\right)^{1/n}$;
(b) $\delta = \left(\frac{L}{B\sin\beta}\right)\left(\frac{P}{2A\sin\beta}\right)^n$

10.7-1 $\delta = P^nL/[2^nA^nB(\sin\beta)^{n+1}]$

10.7-2 $\delta_h = P^2L/A^2B^2$

10.7-3 $\delta = \gamma^nL^{n+1}/B(n+1)$

10.8-1 $\delta_c = Pb^2(L+b)/3EI$ **10.8-2** $\delta_c = 3PL^3/256EI$

10.8-3 $\delta_c = Pb^2(3L+b)/3EI$

10.8-4 $\delta_b = PL(2\sqrt{2}+1)/EA$

10.8-5 $\delta_b = PL/4EA\cos^3\theta$ **10.8-6** $\delta_b = \pi PR^3/2EI$

10.8-7 $\delta_b = L^3(2P_1 + 5P_2)/48EI$

10.8-8 $\theta_b = -7qL^3/384EI + M_0L/3EI$ (positive is clockwise)

10.8-9 $\delta_{max} = 5qL^4/384EI$ **10.8-10** $\theta_a = 7qL^3/48EI$

10.8-11 $\delta_c = 2\sqrt{2}PL/EA$

APPENDIX C

C.2-2 $\bar{x} = \bar{y} = 5a/12$ **C.2-3** $\bar{y} = 1.1$ in.

C.2-4 $2c^2 = ab$ **C.2-5** $\bar{y} = 13.9$ in.

C.2-6 $\bar{y} = 52.5$ mm **C.2-7** $\bar{x} = 0.99$ in., $\bar{y} = 1.99$ in.

C.2-8 $\bar{x} = 137$ mm, $\bar{y} = 132$ mm

C.3-6 $I_x = 20.2 \times 10^6$ mm^4

C.3-7 $I_x = 36.1$ in.^{4}, $I_y = 10.9$ in.^{4}

C.3-8 $I_x = I_y = 195 \times 10^6$ mm^4, $r_x = r_y = 80.1$ mm

C.3-9 $I_1 = 1480$ in.^{4}, $I_2 = 186$ in.^{4}, $r_1 = 7.10$ in., $r_2 = 2.52$ in.

C.4-1 $I_b = 940$ in.^{4} **C.4-2** $I_c = 11a^4/192$

C.4-3 $I_{x_c} = 7.23$ in.^{4}, $I_{y_c} = 43.3$ in.^{4}

C.4-4 $I_2 = 405 \times 10^3$ mm^4

C.4-5 $I_{x_c} = 6050$ in.^{4}, $I_{y_c} = 475$ in.^{4}

C.4-6 $I_c = 106 \times 10^6$ mm^4

C.4-7 $I_{x_c} = 17.40$ in.^{4}, $I_{y_c} = 6.27$ in.^{4}

C.4-8 $b = 250$ mm **C.5-1** $I_p = bh(b^2 + 12h^2)/48$

C.5-2 $I_{p_c} = \frac{r^4}{18\alpha}(9\alpha^2 - 8\sin^2\alpha)$ **C.5-3** $I_p = 233$ in.^{4}

C.5-4 $I_p = bh(b^2 + h^2)/24$

C.5-5 $I_{p_c} = \frac{r^4}{72}\left(\frac{176 - 84\pi + 9\pi^2}{4-\pi}\right)$

C.5-6 $I_{p_c} = 132 \times 10^6$ mm^4 **C.6-2** $I_{xy} = r^4/24$

C.6-3 $b = 2r$ **C.6-4** $I_{xy} = t^2(2b^2 - t^2)/4$

C.6-5 $I_{12} = -20.5$ in.^{4} **C.6-6** $I_{xy} = 24.3 \times 10^6$ mm^4

C.6-7 $I_{x_cy_c} = -6.079$ in.^{4} **C.6-8** $I_{xy} = 99.9 \times 10^6$ mm^4

C.7-1 $I_{x_1} = I_{y_1} = b^4/12$, $I_{x_1y_1} = 0$

C.7-2 $I_{x_1} = \frac{b^3h^3}{6(b^2+h^2)}$, $I_{y_1} = \frac{bh(b^4+h^4)}{12(b^2+h^2)}$, $I_{x_1y_1} = \frac{b^2h^2(h^2-b^2)}{12(b^2+h^2)}$

C.7-3 $I_d = 159$ in.^{4}

C.7-4 $I_{x_1} = 12.4 \times 10^6$ mm^4, $I_{y_1} = 9.68 \times 10^6$ mm^4, $I_{x_1y_1} = 6.02 \times 10^6$ mm^4

C.7-5 $I_{x_1} = 13.5$ in.^{4}, $I_{y_1} = 3.84$ in.^{4}, $I_{x_1y_1} = 4.76$ in.^{4}

C.7-6 $I_{x_1} = 8.75 \times 10^6$ mm^4, $I_{y_1} = 1.02 \times 10^6$ mm^4, $I_{x_1y_1} = -0.356 \times 10^6$ mm^4

C.8-1 (a) $c = \frac{\sqrt{a^2-b^2}}{2}$; (b) $\frac{a}{b} = \sqrt{5}$; (c) $1 < \frac{a}{b} < \sqrt{5}$

C.8-2 Show that two different sets of principal axes exist at points P_1 and P_2.

C.8-3 $\theta_{p_1} = -29.9°$, $\theta_{p_2} = 60.1°$, $I_1 = 311.1$ in.^{4}, $I_2 = 88.9$ in.^{4}

C.8-4 $\theta_{p_1} = -8.54°$, $\theta_{p_2} = 81.46°$, $I_1 = 17.24 \times 10^6$ mm^4, $I_2 = 4.88 \times 10^6$ mm^4

C.8-5 $\theta_{p_1} = 37.7°$, $\theta_{p_2} = 127.7°$, $I_1 = 15.5$ in.^{4}, $I_2 = 1.9$ in.^{4}

C.8-6 $\theta_{p_1} = 32.6°$, $\theta_{p_2} = 122.6°$, $I_1 = 8.76 \times 10^6$ mm^4, $I_2 = 1.00 \times 10^6$ mm^4

C.8-7 $\theta_{p_1} = 16.8°$, $\theta_{p_2} = 106.8°$, $I_1 = 0.2390b^4$, $I_2 = 0.03873b^4$

C.8-8 $\theta_{p_1} = -24.4°$, $\theta_{p_2} = 65.6°$, $I_1 = 11.46 \times 10^6$ mm^4, $I_2 = 1.05 \times 10^6$ mm^4

C.8-9 $\theta_{p_1} = -22.4°$, $\theta_{p_2} = 67.6°$, $I_1 = 25.8$ in.^{4}, $I_2 = 2.2$ in.^{4}

C.8-10 $\theta_{p_1} = 74.1°$, $\theta_{p_2} = -15.9°$, $I_1 = 8.29 \times 10^6$ mm^4, $I_2 = 1.00 \times 10^6$ mm^4

C.8-11 $\theta_{p_1} = 75.7°$, $\theta_{p_2} = -14.3°$, $I_1 = 20.1$ in.^{4}, $I_2 = 2.12$ in.^{4}

C.8-12 $x_1 = -0.0950b$, $y_1 = 0.537b$, $x_2 = 0.762b$, $y_2 = 0.796b$, $I = 0.2390b^4$

Name Index

* Page numbers with an asterisk indicate biographical information

Subject Index